Instructor's Solutions Manual

for

Serway and Jewett's

Principles of Physics
A Calculus-Based Text

Fourth Edition

Ralph V. McGrew
State University of New York

THOMSON
BROOKS/COLE
Australia • Canada • Mexico • Singapore • Spain • United Kingdom • United States

Printed in the United States of America
1 2 3 4 5 6 7 09 08 07 06 05

Printer: Thomson/West

0-534-46482-3

Thomson Higher Education
10 Davis Drive
Belmont, CA 94002-3098
USA

Asia (including India)
Thomson Learning
5 Shenton Way
#01-01 UIC Building
Singapore 068808

Australia/New Zealand
Thomson Learning Australia
102 Dodds Street
Southbank, Victoria 3006
Australia

Canada
Thomson Nelson
1120 Birchmount Road
Toronto, Ontario M1K 5G4
Canada

UK/Europe/Middle East/Africa
Thomson Learning
High Holborn House
50–51 Bedford Row
London WC1R 4LR
United Kingdom

Latin America
Thomson Learning
Seneca, 53
Colonia Polanco
11560 Mexico
D.F. Mexico

Spain (including Portugal)
Thomson Paraninfo
Calle Magallanes, 25
28015 Madrid, Spain

CONTENTS

PREFACE

This *Instructor's Solutions Manual* accompanies the fourth edition of the textbook *Principles of Physics* by Raymond A. Serway and John W. Jewett, Jr. It contains instructor's solutions for all of the end-of-chapter problems, with answers boxed; a list of answers to the even-numbered problems; and answers to the end-of-chapter questions. Problems new to this edition have asterisks next to their solutions.

The first word is *Instructor's*. Please do not give students access to this manual. Bear the following in mind:

- The publisher provides you with this copyrighted manual for your convenience, free of charge. This manual is not to be sold.

- Many teachers at many colleges use the textbook. Some use problems from it in graded homework and in examinations.

- Solutions in this manual are written for efficient communication with instructors, not for maximally effective help to students.

- Students never need to see solutions to all of the problems. The textbook is rich with examples. The *Student Solutions Manual and Study Guide* contains detailed solutions to about twelve problems and several questions per chapter—those with their numbers boxed in the textbook. Of these, five per chapter are on the textbook's website.

- The useful life of this edition can continue only so long as the users maintain the integrity and security of the problem set and solutions.

You should take precautions to keep the manual as secure as possible. Within the context of a course using the textbook, the teacher may use this manual to make answers and solutions to selected questions and problems available for a limited time to students in the course. We ask you not to make the whole manual available to a student, not to make any selections from it available at times when you are not teaching a course using the textbook, and not to make any selections from it available to the public, either physically or electronically.

As novice problem solvers are different from expert problem solvers, so the solutions in this manual are very different from solutions (such as those in the textbook examples) addressed to students. With the goals of communicating efficiently with you and of limiting the bulk of this manual, we have often omitted problem restatements, commentary, intermediate steps, and even initial steps that would make the solutions more clear to students. We generally answer questions briefly, without recitation of reasoning. Students profit from explicit help in understanding a problem, in representing it in different ways, in reasoning about it qualitatively, in planning a solution, and in evaluating a result for reasonableness. They will not get that help from this manual. Students need to see a problem situation in a context of direct experience with observed reality. They must understand the principles used in a problem solution as drawn from a remarkably small set of physical laws. Finally, their solutions must use techniques drawn from a comprehensible set of analysis models. If you provide students with solutions drawn from this manual, try including your own ideas in order to communicate most effectively with your students, and to reveal more of what really goes into solving a problem. We recommend following the *Conceptualize—Categorize—Analyze—Finalize* format introduced in the textbook. The classic book *How to Solve It* by George Polya [Princeton University Press] is especially good at explaining how a teacher can help a student who is actively working a problem, and how the student can learn a lot at the *Finalize* step of problem-solving by thinking critically about her solution.

Problems in the text are varied in subject matter and in what they ask the student to do. Your choice of the questions and problems you assign strongly influences your students' perception of the course, what they learn, and how much they learn. Use of this manual may help you to make a thoughtful individual choice of questions and problems to assign. Problems requiring *computer use*, problems involving *biological or medical applications*, and *review problems* are marked with icons or headings in the textbook. Five other sorts of problems deserve further explanation:

- *"Fermi problems."* Each chapter contains at least one problem asking for estimates of data and an order-of-magnitude calculation. Solutions here look simple. The statement of an order-of-magnitude problem usually assigns the student to decide on what quantities should be taken as data. Be prepared for students who have difficulty in deciding what quantities are relevant, in making estimates, and in believing that order-of-magnitude estimates have value. If you choose to include estimation problems in your course, I suggest examples, encouragement, working in pairs or groups, time to think, regularly repeated practice starting early in the course, and acceptance of different approaches. You may find that explicit study of Section 1.4 and assignment of Problem 1.58 [Problem 58 in Chapter 1] and Problem 30.58 are necessary. Here is a list of the order-of-magnitude problems:

1.19	1.20	1.21	1.22	1.65	2.52	3.56	4.15	5.32	6.40
7.45	8.2	9.49	10.8	11.55	12.39	13.42	14.32	15.22	16.26
17.37	18.33	19.3	20.56	21.24	22.59	23.51	23.61	24.34	24.56
25.21	26.1	27.30	27.44	27.54	28.23	29.50	29.58	29.61	30.1
30.58	31.2	31.7	31.10	31.31	31.45	31.49			

- *"Paired problems."* While every chapter contains problems requiring symbolic solutions, Chapters 1 through 4, 9, 16 through 21, and 25 contain pairs of adjacent problems stated in identical language, the first giving numerical data and asking for a numerical answer and the second asking for a symbolic solution. The pairs are identified by a tan background. You can use either problem by itself. At any point in your course, for students who have learned to do numerical calculations but find difficulty with symbols, you can use both problems together. With your help, students will see that the same steps solve both problems, and that the symbolic problem can be much more instructive. For students who rely too much on an equation-solving calculator, you can assign both problems. With your help, students will see the need to do their algebra themselves.

- *"Jeopardy problems."* Following on the suggestion of Alan Van Heuvelen and David Maloney, *American Journal of Physics* **67**(3) 252–256 (March 1999), problems in this list give students a mathematical representation of a situation and ask for a description in ordinary language or for a diagram, as well as asking for a solution. These problems therefore give students extra practice in switching between different representations:

1.56	2.46	3.18	5.46	7.20	8.50	10.74	20.14	21.38	26.36

- *"Design problems."* Many problems describe a practical device and require calculation of results of its use. Problems in the following list require identifying and combining physical principles describing the operation of a practical device, and determining design parameters necessary to achieve a certain result. A workshop led by Kathleen Harper suggested the idea. These problems can be assigned to any students, but they may be useful as one motivational tool for students studying engineering design. They may be useful in displaying, in one simple and tangible way, how the physics course supports an engineering curriculum.

2.50	3.42	4.23	5.47	6.18	6.49	7.18	8.43	10.18	12.20
12.45	13.16	14.24	14.44	15.57	15.64	16.46	16.48	16.62	17.4
17.57	18.42	19.56	20.45	20.73	21.22	21.33	22.47	23.51	24.68
25.32	26.41	26.54	27.60	28.55	30.63				

- *"Not-just-a-number problems."* The distinction between questions and problems is fuzzy, not sharp. I recommend regularly assigning items listed as questions along with problems, and including questions on tests. The textbook contains many "Thinking Physics" example questions to support this approach; the *Student Manual and Study Guide* contains answers to some chapter-end questions and this Instructor's Manual contains brief answers to all. The group of students who are very good at answering questions may be different from the group who are very good at solving numerical problems. Students may see answering well-chosen questions as a more creative and more enjoyable learning experience than solving numerical problems. Answering some chapter-end questions requires constructing chains of reasoning that are as instructive as any problem solutions. Conversely, some items listed as problems require answers in prose for some parts. Examples among the problems listed for Chapter 4 include 4.21, 4.42, and 4.43. If you assign students to explain their reasoning in problem solutions, all problems may qualify as not-just-a-number problems. I do not attempt a comprehensive list, because the classification itself is fuzzy; but I heartily suggest that you teach students to take seriously the qualitative parts of tasks you assign.

I invite your comments and those of your students on the instructional value of individual questions and problems, and on the value of the eight classes of problems listed above. However we impose classifications on them, some sorts of problems are unique to Serway textbooks, and some are new to this edition.

When a problem quotes data to three significant digits, I quote answers to three significant digits. The last digit is uncertain, often depending, for example, on the precision of the values assumed for physical constants. If the calculation involves a chain of steps, I carry forward many digits in intermediate results, even though I write down only three digits. I "round off" only at the end of the calculation, never anywhere in the middle. The regularity of three significant digits in nearly all textbook problems can help students focus on physical concepts without distraction. I recognize that students may get the false impression that every physics problem should have values precise to three digits. You may assign your students to work more thoughtfully with laboratory measurements.

The tables of analysis models and structural models appearing on pages ix–xi do not summarize the content of the book. For example, Newton's law of gravitation and Maxwell's equations are not included. The tables are not necessarily intended for students, but rather to display to the interested teacher how systematically the textbook uses the idea of modeling.

If you find errors in the textbook problems or solutions, please notify me at the address below. I will reply and will correct errors in future printings. I thank those who have helped to clarify and correct solutions since the publication of the last edition: the student David Faunce and the teachers Erika Hermon, Ken Menningen, Chris Church, David Tamres, Kevin Kilty, and Daniel Fernandez.

I am grateful to Mary Toscano and the staff at M and N Toscano of Somerville, MA, for assembling and typing this manual and preparing diagrams and page layouts. Richard Miers checked the manual for accuracy. Their prompt and exacting work has contributed greatly to the book's correctness, clarity, and usefulness. Various solutions in this manual were written originally by Louis Cadwell, John R. Gordon,

Lawrence Hmurcik, Henry Leap, Steve Van Wyk, Laurent Hodges, N. John DiNardo, Richard Cohen, Ronald Bieniek, Charles Teague, Duane Deardorff, Richard McGrew, Michael Hones, Edward Adelson, Michael Browne, Andrew Duffy, Perry Ganas, Boris Korsunsky, Edwin Lo, Clement Moses, and Raymond Serway; and by Robert Forsythe, John Jewett, Randall Jones, and Jerzy Wrobel, who contributed new problems for this edition of the text. I thank Jay Campbell and Sarah Lowe, Developmental Editors at Brooks/Cole Thomson Learning, who coordinated this project and provided resources for it.

Ralph McGrew, Engineering Science and Physics Department,
Broome Community College, Binghamton, NY 13902-1017
mcgrew_r@sunybroome.edu

Structural Models

Scale	Category	Structural Model
Macroscopic	Solar system	geocentric model
		heliocentric model
		Keplerian model
	Universe	Big Bang model
Microscopic	Solids	atom–spring model
		free electron model for current
	Liquids	ideal fluid in streamline flow model
	Gases	kinetic theory model
	Atoms	Thomson model
		Rutherford model
		Bohr model
		quantum model
	Nuclei	nucleon model
		quantum model
	Nucleons	quark model
Macroscopic/microscopic		Standard model

Analysis Models Used in Problem Solving

Simplification Model	Analysis Model	Primary or Sample Equations and/or Principles	Chapter in which model is first introduced
Particle	particle under constant speed	distance $= v\Delta t$	2
	particle under constant velocity	$x_f = x_i + vt$	2
	particle under constant acceleration	$v_{xf} = v_{xi} + a_x t$	2
		$x_f = x_i + \frac{1}{2}(v_{xi} + v_{xf})t$	
		$x_f = x_i + v_{xi}t + \frac{1}{2}a_x t^2$	
		$v_{xf}^2 = v_{xi}^2 + 2a_x(x_f - x_i)$	
	particle in uniform circular motion	$a_c = \frac{v^2}{r}$	3
		$T = \frac{2\pi r}{v}$	
	particle under a net force	$\Sigma\vec{\mathbf{F}} = m\vec{\mathbf{a}}$	4
	particle in equilibrium	$\Sigma\vec{\mathbf{F}} = 0$	4
	particle in a field		
	gravitational	$\vec{\mathbf{F}}_g = m\vec{\mathbf{g}}$	5
	electric	$\vec{\mathbf{F}}_e = q\vec{\mathbf{E}}$	19
	magnetic	$\vec{\mathbf{F}}_B = q\vec{\mathbf{v}} \times \vec{\mathbf{B}}$	22
	simple harmonic motion	$\vec{\mathbf{F}}_s = -k\vec{\mathbf{x}}$	12
		$x = A\cos(\omega t + \phi)$	
	quantum particle	Principle of superposition	28
		$\Delta p_x \Delta x \geq \frac{\hbar}{2}$	
	quantum particle under boundary conditions	Schrödinger equation	28
System	isolated system		
	energy	$\Delta E = 0$	7
	momentum	$\Delta\vec{\mathbf{p}} = 0$	8
	angular momentum	$\Delta\vec{\mathbf{L}} = 0$	10
	charge		19

Analysis Models Used in Problem Solving

Simplification Model	Analysis Model	Primary or Sample Equations and/or Principles	Chapter in which model is first introduced
System (cont.)	non-isolated system		
	energy	$\Delta E = \Sigma T$	6
	momentum	$\Sigma \vec{\mathbf{F}} = \frac{d\vec{\mathbf{p}}}{dt}$	8
	angular momentum	$\Sigma \vec{\boldsymbol{\tau}} = \frac{d\vec{\mathbf{L}}}{dt}$	10
	non-isolated system in steady state		
	energy	$0 = \Sigma T$	7
		Kirchhoff's loop rule	21
	momentum	$\Sigma \vec{\mathbf{F}} = 0$	8
	angular momentum	$\Sigma \vec{\boldsymbol{\tau}} = 0$	10
	charge	Kirchhoff's junction rule	21
Rigid body	rigid body under constant angular speed	$\theta_f = \theta_i + \omega t$	10
	rigid body under constant angular acceleration	$\omega_f = \omega_i + \alpha t$	10
		$\theta_f = \theta_i + \frac{1}{2}(\omega_i + \omega_f)t$	
		$\theta_f = \theta_i + \omega_i t + \frac{1}{2}\alpha t^2$	
		$\omega_f^2 = \omega_i^2 + 2\alpha(\theta_f - \theta_i)$	
	rigid body under a net torque	$\Sigma \vec{\boldsymbol{\tau}} = I\vec{\boldsymbol{\alpha}}$	10
	rigid body in equilibrium	$\Sigma \vec{\boldsymbol{\tau}} = 0$	10
Wave	traveling wave	$v = \lambda f$	13
		$y = A\sin(kx - \omega t)$	
	wave under boundary conditions	distance between nodes $= \frac{\lambda}{2}$	14
		distance between antinodes $= \frac{\lambda}{2}$	
	wave under reflection	$\theta_1' = \theta_1$	25
	wave under refraction	$\left(\frac{1}{v_1}\right)\sin\theta_1 = \left(\frac{1}{v_2}\right)\sin\theta_2$	25
	waves in interference	Principle of superposition	14
		Constructive: $\delta = m\lambda$	27
		Destructive: $\delta = \left(m + \frac{1}{2}\right)\lambda$	

Chapter 1

Introduction and Vectors

CHAPTER OUTLINE

ANSWERS TO QUESTIONS

Q1.1 Atomic clocks are based on electromagnetic waves which atoms emit. Also, pulsars are highly regular astronomical clocks.

Q1.2 Density varies with temperature and pressure. It would be necessary to measure both mass and volume very accurately in order to use the density of water as a standard.

Q1.3 (a) 0.3 millimeters (b) 50 microseconds (c) 7.2 kilograms

Q1.4 (b) and (d). You cannot add or subtract quantities of different dimension.

Q1.5 A dimensionally correct equation need not be true. Example: 1 chimpanzee = 2 chimpanzee is dimensionally correct. If an equation is not dimensionally correct, it cannot be correct.

Q1.6 On February 7, 2005, I am 59 years and 39 days old.

$$59\text{ yr}\left(\frac{365.25\text{ d}}{1\text{ yr}}\right)+39\text{ d}=21\,589\text{ d}\left(\frac{86\,400\text{ s}}{1\text{ d}}\right)=1.87\times10^{9}\text{ s}\sim10^{9}\text{ s}.$$

Many college students are just approaching 1 Gs.

Q1.7 Zero significant digits. An order-of-magnitude calculation is accurate only within a factor of 10.

Q1.8 With one datum known to one significant digit, we have 80 million yr + 24 yr = 80 million yr.

Q1.9 No, the magnitude of the displacement is always less than or equal to the distance traveled. If two displacements in the same direction are added, then the magnitude of their sum will be equal to the distance traveled. Two vectors in any other orientation will give a displacement less than the distance traveled. If you first walk 3 meters east, and then 4 meters south, you will have walked a total distance of 7 meters, but you will only be 5 meters from your starting point.

Q1.10 Only force and velocity are vectors. None of the other quantities requires a direction to be described.

Q1.11 If the direction-angle of $\vec{\mathbf{A}}$ is between 180 degrees and 270 degrees, its components are both negative. If a vector is in the second quadrant or the fourth quadrant, its components have opposite signs.

Q1.12 The book's displacement is zero, as it ends up at the point from which it started. The distance traveled is 6.0 meters.

Q1.13 85 miles. The magnitude of the displacement is the distance from the starting point, the 260-mile mark, to the ending point, the 175-mile mark.

Q1.14 Vectors $\vec{\mathbf{A}}$ and $\vec{\mathbf{B}}$ are perpendicular to each other.

Q1.15 No, the magnitude of a vector is always positive. A minus sign in a vector only indicates direction, not magnitude.

Q1.16 Any vector that points along a line at 45° to the x and y axes has components equal in magnitude.

Q1.17 Addition of a vector to a scalar is not defined. Think of apples and oranges.

Q1.18 If the direction of a vector is specified by giving the angle of the vector measured clockwise from the positive y-axis, then the x-component of the vector is equal to the sine of the angle multiplied by the magnitude of the vector.

Q1.19 (a) Rotation does not noticeably affect revolution: simplification.
(b) Geometric (c) Structural (d) Simplification
(e) Geometric (f) Simplification (g) Structural

SOLUTIONS TO PROBLEMS

Section 1.1 Standards of Length, Mass, and Time

***P1.1** Modeling the Earth as a sphere, we find its volume as $\frac{4}{3}\pi r^3 = \frac{4}{3}\pi\left(6.37\times10^6 \text{ m}\right)^3 = 1.08\times10^{21} \text{ m}^3$. Its density is then $\rho = \frac{m}{V} = \frac{5.98\times10^{24} \text{ kg}}{1.08\times10^{21} \text{ m}^3} = \boxed{5.52\times10^3 \text{ kg/m}^3}$. This value is intermediate between the tabulated densities of aluminum and iron. Typical rocks have densities around 2 000 to 3 000 kg/m^3. The average density of the Earth is significantly higher, so higher-density material must be down below the surface.

P1.2 Let V represent the volume of the model, the same in $\rho = \frac{m}{V}$ for both. Then $\rho_{\text{iron}} = 9.35 \text{ kg}/V$ and $\rho_{\text{gold}} = \frac{m_{\text{gold}}}{V}$. Next, $\frac{\rho_{\text{gold}}}{\rho_{\text{iron}}} = \frac{m_{\text{gold}}}{9.35 \text{ kg}}$ and $m_{\text{gold}} = 9.35 \text{ kg}\left(\frac{19.3\times10^3 \text{ kg/m}^3}{7.86\times10^3 \text{ kg/m}^3}\right) = \boxed{23.0 \text{ kg}}$.

P1.3

$$V = V_o - V_i = \frac{4}{3}\pi\left(r_2^3 - r_1^3\right)$$

$$\rho = \frac{m}{V}, \text{ so } m = \rho V = \rho\left(\frac{4}{3}\pi\right)\left(r_2^3 - r_1^3\right) = \boxed{\frac{4\pi\rho\left(r_2^3 - r_1^3\right)}{3}}.$$

***P1.4** For either sphere the volume is $V=\frac{4}{3}\pi r^3$ and the mass is $m=\rho V=\rho\frac{4}{3}\pi r^3$. We divide this equation for the larger sphere by the same equation for the smaller:

$$\frac{m_\ell}{m_s}=\frac{\rho 4\pi r_\ell^3 3}{\rho 4\pi r_s^3 3}=\frac{r_\ell^3}{r_s^3}=5.$$

Then $r_\ell=r_s\sqrt[3]{5}=4.50\text{ cm}(1.71)=\boxed{7.69\text{ cm}}$.

Section 1.2 Dimensional Analysis

P1.5 The term x has dimensions of L, a has dimensions of LT^{-2}, and t has dimensions of T. Therefore, the equation $x=ka^m t^n$ has dimensions of

$$L=\left(LT^{-2}\right)^m(T)^n \text{ or } L^1T^0=L^mT^{n-2m}.$$

The powers of L and T must be the same on each side of the equation. Therefore,

$$L^1=L^m \text{ and } \boxed{m=1}.$$

Likewise, equating terms in T, we see that $n-2m$ must equal 0. Thus, $\boxed{n=2}$. The value of k, a dimensionless constant, $\boxed{\text{cannot be obtained by dimensional analysis}}$.

P1.6 (a) Circumference has dimensions of L.

(b) Volume has dimensions of L^3.

(c) Area has dimensions of L^2.

Expression (i) has dimension $L\left(L^2\right)^{1/2}=L^2$, so this must be area (c).
Expression (ii) has dimension L, so it is (a).
Expression (iii) has dimension $L\left(L^2\right)=L^3$, so it is (b). Thus, $\boxed{\text{(a)}=\text{ii};\quad \text{(b)}=\text{iii},\ \text{(c)}=\text{i}}$.

P1.7 (a) $\boxed{\text{This is incorrect}}$ since the units of $[ax]$ are m^2/s^2, while the units of $[v]$ are m/s.

(b) $\boxed{\text{This is correct}}$ since the units of $[y]$ are m, and $\cos(kx)$ is dimensionless if $[k]$ is in m^{-1}.

Section 1.3 Conversion of Units

P1.8 Apply the following conversion factors:

$$1\text{ in}=2.54\text{ cm},\ 1\text{ d}=86\,400\text{ s},\ 100\text{ cm}=1\text{ m},\text{ and } 10^9\text{ nm}=1\text{ m}$$

$$\left(\frac{1}{32}\text{ in/day}\right)\frac{(2.54\text{ cm/in})\left(10^{-2}\text{ m/cm}\right)\left(10^9\text{ nm/m}\right)}{86\,400\text{ s/day}}=\boxed{9.19\text{ nm/s}}.$$

This means the proteins are assembled at a rate of many layers of atoms each second.

P1.9 (a) Seven minutes is 420 seconds, so the rate is

$$r = \frac{30.0 \text{ gal}}{420 \text{ s}} = \boxed{7.14\times10^{-2} \text{ gal/s}}.$$

(b) Converting gallons first to liters, then to m^3,

$$r = \left(7.14\times10^{-2} \text{ gal/s}\right)\left(\frac{3.786 \text{ L}}{1 \text{ gal}}\right)\left(\frac{10^{-3} \text{ m}^3}{1 \text{ L}}\right)$$

$$r = \boxed{2.70\times10^{-4} \text{ m}^3/\text{s}}.$$

(c) At that rate, to fill a 1-m^3 tank would take

$$t = \left(\frac{1 \text{ m}^3}{2.70\times10^{-4} \text{ m}^3/\text{s}}\right)\left(\frac{1 \text{ h}}{3\,600}\right) = \boxed{1.03 \text{ h}}.$$

P1.10 It is often useful to remember that the 1 600-m race at track and field events is approximately 1 mile in length. To be precise, there are 1 609 meters in a mile. Thus, 1 acre is equal in area to

$$(1 \text{ acre})\left(\frac{1 \text{ mi}^2}{640 \text{ acres}}\right)\left(\frac{1\,609 \text{ m}}{\text{mi}}\right)^2 = \boxed{4.05\times10^3 \text{ m}^2}.$$

P1.11 The weight flow rate is $1\,200 \dfrac{\text{ton}}{\text{h}}\left(\dfrac{2\,000 \text{ lb}}{\text{ton}}\right)\left(\dfrac{1 \text{ h}}{60 \text{ min}}\right)\left(\dfrac{1 \text{ min}}{60 \text{ s}}\right) = \boxed{667 \text{ lb/s}}$.

P1.12 (a) $$\left(\frac{7\times10^{12} \text{ \$}}{1\,000 \text{ \$/s}}\right)\left(\frac{1 \text{ h}}{3\,600 \text{ s}}\right)\left(\frac{1 \text{ day}}{24 \text{ h}}\right)\left(\frac{1 \text{ yr}}{365 \text{ days}}\right) = \boxed{220 \text{ years}}$$

(b) The circumference of the Earth at the equator is $2\pi\left(6.378\times10^6 \text{ m}\right) = 4.01\times10^7 \text{ m}$. The length of one dollar bill is 0.155 m so that the length of 7 trillion bills is 1.08×10^{12} m. Thus, the 7 trillion dollars would encircle the Earth

$$\frac{1.08\times10^{12} \text{ m}}{4.01\times10^7 \text{ m}} = \boxed{2.71\times10^4 \text{ times}}.$$

P1.13 $V = At$ so $t = \dfrac{V}{A} = \dfrac{3.78\times10^{-3} \text{ m}^3}{25.0 \text{ m}^2} = \boxed{1.51\times10^{-4} \text{ m (or 151 } \mu\text{m)}}$

***P1.14** $N_{\text{atoms}} = \dfrac{m_{\text{Sun}}}{m_{\text{atom}}} = \dfrac{1.99\times10^{30} \text{ kg}}{1.67\times10^{-27} \text{ kg}} = \boxed{1.19\times10^{57} \text{ atoms}}$

P1.15 To balance, $m_{\text{Fe}} = m_{\text{Al}}$ or $\rho_{\text{Fe}}V_{\text{Fe}} = \rho_{\text{Al}}V_{\text{Al}}$

$$\rho_{\text{Fe}}\left(\frac{4}{3}\right)\pi r_{\text{Fe}}^3 = \rho_{\text{Al}}\left(\frac{4}{3}\right)\pi r_{\text{Al}}^3$$

$$r_{\text{Al}} = r_{\text{Fe}}\left(\frac{\rho_{\text{Fe}}}{\rho_{\text{Al}}}\right)^{1/3} = (2.00 \text{ cm})\left(\frac{7.86}{2.70}\right)^{1/3} = \boxed{2.86 \text{ cm}}.$$

P1.16 The mass of each sphere is $m_{\text{Al}} = \rho_{\text{Al}} V_{\text{Al}} = \dfrac{4\pi \rho_{\text{Al}} r_{\text{Al}}^3}{3}$ and $m_{\text{Fe}} = \rho_{\text{Fe}} V_{\text{Fe}} = \dfrac{4\pi \rho_{\text{Fe}} r_{\text{Fe}}^3}{3}$. Setting these masses equal,

$$\frac{4\pi \rho_{\text{Al}} r_{\text{Al}}^3}{3} = \frac{4\pi \rho_{\text{Fe}} r_{\text{Fe}}^3}{3} \text{ and } \boxed{r_{\text{Al}} = r_{\text{Fe}} \sqrt[3]{\frac{\rho_{\text{Fe}}}{\rho_{\text{Al}}}}}.$$

The resulting expression shows that the radius of the aluminum sphere is directly proportional to the radius of the balancing iron sphere. The fraction $\dfrac{\rho_{\text{Fe}}}{\rho_{\text{Al}}}$ is the factor of change between the densities. Its cube root is a number much closer to 1. The relatively small change in radius implies a change in volume sufficient to compensate for the change in density. The sphere of smaller density has larger radius.

P1.17 (a) $$d_{\text{nucleus, scale}} = d_{\text{nucleus, real}}\left(\frac{d_{\text{atom, scale}}}{d_{\text{atom, real}}}\right) = \left(2.40\times10^{-15}\text{ m}\right)\left(\frac{300\text{ ft}}{1.06\times10^{-10}\text{ m}}\right) = 6.79\times10^{-3}\text{ ft, or}$$

$$d_{\text{nucleus, scale}} = \left(6.79\times10^{-3}\text{ ft}\right)\left(304.8\text{ mm/1 ft}\right) = \boxed{2.07\text{ mm}}$$

(b) $$\frac{V_{\text{atom}}}{V_{\text{nucleus}}} = \frac{\frac{4\pi r_{\text{atom}}^3}{3}}{\frac{4\pi r_{\text{nucleus}}^3}{3}} = \left(\frac{r_{\text{atom}}}{r_{\text{nucleus}}}\right)^3 = \left(\frac{d_{\text{atom}}}{d_{\text{nucleus}}}\right)^3 = \left(\frac{1.06\times10^{-10}\text{ m}}{2.40\times10^{-15}\text{ m}}\right)^3$$

$$= \boxed{8.62\times10^{13}\text{ times as large}}$$

Section 1.4 Order-of-Magnitude Calculations

P1.18 A reasonable guess for the diameter of a tire might be 2.5 ft, with a circumference of about 8 ft. Thus, the tire would make $(50\,000\text{ mi})(5\,280\text{ ft/mi})(1\text{ rev/8 ft}) = 3\times10^7\text{ rev} \sim \boxed{10^7\text{ rev}}$.

P1.19 Model the room as a rectangular solid with dimensions 4 m by 4 m by 3 m, and each ping-pong ball as a sphere of diameter 0.038 m. The volume of the room is $4\times4\times3 = 48\text{ m}^3$, while the volume of one ball is

$$\frac{4\pi}{3}\left(\frac{0.038\text{ m}}{2}\right)^3 = 2.87\times10^{-5}\text{ m}^3.$$

Therefore, one can fit about $\dfrac{48}{2.87\times10^{-5}} \sim \boxed{10^6}$ ping-pong balls in the room.

As an aside, the actual number is smaller than this because there will be a lot of space in the room that cannot be covered by balls. In fact, even in the best arrangement, the so-called "best packing fraction" is $\frac{1}{6}\pi\sqrt{2} = 0.74$ so that at least 26% of the space will be empty. Therefore, the above estimate reduces to $1.67\times10^6\times0.740 \sim 10^6$.

P1.20 Assume the tub measures 1.3 m by 0.5 m by 0.3 m. One-half of its volume is then

$$V=(0.5)(1.3\text{ m})(0.5\text{ m})(0.3\text{ m})=0.10\text{ m}^3.$$

The mass of this volume of water is

$$m_{\text{water}}=\rho_{\text{water}}V=\left(1\,000\text{ kg/m}^3\right)\left(0.10\text{ m}^3\right)=100\text{ kg}\boxed{\sim10^2\text{ kg}}.$$

Pennies are now mostly zinc, but consider copper pennies filling 50% of the volume of the tub. The mass of copper required is

$$m_{\text{copper}}=\rho_{\text{copper}}V=\left(8\,920\text{ kg/m}^3\right)\left(0.10\text{ m}^3\right)=892\text{ kg}\boxed{\sim10^3\text{ kg}}.$$

***P1.21** Assume: Total population $=10^7$; one out of every 100 people has a piano; one tuner can serve about 1 000 pianos (about 4 per day for 250 weekdays, assuming each piano is tuned once per year). Therefore,

$$\#\text{ tuners}\sim\left(\frac{1\text{ tuner}}{1\,000\text{ pianos}}\right)\left(\frac{1\text{ piano}}{100\text{ people}}\right)\left(10^7\text{ people}\right)=\boxed{100}.$$

P1.22 The typical person probably drinks 2 to 3 soft drinks daily. Perhaps half of these were in aluminum cans. Thus, we will estimate 1 aluminum can disposal per person per day. In the U.S. there are ~250 million people, and 365 days in a year, so

$$\left(250\times10^6\text{ cans/day}\right)(365\text{ days/year})\sim\boxed{10^{11}\text{ cans}}$$

are thrown away or recycled each year. Guessing that each can weighs around $\frac{1}{10}$ of an ounce, we estimate this represents

$$\left(10^{11}\text{ cans}\right)(0.1\text{ oz/can})(1\text{ lb/16 oz})(1\text{ ton/2 000 lb})\approx3.1\times10^5\text{ tons/year}.\ \boxed{\sim10^5\text{ tons}}$$

Section 1.5 Significant Figures

P1.23 (a) $\boxed{3}$ (b) $\boxed{4}$ (c) $\boxed{3}$ (d) $\boxed{2}$

P1.24 (a)

$$\begin{array}{r}756.??\\37.2?\\0.83\\+\;2.5?\\\hline796.\not{5}\not{3}\end{array}=\boxed{797}$$

(b) $0.003\,2(2\text{ s.f.})\times356.3(4\text{ s.f.})=1.140\,16=(2\text{ s.f.})\ \boxed{1.1}$

(c) $5.620(4\text{ s.f.})\times\pi(>4\text{ s.f.})=17.656=(4\text{ s.f.})\ \boxed{17.66}$

P1.25 We work to nine significant digits:

$$1\text{ yr} = 1\text{ yr}\left(\frac{365.242\,199\text{ d}}{1\text{ yr}}\right)\left(\frac{24\text{ h}}{1\text{ d}}\right)\left(\frac{60\text{ min}}{1\text{ h}}\right)\left(\frac{60\text{ s}}{1\text{ min}}\right) = \boxed{31\,556\,926.0\text{ s}}.$$

***P1.26** $r = (6.50 \pm 0.20)\text{ cm} = (6.50 \pm 0.20)\times 10^{-2}\text{ m}$

$m = (1.85 \pm 0.02)\text{ kg}$

$$\rho = \frac{m}{\left(\frac{4}{3}\right)\pi r^3}$$

also, $\dfrac{\delta\rho}{\rho} = \dfrac{\delta m}{m} + \dfrac{3\delta r}{r}$. In other words, the percentages of uncertainty are cumulative. Therefore,

$$\frac{\delta\rho}{\rho} = \frac{0.02}{1.85} + \frac{3(0.20)}{6.50} = 0.103\,,$$

$$\rho = \frac{1.85}{\left(\frac{4}{3}\right)\pi\left(6.5\times 10^{-2}\text{ m}\right)^3} = \boxed{1.61\times 10^3\text{ kg/m}^3}$$

and

$$\rho \pm \delta\rho = \boxed{(1.61 \pm 0.17)\times 10^3\text{ kg/m}^3} = (1.6 \pm 0.2)\times 10^3\text{ kg/m}^3\,.$$

P1.27 $V = 2V_1 + 2V_2 = 2(V_1 + V_2)$

$V_1 = (17.0\text{ m} + 1.0\text{ m} + 1.0\text{ m})(1.0\text{ m})(0.09\text{ m}) = 1.70\text{ m}^3$

$V_2 = (10.0\text{ m})(1.0\text{ m})(0.090\text{ m}) = 0.900\text{ m}^3$

$V = 2\left(1.70\text{ m}^3 + 0.900\text{ m}^3\right) = \boxed{5.2\text{ m}^3}$

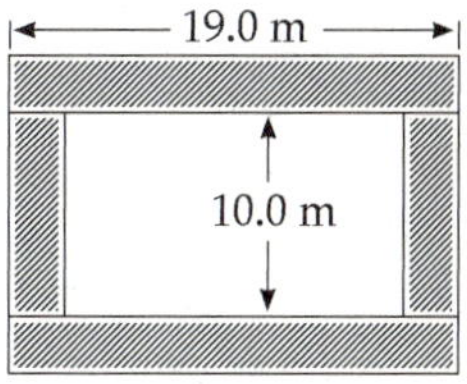

FIG. P1.27

$$\left.\begin{array}{l} \dfrac{\delta \ell_1}{\ell_1} = \dfrac{0.12\text{ m}}{19.0\text{ m}} = 0.0063 \\ \dfrac{\delta w_1}{w_1} = \dfrac{0.01\text{ m}}{1.0\text{ m}} = 0.010 \\ \dfrac{\delta t_1}{t_1} = \dfrac{0.1\text{ cm}}{9.0\text{ cm}} = 0.011 \end{array}\right\} \dfrac{\delta V}{V} = 0.006 + 0.010 + 0.011 = 0.027 = \boxed{3\%}$$

***P1.28** The equation $2x^4 - 3x^3 + 5x - 70 = 0$ is quartic, so we do not attempt to solve it with algebra. To find how many solutions the equation has and to estimate them, we graph the expression:

x	–3	–2	–1	0	1	2	3	4
$y = 2x^4 - 3x^3 + 5x - 70$	158	–24	–70	–70	–66	–52	26	270

We see that the equation $y = 0$ has two roots, one around $x = -2.2$ and the other near $x = +2.7$. To home in on the first of these solutions we compute in sequence: When $x = -2.2$, $y = -2.20$. The root must be between $x = -2.2$ and $x = -3$. When $x = -2.3$, $y = 11.0$. The root is between $x = -2.2$ and $x = -2.3$. When $x = -2.23$, $y = 1.58$. The root is between $x = -2.20$ and $x = -2.23$. When $x = -2.22$, $y = 0.301$. The root is between $x = -2.20$ and –2.22. When $x = -2.215$, $y = -0.331$. The root is between $x = -2.215$ and –2.22. We could next try $x = -2.218$, but we already know to three-digit precision that the root is $x = -2.22$.

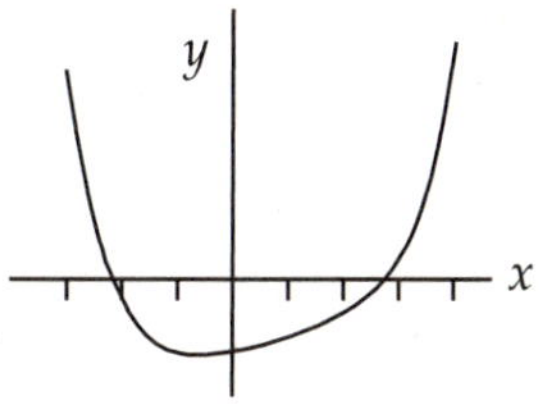

FIG. P1.28

***P1.29** We require $\sin\theta = -3\cos\theta$, or $\dfrac{\sin\theta}{\cos\theta} = -3$, or $\tan\theta = -3$. For $\tan^{-1}(-3) = \arctan(-3)$, your calculator may return –71.6°, but this angle is not between 0° and 360° as the problem requires. The tangent function is negative in the second quadrant (between 90° and 180°) and in the fourth quadrant (from 270° to 360°). The solutions to the equation are then $360° - 71.6° = \boxed{288°}$ and $180° - 71.6 = \boxed{108°}$.

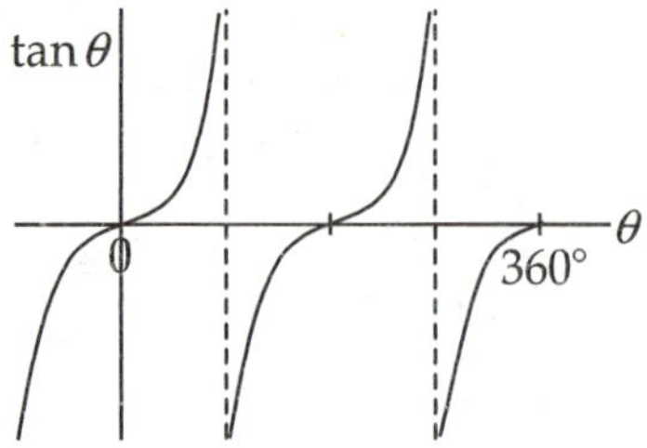

FIG. P1.29

***P1.30** We draw the radius to the initial point and the radius to the final point. The angle θ between these two radii has its sides perpendicular, right side to right side and left side to left side, to the 35° angle between the original and final tangential directions of travel. A most useful theorem from geometry then identifies these angles as equal: $\theta = 35°$. The whole circumference of a 360° circle of the same radius is $2\pi R$. By proportion, then $\dfrac{2\pi R}{360°} = \dfrac{840\text{ m}}{35°}$.

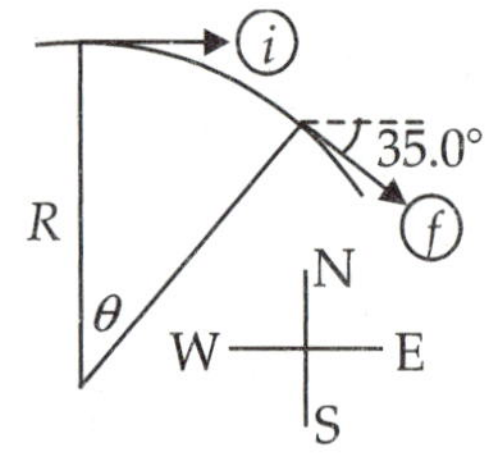

FIG. P1.30

$$R = \frac{360°}{2\pi}\frac{840\text{ m}}{35°} = \frac{840\text{ m}}{0.611} = \boxed{1.38\times10^3\text{ m}}.$$

We could equally well say that the measure of the angle in radians is

$$\theta = 35° = 35°\left(\frac{2\pi\,\text{radians}}{360°}\right) = 0.611\text{ rad} = \frac{840\text{ m}}{R}.$$

$R = 1.38\text{ km}$.

***P1.31** We use substitution, as the most generally applicable method for solving simultaneous equations. We substitute $p = 3q$ into each of the other two equations to eliminate p:

$$\begin{cases} 3qr = qs \\ \frac{1}{2}3qr^2 + \frac{1}{2}qs^2 = \frac{1}{2}qt^2 \end{cases}$$

These simplify to $\begin{cases} 3r = s \\ 3r^2 + s^2 = t^2 \end{cases}$. We substitute to eliminate s: $\begin{array}{l} 3r^2 + (3r)^2 = t^2 \\ 12r^2 = t^2 \end{array}$. We solve for the combination $\dfrac{t}{r}$: $\dfrac{t^2}{r^2} = 12$.

$$\frac{t}{r} = \boxed{\text{either } 3.46 \text{ or } -3.46}$$

Section 1.6 Coordinate Systems

P1.32 $x = r\cos\theta = (5.50\text{ m})\cos 240° = (5.50\text{ m})(-0.5) = \boxed{-2.75\text{ m}}$

$y = r\sin\theta = (5.50\text{ m})\sin 240° = (5.50\text{ m})(-0.866) = \boxed{-4.76\text{ m}}$

P1.33 The x distance out to the fly is 2.00 m and the y distance up to the fly is 1.00 m.

(a) We can use the Pythagorean theorem to find the distance from the origin to the fly.

$$\text{distance} = \sqrt{x^2+y^2} = \sqrt{(2.00\text{ m})^2+(1.00\text{ m})^2} = \sqrt{5.00\text{ m}^2} = \boxed{2.24\text{ m}}$$

(b) $\theta = \tan^{-1}\left(\frac{1}{2}\right) = 26.6°$; $\vec{\mathbf{r}} = \boxed{2.24\text{ m}, 26.6°}$

P1.34 (a) $d = \sqrt{(x_2-x_1)^2+(y_2-y_1)^2} = \sqrt{(2.00-[-3.00])^2+(-4.00-3.00)^2}$

$d = \sqrt{25.0+49.0} = \boxed{8.60\text{ m}}$

(b) $r_1 = \sqrt{(2.00)^2+(-4.00)^2} = \sqrt{20.0} = \boxed{4.47\text{ m}}$

$\theta_1 = \tan^{-1}\left(-\frac{4.00}{2.00}\right) = \boxed{-63.4°}$

$r_2 = \sqrt{(-3.00)^2+(3.00)^2} = \sqrt{18.0} = \boxed{4.24\text{ m}}$

$\theta_2 = \boxed{135°}$ measured from the $+x$ axis.

P1.35 We have $r = \sqrt{x^2+y^2}$ and $\theta = \tan^{-1}\left(\frac{y}{x}\right)$.

(a) The radius for this new point is

$$\sqrt{(-x)^2+y^2} = \sqrt{x^2+y^2} = \boxed{r}$$

and its angle is

$$\tan^{-1}\left(\frac{y}{-x}\right) = \boxed{180°-\theta}.$$

(b) $\sqrt{(-2x)^2+(-2y)^2} = \boxed{2r}$. This point is in the third quadrant if (x, y) is in the first quadrant or in the fourth quadrant if (x, y) is in the second quadrant. It is at an angle of $\boxed{180°+\theta}$.

(c) $\sqrt{(3x)^2+(-3y)^2} = \boxed{3r}$. This point is in the fourth quadrant if (x, y) is in the first quadrant or in the third quadrant if (x, y) is in the second quadrant. It is at an angle of $\boxed{-\theta}$.

Section 1.7 Vectors and Scalars

Section 1.8 Some Properties of Vectors

P1.36 $-\vec{\mathbf{R}} = \boxed{310 \text{ km at } 57° \text{ S of W}}$

(Scale: 1 unit = 20 km)

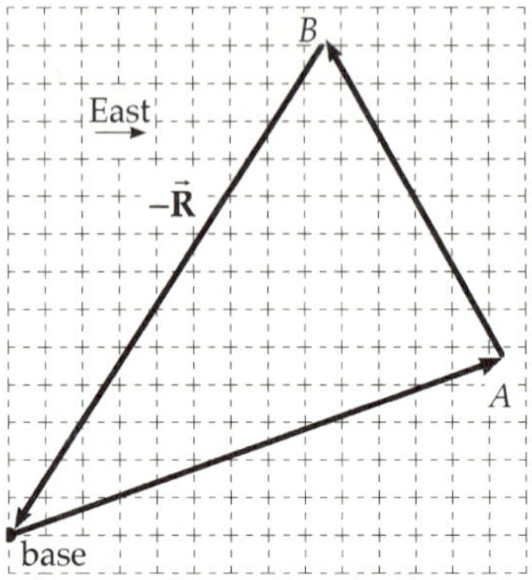

FIG. P1.36

P1.37 (a) $\left|\vec{\mathbf{d}}\right| = \left|-10.0\hat{\mathbf{i}}\right| = \boxed{10.0 \text{ m}}$ since the displacement is in a straight line from point *A* to point *B*.

(b) The actual distance skated is not equal to the straight-line displacement. The distance follows the curved path of the semi-circle (*ACB*).

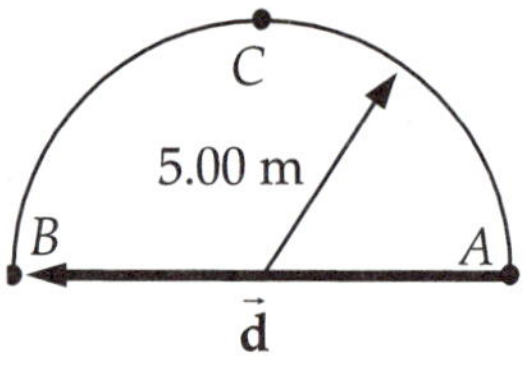

FIG. P1.37

$$s = \frac{1}{2}(2\pi r) = 5\pi = \boxed{15.7 \text{ m}}$$

(c) If the circle is complete, $\vec{\mathbf{d}}$ begins and ends at point *A*. Hence, $\left|\vec{\mathbf{d}}\right| = \boxed{0}$.

P1.38 To find these vector expressions graphically, we draw each set of vectors. Measurements of the results are taken using a ruler and protractor. (Scale: 1 unit = 0.5 m)

(a) $\vec{\mathbf{A}} + \vec{\mathbf{B}} = \boxed{5.2 \text{ m at } 60°}$

(b) $\vec{\mathbf{A}} - \vec{\mathbf{B}} = \boxed{3.0 \text{ m at } 330°}$

(c) $\vec{\mathbf{B}} - \vec{\mathbf{A}} = \boxed{3.0 \text{ m at } 150°}$

(d) $\vec{\mathbf{A}} - 2\vec{\mathbf{B}} = \boxed{5.2 \text{ m at } 300°}$.

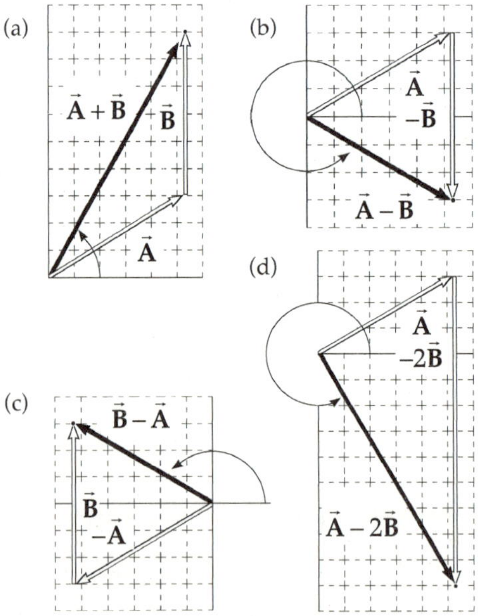

FIG. P1.38

P1.39 The scale drawing for the graphical solution should be similar to the figure to the right. The magnitude and direction of the final displacement from the starting point are obtained by measuring d and θ on the drawing and applying the scale factor used in making the drawing. The results should be

$$\boxed{d = 420 \text{ ft and } \theta = -3°}$$

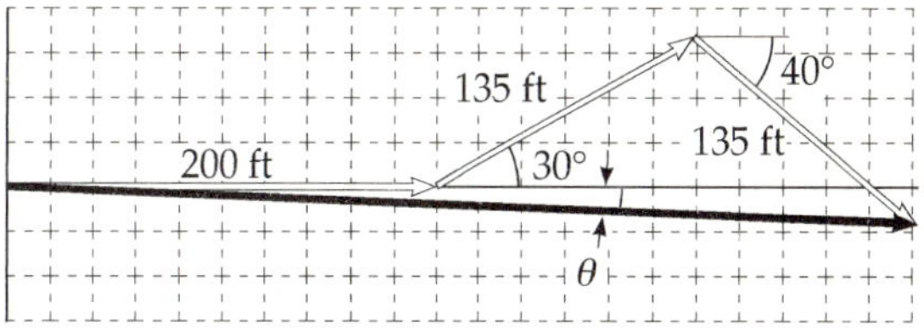

(Scale: 1 unit = 20 ft)

FIG. P1.39

Section 1.9 Components of a Vector and Unit Vectors

P1.40 Coordinates of the super-hero are:

$$x = (100 \text{ m})\cos(-30.0°) = \boxed{86.6 \text{ m}}$$

$$y = (100 \text{ m})\sin(-30.0°) = \boxed{-50.0 \text{ m}}$$

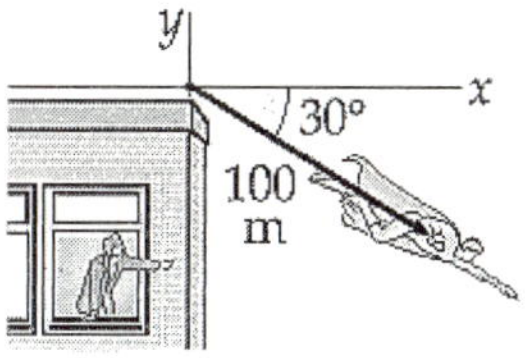

FIG. P1.40

P1.41 $A_x = -25.0$

$A_y = 40.0$

$$A = \sqrt{A_x^2 + A_y^2} = \sqrt{(-25.0)^2 + (40.0)^2} = \boxed{47.2 \text{ units}}$$

We observe that $\tan\phi = \dfrac{|A_y|}{|A_x|}$. So

$$\phi = \tan^{-1}\left(\frac{A_y}{|A_x|}\right) = \tan\frac{40.0}{25.0} = \tan^{-1}(1.60) = 58.0°.$$

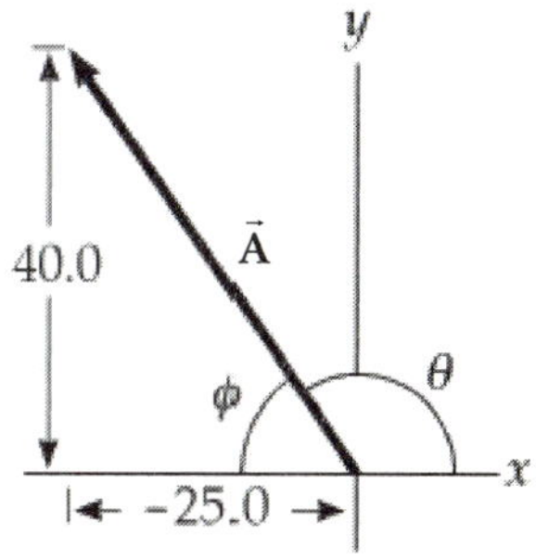

FIG. P1.41

The diagram shows that the angle from the $+x$ axis can be found by subtracting from 180°:

$$\theta = 180° - 58° = \boxed{122°}.$$

P1.42 (a) See figure to the right.

(b) $\vec{\mathbf{C}} = \vec{\mathbf{A}} + \vec{\mathbf{B}} = 2.00\hat{\mathbf{i}} + 6.00\hat{\mathbf{j}} + 3.00\hat{\mathbf{i}} - 2.00\hat{\mathbf{j}} = \boxed{5.00\hat{\mathbf{i}} + 4.00\hat{\mathbf{j}}}$

$\vec{\mathbf{C}} = \sqrt{25.0 + 16.0}$ at $\tan^{-1}\left(\frac{4}{5}\right) = \boxed{6.40 \text{ at } 38.7°}$

$\vec{\mathbf{D}} = \vec{\mathbf{A}} - \vec{\mathbf{B}} = 2.00\hat{\mathbf{i}} + 6.00\hat{\mathbf{j}} - 3.00\hat{\mathbf{i}} + 2.00\hat{\mathbf{j}} = \boxed{-1.00\hat{\mathbf{i}} + 8.00\hat{\mathbf{j}}}$

$\vec{\mathbf{D}} = \sqrt{(-1.00)^2 + (8.00)^2}$ at $\tan^{-1}\left(\frac{8.00}{-1.00}\right)$

$\vec{\mathbf{D}} = 8.06$ at $(180° - 82.9°) = \boxed{8.06 \text{ at } 97.2°}$

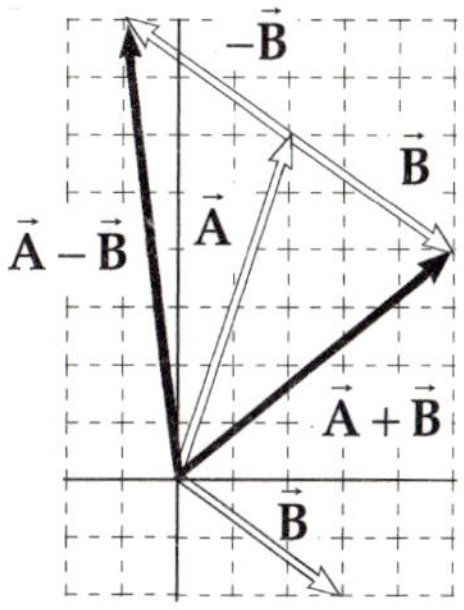

FIG. P1.42

P1.43 We have $\vec{\mathbf{B}} = \vec{\mathbf{R}} - \vec{\mathbf{A}}$:

$$A_x = 150\cos 120° = -75.0 \text{ cm}$$
$$A_y = 150\sin 120° = 130 \text{ cm}$$
$$R_x = 140\cos 35.0° = 115 \text{ cm}$$
$$R_y = 140\sin 35.0° = 80.3 \text{ cm}$$

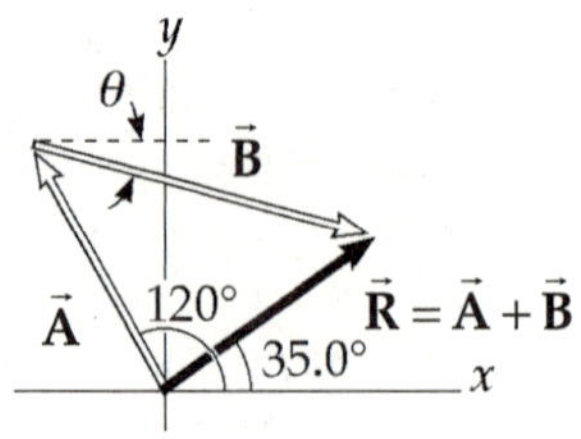

FIG. P1.43

Therefore,

$$\vec{\mathbf{B}} = [115-(-75)]\hat{\mathbf{i}} + [80.3-130]\hat{\mathbf{j}} = (190\hat{\mathbf{i}} - 49.7\hat{\mathbf{j}}) \text{ cm}$$
$$\left|\vec{\mathbf{B}}\right| = \sqrt{190^2 + 49.7^2} = \boxed{196 \text{ cm}}$$
$$\theta = \tan^{-1}\left(-\frac{49.7}{190}\right) = \boxed{-14.7°}.$$

P1.44 $\vec{\mathbf{A}} = -8.70\hat{\mathbf{i}} + 15.0\hat{\mathbf{j}}$ and $\vec{\mathbf{B}} = 13.2\hat{\mathbf{i}} - 6.60\hat{\mathbf{j}}$

$\vec{\mathbf{A}} - \vec{\mathbf{B}} + 3\vec{\mathbf{C}} = 0$:

$$3\vec{\mathbf{C}} = \vec{\mathbf{B}} - \vec{\mathbf{A}} = 21.9\hat{\mathbf{i}} - 21.6\hat{\mathbf{j}}$$
$$\vec{\mathbf{C}} = 7.30\hat{\mathbf{i}} - 7.20\hat{\mathbf{j}}$$

or

$$C_x = \boxed{7.30 \text{ cm}};\ C_y = \boxed{-7.20 \text{ cm}}$$

P1.45 (a) $\left(\vec{\mathbf{A}} + \vec{\mathbf{B}}\right) = \left(3\hat{\mathbf{i}} - 2\hat{\mathbf{j}}\right) + \left(-\hat{\mathbf{i}} - 4\hat{\mathbf{j}}\right) = \boxed{2\hat{\mathbf{i}} - 6\hat{\mathbf{j}}}$

(b) $\left(\vec{\mathbf{A}} - \vec{\mathbf{B}}\right) = \left(3\hat{\mathbf{i}} - 2\hat{\mathbf{j}}\right) - \left(-\hat{\mathbf{i}} - 4\hat{\mathbf{j}}\right) = \boxed{4\hat{\mathbf{i}} + 2\hat{\mathbf{j}}}$

(c) $\left|\vec{\mathbf{A}} + \vec{\mathbf{B}}\right| = \sqrt{2^2 + 6^2} = \boxed{6.32}$

(d) $\left|\vec{\mathbf{A}} - \vec{\mathbf{B}}\right| = \sqrt{4^2 + 2^2} = \boxed{4.47}$

(e) $\theta_{\left|\vec{\mathbf{A}}+\vec{\mathbf{B}}\right|} = \tan^{-1}\left(-\frac{6}{2}\right) = -71.6° = \boxed{288°}$

$\theta_{\left|\vec{\mathbf{A}}-\vec{\mathbf{B}}\right|} = \tan^{-1}\left(\frac{2}{4}\right) = \boxed{26.6°}$

P1.46 (a) $\vec{\mathbf{D}} = \vec{\mathbf{A}} + \vec{\mathbf{B}} + \vec{\mathbf{C}} = 2\hat{\mathbf{i}} + 4\hat{\mathbf{j}}$

$\left|\vec{\mathbf{D}}\right| = \sqrt{2^2 + 4^2} = \boxed{4.47 \text{ m at } \theta = 63.4°}$

(b) $\vec{\mathbf{E}} = -\vec{\mathbf{A}} - \vec{\mathbf{B}} + \vec{\mathbf{C}} = -6\hat{\mathbf{i}} + 6\hat{\mathbf{j}}$

$\left|\vec{\mathbf{E}}\right| = \sqrt{6^2 + 6^2} = \boxed{8.49 \text{ m at } \theta = 135°}$

P1.47 $\vec{\mathbf{d}}_1 = 100\hat{\mathbf{i}}$

$\vec{\mathbf{d}}_2 = -300\hat{\mathbf{j}}$

$\vec{\mathbf{d}}_3 = -150\cos(30.0°)\hat{\mathbf{i}} - 150\sin(30.0°)\hat{\mathbf{j}} = -130\hat{\mathbf{i}} - 75.0\hat{\mathbf{j}}$

$\vec{\mathbf{d}}_4 = -200\cos(60.0°)\hat{\mathbf{i}} + 200\sin(60.0°)\hat{\mathbf{j}} = -100\hat{\mathbf{i}} + 173\hat{\mathbf{j}}$

$\vec{\mathbf{R}} = \vec{\mathbf{d}}_1 + \vec{\mathbf{d}}_2 + \vec{\mathbf{d}}_3 + \vec{\mathbf{d}}_4 = \boxed{\left(-130\hat{\mathbf{i}} - 202\hat{\mathbf{j}}\right)\text{ m}}$

$\left|\vec{\mathbf{R}}\right| = \sqrt{(-130)^2 + (-202)^2} = \boxed{240\text{ m}}$

$\phi = \tan^{-1}\left(\dfrac{202}{130}\right) = 57.2°$

$\theta = 180 + \phi = \boxed{237°}$

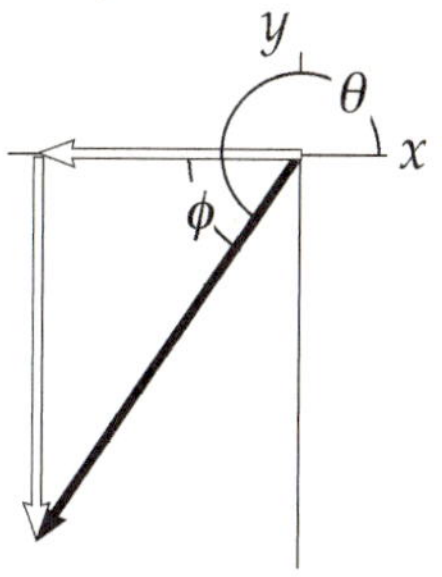

FIG. P1.47

P1.48 $\vec{\mathbf{A}} = 3.00\text{ m},\ \theta_A = 30.0°$ $\vec{\mathbf{B}} = 3.00\text{ m},\ \theta_B = 90.0°$

$A_x = A\cos\theta_A = 3.00\cos 30.0° = 2.60\text{ m}$ $A_y = A\sin\theta_A = 3.00\sin 30.0° = 1.50\text{ m}$

$\vec{\mathbf{A}} = A_x\hat{\mathbf{i}} + A_y\hat{\mathbf{j}} = \left(2.60\hat{\mathbf{i}} + 1.50\hat{\mathbf{j}}\right)\text{ m}$

$B_x = 0,\ B_y = 3.00\text{ m}$ so $\vec{\mathbf{B}} = 3.00\hat{\mathbf{j}}\text{ m}$

$\vec{\mathbf{A}} + \vec{\mathbf{B}} = \left(2.60\hat{\mathbf{i}} + 1.50\hat{\mathbf{j}}\right) + 3.00\hat{\mathbf{j}} = \boxed{\left(2.60\hat{\mathbf{i}} + 4.50\hat{\mathbf{j}}\right)\text{ m}}$

P1.49 Let the positive x-direction be eastward, the positive y-direction be vertically upward, and the positive z-direction be southward. The total displacement is then

$$\vec{\mathbf{d}} = \left(4.80\hat{\mathbf{i}} + 4.80\hat{\mathbf{j}}\right)\text{ cm} + \left(3.70\hat{\mathbf{j}} - 3.70\hat{\mathbf{k}}\right)\text{ cm} = \left(4.80\hat{\mathbf{i}} + 8.50\hat{\mathbf{j}} - 3.70\hat{\mathbf{k}}\right)\text{ cm}.$$

(a) The magnitude is $d = \sqrt{(4.80)^2 + (8.50)^2 + (-3.70)^2}\text{ cm} = \boxed{10.4\text{ cm}}$.

(b) Its angle with the y-axis follows from $\cos\theta = \dfrac{8.50}{10.4}$, giving $\boxed{\theta = 35.5°}$.

P1.50 $\vec{\mathbf{B}} = B_x\hat{\mathbf{i}} + B_y\hat{\mathbf{j}} + B_z\hat{\mathbf{k}} = 4.00\hat{\mathbf{i}} + 6.00\hat{\mathbf{j}} + 3.00\hat{\mathbf{k}}$

$\left|\vec{\mathbf{B}}\right| = \sqrt{4.00^2 + 6.00^2 + 3.00^2} = \boxed{7.81}$

$\theta_x = \cos^{-1}\left(\dfrac{4.00}{7.81}\right) = \boxed{59.2°}$ is the angle with the x axis

$\theta_y = \cos^{-1}\left(\dfrac{6.00}{7.81}\right) = \boxed{39.8°}$

$\theta_z = \cos^{-1}\left(\dfrac{3.00}{7.81}\right) = \boxed{67.4°}$

P1.51 (a) $\vec{\mathbf{A}} = \boxed{8.00\hat{\mathbf{i}} + 12.0\hat{\mathbf{j}} - 4.00\hat{\mathbf{k}}}$

(b) $\vec{\mathbf{B}} = \frac{\vec{\mathbf{A}}}{4} = \boxed{2.00\hat{\mathbf{i}} + 3.00\hat{\mathbf{j}} - 1.00\hat{\mathbf{k}}}$

(c) $\vec{\mathbf{C}} = -3\vec{\mathbf{A}} = \boxed{-24.0\hat{\mathbf{i}} - 36.0\hat{\mathbf{j}} + 12.0\hat{\mathbf{k}}}$

P1.52 (a) $\vec{\mathbf{E}} = (17.0 \text{ cm})\cos 27.0°\,\hat{\mathbf{i}} + (17.0 \text{ cm})\sin 27.0°\,\hat{\mathbf{j}}$

$\vec{\mathbf{E}} = \boxed{\left(15.1\hat{\mathbf{i}} + 7.72\hat{\mathbf{j}}\right) \text{ cm}}$

(b) $\vec{\mathbf{F}} = -(17.0 \text{ cm})\sin 27.0°\,\hat{\mathbf{i}} + (17.0 \text{ cm})\cos 27.0°\,\hat{\mathbf{j}}$

$\vec{\mathbf{F}} = \boxed{\left(-7.72\hat{\mathbf{i}} + 15.1\hat{\mathbf{j}}\right) \text{ cm}}$

(c) $\vec{\mathbf{G}} = -(17.0 \text{ cm})\sin 27.0°\,\hat{\mathbf{i}} - (17.0 \text{ cm})\cos 27.0°\,\hat{\mathbf{j}}$

$\vec{\mathbf{G}} = \boxed{\left(-7.72\hat{\mathbf{i}} - 15.1\hat{\mathbf{j}}\right) \text{ cm}}$

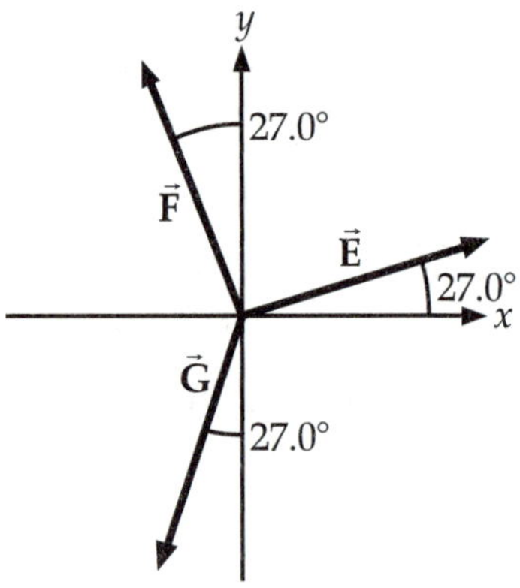

FIG. P1.52

P1.53 (a) $R_x = 40.0\cos 45.0° + 30.0\cos 45.0° = 49.5$
$R_y = 40.0\sin 45.0° - 30.0\sin 45.0° + 20.0 = 27.1$

$\vec{\mathbf{R}} = \boxed{49.5\hat{\mathbf{i}} + 27.1\hat{\mathbf{j}}}$

(b) $\left|\vec{\mathbf{R}}\right| = \sqrt{(49.5)^2 + (27.1)^2} = \boxed{56.4}$

$\theta = \tan^{-1}\left(\frac{27.1}{49.5}\right) = \boxed{28.7°}$

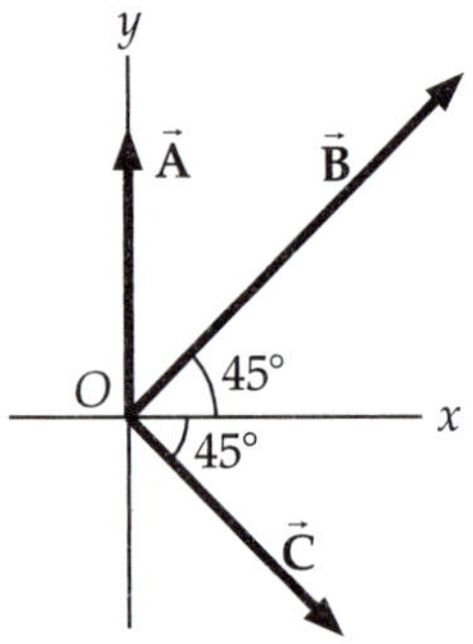

FIG. P1.53

P1.54 Taking components along $\hat{\mathbf{i}}$ and $\hat{\mathbf{j}}$, we get two equations: $6.00a - 8.00b + 26.0 = 0$ and $-8.00a + 3.00b + 19.0 = 0$. Solving simultaneously, $a = 1.33b - 4.33$; $-8(1.33b - 4.33) + 3b + 19 = 0$

$$\boxed{a = 5.00,\ b = 7.00}.$$

Therefore, $5.00\vec{\mathbf{A}} + 7.00\vec{\mathbf{B}} + \vec{\mathbf{C}} = 0$.

Section 1.10 Modeling, Alternative Representations, and Problem-Solving Strategy

P1.55 $\tan 35.0° = \dfrac{x}{100 \text{ m}}$

$$x = (100 \text{ m})\tan 35.0° = \boxed{70.0 \text{ m}}$$

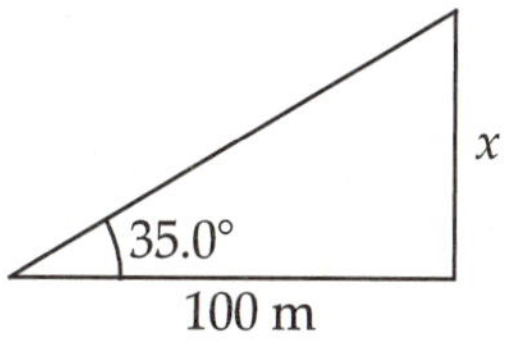

FIG. P1.55

***P1.56** We note that $-\hat{\mathbf{i}} = \text{west}$ and $-\hat{\mathbf{j}} = \text{south}$. The given mathematical representation of the trip can be written as $6.3b$ west $+ 4b$ at $40°$ south of west $+ 3b$ at $50°$ south of east $+ 5b$ south.

(a)

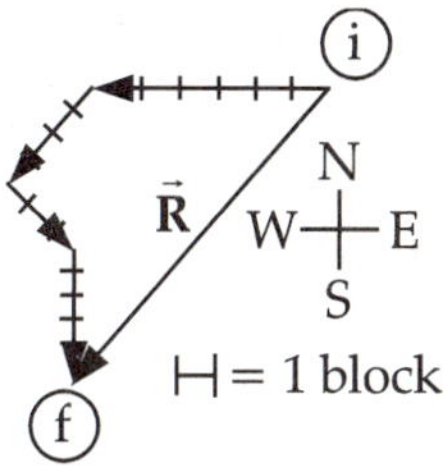

(b) The total odometer distance is the sum of the magnitudes of the four displacements: $6.3b + 4b + 3b + 5b = \boxed{18.3b}$.

(c)
$$\begin{aligned}\vec{\mathbf{R}} &= (-6.3 - 3.06 + 1.93)b\,\hat{\mathbf{i}} + (-2.57 - 2.30 - 5)b\,\hat{\mathbf{j}} \\ &= -7.44b\,\hat{\mathbf{i}} - 9.87b\,\hat{\mathbf{j}} \\ &= \sqrt{(7.44b)^2 + (9.87b)^2} \text{ at } \tan^{-1}\frac{9.87}{7.44} \text{ south of west} \\ &= 12.4b \text{ at } 53.0° \text{ south of west} \\ &= \boxed{12.4b \text{ at } 233° \text{ counterclockwise from east}}\end{aligned}$$

P1.57 From the figure, we may see that the spacing between diagonal planes is half the distance between diagonally adjacent atoms on a flat plane. This diagonal distance may be obtained from the Pythagorean theorem, $L_{\text{diag}} = \sqrt{L^2 + L^2}$. Thus, since the atoms are separated by a distance $L = 0.200$ nm, the diagonal planes are separated by $\frac{1}{2}\sqrt{L^2 + L^2} = \boxed{0.141 \text{ nm}}$.

Additional Problems

P1.58 It is desired to find the distance x such that

$$\frac{x}{100 \text{ m}} = \frac{1\,000 \text{ m}}{x}$$

(i.e., such that x is the same multiple of 100 m as the multiple that 1 000 m is of x). Thus, it is seen that

$$x^2 = (100 \text{ m})(1\,000 \text{ m}) = 1.00 \times 10^5 \text{ m}^2$$

and therefore

$$x = \sqrt{1.00 \times 10^5 \text{ m}^2} = \boxed{316 \text{ m}}.$$

P1.59 $$A_{\text{total}} = (N)(A_{\text{drop}}) = \left(\frac{V_{\text{total}}}{V_{\text{drop}}}\right)(A_{\text{drop}}) = \left(\frac{V_{\text{total}}}{\frac{4\pi r^3}{3}}\right)(4\pi r^2)$$

$$A_{\text{total}} = \left(\frac{3V_{\text{total}}}{r}\right) = 3\left(\frac{30.0\times10^{-6}\ \text{m}^3}{2.00\times10^{-5}\ \text{m}}\right) = \boxed{4.50\ \text{m}^2}$$

P1.60 One month is

$$1\ \text{mo} = (30\ \text{day})(24\ \text{h/day})(3\,600\ \text{s/h}) = 2.592\times10^6\ \text{s}.$$

Applying units to the equation,

$$V = \left(1.50\ \text{Mft}^3/\text{mo}\right)t + \left(0.008\,00\ \text{Mft}^3/\text{mo}^2\right)t^2.$$

Since $1\ \text{Mft}^3 = 10^6\ \text{ft}^3$,

$$V = \left(1.50\times10^6\ \text{ft}^3/\text{mo}\right)t + \left(0.008\,00\times10^6\ \text{ft}^3/\text{mo}^2\right)t^2.$$

Converting months to seconds,

$$V = \frac{1.50\times10^6\ \text{ft}^3/\text{mo}}{2.592\times10^6\ \text{s/mo}}t + \frac{0.008\,00\times10^6\ \text{ft}^3/\text{mo}^2}{\left(2.592\times10^6\ \text{s/mo}\right)^2}t^2.$$

Thus, $\boxed{V\,[\text{ft}^3] = \left(0.579\ \text{ft}^3/\text{s}\right)t + \left(1.19\times10^{-9}\ \text{ft}^3/\text{s}^2\right)t^2}$.

P1.61 The actual number of seconds in a year is

$$(86\,400\ \text{s/day})(365.25\ \text{day/yr}) = 31\,557\,600\ \text{s/yr}.$$

The percent error in the approximation is

$$\frac{\left|\left(\pi\times10^7\ \text{s/yr}\right)-(31\,557\,600\ \text{s/yr})\right|}{31\,557\,600\ \text{s/yr}}\times100\% = \boxed{0.449\%}.$$

P1.62

α'(deg)	α(rad)	$\tan(\alpha)$	$\sin(\alpha)$	difference between α and $\tan\alpha$
15.0	0.262	0.268	0.259	2.30%
20.0	0.349	0.364	0.342	4.09%
30.0	0.524	0.577	0.500	9.32%
33.0	0.576	0.649	0.545	11.3%
31.0	0.541	0.601	0.515	9.95%
31.1	0.543	0.603	0.516	10.02%

Thus $\boxed{31.0°}$ is the largest angle for which $\frac{\tan\alpha-\alpha}{\tan\alpha} < 0.1$.

P1.63 (a) The speed of rise may be found from

$$v = \frac{(\text{Vol rate of flow})}{(\text{Area: } \frac{\pi D^2}{4})} = \frac{16.5 \text{ cm}^3/\text{s}}{\frac{\pi (6.30 \text{ cm})^2}{4}} = \boxed{0.529 \text{ cm/s}}.$$

(b) Likewise, at a 1.35 cm diameter,

$$v = \frac{16.5 \text{ cm}^3/\text{s}}{\frac{\pi (1.35 \text{ cm})^2}{4}} = \boxed{11.5 \text{ cm/s}}.$$

***P1.64** (a) 1 cubic meter of water has a mass

$$m = \rho V = \left(1.00 \times 10^{-3} \text{ kg/cm}^3\right)\left(1.00 \text{ m}^3\right)\left(10^2 \text{ cm/m}\right)^3 = \boxed{1\,000 \text{ kg}}.$$

(b) As a rough calculation, we treat each item as if it were 100% water.

cell: $$m = \rho V = \rho\left(\frac{4}{3}\pi R^3\right) = \rho\left(\frac{1}{6}\pi D^3\right) = \left(1\,000 \text{ kg/m}^3\right)\left(\frac{1}{6}\pi\right)\left(1.0 \times 10^{-6} \text{ m}\right)^3$$
$$= \boxed{5.24 \times 10^{-16} \text{ kg}}$$

kidney: $$m = \rho V = \rho\left(\frac{4}{3}\pi R^3\right) = \left(1.00 \times 10^{-3} \text{ kg/cm}^3\right)\left(\frac{4}{3}\pi\right)(4.0 \text{ cm})^3$$
$$= \boxed{0.268 \text{ kg}}$$

fly: $$m = \rho\left(\frac{\pi}{4}D^2 h\right) = \left(1 \times 10^{-3} \text{ kg/cm}^3\right)\left(\frac{\pi}{4}\right)(2.0 \text{ mm})^2(4.0 \text{ mm})\left(10^{-1} \text{ cm/mm}\right)^3$$
$$= \boxed{1.26 \times 10^{-5} \text{ kg}}$$

P1.65 The volume of the galaxy is

$$\pi r^2 t = \pi\left(10^{21} \text{ m}\right)^2\left(10^{19} \text{ m}\right) \sim 10^{61} \text{ m}^3.$$

If the distance between stars is 4×10^{16} m, then there is one star in a volume on the order of

$$\left(4 \times 10^{16} \text{ m}\right)^3 \sim 10^{50} \text{ m}^3.$$

The number of stars is about $\dfrac{10^{61} \text{ m}^3}{10^{50} \text{ m}^3/\text{star}} \sim \boxed{10^{11} \text{ stars}}$.

P1.66 Let θ represent the angle between the directions of $\vec{A}$ and $\vec{B}$. Since $\vec{A}$ and $\vec{B}$ have the same magnitudes, $\vec{A}$, $\vec{B}$, and $\vec{R} = \vec{A} + \vec{B}$ form an isosceles triangle in which the angles are $180° - \theta$, $\frac{\theta}{2}$, and $\frac{\theta}{2}$. The magnitude of $\vec{R}$ is then $R = 2A\cos\left(\frac{\theta}{2}\right)$. [*Hint*: apply the law of cosines to the isosceles triangle and use the fact that $B = A$.]

Again, $\vec{A}$, $-\vec{B}$, and $\vec{D} = \vec{A} - \vec{B}$ form an isosceles triangle with apex angle θ. Applying the law of cosines and the identity

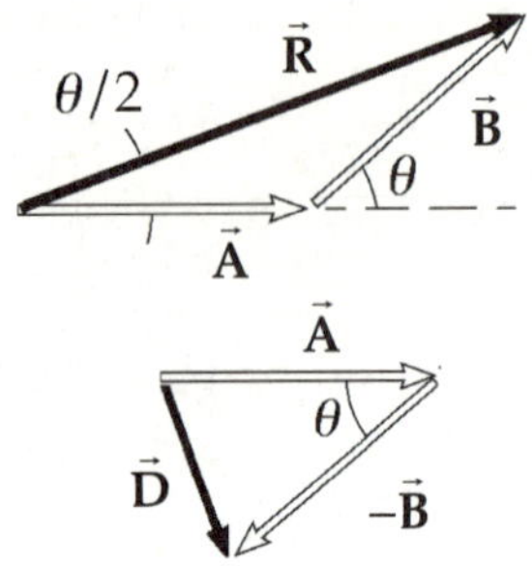

FIG. P1.66

$$(1 - \cos\theta) = 2\sin^2\left(\frac{\theta}{2}\right)$$

gives the magnitude of $\vec{D}$ as $D = 2A\sin\left(\frac{\theta}{2}\right)$. The problem requires that $R = nD$ or $\cos\left(\frac{\theta}{2}\right) = n\sin\left(\frac{\theta}{2}\right)$

giving $\boxed{\theta = 2\tan^{-1}\left(\frac{1}{n}\right)}$.

P1.67 (a)

$$\vec{F} = \vec{F}_1 + \vec{F}_2$$

$$\vec{F} = 120\cos(60.0°)\hat{i} + 120\sin(60.0°)\hat{j} - 80.0\cos(75.0°)\hat{i} + 80.0\sin(75.0°)\hat{j}$$

$$\vec{F} = 60.0\hat{i} + 104\hat{j} - 20.7\hat{i} + 77.3\hat{j} = \left(39.3\hat{i} + 181\hat{j}\right)\text{ N}$$

$$\left|\vec{F}\right| = \sqrt{39.3^2 + 181^2} = \boxed{185\text{ N}}$$

$$\theta = \tan^{-1}\left(\frac{181}{39.3}\right) = \boxed{77.8°}$$

(b)

$$\vec{F}_3 = -\vec{F} = \boxed{\left(-39.3\hat{i} - 181\hat{j}\right)\text{ N}}$$

P1.68 The position vector from the ground under the controller of the first airplane is

$$\vec{r}_1 = (19.2\text{ km})(\cos 25°)\hat{i} + (19.2\text{ km})(\sin 25°)\hat{j} + (0.8\text{ km})\hat{k}$$
$$= \left(17.4\hat{i} + 8.11\hat{j} + 0.8\hat{k}\right)\text{ km}.$$

The second is at

$$\vec{r}_2 = (17.6\text{ km})(\cos 20°)\hat{i} + (17.6\text{ km})(\sin 20°)\hat{j} + (1.1\text{ km})\hat{k}$$
$$= \left(16.5\hat{i} + 6.02\hat{j} + 1.1\hat{k}\right)\text{ km}.$$

Now the displacement from the first plane to the second is

$$\vec{r}_2 - \vec{r}_1 = \left(-0.863\hat{i} - 2.09\hat{j} + 0.3\hat{k}\right)\text{ km}$$

with magnitude

$$\sqrt{(0.863)^2 + (2.09)^2 + (0.3)^2} = \boxed{2.29\text{ km}}.$$

P1.69 (a) You start at point A: $\vec{\mathbf{r}}_1 = \vec{\mathbf{r}}_A = \left(30.0\hat{\mathbf{i}} - 20.0\hat{\mathbf{j}}\right)$ m.

The displacement to B is

$$\vec{\mathbf{r}}_B - \vec{\mathbf{r}}_A = 60.0\hat{\mathbf{i}} + 80.0\hat{\mathbf{j}} - 30.0\hat{\mathbf{i}} + 20.0\hat{\mathbf{j}} = 30.0\hat{\mathbf{i}} + 100\hat{\mathbf{j}}.$$

You cover half of this, $\left(15.0\hat{\mathbf{i}} + 50.0\hat{\mathbf{j}}\right)$ to move to

$\vec{\mathbf{r}}_2 = 30.0\hat{\mathbf{i}} - 20.0\hat{\mathbf{j}} + 15.0\hat{\mathbf{i}} + 50.0\hat{\mathbf{j}} = 45.0\hat{\mathbf{i}} + 30.0\hat{\mathbf{j}}$.

Now the displacement from your current position to C is

$$\vec{\mathbf{r}}_C - \vec{\mathbf{r}}_2 = -10.0\hat{\mathbf{i}} - 10.0\hat{\mathbf{j}} - 45.0\hat{\mathbf{i}} - 30.0\hat{\mathbf{j}} = -55.0\hat{\mathbf{i}} - 40.0\hat{\mathbf{j}}.$$

You cover one-third, moving to

$$\vec{\mathbf{r}}_3 = \vec{\mathbf{r}}_2 + \Delta\vec{\mathbf{r}}_{23} = 45.0\hat{\mathbf{i}} + 30.0\hat{\mathbf{j}} + \frac{1}{3}\left(-55.0\hat{\mathbf{i}} - 40.0\hat{\mathbf{j}}\right) = 26.7\hat{\mathbf{i}} + 16.7\hat{\mathbf{j}}.$$

The displacement from where you are to D is

$$\vec{\mathbf{r}}_D - \vec{\mathbf{r}}_3 = 40.0\hat{\mathbf{i}} - 30.0\hat{\mathbf{j}} - 26.7\hat{\mathbf{i}} - 16.7\hat{\mathbf{j}} = 13.3\hat{\mathbf{i}} - 46.7\hat{\mathbf{j}}.$$

You traverse one-quarter of it, moving to

$$\vec{\mathbf{r}}_4 = \vec{\mathbf{r}}_3 + \frac{1}{4}\left(\vec{\mathbf{r}}_D - \vec{\mathbf{r}}_3\right) = 26.7\hat{\mathbf{i}} + 16.7\hat{\mathbf{j}} + \frac{1}{4}\left(13.3\hat{\mathbf{i}} - 46.7\hat{\mathbf{j}}\right) = 30.0\hat{\mathbf{i}} + 5.00\hat{\mathbf{j}}.$$

The displacement from your new location to E is

$$\vec{\mathbf{r}}_E - \vec{\mathbf{r}}_4 = -70.0\hat{\mathbf{i}} + 60.0\hat{\mathbf{j}} - 30.0\hat{\mathbf{i}} - 5.00\hat{\mathbf{j}} = -100\hat{\mathbf{i}} + 55.0\hat{\mathbf{j}}$$

of which you cover one-fifth the distance, $-20.0\hat{\mathbf{i}} + 11.0\hat{\mathbf{j}}$, moving to

$$\vec{\mathbf{r}}_4 + \Delta\vec{\mathbf{r}}_{45} = 30.0\hat{\mathbf{i}} + 5.00\hat{\mathbf{j}} - 20.0\hat{\mathbf{i}} + 11.0\hat{\mathbf{j}} = 10.0\hat{\mathbf{i}} + 16.0\hat{\mathbf{j}}.$$

The treasure is at $\boxed{(10.0\text{ m}, 16.0\text{ m})}$.

(b) Following the directions brings you to the average position of the trees. The steps we took numerically in part (a) bring you to

$$\vec{\mathbf{r}}_A + \frac{1}{2}\left(\vec{\mathbf{r}}_B - \vec{\mathbf{r}}_A\right) = \left(\frac{\vec{\mathbf{r}}_A + \vec{\mathbf{r}}_B}{2}\right)$$

then to $\dfrac{\left(\vec{\mathbf{r}}_A + \vec{\mathbf{r}}_B\right)}{2} + \dfrac{\vec{\mathbf{r}}_C - \frac{\left(\vec{\mathbf{r}}_A + \vec{\mathbf{r}}_B\right)}{2}}{3} = \dfrac{\vec{\mathbf{r}}_A + \vec{\mathbf{r}}_B + \vec{\mathbf{r}}_C}{3}$

then to $\dfrac{\left(\vec{\mathbf{r}}_A + \vec{\mathbf{r}}_B + \vec{\mathbf{r}}_C\right)}{3} + \dfrac{\vec{\mathbf{r}}_D - \frac{\left(\vec{\mathbf{r}}_A + \vec{\mathbf{r}}_B + \vec{\mathbf{r}}_C\right)}{3}}{4} = \dfrac{\vec{\mathbf{r}}_A + \vec{\mathbf{r}}_B + \vec{\mathbf{r}}_C + \vec{\mathbf{r}}_D}{4}$

and last to $\dfrac{\left(\vec{\mathbf{r}}_A + \vec{\mathbf{r}}_B + \vec{\mathbf{r}}_C + \vec{\mathbf{r}}_D\right)}{4} + \dfrac{\vec{\mathbf{r}}_E - \frac{\left(\vec{\mathbf{r}}_A + \vec{\mathbf{r}}_B + \vec{\mathbf{r}}_C + \vec{\mathbf{r}}_D\right)}{4}}{5} = \dfrac{\vec{\mathbf{r}}_A + \vec{\mathbf{r}}_B + \vec{\mathbf{r}}_C + \vec{\mathbf{r}}_D + \vec{\mathbf{r}}_E}{5}$.

This center of mass of the tree distribution is the same location whatever order we take the trees in.

P1.70 (a) Let T represent the force exerted by each child. The x-component of the resultant force is

$$T\cos 0 + T\cos 120° + T\cos 240° = T(1) + T(-0.5) + T(-0.5) = 0.$$

The y-component is

$$T\sin 0 + T\sin 120 + T\sin 240 = 0 + 0.866T - 0.866T = 0.$$

Thus,

$$\sum \vec{\mathbf{F}} = 0.$$

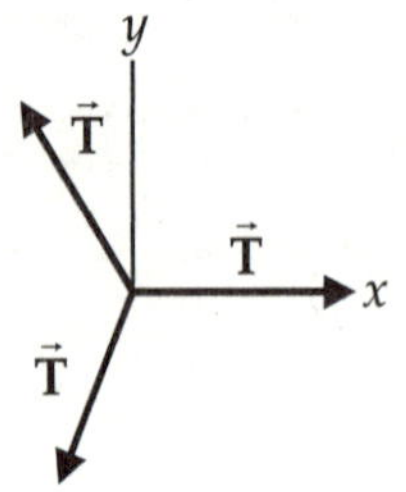

FIG. P1.70

(b) If the total force is not zero, it must point in some direction. When each child moves one space clockwise, the total must turn clockwise by that angle, $\frac{360°}{N}$. Since each child exerts the same force, the new situation is identical to the old and the net force on the tire must still point in the original direction. The contradiction indicates that we were wrong in supposing that the total force is not zero. The total force *must* be zero.

P1.71 (a) From the picture, $\vec{\mathbf{R}}_1 = a\hat{\mathbf{i}} + b\hat{\mathbf{j}}$ and $\left|\vec{\mathbf{R}}_1\right| = \sqrt{a^2 + b^2}$.

(b) $\vec{\mathbf{R}}_2 = a\hat{\mathbf{i}} + b\hat{\mathbf{j}} + c\hat{\mathbf{k}}$; its magnitude is

$$\sqrt{\left|\vec{\mathbf{R}}_1\right|^2 + c^2} = \sqrt{a^2 + b^2 + c^2}.$$

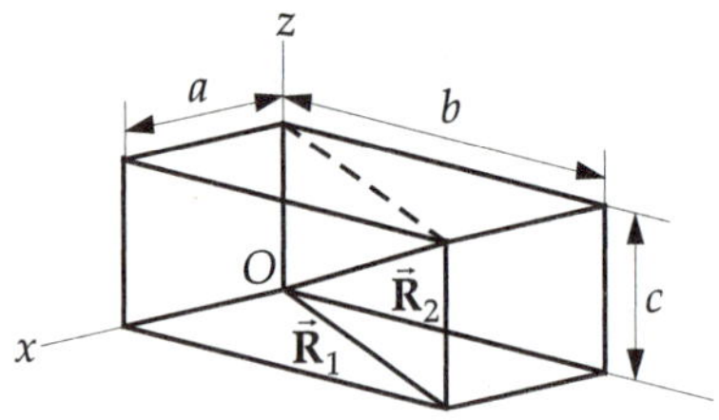

FIG. P1.71

P1.72 Since

$$\vec{\mathbf{A}} + \vec{\mathbf{B}} = 6.00\hat{\mathbf{j}},$$

we have

$$(A_x + B_x)\hat{\mathbf{i}} + (A_y + B_y)\hat{\mathbf{j}} = 0\hat{\mathbf{i}} + 6.00\hat{\mathbf{j}}$$

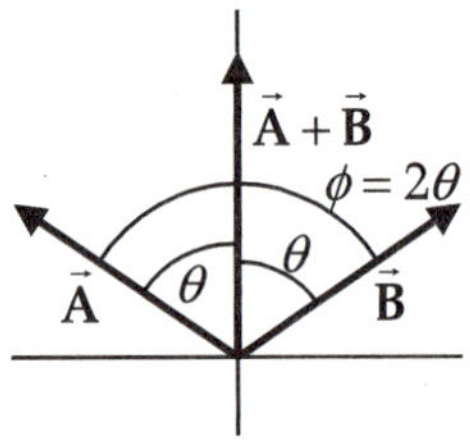

FIG. P1.72

giving

$$A_x + B_x = 0 \text{ or } A_x = -B_x \qquad [1]$$

and

$$A_y + B_y = 6.00. \qquad [2]$$

Since both vectors have a magnitude of 5.00, we also have

$$A_x^2 + A_y^2 = B_x^2 + B_y^2 = 5.00^2.$$

From $A_x = -B_x$, it is seen that

$$A_x^2 = B_x^2.$$

Therefore, $A_x^2 + A_y^2 = B_x^2 + B_y^2$ gives

$$A_y^2 = B_y^2.$$

continued on next page

Then, $A_y = B_y$ and Eq. [2] gives

$$A_y = B_y = 3.00.$$

Defining θ as the angle between either $\vec{\mathbf{A}}$ or $\vec{\mathbf{B}}$ and the y axis, it is seen that

$$\cos\theta = \frac{A_y}{A} = \frac{B_y}{B} = \frac{3.00}{5.00} = 0.600 \text{ and } \theta = 53.1°.$$

The angle between $\vec{\mathbf{A}}$ and $\vec{\mathbf{B}}$ is then $\boxed{\phi = 2\theta = 106°}$.

ANSWERS TO EVEN PROBLEMS

P1.2 23.0 kg

P1.4 7.69 cm

P1.6 (a) ii; (b) iii; (c) i

P1.8 $9.19\ \text{nm/s}$

P1.10 $4.05 \times 10^3\ \text{m}^2$

P1.12 (a) 220 yr; (b) 2.71×10^4 times

P1.14 1.19×10^{57} atoms

P1.16 $r_{\text{Al}} = r_{\text{Fe}} \sqrt[3]{\dfrac{\rho_{\text{Fe}}}{\rho_{\text{Al}}}}$

P1.18 $\sim 10^7$ rev or 10^8 rev

P1.20 $\sim 10^2$ kg; $\sim 10^3$ kg

P1.22 $\sim 10^{11}$ cans; $\sim 10^5$ tons

P1.24 (a) 797; (b) 1.1; (c) 17.66

P1.26 $(1.61 \pm 0.17) \times 10^3\ \text{kg/m}^3$

P1.28 see the solution

P1.30 1.38 km

P1.32 (–2.75, –4.76) m

P1.34 (a) 8.60 m;
(b) 4.47 m at –63.4°, 4.24 m at 135°

P1.36 310 km at 57° S of W

P1.38 (a) 5.2 m at 60°; (b) 3.0 m at 330°;
(c) 3.0 m at 150°; (d) 5.2 m at 300°

P1.40 86.6 m and –50.0 m

P1.42 (a) see the solution;
(b) $5.00\hat{\mathbf{i}} + 4.00\hat{\mathbf{j}}$, 6.40 at 38.7°,
$-1.00\hat{\mathbf{i}} + 8.00\hat{\mathbf{j}}$, 8.06 at 97.2°

P1.44 $C_x = 7.30$ cm, $C_y = -7.20$ cm

P1.46 (a) 4.47 m at 63.4°; (b) 8.49 m at 135°

P1.48 $\vec{\mathbf{A}} + \vec{\mathbf{B}} = \left(2.60\hat{\mathbf{i}} + 4.50\hat{\mathbf{j}}\right)$ m

P1.50 $\left|\vec{\mathbf{B}}\right| = 7.81$, $\theta_x = 59.2°$, $\theta_y = 39.8°$,
$\theta_z = 67.4°$

P1.52 (a) $\left(15.1\hat{\mathbf{i}} + 7.72\hat{\mathbf{j}}\right)$ cm;
(b) $\left(-7.72\hat{\mathbf{i}} + 15.1\hat{\mathbf{j}}\right)$ cm;
(c) $\left(-7.72\hat{\mathbf{i}} - 15.1\hat{\mathbf{j}}\right)$ cm

P1.54 $a = 5.00$, $b = 7.00$

P1.56 (a) see the solution ;(b) 18.3 b;
(c) 12.4 b at 233° counterclockwise from east

P1.58 316 m

P1.60 $0.579t\ \text{ft}^3/\text{s} + 1.19\times10^{-9}t^2\ \text{ft}^3/\text{s}^2$

P1.62 0.541 rad

P1.64 (a) 1 000 kg; (b) 5.24×10^{-16} kg, 0.268 kg, 1.26×10^{-5} kg

P1.66 $2\tan^{-1}\left(\frac{1}{n}\right)$

P1.68 2.29 km

P1.70 (a) zero; (b) zero

P1.72 106°

Chapter 2

Motion in One Dimension

CHAPTER OUTLINE

ANSWERS TO QUESTIONS

Q2.1 If I count 5.0 s between lightning and thunder, the sound has traveled $(331\ \text{m/s})(5.0\ \text{s}) = 1.7\ \text{km}$. The transit time for the light is smaller by

$$\frac{3.00\times10^{8}\ \text{m/s}}{331\ \text{m/s}} = 9.06\times10^{5}\ \text{times},$$

so it is negligible in comparison.

Q2.2 Yes. Yes, if the particle winds up in the $+x$ region at the end.

Q2.3 Zero.

Q2.4 Yes. Yes.

Q2.5 No. Consider a sprinter running a straight-line race. His average velocity would simply be the length of the race divided by the time it took for him to complete the race. If he stops along the way to tie his shoe, then his instantaneous velocity at that point would be zero.

Q2.6 We assume the object moves along a straight line. If its average velocity is zero, then the displacement must be zero over the time interval, according to Equation 2.2. The object might be stationary throughout the interval. If it is moving to the right at first, it must later move to the left to return to its starting point. Its velocity must be zero as it turns around. The graph of the motion shown to the right represents such motion, as the initial and final positions are the same. In an x vs. t graph, the instantaneous velocity at any time t is the slope of the curve at that point. At t_0 in the graph, the slope of the curve is zero, and thus the instantaneous velocity at that time is also zero.

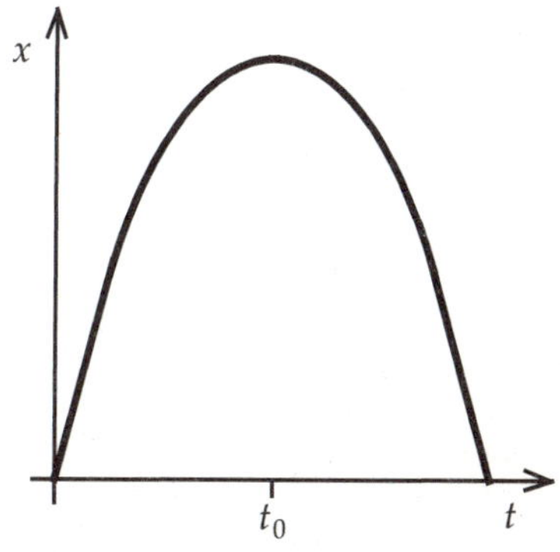

FIG. Q2.6

Q2.7 Yes. If the velocity of the particle is nonzero, the particle is in motion. If the acceleration is zero, the velocity of the particle is unchanging, or is a constant.

Q2.8 Yes. If you drop a doughnut from rest $(v = 0)$, then its acceleration is not zero. A common misconception is that immediately after the doughnut is released, both the velocity and acceleration are zero. If the acceleration were zero, then the velocity would not change, leaving the doughnut floating at rest in mid-air.

Q2.9 No: Car A might have greater acceleration than B, but they might both have zero acceleration, or otherwise equal accelerations; or the driver of B might have tramped hard on the gas pedal in the recent past.

Q2.10 Yes. Consider throwing a ball straight up. As the ball goes up, its velocity is upward $(v > 0)$, and its acceleration is directed down $(a < 0)$. A graph of v vs. t for this situation would look like the figure to the right. The acceleration is the slope of a v vs. t graph, and is always negative in this case, even when the velocity is positive.

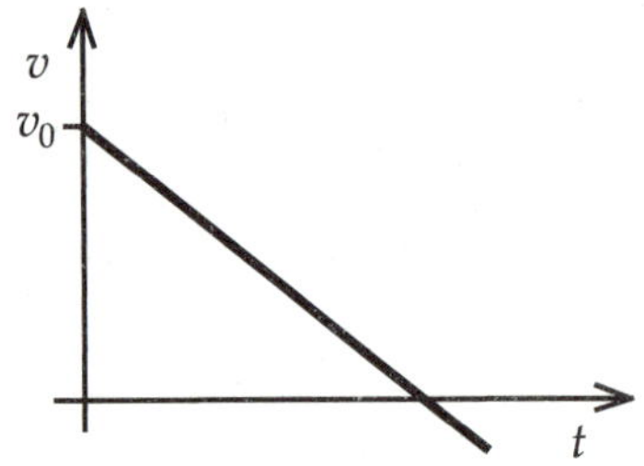

FIG. Q2.10

Q2.11 (a) Speeding up as it moves east (b) Braking East

(c) Cruising East (d) Braking West

(e) Speeding up as it moves west (f) Cruising West

(g) Stopped but starting to move East (h) Stopped but starting to move West

Q2.12 No. Constant acceleration only. Yes. Zero is a constant.

Q2.13 Once the objects leave the hand, both are in free fall, and both experience the same downward acceleration equal to the free-fall acceleration, $-g$.

Q2.14 With $h = \frac{1}{2}gt^2$,

(a) $0.5h = \frac{1}{2}g(0.707t)^2$. The time is later than $0.5t$.

(b) The distance fallen is $0.25h = \frac{1}{2}g(0.5t)^2$. The elevation is $0.75h$, greater than $0.5h$.

Q2.15 They are the same. After the first ball reaches its apex and falls back downward past the student, it will have a downward velocity equal to v_i. This velocity is the same as the velocity of the second ball, so after they fall through equal heights their impact speeds will also be the same.

Q2.16 Above. Your ball has zero initial speed and smaller average speed during the time of flight to the passing point.

SOLUTIONS TO PROBLEMS

Section 2.1 Average Velocity

P2.1 (a) $\bar{v} = \boxed{2.30 \text{ m/s}}$

(b) $\bar{v} = \dfrac{\Delta x}{\Delta t} = \dfrac{57.5 \text{ m} - 9.20 \text{ m}}{3.00 \text{ s}} = \boxed{16.1 \text{ m/s}}$

(c) $\bar{v} = \dfrac{\Delta x}{\Delta t} = \dfrac{57.5 \text{ m} - 0 \text{ m}}{5.00 \text{ s}} = \boxed{11.5 \text{ m/s}}$

P2.2 $x = 10t^2$: For $\begin{cases} t(\text{s}) & = & 2.0 & 2.1 & 3.0 \\ x(\text{m}) & = & 40 & 44.1 & 90 \end{cases}$

(a) $\bar{v} = \dfrac{\Delta x}{\Delta t} = \dfrac{50 \text{ m}}{1.0 \text{ s}} = \boxed{50.0 \text{ m/s}}$

(b) $\bar{v} = \dfrac{\Delta x}{\Delta t} = \dfrac{4.1 \text{ m}}{0.1 \text{ s}} = \boxed{41.0 \text{ m/s}}$

P2.3 (a) $\bar{v} = \dfrac{\Delta x}{\Delta t} = \dfrac{10 \text{ m}}{2 \text{ s}} = \boxed{5 \text{ m/s}}$

(b) $\bar{v} = \dfrac{5 \text{ m}}{4 \text{ s}} = \boxed{1.2 \text{ m/s}}$

(c) $\bar{v} = \dfrac{x_2 - x_1}{t_2 - t_1} = \dfrac{5 \text{ m} - 10 \text{ m}}{4 \text{ s} - 2 \text{ s}} = \boxed{-2.5 \text{ m/s}}$

(d) $\bar{v} = \dfrac{x_2 - x_1}{t_2 - t_1} = \dfrac{-5 \text{ m} - 5 \text{ m}}{7 \text{ s} - 4 \text{ s}} = \boxed{-3.3 \text{ m/s}}$

(e) $\bar{v} = \dfrac{x_2 - x_1}{t_2 - t_1} = \dfrac{0 - 0}{8 - 0} = \boxed{0 \text{ m/s}}$

***P2.4** (a) Let d represent the distance between A and B. Let t_1 be the time for which the walker has the higher speed in $5.00 \text{ m/s} = \dfrac{d}{t_1}$. Let t_2 represent the longer time for the return trip in $-3.00 \text{ m/s} = -\dfrac{d}{t_2}$. Then the times are $t_1 = \dfrac{d}{(5.00 \text{ m/s})}$ and $t_2 = \dfrac{d}{(3.00 \text{ m/s})}$. The average speed is:

continued on next page

$$\bar{v} = \frac{\text{Total distance}}{\text{Total time}} = \frac{d+d}{\frac{d}{(5.00\text{ m/s})} + \frac{d}{(3.00\text{ m/s})}} = \frac{2d}{\frac{(8.00\text{ m/s})d}{(15.0\text{ m}^2/\text{s}^2)}}$$

$$\bar{v} = \frac{2(15.0\text{ m}^2/\text{s}^2)}{8.00\text{ m/s}} = \boxed{3.75\text{ m/s}}$$

(b) She starts and finishes at the same point A. With total displacement = 0, average velocity $= \boxed{0}$.

Section 2.2 Instantaneous Velocity

P2.5 (a) at $t_i = 1.5$ s, $x_i = 8.0$ m (Point A)

at $t_f = 4.0$ s, $x_f = 2.0$ m (Point B)

$$\bar{\mathbf{v}} = \frac{x_f - x_i}{t_f - t_i} = \frac{(2.0 - 8.0)\text{ m}}{(4 - 1.5)\text{ s}} = -\frac{6.0\text{ m}}{2.5\text{ s}} = \boxed{-2.4\text{ m/s}}$$

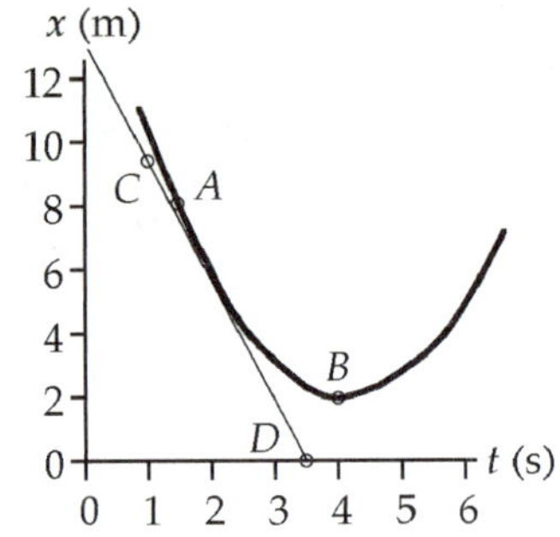

FIG. P2.5

(b) The slope of the tangent line can be found from Points C and D. $(t_C = 1.0\text{ s}, x_C = 9.5\text{ m})$ and $(t_D = 3.5\text{ s}, x_D = 0)$,

$$v \approx \boxed{-3.8\text{ m/s}}.$$

(c) The velocity is zero when x is a minimum. This is at $t \approx \boxed{4\text{ s}}$.

P2.6 (a) At any time, t, the position is given by $x = (3.00\text{ m/s}^2)t^2$.

Thus, at $t_i = 3.00$ s: $x_i = (3.00\text{ m/s}^2)(3.00\text{ s})^2 = \boxed{27.0\text{ m}}$.

(b) At $t_f = 3.00\text{ s} + \Delta t$: $x_f = (3.00\text{ m/s}^2)(3.00\text{ s} + \Delta t)^2$, or

$$x_f = \boxed{27.0\text{ m} + (18.0\text{ m/s})\Delta t + (3.00\text{ m/s}^2)(\Delta t)^2}.$$

(c) The instantaneous velocity at $t = 3.00$ s is:

$$v = \lim_{\Delta t \to 0}\left(\frac{x_f - x_i}{\Delta t}\right) = \lim_{\Delta t \to 0}\left(18.0\text{ m/s} + (3.00\text{ m/s}^2)\Delta t\right) = \boxed{18.0\text{ m/s}}.$$

P2.7 (a)

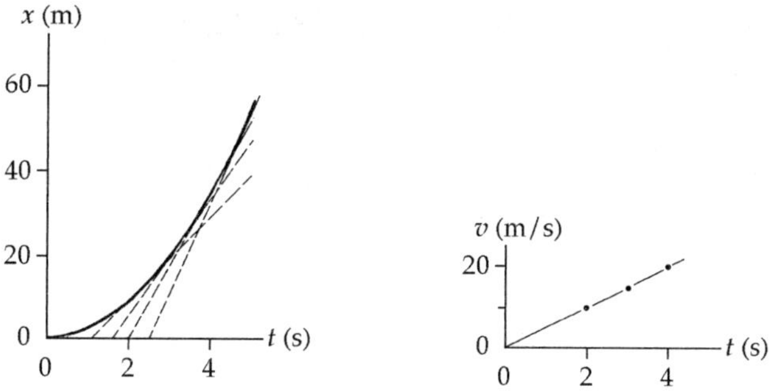

FIG. P2.7(a)

(b) At $t = 5.0$ s, the slope is $v \approx \dfrac{58\text{ m}}{2.5\text{ s}} \approx \boxed{23\text{ m/s}}$.

At $t = 4.0$ s, the slope is $v \approx \dfrac{54\text{ m}}{3\text{ s}} \approx \boxed{18\text{ m/s}}$.

At $t = 3.0$ s, the slope is $v \approx \dfrac{49\text{ m}}{3.4\text{ s}} \approx \boxed{14\text{ m/s}}$.

At $t = 2.0$ s, the slope is $v \approx \dfrac{36\text{ m}}{4.0\text{ s}} \approx \boxed{9.0\text{ m/s}}$.

(c) $\bar{a} = \dfrac{\Delta v}{\Delta t} \approx \dfrac{23\text{ m/s}}{5.0\text{ s}} \approx \boxed{4.6\text{ m/s}^2}$

(d) Initial velocity of the car was $\boxed{\text{zero}}$.

P2.8 (a) $v = \dfrac{(5-0)\text{ m}}{(1-0)\text{ s}} = \boxed{5\text{ m/s}}$

(b) $v = \dfrac{(5-10)\text{ m}}{(4-2)\text{ s}} = \boxed{-2.5\text{ m/s}}$

(c) $v = \dfrac{(5\text{ m} - 5\text{ m})}{(5\text{ s} - 4\text{ s})} = \boxed{0}$

(d) $v = \dfrac{0-(-5\text{ m})}{(8\text{ s} - 7\text{ s})} = \boxed{+5\text{ m/s}}$

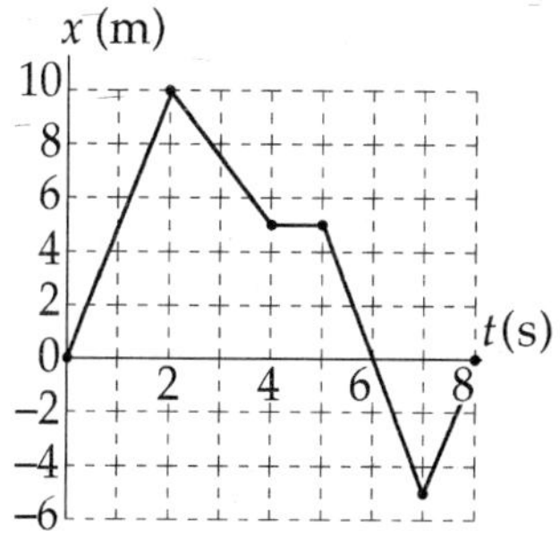

FIG. P2.8

Section 2.3 Analysis Models—The Particle Under Constant Velocity

P2.9 Once it resumes the race, the hare will run for a time of

$$t=\frac{x_f-x_i}{v_x}=\frac{1\,000\ \text{m}-800\ \text{m}}{8\ \text{m/s}}=25\ \text{s}.$$

In this time, the tortoise can crawl a distance

$$x_f-x_i=(0.2\ \text{m/s})(25\ \text{s})=\boxed{5.00\ \text{m}}.$$

Section 2.4 Acceleration

P2.10 Choose the positive direction to be the outward direction, perpendicular to the wall.

$$v_f=v_i+at:\ a=\frac{\Delta v}{\Delta t}=\frac{22.0\ \text{m/s}-(-25.0\ \text{m/s})}{3.50\times10^{-3}\ \text{s}}=\boxed{1.34\times10^4\ \text{m/s}^2}.$$

P2.11 (a) Acceleration is constant over the first ten seconds, so at the end,

$$v_f=v_i+at=0+\left(2.00\ \text{m/s}^2\right)(10.0\ \text{s})=\boxed{20.0\ \text{m/s}}.$$

Then $a=0$ so v is constant from $t=10.0$ s to $t=15.0$ s. And over the last five seconds the velocity changes to

$$v_f=v_i+at=20.0\ \text{m/s}+\left(-3.00\ \text{m/s}^2\right)(5.00\ \text{s})=\boxed{5.00\ \text{m/s}}.$$

(b) In the first ten seconds,

$$x_f=x_i+v_it+\frac{1}{2}at^2=0+0+\frac{1}{2}\left(2.00\ \text{m/s}^2\right)(10.0\ \text{s})^2=100\ \text{m}.$$

Over the next five seconds the position changes to

$$x_f=x_i+v_it+\frac{1}{2}at^2=100\ \text{m}+(20.0\ \text{m/s})(5.00\ \text{s})+0=200\ \text{m}.$$

And at $t=20.0$ s,

$$x_f=x_i+v_it+\frac{1}{2}at^2=200\ \text{m}+(20.0\ \text{m/s})(5.00\ \text{s})+\frac{1}{2}\left(-3.00\ \text{m/s}^2\right)(5.00\ \text{s})^2=\boxed{262\ \text{m}}.$$

P2.12 (a) At $t = 2.00$ s, $x = \left[3.00(2.00)^2 - 2.00(2.00) + 3.00\right]$ m $= 11.0$ m.

At $t = 3.00$ s, $x = \left[3.00(9.00)^2 - 2.00(3.00) + 3.00\right]$ m $= 24.0$ m

so

$$\bar{v} = \frac{\Delta x}{\Delta t} = \frac{24.0 \text{ m} - 11.0 \text{ m}}{3.00 \text{ s} - 2.00 \text{ s}} = \boxed{13.0 \text{ m/s}}.$$

(b) At all times the instantaneous velocity is

$$v = \frac{d}{dt}\left(3.00t^2 - 2.00t + 3.00\right) = (6.00t - 2.00) \text{ m/s}$$

At $t = 2.00$ s, $v = \left[6.00(2.00) - 2.00\right]$ m/s $= \boxed{10.0 \text{ m/s}}$.

At $t = 3.00$ s, $v = \left[6.00(3.00) - 2.00\right]$ m/s $= \boxed{16.0 \text{ m/s}}$.

(c) $\bar{a} = \dfrac{\Delta v}{\Delta t} = \dfrac{16.0 \text{ m/s} - 10.0 \text{ m/s}}{3.00 \text{ s} - 2.00 \text{ s}} = \boxed{6.00 \text{ m/s}^2}$

(d) At all times $a = \dfrac{d}{dt}(6.00t - 2.00) = \boxed{6.00 \text{ m/s}^2}$. (This includes both $t = 2.00$ s and $t = 3.00$ s).

P2.13 $x = 2.00 + 3.00t - t^2$, $v = \dfrac{dx}{dt} = 3.00 - 2.00t$, $a = \dfrac{dv}{dt} = -2.00$

At $t = 3.00$ s:

(a) $x = (2.00 + 9.00 - 9.00)$ m $= \boxed{2.00 \text{ m}}$

(b) $v = (3.00 - 6.00)$ m/s $= \boxed{-3.00 \text{ m/s}}$

(c) $a = \boxed{-2.00 \text{ m/s}^2}$

P2.14 (a) See the graphs at the right.

Choose $x = 0$ at $t = 0$.

At $t = 3$ s, $x = \frac{1}{2}(8 \text{ m/s})(3 \text{ s}) = 12$ m.

At $t = 5$ s, $x = 12 \text{ m} + (8 \text{ m/s})(2 \text{ s}) = 28$ m.

At $t = 7$ s, $x = 28 \text{ m} + \frac{1}{2}(8 \text{ m/s})(2 \text{ s}) = 36$ m.

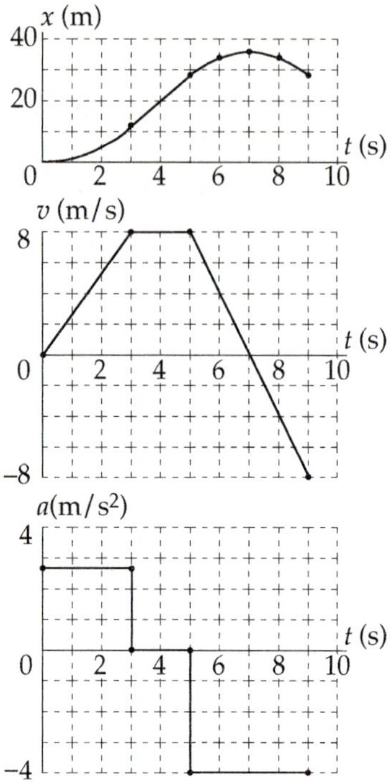

FIG. P2.14

(b) For $0 < t < 3$ s, $a = \frac{8 \text{ m/s}}{3 \text{ s}} = 2.67 \text{ m/s}^2$.

For $3 < t < 5$ s, $a = 0$.

(c) For $5 \text{ s} < t < 9$ s, $a = -\frac{16 \text{ m/s}}{4 \text{ s}} = \boxed{-4 \text{ m/s}^2}$.

(d) At $t = 6$ s, $x = 28 \text{ m} + (6 \text{ m/s})(1 \text{ s}) = \boxed{34 \text{ m}}$.

(e) At $t = 9$ s, $x = 36 \text{ m} + \frac{1}{2}(-8 \text{ m/s})(2 \text{ s}) = \boxed{28 \text{ m}}$.

P2.15 (a) $a = \frac{\Delta v}{\Delta t} = \frac{8.00 \text{ m/s}}{6.00 \text{ s}} = \boxed{1.3 \text{ m/s}^2}$

(b) Maximum positive acceleration is at $t = 3$ s, and is approximately $\boxed{2 \text{ m/s}^2}$.

(c) $a = 0$, at $\boxed{t = 6 \text{ s}}$, and also for $\boxed{t > 10 \text{ s}}$.

(d) Maximum negative acceleration is at $t = 8$ s, and is approximately $\boxed{-1.5 \text{ m/s}^2}$.

Section 2.5 Motion Diagrams

P2.16 (a)

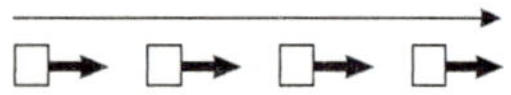

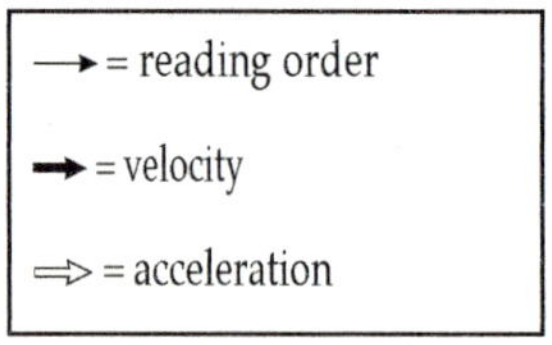

(b)

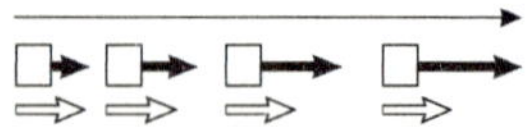

(c)

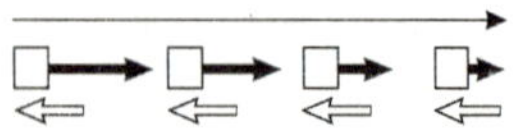

(d)

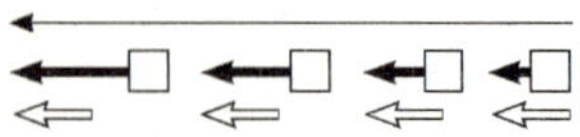

continued on next page

(e) 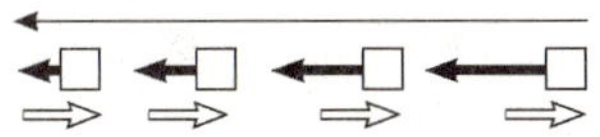

(f) One way of phrasing the answer: The spacing of the successive positions would change with less regularity.

Another way: The object would move with some combination of the kinds of motion shown in (a) through (e). Within one drawing, the accelerations vectors would vary in magnitude and direction.

Section 2.6 The Particle Under Constant Acceleration

***P2.17** (a) $x_f - x_i = \frac{1}{2}(v_i + v_f)t$ becomes $40\text{ m} = \frac{1}{2}(v_i + 2.80\text{ m/s})(8.50\text{ s})$ which yields $v_i = \boxed{6.61\text{ m/s}}$.

(b) $$a = \frac{v_f - v_i}{t} = \frac{2.80\text{ m/s} - 6.61\text{ m/s}}{8.50\text{ s}} = \boxed{-0.448\text{ m/s}^2}$$

P2.18 Method One

Suppose the unknown acceleration is constant as a car moving at $v_{i1} = 35.0\text{ mi/h}$ comes to a $v_f = 0$ stop in $x_{f1} - x_i = 40.0\text{ ft}$. We find its acceleration from $v_{f1}{}^2 = v_{i1}{}^2 + 2a(x_{f1} - x_i)$:

$$a = \frac{v_{f1}{}^2 - v_i{}^2}{2(x_{f1} - x_i)} = \frac{0 - (35.0\text{ mi/h})^2}{2(40.0\text{ ft})}\left(\frac{5280\text{ ft}}{\text{mi}}\right)^2\left(\frac{1\text{ h}}{3600\text{ s}}\right)^2 = -32.9\text{ ft/s}^2$$

Now consider a car moving at $v_{i2} = 70.0\text{ mi/h}$ and stopping to $v_f = 0$ with $a = -32.9\text{ ft/s}^2$. From the same equation its stopping distance is

$$x_{f2} - x_i = \frac{v_{f2}{}^2 - v_i{}^2}{2a} = \frac{0 - (70.0\text{ mi/h})^2}{2(-32.9\text{ ft/s}^2)}\left(\frac{5280\text{ ft}}{1\text{ mi}}\right)^2\left(\frac{1\text{ h}}{3600\text{ s}}\right)^2 = \boxed{160\text{ ft}}$$

Method Two

For the process of stopping from the lower speed v_{i1} we have $v_f^2 = v_{i1}^2 + 2a(x_{f1} - x_i)$; $0 = v_{i1}^2 + 2ax_{f1}$; $v_{i1}^2 = -2ax_{f1}$. For stopping from $v_{i2} = 2v_{i1}$, similarly $0 = v_{i2}^2 + 2a(x_{f2} - 0)$; $v_{i2}^2 = -2ax_{f2}$. Dividing gives $\frac{v_{i2}^2}{v_{i1}^2} = \frac{x_{f2}}{x_{i1}}$; $x_{f2} = 40\text{ ft} \times 2^2 = \boxed{160\text{ ft}}$.

P2.19 Given $v_i = 12.0\text{ cm/s}$ when $x_i = 3.00\text{ cm}(t = 0)$, and at $t = 2.00\text{ s}$, $x_f = -5.00\text{ cm}$,

$$x_f - x_i = v_i t + \frac{1}{2}at^2: -5.00 - 3.00 = 12.0(2.00) + \frac{1}{2}a(2.00)^2$$

$$-8.00 = 24.0 + 2a \qquad a = -\frac{32.0}{2} = \boxed{-16.0\text{ cm/s}^2}.$$

P2.20 (a) Choose the initial point where the pilot reduces the throttle and the final point where the boat passes the buoy:

$$x_i = 0,\ x_f = 100\ \text{m},\ v_{xi} = 30\ \text{m/s},\ v_{xf} = ?,\ a_x = -3.5\ \text{m/s}^2,\ t = ?$$

$x_f = x_i + v_{xi}t + \frac{1}{2}a_x t^2$:

$$100\ \text{m} = 0 + (30\ \text{m/s})t + \frac{1}{2}(-3.5\ \text{m/s}^2)t^2$$

$$(1.75\ \text{m/s}^2)t^2 - (30\ \text{m/s})t + 100\ \text{m} = 0.$$

We use the quadratic formula:

$$t = \frac{-b \pm \sqrt{b^2 - 4ac}}{2a}$$

$$t = \frac{30\ \text{m/s} \pm \sqrt{900\ \text{m}^2/\text{s}^2 - 4(1.75\ \text{m/s}^2)(100\ \text{m})}}{2(1.75\ \text{m/s}^2)} = \frac{30\ \text{m/s} \pm 14.1\ \text{m/s}}{3.5\ \text{m/s}^2} = 12.6\ \text{s or } \boxed{4.53\ \text{s}}.$$

The smaller value is the physical answer. If the boat kept moving with the same acceleration, it would stop and move backward, then gain speed, and pass the buoy again at 12.6 s.

(b) $v_{xf} = v_{xi} + a_x t = 30\ \text{m/s} - (3.5\ \text{m/s}^2)4.53\ \text{s} = \boxed{14.1\ \text{m/s}}$

P2.21 (a) $v_i = 100\ \text{m/s},\ a = -5.00\ \text{m/s}^2,\ v_f = v_i + at$ so $0 = 100 - 5t,\ v_f^2 = v_i^2 + 2a(x_f - x_i)$ so $0 = (100)^2 - 2(5.00)(x_f - 0)$. Thus $x_f = 1\,000\ \text{m}$ and $t = \boxed{20.0\ \text{s}}$.

(b) At this acceleration the plane would overshoot the runway: $\boxed{\text{No}}$.

P2.22 (a) Compare the position equation $x = 2.00 + 3.00t - 4.00t^2$ to the general form

$$x_f = x_i + v_i t + \frac{1}{2}at^2$$

to recognize that $x_i = 2.00\ \text{m}$, $v_i = 3.00\ \text{m/s}$, and $a = -8.00\ \text{m/s}^2$. The velocity equation, $v_f = v_i + at$, is then

$$v_f = 3.00\ \text{m/s} - (8.00\ \text{m/s}^2)t.$$

The particle changes direction when $v_f = 0$, which occurs at $t = \frac{3}{8}$ s. The position at this time is:

$$x = 2.00\ \text{m} + (3.00\ \text{m/s})\left(\frac{3}{8}\ \text{s}\right) - (4.00\ \text{m/s}^2)\left(\frac{3}{8}\ \text{s}\right)^2 = \boxed{2.56\ \text{m}}.$$

continued on next page

(b) From $x_f = x_i + v_i t + \frac{1}{2}at^2$, observe that when $x_f = x_i$, the time is given by $t = -\frac{2v_i}{a}$. Thus, when the particle returns to its initial position, the time is

$$t = \frac{-2(3.00 \text{ m/s})}{-8.00 \text{ m/s}^2} = \frac{3}{4} \text{ s}$$

and the velocity is $v_f = 3.00 \text{ m/s} - (8.00 \text{ m/s}^2)\left(\frac{3}{4} \text{ s}\right) = \boxed{-3.00 \text{ m/s}}$.

P2.23 In the simultaneous equations:

$$\left\{\begin{array}{c} v_{xf} = v_{xi} + a_x t \\ x_f - x_i = \frac{1}{2}(v_{xi} + v_{xf})t \end{array}\right\} \text{ we have } \left\{\begin{array}{c} v_{xf} = v_{xi} - (5.60 \text{ m/s}^2)(4.20 \text{ s}) \\ 62.4 \text{ m} = \frac{1}{2}(v_{xi} + v_{xf})(4.20 \text{ s}) \end{array}\right\}.$$

So substituting for v_{xi} gives $62.4 \text{ m} = \frac{1}{2}\left[v_{xf} + (5.60 \text{ m/s}^2)(4.20 \text{ s}) + v_{xf}\right](4.20 \text{ s})$

$$14.9 \text{ m/s} = v_{xf} + \frac{1}{2}(5.60 \text{ m/s}^2)(4.20 \text{ s}).$$

Thus

$$v_{xf} = \boxed{3.10 \text{ m/s}}.$$

P2.24 Take any two of the standard four equations, such as $\left\{\begin{array}{c} v_{xf} = v_{xi} + a_x t \\ x_f - x_i = \frac{1}{2}(v_{xi} + v_{xf})t \end{array}\right\}$. Solve one for v_{xi}, and substitute into the other: $v_{xi} = v_{xf} - a_x t$

$$x_f - x_i = \frac{1}{2}(v_{xf} - a_x t + v_{xf})t.$$

Thus

$$\boxed{x_f - x_i = v_{xf} t - \frac{1}{2}a_x t^2}.$$

We note that the equation is dimensionally correct. The units are units of length in each term. Like the standard equation $x_f - x_i = v_{xi}t + \frac{1}{2}a_x t^2$, this equation represents that displacement is a quadratic function of time. Back in problem 23, $62.4 \text{ m} = v_{xf}(4.20 \text{ s}) - \frac{1}{2}(-5.60 \text{ m/s}^2)(4.20 \text{ s})^2$

$$v_{xf} = \frac{62.4 \text{ m} - 49.4 \text{ m}}{4.20 \text{ s}} = \boxed{3.10 \text{ m/s}}.$$

P2.25 (a) The time it takes the truck to reach 20.0 m/s is found from $v_f = v_i + at$. Solving for t yields

$$t = \frac{v_f - v_i}{a} = \frac{20.0\ \text{m/s} - 0\ \text{m/s}}{2.00\ \text{m/s}^2} = 10.0\ \text{s}.$$

The total time is thus

$$10.0\ \text{s} + 20.0\ \text{s} + 5.00\ \text{s} = \boxed{35.0\ \text{s}}.$$

(b) The average velocity is the total distance traveled divided by the total time taken. The distance traveled during the first 10.0 s is

$$x_1 = \bar{v}t = \left(\frac{0 + 20.0}{2}\right)(10.0) = 100\ \text{m}.$$

With a being 0 for this interval, the distance traveled during the next 20.0 s is

$$x_2 = v_i t + \frac{1}{2}at^2 = (20.0)(20.0) + 0 = 400\ \text{m}.$$

The distance traveled in the last 5.00 s is

$$x_3 = \bar{v}t = \left(\frac{20.0 + 0}{2}\right)(5.00) = 50.0\ \text{m}.$$

The total distance $x = x_1 + x_2 + x_3 = 100 + 400 + 50 = 550\ \text{m}$, and the average velocity is given by $\bar{v} = \frac{x}{t} = \frac{550}{35.0} = \boxed{15.7\ \text{m/s}}$.

***P2.26** We have $v_i = 2.00 \times 10^4\ \text{m/s}$, $v_f = 6.00 \times 10^6\ \text{m/s}$, $x_f - x_i = 1.50 \times 10^{-2}\ \text{m}$.

(a) $$x_f - x_i = \frac{1}{2}(v_i + v_f)t:\ t = \frac{2(x_f - x_i)}{v_i + v_f} = \frac{2(1.50 \times 10^{-2}\ \text{m})}{2.00 \times 10^4\ \text{m/s} + 6.00 \times 10^6\ \text{m/s}} = \boxed{4.98 \times 10^{-9}\ \text{s}}$$

(b) $v_f^2 = v_i^2 + 2a_x(x_f - x_i)$:

$$a_x = \frac{v_f^2 - v_i^2}{2(x_f - x_i)} = \frac{(6.00 \times 10^6\ \text{m/s})^2 - (2.00 \times 10^4\ \text{m/s})^2}{2(1.50 \times 10^{-2}\ \text{m})} = \boxed{1.20 \times 10^{15}\ \text{m/s}^2}$$

P2.27 Take the original point to be when Sue notices the van. Choose the origin of the x-axis at Sue's car. For her we have $x_{is}=0$, $v_{is}=30.0\ \text{m/s}$, $a_s=-2.00\ \text{m/s}^2$ so her position is given by

$$x_s(t)=x_{is}+v_{is}t+\frac{1}{2}a_st^2=(30.0\ \text{m/s})t+\frac{1}{2}\left(-2.00\ \text{m/s}^2\right)t^2.$$

For the van, $x_{iv}=155\ \text{m}$, $v_{iv}=5.00\ \text{m/s}$, $a_v=0$ and

$$x_v(t)=x_{iv}+v_{iv}t+\frac{1}{2}a_vt^2=155+(5.00\ \text{m/s})t+0.$$

To test for a collision, we look for an instant t_c when both are at the same place:

$$30.0t_c-t_c^2=155+5.00t_c$$
$$0=t_c^2-25.0t_c+155.$$

From the quadratic formula

$$t_c=\frac{25.0\pm\sqrt{(25.0)^2-4(155)}}{2}=13.6\ \text{s or } \boxed{11.4\ \text{s}}.$$

The smaller value is the collision time. (The larger value tells when the van would pull ahead again if the vehicles could move through each other). The wreck happens at position

$$155\ \text{m}+(5.00\ \text{m/s})(11.4\ \text{s})=\boxed{212\ \text{m}}.$$

Section 2.7 Freely Falling Objects

***P2.28** Choose the origin ($y=0$, $t=0$) at the starting point of the cat and take upward as positive. Then $y_i=0$, $v_i=0$, and $a=-g=-9.80\ \text{m/s}^2$. The position and the velocity at time t become:

$$y_f-y_i=v_it+\frac{1}{2}at^2:\ y_f=-\frac{1}{2}gt^2=-\frac{1}{2}\left(9.80\ \text{m/s}^2\right)t^2$$

and

$$v_f=v_i+at:\ v_f=-gt=-\left(9.80\ \text{m/s}^2\right)t.$$

(a) at $t=0.1\ \text{s}$: $y_f=-\frac{1}{2}\left(9.80\ \text{m/s}^2\right)(0.1\ \text{s})^2=\boxed{-0.0490\ \text{m}}$

(b) at $t=0.2\ \text{s}$: $y_f=-\frac{1}{2}\left(9.80\ \text{m/s}^2\right)(0.2\ \text{s})^2=\boxed{-0.196\ \text{m}}$

(c) at $t=0.3\ \text{s}$: $y_f=-\frac{1}{2}\left(9.80\ \text{m/s}^2\right)(0.3\ \text{s})^2=\boxed{-0.441\ \text{m}}$

continued on next page

(a) at $t = 0.1\text{ s}$: $v_f = -(9.80\text{ m/s}^2)(0.1\text{ s}) = \boxed{-0.980\text{ m/s}}$

(b) at $t = 0.2\text{ s}$: $v_f = -(9.80\text{ m/s}^2)(0.2\text{ s}) = \boxed{-1.96\text{ m/s}}$

(c) at $t = 0.3\text{ s}$: $v_f = -(9.80\text{ m/s}^2)(0.3\text{ s}) = \boxed{-2.94\text{ m/s}}$

P2.29 (a) $v_f = v_i - gt$: $v_f = 0$ when $t = 3.00\text{ s}$, $g = 9.80\text{ m/s}^2$. Therefore,

$$v_i = gt = (9.80\text{ m/s}^2)(3.00\text{ s}) = \boxed{29.4\text{ m/s}}.$$

(b) $y_f - y_i = \frac{1}{2}(v_f + v_i)t$

$$y_f - y_i = \frac{1}{2}(29.4\text{ m/s})(3.00\text{ s}) = \boxed{44.1\text{ m}}$$

P2.30 Assume that air resistance may be neglected. Then, the acceleration at all times during the flight is that due to gravity, $a = -g = -9.80\text{ m/s}^2$. During the flight, Goff went 1 mile (1 609 m) up and then 1 mile back down. Determine his speed just after launch by considering his upward flight:

$$v_f^2 = v_i^2 + 2a(y_f - y_i): \quad 0 = v_i^2 - 2(9.80\text{ m/s}^2)(1\,609\text{ m})$$
$$v_i = 178\text{ m/s}.$$

His time in the air may be found by considering his motion from just after launch to just before impact:

$$y_f - y_i = v_i t + \frac{1}{2}at^2: 0 = (178\text{ m/s})t - \frac{1}{2}(-9.80\text{ m/s}^2)t^2.$$

The root $t = 0$ describes launch; the other root, $t = 36.2\text{ s}$, describes his flight time. His rate of pay may then be found from

$$\text{pay rate} = \frac{\$1.00}{36.2\text{ s}} = (0.027\,6\text{ \$/s})(3\,600\text{ s/h}) = \boxed{\$99.3/\text{h}}.$$

We have assumed that the workman's flight time, "a mile", and "a dollar", were measured to three-digit precision. We have interpreted "up in the sky" as referring to the free fall time, not to the launch and landing times. Both the takeoff and landing times must be several seconds away from the job, in order for Goff to survive to resume work.

P2.31 (a) $y_f - y_i = v_i t + \frac{1}{2}at^2$: $4.00 = (1.50)v_i - (4.90)(1.50)^2$ and $v_i = \boxed{10.0\text{ m/s upward}}$.

(b) $v_f = v_i + at = 10.0 - (9.80)(1.50) = -4.68\text{ m/s}$

$$v_f = \boxed{4.68\text{ m/s downward}}$$

P2.32 We have $y_f = -\frac{1}{2}gt^2 + v_i t + y_i$

$$0 = -\left(4.90 \text{ m/s}^2\right)t^2 - (8.00 \text{ m/s})t + 30.0 \text{ m}.$$

Solving for t,

$$t = \frac{8.00 \pm \sqrt{64.0 + 588}}{-9.80}.$$

Using only the positive value for t, we find that $t = \boxed{1.79 \text{ s}}$.

P2.33 Time to fall 3.00 m is found from Eq. 2.12 with $v_i = 0$, $3.00 \text{ m} = \frac{1}{2}\left(9.80 \text{ m/s}^2\right)t^2$, $t = 0.782$ s.

(a) With the horse galloping at 10.0 m/s, the horizontal distance is $vt = \boxed{7.82 \text{ m}}$.

(b) $t = \boxed{0.782 \text{ s}}$

***P2.34** (a) Consider the upward flight of the arrow.

$$v_{yf}^2 = v_{yi}^2 + 2a_y\left(y_f - y_i\right)$$

$$0 = (100 \text{ m/s})^2 + 2\left(-9.8 \text{ m/s}^2\right)\Delta y$$

$$\Delta y = \frac{10\,000 \text{ m}^2/\text{s}^2}{19.6 \text{ m/s}^2} = \boxed{510 \text{ m}}$$

(b) Consider the whole flight of the arrow.

$$y_f = y_i + v_{yi}t + \frac{1}{2}a_y t^2$$

$$0 = 0 + (100 \text{ m/s})t + \frac{1}{2}\left(-9.8 \text{ m/s}^2\right)t^2$$

The root $t = 0$ refers to the starting point. The time of flight is given by

$$t = \frac{100 \text{ m/s}}{4.9 \text{ m/s}^2} = \boxed{20.4 \text{ s}}.$$

Section 2.8 Context Connection—Acceleration Required by Consumers

*P2.35 (a) The Lamborghini requires the shortest time to speed up from 0 to 60 mi/h, so it has the greatest average acceleration:

$$v_{xf} = v_{xi} + a_x t$$

$$a_x = \frac{v_{xf} - v_{xi}}{t} = \frac{60\ \text{mi/h} - 0}{3.6\ \text{s}} = 16.7\ \text{mi/h}\cdot\text{s}, \text{ in agreement}$$

with the value listed. For the Toyota Prius,

$$a_x = \frac{v_{xf} - v_{xi}}{t} = \frac{60\ \text{mi/h} - 0}{12.7\ \text{s}} = 4.72\ \text{mi/h}\cdot\text{s}, \text{ in agreement}$$

with $4.7\ \text{mi/h}\cdot\text{s}$ listed.

(b) Lamborghini: $\frac{60\ \text{mi/h}}{3.6\ \text{s}}\left(\frac{1\,609\ \text{m}}{1\ \text{mi}}\right)\left(\frac{1\ \text{h}}{3\,600\ \text{s}}\right) = \boxed{7.4\ \text{m/s}^2}$ to two significant digits.

Prius: $\frac{60\ \text{mi/h}}{12.7\ \text{s}}\left(\frac{1\,609\ \text{m}}{1\ \text{mi}}\right)\left(\frac{1\ \text{h}}{3\,600\ \text{s}}\right) = \boxed{2.1\ \text{m/s}^2}$

(c) $x_f - x_i = \frac{1}{2}\left(v_{xi} + v_{xf}\right)t$

Lamborghini: $x_f - x_i = \frac{1}{2}(0 + 60\ \text{mi/h})3.6\ \text{s}\left(\frac{1\,609\ \text{m}}{1\ \text{mi}}\right)\left(\frac{1\ \text{h}}{3\,600\ \text{s}}\right) = \boxed{48\ \text{m}}$

Prius: $x_f - x_i = \frac{1}{2}(60\ \text{mi/h})12.7\ \text{s}\left(\frac{1\,609\ \text{m}}{1\ \text{mi}}\right)\left(\frac{1\ \text{h}}{3\,600\ \text{s}}\right) = \boxed{170\ \text{m}}$

(d) $v_{xf} = v_{xi} + a_x t$

$$t = \frac{v_{xf} - v_{xi}}{a_x} = \frac{60\ \text{mi/h} - 0}{9.80\ \text{m/s}^2}\left(\frac{1\,609\ \text{mi}}{1\ \text{mi}}\right)\left(\frac{1\ \text{h}}{3\,600\ \text{s}}\right) = \boxed{2.74\ \text{s}}$$

Building a car capable of such a large acceleration would be difficult. So would steering it as it is speeding up.

P2.36 (a) Assuming a constant acceleration: $a = \frac{v_f - v_i}{t} = \frac{42.0\ \text{m/s}}{8.00\ \text{s}} = \boxed{5.25\ \text{m/s}^2}$

(b) Taking the origin at the original position of the car,

$$x_f = \frac{1}{2}\left(v_i + v_f\right)t = \frac{1}{2}(42.0\ \text{m/s})(8.00\ \text{s}) = \boxed{168\ \text{m}}$$

(c) From $v_f = v_i + at$, the velocity 10.0 s after the car starts from rest is:

$$v_f = 0 + \left(5.25\ \text{m/s}^2\right)(10.0\ \text{s}) = \boxed{52.5\ \text{m/s}}$$

***P2.37** (a) $v_{xf} = v_{xi} + a_x t$

$$a_x = \frac{v_{xf} - v_{xi}}{t} = \frac{175\ \text{mi/h} - 0}{2.50\ \text{s}} = 70.0\ \text{mi/h}\cdot\text{s} = 70.0\ \text{mi/h}\cdot\text{s}\left(\frac{1\,609\ \text{mi}}{1\ \text{m}}\right)\left(\frac{1\ \text{h}}{3\,600\ \text{s}}\right)$$

$$= \boxed{31.3\ \text{m/s}^2} = 31.3\ \text{m/s}^2\left(\frac{g}{9.8\ \text{m/s}^2}\right) = 3.19g$$

(b) $$x_f - x_i = \frac{1}{2}(v_{xi} + v_{xf})t = \frac{1}{2}(0 + 175\ \text{mi/h})2.5\ \text{s}\left(\frac{1\,609\ \text{mi}}{1\ \text{mi}}\right)\left(\frac{1\ \text{h}}{3\,600\ \text{s}}\right) = \boxed{97.8\ \text{m}}$$

***P2.38** (a) Starting from rest and accelerating at $a_b = 13.0\ \text{mi/h}\cdot\text{s}$, the bicycle reaches its maximum speed of $v_{b,\max} = 20.0\ \text{mi/h}$ in a time

$$t_{b,1} = \frac{v_{b,\max} - 0}{a_b} = \frac{20.0\ \text{mi/h}}{13.0\ \text{mi/h}\cdot\text{s}} = 1.54\ \text{s}$$

Since the acceleration a_c of the car is less than that of the bicycle, the car cannot catch the bicycle until some time $t > t_{b,1}$ (that is, until the bicycle is at its maximum speed and coasting). The total displacement of the bicycle at time *t* is

$$\Delta x_b = \frac{1}{2}a_b t_{b,1}^2 + v_{b,\max}(t - t_{b,1})$$
$$= \left(\frac{1.47\ \text{ft/s}}{1\ \text{mi/h}}\right)\left[\frac{1}{2}\left(13.0\ \frac{\text{mi/h}}{\text{s}}\right)(1.54\ \text{s})^2 + (20.0\ \text{mi/h})(t - 1.54\ \text{s})\right]$$
$$= (29.4\ \text{ft/s})t - 22.6\ \text{ft}$$

The total displacement of the car at this time is

$$\Delta x_c = \frac{1}{2}a_c t^2 = \left(\frac{1.47\ \text{ft/s}}{1\ \text{mi/h}}\right)\left[\frac{1}{2}\left(9.00\ \frac{\text{mi/h}}{\text{s}}\right)t^2\right] = (6.62\ \text{ft/s})t^2$$

At the time the car catches the bicycle $\Delta x_c = \Delta x_b$. This gives

$$\left(6.62\ \text{ft/s}^2\right)t^2 = (29.4\ \text{ft/s})t - 22.6\ \text{ft} \qquad \text{or} \qquad t^2 - (4.44\ \text{s})t + 3.42\ \text{s}^2 = 0$$

that has only one acceptable solution $t > t_{b,1}$. This solution gives the total time the bicycle leads the car and is $t = \boxed{3.45\ \text{s}}$.

(b) The lead the bicycle has over the car continues to increase as long as the bicycle is moving faster than the car. This means until the car attains a speed of $v_c = v_{b,\max} = 20.0\ \text{mi/h}$. Thus, the elapsed time when the bicycle's lead ceases to increase is

$$t = \frac{v_{b,\max}}{a_c} = \frac{20.0\ \text{mi/h}}{9.00\ \text{mi/h}\cdot\text{s}} = 2.22\ \text{s}$$

At this time, the lead is

$$(\Delta x_b - \Delta x_c)_{\max} = (\Delta x_b - \Delta x_c)\big|_{t=2.22\ \text{s}} = \left[(29.4\ \text{ft/s})(2.22\ \text{s}) - 22.6\ \text{ft}\right] - \left[\left(6.62\ \text{ft/s}^2\right)(2.22\ \text{s})^2\right]$$

or $(\Delta x_b - \Delta x_c)_{\max} = \boxed{10.0\ \text{ft}}$.

Additional Problems

P2.39 (a) $a = \frac{v_f - v_i}{t} = \frac{632\left(\frac{5\,280}{3\,600}\right)}{1.40} = \boxed{-662 \text{ ft/s}^2} = -202 \text{ m/s}^2 = -20.6g$

(b) $x_f = v_i t + \frac{1}{2}at^2 = (632)\left(\frac{5\,280}{3\,600}\right)(1.40) - \frac{1}{2}(662)(1.40)^2 = \boxed{649 \text{ ft}} = \boxed{198 \text{ m}}$

***P2.40** Take downward as the positive y direction.

(a) While the woman was in free fall,

$$\Delta y = 144 \text{ ft},\ v_i = 0, \text{ and we take } a = g = 32.0 \text{ ft/s}^2.$$

Thus, $\Delta y = v_i t + \frac{1}{2}at^2 \rightarrow 144 \text{ ft} = 0 + (16.0 \text{ ft/s}^2)t^2$ giving $t_{\text{fall}} = 3.00$ s. Her velocity just before impact is:

$$v_f = v_i + gt = 0 + (32.0 \text{ ft/s}^2)(3.00 \text{ s}) = \boxed{96.0 \text{ ft/s}}.$$

(b) While crushing the box, $v_i = 96.0$ ft/s, $v_f = 0$, and $\Delta y = 18.0 \text{ in.} = 1.50$ ft. Therefore,

$$a = \frac{v_f^2 - v_i^2}{2(\Delta y)} = \frac{0 - (96.0 \text{ ft/s})^2}{2(1.50 \text{ ft})} = -3.07 \times 10^3 \text{ ft/s}^2, \text{ or } \boxed{a = 3.07 \times 10^3 \text{ ft/s}^2 \text{ upward}} = 96.0g.$$

(c) Time to crush box: $\Delta t = \frac{\Delta y}{\bar{v}} = \frac{\Delta y}{\frac{v_f + v_i}{2}} = \frac{2(1.50 \text{ ft})}{0 + 96.0 \text{ ft/s}}$ or $\boxed{\Delta t = 3.13 \times 10^{-2} \text{ s}}$.

P2.41 From $v_f^2 = v_i^2 + 2ax$, we have $(10.97 \times 10^3 \text{ m/s})^2 = 0 + 2a(220 \text{ m})$, so that $\boxed{a = 2.74 \times 10^5 \text{ m/s}^2}$ which is $\boxed{a = 2.79 \times 10^4 \text{ times } g}$.

***P2.42** Take the x-axis along the tail section of the snake. The displacement from tail to head is

$$240 \text{ m}\hat{\mathbf{i}} + (420 - 240) \text{ m} \cos(180° - 105°)\hat{\mathbf{i}} - 180 \text{ m} \sin 75° \hat{\mathbf{j}} = 287 \text{ m}\hat{\mathbf{i}} - 174 \text{ m}\hat{\mathbf{j}}.$$

Its magnitude is $\sqrt{(287)^2 + (174)^2}$ m $= 335$ m. From $v = \frac{\text{distance}}{\Delta t}$, the time for each child's run is

$$\text{Inge: } \Delta t = \frac{\text{distance}}{v} = \frac{335 \text{ m(h)(1 km)(3 600 s)}}{(12 \text{ km})(1\,000 \text{ m})(1 \text{ h})} = 101 \text{ s}$$

$$\text{Olaf: } \Delta t = \frac{420 \text{ m} \cdot \text{s}}{3.33 \text{ m}} = 126 \text{ s}.$$

Inge wins by $126 - 101 = \boxed{25.4 \text{ s}}$.

P2.43 (a) Take initial and final points at top and bottom of the incline. If the ball starts from rest,

$$v_i = 0,\ a = 0.500\ \text{m/s}^2,\ x_f - x_i = 9.00\ \text{m}.$$

Then

$$v_f^2 = v_i^2 + 2a(x_f - x_i) = 0^2 + 2(0.500\ \text{m/s}^2)(9.00\ \text{m})$$

$$v_f = \boxed{3.00\ \text{m/s}}.$$

(b) $x_f - x_i = v_i t + \frac{1}{2}at^2$

$$9.00 = 0 + \frac{1}{2}(0.500\ \text{m/s}^2)t^2$$

$$t = \boxed{6.00\ \text{s}}$$

(c) Take initial and final points at the bottom of the planes and the top of the second plane, respectively:

$$v_i = 3.00\ \text{m/s},\ v_f = 0,\ x_f - x_i = 15.00\ \text{m}.$$

$v_f^2 = v_i^2 + 2a(x_f - x_i)$ gives

$$a = \frac{v_f^2 - v_i^2}{2(x_f - x_i)} = \frac{\left[0 - (3.00\ \text{m/s})^2\right]}{2(15.0\ \text{m})} = \boxed{-0.300\ \text{m/s}^2}.$$

(d) Take the initial point at the bottom of the planes and the final point 8.00 m along the second: $v_i = 3.00\ \text{m/s},\ x_f - x_i = 8.00\ \text{m},\ a = -0.300\ \text{m/s}^2$

$$v_f^2 = v_i^2 + 2a(x_f - x_i) = (3.00\ \text{m/s})^2 + 2(-0.300\ \text{m/s}^2)(8.00\ \text{m}) = 4.20\ \text{m}^2/\text{s}^2$$

$$v_f = \boxed{2.05\ \text{m/s}}.$$

***P2.44** Let the glider enter the photogate with velocity v_i and move with constant acceleration a. For its motion from entry to exit,

$$x_f = x_i + v_{xi}t + \frac{1}{2}a_x t^2$$

$$\ell = 0 + v_i\Delta t_d + \frac{1}{2}a\Delta t_d^2 = v_d\Delta t_d$$

$$v_d = v_i + \frac{1}{2}a\Delta t_d$$

(a) The speed halfway through the photogate in space is given by

$$v_{hs}^2 = v_i^2 + 2a\left(\frac{\ell}{2}\right) = v_i^2 + av_d\Delta t_d.$$

$v_{hs} = \sqrt{v_i^2 + av_d\Delta t_d}$ and this is not equal to v_d unless $a = 0$.

(b) The speed halfway through the photogate in time is given by $v_{ht} = v_i + a\left(\frac{\Delta t_d}{2}\right)$ and this is equal to v_d as determined above.

P2.45 Average speed of every point on the train as the first car passes Liz:

$$\frac{\Delta x}{\Delta t} = \frac{8.60 \text{ m}}{1.50 \text{ s}} = 5.73 \text{ m/s}.$$

The train has this as its instantaneous speed halfway through the 1.50 s time. Similarly, halfway through the next 1.10 s, the speed of the train is $\frac{8.60 \text{ m}}{1.10 \text{ s}} = 7.82 \text{ m/s}$. The time required for the speed to change from 5.73 m/s to 7.82 m/s is

$$\frac{1}{2}(1.50 \text{ s}) + \frac{1}{2}(1.10 \text{ s}) = 1.30 \text{ s}$$

so the acceleration is: $a_x = \frac{\Delta v_x}{\Delta t} = \frac{7.82 \text{ m/s} - 5.73 \text{ m/s}}{1.30 \text{ s}} = \boxed{1.60 \text{ m/s}^2}$.

***P2.46** (a) As we see from the graph, from about –50 s to 50 s Acela is cruising at a constant positive velocity in the $+x$ direction. From 50 s to 200 s, Acela accelerates in the $+x$ direction reaching a top speed of about 170 mi/h. Around 200 s, the engineer applies the brakes, and the train, still traveling in the $+x$ direction, slows down and then stops at 350 s. Just after 350 s, Acela reverses direction (v becomes negative) and steadily gains speed in the $-x$ direction.

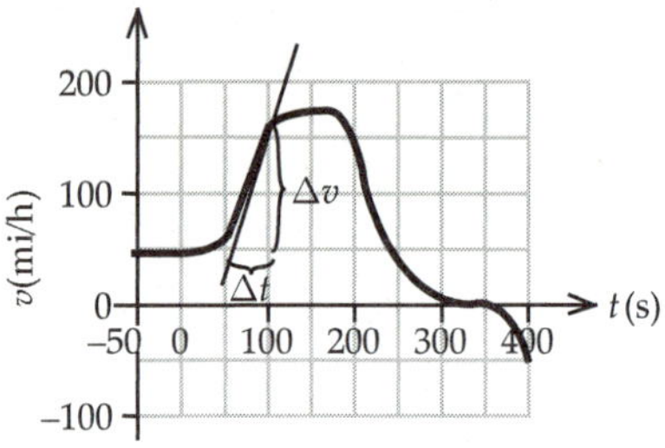

FIG. P2.46(a)

(b) The peak acceleration between 45 and 170 mi/h is given by the slope of the steepest tangent to the v versus t curve in this interval. From the tangent line shown, we find

$$a = \text{slope} = \frac{\Delta v}{\Delta t} = \frac{(155 - 45) \text{ mi/h}}{(100 - 50) \text{ s}} = \boxed{2.2 \text{ (mi/h)/s}} = 0.98 \text{ m/s}^2.$$

(c) Let us use the fact that the area under the v versus t curve equals the displacement. The train's displacement between 0 and 200 s is equal to the area of the gray shaded region, which we have approximated with a series of triangles and rectangles.

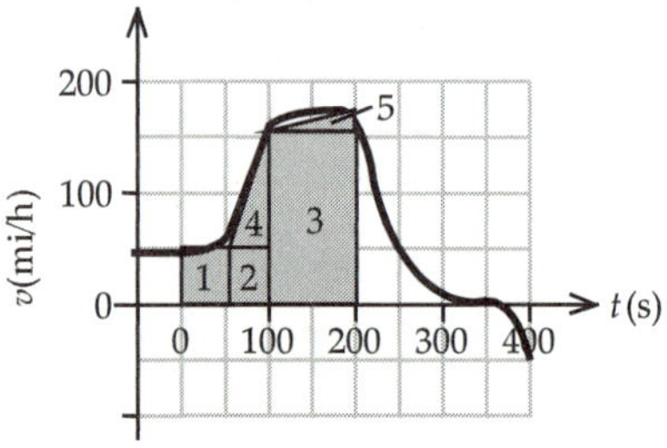

FIG. P2.46(c)

$$\begin{aligned}\Delta x_{0\to 200 \text{ s}} &= \text{area}_1 + \text{area}_2 + \text{area}_3 + \text{area}_4 + \text{area}_5 \\ &\approx (50 \text{ mi/h})(50 \text{ s}) + (50 \text{ mi/h})(50 \text{ s}) \\ &+ (160 \text{ mi/h})(100 \text{ s}) + \frac{1}{2}(50 \text{ s})(100 \text{ mi/h}) \\ &+ \frac{1}{2}(100 \text{ s})(170 \text{ mi/h} - 160 \text{ mi/h}) \\ &= 24\,000 (\text{mi/h})(\text{s})\end{aligned}$$

Now, at the end of our calculation, we can find the displacement in miles by converting hours to seconds. As $1 \text{ h} = 3\,600 \text{ s}$,

$$\Delta x_{0\to 200 \text{ s}} \approx \left(\frac{24\,000 \text{ mi}}{3\,600 \text{ s}}\right)(\text{s}) = \boxed{6.7 \text{ mi}}.$$

P2.47 Let point 0 be at ground level and point 1 be at the end of the engine burn. Let point 2 be the highest point the rocket reaches and point 3 be just before impact. The data in the table are found for each phase of the rocket's motion.

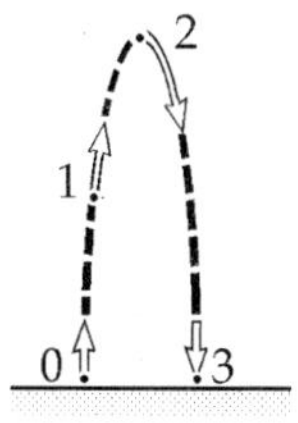

FIG. P2.47

(0 to 1) $v_f^2 - (80.0)^2 = 2(4.00)(1\,000)$ so $v_f = 120$ m/s

$120 = 80.0 + (4.00)t$ giving $t = 10.0$ s

(1 to 2) $0 - (120)^2 = 2(-9.80)(x_f - x_i)$ giving

$x_f - x_i = 735$ m

$0 - 120 = -9.80t$ giving $t = 12.2$ s

This is the time of maximum height of the rocket.

(2 to 3) $v_f^2 - 0 = 2(-9.80)(-1\,735)$

$v_f = -184 = (-9.80)t$ giving $t = 18.8$ s

(a) $t_{\text{total}} = 10 + 12.2 + 18.8 = \boxed{41.0 \text{ s}}$

(b) $(x_f - x_i)_{\text{total}} = \boxed{1.73 \text{ km}}$

(c) $v_{\text{final}} = \boxed{-184 \text{ m/s}}$

		t	x	v	a
0		0.0	0	80	+4.00
#1	End Thrust	10.0	1 000	120	+4.00
#2	Rise Upwards	22.2	1 735	0	–9.80
#3	Fall to Earth	41.0	0	–184	–9.80

P2.48 Distance traveled by motorist $= (15.0 \text{ m/s})t$

Distance traveled by policeman $= \frac{1}{2}(2.00 \text{ m/s}^2)t^2$

(a) intercept occurs when $15.0t = t^2$, or $t = \boxed{15.0 \text{ s}}$

(b) $v(\text{officer}) = (2.00 \text{ m/s}^2)t = \boxed{30.0 \text{ m/s}}$

(c) $x(\text{officer}) = \frac{1}{2}(2.00 \text{ m/s}^2)t^2 = \boxed{225 \text{ m}}$

P2.49 (a) Let x be the distance traveled at acceleration a until maximum speed v is reached. If this is achieved in time t_1 we can use the following three equations:

$$x=\frac{1}{2}(v+v_i)t_1, \qquad 100-x=v(10.2-t_1) \qquad \text{and } v=v_i+at_1.$$

The first two give

$$100=\left(10.2-\frac{1}{2}t_1\right)v=\left(10.2-\frac{1}{2}t_1\right)at_1$$

$$a=\frac{200}{(20.4-t_1)t_1}.$$

$$\text{For Maggie: } a=\frac{200}{(18.4)(2.00)}=\boxed{5.43\ \text{m/s}^2}$$

$$\text{For Judy: } a=\frac{200}{(17.4)(3.00)}=\boxed{3.83\ \text{m/s}^2}$$

(b) $v=at_1$

$$\text{Maggie: } v=(5.43)(2.00)=\boxed{10.9\ \text{m/s}}$$

$$\text{Judy: } v=(3.83)(3.00)=\boxed{11.5\ \text{m/s}}$$

(c) At the six-second mark

$$x=\frac{1}{2}at_1^2+v(6.00-t_1)$$

$$\text{Maggie: } x=\frac{1}{2}(5.43)(2.00)^2+(10.9)(4.00)=54.3\ \text{m}$$

$$\text{Judy: } x=\frac{1}{2}(3.83)(3.00)^2+(11.5)(3.00)=51.7\ \text{m}$$

Maggie is ahead by $\boxed{2.62\ \text{m}}$.

P2.50 $a_1 = 0.100\ \text{m/s}^2$ $\qquad a_2 = -0.500\ \text{m/s}^2$

$$x = 1\,000\ \text{m} = \frac{1}{2}a_1 t_1^2 + v_1 t_2 + \frac{1}{2}a_2 t_2^2 \qquad t = t_1 + t_2 \text{ and } v_1 = a_1 t_1 = -a_2 t_2$$

$$1\,000 = \frac{1}{2}a_1 t_1^2 + a_1 t_1\left(-\frac{a_1 t_1}{a_2}\right) + \frac{1}{2}a_2\left(\frac{a_1 t_1}{a_2}\right)^2 \qquad 1\,000 = \frac{1}{2}a_1\left(1 - \frac{a_1}{a_2}\right)t_1^2$$

$$t_1 = \sqrt{\frac{20\,000}{1.20}} = \boxed{129\ \text{s}}$$

$$t_2 = \frac{a_1 t_1}{-a_2} = \frac{12.9}{0.500} \approx 26\ \text{s} \qquad \text{Total time} = t = \boxed{155\ \text{s}}$$

P2.51 (a) $$y_f = v_{i1}t + \frac{1}{2}at^2 = 50.0 = 2.00t + \frac{1}{2}(9.80)t^2,$$

$$4.90t^2 + 2.00t - 50.0 = 0$$

$$t = \frac{-2.00 + \sqrt{2.00^2 - 4(4.90)(-50.0)}}{2(4.90)}$$

Only the positive root is physically meaningful:

$t = \boxed{3.00\ \text{s}}$ after the first stone is thrown.

(b) $$y_f = v_{i2}t + \frac{1}{2}at^2 \text{ and } t = 3.00 - 1.00 = 2.00\ \text{s}$$

substitute $50.0 = v_{i2}(2.00) + \frac{1}{2}(9.80)(2.00)^2$:

$v_{i2} = \boxed{15.3\ \text{m/s}}$ downward

(c) $v_{1f} = v_{i1} + at = 2.00 + (9.80)(3.00) = \boxed{31.4\ \text{m/s}}$ downward

$v_{2f} = v_{i2} + at = 15.3 + (9.80)(2.00) = \boxed{34.8\ \text{m/s}}$ downward

P2.52 Let the ball fall 1.50 m. It strikes at speed given by

$$v_{xf}^2 = v_{xi}^2 + 2a(x_f - x_i):$$

$$v_{xf}^2 = 0 + 2(-9.80 \text{ m/s}^2)(-1.50 \text{ m})$$

$$v_{xf} = -5.42 \text{ m/s}$$

and its stopping is described by

$$v_{xf}^2 = v_{xi}^2 + 2a_x(x_f - x_i)$$

$$0 = (-5.42 \text{ m/s})^2 + 2a_x(-10^{-2} \text{ m})$$

$$a_x = \frac{-29.4 \text{ m}^2/\text{s}^2}{-2.00 \times 10^{-2} \text{ m}} = +1.47 \times 10^3 \text{ m/s}^2.$$

Its maximum acceleration will be larger than the average acceleration we estimate by imagining constant acceleration, but will still be of order of magnitude $\boxed{\sim 10^3 \text{ m/s}^2}$.

P2.53 (a) In walking a distance Δx, in a time Δt, the length of rope ℓ is only increased by $\Delta x \sin\theta$.

$\therefore$ The pack lifts at a rate $\frac{\Delta x}{\Delta t}\sin\theta$.

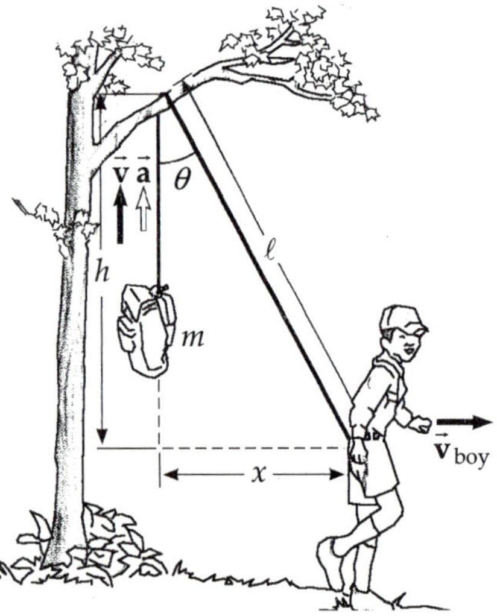

FIG. P2.53

$$v = \frac{\Delta x}{\Delta t}\sin\theta = v_{\text{boy}}\frac{x}{\ell} = \boxed{v_{\text{boy}}\frac{x}{\sqrt{x^2+h^2}}}$$

(b)
$$a = \frac{dv}{dt} = \frac{v_{\text{boy}}}{\ell}\frac{dx}{dt} + v_{\text{boy}}x\frac{d}{dt}\left(\frac{1}{\ell}\right)$$

$$a = v_{\text{boy}}\frac{v_{\text{boy}}}{\ell} - \frac{v_{\text{boy}}x}{\ell^2}\frac{d\ell}{dt}, \text{ but } \frac{d\ell}{dt} = v = v_{\text{boy}}\frac{x}{\ell}$$

$$\therefore a = \frac{v_{\text{boy}}^2}{\ell}\left(1 - \frac{x^2}{\ell^2}\right) = \frac{v_{\text{boy}}^2}{\ell}\frac{h^2}{\ell^2} = \boxed{\frac{h^2 v_{\text{boy}}^2}{(x^2+h^2)^{3/2}}}$$

(c) $\frac{v_{\text{boy}}^2}{h}$, 0

(d) v_{boy}, 0

P2.54 $h = 6.00\text{ m},\ v_{\text{boy}} = 2.00\text{ m/s}\ \ v = \dfrac{\Delta x}{\Delta t}\sin\theta = v_{\text{boy}}\dfrac{x}{\ell} = \dfrac{v_{\text{boy}}x}{\left(x^2 + h^2\right)^{1/2}}$.

However, $x = v_{\text{boy}}t$: $\therefore v = \dfrac{v_{\text{boy}}^2 t}{\left(v_{\text{boy}}^2 t^2 + h^2\right)^{1/2}} = \dfrac{4t}{\left(4t^2 + 36\right)^{1/2}}$.

(a)

t(s)	v(m/s)
0	0
0.5	0.33
1	0.63
1.5	0.89
2	1.11
2.5	1.28
3	1.41
3.5	1.52
4	1.60
4.5	1.66
5	1.71

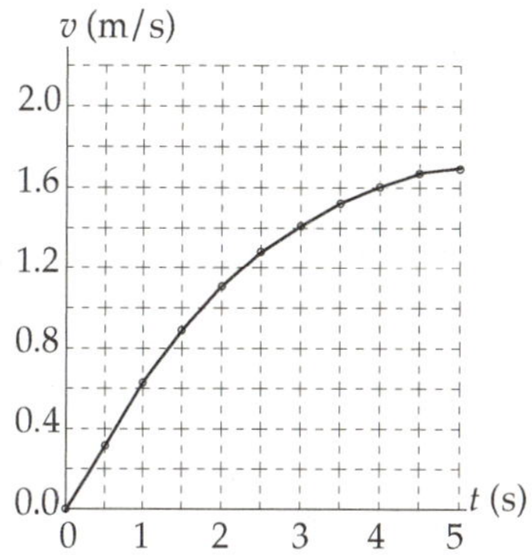

FIG. P2.54(a)

(b) From problem 2.53 above, $a = \dfrac{h^2 v_{\text{boy}}^2}{\left(x^2 + h^2\right)^{3/2}} = \dfrac{h^2 v_{\text{boy}}^2}{\left(v_{\text{boy}}^2 t^2 + h^2\right)^{3/2}} = \dfrac{144}{\left(4t^2 + 36\right)^{3/2}}$.

t(s)	$a\left(\text{m/s}^2\right)$
0	0.67
0.5	0.64
1	0.57
1.5	0.48
2	0.38
2.5	0.30
3.	0.24
3.5	0.18
4.	0.14
4.5	0.11
5	0.09

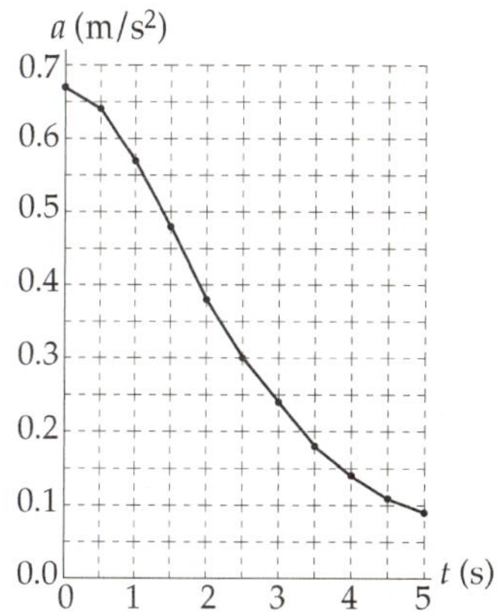

FIG. P2.54(b)

P2.55 (a) $d = \dfrac{1}{2}(9.80)t_1^2$ $\qquad d = 336t_2$

$t_1 + t_2 = 2.40$ $\qquad 336t_2 = 4.90(2.40 - t_2)^2$

$4.90t_2^2 - 359.5t_2 + 28.22 = 0$ $\qquad t_2 = \dfrac{359.5 \pm \sqrt{359.5^2 - 4(4.90)(28.22)}}{9.80}$

$t_2 = \dfrac{359.5 \pm 358.75}{9.80} = 0.076\,5\text{ s}$ so $d = 336t_2 = \boxed{26.4\text{ m}}$

(b) Ignoring the sound travel time, $d = \dfrac{1}{2}(9.80)(2.40)^2 = 28.2\text{ m}$, an error of $\boxed{6.82\%}$.

P2.56 (a)

Time t (s)	Height h (m)	Δh (m)	Δt (s)	$\overline{v}$ (m/s)	midpoint time t (s)
0.00	5.00				
		0.75	0.25	3.00	0.13
0.25	5.75				
		0.65	0.25	2.60	0.38
0.50	6.40				
		0.54	0.25	2.16	0.63
0.75	6.94				
		0.44	0.25	1.76	0.88
1.00	7.38				
		0.34	0.25	1.36	1.13
1.25	7.72				
		0.24	0.25	0.96	1.38
1.50	7.96				
		0.14	0.25	0.56	1.63
1.75	8.10				
		0.03	0.25	0.12	1.88
2.00	8.13				
		–0.06	0.25	–0.24	2.13
2.25	8.07				
		–0.17	0.25	–0.68	2.38
2.50	7.90				
		–0.28	0.25	–1.12	2.63
2.75	7.62				
		–0.37	0.25	–1.48	2.88
3.00	7.25				
		–0.48	0.25	–1.92	3.13
3.25	6.77				
		–0.57	0.25	–2.28	3.38
3.50	6.20				
		–0.68	0.25	–2.72	3.63
3.75	5.52				
		–0.79	0.25	–3.16	3.88
4.00	4.73				
		–0.88	0.25	–3.52	4.13
4.25	3.85				
		–0.99	0.25	–3.96	4.38
4.50	2.86				
		–1.09	0.25	–4.36	4.63
4.75	1.77				
		–1.19	0.25	–4.76	4.88
5.00	0.58				

TABLE P2.56

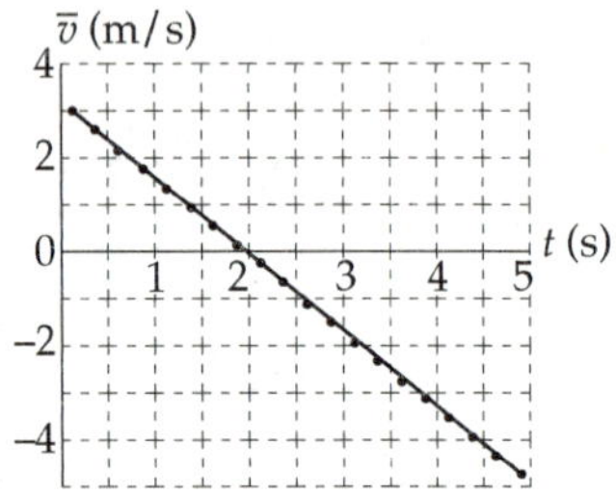

FIG. P2.56(b)

acceleration = slope of line is constant.

$\overline{a} = -1.63\ \text{m/s}^2 = \boxed{1.63\ \text{m/s}^2 \text{ downward}}$

P2.57 The distance x and y are always related by $x^2+y^2=L^2$. Differentiating this equation with respect to time, we have

$$2x\frac{dx}{dt}+2y\frac{dy}{dt}=0$$

Now $\frac{dy}{dt}$ is v_B, the unknown velocity of B; and $\frac{dx}{dt}=-v$.
From the equation resulting from differentiation, we have

$$\frac{dy}{dt}=-\frac{x}{y}\left(\frac{dx}{dt}\right)=-\frac{x}{y}(-v).$$

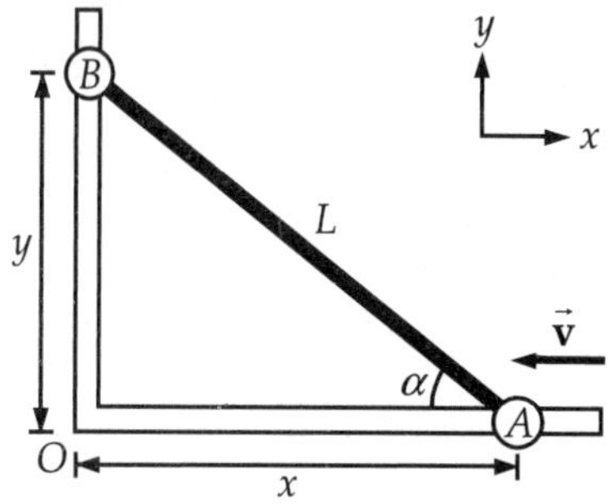

FIG. P2.57

But $\frac{y}{x}=\tan\alpha$ so $v_B=\left(\frac{1}{\tan\alpha}\right)v$. When $\alpha=60.0°$, $v_B=\frac{v}{\tan 60.0°}=\frac{v\sqrt{3}}{3}=\boxed{0.577v}$.

ANSWERS TO EVEN PROBLEMS

P2.2 (a) 50.0 m/s; (b) 41.0 m/s

P2.4 (a) 3.75 m/s; (b) 0

P2.6 (a) 27.0 m;
(b) $x_f=27.0\text{ m}+(18.0\text{ m/s})\Delta t+(3.00\text{ m/s}^2)(\Delta t)^2$;
(c) 18.0 m/s

P2.8 (a) 5.0 m/s; (b) –2.5 m/s; (c) 0; (d) 5.0 m/s

P2.10 $1.34\times10^4\text{ m/s}^2$

P2.12 (a) 13.0 m/s; (b) 10.0 m/s, 16.0 m/s; (c) 6.00 m/s^2; (d) 6.00 m/s^2

P2.14 (a) see the solution; (b) see the solution; (c) –4 m/s^2; (d) 34 m; (e) 28 m

P2.16 (a) see the solution; (b) see the solution; (c) see the solution; (d) see the solution; (e) see the solution; (f) The spacing of the successive positions would change with less regularity.

P2.18 160 ft

P2.20 (a) 4.53 s; (b) 14.1 m/s

P2.22 (a) 2.56 m; (b) –3.00 m/s

P2.24 $x_f-x_i=v_{xf}t-\frac{1}{2}a_xt^2$; 3.10 m/s

P2.26 (a) 4.98×10^{-9} s; (b) $1.20\times10^{15}\text{ m/s}^2$

P2.28 (a) 4.90 cm below the top of the TV, 0.980 m/s downward;
(b) 19.6 cm down, 1.96 m/s downward;
(c) 44.1 cm down, 2.94 m/s downward

P2.30 \$99.3/h

P2.32 1.79 s

P2.34 (a) 510 m; (b) 20.4 s

P2.36 (a) 5.25 m/s^2; (b) 168 m; (c) 52.5 m/s

P2.38 (a) 3.45 s; (b) 10.0 ft

P2.40 (a) 96.0 ft/s;
(b) $a=3.07\times10^3\text{ ft/s}^2$ upward
(c) $\Delta t=3.13\times10^{-2}$ s

P2.42 25.4 s

P2.44 (a) false unless the acceleration is zero; see the solution;
(b) true

P2.46 (a) see the solution; (b) 2.2 (mi/h)/s; (c) 6.7 mi

P2.48 (a) 15.0 s; (b) 30.0 m/s; (c) 225 m

P2.50 155 s, 129 s

P2.52 $\sim 10^3 \text{ m/s}^2$

P2.54 see the solution

P2.56 see the solution, $a = 1.63 \text{ m/s}^2$ downward

Chapter 3

Motion in Two Dimensions

CHAPTER OUTLINE

ANSWERS TO QUESTIONS

Q3.1 No, you cannot determine the instantaneous velocity. Yes, you can determine the average velocity. The points could be widely separated. In this case, you can only determine the average velocity, which is

$$\vec{\mathbf{v}} = \frac{\Delta \vec{\mathbf{x}}}{\Delta t}.$$

Q3.2 (a)

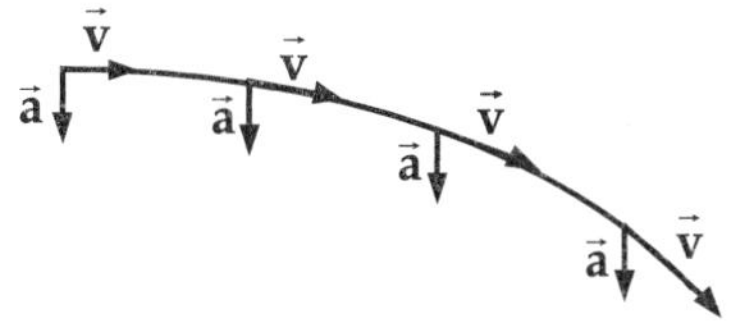

FIG. Q3.2(a)

(b)

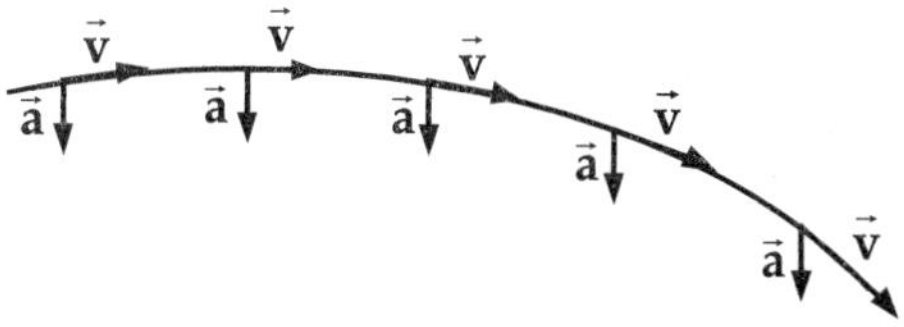

FIG. Q3.2(b)

Q3.3 The easiest way to approach this problem is to determine acceleration first, velocity second and finally position. Determine

(c) acceleration first,

(b) velocity second, and

(a) finally position.

Vertical: In free flight, $a_y = -g$. At the top of a projectile's trajectory, $v_y = 0$. Using this, the maximum height can be found using $v_{fy}^2 = v_{iy}^2 + 2a_y(y_f - y_i)$.

Horizontal: $a_x = 0$, so v_x is always the same. To find the horizontal position at maximum height, one needs the flight time, t. Using the vertical information found previously, the flight time can be found using $v_{fy} = v_{iy} + a_y t$. The horizontal position is $x_f = v_{ix} t$.

If air resistance were taken into account, then the acceleration in both the x and y-directions would have an additional term due to the drag.

Q3.4 The balls will be closest together as the second ball is thrown. Yes, the first ball will always be moving faster, since its flight time is larger, and thus the vertical component of the velocity is larger. The time interval will be one second. No, since the vertical component of the motion determines the flight time.

Q3.5 A parabola.

Q3.6 (a) no (b) yes (c) yes (d) no

Q3.7 The projectile is in free fall. Its vertical component of acceleration is the downward acceleration of gravity. Its horizontal component of acceleration is zero.

Q3.8 The optimal angle would be less than 45°. The longer the projectile is in the air, the more that air resistance will change the components of the velocity. Since the vertical component of the motion determines the flight time, an angle less than 45° would increase range.

Q3.9 The projectile on the moon would have both the larger range and the greater altitude. *Apollo* astronauts performed the experiment with golf balls.

Q3.10 The racing car rounds the turn at a constant *speed* of 90 miles per hour.

Q3.11 (a) no (b) yes

In the second case, the particle is continuously changing the direction of its velocity vector.

Q3.12 (a) The velocity is not constant because the object is constantly changing the direction of its motion.

(b) The acceleration is not constant because the acceleration always points towards the center of the circle. The magnitude of the acceleration is constant, but not the direction.

Q3.13 (a) straight ahead (b) in a circle or straight ahead

Q3.14

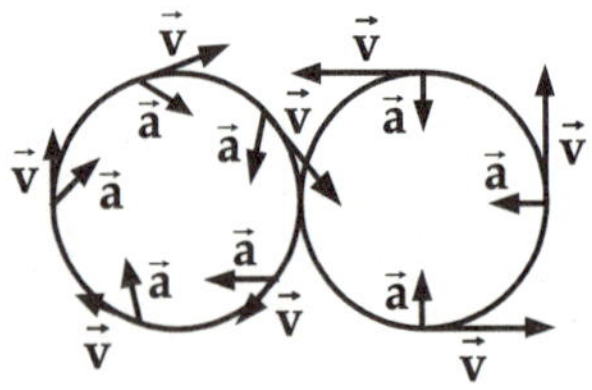

FIG. Q3.14

Q3.15 The wrench will hit at the base of the mast. If air resistance is a factor, it will hit slightly leeward of the base of the mast, displaced in the direction in which air is moving relative to the deck. If the boat is scudding before the wind, for example, the wrench's impact point can be in front of the mast.

Q3.16 (a) The ball would move straight up and down as observed by the passenger. The ball would move in a parabolic trajectory as seen by the ground observer.

(b) Both the passenger and the ground observer would see the ball move in a parabolic trajectory, although the two observed paths would not be the same.

SOLUTIONS TO PROBLEMS

Section 3.1 The Position, Velocity, and Acceleration Vectors

P3.1

$x(m)$	$y(m)$
0	−3 600
−3 000	0
−1 270	1 270
−4 270 m	−2 330 m

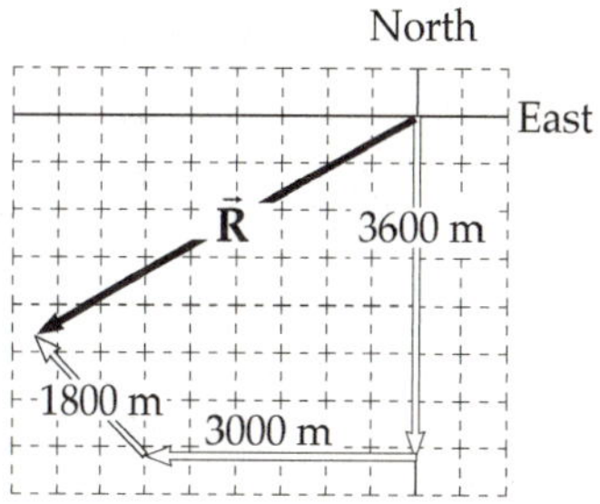

FIG. P3.1

(a) Net displacement $=\sqrt{x^2+y^2}$

$$= \boxed{4.87 \text{ km at } 28.6° \text{ S of W}}$$

(b) $$\text{Average speed} = \frac{(20.0 \text{ m/s})(180 \text{ s})+(25.0 \text{ m/s})(120 \text{ s})+(30.0 \text{ m/s})(60.0 \text{ s})}{180 \text{ s}+120 \text{ s}+60.0 \text{ s}} = \boxed{23.3 \text{ m/s}}$$

(c) $$\text{Average velocity} = \frac{4.87\times 10^3 \text{ m}}{360 \text{ s}} = \boxed{13.5 \text{ m/s along } \vec{\mathbf{R}}}$$

P3.2 (a) For the average velocity, we have

$$\vec{\mathbf{v}}_{av} = \left(\frac{x(4.00)-x(2.00)}{4.00 \text{ s}-2.00 \text{ s}}\right)\hat{\mathbf{i}} + \left(\frac{y(4.00)-y(2.00)}{4.00 \text{ s}-2.00 \text{ s}}\right)\hat{\mathbf{j}} = \left(\frac{5.00 \text{ m}-3.00 \text{ m}}{2.00 \text{ s}}\right)\hat{\mathbf{i}} + \left(\frac{3.00 \text{ m}-1.50 \text{ m}}{2.00 \text{ s}}\right)\hat{\mathbf{j}}$$

$$\vec{\mathbf{v}}_{av} = \boxed{\left(1.00\hat{\mathbf{i}}+0.750\hat{\mathbf{j}}\right) \text{ m/s}}$$

(b) For the velocity components,

we have $$v_x = \frac{dx}{dt} = a = 1.00 \text{ m/s}$$

and $$v_y = \frac{dy}{dt} = 2ct = \left(0.250 \text{ m/s}^2\right)t$$

Therefore, $$\vec{\mathbf{v}} = v_x\hat{\mathbf{i}} + v_y\hat{\mathbf{j}} = (1.00 \text{ m/s})\hat{\mathbf{i}} + \left(0.250 \text{ m/s}^2\right)t\,\hat{\mathbf{j}}$$

$$\boxed{\vec{\mathbf{v}}(t=2.00 \text{ s}) = (1.00 \text{ m/s})\hat{\mathbf{i}} + (0.500 \text{ m/s})\hat{\mathbf{j}}}$$

and the speed is $$\left|\vec{\mathbf{v}}(t=2.00 \text{ s})\right| = \sqrt{(1.00 \text{ m/s})^2+(0.500 \text{ m/s})^2} = \boxed{1.12 \text{ m/s}}$$

Section 3.2 Two-Dimensional Motion with Constant Acceleration

P3.3 $\vec{\mathbf{v}}_i = (4.00\hat{\mathbf{i}} + 1.00\hat{\mathbf{j}})$ m/s and $\vec{\mathbf{v}}(20.0) = (20.0\hat{\mathbf{i}} - 5.00\hat{\mathbf{j}})$ m/s

(a) $$a_x = \frac{\Delta v_x}{\Delta t} = \frac{20.0 - 4.00}{20.0} \text{ m/s}^2 = \boxed{0.800 \text{ m/s}^2}$$

$$a_y = \frac{\Delta v_y}{\Delta t} = \frac{-5.00 - 1.00}{20.0} \text{ m/s}^2 = \boxed{-0.300 \text{ m/s}^2}$$

(b) $$\theta = \tan^{-1}\left(\frac{-0.300}{0.800}\right) = -20.6° = \boxed{339° \text{ from } +x \text{ axis}}$$

(c) At $t = 25.0$ s

$$x_f = x_i + v_{xi}t + \frac{1}{2}a_x t^2 = 10.0 + 4.00(25.0) + \frac{1}{2}(0.800)(25.0)^2 = \boxed{360 \text{ m}}$$

$$y_f = y_i + v_{yi}t + \frac{1}{2}a_y t^2 = -4.00 + 1.00(25.0) + \frac{1}{2}(-0.300)(25.0)^2 = \boxed{-72.7 \text{ m}}$$

$$v_{xf} = v_{xi} + a_x t = 4 + 0.8(25) = 24 \text{ m/s}$$

$$v_{yf} = v_{yi} + a_y t = 1 - 0.3(25) = -6.5 \text{ m/s}$$

$$\theta = \tan^{-1}\left(\frac{v_y}{v_x}\right) = \tan^{-1}\left(\frac{-6.50}{24.0}\right) = \boxed{-15.2°}$$

P3.4 (a) $\vec{\mathbf{v}}_f = \vec{\mathbf{v}}_i + \vec{\mathbf{a}}t$

$$\vec{\mathbf{a}} = \frac{\vec{\mathbf{v}}_f - \vec{\mathbf{v}}_i}{t} = \frac{(9.00\hat{\mathbf{i}} + 7.00\hat{\mathbf{j}}) - (3.00\hat{\mathbf{i}} - 2.00\hat{\mathbf{j}})}{3.00} = \boxed{(2.00\hat{\mathbf{i}} + 3.00\hat{\mathbf{j}}) \text{ m/s}^2}$$

(b) $$\vec{\mathbf{r}}_f = \vec{\mathbf{r}}_i + \vec{\mathbf{v}}_i t + \frac{1}{2}\vec{\mathbf{a}}t^2 = (3.00\hat{\mathbf{i}} - 2.00\hat{\mathbf{j}})t + \frac{1}{2}(2.00\hat{\mathbf{i}} + 3.00\hat{\mathbf{j}})t^2$$

$$\boxed{x = (3.00t + t^2) \text{ m}} \text{ and } \boxed{y = (1.50t^2 - 2.00t) \text{ m}}$$

P3.5 $\vec{\mathbf{a}} = 3.00\hat{\mathbf{j}}$ m/s^2 ; $\vec{\mathbf{v}}_i = 5.00\hat{\mathbf{i}}$ m/s ; $\vec{\mathbf{r}}_i = 0\hat{\mathbf{i}} + 0\hat{\mathbf{j}}$

(a) $$\vec{\mathbf{r}}_f = \vec{\mathbf{r}}_i + \vec{\mathbf{v}}_i t + \frac{1}{2}\vec{\mathbf{a}}t^2 = \boxed{\left[5.00t\hat{\mathbf{i}} + \frac{1}{2}3.00t^2\hat{\mathbf{j}}\right] \text{ m}}$$

$$\vec{\mathbf{v}}_f = \vec{\mathbf{v}}_i + \vec{\mathbf{a}}t = \boxed{(5.00\hat{\mathbf{i}} + 3.00t\hat{\mathbf{j}}) \text{ m/s}}$$

continued on next page

(b) $t = 2.00\text{ s},\ \vec{\mathbf{r}}_f = 5.00(2.00)\hat{\mathbf{i}} + \frac{1}{2}(3.00)(2.00)^2\hat{\mathbf{j}} = (10.0\hat{\mathbf{i}} + 6.00\hat{\mathbf{j}})\text{ m}$

so $x_f = \boxed{10.0\text{ m}}$, $y_f = \boxed{6.00\text{ m}}$

$$\vec{\mathbf{v}}_f = 5.00\hat{\mathbf{i}} + 3.00(2.00)\hat{\mathbf{j}} = (5.00\hat{\mathbf{i}} + 6.00\hat{\mathbf{j}})\text{ m/s}$$

$$v_f = |\vec{\mathbf{v}}_f| = \sqrt{v_{xf}^2 + v_{yf}^2} = \sqrt{(5.00)^2 + (6.00)^2} = \boxed{7.81\text{ m/s}}$$

P3.6 (a) For the x-component of the motion we have $x_f = x_i + v_{xi}t + \frac{1}{2}a_x t^2$.

$$0.01\text{ m} = 0 + (1.80\times10^7\text{ m/s})t + \frac{1}{2}(8\times10^{14}\text{ m/s}^2)t^2$$

$$(4\times10^{14}\text{ m/s}^2)t^2 + (1.80\times10^7\text{ m/s})t - 10^{-2}\text{ m} = 0$$

$$t = \frac{-1.80\times10^7\text{ m/s} \pm \sqrt{(1.8\times10^7\text{ m/s})^2 - 4(4\times10^{14}\text{ m/s}^2)(-10^{-2}\text{m})}}{2(4\times10^{14}\text{ m/s}^2)} = \frac{-1.8\times10^7 \pm 1.84\times10^7\text{ m/s}}{8\times10^{14}\text{ m/s}^2}$$

We choose the + sign to represent the physical situation

$$t = \frac{4.39\times10^5\text{ m/s}}{8\times10^{14}\text{ m/s}^2} = 5.49\times10^{-10}\text{ s}.$$

Here

$$y_f = y_i + v_{yi}t + \frac{1}{2}a_y t^2 = 0 + 0 + \frac{1}{2}(1.6\times10^{15}\text{ m/s}^2)(5.49\times10^{-10}\text{ s})^2 = 2.41\times10^{-4}\text{ m}.$$

So, $\boxed{\vec{\mathbf{r}}_f = (10.0\,\hat{\mathbf{i}} + 0.241\,\hat{\mathbf{j}})\text{ mm}}$.

(b)
$$\vec{\mathbf{v}}_f = \vec{\mathbf{v}}_i + \vec{\mathbf{a}}t = 1.80\times10^7\text{ m/s}\hat{\mathbf{i}} + (8\times10^{14}\text{ m/s}^2\,\hat{\mathbf{i}} + 1.6\times10^{15}\text{ m/s}^2\,\hat{\mathbf{j}})(5.49\times10^{-10}\text{ s})$$
$$= (1.80\times10^7\text{ m/s})\hat{\mathbf{i}} + (4.39\times10^5\text{ m/s})\hat{\mathbf{i}} + (8.78\times10^5\text{ m/s})\hat{\mathbf{j}}$$
$$= \boxed{(1.84\times10^7\text{ m/s})\hat{\mathbf{i}} + (8.78\times10^5\text{ m/s})\hat{\mathbf{j}}}$$

(c)
$$|\vec{\mathbf{v}}_f| = \sqrt{(1.84\times10^7\text{ m/s})^2 + (8.78\times10^5\text{ m/s})^2} = \boxed{1.85\times10^7\text{ m/s}}$$

(d)
$$\theta = \tan^{-1}\left(\frac{v_y}{v_x}\right) = \tan^{-1}\left(\frac{8.78\times10^5}{1.84\times10^7}\right) = \boxed{2.73^\circ}$$

Section 3.3 Projectile Motion

P3.7 (a) The mug leaves the counter horizontally with a velocity v_{xi} (say). If time t elapses before it hits the ground, then since there is no horizontal acceleration, $x_f = v_{xi}t$, i.e.,

$$t = \frac{x_f}{v_{xi}} = \frac{(1.40 \text{ m})}{v_{xi}}$$

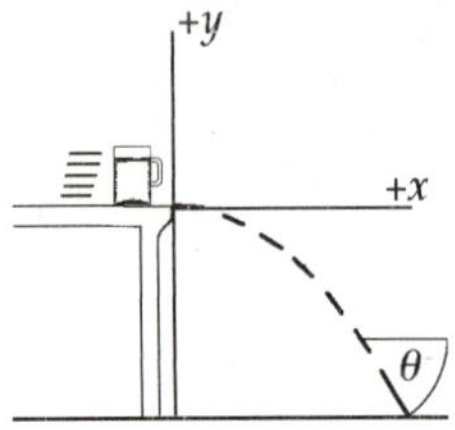

FIG. P3.7

In the same time it falls a distance of 0.860 m with acceleration downward of 9.80 m/s^2. Then

$$y_f = y_i + v_{yi}t + \frac{1}{2}a_y t^2: \ 0 = 0.860 \text{ m} + \frac{1}{2}\left(-9.80 \text{ m/s}^2\right)\left(\frac{1.40 \text{ m}}{v_{xi}}\right)^2.$$

Thus,

$$v_{xi} = \sqrt{\frac{\left(4.90 \text{ m/s}^2\right)\left(1.96 \text{ m}^2\right)}{0.860 \text{ m}}} = \boxed{3.34 \text{ m/s}}.$$

(b) The vertical velocity component with which it hits the floor is

$$v_{yf} = v_{yi} + a_y t = 0 + \left(-9.80 \text{ m/s}^2\right)\left(\frac{1.40 \text{ m}}{3.34 \text{ m/s}}\right) = -4.11 \text{ m/s}.$$

Hence, the angle θ at which the mug strikes the floor is given by

$$\theta = \tan^{-1}\left(\frac{v_{yf}}{v_{xf}}\right) = \tan^{-1}\left(\frac{-4.11}{3.34}\right) = \boxed{-50.9°}.$$

P3.8 The mug is a projectile from just after leaving the counter until just before it reaches the floor. Taking the origin at the point where the mug leaves the bar, the coordinates of the mug at any time are

$$x_f = v_{xi}t + \frac{1}{2}a_x t^2 = v_{xi}t + 0 \text{ and } y_f = v_{yi}t + \frac{1}{2}a_y t^2 = 0 - \frac{1}{2}gt^2.$$

When the mug reaches the floor, $y_f = -h$ so

$$-h = -\frac{1}{2}gt^2$$

which gives the time of impact as $t = \sqrt{\frac{2h}{g}}$.

continued on next page

(a) Since $x_f = d$ when the mug reaches the floor, $x_f = v_{xi} t$ becomes $d = v_{xi}\sqrt{\frac{2h}{g}}$ giving the initial velocity as $\boxed{v_{xi} = d\sqrt{\frac{g}{2h}}}$.

(b) Just before impact, the x-component of velocity is still

$$v_{xf} = v_{xi}$$

while the y-component is $v_{yf} = v_{yi} + a_y t = 0 - g\sqrt{\frac{2h}{g}}$.

Then the direction of motion just before impact is below the horizontal at an angle of

$$\theta = \tan^{-1}\left(\frac{|v_{yf}|}{v_{xf}}\right) = \tan^{-1}\left(\frac{g\sqrt{\frac{2h}{g}}}{d\sqrt{\frac{g}{2h}}}\right) = \boxed{\tan^{-1}\left(\frac{2h}{d}\right)}.$$

The answer for v_{xi} indicates that a larger measured value for d would imply larger takeoff speed in direct proportion. A tape measure lying on the floor could be calibrated as a speedometer. A larger value for h would imply a smaller value for speed by an inverse proportionality to the square root of h. That is, if h were 9 times larger, v_{xi} would be 3 times smaller. The answer for θ shows that the impact velocity makes an angle with the horizontal whose tangent is just twice as large as that of the elevation angle α of the edge of the table as seen from the impact point.

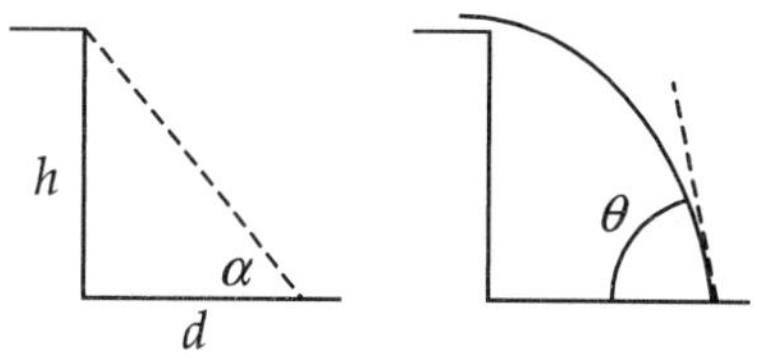

FIG. P3.8

***P3.9** At the maximum height $v_y = 0$, and the time to reach this height is found from

$$v_{yf} = v_{iy} + a_y t \text{ as } t = \frac{v_{yf} - v_{iy}}{a_y} = \frac{0 - v_{iy}}{-g} = \frac{v_{iy}}{g}.$$

The vertical displacement that has occurred during this time is

$$(\Delta y)_{\max} = v_{y,\text{avg}} t = \left(\frac{v_{yf} + v_{iy}}{2}\right)t = \left(\frac{0 + v_{iy}}{2}\right)\left(\frac{v_{iy}}{g}\right) = \frac{v_{iy}^2}{2g}.$$

Thus, if $(\Delta y)_{\max} = 12 \text{ ft}\left(\frac{1 \text{ m}}{3.281 \text{ ft}}\right) = 3.66 \text{ m}$, then

$$v_{iy} = \sqrt{2g(\Delta y)_{\max}} = \sqrt{2(9.80 \text{ m/s}^2)(3.66 \text{ m})} = 8.47 \text{ m/s},$$

continued on next page

and if the angle of projection is $\theta = 45°$, the launch speed is

$$v_i = \frac{v_{iy}}{\sin\theta} = \frac{8.47\ \text{m/s}}{\sin 45°} = \boxed{12.0\ \text{m/s}}.$$

P3.10 From Equation 3.16 with $R = 15.0$ m, $v_i = 3.00$ m/s, $\theta_{max} = 45.0°$

$$\therefore g = \frac{v_i^2}{R} = \frac{9.00}{15.0} = \boxed{0.600\ \text{m/s}^2}$$

P3.11 Take the origin at the mouth of the cannon.

$x_f = v_{xi}t$ $\qquad$ $2\,000\ \text{m} = (1\,000\ \text{m/s})\cos\theta_i t$

Therefore, $\qquad$ $t = \dfrac{2.00\ \text{s}}{\cos\theta_i}$

$y_f = v_{yi}t + \frac{1}{2}a_y t^2$: $\qquad$ $800\ \text{m} = (1\,000\ \text{m/s})\sin\theta_i t + \frac{1}{2}(-9.80\ \text{m/s}^2)t^2$

$$800\ \text{m} = (1\,000\ \text{m/s})\sin\theta_i\left(\frac{2.00\ \text{s}}{\cos\theta_i}\right) - \frac{1}{2}(9.80\ \text{m/s}^2)\left(\frac{2.00\ \text{s}}{\cos\theta_i}\right)^2$$

$$800\ \text{m}(\cos^2\theta_i) = 2\,000\ \text{m}(\sin\theta_i\cos\theta_i) - 19.6\ \text{m}$$

$$19.6\ \text{m} + 800\ \text{m}(\cos^2\theta_i) = 2\,000\ \text{m}\sqrt{1-\cos^2\theta_i}(\cos\theta_i)$$

$$384 + (31\,360)\cos^2\theta_i + (640\,000)\cos^4\theta_i = (4\,000\,000)\cos^2\theta_i - (4\,000\,000)\cos^4\theta_i$$

$$4\,640\,000\cos^4\theta_i - 3\,968\,640\cos^2\theta_i + 384 = 0$$

$$\cos^2\theta_i = \frac{3\,968\,640 \pm \sqrt{(3\,968\,640)^2 - 4(4\,640\,000)(384)}}{9\,280\,000}$$

$\cos\theta_i = 0.925$ $\qquad$ or $\qquad$ 0.00984

$\theta_i = \boxed{22.4° \text{ or } 89.4°}$ $\qquad$ (Both solutions are valid.)

P3.12 (a) $x_f = v_{xi}t = 8.00\cos 20.0°(3.00) = \boxed{22.6\ \text{m}}$

(b) Taking y positive downwards, and the final point just before ground impact,

$$y_f = v_{yi}t + \frac{1}{2}gt^2$$

$$y_f = 8.00\sin 20.0°(3.00) + \frac{1}{2}(9.80)(3.00)^2 = \boxed{52.3\ \text{m}}.$$

continued on next page

(c) Now take the final point 10 m below the window.

$$10.0 = 8.00(\sin 20.0°)t + \frac{1}{2}(9.80)t^2$$

$$4.90t^2 + 2.74t - 10.0 = 0$$

$$t = \frac{-2.74 \pm \sqrt{(2.74)^2 + 196}}{9.80} = \boxed{1.18 \text{ s}}$$

P3.13 Consider the motion from original zero height to maximum height h:

$v_{yf}^2 = v_{yi}^2 + 2a_y(y_f - y_i)$ gives $0 = v_{yi}^2 - 2g(h - 0)$ or $v_{yi} = \sqrt{2gh}$

Now consider the motion from the original point to half the maximum height:

$v_{yf}^2 = v_{yi}^2 + 2a_y(y_f - y_i)$ gives $v_{yh}^2 = 2gh + 2(-g)\left(\frac{1}{2}h - 0\right)$ so $v_{yh} = \sqrt{gh}$

At maximum height, the speed is $v_x = \frac{1}{2}\sqrt{v_x^2 + v_{yh}^2} = \frac{1}{2}\sqrt{v_x^2 + gh}$

Solving, $v_x = \sqrt{\frac{gh}{3}}$

Now the projection angle is $\theta_i = \tan^{-1}\frac{v_{yi}}{v_x} = \tan^{-1}\frac{\sqrt{2gh}}{\sqrt{gh/3}} = \tan^{-1}\sqrt{6} = \boxed{67.8°}$.

P3.14 We interpret the problem to mean that the displacement from fish to bug is

$$2.00 \text{ m at } 30° = (2.00 \text{ m})\cos 30°\hat{\mathbf{i}} + (2.00 \text{ m})\sin 30°\hat{\mathbf{j}} = (1.73 \text{ m})\hat{\mathbf{i}} + (1.00 \text{ m})\hat{\mathbf{j}}.$$

If the water should drop 0.03 m during its flight, then the fish must aim at a point 0.03 m above the bug. The initial velocity of the water then is directed through the point with displacement

$$(1.73 \text{ m})\hat{\mathbf{i}} + (1.03 \text{ m})\hat{\mathbf{j}} = 2.015 \text{ m at } 30.7°.$$

For the time of flight of a water drop we have $x_f = x_i + v_{xi}t + \frac{1}{2}a_x t^2$

$$1.73 \text{ m} = 0 + (v_i \cos 30.7°)t + 0$$

so $t = \frac{1.73 \text{ m}}{v_i \cos 30.7°}$. The vertical motion is described by $y_f = y_i + v_{yi}t + \frac{1}{2}a_y t^2$. The "drop on its path" is $-3.00 \text{ cm} = \frac{1}{2}(-9.80 \text{ m/s}^2)\left(\frac{1.73 \text{ m}}{v_i \cos 30.7°}\right)^2$. Thus,

$$v_i = \frac{1.73 \text{ m}}{\cos 30.7°}\sqrt{\frac{9.80 \text{ m/s}^2}{2 \times 0.03 \text{ m}}} = 2.015 \text{ m}\left(12.8 \text{ s}^{-1}\right) = \boxed{25.8 \text{ m/s}}.$$

P3.15 (a) We use the trajectory equation: $y_f = x_f \tan\theta_i - \dfrac{g x_f^2}{2v_i^2\cos^2\theta_i}$. With $x_f = 36.0\text{ m}$, $v_i = 20.0\text{ m/s}$, and $\theta = 53.0°$ we find $y_f = (36.0\text{ m})\tan 53.0° - \dfrac{(9.80\text{ m/s}^2)(36.0\text{ m})^2}{2(20.0\text{ m/s})^2\cos^2(53.0°)} = 3.94\text{ m}$. The ball clears the bar by $(3.94 - 3.05)\text{ m} = \boxed{0.889\text{ m}}$.

(b) The time the ball takes to reach the maximum height is

$$t_1 = \frac{v_i \sin\theta_i}{g} = \frac{(20.0\text{ m/s})(\sin 53.0°)}{9.80\text{ m/s}^2} = 1.63\text{ s}.$$

The time to travel 36.0 m horizontally is $t_2 = \dfrac{x_f}{v_{ix}}$

$$t_2 = \frac{36.0\text{ m}}{(20.0\text{ m/s})(\cos 53.0°)} = 2.99\text{ s}.$$

Since $t_2 > t_1$ $\boxed{\text{the ball clears the goal on its way down}}$.

P3.16 The horizontal component of displacement is $x_f = v_{xi}t = (v_i\cos\theta_i)t$. Therefore, the time required to reach the building a distance d away is $t = \dfrac{d}{v_i\cos\theta_i}$. At this time, the altitude of the water is

$$y_f = v_{yi}t + \frac{1}{2}a_y t^2 = v_i\sin\theta_i\left(\frac{d}{v_i\cos\theta_i}\right) - \frac{g}{2}\left(\frac{d}{v_i\cos\theta_i}\right)^2.$$

Therefore the water strikes the building at a height h above ground level of

$$h = y_f = \boxed{d\tan\theta_i - \frac{gd^2}{2v_i^2\cos^2\theta_i}}.$$

P3.17 (a) For the horizontal motion, we have

$$x_f = x_i + v_{xi}t + \frac{1}{2}a_x t^2$$
$$24\text{ m} = 0 + v_i(\cos 53°)(2.2\text{ s}) + 0$$
$$v_i = \boxed{18.1\text{ m/s}}.$$

(b) As it passes over the wall, the ball is above the street by $y_f = y_i + v_{yi}t + \frac{1}{2}a_y t^2$

$$y_f = 0 + (18.1\text{ m/s})(\sin 53°)(2.2\text{ s}) + \frac{1}{2}\left(-9.8\text{ m/s}^2\right)(2.2\text{ s})^2 = 8.13\text{ m}.$$

So it clears the parapet by $8.13\text{ m} - 7\text{ m} = \boxed{1.13\text{ m}}$.

continued on next page

(c) Note that the highest point of the ball's trajectory is not directly above the wall. For the whole flight, we have from the trajectory equation $y_f = (\tan\theta_i)x_f - \left(\frac{g}{2v_i^2\cos^2\theta_i}\right)x_f^2$ or $6\text{ m} = (\tan 53°)x_f - \left(\frac{9.8\text{ m/s}^2}{2(18.1\text{ m/s})^2\cos^2 53°}\right)x_f^2$. Solving, $\left(0.041\,2\text{ m}^{-1}\right)x_f^2 - 1.33x_f + 6\text{ m} = 0$ and $x_f = \frac{1.33 \pm \sqrt{1.33^2 - 4(0.041\,2)(6)}}{2\left(0.041\,2\text{ m}^{-1}\right)}$. This yields two results: $x_f = 26.8$ m or 5.44 m. The ball passes twice through the level of the roof. It hits the roof at distance from the wall $26.8\text{ m} - 24\text{ m} = \boxed{2.79\text{ m}}$.

***P3.18** We match the given equations

$$x_f = 0 + (11.2\text{ m/s})\cos 18.5°\,t$$
$$0.360\text{ m} = 0.840\text{ m} + (11.2\text{ m/s})\sin 18.5°\,t - \frac{1}{2}\left(9.80\text{ m/s}^2\right)t^2$$

to the equations for the coordinates of the final position of a projectile

$$x_f = x_i + v_{xi}t$$
$$y_f = y_i + v_{yi}t - \frac{1}{2}gt^2$$

For the equations to represent the same functions of time, all coefficients must agree: $x_i = 0$, $y_i = 0.840\text{ m}$, $v_{xi} = (11.2\text{ m/s})\cos 18.5°$, $v_{yi} = (11.2\text{ m/s})\sin 18.5°$ and $g = 9.80\text{ m/s}^2$.

(a) Then the original position of the athlete's center of mass is the point with coordinates $(x_i,\, y_i) = \boxed{(0,\ 0.840\text{ m})}$. That is, his original position has position vector $\vec{\mathbf{r}} = 0\hat{\mathbf{i}} + 0.840\text{ m}\hat{\mathbf{j}}$.

(b) His original velocity is $\vec{\mathbf{v}}_i = (11.2\text{ m/s})\cos 18.5°\,\hat{\mathbf{i}} + (11.2\text{ m/s})\sin 18.5°\,\hat{\mathbf{j}} = \boxed{11.2\text{ m/s at }18.5°}$ above the x-axis.

(c) From $4.90\text{ m/s}^2\,t^2 - 3.55\text{ m/s}\,t - 0.48\text{ m} = 0$ we find the time of flight, which must be positive $t = \frac{+3.55\text{ m/s} + \sqrt{(3.55\text{ m/s})^2 - 4\left(4.9\text{ m/s}^2\right)(-0.48\text{ m})}}{2\left(4.9\text{ m/s}^2\right)} = 0.842\text{ s}$. Then $x_f = (11.2\text{ m/s})\cos 18.5°(0.8425) = \boxed{8.94\text{ m}}$.

(d)

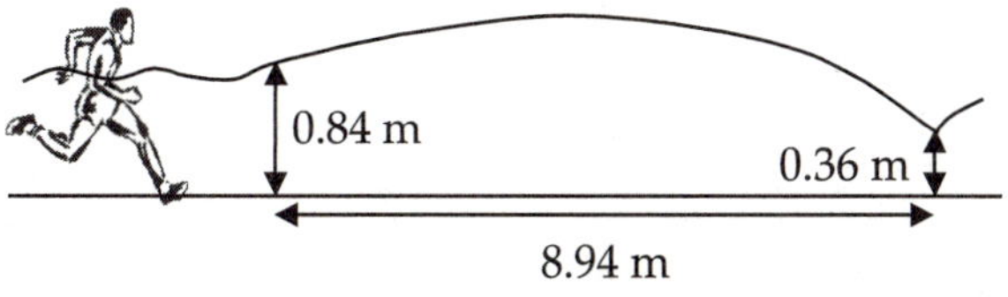

FIG. P3.18

P3.19 The horizontal kick gives zero vertical velocity to the rock. Then its time of flight follows from

$$y_f = y_i + v_{yi}t + \frac{1}{2}a_y t^2$$
$$-40.0 \text{ m} = 0 + 0 + \frac{1}{2}\left(-9.80 \text{ m/s}^2\right)t^2$$
$$t = 2.86 \text{ s}.$$

The extra time $3.00 \text{ s} - 2.86 \text{ s} = 0.143 \text{ s}$ is the time required for the sound she hears to travel straight back to the player.

It covers distance

$$(343 \text{ m/s})0.143 \text{ s} = 49.0 \text{ m} = \sqrt{x^2 + (40.0 \text{ m})^2}$$

where x represents the horizontal distance the rock travels.

$$x = 28.3 \text{ m} = v_{xi}t + 0t^2$$
$$\therefore v_{xi} = \frac{28.3 \text{ m}}{2.86 \text{ s}} = \boxed{9.91 \text{ m/s}}$$

P3.20 From the instant he leaves the floor until just before he lands, the basketball star is a projectile. His vertical velocity and vertical displacement are related by the equation $v_{yf}^2 = v_{yi}^2 + 2a_y\left(y_f - y_i\right)$. Applying this to the upward part of his flight gives $0 = v_{yi}^2 + 2\left(-9.80 \text{ m/s}^2\right)(1.85 - 1.02) \text{ m}$. From this, $v_{yi} = 4.03 \text{ m/s}$. [Note that this is the answer to part (c) of this problem.]

For the downward part of the flight, the equation gives $v_{yf}^2 = 0 + 2\left(-9.80 \text{ m/s}^2\right)(0.900 - 1.85) \text{ m}$. Thus the vertical velocity just before he lands is

$$v_{yf} = -4.32 \text{ m/s}.$$

(a) His hang time may then be found from $v_{yf} = v_{yi} + a_y t$:

$$-4.32 \text{ m/s} = 4.03 \text{ m/s} + \left(-9.80 \text{ m/s}^2\right)t$$

or $t = \boxed{0.852 \text{ s}}$.

(b) Looking at the total horizontal displacement during the leap, $x = v_{xi}t$ becomes

$$2.80 \text{ m} = v_{xi}(0.852 \text{ s})$$

which yields $v_{xi} = \boxed{3.29 \text{ m/s}}$.

(c) $v_{yi} = \boxed{4.03 \text{ m/s}}$. See above for proof.

(d) The takeoff angle is: $\theta = \tan^{-1}\left(\frac{v_{yi}}{v_{xi}}\right) = \tan^{-1}\left(\frac{4.03 \text{ m/s}}{3.29 \text{ m/s}}\right) = \boxed{50.8°}$.

continued on next page

(e) Similarly for the deer, the upward part of the flight gives

$v_{yf}^2 = v_{yi}^2 + 2a_y(y_f - y_i)$:

$$0 = v_{yi}^2 + 2(-9.80 \text{ m/s}^2)(2.50 - 1.20) \text{ m}$$

so $v_{yi} = 5.04$ m/s.

For the downward part, $v_{yf}^2 = v_{yi}^2 + 2a_y(y_f - y_i)$ yields $v_{yf}^2 = 0 + 2(-9.80 \text{ m/s}^2)(0.700 - 2.50)$ m and $v_{yf} = -5.94$ m/s.

The hang time is then found as $v_{yf} = v_{yi} + a_y t$: $-5.94 \text{ m/s} = 5.04 \text{ m/s} + (-9.80 \text{ m/s}^2)t$ and

$$\boxed{t = 1.12 \text{ s}}.$$

P3.21 For the smallest impact angle

$$\theta = \tan^{-1}\left(\frac{v_{yf}}{v_{xf}}\right),$$

we want to minimize v_{yf} and maximize $v_{xf} = v_{xi}$. The final *y*-component of velocity is related to v_{yi} by $v_{yf}^2 = v_{yi}^2 + 2gh$, so we want to minimize v_{yi} and maximize v_{xi}. Both are accomplished by making the initial velocity horizontal. Then $v_{xi} = v$, $v_{yi} = 0$, and $v_{yf} = \sqrt{2gh}$. At last, the impact angle is

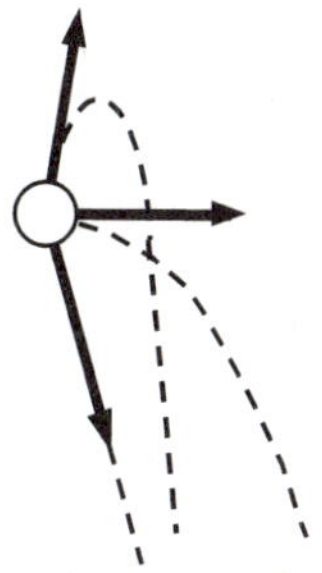

FIG. P3.21

$$\theta = \tan^{-1}\left(\frac{v_{yf}}{v_{xf}}\right) = \boxed{\tan^{-1}\left(\frac{\sqrt{2gh}}{v}\right)}.$$

Section 3.4 The Particle in Uniform Circular Motion

P3.22 $a = \dfrac{v^2}{R}$, $T = 24 \text{ h}(3\,600 \text{ s/h}) = 86\,400$ s

$$v = \frac{2\pi R}{T} = \frac{2\pi(6.37\times10^6 \text{ m})}{86\,400 \text{ s}} = 463 \text{ m/s}$$

$$a = \frac{(463 \text{ m/s})^2}{6.37\times10^6 \text{ m}} = \boxed{0.033\,7 \text{ m/s}^2 \text{ directed toward the center of Earth}}.$$

P3.23 $a_c = \dfrac{v^2}{r} = \dfrac{(20.0 \text{ m/s})^2}{1.06 \text{ m}} = \boxed{377 \text{ m/s}^2}$

The mass is unnecessary information.

***P3.24** $a_c = \frac{v^2}{r}$. Let f represent the rotation rate. Each revolution carries each bit of metal through distance $2\pi r$, so $v = 2\pi r f$ and

$$a_c = \frac{v^2}{r} = 4\pi^2 r f^2 = 100\,g.$$

A smaller radius implies smaller acceleration. To meet the criterion for each bit of metal we consider the minimum radius:

$$f = \left(\frac{100\,g}{4\pi^2 r}\right)^{1/2} = \left(\frac{100 \cdot 9.8\ \text{m/s}^2}{4\pi^2(0.021\ \text{m})}\right)^{1/2} = 34.4\ \frac{1}{\text{s}}\left(\frac{60\ \text{s}}{1\ \text{min}}\right) = \boxed{2.06 \times 10^3\ \text{rev/min}}.$$

P3.25 $r = 0.500$ m;

$$v_t = \frac{2\pi r}{T} = \frac{2\pi(0.500\ \text{m})}{\frac{60.0\ \text{s}}{200\ \text{rev}}} = 10.47\ \text{m/s} = \boxed{10.5\ \text{m/s}}$$

$$a = \frac{v^2}{R} = \frac{(10.47)^2}{0.5} = \boxed{219\ \text{m/s}^2\ \text{inward}}$$

P3.26 $a_c = \frac{v^2}{r}$

$$v = \sqrt{a_c r} = \sqrt{3(9.8\ \text{m/s}^2)(9.45\ \text{m})} = 16.7\ \text{m/s}$$

Each revolution carries the astronaut over a distance of $2\pi r = 2\pi(9.45\ \text{m}) = 59.4$ m. Then the rotation rate is

$$16.7\ \text{m/s}\left(\frac{1\ \text{rev}}{59.4\ \text{m}}\right) = \boxed{0.281\ \text{rev/s}}.$$

P3.27 The satellite is in free fall. Its acceleration is due to gravity and is by effect a centripetal acceleration.

$$a_c = g$$

so $\frac{v^2}{r} = g$. Solving for the velocity, $v = \sqrt{rg} = \sqrt{(6{,}400 + 600)(10^3\ \text{m})(8.21\ \text{m/s}^2)} = \boxed{7.58 \times 10^3\ \text{m/s}}$.

$v = \frac{2\pi r}{T}$ and

$$T = \frac{2\pi r}{v} = \frac{2\pi(7{,}000 \times 10^3\ \text{m})}{7.58 \times 10^3\ \text{m/s}} = \boxed{5.80 \times 10^3\ \text{s}}$$

$$T = 5.80 \times 10^3\ \text{s}\left(\frac{1\ \text{min}}{60\ \text{s}}\right) = 96.7\ \text{min}.$$

Section 3.5 Tangential and Radial Acceleration

P3.28 (b) We do part (b) first. The tangential speed is described by $v_f = v_i + a_t t$

$0.7\ \text{m/s} = 0 + a_t(1.75\ \text{s})$ so $\boxed{a_t = 0.400\ \text{m/s}^2\ \text{forward}}$

(a) Now at $t = 1.25\ \text{s}$, $v_f = v_i + a_t t = 0 + \left(0.4\ \text{m/s}^2\right)1.25\ \text{s}$

$v_f = 0.5\ \text{m/s}$

so $a_c = \dfrac{v^2}{r} = \dfrac{(0.5\ \text{m/s})^2}{0.2\ \text{m}} = \boxed{1.25\ \text{m/s}^2\ \text{toward the center}}$

(c) $\vec{\mathbf{a}} = \vec{\mathbf{a}}_r + \vec{\mathbf{a}}_t = 0.4\ \text{m/s}^2\ \text{forward} + 1.25\ \text{m/s}^2\ \text{inward}$

$\vec{\mathbf{a}} = \sqrt{0.4^2 + 1.25^2}$ forward and inward at $\theta = \tan^{-1}\left(\dfrac{1.25}{0.4}\right)$

$\vec{\mathbf{a}} = \boxed{1.31\ \text{m/s}^2\ \text{forward and } 72.3°\ \text{inward}}$

P3.29 We assume the train is still slowing down at the instant in question.

$$a_c = \frac{v^2}{r} = 1.29\ \text{m/s}^2$$

$$a_t = \frac{\Delta v}{\Delta t} = \frac{(-40.0\ \text{km/h})\left(10^3\ \text{m/km}\right)\left(\frac{1\ \text{h}}{3\,600\ \text{s}}\right)}{15.0\ \text{s}} = -0.741\ \text{m/s}^2$$

$$a = \sqrt{a_c^2 + a_t^2} = \sqrt{\left(1.29\ \text{m/s}^2\right)^2 + \left(-0.741\ \text{m/s}^2\right)^2}$$

at an angle of $\tan^{-1}\left(\dfrac{|a_t|}{a_c}\right) = \tan^{-1}\left(\dfrac{0.741}{1.29}\right)$

$\vec{\mathbf{a}} = \boxed{1.48\ \text{m/s}^2\ \text{inward and } 29.9°\ \text{backward}}$

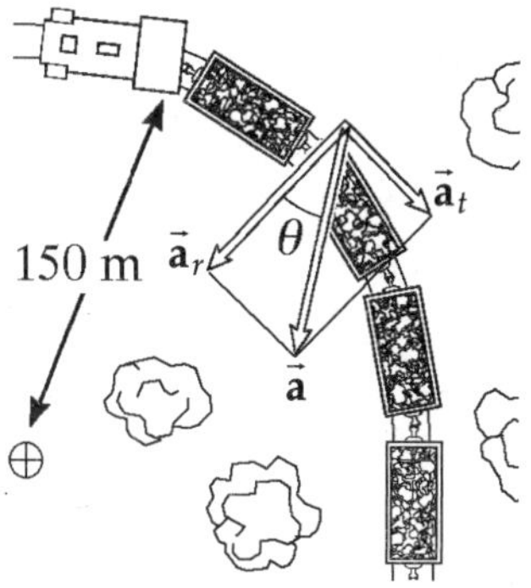

FIG. P3.29

P3.30 (a) See figure to the right.

(b) The components of the 20.2 and the 22.5 m/s^2 along the rope together constitute the centripetal acceleration:

$$a_c = \left(22.5\ \text{m/s}^2\right)\cos(90.0° - 36.9°) + \left(20.2\ \text{m/s}^2\right)\cos 36.9° = \boxed{29.7\ \text{m/s}^2}$$

(c) $a_c = \dfrac{v^2}{r}$ so $v = \sqrt{a_c r} = \sqrt{29.7\ \text{m/s}^2\,(1.50\ \text{m})} = 6.67\ \text{m/s}$ tangent to circle

$\vec{\mathbf{v}} = \boxed{6.67\ \text{m/s at } 36.9°\ \text{above the horizontal}}$

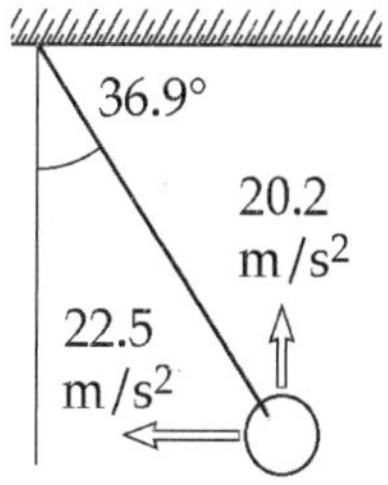

FIG. P3.30

P3.31 $r = 2.50\text{ m}$, $a = 15.0\text{ m/s}^2$

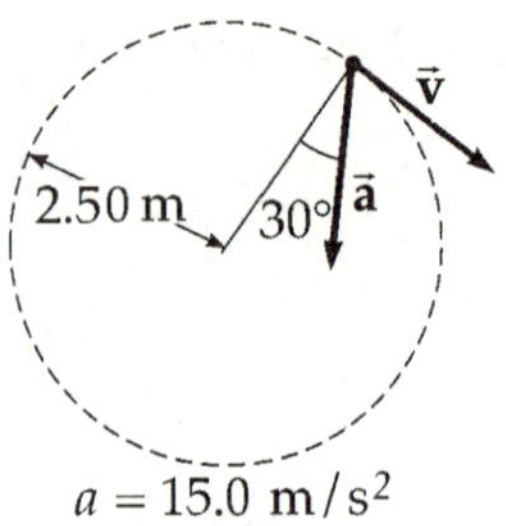

FIG. P3.31

(a) $a_c = a\cos 30.0° = (15.0\text{ m/s}^2)(\cos 30°) = \boxed{13.0\text{ m/s}^2}$

(b) $a_c = \dfrac{v^2}{r}$

so $v^2 = ra_c = 2.50\text{ m}(13.0\text{ m/s}^2) = 32.5\text{ m}^2/\text{s}^2$

$v = \sqrt{32.5}\text{ m/s} = \boxed{5.70\text{ m/s}}$

(c) $a^2 = a_t^2 + a_r^2$

so $a_t = \sqrt{a^2 - a_r^2} = \sqrt{(15.0\text{ m/s}^2)^2 - (13.0\text{m/s}^2)} = \boxed{7.50\text{ m/s}^2}$

Section 3.6 Relative Velocity

P3.32 The bumpers are initially $100\text{ m} = 0.100\text{ km}$ apart. After time t the bumper of the leading car travels $40.0\,t$, while the bumper of the chasing car travels $60.0t$.

Since the cars are side by side at time t, we have

$$0.100 + 40.0t = 60.0t,$$

yielding

$$t = 5.00 \times 10^{-3}\text{ h} = \boxed{18.0\text{ s}}.$$

P3.33 Total time in still water $t = \dfrac{d}{v} = \dfrac{2\,000}{1.20} = 1.67 \times 10^3\text{ s}$.

Total time = time upstream plus time downstream:

$$t_{\text{up}} = \frac{1\,000}{(1.20 - 0.500)} = 1.43 \times 10^3\text{ s}$$

$$t_{\text{down}} = \frac{1\,000}{1.20 + 0.500} = 588\text{ s}.$$

Therefore, $t_{\text{total}} = 1.43 \times 10^3 + 588 = \boxed{2.02 \times 10^3\text{ s, 21.0\% more than if the water were still.}}$.

***P3.34** $\vec{\mathbf{v}}_{ce}$ = the velocity of the car relative to the earth.

$\vec{\mathbf{v}}_{wc}$ = the velocity of the water relative to the car.

$\vec{\mathbf{v}}_{we}$ = the velocity of the water relative to the earth.

These velocities are related as shown in the diagram at the right.

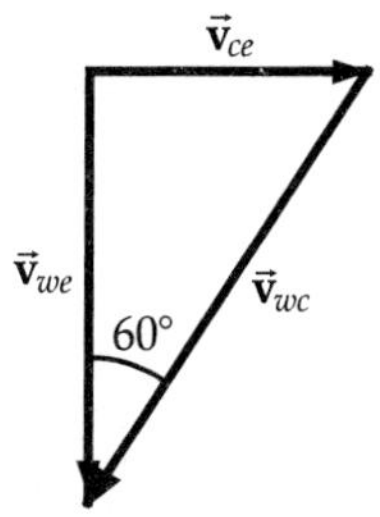

$\vec{\mathbf{v}}_{we} = \vec{\mathbf{v}}_{ce} + \vec{\mathbf{v}}_{wc}$

FIG. P3.34

(a) Since $\vec{\mathbf{v}}_{we}$ is vertical, $v_{wc}\sin 60.0° = v_{ce} = 50.0\text{ km/h}$ or

$\vec{\mathbf{v}}_{wc} = \boxed{57.7\text{ km/h at }60.0°\text{ west of vertical}}$.

(b) Since $\vec{\mathbf{v}}_{ce}$ has zero vertical component,

$$v_{we} = v_{wc}\cos 60.0° = (57.7\text{ km/h})\cos 60.0° = \boxed{28.9\text{ km/h downward}}.$$

P3.35 $v = \sqrt{150^2 + 30.0^2} = \boxed{153\text{ km/h}}$

$\theta = \tan^{-1}\left(\dfrac{30.0}{150}\right) = \boxed{11.3°\text{ north of west}}$

P3.36 For Alan, his speed downstream is $c + v$, while his speed upstream is $c - v$.

Therefore, the total time for Alan is

$$t_1 = \frac{L}{c+v} + \frac{L}{c-v} = \frac{\frac{2L}{c}}{1-\frac{v^2}{c^2}}.$$

For Beth, her cross-stream speed (both ways) is

$$\sqrt{c^2 - v^2}.$$

Thus, the total time for Beth is $t_2 = \dfrac{2L}{\sqrt{c^2 - v^2}} = \dfrac{\frac{2L}{c}}{\sqrt{1-\frac{v^2}{c^2}}}$.

Since $1 - \dfrac{v^2}{c^2} < 1$, $t_1 > t_2$, or $\boxed{\text{Beth, who swims cross-stream, returns first}}$.

P3.37 Identify the student as the S' observer and the professor as the S observer. For the initial motion in S', we have

$$\frac{v'_y}{v'_x} = \tan 60.0° = \sqrt{3}.$$

Let u represent the speed of S' relative to S. Then because there is no x-motion in S, we can write $v_x = v'_x + u = 0$ so that $v'_x = -u = -10.0$ m/s. Hence the ball is thrown backwards in S'. Then,

$$v_y = v'_y = \sqrt{3}\left|v'_x\right| = 10.0\sqrt{3} \text{ m/s}.$$

Using $v_y^2 = 2gh$ we find

$$h = \frac{\left(10.0\sqrt{3} \text{ m/s}\right)^2}{2\left(9.80 \text{ m/s}^2\right)} = \boxed{15.3 \text{ m}}.$$

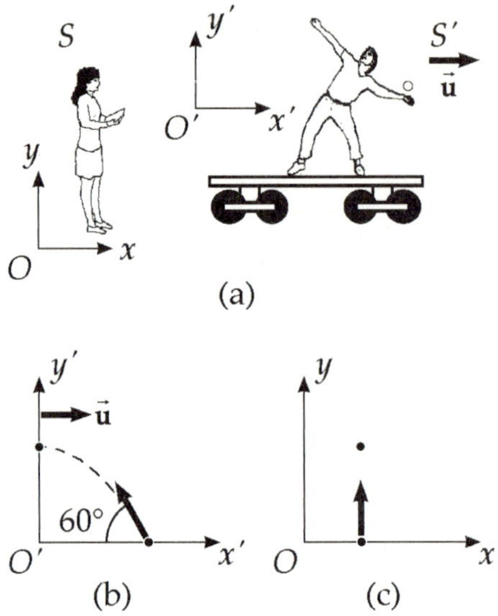

FIG. P3.37

The motion of the ball as seen by the student in S' is shown in diagram (b). The view of the professor in S is shown in diagram (c).

***P3.38** Choose the x-axis along the 20-km distance. The y-components of the displacements of the ship and the speedboat must agree:

$$(26 \text{ km/h})t \sin(40° - 15°) = (50 \text{ km/h})t \sin\alpha$$

$$\alpha = \sin^{-1}\frac{11.0}{50} = 12.7°.$$

The speedboat should head

$$15° + 12.7° = \boxed{27.7° \text{ east of north}}.$$

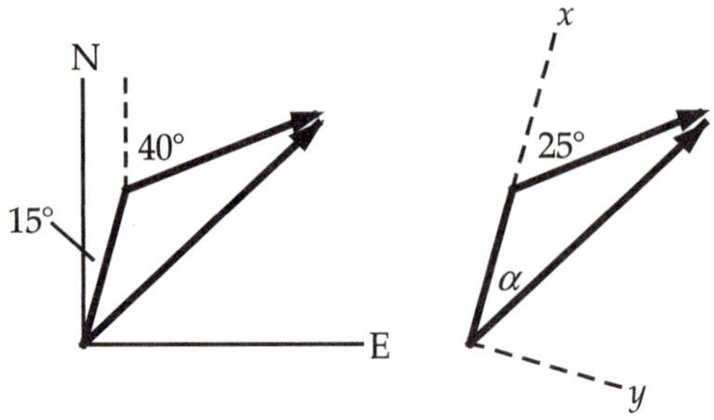

FIG. P3.38

Section 3.7 Context Connection—Lateral Acceleration of Automobiles

***P3.39** $$a = \frac{v^2}{r} = \frac{\left[139 \text{ km/h}(1 \text{ h}/3\,600 \text{ s})(1\,000 \text{ m}/1 \text{ km})\right]^2}{156 \text{ m}} = 9.56 \text{ m/s}^2 = 9.56 \text{ m/s}^2\left(\frac{1g}{9.80 \text{ m/s}^2}\right) = \boxed{0.975g}$$

This agrees to two digits with the tabulated $0.98g$.

***P3.40** We assume that in both cornering processes the vehicle can have the same maximum acceleration.

Then $a = \dfrac{v_1^2}{r_1} = \dfrac{v_2^2}{r_2}$, $\quad \dfrac{(32 \text{ m/s})^2}{150 \text{ m}} = \dfrac{v_2^2}{75 \text{ m}}$ $\quad$ and $\quad v_2 = \sqrt{0.5}(32 \text{ m/s}) = \boxed{22.6 \text{ m/s}}$.

Additional Problems

***P3.41** (a) The speed at the top is $v_x = v_i \cos\theta_i = (143 \text{ m/s})\cos 45° = \boxed{101 \text{ m/s}}$.

(b) In free fall the plane reaches altitude given by

$$v_{yf}^2 = v_{yi}^2 + 2a_y(y_f - y_i)$$
$$0 = (143 \text{ m/s} \sin 45°)^2 + 2(-9.8 \text{ m/s}^2)(y_f - 31\,000 \text{ ft})$$
$$y_f = 31\,000 \text{ ft} + 522 \text{ m}\left(\frac{3.28 \text{ ft}}{1 \text{ m}}\right) = \boxed{3.27 \times 10^4 \text{ ft}}.$$

(c) For the whole free fall motion $v_{yf} = v_{yi} + a_y t$

$$-101 \text{ m/s} = +101 \text{ m/s} - (9.8 \text{ m/s}^2)t$$
$$t = \boxed{20.6 \text{ s}}$$

(d) $a_c = \frac{v^2}{r}$

$$v = \sqrt{a_c r} = \sqrt{0.8(9.8 \text{ m/s}^2)4{,}130 \text{ m}} = \boxed{180 \text{ m/s}}$$

***P3.42** The time of flight of a water drop is given by $y_f = y_i + v_{yi}t + \frac{1}{2}a_y t^2$.

$$0 = 2.35 \text{ m} + 0 - \frac{1}{2}(9.8 \text{ m/s}^2)t_1^2$$

For $t_1 > 0$, the root is $t_1 = \sqrt{\frac{2(2.35 \text{ m})}{9.8 \text{ m/s}^2}} = 0.693 \text{ s}$.

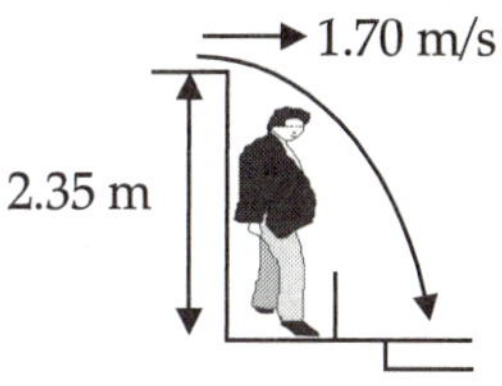

FIG. P3.42

(a) The horizontal range of the font is

$$x_{f1} = x_i + v_{xi}t + \frac{1}{2}a_x t^2$$
$$= 0 + 1.70 \text{ m/s}(0.693 \text{ s}) + 0 = 1.18 \text{ m}.$$

This is about the width of a town sidewalk, so $\boxed{\text{there is}}$ space for a walkway behind the waterfall. Unless the lip of the channel is well designed, water may drip on the visitors. A tall or inattentive person may get his head wet.

(b) Now the flight time t_2 is given by $0 = y_2 + 0 - \frac{1}{2}g t_2^2$. $t_2 = \sqrt{\frac{2y_2}{g}} = \sqrt{\frac{2y_1}{g(12)}} = \frac{1}{\sqrt{12}}\sqrt{\frac{2y_1}{g}} = \frac{t_1}{\sqrt{12}}$.

From the same equation as in part (a) for horizontal range, $x_2 = v_2 t_2$.

$$\frac{x_1}{12} = v_2 \frac{t_1}{\sqrt{12}} \qquad v_2 = \frac{x_1}{t_1\sqrt{12}} = \frac{v_1}{\sqrt{12}} = \frac{1.70 \text{ m/s}}{\sqrt{12}} = \boxed{0.491 \text{ m/s}}$$

The rule that the scale factor for speed is the square root of the scale factor for distance is Froude's law, published in 1870.

P3.43 After the string breaks the ball is a projectile, and reaches the ground at time t: $y_f = v_{yi}t + \frac{1}{2}a_y t^2$

$$-1.20 \text{ m} = 0 + \frac{1}{2}\left(-9.80 \text{ m/s}^2\right)t^2$$

so $t = 0.495$ s.

Its constant horizontal speed is $v_x = \frac{x}{t} = \frac{2.00 \text{ m}}{0.495 \text{ s}} = 4.04 \text{ m/s}$

so before the string breaks $a_c = \frac{v_x^2}{r} = \frac{(4.04 \text{ m/s})^2}{0.300 \text{ m}} = \boxed{54.4 \text{ m/s}^2}$.

P3.44 (a) $y_f = (\tan\theta_i)(x_f) - \frac{g}{2v_i^2\cos^2\theta_i}x_f^2$

Setting $x_f = d\cos\phi$, and $y_f = d\sin\phi$, we have

$$d\sin\phi = (\tan\theta_i)(d\cos\phi) - \frac{g}{2v_i^2\cos^2\theta_i}(d\cos\phi)^2.$$

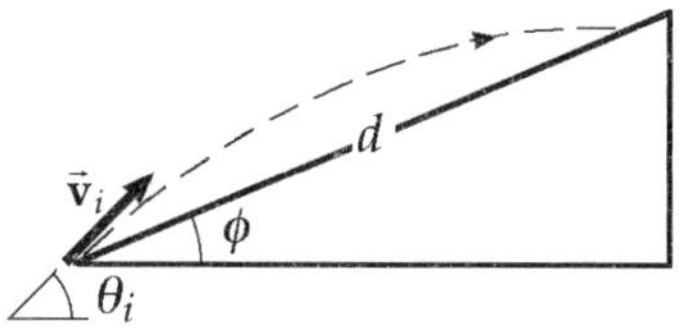

FIG. P3.44

Solving for d yields, $d = \frac{2v_i^2\cos\theta_i[\sin\theta_i\cos\phi - \sin\phi\cos\theta_i]}{g\cos^2\phi}$

or $d = \boxed{\frac{2v_i^2\cos\theta_i\sin(\theta_i - \phi)}{g\cos^2\phi}}$.

(b) Setting $\frac{dd}{d\theta_i} = 0$ leads to $\boxed{\theta_i = 45° + \frac{\phi}{2}}$ and $\boxed{d_{\max} = \frac{v_i^2(1-\sin\phi)}{g\cos^2\phi}}$.

P3.45 Refer to the sketch:

(b) $\Delta x = v_{xi}t$; substitution yields $130 = (v_i\cos 35.0°)t$.

$\Delta y = v_{yi}t + \frac{1}{2}at^2$; substitution yields

$$20.0 = (v_i\sin 35.0°)t + \frac{1}{2}(-9.80)t^2.$$

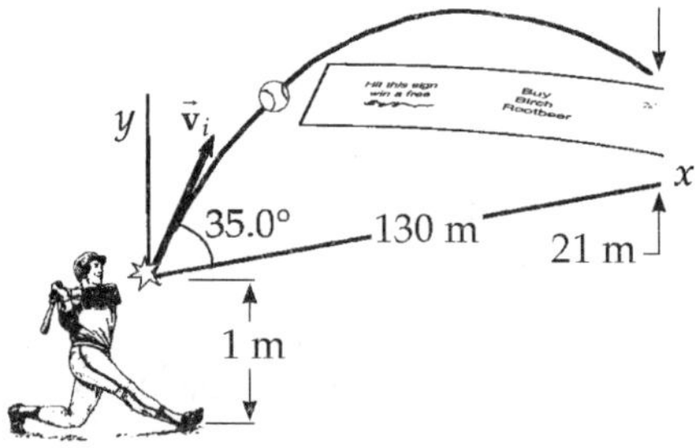

Solving gives $t^2 = \frac{130\tan 35° - 20}{4.9} \Rightarrow t = \boxed{3.81 \text{ s}}$.

FIG. P3.45

(a) $v_i = \boxed{41.7 \text{ m/s}}$

continued on next page

(c) $v_{yf} = v_i \sin\theta_i - gt$, $v_x = v_i \cos\theta_i$

At $t = 3.81$ s, $v_{yf} = 41.7\sin 35.0° - (9.80)(3.81) = \boxed{-13.4 \text{ m/s}}$

$$v_x = (41.7\cos 35.0°) = \boxed{34.1 \text{ m/s}}$$

$$v_f = \sqrt{v_x^2 + v_{yf}^2} = \boxed{36.7 \text{ m/s}}.$$

P3.46 (a) The moon's gravitational acceleration is the probe's centripetal acceleration: (For the moon's radius, see end papers of text.)

$$a = \frac{v^2}{r}$$

$$\frac{1}{6}\left(9.80 \text{ m/s}^2\right) = \frac{v^2}{1.74 \times 10^6 \text{ m}}$$

$$v = \sqrt{2.84 \times 10^6 \text{ m}^2/\text{s}^2} = \boxed{1.69 \text{ km/s}}$$

(b) $v = \dfrac{2\pi r}{T}$

$$T = \frac{2\pi r}{v} = \frac{2\pi(1.74 \times 10^6 \text{ m})}{1.69 \times 10^3 \text{ m/s}} = 6.47 \times 10^3 \text{ s} = \boxed{1.80 \text{ h}}$$

P3.47 $x_f = v_{ix}t = v_i t \cos 40.0°$

Thus, when $x_f = 10.0$ m, $t = \dfrac{10.0 \text{ m}}{v_i \cos 40.0°}$.

At this time, y_f should be $3.05 \text{ m} - 2.00 \text{ m} = 1.05 \text{ m}$.

Thus, $1.05 \text{ m} = \dfrac{(v_i \sin 40.0°)10.0 \text{ m}}{v_i \cos 40.0°} + \dfrac{1}{2}\left(-9.80 \text{ m/s}^2\right)\left[\dfrac{10.0 \text{ m}}{v_i \cos 40.0°}\right]^2$.

From this, $v_i = \boxed{10.7 \text{ m/s}}$.

P3.48 The special conditions allowing use of the horizontal range equation apply.
For the ball thrown at 45°,

$$D = R_{45} = \frac{v_i^2 \sin 90}{g}.$$

For the bouncing ball,

$$D = R_1 + R_2 = \frac{v_i^2 \sin 2\theta}{g} + \frac{\left(\frac{v_i}{2}\right)^2 \sin 2\theta}{g}$$

where θ is the angle it makes with the ground when thrown and when bouncing.

continued on next page

(a) We require:

$$\frac{v_i^2}{g}=\frac{v_i^2\sin 2\theta}{g}+\frac{v_i^2\sin 2\theta}{4g}$$

$$\sin 2\theta=\frac{4}{5}$$

$$\boxed{\theta=26.6°}$$

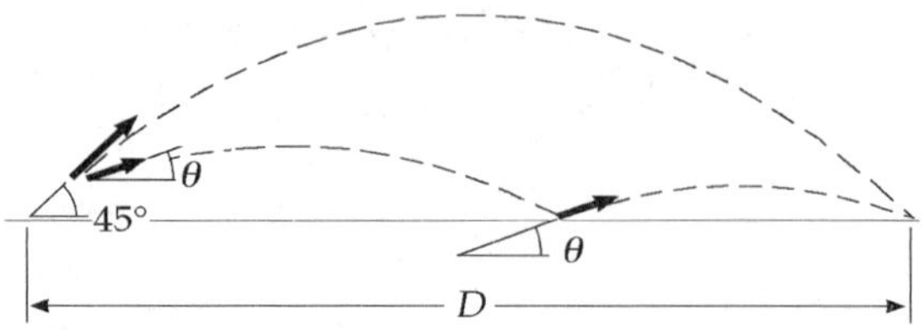

FIG. P3.48

(b) The time for any symmetric parabolic flight is given by

$$y_f=v_{yi}t-\frac{1}{2}gt^2$$
$$0=v_i\sin\theta_i t-\frac{1}{2}gt^2.$$

If $t=0$ is the time the ball is thrown, then $t=\frac{2v_i\sin\theta_i}{g}$ is the time at landing.

So for the ball thrown at 45.0°

$$t_{45}=\frac{2v_i\sin 45.0°}{g}.$$

For the bouncing ball,

$$t=t_1+t_2=\frac{2v_i\sin 26.6°}{g}+\frac{2\left(\frac{v_i}{2}\right)\sin 26.6°}{g}=\frac{3v_i\sin 26.6°}{g}.$$

The ratio of this time to that for no bounce is

$$\frac{\frac{3v_i\sin 26.6°}{g}}{\frac{2v_i\sin 45.0°}{g}}=\frac{1.34}{1.41}=\boxed{0.949}.$$

P3.49 (a) $\Delta y=-\frac{1}{2}gt^2$; $\Delta x=v_i t$

Combine the equations eliminating t:

$$\Delta y=-\frac{1}{2}g\left(\frac{\Delta x}{v_i}\right)^2.$$

From this, $(\Delta x)^2=\left(\frac{-2\Delta y}{g}\right)v_i^2$

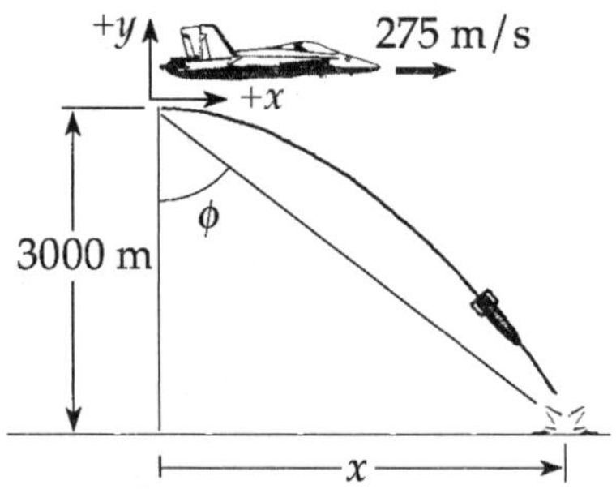

FIG. P3.49

thus $\Delta x=v_i\sqrt{\frac{-2\Delta y}{g}}=275\sqrt{\frac{-2(-3\,000)}{9.80}}=6.80\times10^3=\boxed{6.80\text{ km}}$.

continued on next page

(b) The plane has the same velocity as the bomb in the x direction. Therefore, the plane will be $\boxed{3\,000 \text{ m directly above the bomb}}$ when it hits the ground.

(c) When ϕ is measured from the vertical, $\tan\phi = \dfrac{\Delta x}{\Delta y}$

therefore, $\phi = \tan^{-1}\left(\dfrac{\Delta x}{\Delta y}\right) = \tan^{-1}\left(\dfrac{6\,800}{3\,000}\right) = \boxed{66.2°}$.

P3.50 Measure heights above the level ground. The elevation y_b of the ball follows

$$y_b = R + 0 - \frac{1}{2}gt^2$$

with $x = v_i t$ so $y_b = R - \dfrac{gx^2}{2v_i^2}$.

(a) The elevation y_r of points on the rock is described by

$$y_r^2 + x^2 = R^2.$$

We will have $y_b = y_r$ at $x = 0$, but for all other x we require the ball to be above the rock surface as in $y_b > y_r$. Then $y_b^2 + x^2 > R^2$

$$\left(R - \frac{gx^2}{2v_i^2}\right)^2 + x^2 > R^2$$

$$R^2 - \frac{gx^2R}{v_i^2} + \frac{g^2x^4}{4v_i^4} + x^2 > R^2$$

$$\frac{g^2x^4}{4v_i^4} + x^2 > \frac{gx^2R}{v_i^2}.$$

If this inequality is satisfied for x approaching zero, it will be true for all x. If the ball's parabolic trajectory has large enough radius of curvature at the start, the ball will clear the whole rock: $1 > \dfrac{gR}{v_i^2}$

$$\boxed{v_i > \sqrt{gR}}.$$

(b) With $v_i = \sqrt{gR}$ and $y_b = 0$, we have $0 = R - \dfrac{gx^2}{2gR}$

or $x = R\sqrt{2}$.

The distance from the rock's base is

$$x - R = \boxed{\left(\sqrt{2} - 1\right)R}.$$

P3.51 (a) While on the incline

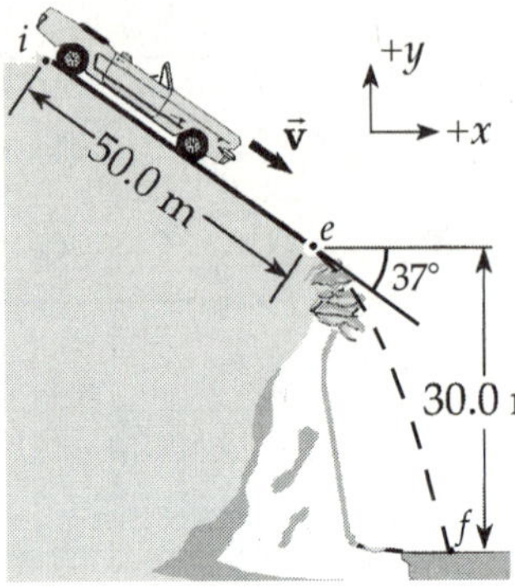

FIG. P3.51

$$v_f^2 - v_i^2 = 2a\Delta x$$
$$v_f - v_i = at$$
$$v_f^2 - 0 = 2(4.00)(50.0)$$
$$20.0 - 0 = 4.00t$$
$$v_f = \boxed{20.0 \text{ m/s}}$$
$$t = \boxed{5.00 \text{ s}}$$

(b) Initial free-flight conditions give us

$$v_{xi} = 20.0\cos 37.0° = 16.0 \text{ m/s}$$

and

$$v_{yi} = -20.0\sin 37.0° = -12.0 \text{ m/s}$$

$$v_{xf} = v_{xi} \text{ since } a_x = 0$$

$$v_{yf} = -\sqrt{2a_y\Delta y + v_{yi}^2} = -\sqrt{2(-9.80)(-30.0)+(-12.0)^2} = -27.1 \text{ m/s}$$

$$v_f = \sqrt{v_{xf}^2 + v_{yf}^2} = \sqrt{(16.0)^2 + (-27.1)^2} = \boxed{31.5 \text{ m/s at } 59.4° \text{ below the horizontal}}$$

(c) $t_1 = 5 \text{ s};\ t_2 = \dfrac{v_{yf} - v_{yi}}{a_y} = \dfrac{-27.1 + 12.0}{-9.80} = 1.53 \text{ s}$

$$t = t_1 + t_2 = \boxed{6.53 \text{ s}}$$

(d) $\Delta x = v_{xi}t_2 = 16.0(1.53) = \boxed{24.5 \text{ m}}$

P3.52 Equation of bank: $y^2 = 16x$ (1)

Equations of motion: $x = v_i t$ (2)

$y = -\dfrac{1}{2}gt^2$ (3)

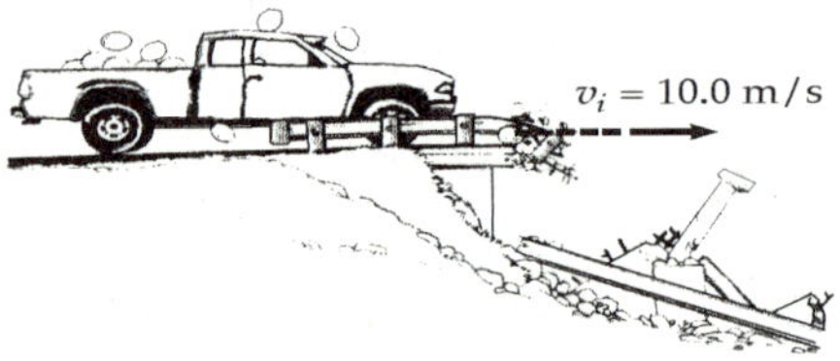

FIG. P3.52

Substitute for t from (2) into (3) $y = -\dfrac{1}{2}g\left(\dfrac{x^2}{v_i^2}\right)$. Equate y from the bank equation to y from the equations of motion:

$$16x = \left[-\frac{1}{2}g\left(\frac{x^2}{v_i^2}\right)\right]^2 \Rightarrow \frac{g^2x^4}{4v_i^4} - 16x = x\left(\frac{g^2x^3}{4v_i^4} - 16\right) = 0.$$

continued on next page

From this, $x = 0$ or $x^3 = \dfrac{64v_i^4}{g^2}$ and $x = 4\left(\dfrac{10^4}{9.80^2}\right)^{1/3} = \boxed{18.8\text{ m}}$. Also,

$$y = -\frac{1}{2}g\left(\frac{x^2}{v_i^2}\right) = -\frac{1}{2}\frac{(9.80)(18.8)^2}{(10.0)^2} = \boxed{-17.3\text{ m}}.$$

P3.53 (a) Coyote: $\Delta x = \frac{1}{2}at^2$; $70.0 = \frac{1}{2}(15.0)t^2$

Roadrunner: $\Delta x = v_i t$; $70.0 = v_i t$

Solving the above, we get $v_i = \boxed{22.9\text{ m/s}}$ and $t = 3.06$ s.

(b) At the edge of the cliff, $v_{xi} = at = (15.0)(3.06) = 45.8$ m/s.

Substituting into $\Delta y = \frac{1}{2}a_y t^2$, we find

$$-100 = \frac{1}{2}(-9.80)t^2$$
$$t = 4.52\text{ s}$$
$$\Delta x = v_{xi}t + \frac{1}{2}a_x t^2 = (45.8)(4.52\text{ s}) + \frac{1}{2}(15.0)(4.52\text{ s})^2.$$

Solving, $\Delta x = \boxed{360\text{ m}}$.

(c) For the Coyote's motion through the air

$$v_{xf} = v_{xi} + a_x t = 45.8 + 15(4.52) = \boxed{114\text{ m/s}}$$
$$v_{yf} = v_{yi} + a_y t = 0 - 9.80(4.52) = \boxed{-44.3\text{ m/s}}.$$

P3.54 Equation 3.15: $h = \dfrac{v_i^2 \sin^2\theta_i}{2g}$

Equation 3.16: $R = \dfrac{v_i^2 \sin 2\theta_i}{g} = \dfrac{2v_i^2 \sin\theta_i \cos\theta_i}{g}$

If $h = \dfrac{R}{6}$, Equation 3.15 yields $v_i \sin\theta_i = \sqrt{\dfrac{gR}{3}}$ (1)

Substituting Equation (1) above into Equation 3.16 gives $R = \dfrac{2\left(\sqrt{gR/3}\right)v_i \cos\theta_i}{g}$

which reduces to $v_i \cos\theta_i = \dfrac{1}{2}\sqrt{3gR}$ (2)

continued on next page

(a) From $v_{yf} = v_{yi} + a_y t$, the time to reach the peak of the path (where $v_{yf} = 0$) is found to be

$$t_{peak} = v_i \sin\theta_i / g$$

Using Equation (1), this gives $$t_{peak} = \sqrt{\frac{R}{3g}}$$

The total time of the ball's flight is then $$\boxed{t_{flight} = 2t_{peak} = 2\sqrt{\frac{R}{3g}}}$$

(b) At the path's peak, the ball moves horizontally with speed $$v_{peak} = v_{xi} = v_i \cos\theta_i$$

Using Equation (1), this becomes $$v_{peak} = \boxed{\frac{1}{2}\sqrt{3gR}}$$

(c) The initial vertical component of velocity is $$v_{yi} = v_i \sin\theta_i$$

From Equation (1), $$v_{yi} = \boxed{\sqrt{\frac{gR}{3}}}$$

(d) Squaring Eq. (1) and (2) and adding the results, $$v_i^2\left(\sin^2\theta_i + \cos^2\theta_i\right) = \frac{gR}{3} + \frac{3gR}{4} = \frac{13gR}{12}$$

Thus, the initial speed is $$v_i = \boxed{\sqrt{\frac{13gR}{12}}}$$

(e) Dividing Equation (1) by (2) yields $$\tan\theta_i = \frac{v_i \sin\theta_i}{v_i \cos\theta_i} = \left[\frac{\left(\sqrt{gR/3}\right)}{\left(\frac{1}{2}\sqrt{3gR}\right)}\right] = \frac{2}{3}$$

Therefore, $$\theta_i = \tan^{-1}\left(\frac{2}{3}\right) = \boxed{33.7°}$$

(f) For a given initial speed, the projection angle yielding maximum peak height is $\theta_i = 90.0°$. With the speed found in (d), Equation 3.15 then yields

$$h_{max} = \frac{(13gR/12)\sin^2 90.0°}{2g} = \boxed{\frac{13}{24}R}$$

(g) For a given initial speed, the projection angle yielding maximum range is $\theta_i = 45.0°$. With the speed found in (d), Equation 3.16 then gives

$$R_{max} = \frac{(13gR/12)\sin 90.0°}{g} = \boxed{\frac{13}{12}R}$$

P3.55 Consider the rocket's trajectory in 3 parts as shown in the sketch. Our initial conditions give:

$$a_y = 30.0 \sin 53.0° = 24.0 \text{ m/s}^2$$

$$a_x = 30.0 \cos 53.0° = 18.1 \text{ m/s}^2$$

$$v_{yi} = 100 \sin 53.0° = 79.9 \text{ m/s}$$

$$v_{xi} = 100 \cos 53.0° = 60.2 \text{ m/s}$$

The distances traveled during each phase of the motion are given in the table.

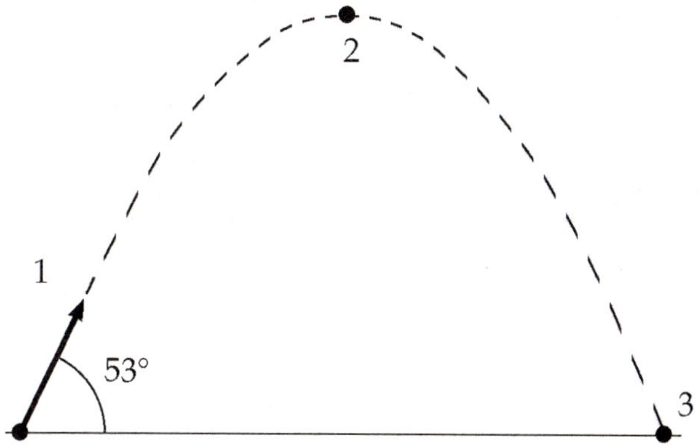

FIG. P3.55

Path #1: $v_{yf} - 79.9 = 24.0(3.00)$ or $v_{yf} = 152 \text{ m/s}$

$v_{xf} - 60.2 = 18.1(3.00)$ or $v_{xf} = 114 \text{ m/s}$

$$\Delta y = 79.9(3.00) + \frac{1}{2}(24.0)(3.00)^2 = 347 \text{ m}$$

$$\Delta x = 60.2(3.00) + \frac{1}{2}(18.1)(3.00)^2 = 262 \text{ m}$$

Path #2: $a_x = 0$, $v_{xf} = v_{xi} = 114 \text{ m/s}$

$0 - 152 = -(9.80)t$ or $t = 15.5 \text{ s}$

$\Delta x = 114(15.5) = 1.77 \times 10^3 \text{ m};$

$$\Delta y = 152(15.5) - \frac{1}{2}(9.80)(15.5)^2 = 1.17 \times 10^3$$

Path #3: $(v_{yf})^2 - 0 = 2(-9.80)(-1.52 \times 10^3)$

$v_{yf} = -173 \text{ m/s}$ $v_{xf} = v_{xi} = 114 \text{ m/s}$, since $a_x = 0$

$-173 - 0 = -(9.80)t$ or $t = 17.6 \text{ s}$

$\Delta x = 114(17.6) = 2.02 \times 10^3 \text{ m}$

continued on next page

(a) $\Delta y(\text{max}) = \boxed{1.52 \times 10^{-3} \text{ m}}$

(b) $t(\text{net}) = 3.00 + 15.5 + 17.6 = \boxed{36.1 \text{ s}}$

(c) $\Delta x(\text{net}) = 262 + 1.77 \times 10^3 + 2.02 \times 10^3$

$\Delta x(\text{net}) = \boxed{4.05 \times 10^3 \text{ m}}$

	Path Part		
	#1	#2	#3
a_y	24.0	–9.80	–9.80
a_x	18.1	0.0	0.00
v_{yf}	152	0.0	–173
v_{xf}	114	114	114
v_{yi}	79.9	152	0.00
v_{xi}	60.2	114	114
Δy	347	1.17×10^3	-1.52×10^3
Δx	262	1.77×10^3	2.02×10^3
t	3.00	15.5	17.6

P3.56 Think of shaking down the mercury in an old fever thermometer. Swing your hand through a circular arc, quickly reversing direction at the bottom end. Suppose your hand moves through one-quarter of a circle of radius 60 cm in 0.1 s. Its speed is

$$\frac{\frac{1}{4}(2\pi)(0.6 \text{ m})}{0.1 \text{ s}} \approx 9 \text{ m/s}$$

and its centripetal acceleration is $\frac{v^2}{r} \approx \frac{(9 \text{ m/s})^2}{0.6 \text{ m}} \boxed{\sim 10^2 \text{ m/s}^2}$.

The tangential acceleration of stopping and reversing the motion will make the total acceleration somewhat larger, but will not affect its order of magnitude.

P3.57 (a) $\Delta x = v_{xi}t,\ \Delta y = v_{yi}t + \frac{1}{2}gt^2$

$$d\cos 50.0° = (10.0\cos 15.0°)t$$

and

$$-d\sin 50.0° = (10.0\sin 15.0°)t + \frac{1}{2}(-9.80)t^2.$$

Solving, $10\cos 15\tan 50 + 10\sin 15 = 4.9t$,

$t = 2.88 \text{ s}$ and $d = \boxed{43.2 \text{ m}}$.

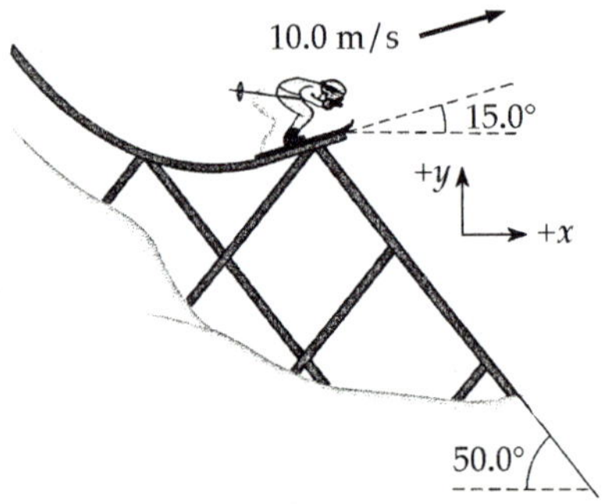

FIG. P3.57

(b) Since $a_x = 0$,

$$v_{xf} = v_{xi} = 10.0\cos 15.0° = \boxed{9.66 \text{ m/s}}$$

$$v_{yf} = v_{yi} + a_y t = 10.0\sin 15.0° - 9.80(2.88) = \boxed{-25.6 \text{ m/s}}.$$

Air resistance would decrease the values of the range and maximum height. As an airfoil, he can get some lift and increase his distance.

P3.58 For one electron, we have

$$y = v_{iy}t,\ D = v_{ix}t + \frac{1}{2}a_x t^2 \approx \frac{1}{2}a_x t^2,\ v_{yf} = v_{yi}, \text{ and } v_{xf} = v_{xi} + a_x t \approx a_x t.$$

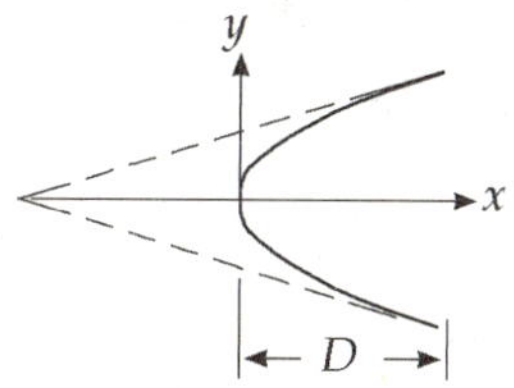

FIG. P3.58

The angle its direction makes with the x-axis is given by

$$\theta = \tan^{-1}\frac{v_{yf}}{v_{xf}} = \tan^{-1}\frac{v_{yi}}{a_x t} = \tan^{-1}\frac{v_{yi}t}{a_x t^2} = \tan^{-1}\frac{y}{2D}.$$

Thus the horizontal distance from the aperture to the virtual source is $2D$. The source is at coordinate $\boxed{x = -D}$.

P3.59 (a) The ice chest floats downstream 2 km in time t, so that $2\text{ km} = v_w t$. The upstream motion of the boat is described by $d = (v - v_w)15\text{ min}$. The downstream motion is described by $d + 2\text{ km} = (v + v_w)(t - 15\text{ min})$. We eliminate $t = \frac{2\text{ km}}{v_w}$ and d by substitution:

$$(v - v_w)15\text{ min} + 2\text{ km} = (v + v_w)\left(\frac{2\text{ km}}{v_w} - 15\text{ min}\right)$$

$$v(15\text{ min}) - v_w(15\text{ min}) + 2\text{ km} = \frac{v}{v_w}2\text{ km} + 2\text{ km} - v(15\text{ min}) - v_w(15\text{ min})$$

$$v(30\text{ min}) = \frac{v}{v_w}2\text{ km}$$

$$v_w = \frac{2\text{ km}}{30\text{ min}} = \boxed{4.00\text{ km/h}}.$$

(b) In the reference frame of the water, the chest is motionless. The boat travels upstream for 15 min at speed v, and then downstream at the same speed, to return to the same point. Thus it travels for 30 min. During this time, the falls approach the chest at speed v_w, traveling 2 km. Thus

$$v_w = \frac{\Delta x}{\Delta t} = \frac{2\text{ km}}{30\text{ min}} = \boxed{4.00\text{ km/h}}.$$

***P3.60** Let the river flow in the x direction.

(a) To minimize time, $\boxed{\text{swim perpendicular to the banks}}$ in the y direction. You are in the water for time t in $\Delta y = v_y t$, $t = \frac{80\text{ m}}{1.5\text{ m/s}} = 53.3\text{ s}$.

(b) The water carries you downstream by $\Delta x = v_x t = (2.50\text{ m/s})53.3\text{ s} = \boxed{133\text{ m}}$.

continued on next page

(c)

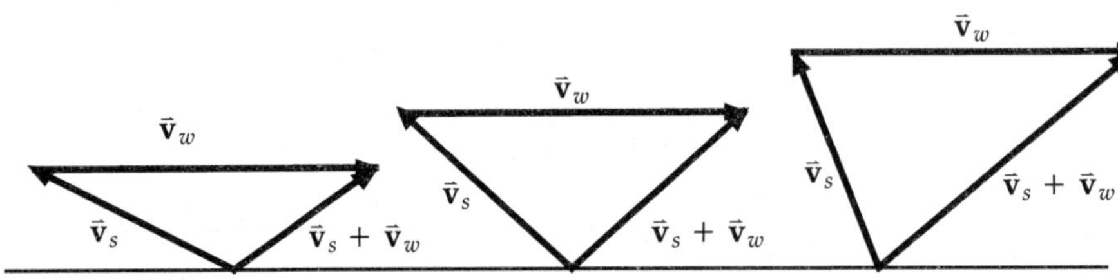

To minimize downstream drift, you should swim so that your resultant velocity $\vec{v}_s + \vec{v}_w$ is perpendicular to your swimming velocity $\vec{v}_s$ relative to the water. This condition is shown in the middle picture. It maximizes the angle between the resultant velocity and the shore. The angle between $\vec{v}_s$ and the shore is given by $\cos\theta = \frac{1.5 \text{ m/s}}{2.5 \text{ m/s}}$, $\boxed{\theta = 53.1°}$.

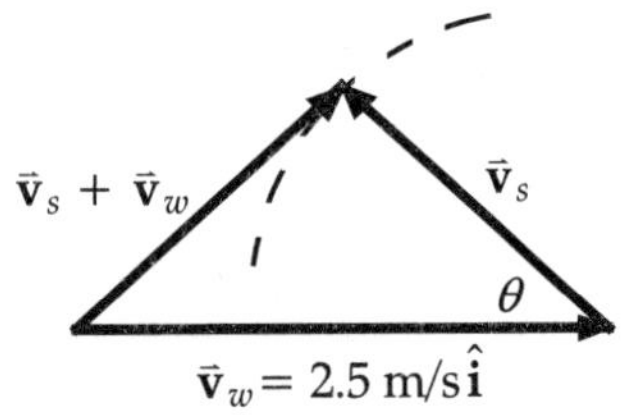

(d) Now $v_y = v_s \sin\theta = 1.5 \text{ m/s} \sin 53.1° = 1.20 \text{ m/s}$

$$t = \frac{\Delta y}{v_y} = \frac{80 \text{ m}}{1.2 \text{ m/s}} = 66.7 \text{ s}$$

$$\Delta x = v_x t = (2.5 \text{ m/s} - 1.5 \text{ m/s} \cos 53.1°)66.7 \text{ s} = \boxed{107 \text{ m}}.$$

P3.61 Find the highest firing angle θ_H for which the projectile will clear the mountain peak; this will yield the range of the closest point of bombardment. Next find the lowest firing angle; this will yield the maximum range under these conditions if both θ_H and θ_L are $> 45°$; $x = 2500$ m, $y = 1800$ m, $v_i = 250$ m/s.

$$y_f = v_{yi}t - \frac{1}{2}gt^2 = v_i(\sin\theta)t - \frac{1}{2}gt^2$$
$$x_f = v_{xi}t = v_i(\cos\theta)t$$

Thus $t = \frac{x_f}{v_i \cos\theta}$. Substitute into the expression for y_f

$$y_f = v_i(\sin\theta)\frac{x_f}{v_i\cos\theta} - \frac{1}{2}g\left(\frac{x_f}{v_i\cos\theta}\right)^2 = x_f\tan\theta - \frac{gx_f^2}{2v_i^2\cos^2\theta}$$

but $\frac{1}{\cos^2\theta} = \tan^2\theta + 1$ so $y_f = x_f\tan\theta - \frac{gx_f^2}{2v_i^2}\left(\tan^2\theta + 1\right)$ and $0 = \frac{gx_f^2}{2v_i^2}\tan^2\theta - x_f\tan\theta + \frac{gx_f^2}{2v_i^2} + y_f$.

Substitute values, use the quadratic formula and find

$$\tan\theta = 3.905 \text{ or } 1.197, \text{ which gives } \theta_H = 75.6° \text{ and } \theta_L = 50.1°.$$

$$\text{Range (at } \theta_H) = \frac{v_i^2\sin 2\theta_H}{g} = 3.07\times10^3 \text{ m from enemy ship}$$

continued on next page

$$3.07\times10^3 - 2\,500 - 300 = 270 \text{ m from shore.}$$

$$\text{Range (at } \theta_L) = \frac{v_i^2 \sin 2\theta_L}{g} = 6.28\times10^3 \text{ m from enemy ship}$$

$$6.28\times10^3 - 2\,500 - 300 = 3.48\times10^3 \text{ from shore.}$$

Therefore, safe distance is $\boxed{< 270 \text{ m}}$ or $\boxed{> 3.48\times10^3 \text{ m}}$ from the shore.

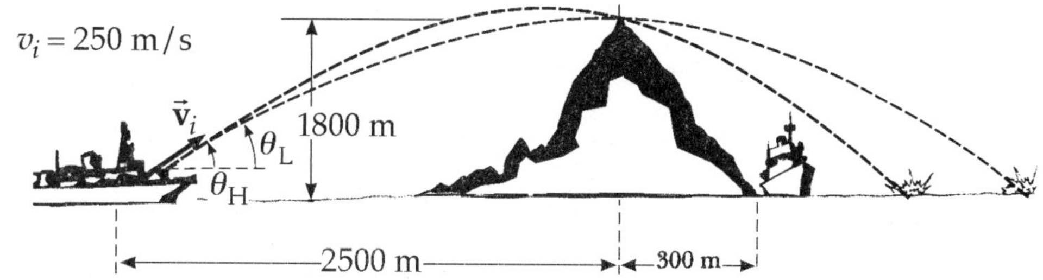

FIG. P3.61

ANSWERS TO EVEN PROBLEMS

P3.2 (a) $(1.00\hat{\mathbf{i}} + 0.750\hat{\mathbf{j}})$ m/s,
(b) $(1.00\hat{\mathbf{i}} + 0.500\hat{\mathbf{j}})$ m/s, 1.12 m/s

P3.4 (a) $(2.00\hat{\mathbf{i}} + 3.00\hat{\mathbf{j}})$ m/s^2;
(b) $(3.00t + t^2)\hat{\mathbf{i}}$ m $+ (1.50t^2 - 2.00t)\hat{\mathbf{j}}$ m

P3.6 (a) $\vec{\mathbf{r}}_f = (10.0\,\hat{\mathbf{i}} + 0.241\,\hat{\mathbf{j}})$ mm;
(b) $(1.84\times10^7 \text{ m/s})\hat{\mathbf{i}} + (8.78\times10^5 \text{ m/s})\hat{\mathbf{j}}$;
(c) 1.85×10^7 m/s; (d) 2.73°

P3.8 (a) $d\sqrt{\dfrac{g}{2h}}$ horizontally;
(b) $\tan^{-1}\left(\dfrac{2h}{d}\right)$ below the horizontal

P3.10 0.600 m/s^2 down

P3.12 (a) 22.6 m; (b) 52.3 m; (c) 1.18 s

P3.14 25.8 m/s

P3.16 $d\tan\theta_i - \dfrac{gd^2}{2v_i^2\cos^2\theta_i}$

P3.18 (a) (0, 0.840 m); (b) 11.2 m/s at 18.5°;
(c) 8.94 m; (d) see the solution

P3.20 (a) 0.852 s; (b) 3.29 m/s; (c) 4.03 m/s;
(d) 50.8°; (e) 1.12 s

P3.22 0.033 7 m/s^2 toward the center of the Earth

P3.24 2.06×10^3 rev/min

P3.26 0.281 rev/s

P3.28 (a) 1.25 m/s^2 toward the center;
(b) 0.400 m/s^2 forward;
(c) 1.31 m/s^2 forward and 72.3° inward

P3.30 (a) see the solution; (b) $29.7\ \text{m/s}^2$; (c) 6.67 m/s at 36.9° above the horizontal

P3.32 18.0 s

P3.34 (a) 57.7 km/h at 60.0° west of vertical; (b) 28.9 km/h downward

P3.36 Beth returns first

P3.38 27.7° east of north

P3.40 22.6 m/s

P3.42 (a) yes; (b) $0.491\ \text{m/s}$

P3.44 (a) $\dfrac{2v_i^2\cos\theta_i\sin(\theta_i-\phi)}{g\cos^2\phi}$; (b) $45°+\dfrac{\phi}{2}$, $\dfrac{v_i^2(1-\sin\phi)}{g\cos^2\phi}$

P3.46 (a) 1.69 km/s; (b) 1.80 h

P3.48 (a) 26.6°; (b) 0.949

P3.50 (a) $v_i > \sqrt{gR}$; (b) $\left(\sqrt{2}-1\right)R$

P3.52 (18.8, –17.3) m

P3.54 (a) $2\sqrt{\dfrac{R}{3g}}$; (b) $\dfrac{1}{2}\sqrt{3gR}$; (c) $\sqrt{\dfrac{gR}{3}}$; (d) $\sqrt{\dfrac{13gR}{12}}$; (e) $33.7°$; (f) $\dfrac{13}{24}R$; (g) $\dfrac{13}{12}R$

P3.56 see the solution, $\sim 10^2\ \text{m/s}^2$

P3.58 $x=-D$

P3.60 (a) swim perpendicular to the banks; (b) 133 m (c) upstream at $53.1°$ to the bank; (d) 107 m

Chapter 4

The Laws of Motion

CHAPTER OUTLINE

ANSWERS TO QUESTIONS

Q4.1 (a) The force due to gravity of the earth pulling down on the ball—the reaction force is the force due to gravity of the ball pulling up on the earth. The force of the hand pushing up on the ball—reaction force is ball pushing down on the hand.

(b) The only force acting on the ball in free-fall is the gravity due to the earth -the reaction force is the gravity due to the ball pulling on the earth.

Q4.2 Mistake one: The car might be momentarily at rest, in the process of (suddenly) reversing forward into backward motion. In this case, the forces on it add to a (large) backward resultant.

Mistake two: There are no cars in interstellar space. If the car is remaining at rest, there are some large forces on it, including its weight and some force or forces of support.

Mistake three: The statement reverses cause and effect, like a politician who thinks that his getting elected was the reason for people to vote for him.

Q4.3 When the bus starts moving, the mass of Claudette is accelerated by the force of the back of the seat on her body. Clark is standing, however, and the only force on him is the friction between his shoes and the floor of the bus. Thus, when the bus starts moving, his feet start accelerating forward, but the rest of his body experiences almost no accelerating force (only that due to his being attached to his accelerating feet!). As a consequence, his body tends to stay almost at rest, according to Newton's first law, relative to the ground. Relative to Claudette, however, he is moving toward her and falls into her lap. (Both performers won Academy Awards.)

Q4.4 No. The reaction to your weight is an upward force you exert on the Earth.

Q4.5 First ask, "Was the bus moving forward or backing up?" If it was moving forward, the passenger is lying. A fast stop would make the suitcase fly toward the front of the bus, not toward the rear. If the bus was backing up at any reasonable speed, a sudden stop could not make a suitcase fly far. Fine her for malicious litigiousness.

Q4.6 It would be smart for the explorer to gently push the rock back into the storage compartment. Newton's 3rd law states that the rock will apply the same size force on her that she applies on it. The harder she pushes on the rock, the larger her resulting acceleration.

Q4.7 The molecules of the floor resist the ball on impact and push the ball back, upward. The actual force acting is due to the forces between molecules that allow the floor to keep its integrity and to prevent the ball from passing through. Notice that for a ball passing through a window, the molecular forces weren't strong enough.

Q4.8 While a football is in flight, the force of gravity and air resistance act on it. When a football is in the process of being kicked, the foot pushes forward on the ball and the ball pushes backward on the foot. At this time and while the ball is in flight, the Earth pulls down on the ball (gravity) and the ball pulls up on the Earth. The moving ball pushes forward on the air and the air backward on the ball.

Q4.9 A particular nugget weighs less in Denver, at greater distance from the center of the earth, so you will pay less for it there, if it is sold by weight. If it is sold by mass, you can buy gold in either place equally well.

Q4.10 A portion of each leaf of grass extends above the metal bar. This portion must accelerate in order for the leaf to bend out of the way. The leaf's mass is small, but when its acceleration is very large, the force exerted by the bar on the leaf puts the leaf under tension large enough to shear it off.

Q4.11 As the barbell goes through the bottom of a cycle, the lifter exerts an upward force on it, and the scale reads the larger upward force that the floor exerts on them together. Around the top of the weight's motion, the scale reads less than average. If the iron is moving upward, the lifter can declare that she has thrown it, just by letting go of it for a moment, so our answer applies also to this case.

Q4.12 It is impossible to string a horizontal cable without its sagging a bit. Since the cable has a mass, gravity pulls it downward. A vertical component of the tension must balance the weight for the cable to be in equilibrium. If the cable were completely horizontal, then there would be no vertical component of the tension to balance the weight.

Some physics teachers demonstrate this by asking a beefy student to pull on the ends of a cord supporting a can of soup at its center. Some get two burly young men to pull on opposite ends of a strong rope, while the smallest person in class gleefully mashes the center of the rope down to the table. Point out the beauty of sagging suspension-bridge cables. With a laser and an optical lever, demonstrate that the mayor makes the courtroom table sag when he sits on it, and the judge bends the bench. Give them "I make the floor sag" buttons, available to instructors using this manual. Estimate the cost of an infinitely strong cable, and the truth will always win.

Q4.13 As the sand leaks out, the acceleration increases. With the same driving force, a decrease in the mass causes an increase in the acceleration.

Q4.14 As the rocket takes off, it burns fuel, pushing the gases from the combustion out the back of the rocket. Since the gases have mass, the total remaining mass of the rocket, fuel, and oxidizer decreases. With a constant thrust, a decrease in the mass results in an increasing acceleration.

Q4.15 The tension in the rope when pulling the car is twice that in the tug-of-war. One could consider the car as behaving like another team of twenty more people.

Q4.16 This statement contradicts Newton's 3rd law. The force that the locomotive exerted on the wall is the same as that exerted by the wall on the locomotive. The wall temporarily exerted on the locomotive a force greater than the force that the wall could exert without breaking.

Q4.17 The sack of sand moves up with the athlete, regardless of how quickly the athlete climbs. Since the athlete and the sack of sand have the same weight, the acceleration of the system must be zero.

Q4.18 The resultant force doesn't always add to zero. If it did, nothing could ever accelerate. If we choose a single object as our system, action and reaction forces can never add to zero, as they act on different objects.

Q4.19 An object cannot exert a force on itself. If it could, then objects would be able to accelerate themselves, without interacting with the environment. You cannot lift yourself by tugging on your bootstraps.

SOLUTIONS TO PROBLEMS

Section 4.1 The Concept of Force

No problems in this section

Section 4.2 Newton's First Law

No problems in this section

Section 4.3 Mass

P4.1 For the same force F, acting on different masses $F = m_1 a_1$

and $F = m_2 a_2$

so $m_1 a_1 = m_2 a_2$.

(a) $$\frac{m_1}{m_2} = \frac{a_2}{a_1} = \boxed{\frac{1}{3}}$$

$$m_2 = 3m_1$$

(b) $$F = (m_1 + m_2)a = 4m_1 a = m_1\left(3.00 \text{ m/s}^2\right)$$

$$a = \boxed{0.750 \text{ m/s}^2}$$

P4.2 Since the car is moving with constant speed and in a straight line, the resultant force on it must be $\boxed{\text{zero}}$ regardless of whether it is moving

(a) toward the right or

(b) the left.

Section 4.4 Newton's Second Law—The Particle Under a Net Force

P4.3 $m = 3.00$ kg

$$\vec{\mathbf{a}} = \left(2.00\hat{\mathbf{i}} + 5.00\hat{\mathbf{j}}\right) \text{ m/s}^2$$

$$\sum \vec{\mathbf{F}} = m\vec{\mathbf{a}} = \boxed{\left(6.00\hat{\mathbf{i}} + 15.0\hat{\mathbf{j}}\right) \text{ N}}$$

$$\left|\sum \vec{\mathbf{F}}\right| = \sqrt{(6.00)^2 + (15.0)^2} \text{ N} = \boxed{16.2 \text{ N}}$$

P4.4 (a) $\Sigma\vec{\mathbf{F}} = \vec{\mathbf{F}}_1 + \vec{\mathbf{F}}_2 = \left(-9.00\hat{\mathbf{i}} + 3.00\hat{\mathbf{j}}\right)\text{ N}$

Acceleration $\vec{\mathbf{a}} = a_x\hat{\mathbf{i}} + a_y\hat{\mathbf{j}} = \dfrac{\Sigma\vec{\mathbf{F}}}{m} = \dfrac{\left(-9.00\hat{\mathbf{i}} + 3.00\hat{\mathbf{j}}\right)\text{ N}}{2.00\text{ kg}} = \left(-4.50\hat{\mathbf{i}} + 1.50\hat{\mathbf{j}}\right)\text{ m/s}^2$

Velocity $\vec{\mathbf{v}}_f = v_x\hat{\mathbf{i}} + v_y\hat{\mathbf{j}} = \vec{\mathbf{v}}_i + \vec{\mathbf{a}}t = \vec{\mathbf{a}}t$

$$\vec{\mathbf{v}}_f = \left(\left(-4.50\hat{\mathbf{i}} + 1.50\hat{\mathbf{j}}\right)\text{m/s}^2\right)(10\text{ s}) = \boxed{\left(-45.0\hat{\mathbf{i}} + 15.0\hat{\mathbf{j}}\right)\text{ m/s}}$$

(b) The direction of motion makes angle θ with the x-direction.

$$\theta = \tan^{-1}\left(\frac{v_y}{v_x}\right) = \tan^{-1}\left(-\frac{15.0\text{ m/s}}{45.0\text{ m/s}}\right)$$

$$\theta = -18.4° + 180° = \boxed{162°\text{ from }+x\text{-axis}}$$

(c) Displacement: $x\text{-displacement} = x_f - x_i = v_{xi}t + \frac{1}{2}a_xt^2 = \frac{1}{2}\left(-4.50\text{ m/s}^2\right)(10.0\text{ s})^2 = -225\text{ m}$

$$y\text{-displacement} = y_f - y_i = v_{yi}t + \frac{1}{2}a_yt^2 = \frac{1}{2}\left(+1.50\text{ m/s}^2\right)(10.0\text{ s})^2 = +75.0\text{ m}$$

$$\Delta\vec{\mathbf{r}} = \boxed{\left(-225\hat{\mathbf{i}} + 75.0\hat{\mathbf{j}}\right)\text{ m}}$$

(d) Position: $\vec{\mathbf{r}}_f = \vec{\mathbf{r}}_i + \Delta\vec{\mathbf{r}}$

$$\vec{\mathbf{r}}_f = \left(-2.00\hat{\mathbf{i}} + 4.00\hat{\mathbf{j}}\right) + \left(-225\hat{\mathbf{i}} + 75.0\hat{\mathbf{j}}\right) = \boxed{\left(-227\hat{\mathbf{i}} + 79.0\hat{\mathbf{j}}\right)\text{ m}}$$

P4.5 $m = 4.00\text{ kg},\ \vec{\mathbf{v}}_i = 3.00\hat{\mathbf{i}}\text{ m/s},\ \vec{\mathbf{v}}_8 = \left(8.00\hat{\mathbf{i}} + 10.0\hat{\mathbf{j}}\right)\text{ m/s},\ t = 8.00\text{ s}$

$$\vec{\mathbf{a}} = \frac{\Delta\vec{\mathbf{v}}}{t} = \frac{5.00\hat{\mathbf{i}} + 10.0\hat{\mathbf{j}}}{8.00}\text{ m/s}^2$$

(a) $\vec{\mathbf{F}} = m\vec{\mathbf{a}} = \boxed{\left(2.50\hat{\mathbf{i}} + 5.00\hat{\mathbf{j}}\right)\text{ N}}$

(b) $F = \sqrt{(2.50)^2 + (5.00)^2} = \boxed{5.59\text{ N}}$

P4.6 $v_x = \dfrac{dx}{dt} = 10t,\ v_y = \dfrac{dy}{dt} = 9t^2$

$a_x = \dfrac{dv_x}{dt} = 10,\ a_y = \dfrac{dv_y}{dt} = 18t$

At $t = 2.00\text{ s},\ a_x = 10.0\text{ m/s}^2,\ a_y = 36.0\text{ m/s}^2$

$\sum F_x = ma_x:\ 3.00\text{ kg}\left(10.0\text{ m/s}^2\right) = 30.0\text{ N}$

$\sum F_y = ma_y:\ 3.00\text{ kg}\left(36.0\text{ m/s}^2\right) = 108\text{ N}$

$$\sum F = \sqrt{F_x^2 + F_y^2} = \boxed{112\text{ N}}$$

P4.7 (a) $\sum \vec{\mathbf{F}} = \vec{\mathbf{F}}_1 + \vec{\mathbf{F}}_2 = \left(20.0\hat{\mathbf{i}} + 15.0\hat{\mathbf{j}}\right)\text{ N}$

$$\sum \vec{\mathbf{F}} = m\vec{\mathbf{a}}: \quad 20.0\hat{\mathbf{i}} + 15.0\hat{\mathbf{j}} = 5.00\vec{\mathbf{a}}$$

$$\vec{\mathbf{a}} = \left(4.00\hat{\mathbf{i}} + 3.00\hat{\mathbf{j}}\right)\text{ m/s}^2$$

or

$$\boxed{a = 5.00\text{ m/s}^2 \text{ at } \theta = 36.9°}$$

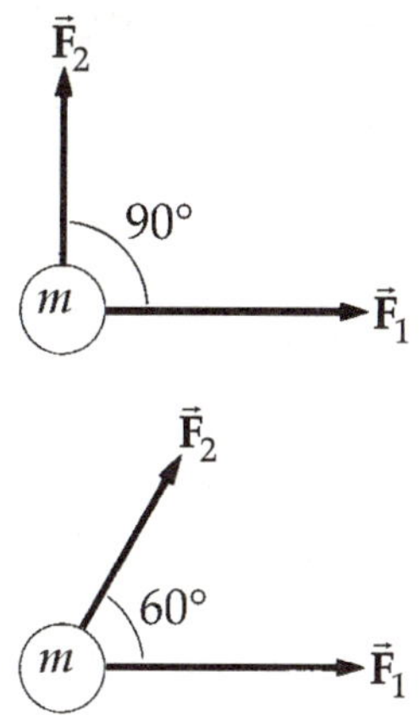

FIG. P4.7

(b) $F_{2x} = 15.0\cos 60.0° = 7.50\text{ N}$

$F_{2y} = 15.0\sin 60.0° = 13.0\text{ N}$

$$\vec{\mathbf{F}}_2 = \left(7.50\hat{\mathbf{i}} + 13.0\hat{\mathbf{j}}\right)\text{ N}$$

$$\sum \vec{\mathbf{F}} = \vec{\mathbf{F}}_1 + \vec{\mathbf{F}}_2 = \left(27.5\hat{\mathbf{i}} + 13.0\hat{\mathbf{j}}\right)\text{ N} = m\vec{\mathbf{a}} = 5.00\vec{\mathbf{a}}$$

$$\vec{\mathbf{a}} = \boxed{\left(5.50\hat{\mathbf{i}} + 2.60\hat{\mathbf{j}}\right)\text{ m/s}^2 = 6.08\text{ m/s}^2 \text{ at } 25.3°}$$

P4.8 $\sum \vec{\mathbf{F}} = m\vec{\mathbf{a}}$ reads

$$\left(-2.00\hat{\mathbf{i}} + 2.00\hat{\mathbf{j}} + 5.00\hat{\mathbf{i}} - 3.00\hat{\mathbf{j}} - 45.0\hat{\mathbf{i}}\right)\text{ N} = m\left(3.75\text{ m/s}^2\right)\hat{\mathbf{a}}$$

where $\hat{\mathbf{a}}$ represents the direction of $\vec{\mathbf{a}}$

$$\left(-42.0\hat{\mathbf{i}} - 1.00\hat{\mathbf{j}}\right)\text{ N} = m\left(3.75\text{ m/s}^2\right)\hat{\mathbf{a}}$$

$$\sum \vec{\mathbf{F}} = \sqrt{(42.0)^2 + (1.00)^2}\text{ N at } \tan^{-1}\left(\frac{1.00}{42.0}\right) \text{ below the } -x\text{-axis}$$

$$\sum \vec{\mathbf{F}} = 42.0\text{ N at } 181° = m\left(3.75\text{ m/s}^2\right)\hat{\mathbf{a}}.$$

For the vectors to be equal, their magnitudes and their directions must be equal.

(a) $\therefore \boxed{\hat{\mathbf{a}} \text{ is at } 181°}$ counterclockwise from the x-axis

(b) $m = \dfrac{42.0\text{ N}}{3.75\text{ m/s}^2} = \boxed{11.2\text{ kg}}$

(d) $\vec{\mathbf{v}}_f = \vec{\mathbf{v}}_i + \vec{\mathbf{a}}t = 0 + \left(3.75\text{ m/s}^2 \text{ at } 181°\right)10.0\text{ s}$ so $\vec{\mathbf{v}}_f = 37.5\text{ m/s at } 181°$

$\vec{\mathbf{v}}_f = 37.5\text{ m/s } \cos 181°\,\hat{\mathbf{i}} + 37.5\text{ m/s } \sin 181°\,\hat{\mathbf{j}}$ so $\vec{\mathbf{v}}_f = \boxed{\left(-37.5\hat{\mathbf{i}} - 0.893\hat{\mathbf{j}}\right)\text{ m/s}}$

(c) $\left|\vec{\mathbf{v}}_f\right| = \sqrt{37.5^2 + 0.893^2}\text{ m/s} = \boxed{37.5\text{ m/s}}$

Section 4.5 The Gravitational Force and Weight

P4.9 (a) $F_g = mg = 120 \text{ lb} = (4.448 \text{ N/lb})(120 \text{ lb}) = \boxed{534 \text{ N}}$

(b) $m = \dfrac{F_g}{g} = \dfrac{534 \text{ N}}{9.80 \text{ m/s}^2} = \boxed{54.5 \text{ kg}}$

P4.10 $F_g = mg = 900 \text{ N}$, $m = \dfrac{900 \text{ N}}{9.80 \text{ m/s}^2} = 91.8 \text{ kg}$

$$\left(F_g\right)_{\text{on Jupiter}} = 91.8 \text{ kg}\left(25.9 \text{ m/s}^2\right) = \boxed{2.38 \text{ kN}}$$

P4.11 Imagine a quick trip by jet, on which you do not visit the rest room and your perspiration is just canceled out by a glass of tomato juice. By subtraction, $\left(F_g\right)_p = mg_p$ and $\left(F_g\right)_C = mg_C$ give

$$\Delta F_g = m\left(g_p - g_C\right).$$

For a person whose mass is 88.7 kg, the change in weight is

$$\Delta F_g = 88.7 \text{ kg}\left(9.809\,5 - 9.780\,8\right) = \boxed{2.55 \text{ N}}.$$

A precise balance scale, as in a doctor's office, reads the same in different locations because it compares you with the standard masses on its beams. A typical bathroom scale is not precise enough to reveal this difference.

P4.12 $F_g = \text{weight of ball} = mg$

$v_{\text{release}} = v$ and time to accelerate $= t$: $\quad \vec{\mathbf{a}} = \dfrac{\Delta \vec{\mathbf{v}}}{\Delta t} = \dfrac{\vec{\mathbf{v}}}{t} = \dfrac{v}{t}\hat{\mathbf{i}}$

(a) Distance $x = \bar{v}t$: $x = \left(\dfrac{v}{2}\right)t = \boxed{\dfrac{vt}{2}}$

(b) $\vec{\mathbf{F}}_p - F_g\hat{\mathbf{j}} = \dfrac{F_g v}{gt}\hat{\mathbf{i}}$

$$\vec{\mathbf{F}}_p = \boxed{\frac{F_g v}{gt}\hat{\mathbf{i}} + F_g\hat{\mathbf{j}}}$$

P4.13 (a) $\sum F = ma$ and $v_f^2 = v_i^2 + 2ax_f$ or $a = \dfrac{v_f^2 - v_i^2}{2x_f}$.

Therefore,

$$\sum F = m\frac{\left(v_f^2 - v_i^2\right)}{2x_f}$$

$$\sum F = 9.11\times10^{-31}\ \text{kg}\frac{\left[\left(7.00\times10^5\ \text{m/s}^2\right)^2 - \left(3.00\times10^5\ \text{m/s}^2\right)^2\right]}{2(0.050\,0\ \text{m})} = \boxed{3.64\times10^{-18}\ \text{N}}.$$

(b) The weight of the electron is

$$F_g = mg = \left(9.11\times10^{-31}\ \text{kg}\right)\left(9.80\ \text{m/s}^2\right) = 8.93\times10^{-30}\ \text{N}$$

The accelerating force is $\boxed{4.08\times10^{11}\ \text{times the weight of the electron.}}$

P4.14 We find acceleration:

$$\vec{\mathbf{r}}_f - \vec{\mathbf{r}}_i = \vec{\mathbf{v}}_i t + \frac{1}{2}\vec{\mathbf{a}}t^2$$

$$4.20\ \text{m}\hat{\mathbf{i}} - 3.30\ \text{m}\hat{\mathbf{j}} = 0 + \frac{1}{2}\vec{\mathbf{a}}(1.20\ \text{s})^2 = 0.720\ \text{s}^2\vec{\mathbf{a}}$$

$$\vec{\mathbf{a}} = \left(5.83\hat{\mathbf{i}} - 4.58\hat{\mathbf{j}}\right)\ \text{m/s}^2.$$

Now $\sum\vec{\mathbf{F}} = m\vec{\mathbf{a}}$ becomes

$$\vec{\mathbf{F}}_g + \vec{\mathbf{F}}_2 = m\vec{\mathbf{a}}$$

$$\vec{\mathbf{F}}_2 = 2.80\ \text{kg}\left(5.83\hat{\mathbf{i}} - 4.58\hat{\mathbf{j}}\right)\ \text{m/s}^2 + (2.80\ \text{kg})\left(9.80\ \text{m/s}^2\right)\hat{\mathbf{j}}$$

$$\vec{\mathbf{F}}_2 = \boxed{\left(16.3\hat{\mathbf{i}} + 14.6\hat{\mathbf{j}}\right)\ \text{N}}.$$

Section 4.6 Newton's Third Law

P4.15 (a) You and the earth exert equal forces on each other: $m_y g = M_e a_e$. If your mass is 70.0 kg,

$$a_e = \frac{(70.0\ \text{kg})\left(9.80\ \text{m/s}^2\right)}{5.98\times10^{24}\ \text{kg}} = \boxed{\sim10^{-22}\ \text{m/s}^2}.$$

(b) You and the planet move for equal times intervals according to $x = \frac{1}{2}at^2$. If the seat is 50.0 cm high,

$$\sqrt{\frac{2x_y}{a_y}} = \sqrt{\frac{2x_e}{a_e}}$$

$$x_e = \frac{a_e}{a_y}x_y = \frac{m_y}{m_e}x_y = \frac{70.0\ \text{kg}(0.500\ \text{m})}{5.98\times10^{24}\ \text{kg}}\boxed{\sim10^{-23}\ \text{m}}.$$

P4.16 (a) Let the x-axis be in the original direction of the molecule's motion.

$$v_f = v_i + at: \quad -670 \text{ m/s} = 670 \text{ m/s} + a(3.00 \times 10^{-13} \text{ s})$$

$$a = \boxed{-4.47 \times 10^{15} \text{ m/s}^2}$$

(b) For the molecule, $\sum \vec{\mathbf{F}} = m\vec{\mathbf{a}}$. Its weight is negligible.

$$\vec{\mathbf{F}}_{\text{wall on molecule}} = 4.68 \times 10^{-26} \text{ kg}(-4.47 \times 10^{15} \text{ m/s}^2) = -2.09 \times 10^{-10} \text{ N}$$

$$\vec{\mathbf{F}}_{\text{molecule on wall}} = \boxed{+2.09 \times 10^{-10} \text{ N}}$$

P4.17 (a) $\boxed{15.0 \text{ lb up}}$

(b) $\boxed{5.00 \text{ lb up}}$

(c) $\boxed{0}$

Section 4.7 Applications of Newton's Laws

P4.18 $T_3 = F_g$ (1)

$T_1 \sin\theta_1 + T_2 \sin\theta_2 = F_g$ (2)

$T_1 \cos\theta_1 = T_2 \cos\theta_2$ (3)

Eliminate T_2 and solve for T_1

$$T_1 = \frac{F_g \cos\theta_2}{(\sin\theta_1 \cos\theta_2 + \cos\theta_1 \sin\theta_2)} = \frac{F_g \cos\theta_2}{\sin(\theta_1 + \theta_2)}$$

$$T_3 = F_g = \boxed{325 \text{ N}}$$

$$T_1 = F_g\left(\frac{\cos 25.0°}{\sin 85.0°}\right) = \boxed{296 \text{ N}}$$

$$T_2 = T_1\left(\frac{\cos\theta_1}{\cos\theta_2}\right) = 296 \text{ N}\left(\frac{\cos 60.0°}{\cos 25.0°}\right) = \boxed{163 \text{ N}}$$

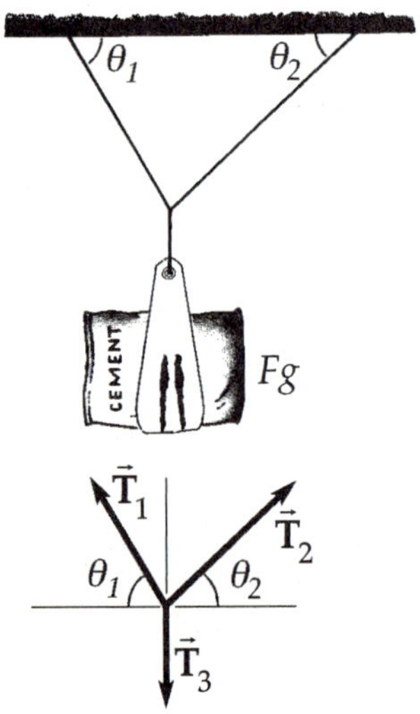

FIG. P4.18

P4.19 See the solution for T_1 in Problem 4.18. The equation indicates that the tension is directly proportional to F_g. As $\theta_1 + \theta_2$ approaches zero (as the angle between the two upper ropes approaches 180°) the tension goes to infinity. Making the right-hand rope horizontal maximizes the tension in the left-hand rope, according to the proportionality of T_1 to $\cos\theta_2$.

***P4.20** As the worker through the pole exerts on the lake bottom a force of 240 N downward at 35° behind the vertical, the lake bottom through the pole exerts a force of 240 N upward at 35° ahead of the vertical. With the x-axis horizontally forward, the pole force on the boat is

$$240\ \text{N}\cos 35°\,\hat{\mathbf{j}} + 240\ \text{N}\sin 35°\,\hat{\mathbf{i}} = 138\ \text{N}\hat{\mathbf{i}} + 197\ \text{N}\hat{\mathbf{j}}.$$

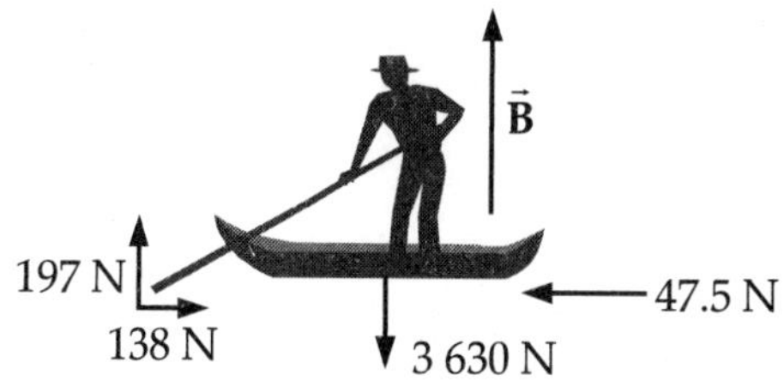

FIG. P4.20

The gravitational force of the whole Earth on boat and worker is $F_g = mg = 370\ \text{kg}\left(9.8\ \text{m/s}^2\right) = 3630\ \text{N down}$.

The acceleration of the boat is purely horizontal, so $\sum F_y = ma_y$ gives $+B + 197\ \text{N} - 3630\ \text{N} = 0$.

(a) The buoyant force is $B = \boxed{3.43\times 10^3\ \text{N}}$.

(b) The acceleration is given by $\sum F_x = ma_x$: $\quad +138\ \text{N} - 47.5\ \text{N} = (370\ \text{kg})a$;
$a = \dfrac{90.2\ \text{N}}{370\ \text{kg}} = 0.244\ \text{m/s}^2$. According to the constant-acceleration model,

$$v_{xf} = v_{xi} + a_x t = 0.857\ \text{m/s} + \left(0.244\ \text{m/s}^2\right)(0.450\ \text{s}) = 0.967\ \text{m/s}$$

$$\vec{\mathbf{v}}_f = \boxed{0.967\hat{\mathbf{i}}\ \text{m/s}}$$

P4.21 (a) An explanation proceeding from fundamental physical principles will be best for the parents and for you. Consider forces on the bit of string touching the weight hanger as shown in the free-body diagram:

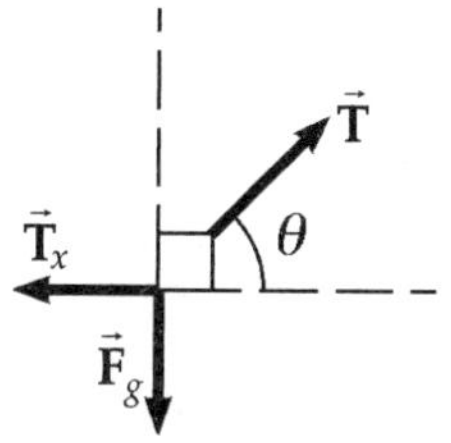

FIG. P4.21

Horizontal Forces: $\sum F_x = ma_x$: $-T_x + T\cos\theta = 0$

Vertical Forces: $\sum F_y = ma_y$: $-F_g + T\sin\theta = 0$

You need only the equation for the vertical forces to find that the tension in the string is given by $T = \boxed{\dfrac{F_g}{\sin\theta}}$. The force the child feels gets smaller, changing from T to $T\cos\theta$, while the counterweight hangs on the string. On the other hand, the kite does not notice what you are doing and the tension in the main part of the string stays constant. You do not need a level, since you learned in physics lab to sight to a horizontal line in a building. Share with the parents your estimate of the experimental uncertainty, which you make by thinking critically about the measurement, by repeating trials, practicing in advance and looking for variations and improvements in technique, including using other observers. You will then be glad to have the parents themselves repeat your measurements.

(b) $$T = \frac{F_g}{\sin\theta} = \frac{0.132\ \text{kg}\left(9.80\ \text{m/s}^2\right)}{\sin 46.3°} = \boxed{1.79\ \text{N}}$$

P4.22 (a) Isolate either mass

$$T - mg = ma = 0$$
$$|T| = |mg|.$$

The scale reads the tension T,

so

$$T = mg = 5.00 \text{ kg}\left(9.80 \text{ m/s}^2\right) = \boxed{49.0 \text{ N}}.$$

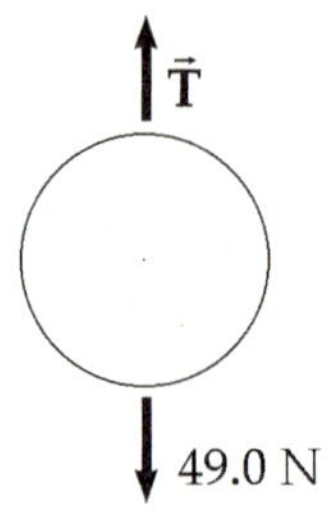

FIG. P4.22(a)

(b) Isolate the pulley

$$\vec{\mathbf{T}}_2 + 2\vec{\mathbf{T}}_1 = 0$$
$$T_2 = 2|T_1| = 2mg = \boxed{98.0 \text{ N}}.$$

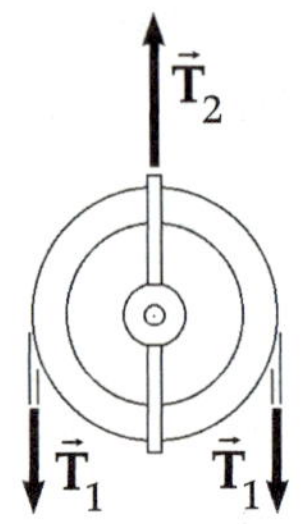

FIG. P4.22(b)

(c) $\sum \vec{\mathbf{F}} = \vec{\mathbf{n}} + \vec{\mathbf{T}} + m\vec{\mathbf{g}} = 0$

Take the component along the incline

$$\vec{\mathbf{n}}_x + \vec{\mathbf{T}}_x + m\vec{\mathbf{g}}_x = 0$$

or

$$0 + T - mg\sin 30.0° = 0$$

$$T = mg\sin 30.0° = \frac{mg}{2} = \frac{5.00(9.80)}{2}$$
$$= \boxed{24.5 \text{ N}}.$$

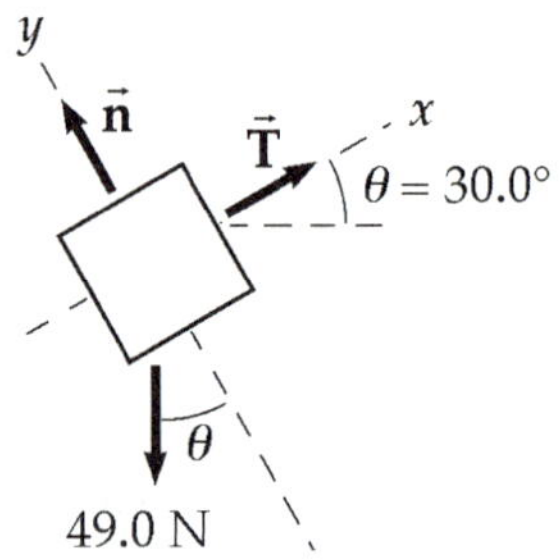

FIG. P4.22(c)

P4.23 (a) Assume the car and mass accelerate horizontally. We consider the forces on the suspended object.

$$\Sigma F_y = ma_y: \quad +T\cos\theta - mg = 0$$

$$\Sigma F_x = ma_x: \quad +T\sin\theta = ma$$

Substitute $T = \dfrac{mg}{\cos\theta}$: $\quad \dfrac{mg\sin\theta}{\cos\theta} = mg\tan\theta = ma$

$$\boxed{a = g\tan\theta}$$

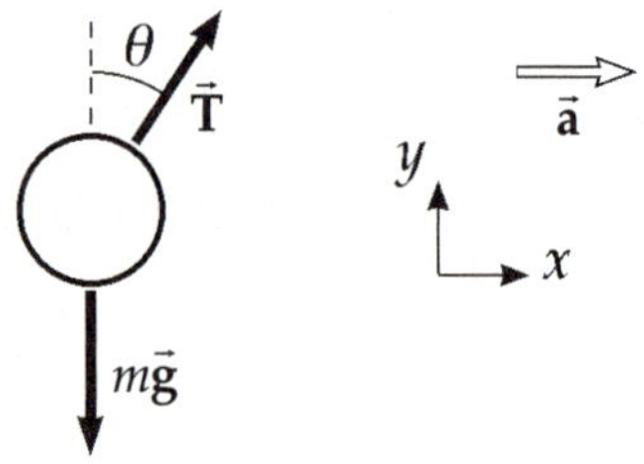

FIG. P4.23

(b) $a = \left(9.80 \text{ m/s}^2\right)\tan 23.0°$ $\quad a = \boxed{4.16 \text{ m/s}^2}$

P4.24 (a) First construct a free body diagram for the 5 kg mass as shown in the Figure 4.24a. Since the mass is in equilibrium, we can require $T_3 - 49\text{ N} = 0$ or $T_3 = 49\text{ N}$. Next, construct a free body diagram for the knot as shown in Figure 4.24a. Again, since the system is moving at constant velocity, $a = 0$ and applying Newton's second law in component form gives

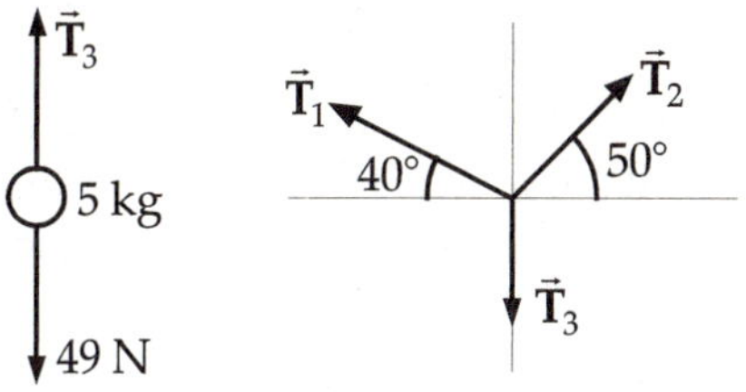

FIG. P4.24(a)

$$\sum F_x = T_2 \cos 50° - T_1 \cos 40° = 0$$
$$\sum F_y = T_2 \sin 50° + T_1 \sin 40° - 49\text{ N} = 0$$

Solving the above equations simultaneously for T_1 and T_2 gives $\boxed{T_1 = 31.5\text{ N}}$ and $\boxed{T_2 = 37.5\text{ N}}$ and above we found $\boxed{T_3 = 49.0\text{ N}}$.

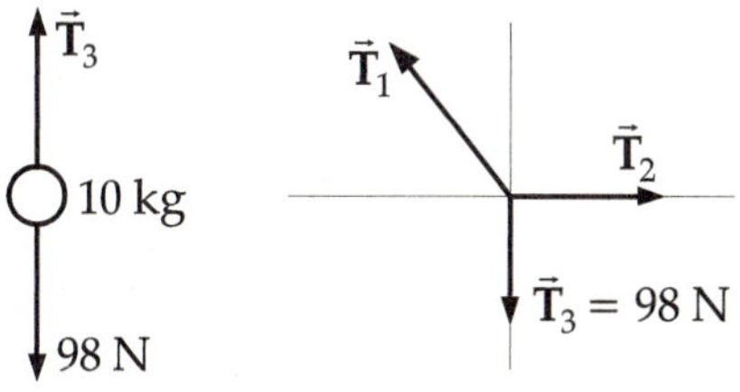

FIG. P4.24(b)

(b) Proceed as in part (a) and construct a free body diagram for the mass and for the knot as shown in Figure 4.24b. Applying Newton's second law in each case (for a constant-velocity system) we find:

$$T_3 - 98\text{ N} = 0$$
$$T_2 - T_1 \cos 60° = 0$$
$$T_1 \sin 60° - T_3 = 0$$

Solving this set of equations we find:

$$\boxed{T_1 = 113\text{ N}} \quad \boxed{T_2 = 56.6\text{ N}}$$

and $\boxed{T_3 = 98.0\text{ N}}$

P4.25 Let us call the forces exerted by each person F_1 and F_2. Thus, for pulling in the same direction,

Newton's second law becomes $F_1 + F_2 = 200\text{ kg}\left(1.52\text{ m/s}^2\right)$

or $F_1 + F_2 = 304\text{ N}$ (1)

When pulling in opposite directions, $F_1 - F_2 = 200\text{ kg}\left(-0.518\text{ m/s}^2\right)$

or $F_1 - F_2 = -104\text{ N}$ (2)

Solving simultaneously, we find $F_1 = \boxed{100\text{ N}}$, and $F_2 = \boxed{204\text{ N}}$

P4.26 The two forces acting on the block are the normal force, n, and the weight, mg. If the block is considered to be a point mass and the x-axis is chosen to be parallel to the plane, then the free body diagram will be as shown in the figure to the right. The angle θ is the angle of inclination of the plane. Applying Newton's second law for the accelerating system (and taking the direction up the plane as the positive x direction) we have

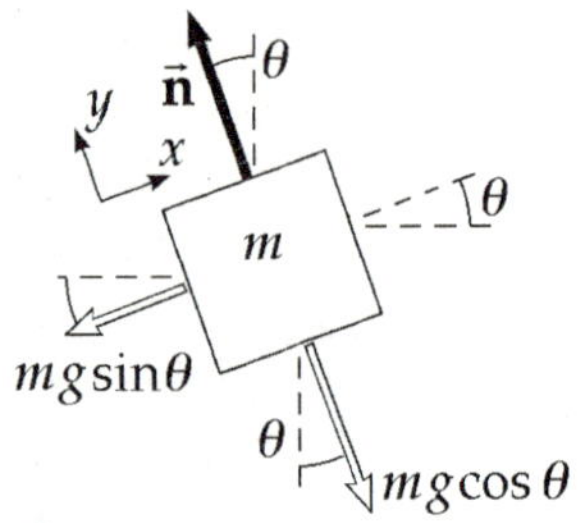

FIG. P4.26

$$\sum F_y = n - mg\cos\theta = 0: \qquad n = mg\cos\theta$$
$$\sum F_x = -mg\sin\theta = ma: \qquad a = -g\sin\theta$$

(a) When $\theta = 15.0°$

$$a = \boxed{-2.54\ \text{m/s}^2}$$

(b) Starting from rest

$$v_f^2 = v_i^2 + 2a(x_f - x_i) = 2ax_f$$
$$|v_f| = \sqrt{2ax_f} = \sqrt{2(-2.54\ \text{m/s}^2)(-2.00\ \text{m})} = \boxed{3.18\ \text{m/s}}$$

P4.27 Choose a coordinate system with $\hat{\mathbf{i}}$ East and $\hat{\mathbf{j}}$ North.

$$\sum \vec{\mathbf{F}} = m\vec{\mathbf{a}} = 1.00\ \text{kg}(10.0\ \text{m/s}^2)\ \text{at}\ 30.0°$$
$$(5.00\ \text{N})\hat{\mathbf{j}} + \vec{\mathbf{F}}_1 = (10.0\ \text{N})\angle 30.0° = (5.00\ \text{N})\hat{\mathbf{j}} + (8.66\ \text{N})\hat{\mathbf{i}}$$
$$\therefore F_1 = \boxed{8.66\ \text{N (East)}}$$

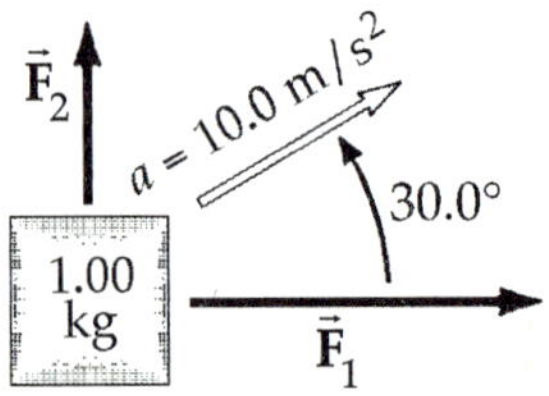

FIG. P4.27

P4.28 First, consider the block moving along the horizontal. The only force in the direction of movement is T. Thus, $\sum F_x = ma$

$$T = (5\ \text{kg})a \qquad (1)$$

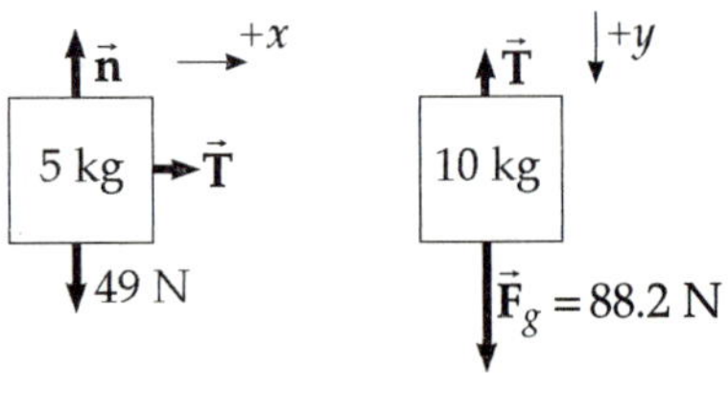

FIG. P4.28

Next consider the block that moves vertically. The forces on it are the tension T and its weight, 88.2 N.

We have $\sum F_y = ma$

$$88.2\ \text{N} - T = (9\ \text{kg})a \qquad (2)$$

Note that both blocks must have the same magnitude of acceleration. Equations (1) and (2) can be added to give $88.2\ \text{N} = (14\ \text{kg})a$. Then

$$\boxed{a = 6.30\ \text{m/s}^2\ \text{and}\ T = 31.5\ \text{N}}.$$

P4.29 After it leaves your hand, the block's speed changes only because of one component of its weight:

$$\sum F_x = ma_x \quad -mg\sin 20.0° = ma$$
$$v_f^2 = v_i^2 + 2a\left(x_f - x_i\right).$$

Taking $v_f = 0$, $v_i = 5.00$ m/s, and $a = -g\sin(20.0°)$ gives

$$0 = (5.00)^2 - 2(9.80)\sin(20.0°)(x_f - 0)$$

or

$$x_f = \frac{25.0}{2(9.80)\sin(20.0°)} = \boxed{3.73 \text{ m}}.$$

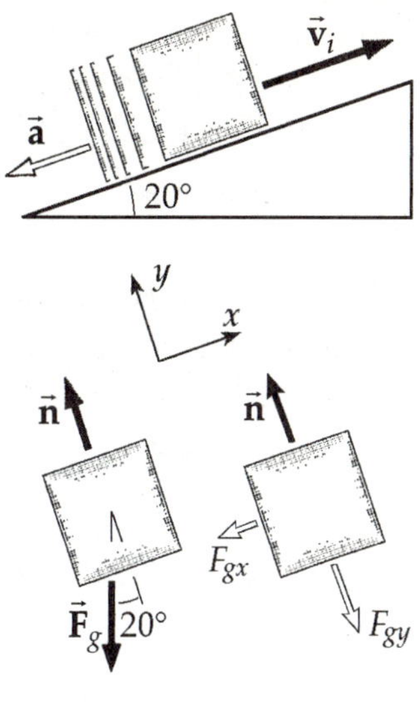

FIG. P4.29

P4.30 $m_1 = 2.00$ kg, $m_2 = 6.00$ kg, $\theta = 55.0°$

(a) $\sum F_x = m_2 g\sin\theta - T = m_2 a$

and

$$T - m_1 g = m_1 a$$
$$a = \frac{m_2 g\sin\theta - m_1 g}{m_1 + m_2} = \boxed{3.57 \text{ m/s}^2}$$

(b) $T = m_1(a+g) = \boxed{26.7 \text{ N}}$

(c) Since $v_i = 0$, $v_f = at = \left(3.57 \text{ m/s}^2\right)(2.00 \text{ s}) = \boxed{7.14 \text{ m/s}}$.

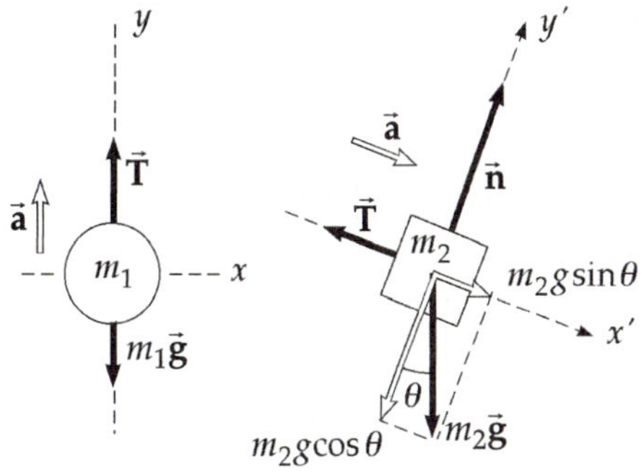

FIG. P4.30

***P4.31** We assume the vertical bar is in compression, pushing up on the pin with force A, and the tilted bar is in tension, exerting force B on the pin at $-50°$.

$$\sum F_x = 0: \quad -2\,500 \text{ N}\cos 30° + B\cos 50° = 0$$
$$\boxed{B = 3.37\times10^3 \text{ N}}$$
$$\sum F_y = 0: \quad -2\,500 \text{ N}\sin 30° + A - 3.37\times10^3 \text{ N}\sin 50° = 0$$
$$\boxed{A = 3.83\times10^3 \text{ N}}$$

Positive answers confirm that

$$\boxed{B \text{ is in tension and } A \text{ is in compression.}}$$

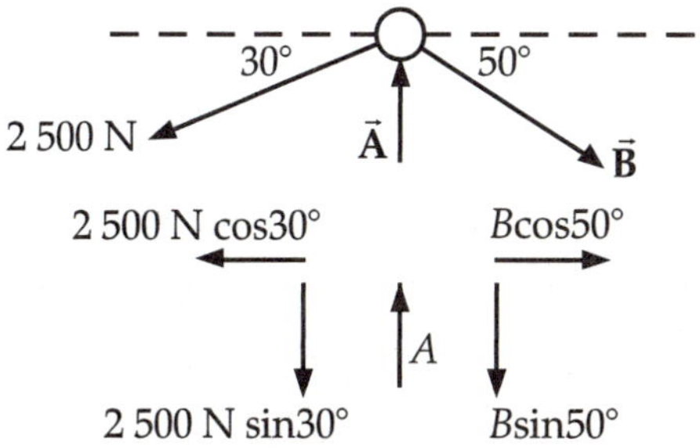

FIG. P4.31

P4.32 First, consider the 3.00 kg rising mass. The forces on it are the tension, T, and its weight, 29.4 N. With the upward direction as positive, the second law becomes

$$\sum F_y = ma_y: T - 29.4\text{ N} = (3.00\text{ kg})a \qquad (1)$$

The forces on the falling 5.00 kg mass are its weight and T, and its acceleration is the same as that of the rising mass. Calling the positive direction down for this mass, we have

$$\sum F_y = ma_y: 49\text{ N} - T = (5.00\text{ kg})a \qquad (2)$$

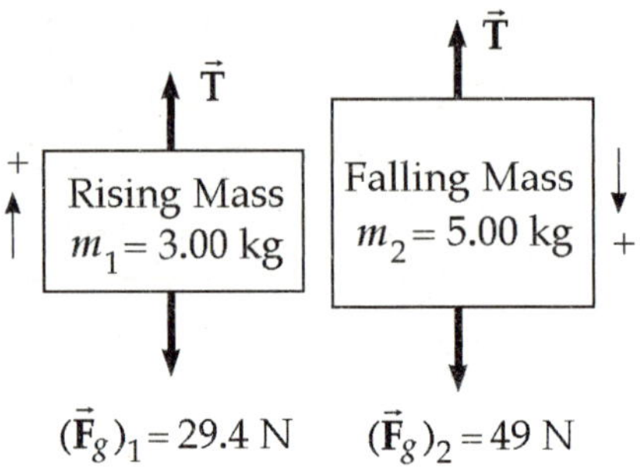

FIG. P4.32

Equations (1) and (2) can be solved simultaneously by adding them:

$$T - 29.4\text{ N} + 49.0\text{ N} - T = (3.00\text{ kg})a + (5.00\text{ kg})a$$

(b) This gives the acceleration as

$$a = \frac{19.6\text{ N}}{8.00\text{ kg}} = \boxed{2.45\text{ m/s}^2}.$$

(a) Then

$$T - 29.4\text{ N} = (3.00\text{ kg})(2.45\text{ m/s}^2) = 7.35\text{ N}.$$

The tension is

$$T = \boxed{36.8\text{ N}}.$$

(c) Consider either mass. We have

$$y = v_i t + \frac{1}{2}at^2 = 0 + \frac{1}{2}(2.45\text{ m/s}^2)(1.00\text{ s})^2 = \boxed{1.23\text{ m}}.$$

***P4.33** As the man rises steadily the pulley turns steadily and the tension in the rope is the same on both sides of the pulley. Choose man-pulley-and-platform as the system:

$$\sum F_y = ma_y$$
$$+T - 950\text{ N} = 0 \qquad T = 950\text{ N}.$$

The worker must pull on the rope with force $\boxed{950\text{ N}}$.

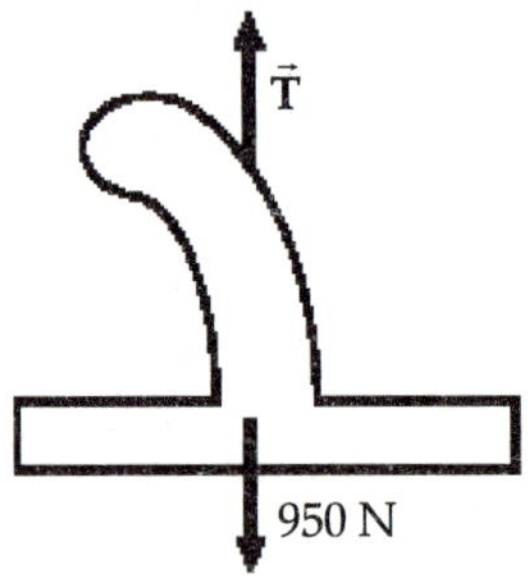

FIG. P4.33

P4.34 Both blocks move with acceleration $a = \left(\frac{m_2 - m_1}{m_2 + m_1}\right) g$:

$$a = \left(\frac{7\text{ kg} - 2\text{ kg}}{7\text{ kg} + 2\text{ kg}}\right) 9.8\text{ m/s}^2 = 5.44\text{ m/s}^2.$$

(a) Take the upward direction as positive for m_1.

$$v_{xf}^2 = v_{xi}^2 + 2a_x(x_f - x_i): \quad 0 = (-2.4\text{ m/s})^2 + 2(5.44\text{ m/s}^2)(x_f - 0)$$

$$x_f = -\frac{5.76\text{ m}^2/\text{s}^2}{2(5.44\text{ m/s}^2)} = -0.529\text{ m}$$

$$x_f = \boxed{0.529\text{ m below its initial level}}$$

(b) $$v_{xf} = v_{xi} + a_x t: \quad v_{xf} = -2.40\text{ m/s} + (5.44\text{ m/s}^2)(1.80\text{ s})$$

$$v_{xf} = \boxed{7.40\text{ m/s upward}}$$

P4.35 Forces acting on 2.00 kg block:

$$T - m_1 g = m_1 a \qquad (1)$$

Forces acting on 8.00 kg block:

$$F_x - T = m_2 a \qquad (2)$$

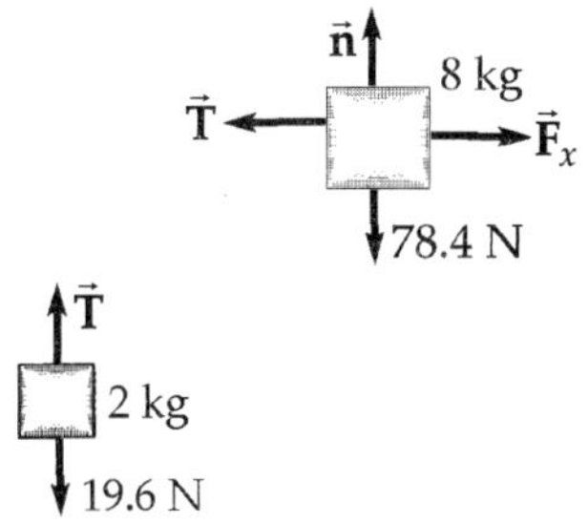

(a) Eliminate T and solve for a:

$$a = \frac{F_x - m_1 g}{m_1 + m_2}$$

$$\boxed{a > 0 \text{ for } F_x > m_1 g = 19.6\text{ N}}.$$

(b) Eliminate a and solve for T:

$$T = \frac{m_1}{m_1 + m_2}(F_x + m_2 g)$$

$$\boxed{T = 0 \text{ for } F_x \le -m_2 g = -78.4\text{ N}}.$$

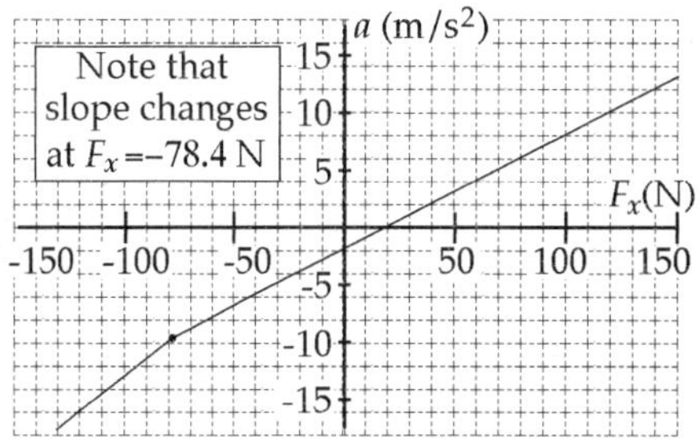

FIG. P4.35

(c)

F_x, N	–100	–78.4	–50.0	0	50.0	100
a_x, m/s^2	–12.5	–9.80	–6.96	–1.96	3.04	8.04

P4.36 (a) For force components along the incline, with the upward direction taken as positive,

$$\sum F_x = ma_x: \quad -mg\sin\theta = ma_x$$

$$a_x = -g\sin\theta = -\left(9.8\ \text{m/s}^2\right)\sin 35° = -5.62\ \text{m/s}^2.$$

For the upward motion,

$$v_{xf}^2 = v_{xi}^2 + 2a_x\left(x_f - x_i\right)$$

$$0 = (5\ \text{m/s})^2 + 2\left(-5.62\ \text{m/s}^2\right)\left(x_f - 0\right)$$

$$x_f = \frac{25\ \text{m}^2/\text{s}^2}{2\left(5.62\ \text{m/s}^2\right)} = \boxed{2.22\ \text{m}}.$$

(b) The time to slide down is given by

$$x_f = x_i + v_{xi}t + \frac{1}{2}a_x t^2$$

$$0 = 2.22\ \text{m} + 0 + \frac{1}{2}\left(-5.62\ \text{m/s}^2\right)t^2$$

$$t = \sqrt{\frac{2(2.22\ \text{m})}{5.62\ \text{m/s}^2}} = 0.890\ \text{s}.$$

For the second particle,

$$x_f = x_i + v_{xi}t + \frac{1}{2}a_x t^2$$

$$0 = 10\ \text{m} + v_{xi}(0.890\ \text{s}) + \frac{1}{2}\left(-5.62\ \text{m/s}^2\right)(0.890\ \text{s})^2$$

$$v_{xi} = \frac{-10\ \text{m} + 2.22\ \text{m}}{0.890\ \text{s}} = -8.74\ \text{m/s}$$

$$\text{speed} = \boxed{8.74\ \text{m/s}}.$$

P4.37 First, we will compute the needed accelerations:

(1) Before it starts to move: $a_y = 0$

(2) During the first 0.800 s: $a_y = \dfrac{v_{yf} - v_{yi}}{t} = \dfrac{1.20 \text{ m/s} - 0}{0.800 \text{ s}} = 1.50 \text{ m/s}^2$

(3) While moving at constant velocity: $a_y = 0$

(4) During the last 1.50 s: $a_y = \dfrac{v_{yf} - v_{yi}}{t} = \dfrac{0 - 1.20 \text{ m/s}}{1.50 \text{ s}} = -0.800 \text{ m/s}^2$

FIG. P4.37

Newton's second law is: $\sum F_y = ma_y$

$$+S - (72.0 \text{ kg})(9.80 \text{ m/s}^2) = (72.0 \text{ kg})a_y$$

$$S = 706 \text{ N} + (72.0 \text{ kg})a_y.$$

(a) When $a_y = 0$, $S = \boxed{706 \text{ N}}$.

(b) When $a_y = 1.50 \text{ m/s}^2$, $S = \boxed{814 \text{ N}}$.

(c) When $a_y = 0$, $S = \boxed{706 \text{ N}}$.

(d) When $a_y = -0.800 \text{ m/s}^2$, $S = \boxed{648 \text{ N}}$.

P4.38 (a) Pulley P_1 has acceleration a_2.
Since m_1 moves *twice* the distance P_1 moves in the same time, m_1 has twice the acceleration of P_1, i.e., $\boxed{a_1 = 2a_2}$.

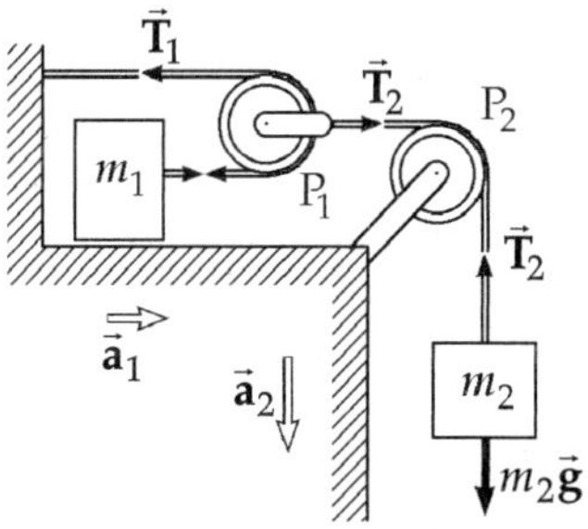

FIG. P4.38

(b) From the figure, and using

$$\sum F = ma: \quad m_2 g - T_2 = m_2 a_2 \qquad (1)$$

$$T_1 = m_1 a_1 = 2m_1 a_2 \qquad (2)$$

$$T_2 - 2T_1 = 0 \qquad (3)$$

Equation (1) becomes $m_2 g - 2T_1 = m_2 a_2$. This equation combined with Equation (2) yields

$$\frac{T_1}{m_1}\left(2m_1 + \frac{m_2}{2}\right) = m_2 g$$

$$\boxed{T_1 = \frac{m_1 m_2}{2m_1 + \frac{1}{2}m_2} g} \text{ and } \boxed{T_2 = \frac{m_1 m_2}{m_1 + \frac{1}{4}m_2} g}.$$

(c) From the values of T_1 and T_2 we find that

$$a_1 = \frac{T_1}{m_1} = \boxed{\frac{m_2 g}{2m_1 + \frac{1}{2}m_2}} \text{ and } a_2 = \frac{1}{2}a_1 = \boxed{\frac{m_2 g}{4m_1 + m_2}}.$$

Section 4.8 Context Connection—Forces on Automobiles

***P4.39** Let situation 1 be the original situation, with $\sum F_1 = m_1 a_1 = m_1(8.40\ \text{mi/h}\cdot\text{s})$. Let situation 2 be the case with larger force $\sum F_2 = (1+0.24)\sum F_1 = m_1 a_2 = 1.24 m_1 a_1$, so $a_2 = 1.24 a_1$. Let situation 3 be the case with the original force but with smaller mass:

$$\sum F_3 = \sum F_1 = m_3 a_3 = (1-0.24) m_1 a_3$$

$$a_3 = \frac{\sum F_1}{0.76 m_1} = 1.32 a_1$$

(a) With $1.32a$, greater than $1.24a_1$, $\boxed{\text{reducing the mass}}$ gives a larger increase in acceleration.

(b) Now with both changes $\sum F_4 = m_4 a_4$

$$1.24 \sum F_1 = 0.76 m_1 a_4$$

$$a_4 = \frac{1.24}{0.76}\frac{\sum F_1}{m_1} = \frac{1.24}{0.76}(8.40\ \text{mi/h}\cdot\text{s}) = \boxed{13.7\ \text{mi/h}\cdot\text{s}}$$

***P4.40** Choose the $+x$-direction to be horizontal and forward with the $+y$ vertical and upward. The common acceleration of the car and trailer then has components of $a_x = +2.15\ \text{m/s}^2$ and $a_y = 0$.

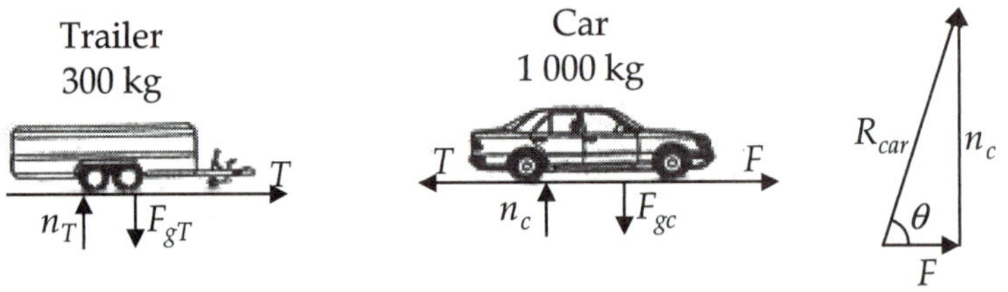

FIG. P4.40

(a) The net force on the car is horizontal and given by

$$\left(\sum F_x\right)_{car} = F - T = m_{car} a_x = (1\,000\ \text{kg})(2.15\ \text{m/s}^2) = \boxed{2.15\times 10^3\ \text{N forward}}.$$

(b) The net force on the trailer is also horizontal and given by

$$\left(\sum F_x\right)_{trailer} = +T = m_{trailer} a_x = (300\ \text{kg})(2.15\ \text{m/s}^2) = \boxed{645\ \text{N forward}}.$$

(c) Consider the free-body diagrams of the car and trailer. The only horizontal force acting on the trailer is $T = 645$ N forward, and this is exerted on the trailer by the car. Newton's third law then states that the force the trailer exerts on the car is $\boxed{\text{645 N toward the rear}}$.

(d) The road exerts two forces on the car. These are F and n_c shown in the free-body diagram of the car. From part (a), $F = T + 2.15\times 10^3\ \text{N} = +2.80\times 10^3\ \text{N}$. Also, $\left(\sum F_y\right)_{car} = n_c - F_{gc} = m_{car} a_y = 0$, so $n_c = F_{gc} = m_{car} g = 9.80\times 10^3\ \text{N}$. The resultant force exerted on the car by the road is then

continued on next page

$$R_{car} = \sqrt{F^2 + n_c^2} = \sqrt{\left(2.80\times10^3\text{ N}\right)^2 + \left(9.80\times10^3\text{ N}\right)^2} = 1.02\times10^4\text{ N}$$

at $\theta = \tan^{-1}\left(\frac{n_c}{F}\right) = \tan^{-1}(3.51) = 74.1°$ above the horizontal and forward. Newton's third law then states that the resultant force exerted on the road by the car is $\boxed{1.02\times10^4\text{ N at }74.1°\text{ below the horizontal and rearward}}$.

Additional Problems

P4.41 (a) See figure to the right.

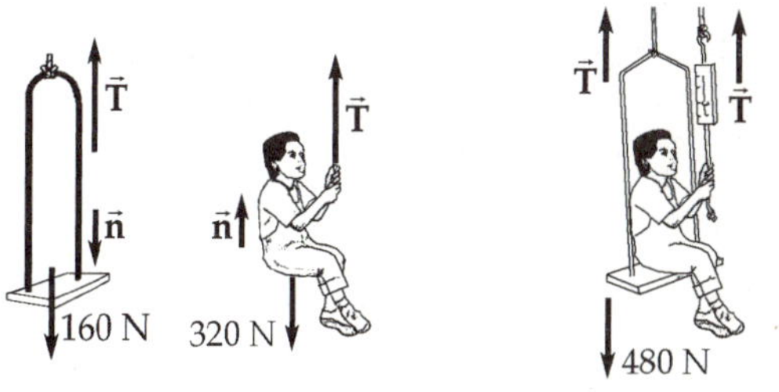

FIG. P4.41

(b) First consider Pat and the chair as the system. Note that *two* ropes support the system, and $T = 250$ N in each rope. Applying $\sum F = ma$

$$2T - 480 = ma\text{, where } m = \frac{480}{9.80} = 49.0\text{ kg}.$$

Solving for a gives

$$a = \frac{500 - 480}{49.0} = \boxed{0.408\text{ m/s}^2}.$$

(c) $\sum F = ma$ on Pat:

$$\sum F = n + T - 320 = ma\text{, where } m = \frac{320}{9.80} = 32.7\text{ kg}$$

$$n = ma + 320 - T = 32.7(0.408) + 320 - 250 = \boxed{83.3\text{ N}}.$$

***P4.42** (a) As soon as Pat passes the rope to the other child,

Pat and the seat, with total weight 480 N, will accelerate down and the other child, only 440 N, will accelerate up.

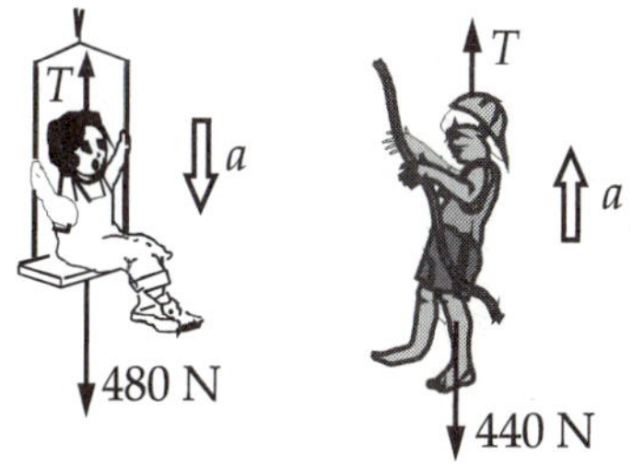

FIG. P4.42

We have $+480\text{ N} - T = \frac{480\text{ N}}{9.8\text{ m/s}^2}a$ and $+T - 440\text{ N} = \frac{440\text{ N}}{9.8\text{ m/s}^2}a$.

Adding,

continued on next page

$$+480\text{ N}-T+T-440\text{ N}=(49.0\text{ kg}+44.9\text{ kg})a$$

$$a=\frac{40\text{ N}}{93.9\text{ kg}}=\boxed{0.426\text{ m/s}^2=a}$$

The rope tension is $T=440\text{ N}+44.9\text{ kg}(0.426\text{ m/s}^2)=459\text{ N}$.

(b) In problem 41, a rope tension of 250 N does not make the rope break. In part (a), the rope is strong enough to support tension 459 N. But now the tension everywhere in the rope is 480 N, so it can exceed the breaking strength of the rope. The tension in the chain supporting the pulley is $480+480\text{ N}=960\text{ N}$, so that chain may break first.

P4.43 (a) $18\text{ N}-P=(2\text{ kg})a$

$$P-Q=(3\text{ kg})a$$

$$Q=(4\text{ kg})a$$

Adding gives $18\text{ N}=(9\text{ kg})a$ so

$$a=\boxed{2.00\text{ m/s}^2}.$$

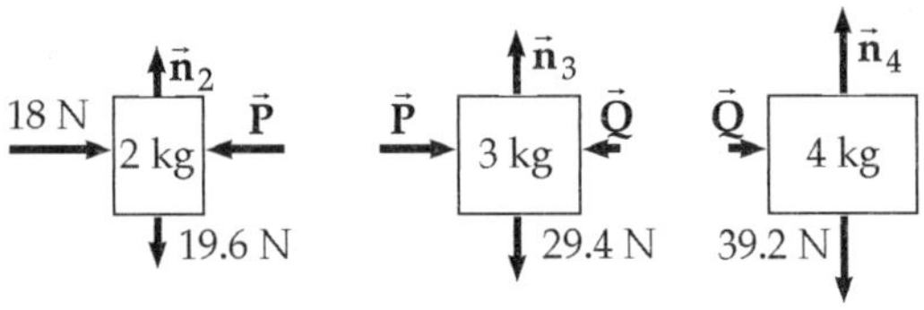

FIG. P4.43

(b) $Q=4\text{ kg}(2\text{ m/s}^2)=\boxed{8.00\text{ N net force on the 4 kg}}$

$P-8\text{ N}=3\text{ kg}(2\text{ m/s}^2)=\boxed{6.00\text{ N net force on the 3 kg}}$ and $P=14\text{ N}$

$18\text{ N}-14\text{ N}=2\text{ kg}(2\text{ m/s}^2)=\boxed{4.00\text{ N net force on the 2 kg}}$

(c) From above, $Q=\boxed{8.00\text{ N}}$ and $P=\boxed{14.0\text{ N}}$.

(d) The 3-kg block models the heavy block of wood. The contact force on your back is represented by Q, which is much less than the force F. The difference between F and Q is the net force causing acceleration of the 5-kg pair of objects. The acceleration is real and nonzero, but lasts for so short a time that it never is associated with a large velocity. The frame of the building and your legs exert forces, small relative to the hammer blow, to bring the partition, block, and you to rest again over a time large relative to the hammer blow. This problem lends itself to interesting lecture demonstrations. One person can hold a lead brick in one hand while another hits the brick with a hammer.

P4.44 (a) Following the in-chapter Example about a block on a frictionless incline, we have

$$a=g\sin\theta=(9.80\text{ m/s}^2)\sin 30.0°$$

$$\boxed{a=4.90\text{ m/s}^2}$$

(b) The block slides distance x on the incline, with $\sin 30.0°=\dfrac{0.500\text{ m}}{x}$

$x=1.00\text{ m}$: $\quad v_f^2=v_i^2+2a(x_f-x_i)=0+2(4.90\text{ m/s}^2)(1.00\text{ m})$

$v_f=\boxed{3.13\text{ m/s}}$ after time $t_s=\dfrac{2x_f}{v_f}=\dfrac{2(1.00\text{ m})}{3.13\text{ m/s}}=0.639\text{ s}$.

continued on next page

(c) Now in free fall $y_f - y_i = v_{yi}t + \frac{1}{2}a_y t^2$:

$$-2.00 = (-3.13\ \text{m/s})\sin 30.0°\, t - \frac{1}{2}\left(9.80\ \text{m/s}^2\right)t^2$$

$$\left(4.90\ \text{m/s}^2\right)t^2 + (1.56\ \text{m/s})t - 2.00\ \text{m} = 0$$

$$t = \frac{-1.56\ \text{m/s} \pm \sqrt{(1.56\ \text{m/s})^2 - 4\left(4.90\ \text{m/s}^2\right)(-2.00\ \text{m})}}{9.80\ \text{m/s}^2}$$

Only one root is physical

$$t = 0.499\ \text{s}$$

$$x_f = v_x t = \left[(3.13\ \text{m/s})\cos 30.0°\right](0.499\ \text{s}) = \boxed{1.35\ \text{m}}$$

(d) total time $= t_s + t = 0.639\ \text{s} + 0.499\ \text{s} = \boxed{1.14\ \text{s}}$

(e) The mass of the block makes no difference.

P4.45 (a) First, we note that $F = T_1$. Next, we focus on the mass M and write $T_5 = Mg$. Next, we focus on the bottom pulley and write $T_5 = T_2 + T_3$. Finally, we focus on the top pulley and write $T_4 = T_1 + T_2 + T_3$.

Since the pulleys are not starting to rotate and are frictionless, $T_1 = T_3$, and $T_2 = T_3$. From this information, we have $T_5 = 2T_2$, so $T_2 = \frac{Mg}{2}$.

Then $\boxed{T_1 = T_2 = T_3 = \frac{Mg}{2}}$, and $\boxed{T_4 = \frac{3Mg}{2}}$, and $\boxed{T_5 = Mg}$.

(b) Since $F = T_1$, we have $\boxed{F = \frac{Mg}{2}}$.

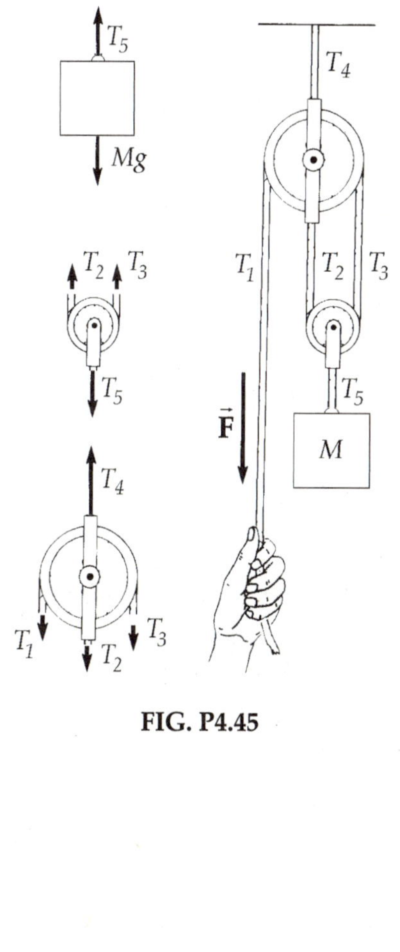

FIG. P4.45

P4.46

t(s)	$t^2\left(s^2\right)$	x(m)
0	0	0
1.02	1.040	0.100
1.53	2.341	0.200
2.01	4.040	0.350
2.64	6.970	0.500
3.30	10.89	0.750
3.75	14.06	1.00

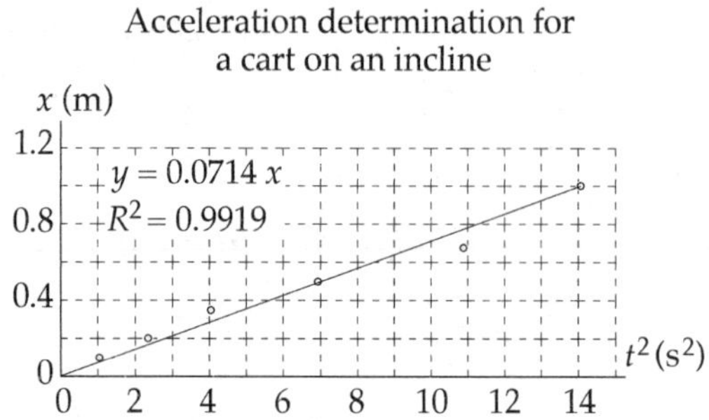

FIG. P4.46

From $x=\frac{1}{2}at^2$ the slope of a graph of x versus t^2 is $\frac{1}{2}a$, and

$$a = 2\times\text{slope} = 2\left(0.0714\ \text{m/s}^2\right) = \boxed{0.143\ \text{m/s}^2}.$$

From $a' = g\sin\theta$,

$$a' = 9.80\ \text{m/s}^2\left(\frac{1.774}{127.1}\right) = 0.137\ \text{m/s}^2\text{, different by 4\%.}$$

The difference is accounted for by the uncertainty in the data, which we may estimate from the third point as

$$\frac{0.350-(0.0714)(4.04)}{0.350} = 18\%.$$

P4.47 $\sum F = ma$

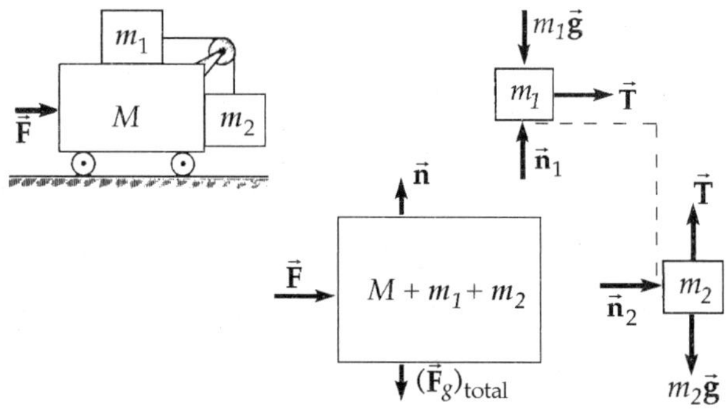

FIG. P4.47

For m_1: $T = m_1 a$

For m_2: $T - m_2 g = 0$

Eliminating T,

$$a = \frac{m_2 g}{m_1}$$

For all 3 blocks:

$$F = (M + m_1 + m_2)a = \boxed{(M+m_1+m_2)\left(\frac{m_2 g}{m_1}\right)}$$

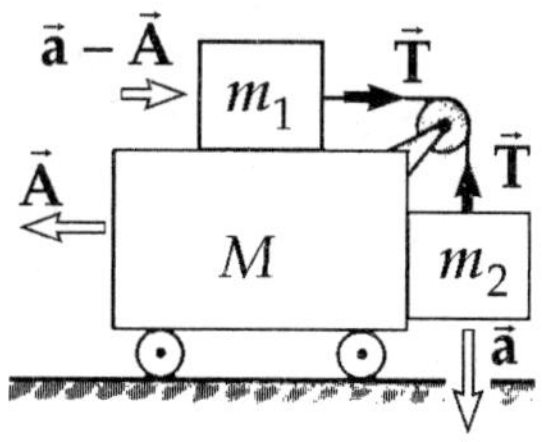

FIG. P4.48

P4.48 (1) $m_1(a-A)=T \Rightarrow \quad a=\dfrac{T}{m_1}+A$

(2) $MA=R_x=T \Rightarrow \quad A=\dfrac{T}{M}$

(3) $m_2 a=m_2 g-T \Rightarrow \quad T=m_2(g-a)$

(a) Substitute the value for a from (1) into (3) and solve for T:

$$T=m_2\left[g-\left(\frac{T}{m_1}+A\right)\right].$$

Substitute for A from (2):

$$T=m_2\left[g-\left(\frac{T}{m_1}+\frac{T}{M}\right)\right]=\boxed{m_2 g\left[\frac{m_1 M}{m_1 M+m_2(m_1+M)}\right]}.$$

(b) Solve (3) for a and substitute value of T:

$$\boxed{a=\frac{m_2 g(m_1+M)}{m_1 M+m_2(M+m_1)}}.$$

(c) From (2), $A=\dfrac{T}{M}$, Substitute the value of T:

$$\boxed{A=\frac{m_1 m_2 g}{m_1 M+m_2(m_1+M)}}.$$

(d) $\boxed{a-A=\dfrac{M m_2 g}{m_1 M+m_2(m_1+M)}}$

***P4.49** (a) Let x represent the position of the glider along the air track. Then $z^2=x^2+h_0^2$, $x=\left(z^2-h_0^2\right)^{1/2}$, $v_x=\dfrac{dx}{dt}=\dfrac{1}{2}\left(z^2-h_0^2\right)^{-1/2}(2z)\dfrac{dz}{dt}$. Now $\dfrac{dz}{dt}$ is the rate at which string passes over the pulley, so it is equal to v_y of the counterweight.

$$v_x=z\left(z^2-h_0^2\right)^{-1/2}v_y=uv_y$$

(b) $a_x=\dfrac{dv_x}{dt}=\dfrac{d}{dt}uv_y=u\dfrac{dv_y}{dt}+v_y\dfrac{du}{dt}$ at release from rest, $v_y=0$ and $a_x=ua_y$.

continued on next page

(c) $\sin 30.0° = \dfrac{80.0 \text{ cm}}{z}$, $z = 1.60$ m, $u = \left(z^2 - h_0^2\right)^{-1/2} z = \left(1.6^2 - 0.8^2\right)^{-1/2}(1.6) = 1.15$.

For the counterweight $\sum F_y = ma_y$: $T - 0.5 \text{ kg } 9.8 \text{ m/s}^2 = -0.5 \text{ kg} a_y$

$$a_y = -2T + 9.8$$

For the glider $\sum F_x = ma_x$: $T\cos 30° = 1.00 \text{ kg } a_x = 1.15 a_y = 1.15(-2T + 9.8) = -2.31T + 11.3 \text{ N}$

$$3.18T = 11.3 \text{ N}$$

$$T = \boxed{3.56 \text{ N}}$$

P4.50 The upward acceleration of the rod is described by

$$y_f = y_i + v_{yi}t + \frac{1}{2}a_y t^2$$

$$1 \times 10^{-3} \text{ m} = 0 + 0 + \frac{1}{2}a_y\left(8 \times 10^{-3} \text{ s}\right)^2$$

$$a_y = 31.2 \text{ m/s}^2$$

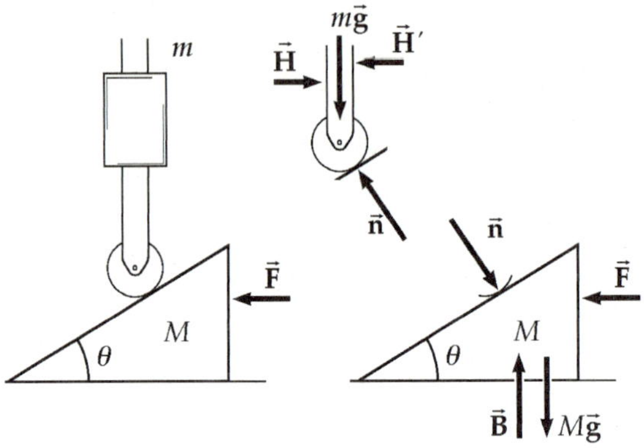

FIG. P4.50

The distance y moved by the rod and the distance x moved by the wedge in the same time are related by $\tan 15° = \dfrac{y}{x} \Rightarrow x = \dfrac{y}{\tan 15°}$. Then their speeds and accelerations are related by

$$\frac{dx}{dt} = \frac{1}{\tan 15°}\frac{dy}{dt}$$

and

$$\frac{d^2x}{dt^2} = \frac{1}{\tan 15°}\frac{d^2y}{dt^2} = \left(\frac{1}{\tan 15°}\right)31.2 \text{ m/s}^2 = 117 \text{ m/s}^2.$$

The free body diagram for the rod is shown. Here H and H' are forces exerted by the guide.

$\sum F_y = ma_y$: $n\cos 15° - mg = ma_y$

$$n\cos 15° - 0.250 \text{ kg}\left(9.8 \text{ m/s}^2\right) = 0.250 \text{ kg}\left(31.2 \text{ m/s}^2\right)$$

$$n = \frac{10.3 \text{ N}}{\cos 15°} = 10.6 \text{ N}$$

For the wedge,

$$\sum F_x = Ma_x: \quad -n\sin 15° + F = 0.5 \text{ kg}\left(117 \text{ m/s}^2\right)$$

$$F = (10.6 \text{ N})\sin 15° + 58.3 \text{ N} = \boxed{61.1 \text{ N}}$$

P4.51 Take the x-axis vertically upward. Your impact speed is given by

$$v_{xf}^2 = v_{xi}^2 + 2a_x(x_f - x_i) = 0 + 2(-9.8\ \text{m/s}^2)(-1.00\ \text{m})$$

$$v_{xf} = -\sqrt{19.6\ \text{m}^2/\text{s}^2} = -4.43\ \text{m/s}$$

In stopping, $\Sigma F_x = ma_x$: $\quad +5.12\times10^4\ \text{N} - 60\ \text{kg}(9.8\ \text{m/s}^2) = (60\ \text{kg})a_x$

$$a_x = \frac{5.06\times10^4\ \text{N}}{60\ \text{kg}} = 844\ \text{m/s}^2$$

$v_{xf}^2 = v_{xi}^2 + 2a_x(x_f - x_i)$ $\quad 0 = (-4.43\ \text{m/s})^2 + 2(844\ \text{m/s}^2)(-d)$

$$d = \frac{19.6\ \text{m}^2/\text{s}^2}{2(844\ \text{m/s}^2)} = \boxed{0.011\ 6\ \text{m}}$$

P4.52 (a) Consider forces on the midpoint of the rope. It is nearly in equilibrium just before the car begins to move. Take the y-axis in the direction of the force you exert:

$$\sum F_y = ma_y:\quad -T\sin\theta + f - T\sin\theta = 0$$

$$\boxed{T = \frac{f}{2\sin\theta}}.$$

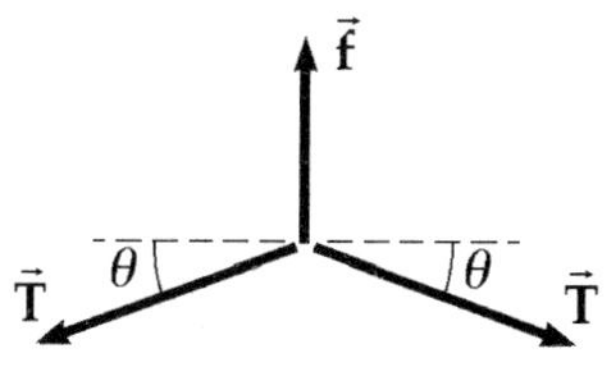

FIG. P4.52

(b) $T = \dfrac{100\ \text{N}}{2\sin 7°} = \boxed{410\ \text{N}}$

P4.53 Choose the x-axis pointing down the slope.

$$v_f = v_i + at:\quad 30.0\ \text{m/s} = 0 + a(6.00\ \text{s})$$

$$a = 5.00\ \text{m/s}^2.$$

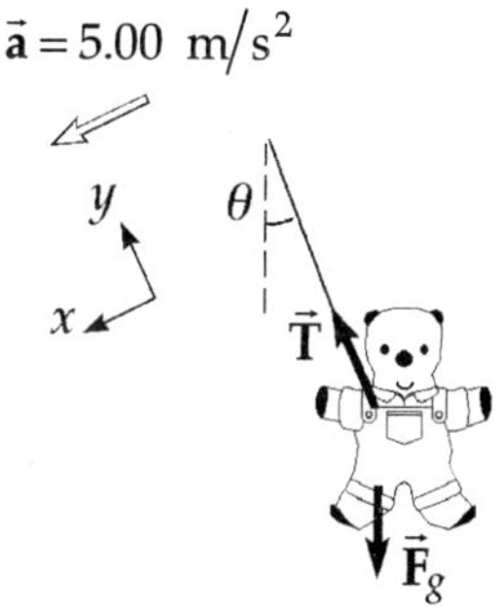

FIG. P4.53

Consider forces on the toy.

(a) $\sum F_x = ma_x:\quad mg\sin\theta = m(5.00\ \text{m/s}^2)$

$$\theta = \boxed{30.7°}$$

(b) $\sum F_y = ma_y:\quad -mg\cos\theta + T = 0$

$$T = mg\cos\theta = (0.100)(9.80)\cos 30.7°$$

$$T = \boxed{0.843\ \text{N}}$$

P4.54 Since it has a larger mass, we expect the 8.00-kg block to move down the plane. The acceleration for both blocks should have the same magnitude since they are joined together by a non-stretching string. Define up the left hand plane as positive for the 3.50-kg object and down the right hand plane as positive for the 8.00-kg object.

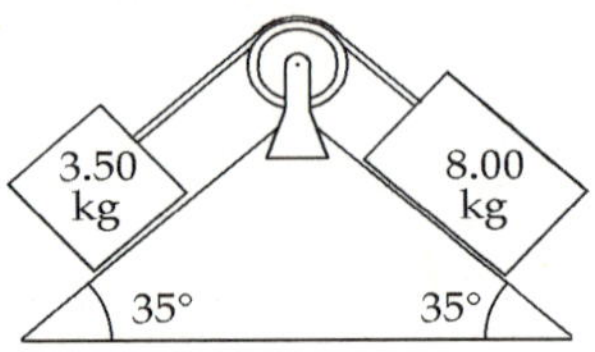

FIG. P4.54

$$\sum F_1 = m_1 a_1: \quad -m_1 g \sin 35.0° + T = m_1 a$$
$$\sum F_2 = m_2 a_2: \quad m_2 g \sin 35.0° - T = m_2 a$$

and

$$-(3.50)(9.80)\sin 35.0° + T = 3.50a$$
$$(8.00)(9.80)\sin 35.0° - T = 8.00a.$$

Adding, we obtain $+45.0 \text{ N} - 19.7 \text{ N} = (11.5 \text{ kg})a$.

(a) Thus the acceleration is $\boxed{a = 2.20 \text{ m/s}^2}$. By substitution,

$$-19.7 \text{ N} + T = (3.50 \text{ kg})(2.20 \text{ m/s}^2) = 7.70 \text{ N}.$$

(b) The tension is

$$\boxed{T = 27.4 \text{ N}}.$$

P4.55 Throughout its up and down motion after release the block has

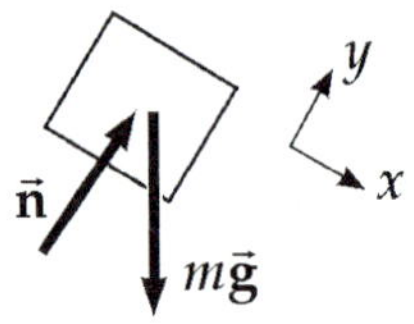

$$\sum F_y = ma_y: \quad +n - mg\cos\theta = 0$$
$$n = mg\cos\theta.$$

Let $\vec{\mathbf{R}} = R_x\hat{\mathbf{i}} + R_y\hat{\mathbf{j}}$ represent the force of table on incline. We have

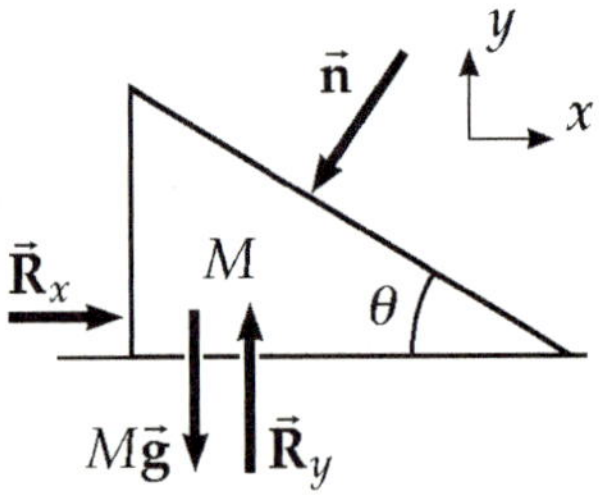

FIG. P4.55

$$\sum F_x = ma_x: \quad +R_x - n\sin\theta = 0$$
$$R_x = mg\cos\theta\sin\theta$$
$$\sum F_y = ma_y: \quad -Mg - n\cos\theta + R_y = 0$$
$$R_y = Mg + mg\cos^2\theta.$$

$$\boxed{\vec{\mathbf{R}} = mg\cos\theta\sin\theta \text{ to the right} + \left(M + m\cos^2\theta\right)g \text{ upward}}$$

P4.56 $\sum F_y = ma_y: n - mg\cos\theta = 0$

or

$$n = 8.40(9.80)\cos\theta$$
$$n = (82.3\text{ N})\cos\theta$$

$\sum F_x = ma_x: mg\sin\theta = ma$

or

$$a = g\sin\theta$$
$$a = (9.80\text{ m/s}^2)\sin\theta$$

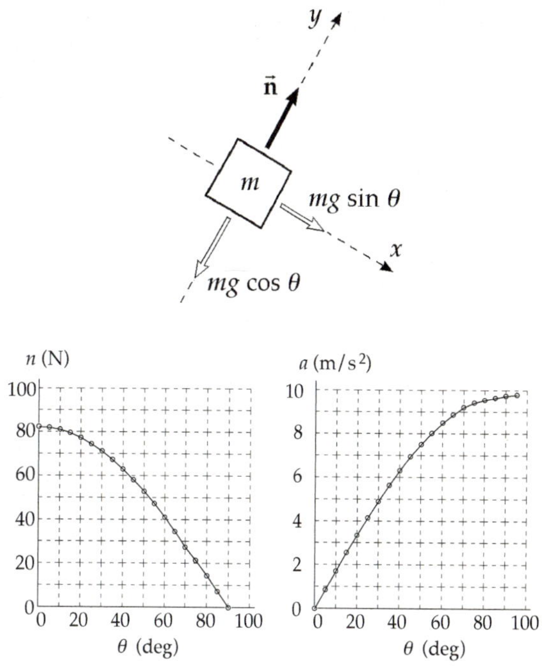

FIG. P4.56

θ, deg	n, N	a, m/s^2
0.00	82.3	0.00
5.00	82.0	0.854
10.0	81.1	1.70
15.0	79.5	2.54
20.0	77.4	3.35
25.0	74.6	4.14
30.0	71.3	4.90
35.0	67.4	5.62
40.0	63.1	6.30
45.0	58.2	6.93
50.0	52.9	7.51
55.0	47.2	8.03
60.0	41.2	8.49
65.0	34.8	8.88
70.0	28.2	9.21
75.0	21.3	9.47
80.0	14.3	9.65
85.0	7.17	9.76
90.0	0.00	9.80

At 0°, the normal force is the full weight and the acceleration is zero. At 90°, the mass is in free fall next to the vertical incline.

P4.57 (a) Apply Newton's second law to two points where butterflies are attached on either half of mobile (other half the same, by symmetry)

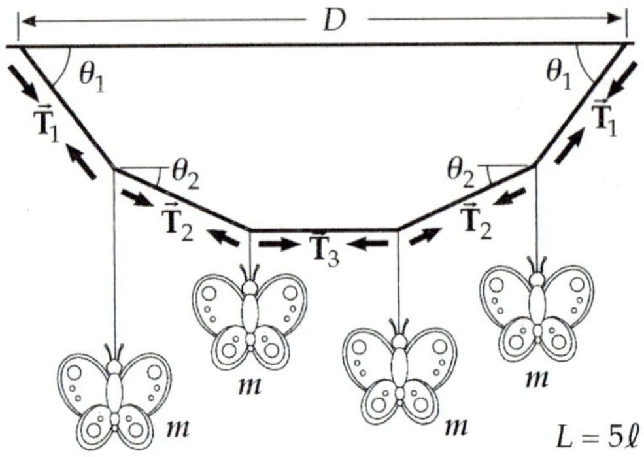

FIG. P4.57

(1) $T_2 \cos\theta_2 - T_1 \cos\theta_1 = 0$

(2) $T_1 \sin\theta_1 - T_2 \sin\theta_2 - mg = 0$

(3) $T_2 \cos\theta_2 - T_3 = 0$

(4) $T_2 \sin\theta_2 - mg = 0$

Substituting (4) into (2) for $T_2 \sin\theta_2$,

$$T_1 \sin\theta_1 - mg - mg = 0.$$

Then

$$\boxed{T_1 = \frac{2mg}{\sin\theta_1}}.$$

Substitute (3) into (1) for $T_2 \cos\theta_2$:

$$T_3 - T_1 \cos\theta_1 = 0, \qquad T_3 = T_1 \cos\theta_1$$

Substitute value of T_1:

$$T_3 = 2mg\frac{\cos\theta_1}{\sin\theta_1} = \boxed{\frac{2mg}{\tan\theta_1} = T_3}.$$

From Equation (4),

$$T_2 = \frac{mg}{\sin\theta_2}.$$

(b) Divide (4) by (3):

$$\frac{T_2 \sin\theta_2}{T_2 \cos\theta_2} = \frac{mg}{T_3}.$$

Substitute value of T_3:

$$\tan\theta_2 = \frac{mg\tan\theta_1}{2mg}, \quad \boxed{\theta_2 = \tan^{-1}\left(\frac{\tan\theta_1}{2}\right)}.$$

Then we can finish answering part (a):

$$\boxed{T_2 = \frac{mg}{\sin\left[\tan^{-1}\left(\frac{1}{2}\tan\theta_1\right)\right]}}.$$

continued on next page

(c) D is the horizontal distance between the points at which the two ends of the string are attached to the ceiling.

$$D = 2\ell\cos\theta_1 + 2\ell\cos\theta_2 + \ell \text{ and } L = 5\ell$$

$$D = \frac{L}{5}\left\{2\cos\theta_1 + 2\cos\left[\tan^{-1}\left(\frac{1}{2}\tan\theta_1\right)\right] + 1\right\}$$

ANSWERS TO EVEN PROBLEMS

P4.2 (a) zero; (b) zero

P4.4 (a) $(-45.0\hat{\mathbf{i}} + 15.0\hat{\mathbf{j}})$ m/s; (b) 162° from the $+x$ axis; (c) $(-225\hat{\mathbf{i}} + 75.0\hat{\mathbf{j}})$ m; (d) $(-227\hat{\mathbf{i}} + 79.0\hat{\mathbf{j}})$ m

P4.6 112 N

P4.8 (a) 181° counterclockwise from x-axis; (b) 11.2 kg; (c) 37.5 m/s; (d) $(-37.5\hat{\mathbf{i}} - 0.893\hat{\mathbf{j}})$ m/s

P4.10 2.38 kN

P4.12 (a) $x = \frac{vt}{2}$; (b) $\frac{F_g v\hat{\mathbf{i}}}{gt} + F_g\hat{\mathbf{j}}$

P4.14 $(16.3\hat{\mathbf{i}} + 14.6\hat{\mathbf{j}})$ N

P4.16 (a) -4.47×10^{15} m/s^2; (b) $+2.09 \times 10^{-10}$ N

P4.18 $T_1 = 296$ N, $T_2 = 163$ N, $T_3 = 325$ N

P4.20 (a) 3.43×10^3 N; (b) $0.967\hat{\mathbf{i}}$ m/s

P4.22 (a) 49.0 N; (b) 98.0 N; (c) 24.5 N

P4.24 (a) 31.5 N, 37.5 N, and 49.0 N; (b) 113 N, 56.6 N, and 98.0 N

P4.26 see the solution (a) -2.54 m/s^2 down the incline; (b) 3.18 m/s

P4.28 see the solution $a = 6.30$ m/s^2, $T = 31.5$ N

P4.30 (a) 3.57 m/s^2; (b) 26.7 N; (c) 7.14 m/s, see the solution

P4.32 (a) 36.8 N; (b) 2.45 m/s^2; (c) 1.23 m

P4.34 (a) 0.529 m; (b) 7.40 m/s upward

P4.36 (a) 2.22 m; (b) 8.74 m/s

P4.38 (a) $a_1 = 2a_2$;

(b) $T_1 = \frac{m_1 m_2}{m_1 + \frac{1}{2}m_2}g$, $T_2 = \frac{m_1 m_2}{m_1 + \frac{1}{4}m_2}g$;

(c) $a_1 = \frac{m_2 g}{2m_1 + \frac{1}{2}m_2}$, $a_2 = \frac{m_2 g}{4m_1 + m_2}$

P4.40 (a) 2.15×10^3 N forward; (b) 645 N forward; (c) 645 N toward the rear; (d) 1.02×10^4 N at 74.1° below the horizontal toward the rear.

P4.42 (a) Pat and the chair accelerate down at 0.426 m/s^2. The other child accelerates up off the ground at the same rate. (b) The tension throughout the rope becomes 480 N, larger than 459 N.

P4.44 (a) 4.90 m/s^2; (b) 3.13 m/s; (c) 1.35 m; (d) 1.14 s; (e) no

P4.46 see the solution $a = 0.143$ m/s^2, approximately 4% high

P4.48 (a) $m_2 g\left[\dfrac{m_1 M}{m_1 M + m_2(m_1 + M)}\right]$;

(b) $\dfrac{m_2 g(M + m_1)}{m_1 M + m_2(m_1 + M)}$;

(c) $\dfrac{m_1 m_2 g}{m_1 M + m_2(m_1 + M)}$;

(d) $\dfrac{M m_2 g}{m_1 M + m_2(m_1 + M)}$

P4.50 61.1 N

P4.52 (a) $T = \dfrac{f}{2\sin\theta}$; (b) 410 N

P4.54 (a) $2.20\ \text{m/s}^2$; (b) 27.4 N

P4.56 see the solution $n = (82.3\ \text{N})\cos\theta$, $a = (9.80\ \text{m/s}^2)\sin\theta$, yes

Chapter 5

More Applications of Newton's Laws

CHAPTER OUTLINE

ANSWERS TO QUESTIONS

Q5.1 (a) $m\vec{a} = \vec{R} + m\vec{g}$ (b) $ma = T - mg$ (c) $ma = f - R$

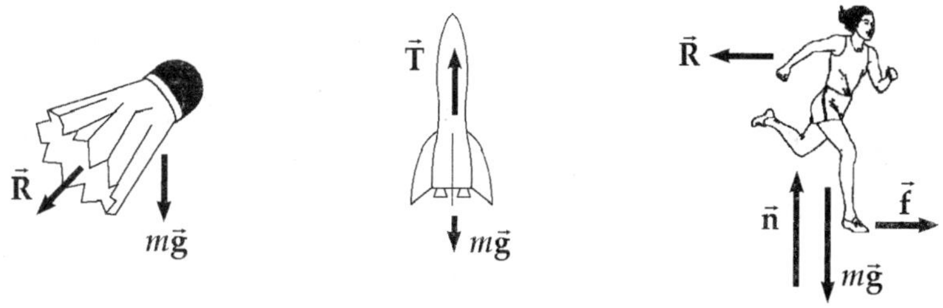

FIG. Q5.1

Q5.2 (a) The friction of the road pushing on the tires of a car causes an automobile to move.

(b) The push of the air on the propeller moves the airplane.

(c) The push of the water on the oars causes the rowboat to move.

Q5.3 As a man takes a step, the action is the force his foot exerts on the Earth; the reaction is the force of the Earth on his foot. In the second case, the action is the force exerted on the girl's back by the snowball; the reaction is the force exerted on the snowball by the girl's back. The third action is the force of the glove on the ball; the reaction is the force of the ball on the glove. The fourth action is the force exerted on the window by the air molecules; the reaction is the force on the air molecules exerted by the window. We could in each case interchange the terms 'action' and 'reaction.'

Q5.4 The tension in the rope must be 9 200 N. Since the rope is moving at a constant speed, then the resultant force on it must be zero. The 49ers are pulling with a force of 9 200 N. If the 49ers were winning with the rope steadily moving in their direction or if the contest was even, then the tension would still be 9 200 N. In all of these cases, the acceleration is zero, and so must be the resultant force on the rope. To win the tug-of-war, a team must exert a larger force on the ground than their opponents do.

Q5.5 If you slam on the brakes, your tires will skid on the road. The force of kinetic friction between the tires and the road is less than the maximum static friction force. Anti-lock brakes work by "pumping" the brakes (much more rapidly that you can) to minimize skidding of the tires on the road.

Q5.6 With friction, it takes longer to come down than to go up. On the way up, the frictional force and the component of the weight down the plane are in the same direction, giving a large acceleration. On the way down, the forces are in opposite directions, giving a relatively smaller acceleration. If the incline is frictionless, it takes the same amount of time to go up as it does to come down.

Q5.7 As you pull away from a stoplight, friction is the force that accelerates forward a box of tissues on the level floor of the car. At the same time, friction of the ground on the tires of the car accelerates the car forward. Drop a stick into a running stream and fluid friction sets the stick into horizontal motion.

Q5.8 The speed changes. The tangential force component causes tangential acceleration.

Q5.9 A torque is exerted by the thrust force of the water times the distance between the nozzles.

Q5.10 This is the same principle as the centrifuge. All the material inside the cylinder tends to move along a straight-line path, but the walls of the cylinder exert an inward force to keep everything moving around in a circular path.

Q5.11 The water has inertia. The water tends to move along a straight line, but the bucket pulls it in and around in a circle.

Q5.12 Blood pressure cannot supply the force necessary both to balance the gravitational force and to provide the centripetal acceleration, to keep blood flowing up to the pilot's brain.

Q5.13 I would not accept that statement for two reasons. First, to be "beyond the pull of gravity," one would have to be infinitely far away from all other matter. Second, astronauts in orbit are moving in a circular path. It is the gravitational pull of Earth on the astronauts that keeps them in orbit. In the space shuttle, just above the atmosphere, gravity is only slightly weaker than at the Earth's surface. Gravity does its job most clearly on an orbiting spacecraft, because the craft feels no other forces and is in free fall.

Q5.14 From the proportionality of the drag force to the speed squared and from Newton's second law, we derive the equation that describes the motion of the skydiver:

$$m\frac{dv_y}{dt} = mg - \frac{D\rho A}{2}v_y^2$$

where D is the coefficient of drag of the parachutist, and A is the projected area of the parachutist's body. At terminal speed,

$$a_y = \frac{dv_y}{dt} = 0 \text{ and } v_T = \left(\frac{2mg}{D\rho A}\right)^{1/2}.$$

When the parachute opens, the coefficient of drag D and the effective area A both increase, thus reducing the speed of the skydiver. Modern parachutes also add a third term, lift, to change the equation to

$$m\frac{dv_y}{dt} = mg - \frac{D\rho A}{2}v_y^2 - \frac{L\rho A}{2}v_x^2$$

where v_y is the vertical velocity, and v_x is the horizontal velocity. The effect of lift is clearly seen in the "paraplane," an ultralight airplane made from a fan, a chair, and a parachute.

Q5.15 Lower air density reduces air resistance, so a tank-truck-load of fuel takes you farther.

Q5.16 The larger drop has higher terminal speed. In the case of spheres, the text demonstrates that terminal speed is proportional to the square root of radius. When moving with terminal speed, an object is in equilibrium and has zero acceleration.

Q5.17 The thesis is false. The moment of decay of a radioactive atomic nucleus (for example) cannot be predicted. Quantum mechanics implies that the future is indeterminate. On the other hand, our sense of free will, of being able to make choices for ourselves that can appear to be random, may be an illusion. It may have nothing to do with the subatomic randomness described by quantum mechanics.

SOLUTIONS TO PROBLEMS

Section 5.1 Forces of Friction

P5.1 For equilibrium: $f = F$ and $n = F_g$. Also, $f = \mu n$ i.e.,

$$\mu = \frac{f}{n} = \frac{F}{F_g}$$

$$\mu_s = \frac{75.0 \text{ N}}{25.0(9.80) \text{ N}} = \boxed{0.306}$$

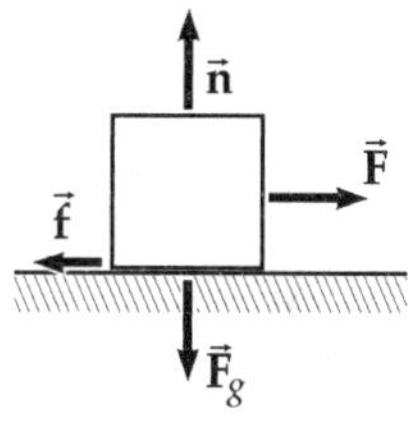

FIG. P5.1

and $\mu_k = \dfrac{60.0 \text{ N}}{25.0(9.80) \text{ N}} = \boxed{0.245}$.

P5.2 $\sum F_y = ma_y$: $+n - mg = 0$

$$f_s \le \mu_s n = \mu_s mg$$

This maximum magnitude of static friction acts so long as the tires roll without skidding.

$$\sum F_x = ma_x: \quad -f_s = ma$$

The maximum acceleration is $a = -\mu_s g$. The initial and final conditions are: $x_i = 0$, $v_i = 50.0 \text{ mi/h} = 22.4 \text{ m/s}$, $v_f = 0$, $v_f^2 = v_i^2 + 2a(x_f - x_i)$: $-v_i^2 = -2\mu_s g x_f$

(a) $$x_f = \frac{v_i^2}{2\mu g}$$

$$x_f = \frac{(22.4 \text{ m/s})^2}{2(0.100)(9.80 \text{ m/s}^2)} = \boxed{256 \text{ m}}$$

(b) $$x_f = \frac{v_i^2}{2\mu g}$$

$$x_f = \frac{(22.4 \text{ m/s})^2}{2(0.600)(9.80 \text{ m/s}^2)} = \boxed{42.7 \text{ m}}$$

P5.3 If all the weight is on the rear wheels,

(a) $F = ma$: $\mu_s mg = ma$

But $\Delta x = \dfrac{at^2}{2} = \dfrac{\mu_s g t^2}{2}$ so $\mu_s = \dfrac{2\Delta x}{gt^2}$: $\mu_s = \dfrac{2(0.250\text{ mi})(1\,609\text{ m/mi})}{(9.80\text{ m/s}^2)(4.96\text{ s})^2} = \boxed{3.34}$.

(b) Time would increase, as the wheels would skid and only kinetic friction would act; or perhaps the car would flip over.

***P5.4**

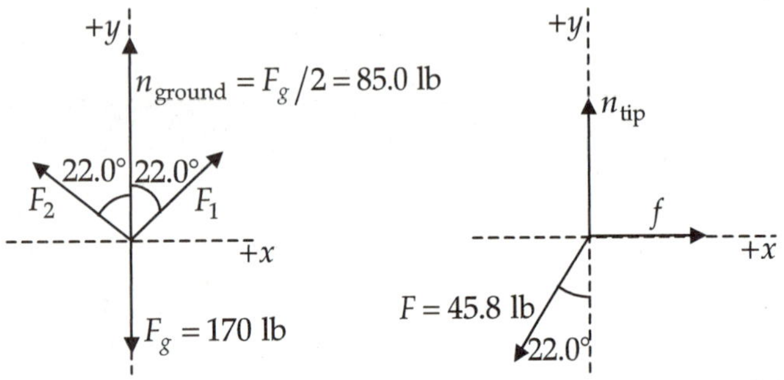

Free-Body Diagram of Person   Free-Body Diagram of Crutch Tip

FIG. P5.4

From the free-body diagram of the person, $\sum F_x = F_1 \sin(22.0°) - F_2 \sin(22.0°) = 0$, which gives $F_1 = F_2 = F$. Then, $\sum F_y = 2F\cos 22.0° + 85.0\text{ lbs} - 170\text{ lbs} = 0$ yields $F = 45.8$ lb.

(a) Now consider the free-body diagram of a crutch tip.

$$\sum F_x = f - (45.8\text{ lb})\sin 22.0° = 0,$$

or

$$f = 17.2\text{ lb}.$$

$$\sum F_y = n_{\text{tip}} - (45.8\text{ lb})\cos 22.0° = 0,$$

which gives

$$n_{\text{tip}} = 42.5\text{ lb}.$$

For minimum coefficient of friction, the crutch tip will be on the verge of slipping, so $f = (f_s)_{\max} = \mu_s n_{\text{tip}}$ and $\mu_s = \dfrac{f}{n_{\text{tip}}} = \dfrac{17.2\text{ lb}}{42.5\text{ lb}} = \boxed{0.404}$.

(b) As found above, the compression force in each crutch is

$$F_1 = F_2 = F = \boxed{45.8\text{ lb}}.$$

P5.5 (a) The person pushes backward on the floor. The floor pushes forward on the person with a force of friction. This is the only horizontal force on the person. If the person's shoe is on the point of slipping the static friction force has its maximum value.

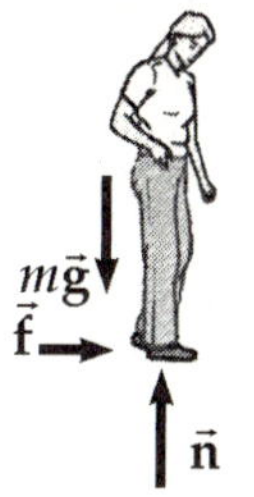

FIG. P5.5

$$\sum F_x = ma_x: \qquad f = \mu_s n = ma_x$$

$$\sum F_y = ma_y: \qquad n - mg = 0$$

$$ma_x = \mu_s mg \qquad a_x = \mu_s g = 0.5\left(9.8\text{ m/s}^2\right) = 4.9\text{ m/s}^2$$

$$x_f = x_i + v_{xi}t + \frac{1}{2}a_x t^2 \qquad 3\text{ m} = 0 + 0 + \frac{1}{2}\left(4.9\text{ m/s}^2\right)t^2$$

$$t = \boxed{1.11\text{ s}}$$

(b) $$x_f = \frac{1}{2}\mu_s g t^2,\; t = \sqrt{\frac{2x_f}{\mu_s g}} = \sqrt{\frac{2(3\text{ m})}{(0.8)\left(9.8\text{ m/s}^2\right)}} = \boxed{0.875\text{ s}}$$

P5.6 If the load is on the point of sliding forward on the bed of the slowing truck, static friction acts backward on the load with its maximum value, to give it the same acceleration as the truck

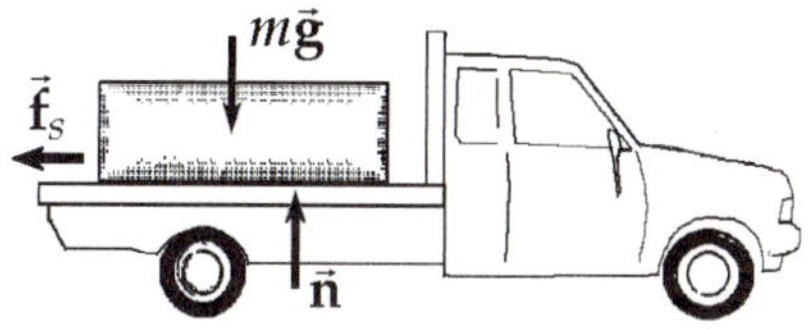

FIG. P5.6

$$\Sigma F_x = ma_x: \qquad -f = m_{\text{load}}a_x$$

$$\Sigma F_y = ma_y: \qquad n - m_{\text{load}}g = 0$$

$$-\mu_s mg = ma_x \qquad a_x = -\mu_s g$$

$$v_{xf}^2 = v_{xi}^2 + 2a_x\left(x_f - x_i\right) \qquad 0 = v_{xi}^2 + 2(-\mu_s g)\left(x_f - 0\right)$$

(a) $$x_f = \frac{v_{xi}^2}{2\mu_s g} = \frac{(12\text{ m/s})^2}{2(0.5)\left(9.8\text{ m/s}^2\right)} = \boxed{14.7\text{ m}}$$

(b) From the expression $x_f = \dfrac{v_{xi}^2}{2\mu_s g}$, $\boxed{\text{neither mass affects the answer}}$.

P5.7 $-f + mg\sin\theta = 0$ and $+n - mg\cos\theta = 0$ with $f = \mu n$ yield $\mu_s = \tan\theta_c = \tan 36.0° = \boxed{0.727}$

$$\mu_k = \tan\theta_c = \tan 30.0° = \boxed{0.577}$$

P5.8 $m_{\text{suitcase}} = 20.0\text{ kg}$, $F = 35.0\text{ N}$

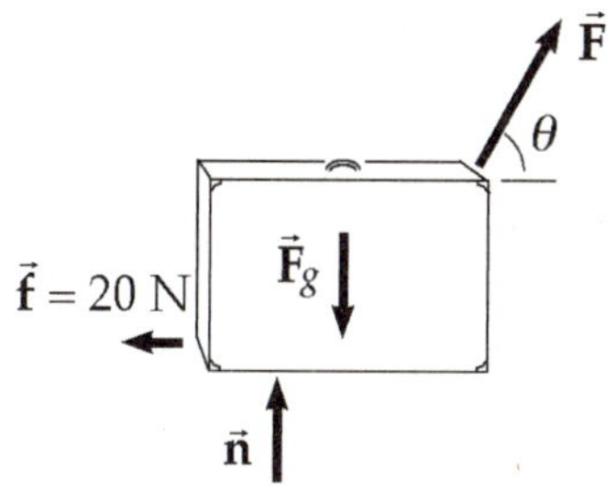

FIG. P5.8

$$\sum F_x = ma_x: \quad -20.0\text{ N} + F\cos\theta = 0$$

$$\sum F_y = ma_y: \quad +n + F\sin\theta - F_g = 0$$

(a) $$F\cos\theta = 20.0\text{ N}$$

$$\cos\theta = \frac{20.0\text{ N}}{35.0\text{ N}} = 0.571$$

$$\boxed{\theta = 55.2°}$$

(b) $$n = F_g - F\sin\theta = \left[196 - 35.0(0.821)\right]\text{ N}$$

$$\boxed{n = 167\text{ N}}$$

P5.9 $m = 3.00\text{ kg}$, $\theta = 30.0°$, $x_f = 2.00\text{ m}$, $t = 1.50\text{ s}$, $x_i = 0$, $v_{xi} = 0$

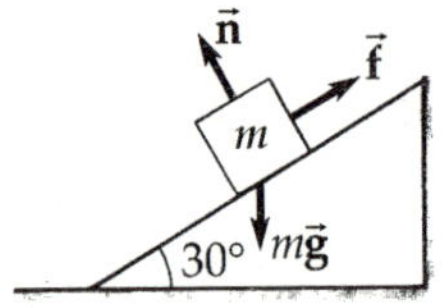

FIG. P5.9

(a) $x_f = \frac{1}{2}at^2$:

$$2.00\text{ m} = \frac{1}{2}a(1.50\text{ s})^2$$

$$a = \frac{4.00}{(1.50)^2} = \boxed{1.78\text{ m/s}^2}$$

$$\sum \vec{\mathbf{F}} = \vec{\mathbf{n}} + \vec{\mathbf{f}} + m\vec{\mathbf{g}} = m\vec{\mathbf{a}}:$$

Along x: $0 - f + mg\sin 30.0° = ma$

$f = m(g\sin 30.0° - a)$

Along y: $n + 0 - mg\cos 30.0° = 0$

$n = mg\cos 30.0°$

(b) $\mu_k = \frac{f}{n} = \frac{m(g\sin 30.0° - a)}{mg\cos 30.0°}$, $\mu_k = \tan 30.0° - \frac{a}{g\cos 30.0°} = \boxed{0.368}$

(c) $f = m(g\sin 30.0° - a)$, $f = 3.00(9.80\sin 30.0° - 1.78) = \boxed{9.37\text{ N}}$

(d) $v_f^2 = v_i^2 + 2a(x_f - x_i)$ where $x_f - x_i = 2.00\text{ m}$

$$v_f^2 = 0 + 2(1.78)(2.00) = 7.11\text{ m}^2/\text{s}^2$$

$$v_f = \sqrt{7.11\text{ m}^2/\text{s}^2} = \boxed{2.67\text{ m/s}}$$

P5.10 $T - f_k = 5.00a$ (for 5.00 kg mass)

$9.00g - T = 9.00a$ (for 9.00 kg mass)

Adding these two equations gives:

$$9.00(9.80) - 0.200(5.00)(9.80) = 14.0a$$

$$a = 5.60\text{ m/s}^2$$

$$\therefore T = 5.00(5.60) + 0.200(5.00)(9.80)$$

$$= \boxed{37.8\text{ N}}$$

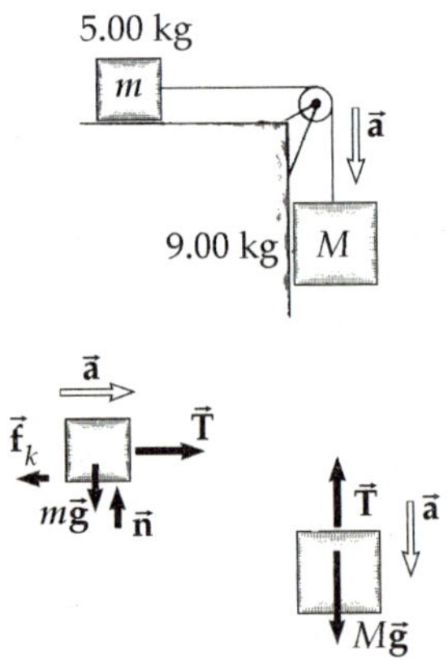

FIG. P5.10

P5.11 (a) See Figure to the right.

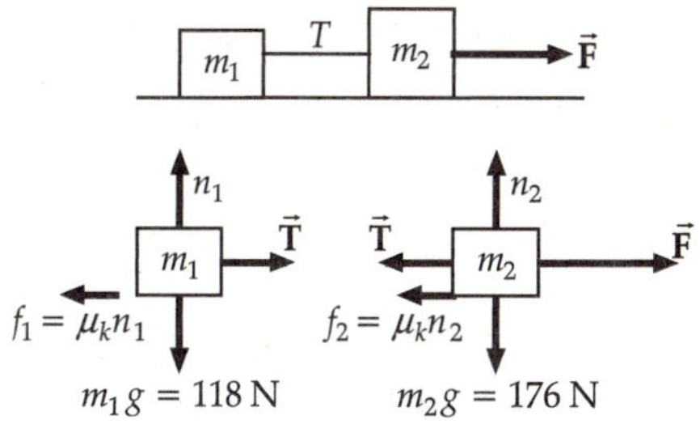

FIG. P5.11

(b)

$$68.0 - T - \mu m_2 g = m_2 a \quad \text{(Block \#2)}$$
$$T - \mu m_1 g = m_1 a \quad \text{(Block \#1)}$$

Adding,

$$68.0 - \mu(m_1 + m_2)g = (m_1 + m_2)a$$
$$a = \frac{68.0}{(m_1 + m_2)} - \mu g = \boxed{1.29 \text{ m/s}^2}$$
$$T = m_1 a + \mu m_1 g = \boxed{27.2 \text{ N}}$$

***P5.12** Let a represent the positive magnitude of the acceleration $-a\hat{\mathbf{j}}$ of m_1, of the acceleration $-a\hat{\mathbf{i}}$ of m_2, and of the acceleration $+a\hat{\mathbf{j}}$ of m_3. Call T_{12} the tension in the left rope and T_{23} the tension in the cord on the right.

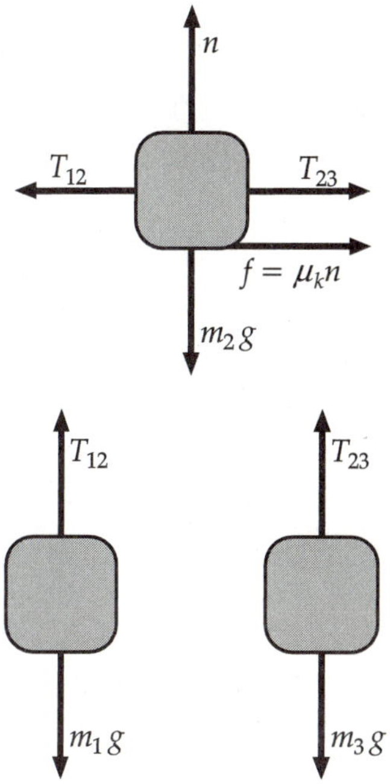

FIG. P5.12

For m_1, $\sum F_y = ma_y \qquad +T_{12} - m_1 g = -m_1 a$

For m_2, $\sum F_x = ma_x \qquad -T_{12} + \mu_k n + T_{23} = -m_2 a$

and $\sum F_y = ma_y \qquad n - m_2 g = 0$

for m_3, $\sum F_y = ma_y \qquad T_{23} - m_3 g = +m_3 a$

we have three simultaneous equations

$$-T_{12} + 39.2 \text{ N} = (4.00 \text{ kg})a$$
$$+T_{12} - 0.350(9.80 \text{ N}) - T_{23} = (1.00 \text{ kg})a$$
$$+T_{23} - 19.6 \text{ N} = (2.00 \text{ kg})a.$$

(a) Add them up:

$$+39.2 \text{ N} - 3.43 \text{ N} - 19.6 \text{ N} = (7.00 \text{ kg})a$$

$$a = \boxed{2.31 \text{ m/s}^2\text{, down for } m_1\text{, left for } m_2\text{, and up for } m_3}.$$

(b) Now $-T_{12} + 39.2 \text{ N} = (4.00 \text{ kg})(2.31 \text{ m/s}^2)$

$$\boxed{T_{12} = 30.0 \text{ N}}$$

and $T_{23} - 19.6 \text{ N} = (2.00 \text{ kg})(2.31 \text{ m/s}^2)$

$$\boxed{T_{23} = 24.2 \text{ N}}.$$

P5.13 (Case 1, impending upward motion)
Setting

$$\sum F_x = 0: \quad P\cos 50.0° - n = 0$$
$$f_{s,\,\max} = \mu_s n: \quad f_{s,\,\max} = \mu_s P\cos 50.0° = 0.250(0.643)P = 0.161P$$

Setting

$$\sum F_y = 0: \quad P\sin 50.0° - 0.161P - 3.00(9.80) = 0$$
$$P_{\max} = \boxed{48.6 \text{ N}}$$

(Case 2, impending downward motion)
As in Case 1,

$$f_{s,\,\max} = 0.161P$$

Setting

$$\sum F_y = 0: \quad P\sin 50.0° + 0.161P - 3.00(9.80) = 0$$
$$P_{\min} = \boxed{31.7 \text{ N}}$$

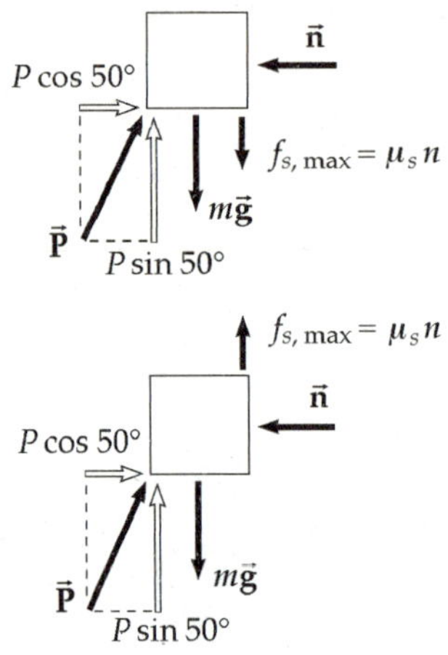

FIG. P5.13

P5.14 We must consider separately the disk when it is in contact with the roof and when it has gone over the top into free fall. In the first case, we take x and y as parallel and perpendicular to the surface of the roof:

$$\sum F_y = ma_y: \quad +n - mg\cos\theta = 0$$
$$n = mg\cos\theta$$

then friction is $f_k = \mu_k n = \mu_k mg\cos\theta$

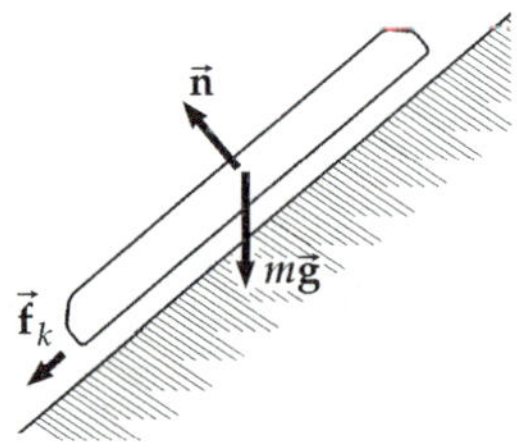

FIG. P5.14

$$\sum F_x = ma_x: \quad -f_k - mg\sin\theta = ma_x$$
$$a_x = -\mu_k g\cos\theta - g\sin\theta = (-0.4\cos 37° - \sin 37°)9.8 \text{ m/s}^2 = -9.03 \text{ m/s}^2$$

The Frisbee goes ballistic with speed given by

$$v_{xf}^2 = v_{xi}^2 + 2a_x(x_f - x_i) = (15 \text{ m/s})^2 + 2(-9.03 \text{ m/s}^2)(10 \text{ m} - 0) = 44.4 \text{ m}^2/\text{s}^2$$
$$v_{xf} = 6.67 \text{ m/s}$$

For the free fall, we take x and y horizontal and vertical:

$$v_{yf}^2 = v_{yi}^2 + 2a_y(y_f - y_i)$$
$$0 = (6.67 \text{ m/s } \sin 37°)^2 + 2(-9.8 \text{ m/s}^2)(y_f - 10 \text{ m } \sin 37°)$$
$$y_f = 6.02 \text{ m} + \frac{(4.01 \text{ m/s})^2}{19.6 \text{ m/s}^2} = \boxed{6.84 \text{ m}}$$

Section 5.2 Newton's Second Law Applied to a Particle in Uniform Circular Motion

P5.15 $m = 3.00\text{ kg}$, $r = 0.800\text{ m}$. The string will break if the tension exceeds the weight corresponding to 25.0 kg, so

$$T_{max} = Mg = 25.0(9.80) = 245\text{ N}.$$

When the 3.00 kg mass rotates in a horizontal circle, the tension causes the centripetal acceleration,

so $$T = \frac{mv^2}{r} = \frac{(3.00)v^2}{0.800}.$$

Then $$v^2 = \frac{rT}{m} = \frac{(0.800)T}{3.00} \le \frac{(0.800)T_{max}}{3.00} = \frac{0.800(245)}{3.00} = 65.3\text{ m}^2/\text{s}^2$$

and $$0 \le v \le \sqrt{65.3}$$

or $$\boxed{0 \le v \le 8.08\text{ m/s}}.$$

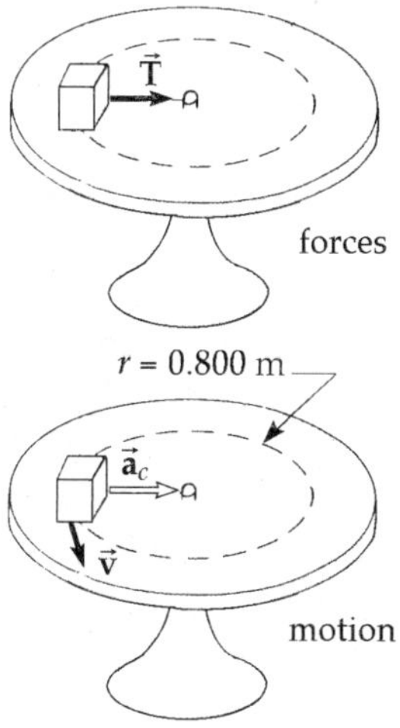

FIG. P5.15

P5.16 (a) $$F = \frac{mv^2}{r} = \frac{(9.11\times10^{-31}\text{ kg})(2.20\times10^{6}\text{ m/s})^2}{0.530\times10^{-10}\text{ m}} = \boxed{8.32\times10^{-8}\text{ N inward}}$$

(b) $$a = \frac{v^2}{r} = \frac{(2.20\times10^{6}\text{ m/s})^2}{0.530\times10^{-10}\text{ m}} = \boxed{9.13\times10^{22}\text{ m/s}^2\text{ inward}}$$

P5.17 $n = mg$ since $a_y = 0$

The force causing the centripetal acceleration is the frictional force f.

From Newton's second law $f = ma_c = \dfrac{mv^2}{r}$.

But the friction condition is $f \le \mu_s n$

i.e., $\dfrac{mv^2}{r} \le \mu_s mg$

$$v \le \sqrt{\mu_s rg} = \sqrt{0.600(35.0\text{ m})(9.80\text{ m/s}^2)} \qquad v \le \boxed{14.3\text{ m/s}}$$

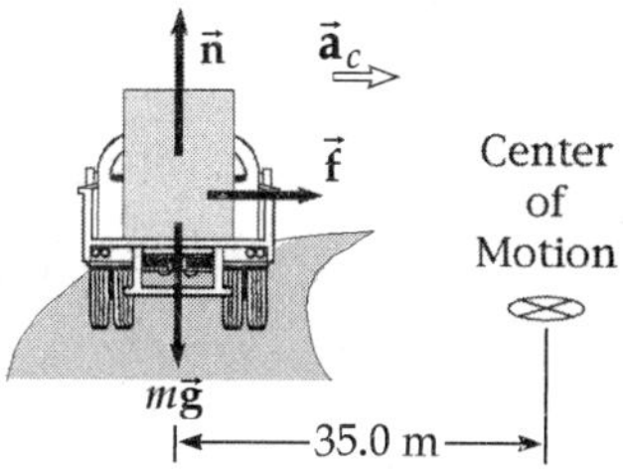

FIG. P5.17

P5.18 (a) $\sum F_y = ma_y$, $\qquad mg_{moon}\text{ down} = \dfrac{mv^2}{r}\text{ down}$

$$v = \sqrt{g_{moon}r} = \sqrt{(1.52\text{ m/s}^2)(1.7\times10^{6}\text{ m} + 100\times10^{3}\text{ m})} = \boxed{1.65\times10^{3}\text{ m/s}}$$

(b) $v = \dfrac{2\pi r}{T}$, $\qquad T = \dfrac{2\pi(1.8\times10^{6}\text{ m})}{1.65\times10^{3}\text{ m/s}} = \boxed{6.84\times10^{3}\text{ s}} = 1.90\text{ h}$

P5.19 $T\cos 5.00° = mg = (80.0\text{ kg})(9.80\text{ m/s}^2)$

(a) $T = 787\text{ N}: \vec{\mathbf{T}} = \boxed{(68.6\text{ N})\hat{\mathbf{i}} + (784\text{ N})\hat{\mathbf{j}}}$

(b) $T\sin 5.00° = ma_c$: $\boxed{a_c = 0.857\text{ m/s}^2}$ toward the center of the circle.

The length of the wire is unnecessary information. We could, on the other hand, use it to find the radius of the circle, the speed of the bob, and the period of the motion.

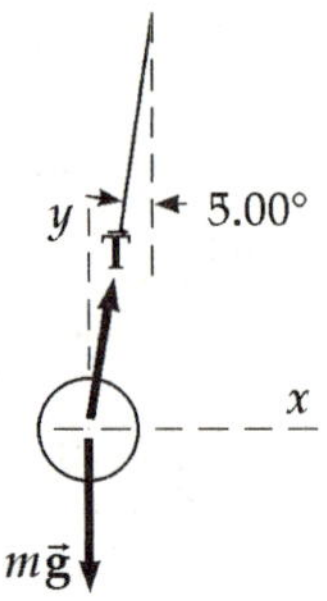

FIG. P5.19

***P5.20** $F_g = mg = (4\text{ kg})(9.8\text{ m/s}^2) = 39.2\text{ N}$

$$\sin\theta = \frac{1.5\text{ m}}{2\text{ m}}$$

$\theta = 48.6°$

$r = (2\text{ m})\cos 48.6° = 1.32\text{ m}$

$$\sum F_x = ma_x = \frac{mv^2}{r}$$

$$T_a\cos 48.6° + T_b\cos 48.6° = \frac{(4\text{ kg})(6\text{ m/s})^2}{1.32\text{ m}}$$

$$T_a + T_b = \frac{109\text{ N}}{\cos 48.6°} = 165\text{ N}$$

$$\sum F_y = ma_y$$

$$+T_a\sin 48.6° - T_b\sin 48.6° - 39.2\text{ N} = 0$$

$$T_a - T_b = \frac{39.2\text{ N}}{\sin 48.6°} = 52.3\text{ N}$$

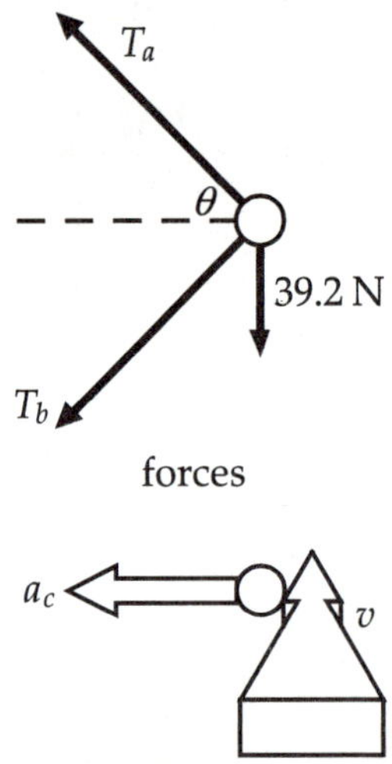

FIG. P5.20

(a) To solve simultaneously, we add the equations in T_a and T_b:

$$T_a + T_b + T_a - T_b = 165\text{ N} + 52.3\text{ N}$$

$$T_a = \frac{217\text{ N}}{2} = \boxed{108\text{ N}}$$

(b) $T_b = 165\text{ N} - T_a = 165\text{ N} - 108\text{ N} = \boxed{56.2\text{ N}}$

Section 5.3 Nonuniform C

P5.21 Let the tension at the low

$\sum F = ma$: $T - mg = m$

$T = m\left(g + \right.$

$T = (85.0$ kg

He doesn't make it acros

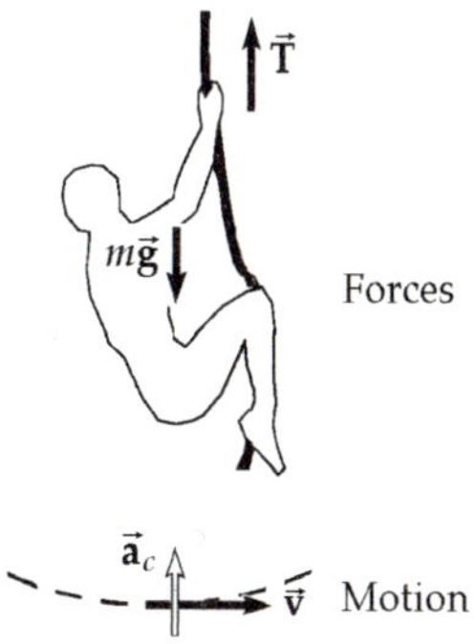

FIG. P5.21

P5.22 (a) $\sum F_y = ma_y = \dfrac{mv^2}{R}$

$mg - n = \dfrac{mv^2}{R}$ $\quad n = \boxed{mg - \dfrac{mv^2}{R}}$

(b) When $n = 0$, $\quad mg = \dfrac{mv^2}{R}$

Then, $\quad v = \boxed{\sqrt{gR}}$.

P5.23 $\sum F_y = \dfrac{mv^2}{r} = mg + n$

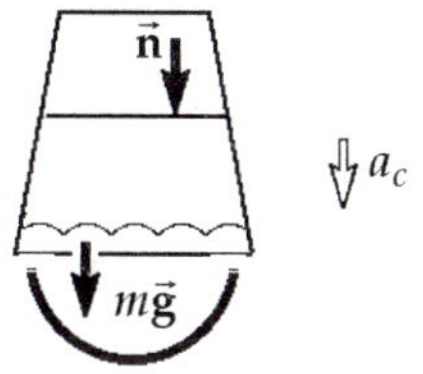

FIG. P5.23

But $n = 0$ at this minimum speed condition, so

$$\frac{mv^2}{r} = mg \Rightarrow v = \sqrt{gr} = \sqrt{(9.80\ \text{m/s}^2)(1.00\ \text{m})} = \boxed{3.13\ \text{m/s}}.$$

P5.24 (a) $a_c = \dfrac{v^2}{r}$ $\quad r = \dfrac{v^2}{a_c} = \dfrac{(13.0\ \text{m/s})^2}{2(9.80\ \text{m/s}^2)} = \boxed{8.62\ \text{m}}$

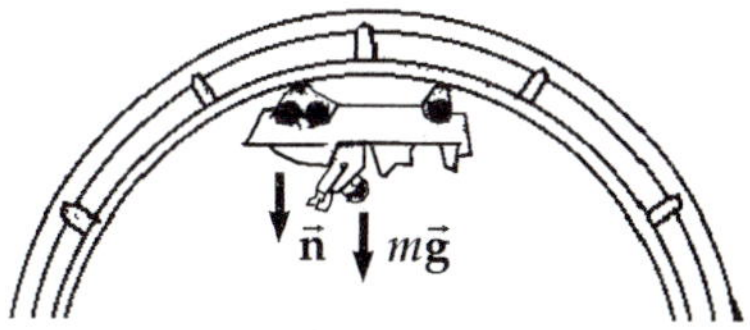

FIG. P5.24

(b) Let n be the force exerted by the rail.

Newton's law gives

$$Mg + n = \frac{Mv^2}{r}$$

$$n = M\left(\frac{v^2}{r} - g\right) = M(2g - g) = \boxed{Mg,\ \text{downward}}$$

continued on next page

(c) $a_c = \dfrac{v^2}{r}$ $\qquad a_c = \dfrac{(13.0 \text{ m/s})^2}{20.0 \text{ m}} = \boxed{8.45 \text{ m/s}^2}$

If the force exerted by the rail is n_1

then $\qquad n_1 + Mg = \dfrac{Mv^2}{r} = Ma_c$

$$n_1 = M(a_c - g) \text{ which is } < 0 \text{ since } a_c = 8.45 \text{ m/s}^2$$

Thus, the normal force would have to point away from the center of the curve. Unless they have belts, the riders will fall from the cars. To be safe we must require n_1 to be positive. Then $a_c > g$. We need $\dfrac{v^2}{r} > g$ or $v > \sqrt{rg} = \sqrt{(20.0 \text{ m})(9.80 \text{ m/s}^2)}$, $v > 14.0$ m/s.

Section 5.4 Motion in the Presence of Velocity-Dependent Resistive Forces

P5.25 (a) $a = g - bv$

When $v = v_T$, $\qquad a = 0$ and $g = bv_T \qquad b = \dfrac{g}{v_T}$

The Styrofoam falls 1.50 m at constant speed v_T in 5.00 s.

Thus, $\qquad v_T = \dfrac{y}{t} = \dfrac{1.50 \text{ m}}{5.00 \text{ s}} = 0.300 \text{ m/s}$

Then $\qquad b = \dfrac{9.80 \text{ m/s}^2}{0.300 \text{ m/s}} = \boxed{32.7 \text{ s}^{-1}}$

(b) At $t = 0$, $v = 0$ and $\qquad a = g = \boxed{9.80 \text{ m/s}^2}$ down

(c) When $v = 0.150$ m/s, $a = g - bv = 9.80 \text{ m/s}^2 - (32.7 \text{ s}^{-1})(0.150 \text{ m/s}) = \boxed{4.90 \text{ m/s}^2}$ down

P5.26 (a) $\rho = \dfrac{m}{V}$, $A = 0.0201 \text{ m}^2$, $R = \dfrac{1}{2}\rho_{\text{air}} A D v_T^2 = mg$

$$m = \rho_{\text{bead}} V = 0.830 \text{ g/cm}^3 \left[\frac{4}{3}\pi (8.00 \text{ cm})^3\right] = 1.78 \text{ kg}$$

Assuming a drag coefficient of $D = 0.500$ for this spherical object, and taking the density of air at 20°C from the endpapers, we have

$$v_T = \sqrt{\frac{2(1.78 \text{ kg})(9.80 \text{ m/s}^2)}{0.500(1.20 \text{ kg/m}^3)(0.0201 \text{ m}^2)}} = \boxed{53.8 \text{ m/s}}$$

(b) $v_f^2 = v_i^2 + 2gh = 0 + 2gh$: $h = \dfrac{v_f^2}{2g} = \dfrac{(53.8 \text{ m/s})^2}{2(9.80 \text{ m/s}^2)} = \boxed{148 \text{ m}}$

P5.27 (a) At terminal velocity, $R = v_T b = mg$

$$\therefore b = \frac{mg}{v_T} = \frac{(3.00\times10^{-3}\text{ kg})(9.80\text{ m/s}^2)}{2.00\times10^{-2}\text{ m/s}} = \boxed{1.47\text{ N}\cdot\text{s/m}}$$

(b) In the equation describing the time variation of the velocity, we have

$v = v_T\left(1-e^{-bt/m}\right)$ $\quad v = 0.632 v_T$ when $e^{-bt/m} = 0.368$

or at time $t = -\left(\frac{m}{b}\right)\ln(0.368) = \boxed{2.04\times10^{-3}\text{ s}}$

(c) At terminal velocity, $R = v_T b = mg = \boxed{2.94\times10^{-2}\text{ N}}$

P5.28 $v = \left(\frac{mg}{b}\right)\left[1-\exp\left(\frac{-bt}{m}\right)\right]$ where $\exp(x) = e^x$ is the exponential function.

At $t\to\infty$, $\quad v \to v_T = \frac{mg}{b}$

At $t = 5.54$ s $\quad 0.500 v_T = v_T\left[1-\exp\left(\frac{-b(5.54\text{ s})}{9.00\text{ kg}}\right)\right]$

$$\exp\left(\frac{-b(5.54\text{ s})}{9.00\text{ kg}}\right) = 0.500;$$

$$\frac{-b(5.54\text{ s})}{9.00\text{ kg}} = \ln 0.500 = -0.693;$$

$$b = \frac{(9.00\text{ kg})(0.693)}{5.54\text{ s}} = 1.13\text{ kg/s}$$

(a) $v_T = \frac{mg}{b}$ $\quad v_T = \frac{(9.00\text{ kg})(9.80\text{ m/s}^2)}{1.13\text{ kg/s}} = \boxed{78.3\text{ m/s}}$

(b) $0.750 v_T = v_T\left[1-\exp\left(\frac{-1.13t}{9.00\text{ s}}\right)\right]$ $\quad \exp\left(\frac{-1.13t}{9.00\text{ s}}\right) = 0.250$

$$t = \frac{9.00(\ln 0.250)}{-1.13}\text{ s} = \boxed{11.1\text{ s}}$$

(c) $\frac{dx}{dt} = \left(\frac{mg}{b}\right)\left[1-\exp\left(-\frac{bt}{m}\right)\right]$; $\quad \int_{x_0}^{x} dx = \int_0^t \left(\frac{mg}{b}\right)\left[1-\exp\left(\frac{-bt}{m}\right)\right]dt$

$$x - x_0 = \frac{mgt}{b} + \left(\frac{m^2 g}{b^2}\right)\exp\left(\frac{-bt}{m}\right)\Bigg|_0^t = \frac{mgt}{b} + \left(\frac{m^2 g}{b^2}\right)\left[\exp\left(\frac{-bt}{m}\right)-1\right]$$

At $t = 5.54$ s, $\quad x = 9.00\text{ kg}(9.80\text{ m/s}^2)\frac{5.54\text{ s}}{1.13\text{ kg/s}} + \left(\frac{(9.00\text{ kg})^2(9.80\text{ m/s}^2)}{(1.13\text{ kg/s})^2}\right)[\exp(-0.693)-1]$

$$x = 434\text{ m} + 626\text{ m}(-0.500) = \boxed{121\text{ m}}$$

P5.29 (a) $v(t)=v_i e^{-ct}$ $\qquad v(20.0\text{ s})=5.00=v_i e^{-20.0c}$, $v_i=10.0\text{ m/s}$.

So $5.00=10.0e^{-20.0c}$ and $-20.0c=\ln\left(\frac{1}{2}\right)$ $\qquad c=-\dfrac{\ln\left(\frac{1}{2}\right)}{20.0}=\boxed{3.47\times10^{-2}\text{ s}^{-1}}$

(b) At $t=40.0\text{ s}$ $\qquad v=(10.0\text{ m/s})e^{-40.0c}=(10.0\text{ m/s})(0.250)=\boxed{2.50\text{ m/s}}$

(c) $v=v_i e^{-ct}$ $\qquad a=\dfrac{dv}{dt}=-cv_i e^{-ct}=\boxed{-cv}$

P5.30

$$\sum F=ma$$

$$-kmv^2=m\frac{dv}{dt}$$

$$-kdt=\frac{dv}{v^2}$$

$$-k\int_0^t dt=\int_{v_0}^{v} v^{-2}dv$$

$$-k(t-0)=\left.\frac{v^{-1}}{-1}\right|_{v_0}^{v}=-\frac{1}{v}+\frac{1}{v_0}$$

$$\frac{1}{v}=\frac{1}{v_0}+kt=\frac{1+v_0kt}{v_0}$$

$$v=\frac{v_0}{1+v_0kt}$$

Section 5.5 The Fundamental Forces of Nature

P5.31 $F=\dfrac{Gm_1m_2}{r^2}=\dfrac{\left(6.672\times10^{-11}\right)(2)(2)}{(0.30)^2}=\boxed{2.97\times10^{-9}\text{ N}}$

P5.32 For two 70–kg persons, modeled as spheres,

$$F_g=\frac{Gm_1m_2}{r^2}=\frac{\left(6.67\times10^{-11}\text{ N}\cdot\text{m}^2/\text{kg}^2\right)(70\text{ kg})(70\text{ kg})}{(2\text{ m})^2}\boxed{\sim10^{-7}\text{ N}}$$

P5.33 $a=\dfrac{MG}{(4R_E)^2}=\dfrac{9.8\text{ m/s}^2}{16}=\boxed{0.613\text{ m/s}^2}$ toward the earth

P5.34 $F=k_e\dfrac{q_1q_2}{(r_{12})^2}=\left(8.99\times10^9\right)\dfrac{(+40)(-40)}{(2000)^2}=-3.60\times10^6\text{ N (attractive)}=\boxed{3.60\times10^6\text{ N downward}}$

Section 5.6 Context Connection—Drag Coefficients of Automobiles

***P5.35** The resistive force is

$$R=\frac{1}{2}D\rho Av^2=\frac{1}{2}(0.250)\left(1.20\ \text{kg/m}^3\right)\left(2.20\ \text{m}^2\right)(27.8\ \text{m/s})^2$$
$$R=255\ \text{N}$$
$$a=-\frac{R}{m}=-\frac{255\ \text{N}}{1\,200\ \text{kg}}=\boxed{-0.212\ \text{m/s}^2}$$

***P5.36** (a) The drag force on this car body at this speed is

$$R=\frac{1}{2}D\rho Av^2=\frac{1}{2}(0.34)\left(1.2\ \text{kg/m}^3\right)\left(2.6\ \text{m}^2\right)(10\ \text{m/s})^2=53.0\ \text{N}.$$

Now Newton's second law is $\sum F_x = ma_x$

$$+f_s-53.0\ \text{N}=1\,300\ \text{kg}\left(3\ \text{m/s}^2\right)$$
$$f_s=\boxed{3.95\times10^3\ \text{N forward}}$$

(b) Newton's second law changes to

$$+3\,950\ \text{N}-\frac{1}{2}(0.2)(1.2)(2.6)(10)^2\ \text{N}=(1\,300\ \text{kg})a$$
$$a=\frac{3\,950-31.2\ \text{N}}{1\,300\ \text{kg}}=\boxed{3.02\ \text{m/s}^2\ \text{forward}}$$

Streamlining to reduce drag makes little difference to the acceleration at this low speed.

(c) $\sum F_x = ma_x$ now reads

$$+3\,950\ \text{N}-\frac{1}{2}(0.34)(1.2)(2.6)\frac{\text{kg}}{\text{m}}v^2=0$$
$$v=\sqrt{\frac{3\,950}{0.530}\ \frac{\text{kg}\cdot\text{m}\cdot\text{m}}{\text{s}^2\cdot\text{kg}}}=\boxed{86.3\ \text{m/s}}$$

(d) Still again, $+3\,950\ \text{N}-\frac{1}{2}(0.2)(1.2)(2.6)\frac{\text{kg}}{\text{m}}v^2=0$

$$v=\sqrt{\frac{3\,950}{0.312}\ \frac{\text{m}^2}{\text{s}^2}}=\boxed{113\ \text{m/s}}$$

According to this model, drag reduction significantly affects maximum speed.

Additional Problems

P5.37 Applying Newton's second law to each object gives:

(1) $T_1 = f_1 + 2m(g\sin\theta + a)$

(2) $T_2 - T_1 = f_2 + m(g\sin\theta + a)$

(3) $T_2 = M(g - a)$

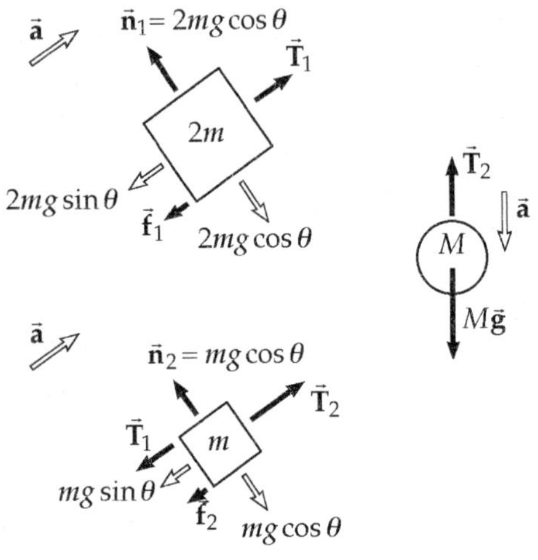

FIG. P5.37

(a), (b) Equilibrium $(a = 0)$

and frictionless incline $(f_1 = f_2 = 0)$

Under these conditions, the equations reduce to

(1′) $\boxed{T_1 = 2mg\sin\theta}$

(2′) $T_2 - T_1 = mg\sin\theta$

(3′) $T_2 = Mg$

Substituting (1′) and (3′) into equation (2′) then gives $\boxed{M = 3m\sin\theta}$

so equation (3′) becomes $\boxed{T_2 = 3mg\sin\theta}$

(c), (d) $M = 6m\sin\theta$ (double the value found above), and $f_1 = f_2 = 0$. With these conditions present, the equations become $T_1 = 2m(g\sin\theta + a)$, $T_2 - T_1 = m(g\sin\theta + a)$ and $T_2 = 6m\sin\theta(g - a)$. Solved simultaneously, these yield

$$\boxed{a = \frac{g\sin\theta}{1 + 2\sin\theta}},\quad \boxed{T_1 = 4mg\sin\theta\left(\frac{1 + \sin\theta}{1 + 2\sin\theta}\right)} \text{ and } \boxed{T_2 = 6mg\sin\theta\left(\frac{1 + \sin\theta}{1 + 2\sin\theta}\right)}$$

(e) Equilibrium $(a = 0)$ and impending motion up the incline so $M = M_{max}$ while $f_1 = 2\mu_s mg\cos\theta$ and $f_2 = \mu_s mg\cos\theta$, both directed down the incline. Under these conditions, the equations become $T_1 = 2mg(\sin\theta + \mu_s\cos\theta)$, $T_2 - T_1 = mg(\sin\theta + \mu_s\cos\theta)$, and $T_2 = M_{max}g$, which yield $\boxed{M_{max} = 3m(\sin\theta + \mu_s\cos\theta)}$

(f) Equilibrium $(a = 0)$ and impending motion **down** the incline so $M = M_{min}$, while $f_1 = 2\mu_s mg\cos\theta$ and $f_2 = \mu_s mg\cos\theta$, both directed **up** the incline. Under these conditions, the equations are $T_1 = 2mg(\sin\theta - \mu_s\cos\theta)$, $T_2 - T_1 = mg(\sin\theta - \mu_s\cos\theta)$, and $T_2 = M_{min}g$, which yield $\boxed{M_{min} = 3m(\sin\theta - \mu_s\cos\theta)}$. When this expression gives a negative value, it corresponds physically to a mass M hanging from a cord over a pulley at the bottom end of the incline.

(g) $\boxed{T_{2,max} - T_{2,min} = M_{max}g - M_{min}g = 6\mu_s mg\cos\theta}$

P5.38 For the system to start to move when released, the force tending to move m_2 down the incline, $m_2 g \sin\theta$, must exceed the maximum friction force which can retard the motion:

$$f_{max} = f_{1,max} + f_{2,max} = \mu_{s,1} n_1 + \mu_{s,2} n_2$$
$$f_{max} = \mu_{s,1} m_1 g + \mu_{s,2} m_2 g \cos\theta$$

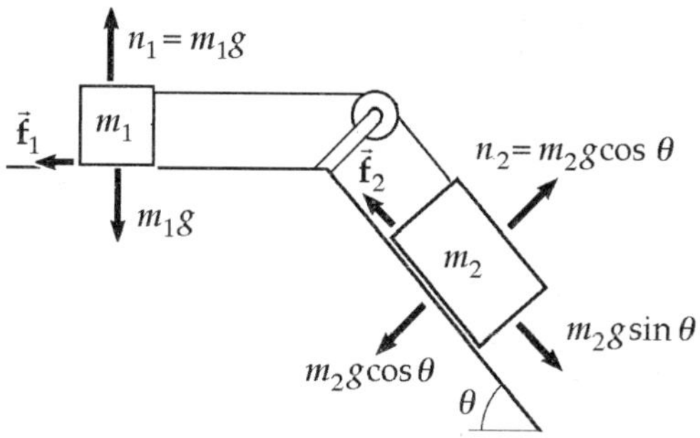

FIG. P5.38

From Table 5.1, $\mu_{s,1} = 0.610$ (aluminum on steel) and $\mu_{s,2} = 0.530$ (copper on steel). With

$$m_1 = 2.00 \text{ kg}, \; m_2 = 6.00 \text{ kg}, \; \theta = 30.0°,$$

the maximum friction force is found to be $f_{max} = 38.9$ N. This exceeds the force tending to cause the system to move, $m_2 g \sin\theta = 6.00 \text{ kg}(9.80 \text{ m/s}^2)\sin 30° = 29.4$ N. Hence,

$$\boxed{\text{the system will not start to move when released}}.$$

The friction forces increase in magnitude until the total friction force retarding the motion, $f = f_1 + f_2$, equals the force tending to set the system in motion. That is, until

$$\boxed{f = m_2 g \sin\theta = 29.4 \text{ N}}.$$

P5.39 (a) The crate is in equilibrium, just before it starts to move. Let the normal force acting on it be n and the friction force, f_s. Resolving vertically: $n = F_g + P\sin\theta$. Horizontally: $P\cos\theta = f_s$. But, $f_s \le \mu_s n$ i.e., $P\cos\theta \le \mu_s(F_g + P\sin\theta)$ or $P(\cos\theta - \mu_s \sin\theta) \le \mu_s F_g$. Divide by $\cos\theta$: $P(1 - \mu_s \tan\theta) \le \mu_s F_g \sec\theta$. Then

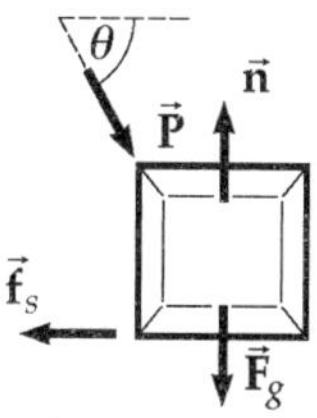

FIG. P5.39

$$\boxed{P_{minimum} = \frac{\mu_s F_g \sec\theta}{1 - \mu_s \tan\theta}}.$$

(b) $$P = \frac{0.400(100 \text{ N})\sec\theta}{1 - 0.400\tan\theta}$$

θ(deg)	0.00	15.0	30.0	45.0	60.0
P(N)	40.0	46.4	60.1	94.3	260

If the angle were 68.2° or more, the expression for P would go to infinity and motion would become impossible.

P5.40 With motion impending,

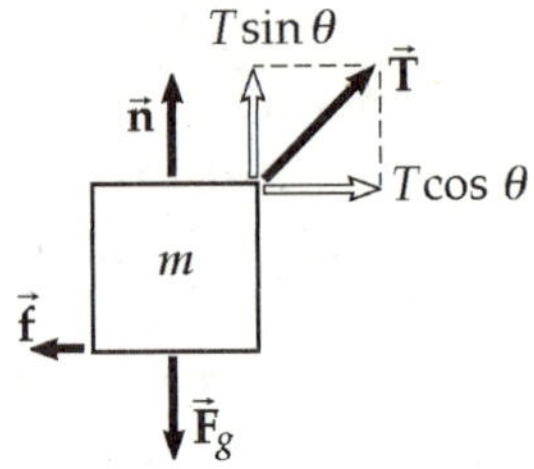

FIG. P5.40

$$n + T\sin\theta - mg = 0$$

$$f = \mu_s(mg - T\sin\theta)$$

and

$$T\cos\theta - \mu_s mg + \mu_s T\sin\theta = 0$$

so

$$T = \frac{\mu_s mg}{\cos\theta + \mu_s\sin\theta}.$$

To minimize T, we maximize $\cos\theta + \mu_s\sin\theta$

$$\frac{d}{d\theta}(\cos\theta + \mu_s\sin\theta) = 0 = -\sin\theta + \mu_s\cos\theta.$$

(a) $\theta = \tan^{-1}\mu_s = \tan^{-1}0.350 = \boxed{19.3°}$

(b) $T = \dfrac{0.350(1.30\text{ kg})(9.80\text{ m/s}^2)}{\cos 19.3° + 0.350\sin 19.3°} = \boxed{4.21\text{ N}}$

P5.41 $\Sigma F_1 = m_1 a$: $\quad -m_1 g\sin 35.0° - f_{k,1} + T = m_1 a$

$$-(3.50)(9.80)\sin 35.0° - \mu_s(3.50)(9.80)\cos 35.0° + T = 3.50(1.50) \qquad (1)$$

$\Sigma F_2 = m_2 a$: $\quad +m_2 g\sin 35.0° - f_{k,2} - T = m_2 a$

$$+(8.00)(9.80)\sin 35.0° - \mu_s(8.00)(9.80)\cos 35.0° - T = 8.00(1.50) \qquad (2)$$

Solving equations (1) and (2) simultaneously gives

(a) $\boxed{\mu_k = 0.0871}$

(b) $\boxed{T = 27.4\text{ N}}$

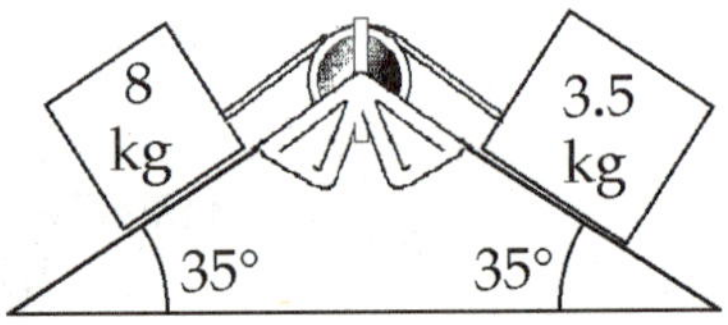

FIG. P5.41

P5.42 (a) See Figure (a) to the right.

(b) See Figure (b) to the right.

(c) For the pin,

$$m\vec{g} = (36.4\text{ kg})(9.8\text{ m/s}^2) = 357\text{ N}$$

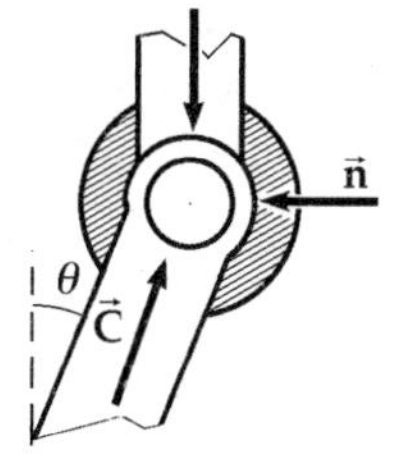

FIG. P5.42(a)

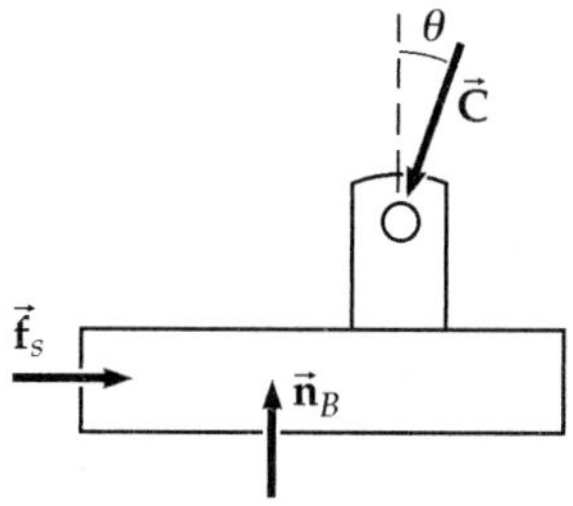

FIG. P5.42(b)

$$\sum F_y = ma_y: \quad C\cos\theta - 357\text{ N} = 0$$
$$C = \frac{357\text{ N}}{\cos\theta}.$$

For the foot,

$$\sum F_y = ma_y: \quad +n_B - C\cos\theta = 0$$
$$n_B = \boxed{357\text{ N}}.$$

(d) For the foot with motion impending,

$$\sum F_x = ma_x: \quad +f_s - C\sin\theta_s = 0$$
$$\mu_s n_B = C\sin\theta_s$$
$$\mu_s = \frac{C\sin\theta_s}{n_B} = \frac{(357\text{ N}/\cos\theta_s)\sin\theta_s}{357\text{ N}} = \tan\theta_s.$$

(e) The maximum coefficient is

$$\mu_s = \tan\theta_s = \tan 50.2° = \boxed{1.20}.$$

P5.43 (a) First, draw a free-body diagram, (top figure) of the top block. Since $a_y = 0$, $n_1 = 19.6\text{ N}$. And $f_k = \mu_k n_1 = 0.300(19.6\text{ N}) = 5.88\text{ N}$. $\Sigma F_x = ma_T$

$$10.0\text{ N} - 5.88\text{ N} = (2.00\text{ kg})a_T$$

or $a_T = 2.06\text{ m/s}^2$ (for top block). Now draw a free-body diagram (middle figure) of the bottom block and observe that $\Sigma F_x = Ma_B$ gives $f = 5.88\text{ N} = (8.00\text{ kg})a_B$ or $a_B = 0.735\text{ m/s}^2$ (for the bottom block). In time t, the distance each block moves (starting from rest) is $d_T = \frac{1}{2}a_T t^2$ and $d_B = \frac{1}{2}a_B t^2$. For the top block to reach the right edge of the bottom block, (see bottom figure) it is necessary that $d_T = d_B + L$ or

$$\frac{1}{2}\left(2.06\text{ m/s}^2\right)t^2 = \frac{1}{2}\left(0.735\text{ m/s}^2\right)t^2 + 3.00\text{ m}$$

which gives: $t = \boxed{2.13\text{ s}}$.

(b) From above, $d_B = \frac{1}{2}\left(0.735\text{ m/s}^2\right)(2.13\text{ s})^2 = \boxed{1.67\text{ m}}$.

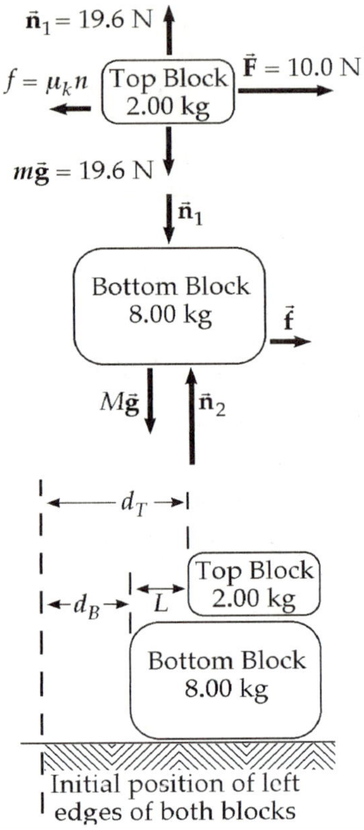

FIG. P5.43

P5.44 (a)

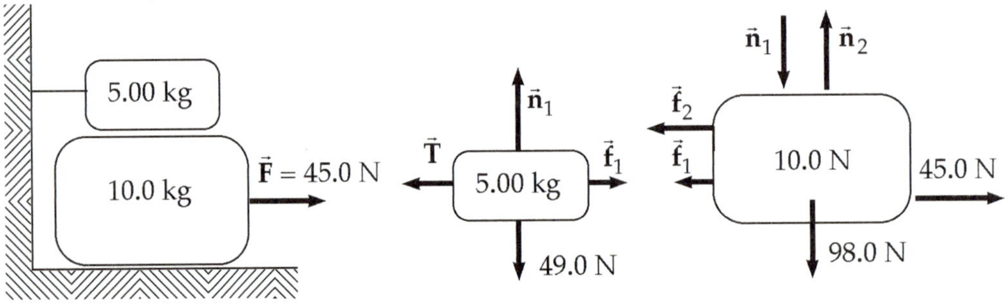

FIG. P5.44

f_1 and n_1 appear in both diagrams as action-reaction pairs

(b) 5.00 kg: $\Sigma F_x = ma$: $n_1 = m_1 g = 5.00(9.80) = 49.0\text{ N}$ $f_1 - T = 0$

$T = f_1 = \mu mg = 0.200(5.00)(9.80) = \boxed{9.80\text{ N}}$

10.0 kg: $\Sigma F_x = ma$: $45.0 - f_1 - f_2 = 10.0a$

$\Sigma F_y = 0$: $n_2 - n_1 - 98.0 = 0$

continued on next page

$$f_2 = \mu n_2 = \mu(n_1 + 98.0) = 0.20(49.0 + 98.0) = 29.4 \text{ N}$$

$$45.0 - 9.80 - 29.4 = 10.0a \qquad a = \boxed{0.580 \text{ m/s}^2}$$

***P5.45** (a) If the car is about to slip *down* the incline, f is directed up the incline.

$\sum F_y = n\cos\theta + f\sin\theta - mg = 0$ where $f = \mu_s n$ gives

$$n = \frac{mg}{\cos\theta(1+\mu_s\tan\theta)} \text{ and } f = \frac{\mu_s mg}{\cos\theta(1+\mu_s\tan\theta)}.$$

Then, $\sum F_x = n\sin\theta - f\cos\theta = m\frac{v_{\text{min}}^2}{R}$ yields

$$\boxed{v_{\text{min}} = \sqrt{\frac{Rg(\tan\theta - \mu_s)}{1+\mu_s\tan\theta}}}.$$

When the car is about to slip *up* the incline, f is directed down the incline. Then, $\sum F_y = n\cos\theta - f\sin\theta - mg = 0$ with $f = \mu_s n$ yields

$$n = \frac{mg}{\cos\theta(1-\mu_s\tan\theta)} \text{ and } f = \frac{\mu_s mg}{\cos\theta(1-\mu_s\tan\theta)}.$$

In this case, $\sum F_x = n\sin\theta + f\cos\theta = m\frac{v_{\text{max}}^2}{R}$, which gives

$$\boxed{v_{\text{max}} = \sqrt{\frac{Rg(\tan\theta + \mu_s)}{1-\mu_s\tan\theta}}}.$$

(b) If $v_{\text{min}} = \sqrt{\frac{Rg(\tan\theta - \mu_s)}{1+\mu_s\tan\theta}} = 0$, then $\boxed{\mu_s = \tan\theta}$.

(c)

$$v_{\text{min}} = \sqrt{\frac{(100 \text{ m})(9.80 \text{ m/s}^2)(\tan 10.0° - 0.100)}{1+(0.100)\tan 10.0°}} = \boxed{8.57 \text{ m/s}}$$

$$v_{\text{max}} = \sqrt{\frac{(100 \text{ m})(9.80 \text{ m/s}^2)(\tan 10.0° + 0.100)}{1-(0.100)\tan 10.0°}} = \boxed{16.6 \text{ m/s}}$$

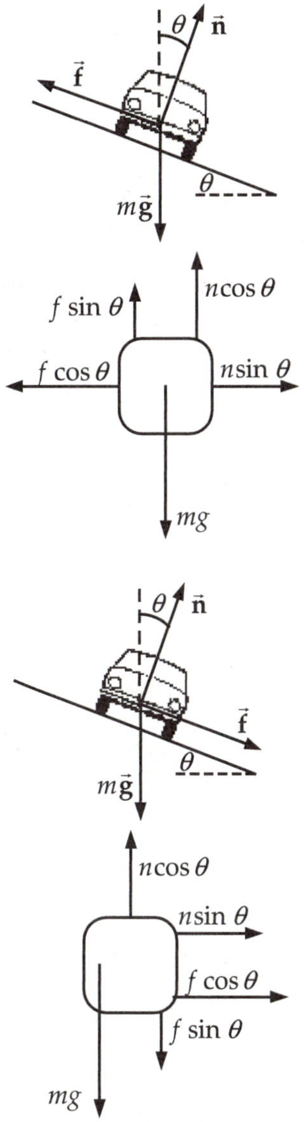

FIG. 5.45

***P5.46** (a) Directly $n = 63.7\text{ N}\cos 13° = 62.1\text{ N}$

$$f_k = 0.36(62.1\text{ N}) = 22.3\text{ N}.$$

Now adding $+T + 14.3\text{ N} - 22.3\text{ N} = (6.5\text{ kg})a$ and $-T + 37.2\text{ N} = (3.8\text{ kg})a$ gives
$37.2\text{ N} - 8.01\text{ N} = (10.3\text{ kg})a$

$$\boxed{a = 2.84\text{ m/s}^2}.$$

Then $T = 37.2\text{ N} - 3.8\text{ kg}(2.84\text{ m/s}^2) = \boxed{26.5\text{ N}}$.

(b) We recognize the equations are describing a 6.5-kg block on an incline at 13° with the horizontal. It has coefficient of friction 0.36 with the incline. It is pulled forward, which is down the incline, by the tension in a cord running to a hanging 3.8-kg object.

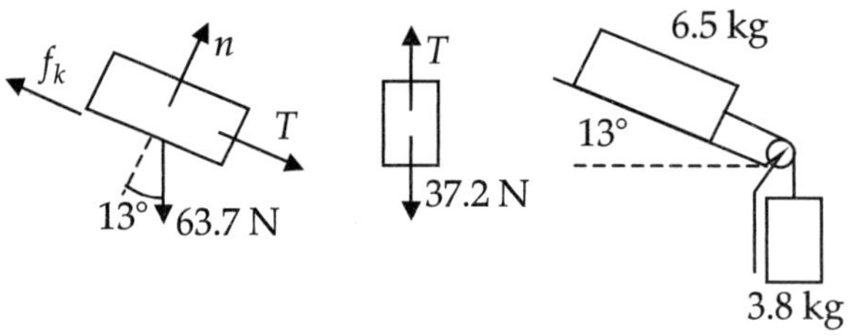

FIG. P5.46

***P5.47** When the cloth is at a lower angle θ, the radial component of $\sum F = ma$ reads

$$n + mg\sin\theta = \frac{mv^2}{r}.$$

At $\theta = 68.0°$, the normal force drops to zero and $g\sin 68° = \frac{v^2}{r}$.

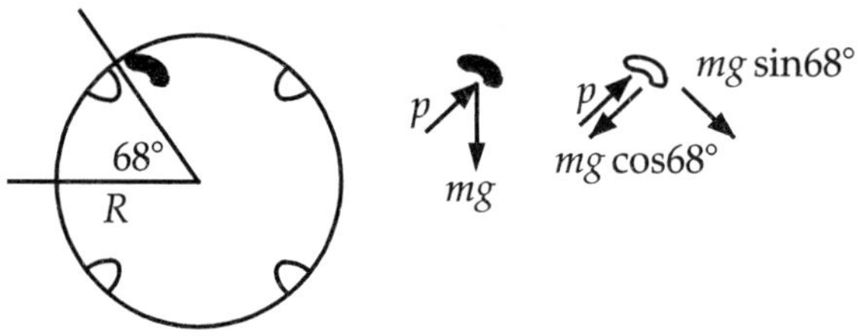

FIG. P5.47

$$v = \sqrt{rg\sin 68°} = \sqrt{(0.33\text{ m})(9.8\text{ m/s}^2)\sin 68°} = 1.73\text{ m/s}$$

The rate of revolution is

$$\text{angular speed} = (1.73\text{ m/s})\left(\frac{1\text{ rev}}{2\pi r}\right)\left(\frac{2\pi r}{2\pi(0.33\text{ m})}\right) = \boxed{0.835\text{ rev/s}} = 50.1\text{ rev/min}.$$

P5.48 (a) While the car negotiates the curve, the accelerometer is at the angle θ.

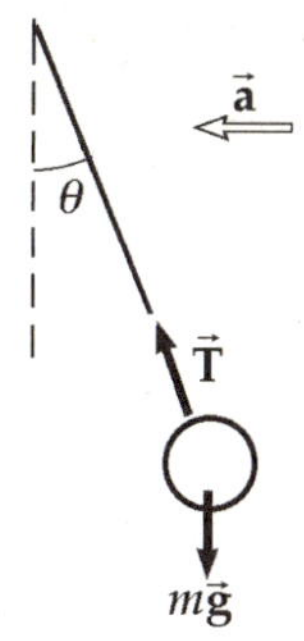

FIG. P5.48

Horizontally: $T\sin\theta = \frac{mv^2}{r}$

Vertically: $T\cos\theta = mg$

where r is the radius of the curve, and v is the speed of the car.

By division, $\tan\theta = \frac{v^2}{rg}$

Then $a_c = \frac{v^2}{r} = g\tan\theta$: $a_c = (9.80\ \text{m/s}^2)\tan 15.0°$

$$a_c = \boxed{2.63\ \text{m/s}^2}$$

(b) $r = \frac{v^2}{a_c}$ $\quad r = \frac{(23.0\ \text{m/s})^2}{2.63\ \text{m/s}^2} = \boxed{201\ \text{m}}$

(c) $v^2 = rg\tan\theta = (201\ \text{m})(9.80\ \text{m/s}^2)\tan 9.00°$ $\quad v = \boxed{17.7\ \text{m/s}}$

P5.49 (a) Since the centripetal acceleration of a person is downward (toward the axis of the earth), it is equivalent to the effect of a falling elevator. Therefore,

$$F_g' = F_g - \frac{mv^2}{r} \text{ or } \boxed{F_g > F_g'}$$

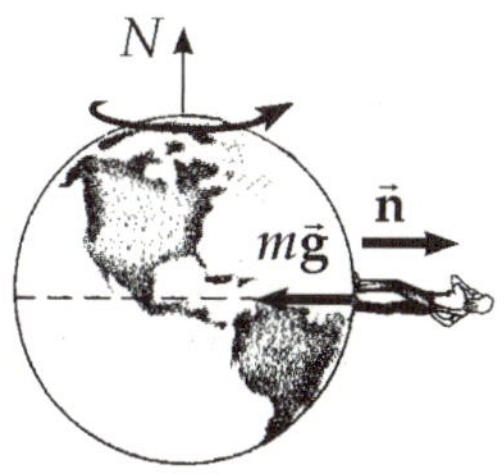

FIG. P5.49

(b) At the poles $v = 0$ and $F_g' = F_g = mg = 75.0(9.80) = \boxed{735\ \text{N}}$ down.

At the equator, $F_g' = F_g - ma_c = 735\ \text{N} - 75.0(0.033\,7)\ \text{N} = \boxed{732\ \text{N}}$ down.

P5.50 (a) Since the object of mass m_2 is in equilibrium, $\sum F_y = T - m_2 g = 0$

or $T = \boxed{m_2 g}$.

(b) The tension in the string provides the required centripetal acceleration of the puck.

Thus, $F_c = T = \boxed{m_2 g}$.

(c) From $F_c = \frac{m_1 v^2}{R}$

we have $v = \sqrt{\frac{RF_c}{m_1}} = \boxed{\sqrt{\left(\frac{m_2}{m_1}\right)gR}}$.

***P5.51** $v = \frac{2\pi r}{T} = \frac{2\pi(9.00\text{ m})}{(15.0\text{ s})} = 3.77\text{ m/s}$

(a) $a_r = \frac{v^2}{r} = \boxed{1.58\text{ m/s}^2}$

(b) $F_{\text{low}} = m(g + a_r) = \boxed{455\text{ N}}$

(c) $F_{\text{high}} = m(g - a_r) = \boxed{329\text{ N}}$

(d) $F_{\text{mid}} = m\sqrt{g^2 + a_r^2} = \boxed{397\text{ N upward and}}$ at $\theta = \tan^{-1}\frac{a_r}{g} = \tan^{-1}\frac{1.58}{9.8} = \boxed{9.15°\text{ inward}}$.

***P5.52** (a) The mass at the end of the chain is in vertical equilibrium. Thus $T\cos\theta = mg$.

Horizontally $T\sin\theta = ma_r = \frac{mv^2}{r}$

$$r = (2.50\sin\theta + 4.00)\text{ m}$$
$$r = (2.50\sin 28.0° + 4.00)\text{ m} = 5.17\text{ m}$$

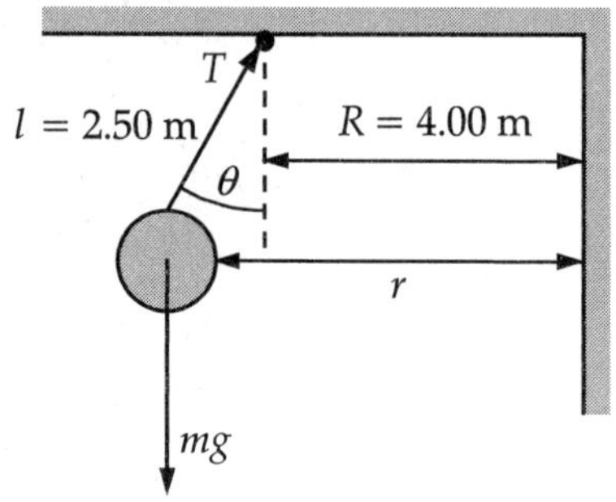

FIG. P5.52

Then $a_r = \frac{v^2}{5.17\text{ m}}$.

By division $\tan\theta = \frac{a_r}{g} = \frac{v^2}{5.17g}$

$$v^2 = 5.17g\tan\theta = (5.17)(9.80)(\tan 28.0°)\text{ m}^2/\text{s}^2$$
$$v = \boxed{5.19\text{ m/s}}$$

(b) $T\cos\theta = mg$

$$T = \frac{mg}{\cos\theta} = \frac{(50.0\text{ kg})(9.80\text{ m/s}^2)}{\cos 28.0°} = \boxed{555\text{ N}}$$

P5.53 Standing on the inner surface of the rim, and moving with it, each person will feel a normal force exerted by the rim. This inward force causes the 3.00 m/s^2 centripetal acceleration:

$a_c = \frac{v^2}{r}$: $\quad v = \sqrt{a_c r} = \sqrt{(3.00\text{ m/s}^2)(60.0\text{ m})} = 13.4\text{ m/s}$

The period of rotation comes from $v = \frac{2\pi r}{T}$: $\quad T = \frac{2\pi r}{v} = \frac{2\pi(60.0\text{ m})}{13.4\text{ m/s}} = 28.1\text{ s}$

so the frequency of rotation is $\quad f = \frac{1}{T} = \frac{1}{28.1\text{ s}} = \frac{1}{28.1\text{ s}}\left(\frac{60\text{ s}}{1\text{ min}}\right) = \boxed{2.14\text{ rev/min}}$.

***P5.54** The volume of the grain is $V = \frac{4}{3}\pi r^3 = \frac{4}{3}\pi\left(5\times10^{-7}\text{ m}\right)^3 = 5.24\times10^{-19}\text{ m}^3$. Its weight is

$$F_g = mg = \rho Vg = \left(19.3\times10^3\text{ kg/m}^3\right)\left(5.24\times10^{-19}\text{ m}^3\right)\left(9.8\text{ m/s}^2\right)$$
$$= 9.90\times10^{-14}\text{ N down}$$

(a) As it settles at terminal speed, $\sum F_y = ma_y$

$$-9.90\times10^{-14}\text{ N} + 0.018\,8\ \frac{\text{N}\cdot\text{s}}{\text{m}^2}\left(5\times10^{-7}\text{ m}\right)v = 0$$
$$v = \boxed{1.05\times10^{-5}\text{ m/s}}$$

(b) $\Delta y = v_y t$

$$t = \frac{0.08\text{ m}}{1.05\times10^{-5}\text{ m/s}} = \boxed{7.59\times10^3\text{ s}} = 2.11\text{ h}$$

(c) The speed of the middle of the tube is

$$3\,000\text{ rev/min}\left(\frac{2\pi(0.09\text{ m})}{1\text{ rev}}\right)\left(\frac{1\text{ min}}{60\text{ s}}\right) = 314\text{ radian/s}(0.09\text{ m}) = 28.3\text{ m/s}.$$

Now $a_c = \frac{v^2}{r} = \frac{(28.3\text{ m/s})^2}{0.09\text{ m}} = \boxed{8.88\times10^3\text{ m/s}^2}$.

(d) In place of part (a) we have $\sum F_x = ma_x$

$$0.018\,8\text{ N}\cdot\text{s/m}^2\left(5\times10^{-7}\text{ m}\right)v = 1.01\times10^{-14}\text{ kg}\left(8.88\times10^3\text{ m/s}^2\right)$$
$$v = 9.55\times10^{-3}\text{ m/s} = \frac{\Delta x}{t}$$
$$t = \frac{0.08\text{ m}\cdot\text{s}}{9.55\times10^{-3}\text{ m}} = \boxed{8.38\text{ s}}$$

P5.55 (a) $n = \dfrac{mv^2}{R}$ $\qquad f - mg = 0$

$f = \mu_s n \qquad v = \dfrac{2\pi R}{T}$

$$T = \boxed{\sqrt{\dfrac{4\pi^2 R\mu_s}{g}}}$$

(b) $T = \boxed{2.54\ \text{s}}$

$$\#\frac{\text{rev}}{\text{min}} = \frac{1\ \text{rev}}{2.54\ \text{s}}\left(\frac{60\ \text{s}}{\text{min}}\right) = \boxed{23.6\ \frac{\text{rev}}{\text{min}}}$$

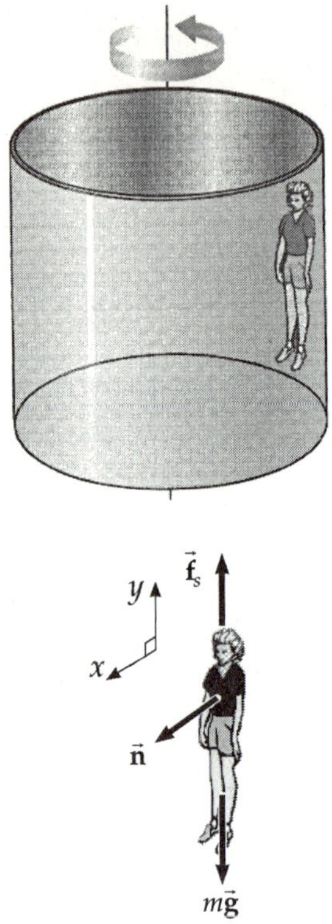

FIG. P5.55

P5.56 (a) The bead moves in a circle with radius $v = R\sin\theta$ at a speed of

$$v = \frac{2\pi r}{T} = \frac{2\pi R\sin\theta}{T}$$

The normal force has

an inward radial component of $n\sin\theta$

and an upward component of $n\cos\theta$

$$\sum F_y = ma_y:\quad n\cos\theta - mg = 0$$

or

$$n = \frac{mg}{\cos\theta}$$

Then $\sum F_x = n\sin\theta = m\dfrac{v^2}{r}$ becomes $\qquad \left(\dfrac{mg}{\cos\theta}\right)\sin\theta = \dfrac{m}{R\sin\theta}\left(\dfrac{2\pi R\sin\theta}{T}\right)^2$

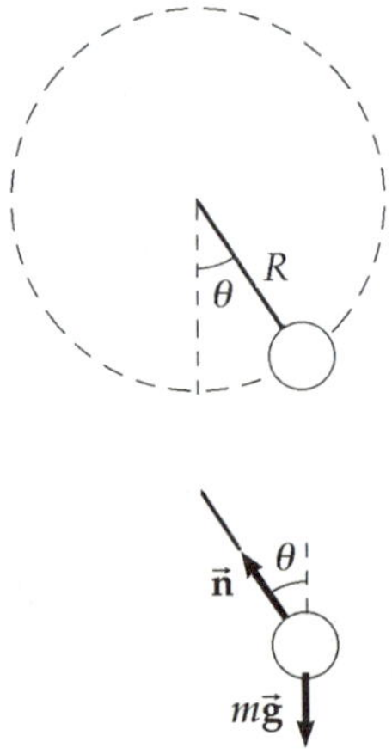

FIG. P5.56

continued on next page

which reduces to $$\frac{g\sin\theta}{\cos\theta}=\frac{4\pi^2R\sin\theta}{T^2}$$

This has two solutions: $$\sin\theta=0\Rightarrow\theta=0° \qquad (1)$$

and $$\cos\theta=\frac{gT^2}{4\pi^2R} \qquad (2)$$

If $R=15.0$ cm and $T=0.450$ s, the second solution yields

$$\cos\theta=\frac{(9.80\text{ m/s}^2)(0.450\text{ s})^2}{4\pi^2(0.150\text{ m})}=0.335\text{ and }\theta=70.4°$$

Thus, in this case, the bead can ride at two positions $\boxed{\theta=70.4°}$ and $\boxed{\theta=0°}$.

(b) At this slower rotation, solution (2) above becomes

$$\cos\theta=\frac{(9.80\text{ m/s}^2)(0.850\text{ s})^2}{4\pi^2(0.150\text{ m})}=1.20\text{, which is impossible.}$$

In this case, the bead can ride only at the bottom of the loop, $\boxed{\theta=0°}$. The loop's rotation must be faster than a certain threshold value in order for the bead to move away from the lowest position.

P5.57 At terminal velocity, the accelerating force of gravity is balanced by frictional drag: $mg=arv+br^2v^2$

(a) $mg=(3.10\times10^{-9})v+(0.870\times10^{-10})v^2$

For water, $$m=\rho V=1\,000\text{ kg/m}^3\left[\frac{4}{3}\pi(10^{-5}\text{ m})^3\right]$$

$$4.11\times10^{-11}=(3.10\times10^{-9})v+(0.870\times10^{-10})v^2$$

Assuming v is small, ignore the second term on the right hand side: $\boxed{v=0.013\,2\text{ m/s}}$.

(b) $mg=(3.10\times10^{-8})v+(0.870\times10^{-8})v^2$ Here we cannot ignore the second term because the coefficients are of nearly equal magnitude.

$$4.11\times10^{-8}=(3.10\times10^{-8})v+(0.870\times10^{-8})v^2$$

$$v=\frac{-3.10\pm\sqrt{(3.10)^2+4(0.870)(4.11)}}{2(0.870)}=\boxed{1.03\text{ m/s}}$$

continued on next page

(c) $mg = (3.10 \times 10^{-7})v + (0.870 \times 10^{-6})v^2$

Assuming $v > 1$ m/s, and ignoring the first term: $4.11 \times 10^{-5} = (0.870 \times 10^{-6})v^2$

$v = \boxed{6.87 \text{ m/s}}$

*P5.58 (a)

t(s)	d(m)
1.00	4.88
2.00	18.9
3.00	42.1
4.00	73.8
5.00	112
6.00	154
7.00	199
8.00	246
9.00	296
10.0	347
11.0	399
12.0	452
13.0	505
14.0	558
15.0	611
16.0	664
17.0	717
18.0	770
19.0	823
20.0	876

(b)

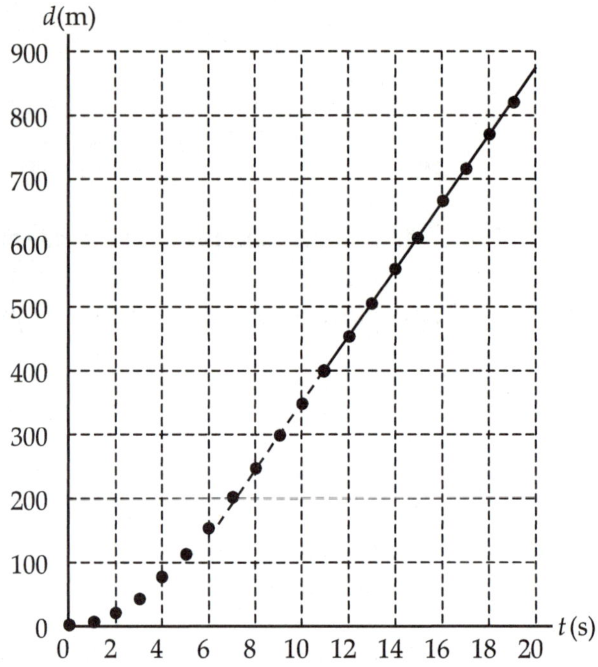

FIG. P5.58(b)

(c) A straight line fits the points from $t = 11.0$ s to 20.0 s quite precisely. Its slope is the terminal speed.

$$v_T = \text{slope} = \frac{876 \text{ m} - 399 \text{ m}}{20.0 \text{ s} - 11.0 \text{ s}} = \boxed{53.0 \text{ m/s}}$$

P5.59 $\sum F_y = L_y - T_y - mg = L\cos 20.0° - T\sin 20.0° - 7.35 \text{ N} = ma_y = 0$

$$\sum F_x = L_x + T_x = L\sin 20.0° + T\cos 20.0° = m\frac{v^2}{r}$$

$$m\frac{v^2}{r} = 0.750 \text{ kg}\frac{(35.0 \text{ m/s})^2}{(60.0 \text{ m})\cos 20.0°} = 16.3 \text{ N}$$

$$\therefore L\sin 20.0° + T\cos 20.0° = 16.3 \text{ N}$$

$$L\cos 20.0° - T\sin 20.0° = 7.35 \text{ N}$$

$$L + T\frac{\cos 20.0°}{\sin 20.0°} = \frac{16.3 \text{ N}}{\sin 20.0°}$$

$$L - T\frac{\sin 20.0°}{\cos 20.0°} = \frac{7.35 \text{ N}}{\cos 20.0°}$$

$$T(\cot 20.0° + \tan 20.0°) = \frac{16.3 \text{ N}}{\sin 20.0°} - \frac{7.35 \text{ N}}{\cos 20.0°}$$

$$T(3.11) = 39.8 \text{ N}$$

$$T = \boxed{12.8 \text{ N}}$$

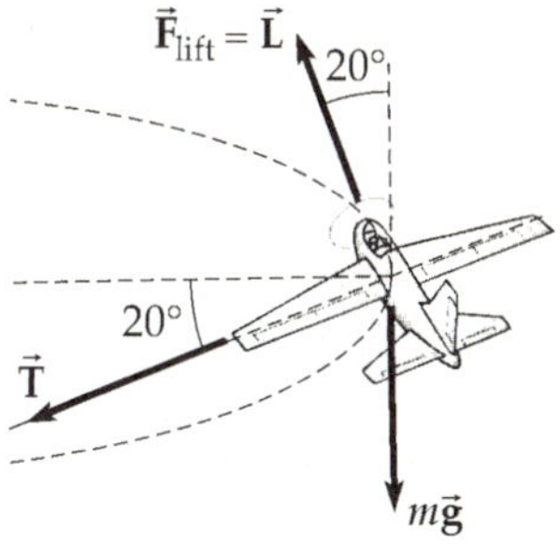

FIG. P5.59

P5.60 $v = v_i - kx$ implies the acceleration is $a = \frac{dv}{dt} = 0 - k\frac{dx}{dt} = -kv$

Then the total force is $\sum F = ma = m(-kv)$

The resistive force is opposite to the velocity: $\boxed{\sum \vec{F} = -km\vec{v}}$.

ANSWERS TO EVEN PROBLEMS

P5.2 (a) 256 m; (b) 42.7 m

P5.4 (a) 0.404; (b) 45.8 lb

P5.6 (a) 14.7 m; (b) neither mass is necessary

P5.8 see the solution (a) 55.2°; (b) 167 N

P5.10 37.8 N

P5.12 (a) 2.31 m/s^2, down for m_1, left for m_2, and up for m_3; (b) 30.0 N and 24.2 N

P5.14 6.84 m

P5.16 (a) 8.32×10^{-8} N inward; (b) $9.13\times10^{22} \text{ m/s}^2$ inward

P5.18 (a) 1.65×10^3 m/s; (b) 6.84×10^3 s

P5.20 (a) 108 N; (b) 56.2 N

P5.22 (a) $mg - \frac{mv^2}{R}$ upward; (b) $\sqrt{gR}$

P5.24 (a) 8.62 m; (b) Mg downward (c) 8.45 m/s^2 unless they have belts, the riders will fall from the cars

P5.26 (a) 53.8 m/s; (b) 148 m

P5.28 (a) 78.3 m/s; (b) 11.1 s; (c) 121 m

P5.30 see the solution

P5.32 $\sim 10^{-7}$ N toward you

P5.34 3.60×10^6 N downward

P5.36 (a) 3.95×10^{3} N forward; (b) $3.02\ \text{m/s}^2$ forward, (c) 86.3 m/s (d) 113 m/s

P5.38 they do not, 29.4 N

P5.40 (a) 19.3°; (b) 4.21 N

P5.42 (a) see the solution; (b) see the solution; (c) 357 N; (d) see the solution; (e) 1.20

P5.44 (a) see the solution; (b) 9.80 N, $0.580\ \text{m/s}^2$

P5.46 (a) $a = 2.84\ \text{m/s}^2$, 26.5 N; (b) see the solution

P5.48 (a) $2.63\ \text{m/s}^2$; (b) 201 m; (c) 17.7 m/s

P5.50 (a) $m_2 g$; (b) $m_2 g$; (c) $\sqrt{\left(\frac{m_2}{m_1}\right)gR}$

P5.52 (a) 5.19 m/s; (b) see the solution, 555 N

P5.54 (a) 1.05×10^{-5} m/s; (b) 7.59×10^{3} s; (c) $8.88\times10^{3}\ \text{m/s}^2$; (d) 8.38 s

P5.56 (a) either 70.4° or 0°; (b) 0°

P5.58 (a) see the solution; (b) see the solution; (c) 53.0 m/s

P5.60 $\sum\vec{\mathbf{F}} = -km\vec{\mathbf{v}}$

Chapter 6

Energy and Energy Transfer

CHAPTER OUTLINE

ANSWERS TO QUESTIONS

Q6.1 The force is perpendicular to every increment of displacement. Therefore, $\vec{\mathbf{F}}\cdot\Delta\vec{\mathbf{r}}=0$.

Q6.2 Yes. Force times distance over which the toe is in contact with the ball. No, he is no longer applying a force. Yes, both air friction and gravity do work.

Q6.3 Force of tension on a ball rotating on the end of a string. Normal force and gravitational force on an object at rest or moving across a level floor.

Q6.4 Work is only done in accelerating the ball from rest. The work is done over the effective length of the pitcher's arm—the distance his hand moves through windup and until release.

Q6.5 (a) Tension (b) Air resistance

(c) Positive in increasing velocity on the downswing.
Negative in decreasing velocity on the upswing.

Q6.6 No. The vectors might be in the third and fourth quadrants, but if the angle between them is less than 90° their dot product is positive.

Q6.7 The scalar product of two vectors is positive if the angle between them is between 0 and 90°. The scalar product is negative when $90°<\theta<180°$.

Q6.8 $k'=2k$. To stretch the smaller piece one meter, each coil would have to stretch twice as much as one coil in the original long spring, since there would be half as many coils. Assuming that the spring is ideal, twice the stretch requires twice the force.

Q6.9 Kinetic energy is always positive. Mass and squared speed are both positive. A moving object can always do positive work in striking another object and causing it to move along the same direction of motion.

Q6.10 The longer barrel will have the higher muzzle speed. Since the accelerating force acts over a longer distance, the change in kinetic energy will be larger.

Q6.11 Kinetic energy is proportional to mass. The first bullet has twice as much kinetic energy.

Q6.12 No violation. Choose the book as the system. You did work and the earth did work on the book. The average force you exerted just counterbalanced the weight of the book. The total work on the book is zero, and is equal to its overall change in kinetic energy.

Q6.13 (a) Kinetic energy is proportional to squared speed. Doubling the speed makes an object's kinetic energy four times larger.

(b) If the total work on an object is zero in some process, its speed must be the same at the final point as it was at the initial point.

Q6.14 The larger engine is unnecessary. Consider a 30-minute commute. If you travel the same speed in each car, it will take the same amount of time, expending the same amount of energy. The extra power available from the larger engine isn't used.

Q6.15 If the instantaneous power output by some agent changes continuously, its average power in a process must be equal to its instantaneous power at least one instant. If its power output is constant, its instantaneous power is always equal to its average power.

Q6.16 The rock increases in speed. The farther it has fallen, the more force it might exert on the sand at the bottom; but it might instead make a deeper crater with an equal-size average force. The farther it falls, the more work it will do in stopping. Its kinetic energy is increasing due to the work that the gravitational force does on it.

Q6.17 The normal force does no work because the angle between the normal force and the direction of motion is usually 90°. Static friction usually does no work because there is no distance through which the force is applied.

Q6.18 An argument for: As a glider moves along an airtrack, the only force that the track applies on the glider is the normal force. Since the angle between the direction of motion and the normal force is 90°, the work done must be zero, even if the track is not level.

Against: An airtrack has bumpers. When a glider bounces from the bumper at the end of the airtrack, it loses a bit of energy, as evidenced by a decreased speed. The airtrack does negative work.

SOLUTIONS TO PROBLEMS

Section 6.1 Systems and Environments

No problems in this section

Section 6.2 Work Done by a Constant Force

P6.1 (a) $W = F\Delta r\cos\theta = (16.0\text{ N})(2.20\text{ m})\cos 25.0° = \boxed{31.9\text{ J}}$

(b), (c) The normal force and the weight are both at 90° to the displacement in any time interval. Both do $\boxed{0}$ work.

(d) $\sum W = 31.9\text{ J} + 0 + 0 = \boxed{31.9\text{ J}}$

P6.2 The component of force along the direction of motion is

$$F\cos\theta = (35.0\text{ N})\cos 25.0° = 31.7\text{ N}.$$

The work done by this force is

$$W = (F\cos\theta)\Delta r = (31.7\text{ N})(50.0\text{ m}) = \boxed{1.59\times 10^3\text{ J}}.$$

P6.3 **Method One.**

Let ϕ represent the instantaneous angle the rope makes with the vertical as it is swinging up from $\phi_i = 0$ to $\phi_f = 60°$. In an incremental bit of motion from angle ϕ to $\phi + d\phi$, the definition of radian measure implies that $\Delta r = (12\text{ m})d\phi$. The angle θ between the incremental displacement and the force of gravity is $\theta = 90° + \phi$. Then $\cos\theta = \cos(90° + \phi) = -\sin\phi$. The work done by the gravitational force on Batman is

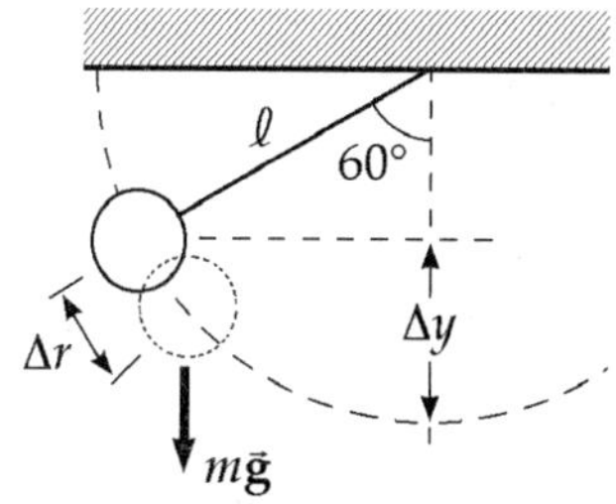

FIG. P6.3

$$W = \int_i^f F\cos\theta dr = \int_{\phi=0}^{\phi=60°} mg(-\sin\phi)(12\text{ m})d\phi$$

$$= -mg(12\text{ m})\int_0^{60°}\sin\phi\, d\phi = (-80\text{ kg})(9.8\text{ m/s}^2)(12\text{ m})(-\cos\phi)\Big|_0^{60°}$$

$$= (-784\text{ N})(12\text{ m})(-\cos 60° + 1) = \boxed{-4.70\times 10^3\text{ J}}$$

Method Two.

The force of gravity on Batman is $mg = (80\text{ kg})(9.8\text{ m/s}^2) = 784\text{ N}$ down. Only his vertical displacement contributes to the work gravity does. His original y-coordinate below the tree limb is -12 m. His final y-coordinate is $(-12\text{ m})\cos 60° = -6\text{ m}$. His change in elevation is $-6\text{ m} - (-12\text{ m}) = 6\text{ m}$. The work done by gravity is

$$W = F\Delta r\cos\theta = (784\text{ N})(6\text{ m})\cos 180° = \boxed{-4.70\text{ kJ}}.$$

P6.4 (a) $W = mgh = (3.35\times 10^{-5})(9.80)(100)\text{ J} = \boxed{3.28\times 10^{-2}\text{ J}}$

(b) Since $R = mg$, $W_{\text{air resistance}} = \boxed{-3.28\times 10^{-2}\text{ J}}$

Section 6.3 The Scalar Product of Two Vectors

P6.5 We must first find the angle between the two vectors. It is:

$$\theta = 360° - 118° - 90.0° - 132° = 20.0°$$

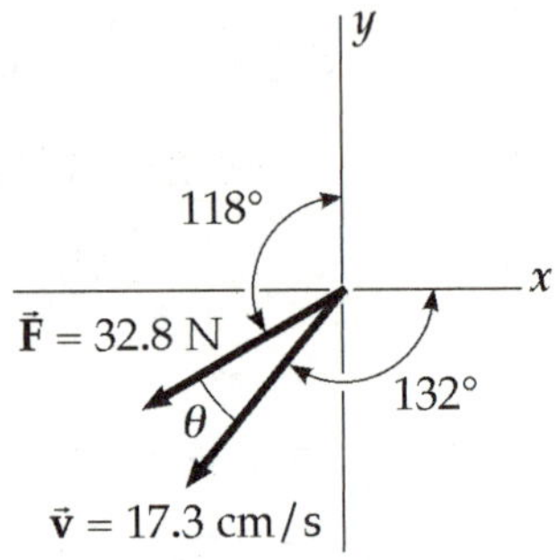

FIG. P6.5

Then

$$\vec{F}\cdot\vec{v} = Fv\cos\theta = (32.8\text{ N})(0.173\text{ m/s})\cos 20.0°$$

or $\vec{F}\cdot\vec{v} = 5.33\dfrac{\text{N}\cdot\text{m}}{\text{s}} = 5.33\dfrac{\text{J}}{\text{s}} = \boxed{5.33\text{ W}}$

P6.6
$$\vec{A}\cdot\vec{B} = \left(A_x\hat{i} + A_y\hat{j} + A_z\hat{k}\right)\cdot\left(B_x\hat{i} + B_y\hat{j} + B_z\hat{k}\right)$$
$$\vec{A}\cdot\vec{B} = A_xB_x\left(\hat{i}\cdot\hat{i}\right) + A_xB_y\left(\hat{i}\cdot\hat{j}\right) + A_xB_z\left(\hat{i}\cdot\hat{k}\right)$$
$$+A_yB_x\left(\hat{j}\cdot\hat{i}\right) + A_yB_y\left(\hat{j}\cdot\hat{j}\right) + A_yB_z\left(\hat{j}\cdot\hat{k}\right)$$
$$+A_zB_x\left(\hat{k}\cdot\hat{i}\right) + A_zB_y\left(\hat{k}\cdot\hat{j}\right) + A_zB_z\left(\hat{k}\cdot\hat{k}\right)$$
$$\vec{A}\cdot\vec{B} = \boxed{A_xB_x + A_yB_y + A_zB_z}$$

P6.7 (a) $W = \vec{F}\cdot\Delta\vec{r} = F_x x + F_y y = (6.00)(3.00)\text{ N}\cdot\text{m} + (-2.00)(1.00)\text{ N}\cdot\text{m} = \boxed{16.0\text{ J}}$

(b) $$\theta = \cos^{-1}\left(\frac{\vec{F}\cdot\Delta\vec{r}}{F\Delta r}\right) = \cos^{-1}\frac{16}{\sqrt{\left((6.00)^2 + (-2.00)^2\right)\left((3.00)^2 + (1.00)^2\right)}} = \boxed{36.9°}$$

P6.8
$$\vec{A} - \vec{B} = \left(3.00\hat{i} + \hat{j} - \hat{k}\right) - \left(-\hat{i} + 2.00\hat{j} + 5.00\hat{k}\right)$$
$$\vec{A} - \vec{B} = 4.00\hat{i} - \hat{j} - 6.00\hat{k}$$
$$\vec{C}\cdot\left(\vec{A} - \vec{B}\right) = \left(2.00\hat{j} - 3.00\hat{k}\right)\cdot\left(4.00\hat{i} - \hat{j} - 6.00\hat{k}\right) = 0 + (-2.00) + (+18.0) = \boxed{16.0}$$

P6.9 (a) $\vec{A} = 3.00\hat{i} - 2.00\hat{j}$

$\vec{B} = 4.00\hat{i} - 4.00\hat{j}$ $\qquad \theta = \cos^{-1}\dfrac{\vec{A}\cdot\vec{B}}{AB} = \cos^{-1}\dfrac{12.0 + 8.00}{\sqrt{(13.0)(32.0)}} = \boxed{11.3°}$

(b) $\vec{B} = 3.00\hat{i} - 4.00\hat{j} + 2.00\hat{k}$

$\vec{A} = -2.00\hat{i} + 4.00\hat{j}$ $\qquad \cos\theta = \dfrac{\vec{A}\cdot\vec{B}}{AB} = \dfrac{-6.00 - 16.0}{\sqrt{(20.0)(29.0)}}$ $\qquad \theta = \boxed{156°}$

(c) $\vec{A} = \hat{i} - 2.00\hat{j} + 2.00\hat{k}$

$\vec{B} = 3.00\hat{j} + 4.00\hat{k}$ $\qquad \theta = \cos^{-1}\left(\dfrac{\vec{A}\cdot\vec{B}}{AB}\right) = \cos^{-1}\left(\dfrac{-6.00 + 8.00}{\sqrt{9.00}\cdot\sqrt{25.0}}\right) = \boxed{82.3°}$

Section 6.4 Work Done by a Varying Force

P6.10 $F_x = (8x - 16)$ N

(a) See figure to the right.

(b) $W_{\text{net}} = \dfrac{-(2.00\text{ m})(16.0\text{ N})}{2} + \dfrac{(1.00\text{ m})(8.00\text{ N})}{2} = \boxed{-12.0\text{ J}}$

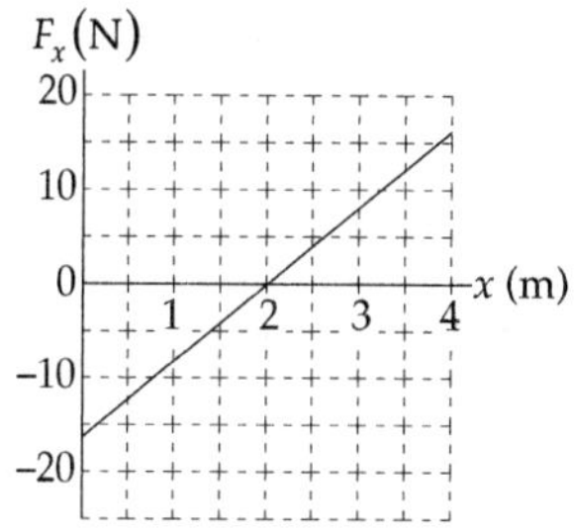

FIG. P6.10

P6.11 $W = \int F_x dx$

and W equals the area under the Force-Displacement curve

(a) For the region $0 \le x \le 5.00$ m,

$$W = \frac{(3.00\text{ N})(5.00\text{ m})}{2} = \boxed{7.50\text{ J}}$$

(b) For the region $5.00 \le x \le 10.0$,

$$W = (3.00\text{ N})(5.00\text{ m}) = \boxed{15.0\text{ J}}$$

(c) For the region $10.0 \le x \le 15.0$,

$$W = \frac{(3.00\text{ N})(5.00\text{ m})}{2} = \boxed{7.50\text{ J}}$$

(d) For the region $0 \le x \le 15.0$

$$W = (7.50 + 7.50 + 15.0)\text{ J} = \boxed{30.0\text{ J}}$$

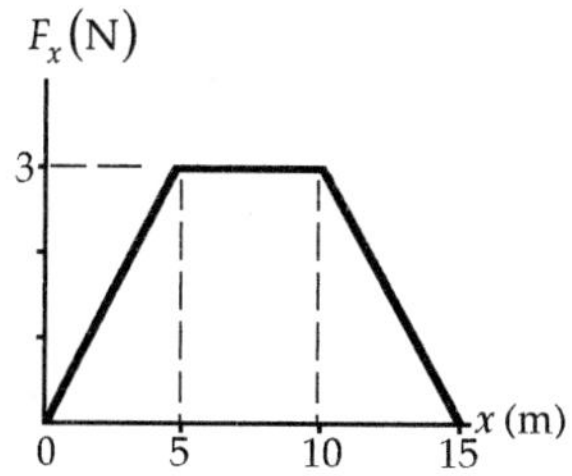

FIG. P6.11

P6.12 Compare an initial picture of the rolling car with a final picture with both springs compressed $K_i + \sum W = K_f$. Work by both springs changes the car's kinetic energy

$$K_i + \frac{1}{2}k_1\left(x_{1i}^2 - x_{1f}^2\right) + \frac{1}{2}k_2\left(x_{2i}^2 - x_{2f}^2\right) = K_f$$

$$\frac{1}{2}mv_i^2 + 0 - \frac{1}{2}(1600 \text{ N/m})(0.500 \text{ m})^2$$

$$+0 - \frac{1}{2}(3400 \text{ N/m})(0.200 \text{ m})^2 = 0$$

$$\frac{1}{2}(6\,000 \text{ kg})v_i^2 - 200 \text{ J} - 68.0 \text{ J} = 0$$

$$v_i = \sqrt{\frac{2(268 \text{ J})}{6\,000 \text{ kg}}} = \boxed{0.299 \text{ m/s}}$$

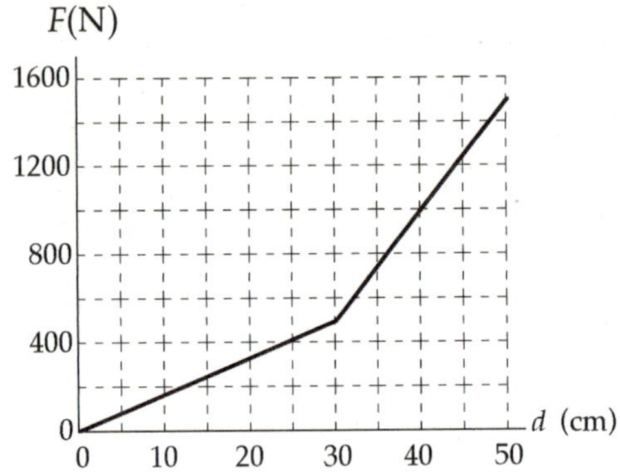

FIG. P6.12

P6.13 $$k = \frac{F}{y} = \frac{Mg}{y} = \frac{(4.00)(9.80) \text{ N}}{2.50 \times 10^{-2} \text{ m}} = 1.57 \times 10^3 \text{ N/m}$$

(a) For 1.50 kg mass $y = \frac{mg}{k} = \frac{(1.50)(9.80)}{1.57 \times 10^3} = \boxed{0.938 \text{ cm}}$

(b) $$\text{Work} = \frac{1}{2}ky^2$$

$$\text{Work} = \frac{1}{2}\left(1.57 \times 10^3 \text{ N} \cdot \text{m}\right)\left(4.00 \times 10^{-2} \text{ m}\right)^2 = \boxed{1.25 \text{ J}}$$

P6.14 $$W = \int_i^f \vec{\mathbf{F}} \cdot d\vec{\mathbf{r}} = \int_0^{5 \text{ m}} \left(4x\hat{\mathbf{i}} + 3y\hat{\mathbf{j}}\right) \text{N} \cdot dx\hat{\mathbf{i}}$$

$$\int_0^{5 \text{ m}} (4 \text{ N/m})x\,dx + 0 = (4 \text{ N/m})\left.\frac{x^2}{2}\right|_0^{5 \text{ m}} = \boxed{50.0 \text{ J}}$$

P6.15 (a) Spring constant is given by $F = kx$

$$k = \frac{F}{x} = \frac{(230 \text{ N})}{(0.400 \text{ m})} = \boxed{575 \text{ N/m}}$$

(b) $$\text{Work} = F_{\text{avg}}x = \frac{1}{2}(230 \text{ N})(0.400 \text{ m}) = \boxed{46.0 \text{ J}}$$

P6.16 (a) $$W=\int_i^f \vec{\mathbf{F}}\cdot d\vec{\mathbf{r}}$$

$$W=\int_0^{0.600\text{ m}}\left(15\,000\text{ N}+10\,000x\text{ N/m}-25\,000x^2\text{ N/m}^2\right)dx\cos 0°$$

$$W=15\,000x+\frac{10\,000x^2}{2}-\frac{25\,000x^3}{3}\bigg|_0^{0.600\text{ m}}$$

$$W=9.00\text{ kJ}+1.80\text{ kJ}-1.80\text{ kJ}=\boxed{9.00\text{ kJ}}$$

(b) Similarly,

$$W=(15.0\text{ kN})(1.00\text{ m})+\frac{(10.0\text{ kN/m})(1.00\text{ m})^2}{2}-\frac{\left(25.0\text{ kN/m}^2\right)(1.00\text{ m})^3}{3}$$

$W=\boxed{11.7\text{ kJ}}$, larger by 29.6%

***P6.17** $$4.00\text{ J}=\frac{1}{2}k(0.100\text{ m})^2$$

$\therefore k=800\text{ N/m}$ and to stretch the spring to 0.200 m requires

$$\Delta W=\frac{1}{2}(800)(0.200)^2-4.00\text{ J}=\boxed{12.0\text{ J}}$$

***P6.18** If the weight of the first tray stretches all four springs by a distance equal to the thickness of the tray, then the proportionality expressed by Hooke's law guarantees that each additional tray will have the same effect, so that the top surface of the top tray will always have the same elevation above the floor. The weight of a tray is $0.580\text{ kg}\left(9.8\text{ m/s}^2\right)=5.68\text{ N}$. The force $\frac{1}{4}(5.68\text{ N})=1.42\text{ N}$ should stretch one spring by 0.450 cm, so its spring constant is $k=\frac{|F_s|}{x}=\frac{1.42\text{ N}}{0.0045\text{ m}}=\boxed{316\text{ N/m}}$. We did not need to know the length or width of the tray.

P6.19 (a) The radius to the object makes angle θ with the horizontal, so its weight makes angle θ with the negative side of the x-axis, when we take the x–axis in the direction of motion tangent to the cylinder.

$$\sum F_x=ma_x$$
$$F-mg\cos\theta=0$$
$$F=\boxed{mg\cos\theta}$$

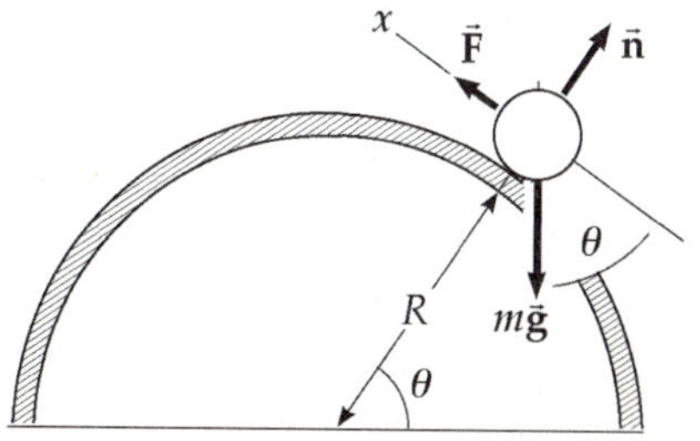

FIG. P6.19

(b) $$W=\int_i^f \vec{\mathbf{F}}\cdot d\vec{\mathbf{r}}$$

continued on next page

We use radian measure to express the next bit of displacement as $dr = Rd\theta$ in terms of the next bit of angle moved through:

$$W = \int_0^{\pi/2} mg\cos\theta R d\theta = mgR\sin\theta\Big|_0^{\pi/2}$$

$$W = mgR(1-0) = \boxed{mgR}$$

***P6.20** The same force makes both light springs stretch.

(a) The hanging mass moves down by

$$x = x_1 + x_2 = \frac{mg}{k_1} + \frac{mg}{k_2} = \boxed{mg\left(\frac{1}{k_1} + \frac{1}{k_2}\right)}$$

(b) We define the effective spring constant as

$$k = \frac{F}{x} = \frac{mg}{mg(1/k_1 + 1/k_2)} = \boxed{\left(\frac{1}{k_1} + \frac{1}{k_2}\right)^{-1}}$$

Section 6.5 Kinetic Energy and the Work-Kinetic Energy Theorem

Section 6.6 The Nonisolated System

P6.21 (a) $K_A = \frac{1}{2}(0.600\text{ kg})(2.00\text{ m/s})^2 = \boxed{1.20\text{ J}}$

(b) $\frac{1}{2}mv_B^2 = K_B$: $\quad v_B = \sqrt{\frac{2K_B}{m}} = \sqrt{\frac{(2)(7.50)}{0.600}} = \boxed{5.00\text{ m/s}}$

(c) $\sum W = \Delta K = K_B - K_A = \frac{1}{2}m\left(v_B^2 - v_A^2\right) = 7.50\text{ J} - 1.20\text{ J} = \boxed{6.30\text{ J}}$

P6.22 (a) $K = \frac{1}{2}mv^2 = \frac{1}{2}(0.300\text{ kg})(15.0\text{ m/s})^2 = \boxed{33.8\text{ J}}$

(b) $K = \frac{1}{2}(0.300)(30.0)^2 = \frac{1}{2}(0.300)(15.0)^2(4) = 4(33.8) = \boxed{135\text{ J}}$

P6.23 $\vec{\mathbf{v}}_i = \left(6.00\hat{\mathbf{i}} - 2.00\hat{\mathbf{j}}\right)\text{ m/s}$

(a) $v_i = \sqrt{v_{ix}^2 + v_{iy}^2} = \sqrt{40.0}\text{ m/s}$

$$K_i = \frac{1}{2}mv_i^2 = \frac{1}{2}(3.00\text{ kg})\left(40.0\text{ m}^2/\text{s}^2\right) = \boxed{60.0\text{ J}}$$

continued on next page

(b) $\vec{\mathbf{v}}_f = 8.00\hat{\mathbf{i}} + 4.00\hat{\mathbf{j}}$

$$v_f^2 = \vec{\mathbf{v}}_f \cdot \vec{\mathbf{v}}_f = 64.0 + 16.0 = 80.0 \text{ m}^2/\text{s}^2$$

$$\Delta K = K_f - K_i = \frac{1}{2}m\left(v_f^2 - v_i^2\right) = \frac{3.00}{2}(80.0) - 60.0 = \boxed{60.0 \text{ J}}$$

P6.24 (a) $\Delta K = K_f - K_i = \frac{1}{2}mv_f^2 - 0 = \sum W =$ (area under curve from $x = 0$ to $x = 5.00$ m)

$$v_f = \sqrt{\frac{2(\text{area})}{m}} = \sqrt{\frac{2(7.50 \text{ J})}{4.00 \text{ kg}}} = \boxed{1.94 \text{ m/s}}$$

(b) $\Delta K = K_f - K_i = \frac{1}{2}mv_f^2 - 0 = \sum W =$ (area under curve from $x = 0$ to $x = 10.0$ m)

$$v_f = \sqrt{\frac{2(\text{area})}{m}} = \sqrt{\frac{2(22.5 \text{ J})}{4.00 \text{ kg}}} = \boxed{3.35 \text{ m/s}}$$

(c) $\Delta K = K_f - K_i = \frac{1}{2}mv_f^2 - 0 = \sum W =$ (area under curve from $x = 0$ to $x = 15.0$ m)

$$v_f = \sqrt{\frac{2(\text{area})}{m}} = \sqrt{\frac{2(30.0 \text{ J})}{4.00 \text{ kg}}} = \boxed{3.87 \text{ m/s}}$$

P6.25 Consider the work done on the pile driver from the time it starts from rest until it comes to rest at the end of the fall. Let $d = 5.00$ m represent the distance over which the driver falls freely, and $h = 0.12$ m the distance it moves the piling.

$$\sum W = \Delta K: \quad W_{\text{gravity}} + W_{\text{beam}} = \frac{1}{2}mv_f^2 - \frac{1}{2}mv_i^2$$

so $$(mg)(h+d)\cos 0° + (\bar{F})(d)\cos 180° = 0 - 0.$$

Thus, $$\bar{F} = \frac{(mg)(h+d)}{d} = \frac{(2\,100 \text{ kg})(9.80 \text{ m/s}^2)(5.12 \text{ m})}{0.120 \text{ m}} = \boxed{8.78 \times 10^5 \text{ N}}$$. The force on the pile driver is $\boxed{\text{upward}}$.

P6.26 (a) $K_i + \sum W = K_f = \frac{1}{2}mv_f^2$

$$0 + \sum W = \frac{1}{2}(15.0 \times 10^{-3} \text{ kg})(780 \text{ m/s})^2 = \boxed{4.56 \text{ kJ}}$$

(b) $$F = \frac{W}{\Delta r \cos\theta} = \frac{4.56 \times 10^3 \text{ J}}{(0.720 \text{ m})\cos 0°} = \boxed{6.34 \text{ kN}}$$

(c) $$a = \frac{v_f^2 - v_i^2}{2x_f} = \frac{(780 \text{ m/s})^2 - 0}{2(0.720 \text{ m})} = \boxed{422 \text{ km/s}^2}$$

(d) $$\sum F = ma = (15 \times 10^{-3} \text{ kg})(422 \times 10^3 \text{ m/s}^2) = \boxed{6.34 \text{ kN}}$$

P6.27 $\sum W = \Delta K = 0$: $\int_0^L mg \sin 35.0°\ dl - \int_0^d kx\ dx = 0$

$$mg \sin 35.0° (L) = \frac{1}{2} kd^2$$

$$d = \sqrt{\frac{2mg \sin 35.0° (L)}{k}}$$

$$d = \sqrt{\frac{2(12.0\text{ kg})(9.80\text{ m/s}^2)(\sin 35.0°)(3.00\text{ m})}{3.00 \times 10^4\text{ N/m}}} = \boxed{0.116\text{ m}}$$

P6.28 (a) $v_f = 0.096\left(3 \times 10^8\text{ m/s}\right) = 2.88 \times 10^7\text{ m/s}$

$$K_f = \frac{1}{2} m v_f^2 = \frac{1}{2}\left(9.11 \times 10^{-31}\text{ kg}\right)\left(2.88 \times 10^7\text{ m/s}\right)^2 = \boxed{3.78 \times 10^{-16}\text{ J}}$$

(b) $K_i + W = K_f$: $0 + F\Delta r \cos\theta = K_f$

$$F(0.028\text{ m})\cos 0° = 3.78 \times 10^{-16}\text{ J}$$

$$F = \boxed{1.35 \times 10^{-14}\text{ N}}$$

(c) $\sum F = ma$; $a = \frac{\sum F}{m} = \frac{1.35 \times 10^{-14}\text{ N}}{9.11 \times 10^{-31}\text{ kg}} = \boxed{1.48 \times 10^{+16}\text{ m/s}^2}$

(d) $v_{xf} = v_{xi} + a_x t$ $2.88 \times 10^7\text{ m/s} = 0 + \left(1.48 \times 10^{16}\text{ m/s}^2\right)t$

$$t = \boxed{1.94 \times 10^{-9}\text{ s}}$$

Check: $x_f = x_i + \frac{1}{2}\left(v_{xi} + v_{xf}\right)t$

$$0.028\text{ m} = 0 + \frac{1}{2}\left(0 + 2.88 \times 10^7\text{ m/s}\right)t$$

$$t = 1.94 \times 10^{-9}\text{ s}$$

Section 6.7 Situations Involving Kinetic Friction

P6.29 $\sum F_y = ma_y$: $n - 392\text{ N} = 0$

$$n = 392\text{ N}$$

$$f_k = \mu_k n = (0.300)(392\text{ N}) = 118\text{ N}$$

(a) $W_F = F\Delta r \cos\theta = (130)(5.00)\cos 0° = \boxed{650\text{ J}}$

(b) $\Delta E_{\text{int}} = f_k d = (118)(5.00) = \boxed{588\text{ J}}$

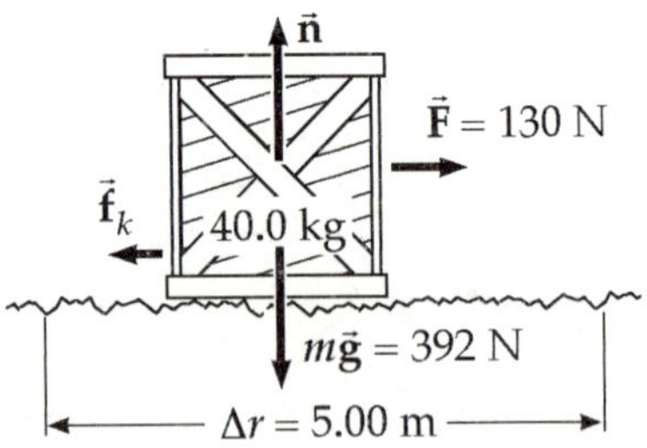

FIG. P6.29

continued on next page

(c) $W_n = n\Delta r\cos\theta = (392)(5.00)\cos 90° = \boxed{0}$

(d) $W_g = mg\Delta r\cos\theta = (392)(5.00)\cos(-90°) = \boxed{0}$

(e) $\Delta K = K_f - K_i = \sum W_{\text{other}} - \Delta E_{\text{int}}$

$$\frac{1}{2}mv_f^2 - 0 = 650\text{ J} - 588\text{ J} + 0 + 0 = \boxed{62.0\text{ J}}$$

(f) $v_f = \sqrt{\frac{2K_f}{m}} = \sqrt{\frac{2(62.0\text{ J})}{40.0\text{ kg}}} = \boxed{1.76\text{ m/s}}$

P6.30 (a) $W_s = \frac{1}{2}kx_i^2 - \frac{1}{2}kx_f^2 = \frac{1}{2}(500)\left(5.00\times10^{-2}\right)^2 - 0 = 0.625\text{ J}$

$$W_s = \frac{1}{2}mv_f^2 - \frac{1}{2}mv_i^2 = \frac{1}{2}mv_f^2 - 0$$

so $v_f = \sqrt{\frac{2(\sum W)}{m}} = \sqrt{\frac{2(0.625)}{2.00}}\text{ m/s} = \boxed{0.791\text{ m/s}}$

(b) $\frac{1}{2}mv_i^2 - f_k d + W_s = \frac{1}{2}mv_f^2$

$$0 - (0.350)(2.00)(9.80)(0.050\,0)\text{ J} + 0.625\text{ J} = \frac{1}{2}mv_f^2$$

$$0.282\text{ J} = \frac{1}{2}(2.00\text{ kg})v_f^2$$

$$v_f = \sqrt{\frac{2(0.282)}{2.00}}\text{ m/s} = \boxed{0.531\text{ m/s}}$$

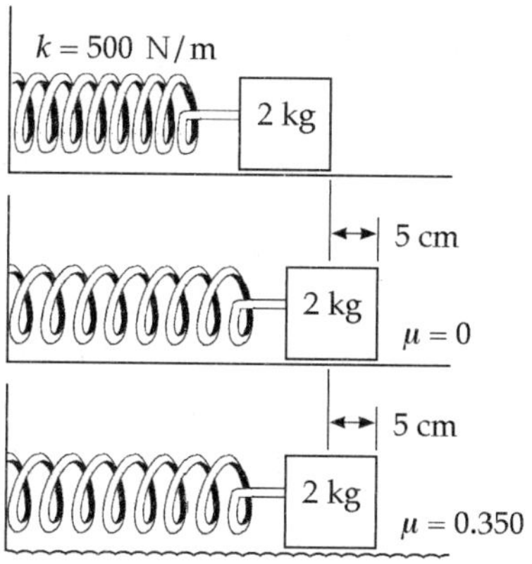

FIG. P6.30

P6.31 (a) $W_g = mg\ell\cos(90.0° + \theta)$

$$W_g = (10.0\text{ kg})\left(9.80\text{ m/s}^2\right)(5.00\text{ m})\cos 110° = \boxed{-168\text{ J}}$$

(b) $f_k = \mu_k n = \mu_k mg\cos\theta$

$$\Delta E_{\text{int}} = \ell f_k = \ell\mu_k mg\cos\theta$$

$$\Delta E_{\text{int}} = (5.00\text{ m})(0.400)(10.0)(9.80)\cos 20.0° = \boxed{184\text{ J}}$$

(c) $W_F = F\ell = (100)(5.00) = \boxed{500\text{ J}}$

(d) $\Delta K = \sum W_{\text{other}} - \Delta E_{\text{int}} = W_F + W_g - \Delta E_{\text{int}} = \boxed{148\text{ J}}$

(e) $\Delta K = \frac{1}{2}mv_f^2 - \frac{1}{2}mv_i^2$

$$v_f = \sqrt{\frac{2(\Delta K)}{m} + v_i^2} = \sqrt{\frac{2(148)}{10.0} + (1.50)^2} = \boxed{5.65\text{ m/s}}$$

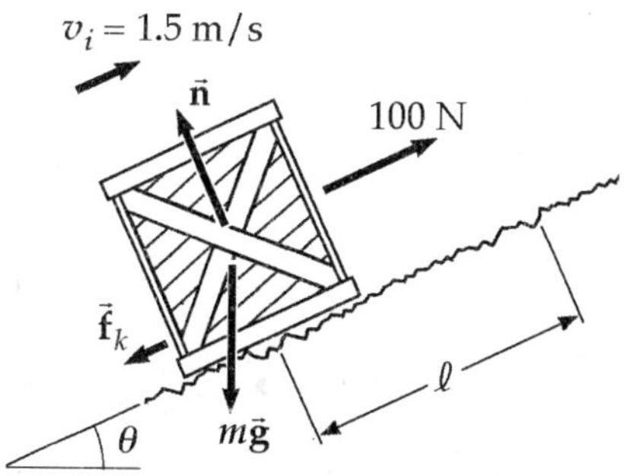

FIG. P6.31

P6.32 $\sum F_y = ma_y$: $n + (70.0\text{ N})\sin 20.0° - 147\text{ N} = 0$

$$n = 123\text{ N}$$

$$f_k = \mu_k n = 0.300(123\text{ N}) = 36.9\text{ N}$$

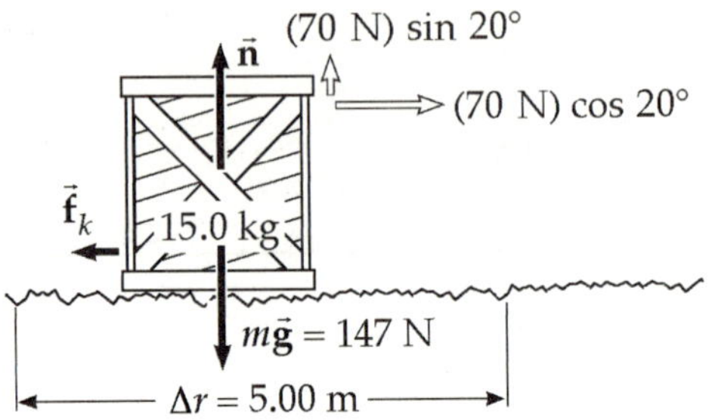

FIG. P6.32

(a) $W = F\Delta r\cos\theta = (70.0\text{ N})(5.00\text{ m})\cos 20.0° = \boxed{329\text{ J}}$

(b) $W = F\Delta r\cos\theta = (123\text{ N})(5.00\text{ m})\cos 90.0° = \boxed{0\text{ J}}$

(c) $W = F\Delta r\cos\theta = (147\text{ N})(5.00\text{ m})\cos 90.0° = \boxed{0}$

(d) $\Delta E_{\text{int}} = f_k d = (36.9\text{ N})(5.00\text{ m}) = \boxed{185\text{ J}}$

(e) $\Delta K = K_f - K_i = \sum W - \Delta E_{\text{int}} = 329\text{ J} - 185\text{ J} = \boxed{+144\text{ J}}$

P6.33 $v_i = 2.00\text{ m/s}$ $\qquad \mu_k = 0.100$

$K_i - f_k d + W_{\text{other}} = K_f$: $\qquad \frac{1}{2}mv_i^2 - f_k d = 0$

$\frac{1}{2}mv_i^2 = \mu_k mgd$ $\qquad d = \frac{v_i^2}{2\mu_k g} = \frac{(2.00\text{ m/s})^2}{2(0.100)(9.80)} = \boxed{2.04\text{ m}}$

Section 6.8 Power

P6.34 (a) The distance moved upward in the first 3.00 s is

$$\Delta y = \bar{v}t = \left[\frac{0 + 1.75\text{ m/s}}{2}\right](3.00\text{ s}) = 2.63\text{ m}.$$

The motor and the earth's gravity do work on the elevator car:

$$\frac{1}{2}mv_i^2 + W_{\text{motor}} + mg\Delta y\cos 180° = \frac{1}{2}mv_f^2$$

$$W_{\text{motor}} = \frac{1}{2}(650\text{ kg})(1.75\text{ m/s})^2 - 0 + (650\text{ kg})g(2.63\text{ m}) = 1.77\times 10^4\text{ J}$$

Also, $W = \bar{\mathcal{P}}\Delta t$ so $\bar{\mathcal{P}} = \frac{W}{\Delta t} = \frac{1.77\times 10^4\text{ J}}{3.00\text{ s}} = \boxed{5.91\times 10^3\text{ W}} = 7.92\text{ hp}.$

(b) When moving upward at constant speed $(v = 1.75\text{ m/s})$ the applied force equals the weight $= (650\text{ kg})(9.80\text{ m/s}^2) = 6.37\times 10^3\text{ N}$. Therefore,

$$\mathcal{P} = Fv = (6.37\times 10^3\text{ N})(1.75\text{ m/s}) = \boxed{1.11\times 10^4\text{ W}} = 14.9\text{ hp}.$$

P6.35 Power $= \dfrac{W}{\Delta t}$ $\qquad \mathscr{P} = \dfrac{mgh}{\Delta t} = \dfrac{(700\text{ N})(10.0\text{ m})}{8.00\text{ s}} = \boxed{875\text{ W}}$

P6.36 (a) $\sum W = \Delta K$, but $\Delta K = 0$ because he moves at constant speed. The skier rises a vertical distance of $(60.0\text{ m})\sin 30.0° = 30.0\text{ m}$. Thus,

$$W_{\text{in}} = -W_g = (70.0\text{ kg})\left(9.8\text{ m/s}^2\right)(30.0\text{ m}) = \boxed{2.06\times10^4\text{ J}} = \boxed{20.6\text{ kJ}}.$$

(b) The time to travel 60.0 m at a constant speed of 2.00 m/s is 30.0 s. Thus,

$$\mathscr{P}_{\text{input}} = \frac{W}{\Delta t} = \frac{2.06\times10^4\text{ J}}{30.0\text{ s}} = \boxed{686\text{ W}} = 0.919\text{ hp}.$$

***P6.37** *energy = power × time*

For the 28.0 W bulb:

$$\text{Energy used} = (28.0\text{ W})\left(1.00\times10^4\text{ h}\right) = 280\text{ kilowatt}\cdot\text{hrs}$$

$$\text{total cost} = \$17.00 + (280\text{ kWh})(\$0.080/\text{kWh}) = \$39.4$$

For the 100 W bulb:

$$\text{Energy used} = (100\text{ W})\left(1.00\times10^4\text{ h}\right) = 1.00\times10^3\text{ kilowatt}\cdot\text{hrs}$$

$$\#\text{ bulb used} = \frac{1.00\times10^4\text{ h}}{750\text{ h/bulb}} = 13.3$$

$$\text{total cost} = 13.3(\$0.420) + \left(1.00\times10^3\text{ kWh}\right)(\$0.080/\text{kWh}) = \$85.6$$

Savings with energy-efficient bulb $= \$85.60 - \$39.40 = \boxed{\$46.2}$

P6.38 (a) Burning 1 lb of fat releases energy $\quad 1\text{ lb}\left(\dfrac{454\text{ g}}{1\text{ lb}}\right)\left(\dfrac{9\text{ kcal}}{1\text{ g}}\right)\left(\dfrac{4\,186\text{ J}}{1\text{ kcal}}\right) = 1.71\times10^7\text{ J}.$

The mechanical energy output is $\quad \left(1.71\times10^7\text{ J}\right)(0.20) = nF\Delta r\cos\theta.$

Then $\quad 3.42\times10^6\text{ J} = nmg\Delta y\cos 0°$

$$3.42\times10^6\text{ J} = n(50\text{ kg})\left(9.8\text{ m/s}^2\right)(80\text{ steps})(0.150\text{ m})$$

$$3.42\times10^6\text{ J} = n\left(5.88\times10^3\text{ J}\right)$$

where the number of times she must climb the steps is $n = \dfrac{3.42\times10^6\text{ J}}{5.88\times10^3\text{ J}} = \boxed{582}$.

This method is impractical compared to limiting food intake.

continued on next page

(b) Her mechanical power output is

$$\mathcal{P} = \frac{W}{\Delta t} = \frac{5.88 \times 10^3 \text{ J}}{65 \text{ s}} = \boxed{90.5 \text{ W}} = 90.5 \text{ W}\left(\frac{1 \text{ hp}}{746 \text{ W}}\right) = \boxed{0.121 \text{ hp}}.$$

P6.39 (a) The fuel economy for walking is $\dfrac{1 \text{ h}}{220 \text{ kcal}}\left(\dfrac{3 \text{ mi}}{\text{h}}\right)\left(\dfrac{1 \text{ kcal}}{4\,186 \text{ J}}\right)\left(\dfrac{1.30 \times 10^8 \text{ J}}{1 \text{ gal}}\right) = \boxed{423 \text{ mi/gal}}$.

(b) For bicycling $\dfrac{1 \text{ h}}{400 \text{ kcal}}\left(\dfrac{10 \text{ mi}}{\text{h}}\right)\left(\dfrac{1 \text{ kcal}}{4\,186 \text{ J}}\right)\left(\dfrac{1.30 \times 10^8 \text{ J}}{1 \text{ gal}}\right) = \boxed{776 \text{ mi/gal}}$.

Section 6.9 Context Connection—Horsepower Ratings of Automobiles

P6.40 A 1 300-kg car speeds up from rest to 55.0 mi/h = 24.6 m/s in 15.0 s. The output work of the engine is equal to its final kinetic energy,

$$\frac{1}{2}(1\,300 \text{ kg})(24.6 \text{ m/s})^2 = 390 \text{ kJ}$$

with power $\mathcal{P} = \dfrac{390\,000 \text{ J}}{15.0 \text{ s}}\ \boxed{\sim 10^4 \text{ W}}$ around 30 horsepower.

P6.41 $\mathcal{P}_a = f_a v$: $\quad f_a = \dfrac{\mathcal{P}_a}{v} = \dfrac{2.24 \times 10^4}{27.0} = \boxed{830 \text{ N}}$

Additional Problems

P6.42 At start, $\vec{\mathbf{v}} = (40.0 \text{ m/s})\cos 30.0°\,\hat{\mathbf{i}} + (40.0 \text{ m/s})\sin 30.0°\,\hat{\mathbf{j}}$

At apex, $\vec{\mathbf{v}} = (40.0 \text{ m/s})\cos 30.0°\,\hat{\mathbf{i}} + 0\hat{\mathbf{j}} = (34.6 \text{ m/s})\hat{\mathbf{i}}$

And $K = \dfrac{1}{2}mv^2 = \dfrac{1}{2}(0.150 \text{ kg})(34.6 \text{ m/s})^2 = \boxed{90.0 \text{ J}}$

***P6.43** Concentration of Energy output $= (0.600 \text{ J/kg}\cdot\text{step})(60.0 \text{ kg})\left(\dfrac{1 \text{ step}}{1.50 \text{ m}}\right) = 24.0 \text{ J/m}$

$$F = (24.0 \text{ J/m})(1 \text{ N}\cdot\text{m/J}) = 24.0 \text{ N}$$
$$\mathcal{P} = Fv$$
$$70.0 \text{ W} = (24.0 \text{ N})v$$
$$v = \boxed{2.92 \text{ m/s}}$$

***P6.44** As it moves at constant speed, the bicycle is in equilibrium. The forward frictional force is equal in magnitude to the air resistance, which we write as av^2 where a is a proportionality constant. The exercising woman exerts the friction force on the ground; by Newton's third law, it is this same size again. The woman's power output is $\mathcal{P} = Fv = av^3 = ch$, where c is another constant and h is her heart rate. We are given $a(22\ \text{km/h})^3 = c(90\ \text{beats/min})$. For her minimum heart rate we have $av_{\text{min}}^3 = c(136\ \text{beats/min})$. By division $\left(\dfrac{v_{\text{min}}}{22\ \text{km/h}}\right)^3 = \dfrac{136}{90}$. $v_{\text{min}} = \left(\dfrac{136}{90}\right)^{1/3}(22\ \text{km/h}) = \boxed{25.2\ \text{km/h}}$.

Similarly, $v_{\text{max}} = \left(\dfrac{166}{90}\right)^{1/3}(22\ \text{km/h}) = \boxed{27.0\ \text{km/h}}$.

P6.45 (a) $x = t + 2.00t^3$

Therefore,

$$v = \frac{dx}{dt} = 1 + 6.00t^2$$

$$K = \frac{1}{2}mv^2 = \frac{1}{2}(4.00)\left(1 + 6.00t^2\right)^2 = \boxed{\left(2.00 + 24.0t^2 + 72.0t^4\right)\text{J}}$$

(b) $a = \dfrac{dv}{dt} = \boxed{(12.0t)\ \text{m/s}^2}$

$F = ma = 4.00(12.0t) = \boxed{(48.0t)\ \text{N}}$

(c) $\mathcal{P} = Fv = (48.0t)\left(1 + 6.00t^2\right) = \boxed{\left(48.0t + 288t^3\right)\text{W}}$

(d) $W = \displaystyle\int_0^{2.00} \mathcal{P}dt = \int_0^{2.00}\left(48.0t + 288t^3\right)dt = \boxed{1\,250\ \text{J}}$

***P6.46** The work done by the applied force is

$$W = \int_i^f F_{\text{applied}}dx = \int_0^{x_{\text{max}}} -\left[-\left(k_1x + k_2x^2\right)\right]dx$$

$$= \int_0^{x_{\text{max}}} k_1x\,dx + \int_0^{x_{\text{max}}} k_2x^2dx = k_1\left.\frac{x^2}{2}\right|_0^{x_{\text{max}}} + k_2\left.\frac{x^3}{3}\right|_0^{x_{\text{max}}}$$

$$= \boxed{k_1\frac{x_{\text{max}}^2}{2} + k_2\frac{x_{\text{max}}^3}{3}}$$

P6.47 (a) The work done by the traveler is mgh_sN where N is the number of steps he climbs during the ride.

$$N = (\text{time on escalator})(n)$$

where
$$(\text{time on escalator}) = \frac{h}{\text{vertical velocity of person}}$$

and
$$\text{vertical velocity of person} = v + nh_s$$

Then,
$$N = \frac{nh}{v + nh_s}$$

and the work done by the person becomes $W_{\text{person}} = \boxed{\dfrac{mgnhh_s}{v + nh_s}}$

(b) The work done by the escalator is

$$W_e = (\text{power})(\text{time}) = [(\text{force exerted})(\text{speed})(\text{time})] = mgvt$$

where
$$t = \frac{h}{v + nh_s} \text{ as above.}$$

Thus,
$$W_e = \boxed{\frac{mgvh}{v + nh_s}}.$$

As a check, the total work done on the person's body must add up to *mgh*, the work an elevator would do in lifting him.

It does add up as follows:
$$\sum W = W_{\text{person}} + W_e = \frac{mgnhh_s}{v + nh_s} + \frac{mgvh}{v + nh_s} = \frac{mgh(nh_s + v)}{v + nh_s} = mgh$$

***P6.48** During its whole motion from $y = 10.0$ m to $y = -3.20$ mm, the force of gravity and the force of the plate do work on the ball. It starts and ends at rest

$$K_i + \sum W = K_f$$
$$0 + F_g \Delta y \cos 0° + F_p \Delta x \cos 180° = 0$$
$$mg(10.003\,2 \text{ m}) - F_p(0.003\,20 \text{ m}) = 0$$
$$F_p = \frac{5 \text{ kg}(9.8 \text{ m/s}^2)(10 \text{ m})}{3.2 \times 10^{-3} \text{ m}} = \boxed{1.53 \times 10^5 \text{ N upward}}$$

P6.49 $\Sigma F_x = ma_x$: $kx = ma$

$$k = \frac{ma}{x} = \frac{(4.70 \times 10^{-3} \text{ kg})0.800(9.80 \text{ m/s}^2)}{0.500 \times 10^{-2} \text{m}} = \boxed{7.37 \text{ N/m}}$$

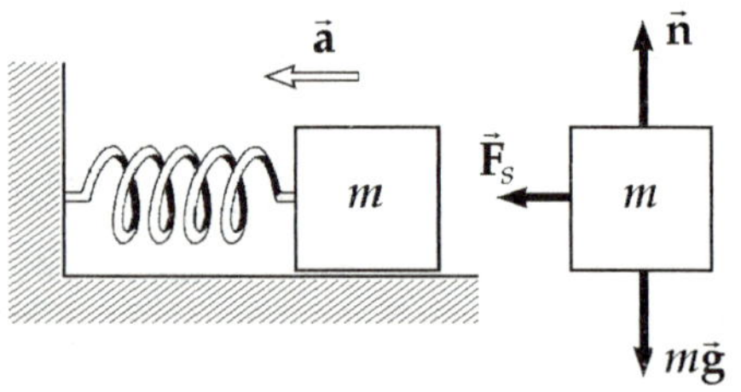

FIG. P6.49

***P6.50** The spring exerts on each block an outward force of magnitude

$$|F_s| = kx = (3.85\ \text{N/m})(0.08\ \text{m}) = 0.308\ \text{N}.$$

For the light block on the left, the vertical forces are given by $F_g = mg = (0.25\ \text{kg})(9.8\ \text{m/s}^2) = 2.45\ \text{N}$, $\sum F_y = 0$, $n - 2.45\ \text{N} = 0$, $n = 2.45\ \text{N}$. Similarly for the heavier block $n = F_g = (0.5\ \text{kg})(9.8\ \text{m/s}^2) = 4.9\ \text{N}$.

(a) For the block on the left, $\sum F_x = ma_x$, $-0.308\ \text{N} = (0.25\ \text{kg})a$, $a = \boxed{-1.23\ \text{m/s}^2}$. For the heavier block, $+0.308\ \text{N} = (0.5\ \text{kg})a$, $a = \boxed{0.616\ \text{m/s}^2}$.

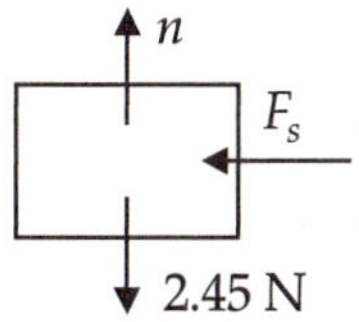

FIG. P6.50

(b) For the block on the left, $f_k = \mu_k n = 0.1(2.45\ \text{N}) = 0.245\ \text{N}$

$$\sum F_x = ma_x$$
$$-0.308\ \text{m/s}^2 + 0.245\ \text{N} = (0.25\ \text{kg})a$$
$$a = \boxed{-0.252\ \text{m/s}^2}$$

if the force of static friction is not too large. For the block on the right, $f_k = \mu_k n = 0.490\ \text{N}$. The maximum force of static friction would be larger, so no motion would begin and the acceleration is $\boxed{\text{zero}}$.

(c) Left block: $f_k = 0.462(2.45\ \text{N}) = 1.13\ \text{N}$. The maximum static friction force would be larger, so the spring force would produce no motion of this block or of the right-hand block, which could feel even more friction force. For both $a = \boxed{0}$.

P6.51 (a) $\mathcal{P} = Fv = F(v_i + at) = F\left(0 + \frac{F}{m}t\right) = \boxed{\left(\frac{F^2}{m}\right)t}$

(b) $\mathcal{P} = \left[\frac{(20.0\ \text{N})^2}{5.00\ \text{kg}}\right](3.00\ \text{s}) = \boxed{240\ \text{W}}$

P6.52 (a) The new length of each spring is $\sqrt{x^2 + L^2}$, so its extension is $\sqrt{x^2 + L^2} - L$ and the force it exerts is $k\left(\sqrt{x^2 + L^2} - L\right)$ toward its fixed end. The y components of the two spring forces add to zero. Their x components add to

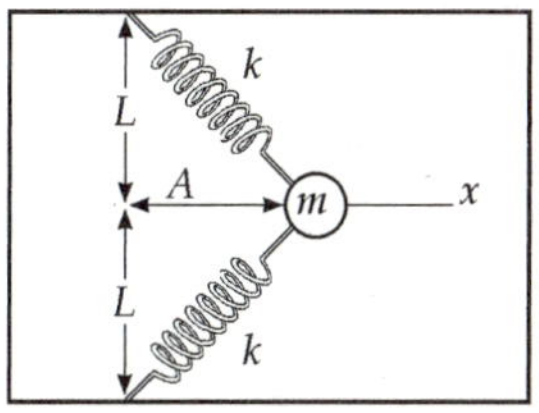

FIG. P6.52

$$\vec{\mathbf{F}} = -2\hat{\mathbf{i}}k\left(\sqrt{x^2 + L^2} - L\right)\frac{x}{\sqrt{x^2 + L^2}} = \boxed{-2kx\hat{\mathbf{i}}\left(1 - \frac{L}{\sqrt{x^2 + L^2}}\right)}.$$

continued on next page

(b) $W = \int_i^f F_x dx$ $\qquad W = \int_A^0 -2kx\left(1 - \frac{L}{\sqrt{x^2+L^2}}\right)dx$

$$W = -2k\int_A^0 x\,dx + kL\int_A^0 \left(x^2+L^2\right)^{-1/2} 2x\,dx \qquad W = -2k\frac{x^2}{2}\Big|_A^0 + kL\frac{\left(x^2+L^2\right)^{1/2}}{(1/2)}\Bigg|_A^0$$

$$W = -0 + kA^2 + 2kL^2 - 2kL\sqrt{A^2+L^2} \qquad W = \boxed{2kL^2 + kA^2 - 2kL\sqrt{A^2+L^2}}$$

P6.53 (a) $\sum W = \Delta K$: $\qquad W_s + W_g = 0$

$$\frac{1}{2}kx_i^2 - 0 + mg\Delta x\cos(90°+60°) = 0$$

$$\frac{1}{2}\left(1.40\times10^3\ \text{N/m}\right)\times(0.100)^2 - (0.200)(9.80)(\sin 60.0°)\Delta x = 0$$

$$\Delta x = \boxed{4.12\ \text{m}}$$

(b) $\sum W = \Delta K + \Delta E_{\text{int}}$: $\qquad W_s + W_g - \Delta E_{\text{int}} = 0$

$$\frac{1}{2}kx_i^2 + mg\Delta x\cos 150° - \mu_k mg\cos 60°\,\Delta x = 0$$

$$\frac{1}{2}\left(1.40\times10^3\ \text{N/m}\right)\times(0.100)^2 - (0.200)(9.80)(\sin 60.0°)\Delta x - (0.200)(9.80)(0.400)(\cos 60.0°)\Delta x = 0$$

$$\Delta x = \boxed{3.35\ \text{m}}$$

P6.54 (a) $\vec{\mathbf{F}}_1 = (25.0\ \text{N})\left(\cos 35.0°\,\hat{\mathbf{i}} + \sin 35.0°\,\hat{\mathbf{j}}\right) = \boxed{\left(20.5\hat{\mathbf{i}} + 14.3\hat{\mathbf{j}}\right)\ \text{N}}$

$$\vec{\mathbf{F}}_2 = (42.0\ \text{N})\left(\cos 150°\,\hat{\mathbf{i}} + \sin 150°\,\hat{\mathbf{j}}\right) = \boxed{\left(-36.4\hat{\mathbf{i}} + 21.0\hat{\mathbf{j}}\right)\ \text{N}}$$

(b) $\sum \vec{\mathbf{F}} = \vec{\mathbf{F}}_1 + \vec{\mathbf{F}}_2 = \boxed{\left(-15.9\hat{\mathbf{i}} + 35.3\hat{\mathbf{j}}\right)\ \text{N}}$

(c) $\vec{\mathbf{a}} = \frac{\sum \vec{\mathbf{F}}}{m} = \boxed{\left(-3.18\hat{\mathbf{i}} + 7.07\hat{\mathbf{j}}\right)\ \text{m/s}^2}$

(d) $\vec{\mathbf{v}}_f = \vec{\mathbf{v}}_i + \vec{\mathbf{a}}t = \left(4.00\hat{\mathbf{i}} + 2.50\hat{\mathbf{j}}\right)\ \text{m/s} + \left(-3.18\hat{\mathbf{i}} + 7.07\hat{\mathbf{j}}\right)\left(\text{m/s}^2\right)(3.00\ \text{s})$

$$\vec{\mathbf{v}}_f = \boxed{\left(-5.54\hat{\mathbf{i}} + 23.7\hat{\mathbf{j}}\right)\ \text{m/s}}$$

(e) $\vec{\mathbf{r}}_f = \vec{\mathbf{r}}_i + \vec{\mathbf{v}}_i t + \frac{1}{2}\vec{\mathbf{a}}t^2$

$$\vec{\mathbf{r}}_f = 0 + \left(4.00\hat{\mathbf{i}} + 2.50\hat{\mathbf{j}}\right)(\text{m/s})(3.00\ \text{s}) + \frac{1}{2}\left(-3.18\hat{\mathbf{i}} + 7.07\hat{\mathbf{j}}\right)\left(\text{m/s}^2\right)(3.00\ \text{s})^2$$

$$\Delta\vec{\mathbf{r}} = \vec{\mathbf{r}}_f = \boxed{\left(-2.30\hat{\mathbf{i}} + 39.3\hat{\mathbf{j}}\right)\ \text{m}}$$

continued on next page

(f) $K_f = \frac{1}{2}mv_f^2 = \frac{1}{2}(5.00\text{ kg})\left[(5.54)^2 + (23.7)^2\right]\left(\text{m/s}^2\right) = \boxed{1.48\text{ kJ}}$

(g) $K_f = \frac{1}{2}mv_i^2 + \sum \vec{\mathbf{F}} \cdot \Delta\vec{\mathbf{r}}$

$$K_f = \frac{1}{2}(5.00\text{ kg})\left[(4.00)^2 + (2.50)^2\right](\text{m/s})^2 + \left[(-15.9\text{ N})(-2.30\text{ m}) + (35.3\text{ N})(39.3\text{ m})\right]$$

$$K_f = 55.6\text{ J} + 1\,426\text{ J} = \boxed{1.48\text{ kJ}}$$

P6.55 $K_i + W_s + W_g = K_f$

$$\frac{1}{2}mv_i^2 + \frac{1}{2}kx_i^2 - \frac{1}{2}kx_f^2 + mg\Delta x\cos\theta = \frac{1}{2}mv_f^2$$

$$0 + \frac{1}{2}kx_i^2 - 0 + mgx_i\cos 100° = \frac{1}{2}mv_f^2$$

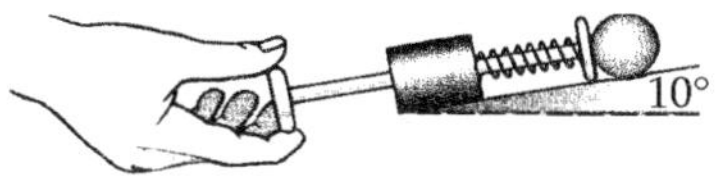

FIG. P6.55

$$\frac{1}{2}(1.20\text{ N/cm})(5.00\text{ cm})(0.050\,0\text{ m}) - (0.100\text{ kg})\left(9.80\text{ m/s}^2\right)(0.050\,0\text{ m})\sin 10.0° = \frac{1}{2}(0.100\text{ kg})v^2$$

$$0.150\text{ J} - 8.51\times10^{-3}\text{ J} = (0.050\,0\text{ kg})v^2$$

$$v = \sqrt{\frac{0.141}{0.050\,0}} = \boxed{1.68\text{ m/s}}$$

P6.56 (a)

F(N)	L(mm)	F(N)	L(mm)
2.00	15.0	14.0	112
4.00	32.0	16.0	126
6.00	49.0	18.0	149
8.00	64.0	20.0	175
10.0	79.0	22.0	190
12.0	98.0		

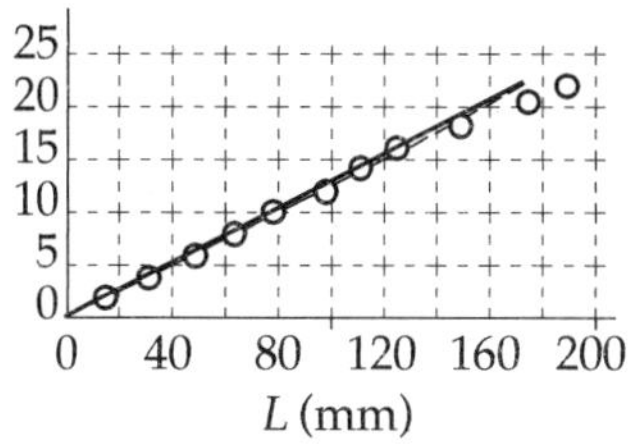

FIG. P6.56

(b) A straight line fits the first eight points, together with the origin. By least-square fitting, its slope is $0.125\text{ N/mm} \pm 2\% = \boxed{125\text{ N/m}} \pm 2\%$. In $F = kx$, the spring constant is $k = \frac{F}{x}$, the same as the slope of the F-versus-x graph.

(c) $F = kx = (125\text{ N/m})(0.105\text{ m}) = \boxed{13.1\text{ N}}$

P6.57 If positive F represents an outward force, (same as direction as r), then

$$W = \int_i^f \vec{\mathbf{F}} \cdot d\vec{\mathbf{r}} = \int_{r_i}^{r_f} \left(2F_0\sigma^{13}r^{-13} - F_0\sigma^7 r^{-7}\right) dr$$

$$W = \left.\frac{2F_0\sigma^{13}r^{-12}}{-12} - \frac{F_0\sigma^7 r^{-6}}{-6}\right|_{r_i}^{r_f}$$

$$W = \frac{-F_0\sigma^{13}\left(r_f^{-12} - r_i^{-12}\right)}{6} + \frac{F_0\sigma^7\left(r_f^{-6} - r_i^{-6}\right)}{6} = \frac{F_0\sigma^7}{6}\left[r_f^{-6} - r_i^{-6}\right] - \frac{F_0\sigma^{13}}{6}\left[r_f^{-12} - r_i^{-12}\right]$$

$$W = 1.03\times10^{-77}\left[r_f^{-6} - r_i^{-6}\right] - 1.89\times10^{-134}\left[r_f^{-12} - r_i^{-12}\right]$$

$$W = 1.03\times10^{-77}\left[1.88\times10^{-6} - 2.44\times10^{-4}\right]10^{60} - 1.89\times10^{-134}\left[3.54\times10^{-12} - 5.96\times10^{-8}\right]10^{120}$$

$$W = -2.49\times10^{-21}\text{ J} + 1.12\times10^{-21}\text{ J} = \boxed{-1.37\times10^{-21}\text{ J}}$$

P6.58 (a) $\Delta E_{\text{int}} = -\Delta K = -\frac{1}{2}m\left(v_f^2 - v_i^2\right)$: $\Delta E_{\text{int}} = -\frac{1}{2}(0.400\text{ kg})\left((6.00)^2 - (8.00)^2\right)(\text{m/s})^2 = \boxed{5.60\text{ J}}$

(b) $\Delta E_{\text{int}} = fd = \mu_k mg(2\pi r)$: $5.60\text{ J} = \mu_k(0.400\text{ kg})\left(9.80\text{ m/s}^2\right)2\pi(1.50\text{ m})$

Thus, $\mu_k = \boxed{0.152}$.

(c) After N revolutions, the object comes to rest and $K_f = 0$.

Thus, $\Delta E_{\text{int}} = -\Delta K = -0 + K_i = \frac{1}{2}mv_i^2$

or $\mu_k mg\left[N(2\pi r)\right] = \frac{1}{2}mv_i^2$.

This gives $N = \dfrac{\frac{1}{2}mv_i^2}{\mu_k mg(2\pi r)} = \dfrac{\frac{1}{2}(8.00\text{ m/s})^2}{(0.152)\left(9.80\text{ m/s}^2\right)2\pi(1.50\text{ m})} = \boxed{2.28\text{ rev}}$.

P6.59 We evaluate $\displaystyle\int_{12.8}^{23.7} \frac{375dx}{x^3 + 3.75x}$ by calculating

$$\frac{375(0.100)}{(12.8)^3 + 3.75(12.8)} + \frac{375(0.100)}{(12.9)^3 + 3.75(12.9)} + \ldots \frac{375(0.100)}{(23.6)^3 + 3.75(23.6)} = 0.806$$

and

$$\frac{375(0.100)}{(12.9)^3 + 3.75(12.9)} + \frac{375(0.100)}{(13.0)^3 + 3.75(13.0)} + \ldots \frac{375(0.100)}{(23.7)^3 + 3.75(23.7)} = 0.791.$$

The answer must be between these two values. We may find it more precisely by using a value for Δx smaller than 0.100. Thus, we find the integral to be $\boxed{0.799\text{ N}\cdot\text{m}}$.

P6.60 $\mathscr{P}\Delta t = W = \Delta K = \dfrac{(\Delta m)v^2}{2}$

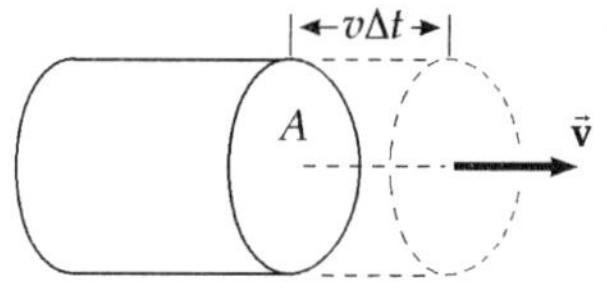

FIG. P6.60

The density is $\rho = \dfrac{\Delta m}{\text{vol}} = \dfrac{\Delta m}{A\Delta x}$.

Substituting this into the first equation and solving for $\mathscr{P}$, since $\dfrac{\Delta x}{\Delta t} = v$,

for a constant speed, we get $\boxed{\mathscr{P} = \dfrac{\rho A v^3}{2}}$.

Also, since $\mathscr{P} = Fv$, $\boxed{F = \dfrac{\rho A v^2}{2}}$.

Our model predicts the same proportionalities as the empirical equation, and gives $D = 1$ for the drag coefficient. Air actually slips around the moving object, instead of accumulating in front of it. For this reason, the drag coefficient is not necessarily unity. It is typically less than one for a streamlined object and can be greater than one if the airflow around the object is complicated.

***P6.61** $\mathscr{P} = \dfrac{1}{2} D\rho \pi r^2 v^3$

(a) $\mathscr{P}_a = \dfrac{1}{2} 1\left(1.20 \text{ kg/m}^3\right)\pi(1.5 \text{ m})^2(8 \text{ m/s})^3 = \boxed{2.17 \times 10^3 \text{ W}}$

(b) $\dfrac{\mathscr{P}_b}{\mathscr{P}_a} = \dfrac{v_b^3}{v_a^3} = \left(\dfrac{24 \text{ m/s}}{8 \text{ m/s}}\right)^3 = 3^3 = 27$

$\mathscr{P}_b = 27\left(2.17 \times 10^3 \text{ W}\right) = \boxed{5.86 \times 10^4 \text{ W}}$

***P6.62** (a) So long as the spring force is greater than the friction force, the block will be gaining speed. The block slows down when the friction force becomes the greater. It has maximum speed when $-kx_a - f_k = ma = 0$.

$-\left(1.0 \times 10^3 \text{ N/m}\right)x_a - 4.0 \text{ N} = 0$ $\boxed{x = -4.0 \times 10^{-3} \text{ m}}$

(b) By the same logic,

$-\left(1.0 \times 10^3 \text{ N/m}\right)x_b - 10.0 \text{ N} = 0$ $\boxed{x = -1.0 \times 10^{-2} \text{ m}}$

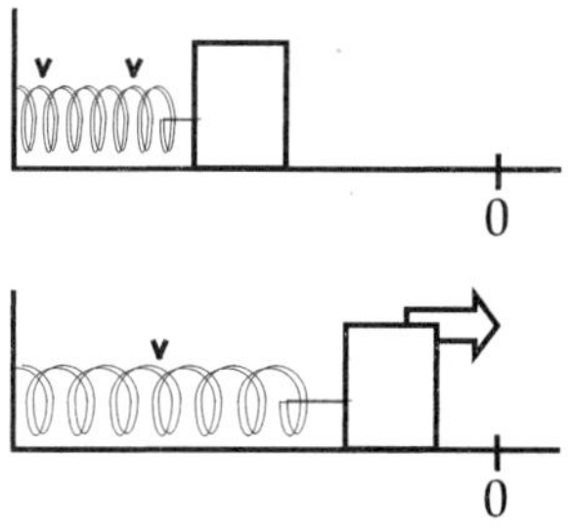

FIG. P6.62

ANSWERS TO EVEN PROBLEMS

P6.2 1.59×10^3 J

P6.4 (a) 32.8 mJ; (b) –32.8 mJ

P6.6 see the solution

P6.8 16.0

P6.10 (a) see the solution; (b) –12.0 J

P6.12 0.299 m/s

P6.14 50.0 J

P6.16 (a) 9.00 kJ; (b) 11.7 kJ, larger by 29.6%

P6.18 see the solution, $316\ \text{N/m}$

P6.20 (a) $mg\left(\frac{1}{k_1}+\frac{1}{k_2}\right)$; (b) $\left(\frac{1}{k_1}+\frac{1}{k_2}\right)^{-1}$

P6.22 (a) 33.8 J; (b) 135 J

P6.24 (a) $1.94\ \text{m/s}$; (b) $3.35\ \text{m/s}$; (c) $3.87\ \text{m/s}$

P6.26 (a) 4.56 kJ; (b) 6.34 kN; (c) $422\ \text{km/s}^2$; (d) 6.34 kN

P6.28 (a) $3.78\times10^{-16}\ \text{J}$; (b) $1.35\times10^{-14}\ \text{N}$; (c) $1.48\times10^{+16}\ \text{m/s}^2$; (d) $1.94\times10^{-9}\ \text{s}$

P6.30 (a) $0.791\ \text{m/s}$; (b) $0.531\ \text{m/s}$

P6.32 (a) 329 J; (b) 0; (c) 0; (d) 185 J; (e) 144 J

P6.34 (a) $5.91\times10^{3}\ \text{W}$; (b) it is 53.0% of $1.11\times10^{4}\ \text{W}$

P6.36 (a) 20.6 kJ; (b) 686 W

P6.38 (a) 582, impractical; (b) $90.5\ \text{W} = 0.121\ \text{hp}$

P6.40 $\sim10^{4}\ \text{W}$

P6.42 90.0 J

P6.44 $25.2\ \text{km/h}$, $27.0\ \text{km/h}$

P6.46 $k_1\frac{x_{\max}^2}{2}+k_2\frac{x_{\max}^3}{3}$

P6.48 $1.53\times10^{5}\ \text{N}$ upward

P6.50 (a) $-1.23\ \text{m/s}^2$ and $0.616\ \text{m/s}^2$; (b) $-0.252\ \text{m/s}^2$ and 0; (c) 0 and 0

P6.52 (a) see the solution; (b) $2kL^2+kA^2-2kL\sqrt{A^2+L^2}$

P6.54 (a) $\vec{\mathbf{F}}_1=(20.5\hat{\mathbf{i}}+14.3\hat{\mathbf{j}})\ \text{N}$, $\vec{\mathbf{F}}_2=(-36.4\hat{\mathbf{i}}+21.0\hat{\mathbf{j}})\ \text{N}$; (b) $(-15.9\hat{\mathbf{i}}+35.3\hat{\mathbf{j}})\ \text{N}$; (c) $(-3.18\hat{\mathbf{i}}+7.07\hat{\mathbf{j}})\ \text{m/s}^2$; (d) $(-5.54\hat{\mathbf{i}}+23.7\hat{\mathbf{j}})\ \text{m/s}$; (e) $(-2.30\hat{\mathbf{i}}+39.3\hat{\mathbf{j}})\ \text{m}$; (f) $1.48\ \text{kJ}$; (g) 1.48 kJ

P6.56 (a) see the solution; (b) $125\ \text{N/m}\pm2\%$; (c) 13.1 N

P6.58 (a) 5.60 J; (b) 0.152; (c) 2.28 rev

P6.60 see the solution

P6.62 (a) $x=-4.0\ \text{mm}$; (b) $-1.0\ \text{cm}$

Chapter 7

Potential Energy

CHAPTER OUTLINE

ANSWERS TO QUESTIONS

Q7.1 The final speed of the children will not depend on the slide length or the presence of bumps if there is no friction. If there is friction, a longer slide will result in a lower final speed. Bumps will have the same effect as they effectively lengthen the distance over which friction can do work, to decrease the total mechanical energy of the children.

Q7.2 Total energy is the sum of kinetic and potential energies. Potential energy can be negative, so the sum of kinetic plus potential can also be negative.

Q7.3 Both agree on the *change* in potential energy, and the kinetic energy. They may disagree on the value of gravitational potential energy, depending on their choice of a zero point.

Q7.4 (a) mgh is provided by the muscles.

(b) No further energy is supplied to the object-Earth system, but some chemical energy must be supplied to the muscles as they keep the weight aloft.

(c) The object loses energy mgh, giving it back to the muscles, where most of it becomes internal energy.

Q7.5 Lift a book from a low shelf to place it on a high shelf. The net change in its kinetic energy is zero, but the book-Earth system increases in gravitational potential energy. Stretch a rubber band to encompass the ends of a ruler. It increases in elastic energy. Rub your hands together or let a pearl drift down at constant speed in a bottle of shampoo. Each system (two hands; pearl and shampoo) increases in internal energy.

Q7.6 Three potential energy terms will appear in the expression of total mechanical energy, one for each conservative force. If you write an equation with initial energy on one side and final energy on the other, the equation contains six potential-energy terms.

Q7.7 (a) It does if it makes the object's speed change, but not if it only makes the direction of the velocity change.

(b) Yes, according to Newton's second law.

Q7.8 The original kinetic energy of the skidding car can be degraded into kinetic energy of random molecular motion in the tires and the road: it is internal energy. If the brakes are used properly, the same energy appears as internal energy in the brake shoes and drums.

Q7.9 All the energy is supplied by foodstuffs that gained their energy from the sun.

Q7.10 Elastic potential energy of plates under stress plus gravitational energy is released when the plates "slip". It is carried away by mechanical waves.

Q7.11 The total energy of the ball-Earth system is conserved. Since the system initially has gravitational energy *mgh* and no kinetic energy, the ball will again have zero kinetic energy when it returns to its original position. Air resistance will cause the ball to come back to a point slightly below its initial position. On the other hand, if anyone gives a forward push to the ball anywhere along its path, the demonstrator will have to duck.

Q7.12 Kinetic energy is greatest at the starting point. Gravitational energy is a maximum at the top of the flight of the ball.

Q7.13 Gravitational energy is proportional to mass, so it doubles.

Q7.14 In stirring cake batter and in weightlifting, your body returns to the same conformation after each stroke. During each stroke chemical energy is irreversibly converted into output work (and internal energy). This observation proves that muscular forces are nonconservative.

Q7.15 Let the gravitational energy be zero at the lowest point in the motion. If you start the vibration by pushing down on the block (2), its kinetic energy becomes extra elastic potential energy in the spring (U_s). After the block starts moving up at its lower turning point (3), this energy becomes both kinetic energy (K) and gravitational potential energy (U_g), and then just gravitational energy when the block is at its greatest height (1). The energy then turns back into kinetic and elastic potential energy, and the cycle repeats.

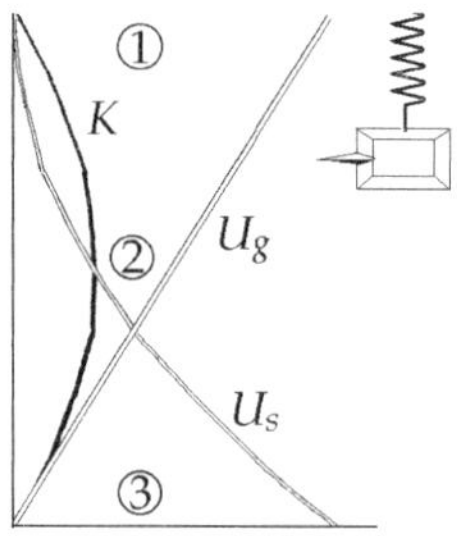

FIG. Q7.15

Q7.16 Chemical energy in the fuel turns into internal energy as the fuel burns. Most of this leaves the car by heat through the walls of the engine and by matter transfer in the exhaust gases. Some leaves the system of fuel by work done to push down the piston. Of this work, a little results in internal energy in the bearings and gears, but most becomes work done on the air to push it aside. The work on the air immediately turns into internal energy in the air. If you use the windshield wipers, you take energy from the crankshaft and turn it into extra internal energy in the glass and wiper blades and wiper-motor coils. If you turn on the air conditioner, your end effect is to put extra energy out into the surroundings. You must apply the brakes at the end of your trip. As soon as the sound of the engine has died away, all you have to show for it is thermal pollution.

Q7.17 A graph of potential energy versus position is a straight horizontal line for a particle in neutral equilibrium. The graph represents a constant function.

Q7.18 The ball is in neutral equilibrium.

SOLUTIONS TO PROBLEMS

Section 7.1 Potential Energy of a System

P7.1 (a) With our choice for the zero level for potential energy when the car is at point B,

$$\boxed{U_B = 0}.$$

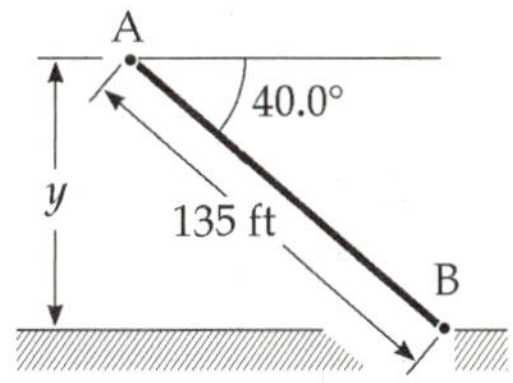

FIG. P7.1

When the car is at point A, the potential energy of the car-Earth system is given by

$$U_A = mgy$$

where y is the vertical height above zero level. With 135 ft $= 41.1$ m, this height is found as:

$$y = (41.1 \text{ m})\sin 40.0° = 26.4 \text{ m}.$$

Thus,

$$U_A = (1\,000 \text{ kg})(9.80 \text{ m/s}^2)(26.4 \text{ m}) = \boxed{2.59 \times 10^5 \text{ J}}.$$

The change in potential energy as the car moves from A to B is

$$U_B - U_A = 0 - 2.59 \times 10^5 \text{ J} = \boxed{-2.59 \times 10^5 \text{ J}}.$$

(b) With our choice of the zero level when the car is at point A, we have $\boxed{U_A = 0}$. The potential energy when the car is at point B is given by $U_B = mgy$ where y is the vertical distance of point B below point A. In part (a), we found the magnitude of this distance to be 26.5 m. Because this distance is now below the zero reference level, it is a negative number.

Thus,

$$U_B = (1\,000 \text{ kg})(9.80 \text{ m/s}^2)(-26.5 \text{ m}) = \boxed{-2.59 \times 10^5 \text{ J}}.$$

The change in potential energy when the car moves from A to B is

$$U_B - U_A = -2.59 \times 10^5 \text{ J} - 0 = \boxed{-2.59 \times 10^5 \text{ J}}.$$

P7.2 (a) We take the zero configuration of system potential energy with the child at the lowest point of the arc. When the string is held horizontal initially, the initial position is 2.00 m above the zero level. Thus,

$$U_g = mgy = (400\ \text{N})(2.00\ \text{m}) = \boxed{800\ \text{J}}.$$

(b) From the sketch, we see that at an angle of 30.0° the child is at a vertical height of $(2.00\ \text{m})(1-\cos 30.0°)$ above the lowest point of the arc. Thus,

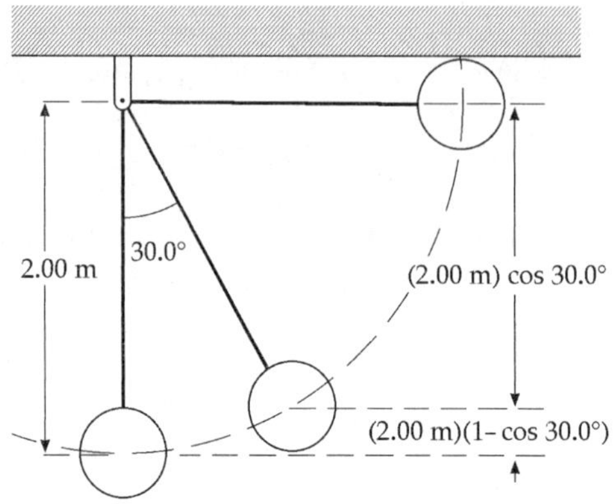

FIG. P7.2

$$U_g = mgy = (400\ \text{N})(2.00\ \text{m})(1-\cos 30.0°) = \boxed{107\ \text{J}}.$$

(c) The zero level has been selected at the lowest point of the arc. Therefore, $\boxed{U_g = 0}$ at this location.

P7.3 The volume flow rate is the volume of water going over the falls each second:

$$3\ \text{m}(0.5\ \text{m})(1.2\ \text{m/s}) = 1.8\ \text{m}^3/\text{s}$$

The mass flow rate is $\frac{m}{t} = \rho\frac{V}{t} = \left(1\,000\ \text{kg/m}^3\right)\left(1.8\ \text{m}^3/\text{s}\right) = 1\,800\ \text{kg/s}$

If the stream has uniform width and depth, the speed of the water below the falls is the same as the speed above the falls. Then no kinetic energy, but only gravitational energy is available for conversion into internal and electric energy.

The input power is $\mathscr{P}_{\text{in}} = \frac{\text{energy}}{\Delta t} = \frac{mgy}{\Delta t} = \frac{m}{\Delta t}gy = (1\,800\ \text{kg/s})\left(9.8\ \text{m/s}^2\right)(5\ \text{m}) = 8.82\times10^4\ \text{J/s}$

The output power is $\mathscr{P}_{\text{useful}} = (\text{efficiency})\mathscr{P}_{\text{in}} = 0.25\left(8.82\times10^4\ \text{W}\right) = \boxed{2.20\times10^4\ \text{W}}$

The efficiency of electric generation at Hoover Dam is about 85% with a head of water (vertical drop) of 174 m. Intensive research is underway to improve the efficiency of low head generators.

Section 7.2 The Isolated System

***P7.4** (a) One child in one jump converts chemical energy into mechanical energy in the amount that her body has as gravitational energy at the top of her jump: $mgy = 36\ \text{kg}\left(9.81\ \text{m/s}^2\right)(0.25\ \text{m}) = 88.3\ \text{J}$. For all of the jumps of the children the energy is $12\left(1.05\times10^6\right)88.3\ \text{J} = \boxed{1.11\times10^9\ \text{J}}$.

(b) The seismic energy is modeled as $E = \frac{0.01}{100}1.11\times10^9\ \text{J} = 1.11\times10^5\ \text{J}$, making the Richter magnitude $\frac{\log E - 4.8}{1.5} = \frac{\log 1.11\times10^5 - 4.8}{1.5} = \frac{5.05-4.8}{1.5} = \boxed{0.2}$.

P7.5 **(a)** $U_i + K_i = U_f + K_f$: $\quad mgh + 0 = mg(2R) + \frac{1}{2}mv^2$

$$g(3.50R) = 2g(R) + \frac{1}{2}v^2$$

$$\boxed{v = \sqrt{3.00gR}}$$

(b) $\sum F = m\frac{v^2}{R}$: $\quad n + mg = m\frac{v^2}{R}$

$$n = m\left[\frac{v^2}{R} - g\right] = m\left[\frac{3.00gR}{R} - g\right] = 2.00mg$$

$$n = 2.00\left(5.00\times10^{-3}\text{ kg}\right)\left(9.80\text{ m/s}^2\right)$$

$$= \boxed{0.098\,0\text{ N downward}}$$

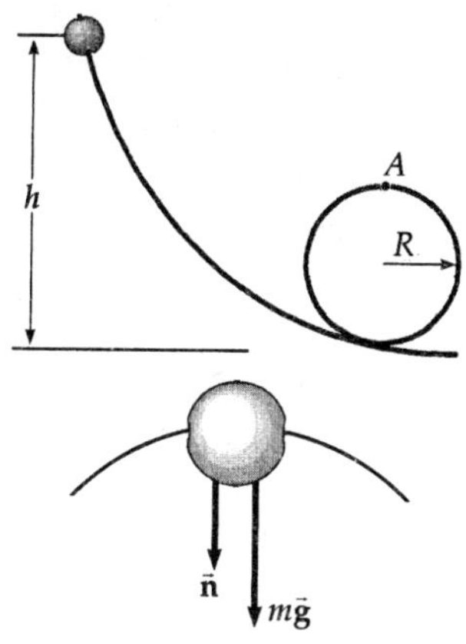

FIG. P7.5

P7.6 (a) Energy of the particle-Earth system is conserved as the particle moves between point P and the apex of the trajectory.

Since the horizontal component of velocity is constant,

$$\frac{1}{2}mv_i^2 = \frac{1}{2}mv_{ix}^2 + \frac{1}{2}mv_{iy}^2$$

$$= \frac{1}{2}mv_{ix}^2 + mgh$$

$$v_{iy} = \sqrt{2(9.80)(20.0)} = \boxed{19.8\text{ m/s}}$$

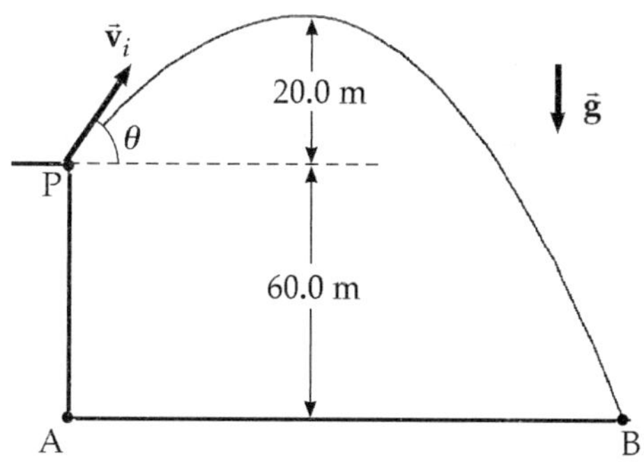

FIG. P7.6

(b) $\Delta K|_{P\to B} = W_g = mg(60.0\text{ m}) = (0.500\text{ kg})\left(9.80\text{ m/s}^2\right)(60.0\text{ m}) = \boxed{294\text{ J}}$

(c) Now let the final point be point B. $\quad v_{xi} = v_{xf} = 30.0\text{ m/s}$

$$\Delta K|_{P\to B} = \frac{1}{2}mv_{yf}^2 - \frac{1}{2}mv_{yi}^2 = 294\text{ J}$$

$$v_{yf}^2 = \frac{2}{m}(294) + v_{yi}^2 = 1\,176 + 392$$

$$v_{yf} = -39.6\text{ m/s}$$

$$\vec{\mathbf{v}}_B = \boxed{(30.0\text{ m/s})\hat{\mathbf{i}} - (39.6\text{ m/s})\hat{\mathbf{j}}}$$

P7.7 From leaving ground to the highest point, $\quad K_i + U_i = K_f + U_f$

$$\frac{1}{2}m(6.00\text{ m/s})^2 + 0 = 0 + m\left(9.80\text{ m/s}^2\right)y$$

The mass makes no difference: $\quad \therefore y = \frac{(6.00\text{ m/s})^2}{(2)\left(9.80\text{ m/s}^2\right)} = \boxed{1.84\text{ m}}$

P7.8 (a) Energy of the object-Earth system is conserved as the object moves between the release point and the lowest point. We choose to measure heights from $y=0$ at the top end of the string.

$$(K+U_g)_i=(K+U_g)_f: \qquad 0+mgy_i=\frac{1}{2}mv_f^2+mgy_f$$

$$(9.8\text{ m/s}^2)(-2\text{ m}\cos 30°)=\frac{1}{2}v_f^2+(9.8\text{ m/s}^2)(-2\text{ m})$$

$$v_f=\sqrt{2(9.8\text{ m/s}^2)(2\text{ m})(1-\cos 30°)}=\boxed{2.29\text{ m/s}}$$

(b) Choose the initial point at $\theta=30°$ and the final point at $\theta=15°$:

$$0+mg(-L\cos 30°)=\frac{1}{2}mv_f^2+mg(-L\cos 15°)$$

$$v_f=\sqrt{2gL(\cos 15°-\cos 30°)}=\sqrt{2(9.8\text{ m/s}^2)(2\text{ m})(\cos 15°-\cos 30°)}=\boxed{1.98\text{ m/s}}$$

P7.9 Using conservation of energy for the system of the Earth and the two objects

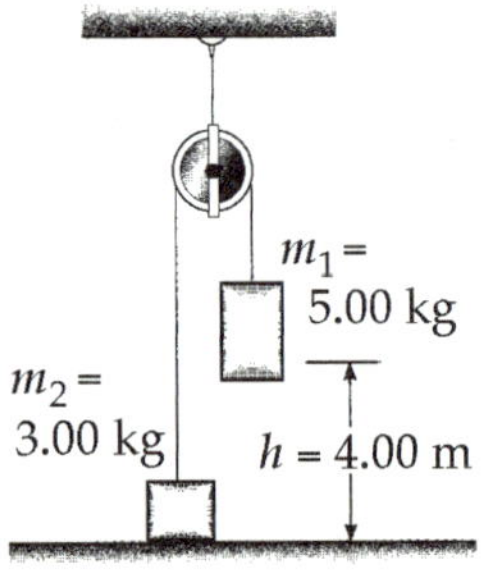

FIG. P7.9

(a) $$(5.00\text{ kg})g(4.00\text{ m})=(3.00\text{ kg})g(4.00\text{ m})+\frac{1}{2}(5.00+3.00)v^2$$

$$v=\sqrt{19.6}=\boxed{4.43\text{ m/s}}$$

(b) Now we apply conservation of energy for the system of the 3.00 kg object and the Earth during the time interval between the instant when the string goes slack and the instant at which the 3.00 kg object reaches its highest position in its free fall.

$$\frac{1}{2}(3.00)v^2=mg\,\Delta y=3.00g\Delta y$$

$$\Delta y=1.00\text{ m}$$

$$y_{\max}=4.00\text{ m}+\Delta y=\boxed{5.00\text{ m}}$$

***P7.10** (a) $$(\Delta K)_{A\to B}=\sum W=W_g=mg\Delta h=mg(5.00-3.20)$$

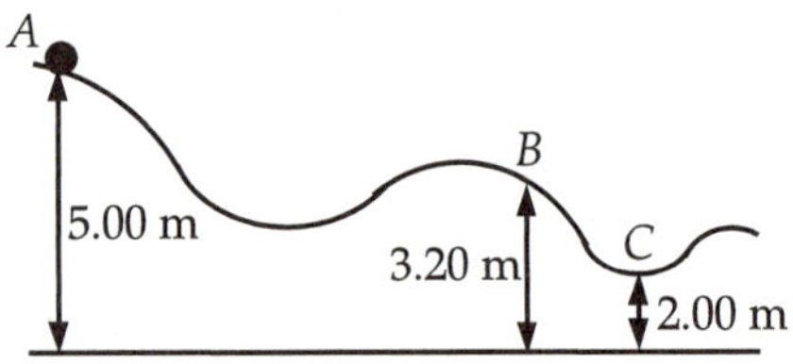

FIG. P7.10

$$\frac{1}{2}mv_B^2-\frac{1}{2}mv_A^2=m(9.80)(1.80)$$

$$v_B=\boxed{5.94\text{ m/s}}$$

Similarly, $v_C=\sqrt{v_A^2+2g(5.00-2.00)}=\boxed{7.67\text{ m/s}}$

(b) $$W_g\big|_{A\to C}=mg(3.00\text{ m})=\boxed{147\text{ J}}$$

P7.11 (a) The force needed to hang on is equal to the force F the trapeze bar exerts on the performer. From the free-body diagram for the performer's body, as shown, $F - mg\cos\theta = m\dfrac{v^2}{\ell}$ or $F = mg\cos\theta + m\dfrac{v^2}{\ell}$. Apply conservation of mechanical energy of the performer-Earth system as the performer moves between the starting point and any later point:

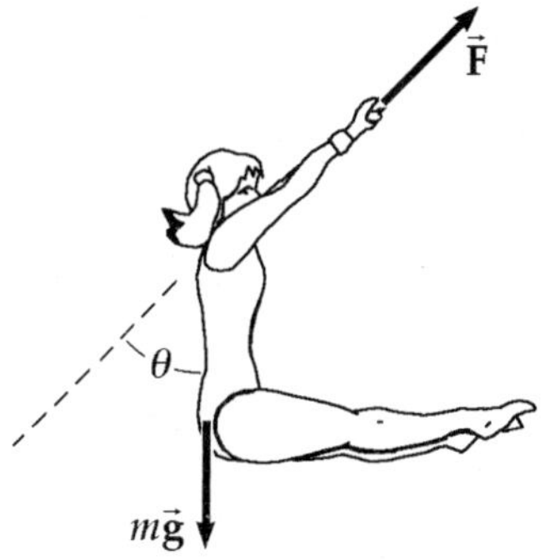

FIG. P7.11

$$mg(\ell - \ell\cos\theta_i) = mg(\ell - \ell\cos\theta) + \frac{1}{2}mv^2$$

Solve for $\dfrac{mv^2}{\ell}$ and substitute into the force equation to obtain $F = \boxed{mg(3\cos\theta - 2\cos\theta_i)}$.

(b) At the bottom of the swing, $\theta = 0°$ so

$$F = mg(3 - 2\cos\theta_i)$$
$$F = 2mg = mg(3 - 2\cos\theta_i)$$

which gives $\theta_i = \boxed{60.0°}$.

P7.12 The force of tension and subsequent force of compression in the rod do no work on the ball, since they are perpendicular to each step of displacement. Consider energy conservation of the ball-Earth system between the instant just after you strike the ball and the instant when it reaches the top. The speed at the top is zero if you hit it just hard enough to get it there.

$K_i + U_{gi} = K_f + U_{gf}$: $\quad \dfrac{1}{2}mv_i^2 + 0 = 0 + mg(2L)$

$$v_i = \sqrt{4gL} = \sqrt{4(9.80)(0.770)}$$
$$v_i = \boxed{5.49 \text{ m/s}}$$

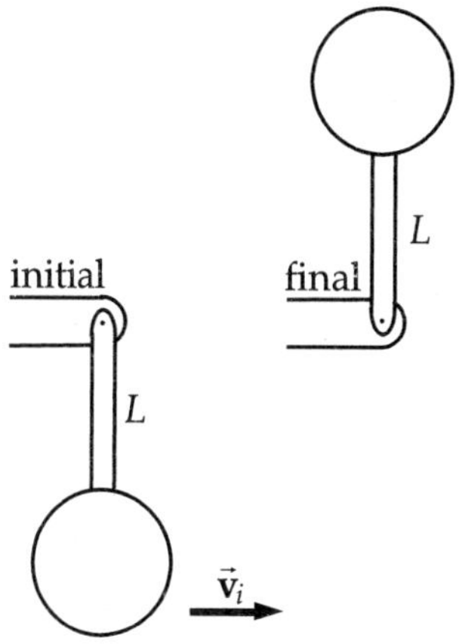

FIG. P7.12

***P7.13** (a) The moving sewage possesses kinetic energy in the same amount as it enters and leaves the pump. The work of the pump increases the gravitational energy of the sewage-Earth system. We take the equation $K_i + U_{gi} + W_{\text{pump}} = K_f + U_{gf}$, subtract out the K terms, and choose $U_{gi} = 0$ at the bottom of the sump, to obtain $W_{\text{pump}} = mgy_f$. Now we differentiate through with respect to time:

$$\mathscr{P}_{\text{pump}} = \frac{\Delta m}{\Delta t} g y_f = \rho \frac{\Delta V}{\Delta t} g y_f$$

$$= 1\,050 \text{ kg/m}^3 \left(1.89 \times 10^6 \text{ L/d}\right)\left(\frac{1 \text{ m}^3}{1\,000 \text{ L}}\right)\left(\frac{1 \text{ d}}{86\,400 \text{ s}}\right)\left(\frac{9.80 \text{ m}}{\text{s}^2}\right)(5.49 \text{ m}) = \boxed{1.24 \times 10^3 \text{ W}}$$

(b) $$\text{efficiency} = \frac{\text{useful output work}}{\text{total input work}} = \frac{\text{useful output work}/\Delta t}{\text{total input work}/\Delta t}$$

$$= \frac{\text{mechanical output power}}{\text{input electric power}} = \frac{1.24 \text{ kW}}{5.90 \text{ kW}} = \boxed{0.209} = 20.9\%$$

The remaining power, $5.90 - 1.24 \text{ kW} = 4.66 \text{ kW}$ is the rate at which internal energy is injected into the sewage and the surroundings of the pump.

Section 7.3 Conservative and Nonconservative Forces

P7.14 (a) $W = \int \vec{\mathbf{F}} \cdot d\vec{\mathbf{r}}$ and if the force is constant, this can be written as

$W = \vec{\mathbf{F}} \cdot \int d\vec{\mathbf{r}} = \boxed{\vec{\mathbf{F}} \cdot \left(\vec{\mathbf{r}}_f - \vec{\mathbf{r}}_i\right), \text{ which depends only on end points, not path.}}$

(b) $$W = \int \vec{\mathbf{F}} \cdot d\vec{\mathbf{r}} = \int \left(3\hat{\mathbf{i}} + 4\hat{\mathbf{j}}\right) \cdot \left(dx\hat{\mathbf{i}} + dy\hat{\mathbf{j}}\right) = (3.00 \text{ N}) \int_0^{5.00 \text{ m}} dx + (4.00 \text{ N}) \int_0^{5.00 \text{ m}} dy$$

$$W = (3.00 \text{ N}) x \Big|_0^{5.00 \text{ m}} + (4.00 \text{ N}) y \Big|_0^{5.00 \text{ m}} = 15.0 \text{ J} + 20.0 \text{ J} = \boxed{35.0 \text{ J}}$$

The same calculation applies for all paths.

P7.15 (a) $$W_{OA} = \int_0^{5.00 \text{ m}} dx\hat{\mathbf{i}} \cdot \left(2y\hat{\mathbf{i}} + x^2\hat{\mathbf{j}}\right) = \int_0^{5.00 \text{ m}} 2y\,dx$$

and since along this path, $y = 0$ $\quad W_{OA} = 0$

$$W_{AC} = \int_0^{5.00 \text{ m}} dy\hat{\mathbf{j}} \cdot \left(2y\hat{\mathbf{i}} + x^2\hat{\mathbf{j}}\right) = \int_0^{5.00 \text{ m}} x^2\,dy$$

For $x = 5.00$ m, $\quad W_{AC} = 125 \text{ J}$

and $\quad W_{OAC} = 0 + 125 = \boxed{125 \text{ J}}$

continued on next page

(b) $$W_{OB} = \int_0^{5.00\text{ m}} dy\hat{\mathbf{j}}\cdot\left(2y\hat{\mathbf{i}}+x^2\hat{\mathbf{j}}\right) = \int_0^{5.00\text{ m}} x^2 dy$$

since along this path, $x=0$, $$W_{OB}=0$$

$$W_{BC} = \int_0^{5.00\text{ m}} dx\hat{\mathbf{i}}\cdot\left(2y\hat{\mathbf{i}}+x^2\hat{\mathbf{j}}\right) = \int_0^{5.00\text{ m}} 2y dx$$

since $y=5.00$ m, $$W_{BC}=50.0\text{ J}$$

$$W_{OBC}=0+50.0=\boxed{50.0\text{ J}}$$

(c) $$W_{OC}=\int\left(dx\hat{\mathbf{i}}+dy\hat{\mathbf{j}}\right)\cdot\left(2y\hat{\mathbf{i}}+x^2\hat{\mathbf{j}}\right)=\int\left(2ydx+x^2dy\right)$$

Since $x=y$ along OC, $$W_{OC}=\int_0^{5.00\text{ m}}\left(2x+x^2\right)dx=\boxed{66.7\text{ J}}$$

(d) F is $\boxed{\text{nonconservative}}$ since the work done is path dependent.

P7.16 Choose the zero point of gravitational potential energy of the object-spring-Earth system as the configuration in which the object comes to rest. Then because the incline is frictionless, we have $E_B=E_A$: $K_B+U_{gB}+U_{sB}=K_A+U_{gA}+U_{sA}$ or $0+mg(d+x)\sin\theta+0=0+0+\frac{1}{2}kx^2$. Solving for d gives $d=\boxed{\dfrac{kx^2}{2mg\sin\theta}-x}$.

P7.17 From conservation of energy for the block-spring-Earth system,

$$U_{gt}=U_{si},$$

or

$$(0.250\text{ kg})\left(9.80\text{ m/s}^2\right)h=\left(\frac{1}{2}\right)(5\,000\text{ N/m})(0.100\text{ m})^2$$

This gives a maximum height $h=\boxed{10.2\text{ m}}$.

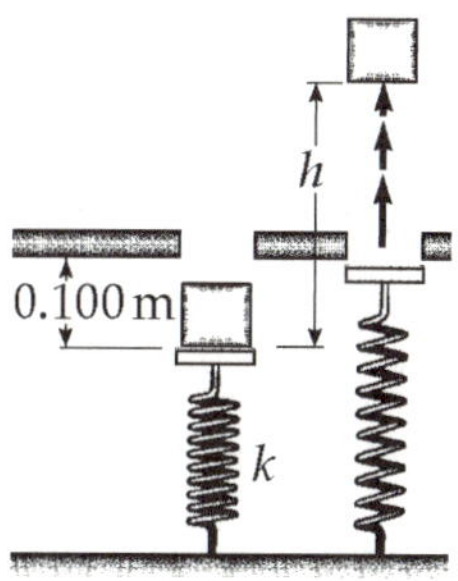

FIG. P7.17

***P7.18** (a) For a 5-m cord the spring constant is described by $F = kx$, $mg = k(1.5 \text{ m})$. For a longer cord of length L the stretch distance is longer so the spring constant is smaller in inverse proportion:

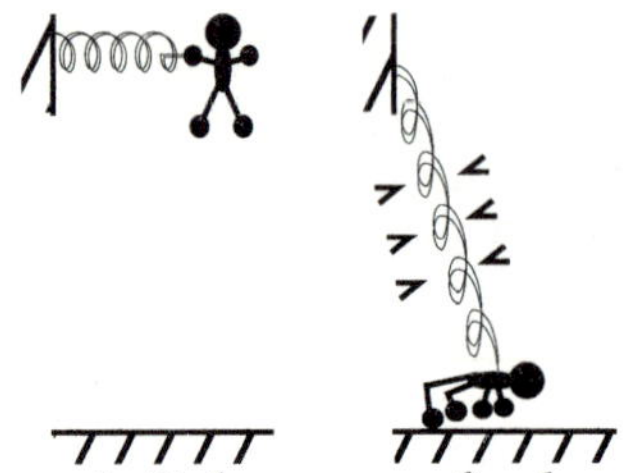

FIG. P7.18

$$k = \frac{5 \text{ m}}{L}\frac{mg}{1.5 \text{ m}} = 3.33\, mg/L$$

$$(K + U_g + U_s)_i = (K + U_g + U_s)_f$$

$$0 + mgy_i + 0 = 0 + mgy_f + \frac{1}{2}kx_f^2$$

$$mg(y_i - y_f) = \frac{1}{2}kx_f^2 = \frac{1}{2}3.33\frac{mg}{L}x_f^2$$

here $y_i - y_f = 55 \text{ m} = L + x_f$

$$55.0 \text{ m}L = \frac{1}{2}3.33(55.0 \text{ m} - L)^2$$

$$55.0 \text{ m}L = 5.04 \times 10^3 \text{ m}^2 - 183 \text{ m}L + 1.67L^2$$

$$0 = 1.67L^2 - 238L + 5.04 \times 10^3 = 0$$

$$L = \frac{238 \pm \sqrt{238^2 - 4(1.67)(5.04 \times 10^3)}}{2(1.67)} = \frac{238 \pm 152}{3.33} = \boxed{25.8 \text{ m}}$$

only the value of L less than 55 m is physical.

(b) $k = 3.33\dfrac{mg}{25.8 \text{ m}}$ $\qquad x_{max} = x_f = 55.0 \text{ m} - 25.8 \text{ m} = 29.2 \text{ m}$

$\sum F = ma$ $\qquad +kx_{max} - mg = ma$

$$3.33\frac{mg}{25.8 \text{ m}}29.2 \text{ m} - mg = ma$$

$$a = 2.77g = \boxed{27.1 \text{ m/s}^2}$$

P7.19 (a) $U_f = K_i - K_f + U_i$ $\qquad U_f = 30.0 - 18.0 + 10.0 = \boxed{22.0 \text{ J}}$

$$\boxed{E = 40.0 \text{ J}}$$

(b) Yes, $\Delta E_{mech} = \Delta K + \Delta U$ is not equal to zero. For conservative forces $\Delta K + \Delta U = 0$.

***P7.20** (a) Simplified, the equation is $0 = (9\,700\ \text{N/m})x^2 - (450.8\ \text{N})x - 1\,395\ \text{N}\cdot\text{m}$. Then

$$x = \frac{-b \pm \sqrt{b^2 - 4ac}}{2a} = \frac{450.8\ \text{N} \pm \sqrt{(450.8\ \text{N})^2 - 4(9\,700\ \text{N/m})(-1\,395\ \text{N}\cdot\text{m})}}{2(9\,700\ \text{N/m})}$$

$$= \frac{450.8\ \text{N} \pm 7\,370\ \text{N}}{19\,400\ \text{N/m}} = \boxed{0.403\ \text{m or } -0.357\ \text{m}}$$

(b) One possible problem statement: From a perch at a height of 2.80 m above the top of the pile of mattresses, a 46.0-kg child jumps nearly straight upward with speed 2.40 m/s. The mattresses behave as a linear spring with force constant 19.4 kN/m. Find the maximum amount by which they are compressed when the child lands on them. Physical meaning: The positive value of x represents the maximum spring compression. The negative value represents the maximum extension of the equivalent spring if the child sticks to the top of the mattress pile as he rebounds upward without friction.

P7.21 The distance traveled by the ball from the top of the arc to the bottom is πR. The work done by the non-conservative force, the force exerted by the pitcher, is $\Delta E = F\Delta r\cos 0° = F(\pi R)$. We shall assign the gravitational energy of the ball-Earth system to be zero with the ball at the bottom of the arc. Then $\Delta E_{\text{mech}} = \frac{1}{2}mv_f^2 - \frac{1}{2}mv_i^2 + mgy_f - mgy_i$ becomes $\frac{1}{2}mv_f^2 = \frac{1}{2}mv_i^2 + mgy_i + F(\pi R)$ or

$$v_f = \sqrt{v_i^2 + 2gy_i + \frac{2F(\pi R)}{m}} = \sqrt{(15.0)^2 + 2(9.80)(1.20) + \frac{2(30.0)\pi(0.600)}{0.250}}$$

$$v_f = \boxed{26.5\ \text{m/s}}$$

***P7.22** (a) Maximum speed occurs after the needle leaves the spring, before it enters the body.

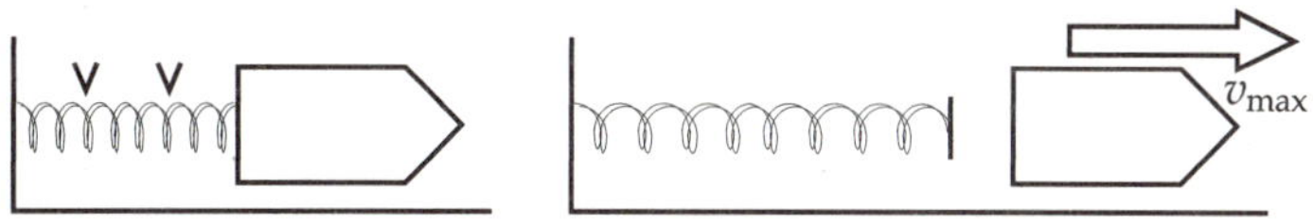

FIG. P7.22(a)

$$K_i + U_i - f_k d = K_f + U_f.$$

We assume the needle is fired horizontally

$$0 + \frac{1}{2}kx^2 - 0 = \frac{1}{2}mv_{\max}^2 + 0$$

$$\frac{1}{2}(375\ \text{N/m})(0.081\ \text{m})^2 = \frac{1}{2}(0.005\,6\ \text{kg})v_{\max}^2$$

$$\left(\frac{2(1.23\ \text{J})}{0.005\,6\ \text{kg}}\right)^{1/2} = v_{\max} = \boxed{21.0\ \text{m/s}}$$

continued on next page

(b) The same energy of 1.23 J as in part (a) now becomes partly internal energy in the soft tissue, partly internal energy in the organ, and partly kinetic energy of the needle just before it runs into the stop. We make up a work-energy equation to describe this process:

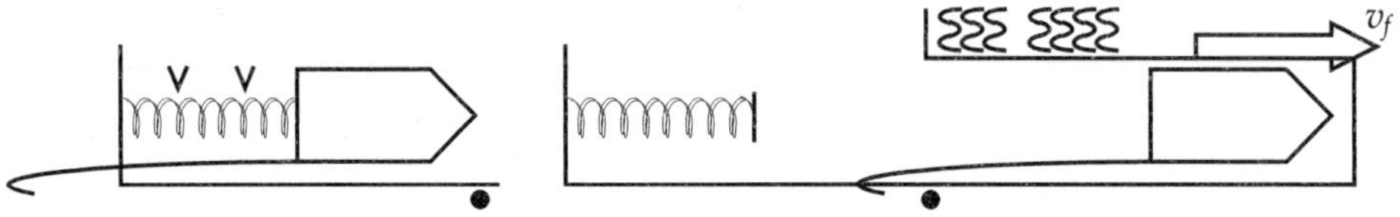

FIG. P7.22(b)

$$K_i + U_i - f_{k1}d_1 - f_{k2}d_2 = K_f + U_f$$

$$1 + \frac{1}{2}kx^2 - f_{k1}d_1 - f_{k2}d_2 = \frac{1}{2}mv_f^2 + 0$$

$$1.23\text{ J} - 7.60\text{ N}(0.024\text{ m}) - 9.20\text{ N}(0.035\text{ m}) = \frac{1}{2}(0.005\,6\text{ kg})v_f^2$$

$$\left(\frac{2(1.23\text{ J} - 0.182\text{ J} - 0.322\text{ J})}{0.005\,6\text{ kg}}\right)^{1/2} = v_f = \boxed{16.1\text{ m/s}}$$

P7.23 $U_i + K_i + \Delta E_{\text{mech}} = U_f + K_f$: $\quad m_2gh - fh = \frac{1}{2}m_1v^2 + \frac{1}{2}m_2v^2$

$$f = \mu n = \mu m_1 g$$

$$m_2gh - \mu m_1 gh = \frac{1}{2}(m_1 + m_2)v^2$$

$$v^2 = \frac{2(m_2 - \mu m_1)(hg)}{m_1 + m_2}$$

$$v = \sqrt{\frac{2(9.80\text{ m/s}^2)(1.50\text{ m})[5.00\text{ kg} - 0.400(3.00\text{ kg})]}{8.00\text{ kg}}} = \boxed{3.74\text{ m/s}}$$

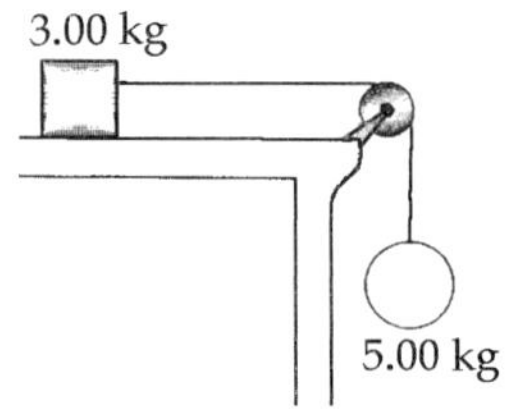

FIG. P7.23

P7.24 We begin with Equation 6.20 for the isolated child-wheelchair-Earth system,

$$\Delta K + \Delta U + \Delta E_{\text{int}} = 0 \qquad (1)$$

We recognize that the change in potential energy is of two types: gravitational and chemical potential energy stored in the body of the child from past meals. We also recognize that the change in internal energy will be due to the friction force (air resistance and rolling resistance) as the child rolls down the hill. Therefore, we write (1) as,

$$\Delta K + \Delta U_g + \Delta U_{\text{body}} + f_k d = 0 \qquad (2)$$

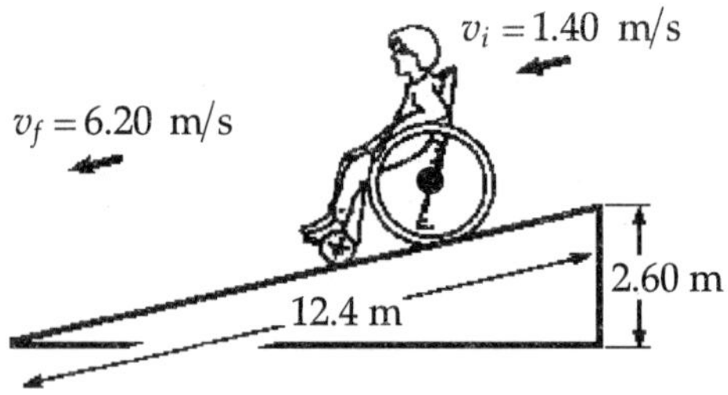

FIG. P7.24

continued on next page

The change in potential energy in the body of the child will be due to work done within the system by the child on the wheels of the chair. Consequently, the requested answer, the work done by the child, is $W_{\text{child}} = -\Delta U_{\text{body}}$. Therefore, (2) can be expressed as,

$$\begin{aligned} W_{\text{child}} &= \Delta K + \Delta U_g + f_k d = \left(\frac{1}{2}mv_f^2 - \frac{1}{2}mv_i^2\right) + \left(mgh_f - mgh_i\right) + f_k d \\ &= \frac{1}{2}m\left(v_f^2 - v_i^2\right) + mg\left(h_f - h_i\right) + f_k d \\ &= \frac{1}{2}(47.0 \text{ kg})\left[(6.20 \text{ m/s})^2 - (1.40 \text{ m/s})^2\right] + (47.0 \text{ kg})\left(9.80 \text{ m/s}^2\right)(0 - 2.60 \text{ m}) + (41.0 \text{ N})(12.4 \text{ m}) \\ &= \boxed{168 \text{ J}} \end{aligned}$$

P7.25 (a) $\Delta K = \frac{1}{2}m\left(v_f^2 - v_i^2\right) = -\frac{1}{2}mv_i^2 = \boxed{-160 \text{ J}}$

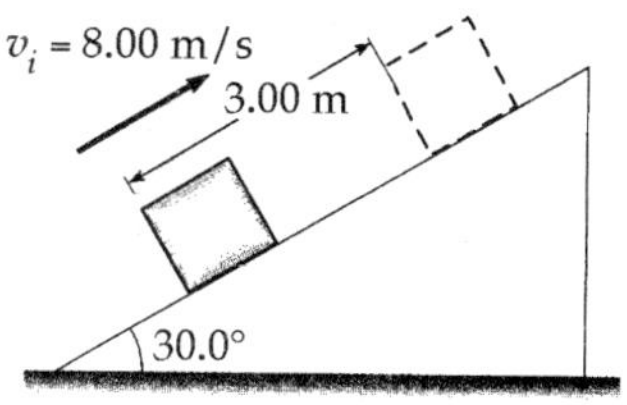

FIG. P7.25

(b) $\Delta U = mg(3.00 \text{ m})\sin 30.0° = \boxed{73.5 \text{ J}}$

(c) The mechanical energy converted due to friction is 86.5 J

$$f = \frac{86.5 \text{ J}}{3.00 \text{ m}} = \boxed{28.8 \text{ N}}$$

(d) $f = \mu_k n = \mu_k mg\cos 30.0° = 28.8 \text{ N}$

$$\mu_k = \frac{28.8 \text{ N}}{(5.00 \text{ kg})\left(9.80 \text{ m/s}^2\right)\cos 30.0°} = \boxed{0.679}$$

P7.26 Consider the whole motion: $K_i + U_i + \Delta E_{\text{mech}} = K_f + U_f$

(a) $0 + mgy_i - f_1\Delta x_1 - f_2\Delta x_2 = \frac{1}{2}mv_f^2 + 0$

$$(80.0 \text{ kg})\left(9.80 \text{ m/s}^2\right)1\,000 \text{ m} - (50.0 \text{ N})(800 \text{ m}) - (3\,600 \text{ N})(200 \text{ m}) = \frac{1}{2}(80.0 \text{ kg})v_f^2$$

$$784\,000 \text{ J} - 40\,000 \text{ J} - 720\,000 \text{ J} = \frac{1}{2}(80.0 \text{ kg})v_f^2$$

$$v_f = \sqrt{\frac{2(24\,000 \text{ J})}{80.0 \text{ kg}}} = \boxed{24.5 \text{ m/s}}$$

(b) $\boxed{\text{Yes}}$ this is too fast for safety.

(c) Now in the same energy equation as in part (a), Δx_2 is unknown, and $\Delta x_1 = 1\,000 \text{ m} - \Delta x_2$:

$$784\,000 \text{ J} - (50.0 \text{ N})(1\,000 \text{ m} - \Delta x_2) - (3\,600 \text{ N})\Delta x_2 = \frac{1}{2}(80.0 \text{ kg})(5.00 \text{ m/s})^2$$

$$784\,000 \text{ J} - 50\,000 \text{ J} - (3\,550 \text{ N})\Delta x_2 = 1\,000 \text{ J}$$

$$\Delta x_2 = \frac{733\,000 \text{ J}}{3\,550 \text{ N}} = \boxed{206 \text{ m}}$$

continued on next page

(d) Really the air drag will depend on the skydiver's speed. It will be larger than her 784 N weight only after the chute is opened. It will be nearly equal to 784 N before she opens the chute and again before she touches down, whenever she moves near terminal speed.

P7.27 (a) $(K+U)_i + \Delta E_{\text{mech}} = (K+U)_f$:

$$0+\frac{1}{2}kx^2 - f\Delta x = \frac{1}{2}mv^2 + 0$$

$$\frac{1}{2}(8.00\text{ N/m})\left(5.00\times10^{-2}\text{ m}\right)^2 - \left(3.20\times10^{-2}\text{ N}\right)(0.150\text{ m}) = \frac{1}{2}\left(5.30\times10^{-3}\text{ kg}\right)v^2$$

$$v = \sqrt{\frac{2\left(5.20\times10^{-3}\text{ J}\right)}{5.30\times10^{-3}\text{ kg}}} = \boxed{1.40\text{ m/s}}$$

(b) When the spring force just equals the friction force, the ball will stop speeding up. Here $\left|\vec{\mathbf{F}}_s\right| = kx$; the spring is compressed by

$$\frac{3.20\times10^{-2}\text{ N}}{8.00\text{ N/m}} = 0.400\text{ cm}$$

and the ball has moved

$$5.00\text{ cm} - 0.400\text{ cm} = \boxed{4.60\text{ cm from the start.}}$$

(c) Between start and maximum speed points,

$$\frac{1}{2}kx_i^2 - f\Delta x = \frac{1}{2}mv^2 + \frac{1}{2}kx_f^2$$

$$\frac{1}{2}8.00\left(5.00\times10^{-2}\right)^2 - \left(3.20\times10^{-2}\right)\left(4.60\times10^{-2}\right) = \frac{1}{2}\left(5.30\times10^{-3}\right)v^2 + \frac{1}{2}8.00\left(4.00\times10^{-3}\right)^2$$

$$v = \boxed{1.79\text{ m/s}}$$

P7.28

$$\sum F_y = n - mg\cos 37.0° = 0$$

$$\therefore n = mg\cos 37.0° = 391\text{ N}$$

$$f = \mu n = 0.250(391\text{ N}) = 97.8\text{ N}$$

$$-f\Delta x = \Delta E_{\text{mech}}$$

$$(-97.8)(20.0) = \Delta U_A + \Delta U_B + \Delta K_A + \Delta K_B$$

$$\Delta U_A = m_A g\left(h_f - h_i\right) = (50.0)(9.80)(20.0\sin 37.0°) = 5.90\times10^3$$

$$\Delta U_B = m_B g\left(h_f - h_i\right) = (100)(9.80)(-20.0) = -1.96\times10^4$$

$$\Delta K_A = \frac{1}{2}m_A\left(v_f^2 - v_i^2\right)$$

$$\Delta K_B = \frac{1}{2}m_B\left(v_f^2 - v_i^2\right) = \frac{m_B}{m_A}\Delta K_A = 2\Delta K_A$$

Adding and solving, $\Delta K_A = \boxed{3.92\text{ kJ}}$.

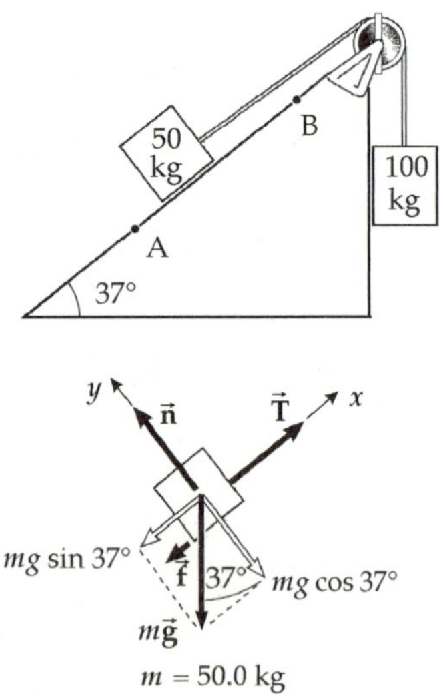

FIG. P7.28

P7.29 (a) The object moved down distance $1.20\text{ m}+x$. Choose $y=0$ at its lower point.

$$K_i+U_{gi}+U_{si}+\Delta E_{\text{mech}}=K_f+U_{gf}+U_{sf}$$
$$0+mgy_i+0+0=0+0+\frac{1}{2}kx^2$$
$$(1.50\text{ kg})(9.80\text{ m/s}^2)(1.20\text{ m}+x)=\frac{1}{2}(320\text{ N/m})x^2$$
$$0=(160\text{ N/m})x^2-(14.7\text{ N})x-17.6\text{ J}$$
$$x=\frac{14.7\text{ N}\pm\sqrt{(-14.7\text{ N})^2-4(160\text{ N/m})(-17.6\text{ N}\cdot\text{m})}}{2(160\text{ N/m})}$$
$$x=\frac{14.7\text{ N}\pm 107\text{ N}}{320\text{ N/m}}$$

The negative root tells how high the object will rebound if it is instantly glued to the spring. We want

$$x=\boxed{0.381\text{ m}}$$

(b) From the same equation,

$$(1.50\text{ kg})(1.63\text{ m/s}^2)(1.20\text{ m}+x)=\frac{1}{2}(320\text{ N/m})x^2$$
$$0=160x^2-2.44x-2.93$$

The positive root is $x=\boxed{0.143\text{ m}}$.

(c) The equation expressing the energy version of the nonisolated system model has one more term:

$$mgy_i-f\Delta x=\frac{1}{2}kx^2$$
$$(1.50\text{ kg})(9.80\text{ m/s}^2)(1.20\text{ m}+x)-0.700\text{ N}(1.20\text{ m}+x)=\frac{1}{2}(320\text{ N/m})x^2$$
$$17.6\text{ J}+14.7\text{ N}x-0.840\text{ J}-0.700\text{ N}x=160\text{ N/m}\,x^2$$
$$160x^2-14.0x-16.8=0$$
$$x=\frac{14.0\pm\sqrt{(14.0)^2-4(160)(-16.8)}}{320}$$
$$x=\boxed{0.371\text{ m}}$$

P7.30 The total mechanical energy of the skysurfer-Earth system is

$$E_{\text{mech}} = K + U_g = \frac{1}{2}mv^2 + mgh.$$

Since the skysurfer has constant speed,

$$\frac{dE_{\text{mech}}}{dt} = mv\frac{dv}{dt} + mg\frac{dh}{dt} = 0 + mg(-v) = -mgv.$$

The rate the system is losing mechanical energy is then

$$\left|\frac{dE_{\text{mech}}}{dt}\right| = mgv = (75.0\text{ kg})(9.80\text{ m/s}^2)(60.0\text{ m/s}) = \boxed{44.1\text{ kW}}.$$

Section 7.4 Conservative Forces and Potential Energy

P7.31 (a) $W = \int F_x dx = \int_1^{5.00\text{ m}} (2x+4)dx = \left(\frac{2x^2}{2} + 4x\right)\Bigg|_1^{5.00\text{ m}} = 25.0 + 20.0 - 1.00 - 4.00 = \boxed{40.0\text{ J}}$

(b) $\Delta K + \Delta U = 0 \qquad \Delta U = -\Delta K = -W = \boxed{-40.0\text{ J}}$

(c) $\Delta K = K_f - \frac{mv_1^2}{2} \qquad K_f = \Delta K + \frac{mv_1^2}{2} = \boxed{62.5\text{ J}}$

P7.32 (a) $U = -\int_0^x \left(-Ax + Bx^2\right)dx = \boxed{\frac{Ax^2}{2} - \frac{Bx^3}{3}}$

(b) $\Delta U = -\int_{2.00\text{ m}}^{3.00\text{ m}} F dx = \frac{A\left[(3.00^2) - (2.00)^2\right]}{2} - \frac{B\left[(3.00)^3 - (2.00)^3\right]}{3} = \boxed{\frac{5.00}{2}A - \frac{19.0}{3}B}$

$\Delta K = \boxed{\left(-\frac{5.00}{2}A + \frac{19.0}{3}B\right)}$

P7.33 $U(r) = \frac{A}{r}$

$F_r = -\frac{\partial U}{\partial r} = -\frac{d}{dr}\left(\frac{A}{r}\right) = \boxed{\frac{A}{r^2}}$. The positive value indicates a force of repulsion if A is positive.

P7.34 $F_x = -\frac{\partial U}{\partial x} = -\frac{\partial\left(3x^3y - 7x\right)}{\partial x} = -\left(9x^2y - 7\right) = 7 - 9x^2y$

$F_y = -\frac{\partial U}{\partial y} = -\frac{\partial\left(3x^3y - 7x\right)}{\partial y} = -\left(3x^3 - 0\right) = -3x^3$

Thus, the force acting at the point (x, y) is $\vec{\mathbf{F}} = F_x\hat{\mathbf{i}} + F_y\hat{\mathbf{j}} = \boxed{\left(7 - 9x^2y\right)\hat{\mathbf{i}} - 3x^3\hat{\mathbf{j}}}$.

Section 7.5 The Nonisolated System in Steady State

No problems in this section

Section 7.6 Potential Energy for Gravitational and Electric Forces

Note: Problems 20.10, 20.13, 20.18, 20.19, 20.20, and 20.62 in Chapter 20 can be assigned with Section 7.6.

P7.35 (a) $$U=-\frac{GM_Em}{r}=-\frac{\left(6.67\times10^{-11}\ \text{N}\cdot\text{m}^2/\text{kg}^2\right)\left(5.98\times10^{24}\ \text{kg}\right)(100\ \text{kg})}{(6.37+2.00)\times10^6\ \text{m}}=\boxed{-4.77\times10^9\ \text{J}}$$

(b), (c) Planet and satellite exert forces of equal magnitude on each other, directed downward on the satellite and upward on the planet.

$$F=\frac{GM_Em}{r^2}=\frac{\left(6.67\times10^{-11}\ \text{N}\cdot\text{m}^2/\text{kg}^2\right)\left(5.98\times10^{24}\ \text{kg}\right)(100\ \text{kg})}{\left(8.37\times10^6\ \text{m}\right)^2}=\boxed{569\ \text{N}}$$

P7.36 $U=-G\dfrac{Mm}{r}$ and $g=\dfrac{GM_E}{R_E{}^2}$ so that

$$\Delta U=-GMm\left(\frac{1}{3R_E}-\frac{1}{R_E}\right)=\frac{2}{3}mgR_E$$

$$\Delta U=\frac{2}{3}(1000\ \text{kg})\left(9.80\ \text{m/s}^2\right)\left(6.37\times10^6\ \text{m}\right)=\boxed{4.17\times10^{10}\ \text{J}}$$

P7.37 The height attained is not small compared to the radius of the Earth, so $U=mgy$ does not apply; $U=-\dfrac{GM_1M_2}{r}$ does. From launch to apogee at height h,

$K_i+U_i+\Delta E_{\text{mech}}=K_f+U_f$:

$$\frac{1}{2}M_pv_i{}^2-\frac{GM_EM_p}{R_E}+0=0-\frac{GM_EM_p}{R_E+h}$$

$$\frac{1}{2}\left(10.0\times10^3\ \text{m/s}\right)^2-\left(6.67\times10^{-11}\ \text{N}\cdot\text{m}^2/\text{kg}^2\right)\left(\frac{5.98\times10^{24}\ \text{kg}}{6.37\times10^6\ \text{m}}\right)$$

$$=-\left(6.67\times10^{-11}\ \text{N}\cdot\text{m}^2/\text{kg}^2\right)\left(\frac{5.98\times10^{24}\ \text{kg}}{6.37\times10^6\ \text{m}+h}\right)$$

$$\left(5.00\times10^7\ \text{m}^2/\text{s}^2\right)-\left(6.26\times10^7\ \text{m}^2/\text{s}^2\right)=\frac{-3.99\times10^{14}\ \text{m}^3/\text{s}^2}{6.37\times10^6\ \text{m}+h}$$

$$6.37\times10^6\ \text{m}+h=\frac{3.99\times10^{14}\ \text{m}^3/\text{s}^2}{1.26\times10^7\ \text{m}^2/\text{s}^2}=3.16\times10^7\ \text{m}$$

$$\boxed{h=2.52\times10^7\ \text{m}}$$

P7.38 (a) $U_{\text{Tot}} = U_{12} + U_{13} + U_{23} = 3U_{12} = 3\left(-\dfrac{Gm_1m_2}{r_{12}}\right)$

$$U_{\text{Tot}} = -\frac{3\left(6.67\times10^{-11}\ \text{N}\cdot\text{m}^2/\text{kg}^2\right)\left(5.00\times10^{-3}\ \text{kg}\right)^2}{0.300\ \text{m}} = \boxed{-1.67\times10^{-14}\ \text{J}}$$

(b) $\boxed{\text{At the center}}$ of the equilateral triangle

---

Section 7.7 Energy Diagrams and Stability of Equilibrium

P7.39 (a) F_x is zero at points A, C and E; F_x is positive at point B and negative at point D.

(b) A and E are unstable, and C is stable.

(c)

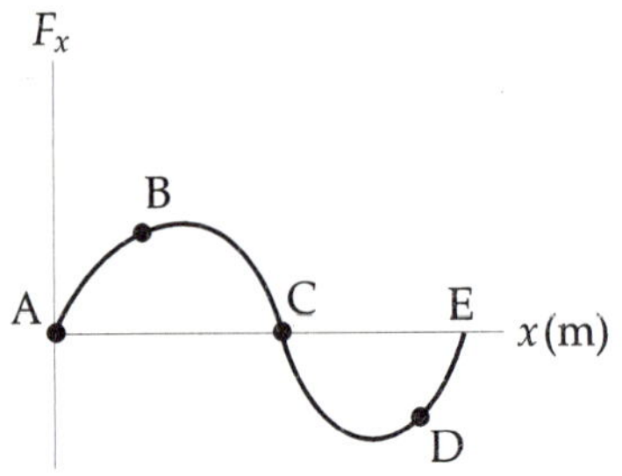

FIG. P7.39(c)

P7.40 (a) There is an equilibrium point wherever the graph of potential energy is horizontal:

At $r = 1.5$ mm and 3.2 mm, the equilibrium is stable.
At $r = 2.3$ mm, the equilibrium is unstable.
A particle moving out toward $r \to \infty$ approaches neutral equilibrium.

(b) The system energy E cannot be less than –5.6 J. The particle is bound if $\boxed{-5.6\ \text{J} \le E < 1\ \text{J}}$.

(c) If the system energy is –3 J, its potential energy must be less than or equal to –3 J. Thus, the particle's position is limited to $\boxed{0.6\ \text{mm} \le r \le 3.6\ \text{mm}}$.

(d) $K + U = E$. Thus, $K_{\text{max}} = E - U_{\text{min}} = -3.0\ \text{J} - (-5.6\ \text{J}) = \boxed{2.6\ \text{J}}$.

(e) Kinetic energy is a maximum when the potential energy is a minimum, at $\boxed{r = 1.5\ \text{mm}}$.

(f) $-3\ \text{J} + W = 1\ \text{J}$. Hence, the binding energy is $W = \boxed{4\ \text{J}}$.

P7.41 (a) When the mass moves distance x, the length of each spring changes from L to $\sqrt{x^2+L^2}$, so each exerts force $k\left(\sqrt{x^2+L^2}-L\right)$ towards its fixed end. The y-components cancel out and the x components add to:

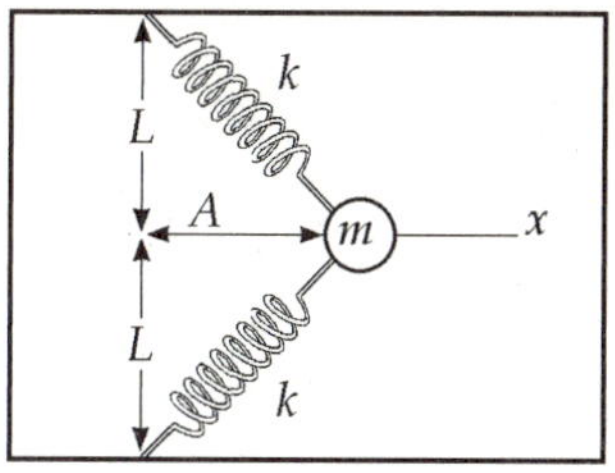

FIG. P7.41(a)

$$F_x = -2k\left(\sqrt{x^2+L^2}-L\right)\left(\frac{x}{\sqrt{x^2+L^2}}\right) = -2kx + \frac{2kLx}{\sqrt{x^2+L^2}}$$

Choose $U=0$ at $x=0$. Then at any point the potential energy of the system is

$$U(x) = -\int_0^x F_x dx = -\int_0^x \left(-2kx + \frac{2kLx}{\sqrt{x^2+L^2}}\right)dx = 2k\int_0^x x\,dx - 2kL\int_0^x \frac{x}{\sqrt{x^2+L^2}}dx$$

$$U(x) = \boxed{kx^2 + 2kL\left(L-\sqrt{x^2+L^2}\right)}$$

(b) $U(x) = 40.0x^2 + 96.0\left(1.20-\sqrt{x^2+1.44}\right)$

For negative x, $U(x)$ has the same value as for positive x. The only equilibrium point (i.e., where $F_x=0$) is $\boxed{x=0}$.

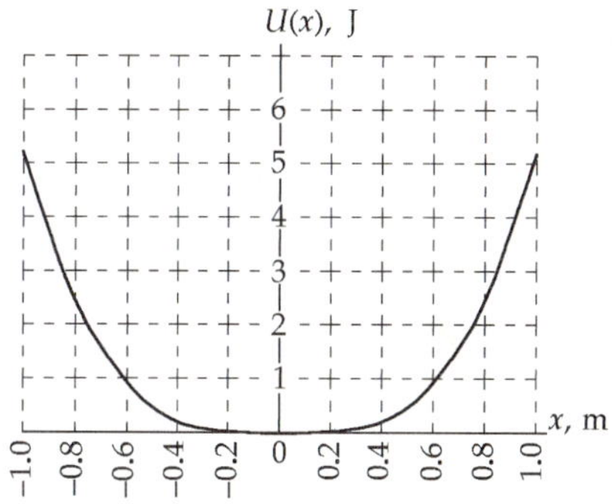

FIG. P7.41(b)

(c) $K_i + U_i + \Delta E_{\text{mech}} = K_f + U_f$

$$0 + 0.400\text{ J} + 0 = \frac{1}{2}(1.18\text{ kg})v_f^2 + 0$$

$$v_f = \boxed{0.823\text{ m/s}}$$

Section 7.8 Context Connection—Potential Energy in Fuels

***P7.42** $v = 100\text{ km/h} = 27.8\text{ m/s}$

The retarding force due to air resistance is

$$R = \frac{1}{2}D\rho Av^2 = \frac{1}{2}(0.330)(1.20\text{ kg/m}^3)(2.50\text{ m}^2)(27.8\text{ m/s})^2 = 382\text{ N}$$

Comparing the energy of the car at two points along the hill,

$$K_i + U_{gi} + \Delta E = K_f + U_{gf}$$

or $$K_i + U_{gi} + \Delta W_e - R(\Delta s) = K_f + U_{gf}$$

where ΔW_e is the work input from the engine. Thus,

$$\Delta W_e = R(\Delta s) + (K_f - K_i) + (U_{gf} - U_{gi})$$

continued on next page

Recognizing that $K_f = K_i$ and dividing by the travel time Δt gives the required power input from the engine as

$$\mathcal{P} = \left(\frac{\Delta W_e}{\Delta t}\right) = R\left(\frac{\Delta s}{\Delta t}\right) + mg\left(\frac{\Delta y}{\Delta t}\right) = Rv + mgv\sin\theta$$

$$\mathcal{P} = (382\text{ N})(27.8\text{ m/s}) + (1\,500\text{ kg})(9.80\text{ m/s}^2)(27.8\text{ m/s})\sin 3.20°$$

$$\mathcal{P} = \boxed{33.4\text{ kW} = 44.8\text{ hp}}$$

***P7.43** The energy per mass values are:

gasoline: 44 MJ/kg

hydrogen: 142 MJ/kg, larger by $\frac{142}{44} = 3.2$ times

hay: 17 MJ/kg, smaller than gasoline by $\frac{44}{17} = 2.6$ times

battery: $\frac{\text{energy}}{\text{mass}} = \frac{\mathcal{P}\Delta t}{m} = \frac{1\,200\text{ W}(3\,600\text{ s})}{16\text{ kg}} = 0.27\text{ MJ/kg}$, smaller than hay by $\frac{17}{0.27} = 63$ times.

Thus in the order requested, the energy densities are

0.27 MJ/kg for a battery
17 MJ/kg for hay is 63 times larger
44 MJ/kg for gasoline is 2.6 times larger still
142 MJ/kg for hydrogen is 3.2 times larger still

***P7.44** (a) $\frac{\mathcal{P}}{\text{area}} = \frac{\text{energy}}{\Delta t\cdot\text{area}} = \frac{600\text{ kWh}}{(30\text{ d})(13\text{ m})(9.5\text{ m})} = \frac{600\times10^3(\text{J/s})\text{h}}{30\text{ d}(123.5\text{ m}^2)}\left(\frac{1\text{ d}}{24\text{ h}}\right) = \boxed{6.75\text{ W/m}^2}$

(b) The car uses gasoline at the rate $55\text{ mi/h}\left(\frac{\text{gal}}{25\text{ mi}}\right)$. Its rate of energy conversion is

$\mathcal{P} = 44\times10^6\text{ J/kg}\left(\frac{2.54\text{ kg}}{1\text{ gal}}\right)(55\text{ mi/h})\left(\frac{\text{gal}}{25\text{ mi}}\right)\left(\frac{1\text{ h}}{3\,600\text{ s}}\right) = 6.83\times10^4\text{ W}$. Its power-per-footprint-area is $\frac{\mathcal{P}}{\text{area}} = \frac{6.83\times10^4\text{ W}}{2.10\text{ m}(4.90\text{ m})} = \boxed{6.64\times10^3\text{ W/m}^2}$.

(c) For an automobile of typical weight and power to run on sunlight, it would have to carry a solar panel huge compared to its own size. Rather than running a conventional car, it is much more natural to use solar energy for agriculture, lighting, space heating, water heating, and small appliances.

Additional Problems

P7.45 At a pace I could keep up for a half-hour exercise period, I climb two stories up, traversing forty steps each 18 cm high, in 20 s. My output work becomes the final gravitational energy of the system of the Earth and me,

$$mgy = (85 \text{ kg})(9.80 \text{ m/s}^2)(40 \times 0.18 \text{ m}) = 6\,000 \text{ J}$$

making my sustainable power $\dfrac{6\,000 \text{ J}}{20 \text{ s}} = \boxed{\sim 10^2 \text{ W}}$.

P7.46 m = mass of pumpkin

R = radius of silo top

$$\sum F_r = ma_r \Rightarrow n - mg\cos\theta = -m\frac{v^2}{R}$$

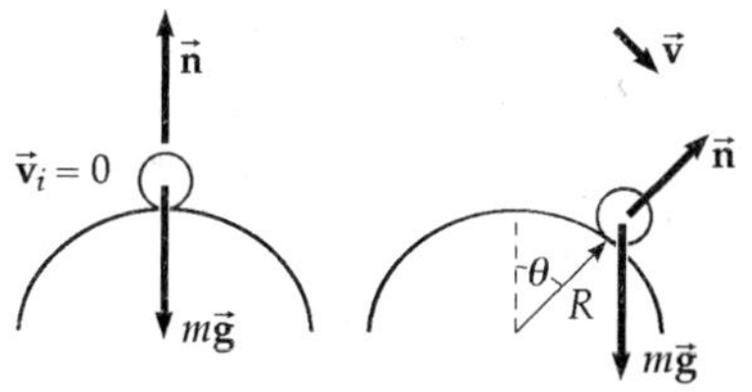

FIG. P7.46

When the pumpkin first loses contact with the surface, $n = 0$. Thus, at the point where it leaves the surface: $v^2 = Rg\cos\theta$.

Choose $U_g = 0$ in the $\theta = 90.0°$ plane. Then applying conservation of energy for the pumpkin-Earth system between the starting point and the point where the pumpkin leaves the surface gives

$$K_f + U_{gf} = K_i + U_{gi}$$

$$\frac{1}{2}mv^2 + mgR\cos\theta = 0 + mgR$$

Using the result from the force analysis, this becomes

$\dfrac{1}{2}mRg\cos\theta + mgR\cos\theta = mgR$, which reduces to

$\cos\theta = \dfrac{2}{3}$, and gives $\theta = \cos^{-1}(2/3) = \boxed{48.2°}$

as the angle at which the pumpkin will lose contact with the surface.

***P7.47** When block B moves up by 1 cm, block A moves down by 2 cm and the separation becomes 3 cm. We then choose the final point to be when B has moved up by $\dfrac{h}{3}$ and has speed $\dfrac{v_A}{2}$. Then A has moved down $\dfrac{2h}{3}$ and has speed v_A:

$$(K_A + K_B + U_g)_i = (K_A + K_B + U_g)_f$$

$$0 + 0 + 0 = \frac{1}{2}mv_A^2 + \frac{1}{2}m\left(\frac{v_A}{2}\right)^2 + \frac{mgh}{3} - \frac{mg2h}{3}$$

$$\frac{mgh}{3} = \frac{5}{8}mv_A^2$$

$$\boxed{v_A = \sqrt{\frac{8gh}{15}}}$$

P7.48 (a) $U_A = mgR = (0.200\text{ kg})(9.80\text{ m/s}^2)(0.300\text{ m}) = \boxed{0.588\text{ J}}$

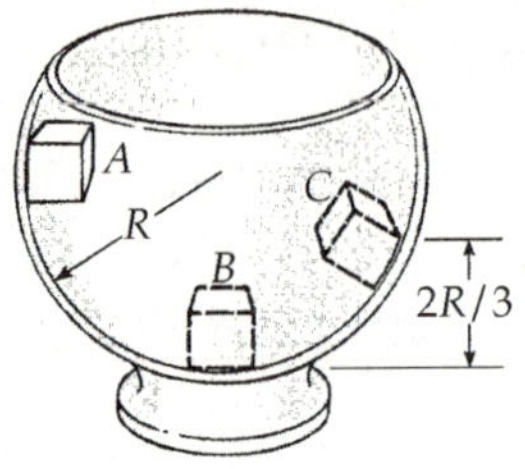

FIG. P7.48

(b) $K_A + U_A = K_B + U_B$

$K_B = K_A + U_A - U_B = mgR = \boxed{0.588\text{ J}}$

(c) $v_B = \sqrt{\dfrac{2K_B}{m}} = \sqrt{\dfrac{2(0.588\text{ J})}{0.200\text{ kg}}} = \boxed{2.42\text{ m/s}}$

(d) $U_C = mgh_C = (0.200\text{ kg})(9.80\text{ m/s}^2)(0.200\text{ m}) = \boxed{0.392\text{ J}}$

$K_C = K_A + U_A - U_C = mg(h_A - h_C)$

$K_C = (0.200\text{ kg})(9.80\text{ m/s}^2)(0.300 - 0.200)\text{ m} = \boxed{0.196\text{ J}}$

P7.49 (a) $K_B = \frac{1}{2}mv_B^2 = \frac{1}{2}(0.200\text{ kg})(1.50\text{ m/s})^2 = \boxed{0.225\text{ J}}$

(b) $$\begin{aligned}\Delta E_{\text{mech}} &= \Delta K + \Delta U = K_B - K_A + U_B - U_A\\ &= K_B + mg(h_B - h_A)\\ &= 0.225\text{ J} + (0.200\text{ kg})(9.80\text{ m/s}^2)(0 - 0.300\text{ m})\\ &= 0.225\text{ J} - 0.588\text{ J} = \boxed{-0.363\text{ J}}\end{aligned}$$

(c) It's possible to estimate an effective coefficient of friction, but not to find the actual value of μ since n and f vary with position.

P7.50 $k = 2.50 \times 10^4\text{ N/m}$, $m = 25.0\text{ kg}$

$x_A = -0.100\text{ m}$, $U_g\big|_{x=0} = U_s\big|_{x=0} = 0$

(a) $E_{\text{mech}} = K_A + U_{gA} + U_{sA}$ $E_{\text{mech}} = 0 + mgx_A + \frac{1}{2}kx_A^2$

$$E_{\text{mech}} = (25.0\text{ kg})(9.80\text{ m/s}^2)(-0.100\text{ m}) + \frac{1}{2}(2.50 \times 10^4\text{ N/m})(-0.100\text{ m})^2$$

$E_{\text{mech}} = -24.5\text{ J} + 125\text{ J} = \boxed{100\text{ J}}$

(b) Since only conservative forces are involved, the total energy of the child-pogo-stick-Earth system at point C is the same as that at point A.

$K_C + U_{gC} + U_{sC} = K_A + U_{gA} + U_{sA}$: $0 + (25.0\text{ kg})(9.80\text{ m/s}^2)x_C + 0 = 0 - 24.5\text{ J} + 125\text{ J}$

$x_C = \boxed{0.410\text{ m}}$

continued on next page

(c) $K_B + U_{gB} + U_{sB} = K_A + U_{gA} + U_{sA}$: $\frac{1}{2}(25.0\text{ kg})v_B^2 + 0 + 0 = 0 + (-24.5\text{ J}) + 125\text{ J}$

$$v_B = \boxed{2.84\text{ m/s}}$$

(d) K and v are at a maximum when $a = \sum F/m = 0$ (i.e., when the magnitude of the upward spring force equals the magnitude of the downward gravitational force).

This occurs at $x < 0$ where $k|x| = mg$

or $$|x| = \frac{(25.0\text{ kg})(9.8\text{ m/s}^2)}{2.50\times10^4\text{ N/m}} = 9.80\times10^{-3}\text{ m}$$

Thus, $K = K_{\max}$ at $x = \boxed{-9.80\text{ mm}}$

(e) $$K_{\max} = K_A + \left(U_{gA} - U_g\big|_{x=-9.80\text{ mm}}\right) + \left(U_{sA} - U_s\big|_{x=-9.80\text{ mm}}\right)$$

or $$\frac{1}{2}(25.0\text{ kg})v_{\max}^2 = (25.0\text{ kg})(9.80\text{ m/s}^2)[(-0.100\text{ m}) - (-0.009\,8\text{ m})]$$
$$+\frac{1}{2}(2.50\times10^4\text{ N/m})\left[(-0.100\text{ m})^2 - (-0.009\,8\text{ m})^2\right]$$

yielding $v_{\max} = \boxed{2.85\text{ m/s}}$

P7.51 $\Delta E_{\text{mech}} = -fd$

$E_f - E_i = -f\cdot d_{BC}$

$\frac{1}{2}kx^2 - mgh = -\mu mgd_{BC}$

$$\mu = \frac{mgh - \frac{1}{2}kx^2}{mgd_{BC}} = \boxed{0.328}$$

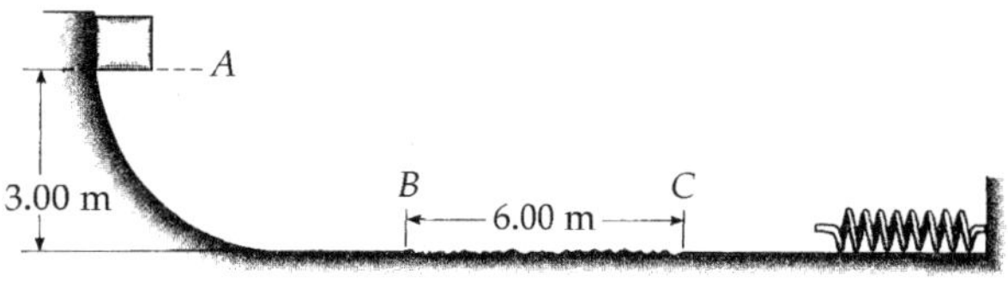

FIG. P7.51

P7.52 (a) $\vec{\mathbf{F}} = -\frac{d}{dx}\left(-x^3 + 2x^2 + 3x\right)\hat{\mathbf{i}} = \boxed{\left(3x^2 - 4x - 3\right)\hat{\mathbf{i}}}$

(b) $F = 0$

when $x = \boxed{1.87\text{ and } -0.535}$

(c) The stable point is at

$x = -0.535$ point of minimum $U(x)$.

The unstable point is at

$x = 1.87$ maximum in $U(x)$.

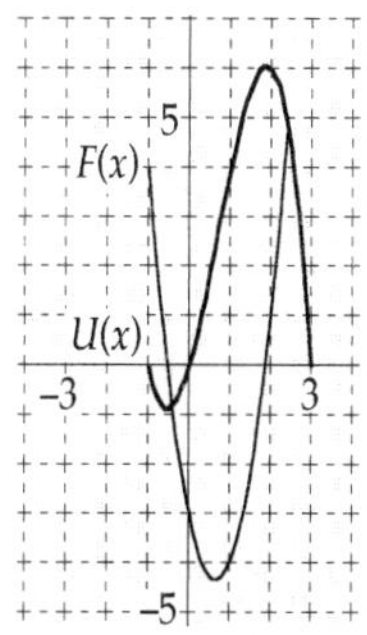

FIG. P7.52

P7.53 $(K+U)_i = (K+U)_f$

$$0+(30.0\text{ kg})(9.80\text{ m/s}^2)(0.200\text{ m})+\frac{1}{2}(250\text{ N/m})(0.200\text{ m})^2$$

$$=\frac{1}{2}(50.0\text{ kg})v^2+(20.0\text{ kg})(9.80\text{ m/s}^2)(0.200\text{ m})\sin 40.0°$$

$$58.8\text{ J}+5.00\text{ J}=(25.0\text{ kg})v^2+25.2\text{ J}$$

$$\boxed{v=1.24\text{ m/s}}$$

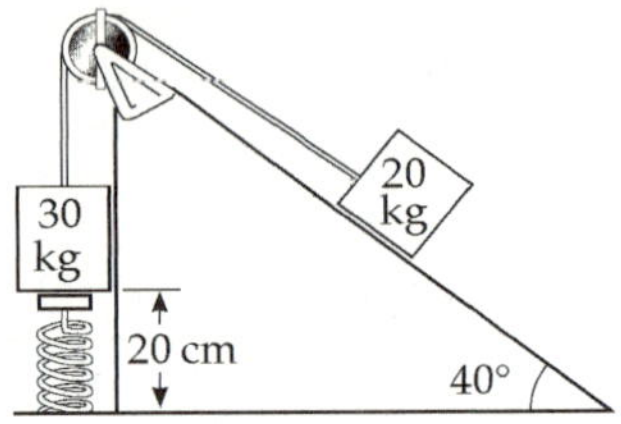

FIG. P7.53

P7.54 (a) Between the second and the third picture, $\Delta E_{\text{mech}} = \Delta K + \Delta U$

$$-\mu mgd = -\frac{1}{2}mv_i^2+\frac{1}{2}kd^2$$

$$\frac{1}{2}(50.0\text{ N/m})d^2+0.250(1.00\text{ kg})(9.80\text{ m/s}^2)d-\frac{1}{2}(1.00\text{ kg})(3.00\text{ m/s})^2=0$$

$$d=\frac{[-2.45\pm 21.35]\text{ N}}{50.0\text{ N/m}}=\boxed{0.378\text{ m}}$$

(b) Between picture two and picture four, $\Delta E_{\text{mech}} = \Delta K + \Delta U$

$$-f(2d)=\frac{1}{2}mv^2-\frac{1}{2}mv_i^2$$

$$v=\sqrt{(3.00\text{ m/s})^2-\frac{2}{(1.00\text{ kg})}(2.45\text{ N})(2)(0.378\text{ m})}$$

$$=\boxed{2.30\text{ m/s}}$$

(c) For the motion from picture two to picture five,
$\Delta E_{\text{mech}} = \Delta K + \Delta U$

$$-f(D+2d)=-\frac{1}{2}(1.00\text{ kg})(3.00\text{ m/s})^2$$

$$D=\frac{9.00\text{ J}}{2(0.250)(1.00\text{ kg})(9.80\text{ m/s}^2)}-2(0.378\text{ m})=\boxed{1.08\text{ m}}$$

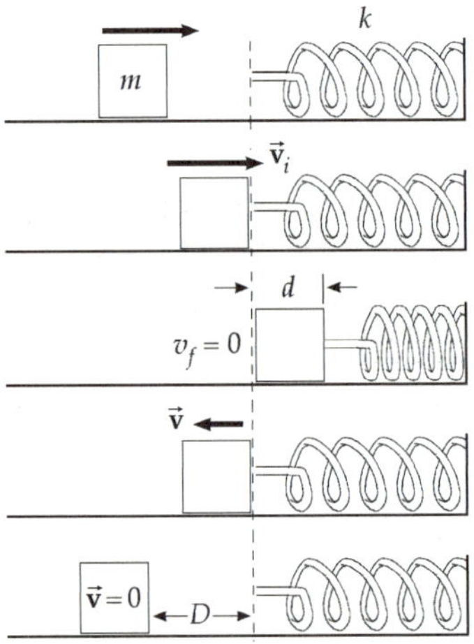

FIG. P7.54

P7.55 (a) Initial compression of spring: $\frac{1}{2}kx^2=\frac{1}{2}mv^2$

$$\frac{1}{2}(450\text{ N/m})(\Delta x)^2=\frac{1}{2}(0.500\text{ kg})(12.0\text{ m/s})^2$$

$$\therefore \Delta x=\boxed{0.400\text{ m}}$$

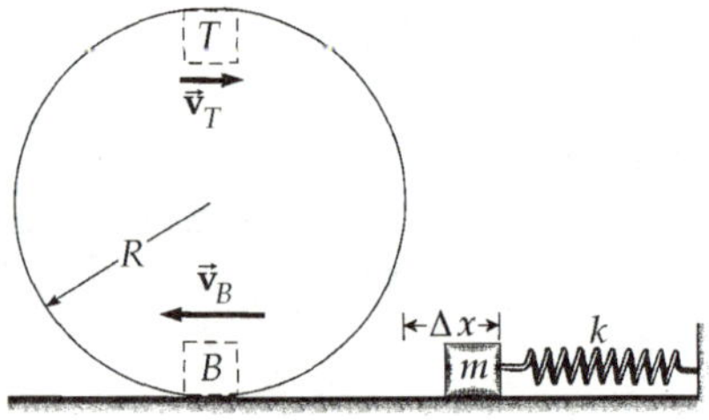

FIG. P7.55

continued on next page

(b) Speed of block at top of track: $\Delta E_{\text{mech}} = -fd$

$$\left(mgh_T + \frac{1}{2}mv_T^2\right) - \left(mgh_B + \frac{1}{2}mv_B^2\right) = -f(\pi R)$$

$$(0.500\text{ kg})(9.80\text{ m/s}^2)(2.00\text{ m}) + \frac{1}{2}(0.500\text{ kg})v_T^2 - \frac{1}{2}(0.500\text{ kg})(12.0\text{ m/s})^2 = -(7.00\text{ N})(\pi)(1.00\text{ m})$$

$$0.250v_T^2 = 4.21$$

$$\therefore v_T = \boxed{4.10\text{ m/s}}$$

(c) Does block fall off at or before top of track? Block falls if $a_c < g$

$$a_c = \frac{v_T^2}{R} = \frac{(4.10)^2}{1.00} = 16.8\text{ m/s}^2$$

Therefore $a_c > g$ and the $\boxed{\text{block stays on the track}}$.

P7.56 Let λ represent the mass of each one meter of the chain and T represent the tension in the chain at the table edge. We imagine the edge to act like a frictionless and massless pulley.

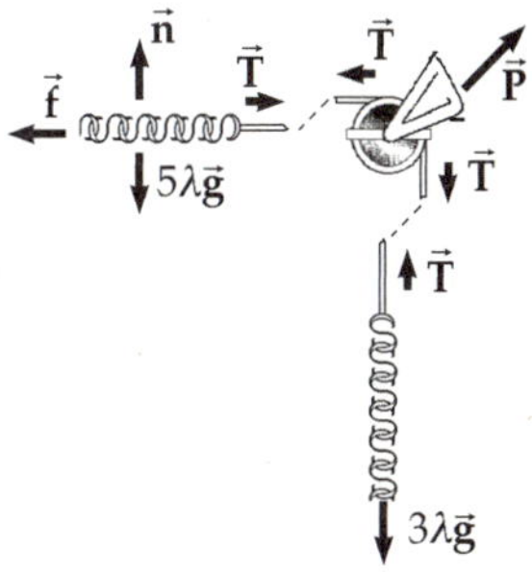

FIG. P7.56

(a) For the five meters on the table with motion impending,

$$\sum F_y = 0: \quad +n - 5\lambda g = 0 \qquad n = 5\lambda g$$

$$f_s \le \mu_s n = 0.6(5\lambda g) = 3\lambda g$$

$$\sum F_x = 0: \quad +T - f_s = 0 \qquad T = f_s \qquad T \le 3\lambda g$$

The maximum value is barely enough to support the hanging segment according to

$$\sum F_y = 0: \quad +T - 3\lambda g = 0 \qquad T = 3\lambda g$$

so it is at this point that the chain starts to slide.

(b) Let x represent the variable distance the chain has slipped since the start.

Then length $(5 - x)$ remains on the table, with now

$$\sum F_y = 0: \quad +n - (5-x)\lambda g = 0 \qquad n = (5-x)\lambda g$$

$$f_k = \mu_k n = 0.4(5-x)\lambda g = 2\lambda g - 0.4x\lambda g$$

Consider energies of the chain-Earth system at the initial moment when the chain starts to slip, and a final moment when $x = 5$, when the last link goes over the brink. Measure heights above the final position of the leading end of the chain. At the moment the final link slips off, the center of the chain is at $y_f = 4$ meters.

Originally, 5 meters of chain is at height 8 m and the middle of the dangling segment is at height $8 - \frac{3}{2} = 6.5$ m.

continued on next page

$$K_i + U_i + \Delta E_{\text{mech}} = K_f + U_f: \qquad 0 + (m_1 g y_1 + m_2 g y_2)_i - \int_i^f f_k dx = \left(\frac{1}{2}mv^2 + mgy\right)_f$$

$$(5\lambda g)8 + (3\lambda g)6.5 - \int_0^5 (2\lambda g - 0.4x\lambda g)dx = \frac{1}{2}(8\lambda)v^2 + (8\lambda g)4$$

$$40.0g + 19.5g - 2.00g\int_0^5 dx + 0.400g\int_0^5 x\,dx = 4.00v^2 + 32.0g$$

$$27.5g - 2.00gx\Big|_0^5 + 0.400g\frac{x^2}{2}\Big|_0^5 = 4.00v^2$$

$$27.5g - 2.00g(5.00) + 0.400g(12.5) = 4.00v^2$$

$$22.5g = 4.00v^2$$

$$v = \sqrt{\frac{(22.5\text{ m})(9.80\text{ m/s}^2)}{4.00}} = \boxed{7.42\text{ m/s}}$$

P7.57 The geometry reveals $D = L\sin\theta + L\sin\phi$, $50.0\text{ m} = 40.0\text{ m}(\sin 50° + \sin\phi)$, $\phi = 28.9°$

(a) From takeoff to alighting for the Jane-Earth system

$$(K + U_g)_i + W_{\text{wind}} = (K + U_g)_f$$

$$\frac{1}{2}mv_i^2 + mg(-L\cos\theta) + FD(-1) = 0 + mg(-L\cos\phi)$$

$$\frac{1}{2}50\text{ kg } v_i^2 + 50\text{ kg}(9.8\text{ m/s}^2)(-40\text{ m}\cos 50°) - 110\text{ N}(50\text{ m}) = 50\text{ kg}(9.8\text{ m/s}^2)(-40\text{ m}\cos 28.9°)$$

$$\frac{1}{2}50\text{ kg } v_i^2 - 1.26\times10^4\text{ J} - 5.5\times10^3\text{ J} = -1.72\times10^4\text{ J}$$

$$v_i = \sqrt{\frac{2(947\text{ J})}{50\text{ kg}}} = \boxed{6.15\text{ m/s}}$$

(b) For the swing back

$$\frac{1}{2}mv_i^2 + mg(-L\cos\phi) + FD(+1) = 0 + mg(-L\cos\theta)$$

$$\frac{1}{2}130\text{ kg } v_i^2 + 130\text{ kg}(9.8\text{ m/s}^2)(-40\text{ m}\cos 28.9°) + 110\text{ N}(50\text{ m})$$
$$= 130\text{ kg}(9.8\text{ m/s}^2)(-40\text{ m}\cos 50°)$$

$$\frac{1}{2}130\text{ kg } v_i^2 - 4.46\times10^4\text{ J} + 5\,500\text{ J} = -3.28\times10^4\text{ J}$$

$$v_i = \sqrt{\frac{2(6\,340\text{ J})}{130\text{ kg}}} = \boxed{9.87\text{ m/s}}$$

P7.58 Case I: Surface is frictionless

$$\frac{1}{2}mv^2 = \frac{1}{2}kx^2$$

$$k = \frac{mv^2}{x^2} = \frac{(5.00\ \text{kg})(1.20\ \text{m/s})^2}{10^{-2}\ \text{m}^2} = 7.20\times 10^2\ \text{N/m}$$

Case II: Surface is rough, $\mu_k = 0.300$

$$\frac{1}{2}mv^2 = \frac{1}{2}kx^2 - \mu_k mgx$$

$$\frac{5.00\ \text{kg}}{2}v^2 = \frac{1}{2}\left(7.20\times 10^2\ \text{N/m}\right)\left(10^{-1}\ \text{m}\right)^2 - (0.300)(5.00\ \text{kg})\left(9.80\ \text{m/s}^2\right)\left(10^{-1}\ \text{m}\right)$$

$$\boxed{v = 0.923\ \text{m/s}}$$

***P7.59** (a) $(K+U_g)_A = (K+U_g)_B$

$$0 + mgy_A = \frac{1}{2}mv_B^2 + 0 \qquad v_B = \sqrt{2gy_A} = \sqrt{2\left(9.8\ \text{m/s}^2\right)6.3\ \text{m}} = \boxed{11.1\ \text{m/s}}$$

(b) $$a_c = \frac{v^2}{r} = \frac{(11.1\ \text{m/s})^2}{6.3\ \text{m}} = \boxed{19.6\ \text{m/s}^2\ \text{up}}$$

(c) $\sum F_y = ma_y \qquad +n_B - mg = ma_c$

$$n_B = 76\ \text{kg}\left(9.8\ \text{m/s}^2 + 19.6\ \text{m/s}^2\right) = \boxed{2.23\times 10^3\ \text{N up}}$$

(d) $W = F\Delta r\cos\theta = 2.23\times 10^3\ \text{N}(0.450\ \text{m})\cos 0° = \boxed{1.01\times 10^3\ \text{J}}$

This quantity of work represents chemical energy, previously in the skateboarder's body, that is converted into extra kinetic energy:

$$\left(K + U_g + U_{\text{chemical}}\right)_B = \left(K + U_g + U_{\text{chemical}}\right)_C$$

$$\frac{1}{2}mv_B^2 + 0 + U_{\text{chemical, }B} - U_{\text{chemical, }C} = K_C + mg(y_C - y_B)$$

$$\frac{1}{2}(76\ \text{kg})(11.1\ \text{m/s})^2 + 1.01\times 10^3\ \text{J} - 76\ \text{kg}\left(9.8\ \text{m/s}^2\right)(0.450\ \text{m}) = K_C$$

(e) $(K+U_g)_C = (K+U_g)_D$

$$5.36\times 10^3\ \text{J} = \frac{1}{2}(76\ \text{kg})v_D^2 + 76\ \text{kg}\left(9.8\ \text{m/s}^2\right)(5.85\ \text{m})$$

$$\sqrt{\frac{\left(5.36\times 10^3\ \text{J} - 4.36\times 10^3\ \text{J}\right)2}{76\ \text{kg}}} = v_D = \boxed{5.14\ \text{m/s}}$$

continued on next page

(f) $(K+U_g)_D = (K+U_g)_E$ where E is the apex of his motion

$$\frac{1}{2}mv_D^2 + 0 = 0 + mg(y_E - y_D) \qquad y_E - y_D = \frac{v_D^2}{2g} = \frac{(5.14 \text{ m/s})^2}{2(9.8 \text{ m/s}^2)} = \boxed{1.35 \text{ m}}$$

(g) Consider the motion with constant acceleration between takeoff and touchdown. The time is the positive root of

$$y_f = y_i + v_{yi}t + \frac{1}{2}a_y t^2$$

$$-2.34 \text{ m} = 0 + 5.14 \text{ m/s}\, t + \frac{1}{2}(-9.8 \text{ m/s}^2)t^2$$

$$4.9t^2 - 5.14t - 2.34 = 0$$

$$t = \frac{5.14 \pm \sqrt{5.14^2 - 4(4.9)(-2.34)}}{9.8} = \boxed{1.39 \text{ s}}$$

P7.60 If the spring is just barely able to lift the lower block from the table, the spring lifts it through no noticeable distance, but exerts on the block a force equal to its weight Mg. The extension of the spring, from $|\vec{\mathbf{F}}_s| = kx$, must be Mg/k. Between an initial point at release and a final point when the moving block first comes to rest, we have

$$K_i + U_{gi} + U_{si} = K_f + U_{gf} + U_{sf}: \qquad 0 + mg\left(-\frac{4mg}{k}\right) + \frac{1}{2}k\left(\frac{4mg}{k}\right)^2 = 0 + mg\left(\frac{Mg}{k}\right) + \frac{1}{2}k\left(\frac{Mg}{k}\right)^2$$

$$-\frac{4m^2g^2}{k} + \frac{8m^2g^2}{k} = \frac{mMg^2}{k} + \frac{M^2g^2}{2k}$$

$$4m^2 = mM + \frac{M^2}{2}$$

$$\frac{M^2}{2} + mM - 4m^2 = 0$$

$$M = \frac{-m \pm \sqrt{m^2 - 4\left(\frac{1}{2}\right)\left(-4m^2\right)}}{2\left(\frac{1}{2}\right)} = -m \pm \sqrt{9m^2}$$

Only a positive mass is physical, so we take $M = m(3-1) = \boxed{2m}$.

***P7.61** (a) Energy is conserved in the swing of the pendulum, and the stationary peg does no work. So the ball's speed does not change when the string hits or leaves the peg, and the ball swings equally high on both sides.

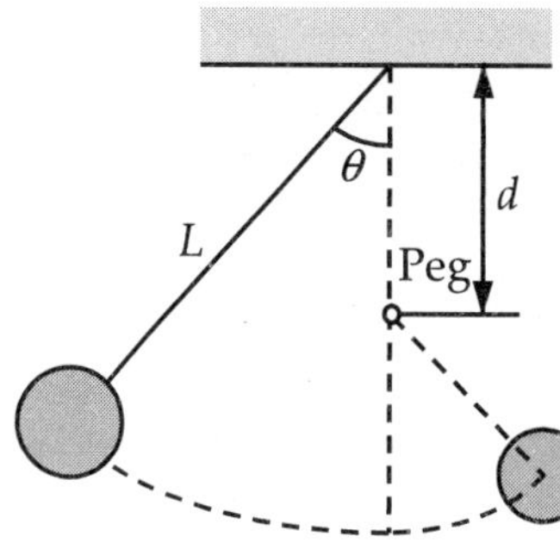

FIG. P7.61

(b) Relative to the point of suspension, $U_i = 0$, $U_f = -mg[d-(L-d)]$. From this we find that $-mg(2d-L)+\frac{1}{2}mv^2 = 0$. Also for centripetal motion, $mg = \frac{mv^2}{R}$ where $R = L-d$. Upon solving, we get $\boxed{d = \frac{3L}{5}}$.

P7.62 (a) At the top of the loop the car and riders are in free fall:

$$\sum F_y = ma_y: \qquad mg \text{ down} = \frac{mv^2}{R} \text{ down}$$

$$v = \sqrt{Rg}$$

Energy of the car-riders-Earth system is conserved between release and top of loop:

$$K_i + U_{gi} = K_f + U_{gf}: \qquad 0 + mgh = \frac{1}{2}mv^2 + mg(2R)$$

$$gh = \frac{1}{2}Rg + g(2R)$$

$$\boxed{h = 2.50R}$$

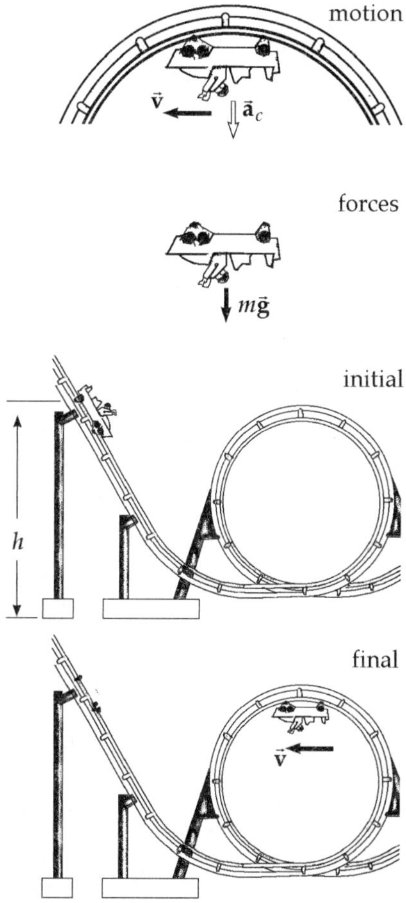

FIG. P7.62

(b) Let h now represent the height $\geq 2.5R$ of the release point. At the bottom of the loop we have

$$mgh = \frac{1}{2}mv_b^2 \qquad \text{or} \qquad v_b^2 = 2gh$$

$$\sum F_y = ma_y: \qquad n_b - mg = \frac{mv_b^2}{R}(\text{up})$$

$$n_b = mg + \frac{m(2gh)}{R}$$

At the top of the loop, $mgh = \frac{1}{2}mv_t^2 + mg(2R)$

$$v_t^2 = 2gh - 4gR$$

continued on next page

$\sum F_y = ma_y$: $\quad -n_t - mg = -\dfrac{mv_t^2}{R}$

$$n_t = -mg + \frac{m}{R}(2gh - 4gR)$$

$$n_t = \frac{m(2gh)}{R} - 5mg$$

Then the normal force at the bottom is larger by

$$n_b - n_t = mg + \frac{m(2gh)}{R} - \frac{m(2gh)}{R} + 5mg = \boxed{6mg}.$$

P7.63 (a) Conservation of energy for the sled-rider-Earth system, between A and C:

$$K_i + U_{gi} = K_f + U_{gf}$$

$$\frac{1}{2}m(2.5\text{ m/s})^2 + m(9.80\text{ m/s}^2)(9.76\text{ m}) = \frac{1}{2}mv_C^2 + 0$$

$$v_C = \sqrt{(2.5\text{ m/s})^2 + 2(9.80\text{ m/s}^2)(9.76\text{ m})} = \boxed{14.1\text{ m/s}}$$

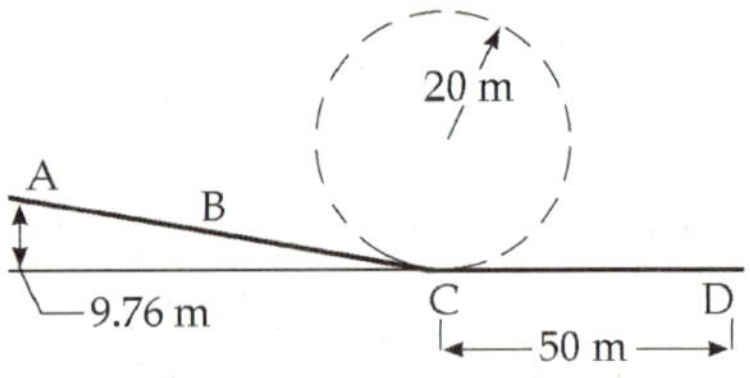

FIG. P7.63(a)

(b) Incorporating the loss of mechanical energy during the portion of the motion in the water, we have, for the entire motion between A and D (the rider's stopping point),

$K_i + U_{gi} - f_k d = K_f + U_{gf}$: $\quad \frac{1}{2}(80\text{ kg})(2.5\text{ m/s})^2 + (80\text{ kg})(9.80\text{ m/s}^2)(9.76\text{ m}) - f_k\Delta x = 0 + 0$

$$-f_k d = \boxed{-7.90 \times 10^3\text{ J}}$$

(c) The water exerts a frictional force $\quad f_k = \dfrac{7.90 \times 10^3\text{ J}}{d} = \dfrac{7.90 \times 10^3\text{ N}\cdot\text{m}}{50\text{ m}} = 158\text{ N}$

and also a normal force of $\quad n = mg = (80\text{ kg})(9.80\text{ m/s}^2) = 784\text{ N}$

The magnitude of the water force is $\quad \sqrt{(158\text{ N})^2 + (784\text{ N})^2} = \boxed{800\text{ N}}$

(d) The angle of the slide is

$$\theta = \sin^{-1}\frac{9.76\text{ m}}{54.3\text{ m}} = 10.4°$$

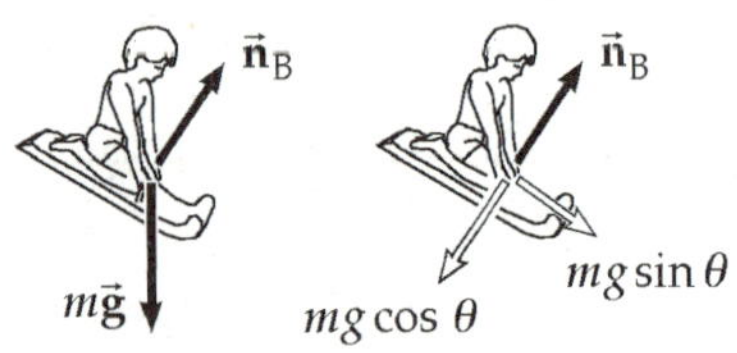

FIG. P7.63(d)

For forces perpendicular to the track at B,

$\sum F_y = ma_y$: $\quad n_B - mg\cos\theta = 0$

$$n_B = (80.0\text{ kg})(9.80\text{ m/s}^2)\cos 10.4° = \boxed{771\text{ N}}$$

continued on next page

(e) $\sum F_y = ma_y$: $+n_C - mg = \dfrac{mv_C^2}{r}$

$$n_C = (80.0 \text{ kg})(9.80 \text{ m/s}^2) + \frac{(80.0 \text{ kg})(14.1 \text{ m/s})^2}{20 \text{ m}}$$

$$n_C = \boxed{1.57 \times 10^3 \text{ N up}}$$

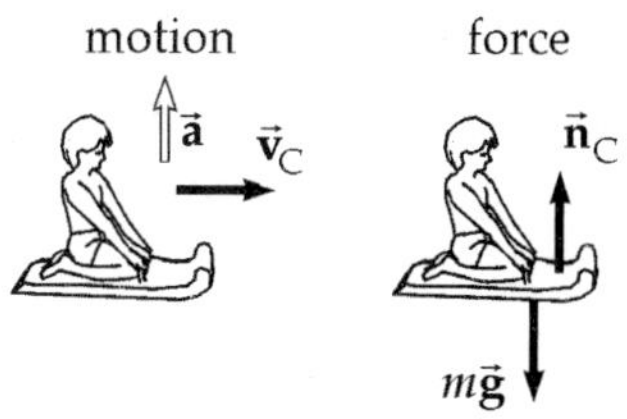

FIG. P7.63(e)

The rider pays for the thrills of a giddy height at A, and a high speed and tremendous splash at C. As a bonus, he gets the quick change in direction and magnitude among the forces we found in parts (d), (e), and (c).

***P7.64** (a) $U_g = mgy = (64 \text{ kg})(9.8 \text{ m/s}^2)y = \boxed{(627 \text{ N})y}$

(b) At the original height and at all heights above $65 \text{ m} - 25.8 \text{ m} = 39.2 \text{ m}$, the cord is unstretched and $\boxed{U_s = 0}$. Below 39.2 m, the cord extension x is given by $x = 39.2 \text{ m} - y$, so the elastic energy is $U_s = \frac{1}{2}kx^2 = \boxed{\frac{1}{2}(81 \text{ N/m})(39.2 \text{ m} - y)^2}$.

(c) For $y > 39.2$ m, $U_g + U_s = \boxed{(627 \text{ N})y}$

For $y \le 39.2$ m,

$$U_g + U_s = (627 \text{ N})y + 40.5 \text{ N/m}\left(1\,537 \text{ m}^2 - (78.4 \text{ m})y + y^2\right)$$

$$= \boxed{(40.5 \text{ N/m})y^2 - (2\,550 \text{ N})y + 62\,200 \text{ J}}$$

(d)

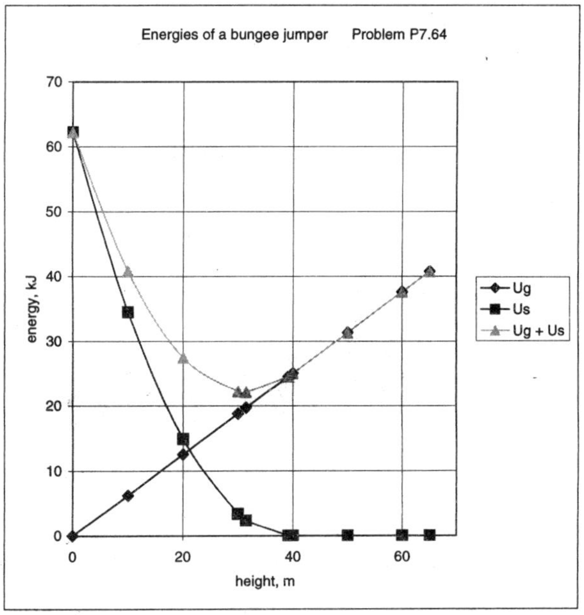

FIG. P7.64(d)

continued on next page

(e) At minimum height, the jumper has zero kinetic energy and the same total energy as at his starting point. $K_i + U_i = K_f + U_f$ becomes

$$627\text{ N}(65\text{ m}) = (40.5\text{ N/m})y_f^2 - (2\,550\text{ N})y_f + 62\,200\text{ J}$$
$$0 = 40.5y_f^2 - 2\,550y_f + 21\,500$$
$$y_f = \boxed{10.0\text{ m}}\ \left[\text{the root 52.9 m is unphysical.}\right]$$

(f) The total potential energy has a minimum, representing a $\boxed{\text{stable equilibrium}}$ position. To find it, we require $\frac{dU}{dy} = 0$:

$$\frac{d}{dy}\left(40.5y^2 - 2\,550y + 62\,200\right) = 0 = 81y - 2\,550$$
$$y = \boxed{31.5\text{ m}}$$

(g) Maximum kinetic energy occurs at minimum potential energy. Between the takeoff point and this location we have $K_i + U_i = K_f + U_f$

$$0 + 40\,800\text{ J} = \frac{1}{2}(64\text{ kg})v_{max}^2 + 40.5(31.5)^2 - 2\,550(31.5) + 62\,200$$
$$v_{max} = \left(\frac{2(40\,800 - 22\,200)}{64}\right)^{1/2} = \boxed{24.1\text{ m/s}}$$

ANSWERS TO EVEN PROBLEMS

P7.2 (a) 800 J; (b) 107 J; (c) 0

P7.4 (a) 1.11×10^9 J; (b) 0.2

P7.6 (a) 19.8 m/s; (b) 294 J; (c) $\left(30.0\hat{\mathbf{i}} - 39.6\hat{\mathbf{j}}\right)$ m/s

P7.8 (a) 2.29 m/s; (b) 1.98 m/s

P7.10 (a) $v_B = 5.94$ m/s, $v_C = 7.67$ m/s; (b) 147 J

P7.12 5.49 m/s

P7.14 (a) see the solution; (b) 35.0 J

P7.16 $d = \dfrac{kx^2}{2mg\sin\theta} - x$

P7.18 (a) 25.8 m; (b) 27.1 m/s^2

P7.20 (a) 0.403 m or −0.357 m; (b) see the solution

P7.22 (a) 21.0 m/s; (b) 16.1 m/s

P7.24 168 J

P7.26 (a) 24.5 m/s; (b) yes; (c) 206 m; (d) unrealistic, see the solution

P7.28 3.92 kJ

P7.30 44.1 kW

P7.32 (a) $\dfrac{Ax^2}{2} - \dfrac{Bx^3}{3}$; (b) $\Delta U = \dfrac{5A}{2} - \dfrac{19B}{3}$; $\Delta K = -\dfrac{5A}{2} + \dfrac{19B}{3}$

P7.34 $\left(7 - 9x^2y\right)\hat{\mathbf{i}} - 3x^3\hat{\mathbf{j}}$

P7.36 4.17×10^{10} J

P7.38 (a) -1.67×10^{-14} J; (b) at the center

P7.40 (a) $r = 1.5$ mm stable, 2.3 mm unstable, 3.2 mm stable, $r \to \infty$ neutral;
(b) $-5.6 \text{ J} \le E < 1 \text{ J}$;
(c) $0.6 \text{ mm} \le r \le 3.6 \text{ mm}$; (d) 2.6 J;
(e) 1.5 mm; (f) 4 J

P7.42 $33.4 \text{ kW} = 44.8 \text{ hp}$

P7.44 (a) 6.75 W/m^2; (b) $6.64 \times 10^3 \text{ W/m}^2$;
(c) see solution

P7.46 48.2°

P7.48 (a) 0.588 J; (b) 0.588 J; (c) 2.42 m/s;
(d) $U_C = 0.392$ J, $K_C = 0.196$ J

P7.50 (a) 100 J; (b) 0.410 m; (c) 2.84 m/s;
(d) –9.80 mm; (e) 2.85 m/s

P7.52 (a) $\left(3x^2 - 4x - 3\right)\hat{\mathbf{i}}$;
(b) $x = 1.87$ and –0.535;
(c) see the solution

P7.54 (a) 0.378 m; (b) 2.30 m/s; (c) 1.08 m

P7.56 (a) see the solution; (b) 7.42 m/s

P7.58 0.923 m/s

P7.60 $2m$

P7.62 (a) $2.5R$; (b) see the solution

P7.64 (a) $(627 \text{ N})y$;
(b) $U_s = 0$ for $y > 39.2$ m and
$U_s = \frac{1}{2}(81 \text{ N/m})(39.2 \text{ m} - y)^2$ for $y \le 39.2$ m;
(c) $U_g + U_s = (627 \text{ N})y$, for $y > 39.2$ m
$U_g + U_s = (40.5 \text{ N/m})y^2 - (2\,550 \text{ N})y + 62\,200 \text{ J}$ for $y \le 39.2$ m;
(d) see the solution; (e) 10.0 m;
(f) yes: stable equilibrium at 31.5 m;
(g) 24.1 m/s

CONTEXT 1 CONCLUSION SOLUTIONS TO PROBLEMS

***CC1.1** (a) $$K = \frac{1}{2}mv^2 = \frac{1}{2}(1\,300 \text{ kg})(22 \text{ m/s})^2 = \boxed{3.15 \times 10^5 \text{ J}}$$

(b) Energy delivered to battery $= 0.70\left(3.15 \times 10^5 \text{ J}\right) = \boxed{2.20 \times 10^5 \text{ J}}$

(c) Energy returned to battery $= 0.85\left(2.20 \times 10^5 \text{ J}\right) = \boxed{1.87 \times 10^5 \text{ J}}$

(d) Kinetic energy produced by motor $= 0.68\left(1.87 \times 10^5 \text{ J}\right) = \boxed{1.27 \times 10^5 \text{ J}}$

(e) $$v = \left(\frac{2K}{m}\right)^{1/2} = \left(\frac{2 \times 1.27 \times 10^5 \text{ J}}{1\,300 \text{ kg}}\right)^{1/2} = \boxed{14.0 \text{ m/s}}$$

(f) $$e = \frac{\text{useful energy output}}{\text{total energy input}} = \frac{1.27 \times 10^5 \text{ J}}{3.15 \times 10^5 \text{ J}} = \boxed{0.405} = 40.5\%$$

continued on next page

(g) The original kinetic energy all becomes either final kinetic energy or internal energy:

$$3.15\times10^5 \text{ J} = 1.27\times10^5 \text{ J} + \Delta E_{\text{int}}$$

$$\Delta E_{\text{int}} = \boxed{1.87\times10^5 \text{ J}}$$

***CC1.2** (a) For the conventional car in the first part of the trip, efficiency $= \dfrac{\text{useful output energy}}{\text{input energy}}$, $0.07 = \dfrac{33 \text{ MJ}}{\text{input}}$, input $= \dfrac{33 \text{ MJ}}{0.07} = 471 \text{ MJ}$. Second part: $0.3 = \dfrac{33 \text{ MJ}}{\text{input}}$, input $= 110 \text{ MJ}$. Total input $= 471 \text{ MJ} + 110 \text{ MJ} = \boxed{581 \text{ MJ}}$. For the hybrid car, $0.3 = \dfrac{66 \text{ MJ}}{\text{input}}$, input $= \boxed{220 \text{ MJ}}$.

(b) conventional car: efficiency $= \dfrac{66 \text{ MJ}}{\text{input}} = 0.114 = \boxed{11.4\%}$. This is much closer to 7% than to 30%. Hybrid car: efficiency $= \boxed{30.0\%}$

ANSWERS TO EVEN CONTEXT 1 CONCLUSION PROBLEMS

CC1.2 (a) 581 MJ, 220 MJ; (b) 11.4%, 30.0%

Chapter 8

Momentum and Collisions

CHAPTER OUTLINE

ANSWERS TO QUESTIONS

Q8.1 No. Impulse, $F\Delta t$, depends on the force and the time for which it is applied.

Q8.2 The momentum doubles since it is proportional to the speed. The kinetic energy quadruples, since it is proportional to the speed-squared.

Q8.3 The momenta of two particles will only be the same if the masses of the particles of the same.

Q8.4 (a) It does not carry force, for if it did, it could accelerate itself.

(b) It cannot deliver more kinetic energy than it possesses. This would violate the law of energy conservation.

(c) It can deliver more momentum in a collision than it possesses in its flight, by bouncing from the object it strikes.

Q8.5 The superhero has finite mass. When he throws the piano, he must recoil in the opposite direction. To hover in midair, he must continuously fly by giving downward momentum to air, rocket exhaust or something that will disturb things below.

Q8.6 No. Only in a precise head-on collision with momenta with equal magnitudes and opposite directions can both objects wind up at rest. Yes. Assume that ball 2, originally at rest, is struck squarely by an equal-mass ball 1. Then ball 2 will take off with the velocity of ball 1, leaving ball 1 at rest.

Q8.7 Interestingly, mutual gravitation brings the ball and the Earth together. As the ball moves downward, the Earth moves upward, although with an acceleration 10^{25} times smaller than that of the ball. The two objects meet, rebound, and separate. Momentum of the ball-Earth system is conserved.

Q8.8 (a) Linear momentum is conserved since there are no external forces acting on the system.

(b) Kinetic energy is not conserved because the chemical potential energy initially in the explosive is converted into kinetic energy of the pieces of the bomb.

Q8.9 Momentum conservation is not violated if we choose as our system the planet along with you. When you receive an impulse forward, the Earth receives the same size impulse backwards. The resulting acceleration of the Earth due to this impulse is significantly smaller than your acceleration forward, but the planet's backward momentum is equal in magnitude to your forward momentum.

Q8.10 Suppose car and truck move along the same line. If one vehicle overtakes the other, the faster-moving one loses more energy than the slower one gains. In a head-on collision, if the speed of the truck is less than $\frac{m_T + 3m_c}{3m_T + m_c}$ times the speed of the car, the car will lose more energy.

Q8.11 The rifle has a much lower speed than the bullet and much less kinetic energy. The butt distributes the recoil force over an area much larger than that of the bullet.

Q8.12 His impact speed is determined by the acceleration of gravity and the distance of fall, in $v_f^2 = v_i^2 - 2g(0 - y_i)$. The force exerted by the pad depends also on the unknown stiffness of the pad.

Q8.13 The product of the mass flow rate and velocity of the water determines the force the firefighters must exert.

Q8.14 The sheet stretches and pulls the two students toward each other. These effects are larger for a faster-moving egg. The time over which the egg stops is extended so that the force stopping it is never too large.

Q8.15 The planet is in motion around the sun, and thus has momentum and kinetic energy of its own. The spacecraft is directed to cross the planet's orbit behind it, so that the planet's gravity has a component pulling forward on the spacecraft. Since this is an elastic collision, and the velocity of the planet remains nearly unchanged, the probe must both increase speed and change direction for both momentum and kinetic energy to be conserved.

Q8.16 Yes. A boomerang, a kitchen stool.

Q8.17 The center of mass of the balls is in free fall, moving up and then down with the acceleration due to gravity, during the 40% of the time when the juggler's hands are empty. During the 60% of the time when the juggler is engaged in catching and tossing, the center of mass must accelerate up with a somewhat smaller average acceleration. The center of mass moves around in a little circle, making three revolutions for every one revolution that one ball makes. Letting T represent the time for one cycle and F_g the weight of one ball, we have $F_J 0.60T = 3F_g T$ and $F_J = 5F_g$. The average force exerted by the juggler is five times the weight of one ball.

Q8.18 Inflate a balloon and release it. The air escaping from the balloon gives the balloon an impulse.

Q8.19 In empty space, the center of mass of a rocket-plus-fuel system does not accelerate during a burn, because no outside force acts on this system. According to the text's 'basic expression for rocket propulsion,' the change in speed of the rocket body will be larger than the speed of the exhaust relative to the rocket, if the final mass is less than 37% of the original mass.

Q8.20 There was a time when the English favored position (a), the Germans position (b), and the French position (c). A Frenchman, Jean D'Alembert, is most responsible for showing that each theory is consistent with the others. All are equally correct. Each is useful for giving a mathematically simple solution for some problems.

SOLUTIONS TO PROBLEMS

Section 8.1 Linear Momentum and Its Conservation

P8.1 $m = 3.00\text{ kg}$, $\vec{\mathbf{v}} = \left(3.00\hat{\mathbf{i}} - 4.00\hat{\mathbf{j}}\right)\text{ m/s}$

(a) $\vec{\mathbf{p}} = m\vec{\mathbf{v}} = \left(9.00\hat{\mathbf{i}} - 12.0\hat{\mathbf{j}}\right)\text{ kg}\cdot\text{m/s}$

Thus, $\boxed{p_x = 9.00\text{ kg}\cdot\text{m/s}}$

and $\boxed{p_y = -12.0\text{ kg}\cdot\text{m/s}}$.

(b) $p = \sqrt{p_x^2 + p_y^2} = \sqrt{(9.00)^2 + (12.0)^2} = \boxed{15.0\text{ kg}\cdot\text{m/s}}$

$$\theta = \tan^{-1}\left(\frac{p_y}{p_x}\right) = \tan^{-1}(-1.33) = \boxed{307°}$$

P8.2 I have mass 85.0 kg and can jump to raise my center of gravity 25.0 cm. I leave the ground with speed given by

$v_f^2 - v_i^2 = 2a\left(x_f - x_i\right)$: $\quad 0 - v_i^2 = 2\left(-9.80\text{ m/s}^2\right)(0.250\text{ m})$

$$v_i = 2.20\text{ m/s}$$

Total momentum of the system of the Earth and me is conserved as I push the earth down and myself up:

$$0 = \left(5.98\times10^{24}\text{ kg}\right)v_e + (85.0\text{ kg})(2.20\text{ m/s})$$

$$v_e \sim \boxed{10^{-23}\text{ m/s}}$$

P8.3 Assume the velocity of the blood is constant over the 0.160 s. Then the patient's body and pallet will have a constant velocity of $\dfrac{6\times10^{-5}\text{ m}}{0.160\text{ s}} = 3.75\times10^{-4}\text{ m/s}$ in the opposite direction. Momentum conservation gives

$\vec{\mathbf{p}}_{1i} + \vec{\mathbf{p}}_{2i} = \vec{\mathbf{p}}_{1f} + \vec{\mathbf{p}}_{2f}$: $\quad 0 = m_{\text{blood}}\,0.5\text{ m/s} + 54\text{ kg}\left(-3.75\times10^{-4}\text{ m/s}\right)$

$$m_{\text{blood}} = 0.0405\text{ kg} = \boxed{40.5\text{ g}}$$

P8.4 (a) The momentum is $p = mv$, so $v = \dfrac{p}{m}$ and the kinetic energy is $K = \dfrac{1}{2}mv^2 = \dfrac{1}{2}m\left(\dfrac{p}{m}\right)^2 = \boxed{\dfrac{p^2}{2m}}$.

(b) $K = \dfrac{1}{2}mv^2$ implies $v = \sqrt{\dfrac{2K}{m}}$, so $p = mv = m\sqrt{\dfrac{2K}{m}} = \boxed{\sqrt{2mK}}$.

P8.5 (a) For the system of two blocks $\Delta p = 0$,

or $p_i = p_f$.

Therefore, $0 = Mv_m + (3M)(2.00\ \text{m/s})$.

Solving gives $v_m = \boxed{-6.00\ \text{m/s}}$ (motion toward the left).

(b) $\frac{1}{2}kx^2 = \frac{1}{2}Mv_M^2 + \frac{1}{2}(3M)v_{3M}^2 = \boxed{8.40\ \text{J}}$

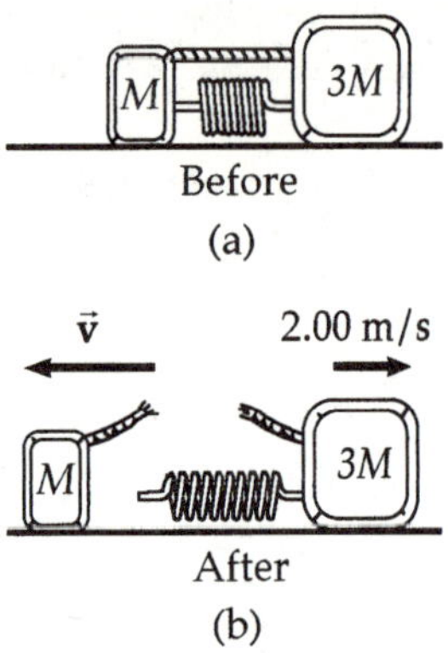

FIG. P8.5

Section 8.2 Impulse and Momentum

***P8.6** From the impulse-momentum theorem, $\bar{F}(\Delta t) = \Delta p = mv_f - mv_i$, the average force required to hold onto the child is

$$\bar{F} = \frac{m(v_f - v_i)}{(\Delta t)} = \frac{(12\ \text{kg})(0 - 60\ \text{mi/h})}{0.050\ \text{s} - 0}\left(\frac{1\ \text{m/s}}{2.237\ \text{mi/h}}\right) = -6.44 \times 10^3\ \text{N}.$$

Therefore, the magnitude of the needed retarding force is $\boxed{6.44 \times 10^3\ \text{N}}$, or 1 400 lb. A person cannot exert a force of this magnitude and a safety device should be used.

P8.7 (a) $I = \int F dt =$ area under curve

$I = \frac{1}{2}(1.50 \times 10^{-3}\ \text{s})(18\,000\ \text{N}) = \boxed{13.5\ \text{N} \cdot \text{s}}$

(b) $F = \frac{13.5\ \text{N} \cdot \text{s}}{1.50 \times 10^{-3}\ \text{s}} = \boxed{9.00\ \text{kN}}$

(c) From the graph, we see that $F_{max} = \boxed{18.0\ \text{kN}}$

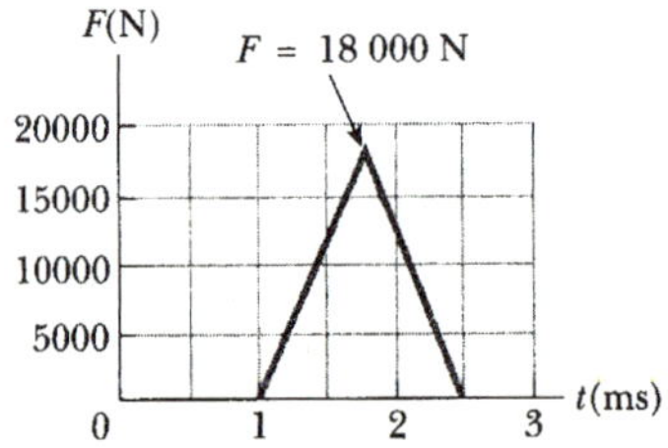

FIG. P8.7

P8.8 Assume the initial direction of the ball in the $-x$ direction.

(a) Impulse, $\vec{\mathbf{I}} = \Delta\vec{\mathbf{p}} = \vec{\mathbf{p}}_f - \vec{\mathbf{p}}_i = (0.060\,0)(40.0)\hat{\mathbf{i}} - (0.060\,0)(50.0)(-\hat{\mathbf{i}}) = \boxed{5.40\hat{\mathbf{i}}\ \text{N} \cdot \text{s}}$

(b) Work $= K_f - K_i = \frac{1}{2}(0.060\,0)\left[(40.0)^2 - (50.0)^2\right] = \boxed{-27.0\ \text{J}}$

P8.9 $\Delta\vec{\mathbf{p}} = \vec{\mathbf{F}}\Delta t$

$$\Delta p_y = m\left(v_{fy} - v_{iy}\right) = m(v\cos 60.0°) - mv\cos 60.0° = 0$$

$$\Delta p_x = m(-v\sin 60.0° - v\sin 60.0°) = -2mv\sin 60.0°$$

$$= -2(3.00 \text{ kg})(10.0 \text{ m/s})(0.866)$$

$$= -52.0 \text{ kg}\cdot\text{m/s}$$

$$F_{\text{ave}} = \frac{\Delta p_x}{\Delta t} = \frac{-52.0 \text{ kg}\cdot\text{m/s}}{0.200 \text{ s}} = \boxed{-260 \text{ N}}$$

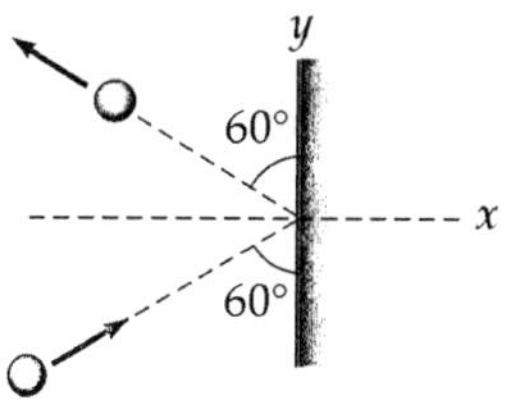

FIG. P8.9

P8.10 Take x-axis toward the pitcher

(a) $p_{ix} + I_x = p_{fx}$: $(0.200 \text{ kg})(15.0 \text{ m/s})(-\cos 45.0°) + I_x = (0.200 \text{ kg})(40.0 \text{ m/s})\cos 30.0°$

$$I_x = 9.05 \text{ N}\cdot\text{s}$$

$p_{iy} + I_y = p_{fy}$: $(0.200 \text{ kg})(15.0 \text{ m/s})(-\sin 45.0°) + I_y = (0.200 \text{ kg})(40.0 \text{ m/s})\sin 30.0°$

$$\vec{\mathbf{I}} = \boxed{\left(9.05\hat{\mathbf{i}} + 6.12\hat{\mathbf{j}}\right) \text{N}\cdot\text{s}}$$

(b) $\vec{\mathbf{I}} = \frac{1}{2}\left(0 + \vec{\mathbf{F}}_m\right)(4.00 \text{ ms}) + \vec{\mathbf{F}}_m(20.0 \text{ ms}) + \frac{1}{2}\vec{\mathbf{F}}_m(4.00 \text{ ms})$

$$\vec{\mathbf{F}}_m \times 24.0 \times 10^{-3} \text{ s} = \left(9.05\hat{\mathbf{i}} + 6.12\hat{\mathbf{j}}\right) \text{N}\cdot\text{s}$$

$$\vec{\mathbf{F}}_m = \boxed{\left(377\hat{\mathbf{i}} + 255\hat{\mathbf{j}}\right) \text{N}}$$

P8.11 The force exerted on the water by the hose is

$$F = \frac{\Delta p_{\text{water}}}{\Delta t} = \frac{mv_f - mv_i}{\Delta t} = \frac{(0.600 \text{ kg})(25.0 \text{ m/s}) - 0}{1.00 \text{ s}} = \boxed{15.0 \text{ N}}.$$

According to Newton's third law, the water exerts a force of equal magnitude back on the hose. Thus, the gardener must apply a 15.0 N force (in the direction of the velocity of the exiting water stream) to hold the hose stationary.

P8.12 (a) Energy is conserved for the spring-mass system:

$K_i + U_{si} = K_f + U_{sf}$: $0 + \frac{1}{2}kx^2 = \frac{1}{2}mv^2 + 0$

$$v = x\sqrt{\frac{k}{m}}$$

(b) From the equation, a $\boxed{\text{smaller}}$ value of m makes $v = x\sqrt{\frac{k}{m}}$ larger.

continued on next page

(c) $I=\left|\vec{\mathbf{p}}_f-\vec{\mathbf{p}}_i\right|=mv_f=0=mx\sqrt{\frac{k}{m}}=x\sqrt{km}$

(d) From the equation, a $\boxed{\text{larger}}$ value of m makes $I=x\sqrt{km}$ larger.

(e) For the glider, $W=K_f-K_i=\frac{1}{2}mv^2-0=\frac{1}{2}kx^2$

The mass makes $\boxed{\text{no difference}}$ to the work.

Section 8.3 Collisions

P8.13 (a) $mv_{1i}+3mv_{2i}=4mv_f$ where $m=2.50\times10^4$ kg

$$v_f=\frac{4.00+3(2.00)}{4}=\boxed{2.50\text{ m/s}}$$

(b) $$K_f-K_i=\frac{1}{2}(4m)v_f^2-\left[\frac{1}{2}mv_{1i}^2+\frac{1}{2}(3m)v_{2i}^2\right]=\left(2.50\times10^4\right)(12.5-8.00-6.00)=\boxed{-3.75\times10^4\text{ J}}$$

P8.14 (a) The internal forces exerted by the actor do not change the total momentum of the system of the four cars and the movie actor

$$(4m)v_i=(3m)(2.00\text{ m/s})+m(4.00\text{ m/s})$$

$$v_i=\frac{6.00\text{ m/s}+4.00\text{ m/s}}{4}=\boxed{2.50\text{ m/s}}$$

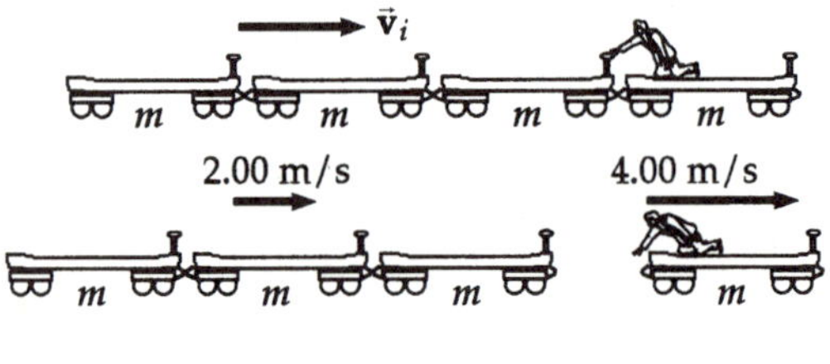

FIG. P8.14

(b) $$W_{\text{actor}}=K_f-K_i=\frac{1}{2}\left[(3m)(2.00\text{ m/s})^2+m(4.00\text{ m/s})^2\right]-\frac{1}{2}(4\text{ m})(2.50\text{ m/s})^2$$

$$W_{\text{actor}}=\frac{\left(2.50\times10^4\text{ kg}\right)}{2}(12.0+16.0-25.0)(\text{m/s})^2=\boxed{37.5\text{ kJ}}$$

(c) **[boxed]** The event considered here is the time reversal of the perfectly inelastic collision in the previous problem. The same momentum conservation equation describes both processes.

P8.15 (a), (b) Let v_g and v_p be the velocity of the girl and the plank relative to the ice surface. Then we may say that $v_g - v_p$ is the velocity of the girl relative to the plank, so that

$$v_g - v_p = 1.50 \qquad (1)$$

But also we must have $m_g v_g + m_p v_p = 0$, since total momentum of the girl-plank system is zero relative to the ice surface. Therefore

$$45.0v_g + 150v_p = 0, \text{ or } v_g = -3.33v_p$$

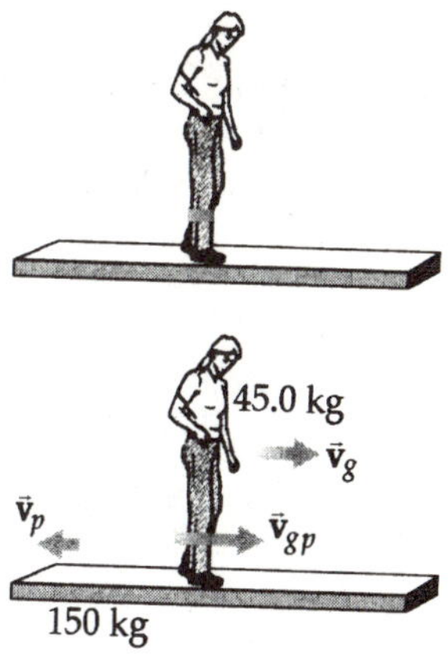

FIG. P8.15

Putting this into the equation (1) above gives $-3.33v_p - v_p = 1.50$ or $v_p = \boxed{-0.346 \text{ m/s}}$. Then $v_g = -3.33(-0.346) = \boxed{1.15 \text{ m/s}}$

P8.16 v_1, speed of m_1 at B before collision.

$$\frac{1}{2}m_1 v_1^2 = m_1 gh$$

$$v_1 = \sqrt{2(9.80)(5.00)} = 9.90 \text{ m/s}$$

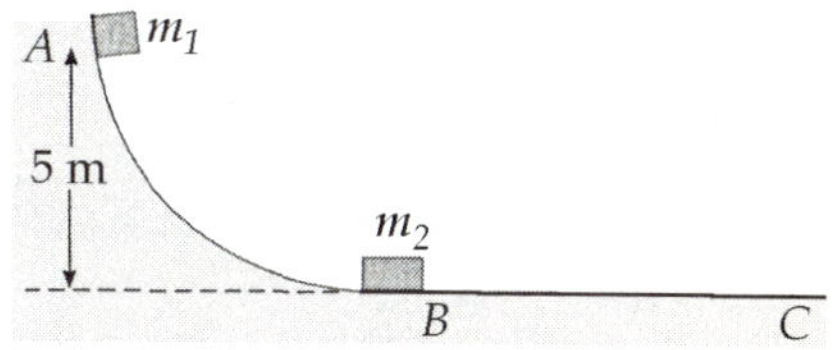

FIG. P8.16

v_{1f}, speed of m_1 at B just after collision.

$$v_{1f} = \frac{m_1 - m_2}{m_1 + m_2}v_1 = -\frac{1}{3}(9.90) \text{ m/s} = -3.30 \text{ m/s}$$

At the highest point (after collision)

$$m_1 g h_{\max} = \frac{1}{2}m_1(-3.30)^2 \qquad h_{\max} = \frac{(-3.30 \text{ m/s})^2}{2(9.80 \text{ m/s}^2)} = \boxed{0.556 \text{ m}}$$

P8.17 For the car-truck-driver-driver system, momentum is conserved:

$$\vec{\mathbf{p}}_{1i} + \vec{\mathbf{p}}_{2i} = \vec{\mathbf{p}}_{1f} + \vec{\mathbf{p}}_{2f}: \qquad (4\,000 \text{ kg})(8 \text{ m/s})\hat{\mathbf{i}} + (800 \text{ kg})(8 \text{ m/s})(-\hat{\mathbf{i}}) = (4\,800 \text{ kg})v_f\hat{\mathbf{i}}$$

$$v_f = \frac{25\,600 \text{ kg}\cdot\text{m/s}}{4\,800 \text{ kg}} = 5.33 \text{ m/s}$$

For the driver of the truck, the impulse-momentum theorem is

$$\vec{\mathbf{F}}\Delta t = \vec{\mathbf{p}}_f - \vec{\mathbf{p}}_i: \qquad \vec{\mathbf{F}}(0.120 \text{ s}) = (80 \text{ kg})(5.33 \text{ m/s})\hat{\mathbf{i}} - (80 \text{ kg})(8 \text{ m/s})\hat{\mathbf{i}}$$

$$\vec{\mathbf{F}} = \boxed{1.78 \times 10^3 \text{ N}(-\hat{\mathbf{i}}) \text{ on the truck driver}}$$

For the driver of the car, $\vec{\mathbf{F}}(0.120 \text{ s}) = (80 \text{ kg})(5.33 \text{ m/s})\hat{\mathbf{i}} - (80 \text{ kg})(8 \text{ m/s})(-\hat{\mathbf{i}})$

$$\vec{\mathbf{F}} = \boxed{8.89 \times 10^3 \text{ N}\hat{\mathbf{i}} \text{ on the car driver}}, \text{ 5 times larger.}$$

P8.18 Energy is conserved for the bob-Earth system between bottom and top of swing. At the top the stiff rod is in compression and the bob nearly at rest.

$$K_i + U_i = K_f + U_f: \qquad \frac{1}{2}Mv_b^2 + 0 = 0 + Mg2\ell$$

$$v_b^2 = g4\ell \text{ so } v_b = 2\sqrt{g\ell}$$

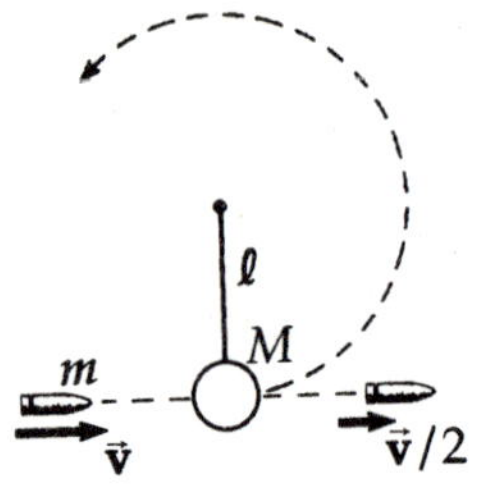

FIG. P8.18

Momentum of the bob-bullet system is conserved in the collision:

$$mv = m\frac{v}{2} + M\left(2\sqrt{g\ell}\right) \qquad \boxed{v = \frac{4M}{m}\sqrt{g\ell}}$$

P8.19 (a) According to the Example in the chapter text, the fraction of total kinetic energy transferred to the moderator is

$$f_2 = \frac{4m_1m_2}{(m_1+m_2)^2}$$

where m_2 is the moderator nucleus and in this case, $m_2 = 12m_1$

$$f_2 = \frac{4m_1(12m_1)}{(13m_1)^2} = \frac{48}{169} = \boxed{0.284 \text{ or } 28.4\%}$$

of the neutron energy is transferred to the carbon nucleus.

(b) $K_C = (0.284)\left(1.6\times10^{-13}\text{ J}\right) = \boxed{4.54\times10^{-14}\text{ J}}$

$K_n = (0.716)\left(1.6\times10^{-13}\text{ J}\right) = \boxed{1.15\times10^{-13}\text{ J}}$

P8.20 We assume equal firing speeds v and equal forces F required for the two bullets to push wood fibers apart. These equal forces act backward on the two bullets.

For the first, $K_i + \Delta E_{\text{mech}} = K_f \qquad \frac{1}{2}\left(7.00\times10^{-3}\text{ kg}\right)v^2 - F\left(8.00\times10^{-2}\text{ m}\right) = 0$

For the second, $p_i = p_f \qquad \left(7.00\times10^{-3}\text{ kg}\right)v = (1.014\text{ kg})v_f$

$$v_f = \frac{\left(7.00\times10^{-3}\right)v}{1.014}$$

Again, $K_i + \Delta E_{\text{mech}} = K_f: \qquad \frac{1}{2}\left(7.00\times10^{-3}\text{ kg}\right)v^2 - Fd = \frac{1}{2}(1.014\text{ kg})v_f^2$

Substituting for v_f, $\frac{1}{2}\left(7.00\times10^{-3}\text{ kg}\right)v^2 - Fd = \frac{1}{2}(1.014\text{ kg})\left(\frac{7.00\times10^{-3}v}{1.014}\right)^2$

$$Fd = \frac{1}{2}\left(7.00\times10^{-3}\right)v^2 - \frac{1}{2}\frac{\left(7.00\times10^{-3}\right)^2}{1.014}v^2$$

Substituting for v, $Fd = F\left(8.00\times10^{-2}\text{ m}\right)\left(1 - \frac{7.00\times10^{-3}}{1.014}\right) \qquad d = \boxed{7.94\text{ cm}}$

P8.21 At impact, momentum of the clay-block system is conserved, so:

$$mv_1 = (m_1 + m_2)v_2$$

After impact, the change in kinetic energy of the clay-block-surface system is equal to the increase in internal energy:

$$\frac{1}{2}(m_1 + m_2)v_2^2 = f_f d = \mu(m_1 + m_2)gd$$

$$\frac{1}{2}(0.112\text{ kg})v_2^2 = 0.650(0.112\text{ kg})(9.80\text{ m/s}^2)(7.50\text{ m})$$

$$v_2^2 = 95.6\text{ m}^2/\text{s}^2 \qquad v_2 = 9.77\text{ m/s}$$

$$(12.0\times10^{-3}\text{ kg})v_1 = (0.112\text{ kg})(9.77\text{ m/s}) \qquad v_1 = \boxed{91.2\text{ m/s}}$$

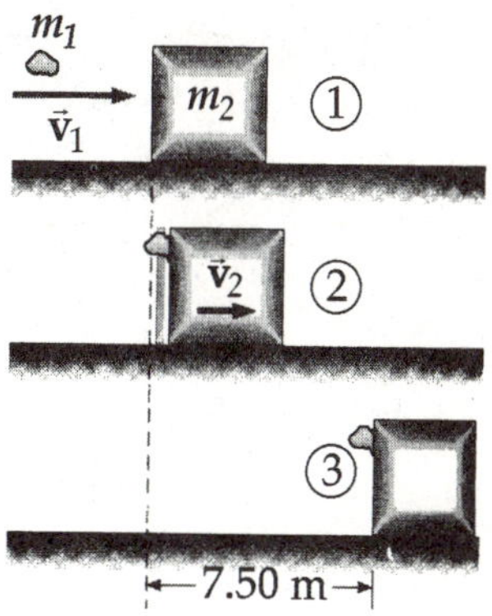

FIG. P8.21

***P8.22** (a) Using conservation of momentum, $\left(\sum\vec{\mathbf{p}}\right)_{\text{after}} = \left(\sum\vec{\mathbf{p}}\right)_{\text{before}}$, gives

$$[(4.0+10+3.0)\text{ kg}]v = (4.0\text{ kg})(5.0\text{ m/s}) + (10\text{ kg})(3.0\text{ m/s}) + (3.0\text{ kg})(-4.0\text{ m/s}).$$

Therefore, $v = +2.24\text{ m/s}$, or $\boxed{2.24\text{ m/s toward the right}}$.

(b) $\boxed{\text{No}}$. For example, if the 10-kg and 3.0-kg mass were to stick together first, they would move with a speed given by solving

$$(13\text{ kg})v_1 = (10\text{ kg})(3.0\text{ m/s}) + (3.0\text{ kg})(-4.0\text{ m/s}),\text{ or } v_1 = +1.38\text{ m/s}.$$

Then when this 13 kg combined mass collides with the 4.0 kg mass, we have

$$(17\text{ kg})v = (13\text{ kg})(1.38\text{ m/s}) + (4.0\text{ kg})(5.0\text{ m/s}),\text{ and } v = +2.24\text{ m/s}$$

just as in part (a). Coupling order makes no difference.

***P8.23** (a) We assume that energy is conserved in the fall of the basketball and the tennis ball. Each reaches its lowest point with a speed given by

$$(K+U_g)_{\text{release}} = (K+U_g)_{\text{bottom}}$$

$$0 + mgy_i = \frac{1}{2}mv_b^2 + 0$$

$$v_b = \sqrt{2gy_i} = \sqrt{2(9.8\text{ m/s}^2)(1.20\text{ m})} = \boxed{4.85\text{ m/s}}$$

continued on next page

(b) The two balls exert no forces on each other as they move down. They collide with each other after the basketball has its velocity reversed by the floor. We choose upward as positive. Momentum conservation:

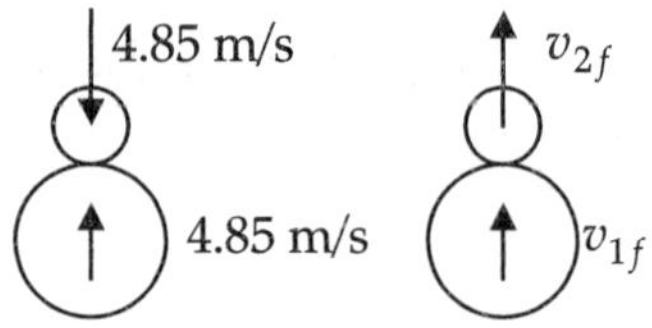

FIG. P8.23(b)

$$(57\text{ g})(-4.85\text{ m/s})+(590\text{ g})(4.85\text{ m/s})=(57\text{ g})v_{2f}+(590\text{ g})v_{1f}$$

To describe the elastic character of the collision, we use the relative velocity equation

$$4.85\text{ m/s}-(-4.85\text{ m/s})=v_{2f}-v_{1f}$$

we solve by substitution

$$v_{1f}=v_{2f}-9.70\text{ m/s}$$
$$2\,580\text{ gm/s}=(57\text{ g})v_{2f}+(590\text{ g})(v_{2f}-9.70\text{ m/s})$$
$$=(57\text{ g})v_{2f}+(590\text{ g})v_{2f}-5\,720\text{ gm/s}$$
$$v_{2f}=\frac{8\,310\text{ m/s}}{647}=12.8\text{ m/s}$$

Now the tennis ball-Earth system keeps constant energy as the ball rises:

$$\frac{1}{2}(57\text{ g})(12.8\text{ m/s})^2=(57\text{ g})(9.8\text{ m/s}^2)y_f$$
$$y_f=\frac{165\text{ m}^2/\text{s}^2}{2(9.8\text{ m/s}^2)}=\boxed{8.41\text{ m}}$$

Section 8.4 Two-Dimensional Collisions

P8.24 (a) First, we conserve momentum for the system of two football players in the x direction (the direction of travel of the fullback).

$$(90.0\text{ kg})(5.00\text{ m/s})+0=(185\text{ kg})V\cos\theta$$

where θ is the angle between the direction of the final velocity V and the x axis. We find

$$V\cos\theta=2.43\text{ m/s} \qquad (1)$$

Now consider conservation of momentum of the system in the y direction (the direction of travel of the opponent).

$$(95.0\text{ kg})(3.00\text{ m/s})+0=(185\text{ kg})(V\sin\theta)$$

which gives,

$$V\sin\theta=1.54\text{ m/s} \qquad (2)$$

continued on next page

Divide equation (2) by (1) $\tan\theta = \dfrac{1.54}{2.43} = 0.633$

From which $\boxed{\theta = 32.3°}$

Then, either (1) or (2) gives $V = \boxed{2.88 \text{ m/s}}$

(b) $K_i = \frac{1}{2}(90.0 \text{ kg})(5.00 \text{ m/s})^2 + \frac{1}{2}(95.0 \text{ kg})(3.00 \text{ m/s})^2 = 1.55\times10^3 \text{ J}$

$K_f = \frac{1}{2}(185 \text{ kg})(2.88 \text{ m/s})^2 = 7.67\times10^2 \text{ J}$

Thus, the kinetic energy lost is $\boxed{\text{783 J into internal energy.}}$

P8.25 $p_{xf} = p_{xi}$: $mv_O\cos\theta + mv_Y\cos(90.0°-\theta) = mv_i$

$$v_O\cos\theta + v_Y\sin\theta = v_i \qquad (1)$$

$p_{yf} = p_{yi}$: $mv_O\sin\theta - mv_Y\sin(90.0°-\theta) = 0$

$$v_O\sin\theta = v_Y\cos\theta \qquad (2)$$

From equation (2),

$$v_O = v_Y\left(\frac{\cos\theta}{\sin\theta}\right) \qquad (3)$$

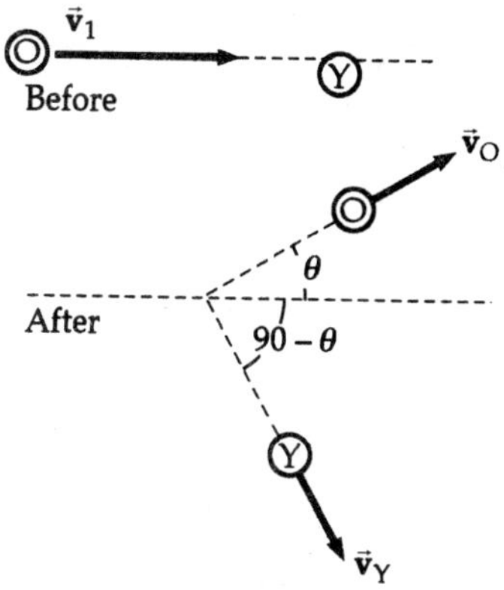

FIG. P8.25

Substituting into equation (1),

$$v_Y\left(\frac{\cos^2\theta}{\sin\theta}\right) + v_Y\sin\theta = v_i$$

so $v_Y\left(\cos^2\theta + \sin^2\theta\right) = v_i\sin\theta$, and $\boxed{v_Y = v_i\sin\theta}$.

Then, from equation (3), $\boxed{v_O = v_i\cos\theta}$.

We did not need to write down an equation expressing conservation of mechanical energy. In the problem situation, the requirement of perpendicular final velocities is equivalent to the condition of elasticity.

P8.26 We use conservation of momentum for the system of two vehicles for both northward and eastward components.

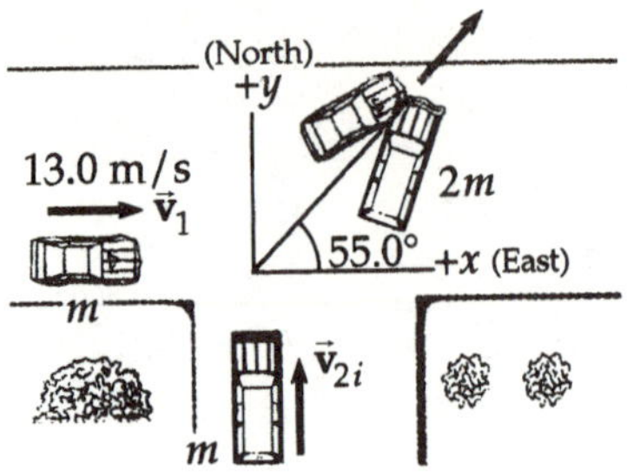

FIG. P8.26

For the eastward direction:

$$M(13.0\ \text{m/s}) = 2MV_f \cos 55.0°$$

For the northward direction:

$$Mv_{2i} = 2MV_f \sin 55.0°$$

Divide the northward equation by the eastward equation to find:

$$v_{2i} = (13.0\ \text{m/s})\tan 55.0° = 18.6\ \text{m/s} = \boxed{41.5\ \text{mi/h}}$$

Thus, the driver of the north bound car was untruthful.

P8.27 By conservation of momentum for the system of the two billiard balls (with all masses equal),

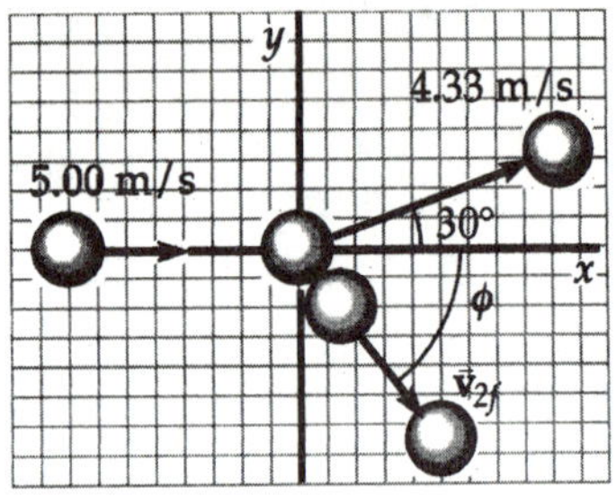

FIG. P8.27

$$5.00\ \text{m/s} + 0 = (4.33\ \text{m/s})\cos 30.0° + v_{2fx}$$
$$v_{2fx} = 1.25\ \text{m/s}$$
$$0 = (4.33\ \text{m/s})\sin 30.0° + v_{2fy}$$
$$v_{2fy} = -2.16\ \text{m/s}$$
$$\vec{\mathbf{v}}_{2f} = \boxed{2.50\ \text{m/s at } -60.0°}$$

Note that we did not need to use the fact that the collision is perfectly elastic.

P8.28 (a) $\vec{\mathbf{p}}_i = \vec{\mathbf{p}}_f$ so $p_{xi} = p_{xf}$

and $p_{yi} = p_{yf}$

$$mv_i = mv\cos\theta + mv\cos\phi \quad (1)$$

$$0 = mv\sin\theta + mv\sin\phi \quad (2)$$

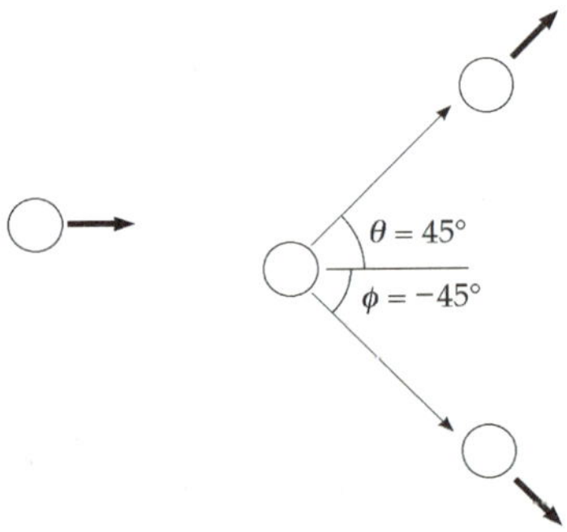

FIG. P8.28

From (2), $\sin\theta = -\sin\phi$

so $\theta = -\phi$

Furthermore, energy conservation for the system of two protons requires

$$\frac{1}{2}mv_i^2 = \frac{1}{2}mv^2 + \frac{1}{2}mv^2$$

so $\boxed{v = \dfrac{v_i}{\sqrt{2}}}$

(b) Hence, (1) gives $v_i = \dfrac{2v_i\cos\theta}{\sqrt{2}}$ $\qquad \theta = \boxed{45.0°}$ $\qquad \phi = \boxed{-45.0°}$

P8.29 $m_1\vec{\mathbf{v}}_{1i}+m_2\vec{\mathbf{v}}_{2i}=(m_1+m_2)\vec{\mathbf{v}}_f$: $\quad 3.00(5.00)\hat{\mathbf{i}}-6.00\hat{\mathbf{j}}=5.00\vec{\mathbf{v}}$

$$\vec{\mathbf{v}}=\boxed{\left(3.00\hat{\mathbf{i}}-1.20\hat{\mathbf{j}}\right)\text{ m/s}}$$

P8.30 (a) We write equations expressing conservation of the x and y components of momentum, with reference to the figures. Let the puck initially at rest be m_2.

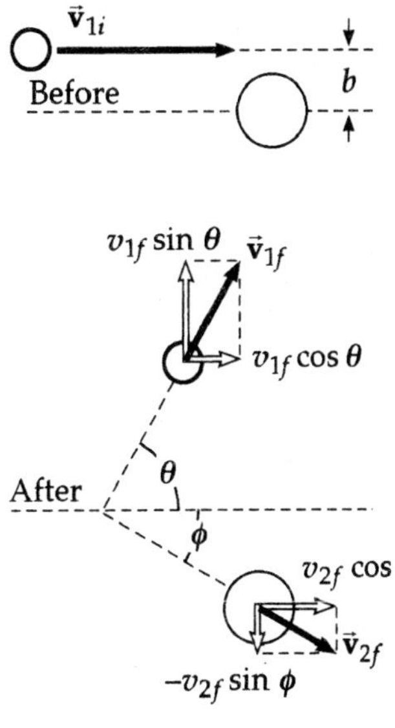

FIG. P8.30

$$m_1v_{1i}=m_1v_{1f}\cos\theta+m_2v_{2f}\cos\phi:$$

$$v_{2f}\cos\phi=\frac{m_1v_{1i}-m_1v_{1f}\cos\theta}{m_2}$$

$$v_{2f}\cos\phi=\frac{(0.200\text{ kg})(2.00\text{ m/s})-(0.200\text{ kg})(1.00\text{ m/s})\cos 53.0°}{0.300\text{ kg}}$$

$$0=m_1v_{1f}\sin\theta-m_2v_{2f}\sin\phi:\qquad v_{2f}\sin\phi=\frac{m_1v_{1f}\sin\theta}{m_2}$$

$$0=(0.200\text{ kg})(1.00\text{ m/s})\sin 53.0°-(0.300\text{ kg})(v_{2f}\sin\phi)$$

From these equations, we find $\tan\phi=\dfrac{\sin\phi}{\cos\phi}=\dfrac{0.160}{0.280}=0.571$

$$\boxed{\phi=29.7°}$$

Then $\quad v_{2f}=\dfrac{(0.160\text{ kg}\cdot\text{m/s})}{(0.300\text{ kg})(\sin 29.7°)}=\boxed{1.07\text{ m/s}}$

(b) $\quad f_{\text{lost}}=\dfrac{\Delta K}{K_i}=\dfrac{K_f-K_i}{K_i}=\boxed{-0.318}$

P8.31 $m_0=17.0\times10^{-27}$ kg $\qquad \vec{\mathbf{v}}_i=0$ (the parent nucleus)

$m_1=5.00\times10^{-27}$ kg $\qquad \vec{\mathbf{v}}_1=6.00\times10^6\,\hat{\mathbf{j}}$ m/s

$m_2=8.40\times10^{-27}$ kg $\qquad \vec{\mathbf{v}}_2=4.00\times10^6\,\hat{\mathbf{i}}$ m/s

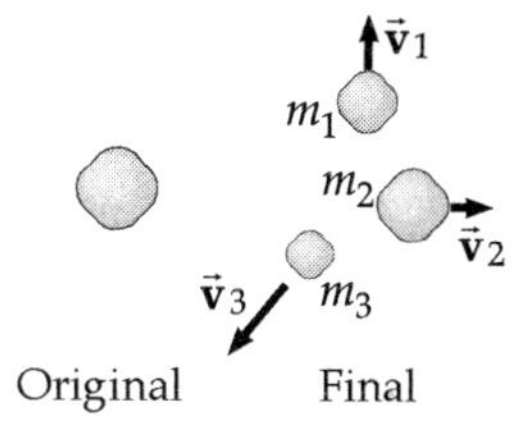

FIG. P8.31

(a) $\quad m_1\vec{\mathbf{v}}_1+m_2\vec{\mathbf{v}}_2+m_3\vec{\mathbf{v}}_3=0$

where $m_3=m_0-m_1-m_2=3.60\times10^{-27}$ kg

$$\left(5.00\times10^{-27}\right)\left(6.00\times10^6\hat{\mathbf{j}}\right)+\left(8.40\times10^{-27}\right)\left(4.00\times10^6\hat{\mathbf{i}}\right)+\left(3.60\times10^{-27}\right)\vec{\mathbf{v}}_3=0$$

$$\vec{\mathbf{v}}_3=\boxed{\left(-9.33\times10^6\hat{\mathbf{i}}-8.33\times10^6\hat{\mathbf{j}}\right)\text{ m/s}}$$

continued on next page

(b) $$E = \frac{1}{2}m_1 v_1^2 + \frac{1}{2}m_2 v_2^2 + \frac{1}{2}m_3 v_3^2$$

$$E = \frac{1}{2}\left[\left(5.00\times10^{-27}\right)\left(6.00\times10^{6}\right)^2 + \left(8.40\times10^{-27}\right)\left(4.00\times10^{6}\right)^2 + \left(3.60\times10^{-27}\right)\left(12.5\times10^{6}\right)^2\right]$$

$$\boxed{E = 4.39\times10^{-13}\text{ J}}$$

Section 8.5 The Center of Mass

P8.32 The x-coordinate of the center of mass is

$$x_{CM} = \frac{\sum m_i x_i}{\sum m_i} = \frac{0+0+0+0}{(2.00\text{ kg} + 3.00\text{ kg} + 2.50\text{ kg} + 4.00\text{ kg})}$$

$$\boxed{x_{CM} = 0}$$

and the y-coordinate of the center of mass is

$$y_{CM} = \frac{\sum m_i y_i}{\sum m_i} = \frac{(2.00\text{ kg})(3.00\text{ m}) + (3.00\text{ kg})(2.50\text{ m}) + (2.50\text{ kg})(0) + (4.00\text{ kg})(-0.500\text{ m})}{2.00\text{ kg} + 3.00\text{ kg} + 2.50\text{ kg} + 4.00\text{ kg}}$$

$$\boxed{y_{CM} = 1.00\text{ m}}$$

P8.33 Let A_1 represent the area of the bottom row of squares, A_2 the middle square, and A_3 the top pair.

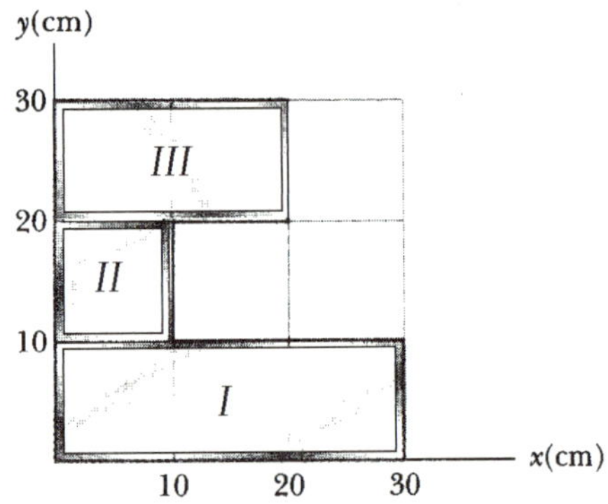

FIG. P8.33

$$A = A_1 + A_2 + A_3$$

$$M = M_1 + M_2 + M_3$$

$$\frac{M_1}{A_1} = \frac{M}{A}$$

$$A_1 = 300\text{ cm}^2,\ A_2 = 100\text{ cm}^2,\ A_3 = 200\text{ cm}^2,\ A = 600\text{ cm}^2$$

$$M_1 = M\left(\frac{A_1}{A}\right) = \frac{300\text{ cm}^2}{600\text{ cm}^2}M = \frac{M}{2}$$

$$M_2 = M\left(\frac{A_2}{A}\right) = \frac{100\text{ cm}^2}{600\text{ cm}^2}M = \frac{M}{6}$$

$$M_3 = M\left(\frac{A_3}{A}\right) = \frac{200\text{ cm}^2}{600\text{ cm}^2}M = \frac{M}{3}$$

$$x_{CM} = \frac{x_1 M_1 + x_2 M_2 + x_3 M_3}{M} = \frac{15.0\text{ cm}\left(\frac{1}{2}M\right) + 5.00\text{ cm}\left(\frac{1}{6}M\right) + 10.0\text{ cm}\left(\frac{1}{3}M\right)}{M}$$

$$x_{CM} = \boxed{11.7\text{ cm}}$$

$$y_{CM} = \frac{\frac{1}{2}M(5.00\text{ cm}) + \frac{1}{6}M(15.0\text{ cm}) + \left(\frac{1}{3}M\right)(25.0\text{ cm})}{M} = 13.3\text{ cm}$$

$$y_{CM} = \boxed{13.3\text{ cm}}$$

P8.34 Take x-axis starting from the oxygen nucleus and pointing toward the middle of the **V**.

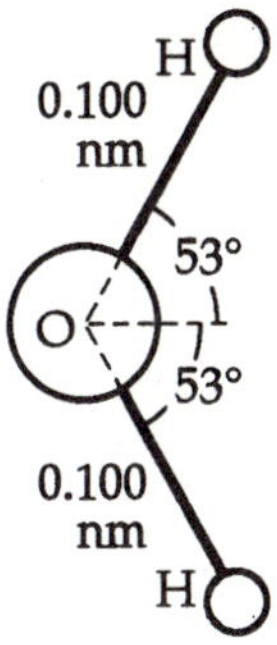

FIG. P8.34

Then $\boxed{y_{CM} = 0}$

and $x_{CM} = \dfrac{\sum m_i x_i}{\sum m_i} =$

$$x_{CM} = \frac{0 + 1.008\text{ u}(0.100\text{ nm})\cos 53.0° + 1.008\text{ u}(0.100\text{ nm})\cos 53.0°}{15.999\text{ u} + 1.008\text{ u} + 1.008\text{ u}}$$

$\boxed{x_{CM} = 0.006\,73\text{ nm from the oxygen nucleus}}$

***P8.35** (a) Represent the height of a particle of mass dm within the object as y. Its contribution to the gravitational energy of the object-Earth system is $(dm)gy$. The total gravitational energy is $U_g = \int_{\text{all mass}} gy\,dm = g\int y\,dm$. For the center of mass we have $y_{CM} = \frac{1}{M}\int y\,dm$, so $U_g = gMy_{CM}$.

(b) The volume of the ramp is $\frac{1}{2}(3.6\text{ m})(15.7\text{ m})(64.8\text{ m}) = 1.83 \times 10^3\text{ m}^3$. Its mass is $\rho V = (3\,800\text{ kg/m}^3)(1.83 \times 10^3\text{ m}^3) = 6.96 \times 10^6\text{ kg}$. Its center of mass is above its base by one-third of its height, $y_{CM} = \frac{1}{3}15.7\text{ m} = 5.23\text{ m}$. Then

$U_g = Mgy_{CM} = 6.96 \times 10^6\text{ kg}(9.8\text{ m/s}^2)5.23\text{ m} = \boxed{3.57 \times 10^8\text{ J}}$.

P8.36 (a) $M = \int_0^{0.300\text{ m}} \lambda dx = \int_0^{0.300\text{ m}} [50.0\text{ g/m} + 20.0x\text{ g/m}^2]dx$

$M = [50.0x\text{ g/m} + 10.0x^2\text{ g/m}^2]_0^{0.300\text{ m}} = \boxed{15.9\text{ g}}$

(b) $x_{CM} = \dfrac{\int_{\text{all mass}} x dm}{M} = \dfrac{1}{M}\int_0^{0.300\text{ m}} \lambda x dx = \dfrac{1}{M}\int_0^{0.300\text{ m}} [50.0x\text{ g/m} + 20.0x^2\text{ g/m}^2]dx$

$$x_{CM} = \frac{1}{15.9\text{ g}}\left[25.0x^2\text{ g/m} + \frac{20x^3\text{ g/m}^2}{3}\right]_0^{0.300\text{ m}} = \boxed{0.153\text{ m}}$$

Section 8.6 Motion of a System of Particles

P8.37 (a) $\vec{\mathbf{v}}_{CM} = \frac{\sum m_i \vec{\mathbf{v}}_i}{M} = \frac{m_1\vec{\mathbf{v}}_1 + m_2\vec{\mathbf{v}}_2}{M}$

$$= \frac{(2.00 \text{ kg})(2.00\hat{\mathbf{i}} \text{ m/s} - 3.00\hat{\mathbf{j}} \text{ m/s}) + (3.00 \text{ kg})(1.00\hat{\mathbf{i}} \text{ m/s} + 6.00\hat{\mathbf{j}} \text{ m/s})}{5.00 \text{ kg}}$$

$$\vec{\mathbf{v}}_{CM} = \boxed{(1.40\hat{\mathbf{i}} + 2.40\hat{\mathbf{j}}) \text{ m/s}}$$

(b) $\vec{\mathbf{p}} = M\vec{\mathbf{v}}_{CM} = (5.00 \text{ kg})(1.40\hat{\mathbf{i}} + 2.40\hat{\mathbf{j}}) \text{ m/s} = \boxed{(7.00\hat{\mathbf{i}} + 12.0\hat{\mathbf{j}}) \text{ kg}\cdot\text{m/s}}$

P8.38 (a) See figure to the right.

(b) Using the definition of the position vector at the center of mass,

$$\vec{\mathbf{r}}_{CM} = \frac{m_1\vec{\mathbf{r}}_1 + m_2\vec{\mathbf{r}}_2}{m_1 + m_2}$$

$$\vec{\mathbf{r}}_{CM} = \frac{(2.00 \text{ kg})(1.00 \text{ m}, 2.00 \text{ m}) + (3.00 \text{ kg})(-4.00 \text{ m}, -3.00 \text{ m})}{2.00 \text{ kg} + 3.00 \text{ kg}}$$

$$\vec{\mathbf{r}}_{CM} = \boxed{(-2.00\hat{\mathbf{i}} - 1.00\hat{\mathbf{j}}) \text{ m}}$$

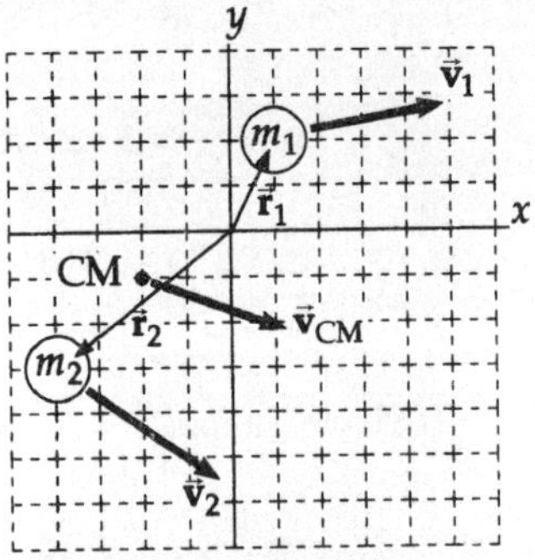

FIG. P8.38

(c) The velocity of the center of mass is

$$\vec{\mathbf{v}}_{CM} = \frac{\vec{\mathbf{P}}}{M} = \frac{m_1\vec{\mathbf{v}}_1 + m_2\vec{\mathbf{v}}_2}{m_1 + m_2} = \frac{(2.00 \text{ kg})(3.00 \text{ m/s}, 0.50 \text{ m/s}) + (3.00 \text{ kg})(3.00 \text{ m/s}, -2.00 \text{ m/s})}{(2.00 \text{ kg} + 3.00 \text{ kg})}$$

$$\vec{\mathbf{v}}_{CM} = \boxed{(3.00\hat{\mathbf{i}} - 1.00\hat{\mathbf{j}}) \text{ m/s}}$$

(d) The total linear momentum of the system can be calculated as $\vec{\mathbf{P}} = M\vec{\mathbf{v}}_{CM}$

or as $\vec{\mathbf{P}} = m_1\vec{\mathbf{v}}_1 + m_2\vec{\mathbf{v}}_2$.

Either gives $\vec{\mathbf{P}} = \boxed{(15.0\hat{\mathbf{i}} - 5.00\hat{\mathbf{j}}) \text{ kg}\cdot\text{m/s}}$.

P8.39 Let $x =$ distance from shore to center of boat

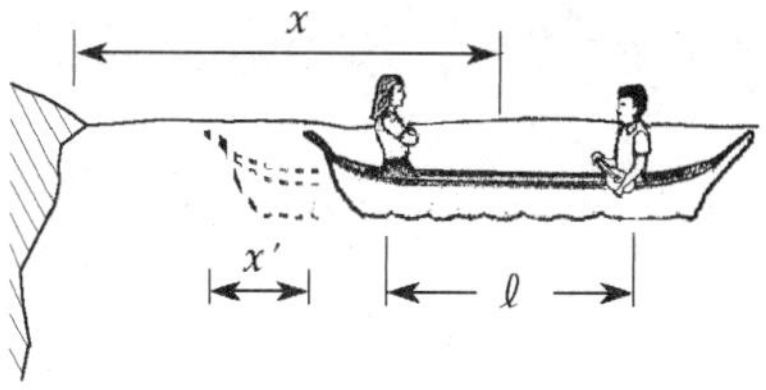

FIG. P8.39

$\ell =$ length of boat

$x' =$ distance boat moves as Juliet moves toward Romeo

The center of mass stays fixed.

Before: $$x_{\text{CM}} = \frac{\left[M_b x + M_J\left(x - \frac{\ell}{2}\right) + M_R\left(x + \frac{\ell}{2}\right)\right]}{\left(M_B + M_J + M_R\right)}$$

After: $$x_{\text{CM}} = \frac{\left[M_B(x - x') + M_J\left(x + \frac{\ell}{2} - x'\right) + M_R\left(x + \frac{\ell}{2} - x'\right)\right]}{\left(M_B + M_J + M_R\right)}$$

$$\ell\left(-\frac{55.0}{2} + \frac{77.0}{2}\right) = x'(-80.0 - 55.0 - 77.0) + \frac{\ell}{2}(55.0 + 77.0)$$

$$x' = \frac{55.0\ell}{212} = \frac{55.0(2.70)}{212} = \boxed{0.700\text{ m}}$$

P8.40 (a) Conservation of momentum for the two-ball system gives us:

$$0.200\text{ kg}(1.50\text{ m/s}) + 0.300\text{ kg}(-0.400\text{ m/s}) = 0.200\text{ kg } v_{1f} + 0.300\text{ kg } v_{2f}$$

Relative velocity equation:

$$v_{2f} - v_{1f} = 1.90\text{ m/s}$$

Then $$0.300 - 0.120 = 0.200 v_{1f} + 0.300\left(1.90 + v_{1f}\right)$$

$$v_{1f} = -0.780\text{ m/s} \qquad v_{2f} = 1.12\text{ m/s}$$

$$\boxed{\vec{\mathbf{v}}_{1f} = -0.780\hat{\mathbf{i}}\text{ m/s}} \qquad \boxed{\vec{\mathbf{v}}_{2f} = 1.12\hat{\mathbf{i}}\text{ m/s}}$$

(b) Before, $$\vec{\mathbf{v}}_{\text{CM}} = \frac{(0.200\text{ kg})(1.50\text{ m/s})\hat{\mathbf{i}} + (0.300\text{ kg})(-0.400\text{ m/s})\hat{\mathbf{i}}}{0.500\text{ kg}}$$

$$\boxed{\vec{\mathbf{v}}_{\text{CM}} = (0.360\text{ m/s})\hat{\mathbf{i}}}$$

Afterwards, the center of mass must move at the same velocity, as momentum of the system is conserved.

Section 8.7 Context Connection—Rocket Propulsion

P8.41 (a) Thrust $= \left| v_e \frac{dM}{dt} \right|$ Thrust $= \left(2.60\times10^3\ \text{m/s}\right)\left(1.50\times10^4\ \text{kg/s}\right) = \boxed{3.90\times10^7\ \text{N}}$

(b) $\sum F_y = \text{Thrust} - Mg = Ma$: $3.90\times10^7 - \left(3.00\times10^6\right)(9.80) = \left(3.00\times10^6\right)a$

$a = \boxed{3.20\ \text{m/s}^2}$

P8.42 (a) The fuel burns at a rate $\frac{dM}{dt} = \frac{12.7\ \text{g}}{1.90\ \text{s}} = 6.68\times10^{-3}\ \text{kg/s}$

Thrust $= v_e \frac{dM}{dt}$: $5.26\ \text{N} = v_e\left(6.68\times10^{-3}\ \text{kg/s}\right)$

$v_e = \boxed{787\ \text{m/s}}$

(b) $v_f - v_i = v_e \ln\left(\frac{M_i}{M_f}\right)$: $v_f - 0 = (787\ \text{m/s})\ln\left(\frac{53.5\ \text{g} + 25.5\ \text{g}}{53.5\ \text{g} + 25.5\ \text{g} - 12.7\ \text{g}}\right)$

$v_f = \boxed{138\ \text{m/s}}$

P8.43 $v = v_e \ln\frac{M_i}{M_f}$

(a) $M_i = e^{v/v_e} M_f$ $M_i = e^5\left(3.00\times10^3\ \text{kg}\right) = 4.45\times10^5\ \text{kg}$

The mass of fuel and oxidizer is $\Delta M = M_i - M_f = (445 - 3.00)\times10^3\ \text{kg} = \boxed{442\ \text{metric tons}}$

(b) $\Delta M = e^2(3.00\ \text{metric tons}) - 3.00\ \text{metric tons} = \boxed{19.2\ \text{metric tons}}$

Because of the exponential, a relatively small increase in fuel and/or engine efficiency causes a large change in the amount of fuel and oxidizer required.

***P8.44** (a) From Equation 8.43, $v - 0 = v_e \ln\left(\frac{M_i}{M_f}\right) = -v_e \ln\left(\frac{M_f}{M_i}\right)$

Now, $M_f = M_i - kt$, so $v = -v_e \ln\left(\frac{M_i - kt}{M_i}\right) = -v_e \ln\left(1 - \frac{k}{M_i}t\right)$

With the definition, $T_p \equiv \frac{M_i}{k}$, this becomes

$$v(t) = \boxed{-v_e \ln\left(1 - \frac{t}{T_p}\right)}$$

continued on next page

(b) With $v_e = 1\,500$ m/s, and $T_p = 144$ s, $v = -(1\,500 \text{ m/s})\ln\left(1 - \frac{t}{144 \text{ s}}\right)$

t(s)	v(m/s)
0	0
20	224
40	488
60	808
80	1220
100	1780
120	2690
132	3730

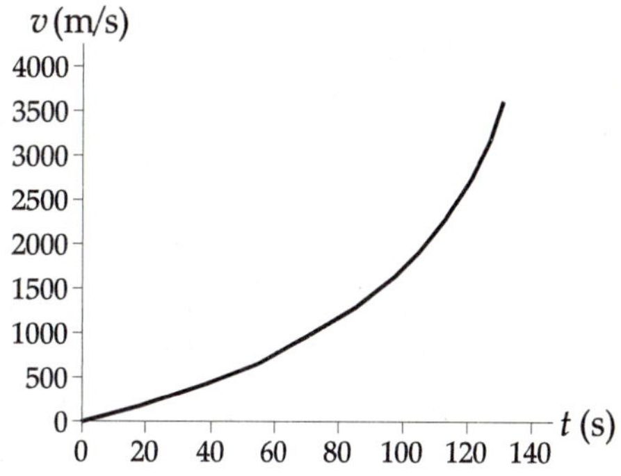

FIG. P8.44(b)

(c) $$a(t) = \frac{dv}{dt} = \frac{d\left[-v_e \ln\left(1 - \frac{t}{T_p}\right)\right]}{dt} = -v_e\left(\frac{1}{1 - \frac{t}{T_p}}\right)\left(-\frac{1}{T_p}\right) = \left(\frac{v_e}{T_p}\right)\left(\frac{1}{1 - \frac{t}{T_p}}\right), \text{ or}$$

$$a(t) = \boxed{\frac{v_e}{T_p - t}}$$

(d) With $v_e = 1\,500$ m/s, and $T_p = 144$ s, $a = \frac{1\,500 \text{ m/s}}{144 \text{ s} - t}$

t(s)	a(m/s^2)
0	10.4
20	12.1
40	14.4
60	17.9
80	23.4
100	34.1
120	62.5
132	125

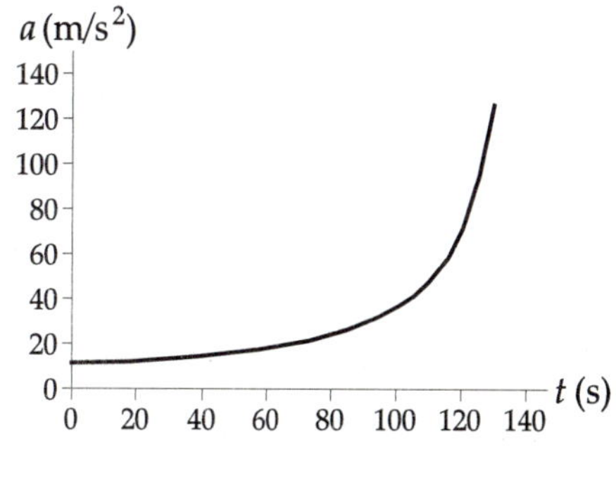

FIG. P8.44(d)

(e) $$x(t) = 0 + \int_0^t v\,dt = \int_0^t \left[-v_e \ln\left(1 - \frac{t}{T_p}\right)\right] dt = v_e T_p \int_0^t \ln\left[1 - \frac{t}{T_p}\right]\left(-\frac{dt}{T_p}\right)$$

$$x(t) = v_e T_p \left[\left(1 - \frac{t}{T_p}\right)\ln\left(1 - \frac{t}{T_p}\right) - \left(1 - \frac{t}{T_p}\right)\right]_0^t$$

$$x(t) = \boxed{v_e\left(T_p - t\right)\ln\left(1 - \frac{t}{T_p}\right) + v_e t}$$

continued on next page

(f) With $v_e = 1\,500 \text{ m/s} = 1.50 \text{ km/s}$, and $T_p = 144 \text{ s}$,

$$x = 1.50(144 - t)\ln\left(1 - \frac{t}{144}\right) + 1.50t$$

t(s)	x(km)
0	0
20	2.19
40	9.23
60	22.1
80	42.2
100	71.7
120	115
132	153

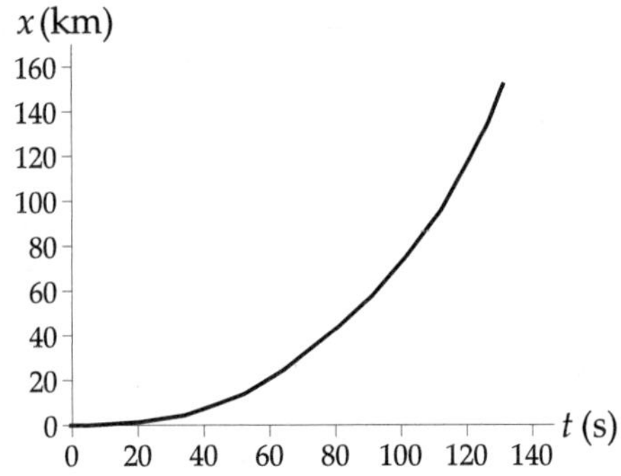

FIG. P8.44(f)

P8.45 The thrust acting on the spacecraft is

$\sum F = ma$: $\quad \sum F = (3\,500 \text{ kg})(2.50 \times 10^{-6})(9.80 \text{ m/s}^2) = 8.58 \times 10^{-2} \text{ N}$

thrust $= \left(\frac{dM}{dt}\right)v_e$: $\quad 8.58 \times 10^{-2} \text{ N} = \left(\frac{\Delta M}{3\,600 \text{ s}}\right)(70 \text{ m/s})$

$$\Delta M = \boxed{4.41 \text{ kg}}$$

Additional Problems

P8.46 (a) When the spring is fully compressed, each cart moves with same velocity $\vec{\mathbf{v}}$. Apply conservation of momentum for the system of two gliders

$\vec{\mathbf{p}}_i = \vec{\mathbf{p}}_f$: $\quad m_1\vec{\mathbf{v}}_1 + m_2\vec{\mathbf{v}}_2 = (m_1 + m_2)\vec{\mathbf{v}} \qquad \boxed{\vec{\mathbf{v}} = \frac{m_1\vec{\mathbf{v}}_1 + m_2\vec{\mathbf{v}}_2}{m_1 + m_2}}$

(b) Only conservative forces act, therefore $\Delta E = 0$. $\frac{1}{2}m_1v_1^2 + \frac{1}{2}m_2v_2^2 = \frac{1}{2}(m_1 + m_2)v^2 + \frac{1}{2}kx_m^2$

Substitute for v from (a) and solve for x_m.

$$x_m^2 = \frac{(m_1 + m_2)m_1v_1^2 + (m_1 + m_2)m_2v_2^2 - (m_1v_1)^2 - (m_2v_2)^2 - 2m_1m_2v_1v_2}{k(m_1 + m_2)}$$

$$x_m = \sqrt{\frac{m_1m_2\left(v_1^2 + v_2^2 - 2v_1v_2\right)}{k(m_1 + m_2)}} = \boxed{(v_1 - v_2)\sqrt{\frac{m_1m_2}{k(m_1 + m_2)}}}$$

continued on next page

(c) $m_1\vec{\mathbf{v}}_1 + m_2\vec{\mathbf{v}}_2 = m_1\vec{\mathbf{v}}_{1f} + m_2\vec{\mathbf{v}}_{2f}$

Conservation of momentum: $m_1\left(\vec{\mathbf{v}}_1 - \vec{\mathbf{v}}_{1f}\right) = m_2\left(\vec{\mathbf{v}}_{2f} - \vec{\mathbf{v}}_2\right)$ (1)

Conservation of energy: $\frac{1}{2}m_1v_1^2 + \frac{1}{2}m_2v_2^2 = \frac{1}{2}m_1v_{1f}^2 + \frac{1}{2}m_2v_{2f}^2$

which simplifies to: $m_1\left(v_1^2 - v_{1f}^2\right) = m_2\left(v_{2f}^2 - v_2^2\right)$

Factoring gives $m_1\left(\vec{\mathbf{v}}_1 - \vec{\mathbf{v}}_{1f}\right)\cdot\left(\vec{\mathbf{v}}_1 + \vec{\mathbf{v}}_{1f}\right) = m_2\left(\vec{\mathbf{v}}_{2f} - \vec{\mathbf{v}}_2\right)\cdot\left(\vec{\mathbf{v}}_{2f} + \vec{\mathbf{v}}_2\right)$

and with the use of the momentum equation (equation (1)),

this reduces to $\left(\vec{\mathbf{v}}_1 + \vec{\mathbf{v}}_{1f}\right) = \left(\vec{\mathbf{v}}_{2f} + \vec{\mathbf{v}}_2\right)$

or $\vec{\mathbf{v}}_{1f} = \vec{\mathbf{v}}_{2f} + \vec{\mathbf{v}}_2 - \vec{\mathbf{v}}_1$ (2)

Substituting equation (2) into equation (1) and simplifying yields:

$$\vec{\mathbf{v}}_{2f} = \boxed{\left(\frac{2m_1}{m_1 + m_2}\right)\vec{\mathbf{v}}_1 + \left(\frac{m_2 - m_1}{m_1 + m_2}\right)\vec{\mathbf{v}}_2}$$

Upon substitution of this expression for $\vec{\mathbf{v}}_{2f}$ into equation 2, one finds

$$\vec{\mathbf{v}}_{1f} = \boxed{\left(\frac{m_1 - m_2}{m_1 + m_2}\right)\vec{\mathbf{v}}_1 + \left(\frac{2m_2}{m_1 + m_2}\right)\vec{\mathbf{v}}_2}$$

Observe that these results are the same as Equations 8.24 and 8.25, which should have been expected since this is a perfectly elastic collision in one dimension.

P8.47 (a) $(60.0 \text{ kg})4.00 \text{ m/s} = (120 + 60.0) \text{ kg}v_f$

$\vec{\mathbf{v}}_f = \boxed{1.33 \text{ m/s}\hat{\mathbf{i}}}$

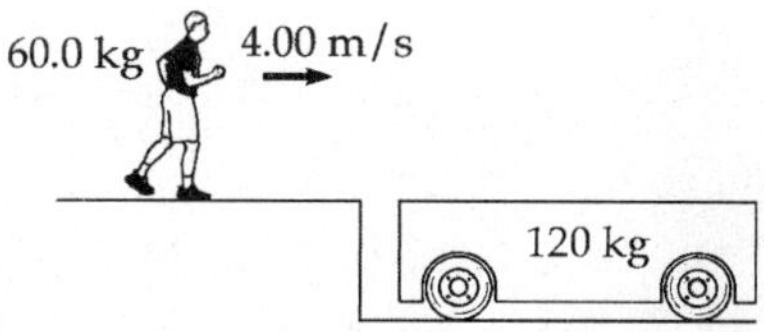

FIG. P8.47

(b) $\sum F_y = 0$: $n - (60.0 \text{ kg})9.80 \text{ m/s}^2 = 0$

$f_k = \mu_k n = 0.400(588 \text{ N}) = 235 \text{ N}$

$\vec{\mathbf{f}}_k = \boxed{-235 \text{ N}\hat{\mathbf{i}}}$

(c) For the person, $p_i + I = p_f$

$mv_i + Ft = mv_f$

$(60.0 \text{ kg})4.00 \text{ m/s} - (235 \text{ N})t = (60.0 \text{ kg})1.33 \text{ m/s}$

$t = \boxed{0.680 \text{ s}}$

continued on next page

(d) person: $m\vec{\mathbf{v}}_f - m\vec{\mathbf{v}}_i = 60.0 \text{ kg}(1.33 - 4.00)\hat{\mathbf{i}} \text{ m/s} = \boxed{-160 \text{ N}\cdot\text{s}\hat{\mathbf{i}}}$

cart: $120 \text{ kg}(1.33 \text{ m/s}) - 0 = \boxed{+160 \text{ N}\cdot\text{s}\hat{\mathbf{i}}}$

(e) $x_f - x_i = \frac{1}{2}(v_i + v_f)t = \frac{1}{2}[(4.00 + 1.33) \text{ m/s}]0.680 \text{ s} = \boxed{1.81 \text{ m}}$

(f) $x_f - x_i = \frac{1}{2}(v_i + v_f)t = \frac{1}{2}(0 + 1.33 \text{ m/s})0.680 \text{ s} = \boxed{0.454 \text{ m}}$

(g) $\frac{1}{2}mv_f^2 - \frac{1}{2}mv_i^2 = \frac{1}{2}60.0 \text{ kg}(1.33 \text{ m/s})^2 - \frac{1}{2}60.0 \text{ kg}(4.00 \text{ m/s})^2 = \boxed{-427 \text{ J}}$

(h) $\frac{1}{2}mv_f^2 - \frac{1}{2}mv_i^2 = \frac{1}{2}120.0 \text{ kg}(1.33 \text{ m/s})^2 - 0 = \boxed{107 \text{ J}}$

(i) The force exerted by the person on the cart must be equal in magnitude and opposite in direction to the force exerted by the cart on the person. The changes in momentum of the two objects must be equal in magnitude and must add to zero. Their changes in kinetic energy are different in magnitude and do not add to zero. The following represent two ways of thinking about 'why.' The distance the cart moves is different from the distance moved by the point of application of the friction force to the cart. The total change in mechanical energy for both objects together, -320 J, becomes $+320$ J of additional internal energy in this perfectly inelastic collision.

***P8.48** Using conservation of momentum from just before to just after the impact of the bullet with the block:

$$mv_i = (M+m)v_f$$

or $$v_i = \left(\frac{M+m}{m}\right)v_f. \qquad (1)$$

The speed of the block and embedded bullet just after impact may be found using kinematic equations:

$$d = v_f t \text{ and } h = \frac{1}{2}gt^2.$$

Thus, $t = \sqrt{\frac{2h}{g}}$ and $v_f = \frac{d}{t} = d\sqrt{\frac{g}{2h}} = \sqrt{\frac{gd^2}{2h}}$.

Substituting into (1) from above gives $v_i = \boxed{\left(\frac{M+m}{m}\right)\sqrt{\frac{gd^2}{2h}}}$.

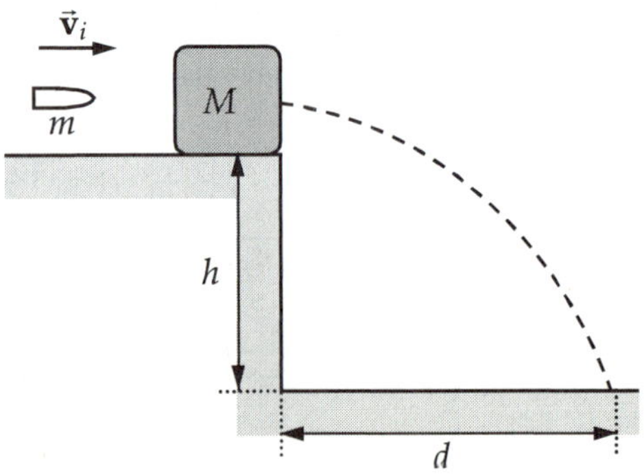

FIG. P8.48

***P8.49** (a) The volume of sea water expelled in one second is $\pi r^2\ell = \pi(0.015\text{ m})^2(16\text{ m}) = 0.011\,3\text{ m}^3$. To say the same thing, the volume flow rate of the jet is $0.011\,3\ \text{m}^3/\text{s}$. Its mass flow rate is $\frac{\Delta m}{\Delta t} = \rho\frac{\Delta V}{\Delta t} = 1\,030\ \text{kg/m}^2\left(0.011\ 3\ \text{m}^3/\text{s}\right) = 11.6\ \text{kg/s}$. The force the squid exerts on the water in the jet is $\frac{\Delta p}{\Delta t} = \frac{\Delta m}{\Delta t}v = 11.6\ \text{kg/s}(16\ \text{m/s}) = 186\text{ N}$. The water in the jet exerts a force of the same size on the squid, to give it initial acceleration $a = \frac{\sum F}{m} = \frac{186\text{ N}}{90\text{ kg}} = \boxed{2.07\ \text{m/s}^2}$.

(b) For the squid in equilibrium at terminal speed,

$$186\text{ N} = R = \frac{1}{2}D\rho A v_T^2$$

$$v_T = \left(\frac{2R}{D\rho A}\right)^{1/2} = \left(\frac{2\cdot 186\text{ N}}{0.3\left(1\,030\ \text{kg/m}^3\right)\left(800\times 10^{-4}\ \text{m}^2\right)}\right)^{1/2} = \boxed{3.88\ \text{m/s}}$$

***P8.50** (a) The mass of the sleigh plus you is 270 kg. Your velocity is 7.50 m/s in the x direction. You unbolt a 15.0-kg seat and throw it back at the ravening wolves, giving it a speed of 8.00 m/s relative to you. Find the velocity of the sleigh after your action, and the velocity of the seat relative to the ground.

(b) We substitute $v_{1f} = 8\ \text{m/s} - v_{2f}$

$$270\ \text{kg}(7.5\ \text{m/s}) = 15\ \text{kg}\left(-8\ \text{m/s} + v_{2f}\right) + (255\ \text{kg})v_{2f}$$

$$2\,025\ \text{kg}\cdot\text{m/s} = -120\ \text{kg}\cdot\text{m/s} + (270\ \text{kg})v_{2f}$$

$$v_{2f} = \frac{2\,145\ \text{m/s}}{270} = \boxed{7.94\ \text{m/s} = v_{2f}}$$

$$v_{1f} = 8\ \text{m/s} - 7.94\ \text{m/s} = \boxed{0.055\,6\ \text{m/s} = v_{1f}}$$

The final velocity of the sleigh is $7.94\ \text{m/s}\,\hat{\mathbf{i}}$. That of the seat is $-0.055\,6\ \text{m/s}\,\hat{\mathbf{i}}$.

(c) You do work on both the sleigh and the seat, to change their kinetic energy according to

$$K_i + W = K_{1f} + K_{2f}$$

$$\frac{1}{2}(270\ \text{kg})(7.5\ \text{m/s})^2 + W = \frac{1}{2}(15\ \text{kg})(0.055\,6\ \text{m/s})^2 + \frac{1}{2}(255\ \text{kg})(7.94\ \text{m/s})^2$$

$$7\,594\text{ J} + W = 0.023\,1\text{ J} + 8\,047\text{ J}$$

$$W = \boxed{453\text{ J}}$$

P8.51 (a) The initial momentum of the system is zero, which remains constant throughout the motion. Therefore, when m_1 leaves the wedge, we must have

$$m_2 v_{\text{wedge}} + m_1 v_{\text{block}} = 0$$

or $$(3.00 \text{ kg})v_{\text{wedge}} + (0.500 \text{ kg})(+4.00 \text{ m/s}) = 0$$

so $$v_{\text{wedge}} = \boxed{-0.667 \text{ m/s}}$$

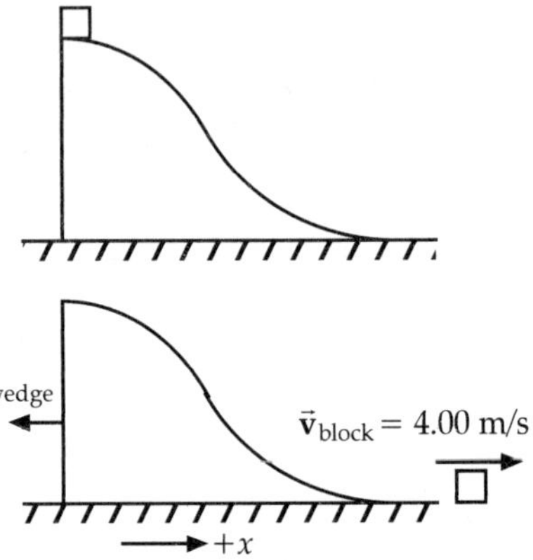

FIG. P8.51

(b) Using conservation of energy for the block-wedge-Earth system as the block slides down the smooth (frictionless) wedge, we have

$$\left[K_{\text{block}} + U_{\text{system}}\right]_i + \left[K_{\text{wedge}}\right]_i = \left[K_{\text{block}} + U_{\text{system}}\right]_f + \left[K_{\text{wedge}}\right]_f$$

or $$\left[0 + m_1 gh\right] + 0 = \left[\frac{1}{2}m_1(4.00)^2 + 0\right] + \frac{1}{2}m_2(-0.667)^2 \text{ which gives } \boxed{h = 0.952 \text{ m}}.$$

P8.52 $$T = \frac{dm}{dt}(\Delta v_{\text{fuel}}) + \frac{dm}{dt}(\Delta v_{\text{air}}): \quad T = (3.00 \text{ kg/s})(600 \text{ m/s}) + (80.0 \text{ kg/s})(600 \text{ m/s} - 223 \text{ m/s})$$

$$T = 1\,800 \text{ N} + 3.02 \times 10^4 \text{ N} = \boxed{3.20 \times 10^4 \text{ N}}$$

The delivered power is the force (or thrust) multiplied by the velocity,

$$\mathcal{P} = Tv = \left(3.20 \times 10^4 \text{ N}\right)(223 \text{ m/s}) = \boxed{7.13 \times 10^6 \text{ W}}$$

***P8.53** It is interesting to compare this problem with Problem P6.50. The force exerted by the spring on each block is in magnitude $|F_s| = kx = (3.85 \text{ N/m})(0.08 \text{ m}) = 0.308 \text{ N}$.

(a) With no friction, the elastic energy in the spring becomes kinetic energy of the blocks, which have momenta of equal magnitude in opposite directions. The blocks move with constant speed after they leave the spring.

$$(K+U)_i = (K+U)_f$$

$$\frac{1}{2}kx^2 = \frac{1}{2}m_1 v_{1f}^2 + \frac{1}{2}m_2 v_{2f}^2$$

$$\frac{1}{2}(3.85 \text{ N/m})(0.08 \text{ m})^2 = \frac{1}{2}(0.25 \text{ kg})v_{1f}^2 + \frac{1}{2}(0.50 \text{ kg})v_{2f}^2$$

$$m_1\vec{v}_{1i} + m_2\vec{v}_{2i} = m_1\vec{v}_{1f} + m_2\vec{v}_{2f}$$

$$0 = (0.25 \text{ kg})v_{1f}\left(-\hat{i}\right) + (0.50 \text{ kg})v_{2f}\hat{i}$$

$$v_{1f} = 2v_{2f}$$

$$0.012\,3 \text{ J} = \frac{1}{2}(0.25 \text{ kg})\left(2v_{2f}\right)^2 + \frac{1}{2}(0.50 \text{ kg})v_{2f}^2 = \frac{1}{2}(1.5 \text{ kg})v_{2f}^2$$

$$v_{2f} = \left(\frac{0.123 \text{ J}}{0.75 \text{ kg}}\right)^{1/2} = 0.128 \text{ m/s} \qquad \boxed{\vec{v}_{2f} = 0.128 \text{ m/s}\,\hat{i}}$$

$$v_{1f} = 2(0.128 \text{ m/s}) = 0.256 \text{ m/s} \qquad \boxed{\vec{v}_{1f} = 0.256 \text{ m/s}\left(-\hat{i}\right)}$$

continued on next page

(b) For the lighter block, $\sum F_y = ma_y$, $n - 0.25\text{ kg}(9.8\text{ m/s}^2) = 0$, $n = 2.45\text{ N}$, $f_k = \mu_k n = 0.1(2.45\text{ N}) = 0.245\text{ N}$. We assume that the maximum force of static friction is a similar size. Since 0.308 N is larger than 0.245 N, this block moves. For the heavier block, the normal force and the frictional force are twice as large: $f_k = 0.490\text{ N}$. Since 0.308 N is less than this, the heavier block stands still. In this case, the frictional forces exerted by the floor change the momentum of the two-block system. The lighter block will gain speed as long as the spring force is larger than the friction force: that is until the spring compresses becomes x_f given by $|F_s| = kx$, $0.245\text{ N} = (3.85\text{ N/m})x_f$, $0.063\,6\text{ m} = x_f$. Now for the energy of the lighter block as it moves to this maximum-speed point we have

$$K_i + U_i - f_k d = K_f + U_f$$

$$0 + 0.012\,3\text{ J} - 0.245\text{ N}(0.08 - 0.063\,6\text{ m}) = \frac{1}{2}(0.25\text{ kg})v_f^2 + \frac{1}{2}(3.85\text{ N/m})(0.063\,6\text{ m})^2$$

$$0.012\,3\text{ J} - 0.004\,01\text{ J} = \frac{1}{2}(0.25\text{ kg})v_f^2 + 0.007\,80\text{ J}$$

$$\left(\frac{2(0.000\,515\text{ J})}{0.25\text{ kg}}\right)^{1/2} = v_f = 0.064\,2\text{ m/s}$$

Thus for the heavier block the maximum velocity is $\boxed{0}$ and for the lighter $\boxed{0.064\,2\text{ m/s}\left(-\hat{\mathbf{i}}\right)}$.

(c) For the lighter block, $f_k = 0.462(2.45\text{ N}) = 1.13\text{ N}$. The force of static friction must be at least as large. The 0.308-N spring force is too small to produce motion of either block. Each has $\boxed{0}$ maximum speed.

P8.54 (a) $\vec{\mathbf{p}}_i + \vec{\mathbf{F}}t = \vec{\mathbf{p}}_f$: $(3.00\text{ kg})(7.00\text{ m/s})\hat{\mathbf{j}} + \left(12.0\text{ N}\hat{\mathbf{i}}\right)(5.00\text{ s}) = (3.00\text{ kg})\vec{\mathbf{v}}_f$

$$\vec{\mathbf{v}}_f = \boxed{\left(20.0\hat{\mathbf{i}} + 7.00\hat{\mathbf{j}}\right)\text{ m/s}}$$

(b) $\vec{\mathbf{a}} = \dfrac{\vec{\mathbf{v}}_f - \vec{\mathbf{v}}_i}{t}$: $\vec{\mathbf{a}} = \dfrac{\left(20.0\hat{\mathbf{i}} + 7.00\hat{\mathbf{j}} - 7.00\hat{\mathbf{j}}\right)\text{ m/s}}{5.00\text{ s}} = \boxed{4.00\hat{\mathbf{i}}\text{ m/s}^2}$

(c) $\vec{\mathbf{a}} = \dfrac{\sum\vec{\mathbf{F}}}{m}$: $\vec{\mathbf{a}} = \dfrac{12.0\text{ N}\hat{\mathbf{i}}}{3.00\text{ kg}} = \boxed{4.00\hat{\mathbf{i}}\text{ m/s}^2}$

(d) $\Delta\vec{\mathbf{r}} = \vec{\mathbf{v}}_i t + \frac{1}{2}\vec{\mathbf{a}}t^2$: $\Delta\vec{\mathbf{r}} = \left(7.00\text{ m/s}\,\hat{\mathbf{j}}\right)(5.00\text{ s}) + \frac{1}{2}\left(4.00\text{ m/s}^2\,\hat{\mathbf{i}}\right)(5.00\text{ s})^2$

$$\Delta\vec{\mathbf{r}} = \boxed{\left(50.0\hat{\mathbf{i}} + 35.0\hat{\mathbf{j}}\right)\text{ m}}$$

(e) $W = \vec{\mathbf{F}}\cdot\Delta\vec{\mathbf{r}}$: $W = \left(12.0\text{ N}\hat{\mathbf{i}}\right)\cdot\left(50.0\text{ m}\hat{\mathbf{i}} + 35.0\text{ m}\hat{\mathbf{j}}\right) = \boxed{600\text{ J}}$

continued on next page

(f) $\frac{1}{2}mv_f^2 = \frac{1}{2}(3.00 \text{ kg})(20.0\hat{\mathbf{i}}+7.00\hat{\mathbf{j}})\cdot(20.0\hat{\mathbf{i}}+7.00\hat{\mathbf{j}}) \text{ m}^2/\text{s}^2$

$$\frac{1}{2}mv_f^2 = (1.50 \text{ kg})(449 \text{ m}^2/\text{s}^2) = \boxed{674 \text{ J}}$$

(g) $\frac{1}{2}mv_i^2 + W = \frac{1}{2}(3.00 \text{ kg})(7.00 \text{ m/s})^2 + 600 \text{ J} = \boxed{674 \text{ J}}$

P8.55 Momentum conservation for the system of the two objects: $3mv_i - mv_i = mv_{1f} + 3mv_{2f}$. The relative velocity equation:

$$v_{1i} - v_{2i} = -v_{1f} + v_{2f} \qquad \begin{cases} -v_i - v_i = -v_{1f} + v_{2f} \\ 2v_i = v_{1f} + 3v_{2f} \end{cases}$$

$$0 = 4v_{2f}$$

$$v_{1f} = \boxed{2v_i} \qquad v_{2f} = \boxed{0}$$

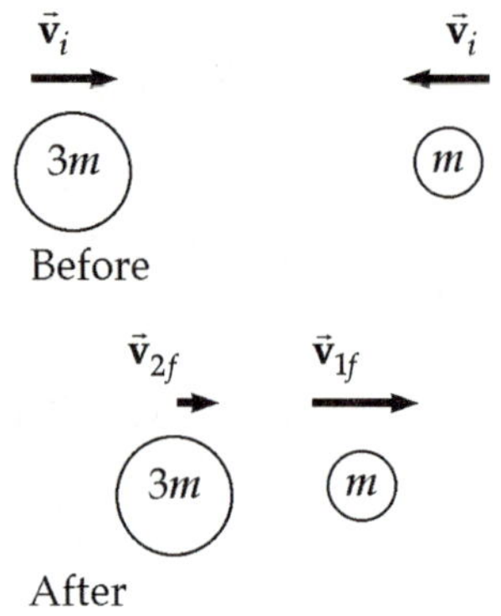

FIG. P8.55

P8.56 x-component of momentum for the system of the two objects:

$p_{1ix} + p_{2ix} = p_{1fx} + p_{2fx}$: $\quad -mv_i + 3mv_i = 0 + 3mv_{2x}$

y-component of momentum of the system: $\quad 0 + 0 = -mv_{1y} + 3mv_{2y}$

by conservation of energy of the system: $\quad +\frac{1}{2}mv_i^2 + \frac{1}{2}3mv_i^2 = \frac{1}{2}mv_{1y}^2 + \frac{1}{2}3m\left(v_{2x}^2 + v_{2y}^2\right)$

we have $\quad v_{2x} = \frac{2v_i}{3}$

also $\quad v_{1y} = 3v_{2y}$

So the energy equation becomes $\quad 4v_i^2 = 9v_{2y}^2 + \frac{4v_i^2}{3} + 3v_{2y}^2$

$$\frac{8v_i^2}{3} = 12v_{2y}^2$$

or $\quad v_{2y} = \frac{\sqrt{2}v_i}{3}$

(a) The object of mass m has final speed $\quad v_{1y} = 3v_{2y} = \boxed{\sqrt{2}v_i}$

and the object of mass 3 m moves at $\quad \sqrt{v_{2x}^2 + v_{2y}^2} = \sqrt{\frac{4v_i^2}{9} + \frac{2v_i^2}{9}}$

$$\sqrt{v_{2x}^2 + v_{2y}^2} = \boxed{\sqrt{\frac{2}{3}}v_i}$$

continued on next page

(b) $\theta = \tan^{-1}\left(\frac{v_{2y}}{v_{2x}}\right)$ $\theta = \tan^{-1}\left(\frac{\sqrt{2}v_i}{3}\frac{3}{2v_i}\right) = \boxed{35.3°}$

***P8.57** (a) Each object swings down according to

$$mgR = \frac{1}{2}mv_1^2 \qquad MgR = \frac{1}{2}Mv_1^2 \qquad v_1 = \sqrt{2gR}$$

The collision: $-mv_1 + Mv_1 = +(m+M)v_2$

$$v_2 = \frac{M-m}{M+m}v_1$$

Swinging up: $\frac{1}{2}(M+m)v_2^2 = (M+m)gR(1-\cos 35°)$

$$v_2 = \sqrt{2gR(1-\cos 35°)}$$

$$\sqrt{2gR(1-\cos 35°)}(M+m) = (M-m)\sqrt{2gR}$$
$$0.425M + 0.425m = M - m$$
$$1.425m = 0.575M$$
$$\boxed{\frac{m}{M} = 0.403}$$

(b) No change is required if the force is different. The nature of the forces within the system of colliding objects does not affect the total momentum of the system. With strong magnetic attraction, the heavier object will be moving somewhat faster and the lighter object faster still. Their extra kinetic energy will all be immediately converted into extra internal energy when the objects latch together. Momentum conservation guarantees that none of the extra kinetic energy remains after the objects join to make them swing higher.

P8.58 (a) Use conservation of the horizontal component of momentum for the system of the shell, the cannon, and the carriage, from just before to just after the cannon firing.

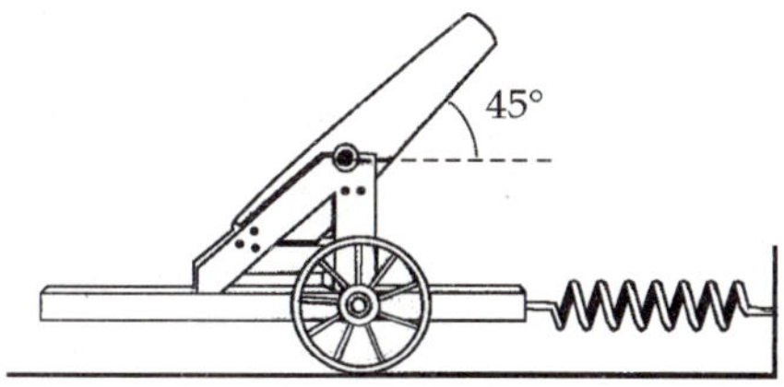

FIG. P8.58

$p_{xf} = p_{xi}$: $\quad m_{shell}v_{shell}\cos 45.0° + m_{cannon}v_{recoil} = 0$

$$(200)(125)\cos 45.0° + (5\,000)v_{recoil} = 0$$

or $\quad v_{recoil} = \boxed{-3.54 \text{ m/s}}$

(b) Use conservation of energy for the system of the cannon, the carriage, and the spring from right after the cannon is fired to the instant when the cannon comes to rest.

$K_f + U_{gf} + U_{sf} = K_i + U_{gi} + U_{si}$: $\quad 0 + 0 + \frac{1}{2}kx_{max}^2 = \frac{1}{2}mv_{recoil}^2 + 0 + 0$

$$x_{max} = \sqrt{\frac{mv_{recoil}^2}{k}} = \sqrt{\frac{(5\,000)(-3.54)^2}{2.00\times 10^4}}\text{ m} = \boxed{1.77 \text{ m}}$$

continued on next page

(c) $\left|F_{s,\max}\right| = kx_{\max}$ $\qquad \left|F_{s,\max}\right| = \left(2.00\times10^4\ \text{N/m}\right)(1.77\ \text{m}) = \boxed{3.54\times10^4\ \text{N}}$

(d) No. The rail exerts a vertical external force (the normal force) on the cannon and prevents it from recoiling vertically. Momentum is not conserved in the vertical direction. The spring does not have time to stretch during the cannon firing. Thus, no external horizontal force is exerted on the system (cannon, carriage, and shell) from just before to just after firing. Momentum of this system is conserved in the horizontal direction during this interval.

P8.59 A picture one second later differs by showing five extra kilograms of sand moving on the belt.

(a) $\dfrac{\Delta p_x}{\Delta t} = \dfrac{(5.00\ \text{kg})(0.750\ \text{m/s})}{1.00\ \text{s}} = \boxed{3.75\ \text{N}}$

(b) The only horizontal force on the sand is belt friction,

so from $\quad p_{xi} + f\Delta t = p_{xf} \quad$ this is $\quad f = \dfrac{\Delta p_x}{\Delta t} = \boxed{3.75\ \text{N}}$

(c) The belt is in equilibrium:

$\sum F_x = ma_x: \quad +F_{\text{ext}} - f = 0 \qquad$ and $\qquad F_{\text{ext}} = \boxed{3.75\ \text{N}}$

(d) $W = F\Delta r\cos\theta = 3.75\ \text{N}(0.750\ \text{m})\cos 0° = \boxed{2.81\ \text{J}}$

(e) $\dfrac{1}{2}(\Delta m)v^2 = \dfrac{1}{2}5.00\ \text{kg}(0.750\ \text{m/s})^2 = \boxed{1.41\ \text{J}}$

(f) $\boxed{\text{Friction between sand and belt converts half of the input work into extra internal energy.}}$

P8.60 The force exerted by the table is equal to the change in momentum of each of the links in the chain.

By the calculus chain rule of derivatives,

$$F_1 = \frac{dp}{dt} = \frac{d(mv)}{dt} = v\frac{dm}{dt} + m\frac{dv}{dt}.$$

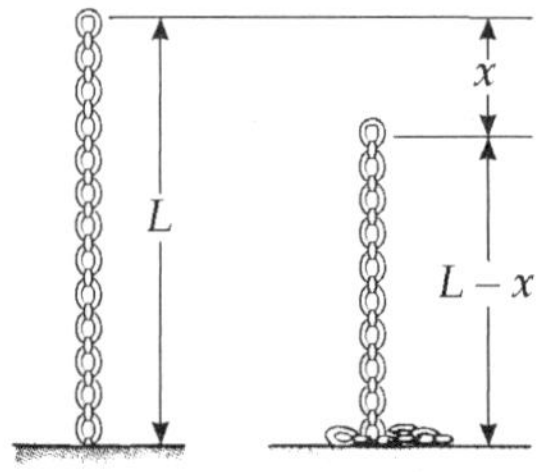

FIG. P8.60

We choose to account for the change in momentum of each link by having it pass from our area of interest just before it hits the table, so that

$$v\frac{dm}{dt} \neq 0 \text{ and } m\frac{dv}{dt} = 0.$$

Since the mass per unit length is uniform, we can express each link of length dx as having a mass dm:

$$dm = \frac{M}{L}dx.$$

continued on next page

The magnitude of the force on the falling chain is the force that will be necessary to stop each of the elements dm.

$$F_1 = v\frac{dm}{dt} = v\left(\frac{M}{L}\right)\frac{dx}{dt} = \left(\frac{M}{L}\right)v^2$$

After falling a distance x, the square of the velocity of each link $v^2 = 2gx$ (from kinematics), hence

$$F_1 = \frac{2Mgx}{L}.$$

The links already on the table have a total length x, and their weight is supported by a force F_2:

$$F_2 = \frac{Mgx}{L}.$$

Hence, the *total* force on the chain is

$$F_{\text{total}} = F_1 + F_2 = \boxed{\frac{3Mgx}{L}}.$$

That is, *the total force is three times the weight of the chain on the table at that instant.*

ANSWERS TO EVEN PROBLEMS

P8.2 $\sim 10^{23}$ m/s

P8.4 (a) $\frac{p^2}{2m}$; (b) $\sqrt{2mK}$

P8.6 The force he would have to exert is 6.44×10^3 N

P8.8 (a) 5.40 N·s in the direction of $\vec{\mathbf{v}}_f$; (b) −27.0 J

P8.10 (a) $(9.05\hat{\mathbf{i}} + 6.12\hat{\mathbf{j}})$ N·s; (b) $(377\hat{\mathbf{i}} + 255\hat{\mathbf{j}})$ N

P8.12 (a) see the solution; (b) small; (c) see the solution; (d) large; (e) no difference

P8.14 (a) 2.50 m/s; (b) 37.5 kJ; (c) see the solution

P8.16 0.556 m

P8.18 $\frac{4M}{m}\sqrt{g\ell}$

P8.20 7.94 cm

P8.22 (a) 2.24 m/s toward the right; (b) No, coupling order makes no difference

P8.24 (a) 2.88 m/s at 32.3° north of east; (b) 783 J into internal energy

P8.26 No, his speed was 41.5 mi/h

P8.28 (a) $\frac{v_i}{\sqrt{2}}$; (b) 45.0° and −45.0°

P8.30 (a) 1.07 m/s at −29.7°; (b) $\frac{\Delta K}{K_i} = -0.318$

P8.32 $x_{CM} = 0$, $y_{CM} = 1.00$ m

P8.34 $y_{CM} = 0$, $x_{CM} = 0.006\,73$ nm from the oxygen nucleus

P8.36 (a) 15.9 g; (b) 0.153 m

P8.38 (a) see the solution; (b) $\left(-2.00\hat{\mathbf{i}}-1.00\hat{\mathbf{j}}\right)$ m; (c) $\left(3.00\hat{\mathbf{i}}-1.00\hat{\mathbf{j}}\right)$ m/s; (d) $\left(15.0\hat{\mathbf{i}}-5.00\hat{\mathbf{j}}\right)$ kg·m/s

P8.40 (a) $\vec{\mathbf{v}}_{1f}=-0.780\hat{\mathbf{i}}$ m/s, $\vec{\mathbf{v}}_{2f}=1.12\hat{\mathbf{i}}$ m/s; (b) $0.360\hat{\mathbf{i}}$ m/s

P8.42 (a) 787 m/s; (b) 138 m/s

P8.44 see the solution

P8.46 (a) $\vec{\mathbf{v}}=\dfrac{m_1\vec{\mathbf{v}}_1+m_2\vec{\mathbf{v}}_2}{m_1+m_2}$

(b) $x_m=(v_1-v_2)\sqrt{\dfrac{m_1m_2}{k(m_1+m_2)}}$

(c) $\vec{\mathbf{v}}_{1f}=\left(\dfrac{m_1-m_2}{m_1+m_2}\right)\vec{\mathbf{v}}_1+\left(\dfrac{2m_2}{m_1+m_2}\right)\vec{\mathbf{v}}_2$,

$\vec{\mathbf{v}}_{2f}=\left(\dfrac{2m_1}{m_1+m_2}\right)\vec{\mathbf{v}}_1+\left(\dfrac{m_2-m_1}{m_1+m_2}\right)\vec{\mathbf{v}}_2$

P8.48 $\left(\dfrac{M+m}{m}\right)\sqrt{\dfrac{gd^2}{2h}}$

P8.50 (a) see the solution;
(b) 0.055 6 m/s in the $-x$ direction, 7.94 m/s in the x direction;
(c) 453 J

P8.52 3.20×10^4 N, 7.13×10^6 W

P8.54 (a) $\left(20.0\hat{\mathbf{i}}+7.00\hat{\mathbf{j}}\right)$ m/s; (b) $4.00\hat{\mathbf{i}}$ m/s^2; (c) $4.00\hat{\mathbf{i}}$ m/s^2; (d) $\left(50.0\hat{\mathbf{i}}+35.0\hat{\mathbf{j}}\right)$ m; (e) 600 J; (f) 674 J; (g) 674 J

P8.56 (a) $\sqrt{2}v_i$ and $\sqrt{\dfrac{2}{3}}v_i$; (b) 35.3°

P8.58 (a) −3.54 m/s; (b) 1.77 m; (c) 3.54×10^4 N; (d) no, the normal force of the rails contributes upward momentum to the system

P8.60 $\dfrac{3Mgx}{L}\hat{\mathbf{j}}$

Chapter 9

Relativity

CHAPTER OUTLINE

ANSWERS TO QUESTIONS

Q9.1 The speed of light c and the speed v of their relative motion.

Q9.2 An ellipsoid. The dimension in the direction of motion would be measured to be scrunched in.

Q9.3 No. The principle of relativity implies that nothing can travel faster than the speed of light in a *vacuum*, which is 300 Mm/s. The electron would emit light in a conical shock wave of Cerenkov radiation.

Q9.4 The clock in orbit runs slower. No, they are not synchronized. Although they both tick at the same rate after return, a time difference has developed between the two clocks.

Q9.5 Suppose a railroad train is moving past you. One way to measure its length is this: You mark the tracks at the cowcatcher forming the front of the moving engine at 9:00:00 AM, while your assistant marks the tracks at the back of the caboose at the same time. Then you find the distance between the marks on the tracks with a tape measure. You and your assistant must make the marks simultaneously in your frame of reference, for otherwise the motion of the train would make its length different from the distance between marks.

Q9.6 (a) Yours does.

(b) His does.

(c) If the velocity of relative motion is constant, both observers have equally valid views.

Q9.7 Get a *Mr. Tompkins* book by George Gamow for a wonderful fictional exploration of this question. Driving home in a hurry, you push on the gas pedal not to increase your speed by very much, but rather to make the blocks get shorter. Big Doppler shifts in wave frequencies make red lights look green as you approach them and make car horns and car radios useless. High-speed transportation is very expensive, requiring huge fuel purchases. And it is dangerous, as a speeding car can knock down a building. Having had breakfast at home, you return hungry for lunch, but you find you have missed dinner. There is a five-day delay in transmission when you watch the Olympics in Australia on live television. It takes ninety-five years for sunlight to reach Earth. We cannot see the Milky Way; the fireball of the Big Bang surrounds us at the distance of Rigel or Deneb.

Q9.8 According to $\vec{\mathbf{p}} = \gamma m\vec{\mathbf{u}}$, doubling the speed u will make the momentum of an object increase by the factor $2\left[\dfrac{c^2 - u^2}{c^2 - 4u^2}\right]^{1/2}$.

Q9.9 As the object approaches the speed of light, its kinetic energy grows without limit. It would take an infinite investment of work to accelerate the object to the speed of light.

Q9.10 There is no upper limit on the momentum of an electron. As more energy E is fed into the object without limit, its speed approaches the speed of light and its momentum approaches $\frac{E}{c}$.

Q9.11 Recall that when a spring of force constant k is compressed or stretched from its relaxed position a distance x, it stores elastic potential energy $U=\frac{1}{2}kx^2$. According to the special theory of relativity, any change in the total energy of the system is equivalent to a change in the mass of the system. Therefore, the mass of a compressed or stretched spring is greater than the mass of a relaxed spring by an amount $\frac{U}{c^2}$. The fractional change is typically unobservably small for a mechanical spring.

Q9.12 You see no change in your reflection at any speed you can attain. You cannot attain the speed of light, for that would take an infinite amount of energy.

Q9.13 Quasar light moves at three hundred million meters per second, just like the light from a firefly at rest.

Q9.14 A photon transports energy. The relativistic equivalence of mass and energy means that is enough to give it momentum.

Q9.15 Any physical theory must agree with experimental measurements within some domain. Newtonian mechanics agrees with experiment for objects moving slowly compared to the speed of light. Relativistic mechanics agrees with experiment for objects at all speeds. Thus the two theories must and do agree with each other for ordinary nonrelativistic objects. Both statements given in the question are formally correct, but the first is clumsily phrased. It seems to suggest that relativistic mechanics applies only to fast-moving objects.

Q9.16 The point of intersection moves to the right. To state the problem precisely, let us assume that each of the two cards moves toward the other parallel to the long dimension of the picture, with velocity of magnitude v. The point of intersection moves to the right at speed $\frac{2v}{\tan\phi}=2v\cot\phi$, where ϕ is the small angle between the cards. As ϕ approaches zero, $\cot\phi$ approaches infinity. Thus the point of intersection can move with a speed faster than c if v is sufficiently large and ϕ sufficiently small. For example, take $v=500$ m/s and $\phi=0.000\,19°$. If you are worried about holding the cards steady enough to be sure of the angle, cut the edge of one card along a curve so that the angle will necessarily be sufficiently small at some place along the edge.

Let us assume the spinning flashlight is at the center of a grain elevator, forming a circular screen of radius R. The linear speed of the spot on the screen is given by $v=\omega R$, where ω is the angular speed of rotation of the flashlight. With sufficiently large ω and R, the speed of the spot moving on the screen can exceed c.

Neither of these examples violates the principle of relativity. Both cases are describing a point of intersection: in the first case, the intersection of two cards and in the second case, the intersection of a light beam with a screen. A point of intersection is not made of matter so it has no mass, and hence no energy. A bug momentarily at the intersection point could yelp, take a bite out of one card, or reflect the light. None of these actions would result in communication reaching another bug so soon as the intersection point reaches him. The second bug would have to wait for sound or light to travel across the distance between the first bug and himself, to get the message.

As a child, the author used an Erector set to build a superluminal speed generator using the intersecting-cards method. Can you get a visible dot to run across a computer screen faster than light? Want'a see it again?

Q9.17 Special relativity describes inertial reference frames: that is, reference frames that are not accelerating. General relativity describes all reference frames.

Q9.18 The downstairs clock runs more slowly because it is closer to the Earth and hence in a stronger gravitational field than the upstairs clock.

SOLUTIONS TO PROBLEMS

Section 9.1 The Principle of Newtonian Relativity

P9.1 The first observer watches some object accelerate under applied forces. Call the instantaneous velocity of the object $\vec{\mathbf{v}}_1$. The second observer has constant velocity $\vec{\mathbf{v}}_{21}$ relative to the first, and measures the object to have velocity $\vec{\mathbf{v}}_2 = \vec{\mathbf{v}}_1 - \vec{\mathbf{v}}_{21}$.

The second observer measures an acceleration of $\vec{\mathbf{a}}_2 = \dfrac{d\vec{\mathbf{v}}_2}{dt} = \dfrac{d\vec{\mathbf{v}}_1}{dt}$.

This is the same as that measured by the first observer. In this nonrelativistic case, they measure the same forces as well. Thus, the second observer also confirms that $\sum \vec{\mathbf{F}} = m\vec{\mathbf{a}}$.

P9.2 The laboratory observer notes Newton's second law to hold: $\vec{\mathbf{F}}_1 = m\vec{\mathbf{a}}_1$
(where the subscript 1 implies the measurement was made in the laboratory frame of reference). The observer in the accelerating frame measures the acceleration of the mass as $\vec{\mathbf{a}}_2 = \vec{\mathbf{a}}_1 - \vec{\mathbf{a}}'$
(where the subscript 2 implies the measurement was made in the accelerating frame of reference, and the primed acceleration term is the acceleration of the accelerated frame with respect to the laboratory frame of reference). If Newton's second law held for the accelerating frame, that observer would then find valid the relation

$$\vec{\mathbf{F}}_2 = m\vec{\mathbf{a}}_2 \qquad \text{or} \qquad \vec{\mathbf{F}}_1 = m\vec{\mathbf{a}}_2$$

(since $\vec{\mathbf{F}}_1 = \vec{\mathbf{F}}_2$ and the mass is unchanged in each). But, instead, the accelerating frame observer will find that $\vec{\mathbf{F}}_2 = m\vec{\mathbf{a}}_2 - m\vec{\mathbf{a}}'$ which is *not* Newton's second law.

P9.3 In the rest frame,

$$p_i = m_1 v_{1i} + m_2 v_{2i} = (2\,000 \text{ kg})(20.0 \text{ m/s}) + (1\,500 \text{ kg})(0 \text{ m/s}) = 4.00 \times 10^4 \text{ kg}\cdot\text{m/s}$$

$$p_f = (m_1 + m_2)v_f = (2\,000 \text{ kg} + 1\,500 \text{ kg})v_f$$

Since $p_i = p_f$, $\qquad v_f = \dfrac{4.00 \times 10^4 \text{ kg}\cdot\text{m/s}}{2\,000 \text{ kg} + 1\,500 \text{ kg}} = 11.429 \text{ m/s}$.

In the moving frame, these velocities are all reduced by +10.0 m/s.

$$v'_{1i} = v_{1i} - v' = 20.0 \text{ m/s} - (+10.0 \text{ m/s}) = 10.0 \text{ m/s}$$

$$v'_{2i} = v_{2i} - v' = 0 \text{ m/s} - (+10.0 \text{ m/s}) = -10.0 \text{ m/s}$$

$$v'_f = 11.429 \text{ m/s} - (+10.0 \text{ m/s}) = 1.429 \text{ m/s}$$

Our initial momentum is then

$$p'_i = m_1 v'_{1i} + m_2 v'_{2i} = (2\,000 \text{ kg})(10.0 \text{ m/s}) + (1\,500 \text{ kg})(-10.0 \text{ m/s}) = 5\,000 \text{ kg}\cdot\text{m/s}$$

and our final momentum is the same:

$$p'_f = (2\,000 \text{ kg} + 1\,500 \text{ kg})v'_f = (3\,500 \text{ kg})(1.429 \text{ m/s}) = 5\,000 \text{ kg}\cdot\text{m/s}.$$

Section 9.2 The Michelson-Morley Experiment

Section 9.3 Einstein's Principle of Relativity

Section 9.4 Consequences of Special Relativity

P9.4 $L = L_p\sqrt{1-\frac{v^2}{c^2}}$ $\qquad v = c\sqrt{1-\left(\frac{L}{L_p}\right)^2}$

Taking $L = \frac{L_p}{2}$ where $L_p = 1.00$ m gives $\qquad v = c\sqrt{1-\left(\frac{L_p/2}{L_p}\right)^2} = c\sqrt{1-\frac{1}{4}} = \boxed{0.866c}$.

P9.5 $\Delta t = \frac{\Delta t_p}{\left[1-(v/c)^2\right]^{1/2}}$ so $v = c\left[1-\left(\frac{\Delta t_p}{\Delta t}\right)^2\right]^{1/2}$.

For $\Delta t = 2\Delta t_p$ $\qquad v = c\left[1-\left(\frac{\Delta t_p}{2\Delta t_p}\right)^2\right]^{1/2} = c\left[1-\frac{1}{4}\right]^{1/2} = \boxed{0.866c}$.

P9.6 (a) $\gamma = \frac{1}{\sqrt{1-(v/c)^2}} = \frac{1}{\sqrt{1-(0.500)^2}} = \frac{2}{\sqrt{3}}$

The time interval between pulses as measured by the Earth observer is

$$\Delta t = \gamma\Delta t_p = \frac{2}{\sqrt{3}}\left(\frac{60.0\text{ s}}{75.0}\right) = 0.924\text{ s}.$$

Thus, the Earth observer records a pulse rate of $\frac{60.0\text{ s/min}}{0.924\text{ s}} = \boxed{64.9/\text{min}}$.

(b) At a relative speed $v = 0.990c$, the relativistic factor γ increases to 7.09 and the pulse rate recorded by the Earth observer decreases to $\boxed{10.6/\text{min}}$. That is, the life span of the astronaut (reckoned by the duration of the total number of his heartbeats) is much longer as measured by an Earth clock than by a clock aboard the space vehicle.

P9.7 (a) The $0.8c$ and the 20 ly are measured in the Earth frame,

so in this frame, $\qquad \Delta t = \frac{x}{v} = \frac{20\text{ ly}}{0.8c} = \frac{20\text{ ly}}{0.8c}\,\frac{1c}{1\text{ ly/yr}} = \boxed{25.0\text{ yr}}$.

(b) We see a clock on the meteoroid moving, so we do not measure proper time; that clock measures proper time.

$\Delta t = \gamma\Delta t_p$: $\qquad \Delta t_p = \frac{\Delta t}{\gamma} = \frac{25.0\text{ yr}}{1/\sqrt{1-v^2/c^2}} = 25.0\text{ yr}\sqrt{1-0.8^2} = 25.0\text{ yr}(0.6) = \boxed{15.0\text{ yr}}$

continued on next page

(c) Method one: We measure the 20 ly on a stick stationary in our frame, so it is proper length. The tourist measures it to be contracted to

$$L=\frac{L_p}{\gamma}=\frac{20\text{ ly}}{1/\sqrt{1-0.8^2}}=\frac{20\text{ ly}}{1.667}=\boxed{12.0\text{ ly}}.$$

Method two: The tourist sees the Earth approaching at $0.8c$

$$(0.8\text{ ly/yr})(15\text{ yr})=\boxed{12.0\text{ ly}}.$$

Not only do distances and times differ between Earth and meteoroid reference frames, but also within the Earth frame apparent distances differ from actual distances. As we have interpreted it, the 20-lightyear actual distance from the Earth to the meteoroid at the time of discovery must be a calculated result, different from the distance measured directly. Because of the finite maximum speed of information transfer, the astronomer sees the meteoroid as it was years previously, when it was much farther away. Call its apparent distance d. The time required for light to reach us from the newly-visible meteoroid is the lookback time $t=\frac{d}{c}$. The astronomer calculates that the meteoroid has approached to be 20 ly away as it moved with constant velocity throughout the lookback time. We can work backwards to reconstruct her calculation:

$$d=20\text{ ly}+0.8ct=20\text{ ly}+\frac{0.8cd}{c}$$

$$0.2d=20\text{ ly}$$

$$d=100\text{ ly}$$

Thus in terms of direct observation, the meteoroid we see covers 100 ly in only 25 years. Such an apparent superluminal velocity is actually observed for some jets of material emanating from quasars, because they happen to be pointed nearly toward the Earth. If we can watch events unfold on the meteoroid, we see them slowed by relativistic time dilation, but also greatly speeded up by the Doppler effect.

P9.8 For $\frac{v}{c}=0.990$, $\gamma=7.09$

(a) The muon's lifetime as measured in the Earth's rest frame is

$$\Delta t=\frac{4.60\text{ km}}{0.990c}$$

and the lifetime measured in the muon's rest frame is

$$\Delta t_p=\frac{\Delta t}{\gamma}=\frac{1}{7.09}\left[\frac{4.60\times10^3\text{ m}}{0.990\left(3.00\times10^8\text{ m/s}\right)}\right]=\boxed{2.18\ \mu\text{s}}.$$

(b) $$L=L_p\sqrt{1-\left(\frac{v}{c}\right)^2}=\frac{L_p}{\gamma}=\frac{4.60\times10^3\text{ m}}{7.09}=\boxed{649\text{ m}}$$

P9.9 $\Delta t = \gamma \Delta t_p = \dfrac{\Delta t_p}{\sqrt{1-v^2/c^2}}$ so $\Delta t_p = \left(\sqrt{1-\dfrac{v^2}{c^2}}\right)\Delta t \approx \left(1-\dfrac{v^2}{2c^2}\right)\Delta t$

and $\Delta t - \Delta t_p = \left(\dfrac{v^2}{2c^2}\right)\Delta t$.

If $v = 1\,000 \text{ km/h} = \dfrac{1.00\times10^6 \text{ m}}{3\,600 \text{ s}} = 277.8 \text{ m/s}$

then $\dfrac{v}{c} = 9.26\times10^{-7}$

and $(\Delta t - \Delta t_p) = (4.28\times10^{-13})(3\,600 \text{ s}) = 1.54\times10^{-9} \text{ s} = \boxed{1.54 \text{ ns}}$.

P9.10 $\gamma = \dfrac{1}{\sqrt{1-(v^2/c^2)}} = 1.01$ so $\boxed{v = 0.140c}$

P9.11 The spaceship is measured by the Earth observer to be length-contracted to

$L = L_p\sqrt{1-\dfrac{v^2}{c^2}}$ or $L^2 = L_p^2\left(1-\dfrac{v^2}{c^2}\right)$.

Also, the contracted length is related to the time required to pass overhead by:

$L = vt$ or $L^2 = v^2t^2 = \dfrac{v^2}{c^2}(ct)^2$.

Equating these two expressions gives $L_p^2 - L_p^2\dfrac{v^2}{c^2} = (ct)^2\dfrac{v^2}{c^2}$

or $\left[L_p^2 + (ct)^2\right]\dfrac{v^2}{c^2} = L_p^2$.

Using the given values: $L_p = 300 \text{ m}$ and $t = 7.50\times10^{-7} \text{ s}$

this becomes $(1.41\times10^5 \text{ m}^2)\dfrac{v^2}{c^2} = 9.00\times10^4 \text{ m}^2$

giving $v = \boxed{0.800c}$.

P9.12 (a) The spaceship is measured by Earth observers to be of length L, where

$$L = L_p\sqrt{1-\frac{v^2}{c^2}} \quad \text{and} \quad L = v\Delta t$$

$$v\Delta t = L_p\sqrt{1-\frac{v^2}{c^2}} \quad \text{and} \quad v^2\Delta t^2 = L_p^2\left(1-\frac{v^2}{c^2}\right).$$

Solving for v, $\quad v^2\left(\Delta t^2 + \frac{L_p^2}{c^2}\right) = L_p^2 \qquad \boxed{v = \frac{cL_p}{\sqrt{c^2\Delta t^2 + L_p^2}}}$.

(b) The tanks move nonrelativistically, so we have $v = \frac{300\text{ m}}{75\text{ s}} = \boxed{4.00\text{ m/s}}$.

(c) For the data in problem 11,

$$v = \frac{c(300\text{ m})}{\sqrt{(3\times10^8\text{ m/s})^2(0.75\times10^{-6}\text{ s})^2 + (300\text{ m})^2}} = \frac{c(300\text{ m})}{\sqrt{225^2+300^2}\text{ m}} = 0.800c$$

in agreement with problem 11. For the data in part (b),

$$v = \frac{c(300\text{ m})}{\sqrt{(3\times10^8\text{ m/s})^2(75\text{ s})^2 + (300\text{ m})^2}} = \frac{c(300\text{ m})}{\sqrt{(2.25\times10^{10})^2+300^2}\text{ m}} = 1.33\times10^{-8}c = 4.00\text{ m/s}$$

in agreement with part (b).

The answer to part (a) shows that the speed determined by the Earth observer is generally proportional to the length of the object (measured by an observer riding on the object.) In part (c) we see one example of the nonrelativistic limit. When $c\Delta t$ is large compared to L_p, the expression in (a) becomes $v \to \frac{cL_p}{\sqrt{c^2\Delta t^2+0}} = \frac{L_p}{\Delta t}$, as in classical physics. The extreme relativistic limit is obtained when Δt is so small that $c\Delta t$ is small compared to L_p. Then the expression in (a) becomes $v \to \frac{cL_p}{\sqrt{0+L_p^2}} = c$. The maximum possible observable speed for an object is the speed of light in vacuum.

P9.13 (a) Since your ship is identical to his, and you are at rest with respect to your own ship, its length is $\boxed{20.0\text{ m}}$.

(b) His ship is in motion relative to you, so you measure its length contracted to $\boxed{19.0\text{ m}}$.

(c) We have $\quad L = L_p\sqrt{1-\frac{v^2}{c^2}}$

from which $\quad \frac{L}{L_p} = \frac{19.0\text{ m}}{20.0\text{ m}} = 0.950 = \sqrt{1-\frac{v^2}{c^2}}$ and $\boxed{v = 0.312c}$.

***P9.14** In the Earth frame, Speedo's trip lasts for a time

$$\Delta t = \frac{\Delta x}{v} = \frac{20.0 \text{ ly}}{0.950 \text{ ly/yr}} = 21.05 \text{ yr}.$$

Speedo's age advances only by the proper time interval

$$\Delta t_p = \frac{\Delta t}{\gamma} = 21.05 \text{ yr}\sqrt{1-0.95^2} = 6.574 \text{ yr during his trip.}$$

Similarly for Goslo,

$$\Delta t_p = \frac{\Delta x}{v}\sqrt{1-\frac{v^2}{c^2}} = \frac{20.0 \text{ ly}}{0.750 \text{ ly/yr}}\sqrt{1-0.75^2} = 17.64 \text{ yr}.$$

While Speedo has landed on Planet X and is waiting for his brother, he ages by

$$\frac{20.0 \text{ ly}}{0.750 \text{ ly/yr}} - \frac{20.0 \text{ ly}}{0.950 \text{ ly/yr}} = 5.614 \text{ yr}.$$

Then $\boxed{\text{Goslo}}$ ends up older by $17.64 \text{ yr} - (6.574 \text{ yr} + 5.614 \text{ yr}) = \boxed{5.45 \text{ yr}}$.

P9.15 (a) $$\Delta t = \gamma \Delta t_p = \frac{\Delta t_p}{\sqrt{1-(v/c)^2}} = \frac{15.0 \text{ yr}}{\sqrt{1-(0.700)^2}} = \boxed{21.0 \text{ yr}}$$

(b) $$d = v(\Delta t) = [0.700c](21.0 \text{ yr}) = [(0.700)(1.00 \text{ ly/yr})](21.0 \text{ yr}) = \boxed{14.7 \text{ ly}}$$

(c) The astronauts see Earth flying out the back window at $0.700c$:

$$d = v(\Delta t_p) = [0.700c](15.0 \text{ yr}) = [(0.700)(1.00 \text{ ly/yr})](15.0 \text{ yr}) = \boxed{10.5 \text{ ly}}$$

(d) Mission control gets signals for 21.0 yr while the battery is operating, and then for 14.7 years after the battery stops powering the transmitter, 14.7 ly away:

$$21.0 \text{ yr} + 14.7 \text{ yr} = \boxed{35.7 \text{ yr}}$$

Section 9.5 The Lorentz Transformation Equations

***P9.16** Let Suzanne be fixed in reference from S and see the two light-emission events with coordinates $x_1 = 0$, $t_1 = 0$, $x_2 = 0$, $t_2 = 3\ \mu s$. Let Mark be fixed in reference frame S′ and give the events coordinate $x_1' = 0$, $t_1' = 0$, $t_2' = 9\ \mu s$.

(a) Then we have

$$t_2' = \gamma\left(t_2 - \frac{v}{c^2}x_2\right)$$

$$9\ \mu s = \frac{1}{\sqrt{1 - v^2/c^2}}(3\ \mu s - 0) \qquad \sqrt{1 - \frac{v^2}{c^2}} = \frac{1}{3}$$

$$\frac{v^2}{c^2} = \frac{8}{9} \qquad \boxed{v = 0.943c}$$

(b) $$x_2' = \gamma(x_2 - vt_2) = 3\left(0 - 0.943c \times 3 \times 10^{-6}\ s\right)\left(\frac{3 \times 10^8\ m/s}{c}\right) = \boxed{2.55 \times 10^3\ m}$$

P9.17 $$\gamma = \frac{1}{\sqrt{1 - v^2/c^2}} = \frac{1}{\sqrt{1 - 0.995^2}} = 10.0$$

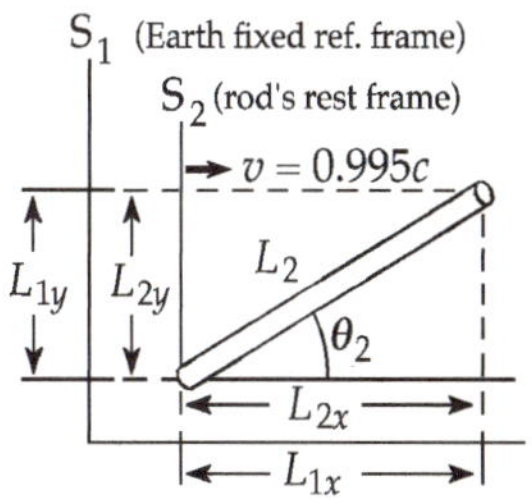

FIG. P9.17

We are also given: $L_1 = 2.00$ m, and $\theta = 30.0°$ (both measured in a reference frame moving relative to the rod).

Thus, $L_{1x} = L_1 \cos\theta_1 = (2.00\ m)(0.867) = 1.73\ m$

and $L_{1y} = L_1 \sin\theta_1 = (2.00\ m)(0.500) = 1.00\ m$

L_{2x} is a proper length, related to L_{1x} by $L_{1x} = \dfrac{L_{2x}}{\gamma}$.

Therefore, $L_{2x} = 10.0 L_{1x} = 17.3\ m$

and $L_{2y} = L_{1y} = 1.00\ m$.

(Lengths perpendicular to the motion are unchanged).

(a) $L_2 = \sqrt{(L_{2x})^2 + (L_{2y})^2}$ gives $\boxed{L_2 = 17.4\ m}$

(b) $\theta_2 = \tan^{-1}\dfrac{L_{2y}}{L_{2x}}$ gives $\boxed{\theta_2 = 3.30°}$

P9.18 Einstein's reasoning about lightning striking the ends of a train shows that the moving observer sees the event toward which she is moving, event B, as occurring first. The S-frame coordinates of the events we may take as $(x=0,\ y=0,\ z=0,\ t=0)$ and $(x=100\text{ m},\ y=0,\ z=0,\ t=0)$. Then the coordinates in S' are given by the Lorentz transformation. Event A is at $(x'=0,\ y'=0,\ z'=0,\ t'=0)$. The time of event B is

$$t'=\gamma\left(t-\frac{v}{c^2}x\right)=\frac{1}{\sqrt{1-0.8^2}}\left(0-\frac{0.8c}{c^2}(100\text{ m})\right)=1.667\left(-\frac{80\text{ m}}{3\times10^{8}\text{ m/s}}\right)=-4.44\times10^{-7}\text{ s}.$$

The time elapsing before A occurs is $\boxed{444\text{ ns}}$.

P9.19 (a) From the Lorentz transformation, the separations between the blue-light and red-light events are described by

$$\Delta x'=\gamma(\Delta x-v\Delta t)\qquad 0=\gamma\left[2.00\text{ m}-v\left(8.00\times10^{-9}\text{ s}\right)\right]$$

$$v=\frac{2.00\text{ m}}{8.00\times10^{-9}\text{ s}}=\boxed{2.50\times10^{8}\text{ m/s}}\qquad \gamma=\frac{1}{\sqrt{1-\left(2.50\times10^{8}\text{ m/s}\right)^2/\left(3.00\times10^{8}\text{ m/s}\right)^2}}=1.81.$$

(b) Again from the Lorentz transformation, $x'=\gamma(x-vt)$:

$$x'=1.81\left[3.00\text{ m}-\left(2.50\times10^{8}\text{ m/s}\right)\left(1.00\times10^{-9}\text{ s}\right)\right]$$

$$x'=\boxed{4.97\text{ m}}.$$

(c) $t'=\gamma\left(t-\dfrac{v}{c^2}x\right)$:

$$t'=1.81\left[1.00\times10^{-9}\text{ s}-\frac{\left(2.50\times10^{8}\text{ m/s}\right)}{\left(3.00\times10^{8}\text{ m/s}\right)^2}(3.00\text{ m})\right]$$

$$t'=\boxed{-1.33\times10^{-8}\text{ s}}$$

P9.20 $u_x=$ Enterprise velocity

$v=$ Klingon velocity

From the Lorentz velocity transformation equation

$$u'_x=\frac{u_x-v}{1-u_xv/c^2}=\frac{0.900c-0.800c}{1-(0.900)(0.800)}=\boxed{0.357c}.$$

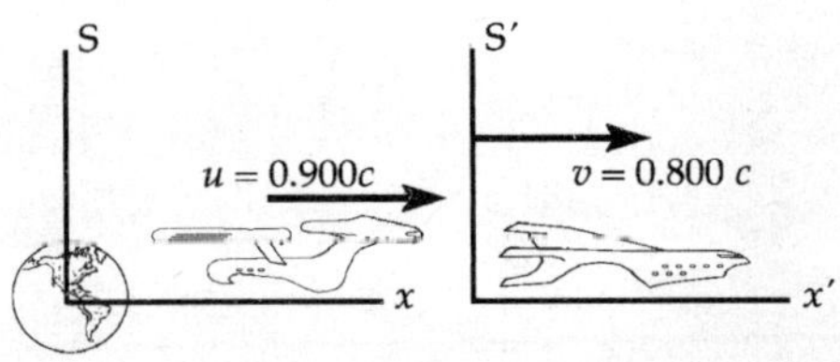

FIG. P9.20

P9.21 Choose the $+x$ axis in the direction of motion of one jet and let v reprsent its velocity.

$$u_x' = \frac{u_x - v}{1 - u_x v/c^2} = \frac{-0.750c - 0.750c}{1 - (-0.750)(0.750)} = \boxed{-0.960c}$$

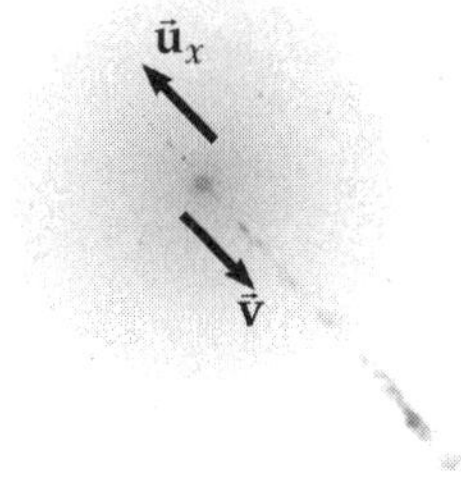

FIG. P9.21

P9.22 Let frame S be the Earth frame of reference. Then $v = -0.7c$.

The components of the velocity of the first spacecraft are $u_x = (0.6c)\cos 50° = 0.386c$

and $u_y = (0.6c)\sin 50° = 0.459c$.

As measured from the S' frame of the second spacecraft,

$$u_x' = \frac{u_x - v}{1 - u_x v/c^2} = \frac{0.386c - (-0.7c)}{1 - \left[(0.386c)(-0.7c)/c^2\right]} = \frac{1.086c}{1.27} = 0.855c$$

$$u_y' = \frac{u_y}{\gamma\left(1 - u_x v/c^2\right)} = \frac{0.459c\sqrt{1-(0.7)^2}}{1-(0.386)(-0.7)} = \frac{0.459c(0.714)}{1.27} = 0.258c$$

The magnitude of $\vec{u}'$ is $\sqrt{(0.855c)^2 + (0.285c)^2} = \boxed{0.893c}$

and its direction is at $\tan^{-1}\dfrac{0.258c}{0.855c} = \boxed{16.8° \text{ above the } x'\text{-axis}}$.

Section 9.6 Relativistic Momentum and the Relativistic Form of Newton's Laws

P9.23 (a) $p = \gamma m u$; for an electron moving at $0.010\,0c$,

$$\gamma = \frac{1}{\sqrt{1-(u/c)^2}} = \frac{1}{\sqrt{1-(0.010\,0)^2}} = 1.000\,05 \approx 1.00.$$

Thus, $p = 1.00\left(9.11\times10^{-31}\text{ kg}\right)(0.010\,0)\left(3.00\times10^{8}\text{ m/s}\right)$

$$p = \boxed{2.73\times10^{-24}\text{ kg}\cdot\text{m/s}}.$$

(b) Following the same steps as used in part (a),

we find at $0.500c$, $\gamma = 1.15$ and $p = \boxed{1.58\times10^{-22}\text{ kg}\cdot\text{m/s}}$.

(c) At $0.900c$, $\gamma = 2.29$ and $p = \boxed{5.64\times10^{-22}\text{ kg}\cdot\text{m/s}}$.

P9.24 Using the relativistic form,
$$p = \frac{mu}{\sqrt{1-(u/c)^2}} = \gamma mu$$

we find the difference Δp from the classical momentum, mu:
$$\Delta p = \gamma mu - mu = (\gamma - 1)mu.$$

(a) The difference is 1.00% when $(\gamma-1)mu = 0.010\,0\gamma mu$:
$$\gamma = \frac{1}{0.990} = \frac{1}{\sqrt{1-(u/c)^2}}$$

thus $1-\left(\frac{u}{c}\right)^2 = (0.990)^2$, and
$$u = \boxed{0.141c}.$$

(b) The difference is 10.0% when $(\gamma-1)mu = 0.100\gamma mu$:
$$\gamma = \frac{1}{0.900} = \frac{1}{\sqrt{1-(u/c)^2}}$$

thus $1-\left(\frac{u}{c}\right)^2 = (0.900)^2$ and
$$u = \boxed{0.436c}.$$

P9.25
$$\frac{p-mu}{mu} = \frac{\gamma mu - mu}{mu} = \gamma - 1: \qquad \gamma - 1 = \frac{1}{\sqrt{1-(u/c)^2}} - 1 \approx 1 + \frac{1}{2}\left(\frac{u}{c}\right)^2 - 1 = \frac{1}{2}\left(\frac{u}{c}\right)^2$$
$$\frac{p-mu}{mu} = \frac{1}{2}\left(\frac{90.0\text{ m/s}}{3.00\times10^8\text{ m/s}}\right)^2 = \boxed{4.50\times10^{-14}}$$

***P9.26** We can express the proportionality of the speeding fine to the excess momentum as $F = d(p - p_\ell)$ where F is the fine, d is a proportionality constant, p is the magnitude of the vehicle's momentum, and p_ℓ is the magnitude of its momentum as it travels at 90 km/h.

(a) METHOD ONE. For low speeds, the classical expression for momentum is accurate. Our equation describes the two cases
$$\$80 = d\left[m(190\text{ km/h}) - m(90\text{ km/h})\right] \text{ and } F_a = d\left[m(1\,090\text{ km/h}) - m(90\text{ km/h})\right].$$

Dividing gives $\frac{F_a}{\$80} = \frac{m(1\,000\text{ km/h})}{m(100\text{ km/h})} \Rightarrow \boxed{F_a = \$800}$.

METHOD TWO. The relativistic momentum expression is always accurate:
$$\$80 = d\left(\frac{m(190\text{ km/h})}{\sqrt{1-(190\text{ km/h})^2/c^2}} - \frac{m(90\text{ km/h})}{\sqrt{1-(90\text{ km/h})^2/c^2}}\right)$$
$$F_a = d\left(\frac{m(1\,090\text{ km/h})}{\sqrt{1-(1\,090\text{ km/h})^2/c^2}} - \frac{m(190\text{ km/h})}{\sqrt{1-(190\text{ km/h})^2/c^2}}\right)$$

To three or even to six digits, the answer is the same: $F_a = \$80(10) = \$800$.

continued on next page

(b) Now the high-speed case must be described relativistically. The speed of light is $3\times10^{8}\ \text{m/s}=1.08\times10^{9}\ \text{km/h}$.

$$\$80=dm(100\ \text{km/h})$$

$$F_b=d\left(\frac{m\left(1.000\,000\,09\times10^{9}\ \text{km/h}\right)}{\sqrt{1-\left(1.000\,000\,09\times10^{9}/1.08\times10^{9}\right)^{2}}}-m(90\ \text{km/h})\right)$$

$$F_b=dm\left(2.65\times10^{9}\ \text{km/h}\right)$$

again dividing, $\frac{F_b}{\$80}=\frac{2.65\times10^{9}}{10^{2}}\Rightarrow F_b=\boxed{\$2.12\times10^{9}}$.

P9.27 Relativistic momentum of the system of fragments must be conserved. For total momentum to be zero after as it was before, we must have, with subscript 2 referring to the heavier fragment, and subscript 1 to the lighter, $p_2=p_1$

or
$$\gamma_2m_2u_2=\gamma_1m_1u_1=\frac{2.50\times10^{-28}\ \text{kg}}{\sqrt{1-(0.893)^{2}}}\times(0.893c)$$

or
$$\frac{\left(1.67\times10^{-27}\ \text{kg}\right)u_2}{\sqrt{1-(u_2/c)^{2}}}=\left(4.960\times10^{-28}\ \text{kg}\right)c.$$

Proceeding to solve, we find
$$\left(\frac{1.67\times10^{-27}}{4.960\times10^{-28}}\frac{u_2}{c}\right)^{2}=1-\frac{u_2^{2}}{c^{2}}$$

$$12.3\frac{u_2^{2}}{c^{2}}=1\ \text{ and }\ u_2=\boxed{0.285c}.$$

P9.28 $p=\frac{mu}{\sqrt{1-(u/c)^{2}}}$ becomes $1-\frac{u^{2}}{c^{2}}=\frac{m^{2}u^{2}}{p^{2}}$

which gives:
$$1=u^{2}\left(\frac{m^{2}}{p^{2}}+\frac{1}{c^{2}}\right)$$

or $c^{2}=u^{2}\left(\frac{m^{2}c^{2}}{p^{2}}+1\right)$ and $\boxed{u=\frac{c}{\sqrt{\left(m^{2}c^{2}/p^{2}\right)+1}}}$.

Section 9.7 Relativistic Energy

P9.29 $\Delta E = (\gamma_1 - \gamma_2)mc^2$

For an electron, $mc^2 = 0.511\text{ MeV}$

(a) $$\Delta E = \left(\sqrt{\frac{1}{(1-0.810)}} - \sqrt{\frac{1}{(1-0.250)}}\right)mc^2 = \boxed{0.582\text{ MeV}}$$

(b) $$\Delta E = \left(\sqrt{\frac{1}{1-(0.990)^2}} - \sqrt{\frac{1}{1-0.810}}\right)mc^2 = \boxed{2.45\text{ MeV}}$$

P9.30 The relativistic kinetic energy of an object of mass m and speed u is $K_r = \left(\frac{1}{\sqrt{1-u^2/c^2}} - 1\right)mc^2$.

For $u = 0.100c$, $$K_r = \left(\frac{1}{\sqrt{1-0.010\,0}} - 1\right)mc^2 = 0.005\,038mc^2.$$

The classical equation $K_c = \frac{1}{2}mu^2$ gives $K_c = \frac{1}{2}m(0.100c)^2 = 0.005\,000mc^2$

different by $$\frac{0.005\,038 - 0.005\,000}{0.005\,038} = 0.751\%.$$

For still smaller speeds the agreement will be still better.

P9.31 (a) $K = E - E_R = 5E_R$

$$E = 6E_R = 6\left(9.11\times10^{-31}\text{ kg}\right)\left(3.00\times10^{8}\text{ m/s}\right)^2 = 4.92\times10^{-13}\text{ J} = \boxed{3.07\text{ MeV}}$$

(b) $E = \gamma mc^2 = \gamma E_R$

Thus $\gamma = \frac{E}{E_R} = 6 = \frac{1}{\sqrt{1-u^2/c^2}}$ which yields $\boxed{u = 0.986c}$

P9.32 (a) Using the classical equation, $K = \frac{1}{2}mv^2 = \frac{1}{2}(78.0\text{ kg})\left(1.06\times10^5\text{ m/s}\right)^2 = \boxed{4.38\times10^{11}\text{ J}}$.

(b) Using the relativistic equation, $K = \left(\frac{1}{\sqrt{1-(v/c)^2}} - 1\right)mc^2$.

$$K = \left[\frac{1}{\sqrt{1-\left(1.06\times10^5/2.998\times10^8\right)^2}} - 1\right](78.0\text{ kg})\left(2.998\times10^8\text{ m/s}\right)^2 = \boxed{4.38\times10^{11}\text{ J}}$$

continued on next page

When $\frac{v}{c} << 1$, the binomial series expansion gives $\left[1-\left(\frac{v}{c}\right)^2\right]^{-1/2} \approx 1+\frac{1}{2}\left(\frac{v}{c}\right)^2$.

Thus, $\left[1-\left(\frac{v}{c}\right)^2\right]^{-1/2} - 1 \approx \frac{1}{2}\left(\frac{v}{c}\right)^2$.

and the relativistic expression for kinetic energy becomes $K \approx \frac{1}{2}\left(\frac{v}{c}\right)^2 mc^2 = \frac{1}{2}mv^2$. That is, in the limit of speeds much smaller than the speed of light, the relativistic and classical expressions yield the same results.

P9.33 (a) $E_R = mc^2 = \left(1.67\times10^{-27}\text{ kg}\right)\left(2.998\times10^8\text{ m/s}\right)^2 = 1.50\times10^{-10}\text{ J} = \boxed{938\text{ MeV}}$

(b) $E = \gamma mc^2 = \dfrac{1.50\times10^{-10}\text{ J}}{\left[1-(0.950c/c)^2\right]^{1/2}} = 4.81\times10^{-10}\text{ J} = \boxed{3.00\times10^3\text{ MeV}}$

(c) $K = E - mc^2 = 4.81\times10^{-10}\text{ J} - 1.50\times10^{-10}\text{ J} = 3.31\times10^{-10}\text{ J} = \boxed{2.07\times10^3\text{ MeV}}$

P9.34 The relativistic density is

$$\frac{E_R}{c^2V} = \frac{mc^2}{c^2V} = \frac{m}{V} = \frac{m}{\left(L_p\right)\left(L_p\right)\left[L_p\sqrt{1-(u/c)^2}\right]} = \frac{8.00\text{ g}}{(1.00\text{ cm})^3\sqrt{1-(0.900)^2}} = \boxed{18.4\text{ g/cm}^3}.$$

***P9.35** From $K = (\gamma-1)mc^2 = \left(\dfrac{1}{\sqrt{1-v^2/c^2}}-1\right)mc^2$ we have $\dfrac{K}{mc^2}+1 = \dfrac{1}{\sqrt{1-v^2/c^2}} = \dfrac{K+mc^2}{mc^2}$.

$$1-\frac{v^2}{c^2} = \frac{m^2c^4}{\left(K+mc^2\right)^2}$$

$$\frac{v^2}{c^2} = 1-\frac{\left(mc^2\right)^2}{\left(K+mc^2\right)^2}$$

$$v = c\left(1-\left(\frac{mc^2}{K+mc^2}\right)^2\right)^{1/2}$$

(a) Electron: $v = c\left(1-\left(\dfrac{0.511}{2.511}\right)^2\right)^{1/2} = \boxed{0.979c}$

(b) Proton: $v = c\left(1-\left(\dfrac{938}{940}\right)^2\right)^{1/2} = \boxed{0.065\,2c}$

continued on next page

(c) $0.979c - 0.065\,2c = \boxed{0.914c} = 2.74\times10^{8}\ \text{m/s}$

In this case the electron is moving relativistically, but the classical expression $\frac{1}{2}mv^2$ is accurate to two digits for the proton.

(d) Electron: $v = c\left(1-\left(\frac{0.511}{2\,000.511}\right)^2\right)^{1/2} = \boxed{0.999\,999\,97c}$

Proton: $v = c\left(1-\left(\frac{938}{2\,938}\right)^2\right)^{1/2} = \boxed{0.948c}$

Excess speed $= 0.999\,999\,97c - 0.948c = \boxed{0.052\,3c} = 1.57\times10^{7}\ \text{m/s}$

As the kinetic energies of both particles become large, the difference in their speeds approaches zero. By contrast, classically the speed difference would become large without any finite limit.

P9.36 We must conserve both energy and relativistic momentum of the system of fragments. With subscript 1 referring to the $0.868c$ particle and subscript 2 to the $0.987c$ particle,

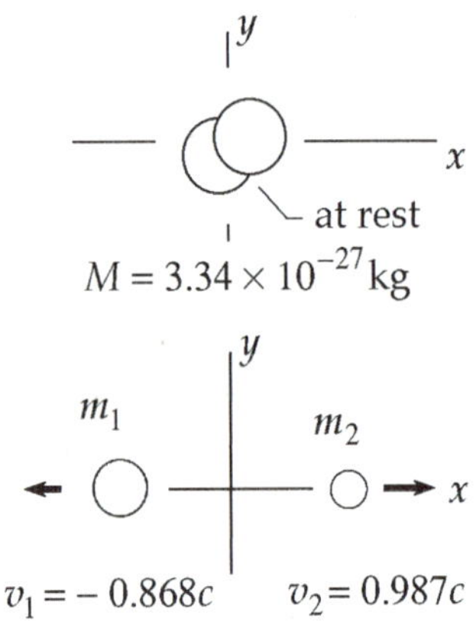

FIG. P9.36

$$\gamma_1 = \frac{1}{\sqrt{1-(0.868)^2}} = 2.01 \text{ and } \gamma_2 = \frac{1}{\sqrt{1-(0.987)^2}} = 6.22.$$

Conservation of energy gives $E_1 + E_2 = E_{\text{total}}$

which is $\gamma_1 m_1 c^2 + \gamma_2 m_2 c^2 = m_{\text{total}} c^2$

or $2.01m_1 + 6.22m_2 = 3.34\times10^{-27}\ \text{kg}.$

This reduces to: $m_1 + 3.09m_2 = 1.66\times10^{-27}\ \text{kg}.$ (1)

Since the final momentum of the system must equal zero, $p_1 = p_2$

gives $\gamma_1 m_1 u_1 = \gamma_2 m_2 u_2$

or $(2.01)(0.868c)m_1 = (6.22)(0.987c)m_2$

which becomes $m_1 = 3.52m_2.$ (2)

Solving (1) and (2) simultaneously, $m_1 = \boxed{8.84\times10^{-28}\ \text{kg}}$ and $m_2 = \boxed{2.51\times10^{-28}\ \text{kg}}$.

P9.37 $E = \gamma mc^2$ $\qquad p = \gamma mu$

$E^2 = \left(\gamma mc^2\right)^2$ $\qquad p^2 = (\gamma mu)^2$

$$E^2 - p^2c^2 = \left(\gamma mc^2\right)^2 - (\gamma mu)^2 c^2 = \gamma^2\left(\left(mc^2\right)^2 - (mc)^2u^2\right) = \left(mc^2\right)^2\left(1-\frac{u^2}{c^2}\right)\left(1-\frac{u^2}{c^2}\right)^{-1} = \left(mc^2\right)^2$$

Q.E.D.

***P9.38** Energy conservation: $\frac{1}{\sqrt{1-0^2}}1\,400\text{ kg}c^2+\frac{900\text{ kg}c^2}{\sqrt{1-0.85^2}}=\frac{Mc^2}{\sqrt{1-v^2/c^2}}$

$$3\,108\text{ kg}\sqrt{1-\frac{v^2}{c^2}}=M.$$

Momentum conservation: $0+\frac{900\text{ kg}(0.85c)}{\sqrt{1-0.85^2}}=\frac{Mv}{\sqrt{1-v^2/c^2}}$

$$1\,452\text{ kg}\sqrt{1-\frac{v^2}{c^2}}=\frac{Mv}{c}.$$

(a) Dividing gives $\frac{v}{c}=\frac{1\,452}{3\,108}=0.467$ $\boxed{v=0.467c}$.

(b) Now by substitution $3\,108\text{ kg}\sqrt{1-0.467^2}=\boxed{M=2.75\times10^3\text{ kg}}$.

P9.39 Conserving total momentum of the decaying particle system, $p_{\text{before decay}}=p_{\text{after decay}}=0$

$$p_\nu=p_\mu=\gamma m_\mu u=\gamma(207m_e)u.$$

Conservation of mass-energy for the system gives $E_\mu+E_\nu=E_\pi$: $\gamma m_\mu c^2+p_\nu c=m_\pi c^2$

$$\gamma(207m_e)+\frac{p_\nu}{c}=273m_e.$$

Substituting from the momentum equation above, $\gamma(207m_e)+\gamma(207m_e)\frac{u}{c}=273m_e$

or $\gamma\left(1+\frac{u}{c}\right)=\frac{273}{207}=1.32$: $\frac{1+u/c}{1-u/c}=1.74$ $\frac{u}{c}=0.270.$

Then, $K_\mu=(\gamma-1)m_\mu c^2=(\gamma-1)207(m_ec^2)$: $K_\mu=\left(\frac{1}{\sqrt{1-(0.270)^2}}-1\right)207(0.511\text{ MeV})$

$$K_\mu=\boxed{4.08\text{ MeV}}.$$

Also, $E_\nu=E_\pi-E_\mu$: $E_\nu=m_\pi c^2-\gamma m_\mu c^2=(273-207\gamma)m_ec^2$

$$E_\nu=\left(273-\frac{207}{\sqrt{1-(0.270)^2}}\right)(0.511\text{ MeV})$$

$$E_\nu=\boxed{29.6\text{ MeV}}$$

Section 9.8 Mass and Energy

P9.40 $E = 2.86 \times 10^5$ J. Also, the mass-energy relation says that $E = mc^2$.

Therefore, $m = \frac{E}{c^2} = \frac{2.86 \times 10^5 \text{ J}}{\left(3.00 \times 10^8 \text{ m/s}\right)^2} = \boxed{3.18 \times 10^{-12} \text{ kg}}$.

No, a mass loss of this magnitude (out of a total of 9.00 g) $\boxed{\text{could not be detected}}$.

P9.41 $\mathcal{P} = \frac{dE}{dt} = \frac{d\left(mc^2\right)}{dt} = c^2 \frac{dm}{dt} = 3.85 \times 10^{26} \text{ W}$

Thus, $\frac{dm}{dt} = \frac{3.85 \times 10^{26} \text{ J/s}}{\left(3.00 \times 10^8 \text{ m/s}\right)^2} = \boxed{4.28 \times 10^9 \text{ kg/s}}$

P9.42 $\Delta m = \frac{E}{c^2} = \frac{\mathcal{P}\Delta t}{c^2} = \frac{0.800\left(1.00 \times 10^9 \text{ J/s}\right)\left(3.00 \text{ yr}\right)\left(3.16 \times 10^7 \text{ s/yr}\right)}{\left(3.00 \times 10^8 \text{ m/s}\right)^2} = \boxed{0.842 \text{ kg}}$

P9.43 $2m_e c^2 = 1.02 \text{ MeV}$ $\qquad E_\gamma \geq \boxed{1.02 \text{ MeV}}$

***P9.44** (a) As the internal energy of the 3 000 kg system increases, its mass increases. Its original kinetic energy of macroscopic motion becomes additional rest energy as its total energy is constant. The speed is small compared to that of light, so we may accurately compute $\Delta E_{R\text{ final}} = K_{1i} + K_{2i} = 2\left(\frac{1}{2}\right)(1\,500 \text{ kg})(20 \text{ m/s})^2 = 6 \times 10^5 \text{ J}$. The mass is $\boxed{\text{greater}}$ according to $\Delta E_R = \Delta mc^2$

$$\Delta m = \frac{6 \times 10^5 \text{ J}}{\left(3 \times 10^8 \text{ m/s}\right)^2} = \boxed{6.67 \times 10^{-12} \text{ kg}}$$

(b) The total original kinetic energy is

$$2(\gamma - 1)mc^2 = 2\left(\frac{1}{\sqrt{1 - \left(2 \times 10^8 / 3 \times 10^8\right)^2}} - 1\right)(1\,500 \text{ kg})\left(3 \times 10^8 \text{ m/s}\right)^2 = 9.22 \times 10^{19} \text{ J}.$$

The final mass is $\boxed{\text{greater}}$ by $\Delta m = \frac{\Delta E_R}{c^2} = \frac{9.22 \times 10^{19} \text{ J}}{9 \times 10^{16} \text{ m}^2/\text{s}^2} = \boxed{1.02 \times 10^3 \text{ kg}}$.

Section 9.9 General Relativity

***P9.45** (a) $v=\frac{2\pi r}{T}=\frac{2\pi\left(2.66\times10^{7}\text{ m}\right)}{43\,080\text{ s}}=\boxed{3.88\times10^{3}\text{ m/s}}$

(b) The small fractional decrease in frequency received is equal in magnitude to the fractional increase in period of the moving oscillator due to time dilation:

$$\text{fractional change in } f=-(\gamma-1)=-\left(\frac{1}{\sqrt{1-\left(3.88\times10^{3}/3\times10^{8}\right)^{2}}}-1\right)$$

$$\approx 1-\left(1-\frac{1}{2}\left[-\left(\frac{3.88\times10^{3}}{3\times10^{8}}\right)^{2}\right]\right)=\boxed{-8.36\times10^{-11}}$$

(c) The orbit altitude is large compared to the radius of the Earth, so we must use $U_g=-\frac{GM_Em}{r}$.

$$\Delta U_g=-\frac{6.67\times10^{-11}\text{ Nm}^2\left(5.98\times10^{24}\text{ kg}\right)m}{\text{kg}^2\ 2.66\times10^{7}\text{ m}}+\frac{6.67\times10^{-11}\text{ Nm}\left(5.98\times10^{24}\text{ kg}\right)m}{\text{kg}\ 6.37\times10^{6}\text{ m}}$$

$$=4.76\times10^{7}\text{ J/kg}\,m$$

$$\frac{\Delta f}{f}=\frac{\Delta U_g}{mc^2}=\frac{4.76\times10^{7}\text{ m}^2/\text{s}^2}{\left(3\times10^{8}\text{ m/s}\right)^2}=\boxed{+5.29\times10^{-10}}$$

(d) $-8.36\times10^{-11}+5.29\times10^{-10}=\boxed{+4.46\times10^{-10}}$

Section 9.10 Context Connection—From Mars to the Stars

P9.46 We find Cooper's speed: $\frac{GMm}{r^2}=\frac{mv^2}{r}$.

Solving, $v=\left[\frac{GM}{(R+h)}\right]^{1/2}=\left[\frac{\left(6.67\times10^{-11}\right)\left(5.98\times10^{24}\right)}{\left(6.37\times10^{6}+0.160\times10^{6}\right)}\right]^{1/2}=7.82\text{ km/s}$.

Then the time period of one orbit, $T=\frac{2\pi(R+h)}{v}=\frac{2\pi\left(6.53\times10^{6}\right)}{7.82\times10^{3}}=5.25\times10^{3}\text{ s}$.

continued on next page

(a) The time difference for 22 orbits is $\Delta t-\Delta t_p=(\gamma-1)\Delta t_p=\left[\left(1-\frac{v^2}{c^2}\right)^{-1/2}-1\right](22T)$

$$\Delta t-\Delta t_p\approx\left(1+\frac{1}{2}\frac{v^2}{c^2}-1\right)(22T)=\frac{1}{2}\left(\frac{7.82\times10^3\text{ m/s}}{3\times10^8\text{ m/s}}\right)^2 22\left(5.25\times10^3\text{ s}\right)=\boxed{39.2\ \mu\text{s}}.$$

(b) For one orbit, $\Delta t-\Delta t_p=\frac{39.2\ \mu\text{s}}{22}=1.78\ \mu\text{s}$. The press report is $\boxed{\text{accurate to one digit}}$.

P9.47 (a) $d_{\text{earth}}=vt_{\text{earth}}=v\gamma t_{\text{astro}}$ so $\left(2.00\times10^6\text{ yr}\right)c=v\frac{1}{\sqrt{1-v^2/c^2}}30.0\text{ yr}$

$$\sqrt{1-\frac{v^2}{c^2}}=\left(\frac{v}{c}\right)\left(1.50\times10^{-5}\right)\qquad 1-\frac{v^2}{c^2}=\frac{v^2\left(2.25\times10^{-10}\right)}{c^2}$$

$$1=\frac{v^2}{c^2}\left(1+2.25\times10^{-10}\right)\quad\text{so}\quad\frac{v}{c}=\left(1+2.25\times10^{-10}\right)^{-1/2}\approx1-\frac{1}{2}\left(2.25\times10^{-10}\right)$$

$$\boxed{\frac{v}{c}=1-1.12\times10^{-10}}$$

(b) $$K=\left(\frac{1}{\sqrt{1-v^2/c^2}}-1\right)mc^2=\left(\frac{2.00\times10^6\text{ yr}}{30\text{ yr}}-1\right)(1\,000)(1\,000\text{ kg})\left(3\times10^8\text{ m/s}\right)^2=\boxed{6.00\times10^{27}\text{ J}}$$

(c) $$6.00\times10^{27}\text{ J}=6.00\times10^{27}\text{ J}\left(\frac{13¢}{\text{kWh}}\right)\left(\frac{\text{k}}{10^3}\right)\left(\frac{\text{W}\cdot\text{s}}{\text{J}}\right)\left(\frac{\text{h}}{3\,600\text{ s}}\right)=\boxed{\$2.17\times10^{20}}$$

Additional Problems

P9.48 (a) When $K_e=K_p$, $\quad m_ec^2(\gamma_e-1)=m_pc^2(\gamma_p-1)$.

In this case, $\quad m_ec^2=0.511\text{ MeV}$, $m_pc^2=938\text{ MeV}$

and $\quad \gamma_e=\left[1-(0.750)^2\right]^{-1/2}=1.511\,9$.

Substituting, $$\gamma_p=1+\frac{m_ec^2(\gamma_e-1)}{m_pc^2}=1+\frac{(0.511\text{ MeV})(1.511\,9-1)}{938\text{ MeV}}=1.000\,279$$

but $$\gamma_p=\frac{1}{\left[1-\left(u_p/c\right)^2\right]^{1/2}}.$$

Therefore, $\quad u_p=c\sqrt{1-\gamma_p^{-2}}=\boxed{0.023\,6c}$.

continued on next page

(b) When $p_e = p_p$ $\quad \gamma_p m_p u_p = \gamma_e m_e u_e$ or $\gamma_p u_p = \dfrac{\gamma_e m_e u_e}{m_p}$.

Thus, $$\gamma_p u_p = \frac{(1.511\,9)\left(0.511 \text{ MeV}/c^2\right)(0.750c)}{938 \text{ MeV}/c^2} = 6.177\,2\times10^{-4}c$$

and $$\frac{u_p}{c} = 6.177\,2\times10^{-4}\sqrt{1-\left(\frac{u_p}{c}\right)^2}$$

which yields $u_p = \boxed{6.18\times10^{-4}c} = 185 \text{ km/s}$.

P9.49 (a) $10^{13}\text{ MeV} = (\gamma-1)m_p c^2$ so $\gamma = 10^{10}$

$v_p \approx c$ $\quad t' = \dfrac{t}{\gamma} = \dfrac{10^5 \text{ yr}}{10^{10}} = 10^{-5}\text{ yr} \sim \boxed{10^2 \text{ s}}$

(b) $d' = ct' \boxed{\sim 10^8 \text{ km}}$

P9.50 (a) Since Mary is in the same reference frame, S', as Ted, she measures the ball to have the same speed Ted observes, namely $|u'_x| = \boxed{0.800c}$.

(b) $$\Delta t' = \frac{L_p}{|u'_x|} = \frac{1.80\times10^{12}\text{ m}}{0.800\left(3.00\times10^8 \text{ m/s}\right)} = \boxed{7.50\times10^3 \text{ s}}$$

(c) $$L = L_p\sqrt{1-\frac{v^2}{c^2}} = \left(1.80\times10^{12}\text{ m}\right)\sqrt{1-\frac{(0.600c)^2}{c^2}} = \boxed{1.44\times10^{12}\text{ m}}$$

Since $v = 0.600c$ and $u'_x = -0.800c$, the velocity Jim measures for the ball is

$$u_x = \frac{u'_x + v}{1+u'_x v/c^2} = \frac{(-0.800c)+(0.600c)}{1+(-0.800)(0.600)} = \boxed{-0.385c}.$$

(d) Jim measures the ball and Mary to be initially separated by 1.44×10^{12} m. Mary's motion at 0.600*c* and the ball's motion at 0.385*c* nibble into this distance from both ends. The gap closes at the rate $0.600c + 0.385c = 0.985c$, so the ball and catcher meet after a time

$$\Delta t = \frac{1.44\times10^{12}\text{ m}}{0.985\left(3.00\times10^8 \text{ m/s}\right)} = \boxed{4.88\times10^3 \text{ s}}.$$

P9.51 $$\frac{\Delta mc^2}{mc^2} = \frac{4(938.78 \text{ MeV}) - 3\,728.4 \text{ MeV}}{4(938.78 \text{ MeV})}\times100\% = \boxed{0.712\%}$$

***P9.52** The energy of the first fragment is given by $E_1^2 = p_1^2c^2 + \left(m_1c^2\right)^2 = (1.75\text{ MeV})^2 + (1.00\text{ MeV})^2$

$E_1 = 2.02\text{ MeV}$.

For the second, $E_2^2 = (2.00\text{ MeV})^2 + (1.50\text{ MeV})^2$ $\qquad E_2 = 2.50\text{ MeV}$.

(a) Energy is conserved, so the unstable object had $E = 4.52\text{ MeV}$. Each component of momentum is conserved, so the original object moved with $p^2 = p_x^2 + p_y^2 = \left(\frac{1.75\text{ MeV}}{c}\right)^2 + \left(\frac{2.00\text{ MeV}}{c}\right)^2$. Then for it

$$(4.52\text{ MeV})^2 = (1.75\text{ MeV})^2 + (2.00\text{ MeV})^2 + \left(mc^2\right)^2 \qquad \boxed{m = \frac{3.65\text{ MeV}}{c^2}}.$$

(b) Now $E = \gamma mc^2$ gives $4.52\text{ MeV} = \frac{1}{\sqrt{1-v^2/c^2}}3.65\text{ MeV}$ $\qquad 1 - \frac{v^2}{c^2} = 0.654$, $\boxed{v = 0.589c}$.

P9.53 (a) Take the spaceship as the primed frame, moving toward the right at $v = +0.600c$.

Then $u_x' = +0.800c$, and $\qquad u_x = \frac{u_x' + v}{1 + (u_x'v)/c^2} = \frac{0.800c + 0.600c}{1 + (0.800)(0.600)} = \boxed{0.946c}$.

(b) $L = \frac{L_p}{\gamma}$: $\qquad L = (0.200\text{ ly})\sqrt{1-(0.600)^2} = \boxed{0.160\text{ ly}}$

(c) The aliens observe the 0.160-ly distance closing because the probe nibbles into it from one end at 0.800*c* and the Earth reduces it at the other end at 0.600*c*.

Thus, $\qquad \text{time} = \frac{0.160\text{ ly}}{0.800c + 0.600c} = \boxed{0.114\text{ yr}}$.

(d) $K = \left(\frac{1}{\sqrt{1-u^2/c^2}} - 1\right)mc^2$: $\qquad K = \left(\frac{1}{\sqrt{1-(0.946)^2}} - 1\right)\left(4.00\times10^5\text{ kg}\right)\left(3.00\times10^8\text{ m/s}\right)^2$

$$K = \boxed{7.50\times10^{22}\text{ J}}$$

P9.54 In this case, the proper time is T_0 (the time measured by the students on a clock at rest relative to them). The dilated time measured by the professor is: $\qquad \Delta t = \gamma T_0$

where $\Delta t = T + t$ Here *T* is the time she waits before sending a signal and *t* is the time required for the signal to reach the students.

Thus, we have: $\qquad T + t = \gamma T_0$. $\qquad$ (1)

To determine the travel time *t*, realize that the distance the students will have moved beyond the professor before the signal reaches them is: $\qquad d = v(T + t)$.

The time required for the signal to travel this distance is: $\qquad t = \frac{d}{c} = \left(\frac{v}{c}\right)(T + t)$.

continued on next page

Solving for t gives: $$t=\frac{(v/c)T}{1-(v/c)}.$$

Substituting this into equation (1) yields: $$T+\frac{(v/c)T}{1-(v/c)}=\gamma T_0$$

or $$\frac{T}{1-v/c}=\gamma T_0.$$

Then $$T=T_0\frac{1-(v/c)}{\sqrt{1-(v^2/c^2)}}=T_0\frac{1-(v/c)}{\sqrt{[1+(v/c)][1-(v/c)]}}=\boxed{T_0\sqrt{\frac{1-(v/c)}{1+(v/c)}}}.$$

P9.55 The observer measures the proper length of the tunnel, 50.0 m, but measures the train contracted to length

$$L=L_p\sqrt{1-\frac{v^2}{c^2}}=100\text{ m}\sqrt{1-(0.950)^2}=31.2\text{ m}$$

shorter than the tunnel by $50.0-31.2=\boxed{18.8\text{ m}}$ so $\boxed{\text{it is completely within the tunnel.}}$.

P9.56 The energy that arrives in one year is

$$E=\mathscr{P}\Delta t=\left(1.79\times10^{17}\text{ J/s}\right)\left(3.16\times10^{7}\text{ s}\right)=5.66\times10^{24}\text{ J}.$$

Thus, $$m=\frac{E}{c^2}=\frac{5.66\times10^{24}\text{ J}}{\left(3.00\times10^{8}\text{ m/s}\right)^2}=\boxed{6.28\times10^{7}\text{ kg}}.$$

P9.57 (a) At any speed, the momentum of the particle is given by

$$p=\gamma mu=\frac{mu}{\sqrt{1-(u/c)^2}}.$$

Since $F=qE=\frac{dp}{dt}$: $$qE=\frac{d}{dt}\left[mu\left(1-\frac{u^2}{c^2}\right)^{-1/2}\right]$$

$$qE=m\left(1-\frac{u^2}{c^2}\right)^{-1/2}\frac{du}{dt}+\frac{1}{2}mu\left(1-\frac{u^2}{c^2}\right)^{-3/2}\left(\frac{2u}{c^2}\right)\frac{du}{dt}.$$

So $$\frac{qE}{m}=\frac{du}{dt}\left[\frac{1-u^2/c^2+u^2/c^2}{\left(1-u^2/c^2\right)^{3/2}}\right]$$

and $$\boxed{a=\frac{du}{dt}=\frac{qE}{m}\left(1-\frac{u^2}{c^2}\right)^{3/2}}.$$

continued on next page

(b) For u small compared to c, the relativistic expression reduces to the classical $a=\frac{qE}{m}$. As u approaches c, the acceleration approaches zero, so that the object can never reach the speed of light.

(c) $$\int_0^u \frac{du}{\left(1-u^2/c^2\right)^{3/2}}=\int_{t=0}^t \frac{qE}{m}dt \qquad \boxed{u=\frac{qEct}{\sqrt{m^2c^2+q^2E^2t^2}}}$$

$$x=\int_0^t u\,dt=qEc\int_0^t \frac{t\,dt}{\sqrt{m^2c^2+q^2E^2t^2}} \qquad \boxed{x=\frac{c}{qE}\left(\sqrt{m^2c^2+q^2E^2t^2}-mc\right)}$$

P9.58 If the energy required to remove a mass m from the surface is equal to its rest energy mc^2,

then $$\frac{GM_s m}{R_g}=mc^2$$

and $$R_g=\frac{GM_s}{c^2}=\frac{\left(6.67\times10^{-11}\text{ N}\cdot\text{m}^2/\text{kg}^2\right)\left(1.99\times10^{30}\text{ kg}\right)}{\left(3.00\times10^{8}\text{ m/s}\right)^2}$$

$$R_g=1.47\times10^{3}\text{ m}=\boxed{1.47\text{ km}}.$$

P9.59 Take the two colliding protons as the system

$E_1=K+mc^2$ $\qquad$ $E_2=mc^2$

$E_1^2=p_1^2c^2+m^2c^4$ $\qquad$ $p_2=0$.

In the final state, $E_f=K_f+Mc^2$: $\quad E_f^2=p_f^2c^2+M^2c^4$.

By energy conservation, $E_1+E_2=E_f$, so

$$E_1^2+2E_1E_2+E_2^2=E_f^2$$

$$p_1^2c^2+m^2c^4+2\left(K+mc^2\right)mc^2+m^2c^4$$
$$=p_f^2c^2+M^2c^4$$

By conservation of momentum, $\quad p_1=p_f$.

Then $$M^2c^4=2Kmc^2+4m^2c^4=\frac{4Km^2c^4}{2mc^2}+4m^2c^4$$

$$Mc^2=2mc^2\sqrt{1+\frac{K}{2mc^2}}.$$

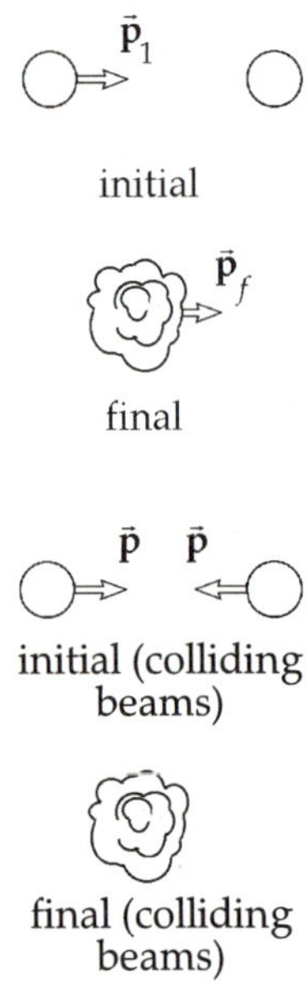

FIG. P9.59

By contrast, for colliding beams we have

In the original state, $\quad E_1=K+mc^2$

$$E_2=K+mc^2.$$

continued on next page

In the final state, $E_f = Mc^2$

$E_1 + E_2 = E_f$: $K + mc^2 + K + mc^2 = Mc^2$

$$Mc^2 = 2mc^2\left(1 + \frac{K}{2mc^2}\right).$$

P9.60 (a) An observer at rest relative to the mirror sees the light travel a distance $D = 2d - x$, where $x = vt_S$ is the distance the ship moves toward the mirror in time t_S. Since this observer agrees that the speed of light is c, the time for it to travel distance D is

$$t_S = \frac{D}{c} = \frac{2d - vt_S}{c} \qquad t_S = \boxed{\frac{2d}{c+v}}.$$

(b) The observer in the rocket measures a length-contracted initial distance to the mirror of

$$L = d\sqrt{1 - \frac{v^2}{c^2}}$$

and the mirror moving toward the ship at speed v. Thus, he measures the distance the light travels as $D = 2(L - y)$ where $y = \frac{vt}{2}$ is the distance the mirror moves toward the ship before the light reflects from it. This observer also measures the speed of light to be c, so the time for it to travel distance D is:

$$t = \frac{D}{c} = \frac{2}{c}\left[d\sqrt{1 - \frac{v^2}{c^2}} - \frac{vt}{2}\right] \text{ so } (c+v)t = \frac{2d}{c}\sqrt{(c+v)(c-v)} \text{ or } t = \boxed{\frac{2d}{c}\sqrt{\frac{c-v}{c+v}}}.$$

***P9.61** (a) $L_0^2 = L_{0x}^2 + L_{0y}^2$ and $L^2 = L_x^2 + L_y^2$.

The motion is in the x direction: $L_y = L_{0y} = L_0 \sin\theta_0$

$$L_x = L_{0x}\sqrt{1 - \left(\frac{v}{c}\right)^2} = (L_0 \cos\theta_0)\sqrt{1 - \left(\frac{v}{c}\right)^2}.$$

Thus, $$L^2 = L_0^2 \cos^2\theta_0\left[1 - \left(\frac{v}{c}\right)^2\right] + L_0^2 \sin^2\theta_0 = L_0^2\left[1 - \left(\frac{v}{c}\right)^2 \cos^2\theta_0\right]$$

or $$\boxed{L = L_0\left[1 - \left(\frac{v}{c}\right)^2 \cos^2\theta_0\right]^{1/2}}.$$

(b) $$\tan\theta = \frac{L_y}{L_x} = \frac{L_{0y}}{L_{0x}\sqrt{1 - (v/c)^2}} = \boxed{\gamma \tan\theta_0}$$

P9.62 Take $m = 1.00\text{ kg}$.

The classical kinetic energy is $$K_c = \frac{1}{2}mu^2 = \frac{1}{2}mc^2\left(\frac{u}{c}\right)^2 = \left(4.50\times10^{16}\text{ J}\right)\left(\frac{u}{c}\right)^2$$

and the actual kinetic energy is $$K_r = \left(\frac{1}{\sqrt{1-(u/c)^2}}-1\right)mc^2 = \left(9.00\times10^{16}\text{ J}\right)\left(\frac{1}{\sqrt{1-(u/c)^2}}-1\right).$$

$\frac{u}{c}$	K_c (J)	K_r (J)
0.000	0.000	0.000
0.100	0.045×10^{16}	0.0453×10^{16}
0.200	0.180×10^{16}	0.186×10^{16}
0.300	0.405×10^{16}	0.435×10^{16}
0.400	0.720×10^{16}	0.820×10^{16}
0.500	1.13×10^{16}	1.39×10^{16}
0.600	1.62×10^{16}	2.25×10^{16}
0.700	2.21×10^{16}	3.60×10^{16}
0.800	2.88×10^{16}	6.00×10^{16}
0.900	3.65×10^{16}	11.6×10^{16}
0.990	4.41×10^{16}	54.8×10^{16}

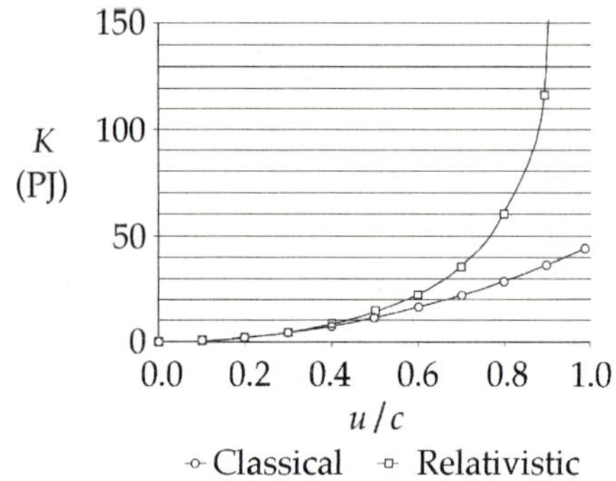

FIG. P9.62

$K_c = 0.990K_r$, when $\frac{1}{2}\left(\frac{u}{c}\right)^2 = 0.990\left[\frac{1}{\sqrt{1-(u/c)^2}}-1\right]$, yielding $u = \boxed{0.115c}$.

Similarly, $K_c = 0.950K_r$ when $u = \boxed{0.257c}$

and $K_c = 0.500K_r$ when $u = \boxed{0.786c}$.

P9.63 (b) Consider a hermit who lives on an asteroid halfway between the Sun and Tau Ceti, stationary with respect to both. Just as our spaceship is passing him, he also sees the blast waves from both explosions. Judging both stars to be stationary, this observer concludes that $\boxed{\text{the two stars blew up simultaneously}}$.

(a) We in the spaceship moving past the hermit do not calculate the explosions to be simultaneous. We measure the distance we have traveled from the Sun as

$$L = L_p\sqrt{1-\left(\frac{v}{c}\right)^2} = (6.00\text{ ly})\sqrt{1-(0.800)^2} = 3.60\text{ ly}.$$

continued on next page

We see the Sun flying away from us at 0.800*c* while the light from the Sun approaches at 1.00*c*. Thus, the gap between the Sun and its blast wave has opened at 1.80*c*, and the time we calculate to have elapsed since the Sun exploded is

$$\frac{3.60\text{ ly}}{1.80c} = 2.00\text{ yr}.$$

We see Tau Ceti as moving toward us at 0.800*c*, while its light approaches at 1.00*c*, only 0.200*c* faster. We measure the gap between that star and its blast wave as 3.60 ly and growing at 0.200*c*. We calculate that it must have been opening for

$$\frac{3.60\text{ ly}}{0.200c} = 18.0\text{ yr}$$

and conclude that $\boxed{\text{Tau Ceti exploded 16.0 years before the Sun}}$.

ANSWERS TO EVEN PROBLEMS

P9.2 see the solution

P9.4 $0.866c$

P9.6 (a) 64.9/min; (b) 10.6/min

P9.8 (a) $2.18\ \mu\text{s}$; (b) 649 m

P9.10 $0.140c$

P9.12 (a) $v = \dfrac{cL_p}{\sqrt{c^2\Delta t^2 + L_p^2}}$; (b) 4.00 m/s; (c) see the solution

P9.14 5.45 yr, Goslo is older

P9.16 (a) $0.943c$; (b) $2.55 \times 10^3\text{ m}$

P9.18 B occurred 444 ns before A

P9.20 $0.357c$

P9.22 $0.893c$ at $16.8°$ above the x'-axis

P9.24 (a) $0.141c$; (b) $0.436c$

P9.26 (a) \$800; (b) $\$2.12 \times 10^9$

P9.28 see the solution

P9.30 see the solution

P9.32 (a) $4.38 \times 10^{11}\text{ J}$; (b) $4.38 \times 10^{11}\text{ J}$

P9.34 18.4 g/cm^3

P9.36 $8.84 \times 10^{-28}\text{ kg}$ and $2.51 \times 10^{-28}\text{ kg}$

P9.38 (a) $0.467c$; (b) $2.75 \times 10^3\text{ kg}$

P9.40 $3.18 \times 10^{-12}\text{ kg}$, could not be detected

P9.42 0.842 kg

P9.44 (a) greater by $6.67 \times 10^{-12}\text{ kg}$; (b) greater by $1.02 \times 10^3\text{ kg}$

P9.46 (a) $39.2\ \mu\text{s}$; (b) Accurate to one digit

P9.48 (a) $0.023\,6c$; (b) $6.18 \times 10^{-4}c$

P9.50 (a) $0.800c$; (b) $7.50 \times 10^3\text{ s}$; (c) 1.44 Tm, $-0.385c$; (d) $4.88 \times 10^3\text{ s}$

P9.52 (a) $\dfrac{3.65\text{ MeV}}{c^2}$; (b) $v = 0.589c$

P9.54 see the solution

P9.56 6.28×10^7 kg

P9.58 1.47 km

P9.60 (a) $\dfrac{2d}{c+v}$; (b) $\dfrac{2d}{c}\sqrt{\dfrac{c-v}{c+v}}$

P9.62 See the solution for the graph.
$K_c = 0.990K_r$ when $u = 0.115c$,
$K_c = 0.950K_r$ when $u = 0.257c$,
$K_c = 0.500K_r$ when $u = 0.786c$.

Chapter 10

Rotational Motion

CHAPTER OUTLINE

ANSWERS TO QUESTIONS

Q10.1 1 rev/min, or $\dfrac{\pi}{30}$ rad/s. Into the wall (clockwise rotation). $\alpha = 0$.

FIG. Q10.1

Q10.2 The speedometer will be inaccurate. The speedometer measures the number of revolutions per second of the tires. A larger tire will travel more distance in one full revolution as $2\pi r$.

Q10.3 The object will start to rotate if the two forces act along different lines. Then the torques of the forces will not be equal in magnitude and opposite in direction.

Q10.4 A quick flip will set the hard–boiled egg spinning faster and more smoothly. The raw egg loses mechanical energy to internal fluid friction.

Q10.5 No, only if its angular momentum changes.

Q10.6 $I_{CM} = MR^2$, $I_{CM} = MR^2$, $I = \frac{1}{3}MR^2$, $I_{CM} = \frac{1}{2}MR^2$

Q10.7 Since the source reel stops almost instantly when the tape stops playing, the friction on the source reel axle must be fairly large. Since the source reel appears to us to rotate at almost constant angular velocity, the angular acceleration must be very small. Therefore, the torque on the source reel due to the tension in the tape must almost exactly balance the frictional torque. In turn, the frictional torque is nearly constant because kinetic friction forces don't depend on velocity, and the radius of the axle where the friction is applied is constant. Thus we conclude that the torque exerted by the tape on the source reel is essentially constant in time as the tape plays.

continued on next page

As the source reel radius R shrinks, the reel's angular speed $\omega = \frac{v}{R}$ must increase to keep the tape speed v constant. But the biggest change is to the reel's moment of inertia. We model the reel as a roll of tape, ignoring any spool or platter carrying the tape. If we think of the roll of tape as a uniform disk, then its moment of inertia is $I = \frac{1}{2}MR^2$. But the roll's mass is proportional to its base area πR^2. Thus, on the whole the moment of inertia is proportional to R^4. The moment of inertia decreases very rapidly as the reel shrinks!

The tension in the tape coming into the read-and-write heads is normally dominated by balancing frictional torque on the source reel, according to $TR \approx \tau_{\text{friction}}$. Therefore, as the tape plays the tension is largest when the reel is smallest. However, in the case of a sudden jerk on the tape, the rotational dynamics of the source reel becomes important. If the source reel is full, then the moment of inertia, proportional to R^4, will be so large that higher tension in the tape will be required to give the source reel its angular acceleration. If the reel is nearly empty, then the same tape acceleration will require a smaller tension. Thus, the tape will be more likely to break when the source reel is nearly full. One sees the same effect in the case of paper towels; it is easier to snap a towel free when the roll is new than when it is nearly empty.

Q10.8 (a) Down–cross–left is away from you: $-\hat{\mathbf{j}} \times (-\hat{\mathbf{i}}) = -\hat{\mathbf{k}}$

(b) Left–cross–down is toward you: $-\hat{\mathbf{i}} \times (-\hat{\mathbf{j}}) = \hat{\mathbf{k}}$

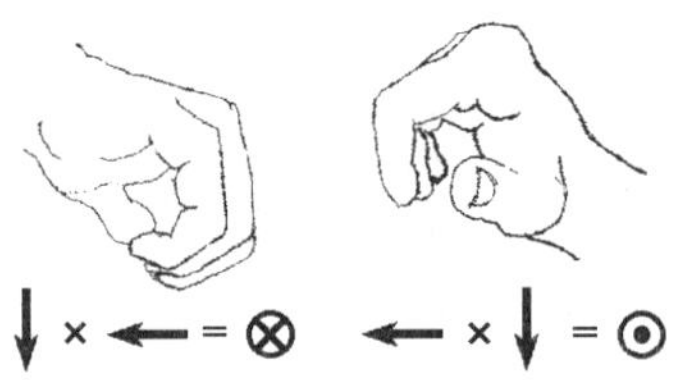

FIG. Q10.8

Q10.9 There must be two rotors to balance the torques on the body of the helicopter. If it had only one rotor, the engine would cause the body of the helicopter to swing around rapidly with angular momentum opposite to the rotor.

Q10.10 The diver leaves the platform with some angular momentum about a horizontal axis through her center of mass. When she draws up her legs, her moment of inertia decreases and her angular speed increases for conservation of angular momentum. Straightening out again slows her rotation.

Q10.11 The long pole has a large moment of inertia about an axis along the rope. An unbalanced torque will then produce only a small angular acceleration of the performer-pole system, to extend the time available for getting back in balance. To keep the center of mass above the rope, the performer can shift the pole left or right, instead of having to bend his body around. The pole sags down at the ends to lower the system center of gravity.

Q10.12 Suppose we look at the motorcycle moving to the right. Its drive wheel is turning clockwise. The wheel speeds up when it leaves the ground. No outside torque about its center of mass acts on the airborne cycle, so its angular momentum is conserved. As the drive wheel's clockwise angular momentum increases, the frame of the cycle acquires counterclockwise angular momentum. The cycle's front end moves up and its back end moves down.

Q10.13 Mass moves away from axis of rotation, so moment of inertia increases, angular speed decreases, and period increases.

Q10.14 Try it first and then explain it! If there is no air resistance and no rolling resistance, Example 10.15 shows that the races will be ties. The race results can reveal a small amount of air resistance.

Q10.15 To win the race, you want to decrease the moment of inertia of the wheels as much as possible. Small, light, solid disk-like wheels would be best!

Q10.16 When you bend over, your center of gravity shifts forward. Once your CG is no longer over your feet, gravity contributes to a nonzero net torque on your body and you begin to rotate.

Q10.17 The free body diagram demonstrates that it is necessary to have friction on the ground to counterbalance the normal force of the wall and to keep the base of the ladder from sliding. Interestingly enough, if there is friction on the floor *and* on the wall, it is not possible to determine whether the ladder will slip from the equilibrium conditions alone.

FIG. Q10.17

Q10.18 (a) Consider pushing up with one hand on one side of a steering wheel and pulling down equally hard with the other hand on the other side. A pair of equal-magnitude oppositely-directed forces applied at different points is called a couple.

(b) An object in free fall has a non-zero net force acting on it, but a net torque of zero about its center of mass.

SOLUTIONS TO PROBLEMS

Section 10.1 Angular Position, Speed, and Acceleration

P10.1 (a) $\theta\big|_{t=0} = \boxed{5.00 \text{ rad}}$

$$\omega\big|_{t=0} = \frac{d\theta}{dt}\bigg|_{t=0} = 10.0 + 4.00t\big|_{t=0} = \boxed{10.0 \text{ rad/s}}$$

$$\alpha_{t=0} = \frac{d\omega}{dt}\bigg|_{t=0} = \boxed{4.00 \text{ rad/s}^2}$$

(b) $\theta\big|_{t=3.00 \text{ s}} = 5.00 + 30.0 + 18.0 = \boxed{53.0 \text{ rad}}$

$$\omega\big|_{t=3.00 \text{ s}} = \frac{d\theta}{dt}\bigg|_{t=3.00 \text{ s}} = 10.0 + 4.00t\big|_{t=3.00 \text{ s}} = \boxed{22.0 \text{ rad/s}}$$

$$\alpha\big|_{t=3.00 \text{ s}} = \frac{d\omega}{dt}\bigg|_{t=3.00 \text{ s}} = \boxed{4.00 \text{ rad/s}^2}$$

Section 10.2 Rotational Kinematics—The Rigid Object Under Constant Angular Acceleration

P10.2 $\omega_f = 2.51 \times 10^4 \text{ rev/min} = 2.63 \times 10^3 \text{ rad/s}$

(a) $$\alpha = \frac{\omega_f - \omega_i}{t} = \frac{2.63 \times 10^3 \text{ rad/s} - 0}{3.2 \text{ s}} = \boxed{8.21 \times 10^2 \text{ rad/s}^2}$$

(b) $$\theta_f = \omega_i t + \frac{1}{2}\alpha t^2 = 0 + \frac{1}{2}\left(8.21 \times 10^2 \text{ rad/s}^2\right)(3.2 \text{ s})^2 = \boxed{4.21 \times 10^3 \text{ rad}}$$

P10.3 $$\omega_i = \frac{100 \text{ rev}}{1.00 \text{ min}}\left(\frac{1 \text{ min}}{60.0 \text{ s}}\right)\left(\frac{2\pi \text{ rad}}{1.00 \text{ rev}}\right) = \frac{10\pi}{3} \text{ rad/s}, \ \omega_f = 0$$

(a) $$t = \frac{\omega_f - \omega_i}{\alpha} = \frac{0 - \frac{10\pi}{3}}{-2.00} \text{ s} = \boxed{5.24 \text{ s}}$$

(b) $$\theta_f = \bar{\omega} t = \left(\frac{\omega_f + \omega_i}{2}\right)t = \left(\frac{10\pi}{6} \text{ rad/s}\right)\left(\frac{10\pi}{6} \text{ s}\right) = \boxed{27.4 \text{ rad}}$$

***P10.4** $\omega_i = 3\,600 \text{ rev/min} = 3.77 \times 10^2 \text{ rad/s}$

$\theta = 50.0 \text{ rev} = 3.14 \times 10^2 \text{ rad and } \omega_f = 0$

$$\omega_f^2 = \omega_i^2 + 2\alpha\theta$$

$$0 = \left(3.77 \times 10^2 \text{ rad/s}\right)^2 + 2\alpha\left(3.14 \times 10^2 \text{ rad}\right)$$

$$\alpha = \boxed{-2.26 \times 10^2 \text{ rad/s}^2}$$

P10.5 $\omega = 5.00 \text{ rev/s} = 10.0\pi \text{ rad/s}$. We will break the motion into two stages: (1) a period during which the tub speeds up and (2) a period during which it slows down.

While speeding up, $\theta_1 = \bar{\omega}t = \dfrac{0 + 10.0\pi \text{ rad/s}}{2}(8.00 \text{ s}) = 40.0\pi \text{ rad}$

While slowing down, $\theta_2 = \bar{\omega}t = \dfrac{10.0\pi \text{ rad/s} + 0}{2}(12.0 \text{ s}) = 60.0\pi \text{ rad}$

So, $\theta_{\text{total}} = \theta_1 + \theta_2 = 100\pi \text{ rad} = \boxed{50.0 \text{ rev}}$

P10.6 $\theta_f - \theta_i = \omega_i t + \frac{1}{2}\alpha t^2$ and $\omega_f = \omega_i + \alpha t$ are two equations in two unknowns ω_i and α

$\omega_i = \omega_f - \alpha t$: $\theta_f - \theta_i = (\omega_f - \alpha t)t + \frac{1}{2}\alpha t^2 = \omega_f t - \frac{1}{2}\alpha t^2$

$$37.0 \text{ rev}\left(\frac{2\pi \text{ rad}}{1 \text{ rev}}\right) = 98.0 \text{ rad/s}(3.00 \text{ s}) - \frac{1}{2}\alpha(3.00 \text{ s})^2$$

$232 \text{ rad} = 294 \text{ rad} - (4.50 \text{ s}^2)\alpha$: $\alpha = \dfrac{61.5 \text{ rad}}{4.50 \text{ s}^2} = \boxed{13.7 \text{ rad/s}^2}$

P10.7 (a) $\omega = \dfrac{\Delta\theta}{\Delta t} = \dfrac{1 \text{ rev}}{1 \text{ day}} = \dfrac{2\pi \text{ rad}}{86\,400 \text{ s}} = \boxed{7.27 \times 10^{-5} \text{ rad/s}}$

(b) $\Delta t = \dfrac{\Delta\theta}{\omega} = \dfrac{107°}{7.27 \times 10^{-5} \text{ rad/s}}\left(\dfrac{2\pi \text{ rad}}{360°}\right) = \boxed{2.57 \times 10^4 \text{ s}}$ or 428 min

---

Section 10.3 Relations Between Rotational and Translational Quantities

P10.8 Estimate the tire's radius at 0.250 m and miles driven as 10 000 per year.

$$\theta = \frac{s}{r} = \frac{1.00 \times 10^4 \text{ mi}}{0.250 \text{ m}}\left(\frac{1\,609 \text{ m}}{1 \text{ mi}}\right) = 6.44 \times 10^7 \text{ rad/yr}$$

$$\theta = 6.44 \times 10^7 \text{ rad/yr}\left(\frac{1 \text{ rev}}{2\pi \text{ rad}}\right) = 1.02 \times 10^7 \text{ rev/yr or } \boxed{\sim 10^7 \text{ rev/yr}}$$

P10.9 (a) $\omega = 2\pi f = \dfrac{2\pi \text{ rad}}{1 \text{ rev}}\left(\dfrac{1\,200 \text{ rev}}{60.0 \text{ s}}\right) = \boxed{126 \text{ rad/s}}$

(b) $v = \omega r = (126 \text{ rad/s})(3.00 \times 10^{-2} \text{ m}) = \boxed{3.77 \text{ m/s}}$

(c) $a_c = \omega^2 r = (126)^2(8.00 \times 10^{-2}) = 1\,260 \text{ m/s}^2$ so $\mathbf{\vec{a}}_r = \boxed{1.26 \text{ km/s}^2 \text{ toward the center}}$

(d) $s = r\theta = \omega r t = (126 \text{ rad/s})(8.00 \times 10^{-2} \text{ m})(2.00 \text{ s}) = \boxed{20.1 \text{ m}}$

P10.10 Given $r = 1.00\text{ m}$, $\alpha = 4.00\text{ rad/s}^2$, $\omega_i = 0$ and $\theta_i = 57.3° = 1.00\text{ rad}$

(a) $\omega_f = \omega_i + \alpha t = 0 + \alpha t$

At $t = 2.00\text{ s}$, $\omega_f = 4.00\text{ rad/s}^2(2.00\text{ s}) = \boxed{8.00\text{ rad/s}}$

(b) $v = r\omega = 1.00\text{ m}(8.00\text{ rad/s}) = \boxed{8.00\text{ m/s}}$

$|a_r| = a_c = r\omega^2 = 1.00\text{ m}(8.00\text{ rad/s})^2 = 64.0\text{ m/s}^2$

$a_t = r\alpha = 1.00\text{ m}(4.00\text{ rad/s}^2) = 4.00\text{ m/s}^2$

The magnitude of the total acceleration is:

$$a = \sqrt{a_r^2 + a_t^2} = \sqrt{(64.0\text{ m/s}^2)^2 + (4.00\text{ m/s}^2)^2} = \boxed{64.1\text{ m/s}^2}$$

The direction of the total acceleration vector makes an angle ϕ with respect to the radius to point P:

$$\phi = \tan^{-1}\left(\frac{a_t}{a_c}\right) = \tan^{-1}\left(\frac{4.00}{64.0}\right) = \boxed{3.58°}$$

(c) $$\theta_f = \theta_i + \omega_i t + \frac{1}{2}\alpha t^2 = (1.00\text{ rad}) + \frac{1}{2}(4.00\text{ rad/s}^2)(2.00\text{ s})^2 = \boxed{9.00\text{ rad}}$$

***P10.11** (a) Consider a tooth on the front sprocket. It gives this speed, relative to the frame, to the link of the chain it engages:

$$v = r\omega = \left(\frac{0.152\text{ m}}{2}\right)76\text{ rev/min}\left(\frac{2\pi\text{ rad}}{1\text{ rev}}\right)\left(\frac{1\text{ min}}{60\text{ s}}\right) = \boxed{0.605\text{ m/s}}$$

(b) Consider the chain link engaging a tooth on the rear sprocket:

$$\omega = \frac{v}{r} = \frac{0.605\text{ m/s}}{\left(\frac{0.07\text{ m}}{2}\right)} = \boxed{17.3\text{ rad/s}}$$

(c) Consider the wheel tread and the road. A thread could be unwinding from the tire with this speed relative to the frame:

$$v = r\omega = \left(\frac{0.673\text{ m}}{2}\right)17.3\text{ rad/s} = \boxed{5.82\text{ m/s}}$$

(d) We did not need to know the length of the pedal cranks, but we could use that information to find the linear speed of the pedals:

$$v = r\omega = 0.175\text{ m }7.96\text{ rad/s}\left(\frac{1}{1\text{ rad}}\right) = 1.39\text{ m/s}$$

P10.12 (a) $\omega_i = \frac{v}{r} = \frac{1.30 \text{ m/s}}{0.023 \text{ m}} = \boxed{56.5 \text{ rad/s}}$

(b) $\omega_f = \frac{1.30 \text{ m/s}}{0.058 \text{ m}} = \boxed{22.4 \text{ rad/s}}$

(c) $\omega_f = \omega_i + \alpha t$: $\alpha = \frac{22.4 \text{ rad/s} - 56.5 \text{ rad/s}}{(74(60)+33) \text{ s}} = \frac{-34.1 \text{ rad/s}}{4473 \text{ s}} = \boxed{-7.63 \times 10^{-3} \text{ rad/s}^2}$

(d) $\theta_f - \theta_i = \frac{1}{2}(\omega_f + \omega_i)t = \frac{1}{2}((56.5 + 22.4) \text{ rad/s})(4473 \text{ s}) = \boxed{1.77 \times 10^5 \text{ rad}}$

(e) $x = vt = (1.30 \text{ m/s})(4\,473 \text{ s}) = \boxed{5.81 \times 10^3 \text{ m}}$

***P10.13** The force of static friction must act forward and then more and more inward on the tires, to produce both tangential and centripetal acceleration. Its tangential component is $m(1.70 \text{ m/s}^2)$. Its radially inward component is $\frac{mv^2}{r}$. This takes the maximum value

$$m\omega_f^2 r = mr(\omega_i^2 + 2\alpha\Delta\theta) = mr\left(0 + 2\alpha\frac{\pi}{2}\right) = m\pi r\alpha = m\pi a_t = m\pi(1.70 \text{ m/s}^2).$$

With skidding impending we have $\sum F_y = ma_y$, $+n - mg = 0$, $n = mg$

$$f_s = \mu_s n = \mu_s mg = \sqrt{m^2(1.70 \text{ m/s}^2)^2 + m^2\pi^2(1.70 \text{ m/s}^2)^2}$$

$$\mu_s = \frac{1.70 \text{ m/s}^2}{g}\sqrt{1+\pi^2} = \boxed{0.572}$$

Section 10.4 Rotational Kinetic Energy

P10.14 $m_1 = 4.00 \text{ kg}$, $r_1 = |y_1| = 3.00 \text{ m}$;

$m_2 = 2.00 \text{ kg}$, $r_2 = |y_2| = 2.00 \text{ m}$;

$m_3 = 3.00 \text{ kg}$, $r_3 = |y_3| = 4.00 \text{ m}$;

$\omega = 2.00 \text{ rad/s}$ about the x-axis

(a) $I_x = m_1 r_1^2 + m_2 r_2^2 + m_3 r_3^2$

$I_x = 4.00(3.00)^2 + 2.00(2.00)^2 + 3.00(4.00)^2 = \boxed{92.0 \text{ kg}\cdot\text{m}^2}$

$K_R = \frac{1}{2}I_x\omega^2 = \frac{1}{2}(92.0)(2.00)^2 = \boxed{184 \text{ J}}$

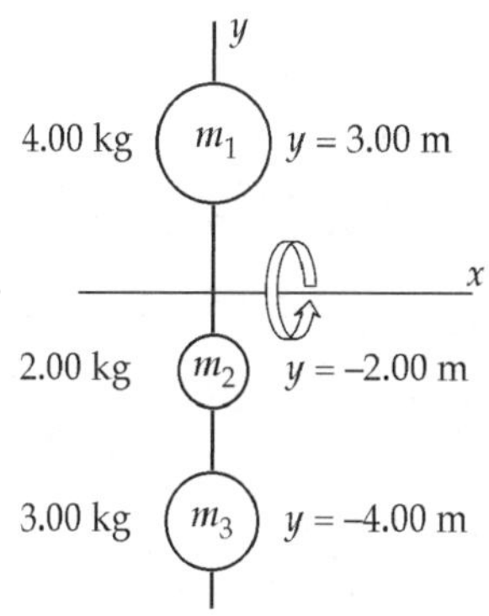

FIG. P10.14

continued on next page

(b) $v_1 = r_1\omega = 3.00(2.00) = \boxed{6.00 \text{ m/s}}$ $\quad K_1 = \frac{1}{2}m_1v_1^2 = \frac{1}{2}(4.00)(6.00)^2 = 72.0 \text{ J}$

$v_2 = r_2\omega = 2.00(2.00) = \boxed{4.00 \text{ m/s}}$ $\quad K_2 = \frac{1}{2}m_2v_2^2 = \frac{1}{2}(2.00)(4.00)^2 = 16.0 \text{ J}$

$v_3 = r_3\omega = 4.00(2.00) = \boxed{8.00 \text{ m/s}}$ $\quad K_3 = \frac{1}{2}m_3v_3^2 = \frac{1}{2}(3.00)(8.00)^2 = 96.0 \text{ J}$

$$K = K_1 + K_2 + K_3 = 72.0 + 16.0 + 96.0 = \boxed{184 \text{ J}} = \frac{1}{2}I_x\omega^2$$

P10.15 From conservation of energy for the object-turntable cylinder-Earth system,

$$\frac{1}{2}I\left(\frac{v}{r}\right)^2 + \frac{1}{2}mv^2 = mgh$$

$$I\frac{v^2}{r^2} = 2mgh - mv^2$$

$$I = \boxed{mr^2\left(\frac{2gh}{v^2} - 1\right)}$$

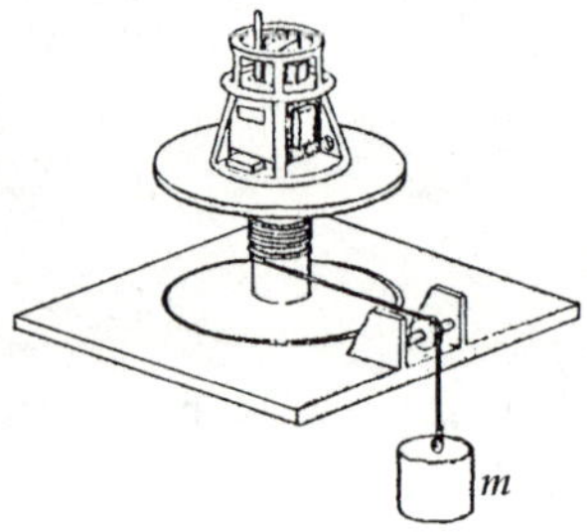

FIG. P10.15

P10.16 The moment of inertia of a thin rod about an axis through one end is $I = \frac{1}{3}ML^2$. The total rotational kinetic energy is given as $K_R = \frac{1}{2}I_h\omega_h^2 + \frac{1}{2}I_m\omega_m^2$

with $\quad I_h = \frac{m_hL_h^2}{3} = \frac{60.0 \text{ kg}(2.70 \text{ m})^2}{3} = 146 \text{ kg}\cdot\text{m}^2$

and $\quad I_m = \frac{m_mL_m^2}{3} = \frac{100 \text{ kg}(4.50 \text{ m})^2}{3} = 675 \text{ kg}\cdot\text{m}^2$

In addition, $\quad \omega_h = \frac{2\pi \text{ rad}}{12 \text{ h}}\left(\frac{1 \text{ h}}{3\,600 \text{ s}}\right) = 1.45 \times 10^{-4} \text{ rad/s}$

while $\quad \omega_m = \frac{2\pi \text{ rad}}{1 \text{ h}}\left(\frac{1 \text{ h}}{3\,600 \text{ s}}\right) = 1.75 \times 10^{-3} \text{ rad/s}$

Therefore, $\quad K_R = \frac{1}{2}(146)\left(1.45 \times 10^{-4}\right)^2 + \frac{1}{2}(675)\left(1.75 \times 10^{-3}\right)^2 = \boxed{1.03 \times 10^{-3} \text{ J}}$

P10.17 Take the two objects, pulley, and Earth as the system. If we neglect friction in the system, then mechanical energy is conserved and we can state that the increase in kinetic energy of the system equals the decrease in potential energy. Since $K_i = 0$ (the system is initially at rest), we have $\Delta K = K_f - K_i = \frac{1}{2}m_1v^2 + \frac{1}{2}m_2v^2 + \frac{1}{2}I\omega^2$ where m_1 and m_2 have a common speed. But $v = R\omega$ so that $\Delta K = \frac{1}{2}\left(m_1 + m_2 + \frac{I}{R^2}\right)v^2$. From Figure P10.17, we see that a loss of gravitational energy is associated with the motion of m_1 and a gain with the motion of m_2. Applying the law of conservation of energy $\Delta K + \Delta U = 0$ gives $\frac{1}{2}\left(m_1 + m_2 + \frac{I}{R^2}\right)v^2 + m_2gh - m_1gh = 0$.

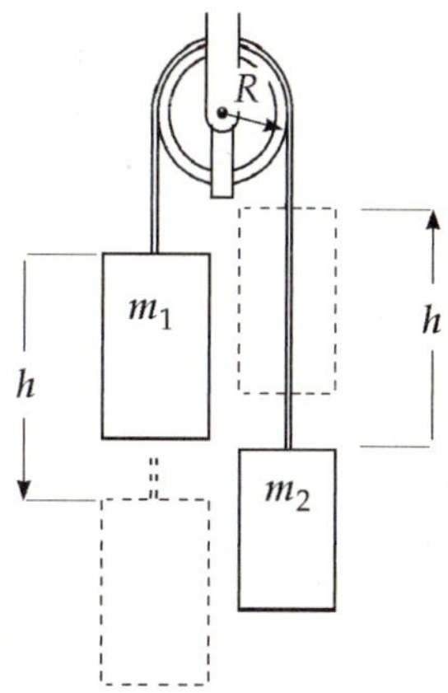

FIG. P10.17

$$\boxed{v = \sqrt{\frac{2(m_1 - m_2)gh}{\left(m_1 + m_2 + \frac{I}{R^2}\right)}}}$$

Since $v = R\omega$, the angular speed of the pulley at this instant is given by

$$\boxed{\omega = \frac{v}{R} = \sqrt{\frac{2(m_1 - m_2)gh}{(m_1R^2 + m_2R^2 + I)}}}$$

***P10.18** For large energy storage at a particular rotation rate, we want a large moment of inertia. To combine this requirement with small mass, we place the mass as far away from the axis as possible.

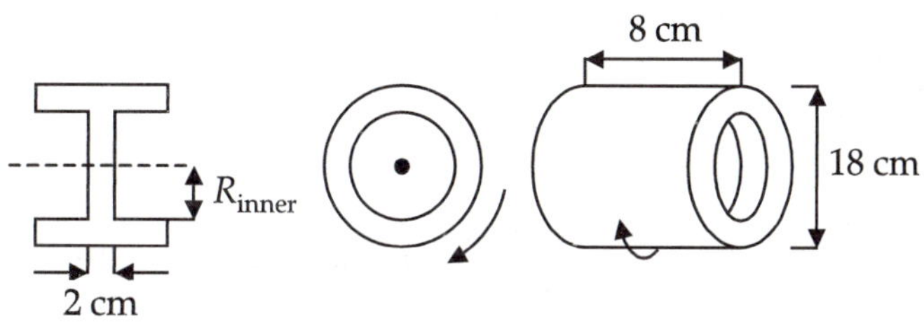

FIG. P10.18

We choose to make the flywheel as a hollow cylinder 18 cm in diameter and 8 cm long. To support this rim, we place a disk across its center. We assume that a disk 2 cm thick will be sturdy enough to support the hollow cylinder securely. The one remaining adjustable parameter is the thickness of the wall of the hollow cylinder. From Table 10.2, the moment of inertia can be written as

continued on next page

$$\begin{aligned}
I_{\text{disk}} + I_{\text{hollow cylinder}} &= \frac{1}{2}M_{\text{disk}}R_{\text{disk}}^2 + \frac{1}{2}M_{\text{wall}}\left(R_{\text{outer}}^2 + R_{\text{inner}}^2\right) \\
&= \frac{1}{2}\rho V_{\text{disk}}R_{\text{outer}}^2 + \frac{1}{2}\rho V_{\text{wall}}\left(R_{\text{outer}}^2 + R_{\text{inner}}^2\right) \\
&= \frac{\rho}{2}\pi R_{\text{outer}}^2(2\text{ cm})R_{\text{outer}}^2 + \frac{\rho}{2}\left[\pi R_{\text{outer}}^2 - \pi R_{\text{inner}}^2\right](6\text{ cm})\left(R_{\text{outer}}^2 + R_{\text{inner}}^2\right) \\
&= \frac{\rho\pi}{2}\left[(9\text{ cm})^4(2\text{ cm}) + (6\text{ cm})\left((9\text{ cm})^2 - R_{\text{inner}}^2\right)\left((9\text{ cm})^2 + R_{\text{inner}}^2\right)\right] \\
&= \rho\pi\left[6\,561\text{ cm}^5 + (3\text{ cm})\left((9\text{ cm})^4 - R_{\text{inner}}^4\right)\right] \\
&= \rho\pi\left[26\,244\text{ cm}^5 - (3\text{ cm})R_{\text{inner}}^4\right]
\end{aligned}$$

For the required energy storage,

$$\frac{1}{2}I\omega_1^2 = \frac{1}{2}I\omega_2^2 + W_{\text{out}}$$

$$\frac{1}{2}I\left[(800\text{ rev/min})\left(\frac{2\pi\text{ rad}}{1\text{ rev}}\right)\left(\frac{1\text{ min}}{60\text{ s}}\right)\right]^2 - \frac{1}{2}I\left[(600)\left(\frac{2\pi\text{ rad}}{60\text{ s}}\right)\right]^2 = 60\text{ J}$$

$$I = \frac{60\text{ J}}{1\,535/\text{s}^2} = \left(7.86\times10^3\text{ kg/m}^3\right)\pi\left[26\,244\text{ cm}^5 - (3\text{ cm})R_{\text{inner}}^4\right]$$

$$1.58\times10^{-6}\text{ m}^5\left(\frac{100\text{ cm}}{1\text{ m}}\right)^5 = 26\,244\text{ cm}^5 - (3\text{ cm})R_{\text{inner}}^4$$

$$R_{\text{inner}} = \left(\frac{26\,244\text{ cm}^4 - 15\,827\text{ cm}^4}{3}\right)^{1/4} = 7.68\text{ cm}$$

The mass of the flywheel is then

$$\begin{aligned}
M_{\text{disk}} + M_{\text{wall}} &= \rho\pi R_{\text{outer}}^2(2\text{ cm}) + \rho\left[\pi R_{\text{outer}}^2 - \pi R_{\text{inner}}^2\right](6\text{ cm}) \\
&= \left(7.86\times10^3\text{ kg/m}^3\right)\pi\left[(0.09\text{ m})^2(0.02\text{ m}) + \left[(0.09\text{ m})^2 - (0.076\,8\text{ m})^2\right](0.06\text{ m})\right] \\
&= 7.27\text{ kg}
\end{aligned}$$

If we made the thickness of the disk somewhat less than 2 cm and the inner radius of the cylindrical wall less than 7.68 cm to compensate, the mass could be a bit less than 7.27 kg.

***P10.19** Note that the torque on the trebuchet is not constant, so its angular acceleration changes in time. At our mathematical level it would be unproductive to calculate values for α on the way to find ω_f. Instead, we consider that gravitational energy of the 60-kg-Earth system becomes gravitational energy of the lighter mass plus kinetic energy of both masses. The maximum speed appears as the rod passes through the vertical. Let v_1 represent the speed of the small-mass particle m_1. Then here the rod is turning at $\omega_1 = \frac{v_1}{2.86\text{ m}}$. The larger-mass particle is moving at

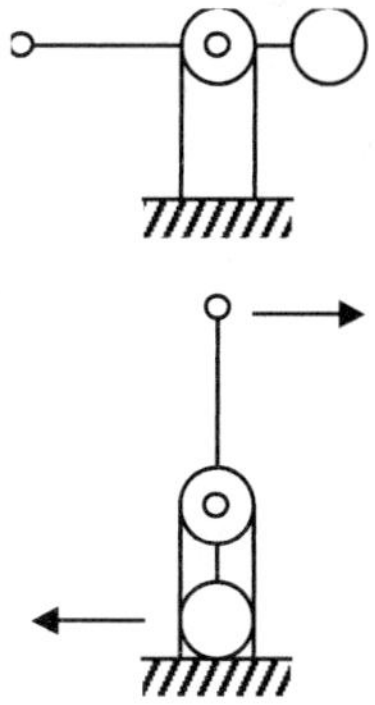

FIG. P10.19

$$v_2 = (0.14\text{ m})\omega_1 = \frac{0.14 v_1}{2.86}.$$

Now the energy-conservation equation becomes

$$\left(K_1 + K_2 + U_{g1} + U_{g2}\right)_i = \left(K_1 + K_2 + U_{g1} + U_{g2}\right)_f$$

$$0 + 0 + 0 + m_2 g y_{2i} = \frac{1}{2}m_1 v_1^2 + \frac{1}{2}m_2 v_2^2 + m_1 g y_{1f} + 0$$

$$(60\text{ kg})\left(9.8\text{ m/s}^2\right)(0.14\text{ m}) = \frac{1}{2}(0.12\text{ kg})v_1^2 + \frac{1}{2}(60\text{ kg})\left(\frac{0.14 v_1}{2.86}\right)^2 + (0.12\text{ kg})\left(9.8\text{ m/s}^2\right)(2.86\text{ m})$$

$$82.32\text{ J} = \frac{1}{2}(0.12\text{ kg})v_1^2 + \frac{1}{2}(0.144\text{ kg})v_1^2 + 3.36\text{ J}$$

$$v_1 = \left(\frac{2(79.0\text{ J})}{0.264\text{ kg}}\right)^{1/2} = \boxed{24.5\text{ m/s}}$$

Section 10.5 Torque and the Vector Product

P10.20 Resolve the 100 N force into components perpendicular to and arallel to the rod, as

$$F_{\text{par}} = (100\text{ N})\cos 57.0° = 54.5\text{ N}$$

and $F_{\text{perp}} = (100\text{ N})\sin 57.0° = 83.9\text{ N}$

The torque of F_{par} is zero since its line of action passes through he pivot point.

The torque of F_{perp} is $\tau = 83.9\text{ N}(2.00\text{ m}) = \boxed{168\text{ N}\cdot\text{m}}$ (clockwise)

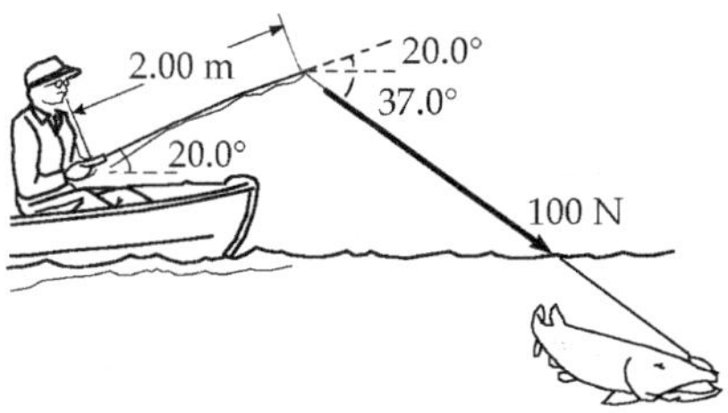

FIG. P10.20

P10.21 $\sum \tau = 0.100\text{ m}(12.0\text{ N}) - 0.250\text{ m}(9.00\text{ N}) - 0.250\text{ m}(10.0\text{ N}) = \boxed{-3.55\text{ N}\cdot\text{m}}$

The thirty-degree angle is unnecessary information.

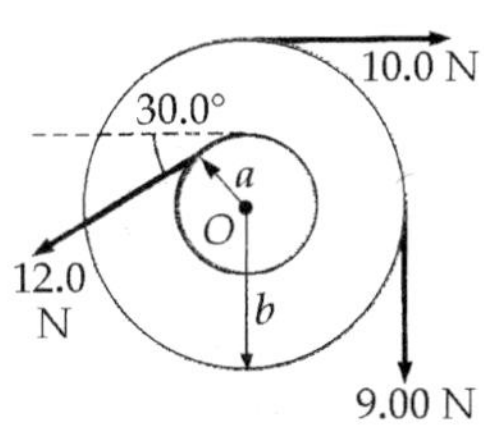

FIG. P10.21

P10.22 $\vec{\mathbf{M}}\times\vec{\mathbf{N}}=\begin{vmatrix}\hat{\mathbf{i}} & \hat{\mathbf{j}} & \hat{\mathbf{k}}\\ 6 & 2 & -1\\ 2 & -1 & -3\end{vmatrix}=\boxed{-7.00\hat{\mathbf{i}}+16.0\hat{\mathbf{j}}-10.0\hat{\mathbf{k}}}$

P10.23 (a) $\vec{\tau}=\vec{\mathbf{r}}\times\vec{\mathbf{F}}=\left(4.00\hat{\mathbf{i}}+5.00\hat{\mathbf{j}}\right)\times\left(2.00\hat{\mathbf{i}}+3.00\hat{\mathbf{j}}\right)$

$$|\vec{\tau}|=\left|12.00\hat{\mathbf{k}}-10.0\hat{\mathbf{k}}\right|=\left|2.00\hat{\mathbf{k}}\right|=\boxed{2.00\ \mathrm{N\cdot m}}$$

(b) The torque vector is in the direction of the unit vector $\hat{\mathbf{k}}$, or in the $\boxed{+z\text{ direction}}$.

P10.24 $\vec{\mathbf{A}}\cdot\vec{\mathbf{B}}=-3.00(6.00)+7.00(-10.0)+(-4.00)(9.00)=-124$

$$AB=\sqrt{(-3.00)^2+(7.00)^2+(-4.00)^2}\cdot\sqrt{(6.00)^2+(-10.0)^2+(9.00)^2}=127$$

(a) $\cos^{-1}\left(\dfrac{\vec{\mathbf{A}}\cdot\vec{\mathbf{B}}}{AB}\right)=\cos^{-1}(-0.979)=\boxed{168°}$

(b) $\vec{\mathbf{A}}\times\vec{\mathbf{B}}=\begin{vmatrix}\hat{\mathbf{i}} & \hat{\mathbf{j}} & \hat{\mathbf{k}}\\ -3.00 & 7.00 & -4.00\\ 6.00 & -10.0 & 9.00\end{vmatrix}=23.0\hat{\mathbf{i}}+3.00\hat{\mathbf{j}}-12.0\hat{\mathbf{k}}$

$$\left|\vec{\mathbf{A}}\times\vec{\mathbf{B}}\right|=\sqrt{(23.0)^2+(3.00)^2+(-12.0)^2}=26.1$$

$$\sin^{-1}\left(\frac{\left|\vec{\mathbf{A}}\times\vec{\mathbf{B}}\right|}{AB}\right)=\sin^{-1}(0.206)=\boxed{11.9°}\text{ or }168°$$

(c) Only $\boxed{\text{the first method}}$ gives the angle between the vectors unambiguously.

P10.25 $\left|\hat{\mathbf{i}}\times\hat{\mathbf{i}}\right|=1\cdot 1\cdot\sin 0°=0$

$\hat{\mathbf{j}}\times\hat{\mathbf{j}}$ and $\hat{\mathbf{k}}\times\hat{\mathbf{k}}$ are zero similarly since the vectors being multiplied are parallel.

$\left|\hat{\mathbf{i}}\times\hat{\mathbf{j}}\right|=1\cdot 1\cdot\sin 90°=1$

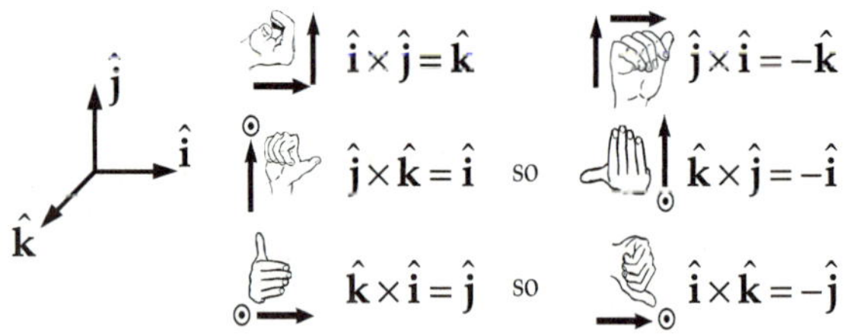

FIG. P10.25

Section 10.6 The Rigid Object in Equilibrium

P10.26 $\sum F_y = 0$: $+380\text{ N} - F_g + 320\text{ N} = 0$

$F_g = 700\text{ N}$

Take torques about her feet:

$\sum \tau = 0$: $-380\text{ N}(2.00\text{ m}) + (700\text{ N})x + (320\text{ N})0 = 0$

$x = \boxed{1.09\text{ m}}$

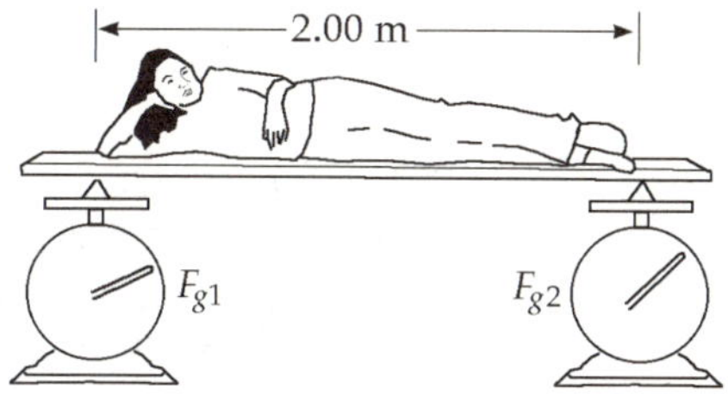

FIG. P10.26

P10.27 Take torques about *P*.

$$\sum \tau_p = -n_0\left[\frac{\ell}{2}+d\right] + m_1 g\left[\frac{\ell}{2}+d\right] + m_b g d - m_2 g x = 0$$

We want to find x for which $n_0 = 0$.

$$x = \frac{(m_1 g + m_b g)d + m_1 g \frac{\ell}{2}}{m_2 g} = \boxed{\frac{(m_1 + m_b)d + m_1\,\ell/2}{m_2}}$$

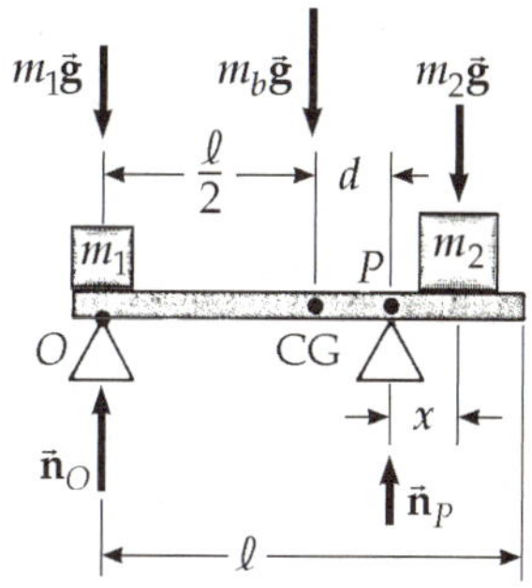

FIG. P10.27

***P10.28** See the free-body diagram at the right.
When the plank is on the verge of tipping about point *P*, the normal force n_1 goes to zero. Then, summing torques about point *P* gives

$$\sum \tau_p = -mgd + Mgx = 0 \qquad \text{or} \qquad x = \left(\frac{m}{M}\right)d.$$

From the dimensions given on the free-body diagram, observe that $d = 1.50\text{ m}$ Thus, when the plank is about to tip,

$$x = \left(\frac{30.0\text{ kg}}{70.0\text{ kg}}\right)(1.50\text{ m}) = \boxed{0.643\text{ m}}.$$

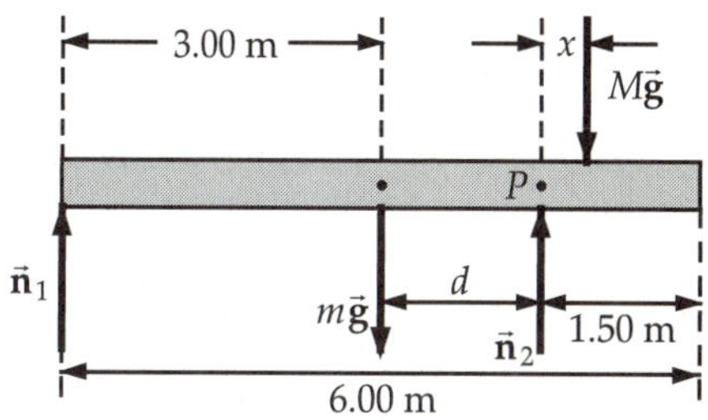

FIG. P10.28

P10.29 (a) Taking moments about P,

$$(R\sin 30.0°)0+(R\cos 30.0°)(5.00\text{ cm})-(150\text{ N})(30.0\text{ cm})=0$$
$$R=1\,039.2\text{ N}=1.04\text{ kN}$$

The force exerted by the hammer on the nail is equal in magnitude and opposite in direction:

$\boxed{1.04\text{ kN at }60°\text{ upward and to the right.}}$

(b) $f=R\sin 30.0°-150\text{ N}=370\text{ N}$

$n=R\cos 30.0°=900\text{ N}$

$\boxed{\vec{\mathbf{F}}_{\text{surface}}=(370\text{ N})\hat{\mathbf{i}}+(900\text{ N})\hat{\mathbf{j}}}$

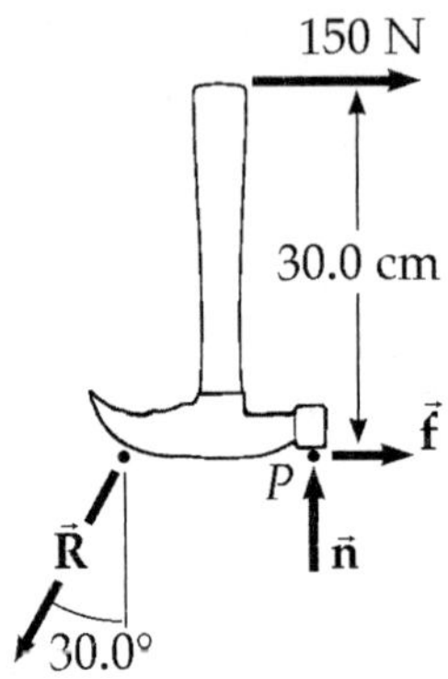

FIG. P10.29

P10.30 (a) $\sum F_x=f-n_w=0$ (1)

$\sum F_y=n_g-m_1g-m_2g=0$ (2)

$$\sum\tau_A=-m_1g\left(\frac{L}{2}\right)\cos\theta-m_2gx\cos\theta+n_wL\sin\theta=0$$

From the torque equation,

$$n_w=\left[\frac{1}{2}m_1g+\left(\frac{x}{L}\right)m_2g\right]\cot\theta$$

Then, from equation (1): $f=n_w=\boxed{\left[\frac{1}{2}m_1g+\left(\frac{x}{L}\right)m_2g\right]\cot\theta}$

and from equation (2): $n_g=\boxed{(m_1+m_2)g}$

(b) If the ladder is on the verge of slipping when $x=d$,

then $\mu=\frac{f|_{x=d}}{n_g}=\boxed{\frac{(m_1/2+m_2d/L)\cot\theta}{m_1+m_2}}$.

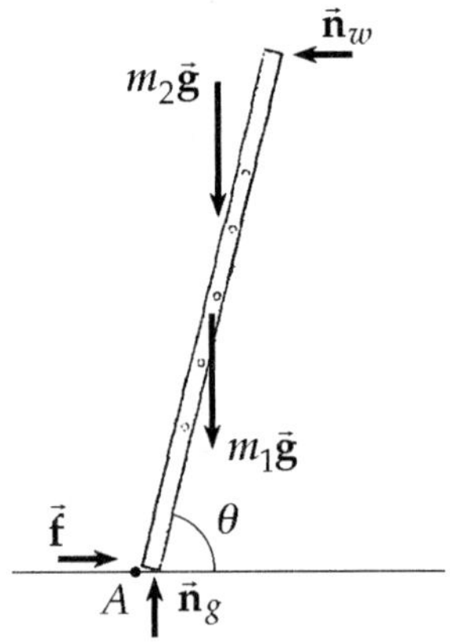

FIG. P10.30

P10.31 Using $\sum F_x = \sum F_y = \sum \tau = 0$, choosing the origin at the left end of the beam, we have (neglecting the weight of the beam)

$$\sum F_x = R_x - T\cos\theta = 0,$$
$$\sum F_y = R_y + T\sin\theta - F_g = 0,$$

and $\sum \tau = -F_g(L+d) + T\sin\theta(2L+d) = 0$.

Solving these equations, we find:

(a) $T = \boxed{\dfrac{F_g(L+d)}{\sin\theta(2L+d)}}$

(b) $R_x = \boxed{\dfrac{F_g(L+d)\cot\theta}{2L+d}}$ $\quad R_y = \boxed{\dfrac{F_g L}{2L+d}}$

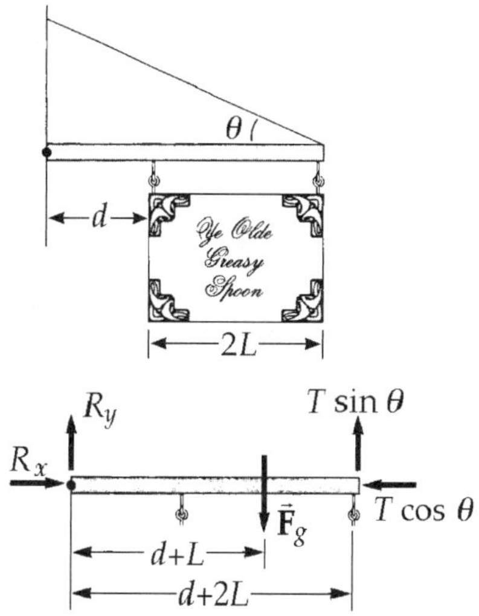

FIG. P10.31

P10.32 We interpret the problem to mean that the support at point B is frictionless. Then the support exerts a force in the x direction and

$$\boxed{F_{By} = 0}$$
$$\sum F_x = F_{Bx} - F_{Ax} = 0$$
$$F_{Ay} - (3\,000 + 10\,000)g = 0$$

and $\sum \tau = -(3\,000g)(2.00) - (10\,000g)(6.00) + F_{Bx}(1.00) = 0$.

These equations combine to give

$$F_{Ax} = F_{Bx} = \boxed{6.47 \times 10^5 \text{ N}}$$
$$F_{Ay} = \boxed{1.27 \times 10^5 \text{ N}}$$

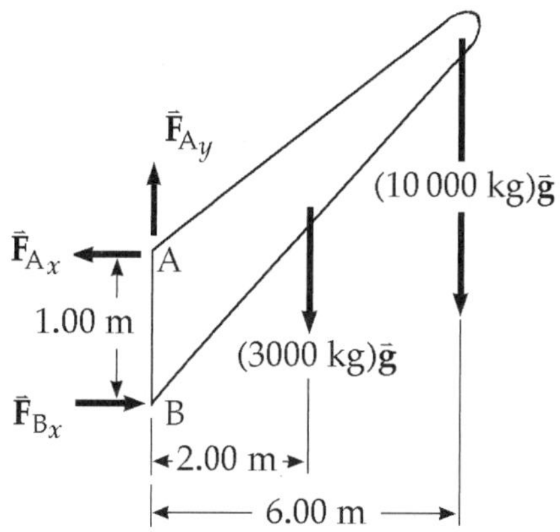

FIG. P10.32

Section 10.7 The Rigid Object Under a Net Torque

P10.33 $\omega_f = \omega_i + \alpha t$: $\quad 10.0 \text{ rad/s} = 0 + \alpha(6.00 \text{ s})$

$$\alpha = \frac{10.00}{6.00} \text{ rad/s}^2 = 1.67 \text{ rad/s}^2$$

(a) $\sum \tau = 36.0 \text{ N}\cdot\text{m} = I\alpha$: $\quad I = \frac{\sum \tau}{\alpha} = \frac{36.0 \text{ N}\cdot\text{m}}{1.67 \text{ rad/s}^2} = \boxed{21.6 \text{ kg}\cdot\text{m}^2}$

(b) $\omega_f = \omega_i + \alpha t$: $\quad 0 = 10.0 + \alpha(60.0)$

$$\alpha = -0.167 \text{ rad/s}^2$$

$$|\tau| = |I\alpha| = (21.6 \text{ kg}\cdot\text{m}^2)(0.167 \text{ rad/s}^2) = \boxed{3.60 \text{ N}\cdot\text{m}}$$

(c) Number of revolutions $\theta_f = \theta_i + \omega_i t + \frac{1}{2}\alpha t^2$

During first 6.00 s $\quad \theta_f = \frac{1}{2}(1.67)(6.00)^2 = 30.1 \text{ rad}$

During next 60.0 s $\quad \theta_f = 10.0(60.0) - \frac{1}{2}(0.167)(60.0)^2 = 299 \text{ rad}$

$$\theta_{\text{total}} = 329 \text{ rad}\left(\frac{1 \text{ rev}}{2\pi \text{ rad}}\right) = \boxed{52.4 \text{ rev}}$$

P10.34 $I = \frac{1}{2}mR^2 = \frac{1}{2}(100 \text{ kg})(0.500 \text{ m})^2 = 12.5 \text{ kg}\cdot\text{m}^2$

$\omega_i = 50.0 \text{ rev/min} = 5.24 \text{ rad/s}$

$$\alpha = \frac{\omega_f - \omega_i}{t} = \frac{0 - 5.24 \text{ rad/s}}{6.00 \text{ s}} = -0.873 \text{ rad/s}^2$$

$$\tau = I\alpha = 12.5 \text{ kg}\cdot\text{m}^2(-0.873 \text{ rad/s}^2) = -10.9 \text{ N}\cdot\text{m}$$

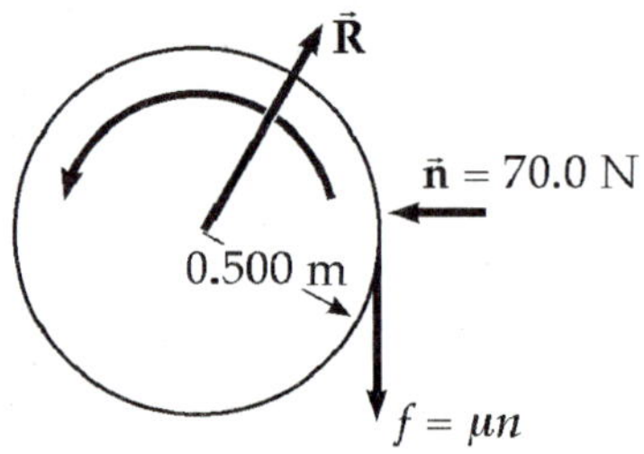

FIG. P10.34

The magnitude of the torque is given by $fR = 10.9 \text{ N}\cdot\text{m}$, where f is the force of friction.

Therefore, $\quad f = \frac{10.9 \text{ N}\cdot\text{m}}{0.500 \text{ m}} \quad$ and $\quad f = \mu_k n$

yields $\quad \mu_k = \frac{f}{n} = \frac{21.8 \text{ N}}{70.0 \text{ N}} = \boxed{0.312}$

P10.35 $\sum \tau = I\alpha = \frac{1}{2}MR^2\alpha$

$$-135 \text{ N}(0.230 \text{ m}) + T(0.230 \text{ m}) = \frac{1}{2}(80 \text{ kg})\left(\frac{1.25}{2} \text{ m}\right)^2(-1.67 \text{ rad/s}^2)$$

$T = \boxed{21.5 \text{ N}}$

P10.36 (a) $I=\frac{1}{2}M\left(R_1^2+R_2^2\right)=\frac{1}{2}(0.35\text{ kg})\left[(0.02\text{ m})^2+(0.03\text{ m})^2\right]=2.28\times10^{-4}\text{ kg}\cdot\text{m}^2$

$$\left(K_1+K_2+K_{\text{rot}}+U_{g2}\right)_i-f_kd=\left(K_1+K_2+K_{\text{rot}}\right)_f$$

$$\frac{1}{2}(0.850\text{ kg})(0.82\text{ m/s})^2+\frac{1}{2}(0.42\text{ kg})(0.82\text{ m/s})^2+\frac{1}{2}\left(2.28\times10^{-4}\text{ kg}\cdot\text{m}^2\right)\left(\frac{0.82\text{ m/s}}{0.03\text{ m}}\right)^2$$

$$+0.42\text{ kg}\left(9.8\text{ m/s}^2\right)(0.7\text{ m})-0.25(0.85\text{ kg})\left(9.8\text{ m/s}^2\right)(0.7\text{ m})$$

$$=\frac{1}{2}(0.85\text{ kg})v_f^2+\frac{1}{2}(0.42\text{ kg})v_f^2+\frac{1}{2}\left(2.28\times10^{-4}\text{ kg}\cdot\text{m}^2\right)\left(\frac{v_f}{0.03\text{ m}}\right)^2$$

$$0.512\text{ J}+2.88\text{ J}-1.46\text{ J}=(0.761\text{ kg})v_f^2$$

$$v_f=\sqrt{\frac{1.94\text{ J}}{0.761\text{ kg}}}=\boxed{1.59\text{ m/s}}$$

(b) $\omega=\frac{v}{r}=\frac{1.59\text{ m/s}}{0.03\text{ m}}=\boxed{53.1\text{ rad/s}}$

P10.37 (a) $m_2g-T_2=m_2a$

$$T_2=m_2(g-a)=20.0\text{ kg}\left(9.80\text{ m/s}^2-2.00\text{ m/s}^2\right)=\boxed{156\text{ N}}$$

$$T_1-m_1g\sin 37.0°=m_1a$$

$$T_1=(15.0\text{ kg})(9.80\sin 37.0°+2.00)\text{ m/s}^2=\boxed{118\text{ N}}$$

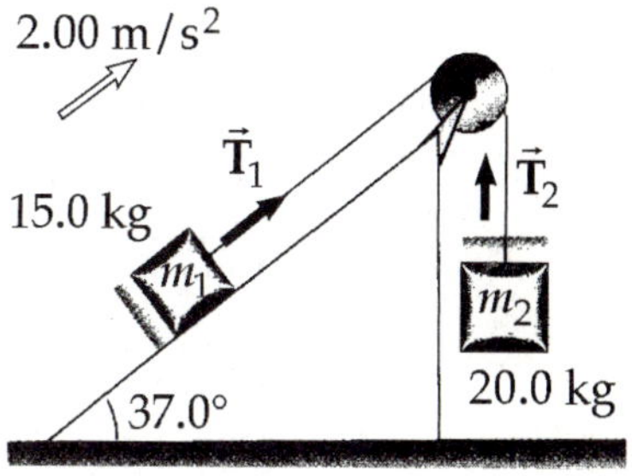

FIG. P10.37

(b) $(T_2-T_1)R=I\alpha=I\left(\frac{a}{R}\right)$

$$I=\frac{(T_2-T_1)R^2}{a}=\frac{(156\text{ N}-118\text{ N})(0.250\text{ m})^2}{2.00\text{ m/s}^2}=\boxed{1.17\text{ kg}\cdot\text{m}^2}$$

P10.38 For the purpose of computing the torque on the rod, we recall that the whole weight Mg can be modeled as acting at the center of mass of the rod. Because the rod is uniform, its center of mass is at its geometric center. The magnitude of the torque due to this force about an axis through the pivot is

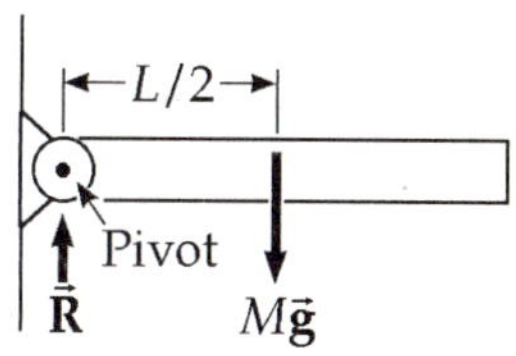

FIG. P10.38

$$\tau=\frac{MgL}{2}$$

The support force at the hinge has zero torque about an axis through the pivot, because this force passes through the axis (hence $r=0$). Since $\tau=I\alpha$, where $I=\frac{1}{3}ML^2$ for this axis of rotation (see Table 10.2), we get $I\alpha=Mg\frac{L}{2}$: $\alpha=\frac{Mg(L/2)}{(1/3)ML^2}=\boxed{\frac{3g}{2L}}$.

The angular acceleration is common to *all* points on the rod.

continued on next page

To find the linear acceleration of the right end of the rod, we use the relation $a_t = R\alpha$ with $R = L$: this gives

$$a_t = L\alpha = \boxed{\frac{3}{2}g}.$$

This result is rather interesting, since $a_t > g$. That is, the end of the rod has an acceleration *greater* than the acceleration due to gravity. Therefore, if a coin were placed at the end of the rod, the end of the rod would fall faster than the coin when released.

Other points on the rod have a linear acceleration less than $\frac{3}{2}g$. For example, the middle of the rod has an acceleration $\frac{3}{4}g$.

P10.39 (a) For the counterweight,

$\sum F_y = ma_y$ becomes: $\quad 50.0 - T = \left(\frac{50.0}{9.80}\right)a$

For the reel $\sum \tau = I\alpha$ reads $TR = I\alpha = I\frac{a}{R}$

where $\quad I = \frac{1}{2}MR^2 = 0.093\,8\text{ kg}\cdot\text{m}^2$

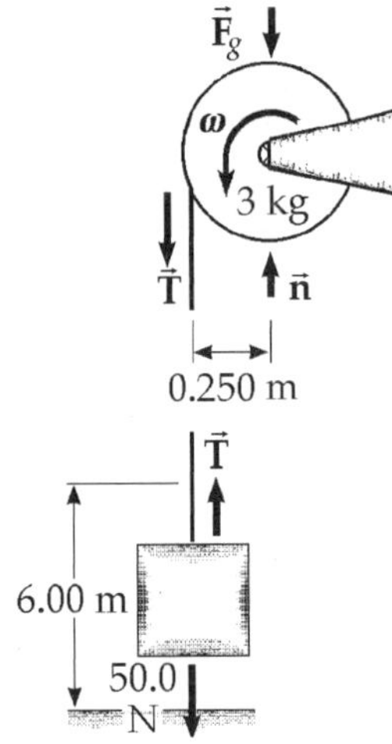

FIG. P10.39

We substitute to eliminate the acceleration:

$$50.0 - T = 5.10\left(\frac{TR^2}{I}\right)$$

$$T = \boxed{11.4\text{ N}}$$

and $\quad a = \frac{50.0 - 11.4}{5.10} = \boxed{7.57\text{ m/s}^2}$

$v_f^2 = v_i^2 + 2a(x_f - x_i)$: $\quad v_f = \sqrt{2(7.57)6.00} = \boxed{9.53\text{ m/s}}$

(b) Use conservation of energy for the system of the object, the reel, and the Earth:

$(K+U)_i = (K+U)_f$: $\quad mgh = \frac{1}{2}mv^2 + \frac{1}{2}I\omega^2$

$$2mgh = mv^2 + I\left(\frac{v^2}{R^2}\right) = v^2\left(m + \frac{I}{R^2}\right)$$

$$v = \sqrt{\frac{2mgh}{m + \frac{I}{R^2}}} = \sqrt{\frac{2(50.0\text{ N})(6.00\text{ m})}{5.10\text{ kg} + \frac{0.093\,8}{(0.250)^2}}} = \boxed{9.53\text{ m/s}}$$

Section 10.8 Angular Momentum

P10.40 Whether we think of the Earth's surface as curved or flat, we interpret the problem to mean that the plane's line of flight extended is precisely tangent to the mountain at its peak, and nearly parallel to the wheat field. Let the positive x direction be eastward, positive y be northward, and positive z be vertically upward.

(a) $\vec{\mathbf{r}} = (4.30\text{ km})\hat{\mathbf{k}} = \left(4.30\times10^{3}\text{ m}\right)\hat{\mathbf{k}}$

$$\vec{\mathbf{p}} = m\vec{\mathbf{v}} = 12\,000\text{ kg}\left(-175\hat{\mathbf{i}}\text{ m/s}\right) = -2.10\times10^{6}\hat{\mathbf{i}}\text{ kg}\cdot\text{m/s}$$

$$\vec{\mathbf{L}} = \vec{\mathbf{r}}\times\vec{\mathbf{p}} = \left(4.30\times10^{3}\hat{\mathbf{k}}\text{ m}\right)\times\left(-2.10\times10^{6}\hat{\mathbf{i}}\text{ kg}\cdot\text{m/s}\right) = \boxed{\left(-9.03\times10^{9}\text{ kg}\cdot\text{m}^2/\text{s}\right)\hat{\mathbf{j}}}$$

(b) $\boxed{\text{No}}$. $L = |\vec{\mathbf{r}}||\vec{\mathbf{p}}|\sin\theta = mv(r\sin\theta)$, and $r\sin\theta$ is the altitude of the plane. Therefore, $L =$ *constant* as the plane moves in level flight with constant velocity.

(c) $\boxed{\text{Zero}}$. The position vector from Pike's Peak to the plane is anti-parallel to the velocity of the plane. That is, it is directed along the same line and opposite in direction.

Thus, $L = mvr\sin 180° = 0$.

P10.41 $\vec{\mathbf{r}} = \left(6.00\hat{\mathbf{i}} + 5.00t\hat{\mathbf{j}}\text{ m}\right)$ $\quad\quad$ $\vec{\mathbf{v}} = \dfrac{d\vec{\mathbf{r}}}{dt} = 5.00\hat{\mathbf{j}}\text{ m/s}$

so $\vec{\mathbf{p}} = m\vec{\mathbf{v}} = 2.00\text{ kg}\left(5.00\hat{\mathbf{j}}\text{ m/s}\right) = 10.0\hat{\mathbf{j}}\text{ kg}\cdot\text{m/s}$

and $\vec{\mathbf{L}} = \vec{\mathbf{r}}\times\vec{\mathbf{p}} = \begin{vmatrix}\hat{\mathbf{i}} & \hat{\mathbf{j}} & \hat{\mathbf{k}}\\ 6.00 & 5.00t & 0\\ 0 & 10.0 & 0\end{vmatrix} = \boxed{\left(60.0\text{ kg}\cdot\text{m}^2/\text{s}\right)\hat{\mathbf{k}}}$

P10.42 The total angular momentum about the center point is given by $L = I_h\omega_h + I_m\omega_m$

with $$I_h = \frac{m_h L_h^2}{3} = \frac{60.0\text{ kg}(2.70\text{ m})^2}{3} = 146\text{ kg}\cdot\text{m}^2$$

and $$I_{m3} = \frac{m_m L_m^2}{3} = \frac{100\text{ kg}(4.50\text{ m})^2}{3} = 675\text{ kg}\cdot\text{m}^2$$

In addition, $$\omega_h = \frac{2\pi\text{ rad}}{12\text{ h}}\left(\frac{1\text{ h}}{3\,600\text{ s}}\right) = 1.45\times10^{-4}\text{ rad/s}$$

while $$\omega_m = \frac{2\pi\text{ rad}}{1\text{ h}}\left(\frac{1\text{ h}}{3\,600\text{ s}}\right) = 1.75\times10^{-3}\text{ rad/s}$$

Thus, $$L = 146\text{ kg}\cdot\text{m}^2\left(1.45\times10^{-4}\text{ rad/s}\right) + 675\text{ kg}\cdot\text{m}^2\left(1.75\times10^{-3}\text{ rad/s}\right)$$

or $$\boxed{L = 1.20\text{ kg}\cdot\text{m}^2/\text{s}}$$

P10.43 (a) $I = \frac{1}{12}m_1L^2 + m_2(0.500)^2 = \frac{1}{12}(0.100)(1.00)^2 + 0.400(0.500)^2 = 0.108\,3\text{ kg}\cdot\text{m}^2$

$$L = I\omega = 0.108\,3(4.00) = \boxed{0.433\text{ kg}\cdot\text{m}^2/\text{s}}$$

(b) $I = \frac{1}{3}m_1L^2 + m_2R^2 = \frac{1}{3}(0.100)(1.00)^2 + 0.400(1.00)^2 = 0.433$

$$L = I\omega = 0.433(4.00) = \boxed{1.73\text{ kg}\cdot\text{m}^2/\text{s}}$$

***P10.44** We require $a_c = g = \frac{v^2}{r} = \omega^2 r$

$$\omega = \sqrt{\frac{g}{r}} = \sqrt{\frac{(9.80\text{ m/s}^2)}{100\text{ m}}} = 0.313\text{ rad/s}$$

$$I = Mr^2 = 5\times10^4\text{ kg}(100\text{ m})^2 = 5\times10^8\text{ kg}\cdot\text{m}^2$$

(a) $L = I\omega = 5\times10^8\text{ kg}\cdot\text{m}^2\,0.313/\text{s} = \boxed{1.57\times10^8\text{ kg}\cdot\text{m}^2/\text{s}}$

(c) $\sum\tau = I\alpha = \frac{I(\omega_f - \omega_i)}{\Delta t}$

$$\sum\tau\Delta t = I\omega_f - I\omega_i = L_f - L_i$$

This is the angular impulse-angular momentum theorem.

(b) $\Delta t = \frac{L_f - 0}{\sum\tau} = \frac{1.57\times10^8\text{ kg}\cdot\text{m}^2/\text{s}}{2(125\text{ N})(100\text{ m})} = \boxed{6.26\times10^3\text{ s}} = 1.74\text{ h}$

Section 10.9 Conservation of Angular Momentum

P10.45 (a) From conservation of angular momentum for the system of two cylinders:

$$(I_1 + I_2)\omega_f = I_1\omega_i \quad\text{or}\quad \omega_f = \boxed{\frac{I_1}{I_1+I_2}\omega_i}$$

(b) $K_f = \frac{1}{2}(I_1 + I_2)\omega_f^2$ and $K_i = \frac{1}{2}I_1\omega_i^2$

so $$\frac{K_f}{K_i} = \frac{\frac{1}{2}(I_1+I_2)}{\frac{1}{2}I_1\omega_i^2}\left(\frac{I_1}{I_1+I_2}\omega_i\right)^2 = \boxed{\frac{I_1}{I_1+I_2}\text{ which is less than 1}}.$$

P10.46 $I_i\omega_i = I_f\omega_f$: $\left(250\text{ kg}\cdot\text{m}^2\right)(10.0\text{ rev/min}) = \left[250\text{ kg}\cdot\text{m}^2 + 25.0\text{ kg}(2.00\text{ m})^2\right]\omega_2$

$$\omega_2 = \boxed{7.14\text{ rev/min}}$$

P10.47 (a) The table turns opposite to the way the woman walks, so its angular momentum cancels that of the woman. From conservation of angular momentum for the system of the woman and the turntable, we have $L_f = L_i = 0$

so, $L_f = I_{\text{woman}}\omega_{\text{woman}} + I_{\text{table}}\omega_{\text{table}} = 0$

and $$\omega_{\text{table}} = \left(-\frac{I_{\text{woman}}}{I_{\text{table}}}\right)\omega_{\text{woman}} = \left(-\frac{m_{\text{woman}}r^2}{I_{\text{table}}}\right)\left(\frac{v_{\text{woman}}}{r}\right) = -\frac{m_{\text{woman}}rv_{\text{woman}}}{I_{\text{table}}}$$

$$\omega_{\text{table}} = -\frac{60.0\text{ kg}(2.00\text{ m})(1.50\text{ m/s})}{500\text{ kg}\cdot\text{m}^2} = -0.360\text{ rad/s}$$

or $$\omega_{\text{table}} = \boxed{0.360\text{ rad/s (counterclockwise)}}$$

(b) $$\text{work done} = \Delta K = K_f - 0 = \frac{1}{2}m_{\text{woman}}v_{\text{woman}}^2 + \frac{1}{2}I\omega_{\text{table}}^2$$

$$W = \frac{1}{2}(60\text{ kg})(1.50\text{ m/s})^2 + \frac{1}{2}\left(500\text{ kg}\cdot\text{m}^2\right)(0.360\text{ rad/s})^2 = \boxed{99.9\text{ J}}$$

P10.48 (a) The total angular momentum of the system of the student, the stool, and the weights about the axis of rotation is given by

$$I_{\text{total}} = I_{\text{weights}} + I_{\text{student}} = 2\left(mr^2\right) + 3.00\text{ kg}\cdot\text{m}^2$$

Before: $r = 1.00\text{ m}$.

Thus, $I_i = 2(3.00\text{ kg})(1.00\text{ m})^2 + 3.00\text{ kg}\cdot\text{m}^2 = 9.00\text{ kg}\cdot\text{m}^2$

After: $r = 0.300\text{ m}$

Thus, $I_f = 2(3.00\text{ kg})(0.300\text{ m})^2 + 3.00\text{ kg}\cdot\text{m}^2 = 3.54\text{ kg}\cdot\text{m}^2$

We now use conservation of angular momentum.

$$I_f\omega_f = I_i\omega_i$$

or $$\omega_f = \left(\frac{I_i}{I_f}\right)\omega_i = \left(\frac{9.00}{3.54}\right)(0.750\text{ rad/s}) = \boxed{1.91\text{ rad/s}}$$

(b) $$K_i = \frac{1}{2}I_i\omega_i^2 = \frac{1}{2}\left(9.00\text{ kg}\cdot\text{m}^2\right)(0.750\text{ rad/s})^2 = \boxed{2.53\text{ J}}$$

$$K_f = \frac{1}{2}I_f\omega_f^2 = \frac{1}{2}\left(3.54\text{ kg}\cdot\text{m}^2\right)(1.91\text{ rad/s})^2 = \boxed{6.44\text{ J}}$$

P10.49 When they touch, the center of mass is distant from the center of the larger puck by

$$y_{CM} = \frac{0 + 80.0\text{ g}(4.00\text{ cm} + 6.00\text{ cm})}{120\text{ g} + 80.0\text{ g}} = 4.00\text{ cm}$$

(a) $$L = r_1 m_1 v_1 + r_2 m_2 v_2 = 0 + \left(6.00 \times 10^{-2}\text{ m}\right)\left(80.0 \times 10^{-3}\text{ kg}\right)(1.50\text{ m/s}) = \boxed{7.20 \times 10^{-3}\text{ kg}\cdot\text{m}^2/\text{s}}$$

(b) The moment of inertia about the CM is

$$I = \left(\frac{1}{2}m_1 r_1^2 + m_1 d_1^2\right) + \left(\frac{1}{2}m_2 r_2^2 + m_2 d_2^2\right)$$

$$I = \frac{1}{2}(0.120\text{ kg})\left(6.00 \times 10^{-2}\text{ m}\right)^2 + (0.120\text{ kg})\left(4.00 \times 10^{-2}\right)^2$$

$$+\frac{1}{2}\left(80.0 \times 10^{-3}\text{ kg}\right)\left(4.00 \times 10^{-2}\text{ m}\right)^2 + \left(80.0 \times 10^{-3}\text{ kg}\right)\left(6.00 \times 10^{-2}\text{ m}\right)^2$$

$$I = 7.60 \times 10^{-4}\text{ kg}\cdot\text{m}^2$$

Angular momentum of the two-puck system is conserved: $L = I\omega$

$$\omega = \frac{L}{I} = \frac{7.20 \times 10^{-3}\text{ kg}\cdot\text{m}^2/\text{s}}{7.60 \times 10^{-4}\text{ kg}\cdot\text{m}^2} = \boxed{9.47\text{ rad/s}}$$

P10.50 For one of the crew,

$\sum F_r = ma_r$: $\quad n = \frac{mv^2}{r} = m\omega_i^2 r$

We require $\quad n = mg$, so $\omega_i = \sqrt{\frac{g}{r}}$

Now, $\quad I_i\omega_i = I_f\omega_f$

$$\left[5.00 \times 10^8\text{ kg}\cdot\text{m}^2 + 150 \times 65.0\text{ kg} \times (100\text{ m})^2\right]\sqrt{\frac{g}{r}} = \left[5.00 \times 10^8\text{ kg}\cdot\text{m}^2 + 50 \times 65.0\text{ kg}(100\text{ m})^2\right]\omega_f$$

$$\left(\frac{5.98 \times 10^8}{5.32 \times 10^8}\right)\sqrt{\frac{g}{r}} = \omega_f = 1.12\sqrt{\frac{g}{r}}$$

Now, $\quad |a_r| = \omega_f^2 r = 1.26g = \boxed{12.3\text{ m/s}^2}$

P10.51 $I_i = mr_i^2 = 0.12\ \text{kg}(0.4\ \text{m})^2 = 1.92\times10^{-2}\ \text{kg}\cdot\text{m}^2$

$$I_f = mr_f^2 = 0.12\ \text{kg}(0.25\ \text{m})^2 = 7.5\times10^{-3}\ \text{kg}\cdot\text{m}^2$$

$$\omega_i = \frac{v_i}{r_i} = \frac{0.8\ \text{m/s}}{0.4\ \text{m}} = 2\ \text{rad/s}$$

Now, use conservation of angular momentum for the system of the ball

$$\omega_f = \omega_i\left(\frac{I_i}{I_f}\right) = (2\ \text{rad/s})\left(\frac{1.92\times10^{-2}\ \text{kg}\cdot\text{m}^2}{7.5\times10^{-3}\ \text{kg}\cdot\text{m}^2}\right) = 5.12\ \text{rad/s}$$

The work done $= \Delta K = \frac{1}{2}I_f\omega_f^2 - \frac{1}{2}I_i\omega_i^2$. Substituting the appropriate values found earlier, we have the work done: $\boxed{5.99\times10^{-2}\ \text{J}}$.

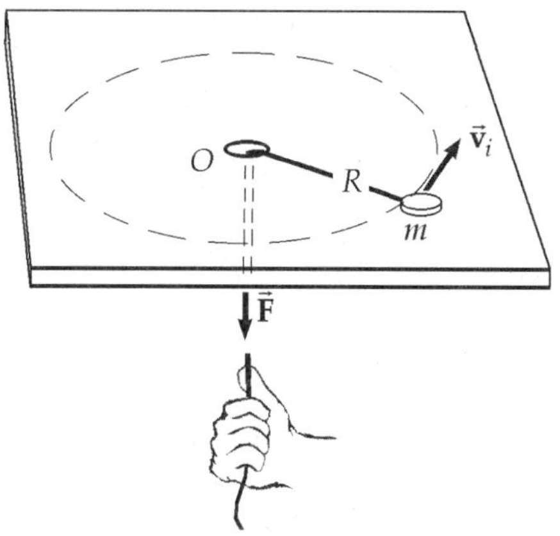

FIG. P10.51

Section 10.10 Precessional Motion of Gyroscopes

P10.52 $I = \frac{2}{5}MR^2 = \frac{2}{5}\left(5.98\times10^{24}\ \text{kg}\right)\left(6.37\times10^6\ \text{m}\right)^2 = 9.71\times10^{37}\ \text{kg}\cdot\text{m}^2$

$$L = I\omega = 9.71\times10^{37}\ \text{kg}\cdot\text{m}^2\left(\frac{2\pi\,\text{rad}}{86\ 400\ \text{s}}\right) = 7.06\times10^{33}\ \text{kg}\cdot\text{m}^2/\text{s}^2$$

$$\tau = L\omega_p = \left(7.06\times10^{33}\ \text{kg}\cdot\text{m}^2/\text{s}\right)\left(\frac{2\pi\,\text{rad}}{2.58\times10^4\ \text{yr}}\right)\left(\frac{1\ \text{yr}}{365.25\ \text{d}}\right)\left(\frac{1\ \text{d}}{86\ 400\ \text{s}}\right) = \boxed{5.45\times10^{22}\ \text{N}\cdot\text{m}}$$

Section 10.11 Rolling Motion of Rigid Objects

P10.53 (a) $K_{\text{trans}} = \frac{1}{2}mv^2 = \frac{1}{2}(10.0\ \text{kg})(10.0\ \text{m/s})^2 = \boxed{500\ \text{J}}$

(b) $K_{\text{rot}} = \frac{1}{2}I\omega^2 = \frac{1}{2}\left(\frac{1}{2}mr^2\right)\left(\frac{v^2}{r^2}\right) = \frac{1}{4}(10.0\ \text{kg})(10.0\ \text{m/s})^2 = \boxed{250\ \text{J}}$

(c) $K_{\text{total}} = K_{\text{trans}} + K_{\text{rot}} = \boxed{750\ \text{J}}$

P10.54 $K=\frac{1}{2}mv^2+\frac{1}{2}I\omega^2=\frac{1}{2}\left[m+\frac{I}{R^2}\right]v^2$ where $\omega=\frac{v}{R}$ since no slipping.

Also, $U_i=mgh,\ U_f=0,$ and $v_i=0$

Therefore, $$\frac{1}{2}\left[m+\frac{I}{R^2}\right]v^2=mgh$$

Thus, $$v^2=\frac{2gh}{\left[1+\left(\frac{I}{mR^2}\right)\right]}$$

For a disk, $$I=\frac{1}{2}mR^2$$

So $v^2=\frac{2gh}{1+\frac{1}{2}}$ or $\boxed{v_{\text{disk}}=\sqrt{\frac{4gh}{3}}}$

For a ring, $I=mR^2$ so $v^2=\frac{2gh}{2}$ or $\boxed{v_{\text{ring}}=\sqrt{gh}}$

Since $v_{\text{disk}}>v_{\text{ring}}$, $\boxed{\text{the disk}}$ reaches the bottom first.

P10.55 (a) Energy conservation for the system of the ball and the Earth between the horizontal section and top of loop:

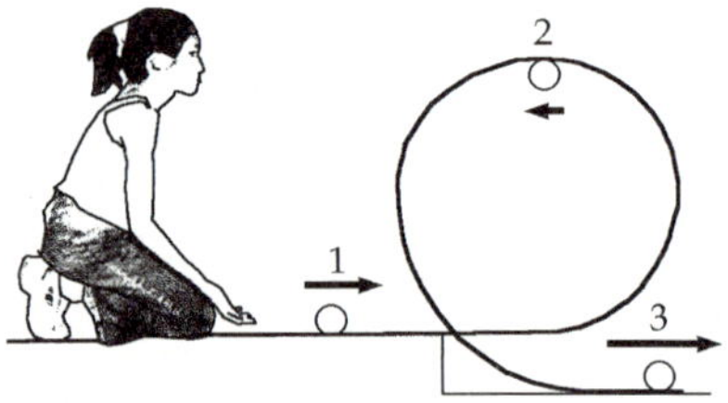

FIG. P10.55

$$\frac{1}{2}mv_2^2+\frac{1}{2}I\omega_2^2+mgy_2=\frac{1}{2}mv_1^2+\frac{1}{2}I\omega_1^2$$

$$\frac{1}{2}mv_2^2+\frac{1}{2}\left(\frac{2}{3}mr^2\right)\left(\frac{v_2}{r}\right)^2+mgy_2$$
$$=\frac{1}{2}mv_1^2+\frac{1}{2}\left(\frac{2}{3}mr^2\right)\left(\frac{v_1}{r}\right)^2$$

$$\frac{5}{6}v_2^2+gy_2=\frac{5}{6}v_1^2$$

$$v_2=\sqrt{v_1^2-\frac{6}{5}gy_2}=\sqrt{(4.03\text{ m/s})^2-\frac{6}{5}(9.80\text{ m/s}^2)(0.900\text{ m})}=\boxed{2.38\text{ m/s}}$$

The centripetal acceleration is $\frac{v_2^2}{r}=\frac{(2.38\text{ m/s})^2}{0.450\text{ m}}=12.6\text{ m/s}^2>g$

Thus, the ball must be in contact with the track, with the track pushing downward on it.

(b) $$\frac{1}{2}mv_3^2+\frac{1}{2}\left(\frac{2}{3}mr^2\right)\left(\frac{v_3}{r}\right)^2+mgy_3=\frac{1}{2}mv_1^2+\frac{1}{2}\left(\frac{2}{3}mr^2\right)\left(\frac{v_1}{r}\right)^2$$

$$v_3=\sqrt{v_1^2-\frac{6}{5}gy_3}=\sqrt{(4.03\text{ m/s})^2-\frac{6}{5}(9.80\text{ m/s}^2)(-0.200\text{ m})}=\boxed{4.31\text{ m/s}}$$

continued on next page

(c) $$\frac{1}{2}mv_2^2 + mgy_2 = \frac{1}{2}mv_1^2$$

$$v_2 = \sqrt{v_1^2 - 2gy_2} = \sqrt{(4.03 \text{ m/s})^2 - 2(9.80 \text{ m/s}^2)(0.900 \text{ m})} = \sqrt{-1.40 \text{ m}^2/\text{s}^2}$$

This result is imaginary. In the case where the ball does not roll, the ball starts with less energy than in part (a) and $\boxed{\text{never makes it to the top}}$ of the loop.

P10.56 $$\bar{v} = \frac{\Delta x}{\Delta t} = \frac{3.00 \text{ m}}{1.50 \text{ s}} = 2.00 \text{ m/s} = \frac{1}{2}(0 + v_f)$$

$$v_f = 4.00 \text{ m/s and } \omega_f = \frac{v_f}{r} = \frac{4.00 \text{ m/s}}{(6.38 \times 10^{-2} \text{ m})/2} = \frac{8.00}{6.38 \times 10^{-2}} \text{ rad/s}$$

We ignore internal friction and suppose the can rolls without slipping.

$$(K_{\text{trans}} + K_{\text{rot}} + U_g)_i + \Delta E_{\text{mech}} = (K_{\text{trans}} + K_{\text{rot}} + U_g)_f$$

$$(0 + 0 + mgy_i) + 0 = \left(\frac{1}{2}mv_f^2 + \frac{1}{2}I\omega_f^2 + 0\right)$$

$$0.215 \text{ kg}(9.80 \text{ m/s}^2)[(3.00 \text{ m})\sin 25.0°] = \frac{1}{2}(0.215 \text{ kg})(4.00 \text{ m/s})^2 + \frac{1}{2}I\left(\frac{8.00}{6.38 \times 10^{-2}} \text{ rad/s}\right)^2$$

$$2.67 \text{ J} = 1.72 \text{ J} + (7\,860 \text{ s}^{-2})I$$

$$I = \frac{0.951 \text{ kg} \cdot \text{m}^2/\text{s}^2}{7\,860 \text{ s}^{-2}} = \boxed{1.21 \times 10^{-4} \text{ kg} \cdot \text{m}^2}$$

The $\boxed{\text{height of the can}}$ is unnecessary data.

Section 10.12 Context Connection—Turning the Spacecraft

P10.57 Angular momentum of the system of the spacecraft and the gyroscope is conserved. The gyroscope and spacecraft turn in opposite directions.

$$0 = I_1\omega_1 + I_2\omega_2: \quad -I_1\omega_1 = I_2\frac{\theta}{t}$$

$$-20 \text{ kg} \cdot \text{m}^2(-100 \text{ rad/s}) = 5 \times 10^5 \text{ kg} \cdot \text{m}^2\left(\frac{30°}{t}\right)\left(\frac{\pi \text{ rad}}{180°}\right)$$

$$t = \frac{2.62 \times 10^5 \text{ s}}{2\,000} = \boxed{131 \text{ s}}$$

Additional Problems

P10.58 The resistive force on each ball is $R = D\rho A v^2$. Here $v = r\omega$, where r is the radius of each ball's path. The resistive torque on each ball is $\tau = rR$, so the total resistive torque on the three ball system is $\tau_{\text{total}} = 3rR$.

The power required to maintain a constant rotation rate is $\mathcal{P} = \tau_{\text{total}}\omega = 3rR\omega$. This required power may be written as

$$\mathcal{P} = \tau_{\text{total}}\omega = 3r\left[D\rho A(r\omega)^2\right]\omega = \left(3r^3 DA\omega^3\right)\rho$$

With $$\omega = \frac{2\pi\,\text{rad}}{1\text{ rev}}\left(\frac{10^3\text{ rev}}{1\text{ min}}\right)\left(\frac{1\text{ min}}{60.0\text{ s}}\right) = \frac{1\,000\pi}{30.0}\text{ rad/s}$$

$$\mathcal{P} = 3(0.100\text{ m})^3(0.600)\left(4.00\times10^{-4}\text{ m}^2\right)\left(\frac{1\,000\pi}{30.0\text{ s}}\right)^3\rho$$

or $\mathcal{P} = \left(0.827\text{ m}^5/\text{s}^3\right)\rho$, where ρ is the density of the resisting medium.

(a) In air, $\rho = 1.20\text{ kg/m}^3$,

and $\mathcal{P} = 0.827\text{ m}^5/\text{s}^3\left(1.20\text{ kg/m}^3\right) = 0.992\text{ N}\cdot\text{m/s} = \boxed{0.992\text{ W}}$

(b) In water, $\rho = 1\,000\text{ kg/m}^3$ and $\mathcal{P} = \boxed{827\text{ W}}$.

P10.59 (a) Since only conservative forces act within the system of the rod and the Earth,

$\Delta E = 0$ so $K_f + U_f = K_i + U_i$

$$\frac{1}{2}I\omega^2 + 0 = 0 + Mg\left(\frac{L}{2}\right)$$

where $I = \frac{1}{3}ML^2$

Therefore, $$\omega = \boxed{\sqrt{\frac{3g}{L}}}$$

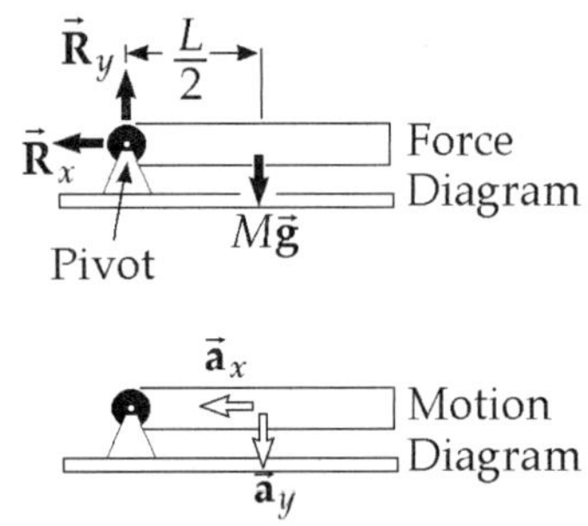

FIG. P10.59

(b) $\sum\tau = I\alpha$, so that in the horizontal orientation,

$$Mg\left(\frac{L}{2}\right) = \frac{ML^2}{3}\alpha$$

$$\alpha = \boxed{\frac{3g}{2L}}$$

(c) $a_x = a_r = -r\omega^2 = -\left(\frac{L}{2}\right)\omega^2 = \boxed{-\frac{3g}{2}}$ $\quad a_y = -a_t = -r\alpha = -\alpha\left(\frac{L}{2}\right) = \boxed{-\frac{3g}{4}}$

continued on next page

(d) Using Newton's second law, we have

$$R_x = Ma_x = \boxed{-\frac{3Mg}{2}}$$

$$R_y - Mg = Ma_y = -\frac{3Mg}{4} \qquad R_y = \boxed{\frac{Mg}{4}}$$

P10.60 τ_f will oppose the torque due to the hanging object:

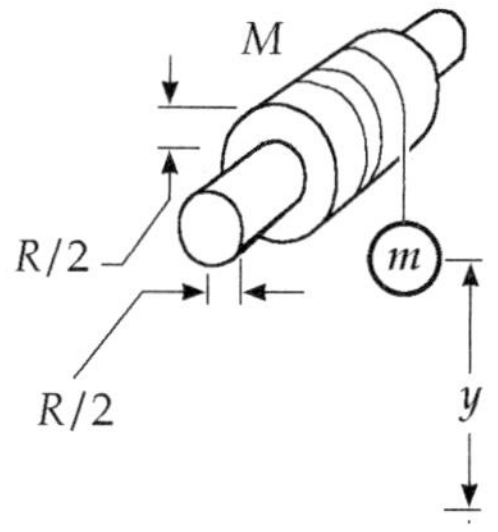

FIG. P10.60

$$\sum \tau = I\alpha = TR - \tau_f: \qquad \tau_f = TR - I\alpha \qquad (1)$$

Now find T, I and α in given or known terms and substitute into equation (1).

$$\sum F_y = T - mg = -ma: \qquad T = m(g-a) \qquad (2)$$

$$\text{also } \Delta y = v_i t + \frac{at^2}{2} \qquad a = \frac{2y}{t^2} \qquad (3)$$

and

$$\alpha = \frac{a}{R} = \frac{2y}{Rt^2}: \qquad (4)$$

$$I = \frac{1}{2}M\left[R^2 + \left(\frac{R}{2}\right)^2\right] = \frac{5}{8}MR^2 \qquad (5)$$

Substituting (2), (3), (4), and (5) into (1),

we find

$$\tau_f = m\left(g - \frac{2y}{t^2}\right)R - \frac{5}{8}\frac{MR^2(2y)}{Rt^2} = \boxed{R\left[m\left(g - \frac{2y}{t^2}\right) - \frac{5}{4}\frac{My}{t^2}\right]}$$

P10.61 (a)

$$W = \Delta K + \Delta U$$

$$W = K_f - K_i + U_f - U_i$$

$$0 = \frac{1}{2}mv^2 + \frac{1}{2}I\omega^2 - mgd\sin\theta - \frac{1}{2}kd^2$$

$$\frac{1}{2}\omega^2\left(I + mR^2\right) = mgd\sin\theta + \frac{1}{2}kd^2$$

$$\boxed{\omega = \sqrt{\frac{2mgd\sin\theta + kd^2}{I + mR^2}}}$$

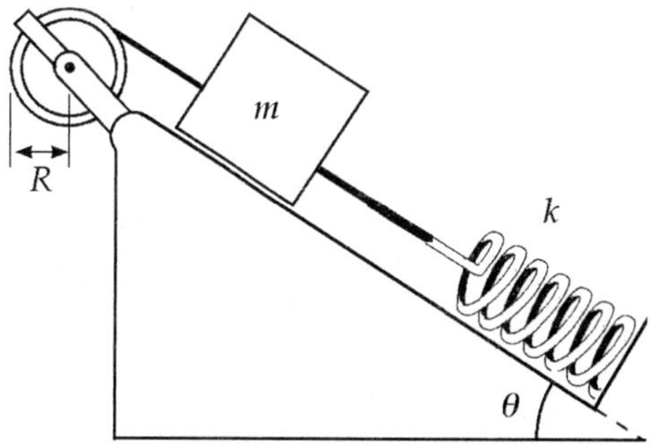

FIG. P10.61

(b)

$$\omega = \sqrt{\frac{2(0.500\text{ kg})\left(9.80\text{ m/s}^2\right)(0.200\text{ m})(\sin 37.0°) + 50.0\text{ N/m}(0.200\text{ m})^2}{1.00\text{ kg}\cdot\text{m}^2 + 0.500\text{ kg}(0.300\text{ m})^2}}$$

$$\omega = \sqrt{\frac{1.18 + 2.00}{1.05}} = \sqrt{3.04} = \boxed{1.74\text{ rad/s}}$$

P10.62 For m_1,

$\sum F_y = ma_y$: $+n - m_1 g = 0$

$n_1 = m_1 g = 19.6 \text{ N}$

$f_{k1} = \mu_k n_1 = 7.06 \text{ N}$

$\sum F_x = ma_x$: $-7.06 \text{ N} + T_1 = (2.00 \text{ kg})a$ (1)

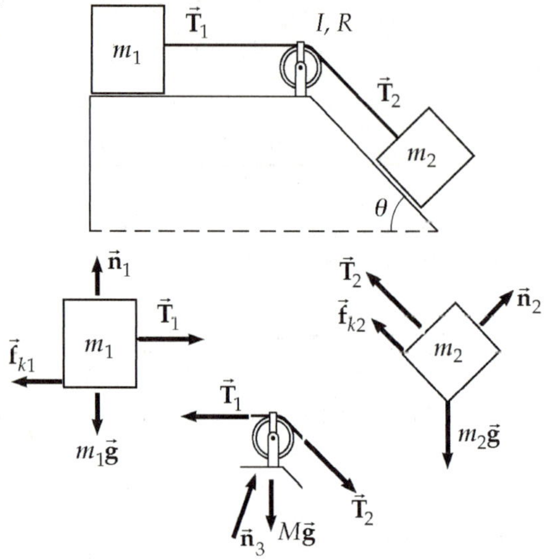

FIG. P10.62

For the pulley,

$\sum \tau = I\alpha$: $-T_1 R + T_2 R = \frac{1}{2}MR^2\left(\frac{a}{R}\right)$

$-T_1 + T_2 = \frac{1}{2}(10.0 \text{ kg})a$

$-T_1 + T_2 = (5.00 \text{ kg})a$ (2)

For m_2, $+n_2 - m_2 g \cos\theta = 0$

$n_2 = 6.00 \text{ kg}(9.80 \text{ m/s}^2)(\cos 30.0°)$

$= 50.9 \text{ N}$

$f_{k2} = \mu_k n_2$

$= 18.3 \text{ N}$: $-18.3 \text{ N} - T_2 + m_2 g \sin\theta = m_2 a$

$-18.3 \text{ N} - T_2 + 29.4 \text{ N} = (6.00 \text{ kg})a$ (3)

(a) Add equations (1), (2), and (3):

$$-7.06 \text{ N} - 18.3 \text{ N} + 29.4 \text{ N} = (13.0 \text{ kg})a$$

$$a = \frac{4.01 \text{ N}}{13.0 \text{ kg}} = \boxed{0.309 \text{ m/s}^2}$$

(b) $T_1 = 2.00 \text{ kg}(0.309 \text{ m/s}^2) + 7.06 \text{ N} = \boxed{7.67 \text{ N}}$

$T_2 = 7.67 \text{ N} + 5.00 \text{ kg}(0.309 \text{ m/s}^2) = \boxed{9.22 \text{ N}}$

P10.63 For the board just starting to move,

$$\sum \tau = I\alpha: \qquad mg\left(\frac{\ell}{2}\right)\cos\theta = \left(\frac{1}{3}m\ell^2\right)\alpha$$

$$\alpha = \frac{3}{2}\left(\frac{g}{\ell}\right)\cos\theta$$

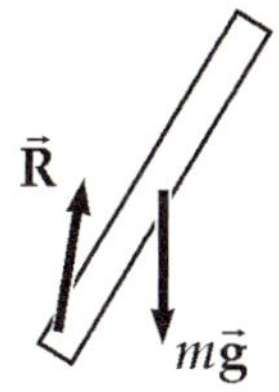

FIG. P10.63

The tangential acceleration of the end is $a_t = \ell\alpha = \frac{3}{2}g\cos\theta$

The vertical component is $a_y = a_t\cos\theta = \frac{3}{2}g\cos^2\theta$

If this is greater than g, the board will pull ahead of the ball falling:

(a) $\frac{3}{2}g\cos^2\theta \geq g$ gives $\cos^2\theta \geq \frac{2}{3}$ so $\cos\theta \geq \sqrt{\frac{2}{3}}$ and $\boxed{\theta \leq 35.3°}$

(b) When $\theta = 35.3°$, the cup will land underneath the release-point of the ball if $r_c = \ell\cos\theta$

When $\ell = 1.00\text{ m}$, and $\theta = 35.3°$ $r_c = 1.00\text{ m}\sqrt{\frac{2}{3}} = 0.816\text{ m}$

so the cup should be $(1.00\text{ m} - 0.816\text{ m}) = \boxed{0.184\text{ m from the moving end}}$

P10.64 Consider the total weight of each hand to act at the center of gravity (mid-point) of that hand. Then the total torque (taking CCW as positive) of these hands about the center of the clock is given by

$$\tau = -m_h g\left(\frac{L_h}{2}\right)\sin\theta_h - m_m g\left(\frac{L_m}{2}\right)\sin\theta_m = -\frac{g}{2}\left(m_h L_h\sin\theta_h + m_m L_m\sin\theta_m\right)$$

If we take $t = 0$ at 12 o'clock, then the angular positions of the hands at time t are

$$\theta_h = \omega_h t,$$

where $\omega_h = \frac{\pi}{6}\text{ rad/h}$

and $\theta_m = \omega_m t,$

where $\omega_m = 2\pi\text{ rad/h}$

Therefore, $\tau = -4.90\text{ m/s}^2\left[60.0\text{ kg}(2.70\text{ m})\sin\left(\frac{\pi t}{6}\right) + 100\text{ kg}(4.50\text{ m})\sin 2\pi t\right]$

or $\tau = -794\text{ N}\cdot\text{m}\left[\sin\left(\frac{\pi t}{6}\right) + 2.78\sin 2\pi t\right]$, where t is in hours.

continued on next page

(a) (i) At 3:00, $t = 3.00$ h,

so $$\tau = -794\text{ N}\cdot\text{m}\left[\sin\left(\frac{\pi}{2}\right) + 2.78\sin 6\pi\right] = \boxed{-794\text{ N}\cdot\text{m}}$$

(ii) At 5:15, $t = 5\text{ h} + \frac{15}{60}\text{ h} = 5.25\text{ h}$, and substitution gives:

$$\tau = \boxed{-2\,510\text{ N}\cdot\text{m}}$$

(iii) At 6:00, $\tau = \boxed{0\text{ N}\cdot\text{m}}$

(iv) At 8:20, $\tau = \boxed{-1\,160\text{ N}\cdot\text{m}}$

(v) At 9:45, $\tau = \boxed{-2\,940\text{ N}\cdot\text{m}}$

(b) The total torque is zero at those times when

$$\sin\left(\frac{\pi t}{6}\right) + 2.78\sin 2\pi t = 0$$

We proceed numerically, to find 0, 0.515 295 5, ..., corresponding to the times

12:00:00	12:30:55	12:58:19	1:32:31	1:57:01
2:33:25	2:56:29	3:33:22	3:56:55	4:32:24
4:58:14	5:30:52	6:00:00	6:29:08	7:01:46
7:27:36	8:03:05	8:26:38	9:03:31	9:26:35
10:02:59	10:27:29	11:01:41	11:29:05	

P10.65 $\sum F = T - Mg = -Ma$: $\sum \tau = TR = I\alpha = \frac{1}{2}MR^2\left(\frac{a}{R}\right)$

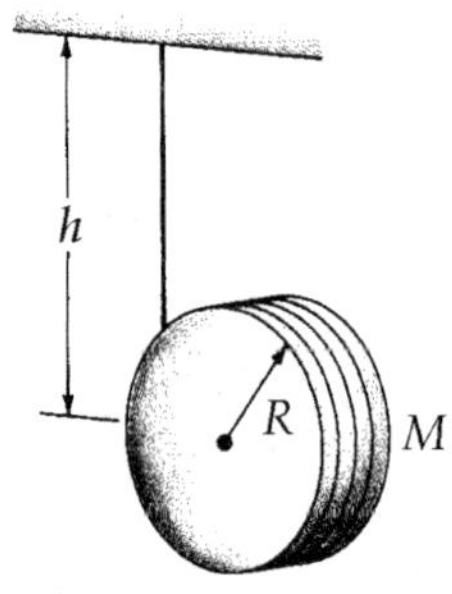

FIG. P10.65

(a) Combining the above two equations we find $T = M(g - a)$ and $a = \frac{2T}{M}$ thus $T = \boxed{\frac{Mg}{3}}$.

(b) $a = \frac{2T}{M} = \frac{2}{M}\left(\frac{Mg}{3}\right) = \boxed{\frac{2}{3}g}$

(c) $v_f^2 = v_i^2 + 2a(x_f - x_i)$ $\qquad v_f^2 = 0 + 2\left(\frac{2}{3}g\right)(h - 0)$

$$v_f = \boxed{\sqrt{\frac{4gh}{3}}}$$

continued on next page

For comparison, from conservation of energy for the system of the disk and the Earth we have

$$U_{gi}+K_{\text{rot}\,i}+K_{\text{trans}\,i}=U_{gf}+K_{\text{rot}\,f}+K_{\text{trans}\,f}: \qquad Mgh+0+0=0+\frac{1}{2}\left(\frac{1}{2}MR^2\right)\left(\frac{v_f}{R}\right)^2+\frac{1}{2}Mv_f^{\,2}$$

$$v_f=\sqrt{\frac{4gh}{3}}$$

***P10.66** Model the stove as a uniform 68 kg box. Its center of mass is at its geometric center, $\frac{28}{2}=14$ inches behind its feet at the front corners. Assume that the light oven door opens to be horizontal and that a person stands on its outer end, $46.375-28=18.375$ inches in front of the front feet. We find the weight F_g of a person standing on the oven door with the stove balanced on its front feet in equilibrium: $\sum\tau=0$

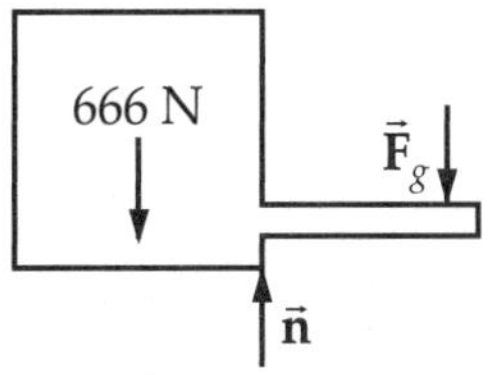

FIG. P10.66

$$(68\text{ kg})\left(9.8\text{ m/s}^2\right)(14\text{ in.})+n(0)-F_g(18.375\text{ in.})=0$$

$$F_g=\boxed{508\text{ N}}$$

If the weight of the person is greater than this, the stove can tip forward. This weight corresponds to mass 51.8 kg, so the person could be a child. If the oven door is heavy (compared to the backsplash) or if the front feet are significantly far behind the front corners, the maximum weight will be significantly less than 508 N.

P10.67 (a) As the bicycle frame moves forward at speed v, the center of each wheel moves forward at the same speed and the wheels turn at angular speed $\omega=\frac{v}{R}$. The total kinetic energy of the bicycle is $K=K_{\text{trans}}+K_{\text{rot}}$ or

$$K=\frac{1}{2}(m_{\text{frame}}+2m_{\text{wheel}})v^2+2\left(\frac{1}{2}I_{\text{wheel}}\omega^2\right)=\frac{1}{2}(m_{\text{frame}}+2m_{\text{wheel}})v^2+\left(\frac{1}{2}m_{\text{wheel}}R^2\right)\left(\frac{v^2}{R^2}\right).$$

This yields

$$K=\frac{1}{2}(m_{\text{frame}}+3m_{\text{wheel}})v^2=\frac{1}{2}\left[8.44\text{ kg}+3(0.820\text{ kg})\right](3.35\text{ m/s})^2=\boxed{61.2\text{ J}}.$$

(b) As the block moves forward with speed v, the top of each trunk moves forward at the same speed and the center of each trunk moves forward at speed $\frac{v}{2}$. The angular speed of each roller is $\omega=\frac{v}{2R}$. As in part (a), we have one object undergoing pure translation and two identical objects rolling without slipping. The total kinetic energy of the system of the stone and the trees is

$$K=K_{\text{trans}}+K_{\text{rot}}$$

continued on next page

or

$$K=\frac{1}{2}m_{\text{stone}}v^2+2\frac{1}{2}m_{\text{tree}}\left(\frac{v}{2}\right)^2+2\left(\frac{1}{2}I_{\text{tree}}\omega^2\right)=\frac{1}{2}\left(m_{\text{stone}}+\frac{1}{2}m_{\text{tree}}\right)v^2+\left(\frac{1}{2}m_{\text{tree}}R^2\right)\left(\frac{v^2}{4R^2}\right).$$

This gives

$$K=\frac{1}{2}\left(m_{\text{stone}}+\frac{3}{4}m_{\text{tree}}\right)v^2=\frac{1}{2}\left[844\text{ kg}+0.75(82.0\text{ kg})\right](0.335\text{ m/s})^2=\boxed{50.8\text{ J}}.$$

***P10.68** (a) $(K+U_s)_A=(K+U_s)_B$

$$0+mgy_A=\frac{1}{2}mv_B^2+0$$

$$v_B=\sqrt{2gy_A}=\sqrt{2\left(9.8\text{ m/s}^2\right)6.30\text{ m}}=\boxed{11.1\text{ m/s}}$$

(b) $L=mvr=76\text{ kg }11.1\text{ m/s }6.3\text{ m}=\boxed{5.32\times10^3\text{ kg}\cdot\text{m}^2/\text{s}}$ toward you along the axis of the channel.

(c) The wheels on his skateboard prevent any tangential force from acting on him. Then no torque about the axis of the channel acts on him and his angular momentum is constant. His legs convert chemical into mechanical energy. They do work to increase his kinetic energy. The normal force acts forward on his body on its rising trajectory, to increase his linear momentum.

(d) $L=mvr$ $\qquad v=\dfrac{5.32\times10^3\text{ kg}\cdot\text{m}^2/\text{s}}{76\text{ kg }5.85\text{ m}}=\boxed{12.0\text{ m/s}}$

(e) For the skateboarder we may write conservation of energy as $\left(K+U_g+U_{\text{chem}}\right)_B=\left(K+U_g\right)_C$ where U_{chem} is excess chemical energy converted by his muscles into mechanical energy between points B and C. This energy converted can also be counted as work, in

$$\left(K+U_g\right)_B+W=\left(K+U_g\right)_C$$

$$\frac{1}{2}76\text{ kg}(11.1\text{ m/s})^2+0+W=\frac{1}{2}76\text{ kg}(12.0\text{ m/s})^2+76\text{ kg }9.8\text{ m/s}^2\ 0.45\text{ m}$$

$$W=5.44\text{ kJ}-4.69\text{ kJ}+335\text{ J}=\boxed{1.08\text{ kJ}}$$

(f) $\left(K+U_g\right)_C=\left(K+U_g\right)_D$

$$\frac{1}{2}76\text{ kg}(12.0\text{ m/s})^2+0=\frac{1}{2}76\text{ kg}v_D^2+76\text{ kg }9.8\text{ m/s}^2\ 5.85\text{ m}$$

$$v_D=\boxed{5.34\text{ m/s}}$$

continued on next page

(g) Let point E be the apex of his flight:

$$\left(K+U_g\right)_D=\left(K+U_g\right)_E$$

$$\frac{1}{2}76\text{ kg}(5.34\text{ m/s})^2+0=0+76\text{ kg}\left(9.8\text{ m/s}^2\right)(y_E-y_D)$$

$$(y_E-y_D)=\boxed{1.46\text{ m}}$$

(h) For the motion between takeoff and touchdown

$$y_f=y_i+v_{yi}t+\frac{1}{2}a_yt^2$$

$$-2.34\text{ m}=0+5.34\text{ m/s}\,t-4.9\text{ m/s}^2\,t^2$$

$$t=\frac{-5.34\pm\sqrt{5.34^2+4(4.9)(2.34)}}{-9.8}=\boxed{1.43\text{ s}}$$

(i) This solution is more accurate. In chapter 7 we modeled the normal force as constant while the skateboarder stands up. Really it increases as the process goes on.

P10.69 (a) $L_i=2\left[Mv\left(\frac{d}{2}\right)\right]=\boxed{Mvd}$

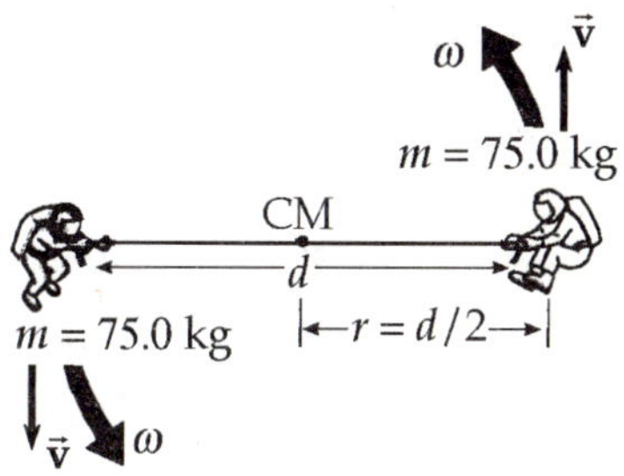

FIG. P10.69

(b) $K=2\left(\frac{1}{2}Mv^2\right)=\boxed{Mv^2}$

(c) $L_f=L_i=\boxed{Mvd}$

(d) $v_f=\frac{L_f}{2Mr_f}=\frac{Mvd}{2M\left(\frac{d}{4}\right)}=\boxed{2v}$

(e) $K_f=2\left(\frac{1}{2}Mv_f^2\right)=M(2v)^2=\boxed{4Mv^2}$

(f) $W=K_f-K_i=\boxed{3Mv^2}$

***P10.70** Choosing the origin at ,$\vec{\mathbf{R}}$

(1) $\sum F_x = +R\sin 15.0° - T\sin\theta = 0$

(2) $\sum F_y = 700 - R\cos 15.0° + T\cos\theta = 0$

(3) $\sum \tau = -700\cos\theta(0.180) + T(0.070\,0) = 0$

Solve the equations for θ

from (3), $T = 1\,800\cos\theta$ from (1), $R = \dfrac{1\,800\sin\theta\cos\theta}{\sin 15.0°}$

Then (2) gives $700 - \dfrac{1\,800\sin\theta\cos\theta\cos 15.0°}{\sin 15.0°} + 1\,800\cos^2\theta = 0$

or $\cos^2\theta + 0.388\,9 - 3.732\sin\theta\cos\theta = 0$

Squaring, $\cos^4\theta - 0.880\,9\cos^2\theta + 0.010\,13 = 0$

Let $u = \cos^2\theta$. Then using the quadratic formula

$u = 0.011\,65$ or $0.869\,3$

Only the second root is physically possible,

$\therefore \theta = \cos^{-1}\sqrt{0.869\,3} = \boxed{21.2°}$

$\therefore T = \boxed{1.68\times 10^3\text{ N}}$ and $R = \boxed{2.34\times 10^3\text{ N}}$

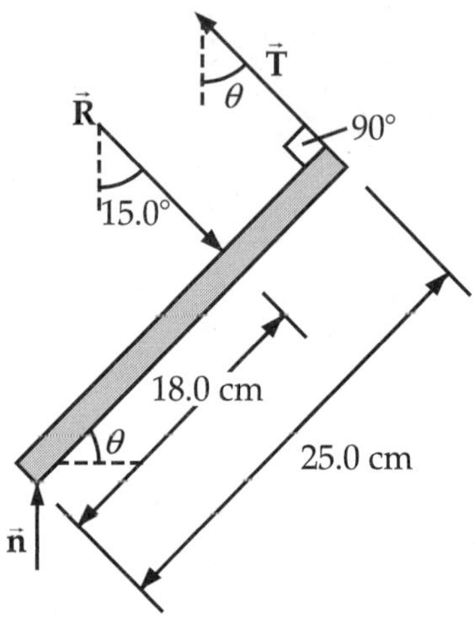

FIG. P10.70

P10.71 Choosing torques about $\vec{\mathbf{R}}$, with $\sum\tau = 0$

$$-\frac{L}{2}(350\text{ N}) + (T\sin 12.0°)\left(\frac{2L}{3}\right) - (200\text{ N})L = 0.$$

From which, $T = \boxed{2.71\text{ kN}}$.

Let R_x = compression force along spine, and from $\sum F_x = 0$

$$R_x = T_x = T\cos 12.0° = \boxed{2.65\text{ kN}}.$$

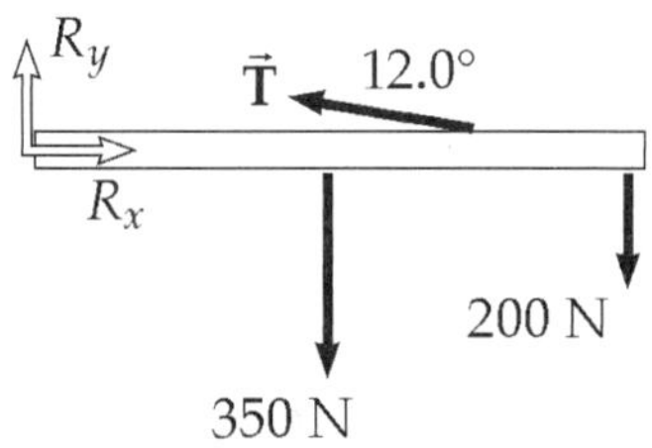

FIG. P10.71

P10.72 (a) Consider the system to consist of the wad of clay and the cylinder. No external forces acting on this system have a torque about the center of the cylinder. Thus, angular momentum of the system is conserved about the axis of the cylinder.

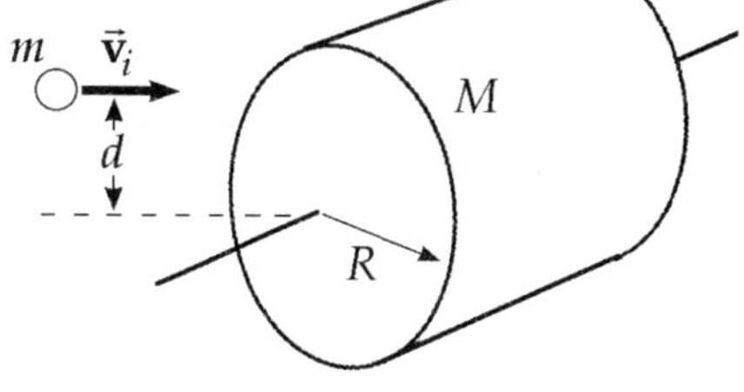

FIG. P10.72

$L_f = L_i$: $\quad I\omega = mv_i d$

or $\quad \left[\frac{1}{2}MR^2 + mR^2\right]\omega = mv_i d$

Thus, $\quad \omega = \boxed{\dfrac{2mv_i d}{(M+2m)R^2}}$.

(b) $\boxed{\text{No}}$. Some mechanical energy changes to internal energy in this perfectly inelastic collision.

P10.73 (a) Locate the origin at the bottom left corner of the cabinet and let $x =$ distance between the *resultant normal force* and the *front of the cabinet*. Then we have

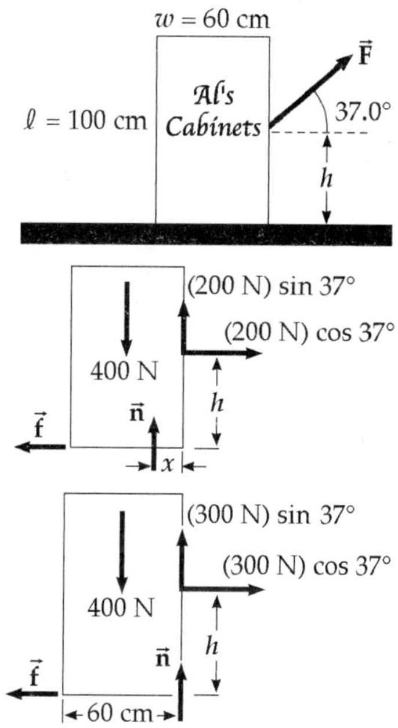

FIG. P10.73

$$\sum F_x = 200\cos 37.0° - \mu n = 0 \quad (1)$$

$$\sum F_y = 200\sin 37.0° + n - 400 = 0 \quad (2)$$

$$\sum \tau = n(0.600 - x) - 400(0.300) + 200\sin 37.0°(0.600)$$

$$-200\cos 37.0°(0.400) = 0 \quad (3)$$

From (2), $\quad n = 400 - 200\sin 37.0° = 280\text{ N}$

From (3), $\quad x = \dfrac{72.2 - 120 + 280(0.600) - 64.0}{280}$

$x = \boxed{20.1\text{ cm}}$ to the left of the front edge

From (1), $\quad \mu_k = \dfrac{200\cos 37.0°}{280} = \boxed{0.571}$

(b) In this case, locate the origin $x = 0$ at the bottom right corner of the cabinet. Since the cabinet is about to tip, we can use $\sum \tau = 0$ to find h:

$$\sum \tau = 400(0.300) - (300\cos 37.0°)h = 0 \qquad h = \frac{120}{300\cos 37.0°} = \boxed{0.501\text{ m}}$$

***P10.74** The 392 N is the weight of the uniform gate, which is 3 m wide. The hinges are 1.8 m apart. They exert horizontal forces A and C. Only one hinge exerts a vertical force. We assume it is the upper hinge.

(a) Free body diagram:

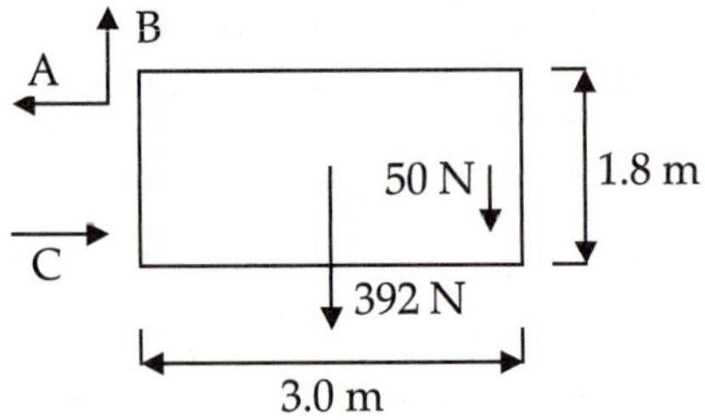

Statement:

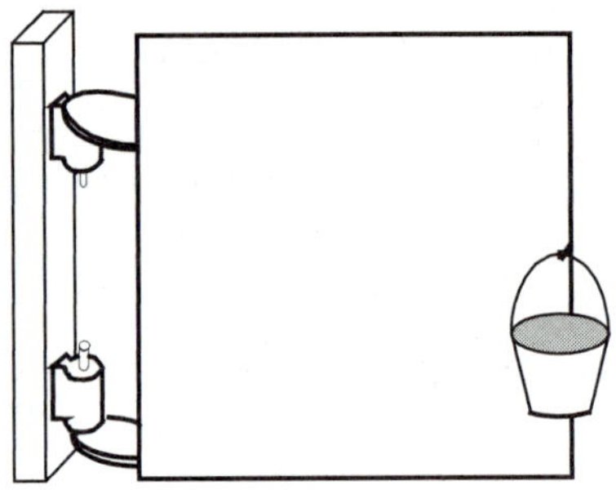

FIG. P10.74

A uniform 40.0-kg farm gate, 3.00 m wide and 1.80 m high, supports a 50.0-N bucket of grain hanging from its latch as shown. The gate is supported by two hinges. Find the force each hinge exerts on the gate.

(b) From the torque equation,

$$C = \frac{738\text{ N}\cdot\text{m}}{1.8\text{ m}} = 410\text{ N}.$$

Then $A = 410\text{ N}$. Also $B = 442\text{ N}$.

The upper hinge exerts 410 N to the left and 442 N up. The lower hinge exerts 410 N to the right.

P10.75 From geometry, observe that

$$\cos\theta = \frac{1}{4} \qquad \text{and} \qquad \theta = 75.5°$$

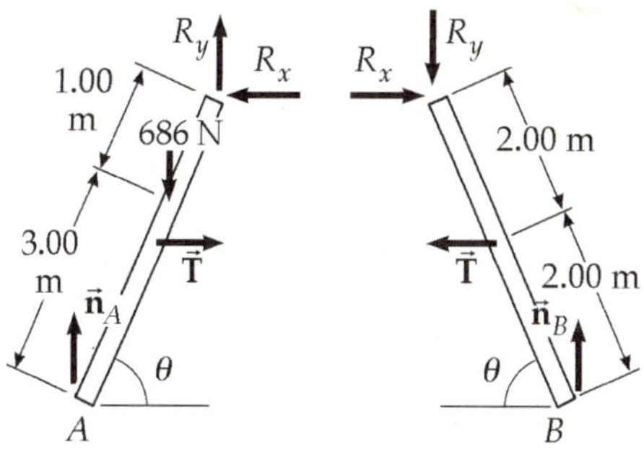

FIG. P10.75

For the left half of the ladder, we have

$$\sum F_x = T - R_x = 0 \qquad (1)$$

$$\sum F_y = R_y + n_A - 686\text{ N} = 0 \qquad (2)$$

$$\sum \tau_{\text{top}} = 686\text{ N}(1.00\cos 75.5°) + T(2.00\sin 75.5°)$$

$$-n_A(4.00\cos 75.5°) = 0 \qquad (3)$$

For the right half of the ladder we have

$$\sum F_x = R_x - T = 0$$

$$\sum F_y = n_B - R_y = 0 \qquad (4)$$

$$\sum \tau_{\text{top}} = n_B(4.00\cos 75.5°) - T(2.00\sin 75.5°) = 0 \qquad (5)$$

Solving equations 1 through 5 simultaneously yields:

(a) $\boxed{T = 133\text{ N}}$

(b) $\boxed{n_A = 429\text{ N}}$ and $\boxed{n_B = 257\text{ N}}$

(c) $\boxed{R_x = 133\text{ N}}$ and $\boxed{R_y = 257\text{ N}}$

The force exerted by the left half of the ladder on the right half is to the right and downward.

***P10.76** (a) $\Delta K_{\text{rot}} + \Delta K_{\text{trans}} + \Delta U = 0$

Note that initially the center of mass of the sphere is a distance $h + r$ above the bottom of the loop; and as the mass reaches the top of the loop, this distance above the reference level is $2R - r$. The conservation of energy requirement gives

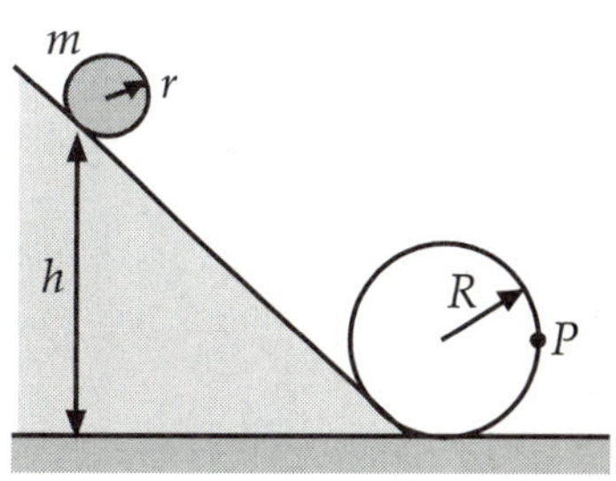

FIG. P10.76

$$mg(h+r) = mg(2R-r) + \frac{1}{2}mv^2 + \frac{1}{2}I\omega^2$$

For the sphere $I = \frac{2}{5}mr^2$ and $v = r\omega$ so that the expression becomes

continued on next page

$$gh + 2gr = 2gR + \frac{7}{10}v^2 \qquad (1)$$

Note that $h = h_{\min}$ when the speed of the sphere at the top of the loop satisfies the condition

$$\sum F = mg = \frac{mv^2}{(R-r)} \text{ or } v^2 = g(R-r)$$

Substituting this into Equation (1) gives

$$h_{\min} = 2(R-r) + 0.700(R-r) \text{ or } \boxed{h_{\min} = 2.70(R-r) \approx 2.70R}$$

(b) When the sphere is initially at $h = 3R$ and finally at point P, the conservation of energy equation gives

$$mg(3R+r) = mgR + \frac{1}{2}mv^2 + \frac{1}{5}mv^2, \text{ or } v^2 = \frac{10}{7}(2R+r)g$$

Turning clockwise as it rolls without slipping past point P, the sphere is slowing down with counterclockwise angular acceleration caused by the torque of an upward force f of static friction. We have $\sum F_y = ma_y$ and $\sum \tau = I\alpha$ becoming $f - mg = -m\alpha r$ and $fr = \left(\frac{2}{5}\right)mr^2\alpha$.

Eliminating f by substitution yields $\alpha = \frac{5g}{7r}$ so that $\sum F_y = \boxed{\frac{-5mg}{7}}$

$$\sum F_x = -n = -\frac{mv^2}{R-r} = -\frac{\left(\frac{10}{7}\right)(2R+r)}{R-r}mg = \boxed{\frac{-20mg}{7}} \text{ (since } R >> r\text{)}$$

P10.77 When it is on the verge of slipping, the cylinder is in equilibrium.

$\sum F_x = 0$: $\quad f_1 = n_2 = \mu_s n_1 \quad$ and $\quad f_2 = \mu_s n_2$

$\sum F_y = 0$: $\quad P + n_1 + f_2 = F_g$

$\sum \tau = 0$: $\quad P = f_1 + f_2$

As P grows so do f_1 and f_2

Therefore, since $\mu_s = \frac{1}{2}$, $\quad f_1 = \frac{n_1}{2} \quad$ and $\quad f_2 = \frac{n_2}{2} = \frac{n_1}{4}$

then $\quad P + n_1 + \frac{n_1}{4} = F_g \quad (1) \quad$ and $\quad P = \frac{n_1}{2} + \frac{n_1}{4} = \frac{3}{4}n_1 \quad (2)$

So $\quad P + \frac{5}{4}n_1 = F_g \quad$ becomes $\quad P + \frac{5}{4}\left(\frac{4}{3}P\right) = F_g \quad$ or $\quad \frac{8}{3}P = F_g$

Therefore, $P = \boxed{\frac{3}{8}F_g}$.

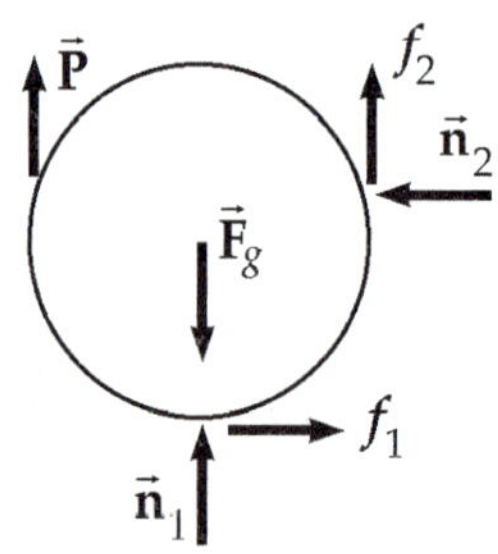

FIG. P10.77

ANSWERS TO EVEN PROBLEMS

P10.2 (a) $8.22\times10^{2}\ \text{rad/s}^2$; (b) 4.21×10^{3} rad

P10.4 $-226\ \text{rad/s}^2$

P10.6 $13.7\ \text{rad/s}^2$

P10.8 $\sim10^{7}\ \text{rev/yr}$

P10.10 (a) 8.00 rad/s;
(b) 8.00 m/s, $64.1\ \text{m/s}^2$ inward at 3.58° ahead of the radius;
(c) 9.00 rad

P10.12 (a) 56.5 rad/s; (b) 22.4 rad/s;
(c) $-7.63\times10^{-3}\ \text{rad/s}^2$; (d) 1.77×10^{5} rad ;
(e) 5.81×10^{3} m

P10.14 (a) $92.0\ \text{kg}\cdot\text{m}^2$, 184 J;
(b) 6.00 m/s, 4.00 m/s, 8.00 m/s, 184 J

P10.16 1.03×10^{-3} J

P10.18 see the solution

P10.20 168 N·m clockwise

P10.22 $-7.00\hat{\mathbf{i}}+16.0\hat{\mathbf{j}}-10.0\hat{\mathbf{k}}$

P10.24 (a) 168°; (b) principal value 11.9°;
(c) the first method is unambiguous

P10.26 1.09 m

P10.28 see the solution, 0.643 m

P10.30 (a) $\left[\dfrac{m_1 g}{2}+\dfrac{x m_2 g}{L}\right]\cot\theta,\ (m_1+m_2)g;$

(b) $\dfrac{\left(\frac{m_1}{2}+\frac{m_2 d}{L}\right)\cot\theta}{m_1+m_2}$

P10.32 $F_{By}=0$, $F_{Ax}=6.47\times10^{5}$ N left,
$F_{Bx}=6.47\times10^{5}$ N right,
$F_{Ay}=1.27\times10^{5}$ N up

P10.34 0.312

P10.36 (a) 1.59 m/s; (b) 53.1 rad/s

P10.38 $\dfrac{3g}{2L};\ -\dfrac{3}{2}g\hat{\mathbf{j}}$

P10.40 (a) $\left(-9.03\times10^{9}\ \text{kg}\cdot\text{m}^2/\text{s}\right)\hat{\mathbf{j}}$; (b) no;
(c) zero

P10.42 $L=1.20\ \text{kg}\cdot\text{m}^2/\text{s}$ perpendicularly into the clock face

P10.44 (a) $1.57\times10^{8}\ \text{kg}\cdot\text{m}^2/\text{s}$; (b) 6.26×10^{3} s ;
(c) see the solution

P10.46 7.14 rev/min

P10.48 (a) 1.91 rad/s; (b) 2.53 J, 6.44 J

P10.50 $12.3\ \text{m/s}^2$

P10.52 5.45×10^{22} N·m

P10.54 $v_{\text{disk}}=\sqrt{\dfrac{4gh}{3}}$, $v_{\text{ring}}=\sqrt{gh}$, the disk

P10.56 $1.21\times10^{-4}\ \text{kg}\cdot\text{m}^2$, height of the can

P10.58 (a) 0.992 W; (b) 827 W

P10.60 see the solution

P10.62 (a) $0.309\ \text{m/s}^2$; (b) 7.67 N, 9.22 N

P10.64 (a) −794 N·m; −2 510 N·m; 0;
−1 160 N·m; −2 940 N·m;
(b) At the following times: 12:00:00 12:30:55 12:58:19 1:32:31 1:57:01 2:33:25 2:56:29 3:33:22 3:56:55 4:32:24 4:58:14 5:30:52 6:00:00 6:29:08 7:01:46 7:27:36 8:03:05 8:26:38 9:03:31 9:26:35 10:02:59 10:27:29 11:01:41 11:29:05

P10.66 The weight must be 508 N or more. The person could be a child, see the solution

P10.68 (a) 11.1 m/s; (b) $5.32\times10^{3}\ \text{kg}\cdot\text{m}^2/\text{s}$; (c) see the solution; (d) 12.0 m/s; (e) 1.08 kJ; (f) 5.34 m/s; (g) 1.46 m; (h) 1.43 s; (i) see the solution

P10.70 $T=(1.68\ \text{kN})$, $R=2.34\ \text{kN}$, $\theta=21.2°$

P10.72 (a) $\omega=\dfrac{2mv_id}{(M+2m)R^2}$; (b) no, some mechanical energy changes into internal energy

P10.74 (a) see the solution; (b) The upper hinge exerts 410 N to the left and 442 N up. The lower hinge exerts 410 N to the right.

P10.76 (a) 2.70 R; (b) $F_x=\dfrac{-20mg}{7}$, $F_y=\dfrac{-5mg}{7}$

Chapter 11

Gravity, Planetary Orbits, and the Hydrogen Atom

CHAPTER OUTLINE

ANSWERS TO QUESTIONS

Q11.1 The Earth creates the same gravitational field g for all objects near the Earth's surface. The larger mass needs a larger force to give it just the same acceleration.

Q11.2 To a good first approximation, your bathroom scale reading is unaffected because you, the Earth, and the scale are all in free fall in the Sun's gravitational field, in orbit around the Sun. To a precise second approximation, you weigh slightly less at noon and at midnight than you do at sunrise or sunset. The Sun's gravitational field is a little weaker at the center of the Earth than at the surface subsolar point, and a little weaker still on the far side of the planet. When the Sun is high in your sky, its gravity pulls up on you a little more strongly than on the Earth as a whole. At midnight the Sun pulls down on you a little less strongly than it does on the Earth below you. So you can have another doughnut with lunch, and your bedsprings will still last a little longer.

Q11.3 Because both the Earth and Moon are moving in orbit about the Sun. As described by $F_{\text{gravitational}} = ma_{\text{centripetal}}$, the gravitational force of the Sun merely keeps the Moon (and Earth) in a nearly circular orbit of radius 150 million kilometers. Because of its velocity, the Moon is kept in its orbit about the Earth by the gravitational force of the Earth. There is no imbalance of these forces, at new moon or full moon.

Q11.4 Air resistance causes a decrease in the energy of the satellite-Earth system. This reduces the diameter of the orbit, bringing the satellite closer to the surface of the Earth. A satellite in a smaller orbit, however, must travel faster. Thus, the effect of air resistance is to speed up the satellite!

Q11.5 The escape speed from the Earth is 11.2 km/s and that from the Moon is 2.3 km/s, smaller by a factor of 5. The energy required—and fuel—would be proportional to v^2, or 25 times more fuel is required to leave the Earth versus leaving the Moon.

Q11.6 In a circular orbit each increment of displacement is perpendicular to the force applied. The dot product of force and displacement is zero. The work done by the gravitational force on a planet in an elliptical orbit speeds up the planet at closest approach, but negative work is done by gravity and the planet slows as it sweeps out to its farthest distance from the Sun. Therefore, net work in one complete orbit is zero.

Q11.7 For a satellite in orbit, one focus of an elliptical orbit, or the center of a circular orbit, must be located at the center of the Earth. If the satellite is over the northern hemisphere for half of its orbit, it must be over the southern hemisphere for the other half. We could share with Easter Island a satellite that would look straight down on Arizona each morning and vertically down on Easter Island each evening.

Q11.8 The gravitational force of the Earth on an extra particle at its center must be zero, not infinite as one interpretation of Equation 11.1 would suggest. All the bits of matter that make up the Earth will pull in different outward directions on the extra particle.

Q11.9 Speed is maximum at closest approach. Speed is minimum at farthest distance.

Q11.10 The gravitational force is conservative. An encounter with a stationary mass cannot permanently speed up a spacecraft. Jupiter is moving. A spacecraft flying across its orbit just behind the planet will gain kinetic energy as the planet's gravity does net positive work on it.

Q11.11 Cavendish determined G. Then from $g = \frac{GM}{R^2}$, one may determine the mass of the Earth.

Q11.12 The spacecraft did not have enough fuel to stop dead in its high-speed course for the Moon.

Q11.13 If an electron moved like a hockey puck, it could have any arbitrary frequency of revolution around an atomic nucleus. If it behaved like a charge in a radio antenna, it would radiate light with frequency equal to its own frequency of oscillation. Thus, the electron in hydrogen atoms would emit a continuous spectrum, electromagnetic waves of all frequencies smeared together.

Q11.14 (a) Yes—provided that the energy of the photon is *precisely* enough to put the electron into one of the allowed energy states. Strangely—more precisely non-classically—enough, if the energy of the photon is not sufficient to put the electron into a particular excited energy level, the photon will not interact with the atom at all!

(b) Yes—a photon of any energy greater than 13.6 eV will ionize the atom. Any "extra" energy will go into kinetic energy of the newly liberated electron.

Q11.15 An atomic electron does not possess enough kinetic energy to escape from its electrical attraction to the nucleus. Positive ionization energy must be injected to pull the electron out to a very large separation from the nucleus, a condition for which we define the energy of the atom to be zero. The atom is a bound system. All this is summarized by saying that the total energy of an atom is negative.

Q11.16 From Equations 11.19, 11.20 and 11.21, we have $-|E| = -\frac{k_e e^2}{2r} = +\frac{k_e e^2}{2r} - \frac{k_e e^2}{r} = K + U_e$. Then $K = |E|$ and $U_e = -2|E|$.

SOLUTIONS TO PROBLEMS

Section 11.1 Newton's Law of Universal Gravitation Revisited

P11.1 $F = m_1 g = \dfrac{Gm_1 m_2}{r^2}$

$$g = \frac{Gm_2}{r^2} = \frac{\left(6.67\times10^{-11}\ \text{N}\cdot\text{m}^2/\text{kg}^2\right)\left(4.00\times10^4\times10^3\ \text{kg}\right)}{(100\ \text{m})^2} = \boxed{2.67\times10^{-7}\ \text{m/s}^2}$$

P11.2 (a) At the midpoint between the two objects, the forces exerted by the 200-kg and 500-kg objects are oppositely directed,

and from $F_g = \dfrac{Gm_1 m_2}{r^2}$

we have $\sum F = \dfrac{G(50.0\ \text{kg})(500\ \text{kg} - 200\ \text{kg})}{(0.200\ \text{m})^2} = \boxed{2.50\times10^{-5}\ \text{N}}$ toward the 500-kg object.

(b) At a point between the two objects at a distance d from the 500-kg objects, the net force on the 50.0-kg object will be zero when

$$\frac{G(50.0\ \text{kg})(200\ \text{kg})}{(0.400\ \text{m} - d)^2} = \frac{G(50.0\ \text{kg})(500\ \text{kg})}{d^2}$$

or $d = \boxed{0.245\ \text{m}}$

P11.3 $F = \dfrac{GMm}{r^2} = \left(6.67\times10^{-11}\ \text{N}\cdot\text{m}^2/\text{kg}^2\right)\dfrac{(1.50\ \text{kg})\left(15.0\times10^{-3}\ \text{kg}\right)}{\left(4.50\times10^{-2}\ \text{m}\right)^2} = \boxed{7.41\times10^{-10}\ \text{N}}$

P11.4 The force exerted on the 4.00-kg mass by the 2.00-kg mass is directed upward and given by

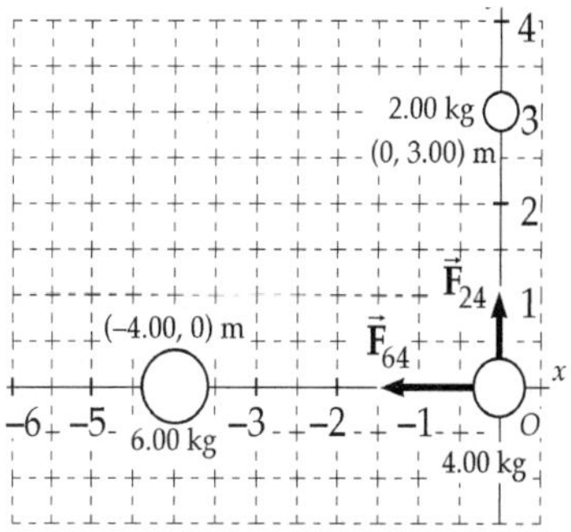

FIG. P11.4

$$\vec{\mathbf{F}}_{24} = G\frac{m_4 m_2}{r_{24}^2}\hat{\mathbf{j}} = \left(6.67\times10^{-11}\ \text{N}\cdot\text{m}^2/\text{kg}^2\right)\frac{(4.00\ \text{kg})(2.00\ \text{kg})}{(3.00\ \text{m})^2}\hat{\mathbf{j}}$$
$$= 5.93\times10^{-11}\,\hat{\mathbf{j}}\text{N}$$

The force exerted on the 4.00-kg mass by the 6.00-kg mass is directed to the left

$$\vec{\mathbf{F}}_{64} = G\frac{m_4 m_6}{r_{64}^2}\left(-\hat{\mathbf{i}}\right) = \left(-6.67\times10^{-11}\ \text{N}\cdot\text{m}^2/\text{kg}^2\right)\frac{(4.00\ \text{kg})(6.00\ \text{kg})}{(4.00\ \text{m})^2}\hat{\mathbf{i}}$$
$$= -10.0\times10^{-11}\,\hat{\mathbf{i}}\text{N}$$

Therefore, the resultant force on the 4.00-kg mass is $\vec{\mathbf{F}}_4 = \vec{\mathbf{F}}_{24} + \vec{\mathbf{F}}_{64} = \boxed{\left(-10.0\hat{\mathbf{i}} + 5.93\hat{\mathbf{j}}\right)\times10^{-11}\ \text{N}}$.

P11.5 (a) The Sun-Earth distance is 1.496×10^{11} m and the Earth-Moon distance is 3.84×10^{8} m, so the distance from the Sun to the Moon during a solar eclipse is

$$1.496 \times 10^{11} \text{ m} - 3.84 \times 10^{8} \text{ m} = 1.492 \times 10^{11} \text{ m}$$

The mass of the Sun, Earth, and Moon are $M_S = 1.99 \times 10^{30}$ kg

$M_E = 5.98 \times 10^{24}$ kg

and $M_M = 7.36 \times 10^{22}$ kg

We have $F_{SM} = \dfrac{Gm_1m_2}{r^2} = \dfrac{(6.67 \times 10^{-11})(1.99 \times 10^{30})(7.36 \times 10^{22})}{(1.492 \times 10^{11})^2} = \boxed{4.39 \times 10^{20} \text{ N}}$

(b) $$F_{EM} = \frac{(6.67 \times 10^{-11} \text{ N} \cdot \text{m}^2/\text{kg}^2)(5.98 \times 10^{24})(7.36 \times 10^{22})}{(3.84 \times 10^{8})^2} = \boxed{1.99 \times 10^{20} \text{ N}}$$

(c) $$F_{SE} = \frac{(6.67 \times 10^{-11} \text{ N} \cdot \text{m}^2/\text{kg}^2)(1.99 \times 10^{30})(5.98 \times 10^{24})}{(1.496 \times 10^{11})^2} = \boxed{3.55 \times 10^{22} \text{ N}}$$

Note that the force exerted by the Sun on the Moon is much stronger than the force of the Earth on the Moon. In a sense, the Moon orbits the Sun more than it orbits the Earth. The Moon's path is everywhere concave toward the Sun. Only by subtracting out the solar orbital motion of the Earth-Moon system do we see the Moon orbiting the center of mass of this system.

P11.6 Let θ represent the angle each cable makes with the vertical, L the cable length, x the distance each ball scrunches in, and $d = 1$ m the original distance between them. Then $r = d - 2x$ is the separation of the balls. We have

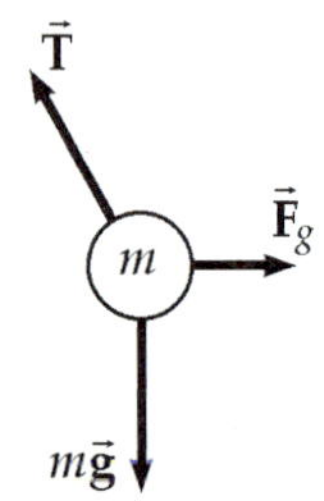

FIG. P11.6

$\sum F_y = 0$: $\quad T\cos\theta - mg = 0$

$\sum F_x = 0$: $\quad T\sin\theta - \dfrac{Gmm}{r^2} = 0$

Then $\quad \tan\theta = \dfrac{Gmm}{r^2 mg} \quad\quad \dfrac{x}{\sqrt{L^2 - x^2}} = \dfrac{Gm}{g(d-2x)^2} \quad\quad x(d-2x)^2 = \dfrac{Gm}{g}\sqrt{L^2 - x^2}$.

The factor $\dfrac{Gm}{g}$ is numerically small. There are two possibilities: either x is small or else $d - 2x$ is small.

Possibility one: We can ignore x in comparison to d and L, obtaining

$$x(1 \text{ m})^2 = \frac{(6.67 \times 10^{-11} \text{ N} \cdot \text{m}^2/\text{kg}^2)(100 \text{ kg})}{(9.8 \text{ m/s}^2)} 45 \text{ m} \qquad x = 3.06 \times 10^{-8} \text{ m}.$$

continued on next page

The separation distance is $r = 1\text{ m} - 2\left(3.06\times10^{-8}\text{ m}\right) = \boxed{1.000\text{ m} - 61.3\text{ nm}}$.

Possibility two: If $d - 2x$ is small, $x \approx 0.5\text{ m}$ and the equation becomes

$$(0.5\text{ m})r^2 = \frac{\left(6.67\times10^{-11}\text{ N}\cdot\text{m}^2/\text{kg}^2\right)(100\text{ kg})}{(9.8\text{ N/kg})}\sqrt{(45\text{ m})^2 - (0.5\text{ m})^2} \qquad r = \boxed{2.74\times10^{-4}\text{ m}}.$$

For this answer to apply, the spheres would have to be compressed to a density like that of the nucleus of atom.

P11.7 $$g = \frac{GM}{R^2} = \frac{G\rho\left(\frac{4\pi R^3}{3}\right)}{R^2} = \frac{4}{3}\pi G\rho R$$

If $$\frac{g_M}{g_E} = \frac{1}{6} = \frac{4\pi G\rho_M R_M}{3}\frac{3}{4\pi G\rho_E R_E}$$

then $$\frac{\rho_M}{\rho_E} = \left(\frac{g_M}{g_E}\right)\left(\frac{R_E}{R_M}\right) = \left(\frac{1}{6}\right)(4) = \boxed{\frac{2}{3}}.$$

P11.8 (a) At the zero-total field point, $\dfrac{GmM_E}{r_E^2} = \dfrac{GmM_M}{r_M^2}$

so $$r_M = r_E\sqrt{\frac{M_M}{M_E}} = r_E\sqrt{\frac{7.36\times10^{22}}{5.98\times10^{24}}} = \frac{r_E}{9.01}$$

$$r_E + r_M = 3.84\times10^8\text{ m} = r_E + \frac{r_E}{9.01}$$

$$r_E = \frac{3.84\times10^8\text{ m}}{1.11} = \boxed{3.46\times10^8\text{ m}}$$

(b) At this distance the acceleration due to the Earth's gravity is

$$g_E = \frac{GM_E}{r_E^2} = \frac{\left(6.67\times10^{-11}\text{ N}\cdot\text{m}^2/\text{kg}^2\right)\left(5.98\times10^{24}\text{ kg}\right)}{\left(3.46\times10^8\text{ m}\right)^2}$$

$$g_E = \boxed{3.34\times10^{-3}\text{ m/s}^2\text{ directed toward the Earth}}$$

***P11.9** (a) For the gravitational force on an object in the neighborhood of Miranda we have

$$m_{\text{obj}}g = \frac{Gm_{\text{obj}}m_{\text{Miranda}}}{r_{\text{Miranda}}^2}$$

$$g = \frac{Gm_{\text{Miranda}}}{r_{\text{Miranda}}^2} = \frac{6.67\times10^{-11}\text{ N}\cdot\text{m}^2\left(6.68\times10^{19}\text{ kg}\right)}{\text{kg}^2\left(242\times10^3\text{ m}\right)^2} = \boxed{0.0761\text{ m/s}^2}$$

continued on next page

(b) We ignore the difference (of about 4%) in g between the lip and the base of the cliff. For the vertical motion of the athlete we have

$$y_f = y_i + v_{yi} + \frac{1}{2}a_y t^2$$

$$-5\,000\text{ m} = 0 + 0 + \frac{1}{2}\left(-0.0761\text{ m/s}^2\right)t^2$$

$$t = \left(\frac{2(5\,000\text{ m})\text{s}^2}{0.0761\text{ m}}\right)^{1/2} = \boxed{363\text{ s}}$$

(c) $x_f = x_i + v_{xi}t + \frac{1}{2}a_x t^2 = 0 + (8.5\text{ m/s})(363\text{ s}) + 0 = \boxed{3.08\times10^3\text{ m}}$

We ignore the curvature of the surface (of about 0.7°) over the athlete's trajectory.

(d) $v_{xf} = v_{xi} = 8.50\text{ m/s}$

$$v_{yf} = v_{yi} + a_y t = 0 - \left(0.0761\text{ m/s}^2\right)(363\text{ s}) = -27.6\text{ m/s}$$

Thus $\vec{\mathbf{v}}_f = \left(8.50\hat{\mathbf{i}} - 27.6\hat{\mathbf{j}}\right)\text{ m/s} = \sqrt{8.5^2 + 27.6^2}\text{ m/s at }\tan^{-1}\frac{27.6}{8.5}$ below the x axis.

$$\boxed{\vec{\mathbf{v}}_f = 28.9\text{ m/s at }72.9°\text{ below the horizontal}}$$

P11.10 (a) $F = \dfrac{GMm}{r^2} = \dfrac{\left(6.67\times10^{-11}\text{ N}\cdot\text{m}^2/\text{kg}^2\right)\left[100\left(1.99\times10^{30}\text{ kg}\right)\left(10^3\text{ kg}\right)\right]}{\left(1.00\times10^4\text{ m} + 50.0\text{ m}\right)^2} = \boxed{1.31\times10^{17}\text{ N}}$

(b) $\Delta F = \dfrac{GMm}{r_{\text{front}}^2} - \dfrac{GMm}{r_{\text{back}}^2}$

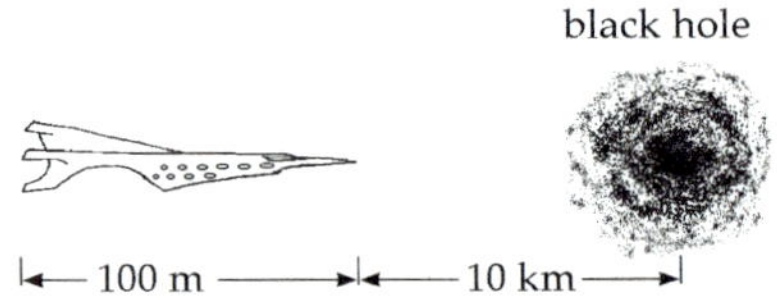

FIG. P11.10

$$\Delta g = \frac{\Delta F}{m} = \frac{GM\left(r_{\text{back}}^2 - r_{\text{front}}^2\right)}{r_{\text{front}}^2 r_{\text{back}}^2}$$

$$\Delta g = \frac{\left(6.67\times10^{-11}\right)\left[100\left(1.99\times10^{30}\right)\right]\left[\left(1.01\times10^4\text{ m}\right)^2 - \left(1.00\times10^4\text{ m}\right)^2\right]}{\left(1.00\times10^4\text{ m}\right)^2\left(1.01\times10^4\text{ m}\right)^2}$$

$$\Delta g = \boxed{2.62\times10^{12}\text{ N/kg}}$$

P11.11 $g_1 = g_2 = \dfrac{MG}{r^2+a^2}$

$g_{1y} = -g_{2y}$ $\qquad g_y = g_{1y} + g_{2y}$

$g_{1x} = g_{2x} = g_2 \cos\theta$ $\qquad \cos\theta = \dfrac{r}{\left(a^2+r^2\right)^{1/2}}$

$$\vec{\mathbf{g}} = 2g_{2x}\left(-\hat{\mathbf{i}}\right)$$

or $\vec{\mathbf{g}} = \boxed{\dfrac{2MGr}{\left(r^2+a^2\right)^{3/2}} \text{ toward the center of mass}}$

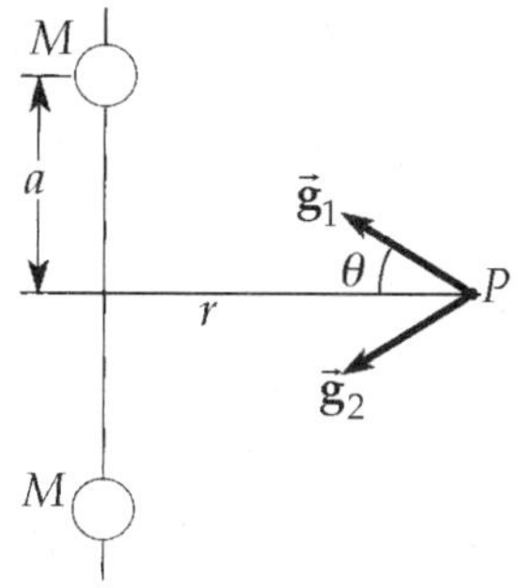

FIG. P11.11

P11.12 (a) We require that $\dfrac{GM_E m}{r^2} = \dfrac{mv^2}{r}$, but $g = \dfrac{M_E G}{R_E{}^2}$. In this case $r = 2R_E$ therefore, $\dfrac{g}{4} = \dfrac{v^2}{2R_E}$ or

$$v = \sqrt{\frac{gR_E}{2}} = \sqrt{\frac{9.80 \text{ m/s}^2\left(6.37\times10^6 \text{ m}\right)}{2}} = \boxed{5.59\times10^3 \text{ m/s}}.$$

(b) $$T = \frac{2\pi r}{v} = \frac{2\pi(2)\left(6.37\times10^6 \text{ m}\right)}{5.59\times10^3 \text{ m/s}} = \boxed{239 \text{ min}}$$

(c) $$F = \frac{GM_E m}{\left(2R_E\right)^2} = \frac{mg}{4} = \frac{(300 \text{ kg})\left(9.80 \text{ m/s}^2\right)}{4} = \boxed{735 \text{ N}}$$

Section 11.2 Structural Models

No problems in this section

Section 11.3 Kepler's Laws

P11.13 (a) For the geosynchronous satellite, $\Sigma F_r = ma_r$ becomes $\dfrac{GmM_E}{r^2} = \dfrac{mv^2}{r}$ and in turn

$\dfrac{GM_E}{r} = \left(\dfrac{2\pi r}{T}\right)^2$ or $r^3 = \dfrac{GM_E T^2}{4\pi^2}$. Thus, the radius of the satellite orbit is

$$r = \left[\frac{\left(6.67\times10^{-11} \text{ N}\cdot\text{m}^2/\text{kg}^2\right)\left(5.98\times10^{24} \text{ kg}\right)\left(86\,400 \text{ s}\right)^2}{4\pi^2}\right]^{1/3} = \boxed{4.23\times10^7 \text{ m}}.$$

continued on next page

(b) The satellite is so far out that its distance from the north pole,
$d=\sqrt{(6.37\times10^6\ \text{m})^2+(4.23\times10^7\ \text{m})^2}=4.27\times10^7\ \text{m}$ is nearly the same as its orbital radius.
The travel time for the radio signal is

$$t=\frac{2d}{c}=\frac{2(4.27\times10^7\ \text{m})}{3.00\times10^8\ \text{m/s}}=\boxed{0.285\ \text{s}}.$$

P11.14 By conservation of angular momentum for the satellite,

$$r_pv_p=r_av_a \qquad \frac{v_p}{v_a}=\frac{r_a}{r_p}=\frac{2\,289\ \text{km}+6.37\times10^3\ \text{km}}{459\ \text{km}+6.37\times10^3\ \text{km}}=\frac{8\,659\ \text{km}}{6\,829\ \text{km}}=\boxed{1.27}.$$

We do not need to know the period.

P11.15 $T^2=\dfrac{4\pi^2a^3}{GM}$ (Kepler's third law with $m<<M$)

$$M=\frac{4\pi^2a^3}{GT^2}=\frac{4\pi^2(4.22\times10^8\ \text{m})^3}{(6.67\times10^{-11}\ \text{N}\cdot\text{m}^2/\text{kg}^2)(1.77\times86\,400\ \text{s})^2}=\boxed{1.90\times10^{27}\ \text{kg}}$$

(Approximately 316 Earth masses)

P11.16 By Kepler's Third Law, $T^2=ka^3$ ($a=$ semi-major axis)

For any object orbiting the Sun, with T in years and a in A.U., $k=1.00$. Therefore, for Comet Halley

$$(75.6)^2=(1.00)\left(\frac{0.570+y}{2}\right)^3$$

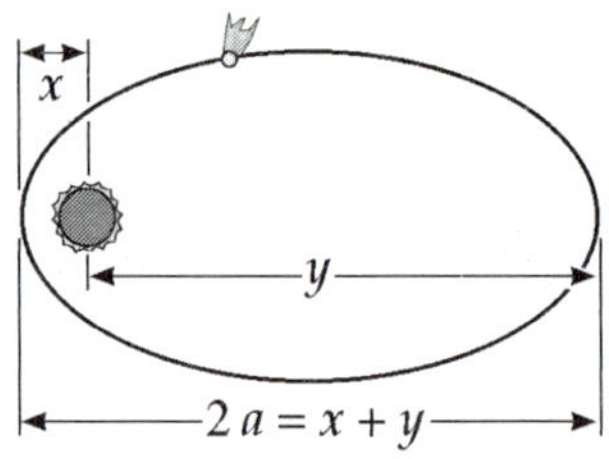

FIG. P11.16

The farthest distance the comet gets from the Sun is

$y=2(75.6)^{2/3}-0.570=\boxed{35.2\ \text{A.U.}}$ (out around the orbit of Pluto)

P11.17 Applying Newton's 2nd Law, $\sum F=ma$ yields $F_g=ma_c$ for each star:

$$\frac{GMM}{(2r)^2}=\frac{Mv^2}{r} \quad \text{or} \quad M=\frac{4v^2r}{G}.$$

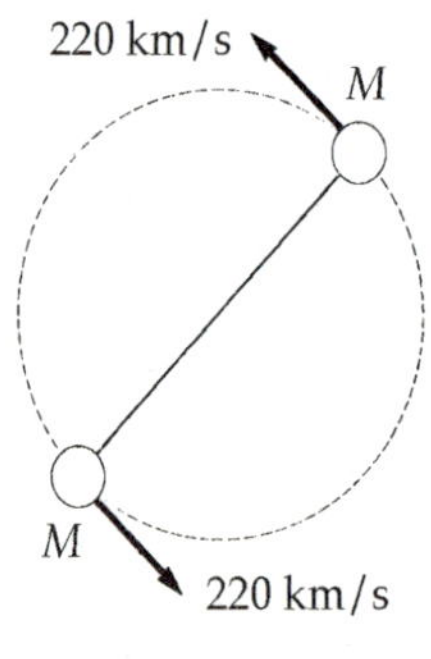

FIG. P11.17

We can write r in terms of the period, T, by considering the time and distance of one complete cycle. The distance traveled in one orbit is the circumference of the stars' common orbit, so $2\pi r=vT$. Therefore

$$M=\frac{4v^2r}{G}=\frac{4v^2}{G}\left(\frac{vT}{2\pi}\right)$$

so, $$M=\frac{2v^3T}{\pi G}=\frac{2(220\times10^3\ \text{m/s})^3(14.4\ \text{d})(86\,400\ \text{s/d})}{\pi(6.67\times10^{-11}\ \text{N}\cdot\text{m}^2/\text{kg}^2)}=\boxed{1.26\times10^{32}\ \text{kg}=63.3\ \text{solar masses}}$$

P11.18 $\sum F = ma$: $\dfrac{Gm_{\text{planet}}M_{\text{star}}}{r^2} = \dfrac{m_{\text{planet}}v^2}{r}$

$$\frac{GM_{\text{star}}}{r} = v^2 = r^2\omega^2$$

$$GM_{\text{star}} = r^3\omega^3 = r_x^3\omega_x^2 = r_y^3\omega_y^2$$

$$\omega_y = \omega_x\left(\frac{r_x}{r_y}\right)^{3/2} \qquad \omega_y = \left(\frac{90.0°}{5.00\text{ yr}}\right)3^{3/2} = \frac{468°}{5.00\text{ yr}}$$

So $\boxed{\text{planet } Y \text{ has turned through 1.30 revolutions}}$.

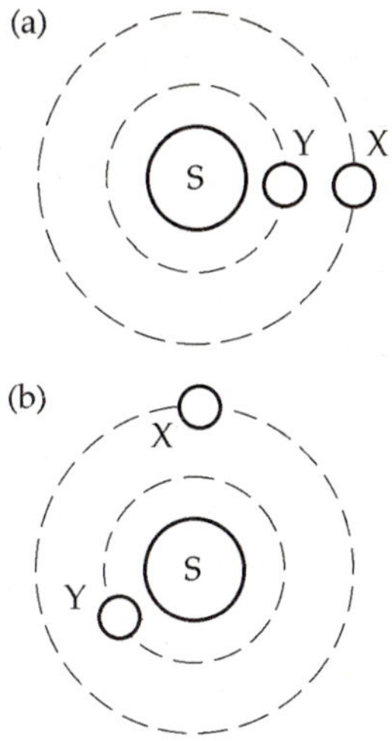

FIG. P11.18

P11.19 The speed of a planet in a circular orbit is given by

$\sum F = ma$: $\dfrac{GM_{\text{sun}}m}{r^2} = \dfrac{mv^2}{r}$ $\qquad v = \sqrt{\dfrac{GM_{\text{sun}}}{r}}$.

For Mercury the speed is $\qquad v_M = \sqrt{\dfrac{(6.67\times10^{-11})(1.99\times10^{30})\text{ m}^2}{(5.79\times10^{10})\text{ s}^2}} = 4.79\times10^4\text{ m/s}$

and for Pluto, $\qquad v_P = \sqrt{\dfrac{(6.67\times10^{-11})(1.99\times10^{30})\text{ m}^2}{(5.91\times10^{12})\text{ s}^2}} = 4.74\times10^3\text{ m/s}$.

With greater speed, Mercury will eventually move farther from the Sun than Pluto. With original distances r_P and r_M perpendicular to their lines of motion, they will be equally far from the Sun after time t where

$$\sqrt{r_P^2 + v_P^2t^2} = \sqrt{r_M^2 + v_M^2t^2}$$

$$r_P^2 - r_M^2 = \left(v_M^2 - v_P^2\right)t^2$$

$$t = \sqrt{\frac{(5.91\times10^{12}\text{ m})^2 - (5.79\times10^{10}\text{ m})^2}{(4.79\times10^4\text{ m/s})^2 - (4.74\times10^3\text{ m/s})^2}} = \sqrt{\frac{3.49\times10^{25}\text{ m}^2}{2.27\times10^9\text{ m}^2/\text{s}^2}} = 1.24\times10^8\text{ s} = \boxed{3.93\text{ yr}}.$$

P11.20 For the Earth, $\sum F = ma$: $\frac{GM_s m}{r^2} = \frac{mv^2}{r} = \frac{m}{r}\left(\frac{2\pi r}{T}\right)^2$.

Then $GM_s T^2 = 4\pi^2 r^3$.

Also the angular momentum $L = mvr = m\frac{2\pi r}{T}r$ is a constant for the Earth.

We eliminate $r = \sqrt{\frac{LT}{2\pi m}}$ between the equations:

$$GM_s T^2 = 4\pi^2\left(\frac{LT}{2\pi m}\right)^{3/2} \qquad GM_s T^{1/2} = 4\pi^2\left(\frac{L}{2\pi m}\right)^{3/2}.$$

Now the rate of change is described by

$$GM_s\left(\frac{1}{2}T^{-1/2}\frac{dT}{dt}\right) + G\left(1\frac{dM_s}{dt}T^{1/2}\right) = 0 \qquad \frac{dT}{dt} = -\frac{dM_s}{dt}\left(2\frac{T}{M_s}\right) \approx \frac{\Delta T}{\Delta t}$$

$$\Delta T \approx -\Delta t\frac{dM_s}{dt}\left(2\frac{T}{M_s}\right) = -5\,000\text{ yr}\left(\frac{3.16\times10^7\text{ s}}{1\text{ yr}}\right)\left(-3.64\times10^9\text{ kg/s}\right)\left(2\frac{1\text{ yr}}{1.991\times10^{30}\text{ kg}}\right)$$

$$\Delta T = \boxed{1.82\times10^{-2}\text{ s}}$$

Section 11.4 Energy Considerations in Planetary and Satellite Motion

P11.21 (a) $\rho = \frac{M_S}{\frac{4}{3}\pi r_E^2} = \frac{3\left(1.99\times10^{30}\text{ kg}\right)}{4\pi\left(6.37\times10^6\text{ m}\right)^3} = \boxed{1.84\times10^9\text{ kg/m}^3}$

(b) $g = \frac{GM_S}{r_E^2} = \frac{\left(6.67\times10^{-11}\text{ N}\cdot\text{m}^2/\text{kg}^2\right)\left(1.99\times10^{30}\text{ kg}\right)}{\left(6.37\times10^6\text{ m}\right)^2} = \boxed{3.27\times10^6\text{ m/s}^2}$

(c) $U_g = -\frac{GM_S m}{r_E} = -\frac{\left(6.67\times10^{-11}\text{ N}\cdot\text{m}^2/\text{kg}^2\right)\left(1.99\times10^{30}\text{ kg}\right)(1.00\text{ kg})}{6.37\times10^6\text{ m}} = \boxed{-2.08\times10^{13}\text{ J}}$

P11.22 $W = -\Delta U = -\left(\frac{-Gm_1m_2}{r} - 0\right)$

$$W = \frac{\left(+6.67\times10^{-11}\text{ N}\cdot\text{m}^2/\text{kg}^2\right)\left(7.36\times10^{22}\text{ kg}\right)\left(1.00\times10^3\text{ kg}\right)}{1.74\times10^6\text{ m}} = \boxed{2.82\times10^9\text{ J}}$$

P11.23 For her jump on earth,

$$\frac{1}{2}mv_i^2 = mgy_f \qquad (1)$$

$$v_i = \sqrt{2gy_f} = \sqrt{2(9.8\ \text{m/s}^2)(0.5\ \text{m})} = 3.13\ \text{m/s}$$

We assume that she has the same takeoff speed on the asteroid. Here

$$\frac{1}{2}mv_i^2 - \frac{GM_Am}{R_A} = 0 + 0 \qquad (2)$$

The equality of densities between planet and asteroid,

$$\rho = \frac{M_E}{\frac{4}{3}\pi R_E^2} = \frac{M_A}{\frac{4}{3}\pi R_A^2} \text{ implies} \qquad M_A = \left(\frac{R_A}{R_E}\right)^3 M_E \qquad (3)$$

Note also at Earth's surface

$$g = \frac{GM_E}{R_E^2} \qquad (4)$$

Combining the equations (2), (1), (3) and (4) by substitution gives

$$\frac{1}{2}v_i^2 = \frac{GM_A}{R_A} \qquad \frac{1}{2}(2gy_f) = \frac{G}{R_A}\left(\frac{R_A}{R_E}\right)^3 M_E$$

$$\frac{GM_E}{R_E^2}y_f = \frac{GM_ER_A^2}{R_E^3} \qquad R_A^2 = y_fR_E = (0.5\ \text{m})(6.37\times10^6\ \text{m})$$

$$R_A = \boxed{1.78\times10^3\ \text{m}}$$

P11.24 (a) The work must provide the increase in gravitational energy

$$\begin{aligned}
W &= \Delta U_g = U_{gf} - U_{gi} \\
&= -\frac{GM_EM_p}{r_f} + \frac{GM_EM_p}{r_i} \\
&= -\frac{GM_EM_p}{R_E + y} + \frac{GM_EM_p}{R_E} \\
&= GM_EM_p\left(\frac{1}{R_E} - \frac{1}{R_E + y}\right) \\
&= \left(\frac{6.67\times10^{-11}\ \text{N}\cdot\text{m}^2}{\text{kg}^2}\right)(5.98\times10^{24}\ \text{kg})(100\ \text{kg})\left(\frac{1}{6.37\times10^6\ \text{m}} - \frac{1}{7.37\times10^6\ \text{m}}\right) \\
W &= \boxed{850\ \text{MJ}}
\end{aligned}$$

continued on next page

(b) In a circular orbit, gravity supplies the centripetal force:

$$\frac{GM_E M_p}{(R_E+y)^2}=\frac{M_p v^2}{(R_E+y)}$$

Then, $\frac{1}{2}M_p v^2=\frac{1}{2}\frac{GM_E M_p}{(R_E+y)}$

So, additional work = kinetic energy required

$$=\frac{1}{2}\frac{(6.67\times10^{-11}\ \text{N}\cdot\text{m}^2)(5.98\times10^{24}\ \text{kg})(100\ \text{kg})}{(\text{kg}^2)(7.37\times10^6\ \text{m})}$$

$$\Delta W=\boxed{2.71\times10^9\ \text{J}}$$

P11.25 $\frac{1}{2}mv_i^2+GM_E m\left(\frac{1}{r_f}-\frac{1}{r_i}\right)=\frac{1}{2}mv_f^2$ $\qquad\frac{1}{2}v_i^2+GM_E\left(0-\frac{1}{R_E}\right)=\frac{1}{2}v_f^2$

or $\qquad v_f^2=v_1^2-\frac{2GM_E}{R_E}$

and $\qquad v_f=\left(v_1^2-\frac{2GM_E}{R_E}\right)^{1/2}$

$$v_f=\left[(2.00\times10^4)^2-1.25\times10^8\right]^{1/2}=\boxed{1.66\times10^4\ \text{m/s}}$$

P11.26 (a) $v_{\text{solar escape}}=\sqrt{\frac{2M_{\text{Sun}}G}{R_{\text{E-Sun}}}}=\boxed{42.1\ \text{km/s}}$

(b) Let $r=R_{E\text{-}S}x$ represent variable distance from the Sun, with x in astronomical units.

$$v=\sqrt{\frac{2M_{\text{Sun}}G}{R_{E\text{-}S}x}}=\frac{42.1}{\sqrt{x}}$$

If $v=\frac{125\,000\ \text{km}}{3\,600\ \text{s}}$, then $x=1.47\ \text{A.U.}=\boxed{2.20\times10^{11}\ \text{m}}$

(at or beyond the orbit of Mars, 125 000 km/h is sufficient for escape).

P11.27 To obtain the orbital velocity, we use

$$\sum F = \frac{mMG}{R^2} = \frac{mv^2}{R}$$

or

$$v = \sqrt{\frac{MG}{R}}$$

We can obtain the escape velocity from

$$\frac{1}{2}mv_{esc}^2 = \frac{mMG}{R}$$

or

$$v_{esc} = \sqrt{\frac{2MG}{R}} = \boxed{\sqrt{2}v}$$

***P11.28** (a) Energy conservation of the object-Earth system from release to radius r:

$$\left(K+U_g\right)_{\text{altitude } h} = \left(K+U_g\right)_{\text{radius } r}$$

$$0 - \frac{GM_E m}{R_E+h} = \frac{1}{2}mv^2 - \frac{GM_E m}{r}$$

$$v = \left(2GM_E\left(\frac{1}{r} - \frac{1}{R_E+h}\right)\right)^{1/2} = -\frac{dr}{dt}$$

(b) $\int_i^f dt = \int_i^f -\frac{dr}{v} = \int_f^i \frac{dr}{v}$. The time of fall is

$$\Delta t = \int_{R_E}^{R_E+h}\left(2GM_E\left(\frac{1}{r} - \frac{1}{R_E+h}\right)\right)^{-1/2} dr$$

$$\Delta t = \int_{6.37\times10^6\text{ m}}^{6.87\times10^6\text{ m}}\left[2\times6.67\times10^{-11}\times5.98\times10^{24}\left(\frac{1}{r} - \frac{1}{6.87\times10^6\text{ m}}\right)\right]^{-1/2} dr$$

We can enter this expression directly into a mathematical calculation program. Alternatively, to save typing we can change variables to $u = \frac{r}{10^6}$. Then

$$\Delta t = \left(7.977\times10^{14}\right)^{-1/2}\int_{6.37}^{6.87}\left(\frac{1}{10^6 u} - \frac{1}{6.87\times10^6}\right)^{-1/2} 10^6\,du = 3.541\times10^{-8}\frac{10^6}{\left(10^6\right)^{-1/2}}\int_{6.37}^{6.87}\left(\frac{1}{u} - \frac{1}{6.87}\right)^{-1/2} du$$

A mathematics program returns the value 9.596 for this integral, giving for the time of fall $\Delta t = 3.541\times10^{-8}\times10^9\times9.596 = 339.8 = \boxed{340\text{ s}}$.

P11.29 $$\frac{v_i^2}{R_E+h}=\frac{GM_E}{(R_E+h)^2}$$

$$K_i=\frac{1}{2}mv_i^2=\frac{1}{2}\left(\frac{GM_Em}{R_E+h}\right)=\frac{1}{2}\left[\frac{(6.67\times10^{-11}\ \text{N}\cdot\text{m}^2/\text{kg}^2)(5.98\times10^{24}\ \text{kg})(500\ \text{kg})}{(6.37\times10^6\ \text{m})+(0.500\times10^6\ \text{m})}\right]=1.45\times10^{10}\ \text{J}$$

The change in gravitational potential energy of the satellite-Earth system is

$$\Delta U=\frac{GM_Em}{R_i}-\frac{GM_Em}{R_f}=GM_Em\left(\frac{1}{R_i}-\frac{1}{R_f}\right)$$
$$=(6.67\times10^{-11}\ \text{N}\cdot\text{m}^2/\text{kg}^2)(5.98\times10^{24}\ \text{kg})(500\ \text{kg})(-1.14\times10^{-8}\ \text{m}^{-1})=-2.27\times10^9\ \text{J}$$

Also, $K_f=\frac{1}{2}mv_f^2=\frac{1}{2}(500\ \text{kg})(2.00\times10^3\ \text{m/s})^2=1.00\times10^9\ \text{J}$.

The energy transformed due to friction is

$$\Delta E_{int}=K_i-K_f-\Delta U=(14.5-1.00+2.27)\times10^9\ \text{J}=\boxed{1.58\times10^{10}\ \text{J}}.$$

P11.30 The gravitational force supplies the needed centripetal acceleration.

Thus, $\frac{GM_Em}{(R_E+h)^2}=\frac{mv^2}{(R_E+h)}$ or $v^2=\frac{GM_E}{R_E+h}$

(a) $T=\frac{2\pi r}{v}=\frac{2\pi(R_E+h)}{\sqrt{\frac{GM_E}{(R_E+h)}}}$ $\quad T=\boxed{2\pi\sqrt{\frac{(R_E+h)^3}{GM_E}}}$

(b) $v=\boxed{\sqrt{\frac{GM_E}{R_E+h}}}$

(c) Minimum energy input is $\Delta E_{min}=(K_f+U_{gf})-(K_i-U_{gi})$.

It is simplest to

launch the satellite from a location on the equator, and launch it toward the east. (boxed)

This choice has the object starting with energy $K_i=\frac{1}{2}mv_i^2$

with $v_i=\frac{2\pi R_E}{1.00\ \text{day}}=\frac{2\pi R_E}{86\,400\ \text{s}}$ and $U_{gi}=-\frac{GM_Em}{R_E}$.

Thus, $$\Delta E_{min}=\frac{1}{2}m\left(\frac{GM_E}{R_E+h}\right)-\frac{GM_Em}{R_E+h}-\frac{1}{2}m\left[\frac{4\pi^2R_E^2}{(86\,400\ \text{s})^2}\right]+\frac{GM_Em}{R_E}$$

or $$\Delta E_{min}=\boxed{GM_Em\left[\frac{R_E+2h}{2R_E(R_E+h)}\right]-\frac{2\pi^2R_E^2m}{(86\,400\ \text{s})^2}}$$

***P11.31** (a) Energy conservation for the object-Earth system from firing to apex:

$$\left(K+U_g\right)_i=\left(K+U_g\right)_f$$
$$\frac{1}{2}mv_i^2-\frac{GmM_E}{R_E}=0-\frac{GmM_E}{R_E+h}$$

where $\frac{1}{2}mv_{esc}^2=\frac{GmM_E}{R_E}$. Then

$$\frac{1}{2}v_i^2-\frac{1}{2}v_{esc}^2=-\frac{1}{2}v_{esc}^2\frac{R_E}{R_E+h}$$
$$v_{esc}^2-v_i^2=\frac{v_{esc}^2R_E}{R_E+h}$$
$$\frac{1}{v_{esc}^2-v_i^2}=\frac{R_E+h}{v_{esc}^2R_E}$$
$$h=\frac{v_{esc}^2R_E}{v_{esc}^2-v_i^2}-R_E=\frac{v_{esc}^2R_E-v_{esc}^2R_E+v_i^2R_E}{v_{esc}^2-v_i^2}$$
$$h=\frac{R_Ev_i^2}{v_{esc}^2-v_i^2}$$

(b) $$h=\frac{6.37\times10^6\text{ m }(8.76)^2}{(11.2)^2-(8.76)^2}=\boxed{1.00\times10^7\text{ m}}$$

(c) The fall of the meteorite is the time-reversal of the upward flight of the projectile, so it is described by the same energy equation

$$v_i^2=v_{esc}^2\left(1-\frac{R_E}{R_E+h}\right)=v_{esc}^2\left(\frac{h}{R_E+h}\right)=\left(11.2\times10^3\text{ m/s}\right)^2\left(\frac{2.51\times10^7\text{ m}}{6.37\times10^6\text{ m}+2.51\times10^7\text{ m}}\right)$$
$$=1.00\times10^8\text{ m}^2/\text{s}^2$$
$$v_i=\boxed{1.00\times10^4\text{ m/s}}$$

(d) With $v_i<<v_{esc}$, $h\approx\frac{R_Ev_i^2}{v_{esc}^2}=\frac{R_Ev_i^2R_E}{2GM_E}$. But $g=\frac{GM_E}{R_E^2}$, so $h=\frac{v_i^2}{2g}$, in agreement with $0^2=v_i^2+2(-g)(h-0)$.

P11.32 For a satellite in an orbit of radius r around the Earth, the total energy of the satellite-Earth system is $E=-\frac{GM_E}{2r}$. Thus, in changing from a circular orbit of radius $r=2R_E$ to one of radius $r=3R_E$, the required work is

$$W=\Delta E=-\frac{GM_Em}{2r_f}+\frac{GM_Em}{2r_i}=GM_Em\left[\frac{1}{4R_E}-\frac{1}{6R_E}\right]=\boxed{\frac{GM_Em}{12R_E}}.$$

P11.33 (a) The major axis of the orbit is $2a = 50.5 \text{ AU}$ so $a = 25.25 \text{ AU}$

Further, in Figure 11.5, $a + c = 50 \text{ AU}$ so $c = 24.75 \text{ AU}$

Then $$e = \frac{c}{a} = \frac{24.75}{25.25} = \boxed{0.980}$$

(b) In $T^2 = K_s a^3$ for objects in solar orbit, the Earth gives us

$$(1 \text{ yr})^2 = K_s (1 \text{ AU})^3 \qquad K_s = \frac{(1 \text{ yr})^2}{(1 \text{ AU})^3}$$

Then $$T^2 = \frac{(1 \text{ yr})^2}{(1 \text{ AU})^3}(25.25 \text{ AU})^3 \qquad T = \boxed{127 \text{ yr}}$$

(c) $$U = -\frac{GMm}{r} = -\frac{\left(6.67 \times 10^{-11} \text{ N} \cdot \text{m}^2/\text{kg}^2\right)\left(1.991 \times 10^{30} \text{ kg}\right)\left(1.2 \times 10^{10} \text{ kg}\right)}{50\left(1.496 \times 10^{11} \text{ m}\right)} = \boxed{-2.13 \times 10^{17} \text{ J}}$$

Section 11.5 Atomic Spectra and the Bohr Theory of Hydrogen

P11.34 (a) The energy of the photon is found as $$E = E_i - E_f = \frac{-13.606 \text{ eV}}{n_i^2} - \frac{(-13.606 \text{ eV})}{n_f^2}$$

$$E = 13.606 \text{ eV}\left(\frac{1}{n_f^2} - \frac{1}{n_i^2}\right)$$

Thus, for $n = 3$ to $n = 2$ transition $$E = 13.606 \text{ eV}\left(\frac{1}{4} - \frac{1}{9}\right) = \boxed{1.89 \text{ eV}}$$

(b) $E = \dfrac{hc}{\lambda}$ and $$\lambda = \frac{6.626 \times 10^{-34} \text{ J} \cdot \text{s}\left(2.998 \times 10^8 \text{ m/s}\right)}{1.89 \text{ eV}\left(1.602 \times 10^{-19} \text{ J/eV}\right)} = \boxed{656 \text{ nm}}$$

(c) $f = \dfrac{c}{\lambda}$ $$f = \frac{3 \times 10^8 \text{ m/s}}{6.56 \times 10^{-7} \text{ m}} = \boxed{4.57 \times 10^{14} \text{ Hz}}$$

P11.35 (a) Lyman series $\frac{1}{\lambda} = R\left(1 - \frac{1}{n_i^2}\right)$ $n_i = 2, 3, 4, \ldots$

$$\frac{1}{\lambda} = \frac{1}{94.96 \times 10^{-9}} = \left(1.097 \times 10^7\right)\left(1 - \frac{1}{n_i^2}\right) \qquad \boxed{n_i = 5}$$

(b) Paschen series: $\frac{1}{\lambda} = R\left(\frac{1}{3^2} - \frac{1}{n_i^2}\right)$ $n_i = 4, 5, 6, \ldots$

The shortest wavelength for this series corresponds to $n_i = \infty$ for ionization

$$\frac{1}{\lambda} = 1.097 \times 10^7\left(\frac{1}{9} - \frac{1}{n_i^2}\right)$$

For $n_i = \infty$, this gives $\lambda = 820$ nm

This is larger than 94.96 nm, so this wave length $\boxed{\text{cannot be associated with the Paschen series}}$.

Balmer series: $\frac{1}{\lambda} = R\left(\frac{1}{2^2} - \frac{1}{n_i^2}\right)$ $n_i = 3, 4, 5, \ldots$

$$\frac{1}{\lambda} = 1.097 \times 10^7\left(\frac{1}{4} - \frac{1}{n_i^2}\right)$$ with $n_i = \infty$ for ionization, $\lambda_{\text{min}} = 365$ nm

Once again the shorter given wavelength $\boxed{\text{cannot be associated with the Balmer series}}$.

P11.36 (a) $v_1 = \sqrt{\frac{k_e e^2}{m_e r_1}}$

where $r_1 = (1)^2 a_0 = 0.005\,29\text{ nm} = 5.29 \times 10^{-11}\text{ m}$

$$v_1 = \sqrt{\frac{\left(8.99 \times 10^9\text{ N}\cdot\text{m}^2/\text{C}^2\right)\left(1.60 \times 10^{-19}\text{ C}\right)^2}{\left(9.11 \times 10^{-31}\text{ kg}\right)\left(5.29 \times 10^{-11}\text{ m}\right)}} = \boxed{2.19 \times 10^6\text{ m/s}}$$

(b) $$K_1 = \frac{1}{2}m_e v_1^2 = \frac{1}{2}\left(9.11 \times 10^{-31}\text{ kg}\right)\left(2.19 \times 10^6\text{ m/s}\right)^2 = 2.18 \times 10^{-18}\text{ J} = \boxed{13.6\text{ eV}}$$

(c) $$U_1 = -\frac{k_e e^2}{r_1} = -\frac{\left(8.99 \times 10^9\text{ N}\cdot\text{m}^2/\text{C}^2\right)\left(1.60 \times 10^{-19}\text{ C}\right)^2}{5.29 \times 10^{-11}\text{ m}} = -4.35 \times 10^{-18}\text{ J} = \boxed{-27.2\text{ eV}}$$

P11.37 $\Delta E = (13.6\text{ eV})\left(\frac{1}{n_i^2} - \frac{1}{n_f^2}\right)$

Where for $\Delta E > 0$ we have absorption and for $\Delta E < 0$ we have emission.

(i) for $n_i = 2$ and $n_f = 5$, $\Delta E = 2.86\text{ eV}$ (absorption)

(ii) for $n_i = 5$ and $n_f = 3$, $\Delta E = -0.967\text{ eV}$ (emission)

(iii) for $n_i = 7$ and $n_f = 4$, $\Delta E = -0.572\text{ eV}$ (emission)

(iv) for $n_i = 4$ and $n_f = 7$, $\Delta E = 0.572\text{ eV}$ (absorption)

(a) $E = \frac{hc}{\lambda}$ so the shortest wavelength is emitted in transition $\boxed{\text{ii}}$.

(b) The atom gains most energy in transition $\boxed{\text{i}}$.

(c) The atom loses energy in transitions $\boxed{\text{ii and iii}}$.

P11.38 We use $E_n = \frac{-13.6\text{ eV}}{n^2}$.

To ionize the atom when the electron is in the nth level,

it is necessary to add an amount of energy given by $E = -E_n = \frac{13.6\text{ eV}}{n^2}$.

(a) Thus, in the ground state where $n = 1$, we have $\boxed{E = 13.6\text{ eV}}$.

(b) In the $n = 3$ level, $E = \frac{13.6\text{ eV}}{9} = \boxed{1.51\text{ eV}}$.

P11.39 (a) $r_2^2 = (0.052\,9\text{ nm})(2)^2 = \boxed{0.212\text{ nm}}$

(b) $m_e v_2 = \sqrt{\frac{m_e k_e e^2}{r_2}} = \sqrt{\frac{(9.11\times10^{-31}\text{ kg})(8.99\times10^9\text{ N}\cdot\text{m}^2/\text{C}^2)(1.60\times10^{-19}\text{ C})^2}{0.212\times10^{-9}\text{ m}}}$

$m_e v_2 = \boxed{9.95\times10^{-25}\text{ kg}\cdot\text{m/s}}$

(c) $L_2 = m_e v_2 r_2 = (9.95\times10^{-25}\text{ kg}\cdot\text{m/s})(0.212\times10^{-9}\text{ m}) = \boxed{2.11\times10^{-34}\text{ kg}\cdot\text{m}^2/\text{s}}$

(d) $K_2 = \frac{1}{2}m_e v_2^2 = \frac{(m_e v_2)^2}{2m_e} = \frac{(9.95\times10^{-25}\text{ kg}\cdot\text{m/s})^2}{2(9.11\times10^{-31}\text{ kg})} = 5.43\times10^{-19}\text{ J} = \boxed{3.40\text{ eV}}$

continued on next page

(e) $$U_2 = -\frac{k_e e^2}{r_2} = -\frac{\left(8.99\times10^9\ \text{N}\cdot\text{m}^2/\text{C}^2\right)\left(1.60\times10^{-19}\ \text{C}\right)^2}{0.212\times10^{-9}\ \text{m}} = -1.09\times10^{-18}\ \text{J} = \boxed{-6.80\ \text{eV}}$$

(f) $$E_2 = K_2 + U_2 = 3.40\ \text{eV} - 6.80\ \text{eV} = \boxed{-3.40\ \text{eV}}$$

P11.40 Starting with $$\frac{1}{2}m_e v^2 = \frac{k_e e^2}{2r}$$

we have $$v^2 = \frac{k_e e^2}{m_e r}$$

and using $$r_n = \frac{n^2\hbar^2}{m_e k_e e^2}$$

gives $$v_n^2 = \frac{k_e e^2}{m_e\left(n^2\hbar^2/m_e k_e e^2\right)}$$

or $$v_n = \frac{k_e e^2}{n\hbar}.$$

P11.41 Each atom gives up its kinetic energy in emitting a photon,

so $$\frac{1}{2}mv^2 = \frac{hc}{\lambda} = \frac{\left(6.626\times10^{-34}\ \text{J}\cdot\text{s}\right)\left(3.00\times10^8\ \text{m/s}\right)}{\left(1.216\times10^{-7}\ \text{m}\right)} = 1.63\times10^{-18}\ \text{J}$$

$$v = \boxed{4.42\times10^4\ \text{m/s}}.$$

Section 11.6 Context Connection—Changing from a Circular to an Elliptical Orbit

P11.42 The original orbit radius is $r = a = 6.37\times10^6\ \text{m} + 500\times10^3\ \text{m} = 6.87\times10^6\ \text{m}$. The original energy is

$$E_i = -\frac{GMm}{2a} = -\frac{\left(6.67\times10^{-11}\ \text{N}\cdot\text{m}^2/\text{kg}^2\right)\left(5.98\times10^{24}\ \text{kg}\right)\left(10^4\ \text{kg}\right)}{2\left(6.87\times10^6\ \text{m}\right)} = -2.90\times10^{11}\ \text{J}.$$

We assume that the perigee distance in the new orbit is $6.87\times10^6\ \text{m}$. Then the major axis is $2a = 6.87\times10^6\ \text{m} + 2.00\times10^7\ \text{m} = 2.69\times10^7\ \text{m}$ and the final energy is

$$E_f = -\frac{GMm}{2a} = -\frac{\left(6.67\times10^{-11}\ \text{N}\cdot\text{m}^2/\text{kg}^2\right)\left(5.98\times10^{24}\ \text{kg}\right)\left(10^4\ \text{kg}\right)}{2.69\times10^7\ \text{m}} = -1.48\times10^{11}\ \text{J}.$$

The energy input required from the engine is $E_f - E_i = -1.48\times10^{11}\ \text{J} - \left(-2.90\times10^{11}\ \text{J}\right) = \boxed{1.42\times10^{11}\ \text{J}}$.

P11.43 (a) Energy of the spacecraft-Mars system is conserved as the spacecraft moves between a very distant point and the point of closest approach:

$$0+0=\frac{1}{2}mv_r^{\ 2}-\frac{GM_{\text{Mars}}m}{r}$$

$$v_r=\sqrt{\frac{2GM_{\text{Mars}}}{r}}$$

After the engine burn, for a circular orbit we have

$$\Sigma F=ma: \quad \frac{GM_{\text{Mars}}m}{r^2}=\frac{mv_0^{\ 2}}{r}$$

$$v_0=\sqrt{\frac{GM_{\text{Mars}}}{r}}$$

The percentage reduction from the original speed is

$$\frac{v_r-v_0}{v_r}=\frac{\sqrt{2}v_0-v_0}{\sqrt{2}v_0}=\frac{\sqrt{2}-1}{\sqrt{2}}\times 100\%=\boxed{29.3\%}$$

(b) The answer to part (a) applies with $\boxed{\text{no changes}}$, as the solution to part (a) shows.

Additional Problems

P11.44 Let m represent the mass of the spacecraft, r_E the radius of the Earth's orbit, and x the distance from Earth to the spacecraft.

The Sun exerts on the spacecraft a radial inward force of $\quad F_s=\frac{GM_s m}{(r_E-x)^2}$

while the Earth exerts on it a radial outward force of $\quad F_E=\frac{GM_E m}{x^2}$

The net force on the spacecraft must produce the correct centripetal acceleration for it to have an orbital period of 1.000 year.

Thus,

$$F_S-F_E=\frac{GM_S m}{(r_E-x)^2}-\frac{GM_E m}{x^2}=\frac{mv^2}{(r_E-x)}=\frac{m}{(r_E-x)}\left[\frac{2\pi(r_E-x)}{T}\right]^2$$

which reduces to

$$\frac{GM_S}{(r_E-x)^2}-\frac{GM_E}{x^2}=\frac{4\pi^2(r_E-x)}{T^2}. \qquad (1)$$

Cleared of fractions, this equation would contain powers of x ranging from the fifth to the zeroth. We do not solve it algebraically. We may test the assertion that x is between 1.47×10^9 m and 1.48×10^9 m by substituting both of these as trial solutions, along with the following data: $M_S=1.991\times 10^{30}$ kg, $M_E=5.983\times 10^{24}$ kg, $r_E=1.496\times 10^{11}$ m, and $T=1.000\text{ yr}=3.156\times 10^7$ s.

continued on next page

With $x = 1.47 \times 10^9$ m substituted into equation (1), we obtain

$$6.052 \times 10^{-3}\ \text{m/s}^2 - 1.85 \times 10^{-3}\ \text{m/s}^2 \approx 5.871 \times 10^{-3}\ \text{m/s}^2$$

or $$5.868 \times 10^{-3}\ \text{m/s}^2 \approx 5.871 \times 10^{-3}\ \text{m/s}^2$$

With $x = 1.48 \times 10^9$ m substituted into the same equation, the result is

$$6.053 \times 10^{-3}\ \text{m/s}^2 - 1.82 \times 10^{-3}\ \text{m/s}^2 \approx 5.870\,8 \times 10^{-3}\ \text{m/s}^2$$

or $$5.870\,9 \times 10^{-3}\ \text{m/s}^2 \approx 5.870\,8 \times 10^{-3}\ \text{m/s}^2 .$$

Since the first trial solution makes the left-hand side of equation (1) slightly less than the right hand side, and the second trial solution does the opposite, the true solution is determined as between the trial values. To three-digit precision, it is 1.48×10^9 m.

As an equation of fifth degree, equation (1) has five roots. The Sun-Earth system has five Lagrange points, all revolving around the Sun synchronously with the Earth. The SOHO and ACE satellites are at one. Another is beyond the far side of the Sun. Another is beyond the night side of the Earth. Two more are on the Earth's orbit, ahead of the planet and behind it by 60°. Plans are under way to gain perspective on the Sun by placing a spacecraft at one of these two co-orbital Lagrange points. The Greek and Trojan asteroids are at the co-orbital Lagrange points of the Jupiter-Sun system.

P11.45 The acceleration of an object at the center of the Earth due to the gravitational force of the Moon is given by

$$a = G\frac{M_{\text{Moon}}}{d^2}$$

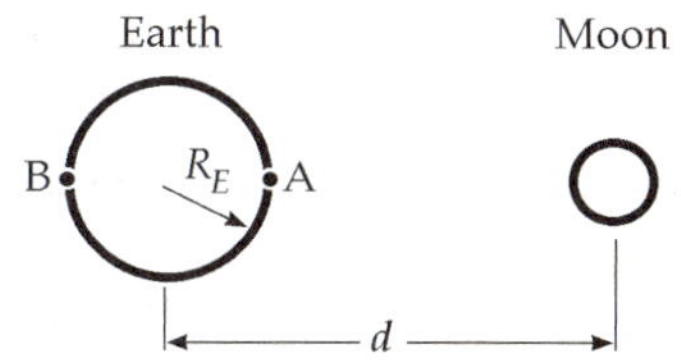

FIG. P11.45

At the point A nearest the Moon, $$a_+ = G\frac{M_M}{(d-r)^2}$$

At the point B farthest from the Moon, $$a_- = G\frac{M_M}{(d+r)^2}$$

$$\Delta a = a_+ - a = GM_M\left[\frac{1}{(d-r)^2} - \frac{1}{d^2}\right]$$

For $d >> r$, $$\Delta a = \frac{2GM_M r}{d^3} = 1.11 \times 10^{-6}\ \text{m/s}^2$$

Across the planet, $$\frac{\Delta g}{g} = \frac{2\Delta a}{g} = \frac{2.22 \times 10^{-6}\ \text{m/s}^2}{9.80\ \text{m/s}^2} = \boxed{2.26 \times 10^{-7}}$$

***P11.46** Energy conservation for the two-sphere system from release to contact:

$$-\frac{Gmm}{R}=-\frac{Gmm}{2r}+\frac{1}{2}mv^2+\frac{1}{2}mv^2$$

$$Gm\left(\frac{1}{2r}-\frac{1}{R}\right)=v^2 \qquad v=\left(Gm\left[\frac{1}{2r}-\frac{1}{R}\right]\right)^{1/2}$$

(a) The injected impulse is the final momentum of each sphere,

$$mv=m^{2/2}\left(Gm\left[\frac{1}{2r}-\frac{1}{R}\right]\right)^{1/2}=\left[Gm^3\left(\frac{1}{2r}-\frac{1}{R}\right)\right]^{1/2}.$$

(b) If they now collide elastically each sphere reverses its velocity to receive impulse

$$mv-(-mv)=2mv=\boxed{2\left[Gm^3\left(\frac{1}{2r}-\frac{1}{R}\right)\right]^{1/2}}$$

P11.47 (a) The free-fall acceleration produced by the Earth is $g=\frac{GM_E}{r^2}=GM_Er^{-2}$ (directed downward)

Its rate of change is $\frac{dg}{dr}=GM_E(-2)r^{-3}=-2GM_Er^{-3}$.

The minus sign indicates that g decreases with increasing height.

At the Earth's surface, $\boxed{\frac{dg}{dr}=-\frac{2GM_E}{R_E^3}}$.

(b) For small differences,

$\frac{|\Delta g|}{\Delta r}=\frac{|\Delta g|}{h}=\frac{2GM_E}{R_E^3}$ Thus, $\boxed{|\Delta g|=\frac{2GM_Eh}{R_E^3}}$

(c) $$|\Delta g|=\frac{2\left(6.67\times10^{-11}\ \text{N}\cdot\text{m}^2/\text{kg}^2\right)\left(5.98\times10^{24}\ \text{kg}\right)(6.00\ \text{m})}{\left(6.37\times10^6\ \text{m}\right)^3}=\boxed{1.85\times10^{-5}\ \text{m/s}^2}$$

***P11.48** (a) Each bit of mass dm in the ring is at the same distance from the object at A. The separate contributions $-\frac{Gmdm}{r}$ to the system energy add up to $-\frac{GmM_{\text{ring}}}{r}$. When the object is at A, this is

$$\frac{-6.67\times10^{-11}\ \text{N}\cdot\text{m}^2 1\,000\ \text{kg}\ 2.36\times10^{20}\ \text{kg}}{\text{kg}^2\sqrt{\left(1\times10^8\ \text{m}\right)^2+\left(2\times10^8\ \text{m}\right)^2}}=\boxed{-7.04\times10^4\ \text{J}}.$$

continued on next page

(b) When the object is at the center of the ring, the potential energy is

$$-\frac{6.67\times10^{-11}\ \text{N}\cdot\text{m}^2\ 1\,000\ \text{kg}\ 2.36\times10^{20}\ \text{kg}}{\text{kg}^2 1\times10^8\ \text{m}}=\boxed{-1.57\times10^5\ \text{J}}.$$

(c) Total energy of the object-ring system is conserved:

$$(K+U_g)_A=(K+U_g)_B$$

$$0-7.04\times10^4\ \text{J}=\frac{1}{2}1\,000\ \text{kg}v_B^2-1.57\times10^5\ \text{J}$$

$$v_B=\left(\frac{2\times8.70\times10^4\ \text{J}}{1\,000\ \text{kg}}\right)^{1/2}=\boxed{13.2\ \text{m/s}}$$

P11.49 To approximate the height of the sulfur, set

$$\frac{mv^2}{2}=mg_{Io}h \qquad h=70\,000\ \text{m} \qquad g_{Io}=\frac{GM}{r^2}=1.79\ \text{m/s}^2$$

$$v=\sqrt{2g_{Io}h} \qquad v=\sqrt{2(1.79)(70\,000)}\approx500\ \text{m/s}\ (\text{over }1\,000\ \text{mi/h})$$

A more precise answer is given by

$$\frac{1}{2}mv^2-\frac{GMm}{r_1}=-\frac{GMm}{r_2}$$

$$\frac{1}{2}v^2=\left(6.67\times10^{-11}\right)\left(8.90\times10^{22}\right)\left(\frac{1}{1.82\times10^6}-\frac{1}{1.89\times10^6}\right) \qquad v=\boxed{492\ \text{m/s}}$$

P11.50 From the walk, $2\pi r=25\,000\ \text{m}$. Thus, the radius of the planet is $r=\dfrac{25\,000\ \text{m}}{2\pi}=3.98\times10^3\ \text{m}$

From the drop: $\Delta y=\dfrac{1}{2}gt^2=\dfrac{1}{2}g(29.2\ \text{s})^2=1.40\ \text{m}$

so, $$g=\frac{2(1.40\ \text{m})}{(29.2\ \text{s})^2}=3.28\times10^{-3}\ \text{m/s}^2=\frac{MG}{r^2} \qquad \therefore M=\boxed{7.79\times10^{14}\ \text{kg}}$$

P11.51 For both circular orbits,

$$\sum F = ma: \qquad \frac{GM_E m}{r^2} = \frac{mv^2}{r}$$

$$v = \sqrt{\frac{GM_E}{r}}$$

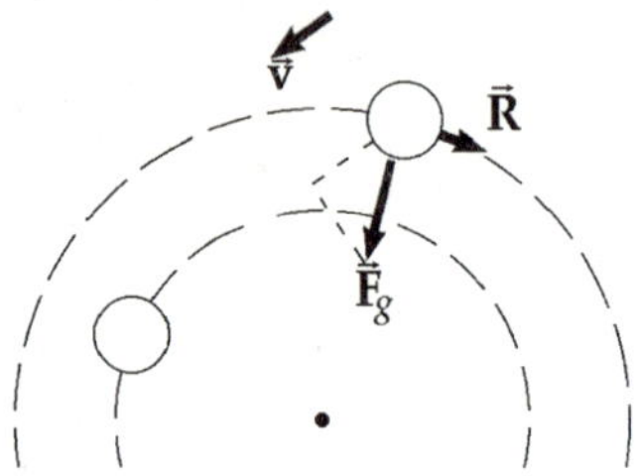

FIG. P11.51

(a) The original speed is $v_i = \sqrt{\frac{(6.67\times10^{-11}\ \text{N}\cdot\text{m}^2/\text{kg}^2)(5.98\times10^{24}\ \text{kg})}{(6.37\times10^6\ \text{m} + 2\times10^5\ \text{m})}} = \boxed{7.79\times10^3\ \text{m/s}}$.

(b) The final speed is $v_i = \sqrt{\frac{(6.67\times10^{-11}\ \text{N}\cdot\text{m}^2/\text{kg}^2)(5.98\times10^{24}\ \text{kg})}{(6.47\times10^6\ \text{m})}} = \boxed{7.85\times10^3\ \text{m/s}}$.

The energy of the satellite-Earth system is

$$K + U_g = \frac{1}{2}mv^2 - \frac{GM_E m}{r} = \frac{1}{2}m\frac{GM_E}{r} - \frac{GM_E}{r} = -\frac{GM_E m}{2r}$$

(c) Originally $E_i = -\frac{(6.67\times10^{-11}\ \text{N}\cdot\text{m}^2/\text{kg}^2)(5.98\times10^{24}\ \text{kg})(100\ \text{kg})}{2(6.57\times10^6\ \text{m})} = \boxed{-3.04\times10^9\ \text{J}}$.

(d) Finally $E_f = -\frac{(6.67\times10^{-11}\ \text{N}\cdot\text{m}^2/\text{kg}^2)(5.98\times10^{24}\ \text{kg})(100\ \text{kg})}{2(6.47\times10^6\ \text{m})} = \boxed{-3.08\times10^9\ \text{J}}$.

(e) Thus the object speeds up as it spirals down to the planet. The loss of gravitational energy is so large that the total energy decreases by

$$E_i - E_f = -3.04\times10^9\ \text{J} - (-3.08\times10^9\ \text{J}) = \boxed{4.69\times10^7\ \text{J}}.$$

(f) The only forces on the object are the backward force of air resistance R, comparatively very small in magnitude, and the force of gravity. Because the spiral path of the satellite is not perpendicular to the gravitational force, $\boxed{\text{one component of the gravitational force}}$ pulls forward on the satellite to do positive work and make its speed increase.

P11.52 (a) The net torque exerted on the Earth is zero. Therefore, the angular momentum of the Earth is conserved;

$$mr_a v_a = mr_p v_p \text{ and } v_a = v_p\left(\frac{r_p}{r_a}\right) = (3.027\times10^4\ \text{m/s})\left(\frac{1.471}{1.521}\right) = \boxed{2.93\times10^4\ \text{m/s}}$$

continued on next page

(b) $K_p = \frac{1}{2}mv_p^2 = \frac{1}{2}(5.98\times10^{24})(3.027\times10^4)^2 = \boxed{2.74\times10^{33}\text{ J}}$

$$U_p = -\frac{GmM}{r_p} = -\frac{(6.673\times10^{-11})(5.98\times10^{24})(1.99\times10^{30})}{1.471\times10^{11}} = \boxed{-5.40\times10^{33}\text{ J}}$$

(c) Using the same form as in part (b), $K_a = \boxed{2.57\times10^{33}\text{ J}}$ and $U_a = \boxed{-5.22\times10^{33}\text{ J}}$.

Compare to find that $K_p + U_p = \boxed{-2.66\times10^{33}\text{ J}}$ and $K_a + U_a = \boxed{-2.65\times10^{33}\text{ J}}$. They agree.

P11.53 (a) At infinite separation $U = 0$ and at rest $K = 0$. Since energy of the two-planet system is conserved we have,

$$0 = \frac{1}{2}m_1v_1^2 + \frac{1}{2}m_2v_2^2 - \frac{Gm_1m_2}{d} \quad (1)$$

The initial momentum of the system is zero and momentum is conserved.

Therefore, $$0 = m_1v_1 - m_2v_2 \quad (2)$$

Combine equations (1) and (2): $\boxed{v_1 = m_2\sqrt{\frac{2G}{d(m_1+m_2)}}}$ and $\boxed{v_2 = m_1\sqrt{\frac{2G}{d(m_1+m_2)}}}$

Relative velocity $$v_r = v_1 - (-v_2) = \boxed{\sqrt{\frac{2G(m_1+m_2)}{d}}}$$

(b) Substitute given numerical values into the equation found for v_1 and v_2 in part (a) to find

$v_1 = 1.03\times10^4$ m/s and $v_2 = 2.58\times10^3$ m/s

Therefore, $K_1 = \frac{1}{2}m_1v_1^2 = \boxed{1.07\times10^{32}\text{ J}}$ and $K_2 = \frac{1}{2}m_2v_2^2 = \boxed{2.67\times10^{31}\text{ J}}$

***P11.54** (a) Let R represent the radius of the asteroid. Then its volume is $\frac{4}{3}\pi R^3$ and its mass is $\rho\frac{4}{3}\pi R^3$.

For your orbital motion, $\sum F = ma$, $\frac{Gm_1m_2}{R^2} = \frac{m_2v^2}{R}$, $\frac{G\rho4\pi R^3}{3R^2} = \frac{v^2}{R}$

$$R = \left(\frac{3v^2}{G\rho4\pi}\right)^{1/2} = \left(\frac{3(8.5\text{ m/s})^2\text{ kg}^2\text{ m}^3}{6.67\times10^{-11}\text{ N}\cdot\text{m}^2(1\,100\text{ kg})4\pi}\right)^{1/2} = \boxed{1.53\times10^4\text{ m}}$$

(b) $\rho\frac{4}{3}\pi R^3 = (1\,100\text{ kg/m}^3)\frac{4}{3}\pi(1.53\times10^4\text{ m})^3 = \boxed{1.66\times10^{16}\text{ kg}}$

(c) $v = \frac{2\pi R}{T}$ $\quad T = \frac{2\pi R}{v} = \frac{2\pi(1.53\times10^4\text{ m})}{8.5\text{ m/s}} = \boxed{1.13\times10^4\text{ s}} = 3.15\text{ h}$

continued on next page

(d) Angular momentum is conserved for the asteroid-you system:

$$\sum L_i = \sum L_f$$

$$0 = m_2 v R - I\omega$$

$$0 = m_2 v R - \frac{2}{5} m_1 R^2 \frac{2\pi}{T_{\text{asteroid}}}$$

$$m_2 v = \frac{4\pi}{5} \frac{m_1 R}{T_{\text{asteroid}}}$$

$$T_{\text{asteroid}} = \frac{4\pi m_1 R}{5 m_2 v} = \frac{4\pi(1.66\times10^{16}\text{ kg})(1.53\times10^4\text{ m})}{5(90\text{ kg})(8.5\text{ m/s})} = \boxed{8.37\times10^{17}\text{ s}} = 26.5\text{ billion years}$$

This problem is realistic. Many asteroids, such as Ida and Eros, are roughly 30 km in diameter. They are typically irregular in shape and not spherical. Satellites such as Phobos (of Mars), Adrastea (of Jupiter), Calypso (of Saturn), and Ophelia (of Uranus) would allow a visitor the same experience of easy orbital motion. So would many Kuiper-belt objects.

P11.55 (a) $$T = \frac{2\pi r}{v} = \frac{2\pi(30\,000\times9.46\times10^{15}\text{ m})}{2.50\times10^5\text{ m/s}} = 7\times10^{15}\text{ s} = \boxed{2\times10^8\text{ yr}}$$

(b) $$M = \frac{4\pi^2 a^3}{GT^2} = \frac{4\pi^2(30\,000\times9.46\times10^{15}\text{ m})^3}{(6.67\times10^{-11}\text{ N}\cdot\text{m}^2/\text{kg}^2)(7.13\times10^{15}\text{ s})^2} = 2.66\times10^{41}\text{ kg}$$

$M = 1.34\times10^{11}$ solar masses $\boxed{\sim10^{11}\text{ solar masses}}$

The number of stars is $\boxed{\text{on the order of }10^{11}}$.

P11.56 (a) From the data about perigee, the energy of the satellite-Earth system is

$$E = \frac{1}{2}mv_p^2 - \frac{GM_E m}{r_p} = \frac{1}{2}(1.60)(8.23\times10^3)^2 - \frac{(6.67\times10^{-11})(5.98\times10^{24})(1.60)}{7.02\times10^6}$$

or $E = \boxed{-3.67\times10^7\text{ J}}$

(b) $$L = mvr\sin\theta = mv_p r_p \sin 90.0° = (1.60\text{ kg})(8.23\times10^3\text{ m/s})(7.02\times10^6\text{ m})$$

$$= \boxed{9.24\times10^{10}\text{ kg}\cdot\text{m}^2/\text{s}}$$

(c) Since both the energy of the satellite-Earth system and the angular momentum of the Earth are conserved,

at apogee we must have $\frac{1}{2}mv_a^2 - \frac{GMm}{r_a} = E$

and $mv_a r_a \sin 90.0° = L$.

continued on next page

Thus, $\frac{1}{2}(1.60)v_a^2 - \frac{(6.67\times10^{-11})(5.98\times10^{24})(1.60)}{r_a} = -3.67\times10^7\ \text{J}$

and $(1.60\ \text{kg})v_a r_a = 9.24\times10^{10}\ \text{kg}\cdot\text{m}^2/\text{s}$.

Solving simultaneously, $\frac{1}{2}(1.60)v_a^2 - \frac{(6.67\times10^{-11})(5.98\times10^{24})(1.60)(1.60)v_a}{9.24\times10^{10}} = -3.67\times10^7$

which reduces to $0.800v_a^2 - 11\,046v_a + 3.672\,3\times10^7 = 0$

so $v_a = \frac{11\,046 \pm \sqrt{(11\,046)^2 - 4(0.800)(3.672\,3\times10^7)}}{2(0.800)}$.

This gives $v_a = 8\,230\ \text{m/s}$ or $\boxed{5\,580\ \text{m/s}}$. The smaller answer refers to the velocity at the apogee while the larger refers to perigee.

Thus, $r_a = \frac{L}{mv_a} = \frac{9.24\times10^{10}\ \text{kg}\cdot\text{m}^2/\text{s}}{(1.60\ \text{kg})(5.58\times10^3\ \text{m/s})} = \boxed{1.04\times10^7\ \text{m}}$.

(d) The major axis is $2a = r_p + r_a$, so the semi-major axis is

$$a = \frac{1}{2}(7.02\times10^6\ \text{m} + 1.04\times10^7\ \text{m}) = \boxed{8.69\times10^6\ \text{m}}$$

(e) $T = \sqrt{\frac{4\pi^2 a^3}{GM_E}} = \sqrt{\frac{4\pi^2(8.69\times10^6\ \text{m})^3}{(6.67\times10^{-11}\ \text{N}\cdot\text{m}^2/\text{kg}^2)(5.98\times10^{24}\ \text{kg})}}$

$T = 8\,060\ \text{s} = \boxed{134\ \text{min}}$

P11.57 Let m represent the mass of the meteoroid and v_i its speed when far away. No torque acts on the meteoroid, so its angular momentum is conserved as it moves between the distant point and the point where it grazes the Earth, moving perpendicular to the radius:

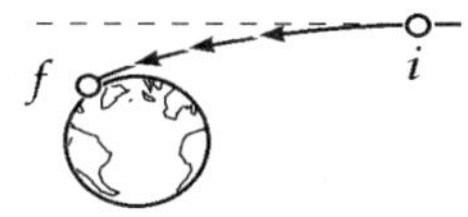

FIG. P11.57

$L_i = L_f$: $\quad m\vec{\mathbf{r}}_i \times \vec{\mathbf{v}}_i = m\vec{\mathbf{r}}_f \times \vec{\mathbf{v}}_f$

$$m(3R_E v_i) = mR_E v_f$$

$$v_f = 3v_i$$

Now energy of the meteoroid-Earth system is also conserved:

$(K + U_g)_i = (K + U_g)_f$: $\quad \frac{1}{2}mv_i^2 + 0 = \frac{1}{2}mv_f^2 - \frac{GM_E m}{R_E}$

$$\frac{1}{2}v_i^2 = \frac{1}{2}(9v_i^2) - \frac{GM_E}{R_E}$$

$\frac{GM_E}{R_E} = 4v_i^2$: $\quad \boxed{v_i = \sqrt{\frac{GM_E}{4R_E}}}$

P11.58 From Kepler's third law, minimum period means minimum orbit size. The "treetop satellite" in Figure P11.27 has minimum period. The radius of the satellite's circular orbit is essentially equal to the radius R of the planet.

$\sum F = ma$:
$$\frac{GMm}{R^2} = \frac{mv^2}{R} = \frac{m}{R}\left(\frac{2\pi R}{T}\right)^2$$

$$G\rho V = \frac{R^2\left(4\pi^2 R^2\right)}{RT^2}$$

$$G\rho\left(\frac{4}{3}\pi R^3\right) = \frac{4\pi^2 R^3}{T^2}$$

The radius divides out: $T^2 G\rho = 3\pi$ $\boxed{T = \sqrt{\frac{3\pi}{G\rho}}}$

P11.59 If we choose the coordinate of the center of mass at the origin, then

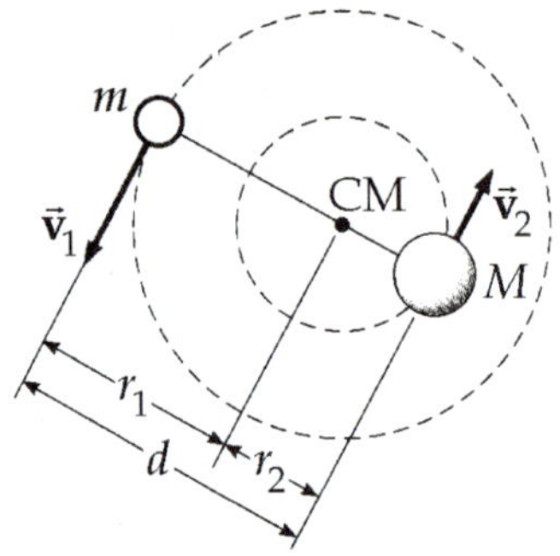

FIG. P11.59

$$0 = \frac{(Mr_2 - mr_1)}{M+m} \quad \text{and} \quad Mr_2 = mr_1$$

(Note: this is equivalent to saying that the net torque must be zero and the two experience no angular acceleration.) For each mass $F = ma$ so

$$mr_1\omega_1^2 = \frac{MGm}{d^2} \quad \text{and} \quad Mr_2\omega_2^2 = \frac{MGm}{d^2}$$

Adding these two equations gives $(r_1 + r_2)\omega^2 = \frac{(M+m)G}{d^2}$

with $\omega_1 = \omega_2 = \omega$. Now using $d = r_1 + r_2$

and $T = \frac{2\pi}{\omega}$

we find $\boxed{T^2 = \frac{4\pi^2 d^3}{G(M+m)}}$.

***P11.60** (a) The gravitational force exerted on m_2 by the Earth (mass m_1) accelerates m_2 according to: $m_2 g_2 = \frac{Gm_1 m_2}{r^2}$. The equal magnitude force exerted on the Earth by m_2 produces negligible acceleration of the Earth. The acceleration of relative approach is then

$$g_2 = \frac{Gm_1}{r^2} = \frac{\left(6.67\times10^{-11}\ \text{N}\cdot\text{m}^2/\text{kg}^2\right)\left(5.98\times10^{24}\ \text{kg}\right)}{\left(1.20\times10^7\ \text{m}\right)^2} = \boxed{2.77\ \text{m/s}^2}.$$

(b) Again, m_2 accelerates toward the center of mass with $g_2 = 2.77\ \text{m/s}^2$. Now the Earth accelerates toward m_2 with an acceleration given as

continued on next page

$$m_1 g_1 = \frac{G m_1 m_2}{r^2}$$

$$g_1 = \frac{G m_2}{r^2} = \frac{\left(6.67\times 10^{-11}\ \text{N}\cdot\text{m}^2/\text{kg}^2\right)\left(2.00\times 10^{24}\ \text{kg}\right)}{\left(1.20\times 10^{7}\ \text{m}\right)^2} = 0.926\ \text{m/s}^2$$

The distance between the masses closes with relative acceleration of

$$g_{\text{rel}} = g_1 + g_2 = 0.926\ \text{m/s}^2 + 2.77\ \text{m/s}^2 = \boxed{3.70\ \text{m/s}^2}.$$

P11.61 Let r represent the distance between the electron and the positron. The two move in a circle of radius $\frac{r}{2}$ around their center of mass with opposite velocities. The total angular momentum of the electron-positron system is quantized according to

$$L_n = \frac{mvr}{2} + \frac{mvr}{2} = n\hbar$$

where $n = 1, 2, 3, \ldots$.

For each particle, $\sum F = ma$ expands to $\frac{k_e e^2}{r^2} = \frac{mv^2}{r/2}$.

We can eliminate $v = \frac{n\hbar}{mr}$ to find $\frac{k_e e^2}{r} = \frac{2mn^2\hbar^2}{m^2 r^2}$.

So the separation distances are $r = \frac{2n^2\hbar^2}{mk_e e^2} = 2a_0 n^2 = \boxed{\left(1.06\times 10^{-10}\ \text{m}\right)n^2}$.

The orbital radii are $\frac{r}{2} = a_0 n^2$, the same as for the electron in hydrogen.

The energy can be calculated from $E = K + U = \frac{1}{2}mv^2 + \frac{1}{2}mv^2 - \frac{k_e e^2}{r}$.

Since $mv^2 = \frac{k_e e^2}{2r}$, $E = \frac{k_e e^2}{2r} - \frac{k_e e^2}{r} = -\frac{k_e e^2}{2r} = \frac{-k_e e^2}{4a_0 n^2} = \boxed{-\frac{6.80\ \text{eV}}{n^2}}$.

ANSWERS TO EVEN PROBLEMS

P11.2 (a) 2.50×10^{-5} N toward the 500-kg mass; (b) between the masses and 0.245 m from the 500 kg mass

P11.4 $\left(-10.0\hat{\mathbf{i}} + 5.93\hat{\mathbf{j}}\right)\times 10^{-11}$ N

P11.6 see the solution, either 1.000 m − 61.3 nm or 2.74×10^{-4} m

P11.8 (a) 3.46×10^{8} m; (b) $3.34\times 10^{-3}\ \text{m/s}^2$ toward the Earth

P11.10 (a) 1.31×10^{17} N toward the black hole; (b) 2.62×10^{12} N/kg

P11.12 (a) 5.59×10^{3} m/s; (b) 239 min; (c) 735 N downward

P11.14 1.27

P11.16 35.2 AU

P11.18 planet Y has turned through 1.30 revolutions

P11.20 18.2 ms

P11.22 2.82×10^{9} J

P11.24 (a) 850 MJ; (b) 2.71×10^{9} J

P11.26 (a) 42.1 km/s; (b) 2.20×10^{11} m

P11.28 (a) see the solution; (b) 340 s

P11.30 (a) $2\pi\sqrt{\frac{(R_E+h)^3}{GM_E}}$; (b) $\sqrt{\frac{GM_E}{R_E+h}}$;
(c) $GM_Em\left[\frac{R_E+2h}{2R_E(R_E+h)}\right]-\frac{2\pi^2R_E^2m}{(86\,400\text{ s})^2}$,
launch the satellite from a location on the equator, and launch it toward the east

P11.32 $\frac{GM_Em}{12R_E}$

P11.34 (a) 1.89 eV; (b) 656 nm; (c) 4.57×10^{14} Hz

P11.36 (a) 2.19×10^{6} m/s; (b) 13.6 eV; (c) –27.2 eV

P11.38 (a) 13.6 eV; (b) 1.51 eV

P11.40 see the solution

P11.42 1.42×10^{11} J

P11.44 see the solution

P11.46 (a) see the solution;
(b) $2\left[Gm^3\left(\frac{1}{2r}-\frac{1}{R}\right)\right]^{1/2}$

P11.48 (a) -7.04×10^{4} J; (b) -1.57×10^{5} J; (c) 13.2 m/s

P11.50 7.79×10^{14} kg

P11.52 (a) 2.93×10^{4} m/s; (b) $K=2.74\times10^{33}$ J, $U=-5.40\times10^{33}$ J;
(c) $K=2.57\times10^{33}$ J, $U=-5.22\times10^{33}$ J, $K_a+U_a=-2.65\times10^{33}$ J, total energy is constant

P11.54 (a) 1.53×10^{4} m; (b) 1.66×10^{16} kg; (c) 1.13×10^{4} s; (d) 8.37×10^{17} s

P11.56 (a) -3.67×10^{7} J; (b) 9.24×10^{10} kg·m²/s;
(c) 5 580 m/s, 1.04×10^{7} m;
(d) 8.69×10^{6} m; (e) 134 min

P11.58 see the solution

P11.60 (a) 2.77 m/s^2; (b) 3.70 m/s^2

CONTEXT 2 CONCLUSION ANSWERS TO QUESTIONS

QCC2.1 The hypothetical anti-Earth will be right here six months from now. Your spacecraft can be here then by leaving the Earth's orbit and slowing down relative to the Sun, to fall into an elliptical orbit around the Sun. This point in the Earth's orbit is the aphelion of the spacecraft's new orbit. For its period to be 1/2 year, we want

$$T^2=\left(\frac{4\pi^2}{GM_S}\right)a^3=\left(2.97\times10^{-19}\text{ s}^2/\text{m}^3\right)a^3$$

$$a^3=\frac{\left(\frac{1}{2}(3.156\times10^{7}\text{ s})\right)^2}{2.97\times10^{-19}\text{ s}^2/\text{m}^3}=8.38\times10^{32}\text{ m}^3$$

$$a=9.43\times10^{10}\text{ m}$$

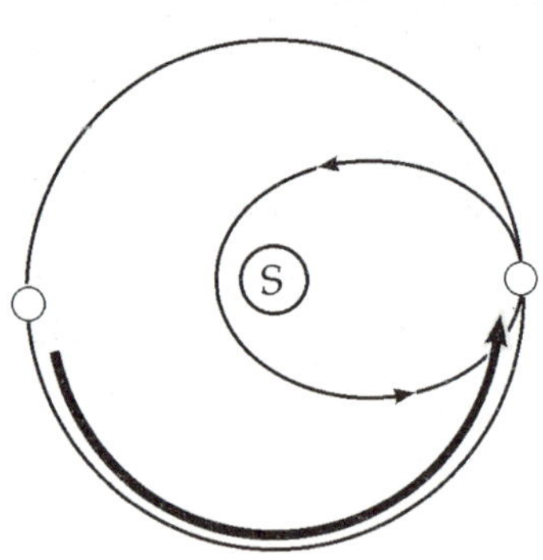

FIG. QCC2.1

continued on next page

Then the major axis is $2a = 1.89\times10^{11}$ m. The perihelion distance is $1.89\times10^{11}\ \text{m} - 1.496\times10^{11}\ \text{m} = 3.9\times10^{10}\ \text{m}$, less than the distance from the Sun to Mercury, so we will want some insulation, air conditioning and a highly reflective surface for the spacecraft. From its current speed in solar orbit,

$$v = \frac{2\pi\left(1.496\times10^{11}\ \text{m}\right)}{3.156\times10^{7}\ \text{s}} = 2.98\times10^{4}\ \text{m/s},$$

the spacecraft needs to drop to a speed given by

$$E = K + U_g$$
$$-\frac{GMm}{2a} = \frac{1}{2}mv^2 - \frac{GMm}{r}$$
$$v^2 = 2GM\left(\frac{1}{r} - \frac{1}{2a}\right)$$
$$v = \sqrt{2\left(6.67\times10^{-11}\ \text{N}\cdot\text{m}^2/\text{kg}^2\right)\left(1.991\times10^{30}\ \text{kg}\right)\left(\frac{1}{1.496\times10^{11}\ \text{m}} - \frac{1}{1.89\times10^{11}\ \text{m}}\right)}$$
$$= 1.92\times10^{4}\ \text{m/s}$$

QCC2.2 If you can survive for an hour, the commander of the other shuttle can most simply give the picnic basket a very slow push (say at 0.5 m/s) tangent to the orbit. Both space shuttles and the basket are in free fall around the Earth. Since 0.5 m/s is negligible in comparison to orbital speed, the basket will maintain nearly the same orbit as the spacecrafts, to bump into yours or pass close to it. More precisely, in this maneuver the speed of the basket would be not quite enough for the circular orbit. It would start at the apogee of a slightly eccentric eclipse and would pass below your spacecraft, perhaps just out of reach. Your benefactor could compensate by doing a calculation and launching the basket with any convenient tangential speed toward you and with a small appropriately chosen radially outward speed component.

CONTEXT 2 CONCLUSION SOLUTIONS TO PROBLEMS

CC2.1 The Hohmann transfer path is one half of an ellipse with major axis:

$$r_{\text{Earth}} + r_{\text{Venus}} = 1.496\times10^{11}\ \text{m} + 1.08\times10^{11}\ \text{m} = 2a$$

Then $a = 1.29\times10^{11}$ m. Now $T^2 = \dfrac{4\pi^2a^3}{GM_S}$ gives for the transfer time

$$\frac{1}{2}T = \frac{1}{2}\sqrt{\frac{4\pi^2\left(1.29\times10^{11}\ \text{m}\right)^3}{\left(6.67\times10^{-11}\ \text{N}\cdot\text{m}^2/\text{kg}^2\right)\left(1.991\times10^{30}\ \text{kg}\right)}}$$
$$= 1.26\times10^{7}\ \text{s} = \boxed{146\ \text{d}}$$

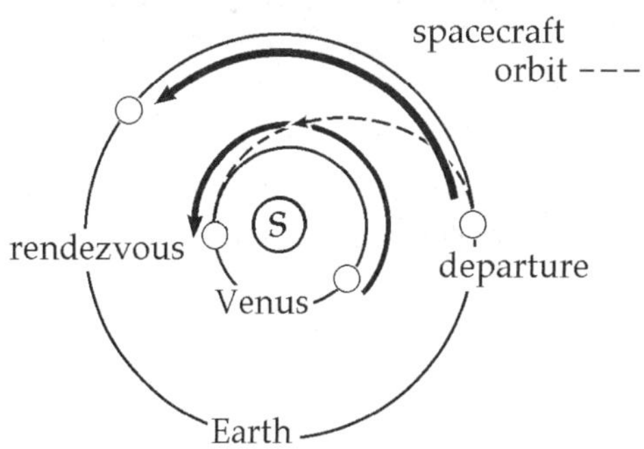

FIG. CC2.1

continued on next page

During this time Venus will move around in its orbit by $\theta = \omega t = 360°\left(\frac{1.26\times10^7\ \text{s}}{1.94\times10^7\ \text{s}}\right) = 234°$. The spacecraft departure and rendezvous points are 180° apart, so the spacecraft should be sent off when $\boxed{\text{Venus is behind}}$ the Earth by $234°-180° = \boxed{53.8°}$.

***CC2.2** The radius of the circular orbit of the space station is

$$r = R_E + 500\ \text{km} = 6.37\times10^6\ \text{m} + 5.00\times10^5\ \text{m} = 6.87\times10^6\ \text{m}.$$

From this, as in Example 11.1, we can find the orbital speed of the space station, which is the initial speed v_i of the golf ball:

$$v_i = \sqrt{\frac{GM_E}{r}} = \sqrt{\frac{\left(6.67\times10^{-11}\ \text{N}\cdot\text{m}^2/\text{kg}^2\right)\left(5.98\times10^{24}\ \text{kg}\right)}{6.87\times10^6\ \text{m}}} = 7.62\times10^3\ \text{m/s}.$$

The initial energy of the golf ball-Earth system, using Equation 11.8, is $E = -\frac{GM_Em}{2r}$ where m is the mass of the golf ball.

After the golf ball is hit, it returns to precisely the same location after the space station has made 2.00 orbits. Thus, the period of the golf ball is 2.00 times that of the space station:

$$\frac{T_b}{T_s} = 2.00 \rightarrow \left(\frac{T_b}{T_s}\right)^2 = 4.00.$$

According to Kepler's third law, the ratio of the squares of the periods of the golf ball and the space station should be the equal to the ratio of the cubes of the semimajor axes:

$$\left(\frac{T_b}{T_s}\right)^2 = \left(\frac{a_b}{a_s}\right)^3 = 4.00 \quad \rightarrow \quad a_b = (4.00)^{1/3}a_s = 1.587a_s = 1.587r$$

where we have identified the semimajor axis of the orbit of the space station as the radius r of its circular orbit. From the semimajor axis of the golf ball orbit, we can find the new energy of the golf ball-Earth system, using Equation 11.9: $E = -\frac{GM_Em}{2a_b}$. Just after the golf ball is hit, the golf ball-Earth system has the same potential energy as it did before the ball was hit, so the difference in energy must be equal to the change in kinetic energy of the golf ball:

$$\Delta E = \Delta K = -\frac{-GM_Em}{2a_b} - \left(-\frac{GM_Em}{2r}\right) = -\frac{GM_Em}{2(1.587r)} + \frac{GM_Em}{2r}$$

$$= \frac{GM_Em}{2r}\left(-\frac{1}{1.587} + 1\right) = 0.185\frac{GM_Em}{r}$$

Setting this equal to $\Delta K = \frac{1}{2}mv_f^2 - \frac{1}{2}mv_i^2$, we find,

continued on next page

$$\frac{1}{2}mv_f^2 - \frac{1}{2}mv_i^2 = 0.185\frac{GM_E m}{r}$$

$$v_f = \sqrt{v_i^2 + 0.370\frac{GM_E}{r}} = \sqrt{v_i^2 + 0.370v_i^2} = 1.17v_i$$

Thus, the speed of the golf ball relative to the space station is,

$$\Delta v = v_f - v_i = 1.17v_i - v_i = 0.17v_i = 0.17\left(7.62\times10^3 \text{ m/s}\right) = \boxed{1.30\times10^3 \text{ m/s}}$$

CC2.3 (a) The speed of an object in a circular orbit at the Earth's distance is given by

$$\Sigma F = ma: \quad \frac{GM_{\text{Sun}}m}{r_E^{\,2}} = \frac{mv_2^{\,2}}{r_E}$$

$$v_2 = \sqrt{\frac{GM_{\text{Sun}}}{r_E}} = \sqrt{\frac{\left(6.67\times10^{-11} \text{ N}\cdot\text{m}^2/\text{kg}^2\right)\left(1.991\times10^{30} \text{ kg}\right)}{1.496\times10^{11} \text{ m}}}$$

$$= 2.98\times10^4 \text{ m/s}$$

For the transfer orbit at perihelion, the speed is described by

$$E = K + U_g \qquad -\frac{GM_{\text{Sun}}m}{2a} = \frac{1}{2}mv_3^{\,2} - \frac{GM_{\text{Sun}}m}{r_E}$$

$$v_3 = \sqrt{2GM_{\text{Sun}}\left(\frac{1}{r_E} - \frac{1}{r_E + r_M}\right)}$$

$$v_3 = \sqrt{2\left(6.67\times10^{-11} \text{ N}\cdot\text{m}^2/\text{kg}^2\right)\left(1.991\times10^{30} \text{ kg}\right)\left(\frac{1}{1.496\times10^{11} \text{ m}} - \frac{1}{\left(1.496\times10^{11} + 2.28\times10^{11}\right)\text{m}}\right)}$$

$$v_3 = 3.27\times10^4 \text{ m/s}$$

The speed change required is $v_3 - v_2 = \boxed{2.95\times10^3 \text{ m/s}}$.

(b) The craft's aphelion speed is $v_4 = \frac{v_3 r_E}{r_M} = \frac{3.27\times10^4 \text{ m/s}\left(1.496\times10^{11} \text{ m}\right)}{\left(2.28\times10^{11} \text{ m}\right)} = 2.15\times10^4 \text{ m/s}$. The speed for a circular orbit at the distance of Mars is

$$v_5 = \sqrt{\frac{GM_{\text{Sun}}}{r_M}} = \sqrt{\frac{\left(6.67\times10^{-11} \text{ N}\cdot\text{m}^2/\text{kg}^2\right)\left(1.991\times10^{30} \text{ kg}\right)}{2.28\times10^{11} \text{ m}}} = 2.41\times10^4 \text{ m/s}.$$

The change in speed required at rendezvous to fly along with Mars is $v_5 - v_4 = \boxed{2.65\times10^3 \text{ m/s}}$.

continued on next page

(c) Consider the Earth as a frame of reference, rather than the Sun. The speed of a launch point on the equator is $v_0 = \frac{2\pi R_E}{1\text{ d}} = \frac{2\pi(6.37\times10^6\text{ m})}{86\,400\text{ s}} = 463\text{ m/s}$. Escape speed from the Earth's gravity is given by

$$\frac{1}{2}mv_1^2 = \frac{GM_E m}{R_E}$$

$$v_1 = \sqrt{\frac{2GM_E}{R_E}} = \sqrt{\frac{2(6.67\times10^{-11}\text{ N}\cdot\text{m}^2/\text{kg}^2)(5.98\times10^{24}\text{ kg})}{6.37\times10^6\text{ m}}} = 1.12\times10^4\text{ m/s}$$

So the launch speed change should be $v_1 - v_0 = \boxed{10.7\times10^3\text{ m/s}}$.

(d) In the reference frame of Mars, imagine a spacecraft at rest with respect to the planet and far from it, then falling in to the surface and making a soft landing at the equator, headed east. The speed it would gain as it falls is given by

$$0+0 = \frac{1}{2}mv_6^2 - \frac{GM_M m}{R_M}$$

$$v_6 = \sqrt{\frac{2GM_M}{R_M}} = \sqrt{\frac{2(6.67\times10^{-11}\text{ N}\cdot\text{m}^2/\text{kg}^2)(6.42\times10^{23}\text{ kg})}{3.37\times10^6\text{ m}}} = 5.04\times10^3\text{ m/s}$$

It should land with speed $v_7 = \frac{2\pi(3.37\times10^6\text{ m})}{24.6\text{ h}(3\,600\text{ s/1 h})} = 239\text{ m/s}$. So it must change speed by $|v_6 - v_7| = \boxed{4.80\times10^3\text{ m/s}}$. Comparison of our answers shows that the most fuel is used in taking off from Earth.

ANSWERS TO EVEN CONTEXT 2 CONCLUSION PROBLEMS

CC2.2 $1.30\times10^3\text{ m/s}$

Chapter 12

Oscillatory Motion

CHAPTER OUTLINE

ANSWERS TO QUESTIONS

Q12.1 Neither are examples of simple harmonic motion, although they are both periodic motion. In neither case is the acceleration proportional to the position. Neither motion is so smooth as SHM. The ball's acceleration is very large when it is in contact with the floor, and the student's when the dismissal bell rings.

Q12.2 The two will be equal if and only if the position of the particle at time zero is its equilibrium position, which we choose as the origin of coordinates.

Q12.3 You can take $\phi = \pi$, or equally well, $\phi = -\pi$. At $t = 0$, the particle is at its turning point on the negative side of equilibrium, at $x = -A$.

Q12.4 No. It is necessary to know both the position and velocity at time zero.

Q12.5 (a) In simple harmonic motion, one-half of the time, the velocity is in the same direction as the displacement away from equilibrium.

(b) Velocity and acceleration are in the same direction half the time.

(c) Acceleration is always opposite to the position vector, and never in the same direction.

Q12.6 We assume that the coils of the spring do not hit one another. The frequency will be higher than f by the factor $\sqrt{2}$. When the spring with two blocks is set into oscillation in space, the coil in the center of the spring does not move. We can imagine clamping the center coil in place without affecting the motion. We can effectively duplicate the motion of each individual block in space by hanging a single block on a half-spring here on Earth. The half-spring with its center coil clamped—or its other half cut off—has twice the spring constant as the original uncut spring, because an applied force of the same size would produce only one-half the extension distance. Thus the oscillation frequency in space is $\left(\frac{1}{2\pi}\right)\left(\frac{2k}{m}\right)^{1/2} = \sqrt{2}f$. The absence of a force required to support the vibrating system in orbital free fall has no effect on the frequency of its vibration.

Q12.7 No; Kinetic, Yes; Potential, No. For constant amplitude, the total energy $\frac{1}{2}kA^2$ stays constant. The kinetic energy $\frac{1}{2}mv^2$ would increase for larger mass if the speed were constant, but here the greater mass causes a decrease in frequency and in the average and maximum speed, so that the kinetic and potential energies at every point are unchanged.

Q12.8 Since the acceleration is not constant in simple harmonic motion, none of the equations in Table 2.2 are valid.

Equation	*Information given by equation*
$x(t)=A\cos(\omega t+\phi)$	position as a function of time
$v(t)=-\omega A\sin(\omega t+\phi)$	velocity as a function of time
$v(x)=\pm\omega\left(A^2-x^2\right)^{1/2}$	velocity as a function of position
$a(t)=-\omega^2 A\cos(\omega t+\phi)$	acceleration as a function of time
$a(x)=-\omega^2 x(t)$	acceleration as a function of position

The angular frequency ω appears in every equation. It is a good idea to figure out the value of angular frequency early in the solution to a problem about vibration, and to store it in calculator memory.

Q12.9 We have $T_i=\sqrt{\frac{L_i}{g}}$ and $T_f=\sqrt{\frac{L_f}{g}}=\sqrt{\frac{2L_i}{g}}=\sqrt{2}T_i$. The period gets larger by $\sqrt{2}$ times. Changing the mass has no effect on the period of a simple pendulum.

Q12.10 Shorten the pendulum to decrease the period between ticks.

Q12.11 No. If the resistive force is greater than the restoring force of the spring (in particular, if $b^2>4mk$), the system will be overdamped and will not oscillate.

Q12.12 Higher frequency. When it supports your weight, the center of the diving board flexes down less than the end does when it supports your weight. Thus the stiffness constant describing the center of the board is greater than the stiffness constant describing the end. And then $f=\left(\frac{1}{2\pi}\right)\sqrt{\frac{k}{m}}$ is greater for you bouncing on the center of the board.

Q12.13 Yes. An oscillator with damping can vibrate at resonance with amplitude that remains constant in time. Without damping, the amplitude would increase without limit at resonance.

Q12.14 An imperceptibly slight breeze may be blowing past the leaves in tiny puffs. As a leaf twists in the wind, the fibers in its stem provide a restoring torque. If the frequency of the breeze matches the natural frequency of vibration of one particular leaf as a torsional pendulum, that leaf can be driven into a large-amplitude resonance vibration. Note that it is not the *size* of the driving force that sets the leaf into resonance, but the *frequency* of the driving force. If the frequency changes, another leaf will be set into resonant oscillation.

SOLUTIONS TO PROBLEMS

Section 12.1 Motion of a Particle Attached to a Spring

P12.1 (a) Since the collision is perfectly elastic, the ball will rebound to the height of 4.00 m and then repeat the motion over and over again. Thus, the $\boxed{\text{motion is periodic}}$.

(b) To determine the period, we use: $x = \frac{1}{2}gt^2$.

The time for the ball to hit the ground is $t = \sqrt{\frac{2x}{g}} = \sqrt{\frac{2(4.00\text{ m})}{9.80\text{ m/s}^2}} = 0.904\text{ s}$

This equals one-half the period, so $T = 2(0.904\text{ s}) = \boxed{1.81\text{ s}}$.

(c) $\boxed{\text{No}}$. The net force acting on the ball is a constant given by $F = -mg$ (except when it is in contact with the ground), which is not in the form of Hooke's law.

Section 12.2 Mathematical Representation of Simple Harmonic Motion

P12.2 (a) $x = (5.00\text{ cm})\cos\left(2t + \frac{\pi}{6}\right)$ At $t = 0$, $x = (5.00\text{ cm})\cos\left(\frac{\pi}{6}\right) = \boxed{4.33\text{ cm}}$

(b) $v = \frac{dx}{dt} = -(10.0\text{ cm/s})\sin\left(2t + \frac{\pi}{6}\right)$ At $t = 0$, $v = \boxed{-5.00\text{ cm/s}}$

(c) $a = \frac{dv}{dt} = -\left(20.0\text{ cm/s}^2\right)\cos\left(2t + \frac{\pi}{6}\right)$ At $t = 0$, $a = \boxed{-17.3\text{ cm/s}^2}$

(d) $A = \boxed{5.00\text{ cm}}$ and $T = \frac{2\pi}{\omega} = \frac{2\pi}{2} = \boxed{3.14\text{ s}}$

P12.3 $x = (4.00\text{ m})\cos(3.00\pi t + \pi)$ compare this with $x = A\cos(\omega t + \phi)$ to find

(a) $\omega = 2\pi f = 3.00\pi$

or $\boxed{f = 1.50\text{ Hz}}$ $T = \frac{1}{f} = \boxed{0.667\text{ s}}$

(b) $A = \boxed{4.00\text{ m}}$

(c) $\phi = \boxed{\pi\,\text{rad}}$

(d) $x(t = 0.250\text{ s}) = (4.00\text{ m})\cos(1.75\pi) = \boxed{2.83\text{ m}}$

P12.4 (a) $\boxed{20.0 \text{ cm}}$

(b) $v_{max} = \omega A = 2\pi f A = \boxed{94.2 \text{ cm/s}}$

This occurs as the particle passes through equilibrium.

(c) $a_{max} = \omega^2 A = (2\pi f)^2 A = \boxed{17.8 \text{ m/s}^2}$

This occurs at maximum excursion from equilibrium.

P12.5 (a) At $t = 0$, $x = 0$ and v is positive (to the right). Therefore, this situation corresponds to $x = A \sin \omega t$

and $\quad v = v_i \cos \omega t$

Since $f = 1.50$ Hz, $\quad \omega = 2\pi f = 3.00\pi$

Also, $A = 2.00$ cm, so that $\quad \boxed{x = (2.00 \text{ cm}) \sin 3.00\pi t}$

(b) $v_{max} = v_i = A\omega = 2.00(3.00\pi) = 6.00\pi \text{ cm/s} = \boxed{18.8 \text{ cm/s}}$

The particle has this speed at $t = 0$ and next at $\quad t = \frac{T}{2} = \boxed{\frac{1}{3} \text{ s}}$

(c) $a_{max} = A\omega^2 = 2.00(3.00\pi)^2 = 18.0\pi^2 \text{ cm/s}^2 = \boxed{178 \text{ cm/s}^2}$

This positive value of acceleration first occurs at $\quad t = \frac{3}{4}T = \boxed{0.500 \text{ s}}$

(d) Since $T = \frac{2}{3}$ s and $A = 2.00$ cm, the particle will travel 8.00 cm in this time.

Hence, in 1.00 s $\left(= \frac{3}{2}T\right)$, the particle will travel $\quad 8.00 \text{ cm} + 4.00 \text{ cm} = \boxed{12.0 \text{ cm}}$.

***P12.6** (a) For constant acceleration, position is given as a function of time by

$$x = x_i + v_{xi}t + \frac{1}{2}a_x t^2$$
$$= 0.27 \text{ m} + (0.14 \text{ m/s})(4.5 \text{ s}) + \frac{1}{2}(-0.32 \text{ m/s}^2)(4.5 \text{ s})^2$$
$$= \boxed{-2.34 \text{ m}}$$

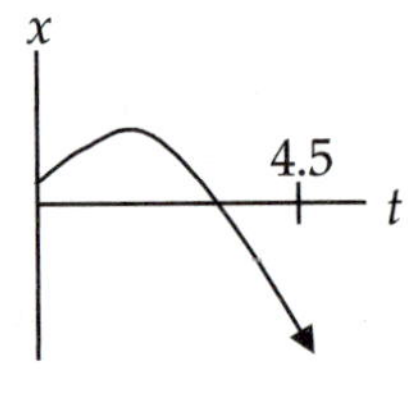

FIG. P12.6(a)

(b) $v_x = v_{xi} + a_x t = 0.14 \text{ m/s} - (0.32 \text{ m/s}^2)(4.5 \text{ s}) = \boxed{-1.30 \text{ m/s}}$

(c) For simple harmonic motion we have instead $x = A\cos(\omega t + \phi)$ and $v = -A\omega \sin(\omega t + \phi)$ where $a = -\omega^2 x$, so that $-0.32 \text{ m/s}^2 = -\omega^2(0.27 \text{ m})$, $\omega = 1.09$ rad/s. At $t = 0$, $0.27 \text{ m} = A\cos\phi$ and $0.14 \text{ m/s} = -A(1.09\text{/s})\sin\phi$. Dividing gives $\frac{0.14 \text{ m/s}}{0.27 \text{ m}} = -(1.09\text{/s})\tan\phi$, $\tan\phi = -0.476$, $\phi = -25.5°$. Still at $t = 0$, $0.27 \text{ m} = A\cos(-25.5°)$, $A = 0.299$ m. Now at $t = 4.5$ s,

continued on next page

$$x=(0.299\text{ m})\cos\left[(1.09\text{ rad/s})(4.5\text{ s})-25.5°\right]=(0.299\text{ m})\cos(4.90\text{ rad}-25.5°)$$
$$=(0.299\text{ m})\cos 255°=\boxed{-0.076\,3\text{ m}}$$

(d) $v=-(0.299\text{ m})(1.09/\text{s})\sin 255°=\boxed{+0.315\text{ m/s}}$

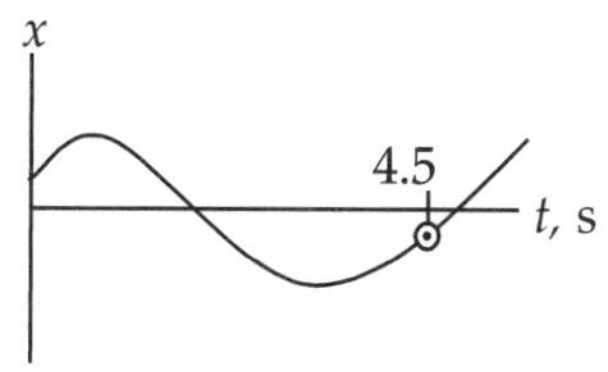

FIG. P12.6(d)

P12.7 The proposed solution $x(t)=x_i\cos\omega t+\left(\frac{v_i}{\omega}\right)\sin\omega t$

implies velocity $v=\frac{dx}{dt}=-x_i\omega\sin\omega t+v_i\cos\omega t$

and acceleration $a=\frac{dv}{dt}=-x_i\omega^2\cos\omega t-v_i\omega\sin\omega t=-\omega^2\left(x_i\cos\omega t+\left(\frac{v_i}{\omega}\right)\sin\omega t\right)=-\omega^2 x$

(a) The acceleration being a negative constant times position means we do have SHM, and its angular frequency is ω At $t=0$ the equations reduce to $x=x_i$ and $v=v_i$ so they satisfy all the requirements.

(b) $v^2-ax=(-x_i\omega\sin\omega t+v_i\cos\omega t)^2-\left(-x_i\omega^2\cos\omega t-v_i\sin\omega t\right)\left(x_i\cos\omega t+\left(\frac{v_i}{\omega}\right)\sin\omega t\right)$

$$v^2-ax=x_i^2\omega^2\sin^2\omega t-2x_iv_i\omega\sin\omega t\cos\omega t+v_i^2\cos^2\omega t$$
$$+x_i^2\omega^2\cos^2\omega t+x_iv_i\omega\cos\omega t\sin\omega t+x_iv_i\omega\sin\omega t\cos\omega t+v_i^2\sin^2\omega t=x_i^2\omega^2+v_i^2$$

So this expression is constant in time. On one hand, it must keep its original value $v_i^2-a_ix_i$. On the other hand, if we evaluate it at a turning point where $v=0$ and $x=A$, it is $A^2\omega^2+0^2=A^2\omega^2$. Thus it is proved.

P12.8 (a) $T=\frac{12.0\text{ s}}{5}=\boxed{2.40\text{ s}}$

(b) $f=\frac{1}{T}=\frac{1}{2.40}=\boxed{0.417\text{ Hz}}$

(c) $\omega=2\pi f=2\pi(0.417)=\boxed{2.62\text{ rad/s}}$

P12.9 $f=\frac{\omega}{2\pi}=\frac{1}{2\pi}\sqrt{\frac{k}{m}}$ or $T=\frac{1}{f}=2\pi\sqrt{\frac{m}{k}}$

Solving for k, $k=\frac{4\pi^2 m}{T^2}=\frac{4\pi^2(7.00\text{ kg})}{(2.60\text{ s})^2}=\boxed{40.9\text{ N/m}}$.

***P12.10** The mass of the cube is

$$m = \rho V = \left(2.7\times10^{3}\ \text{kg/m}^{3}\right)(0.015\ \text{m})^{3} = 9.11\times10^{-3}\ \text{kg}$$

The spring constant of the strip of steel is

$$k = \frac{F}{x} = \frac{14.3\ \text{N}}{0.027\,5\ \text{m}} = 52.0\ \text{N/m}$$

$$f = \frac{\omega}{2\pi} = \frac{1}{2\pi}\sqrt{\frac{k}{m}} = \frac{1}{2\pi}\sqrt{\frac{52\ \text{kg}}{\text{s}^{2}9.11\times10^{-3}\ \text{kg}}} = \boxed{12.0\ \text{Hz}}$$

P12.11 (a) $\omega = \sqrt{\frac{k}{m}} = \sqrt{\frac{8.00\ \text{N/m}}{0.500\ \text{kg}}} = 4.00\ \text{s}^{-1}$ so position is given by $x = 10.0\sin(4.00t)\ \text{cm}$.

From this we find that $v = 40.0\cos(4.00t)\ \text{cm/s}$ $\quad v_{max} = \boxed{40.0\ \text{cm/s}}$

$a = -160\sin(4.00t)\ \text{cm/s}^{2}$ $\quad a_{max} = \boxed{160\ \text{cm/s}^{2}}$.

(b) $t = \left(\frac{1}{4.00}\right)\sin^{-1}\left(\frac{x}{10.0}\right)$ and when $x = 6.00\ \text{cm}$, $t = 0.161\ \text{s}$.

We find $v = 40.0\cos[4.00(0.161)] = \boxed{32.0\ \text{cm/s}}$

$a = -160\sin[4.00(0.161)] = \boxed{-96.0\ \text{cm/s}^{2}}$.

(c) Using $t = \left(\frac{1}{4.00}\right)\sin^{-1}\left(\frac{x}{10.0}\right)$

when $x = 0$, $t = 0$ and when $x = 8.00\ \text{cm}$, $t = 0.232\ \text{s}$.

Therefore, $\Delta t = \boxed{0.232\ \text{s}}$.

P12.12 $m = 1.00\ \text{kg}$, $k = 25.0\ \text{N/m}$, and $A = 3.00\ \text{cm}$. At $t = 0$, $x = -3.00\ \text{cm}$

(a) $\omega = \sqrt{\frac{k}{m}} = \sqrt{\frac{25.0}{1.00}} = 5.00\ \text{rad/s}$

so that, $T = \frac{2\pi}{\omega} = \frac{2\pi}{5.00} = \boxed{1.26\ \text{s}}$

(b) $v_{max} = A\omega = 3.00\times10^{-2}\ \text{m}(5.00\ \text{rad/s}) = \boxed{0.150\ \text{m/s}}$

$a_{max} = A\omega^{2} = 3.00\times10^{-2}\ \text{m}(5.00\ \text{rad/s})^{2} = \boxed{0.750\ \text{m/s}^{2}}$

(c) Because $x = -3.00\ \text{cm}$ and $v = 0$ at $t = 0$, the required solution is $x = -A\cos\omega t$

or $\boxed{x = -3.00\cos(5.00t)\ \text{cm}}$

$$v = \frac{dx}{dt} = \boxed{15.0\sin(5.00t)\ \text{cm/s}}$$

$$a = \frac{dv}{dt} = \boxed{75.0\cos(5.00t)\ \text{cm/s}^{2}}$$

Section 12.3 Energy Considerations in Simple Harmonic Motion

P12.13 Choose the car with its shock-absorbing bumper as the system; by conservation of energy,

$$\frac{1}{2}mv^2 = \frac{1}{2}kx^2: \qquad v = x\sqrt{\frac{k}{m}} = \left(3.16\times10^{-2}\text{ m}\right)\sqrt{\frac{5.00\times10^6}{10^3}} = \boxed{2.23\text{ m/s}}$$

P12.14 $m = 200\text{ g}, T = 0.250\text{ s}, E = 2.00\text{ J}; \ \omega = \frac{2\pi}{T} = \frac{2\pi}{0.250} = 25.1\text{ rad/s}$

(a) $k = m\omega^2 = 0.200\text{ kg}(25.1\text{ rad/s})^2 = \boxed{126\text{ N/m}}$

(b) $E = \frac{kA^2}{2} \Rightarrow A = \sqrt{\frac{2E}{k}} = \sqrt{\frac{2(2.00)}{126}} = \boxed{0.178\text{ m}}$

P12.15 (a) Energy is conserved for the block-spring system between the maximum-displacement and the half-maximum points:

$$(K+U)_i = (K+U)_f \qquad 0 + \frac{1}{2}kA^2 = \frac{1}{2}mv^2 + \frac{1}{2}kx^2$$

$$\frac{1}{2}(6.50\text{ N/m})(0.100\text{ m})^2 = \frac{1}{2}m(0.300\text{ m/s})^2 + \frac{1}{2}(6.50\text{ N/m})\left(5.00\times10^{-2}\text{ m}\right)^2$$

$$32.5\text{ mJ} = \frac{1}{2}m(0.300\text{ m/s})^2 + 8.12\text{ mJ} \qquad m = \frac{2(24.4\text{ mJ})}{9.0\times10^{-2}\text{ m}^2/\text{s}^2} = \boxed{0.542\text{ kg}}$$

(b) $\omega = \sqrt{\frac{k}{m}} = \sqrt{\frac{6.50\text{ N/m}}{0.542\text{ kg}}} = 3.46\text{ rad/s} \qquad \therefore T = \frac{2\pi}{\omega} = \frac{2\pi\text{ rad}}{3.46\text{ rad/s}} = \boxed{1.81\text{ s}}$

(c) $a_{max} = A\omega^2 = 0.100\text{ m}(3.46\text{ rad/s})^2 = \boxed{1.20\text{ m/s}^2}$

P12.16 (a) $E = \frac{kA^2}{2} = \frac{250\text{ N/m}\left(3.50\times10^{-2}\text{ m}\right)^2}{2} = \boxed{0.153\text{ J}}$

(b) $v_{max} = A\omega$ where $\omega = \sqrt{\frac{k}{m}} = \sqrt{\frac{250}{0.500}} = 22.4\text{ s}^{-1}$ $\qquad v_{max} = \boxed{0.784\text{ m/s}}$

(c) $a_{max} = A\omega^2 = 3.50\times10^{-2}\text{ m}\left(22.4\text{ s}^{-1}\right)^2 = \boxed{17.5\text{ m/s}^2}$

P12.17 (a) $E = \frac{1}{2}kA^2 = \frac{1}{2}(35.0\text{ N/m})\left(4.00\times10^{-2}\text{ m}\right)^2 = \boxed{28.0\text{ mJ}}$

(b) $|v| = \omega\sqrt{A^2 - x^2} = \sqrt{\frac{k}{m}}\sqrt{A^2 - x^2}$

$$|v| = \sqrt{\frac{35.0}{50.0\times10^{-3}}}\sqrt{\left(4.00\times10^{-2}\right)^2 - \left(1.00\times10^{-2}\right)^2} = \boxed{1.02\text{ m/s}}$$

(c) $\frac{1}{2}mv^2 = \frac{1}{2}kA^2 - \frac{1}{2}kx^2 = \frac{1}{2}(35.0)\left[\left(4.00\times10^{-2}\right)^2 - \left(3.00\times10^{-2}\right)^2\right] = \boxed{12.2\text{ mJ}}$

(d) $\frac{1}{2}kx^2 = E - \frac{1}{2}mv^2 = \boxed{15.8\text{ mJ}}$

P12.18 (a) $k=\frac{|F|}{x}=\frac{20.0\text{ N}}{0.200\text{ m}}=\boxed{100\text{ N/m}}$

(b) $\omega=\sqrt{\frac{k}{m}}=\sqrt{50.0}\text{ rad/s}$ so $f=\frac{\omega}{2\pi}=\boxed{1.13\text{ Hz}}$

(c) $v_{\max}=\omega A=\sqrt{50.0}(0.200)=\boxed{1.41\text{ m/s}}$ at $x=0$

(d) $a_{\max}=\omega^2 A=50.0(0.200)=\boxed{10.0\text{ m/s}^2}$ at $x=\pm A$

(e) $E=\frac{1}{2}kA^2=\frac{1}{2}(100)(0.200)^2=\boxed{2.00\text{ J}}$

(f) $|v|=\omega\sqrt{A^2-x^2}=\sqrt{50.0}\sqrt{\frac{8}{9}(0.200)^2}=\boxed{1.33\text{ m/s}}$

(g) $|a|=\omega^2 x=50.0\left(\frac{0.200}{3}\right)=\boxed{3.33\text{ m/s}^2}$

***P12.19** (a) $E=\frac{1}{2}kA^2$, so if $A'=2A$, $E'=\frac{1}{2}k(A')^2=\frac{1}{2}k(2A)^2=4E$

Therefore $\boxed{E\text{ increases by factor of 4}}$.

(b) $v_{\max}=\sqrt{\frac{k}{m}}A$, so if A is doubled, $\boxed{v_{\max}\text{ is doubled}}$.

(c) $a_{\max}=\frac{k}{m}A$, so if A is doubled, $\boxed{a_{\max}\text{ also doubles}}$.

(d) $T=2\pi\sqrt{\frac{m}{k}}$ is independent of A, so $\boxed{\text{the period is unchanged}}$.

***P12.20** (a)

$$y_f=y_i+v_{yi}t+\frac{1}{2}a_y t^2$$

$$-11\text{ m}=0+0+\frac{1}{2}\left(-9.8\text{ m/s}^2\right)t^2$$

$$t=\sqrt{\frac{22\text{ m}\cdot\text{s}^2}{9.8\text{ m}}}=\boxed{1.50\text{ s}}$$

(b) Take the initial point where she steps off the bridge and the final point at the bottom of her motion.

$$\left(K+U_g+U_s\right)_i=\left(K+U_g+U_s\right)_f$$

$$0+mgy+0=0+0+\frac{1}{2}kx^2$$

$$65\text{ kg }9.8\text{ m/s}^2\ 36\text{ m}=\frac{1}{2}k(25\text{ m})^2$$

$$k=\boxed{73.4\text{ N/m}}$$

continued on next page

(c) The spring extension at equilibrium is $x=\frac{F}{k}=\frac{65\text{ kg }9.8\text{ m/s}^2}{73.4\text{ N/m}}=8.68\text{ m}$, so this point is $11+8.68\text{ m}=\boxed{19.7\text{ m below the bridge}}$ and the amplitude of her oscillation is $36-19.7=16.3\text{ m}$.

(d) $\omega=\sqrt{\frac{k}{m}}=\sqrt{\frac{73.4\text{ N/m}}{65\text{ kg}}}=\boxed{1.06\text{ rad/s}}$

(e) Take the phase as zero at maximum downward extension. We find what the phase was 25 m higher, where $x=-8.68\text{ m}$:

In $x=A\cos\omega t$, $\quad 16.3\text{ m}=16.3\text{ m}\cos 0$

$-8.68\text{ m}=16.3\text{ m}\cos\left(1.06\frac{t}{\text{s}}\right) \quad 1.06\frac{t}{\text{s}}=-122°=-2.13\text{ rad}$

$t=-2.01\text{ s}$

Then $\boxed{+2.01\text{ s}}$ is the time over which the spring stretches.

(f) total time $=1.50\text{ s}+2.01\text{ s}=\boxed{3.50\text{ s}}$

P12.21 Model the oscillator as a block-spring system.

From energy considerations, $\quad v^2+\omega^2x^2=\omega^2A^2$

$v_{\max}=\omega A$ and $v=\frac{\omega A}{2}$ so $\left(\frac{\omega A}{2}\right)^2+\omega^2x^2=\omega^2A^2$

From this we find $x^2=\frac{3}{4}A^2$ and $x=\frac{\sqrt{3}}{2}A=\boxed{\pm 2.60\text{ cm}}$ where $A=3.00\text{ cm}$

Section 12.4 The Simple Pendulum

Section 12.5 The Physical Pendulum

P12.22 The period in Tokyo is $\quad T_T=2\pi\sqrt{\frac{L_T}{g_T}}$

and the period in Cambridge is $\quad T_C=2\pi\sqrt{\frac{L_C}{g_C}}$

We know $\quad T_T=T_C=2.00\text{ s}$

For which, we see $\quad \frac{L_T}{g_T}=\frac{L_C}{g_C}$

or $\quad \frac{g_C}{g_T}=\frac{L_C}{L_T}=\frac{0.994\,2}{0.992\,7}=\boxed{1.001\,5}$

P12.23 Using the simple harmonic motion model:

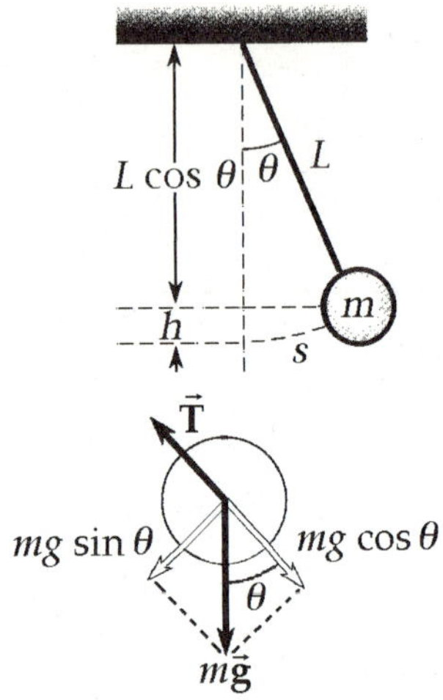

FIG. P12.23

$$A = r\theta = 1 \text{ m } 15° \frac{\pi}{180°} = 0.262 \text{ m}$$

$$\omega = \sqrt{\frac{g}{L}} = \sqrt{\frac{9.8 \text{ m/s}^2}{1 \text{ m}}} = 3.13 \text{ rad/s}$$

(a) $v_{max} = A\omega = 0.262 \text{ m } 3.13/\text{s} = \boxed{0.820 \text{ m/s}}$

(b) $a_{max} = A\omega^2 = 0.262 \text{ m}(3.13/\text{s})^2 = 2.57 \text{ m/s}^2$

$a_{tan} = r\alpha \qquad \alpha = \frac{a_{tan}}{r} = \frac{2.57 \text{ m/s}^2}{1 \text{ m}} = \boxed{2.57 \text{ rad/s}^2}$

(c) $F = ma = 0.25 \text{ kg } 2.57 \text{ m/s}^2 = \boxed{0.641 \text{ N}}$

More precisely,

(a) $mgh = \frac{1}{2}mv_{max}^2$ and $h = L(1-\cos\theta)$

$\therefore v_{max} = \sqrt{2gL(1-\cos\theta)} = \boxed{0.817 \text{ m/s}}$

(b) $I\alpha = mgL\sin\theta$

$\alpha_{max} = \frac{mgL\sin\theta}{mL^2} = \frac{g}{L}\sin\theta_i = \boxed{2.54 \text{ rad/s}^2}$

(c) $F_{max} = mg\sin\theta_i = 0.250(9.80)(\sin 15.0°) = \boxed{0.634 \text{ N}}$

P12.24 $\omega = \frac{2\pi}{T}$: $\quad T = \frac{2\pi}{\omega} = \frac{2\pi}{4.43} = \boxed{1.42 \text{ s}}$

$\omega = \sqrt{\frac{g}{L}}$: $\quad L = \frac{g}{\omega^2} = \frac{9.80}{(4.43)^2} = \boxed{0.499 \text{ m}}$

P12.25 Referring to the sketch we have

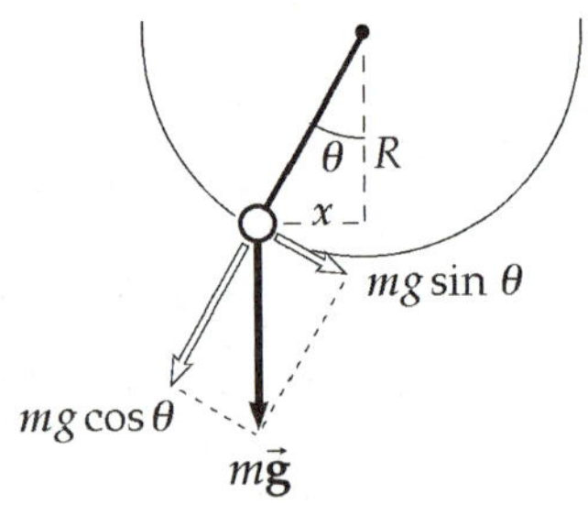

FIG. P12.25

$F = -mg\sin\theta$ and $\tan\theta = \frac{x}{R}$

For small displacements, $\tan\theta \approx \sin\theta$

and $F = -\frac{mg}{R}x = -kx$

Since the restoring force is proportional to the displacement from equilibrium, the motion is simple harmonic motion.

Comparing to $F = -m\omega^2 x$ shows $\boxed{\omega = \sqrt{\frac{k}{m}} = \sqrt{\frac{g}{R}}}$.

***P12.26** (a) $T = \dfrac{\text{total measured time}}{50}$

The measured periods are:

Length, L (m)	1.000	0.750	0.500
Period, T (s)	1.996	1.732	1.422

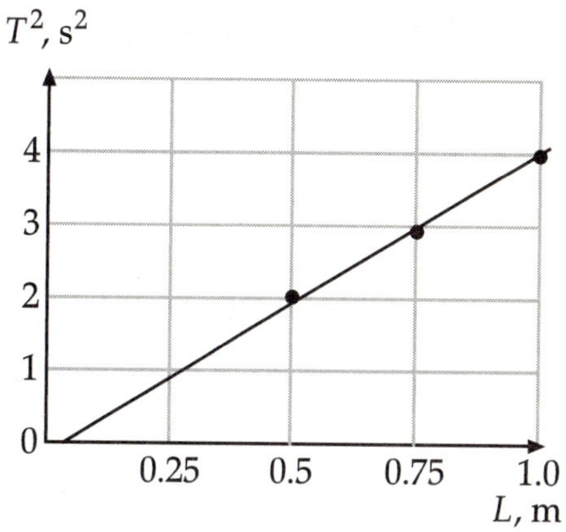

FIG. P12.26

(b) $T = 2\pi\sqrt{\dfrac{L}{g}}$ so $g = \dfrac{4\pi^2 L}{T^2}$

The calculated values for g are:

Period, T (s)	1.996	1.732	1.422
$g\left(\text{m/s}^2\right)$	9.91	9.87	9.76

Thus, $g_{\text{ave}} = \boxed{9.85\ \text{m/s}^2}$ this agrees with the accepted value of $g = 9.80\ \text{m/s}^2$ within 0.5%.

(c) From $T^2 = \left(\dfrac{4\pi^2}{g}\right)L$, the slope of T^2 versus L graph $= \dfrac{4\pi^2}{g} = 4.01\ \text{s}^2/\text{m}$.

Thus, $g = \dfrac{4\pi^2}{\text{slope}} = \boxed{9.85\ \text{m/s}^2}$. This is the same as the value in (b).

P12.27 $f = 0.450$ Hz, $d = 0.350$ m, and $m = 2.20$ kg

$T = \dfrac{1}{f}$;

$T = 2\pi\sqrt{\dfrac{I}{mgd}}$; $T^2 = \dfrac{4\pi^2 I}{mgd}$

$$I = T^2\frac{mgd}{4\pi^2} = \left(\frac{1}{f}\right)^2\frac{mgd}{4\pi^2} = \frac{2.20(9.80)(0.350)}{4\pi^2\left(0.450\ \text{s}^{-1}\right)^2} = \boxed{0.944\ \text{kg}\cdot\text{m}^2}$$

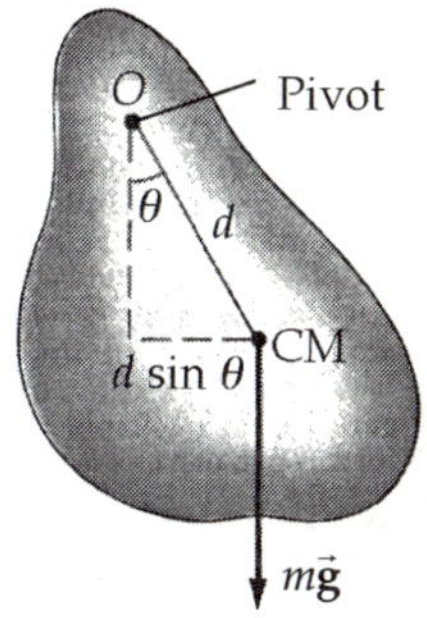

FIG. P12.27

P12.28 (a) The parallel-axis theorem:

$$I = I_{\text{CM}} + Md^2 = \frac{1}{12}ML^2 + Md^2 = \frac{1}{12}M(1.00\ \text{m})^2 + M(1.00\ \text{m})^2$$
$$= M\left(\frac{13}{12}\ \text{m}^2\right)$$

$$T = 2\pi\sqrt{\frac{I}{Mgd}} = 2\pi\sqrt{\frac{M\left(13\ \text{m}^2\right)}{12Mg(1.00\ \text{m})}} = 2\pi\sqrt{\frac{13\ \text{m}}{12\left(9.80\ \text{m/s}^2\right)}} = \boxed{2.09\ \text{s}}$$

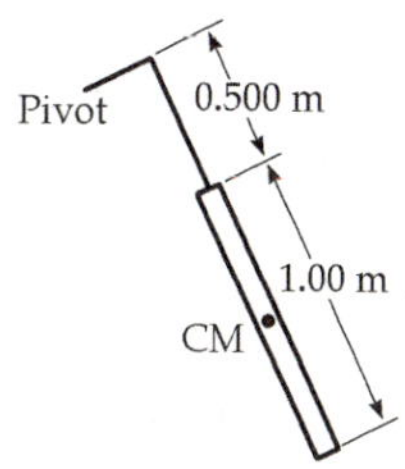

FIG. P12.28

(b) For the simple pendulum

$$T = 2\pi\sqrt{\frac{1.00\ \text{m}}{9.80\ \text{m/s}^2}} = 2.01\ \text{s} \qquad \text{difference} = \frac{2.09\ \text{s} - 2.01\ \text{s}}{2.01\ \text{s}} = \boxed{4.08\%}$$

***P12.29** (a) The parallel axis theorem says directly $I = I_{CM} + md^2$

so $$T = 2\pi\sqrt{\frac{I}{mgd}} = \boxed{2\pi\sqrt{\frac{\left(I_{CM} + md^2\right)}{mgd}}}$$

(b) When d is very large $T \to 2\pi\sqrt{\frac{d}{g}}$ gets large.

When d is very small $T \to 2\pi\sqrt{\frac{I_{CM}}{mgd}}$ gets large.

So there must be a minimum, found by

$$\frac{dT}{dd} = 0 = \frac{d}{dd}2\pi\left(I_{CM} + md^2\right)^{1/2}(mgd)^{-1/2}$$

$$= 2\pi\left(I_{CM} + md^2\right)^{1/2}\left(-\frac{1}{2}\right)(mgd)^{-3/2}mg + 2\pi(mgd)^{-1/2}\left(\frac{1}{2}\right)\left(I_{CM} + md^2\right)^{-1/2}2md$$

$$= \frac{-\pi\left(I_{CM} + md^2\right)mg}{\left(I_{CM} + md^2\right)^{1/2}(mgd)^{3/2}} + \frac{2\pi md\, mgd}{\left(I_{CM} + md^2\right)^{1/2}(mgd)^{3/2}} = 0$$

This requires

$$-I_{CM} - md^2 + 2md^2 = 0$$

or $\boxed{I_{CM} = md^2}$.

Section 12.6 Damped Oscillations

P12.30 The total energy is $E = \frac{1}{2}mv^2 + \frac{1}{2}kx^2$

Taking the time-derivative, $\frac{dE}{dt} = mv\frac{d^2x}{dt^2} + kxv$

Use Equation 12.28: $\frac{md^2x}{dt^2} = -kx - bv$

$$\frac{dE}{dt} = v(-kx - bv) + kvx$$

Thus, $\boxed{\frac{dE}{dt} = -bv^2 < 0}$

P12.31 $\theta_i = 15.0°$ $\theta(t = 1\,000) = 5.50°$

$x = Ae^{-bt/2m}$ $\frac{x_{1\,000}}{x_i} = \frac{Ae^{-bt/2m}}{A} = \frac{5.50}{15.0} = e^{-b(1\,000)/2m}$

$$\ln\left(\frac{5.50}{15.0}\right) = -1.00 = \frac{-b(1\,000)}{2m}$$

$$\therefore \frac{b}{2m} = \boxed{1.00 \times 10^{-3}\ \text{s}^{-1}}$$

P12.32 Show that $x = Ae^{-bt/2m}\cos(\omega t + \phi)$

is a solution of $-kx - b\frac{dx}{dt} = m\frac{d^2x}{dt^2}$ (1)

where $\omega = \sqrt{\frac{k}{m} - \left(\frac{b}{2m}\right)^2}$. (2)

$$x = Ae^{-bt/2m}\cos(\omega t + \phi) \quad (3)$$

$$\frac{dx}{dt} = Ae^{-bt/2m}\left(-\frac{b}{2m}\right)\cos(\omega t + \phi) - Ae^{-bt/2m}\omega\sin(\omega t + \phi) \quad (4)$$

$$\frac{d^2x}{dt^2} = -\frac{b}{2m}\left[Ae^{-bt/2m}\left(-\frac{b}{2m}\right)\cos(\omega t + \phi) - Ae^{-bt/2m}\omega\sin(\omega t + \phi)\right]$$

$$-\left[Ae^{-bt/2m}\left(-\frac{b}{2m}\right)\omega\sin(\omega t + \phi) + Ae^{-bt/2m}\omega^2\cos(\omega t + \phi)\right] \quad (5)$$

Substitute (3), (4) into the left side of (1) and (5) into the right side of (1);

$$-kAe^{-bt/2m}\cos(\omega t + \phi) + \frac{b^2}{2m}Ae^{-bt/2m}\cos(\omega t + \phi) + b\omega Ae^{-bt/2m}\sin(\omega t + \phi)$$

$$= -\frac{b}{2}\left[Ae^{-bt/2m}\left(-\frac{b}{2m}\right)\cos(\omega t + \phi) - Ae^{-bt/2m}\omega\sin(\omega t + \phi)\right]$$

$$+\frac{b}{2}Ae^{-bt/2m}\omega\sin(\omega t + \phi) - m\omega^2 Ae^{-bt/2m}\cos(\omega t + \phi)$$

Compare the coefficients of $Ae^{-bt/2m}\cos(\omega t + \phi)$ and $Ae^{-bt/2m}\sin(\omega t + \phi)$:

cosine-term: $-k + \frac{b^2}{2m} = -\frac{b}{2}\left(-\frac{b}{2m}\right) - m\omega^2 = \frac{b^2}{4m} - m\left(\frac{k}{m} - \frac{b^2}{4m^2}\right) = -k + \frac{b^2}{2m}$

sine-term: $b\omega = +\frac{b}{2}(\omega) + \frac{b}{2}(\omega) = b\omega$

Since the coefficients are equal, $x = Ae^{-bt/2m}\cos(\omega t + \phi)$ is a solution of the equation.

Section 12.7 Forced Oscillations

P12.33 $F = 3.00\sin(2\pi t)$ N and $k = 20.0$ N/m

(a) $\omega = \frac{2\pi}{T} = 2\pi$ rad/s so $T = \boxed{1.00\text{ s}}$

(b) In this case, $\omega_0 = \sqrt{\frac{k}{m}} = \sqrt{\frac{20.0}{2.00}} = 3.16$ rad/s

The equation for the amplitude of a driven oscillator,

with $b = 0$, gives $A = \left(\frac{F_0}{m}\right)\left(\omega^2 - \omega_0^2\right)^{-1} = \frac{3}{2}\left[4\pi^2 - (3.16)^2\right]^{-1}$

Thus $A = 0.050\,9\text{ m} = \boxed{5.09\text{ cm}}$.

P12.34 (a) For resonance, her frequency must match

$$f_0 = \frac{\omega_0}{2\pi} = \frac{1}{2\pi}\sqrt{\frac{k}{m}} = \frac{1}{2\pi}\sqrt{\frac{4.30\times10^3\text{ N/m}}{12.5\text{ kg}}} = \boxed{2.95\text{ Hz}}.$$

(b) From $x = A\cos\omega t$, $v = \frac{dx}{dt} = -A\omega\sin\omega t$, and $a = \frac{dv}{dt} = -A\omega^2\cos\omega t$, the maximum acceleration is $A\omega^2$. When this becomes equal to the acceleration due to gravity, the normal force exerted on her by the mattress will drop to zero at one point in the cycle:

$A\omega^2 = g$ or $A = \frac{g}{\omega^2} = \frac{g}{\frac{k}{m}} = \frac{gm}{k}$ $\quad A = \frac{(9.80\text{ m/s}^2)(12.5\text{ kg})}{4.30\times10^3\text{ N/m}} = \boxed{2.85\text{ cm}}$

P12.35 $F_0\cos\omega t - kx = m\frac{d^2x}{dt^2}$ $\quad \omega_0 = \sqrt{\frac{k}{m}}$ (1)

$x = A\cos(\omega t + \phi)$ (2)

$\frac{dx}{dt} = -A\omega\sin(\omega t + \phi)$ (3)

$\frac{d^2x}{dt^2} = -A\omega^2\cos(\omega t + \phi)$ (4)

Substitute (2) and (4) into (1): $F_0\cos\omega t - kA\cos(\omega t + \phi) = m\left(-A\omega^2\right)\cos(\omega t + \phi)$

Solve for the amplitude: $\left(kA - mA\omega^2\right)\cos(\omega t + \phi) = F_0\cos\omega t$

These will be equal, provided only that ϕ must be zero and $kA - mA\omega^2 = F_0$

Thus, $A = \frac{F_0/m}{(k/m) - \omega^2}$

P12.36 $$A = \frac{F_{\text{ext}}/m}{\sqrt{\left(\omega^2 - \omega_0^2\right)^2 + (b\omega/m)^2}}$$

With $b = 0$, $$A = \frac{F_{\text{ext}}/m}{\sqrt{\left(\omega^2 - \omega_0^2\right)^2}} = \frac{F_{\text{ext}}/m}{\pm\left(\omega^2 - \omega_0^2\right)} = \pm\frac{F_{\text{ext}}/m}{\omega^2 - \omega_0^2}$$

Thus, $$\omega^2 = \omega_0^2 \pm \frac{F_{\text{ext}}/m}{A} = \frac{k}{m} \pm \frac{F_{\text{ext}}}{mA} = \frac{6.30\ \text{N/m}}{0.150\ \text{kg}} \pm \frac{1.70\ \text{N}}{(0.150\ \text{kg})(0.440\ \text{m})}$$

This yields $\omega = 8.23\ \text{rad/s}$ or $\omega = 4.03\ \text{rad/s}$

Then, $f = \dfrac{\omega}{2\pi}$ gives either $f = \boxed{1.31\ \text{Hz}}$ or $f = \boxed{0.641\ \text{Hz}}$

P12.37 The beeper must resonate at the frequency of a simple pendulum of length 8.21 cm:

$$f = \frac{1}{2\pi}\sqrt{\frac{g}{L}} = \frac{1}{2\pi}\sqrt{\frac{9.80\ \text{m/s}^2}{0.0821\ \text{m}}} = \boxed{1.74\ \text{Hz}}.$$

Section 12.8 Context Connection—Resonance in Structures

P12.38 For the resonance vibration with the occupants in the car, we have for the spring constant of the suspension

$$f = \frac{1}{2\pi}\sqrt{\frac{k}{m}} \qquad k = 4\pi^2 f^2 m = 4\pi^2\left(1.8\ \text{s}^{-1}\right)^2\left(1\,130\ \text{kg} + 4(72.4\ \text{kg})\right) = 1.82 \times 10^5\ \text{kg/s}^2$$

Now as the occupants exit $$x = \frac{F}{k} = \frac{4(72.4\ \text{kg})\left(9.8\ \text{m/s}^2\right)}{1.82 \times 10^5\ \text{kg/s}^2} = \boxed{1.56 \times 10^{-2}\ \text{m}}$$

P12.39 Suppose a 100-kg biker compresses the suspension 2.00 cm.

Then, $$k = \frac{F}{x} = \frac{980\ \text{N}}{2.00 \times 10^{-2}\ \text{m}} = 4.90 \times 10^4\ \text{N/m}$$

If total mass of motorcycle and biker is 500 kg, the frequency of free vibration is

$$f = \frac{1}{2\pi}\sqrt{\frac{k}{m}} = \frac{1}{2\pi}\sqrt{\frac{4.90 \times 10^4\ \text{N/m}}{500\ \text{kg}}} = 1.58\ \text{Hz}$$

If he encounters washboard bumps at the same frequency, resonance will make the motorcycle bounce a lot. Assuming a speed of 20.0 m/s, we find these ridges are separated by

$$\frac{20.0\ \text{m/s}}{1.58\ \text{s}^{-1}} = 12.7\ \text{m}\ \boxed{\sim 10^1\ \text{m}}.$$

In addition to this vibration mode of bouncing up and down as one unit, the motorcycle can also vibrate at higher frequencies by rocking back and forth between front and rear wheels, by having just the front wheel bounce inside its fork, or by doing other things. Other spacing of bumps will excite all of these other resonances.

Additional Problems

P12.40 (a) Total energy $= \frac{1}{2}kA^2 = \frac{1}{2}(100 \text{ N/m})(0.200 \text{ m})^2 = 2.00 \text{ J}$

At equilibrium, the total energy is:

$$\frac{1}{2}(m_1 + m_2)v^2 = \frac{1}{2}(16.0 \text{ kg})v^2 = (8.00 \text{ kg})v^2.$$

Therefore,

$$(8.00 \text{ kg})v^2 = 2.00 \text{ J, and } v = \boxed{0.500 \text{ m/s}}.$$

This is the speed of m_1 and m_2 at the equilibrium point. Beyond this point, the mass m_2 moves with the constant speed of 0.500 m/s while mass m_1 starts to slow down due to the restoring force of the spring.

(b) The energy of the m_1-spring system at equilibrium is:

$$\frac{1}{2}m_1 v^2 = \frac{1}{2}(9.00 \text{ kg})(0.500 \text{ m/s})^2 = 1.125 \text{ J}.$$

This is also equal to $\frac{1}{2}k(A')^2$, where A' is the amplitude of the m_1-spring system.

Therefore,

$$\frac{1}{2}(100)(A')^2 = 1.125 \text{ or } A' = 0.150 \text{ m}.$$

The period of the m_1-spring system is $T = 2\pi\sqrt{\frac{m_1}{k}} = 1.885 \text{ s}$

and it takes $\frac{1}{4}T = 0.471 \text{ s}$ after it passes the equilibrium point for the spring to become fully stretched the first time. The distance separating m_1 and m_2 at this time is:

$$D = v\left(\frac{T}{4}\right) - A' = 0.500 \text{ m/s}(0.471 \text{ s}) - 0.150 \text{ m} = 0.0856 = \boxed{8.56 \text{ cm}}.$$

P12.41 $$\left(\frac{d^2x}{dt^2}\right)_{\max} = A\omega^2$$

$$f_{\max} = \mu_s n = \mu_s mg = mA\omega^2$$

$$A = \frac{\mu_s g}{\omega^2} = \boxed{6.62 \text{ cm}}$$

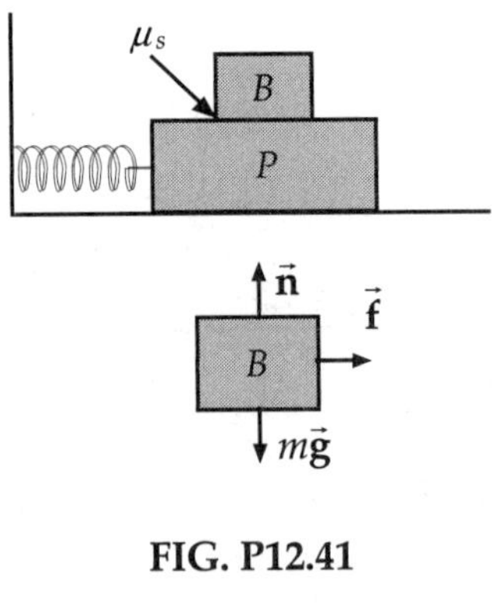

FIG. P12.41

***P12.42** (a) The spring constant of this spring is

$$k = \frac{F}{x} = \frac{0.45\text{ kg}\,9.8\text{ m/s}^2}{0.35\text{ m}} = 12.6\text{ N/m}$$

we take the x-axis pointing downward, so $\phi = 0$

$$x = A\cos\omega t = 18.0\text{ cm}\cos\sqrt{\frac{12.6\text{ kg}}{0.45\text{ kg}\cdot\text{s}^2}}\,84.4\text{ s} = 18.0\text{ cm}\cos 446.6\text{ rad} = \boxed{15.8\text{ cm}}$$

(d) Now $446.6\text{ rad} = 71 \times 2\pi + 0.497\text{ rad}$. In each cycle the object moves $4(18) = 72\text{ cm}$, so it has moved $71(72\text{ cm}) + (18 - 15.8)\text{ cm} = \boxed{51.1\text{ m}}$.

(b) By the same steps, $k = \dfrac{0.44\text{ kg}\,9.8\text{ m/s}^2}{0.355\text{ m}} = 12.1\text{ N/m}$

$$x = A\cos\sqrt{\frac{k}{m}}t = 18.0\text{ cm}\cos\sqrt{\frac{12.1}{0.44}}\,84.4 = 18.0\text{ cm }\cos 443.5\text{ rad} = \boxed{-15.9\text{ cm}}$$

(e) $443.5\text{ rad} = 70(2\pi) + 3.62\text{ rad}$

$$\text{Distance moved} = 70(72\text{ cm}) + 18 + 15.9\text{ cm} = \boxed{50.7\text{ m}}$$

(c) The answers to (d) and (e) are not very different given the difference in the data about the two vibrating systems. But when we ask about details of the future, the imprecision in our knowledge about the present makes it impossible to make precise predictions. The two oscillations start out in phase but get completely out of phase.

P12.43 Deuterium is the isotope of the element hydrogen with atoms having nuclei consisting of one proton and one neutron. For brevity we refer to the molecule formed by two deuterium atoms as D and to the diatomic molecule of hydrogen-1 as H.

$$M_D = 2M_H \qquad \frac{\omega_D}{\omega_H} = \frac{\sqrt{\frac{k}{M_D}}}{\sqrt{\frac{k}{M_H}}} = \sqrt{\frac{M_H}{M_D}} = \sqrt{\frac{1}{2}} \qquad f_D = \frac{f_H}{\sqrt{2}} = \boxed{0.919 \times 10^{14}\text{ Hz}}$$

P12.44 $\omega = \sqrt{\dfrac{k}{m}} = \dfrac{2\pi}{T}$

(a) $k = \omega^2 m = \boxed{\dfrac{4\pi^2 m}{T^2}}$

(b) $m' = \dfrac{k(T')^2}{4\pi^2} = \boxed{m\left(\dfrac{T'}{T}\right)^2}$

***P12.45** (a) The period of the swinging rod is

$$T = 2\pi\sqrt{\frac{I}{mgd}} = 2\pi\sqrt{\frac{(1/3)m\ell^2}{mg\ell/2}} = 2\pi\sqrt{\frac{2\ell}{3g}}$$

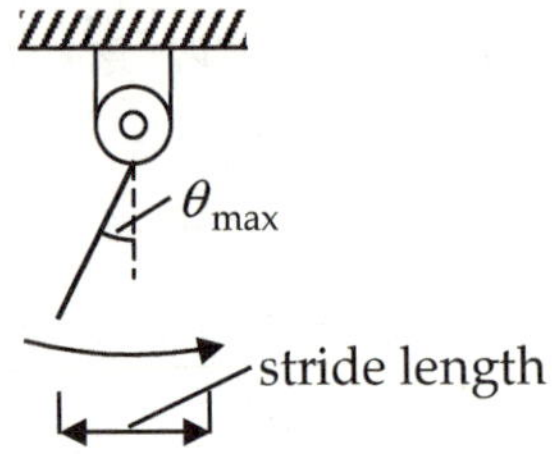

FIG. P12.45(a)

The time for one half a cycle is $\frac{T}{2} = \pi\sqrt{\frac{2\ell}{3g}}$. The distance traveled in this time is the stride length $2\ell\sin\theta_{max}$, so the speed is

$$\frac{d}{t} = \frac{2\ell\sin\theta_{max}}{\pi\sqrt{2\ell/3g}} = \frac{\sqrt{2\ell 3g}\,\sin\theta_{max}}{\pi} = \frac{\sqrt{6g\ell}\,\sin\theta_{max}}{\pi}$$

(b) We use the more precise expression

$$\frac{\sqrt{6g\ell\cos(\theta_{max}/2)}\,\sin\theta_{max}}{\pi} = \frac{\sqrt{6(9.8\text{ m/s}^2)(0.85\text{ m})\cos 14°}\,\sin 28°}{\pi} = \boxed{1.04\text{ m/s}}$$

(c) With

$$v_{old} = \frac{\sqrt{6g\ell_{old}\cos(\theta_{max}/2)}\,\sin\theta_{max}}{\pi}$$

$$v_{new} = \frac{\sqrt{6g\ell_{new}\cos(\theta_{max}/2)}\,\sin\theta_{max}}{\pi}$$

dividing gives

$$\frac{v_{new}}{v_{old}} = \frac{\sqrt{\ell_{new}}}{\sqrt{\ell_{old}}} = 2$$

$$\frac{\ell_{new}}{0.85\text{ m}} = 2^2$$

$$\ell_{new} = \boxed{3.40\text{ m}}$$

***P12.46** (a) Consider the first process of spring compression. It continues as long as glider 1 is moving faster than glider 2. The spring instantaneously has maximum compression when both gliders are moving with the same speed v_a.

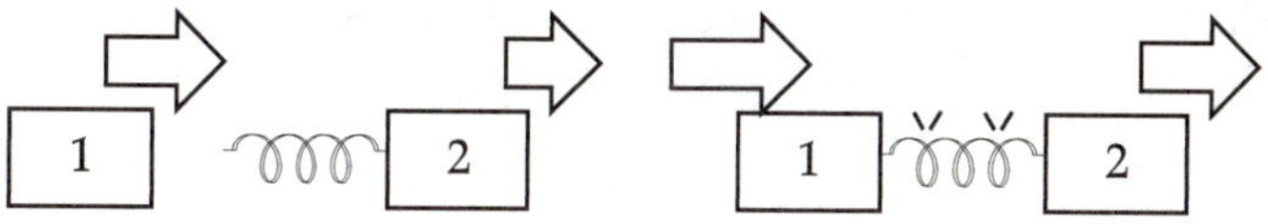

FIG. P12.46(a)

Momentum conservation:

$$m_1 v_{1i} + m_2 v_{2i} = m_1 v_{1f} + m_2 v_f$$

$$(0.24 \text{ kg})(0.74 \text{ m/s}) + (0.36 \text{ kg})(0.12 \text{ m/s}) = (0.24 \text{ kg})v_a + (0.36 \text{ kg})v_a$$

$$\frac{0.220\,8 \text{ kg}\cdot\text{m/s}}{0.60 \text{ kg}} = v_a \qquad \vec{\mathbf{v}}_a = \boxed{0.368 \text{ m/s}\,\hat{\mathbf{i}}}$$

(b) Energy conservation:

$$(K_1 + K_2 + U_s)_i = (K_1 + K_2 + U_s)_f$$

$$\frac{1}{2}m_1 v_{1i}^2 + \frac{1}{2}m_2 v_{2i}^2 + 0 = \frac{1}{2}(m_1 + m_2)v_a^2 + \frac{1}{2}kx^2$$

$$\frac{1}{2}(0.24 \text{ kg})(0.74 \text{ m/s})^2 + \frac{1}{2}(0.36 \text{ kg})(0.12 \text{ m/s})^2 = \frac{1}{2}(0.60 \text{ kg})(0.368 \text{ m/s})^2 + \frac{1}{2}(45 \text{ N/m})x^2$$

$$0.068\,3 \text{ J} = 0.040\,6 \text{ J} + \frac{1}{2}(45 \text{ N/m})x^2$$

$$x = \left(\frac{2(0.027\,7 \text{ J})}{45 \text{ N/m}}\right)^{1/2} = \boxed{0.035\,1 \text{ m}}$$

(c) Conservation of momentum guarantees that the center of mass moves with constant velocity. Imagine viewing the gliders from a reference frame moving with the center of mass. We see the two gliders approach each other with momenta in opposite directions of equal magnitude. Upon colliding they compress the spring and then together bounce, extending and compressing it cyclically.

(d) $$\frac{1}{2}m_{\text{tot}}v_{\text{CM}}^2 = \frac{1}{2}(0.60 \text{ kg})(0.368 \text{ m/s})^2 = \boxed{0.040\,6 \text{ J}}$$

(e) $$\frac{1}{2}kA^2 = \frac{1}{2}(45 \text{ N/m})(0.035\,1 \text{ m})^2 = \boxed{0.027\,7 \text{ J}}$$

P12.47 We draw a free-body diagram of the pendulum. The force $\vec{\mathbf{H}}$ exerted by the hinge causes no torque about the axis of rotation.

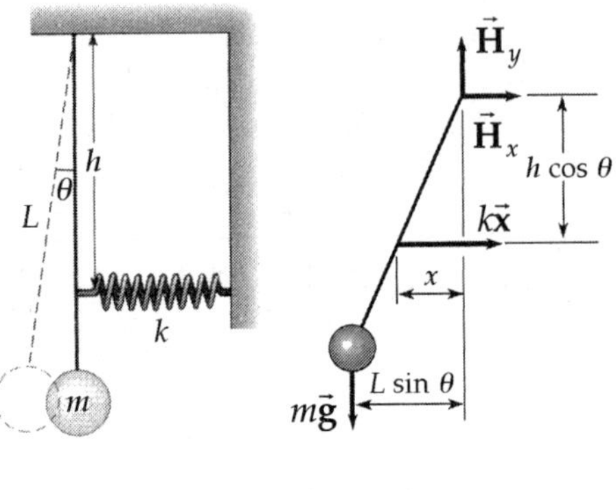

FIG. P12.47

$\tau = I\alpha$ and $\dfrac{d^2\theta}{dt^2} = -\alpha$

The negative sign appears because clockwise θ is counted as positive in the diagram.

$$\tau = MgL\sin\theta + kxh\cos\theta = -I\frac{d^2\theta}{dt^2}$$

For small amplitude vibrations, use the approximations: $\sin\theta \approx \theta$, $\cos\theta \approx 1$, and $x \approx s = h\theta$.

Therefore, $\dfrac{d^2\theta}{dt^2} = -\left(\dfrac{MgL + kh^2}{I}\right)\theta = -\omega^2\theta$ $\qquad \omega = \sqrt{\dfrac{MgL + kh^2}{ML^2}} = 2\pi f$

$$\boxed{f = \frac{1}{2\pi}\sqrt{\frac{MgL + kh^2}{ML^2}}}$$

P12.48 (a) In $x = A\cos(\omega t + \phi)$, $\quad v = -\omega A\sin(\omega t + \phi)$

we have at $t = 0$ $\quad v = -\omega A\sin\phi = -v_{max}$

This requires $\phi = 90°$, so $\quad x = A\cos(\omega t + 90°)$

And this is equivalent to $\quad x = -A\sin\omega t$

Numerically we have $\quad \omega = \sqrt{\dfrac{k}{m}} = \sqrt{\dfrac{50\text{ N/m}}{0.5\text{ kg}}} = 10\text{ s}^{-1}$

and $v_{max} = \omega A$ $\quad 20\text{ m/s} = (10\text{ s}^{-1})A$ $\quad A = 2\text{ m}$

So $\quad \boxed{x = (-2\text{ m})\sin\left[(10\text{ s}^{-1})t\right]}$

(b) In $\dfrac{1}{2}mv^2 + \dfrac{1}{2}kx^2 = \dfrac{1}{2}kA^2$, $\quad \dfrac{1}{2}kx^2 = 3\left(\dfrac{1}{2}mv^2\right)$

implies $\quad \dfrac{1}{3}\dfrac{1}{2}kx^2 + \dfrac{1}{2}kx^2 = \dfrac{1}{2}kA^2$ $\quad \dfrac{4}{3}x^2 = A^2$

$$x = \pm\sqrt{\frac{3}{4}}A = \pm 0.866A = \boxed{\pm 1.73\text{ m}}$$

(c) $\omega = \sqrt{\dfrac{g}{L}}$ $\quad L = \dfrac{g}{\omega^2} = \dfrac{9.8\text{ m/s}^2}{(10\text{ s}^{-1})^2} = \boxed{0.098\,0\text{ m}}$

continued on next page

(d) In $x = (-2\text{ m})\sin\left[\left(10\text{ s}^{-1}\right)t\right]$

the particle is at $x = 0$ at $t = 0$, at $10t = \pi\text{ s}$, and so on.

The particle is at $x = 1\text{ m}$

when $-\frac{1}{2} = \sin\left[\left(10\text{ s}^{-1}\right)t\right]$

with solutions $\left(10\text{ s}^{-1}\right)t = -\frac{\pi}{6}$

$$\left(10\text{ s}^{-1}\right)t = \pi + \frac{\pi}{6}, \text{ and so on.}$$

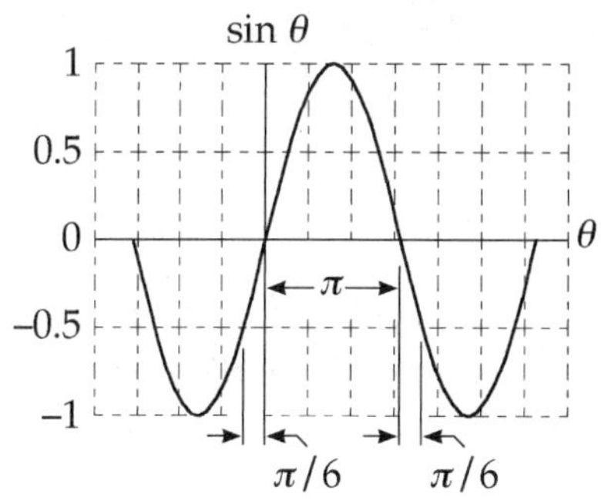

FIG. P12.48(d)

The minimum time for the motion is Δt in

$10\Delta t = \left(\frac{\pi}{6}\right)\text{ s}$

$$\Delta t = \left(\frac{\pi}{60}\right)\text{ s} = \boxed{0.052\ 4\text{ s}}$$

P12.49 (a) At equilibrium, we have

$$\sum \tau = 0 - mg\left(\frac{L}{2}\right) + kx_0L$$

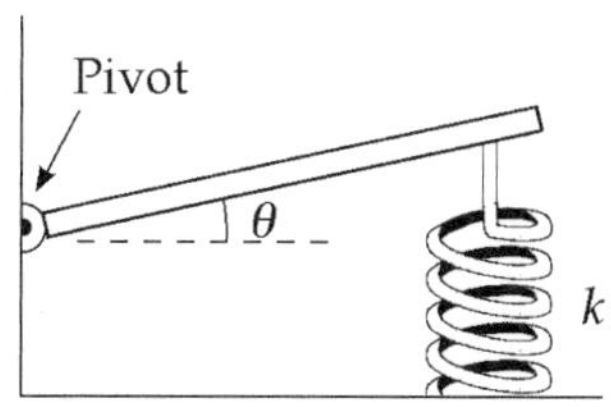

FIG. P12.49

where x_0 is the equilibrium compression.

After displacement by a small angle,

$$\sum \tau = -mg\left(\frac{L}{2}\right) + kxL = -mg\left(\frac{L}{2}\right) + k(x_0 - L\theta)L = -k\theta L^2$$

But, $\sum \tau = I\alpha = \frac{1}{3}mL^2\frac{d^2\theta}{dt^2}$. So $\frac{d^2\theta}{dt^2} = -\frac{3k}{m}\theta$.

The angular acceleration is opposite in direction and proportional to the displacement, so we have simple harmonic motion with $\boxed{\omega^2 = \frac{3k}{m}}$.

(b) $$f = \frac{\omega}{2\pi} = \frac{1}{2\pi}\sqrt{\frac{3k}{m}} = \frac{1}{2\pi}\sqrt{\frac{3(100\text{ N/m})}{5.00\text{ kg}}} = \boxed{1.23\text{ Hz}}$$

P12.50 As it passes through equilibrium, the 4-kg object has speed

$$v_{max} = \omega A = \sqrt{\frac{k}{m}}A = \sqrt{\frac{100\ \text{N/m}}{4\ \text{kg}}}2\ \text{m} = 10.0\ \text{m/s}.$$

In the completely inelastic collision momentum of the two-object system is conserved. So the new 10-kg object starts its oscillation with speed given by

$$4\ \text{kg}(10\ \text{m/s}) + (6\ \text{kg})0 = (10\ \text{kg})v_{max}$$
$$v_{max} = 4.00\ \text{m/s}$$

(a) The new amplitude is given by $\frac{1}{2}mv_{max}^2 = \frac{1}{2}kA^2$

$$10\ \text{kg}(4\ \text{m/s})^2 = (100\ \text{N/m})A^2$$
$$A = 1.26\ \text{m}$$

Thus the amplitude has $\boxed{\text{decreased by}}$ $2.00\ \text{m} - 1.26\ \text{m} = \boxed{0.735\ \text{m}}$

(b) The old period was $T = 2\pi\sqrt{\frac{m}{k}} = 2\pi\sqrt{\frac{4\ \text{kg}}{100\ \text{N/m}}} = 1.26\ \text{s}$

The new period is $T = 2\pi\sqrt{\frac{10}{100}\text{s}^2} = 1.99\ \text{s}$

The period has $\boxed{\text{increased by}}$ $1.99\ \text{m} - 1.26\ \text{m} = \boxed{0.730\ \text{s}}$

(c) The old energy was $\frac{1}{2}mv_{max}^2 = \frac{1}{2}(4\ \text{kg})(10\ \text{m/s})^2 = 200\ \text{J}$

The new mechanical energy is $\frac{1}{2}(10\ \text{kg})(4\ \text{m/s})^2 = 80\ \text{J}$

The energy has $\boxed{\text{decreased by 120 J}}$.

(d) The missing mechanical energy has turned into internal energy in the completely inelastic collision.

P12.51 (a) $T = \frac{2\pi}{\omega} = 2\pi\sqrt{\frac{L}{g}} = \boxed{3.00\ \text{s}}$

(b) $E = \frac{1}{2}mv^2 = \frac{1}{2}(6.74)(2.06)^2 = \boxed{14.3\ \text{J}}$

(c) At maximum angular displacement $mgh = \frac{1}{2}mv^2$ $h = \frac{v^2}{2g} = 0.217\ \text{m}$

$h = L - L\cos\theta = L(1 - \cos\theta)$ $\cos\theta = 1 - \frac{h}{L}$ $\boxed{\theta = 25.5°}$

P12.52 One can write the following equations of motion:

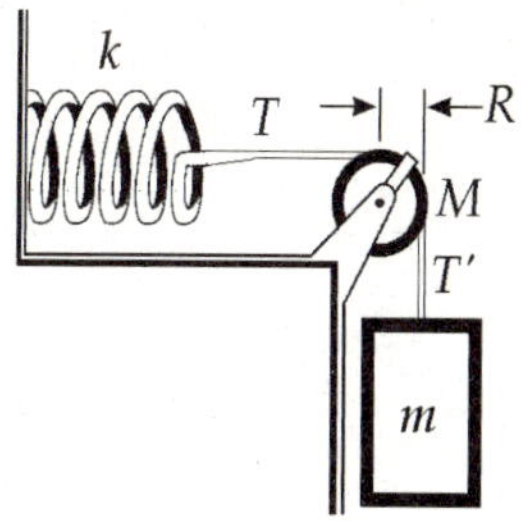

FIG. P12.52

$T - kx = 0$ (describes the spring)

$mg - T' = ma = m\dfrac{d^2x}{dt^2}$ (for the hanging object)

$R(T' - T) = I\dfrac{d^2\theta}{dt^2} = \dfrac{I}{R}\dfrac{d^2x}{dt^2}$ (for the pulley)

with $I = \dfrac{1}{2}MR^2$

Combining these equations gives the equation of motion

$$\left(m + \frac{1}{2}M\right)\frac{d^2x}{dt^2} + kx = mg.$$

The solution is $x(t) = A\sin\omega t + \dfrac{mg}{k}$ (where $\dfrac{mg}{k}$ arises because of the extension of the spring due to the weight of the hanging object), with frequency

$$f = \frac{\omega}{2\pi} = \frac{1}{2\pi}\sqrt{\frac{k}{m + \frac{1}{2}M}} = \frac{1}{2\pi}\sqrt{\frac{100 \text{ N/m}}{0.200 \text{ kg} + \frac{1}{2}M}}.$$

(a) For $M = 0$ $f = \boxed{3.56 \text{ Hz}}$

(b) For $M = 0.250$ kg $f = \boxed{2.79 \text{ Hz}}$

(c) For $M = 0.750$ kg $f = \boxed{2.10 \text{ Hz}}$

P12.53 (a) $\sum \vec{\mathbf{F}} = -2T\sin\theta\,\hat{\mathbf{j}}$ where $\theta = \tan^{-1}\left(\dfrac{y}{L}\right)$

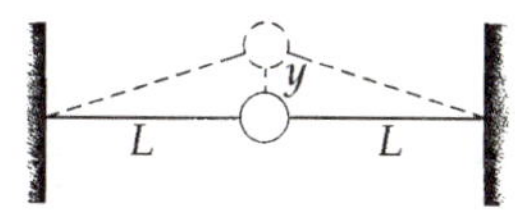

FIG. P12.53

Therefore, for a small displacement

$\sin\theta \approx \tan\theta = \dfrac{y}{L}$ and $\boxed{\sum \vec{\mathbf{F}} = \dfrac{-2Ty}{L}\hat{\mathbf{j}}}$

(b) The total force exerted on the ball is opposite in direction and proportional to its displacement from equilibrium, so the ball moves with simple harmonic motion. For a spring system,

$$\sum \vec{\mathbf{F}} = -k\vec{\mathbf{x}} \quad \text{becomes here} \quad \sum \vec{\mathbf{F}} = -\frac{2T}{L}\vec{\mathbf{y}}.$$

Therefore, the effective spring constant is $\dfrac{2T}{L}$ and $\boxed{\omega = \sqrt{\dfrac{k}{m}} = \sqrt{\dfrac{2T}{mL}}}$.

P12.54 (a) For each segment of the spring

$$dK = \frac{1}{2}(dm)v_x^2.$$

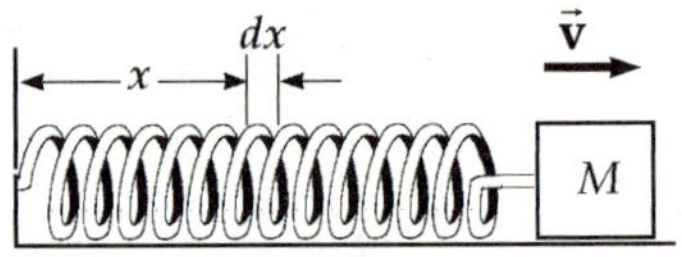

FIG. P12.54

Also, $v_x = \frac{x}{\ell}v$ and $dm = \frac{m}{\ell}dx$.

Therefore, the total kinetic energy of the block-spring system is

$$K = \frac{1}{2}Mv^2 + \frac{1}{2}\int_0^\ell \left(\frac{x^2v^2}{\ell^2}\right)\frac{m}{\ell}dx = \boxed{\frac{1}{2}\left(M + \frac{m}{3}\right)v^2}.$$

(b) $\omega = \sqrt{\frac{k}{m_{eff}}}$ and $\frac{1}{2}m_{eff}v^2 = \frac{1}{2}\left(M + \frac{m}{3}\right)v^2$

Therefore, $T = \frac{2\pi}{\omega} = \boxed{2\pi\sqrt{\frac{M + \frac{m}{3}}{k}}}$.

***P12.55** (a) $\Delta K + \Delta U = 0$

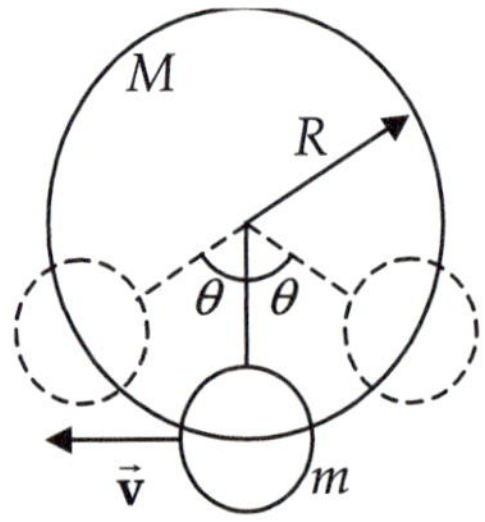

FIG. P12.55

Thus, $K_{top} + U_{top} = K_{bot} + U_{bot}$

where $K_{top} = U_{bot} = 0$

Therefore, $mgh = \frac{1}{2}I\omega^2$, but

$$h = R - R\cos\theta = R(1 - \cos\theta)$$

$$\omega = \frac{v}{R}$$

and $I = \frac{MR^2}{2} + \frac{mr^2}{2} + mR^2$

Substituting we find

$$mgR(1 - \cos\theta) = \frac{1}{2}\left(\frac{MR^2}{2} + \frac{mr^2}{2} + mR^2\right)\frac{v^2}{R^2}$$

$$mgR(1 - \cos\theta) = \left[\frac{M}{4} + \frac{mr^2}{4R^2} + \frac{m}{2}\right]v^2$$

and $v^2 = 4gR\frac{(1 - \cos\theta)}{\left(\frac{M}{m} + \frac{r^2}{R^2} + 2\right)}$

so $\boxed{v = 2\sqrt{\frac{Rg(1 - \cos\theta)}{(M/m) + (r^2/R^2) + 2}}}$

continued on next page

(b) $$T = 2\pi\sqrt{\frac{I}{m_T g d_{CM}}}$$

$$m_T = m + M \qquad d_{CM} = \frac{mR + M(0)}{m+M}$$

$$T = \boxed{2\pi\sqrt{\frac{\frac{1}{2}MR^2 + \frac{1}{2}mr^2 + mR^2}{mgR}}}$$

P12.56 (a) Assuming a Hooke's Law type spring,

$$F = Mg = kx$$

and empirically

$$Mg = 1.74x - 0.113$$

so $k = \boxed{1.74\ \text{N/m} \pm 6\%}$.

M, kg	x, m	Mg, N
0.020 0	0.17	0.196
0.040 0	0.293	0.392
0.050 0	0.353	0.49
0.060 0	0.413	0.588
0.070 0	0.471	0.686
0.080 0	0.493	0.784

(b) We may write the equation as theoretically

$$T^2 = \frac{4\pi^2}{k}M + \frac{4\pi^2}{3k}m_s$$

and empirically

$$T^2 = 21.7M + 0.058\,9$$

so $k = \frac{4\pi^2}{21.7} = \boxed{1.82\ \text{N/m} \pm 3\%}$

Time, s	T, s	M, kg	T^2, s^2
7.03	0.703	0.020 0	0.494
9.62	0.962	0.040 0	0.925
10.67	1.067	0.050 0	1.138
11.67	1.167	0.060 0	1.362
12.52	1.252	0.070 0	1.568
13.41	1.341	0.080 0	1.798

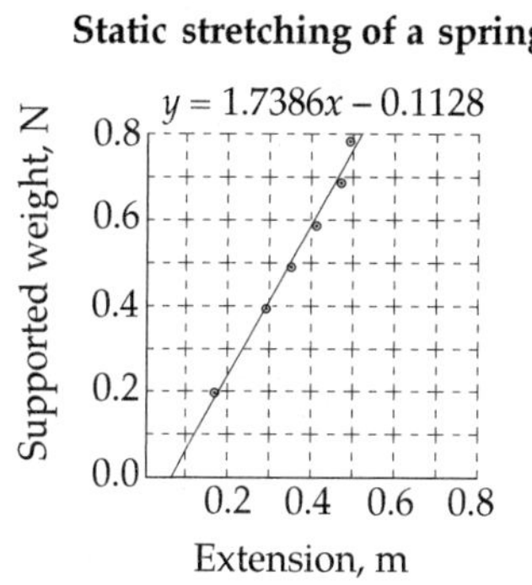

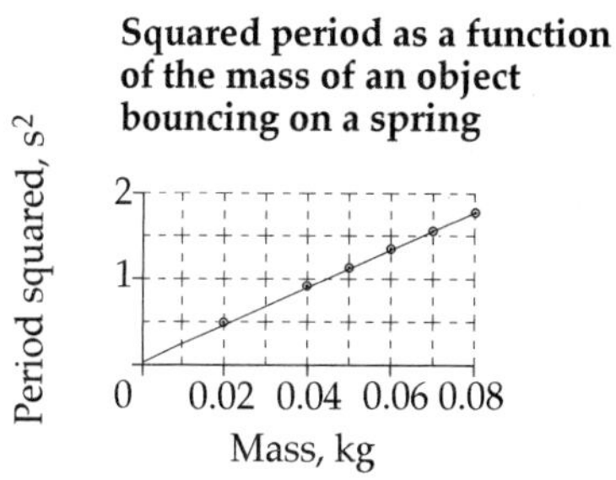

FIG. P12.56

continued on next page

The k values $1.74\ \text{N/m} \pm 6\%$

and $1.82\ \text{N/m} \pm 3\%$ differ by 4%

so $\boxed{\text{they agree.}}$

(c) Utilizing the axis-crossing point, $m_s = 3\left(\dfrac{0.058\,9}{21.7}\right)\text{kg} = \boxed{8\text{ grams} \pm 12\%}$

$\boxed{\text{in agreement}}$ with 7.4 grams.

P12.57 (a) We require $Ae^{-bt/2m} = \dfrac{A}{2}$ $\qquad e^{+bt/2m} = 2$

or $\dfrac{bt}{2m} = \ln 2$ $\qquad$ or $\dfrac{0.100\ \text{kg/s}}{2(0.375\ \text{kg})} t = 0.693$ $\qquad \therefore t = \boxed{5.20\text{ s}}$

The spring constant is irrelevant.

(b) We can evaluate the energy at successive turning points, where

$\cos(\omega t + \phi) = \pm 1$ and the energy is $\dfrac{1}{2}kx^2 = \dfrac{1}{2}kA^2 e^{-2bt/2m}$. We require $\dfrac{1}{2}kA^2 e^{-bt/m} = \dfrac{1}{2}\left(\dfrac{1}{2}kA^2\right)$

or $e^{+bt/m} = 2$ $\qquad \therefore t = \dfrac{m\ln 2}{b} = \dfrac{0.375\ \text{kg}(0.693)}{0.100\ \text{kg/s}} = \boxed{2.60\text{ s}}$.

(c) From $E = \dfrac{1}{2}kA^2$, the fractional rate of change of energy over time is

$$\frac{\frac{dE}{dt}}{E} = \frac{\frac{d}{dt}\left(\frac{1}{2}kA^2\right)}{\frac{1}{2}kA^2} = \frac{\frac{1}{2}k(2A)\frac{dA}{dt}}{\frac{1}{2}kA^2} = 2\frac{\frac{dA}{dt}}{A}$$

two times faster than the fractional rate of change in amplitude.

P12.58 (a) The block moves with the board in what we take as the positive x direction, stretching the spring until the spring force $-kx$ is equal in magnitude to the maximum force of static friction $\mu_s n = \mu_s mg$. This occurs at $x = \dfrac{\mu_s mg}{k}$.

(b) Since v is small, the block is nearly at the rest at this break point. It starts almost immediately to move back to the left, the forces on it being $-kx$ and $+\mu_k mg$. While it is sliding the net force exerted on it can be written as

$$-kx + \mu_k mg = -kx + \frac{k\mu_k mg}{k} = -k\left(x - \frac{\mu_k mg}{k}\right) = -kx_{rel}$$

where x_{rel} is the excursion of the block away from the point $\dfrac{\mu_k mg}{k}$.

continued on next page

Conclusion: the block goes into simple harmonic motion centered about the equilibrium position where the spring is stretched by $\frac{\mu_k mg}{k}$.

(d) The amplitude of its motion is its original displacement, $A = \frac{\mu_s mg}{k} - \frac{\mu_k mg}{k}$. It first comes to rest at spring extension $\frac{\mu_k mg}{k} - A = \frac{(2\mu_k - \mu_s)mg}{k}$. Almost immediately at this point it latches onto the slowly-moving board to move with the board. The board exerts a force of static friction on the block, and the cycle continues.

(c) The graph of the motion looks like this:

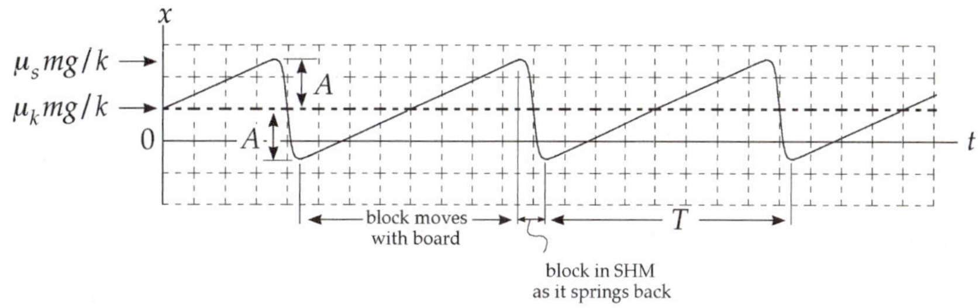

FIG. P12.58(c)

(e) The time during each cycle when the block is moving with the board is $\frac{2A}{v} = \frac{2(\mu_s - \mu_k)mg}{kv}$. The time for which the block is springing back is one half a cycle of simple harmonic motion, $\frac{1}{2}\left(2\pi\sqrt{\frac{m}{k}}\right) = \pi\sqrt{\frac{m}{k}}$. We ignore the times at the end points of the motion when the speed of the block changes from v to 0 and from 0 to v. Since v is small compared to $\frac{2A}{\pi\sqrt{\frac{m}{k}}}$, these times are negligible. Then the period is

$$T = \frac{2(\mu_s - \mu_k)mg}{kv} + \pi\sqrt{\frac{m}{k}}.$$

(f) $$T = \frac{2(0.4 - 0.25)(0.3\text{ kg})(9.8\text{ m/s}^2)}{(0.024\text{ m/s})(12\text{ N/m})} + \pi\sqrt{\frac{0.3\text{ kg}}{12\text{ N/m}}} = 3.06\text{ s} + 0.497\text{ s} = 3.56\text{ s}$$

Then $$f = \frac{1}{T} = \boxed{0.281\text{ Hz}}.$$

(g) $T = \frac{2(\mu_s - \mu_k)mg}{kv} + \pi\sqrt{\frac{m}{k}}$ increases as m increases, so the frequency $\boxed{\text{decreases}}$.

(h) As k increases, T decreases and f $\boxed{\text{increases}}$.

(i) As v increases, T decreases and f $\boxed{\text{increases}}$.

(j) As $(\mu_s - \mu_k)$ increases, T increases and f $\boxed{\text{decreases}}$.

P12.59 (a) When the mass is displaced a distance x from equilibrium, spring 1 is stretched a distance x_1 and spring 2 is stretched a distance x_2.

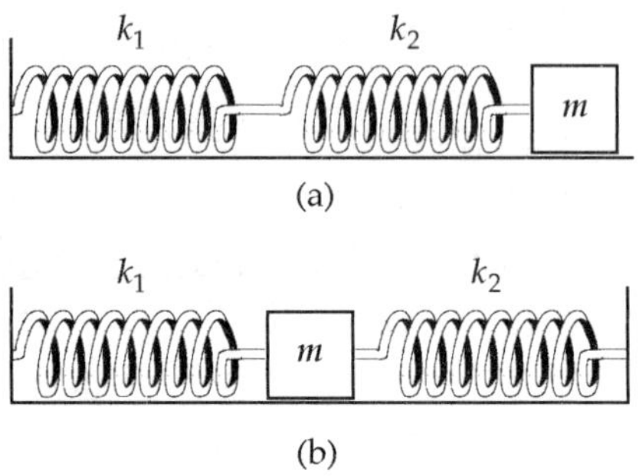

FIG. P12.59

By Newton's third law, we expect

$$k_1x_1 = k_2x_2.$$

When this is combined with the requirement that

$$x = x_1 + x_2,$$

we find $\quad x_1 = \left[\dfrac{k_2}{k_1+k_2}\right]x$

The force on either spring is given by $\quad F_1 = \left[\dfrac{k_1k_2}{k_1+k_2}\right]x = ma$

where a is the acceleration of the mass m.

This is in the form $\quad F = k_{eff}x = ma$

and $\quad T = 2\pi\sqrt{\dfrac{m}{k_{eff}}} = \boxed{2\pi\sqrt{\dfrac{m(k_1+k_2)}{k_1k_2}}}$

(b) In this case each spring is distorted by the distance x which the mass is displaced. Therefore, the restoring force is

$$F = -(k_1+k_2)x \qquad \text{and} \qquad k_{eff} = k_1 + k_2$$

so that $\quad T = \boxed{2\pi\sqrt{\dfrac{m}{(k_1+k_2)}}}$.

P12.60 (a) Newton's law of universal gravitation is $\quad F = -\dfrac{GMm}{r^2} = -\dfrac{Gm}{r^2}\left(\dfrac{4}{3}\pi r^3\right)\rho$

Thus, $\quad F = -\left(\dfrac{4}{3}\pi\rho Gm\right)r$

Which is of Hooke's law form with $\quad k = \dfrac{4}{3}\pi\rho Gm$

(b) The sack of mail moves without friction according to $\quad -\left(\dfrac{4}{3}\right)\pi\rho Gmr = ma$

$$a = -\left(\frac{4}{3}\right)\pi\rho Gr = -\omega^2 r$$

continued on next page

Since acceleration is a negative constant times excursion from equilibrium, it executes SHM with

$$\omega = \sqrt{\frac{4\pi\rho G}{3}} \qquad \text{and period} \qquad T = \frac{2\pi}{\omega} = \sqrt{\frac{3\pi}{\rho G}}$$

The time for a one-way trip through the Earth is $\frac{T}{2} = \sqrt{\frac{3\pi}{4\rho G}}$

We have also $g = \frac{GM_e}{R_e^2} = \frac{G4\pi R_e^3\rho}{3R_e^2} = \frac{4}{3}\pi\rho G R_e$

so $\frac{4\rho G}{3} = \frac{g}{(\pi R_e)}$ and $\frac{T}{2} = \pi\sqrt{\frac{R_e}{g}} = \pi\sqrt{\frac{6.37\times10^6\ \text{m}}{9.8\ \text{m/s}^2}} = 2.53\times10^3\ \text{s} = \boxed{42.2\ \text{min}}$.

(c) For an object moving without friction in a circular orbit just above the Earth's surface, we have $\sum F = ma$ $\qquad \frac{GMm}{R_e^2} = m\frac{v^2}{R_e} = \frac{m}{R_e}\left(\frac{2\pi R_e}{T}\right)^2$

$$mg = \frac{4\pi^2 mR_e}{T^2} \qquad T = 2\pi\sqrt{\frac{R_e}{g}} \qquad \text{so} \qquad \frac{T}{2} = \pi\sqrt{\frac{R_e}{g}}$$

The treetop satellite takes the same time to reach the antipode as the sack of mail. **They arrive at the same time.**

ANSWERS TO EVEN PROBLEMS

P12.2 (a) 4.33 cm; (b) –5.00 cm/s; (c) –17.3 cm/s^2; (d) 5.00 cm; 3.14 s as the particle passes through equilibrium

P12.4 (a) 20.0 cm; (b) 94.2 cm/s as the particle passes through equilibrium; (c) 17.8 m/s^2 at maximum excursion from equilibrium

P12.6 (a) –2.34 m; (b) –1.30 m/s; (c) –0.076 3 m; (d) 0.315 m/s

P12.8 (a) 2.40 s; (b) 0.417 Hz; (c) 2.62 rad/s

P12.10 12.0 Hz

P12.12 (a) 1.26 s; (b) 0.150 m/s; 0.750 m/s^2; (c) $x = -3.00\cos(5.00t)$ cm, $v = 15.0\sin(5.00t)$ cm/s, $a = 75.0\cos(5.00t)$ cm/s^2

P12.14 (a) 126 N/m; (b) 0.178 m

P12.16 (a) 0.153 J; (b) 0.784 m/s; (c) 17.5 m/s^2

P12.18 (a) 100 N/m; (b) 1.13 Hz; (c) 1.41 m/s as the block passes through equilibrium; (d) 10.0 m/s^2 at maximum excursion from equilibrium; (e) 2.00 J; (f) 1.33 m/s; (g) 3.33 m/s^2

P12.20 (a) 1.50 s; (b) 73.4 N/m; (c) 19.7 m below the bridge; (d) 1.06 rad/s; (e) 2.01 s; (f) 3.50 s

P12.22 $\frac{g_{\text{Cambridge}}}{g_{\text{Tokyo}}} = 1.0015$

P12.24 1.42 s; 0.499 m

P12.26 (a) see the solution; (b) 9.85 m/s^2 agreeing with the accepted value within 0.5%; (c) 9.85 m/s^2

P12.28 (a) 2.09 s; (b) 4.08%

P12.30 see the solution

P12.32 see the solution

P12.34 (a) 2.95 Hz; (b) 2.85 cm

P12.36 either 1.31 Hz or 0.641 Hz

P12.38 1.56 cm

P12.40 (a) 0.500 m/s; (b) 8.56 cm

P12.42 (a) 15.8 cm; (b) –15.9 cm; (c) see the solution; (d) 51.1 m; (e) 50.7 m

P12.44 (a) $\frac{4\pi^2 m}{T^2}$; (b) $m\left(\frac{T'}{T}\right)^2$

P12.46 (a) $0.368\hat{\mathbf{i}}$ m/s; (b) 3.51 cm; (c) see the solution; (d) 0.040 6 J; (e) 0.027 7 J

P12.48 (a) $x = (-2\text{ m})\sin\left[\left(10\text{ s}^{-1}\right)t\right]$; (b) at $x = \pm 1.73$ m; (c) 0.098 0 m; (d) 0.052 4 s

P12.50 (a) decreased by 0.735 m; (b) increased by 0.730 s; (c) decreased by 120 J; (d) see the solution

P12.52 (a) 3.56 Hz; (b) 2.79 Hz; (c) 2.10 Hz

P12.54 (a) $\frac{1}{2}\left(M + \frac{m}{3}\right)v^2$; (b) $2\pi\sqrt{\frac{M + \frac{m}{3}}{k}}$

P12.56 (a) $k = 1.74$ N/m $\pm 6\%$; (b) $k = 1.82$ N/m $\pm 3\%$ showing agreement; (c) 8 grams $\pm 12\%$ showing agreement

P12.58 (a) see the solution; (b) see the solution; (c) see the solution; (d) see the solution; (e) see the solution; (f) 0.281 Hz; (g) decreases; (h) increases; (i) increases; (j) decreases

P12.60 (a) see the solution; (b) 42.2 min; (c) they arrive at the same time

Chapter 13

Mechanical Waves

CHAPTER OUTLINE

ANSWERS TO QUESTIONS

Q13.1 To use a slinky to create a longitudinal wave, pull a few coils back and release. For a transverse wave, jostle the end coil side to side.

Q13.2 From $v=\sqrt{\frac{T}{\mu}}$, we must increase the tension by a factor of 4.

Q13.3 It depends on from what the wave reflects. If reflecting from a less dense string, the reflected part of the wave will be right side up.

Q13.4 The section of rope moves up and down in SHM. Its speed is always changing. The wave continues on with constant speed in one direction, setting further sections of the rope into up-and-down motion.

Q13.5 As the source frequency is doubled, the speed of waves on the string stays constant and the wavelength is reduced by one half.

Q13.6 As the source frequency is doubled, the speed of waves on the string stays constant.

Q13.7 Higher tension makes wave speed higher. Greater linear density makes the wave move more slowly.

Q13.8 As the wave passes from the massive string to the less massive string, the wave speed will increase according to $v=\sqrt{\frac{T}{\mu}}$. The frequency will remain unchanged. Since $v=f\lambda$, the wavelength must increase.

Q13.9 Amplitude is increased by a factor of $\sqrt{2}$. The wave speed does not change.

Q13.10 Sound waves are longitudinal because elements of the medium—parcels of air—move parallel and antiparallel to the direction of wave motion.

Q13.11 We assume that a perfect vacuum surrounds the clock. The sound waves require a medium for them to travel to your ear. The hammer on the alarm will strike the bell, and the vibration will spread as sound waves through the body of the clock. If a bone of your skull were in contact with the clock, you would hear the bell. However, in the absence of a surrounding medium like air or water, no sound can be radiated away. A larger-scale example of the same effect: Colossal storms raging on the Sun are deathly still for us.

What happens to the sound energy within the clock? Here is the answer: As the sound wave travels through the steel and plastic, traversing joints and going around corners, its energy is converted into additional internal energy, raising the temperature of the materials. After the sound has died away, the clock will glow very slightly brighter in the infrared portion of the electromagnetic spectrum.

Q13.12 The frequency increases by a factor of 2 because the wave speed, which is dependent only on the medium through which the wave travels, remains constant.

Q13.13 When listening, you are approximately the same distance from all of the members of the group. If different frequencies traveled at different speeds, then you might hear the higher pitched frequencies before you heard the lower ones produced at the same time. Although it might be interesting to think that each listener heard his or her own personal performance depending on where they were seated, a time lag like this could make a Beethoven sonata sound as if it were written by Charles Ives.

Q13.14 He saw the first wave he encountered, light traveling at 3.00×10^8 m/s. At the same moment, infrared as well as visible light began warming his skin, but some time was required to raise the temperature of the outer skin layers before he noticed it. The meteor produced compressional waves in the air and in the ground. The wave in the ground, which can be called either sound or a seismic wave, traveled much faster than the wave in air, since the ground is much stiffer against compression. Our witness received it next and noticed it as a little earthquake. He was no doubt unable to distinguish the P and S waves. The first air-compression wave he received was a shock wave with an amplitude on the order of meters. It transported him off his doorstep. Then he could hear some additional direct sound, reflected sound, and perhaps the sound of the falling trees.

Q13.15 For the sound from a source not to shift in frequency, the radial velocity of the source relative to the observer must be zero; that is, the source must not be moving toward or away from the observer. The source can be moving in a plane perpendicular to the line between it and the observer. Other possibilities: The source and observer might both have zero velocity. They might have equal velocities relative to the medium. The source might be moving around the observer on a sphere of constant radius. Even if the source speeds up on the sphere, slows down, or stops, the frequency heard will be equal to the frequency emitted by the source.

Q13.16 Wind can change a Doppler shift but cannot cause one. Both v_o and v_s in our equations must be interpreted as speeds of observer and source relative to the air. If source and observer are moving relative to each other, the observer will hear one shifted frequency in still air and a different shifted frequency if wind is blowing. If the distance between source and observer is constant, there will never be a Doppler shift.

Q13.17 Let $\Delta t = t_s - t_p$ represent the difference in arrival times of the two waves at a station at distance $d = v_s t_s = v_p t_p$ from the hypocenter. Then $d = \Delta t\left(\frac{1}{v_s} - \frac{1}{v_p}\right)^{-1}$. Knowing the distance from the first station places the hypocenter on a sphere around it. A measurement from a second station limits it to another sphere, which intersects with the first in a circle. Data from a third non-collinear station will generally limit the possibilities to a point.

SOLUTIONS TO PROBLEMS

Section 13.1 Propagation of a Disturbance

P13.1 Replace x by $x - vt = x - 4.5t$

to get $$\boxed{y = \frac{6}{\left[(x-4.5t)^2 + 3\right]}}$$

P13.2

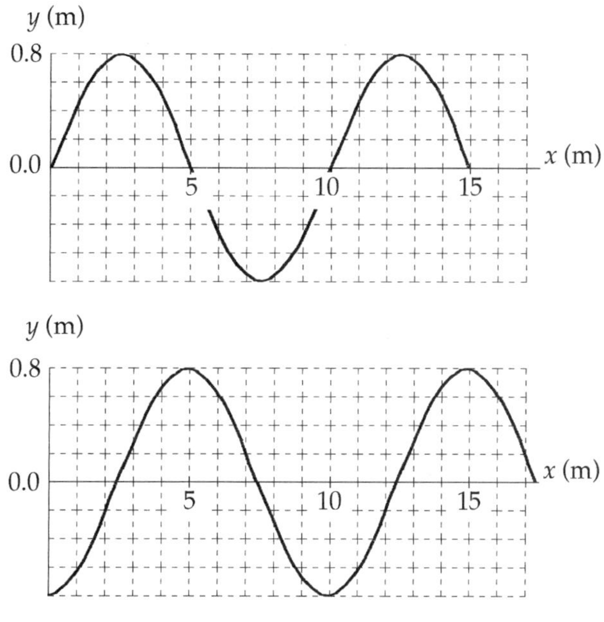

FIG. P13.2

Section 13.2 The Wave Model

Section 13.3 The Traveling Wave

P13.3 $f = \frac{40.0 \text{ vibrations}}{30.0 \text{ s}} = \frac{4}{3} \text{ Hz}$ $\qquad v = \frac{425 \text{ cm}}{10.0 \text{ s}} = 42.5 \text{ cm/s}$

$$\lambda = \frac{v}{f} = \frac{42.5 \text{ cm/s}}{\frac{4}{3} \text{ Hz}} = 31.9 \text{ cm} = \boxed{0.319 \text{ m}}$$

P13.4 Using data from the observations, we have $\lambda = 1.20\text{ m}$ and $f = \dfrac{8.00}{12.0\text{ s}}$. Therefore,

$$v = \lambda f = (1.20\text{ m})\left(\frac{8.00}{12.0\text{ s}}\right) = \boxed{0.800\text{ m/s}}.$$

P13.5 (a) Let $u = 10\pi t - 3\pi x + \dfrac{\pi}{4}$

$$\frac{du}{dt} = 10\pi - 3\pi\frac{dx}{dt} = 0 \text{ at a point of constant phase}$$

$$\frac{dx}{dt} = \frac{10}{3} = \boxed{3.33\text{ m/s}}$$

The velocity is in the $\boxed{\text{positive } x\text{-direction}}$.

(b) $$y(0.100,\ 0) = (0.350\text{ m})\sin\left(-0.300\pi + \frac{\pi}{4}\right) = -0.054\,8\text{ m} = \boxed{-5.48\text{ cm}}$$

(c) $$k = \frac{2\pi}{\lambda} = 3\pi:\ \lambda - \boxed{0.667\text{ m}}$$ $$\omega = 2\pi f = 10\pi:\ f = \boxed{5.00\text{ Hz}}$$

(d) $$v_y = \frac{\partial y}{\partial t} = (0.350)(10\pi)\cos\left(10\pi t - 3\pi x + \frac{\pi}{4}\right)$$ $$v_{y,\,\max} = (10\pi)(0.350) = \boxed{11.0\text{ m/s}}$$

P13.6 $y = (0.020\,0\text{ m})\sin(2.11x - 3.62t)$ in SI units $\quad A = \boxed{2.00\text{ cm}}$

$k = 2.11\text{ rad/m}$ $\quad \lambda = \dfrac{2\pi}{k} = \boxed{2.98\text{ m}}$

$\omega = 3.62\text{ rad/s}$ $\quad f = \dfrac{\omega}{2\pi} = \boxed{0.576\text{ Hz}}$

$$v = f\lambda = \frac{\omega}{2\pi}\frac{2\pi}{k} = \frac{3.62}{2.11} = \boxed{1.72\text{ m/s}}$$

P13.7 (a) $$\omega = 2\pi f = 2\pi\left(5\text{ s}^{-1}\right) = \boxed{31.4\text{ rad/s}}$$

(b) $$\lambda = \frac{v}{f} = \frac{20\text{ m/s}}{5\text{ s}^{-1}} = 4.00\text{ m}$$

$$k = \frac{2\pi}{\lambda} = \frac{2\pi}{4\text{ m}} = \boxed{1.57\text{ rad/m}}$$

(c) In $y = A\sin(kx - \omega t + \phi)$ we take $A = 12\text{ cm}$. At $x = 0$ and $t = 0$ we have $y = (12\text{ cm})\sin\phi$. To make this fit $y = 0$, we take $\phi = 0$. Then $\boxed{y = (12.0\text{ cm})\sin((1.57\text{ rad/m})x - (31.4\text{ rad/s})t)}$

(d) The transverse velocity is $\dfrac{\partial y}{\partial t} = -A\omega\cos(kx - \omega t)$. Its maximum magnitude is $A\omega = 12\text{ cm}(31.4\text{ rad/s}) = \boxed{3.77\text{ m/s}}$

(e) $$a_y = \frac{\partial v_y}{\partial t} = \frac{\partial}{\partial t}(-A\omega\cos(kx - \omega t)) = -A\omega^2\sin(kx - \omega t)$$

The maximum value is $A\omega^2 = (0.12\text{ m})\left(31.4\text{ s}^{-1}\right)^2 = \boxed{118\text{ m/s}^2}$.

***P13.8** At time *t*, the phase of $y = (15.0 \text{ cm})\cos(0.157x - 50.3t)$ at coordinate *x* is $\phi = (0.157 \text{ rad/cm})x - (50.3 \text{ rad/s})t$. Since $60.0° = \frac{\pi}{3}$ rad, the requirement for point *B* is that $\phi_B = \phi_A \pm \frac{\pi}{3}$ rad, or (since $x_A = 0$),

$$(0.157 \text{ rad/cm})x_B - (50.3 \text{ rad/s})t = 0 - (50.3 \text{ rad/s})t \pm \frac{\pi}{3} \text{ rad}.$$

This reduces to $x_B = \frac{\pm\pi \text{ rad}}{3(0.157 \text{ rad/cm})} = \boxed{\pm 6.67 \text{ cm}}$.

P13.9 (a) $A = y_{max} = 8.00 \text{ cm} = 0.080\,0 \text{ m}$ $\qquad k = \frac{2\pi}{\lambda} = \frac{2\pi}{(0.800 \text{ m})} = 7.85 \text{ m}^{-1}$

$$\omega = 2\pi f = 2\pi(3.00) = 6.00\pi \text{ rad/s}$$

Therefore, $y = A\sin(kx + \omega t)$

Or (where $y(0, t) = 0$ at $t = 0$) $\quad \boxed{y = (0.080\,0)\sin(7.85x + 6\pi t) \text{ m}}$

(b) In general, $y = 0.080\,0\sin(7.85x + 6\pi t + \phi)$

Assuming $y(x, 0) = 0$ at $x = 0.100$ m

then we require that $0 = 0.080\,0\sin(0.785 + \phi)$

or $\phi = -0.785$

Therefore, $\boxed{y = 0.080\,0\sin(7.85x + 6\pi t - 0.785) \text{ m}}$

P13.10 $y = (0.120 \text{ m})\sin\left(\frac{\pi}{8}x + 4\pi t\right)$

(a) $v_y = \frac{dy}{dt}$: $\quad v_y = (0.120)(4\pi)\cos\left(\frac{\pi}{8}x + 4\pi t\right)$

$$v_y(0.200 \text{ s}, 1.60 \text{ m}) = \boxed{-1.51 \text{ m/s}}$$

$a_y = \frac{dv}{dt}$: $\quad a_y = (-0.120 \text{ m})(4\pi)^2\sin\left(\frac{\pi}{8}x + 4\pi t\right)$

$$a_y(0.200 \text{ s}, 1.60 \text{ m}) = \boxed{0}$$

(b) $k = \frac{\pi}{8} = \frac{2\pi}{\lambda}$: $\quad \lambda = \boxed{16.0 \text{ m}}$

$\omega = 4\pi = \frac{2\pi}{T}$: $\quad T = \boxed{0.500 \text{ s}}$

$$v = \frac{\lambda}{T} = \frac{16.0 \text{ m}}{0.500 \text{ s}} = \boxed{32.0 \text{ m/s}}$$

P13.11 (a) Let us write the wave function as $y(x,\, t) = A\sin(kx + \omega t + \phi)$

$$y(0,\, 0) = A\sin\phi = 0.020\,0 \text{ m}$$

$$\left.\frac{dy}{dt}\right|_{0,\,0} = A\omega\cos\phi = -2.00 \text{ m/s}$$

Also,
$$\omega = \frac{2\pi}{T} = \frac{2\pi}{0.025\,0 \text{ s}} = 80.0\,\pi/\text{s}$$

$$A^2 = x_i^2 + \left(\frac{v_i}{\omega}\right)^2 = (0.020\,0 \text{ m})^2 + \left(\frac{2.00 \text{ m/s}}{80.0\,\pi/\text{s}}\right)^2$$

$$A = \boxed{0.021\,5 \text{ m}}$$

(b) $\dfrac{A\sin\phi}{A\cos\phi} = \dfrac{0.020\,0}{\frac{-2}{80.0\pi}} = -2.51 = \tan\phi$

Your calculator's answer $\tan^{-1}(-2.51) = -1.19$ rad has a negative sine and positive cosine, just the reverse of what is required. You must look beyond your calculator to find

$\phi = \pi - 1.19 \text{ rad} = \boxed{1.95 \text{ rad}}$

(c) $v_{y,\,\text{max}} = A\omega = 0.021\,5 \text{ m}(80.0\pi/\text{s}) = \boxed{5.41 \text{ m/s}}$

(d) $\lambda = v_x T = (30.0 \text{ m/s})(0.025\,0 \text{ s}) = 0.750 \text{ m}$

$k = \dfrac{2\pi}{\lambda} = \dfrac{2\pi}{0.750 \text{ m}} = 8.38/\text{m}$ $\qquad \omega = 80.0\pi/\text{s}$

$\boxed{y(x,\, t) = (0.021\,5 \text{ m})\sin(8.38x \text{ rad/m} + 80.0\pi t \text{ rad/s} + 1.95 \text{ rad})}$

P13.12 The linear wave equation is $\dfrac{\partial^2 y}{\partial x^2} = \dfrac{1}{v^2}\dfrac{\partial^2 y}{\partial t^2}$

If $y = e^{b(x-vt)}$

then $\dfrac{\partial y}{\partial t} = -bve^{b(x-vt)}$ and $\dfrac{\partial y}{\partial x} = be^{b(x-vt)}$

$\dfrac{\partial^2 y}{\partial t^2} = b^2v^2e^{b(x-vt)}$ and $\dfrac{\partial^2 y}{\partial x^2} = b^2e^{b(x-vt)}$

Therefore, $\dfrac{\partial^2 y}{\partial t^2} = v^2\dfrac{\partial^2 y}{\partial x^2}$, demonstrating that $e^{b(x-vt)}$ is a solution

Section 13.4 The Speed of Transverse Waves on Strings

P13.13 The down and back distance is $4.00\text{ m}+4.00\text{ m}=8.00\text{ m}$.

The speed is then $$v=\frac{d_{\text{total}}}{t}=\frac{4(8.00\text{ m})}{0.800\text{ s}}=40.0\text{ m/s}=\sqrt{\frac{T}{\mu}}$$

Now, $$\mu=\frac{0.200\text{ kg}}{4.00\text{ m}}=5.00\times10^{-2}\text{ kg/m}$$

So $$T=\mu v^2=\left(5.00\times10^{-2}\text{ kg/m}\right)\left(40.0\text{ m/s}\right)^2=\boxed{80.0\text{ N}}$$

P13.14 $T=Mg$ is the tension; $v=\sqrt{\frac{T}{\mu}}=\sqrt{\frac{Mg}{\frac{m}{L}}}=\sqrt{\frac{MgL}{m}}=\frac{L}{t}$ is the wave speed.

Then, $$\frac{MgL}{m}=\frac{L^2}{t^2}$$

and $$g=\frac{Lm}{Mt^2}=\frac{1.60\text{ m}\left(4.00\times10^{-3}\text{ kg}\right)}{3.00\text{ kg}\left(3.61\times10^{-2}\text{ s}\right)^2}=\boxed{1.64\text{ m/s}^2}$$

***P13.15** Since μ is constant, $\mu=\frac{T_2}{v_2^2}=\frac{T_1}{v_1^2}$ and

$$T_2=\left(\frac{v_2}{v_1}\right)^2T_1=\left(\frac{30.0\text{ m/s}}{20.0\text{ m/s}}\right)^2(6.00\text{ N})=\boxed{13.5\text{ N}}.$$

P13.16 From the free-body diagram $mg=2T\sin\theta$

$$T=\frac{mg}{2\sin\theta}$$

The angle θ is found from $$\cos\theta=\frac{\frac{3L}{8}}{\frac{L}{2}}=\frac{3}{4}$$

$$\therefore\theta=41.4°$$

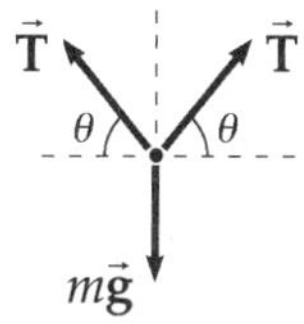

FIG. P13.16

(a) $v=\sqrt{\frac{T}{\mu}}$ $$v=\sqrt{\frac{mg}{2\mu\sin41.4°}}=\left(\sqrt{\frac{9.80\text{ m/s}^2}{2\left(8.00\times10^{-3}\text{ kg/m}\right)\sin41.4°}}\right)\sqrt{m}$$

or $$v=\boxed{\left(30.4\frac{\text{m/s}}{\sqrt{\text{kg}}}\right)\sqrt{m}}$$

(b) $v=60.0=30.4\sqrt{m}$ and $\boxed{m=3.89\text{ kg}}$

P13.17 The total time is the sum of the two times.

In each wire $t = \frac{L}{v} = L\sqrt{\frac{\mu}{T}}$

Let A represent the cross-sectional area of one wire. The mass of one wire can be written both as $m = \rho V = \rho AL$ and also as $m = \mu L$.

Then we have $\mu = \rho A = \frac{\pi \rho d^2}{4}$

Thus, $t = L\left(\frac{\pi \rho d^2}{4T}\right)^{1/2}$

For copper, $t = (20.0)\left[\frac{(\pi)(8\,920)(1.00\times 10^{-3})^2}{(4)(150)}\right]^{1/2} = 0.137 \text{ s}$

For steel, $t = (30.0)\left[\frac{(\pi)(7\,860)(1.00\times 10^{-3})^2}{(4)(150)}\right]^{1/2} = 0.192 \text{ s}$

The total time is $0.137 + 0.192 = \boxed{0.329 \text{ s}}$

Section 13.5 Reflection and Transmission of Waves

P13.18 (a) If the end is fixed, there is inversion of the pulse upon reflection. Thus, when they meet, they cancel and the amplitude is $\boxed{\text{zero}}$.

(b) If the end is free, there is no inversion on reflection. When they meet, the amplitude is $2A = 2(0.150 \text{ m}) = \boxed{0.300 \text{ m}}$.

Section 13.6 Rate of Energy Transfer by Sinusoidal Waves on Strings

P13.19 $f = \frac{v}{\lambda} = \frac{30.0}{0.500} = 60.0 \text{ Hz}$ $\omega = 2\pi f = 120\pi \text{ rad/s}$

$$\mathcal{P} = \frac{1}{2}\mu\omega^2 A^2 v = \frac{1}{2}\left(\frac{0.180}{3.60}\right)(120\pi)^2(0.100)^2(30.0) = \boxed{1.07 \text{ kW}}$$

P13.20 $\mu = 30.0 \text{ g/m} = 30.0\times10^{-3} \text{ kg/m}$

$\lambda = 1.50 \text{ m}$

$f = 50.0 \text{ Hz:} \qquad \omega = 2\pi f = 314 \text{ s}^{-1}$

$2A = 0.150 \text{ m:} \qquad A = 7.50\times10^{-2} \text{ m}$

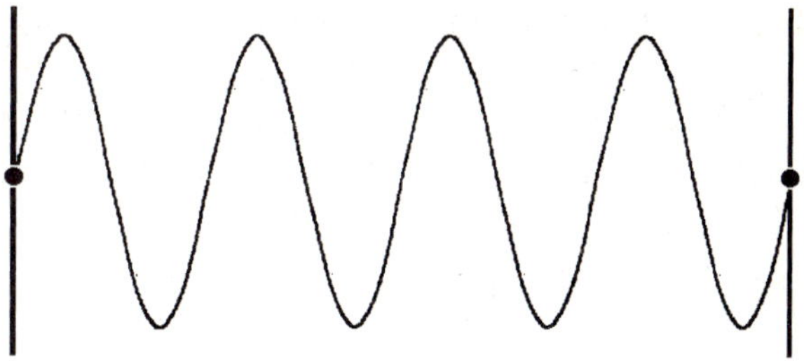

FIG. P13.20

(a) $$y = A\sin\left(\frac{2\pi}{\lambda}x - \omega t\right)$$

$$\boxed{y = \left(7.50\times10^{-2}\right)\sin(4.19x - 314t)}$$

(b) $$\mathcal{P} = \frac{1}{2}\mu\omega^2A^2v = \frac{1}{2}\left(30.0\times10^{-3}\right)(314)^2\left(7.50\times10^{-2}\right)^2\left(\frac{314}{4.19}\right) \text{ W} \qquad \boxed{\mathcal{P} = 625 \text{ W}}$$

P13.21 $A = 5.00\times10^{-2} \text{ m} \qquad \mu = 4.00\times10^{-2} \text{ kg/m} \qquad \mathcal{P} = 300 \text{ W} \qquad T = 100 \text{ N}$

Therefore, $$v = \sqrt{\frac{T}{\mu}} = 50.0 \text{ m/s}$$

$$\mathcal{P} = \frac{1}{2}\mu\omega^2A^2v: \quad \omega^2 = \frac{2\mathcal{P}}{\mu A^2 v} = \frac{2(300)}{\left(4.00\times10^{-2}\right)\left(5.00\times10^{-2}\right)^2(50.0)}$$

$$\omega = 346 \text{ rad/s}$$

$$f = \frac{\omega}{2\pi} = \boxed{55.1 \text{ Hz}}$$

***P13.22** Originally,

$$\mathcal{P}_0 = \frac{1}{2}\mu\omega^2A^2v$$

$$\mathcal{P}_0 = \frac{1}{2}\mu\omega^2A^2\sqrt{\frac{T}{\mu}}$$

$$\mathcal{P}_0 = \frac{1}{2}\omega^2A^2\sqrt{T\mu}$$

The doubled string will have doubled mass-per-length. Presuming that we hold tension constant, it can carry power larger by $\sqrt{2}$ times.

$$\boxed{\sqrt{2}\mathcal{P}_0} = \frac{1}{2}\omega^2A^2\sqrt{T2\mu}$$

Section 13.7 Sound Waves

P13.23 Since $v_{\text{light}} >> v_{\text{sound}}$: $d \approx (343 \text{ m/s})(16.2 \text{ s}) = \boxed{5.56 \text{ km}}$

***P13.24** The sound pulse must travel 150 m before reflection and 150 m after reflection. We have $d = vt$

$$t = \frac{d}{v} = \frac{300\text{ m}}{1\,533\text{ m/s}} = \boxed{0.196\text{ s}}$$

P13.25 (a) $\lambda = \frac{v}{f} = \frac{343\text{ m/s}}{1\,480\text{ s}^{-1}} = \boxed{0.232\text{ m}}$

(b) $\lambda' = \frac{v}{f'} = \frac{343\text{ m/s}}{1\,397\text{ s}^{-1}} = 0.246\text{ m},\ \Delta\lambda = \lambda' - \lambda = \boxed{13.8\text{ mm}}$

P13.26 $\lambda = \frac{v}{f} = \frac{340\text{ m/s}}{60.0 \times 10^{3}\text{ s}^{-1}} = \boxed{5.67\text{ mm}}$

***P13.27** The sound speed is $v = 331\text{ m/s} + 0.6\text{ m/s}\cdot°\text{C}(26°\text{C}) = 347\text{ m/s}$

(a) Let t represent the time for the echo to return. Then

$$d = \frac{1}{2}vt = \frac{1}{2}347\text{ m/s } 24 \times 10^{-3}\text{ s} = \boxed{4.16\text{ m}}.$$

(b) Let Δt represent the duration of the pulse:

$$\Delta t = \frac{10\lambda}{v} = \frac{10\lambda}{f\lambda} = \frac{10}{f} = \frac{10}{22 \times 10^{6}\text{ 1/s}} = \boxed{0.455\ \mu\text{s}}.$$

(c) $L = 10\lambda = \frac{10v}{f} = \frac{10(347\text{ m/s})}{22 \times 10^{6}\text{ 1/s}} = \boxed{0.158\text{ mm}}$

***P13.28** (a) If $f = 2.4\text{ MHz}$, $\lambda = \frac{v}{f} = \frac{1\,500\text{ m/s}}{2.4 \times 10^{6}/\text{s}} = \boxed{0.625\text{ mm}}$

(b) If $f = 1\text{ MHz}$, $\lambda = \frac{v}{f} = \frac{1\,500\text{ m/s}}{10^{6}/\text{s}} = \boxed{1.50\text{ mm}}$

If $f = 20\text{ MHz}$, $\lambda = \frac{1\,500\text{ m/s}}{2 \times 10^{7}/\text{s}} = \boxed{75.0\ \mu\text{m}}$

P13.29 $\Delta P_{\text{max}} = \rho v \omega s_{\text{max}} = \rho v \left(\frac{2\pi v}{\lambda}\right) s_{\text{max}}$

$$\lambda = \frac{2\pi \rho v^{2} s_{\text{max}}}{\Delta P_{\text{max}}} = \frac{2\pi(1.20)(343)^{2}(5.50 \times 10^{-6})}{0.840} = \boxed{5.81\text{ m}}$$

P13.30 (a) $A = \boxed{2.00\ \mu\text{m}}$

$$\lambda = \frac{2\pi}{15.7} = 0.400\ \text{m} = \boxed{40.0\ \text{cm}}$$

$$v = \frac{\omega}{k} = \frac{858}{15.7} = \boxed{54.6\ \text{m/s}}$$

(b) $$s = 2.00\cos\left[(15.7)(0.050\,0) - (858)\left(3.00\times10^{-3}\right)\right] = \boxed{-0.433\ \mu\text{m}}$$

(c) $$v_{\text{max}} = A\omega = (2.00\ \mu\text{m})\left(858\ \text{s}^{-1}\right) = \boxed{1.72\ \text{mm/s}}$$

P13.31 $$k = \frac{2\pi}{\lambda} = \frac{2\pi}{(0.100\ \text{m})} = 62.8\ \text{m}^{-1}$$

$$\omega = \frac{2\pi v}{\lambda} = \frac{2\pi(343\ \text{m/s})}{(0.100\ \text{m})} = 2.16\times10^{4}\ \text{s}^{-1}$$

Therefore, $\boxed{\Delta P = \left(0.200\ \text{N/m}^2\right)\sin\left[62.8\,x/\text{m} - 2.16\times10^{4}\,t/\text{s}\right]}$.

P13.32 $$\Delta P_{\text{max}} = \rho\omega v s_{\text{max}} = \left(1.20\ \text{kg/m}^3\right)\left[2\pi\left(2\,000\ \text{s}^{-1}\right)\right](343\ \text{m/s})\left(2.00\times10^{-8}\ \text{m}\right)$$

$$\Delta P_{\text{max}} = \boxed{0.103\ \text{Pa}}$$

Section 13.8 The Doppler Effect

P13.33 (a) $$f' = \frac{f(v+v_o)}{(v-v_s)}$$

$$f' = 2\,500\frac{(343+25.0)}{(343-40.0)} = \boxed{3.04\ \text{kHz}}$$

(b) $$f' = 2\,500\left(\frac{343+(-25.0)}{343-(-40.0)}\right) = \boxed{2.08\ \text{kHz}}$$

(c) $$f' = 2\,500\left(\frac{343+(-25.0)}{343-40.0}\right) = \boxed{2.62\ \text{kHz}}$$ while police car overtakes

$$f' = 2\,500\left(\frac{343+25.0}{343-(-40.0)}\right) = \boxed{2.40\ \text{kHz}}$$ after police car passes

P13.34 (a) $$\omega = 2\pi f = 2\pi\left(\frac{115/\text{min}}{60.0\ \text{s/min}}\right) = 12.0\ \text{rad/s}$$

$$v_{\max} = \omega A = (12.0\ \text{rad/s})(1.80\times 10^{-3}\ \text{m}) = \boxed{0.0217\ \text{m/s}}$$

(b) The heart wall is a moving observer.

$$f' = f\left(\frac{v+v_O}{v}\right) = (2\,000\,000\ \text{Hz})\left(\frac{1\,500+0.0217}{1\,500}\right) = \boxed{2\,000\,028.9\ \text{Hz}}$$

(c) Now the heart wall is a moving source.

$$f'' = f'\left(\frac{v}{v-v_s}\right) = (2\,000\,029\ \text{Hz})\left(\frac{1\,500}{1\,500-0.0217}\right) = \boxed{2\,000\,057.8\ \text{Hz}}$$

P13.35 Approaching ambulance: $$f' = \frac{f}{(1-v_S/v)}$$

Departing ambulance: $$f'' = \frac{f}{(1-(-v_S/v))}$$

Since $f' = 560$ Hz and $f'' = 480$ Hz $$560\left(1-\frac{v_S}{v}\right) = 480\left(1+\frac{v_S}{v}\right)$$

$$1\,040\frac{v_S}{v} = 80.0$$

$$v_S = \frac{80.0(343)}{1\,040}\ \text{m/s} = \boxed{26.4\ \text{m/s}}$$

P13.36 The maximum speed of the speaker is described by

$$\frac{1}{2}mv_{\max}^2 = \frac{1}{2}kA^2$$

$$v_{\max} = \sqrt{\frac{k}{m}}A = \sqrt{\frac{20.0\ \text{N/m}}{5.00\ \text{kg}}}(0.500\ \text{m}) = 1.00\ \text{m/s}$$

The frequencies heard by the stationary observer range from

$$f'_{\min} = f\left(\frac{v}{v+v_{\max}}\right) \text{ to } f'_{\max} = f\left(\frac{v}{v-v_{\max}}\right)$$

where v is the speed of sound.

$$f'_{\min} = 440\ \text{Hz}\left(\frac{343\ \text{m/s}}{343\ \text{m/s}+1.00\ \text{m/s}}\right) = \boxed{439\ \text{Hz}}$$

$$f'_{\max} = 440\ \text{Hz}\left(\frac{343\ \text{m/s}}{343\ \text{m/s}-1.00\ \text{m/s}}\right) = \boxed{441\ \text{Hz}}$$

P13.37 $f' = f\left(\frac{v}{v - v_s}\right)$ $485 = 512\left(\frac{340}{340 - (-9.80t_{\text{fall}})}\right)$

$$485(340) + (485)(9.80t_f) = (512)(340)$$

$$t_f = \left(\frac{512 - 485}{485}\right)\frac{340}{9.80} = 1.93 \text{ s}$$

$d_1 = \frac{1}{2}gt_f^2 = 18.3 \text{ m}$: $t_{\text{return}} = \frac{18.3}{340} = 0.053\,8 \text{ s}$

The fork continues to fall while the sound returns.

$$t_{\text{total fall}} = t_f + t_{\text{return}} = 1.93 \text{ s} + 0.053\,8 \text{ s} = 1.985 \text{ s}$$

$$d_{\text{total}} = \frac{1}{2}gt_{\text{total fall}}^2 = \boxed{19.3 \text{ m}}$$

P13.38 (a) $v = (331 \text{ m/s}) + 0.6\frac{\text{m}}{\text{s}\cdot{}^\circ\text{C}}(-10^\circ\text{C}) = \boxed{325 \text{ m/s}}$

(b) Approaching the bell, the athlete hears a frequency of $f' = f\left(\frac{v + v_O}{v}\right)$

After passing the bell, she hears a lower frequency of $f'' = f\left(\frac{v + (-v_O)}{v}\right)$

The ratio is $\frac{f''}{f'} = \frac{v - v_O}{v + v_O} = \frac{5}{6}$

which gives $6v - 6v_o = 5v + 5v_o$ or $v_O = \frac{v}{11} = \frac{325 \text{ m/s}}{11} = \boxed{29.5 \text{ m/s}}$

P13.39 (a) Sound moves upwind with speed $(343 - 15)$ m/s. Crests pass a stationary upwind point at frequency 900 Hz.

Then $\lambda = \frac{v}{f} = \frac{328 \text{ m/s}}{900/\text{s}} = \boxed{0.364 \text{ m}}$

(b) By similar logic, $\lambda = \frac{v}{f} = \frac{(343 + 15) \text{ m/s}}{900/\text{s}} = \boxed{0.398 \text{ m}}$

(c) The source is moving through the air at 15 m/s toward the observer. The observer is stationary relative to the air.

$$f' = f\left(\frac{v + v_o}{v - v_s}\right) = 900 \text{ Hz}\left(\frac{343 + 0}{343 - 15}\right) = \boxed{941 \text{ Hz}}$$

(d) The source is moving through the air at 15 m/s away from the downwind firefighter. Her speed relative to the air is 30 m/s toward the source.

$$f' = f\left(\frac{v + v_o}{v - v_s}\right) = 900 \text{ Hz}\left(\frac{343 + 30}{343 - (-15)}\right) = 900 \text{ Hz}\left(\frac{373}{358}\right) = \boxed{938 \text{ Hz}}$$

Section 13.9 Context Connection—Seismic Waves

P13.40 (a) The $\boxed{\text{longitudinal}}$ wave travels a shorter distance and is moving faster, so it will arrive at point B first.

(b) The wave that travels through the Earth must travel

a distance of $2R\sin 30.0° = 2(6.37\times10^6\ \text{m})\sin 30.0° = 6.37\times10^6\ \text{m}$

at a speed of 7 800 m/s

Therefore, it takes $\dfrac{6.37\times10^6\ \text{m}}{7\,800\ \text{m/s}} = 817\ \text{s}$

The wave that travels along the Earth's surface must travel

a distance of $s = R\theta = R\left(\dfrac{\pi}{3}\ \text{rad}\right) = 6.67\times10^6\ \text{m}$

at a speed of 4 500 m/s

Therefore, it takes $\dfrac{6.67\times10^6}{4\,500} = 1\,482\ \text{s}$

The time difference is $\boxed{665\ \text{s}} = 11.1\ \text{min}$

P13.41 The distance the waves have traveled is $d = (7.80\ \text{km/s})t = (4.50\ \text{km/s})(t + 17.3\ \text{s})$

where t is the travel time for the faster wave.

Then, $(7.80 - 4.50)(\text{km/s})t = (4.50\ \text{km/s})(17.3\ \text{s})$

or $t = \dfrac{(4.50\ \text{km/s})(17.3\ \text{s})}{(7.80-4.50)\ \text{km/s}} = 23.6\ \text{s}$

and the distance is $d = (7.80\ \text{km/s})(23.6\ \text{s}) = \boxed{184\ \text{km}}$.

Additional Problems

P13.42 Assume a typical distance between adjacent people $\sim 1\ \text{m}$.

Then the wave speed is $v = \dfrac{\Delta x}{\Delta t} \sim \dfrac{1\ \text{m}}{0.1\ \text{s}} \sim 10\ \text{m/s}$

Model the stadium as a circle with a radius of order 100 m. Then, the time for one circuit around the stadium is

$$T = \frac{2\pi r}{v} \sim \frac{2\pi(10^2)}{10\ \text{m/s}} = 63\ \text{s} \boxed{\sim 1\ \text{min}}.$$

P13.43 Assuming the incline to be frictionless and taking the positive x-direction to be up the incline:

$\sum F_x = T - Mg\sin\theta = 0$ or the tension in the string is $T = Mg\sin\theta$

The speed of transverse waves in the string is then $v = \sqrt{\dfrac{T}{\mu}} = \sqrt{\dfrac{Mg\sin\theta}{m/L}} = \sqrt{\dfrac{MgL\sin\theta}{m}}$

The time interval for a pulse to travel the string's length is $\Delta t = \dfrac{L}{v} = L\sqrt{\dfrac{m}{MgL\sin\theta}} = \boxed{\sqrt{\dfrac{mL}{Mg\sin\theta}}}$

P13.44 $Mgx = \dfrac{1}{2}kx^2$

(a) $T = kx = \boxed{2Mg}$

(b) $L = L_0 + x = \boxed{L_0 + \dfrac{2Mg}{k}}$

(c) $v = \sqrt{\dfrac{T}{\mu}} = \sqrt{\dfrac{TL}{m}} = \boxed{\sqrt{\dfrac{2Mg}{m}\left(L_0 + \dfrac{2Mg}{k}\right)}}$

***P13.45** Let M = mass of block, m = mass of string. For the block, $\sum F = ma$ implies $T = \dfrac{mv_b^2}{r} = m\omega^2 r$. The speed of a wave on the string is then

$$v = \sqrt{\frac{T}{\mu}} = \sqrt{\frac{M\omega^2 r}{\frac{m}{r}}} = r\omega\sqrt{\frac{M}{m}}$$

$$t = \frac{r}{v} = \frac{1}{\omega}\sqrt{\frac{m}{M}}$$

$$\theta = \omega t = \sqrt{\frac{m}{M}} = \sqrt{\frac{0.003\ 2\ \text{kg}}{0.450\ \text{kg}}} = \boxed{0.084\ 3\ \text{rad}}$$

P13.46 (a) Assume the spring is originally stationary throughout, extended to have a length L much greater than its equilibrium length. We start moving one end forward with the speed v at which a wave propagates on the spring. In this way we create a single pulse of compression that moves down the length of the spring. For an increment of spring with length dx and mass dm, just as the pulse swallows it up, $\sum F = ma$ becomes $kdx = adm$ or $\dfrac{k}{\frac{dm}{dx}} = a$. But $\dfrac{dm}{dx} = \mu$ so $a = \dfrac{k}{\mu}$. Also, $a = \dfrac{dv}{dt} = \dfrac{v}{t}$ when $v_i = 0$. But $L = vt$, so $a = \dfrac{v^2}{L}$. Equating the two expressions for a, we have $\dfrac{k}{\mu} = \dfrac{v^2}{L}$ or $\boxed{v = \sqrt{\dfrac{kL}{\mu}}}$.

(b) Using the expression from part (a) $v = \sqrt{\dfrac{kL}{\mu}} = \sqrt{\dfrac{kL^2}{m}} = \sqrt{\dfrac{(100\ \text{N/m})(2.00\ \text{m})^2}{0.400\ \text{kg}}} = \boxed{31.6\ \text{m/s}}$.

P13.47 $v = \sqrt{\frac{T}{\mu}}$ where $T = \mu x g$, the weight of a length x, of rope.

Therefore, $v = \sqrt{gx}$

But $v = \frac{dx}{dt}$, so that $dt = \frac{dx}{\sqrt{gx}}$

and $t = \int_0^L \frac{dx}{\sqrt{gx}} = \frac{1}{\sqrt{g}} \frac{\sqrt{x}}{\frac{1}{2}}\Big|_0^L = \boxed{2\sqrt{\frac{L}{g}}}$

P13.48 At distance x from the bottom, the tension is $T = \left(\frac{mxg}{L}\right) + Mg$, so the wave speed is:

$$v = \sqrt{\frac{T}{\mu}} = \sqrt{\frac{TL}{m}} = \sqrt{xg + \left(\frac{MgL}{m}\right)} = \frac{dx}{dt}.$$

(a) Then $t = \int_0^t dt = \int_0^L \left[xg + \left(\frac{MgL}{m}\right)\right]^{-1/2} dx$ $\qquad t = \frac{1}{g} \frac{\left[xg + (MgL/m)\right]^{1/2}}{\frac{1}{2}}\Bigg|_{x=0}^{x=L}$

$$t = \frac{2}{g}\left[\left(Lg + \frac{MgL}{m}\right)^{1/2} - \left(\frac{MgL}{m}\right)^{1/2}\right] \qquad \boxed{t = 2\sqrt{\frac{L}{g}}\left(\frac{\sqrt{m+M} - \sqrt{M}}{\sqrt{m}}\right)}$$

(b) When $M = 0$, as in the previous problem, $t = 2\sqrt{\frac{L}{g}}\left(\frac{\sqrt{m} - 0}{\sqrt{m}}\right) = \boxed{2\sqrt{\frac{L}{g}}}$

(c) As $m \to 0$ we expand $\sqrt{m+M} = \sqrt{M}\left(1 + \frac{m}{M}\right)^{1/2} = \sqrt{M}\left(1 + \frac{1}{2}\frac{m}{M} - \frac{1}{8}\frac{m^2}{M^2} + \ldots\right)$

to obtain $t = 2\sqrt{\frac{L}{g}}\left(\frac{\sqrt{M} + \frac{1}{2}\left(m/\sqrt{M}\right) - \frac{1}{8}\left(m^2/M^{3/2}\right) + \ldots - \sqrt{M}}{\sqrt{m}}\right)$

$$t \approx 2\sqrt{\frac{L}{g}}\left(\frac{1}{2}\sqrt{\frac{m}{M}}\right) = \boxed{\sqrt{\frac{mL}{Mg}}}$$

P13.49 (a) $\mathcal{P}(x) = \frac{1}{2}\mu\omega^2 A^2 v = \frac{1}{2}\mu\omega^2 A_0^2 e^{-2bx}\left(\frac{\omega}{k}\right) = \boxed{\frac{\mu\omega^3}{2k} A_0^2 e^{-2bx}}$

(b) $\mathcal{P}(0) = \boxed{\frac{\mu\omega^3}{2k} A_0^2}$

(c) $\frac{\mathcal{P}(x)}{\mathcal{P}(0)} = \boxed{e^{-2bx}}$

P13.50 $v = \frac{4\,450 \text{ km}}{9.50 \text{ h}} = 468 \text{ km/h} = \boxed{130 \text{ m/s}}$

$$\bar{d} = \frac{v^2}{g} = \frac{(130 \text{ m/s})^2}{\left(9.80 \text{ m/s}^2\right)} = \boxed{1\,730 \text{ m}}$$

P13.51 (a) $\mu(x)$ is a linear function, so it is of the form $\mu(x) = mx + b$

To have $\mu(0) = \mu_0$ we require $b = \mu_0$. Then $\mu(L) = \mu_L = mL + \mu_0$

so $m = \dfrac{\mu_L - \mu_0}{L}$

Then $\boxed{\mu(x) = \dfrac{(\mu_L - \mu_0)x}{L} + \mu_0}$

(b) From $v = \dfrac{dx}{dt}$, the time required to move from x to $x + dx$ is $\dfrac{dx}{v}$. The time required to move from 0 to L is

$$\Delta t = \int_0^L \frac{dx}{v} = \int_0^L \frac{dx}{\sqrt{\frac{T}{\mu}}} = \frac{1}{\sqrt{T}}\int_0^L \sqrt{\mu(x)}\,dx$$

$$\Delta t = \frac{1}{\sqrt{T}}\int_0^L \left(\frac{(\mu_L - \mu_0)x}{L} + \mu_0\right)^{1/2} \left(\frac{\mu_L - \mu_0}{L}\right)dx\left(\frac{L}{\mu_L - \mu_0}\right)$$

$$\Delta t = \frac{1}{\sqrt{T}}\left(\frac{L}{\mu_L - \mu_0}\right)\left(\frac{(\mu_L - \mu_0)x}{L} + \mu_0\right)^{3/2} \frac{1}{\frac{3}{2}}\Bigg|_0^L$$

$$\Delta t = \frac{2L}{3\sqrt{T}(\mu_L - \mu_0)}\left(\mu_L^{3/2} - \mu_0^{3/2}\right)$$

$$\Delta t = \frac{2L\left(\sqrt{\mu_L} - \sqrt{\mu_0}\right)\left(\mu_L + \sqrt{\mu_L\mu_0} + \mu_0\right)}{3\sqrt{T}\left(\sqrt{\mu_L} - \sqrt{\mu_0}\right)\left(\sqrt{\mu_L} + \sqrt{\mu_0}\right)}$$

$$\Delta t = \frac{2L}{3\sqrt{T}}\left(\frac{\mu_L + \sqrt{\mu_L\mu_0} + \mu_0}{\sqrt{\mu_L} + \sqrt{\mu_0}}\right)$$

P13.52 Sound takes this time to reach the man: $\dfrac{(20.0\text{ m} - 1.75\text{ m})}{343\text{ m/s}} = 5.32 \times 10^{-2}\text{ s}$

so the warning should be shouted no later than before the pot strikes. $0.300\text{ s} + 5.32 \times 10^{-2}\text{ s} = 0.353\text{ s}$

Since the whole time of fall is given by $y = \frac{1}{2}gt^2$: $18.25\text{ m} = \frac{1}{2}(9.80\text{ m/s}^2)t^2$

$t = 1.93\text{ s}$

the warning needs to come $1.93\text{ s} - 0.353\text{ s} = 1.58\text{ s}$

into the fall, when the pot has fallen $\frac{1}{2}(9.80\text{ m/s}^2)(1.58\text{ s})^2 = 12.2\text{ m}$

to be above the ground by $20.0\text{ m} - 12.2\text{ m} = \boxed{7.82\text{ m}}$

P13.53 Since $\cos^2\theta + \sin^2\theta = 1$, $\sin\theta = \pm\sqrt{1 - \cos^2\theta}$ (each sign applying half the time)

$$\Delta P = \Delta P_{max}\sin(kx - \omega t) = \pm\rho v\omega s_{max}\sqrt{1 - \cos^2(kx - \omega t)}$$

Therefore $\Delta P = \pm\rho v\omega\sqrt{s_{max}^2 - s_{max}^2\cos^2(kx - \omega t)} = \pm\rho v\omega\sqrt{s_{max}^2 - s^2}$

P13.54 The trucks form a train analogous to a wave train of crests with speed $v = 19.7\ \text{m/s}$ and unshifted frequency $f = \dfrac{2}{3.00\ \text{min}} = 0.667\ \text{min}^{-1}$.

(a) The cyclist as observer measures a lower Doppler-shifted frequency:

$$f' = f\left(\frac{v + v_o}{v}\right) = \left(0.667\ \text{min}^{-1}\right)\left(\frac{19.7 + (-4.47)}{19.7}\right) = \boxed{0.515/\text{min}}$$

(b) $$f'' = f\left(\frac{v + v_o'}{v}\right) = \left(0.667\ \text{min}^{-1}\right)\left(\frac{19.7 + (-1.56)}{19.7}\right) = \boxed{0.614/\text{min}}$$

The cyclist's speed has decreased very significantly, but there is only a modest increase in the frequency of trucks passing him.

P13.55 $v = \dfrac{2d}{t}$: $d = \dfrac{vt}{2} = \dfrac{1}{2}\left(6.50 \times 10^3\ \text{m/s}\right)(1.85\ \text{s}) = \boxed{6.01\ \text{km}}$

P13.56 (a) $$f' = \frac{fv}{v - u} \qquad f'' = \frac{fv}{v - (-u)} \qquad f' - f'' = fv\left(\frac{1}{v - u} - \frac{1}{v + u}\right)$$

$$\Delta f = \frac{fv(v + u - v + u)}{v^2 - u^2} = \frac{2uvf}{v^2\left(1 - \frac{u^2}{v^2}\right)} = \boxed{\frac{2\frac{u}{v}f}{1 - \frac{u^2}{v^2}}}$$

(b) $130\ \text{km/h} = 36.1\ \text{m/s}$

$$\therefore \Delta f = \frac{2(36.1)(400)}{340\left[1 - \frac{36.1^2}{340^2}\right]} = \boxed{85.9\ \text{Hz}}$$

***P13.57** (a) $$f' = f\frac{v}{\left(v - v_{\text{diver}}\right)}$$

so $$1 - \frac{v_{\text{diver}}}{v} = \frac{f}{f'}$$

$$\Rightarrow v_{\text{diver}} = v\left(1 - \frac{f}{f'}\right)$$

with $v = 343\ \text{m/s}$, $f = 1\,800\ \text{Hz}$ and $f' = 2\,150\ \text{Hz}$

we find

$$v_{\text{diver}} = 343\left(1 - \frac{1\,800}{2\,150}\right) = \boxed{55.8\ \text{m/s}}.$$

(b) If the waves are reflected, and the skydiver is moving into them, we have

$$f'' = f'\frac{\left(v + v_{\text{diver}}\right)}{v} \Rightarrow f'' = f\left[\frac{v}{\left(v - v_{\text{diver}}\right)}\right]\frac{\left(v + v_{\text{diver}}\right)}{v}$$

so $$f'' = 1\,800\frac{(343 + 55.8)}{(343 - 55.8)} = \boxed{2\,500\ \text{Hz}}.$$

***P13.58** (a)

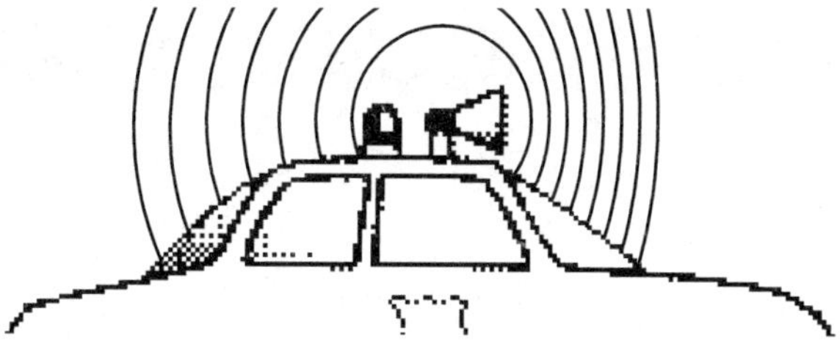

FIG. P13.58(a)

(b) $\lambda = \frac{v}{f} = \frac{343 \text{ m/s}}{1\,000 \text{ s}^{-1}} = \boxed{0.343 \text{ m}}$

(c) $\lambda' = \frac{v}{f'} = \frac{v}{f}\left(\frac{v - v_S}{v}\right) = \frac{(343 - 40.0) \text{ m/s}}{1\,000 \text{ s}^{-1}} = \boxed{0.303 \text{ m}}$

(d) $\lambda'' = \frac{v}{f''} = \frac{v}{f}\left(\frac{v + v_S}{v}\right) = \frac{(343 + 40.0) \text{ m/s}}{1\,000 \text{ s}^{-1}} = \boxed{0.383 \text{ m}}$

(e) $f' = f\left(\frac{v - v_O}{v - v_S}\right) = (1\,000 \text{ Hz})\frac{(343 - 30.0) \text{ m/s}}{(343 - 40.0) \text{ m/s}} = \boxed{1.03 \text{ kHz}}$

***P13.59** Use the Doppler formula, and remember that the bat is a moving source. If the velocity of the insect is v_x,

$$40.4 = 40.0\frac{(340 + 5.00)(340 - v_x)}{(340 - 5.00)(340 + v_x)}.$$

Solving, $v_x = 3.31 \text{ m/s}$. Therefore, $\boxed{\text{the bat is gaining on its prey at 1.69 m/s}}$.

P13.60 (a) If the source and the observer are moving away from each other, we have: $\theta_S - \theta_0 = 180°$, and since $\cos 180° = -1$, we get Equation 13.30 with negative values for both v_O and v_S.

(b) If $v_O = 0$ m/s then $f' = \frac{v}{v - v_S \cos\theta_S} f$. Also, when the train is 40.0 m from the intersection, and the car is 30.0 m from the intersection, $\cos\theta_S = \frac{4}{5}$ so

$$f' = \frac{343 \text{ m/s}}{343 \text{ m/s} - 0.800(25.0 \text{ m/s})}(500 \text{ Hz})$$

or $f' = \boxed{531 \text{ Hz}}$. Note that as the train approaches, passes, and departs from the intersection, θ_S varies from 0° to 180° and the frequency heard by the observer varies from:

$$f'_{\max} = \frac{v}{v - v_S \cos 0°} f = \frac{343 \text{ m/s}}{343 \text{ m/s} - 25.0 \text{ m/s}}(500 \text{ Hz}) = 539 \text{ Hz}$$

$$f'_{\min} = \frac{v}{v - v_S \cos 180°} f = \frac{343 \text{ m/s}}{343 \text{ m/s} + 25.0 \text{ m/s}}(500 \text{ Hz}) = 466 \text{ Hz}$$

ANSWERS TO EVEN PROBLEMS

P13.2 see the solution

P13.4 0.800 m/s

P13.6 2.00 cm, 2.98 m, 0.576 Hz, 1.72 m/s

P13.8 ±6.67 cm

P13.10 (a) –1.51 m/s, 0 m/s^2; (b) 16.0 m, 0.500 s, 32.0 m/s

P13.12 see the solution

P13.14 1.64 m/s^2

P13.16 (a) $v = \left(30.4 \dfrac{\text{m/s}}{\sqrt{\text{kg}}}\right)\sqrt{m}$; (b) 3.89 kg

P13.18 (a) zero; (b) 0.300 m

P13.20 (a) $y = (7.50 \times 10^{-2})\sin(4.19x - 314t)$; (b) 625 W

P13.22 $\sqrt{2}\mathcal{P}_0$

P13.24 0.196 s

P13.26 5.67 mm

P13.28 (a) 0.625 mm; (b) 1.50 mm to 75.0 μm

P13.30 (a) 2.00 μm, 40.0 cm, 54.6 m/s; (b) –0.433 μm; (c) 1.72 mm/s

P13.32 0.103 Pa

P13.34 (a) 0.021 7 m/s; (b) 2 000 028.9 Hz; (c) 2 000 057.8 Hz

P13.36 439 Hz and 441 Hz

P13.38 (a) 325 m/s; (b) 29.5 m/s

P13.40 (a) longitudinal; (b) 665 s

P13.42 ~1 min

P13.44 (a) $2Mg$; (b) $L_0 + \dfrac{2Mg}{k}$; (c) $\sqrt{\dfrac{2Mg}{m}\left(L_0 + \dfrac{2Mg}{k}\right)}$

P13.46 (a) $v = \sqrt{\dfrac{kL}{\mu}}$; (b) 31.6 m/s

P13.48 see the solution

P13.50 130 m/s, 1 730 m

P13.52 7.82 m

P13.54 (a) 0.515/min; (b) 0.614/min

P13.56 (a) $\dfrac{2\frac{u}{v}}{1 - \frac{u^2}{v^2}} f$; (b) 85.9 Hz

P13.58 (a) see the solution; (b) 0.343 m; (c) 0.303 m; (d) 0.383 m; (e) 1.03 kHz

P13.60 (a) see the solution; (b) 531 Hz

Chapter 14

Superposition and Standing Waves

CHAPTER OUTLINE

ANSWERS TO QUESTIONS

Q14.1 No. Waves with other waveforms are also trains of disturbance that add together when waves from different sources move through the same medium at the same time.

Q14.2 No. A wave is not a solid object, but a chain of disturbance. As described by the principle of superposition, the waves move through each other.

Q14.3 They can, wherever the two waves are nearly enough in phase that their displacements will add to create a total displacement greater than the amplitude of either of the two original waves.

When two one-dimensional sinusoidal waves of the same amplitude interfere, this condition is satisfied whenever the absolute value of the phase difference between the two waves is less than 120°.

Q14.4 When the two tubes together are not an efficient transmitter of sound from source to receiver, they are an efficient reflector. The incoming sound is reflected back to the source. The waves reflected by the two tubes separately at the junction interfere constructively.

Q14.5 No. The total energy of the pair of waves remains the same. Energy missing from zones of destructive interference appears in zones of constructive interference.

Q14.6 Damping, and non–linear effects in the vibration turn the energy of vibration into internal energy.

Q14.7 The air in the shower stall can vibrate in standing wave patterns to intensify those frequencies in your voice which correspond to its free vibrations. The hard walls of the bathroom reflect sound very well to make your voice more intense at all frequencies, and giving the room a longer reverberation time. The reverberant sound may help you to stay on key.

Q14.8 The trombone slide and trumpet valves change the length of the air column inside the instrument, to change its resonant frequencies.

Q14.9 In a classical guitar, vibrations of the strings are transferred to the wooden body through the bridge. Because of its large area, the guitar body is a much more efficient radiator of sound than an individual guitar string. Thus, energy associated with the vibration is transferred to the air relatively rapidly by the guitar body, resulting in a more intense sound.

Q14.10 The vibration of the air must have zero amplitude at the closed end. For air in a pipe closed at one end, the diagrams show how resonance vibrations have NA distances that are odd integer submultiples of the NA distance in the fundamental vibration. If the pipe is open, resonance vibrations have NA distances that are all integer submultiples of the NA distance in the fundamental.

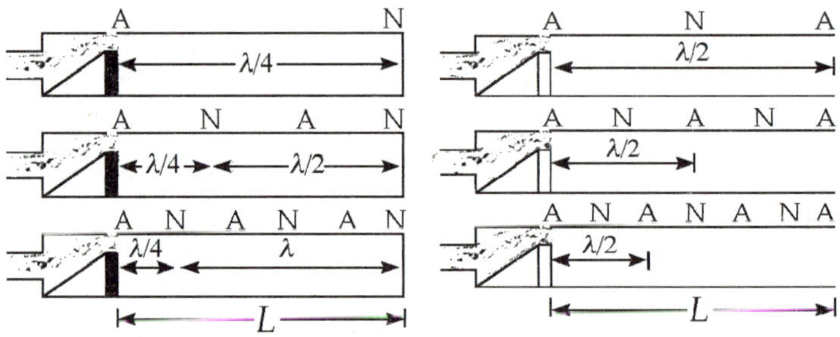

FIG. Q14.10

Q14.11 The bow string is pulled away from equilibrium and released, similar to the way that a guitar string is pulled and released when it is plucked. Thus, standing waves will be excited in the bow string. If the arrow leaves from the exact center of the string, then a series of odd harmonics will be excited. Even harmonies will not be excited because they have a node at the point where the string exhibits its maximum displacement.

Q14.12 What is needed is a tuning fork—or other pure-tone generator—of the desired frequency. Strike the tuning fork and pluck the corresponding string on the piano at the same time. If they are precisely in tune, you will hear a single pitch with no amplitude modulation. If the two pitches are a bit off, you will hear beats. As they vibrate, retune the piano string until the beat frequency goes to zero.

Q14.13 Beats. The propellers are rotating at slightly different frequencies.

Q14.14 Walking makes the person's hand vibrate a little. If the frequency of this motion is equal to the natural frequency of coffee sloshing from side to side in the cup, then a large–amplitude vibration of the coffee will build up in resonance. To get off resonance and back to the normal case of a small-amplitude disturbance producing a small–amplitude result, the person can walk faster, walk slower, or get a larger or smaller cup. Alternatively, even at resonance he can reduce the amplitude by adding damping, as by stirring high–fiber quick–cooking oatmeal into the hot coffee.

Q14.15 Stick a bit of chewing gum to one tine of the second fork. If the beat frequency is then faster than 4 beats per second, the second has a lower frequency than the standard fork. If the beats have slowed down, the second fork has a higher frequency than the standard. Remove the gum, clean the fork, add or subtract 4 Hz according to what you found, and your answer will be the frequency of the second fork.

Q14.16 Instead of just radiating sound very softly into the surrounding air, the tuning fork makes the chalkboard vibrate. With its large area this stiff sounding board radiates sound into the air with higher power. So it drains away the fork's energy of vibration faster and the fork stops vibrating sooner. This process exemplifies conservation of energy, as the energy of vibration of the fork is transferred through the blackboard into energy of vibration of the air.

SOLUTIONS TO PROBLEMS

Section 14.1 The Principle of Superposition

P14.1 $y = y_1 + y_2 = 3.00\cos(4.00x - 1.60t) + 4.00\sin(5.0x - 2.00t)$ evaluated at the given x values.

(a) $x = 1.00,\ t = 1.00$ $\quad y = 3.00\cos(2.40\text{ rad}) + 4.00\sin(+3.00\text{ rad}) = \boxed{-1.65\text{ cm}}$

(b) $x = 1.00,\ t = 0.500$ $\quad y = 3.00\cos(+3.20\text{ rad}) + 4.00\sin(+4.00\text{ rad}) = \boxed{-6.02\text{ cm}}$

(c) $x = 0.500,\ t = 0$ $\quad y = 3.00\cos(+2.00\text{ rad}) + 4.00\sin(+2.50\text{ rad}) = \boxed{1.15\text{ cm}}$

P14.2

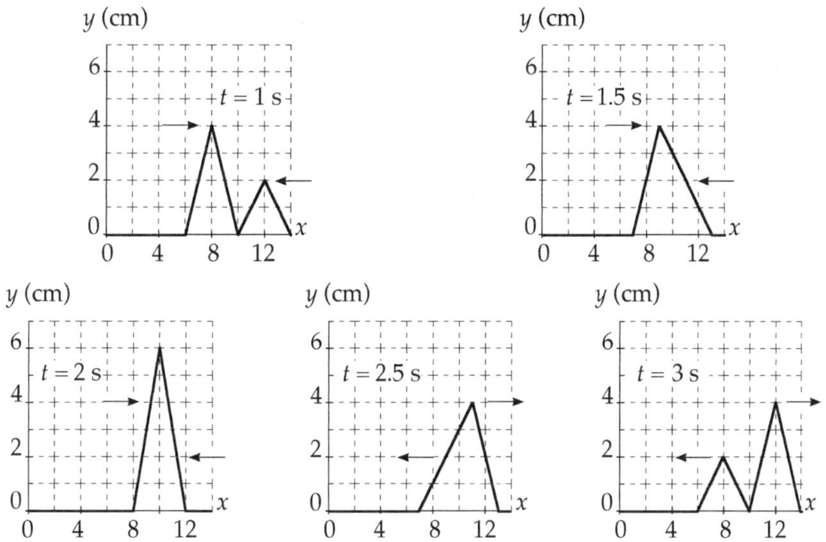

FIG. P14.2

P14.3 (a) $y_1 = f(x - vt)$, so wave 1 travels in the $\boxed{+x\text{ direction}}$

$y_2 = f(x + vt)$, so wave 2 travels in the $\boxed{-x\text{ direction}}$

(b) To cancel, $y_1 + y_2 = 0$: $\quad \dfrac{5}{(3x-4t)^2+2} = \dfrac{+5}{(3x+4t-6)^2+2}$

$$(3x-4t)^2 = (3x+4t-6)^2$$
$$3x - 4t = \pm(3x + 4t - 6)$$

for the positive root, $\quad 8t = 6$ $\quad \boxed{t = 0.750\text{ s}}$

(at $t = 0.750$ s, the waves cancel everywhere)

(c) for the negative root, $\quad 6x = 6$ $\quad \boxed{x = 1.00\text{ m}}$

(at $x = 1.00$ m, the waves cancel always)

Section 14.2 Interference of Waves

P14.4 Suppose the waves are sinusoidal.

The sum is $(4.00\text{ cm})\sin(kx-\omega t)+(4.00\text{ cm})\sin(kx-\omega t+90.0°)$

$$2(4.00\text{ cm})\sin(kx-\omega t+45.0°)\cos 45.0°$$

So the amplitude is $(8.00\text{ cm})\cos 45.0°=\boxed{5.66\text{ cm}}$.

P14.5 The resultant wave function has the form

$$y=2A_0\cos\left(\frac{\phi}{2}\right)\sin\left(kx-\omega t+\frac{\phi}{2}\right)$$

(a) $A=2A_0\cos\left(\frac{\phi}{2}\right)=2(5.00)\cos\left[\frac{-\pi/4}{2}\right]=\boxed{9.24\text{ m}}$

(b) $f=\frac{\omega}{2\pi}=\frac{1\,200\pi}{2\pi}=\boxed{600\text{ Hz}}$

P14.6 $2A_0\cos\left(\frac{\phi}{2}\right)=A_0$ so $\frac{\phi}{2}=\cos^{-1}\left(\frac{1}{2}\right)=60.0°=\frac{\pi}{3}$

Thus, the phase difference is $\phi=120°=\frac{2\pi}{3}$

This phase difference results if the time delay is $\frac{T}{3}=\frac{1}{3f}=\frac{\lambda}{3v}$

$$\text{Time delay}=\frac{3.00\text{ m}}{3(2.00\text{ m/s})}=\boxed{0.500\text{ s}}$$

P14.7 Waves reflecting from the near end travel 28.0 m (14.0 m down and 14.0 m back), while waves reflecting from the far end travel 66.0 m. The path difference for the two waves is:

$$\Delta r=66.0\text{ m}-28.0\text{ m}=38.0\text{ m}$$

Since $\lambda=\frac{v}{f}$

Then $\frac{\Delta r}{\lambda}=\frac{(\Delta r)f}{v}=\frac{(38.0\text{ m})(246\text{ Hz})}{343\text{ m/s}}=27.254$

or, $\Delta r=27.254\lambda$

The phase difference between the two reflected waves is then

$$\phi=0.254(1\text{ cycle})=0.254(2\pi\text{ rad})=\boxed{91.3°}$$

P14.8 (a) $\Delta x = \sqrt{9.00 + 4.00} - 3.00 = \sqrt{13} - 3.00 = 0.606 \text{ m}$

The wavelength is $\lambda = \frac{v}{f} = \frac{343 \text{ m/s}}{300 \text{ Hz}} = 1.14 \text{ m}$

Thus, $\frac{\Delta x}{\lambda} = \frac{0.606}{1.14} = 0.530 \text{ of a wave},$

or $\Delta\phi = 2\pi(0.530) = \boxed{3.33 \text{ rad}}$

(b) For destructive interference, we want $\frac{\Delta x}{\lambda} = 0.500 = f\frac{\Delta x}{v}$

where Δx is a constant in this set up. $f = \frac{v}{2\Delta x} = \frac{343}{2(0.606)} = \boxed{283 \text{ Hz}}$

***P14.9** (a) $\phi_1 = (20.0 \text{ rad/cm})(5.00 \text{ cm}) - (32.0 \text{ rad/s})(2.00 \text{ s}) = 36.0 \text{ rad}$

$\phi_1 = (25.0 \text{ rad/cm})(5.00 \text{ cm}) - (40.0 \text{ rad/s})(2.00 \text{ s}) = 45.0 \text{ rad}$

$\Delta\phi = 9.00 \text{ radians} = 516° = \boxed{156°}$

(b) $\Delta\phi = \left|20.0x - 32.0t - [25.0x - 40.0t]\right| = \left|-5.00x + 8.00t\right|$

At $t = 2.00 \text{ s}$, the requirement is

$$\Delta\phi = \left|-5.00x + 8.00(2.00)\right| = (2n+1)\pi \text{ for any integer } n.$$

For $x < 3.20$, $-5.00x + 16.0$ is positive, so we have

$$-5.00x + 16.0 = (2n+1)\pi, \text{ or}$$

$$x = 3.20 - \frac{(2n+1)\pi}{5.00}$$

The smallest positive value of x occurs for $n = 2$ and is

$$x = 3.20 - \frac{(4+1)\pi}{5.00} = 3.20 - \pi = \boxed{0.058\,4 \text{ cm}}.$$

P14.10 Suppose the man's ears are at the same level as the lower speaker. Sound from the upper speaker is delayed by traveling the extra distance $\Delta r = \sqrt{L^2 + d^2} - L$.

He hears a minimum when $\Delta r = (2n-1)\left(\frac{\lambda}{2}\right)$ with $n = 1, 2, 3, \ldots$

Then, $\sqrt{L^2 + d^2} - L = \left(n - \frac{1}{2}\right)\left(\frac{v}{f}\right)$

$$\sqrt{L^2 + d^2} = \left(n - \frac{1}{2}\right)\left(\frac{v}{f}\right) + L$$

$$L^2 + d^2 = \left(n - \frac{1}{2}\right)^2\left(\frac{v}{f}\right)^2 + 2\left(n - \frac{1}{2}\right)\left(\frac{v}{f}\right)L + L^2$$

continued on next page

$$d^2-\left(n-\frac{1}{2}\right)^2\left(\frac{v}{f}\right)^2=2\left(n-\frac{1}{2}\right)\left(\frac{v}{f}\right)L \qquad (1)$$

Equation 1 gives the distances from the lower speaker at which the man will hear a minimum. The path difference Δr starts from nearly zero when the man is very far away and increases to d when $L=0$.

(a) The number of minima he hears is the greatest integer value for which $L\geq 0$. This is the same as the greatest integer solution to $d\geq\left(n-\frac{1}{2}\right)\left(\frac{v}{f}\right)$, or

$$\boxed{\text{number of minima heard } = n_{\max} = \text{greatest integer } \leq d\left(\frac{f}{v}\right)+\frac{1}{2}}.$$

(b) From equation 1, the distances at which minima occur are given by

$$\boxed{L_n=\frac{d^2-\left(n-\frac{1}{2}\right)^2\left(\frac{v}{f}\right)^2}{2\left(n-\frac{1}{2}\right)\left(\frac{v}{f}\right)} \text{ where } n=1, 2, \ldots, n_{\max}}.$$

P14.11 (a) First we calculate the wavelength: $\lambda=\frac{v}{f}=\frac{344\text{ m/s}}{21.5\text{ Hz}}=16.0\text{ m}$

Then we note that the path difference equals $9.00\text{ m}-1.00\text{ m}=\boxed{\frac{1}{2}\lambda}$

Therefore, the receiver will record a minimum in sound intensity.

(b) We choose the origin at the midpoint between the speakers. If the receiver is located at point (x, y), then we must solve:

$$\sqrt{(x+5.00)^2+y^2}-\sqrt{(x-5.00)^2+y^2}=\frac{1}{2}\lambda$$

Then, $$\sqrt{(x+5.00)^2+y^2}=\sqrt{(x-5.00)^2+y^2}+\frac{1}{2}\lambda$$

Square both sides and simplify to get: $$20.0x-\frac{\lambda^2}{4}=\lambda\sqrt{(x-5.00)^2+y^2}$$

Upon squaring again, this reduces to: $$400x^2-10.0\lambda^2x+\frac{\lambda^4}{16.0}=\lambda^2(x-5.00)^2+\lambda^2y^2$$

Substituting $\lambda=16.0\text{ m}$, and reducing, $$\boxed{9.00x^2-16.0y^2=144}$$

or $$\frac{x^2}{16.0}-\frac{y^2}{9.00}=1$$

(When plotted this yields a curve called a hyperbola.)

Section 14.3 Standing Waves

P14.12 $y = (1.50 \text{ m})\sin(0.400x)\cos(200t) = 2A_0 \sin kx \cos \omega t$

Therefore, $k = \frac{2\pi}{\lambda} = 0.400 \text{ rad/m}$ $\quad \lambda = \frac{2\pi}{0.400 \text{ rad/m}} = \boxed{15.7 \text{ m}}$

and $\omega = 2\pi f$ so $\quad f = \frac{\omega}{2\pi} = \frac{200 \text{ rad/s}}{2\pi \text{rad}} = \boxed{31.8 \text{ Hz}}$

The speed of waves in the medium is $v = \lambda f = \frac{\lambda}{2\pi} 2\pi f = \frac{\omega}{k} = \frac{200 \text{ rad/s}}{0.400 \text{ rad/m}} = \boxed{500 \text{ m/s}}$

P14.13 The facing speakers produce a standing wave in the space between them, with the spacing between nodes being

$$d_{NN} = \frac{\lambda}{2} = \frac{v}{2f} = \frac{343 \text{ m/s}}{2\left(800 \text{ s}^{-1}\right)} = 0.214 \text{ m}$$

If the speakers vibrate in phase, the point halfway between them is an antinode of pressure at a distance from either speaker of

$$\frac{1.25 \text{ m}}{2} = 0.625.$$

Then there isa node at $\quad 0.625 - \frac{0.214}{2} = \boxed{0.518 \text{ m}}$

a node at $\quad 0.518 \text{ m} - 0.214 \text{ m} = \boxed{0.303 \text{ m}}$

a node at $\quad 0.303 \text{ m} - 0.214 \text{ m} = \boxed{0.0891 \text{ m}}$

a node at $\quad 0.518 \text{ m} + 0.214 \text{ m} = \boxed{0.732 \text{ m}}$

a node at $\quad 0.732 \text{ m} + 0.214 \text{ m} = \boxed{0.947 \text{ m}}$

and a node at $\quad 0.947 \text{ m} + 0.214 \text{ m} = \boxed{1.16 \text{ m}}$ from either speaker.

P14.14 $y = 2A_0 \sin kx \cos \omega t$

$$\frac{\partial^2 y}{\partial x^2} = -2A_0 k^2 \sin kx \cos \omega t \qquad \frac{\partial^2 y}{\partial t^2} = -2A_0 \omega^2 \sin kx \cos \omega t$$

Substitution into the wave equation gives $\quad -2A_0 k^2 \sin kx \cos \omega t = \left(\frac{1}{v^2}\right)\left(-2A_0 \omega^2 \sin kx \cos \omega t\right)$

This is satisfied, provided that $\quad v = \frac{\omega}{k}$

P14.15 $y_1 = 3.00\sin[\pi(x+0.600t)]$ cm; $y_2 = 3.00\sin[\pi(x-0.600t)]$ cm

$$y = y_1 + y_2 = [3.00\sin(\pi x)\cos(0.600\pi t) + 3.00\sin(\pi x)\cos(0.600\pi t)] \text{ cm}$$

$$y = (6.00 \text{ cm})\sin(\pi x)\cos(0.600\pi t)$$

(a) We can take $\cos(0.600\pi t) = 1$ to get the maximum y.

At $x = 0.250$ cm, $\quad y_{max} = (6.00 \text{ cm})\sin(0.250\pi) = \boxed{4.24 \text{ cm}}$

(b) At $x = 0.500$ cm, $\quad y_{max} = (6.00 \text{ cm})\sin(0.500\pi) = \boxed{6.00 \text{ cm}}$

(c) Now take $\cos(0.600\pi t) = -1$ to get y_{max}:

At $x = 1.50$ cm, $\quad y_{max} = (6.00 \text{ cm})\sin(1.50\pi)(-1) = \boxed{6.00 \text{ cm}}$

(d) The antinodes occur when $\quad x = \dfrac{n\lambda}{4}$ $(n = 1, 3, 5, \ldots)$

But $k = \dfrac{2\pi}{\lambda} = \pi$, so $\quad \lambda = 2.00$ cm

and $\quad x_1 = \dfrac{\lambda}{4} = \boxed{0.500 \text{ cm}}$ as in (b)

$x_2 = \dfrac{3\lambda}{4} = \boxed{1.50 \text{ cm}}$ as in (c)

$x_3 = \dfrac{5\lambda}{4} = \boxed{2.50 \text{ cm}}$

P14.16 (a) The resultant wave is $\quad y = 2A\sin\left(kx + \dfrac{\phi}{2}\right)\cos\left(\omega t - \dfrac{\phi}{2}\right)$

The nodes are located at $\quad kx + \dfrac{\phi}{2} = n\pi$

so $\quad x = \dfrac{n\pi}{k} - \dfrac{\phi}{2k}$

which means that each node is shifted $\dfrac{\phi}{2k}$ to the left.

(b) The separation of nodes is $\quad \Delta x = \left[(n+1)\dfrac{\pi}{k} - \dfrac{\phi}{2k}\right] - \left[\dfrac{n\pi}{k} - \dfrac{\phi}{2k}\right] \qquad \Delta x = \dfrac{\pi}{k} = \dfrac{\lambda}{2}$

The nodes are still separated by half a wavelength.

Section 14.4 Standing Waves in Strings

P14.17 $L = 30.0\text{ m};\ \mu = 9.00\times10^{-3}\text{ kg/m};\ T = 20.0\text{ N};\ f_1 = \dfrac{v}{2L}$

where $v = \left(\dfrac{T}{\mu}\right)^{1/2} = 47.1\text{ m/s}$

so $f_1 = \dfrac{47.1}{60.0} = \boxed{0.786\text{ Hz}}$ $\quad f_2 = 2f_1 = \boxed{1.57\text{ Hz}}$

$f_3 = 3f_1 = \boxed{2.36\text{ Hz}}$ $\quad f_4 = 4f_1 = \boxed{3.14\text{ Hz}}$

P14.18 $L = 120\text{ cm},\ f = 120\text{ Hz}$

(a) For four segments, $L = 2\lambda$ or $\lambda = 60.0\text{ cm} = \boxed{0.600\text{ m}}$

(b) $v = \lambda f = 72.0\text{ m/s}\ \ f_1 = \dfrac{v}{2L} = \dfrac{72.0}{2(1.20)} = \boxed{30.0\text{ Hz}}$

P14.19 The tension in the string is $T = (4\text{ kg})(9.8\text{ m/s}^2) = 39.2\text{ N}$

Its linear density is $\mu = \dfrac{m}{L} = \dfrac{8\times10^{-3}\text{ kg}}{5\text{ m}} = 1.6\times10^{-3}\text{ kg/m}$

and the wave speed on the string is $v = \sqrt{\dfrac{T}{\mu}} = \sqrt{\dfrac{39.2\text{ N}}{1.6\times10^{-3}\text{ kg/m}}} = 156.5\text{ m/s}$

In its fundamental mode of vibration, we have $\lambda = 2L = 2(5\text{ m}) = 10\text{ m}$

Thus, $f = \dfrac{v}{\lambda} = \dfrac{156.5\text{ m/s}}{10\text{ m}} = \boxed{15.7\text{ Hz}}$

P14.20 (a) Let n be the number of nodes in the standing wave resulting from the 25.0-kg mass. Then $n+1$ is the number of nodes for the standing wave resulting from the 16.0-kg mass. For standing waves, $\lambda = \dfrac{2L}{n}$, and the frequency is $f = \dfrac{v}{\lambda}$.

Thus, $f = \dfrac{n}{2L}\sqrt{\dfrac{T_n}{\mu}}$

and also $f = \dfrac{n+1}{2L}\sqrt{\dfrac{T_{n+1}}{\mu}}$

Thus, $\dfrac{n+1}{n} = \sqrt{\dfrac{T_n}{T_{n+1}}} = \sqrt{\dfrac{(25.0\text{ kg})g}{(16.0\text{ kg})g}} = \dfrac{5}{4}$

Therefore, $4n + 4 = 5n$, or $n = 4$

Then, $f = \dfrac{4}{2(2.00\text{ m})}\sqrt{\dfrac{(25.0\text{ kg})(9.80\text{ m/s}^2)}{0.002\,00\text{ kg/m}}} = \boxed{350\text{ Hz}}$

continued on next page

(b) The largest mass will correspond to a standing wave of 1 loop

$(n = 1)$ so $\quad 350\text{ Hz} = \dfrac{1}{2(2.00\text{ m})}\sqrt{\dfrac{m(9.80\text{ m/s}^2)}{0.002\,00\text{ kg/m}}}$

yielding $\quad m = \boxed{400\text{ kg}}$

***P14.21** $f_1 = \dfrac{v}{2L}$, where $v = \left(\dfrac{T}{\mu}\right)^{1/2}$

(a) If L is doubled, then $f_1 \propto L^{-1}$ will be reduced by a factor $\dfrac{1}{2}$.

(b) If μ is doubled, then $f_1 \propto \mu^{-1/2}$ will be reduced by a factor $\dfrac{1}{\sqrt{2}}$.

(c) If T is doubled, then $f_1 \propto \sqrt{T}$ will increase by a factor of $\sqrt{2}$.

***P14.22** For the whole string vibrating, $d_{NN} = 0.64\text{ m} = \dfrac{\lambda}{2}$; $\lambda = 1.28\text{ m}$. The speed of a pulse on the string is $v = f\lambda = 330\dfrac{1}{s}1.28\text{ m} = 422\text{ m/s}$.

(a) When the string is stopped at the fret, $d_{NN} = \dfrac{2}{3}0.64\text{ m} = \dfrac{\lambda}{2}$;

$\lambda = 0.853\text{ m}$

$$f = \frac{v}{\lambda} = \frac{422\text{ m/s}}{0.853\text{ m}} = \boxed{495\text{ Hz}}.$$

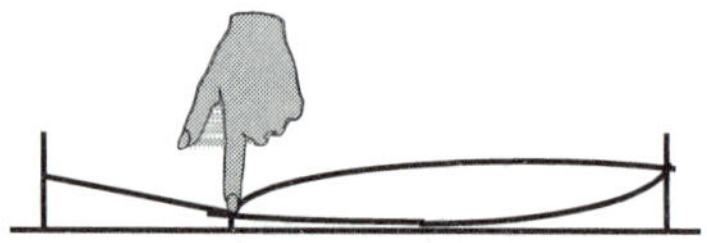

FIG. P14.22(a)

(b) The light touch at a point one third of the way along the string damps out vibration in the two lowest vibration states of the string as a whole. The whole string vibrates in its third resonance possibility: $3d_{NN} = 0.64\text{ m} = 3\dfrac{\lambda}{2}$;

$\lambda = 0.427\text{ m}$

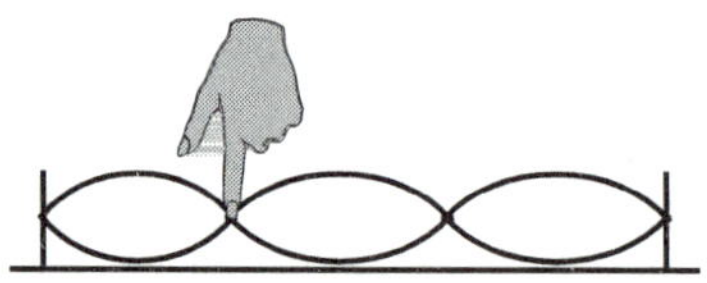

FIG. P14.22(b)

$$f = \frac{v}{\lambda} = \frac{422\text{ m/s}}{0.427\text{ m}} = \boxed{990\text{ Hz}}.$$

P14.23 $d_{NN} = 0.700\text{ m}$

$\lambda = 1.40\text{ m}$

$$f\lambda = v = 308\text{ m/s} = \sqrt{\frac{T}{(1.20\times 10^{-3})/(0.700)}}$$

(a) $T = \boxed{163\text{ N}}$

(b) $f_3 = \boxed{660\text{ Hz}}$

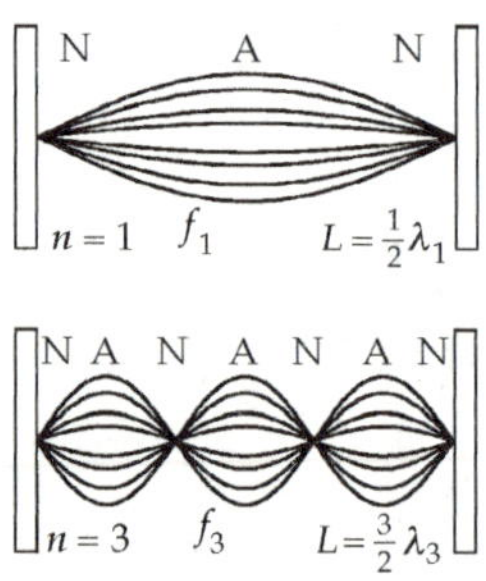

FIG. P14.23

P14.24 $\lambda_G = 2(0.350 \text{ m}) = \dfrac{v}{f_G}$; $\lambda_A = 2L_A = \dfrac{v}{f_A}$

$$L_G - L_A = L_G - \left(\frac{f_G}{f_A}\right)L_G = L_G\left(1 - \frac{f_G}{f_A}\right) = (0.350 \text{ m})\left(1 - \frac{392}{440}\right) = 0.038\,2 \text{ m}$$

Thus, $L_A = L_G - 0.038\,2 \text{ m} = 0.350 \text{ m} - 0.038\,2 \text{ m} = 0.312 \text{ m}$,

or the finger should be placed $\boxed{\text{31.2 cm from the bridge}}$.

$$L_A = \frac{v}{2f_A} = \frac{1}{2f_A}\sqrt{\frac{T}{\mu}};\ dL_A = \frac{dT}{4f_A\sqrt{T\mu}};\ \frac{dL_A}{L_A} = \frac{1}{2}\frac{dT}{T}$$

$$\frac{dT}{T} = 2\frac{dL_A}{L_A} = 2\frac{0.600 \text{ cm}}{(35.0 - 3.82) \text{ cm}} = \boxed{3.84\%}$$

***P14.25** In the fundamental mode, the string above the rod has only two nodes, at A and B, with an anti-node halfway between A and B. Thus,

$$\frac{\lambda}{2} = \overline{AB} = \frac{L}{\cos\theta} \text{ or } \lambda = \frac{2L}{\cos\theta}.$$

Since the fundamental frequency is f, the wave speed in this segment of string is

$$v = \lambda f = \frac{2Lf}{\cos\theta}.$$

Also, $v = \sqrt{\dfrac{T}{\mu}} = \sqrt{\dfrac{T}{m/\overline{AB}}} = \sqrt{\dfrac{TL}{m\cos\theta}}$ where T is the tension in this part of the string. Thus,

$$\frac{2Lf}{\cos\theta} = \sqrt{\frac{TL}{m\cos\theta}} \text{ or } \frac{4L^2f^2}{\cos^2\theta} = \frac{TL}{m\cos\theta}$$

and the mass of string above the rod is:

$$m = \frac{T\cos\theta}{4Lf^2} \qquad \text{[Equation 1]}$$

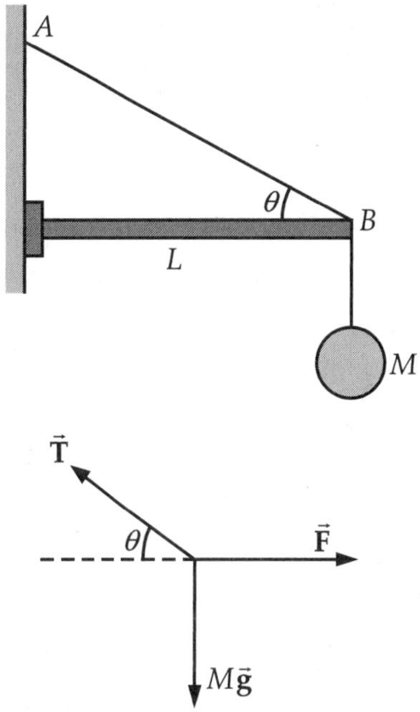

FIG. P14.25

Now, consider the tension in the string. The light rod would rotate about point P if the string exerted any vertical force on it. Therefore, recalling Newton's third law, the rod must exert only a horizontal force on the string. Consider a free-body diagram of the string segment in contact with the end of the rod.

$$\sum F_y = T\sin\theta - Mg = 0 \Rightarrow T = \frac{Mg}{\sin\theta}$$

Then, from Equation 1, the mass of string above the rod is

$$m = \left(\frac{Mg}{\sin\theta}\right)\frac{\cos\theta}{4Lf^2} = \boxed{\frac{Mg}{4Lf^2\tan\theta}}.$$

P14.26 Comparing $y = (0.002\text{ m})\sin\left((\pi\text{ rad/m})x\right)\cos\left((100\pi\text{ rad/s})t\right)$

with $y = 2A\sin kx\cos\omega t$

we find $k = \frac{2\pi}{\lambda} = \pi\,\text{m}^{-1}$, $\lambda = 2.00\text{ m}$, and $\omega = 2\pi f = 100\pi\,\text{s}^{-1}$: $f = 50.0\text{ Hz}$

(a) Then the distance between adjacent nodes is $d_{NN} = \frac{\lambda}{2} = 1.00\text{ m}$

and on the string are $\frac{L}{d_{NN}} = \frac{3.00\text{ m}}{1.00\text{ m}} = \boxed{3\text{ loops}}$

For the speed we have $v = f\lambda = \left(50\text{ s}^{-1}\right)2\text{ m} = 100\text{ m/s}$

(b) In the simplest standing wave vibration, $d_{NN} = 3.00\text{ m} = \frac{\lambda_b}{2}$, $\lambda_b = 6.00\text{ m}$

and $f_b = \frac{v_a}{\lambda_b} = \frac{100\text{ m/s}}{6.00\text{ m}} = \boxed{16.7\text{ Hz}}$

(c) In $v_0 = \sqrt{\frac{T_0}{\mu}}$, if the tension increases to $T_c = 9T_0$ and the string does not stretch, the speed increases to

$$v_c = \sqrt{\frac{9T_0}{\mu}} = 3\sqrt{\frac{T_0}{\mu}} = 3v_0 = 3(100\text{ m/s}) = 300\text{ m/s}$$

Then $\lambda_c = \frac{v_c}{f_a} = \frac{300\text{ m/s}}{50\text{ s}^{-1}} = 6.00\text{ m}$ $\quad d_{NN} = \frac{\lambda_c}{2} = 3.00\text{ m}$

and $\boxed{\text{one}}$ loop fits onto the string.

Section 14.5 Standing Waves in Air Columns

P14.27 (a) For the fundamental mode in a closed pipe, $\lambda = 4L$, as in the diagram.

But $v = f\lambda$, therefore $L = \frac{v}{4f}$

So, $L = \frac{343\text{ m/s}}{4\left(240\text{ s}^{-1}\right)} = \boxed{0.357\text{ m}}$

(b) For an open pipe, $\lambda = 2L$, as in the diagram.

So, $L = \frac{v}{2f} = \frac{343\text{ m/s}}{2\left(240\text{ s}^{-1}\right)} = \boxed{0.715\text{ m}}$

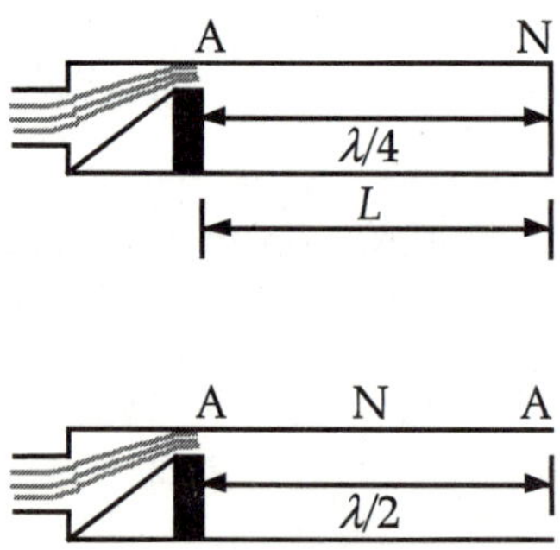

FIG. P14.27

P14.28 $d_{AA} = 0.320\text{ m};\ \lambda = 0.640\text{ m}$

(a) $f = \dfrac{v}{\lambda} = \boxed{531\text{ Hz}}$

(b) $\lambda = 0.085\,0\text{ m};\ d_{AA} = \boxed{42.5\text{ mm}}$

***P14.29** Assuming an air temperature of $T = 37°\text{C} = 310\text{ K}$, the speed of sound inside the pipe is

$$v = 331\text{ m/s} + 0.6\text{ m/s}\cdot\text{C}°(37°\text{C}) = 353\text{ m/s}.$$

In the fundamental resonant mode, the wavelength of sound waves in a pipe closed at one end is $\lambda = 4L$. Thus, for the whooping crane

$$\lambda = 4(5.0\text{ ft}) = 2.0\times10^{1}\text{ ft} \qquad \text{and} \qquad f = \frac{v}{\lambda} = \frac{(353\text{ m/s})}{2.0\times10^{1}\text{ ft}}\left(\frac{3.281\text{ ft}}{1\text{ m}}\right) = \boxed{57.9\text{ Hz}}.$$

P14.30 The wavelength is $\lambda = \dfrac{v}{f} = \dfrac{343\text{ m/s}}{261.6/\text{s}} = 1.31\text{ m}$

so the length of the open pipe vibrating in its simplest (A-N-A) mode is

$$d_{\text{A to A}} = \frac{1}{2}\lambda = \boxed{0.656\text{ m}}$$

A closed pipe has (N-A) for its simplest resonance,

(N-A-N-A) for the second,

and (N-A-N-A-N-A) for the third.

Here, the pipe length is $5d_{\text{N to A}} = \dfrac{5\lambda}{4} = \dfrac{5}{4}(1.31\text{ m}) = \boxed{1.64\text{ m}}$

P14.31 For a closed box, the resonant frequencies will have nodes at both sides, so the permitted wavelengths will be $L = \frac{1}{2}n\lambda,\ (n = 1, 2, 3, \ldots)$.

i.e., $L = \dfrac{n\lambda}{2} = \dfrac{nv}{2f}$ and $f = \dfrac{nv}{2L}$.

Therefore, with $L = 0.860\text{ m}$ and $L' = 2.10\text{ m}$, the resonant frequencies are

$f_n = \boxed{n(206\text{ Hz})}$ for $L = 0.860\text{ m}$ for each n from 1 to 9

and $f_n' = \boxed{n(84.5\text{ Hz})}$ for $L' = 2.10\text{ m}$ for each n from 2 to 23.

P14.32 The air in the auditory canal, about 3 cm long, can vibrate with a node at the closed end and antinode at the open end,

with $d_{\text{N to A}} = 3 \text{ cm} = \frac{\lambda}{4}$

so $\lambda = 0.12 \text{ m}$

and $f = \frac{v}{\lambda} = \frac{343 \text{ m/s}}{0.12 \text{ m}} \approx \boxed{3 \text{ kHz}}$

A small-amplitude external excitation at this frequency can, over time, feed energy into a larger-amplitude resonance vibration of the air in the canal, making it audible.

P14.33 For both open and closed pipes, resonant frequencies are equally spaced as numbers. The set of resonant frequencies then must be 650 Hz, 550 Hz, 450 Hz, 350 Hz, 250 Hz, 150 Hz, 50 Hz. These are odd-integer multipliers of the fundamental frequency of $\boxed{50.0 \text{ Hz}}$. Then the pipe length is

$$d_{NA} = \frac{\lambda}{4} = \frac{v}{4f} = \frac{340 \text{ m/s}}{4(50/\text{s})} = \boxed{1.70 \text{ m}}.$$

P14.34 The wavelength of sound is $\lambda = \frac{v}{f}$

The distance between water levels at resonance is $d = \frac{v}{2f}$ $\quad \therefore Rt = \pi r^2 d = \frac{\pi r^2 v}{2f}$

and $t = \boxed{\frac{\pi r^2 v}{2Rf}}$.

P14.35 $\frac{\lambda}{2} = d_{AA} = \frac{L}{n}$ or $L = \frac{n\lambda}{2}$ for $n = 1, 2, 3, \ldots$

Since $\lambda = \frac{v}{f}$ $\quad L = n\left(\frac{v}{2f}\right)$ for $n = 1, 2, 3, \ldots$

With $v = 343 \text{ m/s}$ and $f = 680 \text{ Hz}$,

$$L = n\left(\frac{343 \text{ m/s}}{2(680 \text{ Hz})}\right) = n(0.252 \text{ m}) \quad \text{for } n = 1, 2, 3, \ldots$$

Possible lengths for resonance are: $L = \boxed{0.252 \text{ m}, 0.504 \text{ m}, 0.757 \text{ m}, \ldots, n(0.252) \text{ m}}$.

P14.36 The length corresponding to the fundamental satisfies $f = \frac{v}{4L}$: $L_1 = \frac{v}{4f} = \frac{343}{4(512)} = 0.167 \text{ m}$.

Since $L > 20.0 \text{ cm}$, the *next* two modes will be observed, corresponding to $f = \frac{3v}{4L_2}$ and $f = \frac{5v}{4L_3}$.

or $L_2 = \frac{3v}{4f} = \boxed{0.502 \text{ m}}$ and $L_3 = \frac{5v}{4f} = \boxed{0.837 \text{ m}}$.

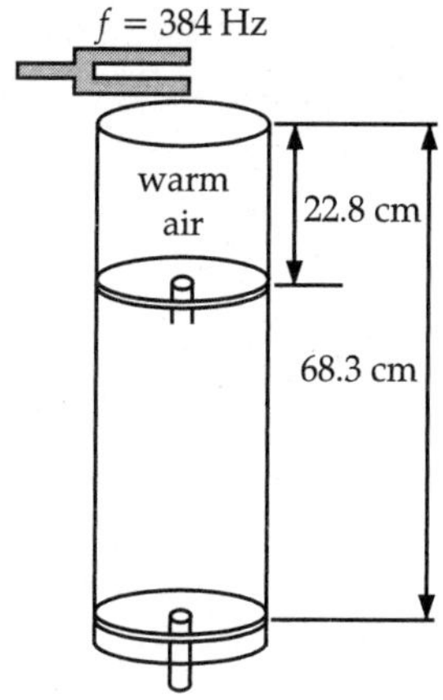

FIG. P14.37

P14.37 For resonance in a narrow tube open at one end,

$$f = n\frac{v}{4L} \quad (n = 1, 3, 5, \ldots).$$

(a) Assuming $n = 1$ and $n = 3$,

$$384 = \frac{v}{4(0.228)} \text{ and } 384 = \frac{3v}{4(0.683)}.$$

In either case, $v = \boxed{350 \text{ m/s}}$.

(b) For the next resonance $n = 5$, and $L = \frac{5v}{4f} = \frac{5(350 \text{ m/s})}{4(384 \text{ s}^{-1})} = \boxed{1.14 \text{ m}}$.

***P14.38** (a) For the fundamental mode of an open tube,

$$L = \frac{\lambda}{2} = \frac{v}{2f} = \frac{343 \text{ m/s}}{2(880 \text{ s}^{-1})} = \boxed{0.195 \text{ m}}.$$

(b) $v = 331 \text{ m/s} + 0.6 \text{ m/s}\cdot{}^\circ\text{C}(-5^\circ\text{C}) = 328 \text{ m/s}$

We ignore the thermal expansion of the metal.

$$f = \frac{v}{\lambda} = \frac{v}{2L} = \frac{328 \text{ m/s}}{2(0.195 \text{ m})} = \boxed{842 \text{ Hz}}$$

The flute is flat by a semitone.

Section 14.6 Beats: Interference in Time

P14.39 $f \propto v \propto \sqrt{T}$ $\qquad f_{\text{new}} = 110\sqrt{\frac{540}{600}} = 104.4 \text{ Hz}$

$\Delta f = \boxed{5.64 \text{ beats/s}}$

P14.40 (a) The string could be tuned to either $\boxed{521 \text{ Hz or } 525 \text{ Hz}}$ from this evidence.

(b) Tightening the string raises the wave speed and frequency. If the frequency were originally 521 Hz, the beats would slow down.

Instead, the frequency must have started at 525 Hz to become $\boxed{526 \text{ Hz}}$.

continued on next page

(c) From $f = \frac{v}{\lambda} = \frac{\sqrt{T/\mu}}{2L} = \frac{1}{2L}\sqrt{\frac{T}{\mu}}$

$$\frac{f_2}{f_1} = \sqrt{\frac{T_2}{T_1}} \text{ and } T_2 = \left(\frac{f_2}{f_1}\right)^2 T_1 = \left(\frac{523 \text{ Hz}}{526 \text{ Hz}}\right)^2 T_1 = 0.989T_1.$$

The fractional change that should be made in the tension is then

$$\text{fractional change} = \frac{T_1 - T_2}{T_1} = 1 - 0.989 = 0.011\,4 = 1.14\% \text{ lower.}$$

The tension should be $\boxed{\text{reduced by } 1.14\%}$.

P14.41 For an echo $f' = f\frac{(v + v_s)}{(v - v_s)}$ the beat frequency is $f_b = |f' - f|$.

Solving for f_b.

gives $f_b = f\frac{(2v_s)}{(v - v_s)}$ when approaching wall.

(a) $f_b = (256)\frac{2(1.33)}{(343 - 1.33)} = \boxed{1.99 \text{ Hz}}$ beat frequency

(b) When he is moving away from the wall, v_s changes sign. Solving for v_s gives

$$v_s = \frac{f_b v}{2f - f_b} = \frac{(5)(343)}{(2)(256) - 5} = \boxed{3.38 \text{ m/s}}.$$

Section 14.7 Nonsinusoidal Wave Patterns

P14.42 We evaluate

$$s = 100\sin\theta + 157\sin 2\theta + 62.9\sin 3\theta + 105\sin 4\theta$$
$$+51.9\sin 5\theta + 29.5\sin 6\theta + 25.3\sin 7\theta$$

where s represents particle displacement in nanometers and θ represents the phase of the wave in radians. As θ advances by 2π, time advances by (1/523) s. Here is the result:

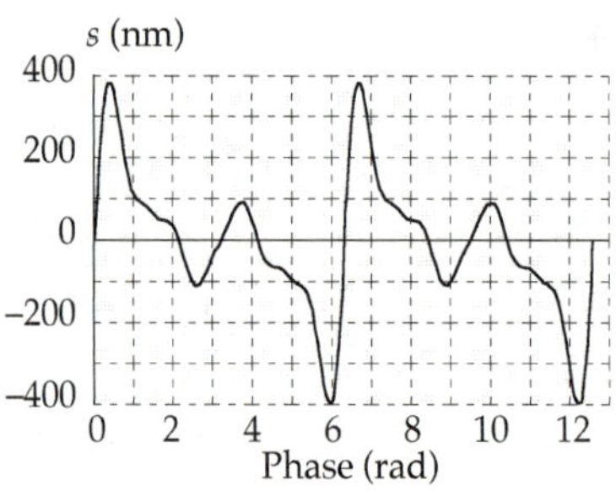

FIG. P14.42

***P14.43** We list the frequencies of the harmonics of each note in Hz:

	Harmonic				
Note	1	2	3	4	5
A	440.00	880.00	1 320.0	1 760.0	2 200.0
C#	554.37	1 108.7	1 663.1	2 217.5	2 771.9
E	659.26	1 318.5	1 977.8	2 637.0	3 296.3

The second harmonic of E is close the the third harmonic of A, and the fourth harmonic of C# is close to the fifth harmonic of A.

Section 14.8 Context Connection—Building on Antinodes

P14.44 (a) The wave speed is $v = \dfrac{9.15\text{ m}}{2.50\text{ s}} = \boxed{3.66\text{ m/s}}$

(b) From the figure, there are antinodes at both ends of the pond, so the distance between adjacent antinodes

is $d_{AA} = \dfrac{\lambda}{2} = 9.15\text{ m},$

and the wavelength is $\lambda = 18.3\text{ m}$

The frequency is then $f = \dfrac{v}{\lambda} = \dfrac{3.66\text{ m/s}}{18.3\text{ m}} = \boxed{0.200\text{ Hz}}$

We have assumed the wave speed is the same for all wavelengths.

P14.45 The wave speed is $v = \sqrt{gd} = \sqrt{(9.80\text{ m/s}^2)(36.1\text{ m})} = 18.8\text{ m/s}$

The bay has one end open and one closed. Its simplest resonance is with a node of horizontal velocity, which is also an antinode of vertical displacement, at the head of the bay and an antinode of velocity, which is a node of displacement, at the mouth. The vibration of the water in the bay is like that in one half of the pond shown in Figure P14.44.

Then, $d_{NA} = 210 \times 10^3\text{ m} = \dfrac{\lambda}{4}$

and $\lambda = 840 \times 10^3\text{ m}$

Therefore, the period is $T = \dfrac{1}{f} = \dfrac{\lambda}{v} = \dfrac{840 \times 10^3\text{ m}}{18.8\text{ m/s}} = 4.47 \times 10^4\text{ s} = \boxed{12\text{ h }24\text{ min}}$

This agrees precisely with the period of the lunar excitation, so we identify the extra-high tides as amplified by resonance.

Additional Problems

P14.46 The distance between adjacent nodes is one-quarter of the circumference.

$$d_{\text{NN}} = d_{\text{AA}} = \frac{\lambda}{2} = \frac{20.0 \text{ cm}}{4} = 5.00 \text{ cm}$$

so $\lambda = 10.0$ cm and $f = \frac{v}{\lambda} = \frac{900 \text{ m/s}}{0.100 \text{ m}} = 9\,000 \text{ Hz} = \boxed{9.00 \text{ kHz}}$.

The singer must match this frequency quite precisely for some interval of time to feed enough energy into the glass to crack it.

P14.47 $f = 87.0$ Hz

speed of sound in air: $v_a = 340 \text{ m/s}$

(a) $\lambda_b = \ell$ $\quad v = f\lambda_b = (87.0 \text{ s}^{-1})(0.400 \text{ m})$

$v = \boxed{34.8 \text{ m/s}}$

(b) $\left.\begin{array}{l}\lambda_a = 4L \\ v_a = \lambda_a f\end{array}\right\}$ $\quad L = \frac{v_a}{4f} = \frac{340 \text{ m/s}}{4(87.0 \text{ s}^{-1})} = \boxed{0.977 \text{ m}}$

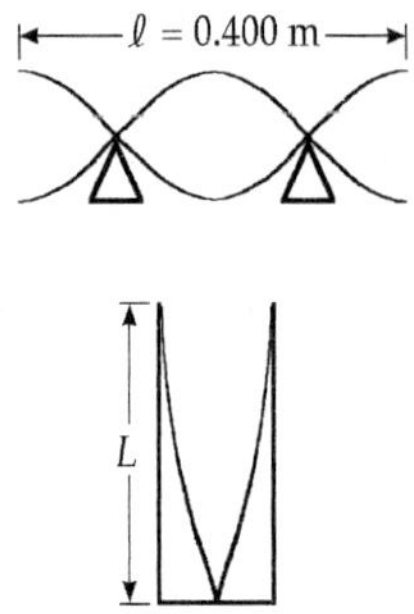

FIG. P14.47

***P14.48**

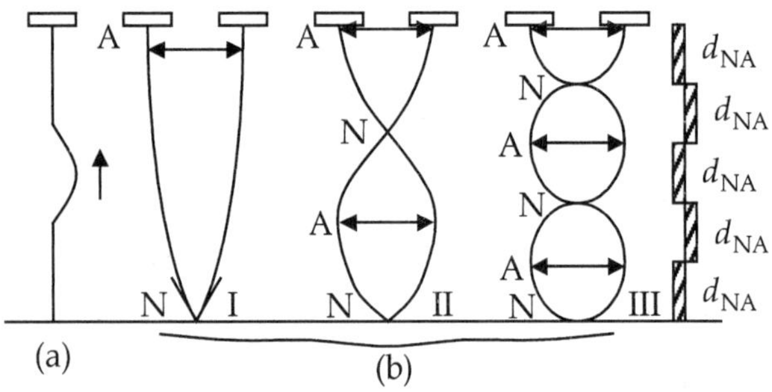

FIG. P14.48

(a) $\mu = \frac{5.5 \times 10^{-3} \text{ kg}}{0.86 \text{ m}} = 6.40 \times 10^{-3} \text{ kg/m}$

$$v = \sqrt{\frac{T}{\mu}} = \sqrt{\frac{1.30 \text{ kg} \cdot \text{m/s}^2}{6.40 \times 10^{-3} \text{ kg/m}}} = \boxed{14.3 \text{ m/s}}$$

(b) In state I, $d_{\text{NA}} = \boxed{0.860 \text{ m}} = \frac{\lambda}{4}$

(c) $\lambda = 3.44$ m $\quad f = \frac{v}{\lambda} = \frac{14.3 \text{ m/s}}{3.44 \text{ m}} = \boxed{4.14 \text{ Hz}}$

In state II, $d_{\text{NA}} = \frac{1}{3}(0.86 \text{ m}) = \boxed{0.287 \text{ m}}$

$\lambda = 4(0.287 \text{ m}) = 1.15 \text{ m}$ $\quad f = \frac{v}{\lambda} = \frac{14.3 \text{ m/s}}{1.15 \text{ m}} = \boxed{12.4 \text{ Hz}}$

continued on next page

In state III, $d_{NA} = \frac{1}{5}(0.86\text{ m}) = \boxed{0.172\text{ m}}$

$$f = \frac{v}{\lambda} = \frac{14.3\text{ m/s}}{4(0.172\text{ m})} = \boxed{20.7\text{ Hz}}$$

P14.49 Moving away from station, frequency is depressed:

$f' = 180 - 2.00 = 178\text{ Hz}$: $\quad 178 = 180\frac{343}{343-(-v)}$

Solving for v gives $\quad v = \frac{(2.00)(343)}{178}$

Therefore, $\quad v = \boxed{3.85\text{ m/s away from station}}$

Moving toward the station, the frequency is enhanced:

$f' = 180 + 2.00 = 182\text{ Hz}$: $\quad 182 = 180\frac{343}{343-v}$

Solving for v gives $\quad 4 = \frac{(2.00)(343)}{182}$

Therefore, $\quad v = \boxed{3.77\text{ m/s toward the station}}$

***P14.50** (a) Use the Doppler formula

$$f' = f\frac{(v \pm v_0)}{(v \mp v_s)}.$$

With f_1' = frequency of the speaker in front of student and

f_2' = frequency of the speaker behind the student.

$$f_1' = (456\text{ Hz})\frac{(343\text{ m/s} + 1.50\text{ m/s})}{(343\text{ m/s} - 0)} = 458\text{ Hz}$$

$$f_2' = (456\text{ Hz})\frac{(343\text{ m/s} - 1.50\text{ m/s})}{(343\text{ m/s} + 0)} = 454\text{ Hz}$$

Therefore, $f_b = f_1' - f_2' = \boxed{3.99\text{ Hz}}$.

(b) The waves broadcast by both speakers have $\lambda = \frac{v}{f} = \frac{343\text{ m/s}}{456/\text{s}} = 0.752\text{ m}$. The standing wave between them has $d_{AA} = \frac{\lambda}{2} = 0.376\text{ m}$. The student walks from one maximum to the next in time $\Delta t = \frac{0.376\text{ m}}{1.50\text{ m/s}} = 0.251\text{ s}$, so the frequency at which she hears maxima is $f = \frac{1}{T} = \boxed{3.99\text{ Hz}}$.

P14.51 Call L the depth of the well and v the speed of sound.

Then for some integer n $$L=(2n-1)\frac{\lambda_1}{4}=(2n-1)\frac{v}{4f_1}=\frac{(2n-1)(343\text{ m/s})}{4\left(51.5\text{ s}^{-1}\right)}$$

and for the next resonance $$L=\left[2(n+1)-1\right]\frac{\lambda_2}{4}=(2n+1)\frac{v}{4f_2}=\frac{(2n+1)(343\text{ m/s})}{4\left(60.0\text{ s}^{-1}\right)}$$

Thus, $$\frac{(2n-1)(343\text{ m/s})}{4\left(51.5\text{ s}^{-1}\right)}=\frac{(2n+1)(343\text{ m/s})}{4\left(60.0\text{ s}^{-1}\right)}$$

and we require an *integer* solution to $$\frac{2n+1}{60.0}=\frac{2n-1}{51.5}$$

The equation gives $n=\frac{111.5}{17}=6.56$, so the best fitting integer is $n=7$.

Then $$L=\frac{\left[2(7)-1\right](343\text{ m/s})}{4\left(51.5\text{ s}^{-1}\right)}=21.6\text{ m}$$

and $$L=\frac{\left[2(7)+1\right](343\text{ m/s})}{4\left(60.0\text{ s}^{-1}\right)}=21.4\text{ m}$$

suggest the best value for the depth of the well is $\boxed{21.5\text{ m}}$.

P14.52 $$v=\sqrt{\frac{(48.0)(2.00)}{4.80\times10^{-3}}}=141\text{ m/s}$$

$$d_{\text{NN}}=1.00\text{ m};\ \lambda=2.00\text{ m};\ f=\frac{v}{\lambda}=70.7\text{ Hz}$$

$$\lambda_a=\frac{v_a}{f}=\frac{343\text{ m/s}}{70.7\text{ Hz}}=\boxed{4.85\text{ m}}$$

P14.53 (a) Since the first node is at the weld, the wavelength in the thin wire is $2L$ or 80.0 cm. The frequency and tension are the same in both sections, so

$$f=\frac{1}{2L}\sqrt{\frac{T}{\mu}}=\frac{1}{2(0.400)}\sqrt{\frac{4.60}{2.00\times10^{-3}}}=\boxed{59.9\text{ Hz}}.$$

(b) As the thick wire is twice the diameter, the linear density is 4 times that of the thin wire.

$\mu'=8.00\text{ g/m}$

so $L'=\frac{1}{2f}\sqrt{\frac{T}{\mu'}}$ $\qquad L'=\left[\frac{1}{(2)(59.9)}\right]\sqrt{\frac{4.60}{8.00\times10^{-3}}}=\boxed{20.0\text{ cm}}$ half the length of the thin wire.

P14.54 The second standing wave mode of the air in the pipe reads ANAN, with $d_{NA} = \frac{\lambda}{4} = \frac{1.75 \text{ m}}{3}$

so $\lambda = 2.33$ m

and $f = \frac{v}{\lambda} = \frac{343 \text{ m/s}}{2.33 \text{ m}} = 147$ Hz

For the string, λ and v are different but f is the same.

$$\frac{\lambda}{2} = d_{NN} = \frac{0.400 \text{ m}}{2}$$

so $\lambda = 0.400$ m

$$v = \lambda f = (0.400 \text{ m})(147 \text{ Hz}) = 58.8 \text{ m/s} = \sqrt{\frac{T}{\mu}}$$

$$T = \mu v^2 = \left(9.00 \times 10^{-3} \text{ kg/m}\right)(58.8 \text{ m/s})^2 = \boxed{31.1 \text{ N}}$$

P14.55 (a) $f = \frac{n}{2L}\sqrt{\frac{T}{\mu}}$

so $\frac{f'}{f} = \frac{L}{L'} = \frac{L}{2L} = \frac{1}{2}$

The $\boxed{\text{frequency should be halved}}$ to get the same number of antinodes for twice the length.

(b) $\frac{n'}{n} = \sqrt{\frac{T}{T'}}$ so $\frac{T'}{T} = \left(\frac{n}{n'}\right)^2 = \left[\frac{n}{n+1}\right]^2$

The tension must be $T' = \boxed{\left[\frac{n}{n+1}\right]^2 T}$

(c) $\frac{f'}{f} = \frac{n'L}{nL'}\sqrt{\frac{T'}{T}}$ so $\frac{T'}{T} = \left(\frac{nfL'}{n'fL}\right)^2$

$\frac{T'}{T} = \left(\frac{3}{2 \cdot 2}\right)^2$ $\boxed{\frac{T'}{T} = \frac{9}{16}}$ to get twice as many antinodes.

P14.56 (a) For the block:

$\sum F_x = T - Mg \sin 30.0° = 0$

so $T = Mg \sin 30.0° = \boxed{\frac{1}{2}Mg}$.

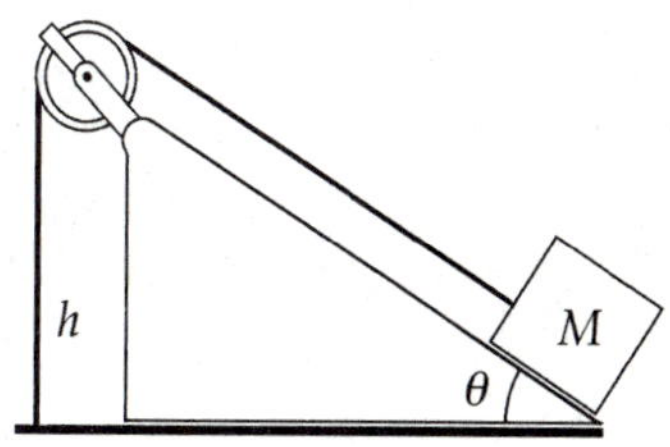

FIG. P14.56

(b) The length of the section of string parallel to the incline is $\frac{h}{\sin 30.0°} = 2h$. The total length of the string is then $\boxed{3h}$.

(c) The mass per unit length of the string is $\mu = \boxed{\frac{m}{3h}}$

continued on next page

(d) The speed of waves in the string is $v=\sqrt{\dfrac{T}{\mu}}=\sqrt{\left(\dfrac{Mg}{2}\right)\left(\dfrac{3h}{m}\right)}=\boxed{\sqrt{\dfrac{3Mgh}{2m}}}$

(e) In the fundamental mode, the segment of length h vibrates as one loop. The distance between adjacent nodes is then $d_{NN}=\dfrac{\lambda}{2}=h$, so the wavelength is $\lambda=2h$.

The frequency is $f=\dfrac{v}{\lambda}=\dfrac{1}{2h}\sqrt{\dfrac{3Mgh}{2m}}=\boxed{\sqrt{\dfrac{3Mg}{8mh}}}$

(g) When the vertical segment of string vibrates with 2 loops (i.e., 3 nodes), then $h=2\left(\dfrac{\lambda}{2}\right)$ and

the wavelength is $\lambda=\boxed{h}$.

(f) The period of the standing wave of 3 nodes (or two loops) is

$$T=\frac{1}{f}=\frac{\lambda}{v}=h\sqrt{\frac{2m}{3Mgh}}=\boxed{\sqrt{\frac{2mh}{3Mg}}}$$

(h) $f_b=1.02f-f=\left(2.00\times10^{-2}\right)f=\boxed{\left(2.00\times10^{-2}\right)\sqrt{\dfrac{3Mg}{8mh}}}$

P14.57 We look for a solution of the form

$$\begin{aligned}5.00\sin(2.00x-10.0t)+10.0\cos(2.00x-10.0t)&=A\sin(2.00x-10.0t+\phi)\\&=A\sin(2.00x-10.0t)\cos\phi+A\cos(2.00x-10.0t)\sin\phi\end{aligned}$$

This will be true if both $5.00=A\cos\phi$ and $10.0=A\sin\phi$,

requiring $(5.00)^2+(10.0)^2=A^2$

$A=11.2$ and $\phi=63.4°$

The resultant wave $\boxed{11.2\sin(2.00x-10.0t+63.4°)}$ is sinusoidal.

P14.58 For the wire, $\mu=\dfrac{0.010\,0\text{ kg}}{2.00\text{ m}}=5.00\times10^{-3}\text{ kg/m}$: $v=\sqrt{\dfrac{T}{\mu}}=\sqrt{\dfrac{\left(200\text{ kg}\cdot\text{m/s}^2\right)}{5.00\times10^{-3}\text{ kg/m}}}$

$v=200\text{ m/s}$

If it vibrates in its simplest state, $d_{NN}=2.00\text{ m}=\dfrac{\lambda}{2}$: $f=\dfrac{v}{\lambda}=\dfrac{(200\text{ m/s})}{4.00\text{ m}}=50.0\text{ Hz}$

(a) The tuning fork can have frequencies $\boxed{45.0\text{ Hz or }55.0\text{ Hz}}$.

continued on next page

(b) If $f = 45.0\text{ Hz}$, $v = f\lambda = (45.0/\text{s})4.00\text{ m} = 180\text{ m/s}$.

Then, $T = v^2\mu = (180\text{ m/s})^2(5.00\times10^{-3}\text{ kg/m}) = \boxed{162\text{ N}}$

$\boxed{\text{or}}$ if $f = 55.0\text{ Hz}$, $T = v^2\mu = f^2\lambda^2\mu = (55.0/\text{s})^2(4.00\text{ m})^2(5.00\times10^{-3}\text{ kg/m}) = \boxed{242\text{ N}}$.

P14.59 (a) Let θ represent the angle each slanted rope makes with the vertical.

In the diagram, observe that:

$$\sin\theta = \frac{1.00\text{ m}}{1.50\text{ m}} = \frac{2}{3}$$

or $\theta = 41.8°$.

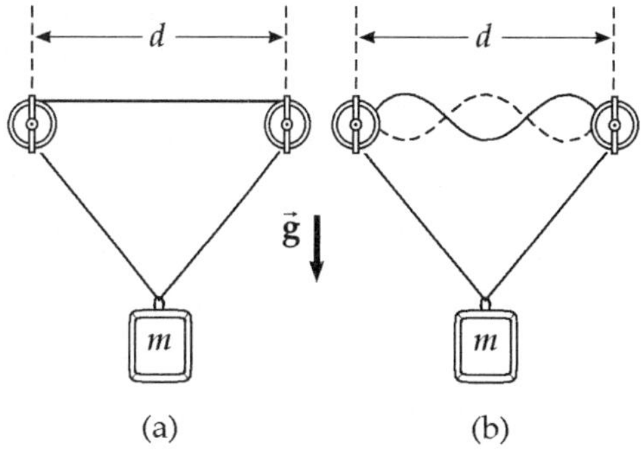

FIG. P14.59

Considering the mass,

$$\sum F_y = 0:\ 2T\cos\theta = mg$$

$$\text{or } T = \frac{(12.0\text{ kg})(9.80\text{ m/s}^2)}{2\cos 41.8°} = \boxed{78.9\text{ N}}$$

(b) The speed of transverse waves in the string is $v = \sqrt{\frac{T}{\mu}} = \sqrt{\frac{78.9\text{ N}}{0.001\,00\text{ kg/m}}} = 281\text{ m/s}$.

For the standing wave pattern shown (3 loops), $d = \frac{3}{2}\lambda$

or $\lambda = \frac{2(2.00\text{ m})}{3} = 1.33\text{ m}$.

Thus, the required frequency is $f = \frac{v}{\lambda} = \frac{281\text{ m/s}}{1.33\text{ m}} = \boxed{211\text{ Hz}}$.

P14.60 $d_{AA} = \frac{\lambda}{2} = 7.05\times10^{-3}\text{ m}$ is the distance between antinodes.

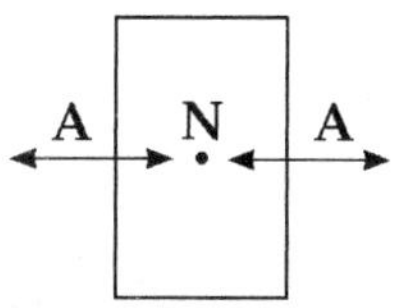

FIG. P14.60

Then $\lambda = 14.1\times10^{-3}\text{ m}$

and $f = \frac{v}{\lambda} = \frac{3.70\times10^3\text{ m/s}}{14.1\times10^{-3}\text{ m}} = \boxed{2.62\times10^5\text{ Hz}}$.

The crystal can be tuned to vibrate at 2^{18} Hz, so that binary counters can derive from it a signal at precisely 1 Hz.

ANSWERS TO EVEN PROBLEMS

P14.2 see the solution

P14.4 5.66 cm

P14.6 0.500 s

P14.8 (a) 3.33 rad; (b) 283 Hz

P14.10 (a) number of minima heard $= n_{max} =$ greatest integer $\le d\left(\frac{f}{v}\right)+\frac{1}{2}$;

(b) $L_n = \dfrac{d^2-\left(n-\frac{1}{2}\right)^2\left(\frac{v}{f}\right)^2}{2\left(n-\frac{1}{2}\right)\left(\frac{v}{f}\right)}$ where $n = 1, 2, \ldots, n_{max}$

P14.12 15.7 m, 31.8 Hz, 500 m/s

P14.14 see the solution

P14.16 (a) see the solution; (b) see the solution

P14.18 (a) 0.600 m; (b) 30.0 Hz

P14.20 (a) 350 Hz; (b) 400 kg

P14.22 (a) 495 Hz; (b) 990 Hz

P14.24 31.2 cm from the bridge, 3.84%

P14.26 (a) 3 loops; (b) 16.7 Hz; (c) 1 loop

P14.28 (a) 531 Hz; (b) 42.5 mm

P14.30 0.656 m, 1.64 m

P14.32 around 3 kHz, A small-amplitude external excitation at this frequency can, over time, feed energy into a larger-amplitude resonance vibration of the air in the canal, making it audible.

P14.34 $\dfrac{\pi r^2 v}{2Rf}$

P14.36 0.502 m, 0.837 m

P14.38 (a) 0.195 m; (b) 842 Hz

P14.40 (a) 521 Hz or 525 Hz; (b) 526 Hz; (c) reduced by 1.14%

P14.42 see the solution

P14.44 (a) 3.66 m/s; (b) 0.200 Hz

P14.46 9.00 kHz

P14.48 (a) 14.3 m/s; (b) 0.860 m, 0.287 m, 0.172 m; (c) 4.14 Hz, 12.4 Hz, 20.7 Hz

P14.50 (a) 3.99 Hz; (b) 3.99 Hz

P14.52 4.85 m

P14.54 31.1 N

P14.56 (a) $\frac{1}{2}Mg$; (b) $3h$; (c) $\frac{m}{3h}$; (d) $\sqrt{\frac{3Mgh}{2m}}$; (e) $\sqrt{\frac{3Mg}{8mh}}$; (f) $\sqrt{\frac{2mh}{3Mg}}$; (g) h; (h) $\left(2.00\times10^{-2}\right)\sqrt{\frac{3Mg}{8mh}}$

P14.58 (a) 45.0 or 55.0 Hz; (b) 162 or 242 N

P14.60 2.62×10^5 Hz

CONTEXT 3 CONCLUSION SOLUTIONS TO PROBLEMS

CC3.1 Let point 1 be $r = 10$ km from the epicenter and point 2 be at 20 km. The intensity is proportional to $\frac{1}{r}$ according to $I = \frac{C}{r}$, where C is some constant. Intensity is defined as the energy a wave carries each second through a unit area of wavefront, so it is proportional to the amplitude squared according to $I = DA^2$, where D is another constant. Then the factors of change are related by

continued on next page

$$\frac{I_2}{I_1} = \frac{DA_2^{\ 2}}{DA_1^{\ 2}} = \frac{r_1 C}{r_2 C}$$

$$\frac{A_2}{A_1} = \sqrt{\frac{r_1}{r_2}}$$

$$A_2 = A_1\sqrt{\frac{r_1}{r_2}} = 5.0\ \text{cm}\sqrt{\frac{10\ \text{km}}{20\ \text{km}}} = \boxed{3.5\ \text{cm}}$$

CC3.2 As in Equation 13.23, the rate of energy transfer in a seismic wave is proportional to the speed and to the amplitude squared. We write $\mathcal{P} = FvA^2$, where F is some constant. If no wave energy is reflected or turns into internal energy, $Fv_{\text{bedrock}}A^2_{\text{bedrock}} = Fv_{\text{mudfill}}A^2_{\text{mudfill}}$

$$\frac{v_{\text{mudfill}}}{v_{\text{bedrock}}} = \left(\frac{A_{\text{bedrock}}}{A_{\text{mudfill}}}\right)^2 = \left(\frac{A_{\text{bedrock}}}{5A_{\text{bedrock}}}\right)^2 = \frac{1}{25}$$

$\boxed{\text{The speed decreases by a factor of 25.}}$

CC3.3 **METHOD ONE**

From the graph, we have for the speed of S waves $v_S = \frac{395\ \text{km}}{100\ \text{s}} = 3.95\ \text{km/s}$, and for the speed of P waves $v_P = \frac{400\ \text{km}}{48\ \text{s}} = 8.33\ \text{km/s}$. From the data of station 1 we can find a value for the time the quake started: $15\ \text{h:}46\ \text{min:}06\ \text{s} - \frac{200\ \text{km}}{8.33\ \text{km/s}} = 15\ \text{h:}45\ \text{min:}66\ \text{s} - 24\ \text{s} = 15\ \text{h:}45\ \text{min:}42\ \text{s}$. Similarly from the data of the other stations, the quake began at $15\text{:}46\text{:}01 - \frac{160\ \text{km}}{8.33\ \text{km/s}} = 15\text{:}45\text{:}41.8$ or $15\text{:}45\text{:}54 - \frac{105\ \text{s}}{8.33} = 15\text{:}45\text{:}41.4$. For the most probable value for the actual time we take the average, $15\text{:}45\text{:}41.7 \pm 0.3\ \text{s}$. Then the S-wave arrival time should be

$$15\text{:}45\text{:}41.7 + \frac{200\ \text{km}}{3.95\ \text{km/s}} = \boxed{15\text{:}46\text{:}32\ \text{for station 1,}}$$

$$15\text{:}45\text{:}41.7 + \frac{160\ \text{s}}{3.95} = \boxed{15\text{:}46\text{:}22\ \text{for station 2,}}$$

$$15\text{:}45\text{:}41.7 + \frac{105\ \text{s}}{3.95} = \boxed{15\text{:}46\text{:}08\ \text{for station 3,}}$$

all with uncertainties of $\boxed{\pm 1\ \text{s}}$

METHOD TWO

With no significant loss of precision, we can use the graph of travel times to read the S wave arrival times almost directly.

For station #1, locate 200 km on the horizontal axis. Vertically above it, read the size of the space between the P and S lines as 27 s. Add this S wave delay time to the P wave arrival time, 15:46:06, to obtain 15:46:33 as the S wave arrival time at station #1.

Similarly for station #2, the S wave should arrive at 21 s + 15:46:01 = 15:46:22.

For station #3, the graph shows that at range 105 km an S wave arrives 14 s after a P wave, placing it at 15:45:54 + 14 = 15:46:08.

Note: The theory developed in the context conclusion can be applied to the following topical problem.

CC3.4 An earthquake or a landslide can produce an ocean wave carrying great energy, called a tsunami, when its wavelength is large compared to the ocean depth d, the speed of a water wave is given approximately by $\sqrt{gd}$.

(a) Explain why the amplitude of the wave increases as the wave approaches shore. What can you consider to be constant in the motion of one wave crest?

(b) Assume that an earthquake, all along a plate boundary running north and south, produces a straight tsunami wave crest moving everywhere to the west. If the wave has amplitude 1.8 m when its speed is 200 m/s, what will be its amplitude where the water is 9.00 m deep?

(c) Explain why the amplitude at the shore should be expected to be still greater, but cannot be meaningfully predicted by your model.

ANSWERS TO EVEN CONTEXT 3 CONCLUSION PROBLEMS

CC3.2 The speed decreases by a factor is 25.

CC3.4 (a) The energy a wave crest carries is constant in the absence of absorption. Then the rate at which energy passes a stationary point, which is the power of the wave, is constant. The power is proportional to the square of the amplitude end to the wave speed. The speed decreases as the wave moves into shallower water near shore, so the amplitude must increase.

(b) For the wave described, with a single direction of energy transport, the intensity is the same at the deep-water location ① and at the place ② with depth 9 m. To express the constant intensity we write

$$A_1^2 v_1 = A_2^2 v_2 = A_2^2 \sqrt{g d_2}$$

$$(1.8\ \text{m})^2\ 200\ \text{m/s} = A_2^2 \sqrt{\left(9.8\ \text{m/s}^2\right) 9\ \text{m}}$$

$$= A_2^2 9.39\ \text{m/s}$$

$$A_2 = 1.8\left(\frac{200\ \text{m/s}}{9.39\ \text{m/s}}\right)^{1/2}$$

$$= \boxed{8.31\ \text{m}}$$

(c) As the water depth goes to zero, our model would predict zero speed and infinite amplitude. In fact the amplitude must be finite as the wave comes ashore. As the speed decreases the wavelength also decreases. When it becomes comparable to the water depth, or smaller, our formula $\sqrt{gd}$ for wave speed no longer applies.

Chapter 15

Fluid Mechanics

CHAPTER OUTLINE

ANSWERS TO QUESTIONS

Q15.1 The weight depends upon the total volume of glass. The pressure depends only on the depth.

Q15.2 Both must be built the same. The force on the back of each dam is the average pressure of the water times the area of the dam. If both reservoirs are equally deep, the force is the same.

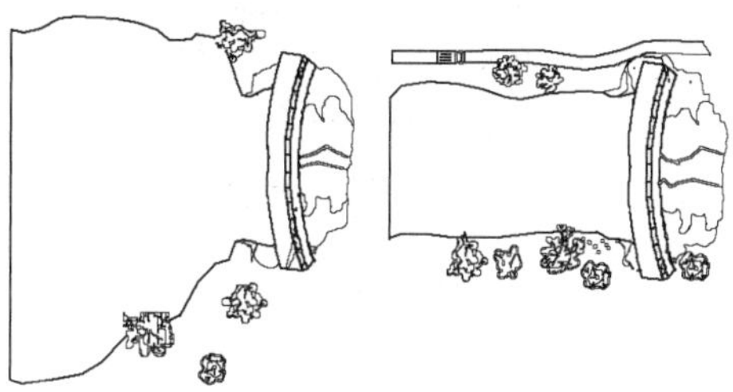

FIG. Q15.2

Q15.3 If the tube were to fill up to the height of several stories of the building, the pressure at the bottom of the depth of the tube of fluid would be very large according to Equation 15.4. This pressure is much larger than that originally exerted by inward elastic forces of the rubber on the water. As a result, water is pushed into the bottle from the tube. As more water is added to the tube, more water continues to enter the bottle, stretching it thin. For a typical bottle, the pressure at the bottom of the tube can become greater than the pressure at which the rubber material will rupture, so the bottle will simply fill with water and expand until it bursts. Blaise Pascal splintered strong barrels by this method.

Q15.4 In the ocean, the ship floats due to the buoyant force from *salt water*. Salt water is denser than fresh water. As the ship is pulled up the river, the buoyant force from the fresh water in the river is not sufficient to support the weight of the ship, and it sinks.

Q15.5 Yes. The propulsive force of the fish on the water causes the scale reading to fluctuate. Its average value will still be equal to the total weight of bucket, water, and fish.

Q15.6 Exactly the same. Buoyancy equals density of water times volume displaced.

Q15.7 The buoyant force is a conservative force. It does positive work on an object displaced upward, negative work on an object displaced downward, and zero work on an object displaced through a closed path. Potential energy is associated with it and it is gravitational potential energy of the object-fluid-Earth system. This system possesses extra energy when you hold a tennis ball at the bottom of a pool of water. As the tennis ball bobs up after it is released, the ball can do work in pushing obstacles out of the way. This work is just the work that can be done by the water sinking down to occupy the space vacated by the ball.

Q15.8 The rapidly moving air above the ball exerts less pressure than the atmospheric pressure below the ball. This can give substantial lift to balance the weight of the ball.

Q15.9 At lower elevation the water pressure is greater because pressure increases with increasing depth below the water surface in the reservoir (or water tower). The penthouse apartment is not so far below the water surface. The pressure behind a closed faucet is weaker there and the flow weaker from an open faucet. Your fire department likely has a record of the precise elevation of every fire hydrant.

Q15.10 The ski–jumper gives her body the shape of an airfoil. She deflects downward the air stream as it rushes past and it deflects her upward by Newton's third law. The air exerts on her a lift force, giving her a higher and longer trajectory. To say it in different words, the pressure on her back is less than the pressure on her front.

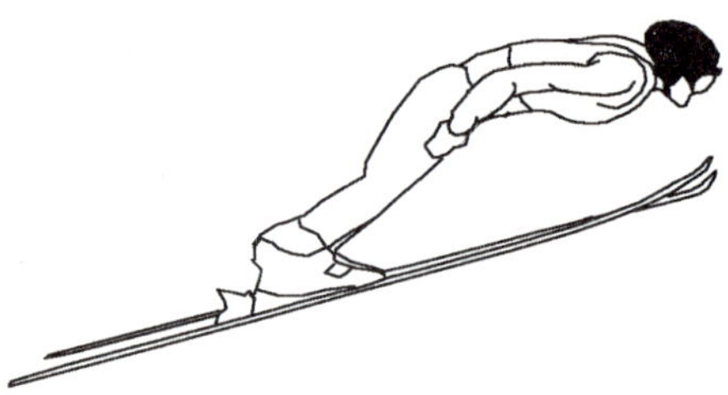

FIG. Q15.10

Q15.11 The glass may have higher density than the liquid, but the air inside has lower density. The total weight of the bottle can be less than the weight of an equal volume of the liquid.

Q15.12 Breathing in makes your volume greater and increases the buoyant force on you. You instinctively take a deep breath if you fall into the lake.

Q15.13 No. The somewhat lighter barge will float higher in the water.

Q15.14 The level of the pond falls. This is because the anchor displaces more water while in the boat. A floating object displaces a volume of water whose weight is equal to the weight of the object. A submerged object displaces a volume of water equal to the volume of the object. Because the density of the anchor is greater than that of water, a volume of water that weighs the same as the anchor will be greater than the volume of the anchor.

Q15.15 The metal is more dense than water. If the metal is sufficiently thin, it can float like a ship, with the lip of the dish above the water line. Most of the volume below the water line is filled with air. The mass of the dish divided by the volume of the part below the water line is just equal to the density of water. Placing a bar of soap into this space to replace the air raises the average density of the compound object and the density can become greater than that of water. The dish sinks with its cargo.

Q15.16 The excess pressure is transmitted undiminished throughout the container. It will compress air inside the wood. The water driven into the wood raises its average density and makes if float lower in the water. Add some thumbtacks to reach neutral buoyancy and you can make the wood sink or rise at will by subtly squeezing a large clear–plastic soft–drink bottle. Bored with graph paper and proving his own existence, René Descartes invented this toy or trick.

Q15.17 The air in your lungs, the blood in your arteries and veins, and the protoplasm in each cell exert nearly the same pressure, so that the wall of your chest can be in equilibrium.

Q15.18 We suppose the compound object floats. In both orientations it displaces its own weight of water, so it displaces equal volumes of water. The water level in the tub will be unchanged when the object is turned over. Now the steel is underwater and the water exerts on the steel a buoyant force that was not present when the steel was on top surrounded by air. Thus, slightly less wood will be below the water line on the block. It will appear to float higher.

Q15.19 Regular cola contains a considerable mass of dissolved sugar. Its density is higher than that of water. Diet cola contains a very small mass of artificial sweetener and has nearly the same density as water. The low–density air in the can has a bigger effect than the thin aluminum shell, so the can of diet cola floats.

Q15.20 The air at the lower hole is in the stagnant layer next to the ground and has nearly zero velocity. The mound entrance is high enough to be in a layer of moving air. A breeze from any direction speeds up to go over the mound and the air pressure drops. Air then flows through the burrow from the lower entrance to the upper entrance.

Q15.21 Clap your shoe or wallet over the hole, or a seat cushion, or your hand. Anything that can sustain a force on the order of 100 N is strong enough to cover the hole and greatly slow down the escape of the cabin air. You need not worry about the air rushing out instantly, or about your body being "sucked" through the hole, or about your blood boiling or your body exploding. If the cabin pressure drops a lot, your ears will pop and the saliva in your mouth may boil—at body temperature—but you will still have a couple of minutes to plug the hole and put on your emergency oxygen mask. Passengers who have been drinking carbonated beverages may find that the carbon dioxide suddenly comes out of solution in their stomachs, distending their vests, making them belch, and all but frothing from their ears; so you might warn them of this effect.

Q15.22 (a) The fluid around the compass needle exerts kinetic friction to take energy out of its oscillation. As the needle's speed approaches zero the friction force goes precisely to zero, so it cannot throw off the final direction of the stationary needle.

(b) When the level is set in a new orientation, the air bubble will move toward a new equilibrium position in the tube. It will tend to oscillate around this position, but fluid friction quickly degrades its energy of oscillation into internal energy. Exerting zero static friction, the water cannot throw off the final position of the bubble. To calibrate a newly constructed level, set it on a known level surface, wait a moment, and scribe two lines on the outer surface of the transparent tube bracketing the ends of the bubble.

(c) The soap will come to rest at the drain. Kinetic friction takes away its kinetic energy. With no static friction, the soap can stand in equilibrium only at the lowest point of the bathtub. Try it! Compare its motion with that of a dense spherical ball released in the tub.

SOLUTIONS TO PROBLEMS

Section 15.1 Pressure

P15.1 $M = \rho_{\text{iron}} V = \left(7\,860 \text{ kg/m}^3\right)\left[\frac{4}{3}\pi(0.015\,0 \text{ m})^3\right]$

$M = \boxed{0.111 \text{ kg}}$

P15.2 Let F_g be its weight. Then each tire supports $\frac{F_g}{4}$,

so $P = \frac{F}{A} = \frac{F_g}{4A}$

yielding $F_g = 4AP = 4\left(0.024\,0 \text{ m}^2\right)\left(200 \times 10^3 \text{ N/m}^2\right) = \boxed{1.92 \times 10^4 \text{ N}}$

P15.3 $P = \frac{F}{A} = \frac{50.0(9.80)}{\pi\left(0.500 \times 10^{-2}\right)^2} = \boxed{6.24 \times 10^6 \text{ N/m}^2}$

P15.4 The Earth's surface area is $4\pi R^2$. The force pushing inward over this area amounts to

$$F = P_0 A = P_0\left(4\pi R^2\right).$$

This force is the weight of the air:

$$F_g = mg = P_0\left(4\pi R^2\right)$$

so the mass of the air is

$$m = \frac{P_0\left(4\pi R^2\right)}{g} = \frac{\left(1.013 \times 10^5 \text{ N/m}^2\right)\left[4\pi\left(6.37 \times 10^6 \text{ m}\right)^2\right]}{9.80 \text{ m/s}^2} = \boxed{5.27 \times 10^{18} \text{ kg}}.$$

Section 15.2 Variation of Pressure with Depth

P15.5 $F_{el} = F_{\text{fluid}}$ or $kx = \rho g h A$

and $h = \frac{kx}{\rho g A}$

$$h = \frac{\left(1\,000 \text{ N/m}^2\right)\left(5.00 \times 10^{-3} \text{ m}\right)}{\left(10^3 \text{ kg/m}^3\right)\left(9.80 \text{ m/s}^2\right)\left[\pi\left(1.00 \times 10^{-2} \text{ m}\right)^2\right]} = \boxed{1.62 \text{ m}}$$

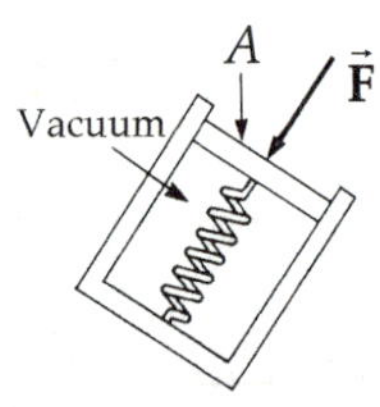

FIG. P15.5

P15.6 (a) $P = P_0 + \rho gh = 1.013 \times 10^5\ \text{Pa} + (1\,024\ \text{kg/m}^3)(9.80\ \text{m/s}^2)(1\,000\ \text{m})$

$$P = \boxed{1.01 \times 10^7\ \text{Pa}}$$

(b) The gauge pressure is the difference in pressure between the water outside and the air inside the submarine, which we suppose is at 1.00 atmosphere.

$$P_{\text{gauge}} = P - P_0 = \rho gh = 1.00 \times 10^7\ \text{Pa}$$

The resultant inward force on the porthole is then

$$F = P_{\text{gauge}} A = 1.00 \times 10^7\ \text{Pa}\left[\pi (0.150\ \text{m})^2\right] = \boxed{7.09 \times 10^5\ \text{N}}.$$

P15.7 $F_g = 80.0\ \text{kg}(9.80\ \text{m/s}^2) = 784\ \text{N}$

When the cup barely supports the student, the normal force of the ceiling is zero and the cup is in equilibrium.

$$F_g = F = PA = (1.013 \times 10^5\ \text{Pa})A$$

$$A = \frac{F_g}{P} = \frac{784}{1.013 \times 10^5} = \boxed{7.74 \times 10^{-3}\ \text{m}^2}$$

FIG. P15.7

P15.8 (a) Suppose the "vacuum cleaner" functions as a high–vacuum pump. The air below the brick will exert on it a lifting force

$$F = PA = 1.013 \times 10^5\ \text{Pa}\left[\pi\left(1.43 \times 10^{-2}\ \text{m}\right)^2\right] = \boxed{65.1\ \text{N}}.$$

(b) The octopus can pull the bottom away from the top shell with a force that could be no larger than

$$F = PA = (P_0 + \rho g h)A = \left[1.013 \times 10^5\ \text{Pa} + (1\,030\ \text{kg/m}^3)(9.80\ \text{m/s}^2)(32.3\ \text{m})\right]\left[\pi\left(1.43 \times 10^{-2}\ \text{m}\right)^2\right]$$

$$F = \boxed{275\ \text{N}}$$

***P15.9** The excess water pressure (over air pressure) halfway down is

$$P_{\text{gauge}} = \rho gh = (1\,000\ \text{kg/m}^3)(9.80\ \text{m/s}^2)(1.20\ \text{m}) = 1.18 \times 10^4\ \text{Pa}.$$

The force on the wall due to the water is

$$F = P_{\text{gauge}} A = (1.18 \times 10^4\ \text{Pa})(2.40\ \text{m})(9.60\ \text{m}) = \boxed{2.71 \times 10^5\ \text{N}}$$

horizontally toward the back of the hole.

P15.10 The pressure on the bottom due to the water is $P_b = \rho g z = 1.96 \times 10^4$ Pa

So, $$F_b = P_b A = \boxed{5.88 \times 10^6 \text{ N}}$$

On each end, $$F = \overline{P}A = 9.80 \times 10^3 \text{ Pa}\left(20.0 \text{ m}^2\right) = \boxed{196 \text{ kN}}$$

On the side, $$F = \overline{P}A = 9.80 \times 10^3 \text{ Pa}\left(60.0 \text{ m}^2\right) = \boxed{588 \text{ kN}}$$

***P15.11** The fluid in the hydraulic jack is originally exerting the same pressure as the air outside. This pressure P_0 results in zero net force on either piston. For the equilibrium of piston 2 we require

$$500 \text{ lb} = (P - P_0)A = (P - P_0)\pi\left(\frac{1.5 \text{ in.}}{2}\right)^2$$

Let F_1 represent the force the lever bar exerts on piston 1. Then similarly

$$F_1 = (P - P_0)\pi\left(\frac{0.25 \text{ in.}}{2}\right)^2$$

We ignore the weights of the pistons, sliding friction, and the slight difference in fluid pressure P due to the height difference between points 1 and 2. By division

$$\frac{F_1}{500 \text{ lb}} = \left(\frac{0.25 \text{ in.}}{1.5 \text{ in.}}\right)^2$$

$$F_1 = \frac{500 \text{ lb}}{36}$$

We say the hydraulic lift has an ideal mechanical advantage of 36. Next for the lever bar we ignore weight and friction, assume equilibrium, and take torques about the fixed hinge.

$$\sum \tau = 0 \qquad F_1(2 \text{ in.}) - F(12 \text{ in.}) = 0 \qquad F = \frac{F_1}{6}$$

The lever has an ideal mechanical advantage of 6. By substitution,

$$F = \frac{500 \text{ lb}}{36 \cdot 6} = \boxed{2.31 \text{ lb}}$$

Section 15.3 Pressure Measurements

P15.12 (a) We imagine the superhero to produce a perfect vacuum in the straw. Take point 1 at the water surface in the basin and point 2 at the water surface in the straw:

$$P_1 + \rho g y_1 = P_2 + \rho g y_2$$

$$1.013 \times 10^5 \text{ N/m}^2 + 0 = 0 + \left(1\,000 \text{ kg/m}^3\right)\left(9.80 \text{ m/s}^2\right)y_2 \qquad y_2 = \boxed{10.3 \text{ m}}$$

(b) No atmosphere can lift the water in the straw through $\boxed{\text{zero}}$ height difference.

P15.13 $P_0 = \rho g h$

$$h = \frac{P_0}{\rho g} = \frac{1.013 \times 10^5 \text{ Pa}}{\left(0.984 \times 10^3 \text{ kg/m}^3\right)\left(9.80 \text{ m/s}^2\right)} = \boxed{10.5 \text{ m}}$$

No. Some alcohol and water will evaporate. The equilibrium vapor pressures of alcohol and water are higher than the vapor pressure of mercury.

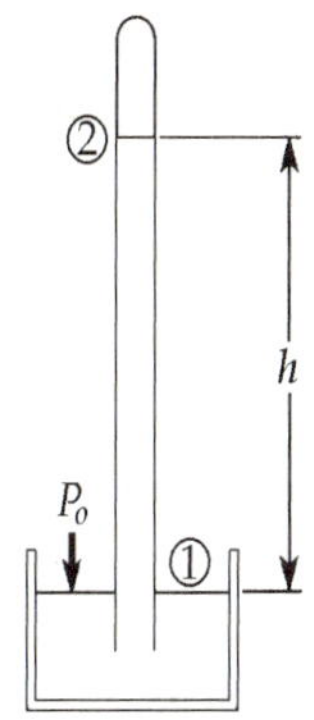

FIG. P15.13

P15.14 (a) Using the definition of density, we have

$$h_w = \frac{m_{\text{water}}}{A_2 \rho_{\text{water}}} = \frac{100 \text{ g}}{5.00 \text{ cm}^2\left(1.00 \text{ g/cm}^3\right)} = \boxed{20.0 \text{ cm}}$$

(b) Sketch (b) at the right represents the situation after the water is added. A volume $(A_2 h_2)$ of mercury has been displaced by water in the right tube. The additional volume of mercury now in the left tube is $A_1 h$. Since the total volume of mercury has not changed,

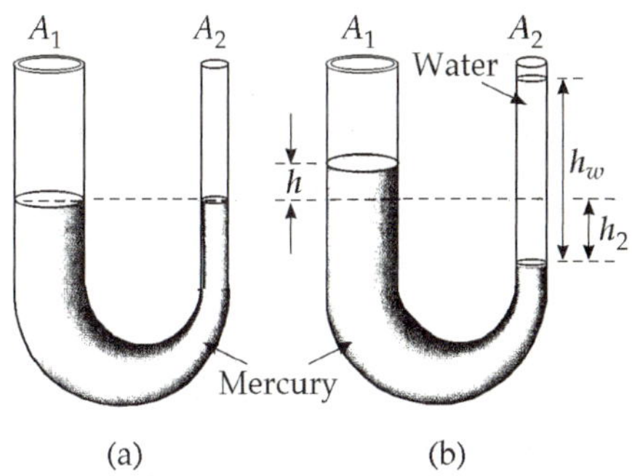

FIG. P15.14

$$A_2 h_2 = A_1 h \qquad \text{or} \qquad h_2 = \frac{A_1}{A_2} h \qquad (1)$$

At the level of the mercury–water interface in the right tube, we may write the absolute pressure as:

$$P = P_0 + \rho_{\text{water}} g h_w$$

The pressure at this same level in the left tube is given by

$$P = P_0 + \rho_{\text{Hg}} g (h + h_2) = P_0 + \rho_{\text{water}} g h_w$$

which, using equation (1) above, reduces to

$$\rho_{\text{Hg}} h \left[1 + \frac{A_1}{A_2}\right] = \rho_{\text{water}} h_w$$

or
$$h = \frac{\rho_{\text{water}} h_w}{\rho_{\text{Hg}}\left(1 + \frac{A_1}{A_2}\right)}.$$

Thus, the level of mercury has risen a distance of $h = \dfrac{\left(1.00 \text{ g/cm}^3\right)(20.0 \text{ cm})}{\left(13.6 \text{ g/cm}^3\right)\left(1 + \frac{10.0}{5.00}\right)} = \boxed{0.490 \text{ cm}}$

above the original level.

P15.15 $\Delta P_0 = \rho g \Delta h = -2.66 \times 10^3 \text{ Pa}: \quad P = P_0 + \Delta P_0 = (1.013 - 0.026\,6) \times 10^5 \text{ Pa} = \boxed{0.986 \times 10^5 \text{ Pa}}$

P15.16 (a) $P = P_0 + \rho g h$

The gauge pressure is

$$P - P_0 = \rho g h = 1\,000 \text{ kg}\left(9.8 \text{ m/s}^2\right)(0.160 \text{ m}) = \boxed{1.57 \text{ kPa}} = 1.57 \times 10^3 \text{ Pa}\left(\frac{1 \text{ atm}}{1.013 \times 10^5 \text{ Pa}}\right)$$

$$= \boxed{0.015\,5 \text{ atm}}.$$

It would lift a mercury column to height

$$h = \frac{P - P_0}{\rho g} = \frac{1\,568 \text{ Pa}}{\left(13\,600 \text{ kg/m}^3\right)\left(9.8 \text{ m/s}^2\right)} = \boxed{11.8 \text{ mm}}.$$

(b) Increased pressure of the cerebrospinal fluid will raise the level of the fluid in the spinal tap.

(c) Blockage of the fluid within the spinal column or between the skull and the spinal column would prevent the fluid level from rising.

***P15.17** (a) The volume of the tube is

$$\pi r^2 \ell = \pi (0.010\,5 \text{ m})^2 (1.6 \text{ m}) = 5.54 \times 10^{-4} \text{ m}^3$$

$$m = \rho V = \left(13.6 \times 10^3 \text{ kg/m}^3\right)\left(5.54 \times 10^{-4} \text{ m}^3\right) = \boxed{7.54 \text{ kg}}$$

(b) We suppose that no air bubbles into the tube. The vacuum above the mercury column exerts no force on the plastic. The mercury column exerts no vertical force on the tube. Since the tube has thin walls, we ignore any buoyant force exerted on the tube by the mercury in the pan. The forces on the tube are the cord tension T upward; its weight $F_g = (0.480 \text{ kg})\left(9.8 \text{ m/s}^2\right) = 4.70 \text{ N}$ downward; and a force exerted downward on its top end by the air above.

$$F = PA = \left(1.013 \times 10^5 \text{ N/m}^2\right)\pi (0.010\,6 \text{ m})^2 = 35.1 \text{ N}$$

Now $\sum F_y = ma_y$, $\quad +T - 4.70 \text{ N} - 35.1 \text{ N} = 0$, $\quad T = \boxed{39.8 \text{ N}}$.

(c) The table supports the weight of all of the mercury and the pan. The surrounding air exerts equal upward and downward forces on most of the footprint area of the mercury, but there is no air above the mercury column. The air vertically below the column exerts an upward force on the pan equal to 35.1 N as computed above. For the equilibrium of the mercury and pan,

$$+n - (7.54 \text{ kg})\left(9.8 \text{ m/s}^2\right) - (0.32 \text{ kg})\left(9.8 \text{ m/s}^2\right) + 35.1 \text{ N} = 0$$

$$n = 77.0 \text{ N} - 35.1 \text{ N} = \boxed{41.9 \text{ N}}$$

continued on next page

(d) In the "Torricellian vacuum" above the mercury column, a barometer reads $\boxed{\text{zero}}$.

(e) The leak slowly raises the air pressure above the mercury column. The height of this column above the mercury in the pan decreases, but this change has no effect on T or n. The air in the space above the mercury column pushes up on the top of the tube to decrease T. It pushes down on the mercury column to increase n.

Section 15.4 Buoyant Forces and Archimedes's Principle

P15.18 $F_g = (m+\rho_s V)g$ must be equal to $F_b = \rho_w V g$

Since $V = Ah$, $m + \rho_s Ah = \rho_w Ah$

and $A = \boxed{\dfrac{m}{(\rho_w - \rho_s)h}}$

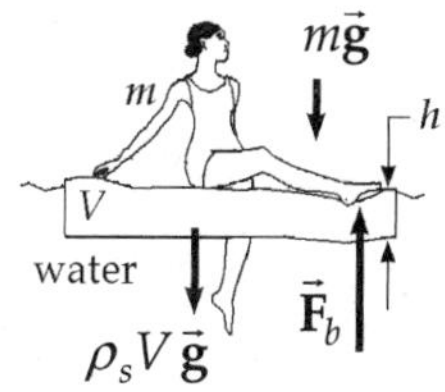

FIG. P15.18

P15.19 At equilibrium $\sum F = 0$ or $F_{app} + mg = B$

where B is the buoyant force.

The applied force, $F_{app} = B - mg$

where $B = \text{Vol}(\rho_{\text{water}})g$

and $m = (\text{Vol})\rho_{\text{ball}}$.

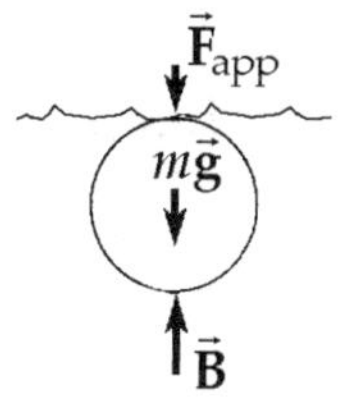

FIG. P15.19

So, $F_{app} = (\text{Vol})g(\rho_{\text{water}} - \rho_{\text{ball}}) = \frac{4}{3}\pi r^3 g(\rho_{\text{water}} - \rho_{\text{ball}})$

$$F_{app} = \frac{4}{3}\pi\left(1.90\times10^{-2}\text{ m}\right)^3\left(9.80\text{ m/s}^2\right)\left(10^3\text{ kg/m}^3 - 84.0\text{ kg/m}^3\right) = \boxed{0.258\text{ N}}$$

***P15.20** (a)

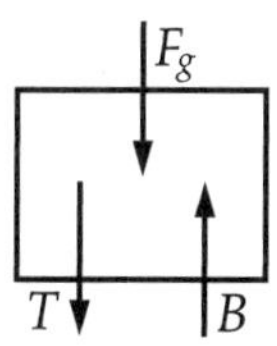

FIG. P15.20(a)

(b) $\sum F_y = 0$: $-15\text{ N} - 10\text{ N} + B = 0$

$\boxed{B = 25.0\text{ N}}$

(c) The oil pushes $\boxed{\text{horizontally inward}}$ on each side of the block.

(d) $\boxed{\text{String tension increases}}$. The oil causes the water below to be under greater pressure, and the water pushes up more strongly on the bottom of the block.

continued on next page

(e) Consider the equilibrium just before the string breaks:

$$-15\text{ N}-60\text{ N}+25\text{ N}+B_{\text{oil}}=0$$
$$B_{\text{oil}}=50\text{ N}$$

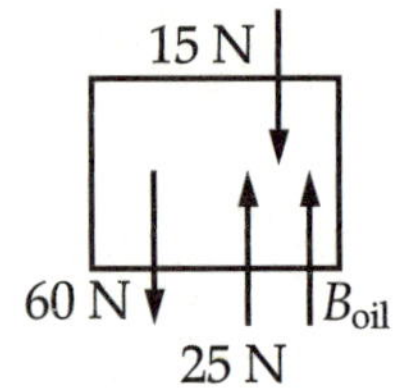

FIG. P15.20(e)

For the buoyant force of the water we have

$$B=\rho Vg \qquad 25\text{ N}=\left(1\,000\text{ kg/m}^3\right)\left(0.25V_{\text{block}}\right)9.8\text{ m/s}^2$$
$$V_{\text{block}}=1.02\times10^{-2}\text{ m}^3$$

For the buoyant force of the oil

$$50\text{ N}=\left(800\text{ kg/m}^3\right)f_e\left(1.02\times10^{-2}\text{ m}^3\right)9.8\text{ m/s}^2$$
$$f_e=0.625=\boxed{62.5\%}$$

(f) $$-15\text{ N}+\left(800\text{ kg/m}^3\right)f_f\left(1.02\times10^{-2}\text{ m}^3\right)9.8\text{ m/s}^2=0$$

$$f_f=0.187=\boxed{18.7\%}$$

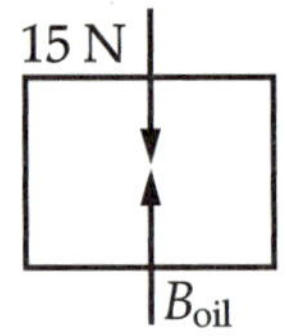

FIG. P15.20(f)

P15.21 (a) $P=P_0+\rho gh$

Taking $P_0=1.013\times10^5\text{ N/m}^2$ and $h=5.00\text{ cm}$

we find $P_{\text{top}}=1.017\,9\times10^5\text{ N/m}^2$

For $h=17.0$ cm, we get $P_{\text{bot}}=1.029\,7\times10^5\text{ N/m}^2$

Since the areas of the top and bottom are $A=(0.100\text{ m})^2=10^{-2}\text{ m}^2$

we find $F_{\text{top}}=P_{\text{top}}A=\boxed{1.017\,9\times10^3\text{ N}}$

and $F_{\text{bot}}=\boxed{1.029\,7\times10^3\text{ N}}$

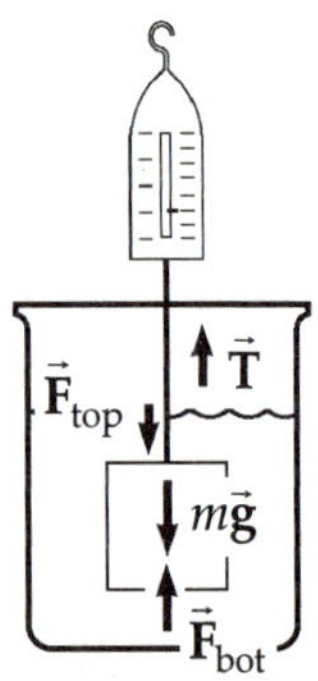

FIG. P15.21

(b) $T+B-Mg=0$

where $B=\rho_w Vg=\left(10^3\text{ kg/m}^3\right)\left(1.20\times10^{-3}\text{ m}^3\right)\left(9.80\text{ m/s}^2\right)=11.8\text{ N}$

and $Mg=10.0(9.80)=98.0\text{ N}$

Therefore, $T=Mg-B=98.0-11.8=\boxed{86.2\text{ N}}$

(c) $F_{\text{bot}}-F_{\text{top}}=(1.029\,7-1.017\,9)\times10^3\text{ N}=\boxed{11.8\text{ N}}$

which is equal to B found in part (b).

P15.22 Consider spherical balloons of radius 12.5 cm containing helium at STP and immersed in air at 0°C and 1 atm. If the rubber envelope has mass 5.00 g, the upward force on each is

$$B - F_{g,\text{He}} - F_{g,env} = \rho_{\text{air}} Vg - \rho_{\text{He}} Vg - m_{env} g$$

$$F_{up} = (\rho_{\text{air}} - \rho_{\text{He}})\left(\frac{4}{3}\pi r^3\right)g - m_{env} g$$

$$F_{up} = \left[(1.29 - 0.179)\ \text{kg/m}^3\right]\left[\frac{4}{3}\pi(0.125\ \text{m})^3\right](9.80\ \text{m/s}^2) - 5.00\times10^{-3}\ \text{kg}(9.80\ \text{m/s}^2) = 0.040\,1\ \text{N}$$

If your weight (including harness, strings, and submarine sandwich) is

$$70.0\ \text{kg}(9.80\ \text{m/s}^2) = 686\ \text{N}$$

you need this many balloons: $\dfrac{686\ \text{N}}{0.040\,1\ \text{N}} = 17\,000 \boxed{\sim 10^4}$.

P15.23 (a) According to Archimedes, $B = \rho_{\text{water}} V_{\text{water}} g = (1.00\ \text{g/cm}^3)[20.0\times 20.0\times(20.0 - h)]g$

But $B = \text{Weight of block} = mg = \rho_{\text{wood}} V_{\text{wood}} g = (0.650\ \text{g/cm}^3)(20.0\ \text{cm})^3 g$

$$0.650(20.0)^3 g = 1.00(20.0)(20.0)(20.0 - h)g$$

$$20.0 - h = 20.0(0.650) \text{ so } h = 20.0(1 - 0.650) = \boxed{7.00\ \text{cm}}$$

(b) $B = F_g + Mg$ where $M = \text{mass of lead}$

$$1.00(20.0)^3 g = 0.650(20.0)^3 g + Mg$$

$$M = (1.00 - 0.650)(20.0)^3 = 0.350(20.0)^3 = 2\,800\ \text{g} = \boxed{2.80\ \text{kg}}$$

P15.24 Let A represent the horizontal cross-sectional area of the rod, which we presume to be constant. The rod is in equilibrium:

$\sum F_y = 0$: $\quad -mg + B = 0 = -\rho_0 V_{\text{whole rod}} g + \rho_{\text{fluid}} V_{\text{immersed}} g$

$$\rho_0 ALg = \rho A(L - h)g$$

The density of the liquid is $\rho = \dfrac{\rho_0 L}{L - h}$.

P15.25 The balloon stops rising when $(\rho_{\text{air}} - \rho_{\text{He}})gV = Mg$ and $(\rho_{\text{air}} - \rho_{\text{He}})V = M$,

Therefore, $V = \dfrac{M}{\rho_{\text{air}} - \rho_{\text{He}}} = \dfrac{400}{1.25e^{-1} - 0.180}$ $\quad V = \boxed{1\,430\ \text{m}^3}$

P15.26 Constant velocity implies zero acceleration, which means that the submersible is in equilibrium under the gravitational force, the upward buoyant force, and the upward resistance force:

$$\sum F_y = ma_y = 0 \qquad -\left(1.20\times10^4 \text{ kg} + m\right)g + \rho_w gV + 1\,100 \text{ N} = 0$$

where m is the mass of the added water and V is the sphere's volume.

$$1.20\times10^4 \text{ kg} + m = 1.03\times10^3\left[\frac{4}{3}\pi(1.50)^3\right] + \frac{1\,100 \text{ N}}{9.8 \text{ m/s}^2}$$

so

$$m = \boxed{2.67\times10^3 \text{ kg}}$$

P15.27 $B = F_g$

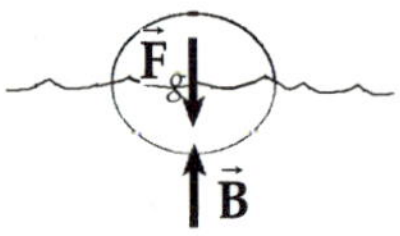

FIG. P15.27

$$\rho_{H_2O}\, g\frac{V}{2} = \rho_{\text{sphere}}\, gV$$

$$\rho_{\text{sphere}} = \frac{1}{2}\rho_{H_2O} = \boxed{500 \text{ kg/m}^3}$$

$$\rho_{\text{glycerin}}\, g\left(\frac{4}{10}V\right) - \rho_{\text{sphere}}\, gV = 0$$

$$\rho_{\text{glycerin}} = \frac{10}{4}\left(500 \text{ kg/m}^3\right) = \boxed{1\,250 \text{ kg/m}^3}$$

P15.28 Let ℓ represent the length below water at equilibrium and M the tube's mass:

$\Sigma F_y = 0$: $\qquad -Mg + \rho\pi r^2 \ell g = 0$

Now with any excursion x from equilibrium: $\qquad -Mg + \rho\pi r^2(\ell - x)g = Ma$

Subtracting the equilibrium equation gives: $\qquad -\rho\pi r^2 gx = Ma$

$$a = -\left(\frac{\rho\pi r^2 g}{M}\right)x = -\omega^2 x$$

The opposite direction and direct proportionality of a to x imply SHM with angular frequency

$$\omega = \sqrt{\frac{\rho\pi r^2 g}{M}}$$

$$T = \frac{2\pi}{\omega} = \boxed{\frac{2}{r}\sqrt{\frac{\pi M}{\rho g}}}$$

***P15.29** The water exerts a buoyant force on the air, given by

$$B=\rho_{\text{fluid}}gV=\left(1\,000\ \text{kg/m}^3\right)\left(9.8\ \text{m/s}^2\right)(10\ \text{L})\left(\frac{1\ \text{m}^3}{10^3\ \text{L}}\right)=98\ \text{N up}$$

The weight of the air is

$$F_g=\rho gV=\left(2.4\ \text{kg/m}^3\right)\left(9.8\ \text{m/s}^2\right)\left(10\times10^{-3}\ \text{m}^3\right)=0.235\ \text{N down}$$

To transport the air down at constant speed requires a downward force D in $+98\ \text{N}-0.235\ \text{N}-D=0$, $D=97.8\ \text{N}$, and work $W=\vec{\mathbf{D}}\cdot\vec{\mathbf{d}}=(97.8\ \text{N})(10.3\ \text{m})\cos 0°=\boxed{1.01\ \text{kJ}}$.

Section 15.5 Fluid Dynamics

Section 15.6 Streamlines and the Continuity Equation for Fluids

Section 15.7 Bernoulli's Equation

P15.30 Volume flow rate $=A_1v_1=A_2v_2$

$$\frac{20.0\ \text{L}}{60.0\ \text{s}}\left(\frac{1\,000\ \text{cm}^3}{1\ \text{L}}\right)=\pi(1.00\ \text{cm})^2v_{\text{hose}}=\pi(0.500\ \text{cm})^2v_{\text{nozzle}}$$

(a) $$v_{\text{hose}}=\frac{333\ \text{cm}^3/\text{s}}{3.14\ \text{cm}^2}=\boxed{106\ \text{cm/s}}$$

(b) $$v_{\text{nozzle}}=\frac{333\ \text{cm}^3/\text{s}}{0.785\ \text{cm}^2}=\boxed{424\ \text{cm/s}}$$

P15.31 By Bernoulli's equation,

$$8.00\times10^4\ \text{N/m}^2+\frac{1}{2}(1\,000)v^2=6.00\times10^4\ \text{N/m}^2+\frac{1}{2}(1\,000)16v^2$$

$$2.00\times10^4\ \text{N/m}^2=\frac{1}{2}(1\,000)15v^2$$

$$v=1.63\ \text{m/s}$$

$$\frac{dm}{dt}=\rho Av=1\,000\pi\left(5.00\times10^{-2}\right)^2(1.63\ \text{m/s})=\boxed{12.8\ \text{kg/s}}$$

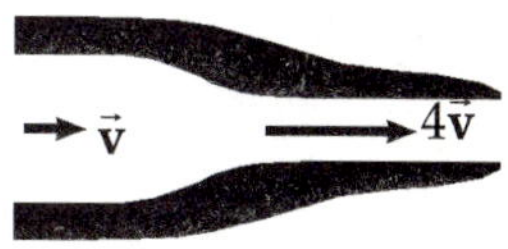

FIG. P15.31

***P15.32** Flow rate $Q=0.012\,0\ \text{m}^3/\text{s}=v_1A_1=v_2A_2$

$$v_2=\frac{Q}{A_2}=\frac{0.012\,0}{A_2}=\boxed{31.6\ \text{m/s}}$$

P15.33 Apply Bernoulli's equation between the top surface and the exiting stream.

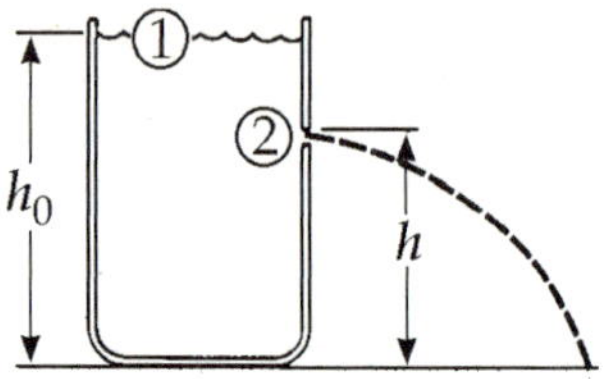

FIG. P15.33

$$P_0+0+\rho gh=P_0+\frac{1}{2}\rho v_x^2+\rho gh$$

$$v_x^2=2g(h_0-h) \qquad \therefore\ v_x=\sqrt{2g(h_0-h)}$$

$$x=v_xt: \qquad y=h=\frac{1}{2}gt^2$$

$$\therefore\ t=\sqrt{\frac{2h}{g}}$$

and $x=v_x\sqrt{\frac{2h}{g}}=\sqrt{2g(h_0-h)}\sqrt{\frac{2h}{g}}$: $\qquad x=\boxed{2\sqrt{h(h_0-h)}}$

***P15.34** Assuming the top is open to the atmosphere, then

$P_1=P_0$.

Note $P_2=P_0$.

$$\text{Flow rate}=2.50\times10^{-3}\ \text{m}^3/\text{min}=4.17\times10^{-5}\ \text{m}^3/\text{s}.$$

(a) $A_1>>A_2$ so $v_1<<v_2$

Assuming $v_1=0$,

$$P_1+\frac{\rho v_1^2}{2}+\rho gy_1=P_2+\frac{\rho v_2^2}{2}+\rho gy_2$$
$$v_2=(2gy_1)^{1/2}=[2(9.80)(16.0)]^{1/2}=\boxed{17.7\ \text{m/s}}$$

(b) $\text{Flow rate}=A_2v_2=\left(\frac{\pi d^2}{4}\right)(17.7)=4.17\times10^{-5}\ \text{m}^3/\text{s}$

$$d=\boxed{1.73\times10^{-3}\ \text{m}}=1.73\ \text{mm}$$

***P15.35** Take point ① at the free surface of the water in the tank and ② inside the nozzle.

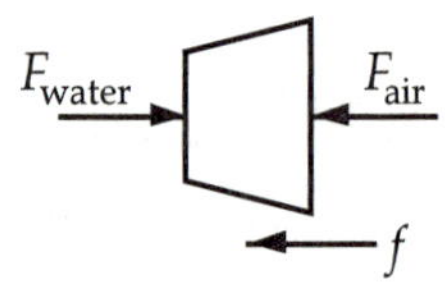

FIG. P15.35

(a) With the cork in place $P_1+\rho gy_1+\frac{1}{2}\rho v_1^2=P_2+\rho gy_2+\frac{1}{2}\rho v_2^2$ becomes $P_0+1\,000\ \text{kg/m}^3\ 9.8\ \text{m/s}^2\ 7.5\ \text{m}+0=P_2+0+0;\ P_2-P_0=7.35\times10^4\ \text{Pa}$.

For the stopper $\sum F_x=0$

$$F_{water}-F_{air}-f=0$$
$$P_2A-P_0A=f$$
$$f=7.35\times10^4\ \text{Pa}\,\pi(0.011\ \text{m})^2=\boxed{27.9\ \text{N}}$$

continued on next page

(b) Now Bernoulli's equation gives

$$P_0 + 7.35\times10^4 \text{ Pa} + 0 = P_0 + 0 + \frac{1}{2}\left(1\,000 \text{ kg/m}^3\right)v_2^2$$

$$v_2 = 12.1 \text{ m/s}$$

The quantity leaving the nozzle in 2 h is

$$\rho V = \rho A v_2 t = \left(1\,000 \text{ kg/m}^3\right)\pi(0.011 \text{ m})^2(12.1 \text{ m/s})7\,200 \text{ s} = \boxed{3.32\times10^4 \text{ kg}}.$$

(c) Take point 1 in the wide hose and 2 just outside the nozzle. Continuity:

$$A_1v_1 = A_2v_2$$

$$\pi\left(\frac{6.6 \text{ cm}}{2}\right)^2 v_1 = \pi\left(\frac{2.2 \text{ cm}}{2}\right)^2 12.1 \text{ m/s}$$

$$v_1 = \frac{12.1 \text{ m/s}}{9} = 1.35 \text{ m/s}$$

$$P_1 + \rho g y_1 + \frac{1}{2}\rho v_1^2 = P_2 + \rho g y_2 + \frac{1}{2}\rho v_2^2$$

$$P_1 + 0 + \frac{1}{2}\left(1\,000 \text{ kg/m}^3\right)(1.35 \text{ m/s})^2 = P_0 + 0 + \frac{1}{2}\left(1\,000 \text{ kg/m}^3\right)(12.1 \text{ m/s})^2$$

$$P_1 - P_0 = 7.35\times10^4 \text{ Pa} - 9.07\times10^2 \text{ Pa} = \boxed{7.26\times10^4 \text{ Pa}}$$

P15.36 (a) $\mathcal{P} = \frac{\Delta E}{\Delta t} = \frac{\Delta mgh}{\Delta t} = \left(\frac{\Delta m}{\Delta t}\right)gh = Rgh$

(b) $\mathcal{P}_{\text{EL}} = 0.85\left(8.5\times10^5\right)(9.8)(87) = \boxed{616 \text{ MW}}$

***P15.37** The volume flow rate is

$$\frac{125 \text{ cm}^3}{16.3 \text{ s}} = Av_1 = \pi\left(\frac{0.96 \text{ cm}}{2}\right)^2 v_1.$$

The speed at the top of the falling column is

$$v_1 = \frac{7.67 \text{ cm}^3/\text{s}}{0.724 \text{ cm}^2} = 10.6 \text{ cm/s}.$$

Take point 2 at 13 cm below:

$$P_1 + \rho g y_1 + \frac{1}{2}\rho v_1^2 = P_2 + \rho g y_2 + \frac{1}{2}\rho v_2^2$$

$$P_0 + \left(1\,000 \text{ kg/m}^3\right)\left(9.8 \text{ m/s}^2\right)0.13 \text{ m} + \frac{1}{2}\left(1\,000 \text{ kg/m}^3\right)(0.106 \text{ m/s})^2$$

$$= P_0 + 0 + \frac{1}{2}\left(1\,000 \text{ kg/m}^3\right)v_2^2$$

$$v_2 = \sqrt{2\left(9.8 \text{ m/s}^2\right)0.13 \text{ m} + (0.106 \text{ m/s})^2} = 1.60 \text{ m/s}$$

continued on next page

The volume flow rate is constant:

$$7.67\ \text{cm}^3/\text{s} = \pi\left(\frac{d}{2}\right)^2 160\ \text{cm/s}$$

$$d = \boxed{0.247\ \text{cm}}$$

P15.38 (a) Between sea surface and clogged hole: $P_1 + \frac{1}{2}\rho v_1^2 + \rho g y_1 = P_2 + \frac{1}{2}\rho v_2^2 + \rho g y_2$

$$1\ \text{atm} + 0 + (1\,030\ \text{kg/m}^3)(9.8\ \text{m/s}^2)(2\ \text{m}) = P_2 + 0 + 0 \quad P_2 = 1\ \text{atm} + 20.2\ \text{kPa}$$

The air on the back of his hand pushes opposite the water, so the net force on his hand is

$$F = PA = (20.2\times10^3\ \text{N/m}^2)\left(\frac{\pi}{4}\right)(1.2\times10^{-2}\ \text{m})^2 \qquad F = \boxed{2.28\ \text{N}}$$

(b) Now, Bernoulli's theorem is

$$1\ \text{atm} + 0 + 20.2\ \text{kPa} = 1\ \text{atm} + \frac{1}{2}(1\,030\ \text{kg/m}^3)v_2^2 + 0 \qquad v_2 = 6.26\ \text{m/s}$$

The volume rate of flow is $A_2 v_2 = \frac{\pi}{4}(1.2\times10^{-2}\ \text{m})^2(6.26\ \text{m/s}) = 7.08\times10^{-4}\ \text{m}^3/\text{s}$

One acre–foot is $4\,047\ \text{m}^2 \times 0.304\,8\ \text{m} = 1\,234\ \text{m}^3$

Requiring $\dfrac{1\,234\ \text{m}^3}{7.08\times10^{-4}\ \text{m}^3/\text{s}} = \boxed{1.74\times10^6\ \text{s}} = 20.2\ \text{days}$

P15.39 (a) Suppose the flow is very slow: $\left(P + \frac{1}{2}\rho v^2 + \rho g y\right)_{\text{river}} = \left(P + \frac{1}{2}\rho v^2 + \rho g y\right)_{\text{rim}}$

$$P + 0 + \rho g(564\ \text{m}) = 1\ \text{atm} + 0 + \rho g(2\,096\ \text{m})$$

$$P = 1\ \text{atm} + (1\,000\ \text{kg/m}^3)(9.8\ \text{m/s}^2)(1\,532\ \text{m}) = \boxed{1\ \text{atm} + 15.0\ \text{MPa}}$$

(b) The volume flow rate is $4\,500\ \text{m}^3/\text{d} = Av = \dfrac{\pi d^2 v}{4}$

$$v = (4\,500\ \text{m}^3/\text{d})\left(\frac{1\ \text{d}}{86\,400\ \text{s}}\right)\left(\frac{4}{\pi(0.150\ \text{m})^2}\right) = \boxed{2.95\ \text{m/s}}$$

(c) Imagine the pressure as applied to stationary water at the bottom of the pipe:

$$\left(P + \frac{1}{2}\rho v^2 + \rho g y\right)_{\text{bottom}} = \left(P + \frac{1}{2}\rho v^2 + \rho g y\right)_{\text{top}}$$

$$P + 0 = 1\ \text{atm} + \frac{1}{2}(1\,000\ \text{kg/m}^3)(2.95\ \text{m/s})^2 + 1\,000\ \text{kg}(9.8\ \text{m/s}^2)(1\,532\ \text{m})$$

$$P = 1\ \text{atm} + 15.0\ \text{MPa} + 4.34\ \text{kPa}$$

The additional pressure is $\boxed{4.34\ \text{kPa}}$.

P15.40 (a) For upward flight of a water-drop projectile from geyser vent to fountain–top,
$v_{yf}^2 = v_{yi}^2 + 2a_y\Delta y$

Then $0 = v_i^2 + 2\left(-9.80 \text{ m/s}^2\right)(+40.0 \text{ m})$ and $\quad v_i = \boxed{28.0 \text{ m/s}}$

(b) Between geyser vent and fountain-top: $\quad P_1 + \frac{1}{2}\rho v_1^2 + \rho g y_1 = P_2 + \frac{1}{2}\rho v_2^2 + \rho g y_2$

Air is so low in density that very nearly $\quad P_1 = P_2 = 1 \text{ atm}$

Then, $\quad \frac{1}{2}v_i^2 + 0 = 0 + \left(9.80 \text{ m/s}^2\right)(40.0 \text{ m})$

$$v_1 = \boxed{28.0 \text{ m/s}}$$

(c) Between the chamber and the fountain-top: $\quad P_1 + \frac{1}{2}\rho v_1^2 + \rho g y_1 = P_2 + \frac{1}{2}\rho v_2^2 + \rho g y_2$

$$P_1 + 0 + \left(1\,000 \text{ kg/m}^3\right)\left(9.80 \text{ m/s}^2\right)(-175 \text{ m}) = P_0 + 0 + \left(1\,000 \text{ kg/m}^3\right)\left(9.80 \text{ m/s}^2\right)(+40.0 \text{ m})$$
$$P_1 - P_0 = \left(1\,000 \text{ kg/m}^3\right)\left(9.80 \text{ m/s}^2\right)(215 \text{ m}) = \boxed{2.11 \text{ MPa}}$$

***P15.41** The assumption of incompressibility is surely unrealistic, but allows an estimate of the speed:

$$P_1 + \rho g y_1 + \frac{1}{2}\rho v_1^2 = P_2 + \rho g y_2 + \frac{1}{2}\rho v_2^2$$
$$1.00 \text{ atm} + 0 + 0 = 0.287 \text{ atm} + 0 + \frac{1}{2}\left(1.20 \text{ kg/m}^3\right)v_2^2$$
$$v_2 = \sqrt{\frac{2(1.00 - 0.287)\left(1.013\times10^5 \text{ N/m}^2\right)}{1.20 \text{ kg/m}^3}} = \boxed{347 \text{ m/s}}$$

Section 15.8 Other Applications of Fluid Dynamics

P15.42 $Mg = (P_1 - P_2)A$ for a balanced condition $\quad \frac{16\,000(9.80)}{A} = 7.00\times10^4 - P_2$

where $\quad A = 80.0 \text{ m}^2 \quad \therefore P_2 = 7.0\times10^4 - 0.196\times10^4 = \boxed{6.80\times10^4 \text{ Pa}}$

P15.43 (a) $P_0 + \rho gh + 0 = P_0 + 0 + \frac{1}{2}\rho v_3^2$ $\quad v_3 = \sqrt{2gh}$

If $h = 1.00$ m, $\quad v_3 = \boxed{4.43 \text{ m/s}}$

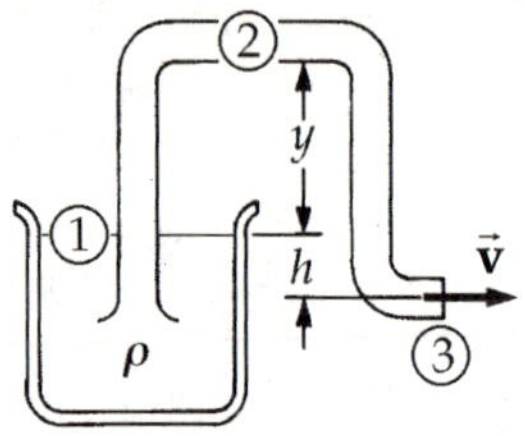

FIG. P15.43

(b) $P + \rho gy + \frac{1}{2}\rho v_2^2 = P_0 + 0 + \frac{1}{2}\rho v_3^2$

Since $v_2 = v_3$, $\quad P = P_0 - \rho gy$

Since $P \geq 0$ $\quad y \leq \frac{P_0}{\rho g} = \frac{1.013 \times 10^5 \text{ Pa}}{(10^3 \text{ kg/m}^3)(9.8 \text{ m/s}^2)} = \boxed{10.3 \text{ m}}$

P15.44 Take points 1 and 2 in the air just inside and outside the window pane.

$$P_1 + \frac{1}{2}\rho v_1^2 + \rho gy_1 = P_2 + \frac{1}{2}\rho v_2^2 + \rho gy_2$$

$$P_0 + 0 = P_2 + \frac{1}{2}(1.30 \text{ kg/m}^3)(11.2 \text{ m/s})^2 \qquad P_2 = P_0 - 81.5 \text{ Pa}$$

(a) The total force exerted by the air is outward,

$$P_1A - P_2A = P_0A - P_0A + (81.5 \text{ N/m}^2)(4 \text{ m})(1.5 \text{ m}) = \boxed{489 \text{ N outward}}$$

(b) $P_1A - P_2A = \frac{1}{2}\rho v_2^2 A = \frac{1}{2}(1.30 \text{ kg/m}^3)(22.4 \text{ m/s})^2(4 \text{ m})(1.5 \text{ m}) = \boxed{1.96 \text{ kN outward}}$

P15.45 In the reservoir, the gauge pressure is $\quad \Delta P = \frac{2.00 \text{ N}}{2.50 \times 10^{-5} \text{ m}^2} = 8.00 \times 10^4 \text{ Pa}$

From the equation of continuity: $\quad A_1v_1 = A_2v_2$

$(2.50 \times 10^{-5} \text{ m}^2)v_1 = (1.00 \times 10^{-8} \text{ m}^2)v_2 \quad v_1 = (4.00 \times 10^{-4})v_2$

Thus, v_1^2 is negligible in comparison to v_2^2.

Then, from Bernoulli's equation: $\quad (P_1 - P_2) + \frac{1}{2}\rho v_1^2 + \rho gy_1 = \frac{1}{2}\rho v_2^2 + \rho gy_2$

$$8.00 \times 10^4 \text{ Pa} + 0 + 0 = 0 + \frac{1}{2}(1\,000 \text{ kg/m}^3)v_2^2$$

$$v_2 = \sqrt{\frac{2(8.00 \times 10^4 \text{ Pa})}{1\,000 \text{ kg/m}^3}} = \boxed{12.6 \text{ m/s}}$$

Section 15.9 Context Connection—A Near Miss Even Before Leaving Southampton

P15.46 (a) The iceberg floats in equilibrium

$\Sigma F_y = 0$: $\quad -F_g + B = 0$

$0 = -m_{\text{iceberg}} g + \rho_{\text{fluid}} V_{\text{immersed}} g \quad \rho_{\text{ice}} V_{\text{iceberg}} = \rho_{\text{fluid}} V_{\text{immersed}}$

$$\frac{V_{\text{immersed}}}{V_{\text{iceberg}}} = \frac{\rho_{\text{ice}}}{\rho_{\text{fluid}}} = \frac{0.917 \times 10^3 \text{ kg/m}^3}{1.03 \times 10^3 \text{ kg/m}^3} = 0.890$$

The fraction of the volume above the water line is $1 - 0.890 = 0.110 = 11.0\%$.

(b) With lower density, more fresh water must be displaced to support the iceberg. A $\boxed{\text{smaller}}$ fraction is above the water line

$$\frac{V_{\text{immersed}}}{V_{\text{iceberg}}} = \frac{\rho_{\text{ice}}}{\rho_{\text{fluid}}} = \frac{0.917 \times 10^3 \text{ kg/m}^3}{1.00 \times 10^3 \text{ kg/m}^3} = 0.917$$

The fraction exposed is $1 - 0.917 = \boxed{8.3\%}$.

P15.47 The weight of the additional water displaced is equal to the weight of the passengers:

$\rho_{\text{fluid}} \Delta V_{\text{immersed}} g = \Delta m g \quad (1.03 \times 10^3 \text{ kg/m}^3)(0.010\,0 \text{ m})A = 2\,205(75 \text{ kg})$

$$A = \frac{1.65 \times 10^5 \text{ kg}}{10.3 \text{ kg/m}^2} = \boxed{1.61 \times 10^4 \text{ m}^2}$$

P15.48 (a) $\theta = \dfrac{s}{r} = \dfrac{15 \text{ m}}{0.25 \text{ m}} = \boxed{60.0 \text{ rad}}$

(b) Let T represent the tension in the rope. Define the positive y axis as pointing down.

For the anchor: $\quad (2\,000 \text{ kg})(9.8 \text{ m/s}^2) - T = (2\,000 \text{ kg})(a)$

For the reel: $\quad \Sigma\tau = I\alpha = \dfrac{1}{2} MR^2 \dfrac{a}{R}$

$T(0.25 \text{ m}) = \dfrac{1}{2}(300 \text{ kg})(0.25 \text{ m})^2 \left(\dfrac{a}{0.25 \text{ m}}\right) \quad T = (150 \text{ kg})a$

Substituting, $\quad 19\,600 \text{ N} - (150 \text{ kg})a = (2\,000 \text{ kg})a$

$$a = \frac{19\,600 \text{ N}}{2\,150 \text{ kg}} = \boxed{9.12 \text{ m/s}^2}$$

continued on next page

(c) The water exerts a buoyant force on the anchor,

$$B=\rho_{\text{fluid}}Vg=\rho_{\text{fluid}}\left(\frac{m}{\rho_{\text{iron}}}\right)g=\left(1.03\times10^{3}\ \text{kg/m}^{3}\right)\left(\frac{2\,000\ \text{kg}}{7.86\times10^{3}\ \text{kg/m}^{3}}\right)\left(9.8\ \text{m/s}^{2}\right)=2\,570\ \text{N}$$

Now for the anchor, $\Sigma F_y=ma_y$: $19\,600\ \text{N}-T-2\,500\ \text{N}-2\,570\ \text{N}=(2\,000\ \text{kg})(a)$ while again for the reel $T=(150\ \text{kg})a$. So $a=\dfrac{19\,600\ \text{N}-2\,500\ \text{N}-2\,570\ \text{N}}{2\,150\ \text{kg}}=\boxed{6.76\ \text{m/s}^{2}}$.

(d) $\tau=Tr=150\ \text{kg}\left(6.76\ \text{m/s}^{2}\right)(0.25\ \text{m})=\boxed{253\ \text{N}\cdot\text{m}}$

Additional Problems

P15.49 The "balanced" condition is one in which the apparent weight of the body equals the apparent weight of the weights. This condition can be written as:

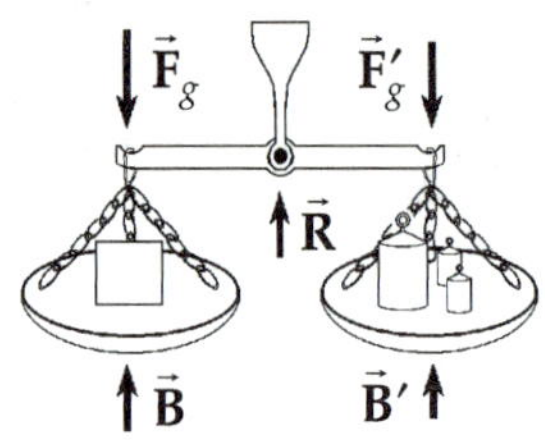

FIG. P15.49

$$F_g-B=F'_g-B'$$

where B and B' are the buoyant forces on the body and weights respectively. The buoyant force experienced by an object of volume V in air equals:

$$\text{Buoyant force}=(\text{Volume of object})\rho_{\text{air}}g$$

so we have $B=V\rho_{\text{air}}g$ and $B'=\left(\dfrac{F'_g}{\rho g}\right)\rho_{\text{air}}g$.

Therefore, $F_g=F'_g+\left(V-\dfrac{F'_g}{\rho g}\right)\rho_{\text{air}}g$.

P15.50 Assume $v_{\text{inside}}\approx0$. From Bernoulli's equation

$$P+0+0=1\ \text{atm}+\frac{1}{2}(1\,000)(30.0)^{2}+1\,000(9.80)(0.500)$$

$$P_{\text{gauge}}=P-1\ \text{atm}=4.50\times10^{5}+4.90\times10^{3}=\boxed{455\ \text{kPa}}$$

P15.51 At equilibrium, $\sum F_y = 0$: $B - F_{\text{spring}} - F_{g,\,\text{He}} - F_{g,\,\text{balloon}} = 0$

giving $F_{\text{spring}} = kL = B - (m_{\text{He}} + m_{\text{balloon}})g$.

But $B = \text{weight of displaced air} = \rho_{\text{air}} Vg$

and $m_{\text{He}} = \rho_{\text{He}} V$.

Therefore, we have: $kL = \rho_{\text{air}} Vg - \rho_{\text{He}} Vg - m_{\text{balloon}} g$

or $$L = \frac{(\rho_{\text{air}} - \rho_{\text{He}})V - m_{\text{balloon}}}{k} g.$$

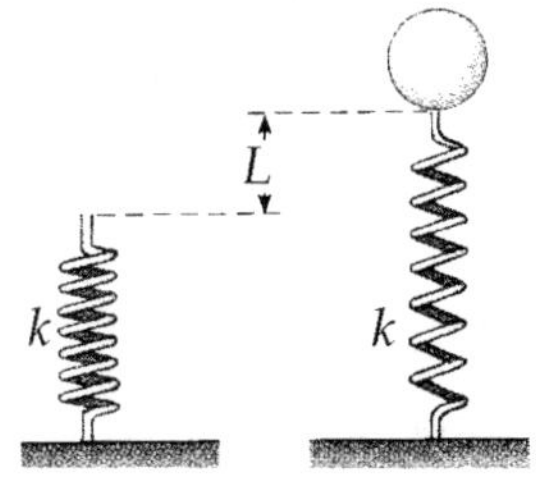

FIG. P15.51

From the data given, $$L = \frac{(1.29\ \text{kg/m}^3 - 0.180\ \text{kg/m}^3)5.00\ \text{m}^3 - 2.00 \times 10^{-3}\ \text{kg}}{90.0\ \text{N/m}}(9.80\ \text{m/s}^2).$$

Thus, this gives $L = \boxed{0.604\ \text{m}}$.

P15.52 $P = \rho g h$ $1.013 \times 10^5 = 1.29(9.80)h$

$h = \boxed{8.01\ \text{km}}$ For Mt. Everest, $29\,300\ \text{ft} = 8.88\ \text{km}$ $\boxed{\text{Yes}}$

P15.53 The torque is $\tau = \int d\tau = \int r\,dF$

From the figure $$\tau = \int_0^H y[\rho g(H - y)w\,dy] = \boxed{\frac{1}{6}\rho g w H^3}$$

The total force is given as $\frac{1}{2}\rho g w H^2$

If this were applied at a height y_{eff} such that the torque remains unchanged, we have

$$\frac{1}{6}\rho g w H^3 = y_{eff}\left[\frac{1}{2}\rho g w H^2\right] \quad \text{and} \quad y_{eff} = \boxed{\frac{1}{3}H}.$$

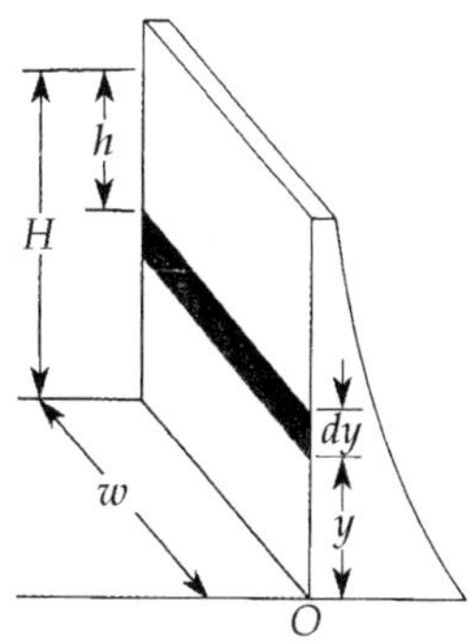

FIG. P15.53

P15.54 (a) The pressure on the surface of the two hemispheres is constant at all points, and the force on each element of surface area is directed along the radius of the hemispheres. The applied force along the axis must balance the force on the "effective" area, which is the projection of the actual surface onto a plane perpendicular to the x axis,

$$A = \pi R^2$$

Therefore, $F = \boxed{(P_0 - P)\pi R^2}$

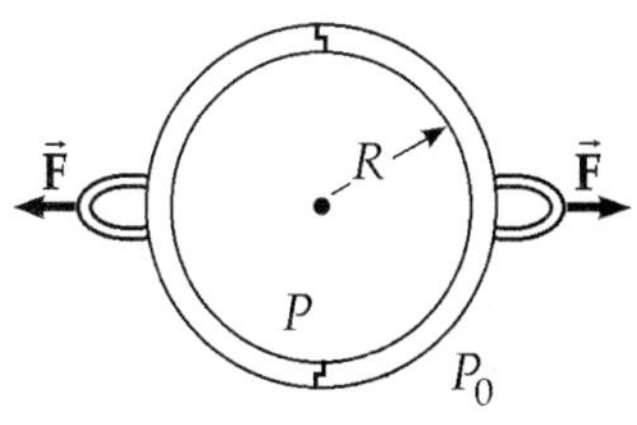

FIG. P15.54

(b) For the values given $F = (P_0 - 0.100P_0)\left[\pi(0.300\ \text{m})^2\right] = 0.254P_0 = \boxed{2.58 \times 10^4\ \text{N}}$

P15.55 Looking at the top scale and the iron block:

$$T_1 + B = F_{g,\text{ Fe}} \qquad \text{where} \qquad B = \rho_0 V_{\text{Fe}} g = \rho_0 \left(\frac{m_{\text{Fe}}}{\rho_{\text{Fe}}}\right) g$$

is the buoyant force exerted on the iron block by the oil.

Thus,
$$T_1 = F_{g,\text{ Fe}} - B = m_{\text{Fe}} g - \rho_0 \left(\frac{m_{\text{Fe}}}{\rho_{\text{Fe}}}\right) g$$

or
$$T_1 = \boxed{\left(1 - \frac{\rho_0}{\rho_{\text{Fe}}}\right) m_{\text{Fe}} g}$$ is the reading on the top scale.

Now, consider the bottom scale, which exerts an upward force of T_2 on the beaker–oil–iron combination.

$\sum F_y = 0$:
$$T_1 + T_2 - F_{g,\text{ beaker}} - F_{g,\text{ oil}} - F_{g,\text{ Fe}} = 0$$

$$T_2 = F_{g,\text{ beaker}} + F_{g,\text{ oil}} + F_{g,\text{ Fe}} - T_1 = (m_b + m_0 + m_{\text{Fe}}) g - \left(1 - \frac{\rho_0}{\rho_{\text{Fe}}}\right) m_{\text{Fe}} g$$

or
$$T_2 = \boxed{\left[m_b + m_0 + \left(\frac{\rho_0}{\rho_{\text{Fe}}}\right) m_{\text{Fe}}\right] g}$$ is the reading on the bottom scale.

P15.56 Let $m = \rho V$ represent the mass of the copper cylinder. The original tension in the wire is $T_1 = mg = \rho V g$. The water exerts a buoyant force $\rho_{\text{water}} \left(\frac{V}{2}\right) g$ on the cylinder, to reduce the tension to

$$T_2 = \rho V g - \rho_{\text{water}} \left(\frac{V}{2}\right) g = \left(\rho - \frac{\rho_{\text{water}}}{2}\right) V g.$$

The speed of a wave on the string changes from $\sqrt{\frac{T_1}{\mu}}$ to $\sqrt{\frac{T_2}{\mu}}$. The frequency changes from

$$f_1 = \frac{v_1}{\lambda} = \sqrt{\frac{T_1}{\mu}} \frac{1}{\lambda} \text{ to } f_2 = \sqrt{\frac{T_2}{\mu}} \frac{1}{\lambda}$$

where we assume $\lambda = 2L$ is constant.

Then
$$\frac{f_2}{f_1} = \sqrt{\frac{T_2}{T_1}} = \sqrt{\frac{\rho - \rho_{\text{water}}/2}{\rho}} = \sqrt{\frac{8.92 - 1.00/2}{8.92}}$$

$$f_2 = 300 \text{ Hz} \sqrt{\frac{8.42}{8.92}} = \boxed{291 \text{ Hz}}.$$

***P15.57** (a) The volume of a cell is $\frac{4}{3}\pi r^3 = \frac{4}{3}\pi\left(8\times10^{-6}\text{ m}\right)^3 = 2.14\times10^{-15}\text{ m}^3$. Its weight is

$$\rho V g = \left(1.02\times10^3\text{ kg/m}^3\right)\left(2.14\times10^{-15}\text{ m}^3\right)\left(9.8\text{ m/s}^2\right) = 2.14\times10^{-11}\text{ N down}$$

The buoyant force is

$$\rho_{\text{fluid}} V g = \left(1.00\times10^3\text{ kg/m}^3\right)\left(2.14\times10^{-15}\text{ m}^3\right)\left(9.8\text{ m/s}^2\right) = 2.10\times10^{-11}\text{ N up}$$

As the cell drifts down at constant speed,

$$\sum F_y = ma_y \quad +2.10\times10^{-11}\text{ N} + \left(0.018\,8\text{ N}\cdot\text{s/m}^2\right)\left(8\times10^{-6}\text{ m}\right)v_T - 2.14\times10^{-11}\text{ N} = 0$$

$$\left(1.50\times10^{-7}\text{ N}\cdot\text{s/m}\right)v_T = 4.20\times10^{-13}\text{ N}$$

$$v_T = \boxed{2.79\times10^{-6}\text{ m/s}}$$

(b) For a cell starting at the top of the tube, $v_T = \frac{x}{t}$,

$$t = \frac{x}{v_T} = \frac{0.08\text{ m}}{2.79\times10^{-6}\text{ m/s}} = 2.86\times10^4\text{ s} = \boxed{7.95\text{ h}}.$$

(c)

$$a_c = \frac{v^2}{r} = \frac{\omega^2 r^2}{r} = \omega^2 r = \left[3\,000\text{ rev/min}\left(\frac{2\pi\text{ rad}}{1\text{ rev}}\right)\left(\frac{1\text{ min}}{60\text{ s}}\right)\right]^2(0.09\text{ m})$$

$$= \boxed{8.88\times10^3\text{ m/s}^2\text{ toward the axis}}$$

(d) Imagine simple sedimentation on a planet where the free-fall acceleration is 8.88 km/s^2. A cell feels forces

$$\rho V g = \left(1.02\times10^3\text{ kg/m}^3\right)\left(2.14\times10^{-15}\text{ m}^3\right)\left(8.88\times10^3\text{ m/s}^2\right) = 1.94\times10^{-8}\text{ N down}$$

$$\rho_{\text{fluid}} V g = \left(1.00\times10^3\text{ kg/m}^3\right)\left(2.14\times10^{-15}\text{ m}^3\right)\left(8.88\times10^3\text{ m/s}^2\right) = 1.91\times10^{-8}\text{ N up}$$

and $\left(0.018\,8\text{ N}\cdot\text{s/m}^2\right)\left(8\times10^{-6}\text{ m}\right)v_T = \left(1.50\times10^{-7}\text{ N}\cdot\text{s/m}\right)v_T$ up. So

$$-1.94\times10^{-8}\text{ N} + 1.91\times10^{-8}\text{ N} + \left(1.50\times10^{-7}\text{ N}\cdot\text{s/m}\right)v_T = 0$$

$$v_T = \frac{3.81\times10^{-10}\text{ N}\cdot\text{m}}{1.50\times10^{-7}\text{ N}\cdot\text{s}} = 2.53\times10^{-3}\text{ m/s} = \frac{0.08\text{ m}}{t}$$

$$t = \boxed{31.6\text{ s}}$$

P15.58 The incremental version of $P - P_0 = \rho g y$ is $dP = -\rho g dy$

We assume that the density of air is proportional to pressure, or $\frac{P}{\rho} = \frac{P_0}{\rho_0}$

Combining these two equations we have $dP = -P\frac{\rho_0}{P_0} g dy$

$$\int_{P_0}^{P} \frac{dP}{P} = -g\frac{\rho_0}{P_0}\int_0^h dy$$

and integrating gives $\ln\left(\frac{P}{P_0}\right) = -\frac{\rho_0 g h}{P_0}$

so where $\alpha = \frac{\rho_0 g}{P_0}$, $P = P_0 e^{-\alpha h}$

P15.59 Energy for the fluid-Earth system is conserved.

$$(K+U)_i + \Delta E_{\text{mech}} = (K+U)_f: \quad 0 + \frac{mgL}{2} + 0 = \frac{1}{2}mv^2 + 0$$

$$v = \sqrt{gL} = \sqrt{2.00 \text{ m}\left(9.8 \text{ m/s}^2\right)} = \boxed{4.43 \text{ m/s}}$$

P15.60 Let s stand for the edge of the cube, h for the depth of immersion, ρ_{ice} stand for the density of the ice, ρ_w stand for density of water, and ρ_a stand for density of the alcohol.

(a) According to Archimedes's principle, at equilibrium we have

$$\rho_{\text{ice}} g s^3 = \rho_w g h s^2 \Rightarrow h = s\frac{\rho_{\text{ice}}}{\rho_w}$$

With $\rho_{\text{ice}} = 0.917 \times 10^3 \text{ kg/m}^3$

$\rho_w = 1.00 \times 10^3 \text{ kg/m}^3$

and $s = 20.0 \text{ mm}$

we get $h = 20.0(0.917) = 18.34 \text{ mm} \approx \boxed{18.3 \text{ mm}}$

(b) We assume that the top of the cube is still above the alcohol surface. Letting h_a stand for the thickness of the alcohol layer, we have

$$\rho_a g s^2 h_a + \rho_w g s^2 h_w = \rho_{\text{ice}} g s^3 \quad \text{so} \quad h_w = \left(\frac{\rho_{\text{ice}}}{\rho_w}\right)s - \left(\frac{\rho_a}{\rho_w}\right)h_a$$

continued on next page

With $\rho_a = 0.806 \times 10^3 \text{ kg/m}^3$

and $h_a = 5.00 \text{ mm}$

we obtain $h_w = 18.34 - 0.806(5.00) = 14.31 \text{ mm} \approx \boxed{14.3 \text{ mm}}$

(c) Here $h'_w = s - h'_a$, so Archimedes's principle gives

$$\rho_a g s^2 h'_a + \rho_w g s^2 (s - h'_a) = \rho_{\text{ice}} g s^3 \Rightarrow \rho_a h'_a + \rho_w (s - h'_a) = \rho_{\text{ice}} s$$

$$h'_a = s\frac{(\rho_w - \rho_{\text{ice}})}{(\rho_w - \rho_a)} = 20.0\frac{(1.000 - 0.917)}{(1.000 - 0.806)} = 8.557 \approx \boxed{8.56 \text{ mm}}$$

P15.61 **Note**: Variation of atmospheric pressure with altitude is included in this solution. Because of the small distances involved, this effect is unimportant in the final answers.

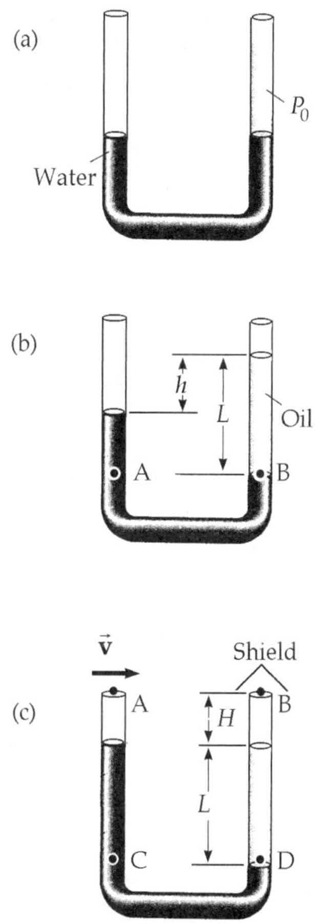

FIG. P15.61

(a) Consider the pressure at points A and B in part (b) of the figure:

Using the left tube: $P_A = P_{\text{atm}} + \rho_a g h + \rho_w g(L - h)$ where the second term is due to the variation of air pressure with altitude.

Using the right tube: $P_B = P_{\text{atm}} + \rho_0 g L$

But Pascal's principle says that $P_A = P_B$.

Therefore, $P_{\text{atm}} + \rho_0 g L = P_{\text{atm}} + \rho_a g h + \rho_w g(L - h)$

or $(\rho_w - \rho_a)h = (\rho_w - \rho_0)L$, giving

$$h = \left(\frac{\rho_w - \rho_0}{\rho_w - \rho_a}\right)L = \left(\frac{1\,000 - 750}{1\,000 - 1.29}\right)5.00 \text{ cm} = \boxed{1.25 \text{ cm}}$$

(b) Consider part (c) of the diagram showing the situation when the air flow over the left tube equalizes the fluid levels in the two tubes. First, apply Bernoulli's equation to points A and B ($y_A = y_B$, $v_A = v$, and $v_B = 0$)

This gives: $P_A + \frac{1}{2}\rho_a v^2 + \rho_a g y_A = P_B + \frac{1}{2}\rho_a (0)^2 + \rho_a g y_B$

and since $y_A = y_B$, this reduces to: $P_B - P_A = \frac{1}{2}\rho_a v^2$ (1)

Now consider points C and D, both at the level of the oil–water interface in the right tube. Using the variation of pressure with depth in static fluids, we have:

$P_C = P_A + \rho_a g H + \rho_w g L$ and $P_D = P_B + \rho_a g H + \rho_0 g L$

continued on next page

But Pascal's principle says that $P_C = P_D$. Equating these two gives:

$$P_B + \rho_a gH + \rho_0 gL = P_A + \rho_a gH + \rho_w gL \quad \text{or} \quad P_B - P_A = (\rho_w - \rho_0)gL \qquad (2)$$

Substitute equation (1) for $P_B - P_A$ into (2) to obtain $\frac{1}{2}\rho_a v^2 = (\rho_w - \rho_0)gL$

or $$v = \sqrt{\frac{2gL(\rho_w - \rho_0)}{\rho_a}} = \sqrt{2(9.80 \text{ m/s}^2)(0.050\,0 \text{ m})\left(\frac{1\,000 - 750}{1.29}\right)}$$

$$v = \boxed{13.8 \text{ m/s}}$$

P15.62 (a) The flow rate, Av, as given may be expressed as follows:

$$\frac{25.0 \text{ liters}}{30.0 \text{ s}} = 0.833 \text{ liters/s} = 833 \text{ cm}^3/\text{s}.$$

The area of the faucet tap is $\pi \text{ cm}^2$, so we can find the velocity as

$$v = \frac{\text{flow rate}}{A} = \frac{833 \text{ cm}^3/\text{s}}{\pi \text{ cm}^2} = 265 \text{ cm/s} = \boxed{2.65 \text{ m/s}}.$$

(b) We choose point 1 to be in the entrance pipe and point 2 to be at the faucet tap. $A_1 v_1 = A_2 v_2$ gives $v_1 = 0.295$ m/s. Bernoulli's equation is:

$$P_1 - P_2 = \frac{1}{2}\rho\left(v_2^2 - v_1^2\right) + \rho g(y_2 - y_1)$$

and gives

$$P_1 - P_2 = \frac{1}{2}\left(10^3 \text{ kg/m}^3\right)\left[(2.65 \text{ m/s})^2 - (0.295 \text{ m/s})^2\right] + \left(10^3 \text{ kg/m}^3\right)\left(9.80 \text{ m/s}^2\right)(2.00 \text{ m})$$

or $$P_{\text{gauge}} = P_1 - P_2 = \boxed{2.31 \times 10^4 \text{ Pa}}.$$

P15.63 (a) Since the upward buoyant force is balanced by the weight of the sphere,

$$m_1 g = \rho V g = \rho\left(\frac{4}{3}\pi R^3\right)g.$$

In this problem, $\rho = 0.789\,45 \text{ g/cm}^3$ at 20.0°C, and $R = 1.00$ cm so we find:

$$m_1 = \rho\left(\frac{4}{3}\pi R^3\right) = \left(0.789\,45 \text{ g/cm}^3\right)\left[\frac{4}{3}\pi(1.00 \text{ cm})^3\right] = \boxed{3.307 \text{ g}}.$$

continued on next page

(b) Following the same procedure as in part (a), with $\rho' = 0.780\,97\ \text{g/cm}^3$ at 30.0°C, we find:

$$m_2 = \rho'\left(\frac{4}{3}\pi R^3\right) = \left(0.780\,97\ \text{g/cm}^3\right)\left[\frac{4}{3}\pi(1.00\ \text{cm})^3\right] = \boxed{3.271\ \text{g}}.$$

(c) When the first sphere is resting on the bottom of the tube,

$n + B = F_{g1} = m_1 g$, where n is the normal force.

Since $B = \rho' V g$

$$n = m_1 g - \rho' V g = \left[3.307\ \text{g} - \left(0.780\,97\ \text{g/cm}^3\right)\frac{4}{3}\pi(1.00\ \text{cm})^3\right]980\ \text{cm/s}^2$$

$$n = 34.8\ \text{g}\cdot\text{cm/s}^2 = \boxed{3.48\times10^{-4}\ \text{N}}$$

***P15.64** (a) For diverging stream lines that pass just above and just below the hydrofoil we have

$$P_t + \rho g y_t + \frac{1}{2}\rho v_t^2 = P_b + \rho g y_b + \frac{1}{2}\rho v_b^2.$$

Ignoring the buoyant force means taking $y_t \approx y_b$

$$P_t + \frac{1}{2}\rho(n v_b)^2 = P_b + \frac{1}{2}\rho v_b^2$$

$$P_b - P_t = \frac{1}{2}\rho v_b^2\left(n^2 - 1\right)$$

The lift force is $(P_b - P_t)A = \frac{1}{2}\rho v_b^2\left(n^2 - 1\right)A$.

(b) For liftoff,

$$\frac{1}{2}\rho v_b^2\left(n^2 - 1\right)A = Mg$$

$$v_b = \left(\frac{2Mg}{\rho\left(n^2 - 1\right)A}\right)^{1/2}$$

The speed of the boat relative to the shore must be nearly equal to this speed of the water below the hydrofoil relative to the boat.

(c) $v^2\left(n^2 - 1\right)A\rho = 2Mg$

$$A = \frac{2(800\ \text{kg})9.8\ \text{m/s}^2}{(9.5\ \text{m/s})^2\left(1.05^2 - 1\right)1\,000\ \text{kg/m}^3} = \boxed{1.70\ \text{m}^2}$$

ANSWERS TO EVEN PROBLEMS

P15.2 1.92×10^4 N

P15.4 5.27×10^{18} kg

P15.6 (a) 1.01×10^7 Pa; (b) 7.09×10^5 N outward

P15.8 (a) 65.1 N; (b) 275 N

P15.10 5.88×10^6 N; 196 kN; 588 kN

P15.12 (a) 10.3 m; (b) 0

P15.14 (a) 20.0 cm; (b) 0.490 cm

P15.16 (a) 1.57 Pa, 0.015 5 atm, 11.8 mmHg;
(b) Increased pressure of the cerebrospinal fluid will raise the level of the fluid in the spinal tap.
(c) Blockage of the fluid within the spinal column or between the skull and the spinal column would prevent the fluid level from rising.

P15.18 $\dfrac{m}{(\rho_w-\rho_s)h}$

P15.20 (a) see the solution; (b) 25.0 N;
(c) horizontally inward;
(d) string tension increases,
see the solution; (e) 62.5%; (f) 18.7%

P15.22 $\sim10^4$ balloons of 25-cm diameter

P15.24 see the solution

P15.26 2.67×10^3 kg

P15.28 see the solution, $T=\dfrac{2}{r}\sqrt{\dfrac{\pi M}{\rho g}}$

P15.30 (a) 106 cm/s; (b) 424 cm/s

P15.32 31.6 m/s

P15.34 a) 17.7 m/s; (b) 1.73×10^{-3} m

P15.36 (a) see the solution; (b) 616 MW

P15.38 (a) 2.28 N toward Holland;
(b) 1.74×10^6 s

P15.40 (a) 28.0 m/s; (b) 28.0 m/s; (c) 2.11 MPa

P15.42 6.80×10^4 Pa

P15.44 (a) 489 N outward; (b) 1.96 kN outward

P15.46 (a) see the solution; (b) smaller, 8.3%

P15.48 (a) 60.0 rad; (b) 9.12 $\mathrm{m/s^2}$; (c) 6.76 $\mathrm{m/s^2}$;
(d) 253 N·m

P15.50 455 kPa

P15.52 8.01 km, yes

P15.54 (a) $(P_0-P)\pi R^2$; (b) 2.58×10^4 N

P15.56 291 Hz

P15.58 see the solution

P15.60 (a) 18.3 mm; (b) 14.3 mm; (c) 8.56 mm

P15.62 (a) 2.65 m/s; (b) 2.31×10^4 Pa

P15.64 (a) see the solution; (b) see the solution;
(c) 1.70 $\mathrm{m^2}$

CONTEXT 4 CONCLUSION SOLUTIONS TO PROBLEMS

CC4.1 The stern half has length $\frac{1}{2}(269\text{ m}) = 134.5\text{ m}$, mass $2.1\times10^7\text{ kg}$, and weight $(2.1\times10^7\text{ kg})(9.8\text{ m/s}^2) = 2.06\times10^8\text{ N}$, which acts one-quarter of the way along the length of the ship. The moment arm of the weight is $\frac{134.5\text{ m}}{2}\cos 45° = 47.6\text{ m}$ so the torque is

$$\tau = Fd = (2.06\times10^8\text{ N})(47.6\text{ m}) = \boxed{9.8\times10^9\text{ N}\cdot\text{m}}$$

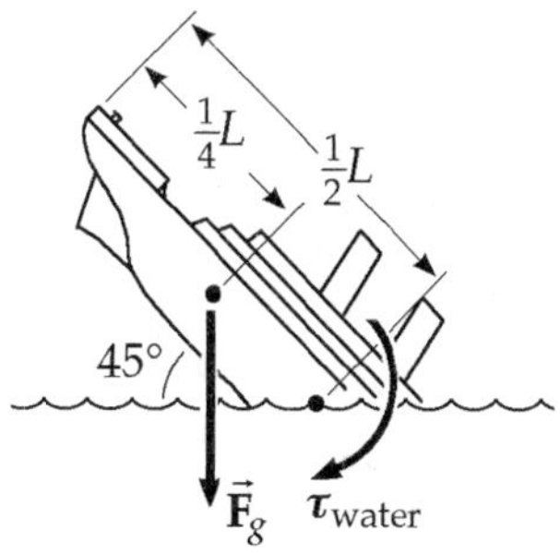

FIG. CC4.1

CC4.2 (a) The absolute pressure is

$$P = P_0 + \rho g h = 1.013\times10^5\text{ Pa} + (1.03\times10^3\text{ kg/m}^3)(9.8\text{ m/s}^2)(119\text{ m}) = \boxed{1.30\times10^6\text{ Pa}}$$

(b) Yes, but only with specialized equipment and techniques.

CC4.3 (a) $v = \frac{\Delta x}{\Delta t} = \frac{3\,784\text{ m}}{2.5\text{ h}}\left(\frac{1\text{ h}}{3\,600\text{ s}}\right) = \boxed{0.42\text{ m/s}}$

(b) The density of the sinking submersible must be slightly $\boxed{\text{higher}}$ than sea water so that it can move down at constant speed with a small upward fluid friction force acting on it.

CC4.4 Momentum is conserved for the two-ship system during the collision:

$$12165\text{ ton}\left((18\text{ knot})\sin 30°\,\hat{\mathbf{i}} - (18\text{ knot})\cos 30°\,\hat{\mathbf{j}}\right)$$
$$+ 29\,100\text{ ton}\left(-(23\text{ knot})\cos 15°\,\hat{\mathbf{i}} - (23\text{ knot})\sin 15°\,\hat{\mathbf{j}}\right)$$
$$= (12165\text{ ton} + 29100\text{ ton})\vec{\mathbf{v}}_f$$
$$(-5.37\times10^5\text{ ton}\cdot\text{knot})\hat{\mathbf{i}} - (3.63\times10^5\text{ ton}\cdot\text{knot})\hat{\mathbf{j}} = (4.13\times10^4\text{ ton})\vec{\mathbf{v}}_f$$

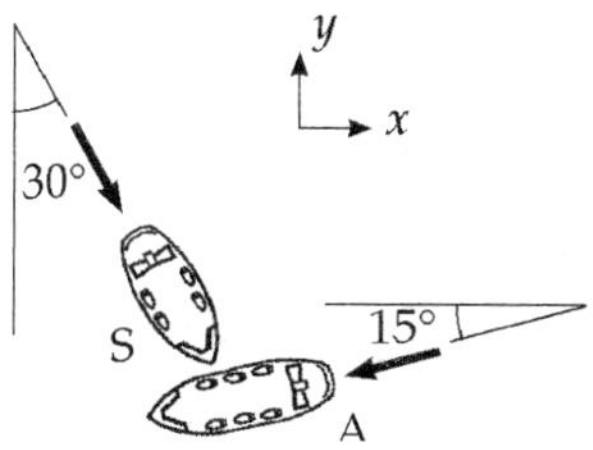

FIG. CC4.4

(a) $\vec{\mathbf{v}}_f = (-13.0\text{ knot})\hat{\mathbf{i}} - (8.79\text{ knot})\hat{\mathbf{j}}$

$$\vec{\mathbf{v}}_f = \sqrt{13.0^2 + 8.79^2}\text{ knot at }\tan^{-1}\left(\frac{8.79}{13.0}\right)\text{ south of west} = \boxed{16\text{ knot at }34°\text{ south of west}}$$

(b) The original kinetic energy of the system is proportional to

$$\frac{1}{2}(12165\text{ ton})(18\text{ knot})^2 + \frac{1}{2}(29100\text{ ton})(23\text{ knot})^2 = 9.67\times10^6\text{ ton}\cdot\text{knot}^2$$

continued on next page

The final kinetic energy is proportional to $\frac{1}{2}(41\,265\text{ ton})(15.7\text{ knot})^2 = 5.09\times10^6\text{ ton}\cdot\text{knot}^2$.

The fraction transformed or transferred is $\frac{9.67-5.09}{9.67} = \boxed{47\%}$.

ANSWERS TO EVEN CONTEXT 4 CONCLUSION PROBLEMS

CC4.2 (a) 1.30×10^6 Pa; (b) Yes, but only with specialized equipment and techniques

CC4.4 (a) 16 knot at 34° south of west; (b) 47%

Chapter 16

Temperature and the Kinetic Theory of Gases

CHAPTER OUTLINE

ANSWERS TO QUESTIONS

Q16.1 The copper's temperature drops and the water temperature rises until both temperatures are the same. Then the metal and the water are in thermal equilibrium.

Q16.2 The astronaut is referring to the temperature of the lunar surface, specifically a 400°F difference. A thermometer would register the temperature of the thermometer liquid. Since there is no atmosphere in the moon, the thermometer will not read a realistic temperature unless it is placed into the lunar soil.

Q16.3 If the amalgam had a larger coefficient of expansion than your tooth, it would expand more than the cavity in your tooth when you take a sip of your ever-beloved coffee, resulting in a broken or cracked tooth! As you ice down your now excruciatingly painful broken tooth, the amalgam would contract more than the cavity in your tooth and fall out, leaving the nerve roots exposed. Isn't it nice that your dentist knows thermodynamics?

Q16.4 The measurements made with the heated steel tape will be too short—but only by a factor of 5×10^{-5} of the measured length.

Q16.5 (a) One mole of H_2 has a mass of 2.016 0 g.

(b) One mole of He has a mass of 4.002 6 g.

(c) One mole of CO has a mass of 28.010 g.

Q16.6 The ideal gas law, $PV = nRT$ predicts zero volume at absolute zero. This is incorrect because the ideal gas law cannot work all the way down to or below the temperature at which gas turns to liquid, or in the case of CO_2, a solid.

Q16.7 Call the process isobaric cooling or isobaric contraction. The rubber wall is easy to stretch. The air inside is nearly at atmospheric pressure originally and stays at atmospheric pressure as the wall moves in, just maintaining equality of pressure outside and inside. The air is nearly an ideal gas to start with, but $PV = nRT$ soon fails. Volume will drop by a larger factor than temperature as the water vapor liquefies and then freezes, as the carbon dioxide turns to snow, as the argon turns to slush, and as the oxygen liquefies. From the outside, you see contraction to a small fraction of the original volume.

Q16.8 Cylinder A must be at lower pressure. If the gas is thin, it will be at one-third the absolute pressure of B.

Q16.9 At high temperature and pressure, the steam inside exerts large forces on the pot and cover. Strong latches hold them together, but they would explode apart if you tried to open the hot cooker.

Q16.10 (a) The water level in the cave rises by a smaller distance than the water outside, as the trapped air is compressed. Air can escape from the cave if the rock is not completely airtight, and also by dissolving in the water.

(b) The ideal cave stays completely full of water at low tide. The water in the cave is supported by atmospheric pressure on the free water surface outside.

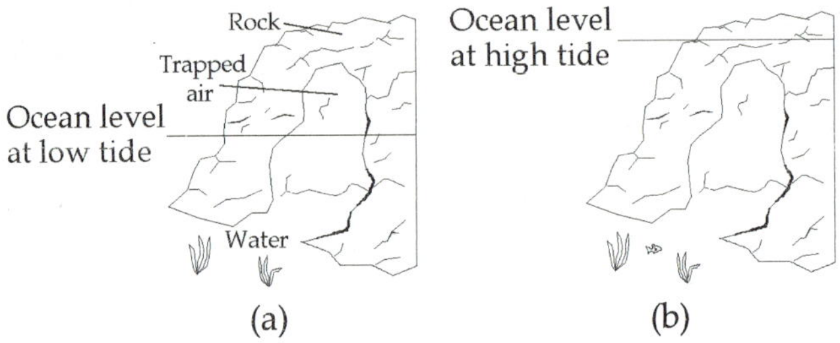

FIG. Q16.10

Q16.11 The sphere expands when heated, so that it no longer fits through the ring. With the sphere still hot, you can separate the sphere and ring by heating the ring. This more surprising result occurs because the thermal expansion of the ring is not like the inflation of a blood-pressure cuff. Rather, it is like a photographic enlargement; every linear dimension, including the hole diameter, increases by the same factor. The reason for this is that the atoms everywhere, including those around the inner circumference, push away from each other. The only way that the atoms can accommodate the greater distances is for the circumference—and corresponding diameter—to grow. This property was once used to fit metal rims to wooden wagon and horse-buggy wheels. If the ring is heated and the sphere left at room temperature, the sphere would pass through the ring with more space to spare.

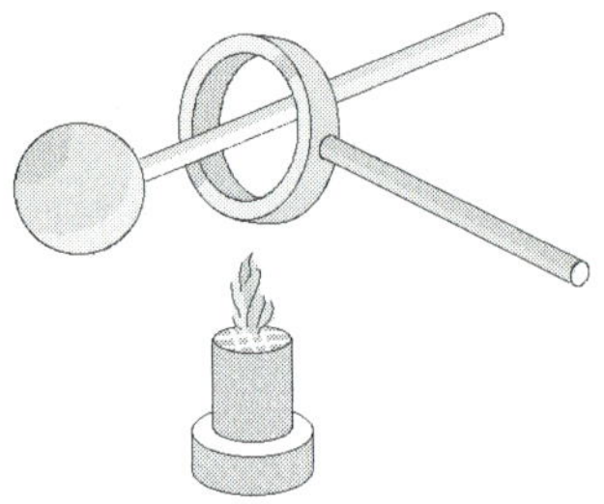

FIG. Q16.11

Q16.12 The coefficient of expansion of metal is larger than that of glass. When hot water is run over the jar, both the glass and the lid expand, but at different rates. Since *all* dimensions expand, there will be a certain temperature at which the inner diameter of the lid has expanded more than the top of the jar, and the lid will be easier to remove.

Q16.13 The alcohol evaporates, absorbing energy from the skin to lower the skin temperature.

Q16.14 The molecules of all different kinds collide with the walls of the container, so molecules of all different kinds exert partial pressures that contribute to the total pressure. The molecules can be so small that they collide with one another relatively rarely and each kind exerts partial pressure as if the other kinds of molecules were absent. If the molecules collide with one another often, the collisions exactly conserve momentum and so do not affect the net force on the walls.

Q16.15 The volume of the balloon will decrease. The pressure inside the balloon is nearly equal to the constant exterior atmospheric pressure. Then from $PV = nRT$, volume must decrease in proportion to the absolute temperature. Call the process isobaric contraction.

Q16.16 The dry air is denser. Since the air and the water vapor are at the same temperature, they have the same kinetic energy per molecule. For a controlled experiment, the humid and dry air are at the same pressure, so the number of molecules per unit volume must be the same for both. The water molecule has a smaller molecular mass (18.0 u) than any of the gases that make up the air, so the humid air must have the smaller mass per unit volume.

Q16.17 Suppose the balloon rises into air uniform in temperature. The air cannot be uniform in pressure because the lower layers support the weight of all the air above them. The rubber in a typical balloon is easy to stretch and stretches or contracts until interior and exterior pressures are nearly equal. So as the balloon rises it expands. This is an isothermal expansion, with P decreasing as V increases by the same factor in $PV = nRT$. If the rubber wall is very strong it will eventually contain the helium at higher pressure than the air outside but at the same density, so that the balloon will stop rising. More likely, the rubber will stretch and break, releasing the helium to keep rising and "boil out" of the Earth's atmosphere.

Q16.18 (a) Average molecular kinetic energy increases by a factor of 3.

(b) The rms speed increases by a factor of $\sqrt{3}$.

(c) Average momentum change increases by $\sqrt{3}$.

(d) Rate of collisions increases by a factor of $\sqrt{3}$ since the mean free path remains unchanged.

(e) Pressure increases by a factor of 3.

SOLUTIONS TO PROBLEMS

Section 16.1 Temperature and the Zeroth Law of Thermodynamics

No problems in this section

Section 16.2 Thermometers and Temperature Scales

P16.1 Since we have a linear graph, the pressure is related to the temperature as $P = A + BT$, where A and B are constants. To find A and B, we use the data

$$0.900 \text{ atm} = A + (-80.0°\text{C})B \qquad (1)$$

$$1.635 \text{ atm} = A + (78.0°\text{C})B \qquad (2)$$

Solving (1) and (2) simultaneously,

we find $A = 1.272 \text{ atm}$

continued on next page

and $B = 4.652 \times 10^{-3}$ atm/°C

Therefore, $P = 1.272 \text{ atm} + \left(4.652 \times 10^{-3} \text{ atm/°C}\right)T$

(a) At absolute zero $P = 0 = 1.272 \text{ atm} + \left(4.652 \times 10^{-3} \text{ atm/°C}\right)T$

which gives $\boxed{T = -273°\text{C}}$.

(b) At the freezing point of water $P = 1.272 \text{ atm} + 0 = \boxed{1.27 \text{ atm}}$.

(c) And at the boiling point $P = 1.272 \text{ atm} + \left(4.652 \times 10^{-3} \text{ atm/°C}\right)(100°\text{C}) = \boxed{1.74 \text{ atm}}$.

P16.2 (a) To convert from Fahrenheit to Celsius, we use $T_C = \frac{5}{9}(T_F - 32.0) = \frac{5}{9}(98.6 - 32.0) = \boxed{37.0°\text{C}}$

and the Kelvin temperature is found as $T = T_C + 273 = \boxed{310 \text{ K}}$

(b) In a fashion identical to that used in (a), we find $T_C = \boxed{-20.6°\text{C}}$

and $T = \boxed{253 \text{ K}}$

P16.3 (a) $T_F = \frac{9}{5}T_C + 32.0°\text{F} = \frac{9}{5}(-195.81) + 32.0 = \boxed{-320°\text{F}}$

(b) $T = T_C + 273.15 = -195.81 + 273.15 = \boxed{77.3 \text{ K}}$

P16.4 (a) $\Delta T = 450°\text{C} = 450°\text{C}\left(\frac{212°\text{F} - 32.0°\text{F}}{100°\text{C} - 0.00°\text{C}}\right) = \boxed{810°\text{F}}$

(b) $\Delta T = 450°\text{C} = \boxed{450 \text{ K}}$

Section 16.3 Thermal Expansion of Solids and Liquids

***P16.5** $\Delta L = \alpha L_i \Delta T = 11 \times 10^{-6} (°\text{C})^{-1} (1\,300 \text{ km})\left[35°\text{C} - (-73°\text{C})\right] = \boxed{1.54 \text{ km}}$

The expansion can be compensated for by mounting the pipeline on rollers and placing Ω-shaped loops between straight sections. They bend as the steel changes length.

P16.6 For the dimensions to increase, $\Delta L = \alpha L_i \Delta T$

$$1.00 \times 10^{-2} \text{ cm} = 1.30 \times 10^{-4} °\text{C}^{-1}(2.20 \text{ cm})(T - 20.0°\text{C})$$

$$T = \boxed{55.0°\text{C}}$$

***P16.7** $\Delta L = \alpha L_i \Delta T = \left(22 \times 10^{-6}/\text{C}°\right)(2.40 \text{ cm})(30°\text{C}) = \boxed{1.58 \times 10^{-3} \text{ cm}}$

***P16.8** We consider the expansion of the horizontal strip of vinyl lying between the snug nailhead and a nail near the far end of the wall, where the siding panel can slide by 1.40 cm under the nailhead:

$$\Delta L = \alpha L_i \Delta T$$

$$0.014 \text{ m} = \alpha (15 \text{ m})[38°\text{C} - (-5°\text{C})]$$

$$\alpha = \boxed{2.17 \times 10^{-5} (°\text{C})^{-1}}$$

P16.9 (a) $\Delta L = \alpha L_i \Delta T = 9.00 \times 10^{-6}\ °\text{C}^{-1}(30.0 \text{ cm})(65.0°\text{C}) = \boxed{0.176 \text{ mm}}$

(b) $\Delta L = \alpha L_i \Delta T = 9.00 \times 10^{-6}\ °\text{C}^{-1}(1.50 \text{ cm})(65.0°\text{C}) = \boxed{8.78 \times 10^{-4} \text{ cm}}$

(c) $\Delta V = 3\alpha V_i \Delta T = 3\left(9.00 \times 10^{-6}\ °\text{C}^{-1}\right)\left(\frac{30.0(\pi)(1.50)^2}{4} \text{ cm}^3\right)(65.0°\text{C}) = \boxed{0.093\,0 \text{ cm}^3}$

P16.10 The horizontal section expands according to $\Delta L = \alpha L_i \Delta T$.

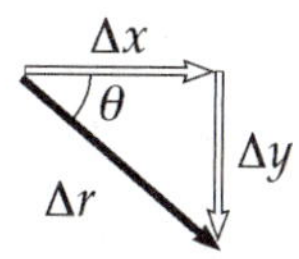

FIG. P16.10

$$\Delta x = \left(17 \times 10^{-6}\ °\text{C}^{-1}\right)(28.0 \text{ cm})(46.5°\text{C} - 18.0°\text{C}) = 1.36 \times 10^{-2} \text{ cm}$$

The vertical section expands similarly by

$$\Delta y = \left(17 \times 10^{-6}\ °\text{C}^{-1}\right)(134 \text{ cm})(28.5°\text{C}) = 6.49 \times 10^{-2} \text{ cm}.$$

The vector displacement of the pipe elbow has magnitude

$$\Delta r = \sqrt{\Delta x^2 + \Delta y^2} = \sqrt{(0.136 \text{ mm})^2 + (0.649 \text{ mm})^2} = 0.663 \text{ mm}$$

and is directed to the right below the horizontal at angle

$$\theta = \tan^{-1}\left(\frac{\Delta y}{\Delta x}\right) = \tan^{-1}\left(\frac{0.649 \text{ mm}}{0.136 \text{ mm}}\right) = 78.2°$$

$$\boxed{\Delta r = 0.663 \text{ mm to the right at } 78.2° \text{ below the horizontal}}$$

P16.11 (a) $\Delta A = 2\alpha A_i \Delta T$: $\Delta A = 2\left(17.0 \times 10^{-6}\ °\text{C}^{-1}\right)(0.080\,0 \text{ m})^2(50.0°\text{C})$

$$\Delta A = 1.09 \times 10^{-5} \text{ m}^2 = \boxed{0.109 \text{ cm}^2}$$

(b) The length of each side of the hole has increased. Thus, this represents an $\boxed{\text{increase}}$ in the area of the hole.

P16.12 $\Delta V = (\beta - 3\alpha)V_i \Delta T = \left(5.81 \times 10^{-4} - 3\left(11.0 \times 10^{-6}\right)\right)(50.0 \text{ gal})(20.0) = \boxed{0.548 \text{ gal}}$

P16.13 (a) $\Delta V = V_t\beta_t\Delta T - V_{\text{Al}}\beta_{\text{Al}}\Delta T = (\beta_t - 3\alpha_{\text{Al}})V_i\Delta T$

$$= (9.00\times10^{-4} - 0.720\times10^{-4})°\text{C}^{-1}(2\,000\text{ cm}^3)(60.0°\text{C})$$

$\Delta V = \boxed{99.4\text{ cm}^3}$ overflows.

(b) The whole new volume of turpentine is

$$2\,000\text{ cm}^3 + 9.00\times10^{-4}\,°\text{C}^{-1}(2\,000\text{ cm}^3)(60.0°\text{C}) = 2\,108\text{ cm}^3$$

so the fraction lost is $\dfrac{99.4\text{ cm}^3}{2\,108\text{ cm}^3} = 4.71\times10^{-2}$

and this fraction of the cylinder's depth will be empty upon cooling:

$$4.71\times10^{-2}(20.0\text{ cm}) = \boxed{0.943\text{ cm}}.$$

P16.14 (a) $L = L_i(1+\alpha\Delta T)$: $5.050\text{ cm} = 5.000\text{ cm}\left[1 + 24.0\times10^{-6}\,°\text{C}^{-1}(T - 20.0°\text{C})\right]$

$T = \boxed{437°\text{C}}$

(b) We must get $L_{\text{Al}} = L_{\text{Brass}}$ for some ΔT, or

$$L_{i,\text{ Al}}(1+\alpha_{\text{Al}}\Delta T) = L_{i,\text{ Brass}}(1+\alpha_{\text{Brass}}\Delta T)$$
$$5.000\text{ cm}\left[1+(24.0\times10^{-6}\,°\text{C}^{-1})\Delta T\right] = 5.050\text{ cm}\left[1+(19.0\times10^{-6}\,°\text{C}^{-1})\Delta T\right]$$

Solving for ΔT, $\Delta T = 2\,080°\text{C}$,

so $\boxed{T = 2\,100°\text{C}}$

This will not work because $\boxed{\text{aluminum melts at }660°\text{C}}$.

Section 16.4 Macroscopic Description of an Ideal Gas

P16.15 Mass of gold abraded: $|\Delta m| = 3.80\text{ g} - 3.35\text{ g} = 0.45\text{ g} = (0.45\text{ g})\left(\dfrac{1\text{ kg}}{10^3\text{ g}}\right) = 4.5\times10^{-4}\text{ kg}$.

Each atom has mass $m_0 = 197\text{ u} = 197\text{ u}\left(\dfrac{1.66\times10^{-27}\text{ kg}}{1\text{ u}}\right) = 3.27\times10^{-25}\text{ kg}$.

Now, $|\Delta m| = |\Delta N|m_0$, and the number of atoms missing is

$$|\Delta N| = \frac{|\Delta m|}{m_0} = \frac{4.5\times10^{-4}\text{ kg}}{3.27\times10^{-25}\text{ kg}} = 1.38\times10^{21}\text{ atoms}.$$

continued on next page

The rate of loss is

$$\frac{|\Delta N|}{\Delta t}=\frac{1.38\times10^{21}\text{ atoms}}{50\text{ yr}}\left(\frac{1\text{ yr}}{365.25\text{ d}}\right)\left(\frac{1\text{ d}}{24\text{ h}}\right)\left(\frac{1\text{ h}}{60\text{ min}}\right)\left(\frac{1\text{ min}}{60\text{ s}}\right)$$

$$\frac{|\Delta N|}{\Delta t}=\boxed{8.72\times10^{11}\text{ atoms/s}}.$$

P16.16 One mole of helium contains Avogadro's number of molecules and has a mass of 4.00 g. Let us call m_0 the mass of one atom, and we have

$$N_A m_0 = 4.00\text{ g/mol}$$

or $$m_0=\frac{4.00\text{ g/mol}}{6.02\times10^{23}\text{ molecules/mol}}=6.64\times10^{-24}\text{ g/molecule}$$

$$m_0=\boxed{6.64\times10^{-27}\text{ kg}}$$

P16.17 (a) Initially, $P_iV_i=n_iRT_i$ $\quad(1.00\text{ atm})V_i=n_iR(10.0+273.15)\text{ K}$

Finally, $P_fV_f=n_fRT_f$ $\quad P_f(0.280V_i)=n_iR(40.0+273.15)\text{ K}$

Dividing these equations, $\quad\dfrac{0.280P_f}{1.00\text{ atm}}=\dfrac{313.15\text{ K}}{283.15\text{ K}}$

giving $\quad P_f=3.95\text{ atm}$

or $\quad P_f=\boxed{4.00\times10^5\text{ Pa(abs.)}}.$

(b) After being driven $\quad P_d(1.02)(0.280V_i)=n_iR(85.0+273.15)\text{ K}$

$$P_d=1.121P_f=\boxed{4.49\times10^5\text{ Pa}}$$

P16.18 $PV=NP'V'=\frac{4}{3}\pi r^3NP'$: $\quad N=\dfrac{3PV}{4\pi r^3P'}=\dfrac{3(150)(0.100)}{4\pi(0.150)^3(1.20)}=884\text{ balloons}$

If we have no special means for squeezing the last 100 L of helium out of the tank, the tank will be full of helium at 1.20 atm when the last balloon is inflated. The number of balloons is then reduced to $884-\dfrac{(0.100\text{ m}^3)3}{4\pi(0.15\text{ m})^3}=\boxed{877}$.

P16.19 The equation of state of an ideal gas is $PV=nRT$ so we need to solve for the number of moles to find N.

$$n=\frac{PV}{RT}=\frac{(1.01\times10^5\text{ N/m}^2)[(10.0\text{ m})(20.0\text{ m})(30.0\text{ m})]}{(8.314\text{ J/mol}\cdot\text{K})(293\text{ K})}=2.49\times10^5\text{ mol}$$

$$N=nN_A=2.49\times10^5\text{ mol}(6.022\times10^{23}\text{ molecules/mol})=\boxed{1.50\times10^{29}\text{ molecules}}$$

***P16.20** Consider the air in the tank during one discharge process. We suppose that the process is slow enough that the temperature remains constant. Then as the pressure drops from 2.40 atm to 1.20 atm, the volume of the air doubles. During the first discharge, the air volume changes from 1 L to 2 L. Just 1 L of water is expelled and 3 L remains. In the second discharge, the air volume changes from 2 L to 4 L and 2 L of water is sprayed out. In the third discharge, only the last 1 L of water comes out. Were it not for male pattern dumbness, each person could more efficiently use his device by starting with the tank half full of water.

P16.21 $\sum F_y = 0$: $\quad \rho_{\text{out}} gV - \rho_{\text{in}} gV - (200\text{ kg})g = 0$

$$(\rho_{\text{out}} - \rho_{\text{in}})(400\text{ m}^3) = 200\text{ kg}$$

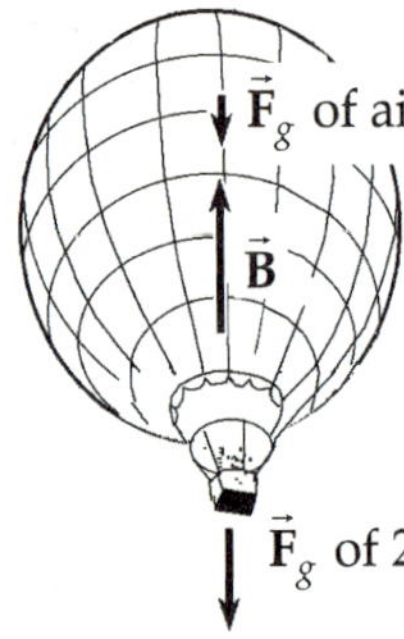

FIG. P16.21

The density of the air outside is 1.25 kg/m^3.

From $PV = nRT$, $\frac{n}{V} = \frac{P}{RT}$

The density is inversely proportional to the temperature, and the density of the hot air is

$$\rho_{\text{in}} = (1.25\text{ kg/m}^3)\left(\frac{283\text{ K}}{T_{\text{in}}}\right)$$

Then
$$(1.25\text{ kg/m}^3)\left(1 - \frac{283\text{ K}}{T_{\text{in}}}\right)(400\text{ m}^3) = 200\text{ kg}$$

$$1 - \frac{283\text{ K}}{T_{\text{in}}} = 0.400$$

$$0.600 = \frac{283\text{ K}}{T_{\text{in}}} \qquad T_{\text{in}} = \boxed{472\text{ K}}$$

P16.22 (a) $PV = nRT \qquad n = \frac{PV}{RT}$

$$m = nM = \frac{PVM}{RT} = \frac{1.013\times10^5\text{ Pa}(0.100\text{ m})^3(28.9\times10^{-3}\text{ kg/mol})}{(8.314\text{ J/mol}\cdot\text{K})(300\text{ K})}$$

$$m = \boxed{1.17\times10^{-3}\text{ kg}}$$

(b) $F_g = mg = 1.17\times10^{-3}\text{ kg}(9.80\text{ m/s}^2) = \boxed{11.5\text{ mN}}$

(c) $F = PA = (1.013\times10^5\text{ N/m}^2)(0.100\text{ m})^2 = \boxed{1.01\text{ kN}}$

(d) The $\boxed{\text{molecules must be moving very fast}}$ to hit the walls hard.

***P16.23** (a) The air in the tube is far from liquification, so it behaves as an ideal gas. At the ocean surface it is described by $P_tV_t = nRT$ where $P_t = 1$ atm, $V_t = A(6.50\text{ cm})$, and A is the cross-sectional area of the interior of the tube. At the bottom of the dive, $P_bV_b = nRT = P_bA(6.50\text{ cm} - 2.70\text{ cm})$. By division,

$$\frac{P_b(3.8\text{ cm})}{(1\text{ atm})(6.5\text{ cm})} = 1$$

$$P_b = 1.013\times10^5\text{ N/m}^2\,\frac{6.5}{3.8} = 1.73\times10^5\text{ N/m}^2$$

The salt water enters the tube until the air pressure is equal to the water pressure at depth, which is described by

$$P_b = P_t + \rho gh$$

$$1.73\times10^5\text{ N/m}^2 = 1.013\times10^5\text{ N/m}^2 + \left(1\,030\text{ kg/m}^3\right)\left(9.8\text{ m/s}^2\right)h$$

$$h = \frac{7.20\times10^4\text{ kg}\cdot\text{m}\cdot\text{m}^2\cdot\text{s}^2}{1.01\times10^4\text{ s}^2\cdot\text{m}^2\cdot\text{kg}} = \boxed{7.13\text{ m}}$$

(b) With a very thin tube, air does not bubble out. At the bottom of the dive, the tube gives a valid reading in any orientation. The open end of the tube should be at the bottom after the bird surfaces, so that the water will drain away as the expanding air pushes it out. Students can make the tubes and dive with them in a swimming pool, to observe how dependably they work.

P16.24 At depth, $P = P_0 + \rho gh$ and $PV_i = nRT_i$

At the surface, $P_0V_f = nRT_f$: $\dfrac{P_0V_f}{(P_0+\rho gh)V_i} = \dfrac{T_f}{T_i}$

Therefore $V_f = V_i\left(\dfrac{T_f}{T_i}\right)\left(\dfrac{P_0+\rho gh}{P_0}\right)$

$$V_f = 1.00\text{ cm}^3\left(\frac{293\text{ K}}{278\text{ K}}\right)\left(\frac{1.013\times10^5\text{ Pa} + \left(1\,025\text{ kg/m}^3\right)\left(9.80\text{ m/s}^2\right)(25.0\text{ m})}{1.013\times10^5\text{ Pa}}\right)$$

$$V_f = \boxed{3.67\text{ cm}^3}$$

P16.25 $PV = nRT$: $\dfrac{m_f}{m_i} = \dfrac{n_f}{n_i} = \dfrac{P_fV_f}{RT_f}\dfrac{RT_i}{P_iV_i} = \dfrac{P_f}{P_i}$

so $m_f = m_i\left(\dfrac{P_f}{P_i}\right)$

$$|\Delta m| = m_i - m_f = m_i\left(\frac{P_i - P_f}{P_i}\right) = 12.0\text{ kg}\left(\frac{41.0\text{ atm} - 26.0\text{ atm}}{41.0\text{ atm}}\right) = \boxed{4.39\text{ kg}}$$

P16.26 My bedroom is 4 m long, 4 m wide, and 2.4 m high, enclosing air at 100 kPa and $20°C = 293\text{ K}$. Think of the air as 80.0% N_2 and 20.0% O_2.

Avogadro's number of molecules has mass

$$(0.800)(28.0\text{ g/mol}) + (0.200)(32.0\text{ g/mol}) = 0.028\,8\text{ kg/mol}$$

Then $$PV = nRT = \left(\frac{m}{M}\right)RT$$

gives $$m = \frac{PVM}{RT} = \frac{(1.00\times10^5\text{ N/m}^2)(38.4\text{ m}^3)(0.028\,8\text{ kg/mol})}{(8.314\text{ J/mol}\cdot\text{K})(293\text{ K})} = 45.4\text{ kg} \boxed{\sim 10^2\text{ kg}}$$

***P16.27** The CO_2 is far from liquefaction, so after it comes out of solution it behaves as an ideal gas. Its molar mass is $M = 12.0\text{ g/mol} + 2(16.0\text{ g/mol}) = 44.0\text{ g/mol}$. The quantity of gas in the cylinder is

$$n = \frac{m_{\text{sample}}}{M} = \frac{6.50\text{ g}}{44.0\text{ g/mol}} = 0.148\text{ mol}$$

Then $$PV = nRT$$

gives $$V = \frac{nRT}{P} = \frac{0.148\text{ mol}(8.314\text{ J/mol}\cdot\text{K})(273\text{ K} + 20\text{ K})}{1.013\times10^5\text{ N/m}^2}\left(\frac{1\text{ N}\cdot\text{m}}{1\text{ J}}\right)\left(\frac{10^3\text{ L}}{1\text{ m}^3}\right) = \boxed{3.55\text{ L}}$$

P16.28 $$N = \frac{PVN_A}{RT} = \frac{(10^{-9}\text{ Pa})(1.00\text{ m}^3)(6.02\times10^{23}\text{ molecules/mol})}{(8.314\text{ J/K}\cdot\text{mol})(300\text{ K})} = \boxed{2.41\times10^{11}\text{ molecules}}$$

P16.29 $$P_0V = n_1RT_1 = \left(\frac{m_1}{M}\right)RT_1$$

$$P_0V = n_2RT_2 = \left(\frac{m_2}{M}\right)RT_2$$

$$\boxed{m_1 - m_2 = \frac{P_0VM}{R}\left(\frac{1}{T_1} - \frac{1}{T_2}\right)}$$

Section 16.5 The Kinetic Theory of Gases

P16.30 Consider the x axis to be perpendicular to the plane of the window. Then, the average force exerted on the window by the hailstones is in magnitude

$$\bar{F} = Nm\left(\frac{\Delta v}{\Delta t}\right) = Nm\left(\frac{v_{xf} - v_{xi}}{t}\right) = Nm\left(\frac{v\sin\theta - (-v\sin\theta)}{t}\right) = \boxed{Nm\left(\frac{2v\sin\theta}{t}\right)}.$$

Thus, the pressure on the windowpane is $P = \frac{\bar{F}}{A} = \boxed{Nm\left(\frac{2v\sin\theta}{At}\right)}$.

P16.31 $$\bar{F} = \frac{(5.00\times10^{23})\left[2(4.68\times10^{-26}\text{ kg})(300\text{ m/s})\right]}{1.00\text{ s}} = 14.0\text{ N}$$

and $P = \frac{\bar{F}}{A} = \frac{14.0\text{ N}}{8.00\times10^{-4}\text{ m}^2} = \boxed{17.6\text{ kPa}}$.

P16.32 (a) $PV = nRT = \frac{Nm_0 v^2}{3}$

The total translational kinetic energy is $\frac{Nm_0 v^2}{2} = E_{\text{trans}}$:

$$E_{\text{trans}} = \frac{3}{2}PV = \frac{3}{2}\left(3.00 \times 1.013 \times 10^5\right)\left(5.00 \times 10^{-3}\right) = \boxed{2.28 \text{ kJ}}$$

(b) $$\frac{m_0 v^2}{2} = \frac{3k_B T}{2} = \frac{3RT}{2N_A} = \frac{3(8.314)(300)}{2\left(6.02 \times 10^{23}\right)} = \boxed{6.21 \times 10^{-21} \text{ J}}$$

P16.33 (a) $$PV = Nk_B T: \quad N = \frac{PV}{k_B T} = \frac{1.013 \times 10^5 \text{ Pa}\left[\frac{4}{3}\pi(0.150 \text{ m})^3\right]}{\left(1.38 \times 10^{-23} \text{ J/K}\right)(293 \text{ K})} = \boxed{3.54 \times 10^{23} \text{ atoms}}$$

(b) $$\overline{K} = \frac{3}{2}k_B T = \frac{3}{2}\left(1.38 \times 10^{-23}\right)(293) \text{ J} = \boxed{6.07 \times 10^{-21} \text{ J}}$$

(c) For helium, the atomic mass is $m_0 = \frac{4.00 \text{ g/mol}}{6.02 \times 10^{23} \text{ molecules/mol}} = 6.64 \times 10^{-24} \text{ g/molecule}$

$$m_0 = 6.64 \times 10^{-27} \text{ kg/molecule}$$

$\frac{1}{2}m_0 \overline{v^2} = \frac{3}{2}k_B T$: $\quad \therefore v_{\text{rms}} = \sqrt{\frac{3k_B T}{m_0}} = \boxed{1.35 \text{ km/s}}$

***P16.34** Let $d = 2r$ represent the diameter of the particle. Its mass is $m = \rho V = \rho\frac{4}{3}\pi r^3 = \rho\frac{4}{3}\pi\left(\frac{d}{2}\right)^3 = \frac{\rho\pi d^3}{6}$.

Then $\frac{1}{2}mv_{\text{rms}}^2 = \frac{3}{2}kT$ gives $\frac{\rho\pi d^3}{6}v_{\text{rms}}^2 = 3kT$

(a) $$v_{\text{rms}} = \left(\frac{18kT}{\rho\pi d^3}\right)^{1/2} = \left(\frac{18\left(1.38 \times 10^{-23} \text{ J/K}\right)(293 \text{ K})}{\left(1\,000 \text{ kg/m}^3\right)\pi\left(3 \times 10^{-6} \text{ m}\right)^3}\right)^{1/2} = \boxed{9.26 \times 10^{-4} \text{ m/s}}$$

(b) $v = \frac{x}{t}$ $\qquad t = \frac{x}{v} = \frac{3 \times 10^{-6} \text{ m}}{9.26 \times 10^{-4} \text{ m/s}} = \boxed{3.24 \text{ ms}}$

(c) $70 \text{ kg} = 1\,000 \text{ kg/m}^3 \frac{\pi d^3}{6}$ $\qquad d = 0.511 \text{ m}$

$$v_{\text{rms}} = \left(\frac{18kT}{\rho\pi d^3}\right)^{1/2} = \left(\frac{18\left(1.38 \times 10^{-23} \text{ J/K}\right)(293 \text{ K})}{\left(1\,000 \text{ kg/m}^3\right)\pi(0.511 \text{ m})^3}\right)^{1/2} = \boxed{1.32 \times 10^{-11} \text{ m/s}}$$

$$t = \frac{0.511 \text{ m}}{1.32 \times 10^{-11} \text{ m/s}} = \boxed{3.88 \times 10^{10} \text{ s}} = 1\,230 \text{ yr}$$

This motion is too slow to observe.

continued on next page

(d) $\left(\dfrac{18kT}{\rho\pi d^3}\right)^{1/2}=\dfrac{d}{1\text{ s}}$ $\qquad \dfrac{18kT}{\rho\pi}=\dfrac{d^5}{1\text{ s}^2}$

$$d=\left(\frac{18\left(1.38\times10^{-23}\text{ J/K}\right)(293\text{ K})\left(1\text{ s}^2\right)}{\left(1\,000\text{ kg/m}^3\right)\pi}\right)^{1/5}=\boxed{2.97\times10^{-5}\text{ m}}$$

Brownian motion is best observed with pollen grains, smoke particles, or latex spheres smaller than this 29.7-μm size. Then they can jitter about convincingly, showing relatively large accelerations several times per second.

P16.35 (a) $\overline{K}=\frac{3}{2}k_BT=\frac{3}{2}\left(1.38\times10^{-23}\text{ J/K}\right)(423\text{ K})=\boxed{8.76\times10^{-21}\text{ J}}$

(b) $\overline{K}=\frac{1}{2}m_0v_{\text{rms}}^2=8.76\times10^{-21}\text{ J}$

so $\qquad v_{rms}=\sqrt{\dfrac{1.75\times10^{-20}\text{ J}}{m_0}}$ (1)

For helium, $\qquad m_0=\dfrac{4.00\text{ g/mol}}{6.02\times10^{23}\text{ molecules/mol}}=6.64\times10^{-24}\text{ g/molecule}$

$m_0=6.64\times10^{-27}\text{ kg/molecule}$

Similarly for argon, $\qquad m_0=\dfrac{39.9\text{ g/mol}}{6.02\times10^{23}\text{ molecules/mol}}=6.63\times10^{-23}\text{ g/molecule}$

$m_0=6.63\times10^{-26}\text{ kg/molecule}$

Substituting in (1) above,

we find for helium, $\qquad \boxed{v_{rms}=1.62\text{ km/s}}$

and for argon, $\qquad \boxed{v_{rms}=514\text{ m/s}}$

Section 16.6 Distribution of Molecular Speeds

P16.36 (a) $v_{\text{av}}=\dfrac{\sum n_iv_i}{N}=\dfrac{1}{15}\left[1(2)+2(3)+3(5)+4(7)+3(9)+2(12)\right]=\boxed{6.80\text{ m/s}}$

(b) $\left(v^2\right)_{\text{av}}=\dfrac{\sum n_iv_i^2}{N}=54.9\text{ m}^2/\text{s}^2$

so $v_{\text{rms}}=\sqrt{\left(v^2\right)_{\text{av}}}=\sqrt{54.9}=\boxed{7.41\text{ m/s}}$

(c) $v_{\text{mp}}=\boxed{7.00\text{ m/s}}$

P16.37 In the Maxwell Boltzmann speed distribution function take $\frac{dN_v}{dv}=0$ to find

$$4\pi N\left(\frac{m_0}{2\pi k_B T}\right)^{3/2}\exp\left(-\frac{m_0 v^2}{2k_B T}\right)\left(2v-\frac{2m_0 v^3}{2k_B T}\right)=0$$

and solve for v to find the most probable speed.

Reject as solutions $v=0$ and $v=\infty$

Retain only $2-\frac{m_0 v^2}{k_B T}=0$

Then $v_{\text{mp}}=\sqrt{\frac{2k_B T}{m_0}}$

P16.38 The most probable speed is $v_{\text{mp}}=\sqrt{\frac{2k_B T}{m_0}}=\sqrt{\frac{2\left(1.38\times10^{-23}\text{ J/K}\right)(4.20\text{ K})}{6.64\times10^{-27}\text{ kg}}}=\boxed{132\text{ m/s}}$.

P16.39 (a) From $v_{\text{av}}=\sqrt{\frac{8k_B T}{\pi m_0}}$

we find the temperature as $T=\frac{\pi\left(6.64\times10^{-27}\text{ kg}\right)\left(1.12\times10^4\text{ m/s}\right)^2}{8\left(1.38\times10^{-23}\text{ J/mol}\cdot\text{K}\right)}=\boxed{2.37\times10^4\text{ K}}$

(b) $T=\frac{\pi\left(6.64\times10^{-27}\text{ kg}\right)\left(2.37\times10^3\text{ m/s}\right)^2}{8\left(1.38\times10^{-23}\text{ J/mol}\cdot\text{K}\right)}=\boxed{1.06\times10^3\text{ K}}$

Section 16.7 Context Connection—The Atmospheric Lapse Rate

P16.40 For a uniform lapse rate, the identity $\frac{\Delta T}{\Delta y}=\frac{T_f-T_i}{\Delta y}$ implies

$$T_f=T_i+\frac{\Delta T}{\Delta y}\Delta y=30°\text{C}-(6.5°\text{C/km})(3.66\text{ km})=\boxed{6.2°\text{C}}$$

P16.41 (a)

$$\frac{dT}{dy}=-\frac{\gamma-1}{\gamma}\frac{gM}{R}=-\frac{0.40}{1.40}\frac{\left(9.8\text{ m/s}^2\right)(28.9\text{ g/mol})}{(8.314\text{ J/moll}\cdot\text{K})}\left(\frac{1\text{ kg}}{1\,000\text{ g}}\right)\left(\frac{1\text{ J}}{1\text{ kg}\cdot\text{m}^2/\text{s}^2}\right)$$
$$=-9.73\times10^{-3}\text{ K/m}=\boxed{-9.73°\text{C/km}}$$

(b) Air contains water vapor. Air does not behave as an ideal gas. As a parcel of air rises in the atmosphere and its temperature drops, its ability to contain water vapor decreases, so water will likely condense out as liquid drops or as ice crystals. (The condensate may or may not be visible as clouds.) The condensate releases its heat of vaporization, raising the air temperature above the value that would be expected according to part (a).

continued on next page

(c) For an object of mass m on Mars, weight = force of planet's gravity: $mg = \frac{GM_{\text{Mars}}m}{r_{\text{Mars}}^2}$ or

$g = \frac{GM_{\text{Mars}}}{r_{\text{Mars}}^2}$

$$g = \frac{(6.67\times10^{-11}\ \text{N}\cdot\text{m}^2/\text{kg}^2)(6.42\times10^{23}\ \text{kg})}{(3.37\times10^6\ \text{m})^2} = 3.77\ \text{m/s}^2$$

$$\frac{dT}{dy} = -\frac{\gamma-1}{\gamma}\frac{gM}{R} = -\frac{0.30}{1.30}\frac{(3.77\ \text{m/s}^2)(0.0440\ \text{kg/mol})}{(8.314\ \text{J/mol}\cdot\text{K})} = -4.60\times10^{-3}\ \text{K/m} = \boxed{-4.60°\text{C/km}}$$

(d) $\frac{\Delta T}{\Delta y} = \frac{dT}{dy}$: $\quad \Delta y = \frac{\Delta T}{\frac{dT}{dy}} = \frac{-60°\text{C} - (-40°\text{C})}{-4.60°\text{C/km}} = \boxed{4.34\ \text{km}}$

(e) The dust in the atmosphere absorbs and scatters energy from the electromagnetic radiation coming through the atmosphere from the sun. The dust contributes energy to the gas molecules high in the atmosphere, resulting in an increase in the internal energy of the atmosphere aloft and a smaller decrease in temperature with height, than in the case where there is no absorption of sunlight. The larger the amount of dust, the more the lapse rate will deviate from the theoretical value in part (c). Thus it was dustier during the *Mariner* flights in 1969.

Additional Problems

P16.42 The excess expansion of the brass is $\Delta L_{\text{rod}} - \Delta L_{\text{tape}} = (\alpha_{\text{brass}} - \alpha_{\text{steel}})L_i\Delta T$

$$\Delta(\Delta L) = (19.0 - 11.0)\times10^{-6}\ (°\text{C})^{-1}(0.950\ \text{m})(35.0°\text{C})$$
$$\Delta(\Delta L) = 2.66\times10^{-4}\ \text{m}$$

(a) The rod contracts more than tape to

a length reading $0.950\,0\ \text{m} - 0.000\,266\ \text{m} = \boxed{0.949\,7\ \text{m}}$

(b) $0.950\,0\ \text{m} + 0.000\,266\ \text{m} = \boxed{0.950\,3\ \text{m}}$

P16.43 At 0°C, 10.0 gallons of gasoline has mass,

from $\quad \rho = \frac{m}{V}$

$$m = \rho V = (730\ \text{kg/m}^3)(10.0\ \text{gal})\left(\frac{0.003\,80\ \text{m}^3}{1.00\ \text{gal}}\right) = 27.7\ \text{kg}$$

The gasoline will expand in volume by

$$\Delta V = \beta V_i\Delta T = 9.60\times10^{-4}\ °\text{C}^{-1}(10.0\ \text{gal})(20.0°\text{C} - 0.0°\text{C}) = 0.192\ \text{gal}$$

continued on next page

At 20.0°C, $10.192\text{ gal} = 27.7\text{ kg}$

$$10.0\text{ gal} = 27.7\text{ kg}\left(\frac{10.0\text{ gal}}{10.192\text{ gal}}\right) = 27.2\text{ kg}$$

The extra mass contained in 10.0 gallons at 0.0°C is

$$27.7\text{ kg} - 27.2\text{ kg} = \boxed{0.523\text{ kg}}.$$

P16.44 (a) The volume of the liquid increases as $\Delta V_\ell = V_i\beta\Delta T$. The volume of the flask increases as $\Delta V_g = 3\alpha V_i\Delta T$. Therefore, the overflow in the capillary is $V_c = V_i\Delta T(\beta - 3\alpha)$; and in the capillary $V_c = A\Delta h$.

Therefore, $\boxed{\Delta h = \frac{V_i}{A}(\beta - 3\alpha)\Delta T}$.

(b) For a mercury thermometer $\beta(\text{Hg}) = 1.82\times10^{-4}\ {}^\circ\text{C}^{-1}$

and for glass, $3\alpha = 3\times3.20\times10^{-6}\ {}^\circ\text{C}^{-1}$

Thus $\beta - 3\alpha \approx \beta$ within 5%

or $\boxed{\alpha \ll \beta}$.

P16.45 (a) $\rho = \frac{m}{V}$ and $d\rho = -\frac{m}{V^2}dV$

For very small changes in V and ρ, this can be expressed as

$$\Delta\rho = -\frac{m}{V}\frac{\Delta V}{V} = -\rho\beta\Delta T.$$

The negative sign means that any increase in temperature causes the density to decrease and vice versa.

(b) For water we have $\beta = \left|\frac{\Delta\rho}{\rho\Delta T}\right| = \left|\frac{1.000\,0\text{ g/cm}^3 - 0.999\,7\text{ g/cm}^3}{(1.000\,0\text{ g/cm}^3)(10.0^\circ\text{C} - 4.0^\circ\text{C})}\right| = \boxed{5\times10^{-5}\ {}^\circ\text{C}^{-1}}$.

P16.46 The astronauts exhale this much CO_2:

$$n = \frac{m_{\text{sample}}}{M} = \frac{1.09\text{ kg}}{\text{astronaut}\cdot\text{day}}\left(\frac{1\,000\text{ g}}{1\text{ kg}}\right)(3\text{ astronauts})(7\text{ days})\left(\frac{1\text{ mol}}{44.0\text{ g}}\right) = 520\text{ mol}.$$

Then 520 mol of methane is generated. It is far from liquefaction and behaves as an ideal gas.

$$P = \frac{nRT}{V} = \frac{520\text{ mol}(8.314\text{ J/mol}\cdot\text{K})(273\text{ K} - 45\text{ K})}{150\times10^{-3}\text{ m}^3} = \boxed{6.57\times10^6\text{ Pa}}$$

P16.47 (a) We assume that air at atmospheric pressure is above the piston.

In equilibrium $P_{gas} = \frac{mg}{A} + P_0$

Therefore, $\frac{nRT}{hA} = \frac{mg}{A} + P_0$

or $\boxed{h = \frac{nRT}{mg + P_0 A}}$

where we have used $V = hA$ as the volume of the gas.

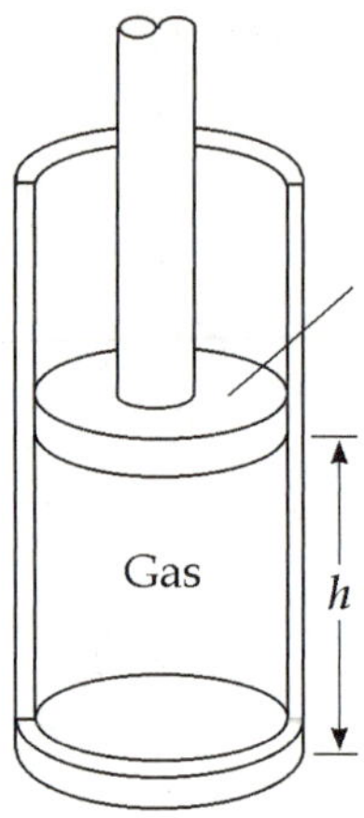

FIG. P16.47

(b) From the data given,

$$h = \frac{0.200 \text{ mol}(8.314 \text{ J/K}\cdot\text{mol})(400 \text{ K})}{20.0 \text{ kg}(9.80 \text{ m/s}^2) + (1.013\times10^5 \text{ N/m}^2)(0.008\,00 \text{ m}^2)}$$
$$= \boxed{0.661 \text{ m}}$$

P16.48 The angle of bending θ, between tangents to the two ends of the strip, is equal to the angle the strip subtends at its center of curvature. (The angles are equal because their sides are perpendicular, right side to the right side and left side to left side.)

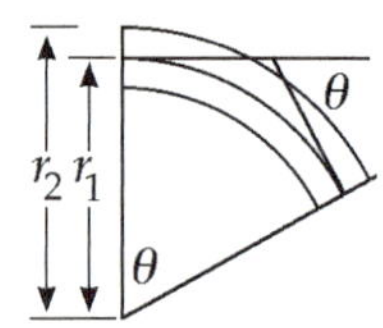

FIG. P16.48

(a) The definition of radian measure gives $L_i + \Delta L_1 = \theta r_1$

and $L_i + \Delta L_2 = \theta r_2$

By subtraction, $\Delta L_2 - \Delta L_1 = \theta(r_2 - r_1)$

$\alpha_2 L_i \Delta T - \alpha_1 L_i \Delta T = \theta \Delta r$

$$\boxed{\theta = \frac{(\alpha_2 - \alpha_1)L_i \Delta T}{\Delta r}}$$

(b) In the expression from part (a), θ is directly proportional to ΔT and also to $(\alpha_2 - \alpha_1)$. Therefore θ is zero when either of these quantities becomes zero.

(c) The material that expands more when heated contracts more when cooled, so the bimetallic strip bends the other way. It is fun to demonstrate this with liquid nitrogen.

(d) $$\theta = \frac{2(\alpha_2 - \alpha_1)L_i \Delta T}{2\Delta r} = \frac{2\left((19\times10^{-6} - 0.9\times10^{-6})^\circ\text{C}^{-1}\right)(200 \text{ mm})(1^\circ\text{C})}{0.500 \text{ mm}}$$
$$= 1.45\times10^{-2} = 1.45\times10^{-2} \text{ rad}\left(\frac{180^\circ}{\pi \text{ rad}}\right) = \boxed{0.830^\circ}$$

P16.49 From the diagram we see that the change in area is

$$\Delta A = \ell \Delta w + w \Delta \ell + \Delta w \Delta \ell .$$

Since $\Delta \ell$ and Δw are each small quantities, the product $\Delta w \Delta \ell$ will be very small. Therefore, we assume $\Delta w \Delta \ell \approx 0$.

Since $\Delta w = w\alpha \Delta T$ and $\Delta \ell = \ell \alpha \Delta T$,

we then have $\Delta A = \ell w \alpha \Delta T + \ell w \alpha \Delta T$

and since $A = \ell w$, $\boxed{\Delta A = 2\alpha A \Delta T}$.

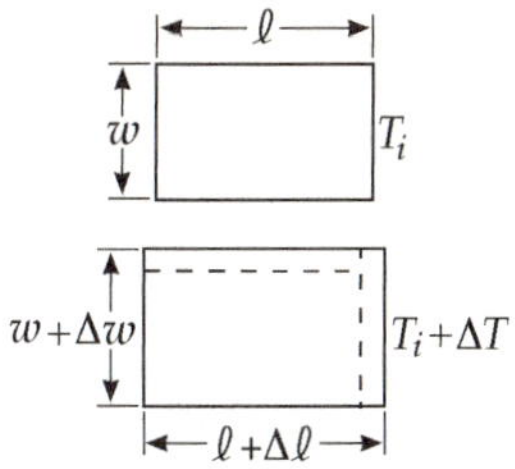

FIG. P16.49

The approximation assumes $\Delta w \Delta \ell \approx 0$, or $\alpha \Delta T \approx 0$. Another way of stating this is $\boxed{\alpha \Delta T << 1}$.

P16.50 (a) $T_i = 2\pi \sqrt{\frac{L_i}{g}}$ so $L_i = \frac{T_i^2 g}{4\pi^2} = \frac{(1.000 \text{ s})^2 (9.80 \text{ m/s}^2)}{4\pi^2} = 0.248\,2 \text{ m}$

$$\Delta L = \alpha L_i \Delta T = 19.0 \times 10^{-6}\ {}^\circ\text{C}^{-1}(0.248\,2 \text{ m})(10.0^\circ\text{C}) = 4.72 \times 10^{-5} \text{ m}$$

$$T_f = 2\pi \sqrt{\frac{L_i + \Delta L}{g}} = 2\pi \sqrt{\frac{0.248\,3 \text{ m}}{9.80 \text{ m/s}^2}} = 1.000\,095\,0 \text{ s}$$

$$\Delta T = \boxed{9.50 \times 10^{-5} \text{ s}}$$

(b) In one week, the time lost is time lost $= 1 \text{ week}\left(9.50 \times 10^{-5} \text{ s lost per second}\right)$

$$\text{time lost} = (7.00 \text{ d/week})\left(\frac{86\,400 \text{ s}}{1.00 \text{ d}}\right)\left(9.50 \times 10^{-5} \frac{\text{s lost}}{\text{s}}\right)$$

$$\text{time lost} = \boxed{57.5 \text{ s lost}}$$

P16.51 $I = \int r^2 dm$ and since $r(T) = r(T_i)(1 + \alpha \Delta T)$

for $\alpha \Delta T << 1$ we find $\frac{I(T)}{I(T_i)} = (1 + \alpha \Delta T)^2 \approx 1 + 2\alpha \Delta T$

thus $\frac{I(T) - I(T_i)}{I(T_i)} \approx 2\alpha \Delta T$

(a) With $\alpha = 17.0 \times 10^{-6}\ {}^\circ\text{C}^{-1}$ and $\Delta T = 100^\circ\text{C}$

we find for Cu: $\frac{\Delta I}{I} = 2\left(17.0 \times 10^{-6}\ {}^\circ\text{C}^{-1}\right)(100^\circ\text{C}) = \boxed{0.340\%}$

(b) With $\alpha = 24.0 \times 10^{-6}\ {}^\circ\text{C}^{-1}$

and $\Delta T = 100^\circ\text{C}$

we find for Al: $\frac{\Delta I}{I} = 2\left(24.0 \times 10^{-6}\ {}^\circ\text{C}^{-1}\right)(100^\circ\text{C}) = \boxed{0.480\%}$

***P16.52** (a) $B = \rho g V'$ $\qquad P' = P_0 + \rho g d \qquad P'V' = P_0 V_i$

$$B = \frac{\rho g P_0 V_i}{P'} = \boxed{\frac{\rho g P_0 V_i}{(P_0 + \rho g d)}}$$

(b) Since d is in the denominator, B must $\boxed{\text{decrease}}$ as the depth increases.

(The volume of the balloon becomes smaller with increasing pressure.)

(c) $$\frac{1}{2} = \frac{B(d)}{B(0)} = \frac{\rho g P_0 V_i/(P_0 + \rho g d)}{\rho g P_0 V_i / P_0} = \frac{P_0}{P_0 + \rho g d}$$

$$P_0 + \rho g d = 2P_0$$

$$d = \frac{P_0}{\rho g} = \frac{1.013 \times 10^5\ \text{N/m}^2}{(1.00 \times 10^3\ \text{kg/m}^3)(9.80\ \text{m/s}^2)} = \boxed{10.3\ \text{m}}$$

P16.53 After expansion, the length of one of the spans is

$$L_f = L_i(1 + \alpha\Delta T) = 125\ \text{m}\left[1 + 12 \times 10^{-6}\ {}^\circ\text{C}^{-1}(20.0^\circ\text{C})\right] = 125.03\ \text{m}.$$

L_f, y, and the original 125 m length of this span form a right triangle with y as the altitude. Using the Pythagorean theorem gives:

$$(125.03\ \text{m})^2 = y^2 + (125\ \text{m})^2$$

yielding $y = \boxed{2.74\ \text{m}}$.

P16.54 After expansion, the length of one of the spans is $L_f = L(1 + \alpha\Delta T)$. L_f, y, and the original length L of this span form a right triangle with y as the altitude. Using the Pythagorean theorem gives

$$L_f^2 = L^2 + y^2, \qquad \text{or} \qquad y = \sqrt{L_f^2 - L^2} = L\sqrt{(1 + \alpha\Delta T)^2 - 1} = L\sqrt{2\alpha\Delta T + (\alpha\Delta T)^2}$$

Since $\alpha\Delta T << 1$, $\qquad y \approx \boxed{L\sqrt{2\alpha\Delta T}}$.

The height of the center of the buckling bridge is directly proportional to the length. A small bridge is geometrically similar to a large one. The height is proportional to the square róot of the temperature increase. Doubling ΔT makes y increase by only 41%. A small value of ΔT can have a surprisingly large effect. In units, the equation reads $\text{m} = \text{m}(^\circ\text{C}/^\circ\text{C})^{1/2}$, so it is dimensionally correct.

***P16.55** (a) No torque acts on the disk so its angular momentum is constant. Its moment of inertia decreases as it contracts so its angular speed must $\boxed{\text{increase}}$.

(b) $$I_i\omega_i = I_f\omega_f = \frac{1}{2}MR_i^2\omega_i = \frac{1}{2}MR_f^2\omega_f = \frac{1}{2}M[R_i + R_i\alpha\Delta T]^2\omega_f = \frac{1}{2}MR_i^2[1 - \alpha|\Delta T|]^2\omega_f$$

$$\omega_f = \omega_i[1 - \alpha|\Delta T|]^{-2} = \frac{25.0\ \text{rad/s}}{\left(1 - (17 \times 10^{-6}\ 1/\text{C}^\circ)830^\circ\text{C}\right)^2} = \frac{25.0\ \text{rad/s}}{0.972} = \boxed{25.7\ \text{rad/s}}$$

P16.56 (a) From $PV = nRT$, the volume is: $V = \left(\frac{nR}{P}\right)T$

Therefore, when pressure is held constant, $\frac{\partial V}{\partial T} = \frac{nR}{P} = \frac{V}{T}$

Thus, $\beta \equiv \left(\frac{1}{V}\right)\frac{\partial V}{\partial T} = \left(\frac{1}{V}\right)\frac{V}{T}$, or $\beta = \boxed{\frac{1}{T}}$

(b) At $T = 0°\text{C} = 273\text{ K}$, this predicts $\beta = \frac{1}{273\text{ K}} = \boxed{3.66 \times 10^{-3}\text{ K}^{-1}}$

Experimental values are: $\beta_{\text{He}} = 3.665 \times 10^{-3}\text{ K}^{-1}$ and $\beta_{\text{air}} = 3.67 \times 10^{-3}\text{ K}^{-1}$

They agree within 0.06% and 0.2%, respectively.

P16.57 (a) Let m represent the sample mass. The number of moles is $n = \frac{m}{M}$ and the density is $\rho = \frac{m}{V}$.

So $PV = nRT$ becomes $PV = \frac{m}{M}RT$ or $PM = \frac{m}{V}RT$.

Then, $\rho = \frac{m}{V} = \boxed{\frac{PM}{RT}}$.

(b) $$\rho = \frac{PM}{RT} = \frac{(1.013 \times 10^5\text{ N/m}^2)(0.0320\text{ kg/mol})}{(8.314\text{ J/mol}\cdot\text{K})(293\text{ K})} = \boxed{1.33\text{ kg/m}^3}$$

P16.58 (a) With piston alone: $T = \text{constant}$, so $PV = P_0V_0$

or $P(Ah_i) = P_0(Ah_0)$

With $A = \text{constant}$, $P = P_0\left(\frac{h_0}{h_i}\right)$

But, $P = P_0 + \frac{m_p g}{A}$

where m_p is the mass of the piston.

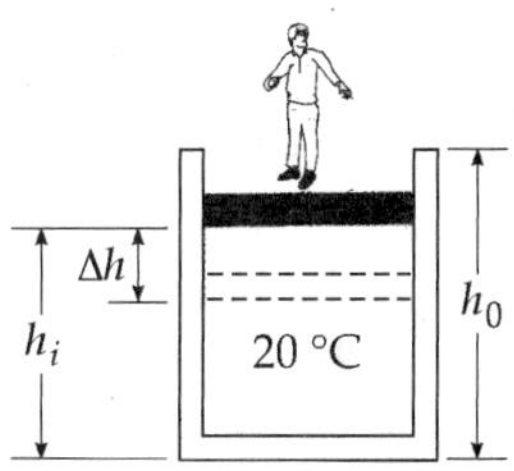

FIG. P16.58

Thus, $P_0 + \frac{m_p g}{A} = P_0\left(\frac{h_0}{h_i}\right)$

which reduces to $$h_i = \frac{h_0}{1 + \frac{m_p g}{P_0 A}} = \frac{50.0\text{ cm}}{1 + \frac{20.0\text{ kg}(9.80\text{ m/s}^2)}{1.013\times10^5\text{ Pa}\left[\pi(0.400\text{ m})^2\right]}} = 49.81\text{ cm}$$

With the man of mass M on the piston, a very similar calculation (replacing m_p by $m_p + M$) gives:

$$h' = \frac{h_0}{1 + \frac{(m_p + M)g}{P_0 A}} = \frac{50.0\text{ cm}}{1 + \frac{95.0\text{ kg}(9.80\text{ m/s}^2)}{1.013\times10^5\text{ Pa}\left[\pi(0.400\text{ m})^2\right]}} = 49.10\text{ cm}$$

continued on next page

Thus, when the man steps on the piston, it moves downward by

$$\Delta h = h_i - h' = 49.81\text{ cm} - 49.10\text{ cm} = 0.706\text{ cm} = \boxed{7.06\text{ mm}}.$$

(b) P = const, so $\dfrac{V}{T} = \dfrac{V'}{T_i}$ or $\dfrac{Ah_i}{T} = \dfrac{Ah'}{T_i}$

giving $T = T_i\left(\dfrac{h_i}{h'}\right) = 293\text{ K}\left(\dfrac{49.81}{49.10}\right) = \boxed{297\text{ K}}$ (or 24°C)

P16.59 Some gas will pass through the porous plug from the reaction chamber 1 to the reservoir 2 as the reaction chamber is heated, but the net quantity of gas stays constant according to $n_{i1} + n_{i2} = n_{f1} + n_{f2}$. Assuming the gas is ideal, we apply $n = \dfrac{PV}{RT}$ to each term:

$$\frac{P_i V_0}{(300\text{ K})R} + \frac{P_i(4V_0)}{(300\text{ K})R} = \frac{P_f V_0}{(673\text{ K})R} + \frac{P_f(4V_0)}{(300\text{ K})R}$$

$$1\text{ atm}\left(\frac{5}{300\text{ K}}\right) = P_f\left(\frac{1}{673\text{ K}} + \frac{4}{300\text{ K}}\right) \qquad \boxed{P_f = 1.12\text{ atm}}$$

P16.60 Let 2θ represent the angle the curved rail subtends. We have

$$L_i + \Delta L = 2\theta R = L_i(1 + \alpha\Delta T)$$

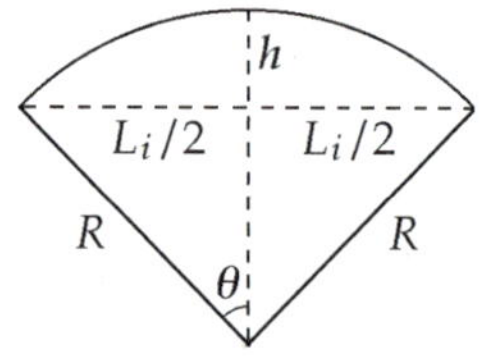

FIG. P16.60

and $\sin\theta = \dfrac{\frac{L_i}{2}}{R} = \dfrac{L_i}{2R}$

Thus, $\theta = \dfrac{L_i}{2R}(1 + \alpha\Delta T) = (1 + \alpha\Delta T)\sin\theta$

and we must solve the transcendental equation $\theta = (1 + \alpha\Delta T)\sin\theta = (1.000\,005\,5)\sin\theta$

Homing in on the non-zero solution gives, to four digits, $\theta = 0.018\,16\text{ rad} = 1.040\,5°$

Now, $h = R - R\cos\theta = \dfrac{L_i(1 - \cos\theta)}{2\sin\theta}$

This yields $\boxed{h = 4.54\text{ m}}$, a remarkably large value compared to $\Delta L = 5.50\text{ cm}$.

***P16.61** (a) Let xL represent the distance of the stationary line below the top edge of the plate. The normal force on the lower part of the plate is $mg(1 - x)\cos\theta$ and the force of kinetic friction on it is $\mu_k mg(1 - x)\cos\theta$ up the roof. Again, $\mu_k mgx\cos\theta$ acts down the roof on the upper part of the plate. The near-equilibrium of the plate requires $\sum F_x = 0$

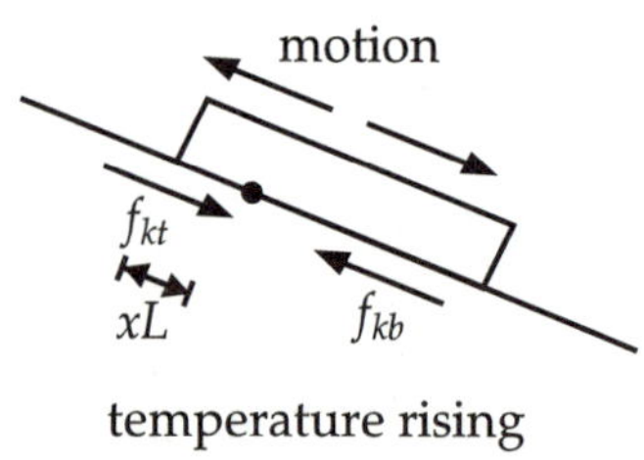

FIG. P16.61(a)

continued on next page

$$-\mu_k mgx\cos\theta+\mu_k mg(1-x)\cos\theta-mg\sin\theta=0$$
$$-2\mu_k mgx\cos\theta=mg\sin\theta-\mu_k mg\cos\theta$$
$$2\mu_k x=\mu_k-\tan\theta$$
$$x=\frac{1}{2}-\frac{\tan\theta}{2\mu_k}$$

and the stationary line is indeed below the top edge by $xL=\dfrac{L}{2}\left(1-\dfrac{\tan\theta}{\mu_k}\right)$.

(b) With the temperature falling, the plate contracts faster than the roof. The upper part slides down and feels an upward frictional force $\mu_k mg(1-x)\cos\theta$. The lower part slides up and feels downward frictional force $\mu_k mgx\cos\theta$. The equation $\sum F_x=0$ is then the same as in part (a) and the stationary line is above the bottom edge by $xL=\dfrac{L}{2}\left(1-\dfrac{\tan\theta}{\mu_k}\right)$.

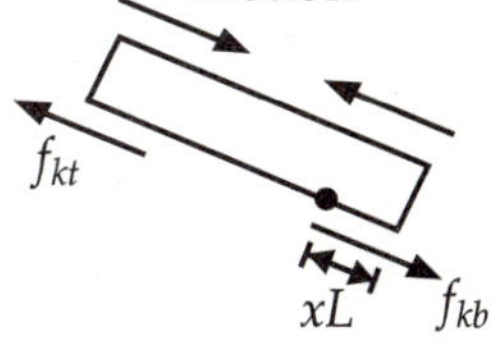

FIG. P16.61(b)

(c) Start thinking about the plate at dawn, as the temperature starts to rise. As in part (a), a line at distance xL below the top edge of the plate stays stationary relative to the roof as long as the temperature rises. The point P on the plate at distance xL above the bottom edge is destined to become the fixed point when the temperature starts falling. As the temperature rises, this point moves down the roof because of the expansion of the central part of the plate. Its displacement for the day is

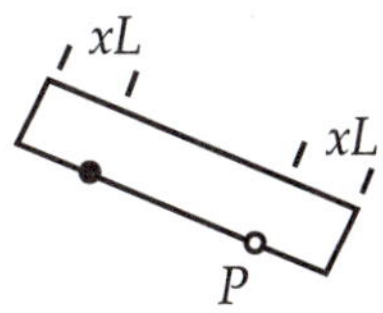

FIG. P16.61(c)

$$\Delta L=(\alpha_2-\alpha_1)(L-xL-xL)\Delta T$$
$$=(\alpha_2-\alpha_1)\left[L-2\frac{L}{2}\left(1-\frac{\tan\theta}{\mu_k}\right)\right](T_h-T_c)$$
$$=(\alpha_2-\alpha_1)\left(\frac{L\tan\theta}{\mu_k}\right)(T_h-T_c).$$

At dawn the next day the point P is farther down the roof by the distance ΔL. It represents the displacement of every other point on the plate.

(d) $$(\alpha_2-\alpha_1)\left(\frac{L\tan\theta}{\mu_k}\right)(T_h-T_c)=\left(24\times10^{-6}\frac{1}{\text{C}^\circ}-15\times10^{-6}\frac{1}{\text{C}^\circ}\right)\frac{1.20\text{ m}\tan 18.5^\circ}{0.42}32^\circ\text{C}=\boxed{0.275\text{ mm}}$$

(e) If $\alpha_2<\alpha_1$, the diagram in part (a) applies to temperature falling and the diagram in part (b) applies to temperature rising. The weight of the plate still pulls it step by step down the roof. The same expression describes how far it moves each day.

P16.62 The pressure of the gas in the lungs of the diver must be the same as the absolute pressure of the water at this depth of 50.0 meters. This is:

$$P = P_0 + \rho g h = 1.00 \text{ atm} + \left(1.03\times10^3 \text{ kg/m}^3\right)\left(9.80 \text{ m/s}^2\right)(50.0 \text{ m})$$

or $$P = 1.00 \text{ atm} + 5.05\times10^5 \text{ Pa}\left(\frac{1.00 \text{ atm}}{1.013\times10^5 \text{ Pa}}\right) = 5.98 \text{ atm}$$

If the partial pressure due to the oxygen in the gas mixture is to be 1.00 atmosphere (or the fraction $\frac{1}{5.98}$ of the total pressure) oxygen molecules should make up only $\frac{1}{5.98}$ of the total number of molecules. This will be true if 1.00 mole of oxygen is used for every 4.98 mole of helium. The ratio by weight is then

$$\frac{(4.98 \text{ mol He})(4.003 \text{ g/mol He})g}{(1.00 \text{ mol } O_2)(2\times15.999 \text{ g/mol } O_2)g} = \boxed{0.623}.$$

P16.63 $$N_v(v) = 4\pi N\left(\frac{m_0}{2\pi k_B T}\right)^{3/2} v^2 \exp\left(\frac{-m_0 v^2}{2k_B T}\right)$$

Note that $$v_{mp} = \left(\frac{2k_B T}{m_0}\right)^{1/2}$$

Thus, $$N_v(v) = 4\pi N\left(\frac{m_0}{2\pi k_B T}\right)^{3/2} v^2 e^{\left(-v^2/v_{mp}^2\right)}$$

And $$\frac{N_v(v)}{N_v(v_{mp})} = \left(\frac{v}{v_{mp}}\right)^2 e^{\left(1-v^2/v_{mp}^2\right)}$$

For $$v = \frac{v_{mp}}{50}$$

$$\frac{N_v(v)}{N_v(v_{mp})} = \left(\frac{1}{50}\right)^2 e^{\left[1-(1/50)^2\right]} = 1.09\times10^{-3}$$

The other values are computed similarly, with the following results:

$\frac{v}{v_{mp}}$	$\frac{N_v(v)}{N_v(v_{mp})}$
$\frac{1}{50}$	1.09×10^{-3}
$\frac{1}{10}$	2.69×10^{-2}
$\frac{1}{2}$	0.529
1	1.00
2	0.199
10	1.01×10^{-41}
50	$1.25\times10^{-1\,082}$

To find the last value, note:

$$(50)^2 e^{1-2\,500} = 2\,500 e^{-2\,499}$$

$$10^{\log 2\,500} e^{(\ln 10)(-2\,499/\ln 10)} = 10^{\log 2\,500} 10^{-2\,499/\ln 10} = 10^{\log 2\,500 - 2\,499/\ln 10} = 10^{-1\,081.904}$$

P16.64 (a) Maxwell's speed distribution function is

$$N_v = 4\pi N\left(\frac{m_0}{2\pi k_B T}\right)^{3/2} v^2 e^{-m_0 v^2/2k_B T}$$

With $N = 1.00\times10^4$,

$$m_0 = \frac{M}{N_A} = \frac{0.032 \text{ kg}}{6.02\times10^{23}} = 5.32\times10^{-26} \text{ kg}$$

$$T = 500 \text{ K}$$

and $k_B = 1.38\times10^{-23}$ J/molecule·K

this becomes $N_v = \left(1.71\times10^{-4}\right)v^2 e^{-\left(3.85\times10^{-6}\right)v^2}$

To the right is a plot of this function for the range $0 \le v \le 1\,500$ m/s.

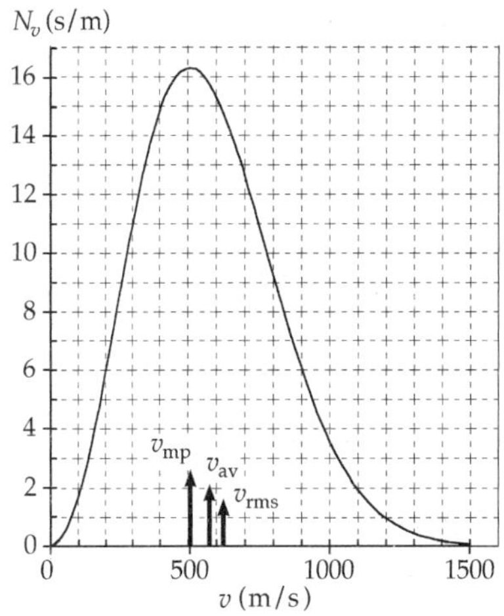

FIG. P16.64(a)

(b) The most probable speed occurs where N_v is a maximum.

From the graph, $\boxed{v_{mp} \approx 510 \text{ m/s}}$

(c) $$v_{av} = \sqrt{\frac{8k_B T}{\pi m_0}} = \sqrt{\frac{8\left(1.38\times10^{-23}\right)(500)}{\pi\left(5.32\times10^{-26}\right)}} = \boxed{575 \text{ m/s}}$$

Also,

$$v_{rms} = \sqrt{\frac{3k_B T}{m_0}} = \sqrt{\frac{3\left(1.38\times10^{-23}\right)(500)}{5.32\times10^{-26}}} = \boxed{624 \text{ m/s}}$$

(d) The fraction of particles in the range $300 \text{ m/s} \le v \le 600 \text{ m/s}$

is $$\frac{\int_{300}^{600} N_v dv}{N}$$

where $N = 10^4$

and the integral of N_v is read from the graph as the area under the curve.

This is approximately 4 400 and the fraction is 0.44 or $\boxed{44\%}$.

ANSWERS TO EVEN PROBLEMS

P16.2 (a) $37.0°\text{C}$, 310 K; (b) $-20.6°\text{C}$, 253 K

P16.4 (a) 810°F; (b) 450 K

P16.6 55.0°C

P16.8 $2.17\times10^{-5}\ (°\text{C})^{-1}$

P16.10 0.663 mm to the right at 78.2° below the horizontal

P16.12 0.548 gal

P16.14 (a) 437°C; (b) 2 100°C, no, aluminum melts at 660°C.

P16.16 6.64×10^{-27} kg

P16.18 877 balloons

P16.20 see the solution

P16.22 (a) 1.17×10^{-3} kg; (b) 11.5 mN; (c) 1.01 kN; (d) molecules must be moving very fast

P16.24 3.67 cm^3

P16.26 $\sim 10^2$ kg

P16.28 2.41×10^{11} molecules

P16.30 $Nm\left(\dfrac{2v\sin\theta}{t}\right)$ perpendicular to the window, $Nm\left(\dfrac{2v\sin\theta}{At}\right)$

P16.32 (a) 2.28 kJ; (b) 6.21×10^{-21} J

P16.34 (a) 9.26×10^{-4} m/s; (b) 3.24 ms; (c) 1.32×10^{-11} m/s, 3.88×10^{10} s; (d) 2.97×10^{-5} m

P16.36 (a) 6.80 m/s; (b) 7.41 m/s; (c) 7.00 m/s

P16.38 132 m/s

P16.40 6.2°C

P16.42 (a) 0.949 7 m; (b) 0.950 3 m

P16.44 (a) $\Delta h = \dfrac{V_i}{A}(\beta - 3\alpha)\Delta T$; (b) $\alpha \ll \beta$

P16.46 6.57×10^6 Pa

P16.48 (a) $\theta = \dfrac{(\alpha_2-\alpha_1)L_i\Delta T}{\Delta r}$; (b) see the solution; (c) see the solution; (d) 0.830°

P16.50 (a) increases by 9.50×10^{-5} s; (b) 57.5 s lost

P16.52 (a) $\dfrac{\rho g P_0 V_i}{(P_0+\rho g d)}$ up; (b) decrease; (c) 10.3 m

P16.54 $y \approx L\sqrt{2\alpha\Delta T}$

P16.56 (a) $\dfrac{1}{T}$; (b) 3.66×10^{-3} K^{-1}, it agrees within 0.2% or better

P16.58 (a) 7.06 mm; (b) 297 K

P16.60 4.54 m

P16.62 0.623

P16.64 (a) see the solution; (b) $v_{mp} \approx 510$ m/s; (c) 575 m/s, 624 m/s; (d) 44%

Chapter 17

Energy in Thermal Processes: The First Law of Thermodynamics

CHAPTER OUTLINE

ANSWERS TO QUESTIONS

Q17.1 Temperature is a measure of molecular motion. Heat is energy in the process of being transferred between objects by random molecular collisions. Internal energy is an object's energy of random molecular motion and molecular interaction.

Q17.2 The ΔT is twice as great in the ethyl alcohol.

Q17.3 The final equilibrium temperature will show no significant increase over the initial temperature of the water.

Q17.4 Some water may boil away. You would have to very precisely measure how much, and very quickly measure the temperature of the steam; it is not necessarily $100°\text{C}$.

Q17.5 Heat is energy being transferred, not energy contained in an object. Further, a large-mass object, or an object made of a material with high specific heat, can contain more internal energy than a higher-temperature object.

Q17.6 There are three properties to consider here: thermal conductivity, specific heat, and mass. With dry aluminum, the thermal conductivity of aluminum is much greater than that of (dry) skin. This means that the internal energy in the aluminum can more readily be transferred to the atmosphere than to your fingers. In essence, your skin acts as a thermal insulator to some degree (pun intended). If the aluminum is wet, it can wet the outer layer of your skin to make it into a good conductor of heat; then more internal energy from the aluminum can get into you. Further, the water itself, with additional mass and with a relatively large specific heat compared to aluminum, can be a significant source of extra energy to burn you. In practical terms, when you let go of a hot, dry piece of aluminum foil, the heat transfer immediately ends. When you let go of a hot *and* wet piece of aluminum foil, the hot water sticks to your skin, continuing the heat transfer, and resulting in more energy transfer to you!

Q17.7 Write $1\,000\text{ kg}(4\,186\text{ J/kg}\cdot{}^\circ\text{C})(1^\circ\text{C}) = V(1.3\text{ kg/m}^3)(1\,000\text{ J/kg}\cdot{}^\circ\text{C})(1^\circ\text{C})$ to find $V = 3.2\times10^3\text{ m}^3$.

Q17.8 (a) and (b) both increase by minuscule amounts.

Q17.9 If the system is isolated, no energy enters or leaves the system by heat, work, or other transfer processes. Within the system energy can change from one form to another, but since energy is conserved these transformations cannot affect the total amount of energy. The total energy is constant.

Q17.10 Yes. If you know the different specific heats of zinc and copper, you can determine the fraction of each by heating a known mass of pennies to a specific initial temperature, say 100°C, and dumping them into a known quantity of water, at say 20°C. The final temperature T will reveal the metal content:

$$m_{\text{pennies}}\left[xc_{\text{Cu}} + (1-x)c_{\text{Zn}}\right](100^\circ\text{C} - T) = m_{\text{H}_2\text{O}}c_{\text{H}_2\text{O}}(T - 20^\circ\text{C}).$$

Since all quantities are known, except x, the fraction of the penny that is copper will be found by putting in the experimental numbers m_{pennies}, $m_{\text{H}_2\text{O}}$, $T(\text{final})$, c_{Zn}, and c_{Cu}.

Q17.11 The tile is a better thermal conductor than carpet. Thus, energy is conducted away from your feet more rapidly by the tile than by the carpeted floor.

Q17.12 The question refers to baking in a conventional oven, not to microwaving. The metal has much higher thermal conductivity than the potato. The metal quickly conducts energy from the hot oven into the center of potato.

Q17.13 In winter the interior of the house is warmer than the air outside. On a summer day we want the interior to stay cooler than the exterior. Heavy draperies over the windows can slow down energy transfer by conduction, by convection, and by radiation, to make it easier to maintain the desired difference in temperature.

Q17.14 In winter the produce is protected from freezing. The heat capacity of the earth is so high that soil freezes only to a depth of a few decimeters in temperate regions. Throughout the year the temperature will stay nearly constant all day and night. Factors to be considered are the insulating properties of soil, the absence of a path for energy to be radiated away from or to the vegetables, and the hindrance to the formation of convection currents in the small, enclosed space.

Q17.15 The high mass and specific heat of the barrel of water and its high heat of fusion mean that a large amount of energy would have to leak out of the cellar before the water and the produce froze solid. Evaporation of the water keeps the relative humidity high to protect foodstuffs from drying out.

Q17.16 The porcelain of the teacup is a thermal insulator. That is, it is a thermal conductor of relatively low conductivity. When you wrap your hands around a cup of hot tea, you make A large and L small in the equation $\mathcal{P} = kA\dfrac{T_h - T_c}{L}$ for the rate of energy transfer by heat from tea into you. When you hold the cup by the handle, you make the rate of energy transfer much smaller by reducing A and increasing L. The air around the cup handle will also reduce the temperature where you are touching it. A paper cup can be fitted into a tubular jacket of corrugated cardboard, with the channels running vertically, for remarkably effective insulation, according to the same principles.

Q17.17 Keep them dry. The air pockets in the pad conduct energy by heat, but only slowly. Wet pads would absorb some energy in warming up themselves, but the pot would still be hot and the water would quickly conduct and convect a lot of energy right into you.

Q17.18 The person should add the cream immediately when the coffee is poured, rather than later when he is ready to drink it. Then the smaller temperature difference between coffee and environment will reduce the rate of energy loss during the several minutes. Also, the lighter color of the coffee with cream may reduce the emissivity to decrease the radiated power.

Q17.19 Convection. The bridge deck loses energy rapidly to the air both above it and below it.

Q17.20 The marshmallow has very small mass compared to the saliva in the teacher's mouth and the surrounding tissues. Mostly air and sugar, the marshmallow also has a low specific heat compared to living matter. Then the marshmallow can zoom up through a large temperature change while causing only a small temperature drop of the teacher's mouth. The marshmallow is a foam with closed cells and it carries very little liquid nitrogen into the mouth. The liquid nitrogen still on the marshmallow comes in contact with the much hotter saliva and immediately boils into cold gaseous nitrogen. This nitrogen gas has very low thermal conductivity. It creates an insulating thermal barrier between the marshmallow and the teacher's mouth (the Leydenfrost effect). A similar effect can be seen when water droplets are put on a hot skillet. Each one dances around as it slowly shrinks, because it is levitated on a thin film of steam. The most extreme demonstration of this effect is pouring liquid nitrogen into one's mouth and blowing out a plume of nitrogen gas. We strongly recommend that you read of Jearl Walker's adventures with this demonstration rather than trying it.

Q17.21 (a) Warm a pot of coffee on a hot stove.

(b) Consider an ice cube at $0°C$ in a glass of lemonade on a picnic table. The ice will absorb energy while melting, but not increase in temperature.

(c) Let a high-pressure gas at room temperature slowly expand by pushing on a piston. Work comes out of the gas in a constant-temperature expansion as the same quantity of heat flows in from the surroundings.

(d) Warm your hands by rubbing them together. Heat your tepid coffee in a microwave oven. Energy input by work, by electromagnetic radiation, or by other means, can all alike produce a temperature increase.

(e) Davy's experiment is an example of this process.

(f) This is not necessarily true. Consider some supercooled liquid water, unstable but with temperature below $0°C$. Drop in a snowflake or a grain of dust to trigger its freezing into ice, and the release of internal energy measured by its latent heat of fusion can actually push its temperature up.

Q17.22 Partially evacuating the container is equivalent to letting the remaining gas expand. This means that the gas does work, making its internal energy and hence its temperature decrease. The liquid in the container will eventually reach thermal equilibrium with the low pressure gas. This effect of an expanding gas decreasing in temperature is a key process in your refrigerator or air conditioner.

SOLUTIONS TO PROBLEMS

Section 17.1 Heat and Internal Energy

P17.1 Imagine a mass m of water contained in a styrofoam box, which prevents any energy flow into or out of it by heat. As the water falls the gravitational energy of the water-Earth system turns into kinetic energy of organized (downward) motion of the water, then into kinetic energy of disorganized (splashing around) macroscopic motion, and at last into kinetic energy of random molecular motion, which is internal energy. There is 100% conversion efficiency from gravitational into internal energy. Heat is not involved in the warming process, but imagine a process in which the mass m of water is placed on a stove and energy mgy is put into it by heat. The temperature change is then specified by

$$mgy = mc\Delta T$$

$$\Delta T = \frac{gy}{c} = \frac{(9.80\ \text{m/s}^2)(120\ \text{m})}{4186\ \text{J/kg}\cdot{}^\circ\text{C}}\left(\frac{1\ \text{J}}{1\ \text{kg}\cdot\text{m}^2/\text{s}^2}\right) = \boxed{0.281^\circ\text{C}}$$

P17.2 The container is thermally insulated, so no energy flows by heat:

$$Q = 0$$

and

$$\Delta E_{\text{int}} = Q + W_{\text{input}} = 0 + W_{\text{input}} = 2mgh$$

The work on the falling weights is equal to the work done on the water in the container by the rotating blades. This work results in an increase in internal energy of the water:

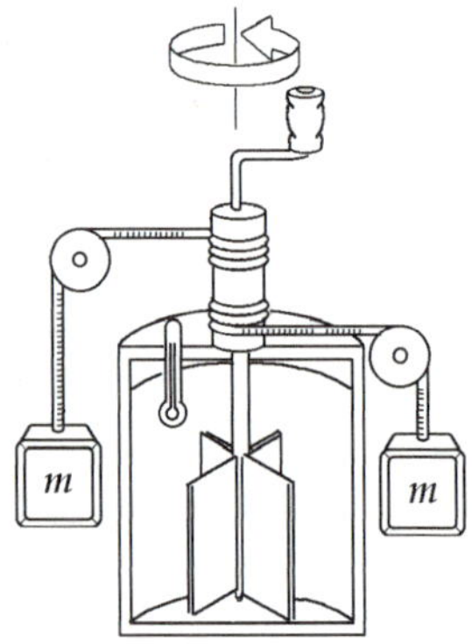

Thermal insulator

FIG. P17.2

$$2mgh = \Delta E_{\text{int}} = m_{\text{water}}c\Delta T$$

$$\Delta T = \frac{2mgh}{m_{\text{water}}c} = \frac{2\times 1.50\ \text{kg}(9.80\ \text{m/s}^2)(3.00\ \text{m})}{0.200\ \text{kg}(4\,186\ \text{J/kg}\cdot{}^\circ\text{C})} = \frac{88.2\ \text{J}}{837\ \text{J/}^\circ\text{C}} = \boxed{0.105^\circ\text{C}}$$

Section 17.2 Specific Heat

P17.3 From $Q = mc\Delta T$

we find $\Delta T = \dfrac{Q}{mc} = \dfrac{1\,200\ \text{J}}{0.050\,0\ \text{kg}(387\ \text{J/kg}\cdot{}^\circ\text{C})} = 62.0^\circ\text{C}$

Thus, the final temperature is $\boxed{87.0^\circ\text{C}}$.

P17.4 We imagine the stone energy reservoir has a large area in contact with air and is always at nearly the same temperature as the air. Its overnight loss of energy is described by

$$\mathcal{P} = \frac{Q}{\Delta t} = \frac{mc\Delta T}{\Delta t}$$

$$m = \frac{\mathcal{P}\Delta t}{c\Delta T} = \frac{(-6\,000\ \text{J/s})(14\ \text{h})(3\,600\ \text{s/h})}{(850\ \text{J/kg}\cdot{}^\circ\text{C})(18^\circ\text{C} - 38^\circ\text{C})} = \frac{3.02\times 10^8\ \text{J}\cdot\text{kg}\cdot{}^\circ\text{C}}{850\ \text{J}(20^\circ\text{C})} = \boxed{1.78\times 10^4\ \text{kg}}$$

P17.5 $Q_{\text{cold}} = -Q_{\text{hot}}$

$$(mc\Delta T)_{\text{water}} = -(mc\Delta T)_{\text{iron}}$$

$$20.0\text{ kg}(4\,186\text{ J/kg}\cdot°\text{C})(T_f - 25.0°\text{C}) = -(1.50\text{ kg})(448\text{ J/kg}\cdot°\text{C})(T_f - 600°\text{C})$$

$$T_f = \boxed{29.6°\text{C}}$$

***P17.6** Let us find the energy transferred in one minute.

$$Q = \left[m_{\text{cup}}c_{\text{cup}} + m_{\text{water}}c_{\text{water}}\right]\Delta T$$

$$Q = \left[(0.200\text{ kg})(900\text{ J/kg}\cdot°\text{C}) + (0.800\text{ kg})(4\,186\text{ J/kg}\cdot°\text{C})\right](-1.50°\text{C}) = -5\,290\text{ J}$$

If this much energy is removed from the system each minute, the rate of removal is

$$\mathcal{P} = \frac{|Q|}{\Delta t} = \frac{5\,290\text{ J}}{60.0\text{ s}} = 88.2\text{ J/s} = \boxed{88.2\text{ W}}.$$

***P17.7** (a) Work that the bit does in deforming the block, breaking chips off, and giving them kinetic energy is not a final destination for energy. All of this work turns entirely into internal energy as soon as the chips stop their macroscopic motion. The amount of energy input to the steel is the work done by the bit:

$$W = \vec{\mathbf{F}}\cdot\vec{\mathbf{d}} = (3.2\text{ N})(40\text{ m/s})(15\text{ s})\cos 0° = 1\,920\text{ J}$$

To evaluate the temperature change produced by this energy we imagine injecting the same quantity of energy as heat from a stove. The bit, chips, and block all undergo the same temperature change. Any difference in temperature between one bit of steel and another would erase itself by causing a heat transfer from the temporarily hotter to the colder region.

$$Q = mc\Delta T$$

$$\Delta T = \frac{Q}{mc} = \frac{1\,920\text{ J}\cdot\text{kg}\cdot°\text{C}}{(0.267\text{ kg})(448\text{ J})} = \boxed{16.1°\text{C}}$$

(b) See part (a).

P17.8 (a) $Q_{\text{cold}} = -Q_{\text{hot}}$

$$(m_w c_w + m_c c_c)(T_f - T_c) = -m_{\text{Cu}}c_{\text{Cu}}(T_f - T_{\text{Cu}}) - m_{unk}c_{unk}(T_f - T_{unk})$$

where w is for water, c the calorimeter, Cu the copper sample, and *unk* the unknown.

$$\left[250\text{ g}(1.00\text{ cal/g}\cdot°\text{C}) + 100\text{ g}(0.215\text{ cal/g}\cdot°\text{C})\right](20.0 - 10.0)°\text{C}$$

$$= -(50.0\text{ g})(0.092\,4\text{ cal/g}\cdot°\text{C})(20.0 - 80.0)°\text{C} - (70.0\text{ g})c_{unk}(20.0 - 100)°\text{C}$$

$$2.44\times10^3\text{ cal} = (5.60\times10^3\text{ g}\cdot°\text{C})c_{unk}$$

or $c_{unk} = \boxed{0.435\text{ cal/g}\cdot°\text{C}}$.

(b) The material of the sample can be $\boxed{\text{beryllium}}$.

***P17.9** We do not know whether the aluminum will rise or drop in temperature. The energy the water can absorb in rising to 26°C is $mc\Delta T = 0.25\text{ kg } 4\,186\dfrac{\text{J}}{\text{kg }^\circ\text{C}}6^\circ\text{C} = 6\,279\text{ J}$. The energy the copper can put out in dropping to 26°C is $mc\Delta T = 0.1\text{ kg } 387\dfrac{\text{J}}{\text{kg }^\circ\text{C}}74^\circ\text{C} = 2\,864\text{ J}$. Since $6\,279\text{ J} > 2\,864\text{ J}$, the final temperature is less than 26°C. We can write $Q_h = -Q_c$ as

$$Q_{\text{water}} + Q_{\text{Al}} + Q_{\text{Cu}} = 0$$

$$0.25\text{ kg } 4\,186\frac{\text{J}}{\text{kg }^\circ\text{C}}\left(T_f - 20^\circ\text{C}\right) + 0.4\text{ kg } 900\frac{\text{J}}{\text{kg }^\circ\text{C}}\left(T_f - 26^\circ\text{C}\right)$$

$$+0.1\text{ kg } 387\frac{\text{J}}{\text{kg }^\circ\text{C}}\left(T_f - 100^\circ\text{C}\right) = 0$$

$$1\,046.5T_f - 20\,930^\circ\text{C} + 360T_f - 9\,360^\circ\text{C} + 38.7T_f - 3\,870^\circ\text{C} = 0$$

$$1\,445.2T_f = 34\,160^\circ\text{C}$$

$$T_f = \boxed{23.6^\circ\text{C}}$$

P17.10 $Q_{\text{cold}} = -Q_{\text{hot}}$

$$m_{\text{Al}}c_{\text{Al}}\left(T_f - T_c\right) + m_c c_w\left(T_f - T_c\right) = -m_h c_w\left(T_f - T_h\right)$$

$$(m_{\text{Al}}c_{\text{Al}} + m_c c_w)T_f - (m_{\text{Al}}c_{\text{Al}} + m_c c_w)T_c = -m_h c_w T_f + m_h c_w T_h$$

$$(m_{\text{Al}}c_{\text{Al}} + m_c c_w + m_h c_w)T_f = (m_{\text{Al}}c_{\text{Al}} + m_c c_w)T_c + m_h c_w T_h$$

$$T_f = \boxed{\frac{(m_{\text{Al}}c_{\text{Al}} + m_c c_w)T_c + m_h c_w T_h}{m_{\text{Al}}c_{\text{Al}} + m_c c_w + m_h c_w}}$$

Section 17.3 Latent Heat and Phase Changes

P17.11 The heat needed is the sum of the following terms:

$$Q_{\text{needed}} = (\text{heat to reach melting point}) + (\text{heat to melt})$$
$$+(\text{heat to reach boiling point}) + (\text{heat to vaporize}) + (\text{heat to reach } 110^\circ\text{C})$$

Thus, we have

$$Q_{\text{needed}} = 0.040\,0\text{ kg}\Big[(2\,090\text{ J/kg}\cdot{}^\circ\text{C})(10.0^\circ\text{C}) + \left(3.33\times10^5\text{ J/kg}\right)$$
$$+(4\,186\text{ J/kg}\cdot{}^\circ\text{C})(100^\circ\text{C}) + \left(2.26\times10^6\text{ J/kg}\right) + (2\,010\text{ J/kg}\cdot{}^\circ\text{C})(10.0^\circ\text{C})\Big]$$

$$Q_{\text{needed}} = \boxed{1.22\times10^5\text{ J}}$$

P17.12 $Q_{\text{cold}} = -Q_{\text{hot}}$

$$(m_w c_w + m_c c_c)(T_f - T_i) = -m_s\left[-L_v + c_w(T_f - 100)\right]$$

$$\left[0.250\text{ kg}(4\,186\text{ J/kg}\cdot{}^\circ\text{C}) + 0.050\,0\text{ kg}(387\text{ J/kg}\cdot{}^\circ\text{C})\right](50.0^\circ\text{C} - 20.0^\circ\text{C})$$

$$= -m_s\left[-2.26\times10^6\text{ J/kg} + (4\,186\text{ J/kg}\cdot{}^\circ\text{C})(50.0^\circ\text{C} - 100^\circ\text{C})\right]$$

$$m_s = \frac{3.20\times10^4\text{ J}}{2.47\times10^6\text{ J/kg}} = 0.012\,9\text{ kg} = \boxed{12.9\text{ g steam}}$$

P17.13 The bullet will not melt all the ice, so its final temperature is 0°C.

Then $\left(\frac{1}{2}mv^2 + mc|\Delta T|\right)_{\text{bullet}} = m_w L_f$

where m_w is the melt water mass

$$m_w = \frac{0.500\left(3.00\times10^{-3}\text{ kg}\right)(240\text{ m/s})^2 + 3.00\times10^{-3}\text{ kg}(128\text{ J/kg}\cdot{}^\circ\text{C})(30.0^\circ\text{C})}{3.33\times10^5\text{ J/kg}}$$

$$m_w = \frac{86.4\text{ J} + 11.5\text{ J}}{333\,000\text{ J/kg}} = \boxed{0.294\text{ g}}$$

P17.14 $Q = m_{\text{Cu}} c_{\text{Cu}} \Delta T = m_{N_2}\left(L_{\text{vap}}\right)_{N_2}$

$$1.00\text{ kg}(0.092\,0\text{ cal/g}\cdot{}^\circ\text{C})(293 - 77.3)^\circ\text{C} = m(48.0\text{ cal/g})$$

$$m = \boxed{0.414\text{ kg}}$$

P17.15 (a) Since the heat required to melt 250 g of ice at 0°C *exceeds* the heat required to cool 600 g of water from 18°C to 0°C, the final temperature of the system (water + ice) must be $\boxed{0^\circ\text{C}}$.

(b) Let m represent the mass of ice that melts before the system reaches equilibrium at 0°C.

$$Q_{\text{cold}} = -Q_{\text{hot}}$$

$$mL_f = -m_w c_w(0^\circ\text{C} - T_i)$$

$$m\left(3.33\times10^5\text{ J/kg}\right) = -(0.600\text{ kg})(4\,186\text{ J/kg}\cdot{}^\circ\text{C})(0^\circ\text{C} - 18.0^\circ\text{C})$$

$$m = 136\text{ g, so the ice remaining} = 250\text{ g} - 136\text{ g} = \boxed{114\text{ g}}$$

P17.16 The original gravitational energy of the hailstone-Earth system changes entirely into additional internal energy in the hailstone, to produce its phase change. No temperature change occurs, either in the hailstone, in the air, or in sidewalk. Then

$$mgy = mL$$

$$y = \frac{L}{g} = \frac{3.33\times10^5\text{ J/kg}}{9.8\text{ m/s}^2}\left(\frac{1\text{ kg}\cdot\text{m}^2/\text{s}^2}{1\text{ J}}\right) = \boxed{3.40\times10^4\text{ m}}$$

***P17.17** The original kinetic energy all becomes thermal energy:

$$\frac{1}{2}mv^2+\frac{1}{2}mv^2=2\left(\frac{1}{2}\right)\left(5.00\times10^{-3}\text{ kg}\right)\left(500\text{ m/s}\right)^2=1.25\text{ kJ}.$$

Raising the temperature to the melting point requires

$$Q=mc\Delta T=10.0\times10^{-3}\text{ kg}(128\text{ J/kg}\cdot{}^\circ\text{C})(327^\circ\text{C}-20.0^\circ\text{C})=393\text{ J}.$$

Since $1\,250\text{ J}>393\text{ J}$, the lead starts to melt. Melting it all requires

$$Q=mL=\left(10.0\times10^{-3}\text{ kg}\right)\left(2.45\times10^4\text{ J/kg}\right)=245\text{ J}.$$

Since $1\,250\text{ J}>393+245\text{ J}$, it all melts. If we assume $\boxed{\text{liquid lead}}$ has the same specific heat as solid lead, the final temperature is given by

$$1.25\times10^3\text{ J}=393\text{ J}+245\text{ J}+10.0\times10^{-3}\text{ kg}(128\text{ J/kg}\cdot{}^\circ\text{C})\left(T_f-327^\circ\text{C}\right)$$

$$\boxed{T_f=805^\circ\text{C}}$$

P17.18 We find the quantity of water vapor in one exhaled breath.

$$PV=nRT:\qquad n=\frac{PV}{RT}=\frac{\left(3.20\times10^3\text{ N/m}^2\right)\left(0.600\times10^{-3}\text{ m}^3\right)}{(8.314\text{ J/mol}\cdot\text{K})(273\text{ K}+37\text{ K})}=7.45\times10^{-4}\text{ mol}$$

The molar mass of water (H_2O) is $M=\left(2(1.00)+16.0\right)\text{ g/mol}=0.018\,0\text{ kg/mol}$. The mass of water vapor exhaled in one breath is $m_{\text{sample}}=nM=7.45\times10^{-4}\text{ mol}(0.0180\text{ kg/mol})=1.34\times10^{-5}\text{ kg}$. The energy absorbed from your body as the water evaporates can be estimate as

$$Q=mL=1.34\times10^{-5}\text{ kg}\left(2.26\times10^6\text{ J/kg}\right)=30.3\text{ J}.$$

Your rate of energy loss is $\mathcal{P}=\frac{Q}{\Delta t}=\frac{30.3\text{ J}}{\text{breath}}\left(\frac{22.0\text{ breath}}{\text{min}}\right)\left(\frac{1\text{ min}}{60\text{ s}}\right)=\boxed{11.1\text{ W}}$. Note that a dog does not perspire. Instead, the dog pants, maximizes energy loss through the pathway considered here.

Section 17.4 Work in Thermodynamic Processes

P17.19 $W_{if} = -\int_i^f P\,dV$

The work done on the gas is the negative of the area under the curve $P = \alpha V^2$ between V_i and V_f.

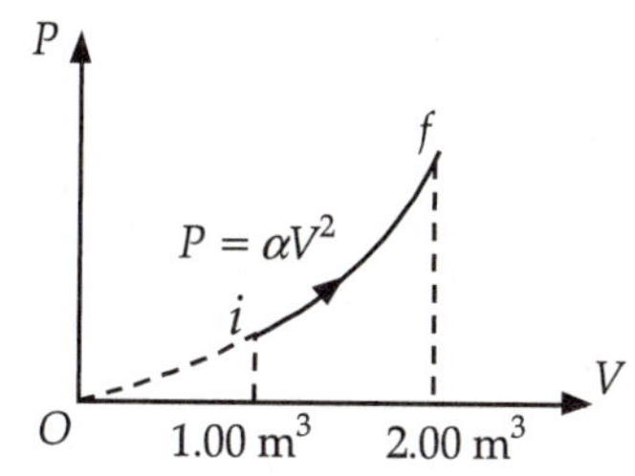

FIG. P17.19

$$W_{if} = -\int_i^f \alpha V^2 dV = -\frac{1}{3}\alpha\left(V_f^3 - V_i^3\right)$$

$$V_f = 2V_i = 2\left(1.00 \text{ m}^3\right) = 2.00 \text{ m}^3$$

$$W_{if} = -\frac{1}{3}\left[\left(5.00 \text{ atm/m}^6\right)\left(1.013\times10^5 \text{ Pa/atm}\right)\right]\left[\left(2.00 \text{ m}^3\right)^3 - \left(1.00 \text{ m}^3\right)^3\right] = \boxed{-1.18 \text{ MJ}}$$

P17.20 (a) $W = -\int P\,dV$

$$W = -\left(6.00\times10^6 \text{ Pa}\right)(2.00-1.00) \text{ m}^3 + -\left(4.00\times10^6 \text{ Pa}\right)(3.00-2.00) \text{ m}^3 + -\left(2.00\times10^6 \text{ Pa}\right)(4.00-3.00) \text{ m}^3$$

$$W_{i\to f} = \boxed{-12.0 \text{ MJ}}$$

(b) $W_{f\to i} = \boxed{+12.0 \text{ MJ}}$

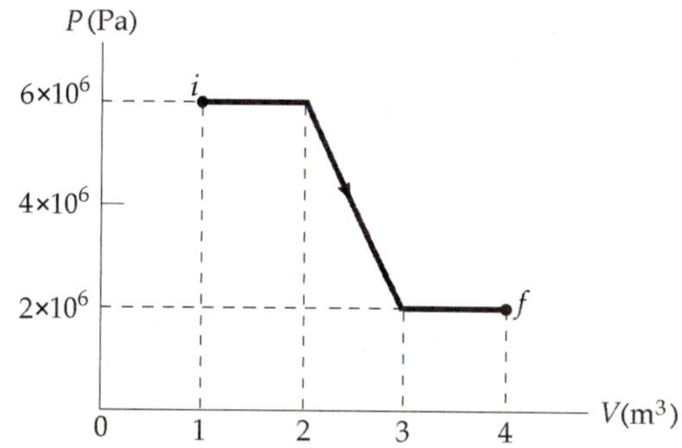

FIG. P17.20

P17.21 $W = -P\Delta V = -P\left(\frac{nR}{P}\right)\left(T_f - T_i\right) = -nR\Delta T = -(0.200)(8.314)(280) = \boxed{-466 \text{ J}}$

P17.22 $W = -\int_i^f P\,dV = -P\int_i^f dV = -P\Delta V = -nR\Delta T = \boxed{-nR(T_2 - T_1)}$ The negative sign indicates that the expanding gas does positive work. The quantity of work is directly proportional to the quantity of gas and to the temperature change.

Section 17.5 The First Law of Thermodynamics

P17.23 $\Delta E_{\text{int}} = Q + W$

$$Q = \Delta E_{\text{int}} - W = -500 \text{ J} - 220 \text{ J} = \boxed{-720 \text{ J}}$$

The negative sign indicates that positive energy is transferred *from* the system by heat.

P17.24 (a) $Q = -W = \text{Area of triangle}$

$$Q = \frac{1}{2}\left(4.00\ \text{m}^3\right)(6.00\ \text{kPa}) = \boxed{12.0\ \text{kJ}}$$

(b) $Q = -W = \boxed{-12.0\ \text{kJ}}$

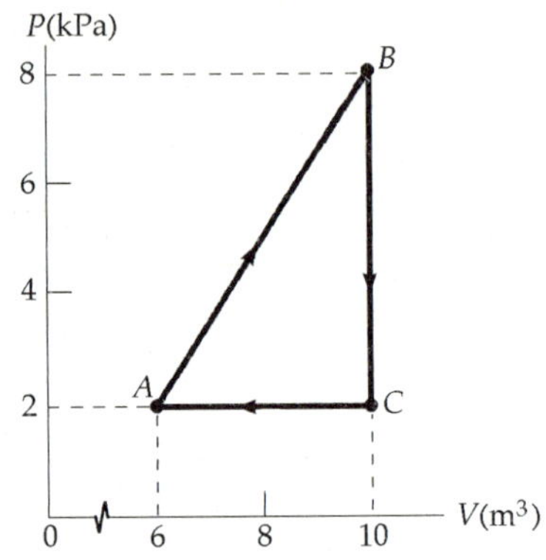

FIG. P17.24

P17.25

	Q	W	ΔE_{int}	
BC	−	0	−	$(Q = \Delta E_{int}$ since $W_{BC} = 0)$
CA	−	+	−	$(\Delta E_{int} < 0$ and $W > 0$, so $Q < 0)$
AB	+	−	+	$(W < 0,\ \Delta E_{int} > 0$ since $\Delta E_{int} < 0$ for $B \to C \to A$; so $Q > 0)$

Section 17.6 Some Applications of the First Law of Thermodynamics

P17.26 (a) $W = -nRT\ln\left(\frac{V_f}{V_i}\right) = -P_f V_f \ln\left(\frac{V_f}{V_i}\right)$

$$\text{so } V_i = V_f \exp\left(+\frac{W}{P_f V_f}\right) = (0.025\,0)\exp\left[\frac{-3\,000}{0.025\,0\left(1.013\times10^5\right)}\right] = \boxed{0.007\,65\ \text{m}^3}$$

(b) $$T_f = \frac{P_f V_f}{nR} = \frac{1.013\times10^5\ \text{Pa}\left(0.025\,0\ \text{m}^3\right)}{1.00\ \text{mol}(8.314\ \text{J/K}\cdot\text{mol})} = \boxed{305\ \text{K}}$$

P17.27 (a) $\Delta E_{int} = Q - P\Delta V = 12.5\ \text{kJ} - 2.50\ \text{kPa}(3.00 - 1.00)\ \text{m}^3 = \boxed{7.50\ \text{kJ}}$

(b) $$\frac{V_1}{T_1} = \frac{V_2}{T_2}$$

$$T_2 = \frac{V_2}{V_1}T_1 = \frac{3.00}{1.00}(300\ \text{K}) = \boxed{900\ \text{K}}$$

P17.28 (a) $W = -P\Delta V = -P[3\alpha V\Delta T]$

$$= -\left(1.013\times10^5\ \text{N/m}^2\right)\left[3\left(24.0\times10^{-6}\,{}^\circ\text{C}^{-1}\right)\left(\frac{1.00\ \text{kg}}{2.70\times10^3\ \text{kg/m}^3}\right)(18.0^\circ\text{C})\right]$$

$$W = \boxed{-48.6\ \text{mJ}}$$

(b) $Q = cm\Delta T = (900\ \text{J/kg}\cdot{}^\circ\text{C})(1.00\ \text{kg})(18.0^\circ\text{C}) = \boxed{16.2\ \text{kJ}}$

(c) $\Delta E_{int} = Q + W = 16.2\ \text{kJ} - 48.6\ \text{mJ} = \boxed{16.2\ \text{kJ}}$

P17.29 $W = -P\Delta V = -P(V_s - V_w) = -\frac{P(nRT)}{P} + P\left[\frac{18.0\text{ g}}{(1.00\text{ g/cm}^3)(10^6\text{ cm}^3/\text{m}^3)}\right]$

$$W = -(1.00\text{ mol})(8.314\text{ J/K}\cdot\text{mol})(373\text{ K}) + (1.013\times 10^5\text{ N/m}^2)\left(\frac{18.0\text{ g}}{10^6\text{ g/m}^3}\right) = \boxed{-3.10\text{ kJ}}$$

$$Q = mL_v = 0.018\,0\text{ kg}(2.26\times 10^6\text{ J/kg}) = 40.7\text{ kJ}$$

$$\Delta E_{\text{int}} = Q + W = \boxed{37.6\text{ kJ}}$$

P17.30 (a) The work done during each step of the cycle equals the negative of the area under that segment of the PV curve.

$$W = W_{DA} + W_{AB} + W_{BC} + W_{CD}$$

$$W = -P_i(V_i - 3V_i) + 0 - 3P_i(3V_i - V_i) + 0 = \boxed{-4P_iV_i}$$

(b) The initial and final values of T for the system are equal. Therefore, $\Delta E_{\text{int}} = 0$ and $Q = -W = \boxed{4P_iV_i}$.

(c) $W = -4P_iV_i = -4nRT_i = -4(1.00)(8.314)(273) = \boxed{-9.08\text{ kJ}}$

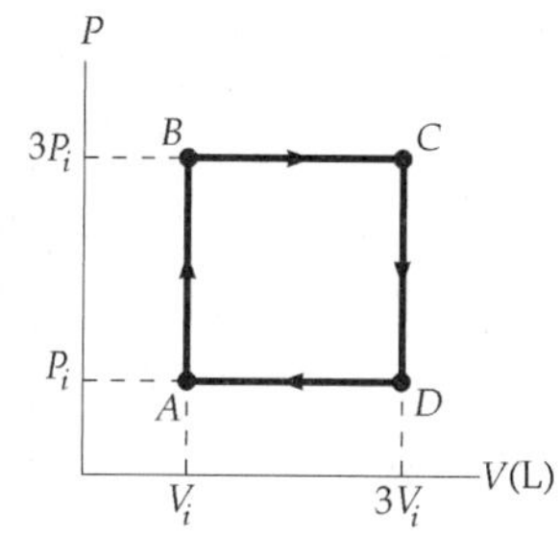

FIG. P17.30

P17.31 (a) $P_iV_i = P_fV_f = nRT = 2.00\text{ mol}(8.314\text{ J/K}\cdot\text{mol})(300\text{ K}) = 4.99\times 10^3\text{ J}$

$$V_i = \frac{nRT}{P_i} = \frac{4.99\times 10^3\text{ J}}{0.400\text{ atm}}$$

$$V_f = \frac{nRT}{P_f} = \frac{4.99\times 10^3\text{ J}}{1.20\text{ atm}} = \frac{1}{3}V_i = \boxed{0.041\,0\text{ m}^3}$$

(b) $W = -\int P\,dV = -nRT\ln\left(\frac{V_f}{V_i}\right) = -(4.99\times 10^3)\ln\left(\frac{1}{3}\right) = \boxed{+5.48\text{ kJ}}$

(c) $\Delta E_{\text{int}} = 0 = Q + W$

$$Q = \boxed{-5.48\text{ kJ}}$$

P17.32 $\Delta E_{\text{int},\,ABC} = \Delta E_{\text{int},\,AC}$ (conservation of energy)

(a) $\Delta E_{\text{int},\,ABC} = Q_{ABC} + W_{ABC}$ (First Law)

$$Q_{ABC} = 800\text{ J} + 500\text{ J} = \boxed{1\,300\text{ J}}$$

(b) $W_{CD} = -P_C\Delta V_{CD}$, $\Delta V_{AB} = -\Delta V_{CD}$, and $P_A = 5P_C$

Then, $W_{CD} = \frac{1}{5}P_A\Delta V_{AB} = -\frac{1}{5}W_{AB} = \boxed{100\text{ J}}$

(+ means that work is done on the system)

(c) $W_{CDA} = W_{CD}$ so that $Q_{CA} = \Delta E_{\text{int},\,CA} - W_{CDA} = -800\text{ J} - 100\text{ J} = \boxed{-900\text{ J}}$

(– means that energy must be removed from the system by heat)

(d) $\Delta E_{\text{int},\,CD} = \Delta E_{\text{int},\,CDA} - \Delta E_{\text{int},\,DA} = -800\text{ J} - 500\text{ J} = -1\,300\text{ J}$

and $Q_{CD} = \Delta E_{\text{int},\,CD} - W_{CD} = -1\,300\text{ J} - 100\text{ J} = \boxed{-1\,400\text{ J}}$

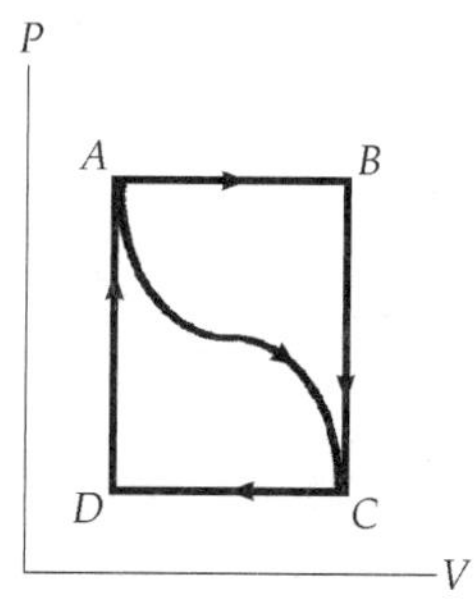

FIG. P17.32

Section 17.7 Molar Specific Heats of Ideal Gases

P17.33 We use the tabulated values for C_p and C_V.

(a) $Q = nC_p\Delta T = 1.00 \text{ mol}(28.8 \text{ J/mol}\cdot\text{K})(420-300)\text{K} = \boxed{3.46 \text{ kJ}}$

(b) $\Delta E_{int} = nC_V\Delta T = 1.00 \text{ mol}(20.4 \text{ J/mol}\cdot\text{K})(120 \text{ K}) = \boxed{2.45 \text{ kJ}}$

(c) $W = -Q + \Delta E_{int} = -3.46 \text{ kJ} + 2.45 \text{ kJ} = \boxed{-1.01 \text{ kJ}}$

P17.34 $E_{int} = \frac{3}{2}nRT$

$$\Delta E_{int} = \frac{3}{2}nR\Delta T = \frac{3}{2}(3.00 \text{ mol})(8.314 \text{ J/mol}\cdot\text{K})(2.00 \text{ K}) = \boxed{74.8 \text{ J}}$$

P17.35 $n = 1.00 \text{ mol}$, $T_i = 300 \text{ K}$

(b) Since $V = \text{constant}$, $W = \boxed{0}$.

(a) $\Delta E_{int} = Q + W = 209 \text{ J} + 0 = \boxed{209 \text{ J}}$

(c) $\Delta E_{int} = nC_V\Delta T = n\left(\frac{3}{2}R\right)\Delta T$ so

$$\Delta T = \frac{2\Delta E_{int}}{3nR} = \frac{2(209 \text{ J})}{3(1.00 \text{ mol})(8.314 \text{ J/mol}\cdot\text{K})} = 16.8 \text{ K}$$

$$T = T_i + \Delta T = 300 \text{ K} + 16.8 \text{ K} = \boxed{317 \text{ K}}$$

P17.36 (a) $C_V = \frac{5}{2}R = \frac{5}{2}(8.314 \text{ J/mol}\cdot\text{K})\left(\frac{1.00 \text{ mol}}{0.0289 \text{ kg}}\right) = 719 \text{ J/kg}\cdot\text{K} = \boxed{0.719 \text{ kJ/kg}\cdot\text{K}}$

(b) $m = Mn = M\left(\frac{PV}{RT}\right)$

$$m = (0.0289 \text{ kg/mol})\left(\frac{200\times10^3 \text{ Pa}\left(0.350 \text{ m}^3\right)}{(8.314 \text{ J/mol}\cdot\text{K})(300 \text{ K})}\right) = \boxed{0.811 \text{ kg}}$$

(c) We consider a constant volume process where no work is done.

$$Q = mC_V\Delta T = 0.811 \text{ kg}(0.719 \text{ kJ/kg}\cdot\text{K})(700 \text{ K} - 300 \text{ K}) = \boxed{233 \text{ kJ}}$$

(d) We now consider a constant pressure process where the internal energy of the gas is increased and work is done.

$$Q = mC_P\Delta T = m(C_V + R)\Delta T = m\left(\frac{7R}{2}\right)\Delta T = m\left(\frac{7C_V}{5}\right)\Delta T$$

$$Q = 0.811 \text{ kg}\left[\frac{7}{5}(0.719 \text{ kJ/kg}\cdot\text{K})\right](400 \text{ K}) = \boxed{327 \text{ kJ}}$$

P17.37 Consider 800 cm^3 of (flavored) water at 90.0 °C mixing with 200 cm^3 of diatomic ideal gas at 20.0°C: $Q_{\text{cold}} = -Q_{\text{hot}}$ or

$$m_{\text{air}} c_{P,\text{air}} \left(T_f - T_{i,\text{air}}\right) = -m_w c_w (\Delta T)_w$$

$$(\Delta T)_w = \frac{-m_{\text{air}} c_{P,\text{air}} (T_f - T_{i,\text{air}})}{m_w c_w} = \frac{-(\rho V)_{\text{air}} c_{P,\text{air}} (90.0°\text{C} - 20.0°\text{C})}{(\rho_w V_w) c_w}$$

where we have anticipated that the final temperature of the mixture will be close to 90.0°C. The molar specific heat of air is $C_{P,\text{ air}} = \frac{7}{2} R$. So the specific heat per gram is

$$c_{P,\text{air}} = \frac{7}{2}\left(\frac{R}{M}\right) = \frac{7}{2}(8.314 \text{ J/mol}\cdot\text{K})\left(\frac{1.00 \text{ mol}}{28.9 \text{ g}}\right) = 1.01 \text{ J/g}\cdot°\text{C}$$

$$(\Delta T)_w = -\frac{[(1.20 \times 10^{-3} \text{ g/cm}^3)(200 \text{ cm}^3)](1.01 \text{ J/g}\cdot°\text{C})(70.0°\text{C})}{[(1.00 \text{ g/cm}^3)(800 \text{ cm}^3)](4.186 \text{ J/kg}\cdot°\text{C})}$$

or $(\Delta T)_w \approx -5.05 \times 10^{-3}\,°\text{C}$. The change of temperature for the water is

$\boxed{\text{between } 10^{-3}\,°\text{C and } 10^{-2}\,°\text{C}}$.

***P17.38** (a) The air is far from liquefaction so it behaves as an ideal gas. From $PV = nRT$ we have $PV = \frac{m}{M} RT$, $PM = \frac{m}{V} RT = \rho RT$. For the samples of air in the balloon at 10°C (cold) and at the elevated temperature (hot) we have $PM = \rho_c R T_c$ and $PM = \rho_h R T_h$. Then $\rho_h T_h = \rho_c T_c$ and $\rho_h = \frac{\rho_c T_c}{T_h}$. For equilibrium of the balloon on the point of rising,

$$\sum F_y = ma_y \qquad +B - F_{g\,\text{hot air}} - F_{g\,\text{cargo}} = 0$$

$$+\rho_c V g - \rho_h V g - mg = 0$$

$$+\rho_c V - \frac{\rho_c T_c}{T_h} V - m = 0$$

$$\left(1.25 \text{ kg/m}^3\right) 400 \text{ m}^3 - \left(1.25 \text{ kg/m}^3\right) \frac{283 \text{ K}}{T_h} 400 \text{ m}^3 - 200 \text{ kg} = 0$$

$$300 \text{ kg} = 500 \text{ kg} \frac{283 \text{ K}}{T_h}$$

$$T_h = \frac{500}{300} 283 \text{ K} = 472 \text{ K}$$

The quantity of air that must be warmed is given by $PV = n_h R T_h$, $n_h = \frac{PV}{RT_h}$. The heat input required is

$$Q = nC_p \Delta T = -\frac{PV}{RT_h} \frac{7}{2} R (T_h - T_c) = \frac{101 \times 10^3 \text{N } 400 \text{ m}^3\ 7(472 \text{ K} - 283 \text{ K})}{\text{m}^2\ 472 \text{ K } 2} = \boxed{5.66 \times 10^7 \text{ J}}$$

(b) $$Q = mH,\ m = \frac{Q}{H} = \frac{5.66 \times 10^7 \text{ J}}{5.03 \times 10^7 \text{ J/kg}} = \boxed{1.12 \text{ kg}}$$

P17.39 $Q = (nC_P\Delta T)_{\text{isobaric}} + (nC_V\Delta T)_{\text{isovolumetric}}$

In the isobaric process, V doubles so T must double, to $2T_i$.

In the isovolumetric process, P triples so T changes from $2T_i$ to $6T_i$.

$$Q = n\left(\frac{7}{2}R\right)(2T_i - T_i) + n\left(\frac{5}{2}R\right)(6T_i - 2T_i) = 13.5nRT_i = \boxed{13.5PV}$$

Section 17.8 Adiabatic Processes for an Ideal Gas

P17.40 (a) $P_iV_i^{\gamma} = P_fV_f^{\gamma}$ so $\dfrac{V_f}{V_i} = \left(\dfrac{P_i}{P_f}\right)^{1/\gamma} = \left(\dfrac{1.00}{20.0}\right)^{5/7} = \boxed{0.118}$

(b) $\dfrac{T_f}{T_i} = \dfrac{P_fV_f}{P_iV_i} = \left(\dfrac{P_f}{P_i}\right)\left(\dfrac{V_f}{V_i}\right) = (20.0)(0.118),\ \dfrac{T_f}{T_i} = \boxed{2.35}$

(c) Since the process is adiabatic, $\boxed{Q = 0}$.

Since $\gamma = 1.40 = \dfrac{C_P}{C_V} = \dfrac{R + C_V}{C_V},\ C_V = \dfrac{5}{2}R$ and $\Delta T = 2.35T_i - T_i = 1.35T_i$

$$\Delta E_{\text{int}} = nC_V\Delta T = (0.0160 \text{ mol})\left(\frac{5}{2}\right)(8.314 \text{ J/mol}\cdot\text{K})[1.35(300 \text{ K})] = \boxed{135 \text{ J}}$$

and $W = -Q + \Delta E_{\text{int}} = 0 + 135 \text{ J} = \boxed{+135 \text{ J}}$.

P17.41 (a) $P_iV_i^{\gamma} = P_fV_f^{\gamma}$

$$P_f = P_i\left(\frac{V_i}{V_f}\right)^{\gamma} = 5.00 \text{ atm}\left(\frac{12.0}{30.0}\right)^{1.40} = \boxed{1.39 \text{ atm}}$$

(b) $$T_i = \frac{P_iV_i}{nR} = \frac{5.00(1.013\times10^5 \text{ Pa})(12.0\times10^{-3} \text{ m}^3)}{2.00 \text{ mol}(8.314 \text{ J/mol}\cdot\text{K})} = \boxed{366 \text{ K}}$$

$$T_f = \frac{P_fV_f}{nR} = \frac{1.39(1.013\times10^5 \text{ Pa})(30.0\times10^{-3} \text{ m}^3)}{2.00 \text{ mol}(8.314 \text{ J/mol}\cdot\text{K})} = \boxed{253 \text{ K}}$$

(c) The process is adiabatic: $\boxed{Q = 0}$

$\gamma = 1.40 = \dfrac{C_P}{C_V} = \dfrac{R + C_V}{C_V},\ C_V = \dfrac{5}{2}R$

$$\Delta E_{\text{int}} = nC_V\Delta T = 2.00 \text{ mol}\left(\frac{5}{2}(8.314 \text{ J/mol}\cdot\text{K})\right)(253 \text{ K} - 365 \text{ K}) = \boxed{-4.66 \text{ kJ}}$$

$$W = \Delta E_{\text{int}} - Q = -4.66 \text{ kJ} - 0 = \boxed{-4.66 \text{ kJ}}$$

P17.42 $\frac{T_f}{T_i} = \left(\frac{V_i}{V_f}\right)^{\gamma-1} = \left(\frac{1}{2}\right)^{0.400}$

If $T_i = 300\text{ K}$, then $T_f = \boxed{227\text{ K}}$.

P17.43 (a) See the diagram at the right.

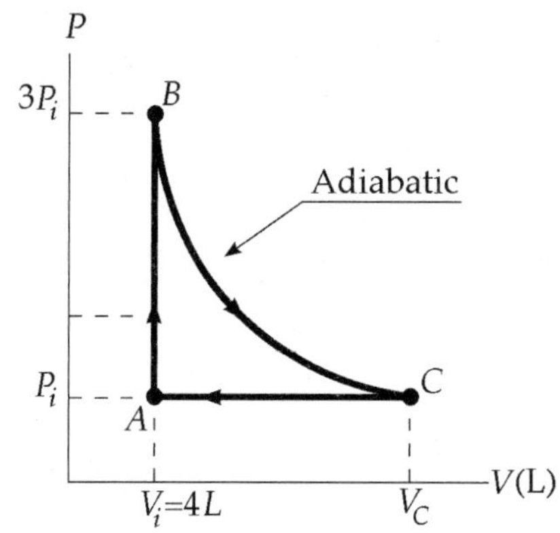

FIG. P17.43

(b)

$$P_B V_B^{\gamma} = P_C V_C^{\gamma}$$
$$3P_i V_i^{\gamma} = P_i V_C^{\gamma}$$
$$V_C = \left(3^{1/\gamma}\right)V_i = \left(3^{5/7}\right)V_i = 2.19V_i$$
$$V_C = 2.19(4.00\text{ L}) = \boxed{8.77\text{ L}}$$

(c)

$$P_B V_B = nRT_B = 3P_i V_i = 3nRT_i$$
$$T_B = 3T_i = 3(300\text{ K}) = \boxed{900\text{ K}}$$

(d) After one whole cycle, $T_A = T_i = \boxed{300\text{ K}}$

(e) In AB, $Q_{AB} = nC_V\Delta V = n\left(\frac{5}{2}R\right)(3T_i - T_i) = (5.00)nRT_i$, $Q_{BC} = 0$ as this process is adiabatic

$P_C V_C = nRT_C = P_i(2.19V_i) = (2.19)nRT_i$ so $T_C = 2.19T_i$,

$$Q_{CA} = nC_P\Delta T = n\left(\frac{7}{2}R\right)(T_i - 2.19T_i) = (-4.17)nRT_i$$

For the whole cycle,

$$Q_{ABCA} = Q_{AB} + Q_{BC} + Q_{CA} = (5.00 - 4.17)nRT_i = (0.829)nRT_i$$
$$(\Delta E_{\text{int}})_{ABCA} = 0 = Q_{ABCA} + W_{ABCA}$$
$$W_{ABCA} = -Q_{ABCA} = -(0.829)nRT_i = -(0.829)P_iV_i$$
$$W_{ABCA} = -(0.829)\left(1.013\times10^5\text{ Pa}\right)\left(4.00\times10^{-3}\text{ m}^3\right) = \boxed{-336\text{ J}}$$

P17.44 (a) The work done *on* the gas is

$$W_{ab} = -\int_{V_a}^{V_b} P\,dV$$

For the isothermal process,

$$W_{ab'} = -nRT_a \int_{V_a}^{V_{b'}} \left(\frac{1}{V}\right) dV$$

$$W_{ab'} = -nRT_a \ln\left(\frac{V_{b'}}{V_a}\right) = nRT \ln\left(\frac{V_a}{V_{b'}}\right)$$

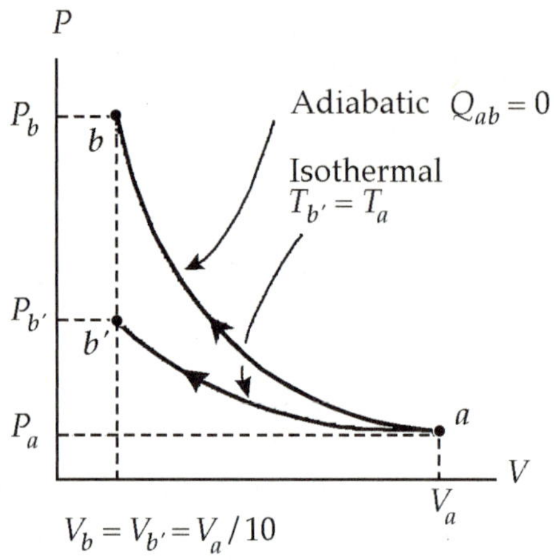

FIG. P17.44

Thus,

$$W_{ab'} = 5.00\text{ mol}(8.314\text{ J/mol}\cdot\text{K})(293\text{ K})\ln(10.0)$$

$$W_{ab'} = \boxed{28.0\text{ kJ}}$$

(b) For the adiabatic process, we must first find the final temperature, T_b. Since air consists primarily of diatomic molecules, we shall use $\gamma_{\text{air}} = 1.40$ and $C_{V,\text{air}} = \frac{5R}{2} = 5\frac{8.314}{2} = 20.8\text{ J/mol}\cdot\text{K}$. Then, for the adiabatic process

$$T_b = T_a\left(\frac{V_a}{V_b}\right)^{\gamma-1} = 293\text{ K}(10.0)^{0.400} = 736\text{ K}.$$

Thus, the work done on the gas during the adiabatic process is

$$W_{ab} = (-Q + \Delta E_{\text{int}})_{ab} = (-0 + nC_V\Delta T)_{ab} = nC_V(T_b - T_a)$$

or $W_{ab} = 5.00\text{ mol}(20.8\text{ J/mol}\cdot\text{K})(736 - 293)\text{ K} = \boxed{46.0\text{ kJ}}$.

(c) For the isothermal process, we have $P_{b'}V_{b'} = P_aV_a$. Thus,

$P_{b'} = P_a\left(\frac{V_{a'}}{V_{b'}}\right) = 1.00\text{ atm}(10.0) = \boxed{10.0\text{ atm}}$. For the adiabatic process, we have $P_bV_b^{\gamma} = P_aV_a^{\gamma}$.

Thus, $P_b = P_a\left(\frac{V_a}{V_b}\right)^{\gamma} = 1.00\text{ atm}(10.0)^{1.40} = \boxed{25.1\text{ atm}}$.

P17.45 We suppose the air plus burnt gasoline behaves like a diatomic ideal gas. We find its final absolute pressure:

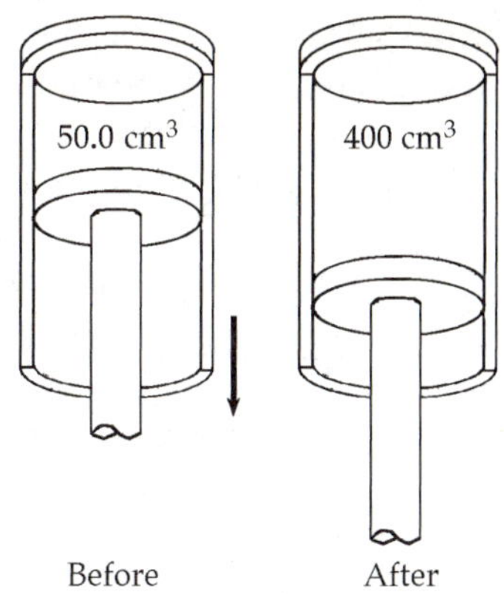

FIG. P17.45

$$21.0 \text{ atm}\left(50.0 \text{ cm}^3\right)^{7/5} = P_f\left(400 \text{ cm}^3\right)^{7/5}$$

$$P_f = 21.0 \text{ atm}\left(\frac{1}{8}\right)^{7/5} = 1.14 \text{ atm}$$

Now $Q = 0$

and $W = \Delta E_{\text{int}} = nC_V\left(T_f - T_i\right)$

$$\therefore W = \frac{5}{2}nRT_f - \frac{5}{2}nRT_i = \frac{5}{2}\left(P_fV_f - P_iV_i\right)$$

$$W = \frac{5}{2}\left[1.14 \text{ atm}\left(400 \text{ cm}^3\right) - 21.0 \text{ atm}\left(50.0 \text{ cm}^3\right)\right]\left(\frac{1.013\times10^5 \text{ N/m}^2}{1 \text{ atm}}\right)\left(10^{-6} \text{ m}^3/\text{cm}^3\right)$$

$$W = -150 \text{ J}$$

The output work is $-W = +150$ J. The time for this stroke is $\frac{1}{4}\left(\frac{1 \text{ min}}{2500}\right)\left(\frac{60 \text{ s}}{1 \text{ min}}\right) = 6.00\times10^{-3}$ s

$$\mathscr{P} = \frac{-W}{\Delta t} = \frac{150 \text{ J}}{6.00\times10^{-3} \text{ s}} = \boxed{25.0 \text{ kW}}$$

Section 17.9 Molar Specific Heats and the Equipartition of Energy

P17.46 (1) $E_{\text{int}} = Nf\left(\frac{k_BT}{2}\right) = f\left(\frac{nRT}{2}\right)$

(2) $C_V = \frac{1}{n}\left(\frac{dE_{\text{int}}}{dT}\right) = \frac{1}{2}fR$

(3) $C_P = C_V + R = \frac{1}{2}(f+2)R$

(4) $\gamma = \frac{C_P}{C_V} = \frac{f+2}{f}$

P17.47 The heat capacity at constant volume is nC_V. An ideal gas of diatomic molecules has three degrees of freedom for translation in the x, y, and z directions. If we take the y axis along the axis of a molecule, then outside forces cannot excite rotation about this axis, since they have no lever arms. Collisions will set the molecule spinning only about the x and z axes.

(a) If the molecules do not vibrate, they have five degrees of freedom. Random collisions put equal amounts of energy $\frac{1}{2}k_BT$ into all five kinds of motion. The average energy of one molecule is $\frac{5}{2}k_BT$. The internal energy of the two-mole sample is

$$N\left(\frac{5}{2}k_BT\right) = nN_A\left(\frac{5}{2}k_BT\right) = n\left(\frac{5}{2}R\right)T = nC_VT.$$

The molar heat capacity is $C_V = \frac{5}{2}R$ and the sample's heat capacity is

$$nC_V = n\left(\frac{5}{2}R\right) = 2\text{ mol}\left(\frac{5}{2}(8.314\text{ J/mol}\cdot\text{K})\right)$$

$$\boxed{nC_V = 41.6\text{ J/K}}$$

For the heat capacity at constant pressure we have

$$nC_P = n(C_V + R) = n\left(\frac{5}{2}R + R\right) = \frac{7}{2}nR = 2\text{ mol}\left(\frac{7}{2}(8.314\text{ J/mol}\cdot\text{K})\right)$$

$$\boxed{nC_P = 58.2\text{ J/K}}$$

(b) In vibration with the center of mass fixed, both atoms are always moving in opposite directions with equal speeds. Vibration adds two more degrees of freedom for two more terms in the molecular energy, for kinetic and for elastic potential energy. We have $nC_V = n\left(\frac{7}{2}R\right) = \boxed{58.2\text{ J/K}}$ and $nC_P = n\left(\frac{9}{2}R\right) = \boxed{74.8\text{ J/K}}$.

Section 17.10 Energy Transfer Mechanisms in Thermal Processes

P17.48 $\mathcal{P} = kA\frac{\Delta T}{L}$

$$k = \frac{\mathcal{P}L}{A\Delta T} = \frac{10.0\text{ W}(0.0400\text{ m})}{1.20\text{ m}^2(15.0°\text{C})} = \boxed{2.22\times10^{-2}\text{ W/m}\cdot°\text{C}}$$

P17.49 In the steady state condition, $\mathcal{P}_{\text{Au}} = \mathcal{P}_{\text{Ag}}$ so that

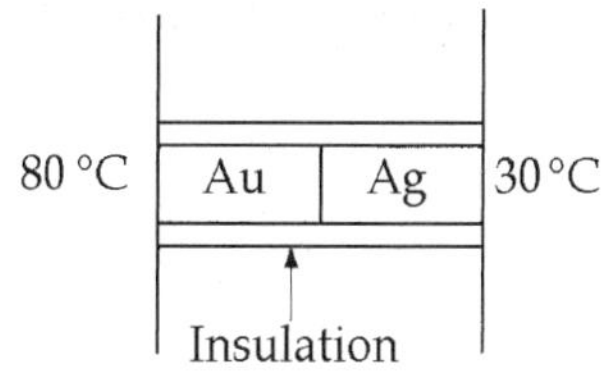

FIG. P17.49

$$k_{\text{Au}} A_{\text{Au}} \left(\frac{\Delta T}{\Delta x} \right)_{\text{Au}} = k_{\text{Ag}} A_{\text{Ag}} \left(\frac{\Delta T}{\Delta x} \right)_{\text{Ag}}$$

In this case

$$\begin{aligned} A_{\text{Au}} &= A_{\text{Ag}} \\ \Delta x_{\text{Au}} &= \Delta x_{\text{Ag}} \\ \Delta T_{\text{Au}} &= (80.0 - T) \\ \Delta T_{\text{Ag}} &= (T - 30.0) \end{aligned}$$

where T is the temperature of the junction. Therefore, $k_{\text{Au}}(80.0 - T) = k_{\text{Ag}}(T - 30.0)$. And $\boxed{T = 51.2°\text{C}}$.

P17.50 We suppose that the area of the transistor is so small that energy flow by heat from the transistor directly to the air is negligible compared to energy conduction through the mica.

$$\mathcal{P} = kA \frac{(T_h - T_c)}{L}$$

$$T_h = T_c + \frac{\mathcal{P} L}{kA} = 35.0°\text{C} + \frac{1.50 \text{ W}\left(0.0852 \times 10^{-3} \text{ m}\right)}{(0.0753 \text{ W/m}\cdot°\text{C})(8.25 \times 6.25)\ 10^{-6} \text{ m}^2} = \boxed{67.9°\text{C}}$$

P17.51 $\mathcal{P} = \sigma A e T^4 = \left(5.67 \times 10^{-8} \text{ W/m}^2 \cdot \text{K}^4\right)\left[4\pi\left(6.96 \times 10^8 \text{ m}\right)^2\right](0.986)(5\ 800 \text{ K})^4$

$\mathcal{P} = \boxed{3.85 \times 10^{26} \text{ W}}$

P17.52 $\mathcal{P} = kA \dfrac{(T_h - T_c)}{L} = (0.210 \text{ W/m}\cdot°\text{C})\left(1.40 \text{ m}^2\right)\left(\dfrac{37.0°\text{C} - 34.0°\text{C}}{0.0250 \text{ m}}\right)$

$= 35.3 \text{ W} = (35.3 \text{ J/s})\left(\dfrac{1 \text{ kcal}}{4186 \text{ J}}\right)\left(\dfrac{3600 \text{ s}}{1 \text{ h}}\right) = \boxed{30.3 \text{ kcal/h}}$

Since this is much less than 240 kcal/h, blood flow is essential to cool the body.

P17.53 The net rate of energy loss from his skin is

$$\mathcal{P}_{\text{net}} = \sigma A e\left(T^4 - T_0^{\ 4}\right) = \left(5.67 \times 10^{-8} \text{ W/m}^2 \cdot \text{K}^4\right)\left(1.50 \text{ m}^2\right)(0.900)\left[(308 \text{ K})^4 - (293 \text{ K})^4\right] = 125 \text{ W}.$$

Note that the temperatures must be in kelvins. The energy loss in ten minutes is

$$Q = \mathcal{P}_{\text{net}} \Delta t = (125 \text{ J/s})(600 \text{ s}) = \boxed{7.48 \times 10^4 \text{ J}}.$$

In the infrared, the person shines brighter than a hundred-watt light bulb.

Section 17.11 Context Connection—Energy Balance for the Earth

P17.54 We suppose the earth below is an insulator. The square meter must radiate in the infrared as much energy as it absorbs, $\mathcal{P} = \sigma AeT^4$. Assuming that $e = 1.00$ for blackbody blacktop:

$$1000 \text{ W} = \left(5.67\times10^{-8} \text{ W/m}^2\cdot\text{K}^4\right)\left(1.00 \text{ m}^2\right)(1.00)T^4$$

$T = \left(1.76\times10^{10} \text{ K}^4\right)^{1/4} = \boxed{364 \text{ K}}$ (You can cook an egg on it.)

P17.55 The sphere of radius R absorbs sunlight over the area of its day hemisphere, projected as a flat circle perpendicular to the light: πR^2. It radiates in all directions, over area $4\pi R^2$. Then, in steady state,

$$\mathcal{P}_{\text{in}} = \mathcal{P}_{\text{out}}$$
$$e\left(1370 \text{ W/m}^2\right)\pi R^2 = e\sigma\left(4\pi R^2\right)T^4$$

The emissivity e, the radius R, and π all cancel. Therefore,

$$T = \left[\frac{1370 \text{ W/m}^2}{4\left(5.67\times10^{-8} \text{ W/m}^2\cdot\text{K}^4\right)}\right]^{1/4} = \boxed{279 \text{ K}} = 6°\text{C}.$$

P17.56 (a) Among atmospheric layers, $PV^\gamma = \text{constant}$ and $\frac{PV}{T} = \text{constant}$. Then $\left(P^{1/\gamma}\right)V = \text{constant}$ and $\frac{PV}{T\left(P^{1/\gamma}\right)V} = \text{constant}$. Then $T\left(P^{(1/\gamma)-1}\right) = T\left(P^{(1-\gamma)/\gamma}\right) = \text{constant}$.

(b) We differentiate the equation $T\left(P^{(1/\gamma)-1}\right) = \text{constant}$ with respect to y:

$$T\left(\frac{1}{\gamma}-1\right)\left(P^{(1/\gamma)-1-1}\right)\frac{dP}{dy} + \left(P^{(1/\gamma)-1}\right)\frac{dT}{dy} = 0$$

$$\frac{T}{P}\left(\frac{1}{\gamma}-1\right)\frac{dP}{dy} + \frac{dT}{dy} = 0$$

$$\frac{dT}{dy} = \frac{T}{P}\left(1-\frac{1}{\gamma}\right)\frac{dP}{dy}$$

(c) Substituting $\frac{dP}{dy} = -\rho g$ gives $\frac{dT}{dy} = \frac{T}{P}\left(1-\frac{1}{\gamma}\right)(-\rho g)$. Now $\rho = \frac{m_{\text{sample}}}{V} = \frac{nM}{V}$. So

$$\frac{dT}{dy} = -\left(1-\frac{1}{\gamma}\right)\frac{TnMg}{PV} = -\left(1-\frac{1}{\gamma}\right)\frac{Mg}{R}$$

Additional Problems

***P17.57** (a) The heat leaving the box during the day is given by $\mathcal{P} = kA\frac{(T_H - T_c)}{L} = \frac{Q}{\Delta t}$

$$Q = 0.012\frac{\text{W}}{\text{m°C}}0.49\ \text{m}^2\frac{37°\text{C} - 23°\text{C}}{0.045\ \text{m}}12\ \text{h}\left(\frac{3\ 600\ \text{s}}{1\ \text{h}}\right) = 7.90 \times 10^4\ \text{J}.$$

The heat lost at night is

$$Q = 0.012\frac{\text{W}}{\text{m°C}}0.49\ \text{m}^2\frac{37°\text{C} - 16°\text{C}}{0.045\ \text{m}}12\ \text{h}\left(\frac{3\ 600\ \text{s}}{1\ \text{h}}\right) = 1.19 \times 10^5\ \text{J}.$$

The total heat is $1.19 \times 10^5\ \text{J} + 7.90 \times 10^4\ \text{J} = 1.98 \times 10^5\ \text{J}$. It must be supplied by the solidifying wax: $Q = mL$

$$m = \frac{Q}{L} = \frac{1.98 \times 10^5\ \text{J}}{205 \times 10^3\ \text{J/kg}} = \boxed{0.964\ \text{kg or more}}$$

(b) The test samples and the inner surface of the insulation can be preheated to 37°C as the box is filled. If this is done, nothing changes in temperature during the test period. Then the masses of the test samples and the insulation make no difference.

P17.58 The increase in internal energy required to melt 1.00 kg of snow is

$$\Delta E_{\text{int}} = (1.00\ \text{kg})(3.33 \times 10^5\ \text{J/kg}) = 3.33 \times 10^5\ \text{J}.$$

The force of friction is

$$f = \mu n = \mu mg = 0.200(75.0\ \text{kg})(9.80\ \text{m/s}^2) = 147\ \text{N}.$$

According to the problem statement, the loss of mechanical energy of the skier is assumed to be equal to the increase in internal energy of the snow. This increase in internal energy is

$$\Delta E_{\text{int}} = f\Delta r = (147\ \text{N})\Delta r = 3.33 \times 10^5\ \text{J}$$

and

$$\Delta r = \boxed{2.27 \times 10^3\ \text{m}}.$$

P17.59 (a) The energy thus far gained by the copper equals the energy loss by the silver. Your down parka is an excellent insulator.

$$Q_{\text{cold}} = -Q_{\text{hot}}$$

or $$m_{\text{Cu}}c_{\text{Cu}}(T_f - T_i)_{\text{Cu}} = -m_{\text{Ag}}c_{\text{Ag}}(T_f - T_i)_{\text{Ag}}$$

$$(9.00\ \text{g})(387\ \text{J/kg}\cdot°\text{C})(16.0°\text{C}) = -(14.0\ \text{g})(234\ \text{J/kg}\cdot°\text{C})(T_f - 30.0°\text{C})_{\text{Ag}}$$

$$(T_f - 30.0°\text{C})_{\text{Ag}} = -17.0°\text{C}$$

so $$T_{f,\text{Ag}} = \boxed{13.0°\text{C}}.$$

continued on next page

(b) Differentiating the energy gain-and-loss equation gives: $m_{Ag}c_{Ag}\left(\frac{dT}{dt}\right)_{Ag} = -m_{Cu}c_{Cu}\left(\frac{dT}{dt}\right)_{Cu}$

$$\left(\frac{dT}{dt}\right)_{Ag} = -\frac{m_{Cu}c_{Cu}}{m_{Ag}c_{Ag}}\left(\frac{dT}{dt}\right)_{Cu} = -\frac{9.00\text{ g}(387\text{ J/kg}\cdot°\text{C})}{14.0\text{ g}(234\text{ J/kg}\cdot°\text{C})}(+0.500°\text{C/s})$$

$$\left(\frac{dT}{dt}\right)_{Ag} = \boxed{-0.532°\text{C/s}} \text{ (negative sign} \Rightarrow \text{decreasing temperature)}$$

P17.60 (a) Before conduction has time to become important, the energy lost by the rod equals the energy gained by the helium. Therefore,

$$(mL_v)_{He} = (mc|\Delta T|)_{Al}$$

or $$(\rho VL_v)_{He} = (\rho Vc|\Delta T|)_{Al}$$

so $$V_{He} = \frac{(\rho Vc|\Delta T|)_{Al}}{(\rho L_v)_{He}}$$

$$V_{He} = \frac{(2.70\text{ g/cm}^3)(62.5\text{ cm}^3)(0.210\text{ cal/g}\cdot°\text{C})(295.8°\text{C})}{(0.125\text{ g/cm}^3)(2.09\times10^4\text{ J/kg})(1.00\text{ cal/4.186 J})(1.00\text{ kg/1 000 g})}$$

$$V_{He} = 1.68\times10^4\text{ cm}^3 = \boxed{16.8\text{ liters}}$$

(b) The rate at which energy is supplied to the rod in order to maintain constant temperatures is given by

$$\mathcal{P} = kA\left(\frac{dT}{dx}\right) = (31.0\text{ J/s}\cdot\text{cm}\cdot\text{K})(2.50\text{ cm}^2)\left(\frac{295.8\text{ K}}{25.0\text{ cm}}\right) = 917\text{ W}$$

This power supplied to the helium will produce a "boil-off" rate of

$$\frac{\mathcal{P}}{\rho L_v} = \frac{(917\text{ W})(10^3\text{ g/kg})}{(0.125\text{ g/cm}^3)(2.09\times10^4\text{ J/kg})} = 351\text{ cm}^3/\text{s} = \boxed{0.351\text{ L/s}}$$

P17.61 $Q = mc\Delta T = (\rho V)c\Delta T$ so that when a constant temperature difference ΔT is maintained,

the rate of adding energy to the liquid is $\mathcal{P} = \frac{dQ}{dt} = \rho\left(\frac{dV}{dt}\right)c\Delta T = \rho Rc\Delta T$

and the specific heat of the liquid is $c = \boxed{\frac{\mathcal{P}}{\rho R\Delta T}}$.

P17.62 (a) Work done by the gas is the negative of the area under the PV curve

$$W = -P_i\left(\frac{V_i}{2} - V_i\right) = \boxed{+\frac{P_iV_i}{2}}.$$

(b) In this case the area under the curve is $W = -\int P\,dV$. Since the process is isothermal,

$$PV = P_iV_i = 4P_i\left(\frac{V_i}{4}\right) = nRT_i$$

and $$W = -\int_{V_i}^{V_i/4}\left(\frac{dV}{V}\right)(P_iV_i) = -P_iV_i\ln\left(\frac{V_i/4}{V_i}\right) = P_iV_i\ln 4$$

$$= \boxed{+1.39P_iV_i}$$

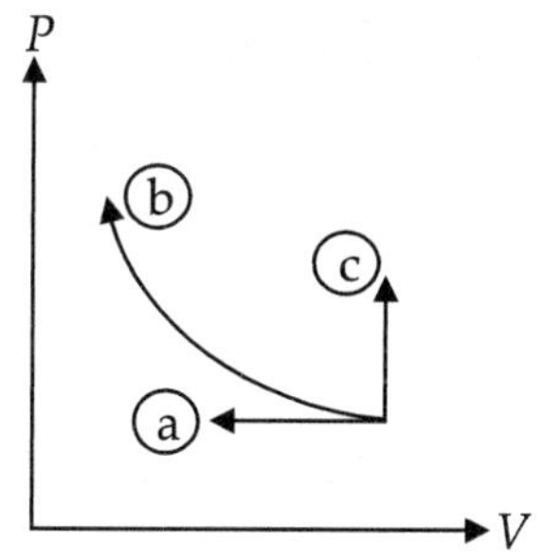

FIG. P17.62

(c) The area under the curve is 0 and $\boxed{W = 0}$.

***P17.63** The initial moment of inertia of the disk is

$$\frac{1}{2}MR^2 = \frac{1}{2}\rho VR^2 = \frac{1}{2}\rho\pi R^2 tR^2 = \frac{1}{2}\left(8\,920\ \text{kg/m}^3\right)\pi(28\ \text{m})^4 1.2\ \text{m} = 1.033\times10^{10}\ \text{kg}\cdot\text{m}^2$$

The rotation speeds up as the disk cools off, according to

$$I_i\omega_i = I_f\omega_f$$

$$\frac{1}{2}MR_i^2\omega_i = \frac{1}{2}MR_f^2\omega_f = \frac{1}{2}MR_i^2(1-\alpha|\Delta T|)^2\omega_f$$

$$\omega_f = \omega_i\frac{1}{(1-\alpha|\Delta T|)^2} = 25\ \text{rad/s}\frac{1}{\left[1-\left(17\times10^{-6}\ 1/°\text{C}\right)830°\text{C}\right]^2} = 25.720\,7\ \text{rad/s}$$

(a) The kinetic energy increases by

$$\frac{1}{2}I_f\omega_f^2 - \frac{1}{2}I_i\omega_i^2 = \frac{1}{2}I_i\omega_i\omega_f - \frac{1}{2}I_i\omega_i^2 = \frac{1}{2}I_i\omega_i(\omega_f - \omega_i)$$

$$= \frac{1}{2}1.033\times10^{10}\ \text{kg}\cdot\text{m}^2(25\ \text{rad/s})0.720\,7\ \text{rad/s} = \boxed{9.31\times10^{10}\ \text{J}}$$

(b) $\Delta E_{\text{int}} = mc\Delta T = 2.64\times10^7\ \text{kg}(387\ \text{J/kg}\cdot°\text{C})(20°\text{C} - 850°\text{C}) = \boxed{-8.47\times10^{12}\ \text{J}}$

(c) As 8.47×10^{12} J leaves the fund of internal energy, 9.31×10^{10} J changes into extra kinetic energy, and the rest, $\boxed{8.38\times10^{12}\ \text{J}}$ is radiated.

***P17.64** The loss of mechanical energy is

$$\frac{1}{2}mv_i^2+\frac{GM_Em}{R_E}=\frac{1}{2}670\text{ kg}\left(1.4\times10^4\text{ m/s}\right)^2+\frac{6.67\times10^{-11}\text{ Nm}^2\,5.98\times10^{24}\text{ kg }670\text{ kg}}{\text{kg}^2\,6.37\times10^6\text{ m}}$$
$$=6.57\times10^{10}\text{ J}+4.20\times10^{10}\text{ J}=1.08\times10^{11}\text{ J}$$

One half becomes extra internal energy in the aluminum: $\Delta E_{\text{int}}=5.38\times10^{10}$ J. To raise the temperature of the aluminum to the melting point requires energy

$$mc\Delta T=670\text{ kg }900\frac{\text{J}}{\text{kg°C}}\left(660-(-15°\text{C})\right)=4.07\times10^8\text{ J}.$$

To melt it, $mL=670\text{ kg }3.97\times10^5\text{ J/kg}=2.66\times10^8\text{ J}$. To raise it to the boiling point, $mc\Delta T=670(1\,170)(2\,450-660)\text{J}=1.40\times10^9\text{J}$. To boil it, $mL=670\text{ kg }1.14\times10^7\text{ J/kg}=7.64\times10^9\text{J}$. Then

$$5.38\times10^{10}\text{ J}=9.71\times10^9\text{ J}+670(1\,170)\left(T_f-2\,450°\text{C}\right)\text{J/°C}$$
$$T_f=\boxed{5.87\times10^4\,°\text{C}}$$

P17.65 The power incident on the solar collector is

$$\mathcal{P}_i=IA=\left(600\text{ W/m}^2\right)\left[\pi(0.300\text{ m})^2\right]=170\text{ W}.$$

For a 40.0% reflector, the collected power is $\mathcal{P}_c=67.9$ W. The total energy required to increase the temperature of the water to the boiling point and to evaporate it is $Q=cm\Delta T+mL_V$:

$$Q=0.500\text{ kg}\left[(4\,186\text{ J/kg·°C})(80.0°\text{C})+2.26\times10^6\text{ J/kg}\right]=1.30\times10^6\text{ J}.$$

The time interval required is $\Delta t=\frac{Q}{\mathcal{P}_c}=\frac{1.30\times10^6\text{ J}}{67.9\text{ W}}=\boxed{5.31\text{ h}}$.

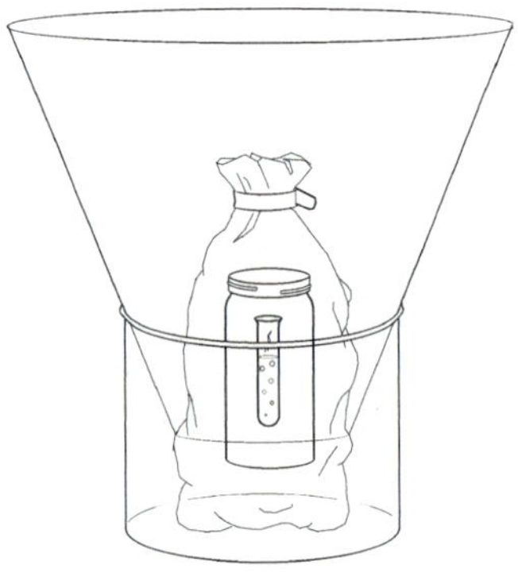

FIG. P17.65

P17.66 (a) $Fv=(50.0\text{ N})(40.0\text{ m/s})=\boxed{2\,000\text{ W}}$

(b) Energy received by each object is $(1\,000\text{ W})(10\text{ s})=10^4\text{ J}=2\,389\text{ cal}$. The specific heat of iron is 0.107 cal/g·°C, so the heat capacity of each object is $5.00\times10^3\times0.107=535.0\text{ cal/°C}$.

$$\Delta T=\frac{2\,389\text{ cal}}{535.0\text{ cal/°C}}=\boxed{4.47°\text{C}}$$

P17.67 (a) The block starts with $K_i = \frac{1}{2}mv_i^2 = \frac{1}{2}(1.60\text{ kg})(2.50\text{ m/s})^2 = 5.00\text{ J}$

All this becomes extra internal energy in ice, melting some according to "Q" $= m_{\text{ice}}L_f$. Thus, the mass of ice that melts is

$$m_{\text{ice}} = \frac{\text{"}Q\text{"}}{L_f} = \frac{K_i}{L_f} = \frac{5.00\text{ J}}{3.33\times10^5\text{ J/kg}} = 1.50\times10^{-5}\text{ kg} = \boxed{15.0\text{ mg}}.$$

For the block: $Q = 0$ (no energy flows by heat since there is no temperature difference)

$W = -5.00\text{ J}$

$\Delta E_{\text{int}} = 0$ (no temperature change)

and $\Delta K = -5.00\text{ J}$

For the ice, $Q = 0$

$W = +5.00\text{ J}$

$\Delta E_{\text{int}} = +5.00\text{ J}$

and $\Delta K = 0$

(b) Again, $K_i = 5.00\text{ J}$ and $m_{\text{ice}} = \boxed{15.0\text{ mg}}$

For the block of ice: $Q = 0$; $\Delta E_{\text{int}} = +5.00\text{ J}$; $\Delta K = -5.00\text{ J}$

so $W = 0$.

For the copper, nothing happens: $Q = \Delta E_{\text{int}} = \Delta K = W = 0$.

(c) Again, $K_i = 5.00\text{ J}$. Both blocks must rise equally in temperature.

"Q" $= mc\Delta T$:

$$\Delta T = \frac{\text{"}Q\text{"}}{mc} = \frac{5.00\text{ J}}{2(1.60\text{ kg})(387\text{ J/kg}\cdot{}^\circ\text{C})} = \boxed{4.04\times10^{-3}\,{}^\circ\text{C}}$$

At any instant, the two blocks are at the same temperature, so for both $Q = 0$.

For the moving block: $\Delta K = -5.00\text{ J}$

and $\Delta E_{\text{int}} = +2.50\text{ J}$

so $W = -2.50\text{ J}$

For the stationary block: $\Delta K = 0$

and $\Delta E_{\text{int}} = +2.50\text{ J}$

so $W = +2.50\text{ J}$

For each object in each situation, the general continuity equation for energy, in the form $\Delta K + \Delta E_{\text{int}} = W + Q$, correctly describes the relationship between energy transfers and changes in the object's energy content.

P17.68 $$\frac{L\rho A dx}{dt}=kA\left(\frac{\Delta T}{x}\right)$$

$$L\rho\int_{4.00}^{8.00}x\,dx=k\Delta T\int_{0}^{\Delta t}dt$$

$$L\rho\left.\frac{x^2}{2}\right|_{4.00}^{8.00}=k\Delta T\Delta t$$

$$\left(3.33\times10^5\ \text{J/kg}\right)\left(917\ \text{kg/m}^3\right)\left(\frac{(0.080\,0\ \text{m})^2-(0.040\,0\ \text{m})^2}{2}\right)=(2.00\ \text{W/m}\cdot{}^\circ\text{C})(10.0^\circ\text{C})\Delta t$$

$$\Delta t=3.66\times10^4\ \text{s}=\boxed{10.2\ \text{h}}$$

P17.69 $$A=A_{\text{end walls}}+A_{\text{ends of attic}}+A_{\text{side walls}}+A_{\text{roof}}$$

$$A=2(8.00\ \text{m}\times5.00\ \text{m})+2\left[2\times\frac{1}{2}\times4.00\ \text{m}\times(4.00\ \text{m})\tan37.0^\circ\right]$$
$$+2(10.0\ \text{m}\times5.00\ \text{m})+2(10.0\ \text{m})\left(\frac{4.00\ \text{m}}{\cos37.0^\circ}\right)$$

$$A=304\ \text{m}^2$$

$$\mathscr{P}=\frac{kA\Delta T}{L}=\frac{\left(4.80\times10^{-4}\ \text{kW/m}\cdot{}^\circ\text{C}\right)\left(304\ \text{m}^2\right)(25.0^\circ\text{C})}{0.210\ \text{m}}=17.4\ \text{kW}=4.15\ \text{kcal/s}$$

Thus, the energy lost per day by heat is $(4.15\ \text{kcal/s})(86\,400\ \text{s})=3.59\times10^5\ \text{kcal/day}$.

The gas needed to replace this loss is $\dfrac{3.59\times10^5\ \text{kcal/day}}{9\,300\ \text{kcal/m}^3}=\boxed{38.6\ \text{m}^3/\text{day}}$.

P17.70 $$Q_{\text{cold}}=-Q_{\text{hot}}$$

or $$Q_{\text{Al}}=-(Q_{\text{water}}+Q_{\text{calo}})$$

$$m_{\text{Al}}c_{\text{Al}}\left(T_f-T_i\right)_{\text{Al}}=-(m_wc_w+m_cc_c)\left(T_f-T_i\right)_w$$

$$(0.200\ \text{kg})c_{\text{Al}}(+39.3^\circ\text{C})=-\left[0.400\ \text{kg}(4\,186\ \text{J/kg}\cdot{}^\circ\text{C})+0.040\,0\ \text{kg}(630\ \text{J/kg}\cdot{}^\circ\text{C})\right](-3.70^\circ\text{C})$$

$$c_{\text{Al}}=\frac{6.29\times10^3\ \text{J}}{7.86\ \text{kg}\cdot{}^\circ\text{C}}=\boxed{800\ \text{J/kg}\cdot{}^\circ\text{C}}$$

P17.71 (a) $$P_f=\boxed{100\ \text{kPa}},\ T_f=\boxed{400\ \text{K}}$$

$$V_f=\frac{nRT_f}{P_f}=\frac{2.00\ \text{mol}(8.314\ \text{J/mol}\cdot\text{K})(400\ \text{K})}{100\times10^3\ \text{Pa}}=0.0665\ \text{m}^3=\boxed{66.5\ \text{L}}$$

$$\Delta E_{\text{int}}=(3.50)nR\Delta T=3.50(2.00\ \text{mol})(8.314\ \text{J/mol}\cdot\text{K})(100\ \text{K})=\boxed{5.82\ \text{kJ}}$$

$$W=-P\Delta V=-nR\Delta T=-(2.00\ \text{mol})(8.314\ \text{J/mol}\cdot\text{K})(100\ \text{K})=\boxed{-1.66\ \text{kJ}}$$

$$Q=\Delta E_{\text{int}}-W=5.82\ \text{kJ}+1.66\ \text{kJ}=\boxed{7.84\ \text{kJ}}$$

continued on next page

(b) $T_f = \boxed{400\text{ K}}$, $V_f = V_i = \dfrac{nRT_i}{P_i} = \dfrac{2.00\text{ mol}(8.314\text{ J/mol}\cdot\text{K})(300\text{ K})}{100\times 10^3\text{ Pa}} = 0.0499\text{ m}^3 = \boxed{49.9\text{ L}}$

$$P_f = P_i\left(\frac{T_f}{T_i}\right) = 100\text{ kPa}\left(\frac{400\text{ K}}{300\text{ K}}\right) = \boxed{133\text{ kPa}},\ W = -\int P\,dV = \boxed{0}\ \text{since } V = \text{constant}$$

$$\Delta E_{\text{int}} = \boxed{5.82\text{ kJ}}\ \text{as in part (a)}\ Q = \Delta E_{\text{int}} - W = 5.82\text{ kJ} - 0 = \boxed{5.82\text{ kJ}}$$

(c) $P_f = \boxed{120\text{ kPa}}$, $T_f = \boxed{300\text{ K}}$, $V_f = V_i\left(\dfrac{P_i}{P_f}\right) = 49.9\text{ L}\left(\dfrac{100\text{ kPa}}{120\text{ kPa}}\right) = \boxed{41.6\text{ L}}$,

$$\Delta E_{\text{int}} = (3.50)nR\Delta T = \boxed{0}\ \text{since } T = \text{constant}$$

$$W = -\int P\,dV = -nRT_i\int_{V_i}^{V_f}\frac{dV}{V} = -nRT_i\ln\left(\frac{V_f}{V_i}\right) = -nRT_i\ln\left(\frac{P_i}{P_f}\right)$$

$$W = -(2.00\text{ mol})(8.314\text{ J/mol}\cdot\text{K})(300\text{ K})\ln\left(\frac{100\text{ kPa}}{120\text{ kPa}}\right) = \boxed{+909\text{ J}}$$

$$Q = \Delta E_{\text{int}} - W = 0 - 909\text{ J} = \boxed{-909\text{ J}}$$

(d) $P_f = \boxed{120\text{ kPa}}$, $\gamma = \dfrac{C_P}{C_V} = \dfrac{C_V + R}{C_V} = \dfrac{3.50R + R}{3.50R} = \dfrac{4.50}{3.50} = \dfrac{9}{7}$

$$P_fV_f^{\gamma} = P_iV_i^{\gamma}:\ \text{so } V_f = V_i\left(\frac{P_i}{P_f}\right)^{1/\gamma} = 49.9\text{ L}\left(\frac{100\text{ kPa}}{120\text{ kPa}}\right)^{7/9} = \boxed{43.3\text{ L}}$$

$$T_f = T_i\left(\frac{P_fV_f}{P_iV_i}\right) = 300\text{ K}\left(\frac{120\text{ kPa}}{100\text{ kPa}}\right)\left(\frac{43.3\text{ L}}{49.9\text{ L}}\right) = \boxed{312\text{ K}}$$

$$\Delta E_{\text{int}} = (3.50)nR\Delta T = 3.50(2.00\text{ mol})(8.314\text{ J/mol}\cdot\text{K})(12.4\text{ K}) = \boxed{722\text{ J}}$$

$$Q = \boxed{0}\ \text{(adiabatic process)}$$

$$W = -Q + \Delta E_{\text{int}} = 0 + 722\text{ J} = \boxed{+722\text{ J}}$$

***P17.72** (a) $PV^{\gamma} = k$. So, $W = -\displaystyle\int_i^f P\,dV = -k\int_i^f \frac{dV}{V^{\gamma}} = \frac{P_fV_f - P_iV_i}{\gamma - 1}$

(b) $dE_{\text{int}} = dQ + dW$ and $dQ = 0$ for an adiabatic process.

Therefore, $W = +\Delta E_{\text{int}} = nC_V\left(T_f - T_i\right)$.

To show consistency between these two equations, consider that $\gamma = \dfrac{C_p}{C_V}$ and $C_p - C_V = R$.

Therefore, $\dfrac{1}{\gamma - 1} = \dfrac{C_V}{R}$.

Using this, result found on part (a) becomes

$$W = \left(P_fV_f - P_iV_i\right)\frac{C_V}{R}.$$

Also, for an ideal gas $\dfrac{PV}{R} = nT$ so that $W = nC_V\left(T_f - T_i\right)$.

***P17.73** (a) $W = nC_V(T_f - T_i)$

$$-2\ 500\ \text{J} = 1\ \text{mol}\frac{3}{2}8.314\ \text{J/mol}\cdot\text{K}(T_f - 500\ \text{K})$$

$$T_f = \boxed{300\ \text{K}}$$

(b) $P_iV_i^\gamma = P_fV_f^\gamma$

$$P_i\left(\frac{nRT_i}{P_i}\right)^\gamma = P_f\left(\frac{nRT_f}{P_f}\right)^\gamma \qquad T_i^\gamma P_i^{1-\gamma} = T_f^\gamma P_f^{1-\gamma}$$

$$\frac{T_i^{\gamma/(\gamma-1)}}{P_i} = \frac{T_f^{\gamma/(\gamma-1)}}{P_f} \qquad P_f = P_i\left(\frac{T_f}{T_i}\right)^{\gamma/(\gamma-1)}$$

$$P_f = P_i\left(\frac{T_f}{T_i}\right)^{(5/3)(3/2)} = 3.60\ \text{atm}\left(\frac{300}{500}\right)^{5/2} = \boxed{1.00\ \text{atm}}$$

P17.74 The ball loses energy $\frac{1}{2}mv_i^2 - \frac{1}{2}mv_f^2 = \frac{1}{2}(0.142\ \text{kg})\left[(47.2)^2 - (42.5)^2\right]\ \text{m}^2/\text{s}^2 = 29.9\ \text{J}$.

The air volume is $V = \pi(0.037\ 0\ \text{m})^2(19.4\ \text{m}) = 0.083\ 4\ \text{m}^3$

and its quantity is $n = \frac{PV}{RT} = \frac{1.013\times10^5\ \text{Pa}(0.083\ 4\ \text{m}^3)}{(8.314\ \text{J/mol}\cdot\text{K})(293\ \text{K})} = 3.47\ \text{mol}$.

The air absorbs energy according to $Q = nC_p\Delta T$. So

$$\Delta T = \frac{Q}{nC_P} = \frac{29.9\ \text{J}}{3.47\ \text{mol}\,\frac{7}{2}(8.314\ \text{J/mol}\cdot\text{K})} = \boxed{0.296°\text{C}}.$$

P17.75 (a) $n = \frac{PV}{RT} = \frac{(1.013\times10^5\ \text{Pa})(5.00\times10^{-3}\ \text{m}^3)}{(8.314\ \text{J/mol}\cdot\text{K})(300\ \text{K})} = \boxed{0.203\ \text{mol}}$

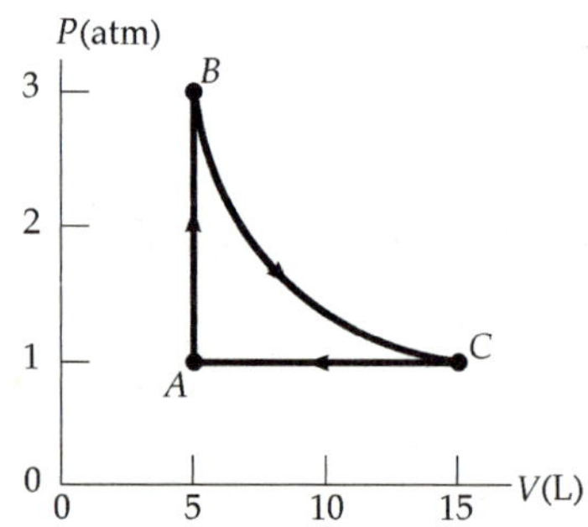

FIG. P17.75

(b) $T_B = T_A\left(\frac{P_B}{P_A}\right) = 300\ \text{K}\left(\frac{3.00}{1.00}\right) = \boxed{900\ \text{K}}$

$$T_C = T_B = \boxed{900\ \text{K}}$$

$$V_C = V_A\left(\frac{T_C}{T_A}\right) = 5.00\ \text{L}\left(\frac{900}{300}\right) = \boxed{15.0\ \text{L}}$$

(c) $E_{\text{int},A} = \frac{3}{2}nRT_A = \frac{3}{2}(0.203\ \text{mol})(8.314\ \text{J/mol}\cdot\text{K})(300\ \text{K})$

$$= \boxed{760\ \text{J}}$$

$$E_{\text{int},B} = E_{\text{int},C} = \frac{3}{2}nRT_B = \frac{3}{2}(0.203\ \text{mol})(8.314\ \text{J/mol}\cdot\text{K})(900\ \text{K}) = \boxed{2.28\ \text{kJ}}$$

continued on next page

(d)

	P (atm)	V(L)	T(K)	E_{int} (kJ)
A	1.00	5.00	300	0.760
B	3.00	5.00	900	2.28
C	1.00	15.00	900	2.28

(e) For the process AB, lock the piston in place and put the cylinder into an oven at 900 K. For BC, keep the sample in the oven while gradually letting the gas expand to lift a gradually decreasing load on the piston as far as it can. For CA, carry the cylinder back into the room at 300 K and let the gas cool without touching the piston yourself.

(f) For AB: $W = \boxed{0}$, $\Delta E_{int} = E_{int,B} - E_{int,A} = (2.28 - 0.760)\text{ kJ} = \boxed{1.52\text{ kJ}}$

$$Q = \Delta E_{int} - W = \boxed{1.52\text{ kJ}}$$

For BC: $\Delta E_{int} = \boxed{0}$, $W = -nRT_B \ln\left(\frac{V_C}{V_B}\right)$

$$W = -(0.203\text{ mol})(8.314\text{ J/mol}\cdot\text{K})(900\text{ K})\ln(3.00) = \boxed{-1.67\text{ kJ}}$$

$$Q = \Delta E_{int} - W = \boxed{1.67\text{ kJ}}$$

For CA: $\Delta E_{int} = E_{int,A} - E_{int,C} = (0.760 - 2.28)\text{ kJ} = \boxed{-1.52\text{ kJ}}$

$$W = -P\Delta V = -nR\Delta T = -(0.203\text{ mol})(8.314\text{ J/mol}\cdot\text{K})(-600\text{ K}) = \boxed{1.01\text{ kJ}}$$

$$Q = \Delta E_{int} - W = -1.52\text{ kJ} - 1.01\text{ kJ} = \boxed{-2.53\text{ kJ}}$$

(g) We add the amounts of energy for each process to find them for the whole cycle.

$$Q_{ABCA} = +1.52\text{ kJ} + 1.67\text{ kJ} - 2.53\text{ kJ} = \boxed{0.656\text{ kJ}}$$

$$W_{ABCA} = 0 - 1.67\text{ kJ} + 1.01\text{ kJ} = \boxed{-0.656\text{ kJ}}$$

$$(\Delta E_{int})_{ABCA} = +1.52\text{ kJ} + 0 - 1.52\text{ kJ} = \boxed{0}$$

***P17.76** (a) The chemical energy input becomes partly work output and partly internal energy. The energy flow each second is described by

$$400\text{ kcal/h} = 60\text{ J/s} + \frac{mL}{\Delta t} = 400\text{ kcal/h}\left(\frac{4\,186\text{ J}}{1\text{ kcal}}\right)\left(\frac{1\text{ h}}{3\,600\text{ s}}\right) = 465\text{ W}$$

$$\frac{m}{\Delta t}L = 465\text{ W} - 60\text{ W} = 405\text{ J/s}$$

$$\frac{m}{\Delta t} = \frac{405\text{ J/s}}{2.26\times10^6\text{ J/kg}}\left(\frac{3\,600\text{ s}}{1\text{ h}}\right) = \boxed{0.645\text{ kg/h}}$$

(b) The rate of fat burning is ideally $\frac{400\text{ kcal/h}}{9\text{ kcal/g}} = 0.044\,4\text{ kg/h}$. The 0.044 4 kg/h of water produced by metabolism is this fraction of the water needed for cooling:
$\frac{0.0\,444\text{ kg/h}}{0.645\text{ kg/h}} = \boxed{0.068\,9} = 6.89\%$. Moral: drink plenty of fruit juice while you exercise.

ANSWERS TO EVEN PROBLEMS

P17.2 $0.105°\text{C}$

P17.4 1.78×10^4 kg

P17.6 88.2 W

P17.8 (a) $0.435\ \text{cal/g}\cdot°\text{C}$; (b) beryllium

P17.10 $\dfrac{(m_{\text{Al}}c_{\text{Al}} + m_c c_w)T_c + m_h c_w T_h}{m_{\text{Al}}c_{\text{Al}} + m_c c_w + m_h c_w}$

P17.12 12.9 g steam

P17.14 0.414 kg

P17.16 3.40×10^4 m

P17.18 11.1 W

P17.20 (a) −12.0 MJ; (b) +12.0 MJ

P17.22 $-nR(T_2 - T_1)$

P17.24 (a) 12.0 kJ; (b) −12.0 kJ

P17.26 (a) $0.007\,65\ \text{m}^3$; (b) 305 K

P17.28 (a) −48.6 mJ; (b) 16.2 kJ; (c) 16.2 kJ

P17.30 (a) $-4P_iV_i$; (b) $4P_iV_i$; (c) −9.08 kJ

P17.32 (a) 1 300 J; (b) 100 J; (c) −900 J; (d) −1 400 J

P17.34 74.8 J

P17.36 (a) $0.719\ \text{kJ/kg}\cdot\text{K}$; (b) 0.811 kg; (c) 233 kJ; (d) 327 kJ

P17.38 (a) 5.66×10^7 J; (b) 1.12 kg

P17.40 (a) 0.118; (b) 2.35; (c) 0, 135 J, +135 J

P17.42 227 K

P17.44 (a) 28.0 kJ; (b) 46.0 kJ; (c) 10.0 atm, 25.1 atm

P17.46 see the solution

P17.48 $2.22 \times 10^{-2}\ \text{W/m}\cdot°\text{C}$

P17.50 67.9°C

P17.52 30.3 kcal/h

P17.54 364 K

P17.56 (a) see the solution; (b) see the solution; (c) see the solution

P17.58 2.27×10^3 m

P17.60 (a) 16.8 liters; (b) 0.351 L/s

P17.62 (a) $+\dfrac{P_iV_i}{2}$; (b) $+1.39P_iV_i$; (c) 0

P17.64 $5.87 \times 10^4\,°\text{C}$

P17.66 (a) 2 000 W; (b) 4.47°C

P17.68 10.2 h

P17.70 $800\ \text{J/kg}\cdot°\text{C}$

P17.72 (a) see the solution; (b) see the solution

P17.74 0.296°C

P17.76 (a) 0.645 kg/h; (b) 0.068 9

Chapter 18

Heat Engines, Entropy, and the Second Law of Thermodynamics

CHAPTER OUTLINE

ANSWERS TO QUESTIONS

Q18.1 First, the efficiency of the automobile engine cannot exceed the Carnot efficiency: it is limited by the temperature of burning fuel and the temperature of the environment into which the exhaust is dumped. Second, the engine block cannot be allowed to go over a certain temperature. Third, any practical engine has friction, incomplete burning of fuel, and limits set by timing and energy transfer by heat.

Q18.2 No. Any heat engine takes in energy by heat and must also put out energy by heat. The energy that is dumped as exhaust into the low-temperature sink will always be thermal pollution in the outside environment. So-called "steady growth" in human energy use cannot continue.

Q18.3 A higher steam temperature means that more energy can be extracted from the steam. For a constant temperature heat sink at T_c, and steam at T_h, the efficiency of the power plant is determined by $\frac{T_h - T_c}{T_h} = 1 - \frac{T_c}{T_h}$ and is maximized for a high T_h.

Q18.4 No. The first law of thermodynamics is a statement about energy conservation, while the second is a statement about stable thermal equilibrium. They are by no means mutually exclusive. For the particular case of a cycling heat engine, the first law implies $|Q_h| = W_{eng} + |Q_c|$, and the second law implies $|Q_c| > 0$.

Q18.5 Take an automobile as an example. According to the first law or the idea of energy conservation, it must take in all the energy it puts out. Its energy source is chemical energy in gasoline. During the combustion process, some of that energy goes into moving the pistons and eventually into the mechanical motion of the car. Clearly much of the energy goes into heat, which, through the cooling system, is transferred into the atmosphere. Moreover, there are numerous places where friction, both mechanical and fluid, turns mechanical energy into heat. In even the most efficient internal combustion engine cars, less than 30% of the energy from the fuel actually goes into moving the car. The rest ends up as useless heat put into the atmosphere.

Q18.6 Suppose the ambient temperature is 20°C. A gas can be heated to the temperature of the bottom of the pond, and allowed to cool as it blows through a turbine. The Carnot efficiency of such an engine is about $e_c = \frac{\Delta T}{T_h} = \frac{80}{373} = 22\%$.

Q18.7 No, because the work done to run the heat pump represents energy transferred into the house by heat.

Q18.8 A slice of hot pizza cools off. Road friction brings a skidding car to a stop. A cup falls to the floor and shatters. Your cat dies. Any process is irreversible if it looks funny or frightening when shown in a videotape running backwards. The free flight of a projectile is nearly reversible.

Q18.9 (a) When the two sides of the semiconductor are at different temperatures, an electric potential (voltage) is generated across the material, which can drive electric current through an external circuit. The two cups at 50°C contain the same amount of internal energy as the pair of hot and cold cups. But no energy flows by heat through the converter bridging between them and no voltage is generated across the semiconductors.

(b) A heat engine must put out exhaust energy by heat. The cold cup provides a sink to absorb output or wasted energy by heat, which has nowhere to go between two cups of equally warm water.

Q18.10 The rest of the Universe must have an entropy change of +8.0 J/K, or more.

Q18.11 (a) For an expanding ideal gas at constant temperature, $\Delta S = \frac{\Delta Q}{T} = nR\ln\left(\frac{V_2}{V_1}\right)$.

(b) For a reversible adiabatic expansion $\Delta Q = 0$, and $\Delta S = 0$. An ideal gas undergoing an irreversible adiabatic expansion can have any positive value for ΔS up to the value given in part (a).

Q18.12 To increase its entropy, raise its temperature. To decrease its entropy, lower its temperature. "Remove energy from it by heat" is not such a good answer, for if you hammer on it or rub it with a blunt file and at the same time remove energy from it by heat into a constant temperature bath, its entropy can stay constant.

Q18.13 An analogy used by Carnot is instructive: A waterfall continuously converts mechanical energy into internal energy. It continuously creates entropy as the organized motion of the falling water turns into disorganized molecular motion. We humans put turbines into the waterfall, diverting some of the energy stream to our use. Water flows spontaneously from high to low elevation and energy spontaneously flows by heat from high to low temperature. Into the great flow of solar radiation from Sun to Earth, living things put themselves. They live on energy flow, more than just on energy. A basking snake diverts energy from a high-temperature source (the Sun) through itself temporarily, before the energy inevitably is radiated from the body of the snake to a low-temperature sink (outer space). A tree builds organized cellulose molecules and we build libraries and babies who look like their grandmothers, all out of a thin diverted stream in the universal flow of energy crashing down to disorder. We do not violate the second law, for we build local reductions in the entropy of one thing within the inexorable increase in the total entropy of the Universe. Your roommate's exercise puts energy into the room by heat.

Q18.14 Either statement can be considered an instructive analogy. We choose to take the first view. All processes require energy, either as energy content or as energy input. The kinetic energy that it possessed at its formation continues to make the Earth go around. Energy released by nuclear reactions in the core of the Sun drives weather on the Earth and essentially all processes in the biosphere. The energy intensity of sunlight controls how lush a forest or jungle can be and how warm a planet is. Continuous energy input is not required for the motion of the planet. Continuous energy input is required for life because energy tends to be continuously degraded, as heat flows into lower-temperature sinks. The continuously increasing entropy of the Universe is the index to energy-transfers completed.

Q18.15 Shaking opens up spaces between jellybeans. The smaller ones more often can fall down into spaces below them. The accumulation of larger candies on top and smaller ones on the bottom implies a small increase in order, a small decrease in one contribution to the total entropy, but the second law is not violated. The total entropy increases as the system warms up, its increase in internal energy coming from the work put into shaking the box and also from a bit of gravitational energy loss as the beans settle compactly together.

SOLUTIONS TO PROBLEMS

Section 18.1 Heat Engines and the Second Law of Thermodynamics

P18.1 (a) $e=\frac{W_{\text{eng}}}{|Q_h|}=\frac{25.0\text{ J}}{360\text{ J}}=\boxed{0.069\,4}$ or $\boxed{6.94\%}$

(b) $|Q_c|=|Q_h|-W_{\text{eng}}=360\text{ J}-25.0\text{ J}=\boxed{335\text{ J}}$

***P18.2** (a) The input energy each hour is

$$\left(7.89\times10^3\text{ J/revolution}\right)\left(2\,500\text{ rev/min}\right)\frac{60\text{ min}}{1\text{ h}}=1.18\times10^9\text{ J/h}$$

implying fuel input $\left(1.18\times10^9\text{ J/h}\right)\left(\frac{1\text{ L}}{4.03\times10^7\text{ J}}\right)=\boxed{29.4\text{ L/h}}$

(b) $Q_h=W_{\text{eng}}+|Q_c|$. For a continuous-transfer process we may divide by time to have

$$\frac{Q_h}{\Delta t}=\frac{W_{\text{eng}}}{\Delta t}+\frac{|Q_c|}{\Delta t}$$

Useful power output $=\frac{W_{\text{eng}}}{\Delta t}=\frac{Q_h}{\Delta t}-\frac{|Q_c|}{\Delta t}$

$$=\left(\frac{7.89\times10^3\text{ J}}{\text{revolution}}-\frac{4.58\times10^3\text{ J}}{\text{revolution}}\right)\frac{2\,500\text{ rev}}{1\text{ min}}\frac{1\text{ min}}{60\text{ s}}=1.38\times10^5\text{ W}$$

$$\mathcal{P}_{\text{eng}}=1.38\times10^5\text{ W}\left(\frac{1\text{ hp}}{746\text{ W}}\right)=\boxed{185\text{ hp}}$$

(c) $\mathcal{P}_{\text{eng}}=\tau\omega\Rightarrow\tau=\frac{\mathcal{P}_{\text{eng}}}{\omega}=\frac{1.38\times10^5\text{ J/s}}{(2\,500\text{ rev/60 s})}\left(\frac{1\text{ rev}}{2\pi\text{ rad}}\right)=\boxed{527\text{ N}\cdot\text{m}}$

(d) $\frac{|Q_c|}{\Delta t}=\frac{4.58\times10^3\text{ J}}{\text{revolution}}\left(\frac{2\,500\text{ rev}}{60\text{ s}}\right)=\boxed{1.91\times10^5\text{ W}}$

P18.3 (a) We have $e=\frac{W_{\text{eng}}}{|Q_h|}=\frac{|Q_h|-|Q_c|}{|Q_h|}=1-\frac{|Q_c|}{|Q_h|}=0.250$

with $|Q_c|=8\,000$ J, we have $|Q_h|=\boxed{10.7\text{ kJ}}$

(b) $W_{\text{eng}}=|Q_h|-|Q_c|=2\,667$ J

and from $\mathcal{P}=\frac{W_{\text{eng}}}{\Delta t}$, we have $\Delta t=\frac{W_{\text{eng}}}{\mathcal{P}}=\frac{2\,667\text{ J}}{5\,000\text{ J/s}}=\boxed{0.533\text{ s}}$.

***P18.4** The engine's output work we identify with the kinetic energy of the bullet:

$$W_{\text{eng}}=K=\frac{1}{2}mv^2=\frac{1}{2}0.002\;4\text{ kg}(320\text{ m/s})^2=123\text{ J}$$

$$e=\frac{W_{\text{eng}}}{Q_h}$$

$$Q_h=\frac{W_{\text{eng}}}{e}=\frac{123\text{ J}}{0.011}=1.12\times10^4\text{ J}$$

$$Q_h=W_{\text{eng}}+|Q_c|$$

The energy exhaust is

$$|Q_c|=Q_h-W_{\text{eng}}=1.12\times10^4\text{ J}-123\text{ J}=1.10\times10^4\text{ J}$$

$$Q=mc\Delta T$$

$$\Delta T=\frac{Q}{mc}=\frac{1.10\times10^4\text{ J kg°C}}{1.80\text{ kg }448\text{ J}}=\boxed{13.7\text{°C}}$$

Section 18.2 Reversible and Irreversible Processes

Section 18.3 The Carnot Engine

P18.5 $T_c=703$ K $\quad T_h=2\,143$ K

(a) $e_c=\frac{\Delta T}{T_h}=\frac{1\,440}{2\,143}=\boxed{67.2\%}$

(b) $|Q_h|=1.40\times10^5$ J, $W_{\text{eng}}=0.420|Q_h|$

$$\mathcal{P}=\frac{W_{\text{eng}}}{\Delta t}=\frac{5.88\times10^4\text{ J}}{1\text{ s}}=\boxed{58.8\text{ kW}}$$

P18.6 When $e = e_c$, $$1 - \frac{T_c}{T_h} = \frac{W_{\text{eng}}}{|Q_h|} \text{ and } \frac{\frac{W_{\text{eng}}}{\Delta t}}{\frac{|Q_h|}{\Delta t}} = 1 - \frac{T_c}{T_h}$$

(a) $$|Q_h| = \frac{\left(\frac{W_{\text{eng}}}{\Delta t}\right)\Delta t}{1 - \frac{T_c}{T_h}} = \frac{(1.50 \times 10^5 \text{ W})(3\,600 \text{ s})}{1 - \frac{293}{773}}$$

$$|Q_h| = 8.70 \times 10^8 \text{ J} = \boxed{870 \text{ MJ}}$$

(b) $$|Q_c| = |Q_h| - \left(\frac{W_{\text{eng}}}{\Delta t}\right)\Delta t = 8.70 \times 10^8 - (1.50 \times 10^5)(3\,600) = 3.30 \times 10^8 \text{ J} = \boxed{330 \text{ MJ}}$$

P18.7 Isothermal expansion at $T_h = 523 \text{ K}$

Isothermal compression at $T_c = 323 \text{ K}$

Gas absorbs 1 200 J during expansion.

(a) $$|Q_c| = |Q_h|\left(\frac{T_c}{T_h}\right) = 1\,200 \text{ J}\left(\frac{323}{523}\right) = \boxed{741 \text{ J}}$$

(b) $$W_{\text{eng}} = |Q_h| - |Q_c| = (1\,200 - 741) \text{ J} = \boxed{459 \text{ J}}$$

P18.8 We use $$e_c = 1 - \frac{T_c}{T_h}$$

as $$0.300 = 1 - \frac{573 \text{ K}}{T_h}$$

From which, $T_h = 819 \text{ K} = \boxed{546°\text{C}}$

P18.9 The Carnot summer efficiency is $$e_{c,s} = 1 - \frac{T_c}{T_h} = 1 - \frac{(273 + 20) \text{ K}}{(273 + 350) \text{ K}} = 0.530$$

And in winter, $$e_{c,w} = 1 - \frac{283}{623} = 0.546$$

Then the actual winter efficiency is $$0.320\left(\frac{0.546}{0.530}\right) = \boxed{0.330} \text{ or } \boxed{33.0\%}$$

P18.10 (a) $e_{\max} = 1 - \frac{T_c}{T_h} = 1 - \frac{278}{293} = 5.12 \times 10^{-2} = \boxed{5.12\%}$

(b) $\mathcal{P} = \frac{W_{\text{eng}}}{\Delta t} = 75.0 \times 10^6 \text{ J/s}$

Therefore, $W_{\text{eng}} = (75.0 \times 10^6 \text{ J/s})(3\,600 \text{ s/h}) = 2.70 \times 10^{11} \text{ J/h}$

From $e = \frac{W_{\text{eng}}}{|Q_h|}$ we find $|Q_h| = \frac{W_{\text{eng}}}{e} = \frac{2.70 \times 10^{11} \text{ J/h}}{5.12 \times 10^{-2}} = 5.27 \times 10^{12} \text{ J/h} = \boxed{5.27 \text{ TJ/h}}$

(c) As fossil-fuel prices rise, this way to use solar energy will become a good buy.

***P18.11** (a) $e = \frac{W_{\text{eng1}} + W_{\text{eng2}}}{Q_{h1}} = \frac{e_1 Q_{1h} + e_2 Q_{2h}}{Q_{h1}}$

Now $Q_{2h} = Q_{1c} = Q_{1h} - W_{\text{eng1}} = Q_{h1} - e_1 Q_{1h}$.

So $e = \frac{e_1 Q_{1h} + e_2 (Q_{1h} - e_1 Q_{1h})}{Q_{1h}} = \boxed{e_1 + e_2 - e_1 e_2}$.

(b) $e = e_1 + e_2 - e_1 e_2 = 1 - \frac{T_i}{T_h} + 1 - \frac{T_c}{T_i} - \left(1 - \frac{T_i}{T_h}\right)\left(1 - \frac{T_c}{T_i}\right) = 2 - \frac{T_i}{T_h} - \frac{T_c}{T_i} - 1 + \frac{T_i}{T_h} + \frac{T_c}{T_i} - \frac{T_c}{T_h} = \boxed{1 - \frac{T_c}{T_h}}$

The combination of reversible engines is itself a reversible engine so it has the Carnot efficiency.

(c) With $W_{\text{eng2}} = W_{\text{eng1}}$, $e = \frac{W_{\text{eng1}} + W_{\text{eng2}}}{Q_{1h}} = \frac{2W_{\text{eng1}}}{Q_{1h}} = 2e_1$

$$1 - \frac{T_c}{T_h} = 2\left(1 - \frac{T_i}{T_h}\right)$$

$$0 - \frac{T_c}{T_h} = 1 - \frac{2T_i}{T_h}$$

$$2T_i = T_h + T_c$$

$$\boxed{T_i = \frac{1}{2}(T_h + T_c)}$$

(d) $e_1 = e_2 = 1 - \frac{T_i}{T_h} = 1 - \frac{T_c}{T_i}$

$$T_i^2 = T_c T_h$$

$$\boxed{T_i = (T_h T_c)^{1/2}}$$

P18.12 (a) First, consider the adiabatic process $D \to A$:

$P_D V_D^\gamma = P_A V_A^\gamma$ so $$P_D = P_A\left(\frac{V_A}{V_D}\right)^\gamma = 1\,400\text{ kPa}\left(\frac{10.0\text{ L}}{15.0\text{ L}}\right)^{5/3} = \boxed{712\text{ kPa}}.$$

Also $$\left(\frac{nRT_D}{V_D}\right)V_D^\gamma = \left(\frac{nRT_A}{V_A}\right)V_A^\gamma$$

or $$T_D = T_A\left(\frac{V_A}{V_D}\right)^{\gamma-1} = 720\text{ K}\left(\frac{10.0}{15.0}\right)^{2/3} = \boxed{549\text{ K}}.$$

Now, consider the isothermal process $C \to D$: $T_C = T_D = \boxed{549\text{ K}}$.

$$P_C = P_D\left(\frac{V_D}{V_C}\right) = \left[P_A\left(\frac{V_A}{V_D}\right)^\gamma\right]\left(\frac{V_D}{V_C}\right) = \frac{P_A V_A^\gamma}{V_C V_D^{\gamma-1}}$$

$$P_C = \frac{1\,400\text{ kPa}(10.0\text{ L})^{5/3}}{24.0\text{ L}(15.0\text{ L})^{2/3}} = \boxed{445\text{ kPa}}$$

Next, consider the adiabatic process $B \to C$: $P_B V_B^\gamma = P_C V_C^\gamma$.

But, $P_C = \dfrac{P_A V_A^\gamma}{V_C V_D^{\gamma-1}}$ from above. Also considering the isothermal process, $P_B = P_A\left(\dfrac{V_A}{V_B}\right)$.

Hence, $P_A\left(\dfrac{V_A}{V_B}\right)V_B^\gamma = \left(\dfrac{P_A V_A^\gamma}{V_C V_D^{\gamma-1}}\right)V_C^\gamma$ which reduces to $V_B = \dfrac{V_A V_C}{V_D} = \dfrac{10.0\text{ L}(24.0\text{ L})}{15.0\text{ L}} = \boxed{16.0\text{ L}}$.

Finally, $P_B = P_A\left(\dfrac{V_A}{V_B}\right) = 1\,400\text{ kPa}\left(\dfrac{10.0\text{ L}}{16.0\text{ L}}\right) = \boxed{875\text{ kPa}}$.

State	P(kPa)	V(L)	T(K)
A	1 400	10.0	720
B	875	16.0	720
C	445	24.0	549
D	712	15.0	549

(b) For the isothermal process $A \to B$: $\Delta E_{\text{int}} = nC_V\Delta T = \boxed{0}$

so $Q = -W = nRT\ln\left(\dfrac{V_B}{V_A}\right) = 2.34\text{ mol}(8.314\text{ J/mol}\cdot\text{K})(720\text{ K})\ln\left(\dfrac{16.0}{10.0}\right) = \boxed{+6.58\text{ kJ}}$.

For the adiabatic process $B \to C$: $Q = \boxed{0}$

$$\Delta E_{\text{int}} = nC_V(T_C - T_B) = 2.34\text{ mol}\left[\frac{3}{2}(8.314\text{ J/mol}\cdot\text{K})\right](549 - 720)\text{ K} = \boxed{-4.98\text{ kJ}}$$

and $W = -Q + \Delta E_{\text{int}} = 0 + (-4.98\text{ kJ}) = \boxed{-4.98\text{ kJ}}$.

continued on next page

For the isothermal process $C \to D$: $\Delta E_{int} = nC_V \Delta T = \boxed{0}$

and $Q = -W = nRT\ln\left(\frac{V_D}{V_C}\right) = 2.34 \text{ mol}(8.314 \text{ J/mol}\cdot\text{K})(549 \text{ K})\ln\left(\frac{15.0}{24.0}\right) = \boxed{-5.02 \text{ kJ}}$.

Finally, for the adiabatic process $D \to A$: $Q = \boxed{0}$

$$\Delta E_{int} = nC_V(T_A - T_D) = 2.34 \text{ mol}\left[\frac{3}{2}(8.314 \text{ J/mol}\cdot\text{K})\right](720 - 549) \text{ K} = \boxed{+4.98 \text{ kJ}}$$

and $W = -Q + \Delta E_{int} = 0 + 4.98 \text{ kJ} = \boxed{+4.98 \text{ kJ}}$.

Process	Q(kJ)	W(kJ)	ΔE_{int} (kJ)
$A \to B$	+6.58	–6.58	0
$B \to C$	0	–4.98	–4.98
$C \to D$	–5.02	+5.02	0
$D \to A$	0	+4.98	+4.98
ABCDA	+1.56	–1.56	0

The work done *by* the engine is the negative of the work input. The output work W_{eng} is given by the work column in the table with all signs reversed.

(c) $$e = \frac{W_{eng}}{|Q_h|} = \frac{-W_{ABCD}}{Q_{A\to B}} = \frac{1.56 \text{ kJ}}{6.58 \text{ kJ}} = 0.237 \text{ or } \boxed{23.7\%}$$

$$e_c = 1 - \frac{T_c}{T_h} = 1 - \frac{549}{720} = 0.237 \text{ or } \boxed{23.7\%}$$

Section 18.4 Heat Pumps and Refrigerators

***P18.13** $\text{COP(refrigerator)} = \frac{Q_c}{W}$

(a) If $Q_c = 120$ J and $\text{COP} = 5.00$, then $\boxed{W = 24.0 \text{ J}}$

(b) Heat expelled = Heat removed + Work done.

$$Q_h = Q_c + W = 120 \text{ J} + 24 \text{ J} = \boxed{144 \text{ J}}$$

***P18.14** $\text{COP} = 3.00 = \frac{Q_c}{W}$. Therefore, $W = \frac{Q_c}{3.00}$.

The heat removed each minute is

$$\frac{Q_C}{t} = (0.030\,0 \text{ kg})(4\,186 \text{ J/kg}\,°\text{C})(22.0°\text{C}) + (0.030\,0 \text{ kg})(3.33\times10^5 \text{ J/kg}) + (0.030\,0 \text{ kg})(2\,090 \text{ J/kg}\,°\text{C})(20.0°\text{C}) = 1.40\times10^4 \text{ J/min}$$

or, $\frac{Q_c}{t} = 233 \text{ J/s}$.

Thus, the work done per sec $= \mathcal{P} = \frac{233 \text{ J/s}}{3.00} = \boxed{77.8 \text{ W}}$.

P18.15 (a) $\left(10.0\,\frac{\text{Btu}}{\text{h}\cdot\text{W}}\right)\left(\frac{1055 \text{ J}}{1 \text{ Btu}}\right)\left(\frac{1 \text{ h}}{3\,600 \text{ s}}\right)\left(\frac{1 \text{ W}}{1 \text{ J/s}}\right) = \boxed{2.93}$

(b) Coefficient of performance for a refrigerator: $\boxed{(\text{COP})_{\text{refrigerator}}}$

(c) With EER 5, $5\,\frac{\text{Btu}}{\text{h}\cdot\text{W}} = \frac{10\,000 \text{ Btu/h}}{\mathcal{P}}$: $\mathcal{P} = \frac{10\,000 \text{ Btu/h}}{\frac{5 \text{ Btu}}{\text{h}\cdot\text{W}}} = 2\,000 \text{ W} = 2.00 \text{ kW}$

Energy purchased is $\mathcal{P}\Delta t = (2.00 \text{ kW})(1\,500 \text{ h}) = 3.00\times10^3 \text{ kWh}$

$\text{Cost} = (3.00\times10^3 \text{ kWh})(0.100 \text{ \$/kWh}) = \$300$

With EER 10, $10\,\frac{\text{Btu}}{\text{h}\cdot\text{W}} = \frac{10\,000 \text{ Btu/h}}{\mathcal{P}}$: $\mathcal{P} = \frac{10\,000 \text{ Btu/h}}{\frac{10 \text{ Btu}}{\text{h}\cdot\text{W}}} = 1\,000 \text{ W} = 1.00 \text{ kW}$

Energy purchased is $\mathcal{P}\Delta t = (1.00 \text{ kW})(1\,500 \text{ h}) = 1.50\times10^3 \text{ kWh}$

$\text{Cost} = (1.50\times10^3 \text{ kWh})(0.100 \text{ \$/kWh}) = \$150$

Thus, the cost for air conditioning is $\boxed{\text{half as much with EER 10}}$

P18.16 $(\text{COP})_{\text{refrig}} = \frac{T_c}{\Delta T} = \frac{270}{30.0} = \boxed{9.00}$

P18.17 (a) For a complete cycle, $\Delta E_{\text{int}} = 0$ and $W = |Q_h| - |Q_c| = |Q_c|\left[\frac{|(Q_h)|}{|Q_c|} - 1\right]$.

We have already shown that for a Carnot cycle (and only for a Carnot cycle) $\frac{|Q_h|}{|Q_c|} = \frac{T_h}{T_c}$.

Therefore, $W = \boxed{|Q_c|\left[\frac{T_h - T_c}{T_c}\right]}$.

continued on next page

(b) We have the definition of the coefficient of performance for a refrigerator, $\text{COP} = \frac{|Q_c|}{W}$.

Using the result from part (a), this becomes $\text{COP} = \boxed{\frac{T_c}{T_h - T_c}}$.

P18.18 $(\text{COP})_{\text{heat pump}} = \frac{|Q_c| + W}{W} = \frac{T_h}{\Delta T} = \frac{295}{25} = \boxed{11.8}$

P18.19 $\text{COP} = 0.100\,\text{COP}_{\text{Carnot cycle}}$

or $\frac{|Q_h|}{W} = 0.100\left(\frac{|Q_h|}{W}\right)_{\text{Carnot cycle}} = 0.100\left(\frac{1}{\text{Carnot efficiency}}\right)$

$\frac{|Q_h|}{W} = 0.100\left(\frac{T_h}{T_h - T_c}\right) = 0.100\left(\frac{293\text{ K}}{293\text{ K} - 268\text{ K}}\right) = 1.17$

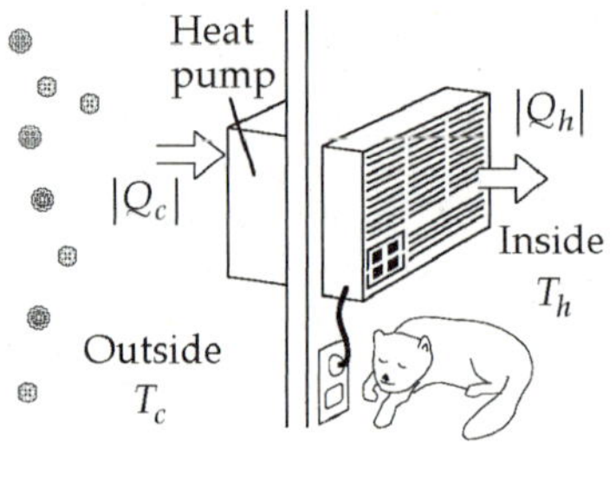

FIG. P18.19

Thus, $\boxed{\text{1.17 joules of energy enter the room by heat for each joule of work done.}}$

***P18.20** $e = \frac{W}{Q_h} = 0.350 \qquad W = 0.350 Q_h$

$Q_h = W + Q_c \qquad Q_c = 0.650 Q_h$

$\text{COP(refrigerator)} = \frac{Q_c}{W} = \frac{0.650 Q_h}{0.350 Q_h} = \boxed{1.86}$

P18.21 $(\text{COP})_{\text{Carnot refrig}} = \frac{T_c}{\Delta T} = \frac{4.00}{289} = 0.013\,8 = \frac{|Q_c|}{W}$

$\therefore W = \boxed{72.2\text{ J}}$ per 1 J energy removed by heat.

***P18.22** (a) The TV set uses energy at the rate $400\text{ J/s}\left(\frac{\$0.15}{\text{kWh}}\right)\left(\frac{1\text{ kWh}}{3.6\times10^{6}\text{ J}}\right)\left(\frac{60\text{ s}}{1\text{ min}}\right) = \boxed{\$1.00\times10^{-3}/\text{min}}$.

(b) An air conditioner is a refrigerator:

$$COP = \frac{|Q_c|}{W} = \frac{\frac{|Q_c|}{\Delta t}}{\frac{W}{\Delta t}}$$

$$\frac{W}{\Delta t} = \frac{\frac{|Q_c|}{\Delta t}}{COP} = \frac{400\text{ J/s}}{4.50} = 88.9\text{ J/s}$$

continued on next page

This is the rate at which extra electric energy is required by the air conditioner to keep the room cool, with cost

$$88.9\text{W}\left(\frac{\$0.15}{\text{kwh}}\right)\left(\frac{\text{k}}{1\,000}\right)\left(\frac{1\text{ h}}{60\text{ min}}\right)=\boxed{\$2.22\times10^{-4}/\text{min}}.$$

(c) Your father blames his extra output of $170\text{ W}-120\text{ W}=50\text{ W}$ on the TV set. It must be pumped out of the room by the air conditioner, with power expenditure

$$\frac{W}{\Delta t}=\frac{|Q_c|\Delta t}{COP}=\frac{50\text{ W}}{4.50}\left(\frac{\$0.15}{\text{kWh}}\right)\left(\frac{\text{k}}{1\,000}\right)\left(\frac{1\text{ h}}{60\text{ min}}\right)=\boxed{\$2.78\times10^{-5}/\text{min}}.$$

Section 18.5 An Alternative Statement of the Second Law

No problems in this section

Section 18.6 Entropy

Section 18.7 Entropy and the Second Law of Thermodynamics

P18.23 $$\Delta S=\int_i^f\frac{dQ}{T}=\int_{T_i}^{T_f}\frac{mc\,dT}{T}=mc\ln\left(\frac{T_f}{T_i}\right)$$

$$\Delta S=250\text{ g}(1.00\text{ cal/g}\cdot{}^\circ\text{C})\ln\left(\frac{353}{293}\right)=46.6\text{ cal/K}=\boxed{195\text{ J/K}}$$

P18.24 For a freezing process,

$$\Delta S=\frac{\Delta Q}{T}=\frac{-(0.500\text{ kg})(3.33\times10^5\text{ J/kg})}{273\text{ K}}=\boxed{-610\text{ J/K}}.$$

***P18.25** (a) The process is $\boxed{\text{isobaric}}$ because it takes place under constant atmospheric pressure. As described by Newton's third law, the stewing syrup must exert the same force on the air as the air exerts on it. The heating process is not adiabatic (energy goes in by heat), isothermal (*T* goes up), isovolumetic (it likely expands a bit), cyclic (it is different at the end), or isentropic (entropy increases). It could be made as nearly reversible as you wish, by not using a kitchen stove but a heater kept always just incrementally higher in temperature than the syrup. The process would then also be eternal, and impractical for food production.

(b) The final temperature is

$$220^\circ\text{F}=212^\circ\text{F}+8^\circ\text{F}=100^\circ\text{C}+8^\circ\text{F}\left(\frac{100-0^\circ\text{C}}{212-32^\circ\text{F}}\right)=104^\circ\text{C}.$$

continued on next page

For the mixture,

$$Q = m_1c_1\Delta T + m_2c_2\Delta T = (900 \text{ g } 1 \text{ cal/g}\cdot°\text{C} + 930 \text{ g } 0.299 \text{ cal/g}\cdot°\text{C})(104.4°\text{C} - 23°\text{C})$$
$$= 9.59\times10^4 \text{ cal} = \boxed{4.02\times10^5 \text{ J}}$$

(c) Consider the reversible heating process described in part (a):

$$\Delta S = \int_i^f \frac{dQ}{T} = \int_i^f \frac{(m_1c_1 + m_2c_2)dT}{T} = (m_1c_1 + m_2c_2)\ln\frac{T_f}{T_i}$$
$$= [900(1) + 930(0.299)](\text{cal}/°\text{C})\left(\frac{4.186 \text{ J}}{1 \text{ cal}}\right)\left(\frac{1°\text{C}}{1 \text{ K}}\right)\ln\left(\frac{273+104}{273+23}\right)$$
$$= (4\,930 \text{ J/K})0.243 = \boxed{1.20\times10^3 \text{ J/K}}$$

P18.26 We take data from Tables 17.1 and 17.2, and we assume a constant specific heat for each phase. As the ice is warmed from –12°C to 0°C, its entropy increases by

$$\Delta S = \int_i^f \frac{dQ}{T} = \int_{261 \text{ K}}^{273 \text{ K}} \frac{mc_{\text{ice}}dT}{T} = mc_{\text{ice}} \int_{261 \text{ K}}^{273 \text{ K}} T^{-1}dT = mc_{\text{ice}} \ln T\Big|_{261 \text{ K}}^{273 \text{ K}}$$

$$\Delta S = 0.027\,9 \text{ kg}(2\,090 \text{ J/kg}\cdot°\text{C})(\ln 273 \text{ K} - \ln 261 \text{ K}) = 0.027\,9 \text{ kg}(2\,090 \text{ J/kg}\cdot°\text{C})\left(\ln\left(\frac{273}{261}\right)\right)$$

$$\Delta S = 2.62 \text{ J/K}$$

As the ice melts its entropy change is

$$\Delta S = \frac{Q}{T} = \frac{mL_f}{T} = \frac{0.027\,9 \text{ kg}(3.33\times10^5 \text{ J/kg})}{273 \text{ K}} = 34.0 \text{ J/K}$$

As liquid water warms from 273 K to 373 K,

$$\Delta S = \int_i^f \frac{mc_{\text{liquid}}dT}{T} = mc_{\text{liquid}} \ln\left(\frac{T_f}{T_i}\right) = 0.027\,9 \text{ kg}(4\,186 \text{ J/kg}\cdot°\text{C})\ln\left(\frac{373}{273}\right) = 36.5 \text{ J/K}$$

As the water boils and the steam warms,

$$\Delta S = \frac{mL_v}{T} + mc_{\text{steam}} \ln\left(\frac{T_f}{T_i}\right)$$

$$\Delta S = \frac{0.027\,9 \text{ kg}(2.26\times10^6 \text{ J/kg})}{373 \text{ K}} + 0.027\,9 \text{ kg}(2\,010 \text{ J/kg}\cdot°\text{C})\ln\left(\frac{388}{373}\right) = 169 \text{ J/K} + 2.21 \text{ J/K}$$

The total entropy change is

$$(2.62 + 34.0 + 36.5 + 169 + 2.21) \text{ J/K} = \boxed{244 \text{ J/K}}.$$

For steam at constant pressure, the molar specific heat in Table 17.3 implies a specific heat of $(35.4 \text{ J/mol}\cdot\text{K})\left(\frac{1 \text{ mol}}{0.018 \text{ kg}}\right) = 1\,970 \text{ J/kg}\cdot\text{K}$, nearly agreeing with $2\,010 \text{ J/kg}\cdot\text{K}$.

P18.27 (a) A 12 can only be obtained $\boxed{\text{one}}$ way: $6+6$

(b) A 7 can be obtained $\boxed{\text{six}}$ ways: $6+1, 5+2, 4+3, 3+4, 2+5, 1+6$

P18.28 (a) The table is shown below. On the basis of the table, the most probable result of a toss is $\boxed{\text{2 heads and 2 tails}}$.

(b) The most ordered state is the least likely state. Thus, on the basis of the table this is $\boxed{\text{either all heads or all tails}}$.

(c) The most disordered is the most likely state. Thus, this is $\boxed{\text{2 heads and 2 tails}}$.

Result	Possible Combinations	Total
All heads	HHHH	1
3H, 1T	THHH, HTHH, HHTH, HHHT	4
2H, 2T	TTHH, THTH, THHT, HTTH, HTHT, HHTT	6
1H, 3T	HTTT, THTT, TTHT, TTTH	4
All tails	TTTT	1

P18.29 (a)

Result	Possible Combinations	Total
All red	RRR	1
2R, 1G	RRG, RGR, GRR	3
1R, 2G	RGG, GRG, GGR	3
All green	GGG	1

(b)

Result	Possible Combinations	Total
All red	RRRRR	1
4R, 1G	RRRRG, RRRGR, RRGRR, RGRRR, GRRRR	5
3R, 2G	RRRGG, RRGRG, RGRRG, GRRRG, RRGGR, RGRGR, GRRGR, RGGRR, GRGRR, GGRRR	10
2R, 3G	GGGRR, GGRGR, GRGGR, RGGGR, GGRRG, GRGRG, RGGRG, GRRGG, RGRGG, RRGGG	10
1R, 4G	RGGGG, GRGGG, GGRGG, GGGRG, GGGGR	5
All green	GGGGG	1

Section 18.8 Entropy Changes in Irreversible Processes

P18.30 $\Delta S = \frac{Q_2}{T_2} - \frac{Q_1}{T_1} = \left(\frac{1\,000}{290} - \frac{1\,000}{5\,700}\right) \text{ J/K} = \boxed{3.27 \text{ J/K}}$

P18.31 The car ends up in the same thermodynamic state as it started, so it undergoes zero changes in entropy. The original kinetic energy of the car is transferred by heat to the surrounding air, adding to the internal energy of the air. Its change in entropy is

$$\Delta S = \frac{\frac{1}{2}mv^2}{T} = \frac{750(20.0)^2}{293} \text{ J/K} = \boxed{1.02 \text{ kJ/K}}.$$

P18.32 $c_{\text{iron}} = 448\ \text{J/kg}\cdot°\text{C};\ c_{\text{water}} = 4\,186\ \text{J/kg}\cdot°\text{C}$

$Q_{\text{cold}} = -Q_{\text{hot}}$: $4.00\ \text{kg}(4\,186\ \text{J/kg}\cdot°\text{C})(T_f - 10.0°\text{C}) = -(1.00\ \text{kg})(448\ \text{J/kg}\cdot°\text{C})(T_f - 900°\text{C})$

which yields $T_f = 33.2°\text{C} = 306.2\ \text{K}$

$$\Delta S = \int_{283\ \text{K}}^{306.2\ \text{K}} \frac{c_{\text{water}} m_{\text{water}}\, dT}{T} + \int_{1\,173\ \text{K}}^{306.2\ \text{K}} \frac{c_{\text{iron}} m_{\text{iron}}\, dT}{T}$$

$$\Delta S = c_{\text{water}} m_{\text{water}} \ln\left(\frac{306.2}{283}\right) + c_{\text{iron}} m_{\text{iron}} \ln\left(\frac{306.2}{1\,173}\right)$$

$$\Delta S = (4\,186\ \text{J/kg}\cdot\text{K})(4.00\ \text{kg})(0.078\,8) + (448\ \text{J/kg}\cdot\text{K})(1.00\ \text{kg})(-1.34)$$

$$\Delta S = \boxed{718\ \text{J/K}}$$

P18.33 Sitting here writing, I convert chemical energy, in ordered molecules in food, into internal energy that leaves my body by heat into the room-temperature surroundings. My rate of energy output is equal to my metabolic rate,

$$2\,500\ \text{kcal/d} = \frac{2\,500 \times 10^3\ \text{cal}}{86\,400\ \text{s}}\left(\frac{4.186\ \text{J}}{1\ \text{cal}}\right) = 120\ \text{W}.$$

My body is in steady state, changing little in entropy, as the environment increases in entropy at the rate

$$\frac{\Delta S}{\Delta t} = \frac{Q/T}{\Delta t} = \frac{Q/\Delta t}{T} = \frac{120\ \text{W}}{293\ \text{K}} = 0.4\ \text{W/K} \sim \boxed{1\ \text{W/K}}.$$

When using powerful appliances or an automobile, my personal contribution to entropy production is much greater than the above estimate, based only on metabolism.

P18.34 $\Delta S = nR\ln\left(\frac{V_f}{V_i}\right) = R\ln 2 = \boxed{5.76\ \text{J/K}}$

There is $\boxed{\text{no change in temperature}}$.

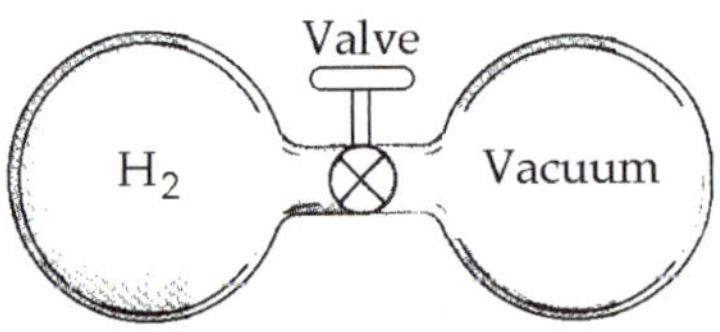

FIG. P18.34

P18.35 $\Delta S = nR\ln\left(\frac{V_f}{V_i}\right) = (0.044\,0)(2)R\ln 2$

$\Delta S = 0.088\,0(8.314)\ln 2 = \boxed{0.507\ \text{J/K}}$

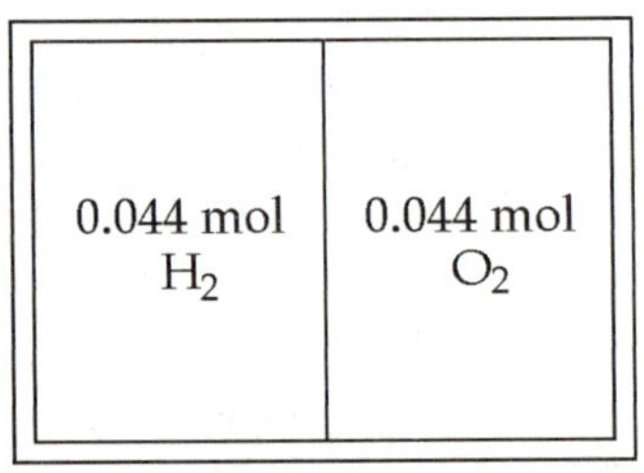

FIG. P18.35

Section 18.9 Context Connection—The Atmosphere as a Heat Engine

P18.36 The Earth presents a circular projected area of $\pi(6.37\times10^6\text{ m})^2$ perpendicular to the direction of energy flow in sunlight. The power incident on the Earth is then

$$\mathcal{P}=IA=(1\ 370\text{ W/m}^2)\left[\pi(6.37\times10^6\text{ m})^2\right]=1.75\times10^{17}\text{ W}.$$

The power absorbed in the atmosphere is $0.64(1.75\times10^{17}\text{ W})=1.1\times10^{17}\text{ W}$. The mechanical wind power is $0.008(1.1\times10^{17}\text{ W})=\boxed{9\times10^{14}\text{ W}}$.

P18.37 (a) The volume of the air is πr^2h and its mass is $m=\rho V=\rho\pi r^2h$. Then its kinetic energy is

$$K=\frac{1}{2}mv^2=\frac{1}{2}\rho\pi r^2hv^2$$

$$K=\frac{1}{2}(1.20\text{ kg/m}^3)\left[\pi(300\ 000\text{ m})^2(11\ 000\text{ m})\right]\left(\frac{60\ 000\text{ m}}{3\ 600\text{ s}}\right)^2$$

$$K=\boxed{5.2\times10^{17}\text{ J}}$$

(b) We can suppose the light shines perpendicularly down onto the hurricane. Its intensity is $I=\frac{\mathcal{P}}{A}=\frac{E}{A\Delta t}$: $\Delta t=\frac{E}{AI}=\frac{5.2\times10^{17}\text{ J}\cdot\text{m}^2\cdot\text{s}}{\pi(3\times10^5\text{ m})^2(10^3\text{ J})}=\boxed{1.8\times10^3\text{ s}}$.

Additional Problems

P18.38 For the Carnot engine, $e_c=1-\frac{T_c}{T_h}=1-\frac{300\text{ K}}{750\text{ K}}=0.600$.

Also, $e_c=\frac{W_{eng}}{|Q_h|}$.

so $|Q_h|=\frac{W_{eng}}{e_c}=\frac{150\text{ J}}{0.600}=250\text{ J}$.

and $|Q_c|=|Q_h|-W_{eng}=250\text{ J}-150\text{ J}=100\text{ J}$.

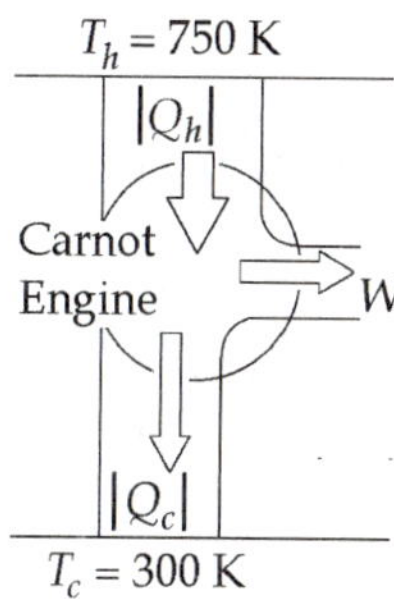

FIG. P18.38

(a) $|Q_h|=\frac{W_{eng}}{e_S}=\frac{150\text{ J}}{0.700}=\boxed{214\text{ J}}$

$|Q_c|=|Q_h|-W_{eng}=214\text{ J}-150\text{ J}=\boxed{64.3\text{ J}}$

(b) $|Q_{h,net}|=214\text{ J}-250\text{ J}=\boxed{-35.7\text{ J}}$

$|Q_{c,net}|=64.3\text{ J}-100\text{ J}=\boxed{-35.7\text{ J}}$

The net flow of energy by heat from the cold to the hot reservoir without work input, is impossible.

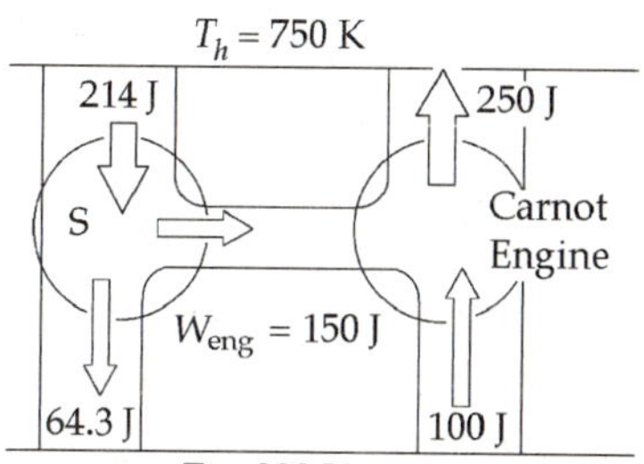

FIG. P18.38(b)

continued on next page

(c) For engine S: $|Q_c| = |Q_h| - W_{\text{eng}} = \dfrac{W_{\text{eng}}}{e_S} - W_{\text{eng}}$.

so $$W_{\text{eng}} = \frac{|Q_c|}{\frac{1}{e_S} - 1} = \frac{100\text{ J}}{\frac{1}{0.700} - 1} = \boxed{233\text{ J}}.$$

and $$|Q_h| = |Q_c| + W_{\text{eng}} = 233\text{ J} + 100\text{ J} = \boxed{333\text{ J}}.$$

(d) $|Q_{h,\text{net}}| = 333\text{ J} - 250\text{ J} = \boxed{83.3\text{ J}}$

$W_{\text{net}} = 233\text{ J} - 150\text{ J} = \boxed{83.3\text{ J}}$

$|Q_{c,\text{net}}| = \boxed{0}$

The output of 83.3 J of energy from the heat engine by work in a cyclic process without any exhaust by heat is impossible.

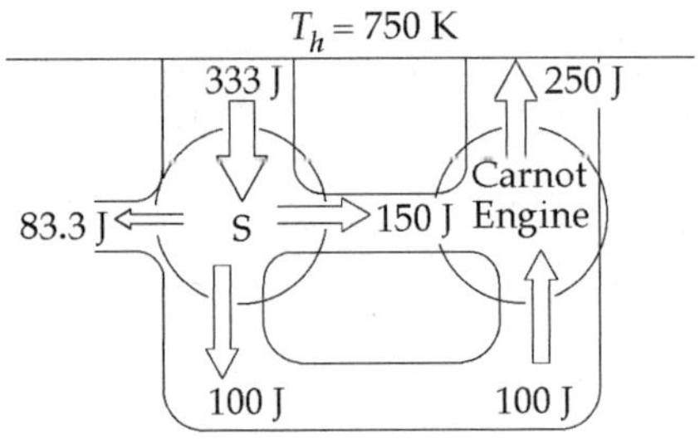

FIG. P18.38(d)

(e) Both engines operate in cycles, so $\Delta S_S = \Delta S_{\text{Carnot}} = 0$.

For the reservoirs, $$\Delta S_h = -\frac{|Q_h|}{T_h} \text{ and } \Delta S_c = +\frac{|Q_c|}{T_c}.$$

Thus, $$\Delta S_{\text{total}} = \Delta S_S + \Delta S_{\text{Carnot}} + \Delta S_h + \Delta S_c = 0 + 0 - \frac{83.3\text{ J}}{750\text{ K}} + \frac{0}{300\text{ K}} = \boxed{-0.111\text{ J/K}}.$$

A decrease in total entropy is impossible.

P18.39 As in Section 6.6, let T_{ET} represent the energy transferred by electrical transmission. Then

(a) $\mathcal{P}_{\text{electric}} = \dfrac{T_{ET}}{\Delta t}$ so if all the electric energy is converted into internal energy, the steady-state condition of the house is described by $T_{ET} = |Q|$.

Therefore, $$\mathcal{P}_{\text{electric}} = \frac{Q}{\Delta t} = \boxed{5\,000\text{ W}}$$

(b) For a heat pump, $$(\text{COP})_{\text{Carnot}} = \frac{T_h}{\Delta T} = \frac{295\text{ K}}{27\text{ K}} = 10.92$$

$$\text{Actual COP} = 0.6(10.92) = 6.55 = \frac{|Q_h|}{W} = \frac{|Q_h|/\Delta t}{W/\Delta t}$$

Therefore, to bring 5 000 W of energy into the house only requires input power

$$\mathcal{P}_{\text{heat pump}} = \frac{W}{\Delta t} = \frac{|Q_h|/\Delta t}{\text{COP}} = \frac{5\,000\text{ W}}{6.56} = \boxed{763\text{ W}}$$

P18.40 The conversion of gravitational potential energy into kinetic energy as the water falls is reversible. But the subsequent conversion into internal energy is not. We imagine arriving at the same final state by adding energy by heat, in amount mgy, to the water from a stove at a temperature infinitesimally above 20.0°C. Then,

$$\Delta S = \int \frac{dQ}{T} = \frac{Q}{T} = \frac{mgy}{T} = \frac{5\,000 \text{ m}^3\left(1\,000 \text{ kg/m}^3\right)\left(9.80 \text{ m/s}^2\right)(50.0 \text{ m})}{293 \text{ K}} = \boxed{8.36\times 10^6 \text{ J/K}}.$$

P18.41 $|Q_c| = mc\Delta T + mL + mc\Delta T =$

$$|Q_c| = 0.500 \text{ kg}(4\,186 \text{ J/kg}\cdot°\text{C})(10°\text{C}) + 0.500 \text{ kg}\left(3.33\times 10^5 \text{ J/kg}\right) + 0.500 \text{ kg}(2\,090 \text{ J/kg}\cdot°\text{C})(20°\text{C})$$

$$|Q_c| = 2.08\times 10^5 \text{ J}$$

$$\frac{|Q_c|}{W} = \text{COP}_c(\text{refrigerator}) = \frac{T_c}{T_h - T_c}$$

$$W = \frac{|Q_c|(T_h - T_c)}{T_c} = \frac{\left(2.08\times 10^5 \text{ J}\right)\left[20.0°\text{C} - (-20.0°\text{C})\right]}{(273 - 20.0) \text{ K}} = \boxed{32.9 \text{ kJ}}$$

***P18.42** (a) Let state i represent the gas before its compression and state f afterwards, $V_f = \frac{V_i}{8}$. For a diatomic ideal gas, $C_v = \frac{5}{2}R$, $C_p = \frac{7}{2}R$, and $\gamma = \frac{C_p}{C_V} = 1.40$. Next,

$$P_iV_i^\gamma = P_fV_f^\gamma$$

$$P_f = P_i\left(\frac{V_i}{V_f}\right)^\gamma = P_i 8^{1.40} = 18.4P_i$$

$$P_iV_i = nRT_i$$

$$P_fV_f = \frac{18.4P_iV_i}{8} = 2.30P_iV_i = 2.30nRT_i = nRT_f$$

so $T_f = 2.30T_i$

$$\Delta E_{\text{int}} = nC_V\Delta T = n\frac{5}{2}R\left(T_f - T_i\right) = \frac{5}{2}nR1.30T_i = \frac{5}{2}1.30P_iV_i$$

$$= \frac{5}{2}1.30\left(1.013\times 10^5 \text{ N/m}^2\right)0.12\times 10^{-3} \text{ m}^3 = 39.4 \text{ J}$$

Since the process is adiabatic, $Q = 0$ and $\Delta E_{\text{int}} = Q + W$ gives $W = \boxed{39.4 \text{ J}}$.

(b) The moment of inertia of the wheel is $I = \frac{1}{2}MR^2 = \frac{1}{2}5.1 \text{ kg}(0.085 \text{ m})^2 = 0.018\,4 \text{ kg}\cdot\text{m}^2$. We want the flywheel to do work 39.4 J, so the work on the flywheel should be –39.4 J:

continued on next page

$$K_{\text{rot}\,i} + W = K_{\text{rot}\,f}$$

$$\frac{1}{2}I\omega_i^2 - 39.4\text{ J} = 0$$

$$\omega_i = \left(\frac{2(39.4\text{ J})}{0.018\ 4\text{ kg}\cdot\text{m}^2}\right)^{1/2} = \boxed{65.4\text{ rad/s}}$$

(c) Now we want $W = 0.05K_{\text{rot}\,i}$

$$39.4J = 0.05\frac{1}{2}0.018\ 4\text{ kg}\cdot\text{m}^2\omega_i^2$$

$$\omega = \left(\frac{2(789\text{ J})}{0.018\ 4\text{ kg}\cdot\text{m}^2}\right)^{1/2} = \boxed{293\text{ rad/s}}$$

P18.43 (a) For an isothermal process, $Q = nRT\ln\left(\frac{V_2}{V_1}\right)$

Therefore, $Q_1 = nR(3T_i)\ln 2$

and $Q_3 = nR(T_i)\ln\left(\frac{1}{2}\right)$

For the constant volume processes, $Q_2 = \Delta E_{\text{int},\,2} = \frac{3}{2}nR(T_i - 3T_i)$

and $Q_4 = \Delta E_{\text{int},\,4} = \frac{3}{2}nR(3T_i - T_i)$

The net energy by heat transferred is then

$$Q = Q_1 + Q_2 + Q_3 + Q_4$$

or $Q = \boxed{2nRT_i\ln 2}$.

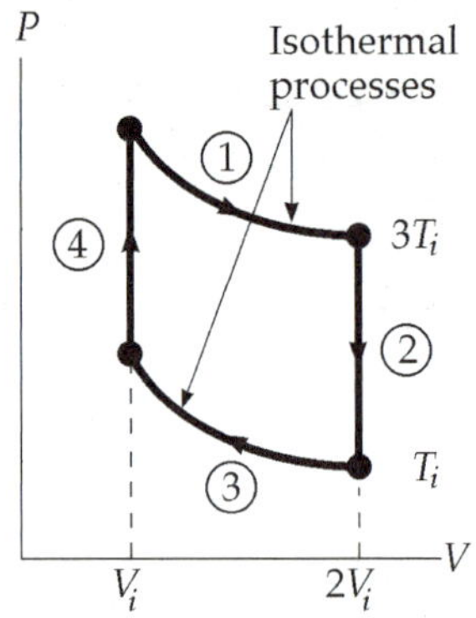

FIG. P18.43

(b) A positive value for heat represents energy transferred into the system.

Therefore, $|Q_h| = Q_1 + Q_4 = 3nRT_i(1 + \ln 2)$

Since the change in temperature for the complete cycle is zero,

$$\Delta E_{\text{int}} = 0 \text{ and } W_{\text{eng}} = Q$$

Therefore, the efficiency is $e_c = \frac{W_{\text{eng}}}{|Q_h|} = \frac{Q}{|Q_h|} = \frac{2\ln 2}{3(1 + \ln 2)} = \boxed{0.273}$

P18.44 $\Delta S_{\text{hot}} = \frac{-1\ 000\text{ J}}{600\text{ K}}$

$\Delta S_{\text{cold}} = \frac{+750\text{ J}}{350\text{ K}}$

(a) $\Delta S_U = \Delta S_{\text{hot}} + \Delta S_{\text{cold}} = \boxed{0.476\text{ J/K}}$

continued on next page

(b) $$e_c = 1 - \frac{T_1}{T_2} = 0.417$$

$$W_{\text{eng}} = e_c|Q_h| = 0.417(1\,000\text{ J}) = \boxed{417\text{ J}}$$

(c) $$W_{\text{net}} = 417\text{ J} - 250\text{ J} = 167\text{ J}$$

$$T_1 \Delta S_U = 350\text{ K}(0.476\text{ J/K}) = \boxed{167\text{ J}}$$

P18.45 $e_c = 1 - \dfrac{T_c}{T_h} = \dfrac{W_{\text{eng}}}{|Q_h|} = \dfrac{\frac{W_{\text{eng}}}{\Delta t}}{\frac{|Q_h|}{\Delta t}}$: $$\frac{|Q_h|}{\Delta t} = \frac{\mathcal{P}}{\left(1 - \frac{T_c}{T_h}\right)} = \frac{\mathcal{P}T_h}{T_h - T_c}$$

$|Q_h| = W_{\text{eng}} + |Q_c|$: $$\frac{|Q_c|}{\Delta t} = \frac{|Q_h|}{\Delta t} - \frac{W_{\text{eng}}}{\Delta t}$$

$$\frac{|Q_c|}{\Delta t} = \frac{\mathcal{P}T_h}{T_h - T_c} - \mathcal{P} = \frac{\mathcal{P}T_c}{T_h - T_c}$$

$|Q_c| = mc\Delta T$: $$\frac{|Q_c|}{\Delta t} = \left(\frac{\Delta m}{\Delta t}\right)c\Delta T = \frac{\mathcal{P}T_c}{T_h - T_c}$$

$$\frac{\Delta m}{\Delta t} = \frac{\mathcal{P}T_c}{(T_h - T_c)c\Delta T}$$

$$\frac{\Delta m}{\Delta t} = \frac{(1.00 \times 10^9\text{ W})(300\text{ K})}{200\text{ K}(4\,186\text{ J/kg}\cdot{}^\circ\text{C})(6.00^\circ\text{C})} = \boxed{5.97 \times 10^4\text{ kg/s}}$$

P18.46 $e_c = 1 - \dfrac{T_c}{T_h} = \dfrac{W_{\text{eng}}}{|Q_h|} = \dfrac{\frac{W_{\text{eng}}}{\Delta t}}{\frac{|Q_h|}{\Delta t}}$ $$\frac{|Q_h|}{\Delta t} = \frac{\mathcal{P}}{\left(1 - \frac{T_c}{T_h}\right)} = \frac{\mathcal{P}T_h}{T_h - T_c}$$

$$\frac{|Q_c|}{\Delta t} = \left(\frac{|Q_h|}{\Delta t}\right) - \mathcal{P} = \frac{\mathcal{P}T_c}{T_h - T_c}$$

$|Q_c| = mc\Delta T$, where c is the specific heat of water.

Therefore, $$\frac{|Q_c|}{\Delta t} = \left(\frac{\Delta m}{\Delta t}\right)c\Delta T = \frac{\mathcal{P}T_c}{T_h - T_c}$$

and $$\frac{\Delta m}{\Delta t} = \boxed{\frac{\mathcal{P}T_c}{(T_h - T_c)c\Delta T}}.$$

We test for dimensional correctness by identifying the units of the right-hand side: $\dfrac{\text{W}\cdot{}^\circ\text{C}}{{}^\circ\text{C}(\text{J/kg}\cdot{}^\circ\text{C}){}^\circ\text{C}} = \dfrac{(\text{J/s})\text{kg}}{\text{J}} = \text{kg/s}$, as on the left hand side. Think of yourself as a power-company engineer arranging to have enough cooling water to carry off your thermal pollution. If the plant power $\mathcal{P}$ increases, the required flow rate increases in direct proportion. If environmental regulations require a smaller temperature change, then the required flow rate increases again, now in inverse proportion. If you can run the reactor core or firebox hotter, since T_h is in the bottom of the fraction, the required flow rate decreases! If the turbines take in steam at higher temperature, they can be made more efficient to reduce waste heat output.

P18.47 We want to evaluate $\frac{dP}{dV}$ for the function implied by $PV = nRT = \text{constant}$, and also for the different function implied by $PV^\gamma = \text{constant}$. We can use implicit differentiation:

From $PV = \text{constant}$ $\quad P\frac{dV}{dV} + V\frac{dP}{dV} = 0 \quad \left(\frac{dP}{dV}\right)_{\text{isotherm}} = -\frac{P}{V}$

From $PV^\gamma = \text{constant}$ $\quad P\gamma V^{\gamma-1} + V^\gamma \frac{dP}{dV} = 0 \quad \left(\frac{dP}{dV}\right)_{\text{adiabat}} = -\frac{\gamma P}{V}$

Therefore, $\left(\frac{dP}{dV}\right)_{\text{adiabat}} = \gamma\left(\frac{dP}{dV}\right)_{\text{isotherm}}$. The theorem is proved.

***P18.48** (a) $35.0°\text{F} = \frac{5}{9}(35.0 - 32.0)°\text{C} = (1.67 + 273.15)\text{ K} = 274.82\text{ K}$

$98.6°\text{F} = \frac{5}{9}(98.6 - 32.0)°\text{C} = (37.0 + 273.15)\text{ K} = 310.15\text{ K}$

$$\Delta S_{\text{ice water}} = \int \frac{dQ}{T} = (453.6\text{ g})(1.00\text{ cal/g}\cdot\text{K}) \times \int_{274.82}^{310.15} \frac{dT}{T} = 453.6\ln\left(\frac{310.15}{274.82}\right) = 54.86\text{ cal/K}$$

$$\Delta S_{\text{body}} = -\frac{|Q|}{T_{\text{body}}} = -(453.6)(1.00)\frac{(310.15 - 274.82)}{310.15} = -51.67\text{ cal/K}$$

$$\Delta S_{\text{system}} = 54.86 - 51.67 = \boxed{3.19\text{ cal/K}}$$

(b) $(453.6)(1)(T_F - 274.82) = (70.0\times 10^3)(1)(310.15 - T_F)$

Thus,

$$(70.0 + 0.453\,6)\times 10^3 T_F = [(70.0)(310.15) + (0.453\,6)(274.82)]\times 10^3$$

and $T_F = 309.92\text{ K} = 36.77°\text{C} = \boxed{98.19°\text{F}}$

$$\Delta S'_{\text{ice water}} = 453.6\ln\left(\frac{309.92}{274.82}\right) = 54.52\text{ cal/K}$$

$$\Delta S'_{\text{body}} = -(70.0\times 10^3)\ln\left(\frac{310.15}{309.92}\right) = -51.93\text{ cal/K}$$

$\Delta S'_{\text{sys}} = 54.52 - 51.93 = \boxed{2.59\text{ cal/K}}$ which is less than the estimate in part (a).

P18.49 (a) For the isothermal process *AB*, the work on the gas is

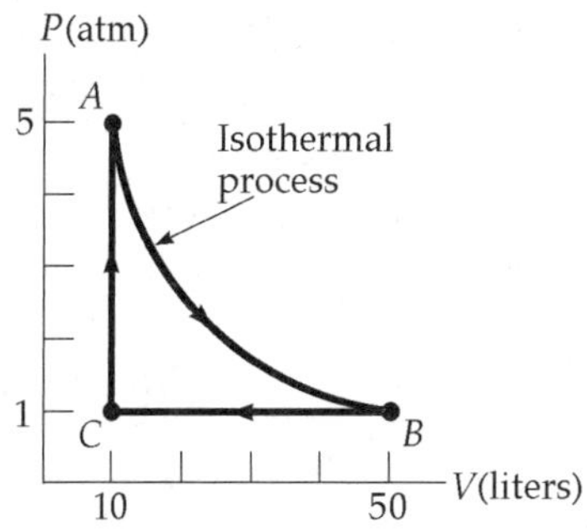

FIG. P18.49

$$W_{AB} = -P_A V_A \ln\left(\frac{V_B}{V_A}\right)$$

$$W_{AB} = -5\left(1.013\times10^5 \text{ Pa}\right)\left(10.0\times10^{-3} \text{ m}^3\right)\ln\left(\frac{50.0}{10.0}\right)$$

$$W_{AB} = -8.15\times10^3 \text{ J}$$

where we have used $1.00 \text{ atm} = 1.013\times10^5 \text{ Pa}$

and $1.00 \text{ L} = 1.00\times10^{-3} \text{ m}^3$

$$W_{BC} = -P_B\Delta V = -\left(1.013\times10^5 \text{ Pa}\right)\left[(10.0-50.0)\times10^{-3}\right]\text{m}^3 = +4.05\times10^3 \text{ J}$$

$$W_{CA} = 0 \text{ and } W_{\text{eng}} = -W_{AB} - W_{BC} = 4.11\times10^3 \text{ J} = \boxed{4.11 \text{ kJ}}$$

(b) Since *AB* is an isothermal process, $\Delta E_{\text{int}, AB} = 0$

and $Q_{AB} = -W_{AB} = 8.15\times10^3 \text{ J}$

For an ideal monatomic gas, $C_V = \frac{3R}{2}$ and $C_P = \frac{5R}{2}$

$$T_B = T_A = \frac{P_B V_B}{nR} = \frac{\left(1.013\times10^5\right)\left(50.0\times10^{-3}\right)}{R} = \frac{5.05\times10^3}{R}$$

Also,
$$T_C = \frac{P_C V_C}{nR} = \frac{\left(1.013\times10^5\right)\left(10.0\times10^{-3}\right)}{R} = \frac{1.01\times10^3}{R}$$

$$Q_{CA} = nC_V\Delta T = 1.00\left(\frac{3}{2}R\right)\left(\frac{5.05\times10^3 - 1.01\times10^3}{R}\right) = 6.08 \text{ kJ}$$

so the total energy absorbed by heat is $Q_{AB} + Q_{CA} = 8.15 \text{ kJ} + 6.08 \text{ kJ} = \boxed{14.2 \text{ kJ}}$.

(c) $$Q_{BC} = nC_P\Delta T = \frac{5}{2}(nR\Delta T) = \frac{5}{2}P_B\Delta V_{BC}$$

$$Q_{BC} = \frac{5}{2}\left(1.013\times10^5\right)\left[(10.0-50.0)\times10^{-3}\right] = -1.01\times10^4 \text{ J} = \boxed{-10.1 \text{ kJ}}$$

(d) $$e = \frac{W_{\text{eng}}}{|Q_h|} = \frac{W_{\text{eng}}}{Q_{AB} + Q_{CA}} = \frac{4.11\times10^3 \text{ J}}{1.42\times10^4 \text{ J}} = 0.289 \text{ or } \boxed{28.9\%}$$

P18.50 Like a refrigerator, an air conditioner has as its purpose the removal of energy by heat from the cold reservoir.

Its ideal COP is $$\text{COP}_{\text{Carnot}} = \frac{T_c}{T_h - T_c} = \frac{280\text{ K}}{20\text{ K}} = 14.0$$

(a) Its actual COP is $$0.400(14.0) = 5.60 = \frac{|Q_c|}{|Q_h| - |Q_c|} = \frac{|Q_c/\Delta t|}{|Q_h/\Delta t| - |Q_c/\Delta t|}$$

$$5.60\left|\frac{Q_h}{\Delta t}\right| - 5.60\left|\frac{Q_c}{\Delta t}\right| = \left|\frac{Q_c}{\Delta t}\right|$$

$$5.60(10.0\text{ kW}) = 6.60\left|\frac{Q_c}{\Delta t}\right| \text{ and } \left|\frac{Q_c}{\Delta t}\right| = \boxed{8.48\text{ kW}}$$

(b) $|Q_h| = W_{\text{eng}} + |Q_c|$: $$\frac{W_{\text{eng}}}{\Delta t} = \left|\frac{Q_h}{\Delta t}\right| - \left|\frac{Q_c}{\Delta t}\right| = 10.0\text{ kW} - 8.48\text{ kW} = \boxed{1.52\text{ kW}}$$

(c) The air conditioner operates in a cycle, so the entropy of the working fluid does not change. The hot reservoir increases in entropy by

$$\frac{|Q_h|}{T_h} = \frac{(10.0\times10^3\text{ J/s})(3\,600\text{ s})}{300\text{ K}} = 1.20\times10^5\text{ J/K}$$

The cold room decreases in entropy by

$$\Delta S = -\frac{|Q_c|}{T_c} = -\frac{(8.48\times10^3\text{ J/s})(3\,600\text{ s})}{280\text{ K}} = -1.09\times10^5\text{ J/K}$$

The net entropy change is positive, as it must be:

$$+1.20\times10^5\text{ J/K} - 1.09\times10^5\text{ J/K} = \boxed{1.09\times10^4\text{ J/K}}$$

(d) The new ideal COP is $$\text{COP}_{\text{Carnot}} = \frac{T_c}{T_h - T_c} = \frac{280\text{ K}}{25\text{ K}} = 11.2$$

We suppose the actual COP is $0.400(11.2) = 4.48$

As a fraction of the original 5.60, this is $\frac{4.48}{5.60} = 0.800$, so the fractional change is to $\boxed{\text{drop by }20.0\%}$.

P18.51 At point A, $P_iV_i = nRT_i$ and $n = 1.00$ mol

At point B, $3P_iV_i = nRT_B$ so $T_B = 3T_i$

At point C, $(3P_i)(2V_i) = nRT_C$ and $T_C = 6T_i$

At point D, $P_i(2V_i) = nRT_D$ so $T_D = 2T_i$

The heat for each step in the cycle is found using $C_V = \frac{3R}{2}$ and $C_P = \frac{5R}{2}$:

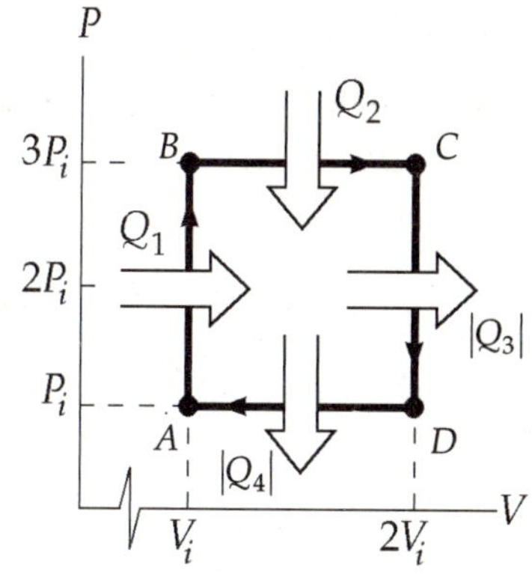

FIG. P18.51

$$Q_{AB} = nC_V(3T_i - T_i) = 3nRT_i$$
$$Q_{BC} = nC_P(6T_i - 3T_i) = 7.50nRT_i$$
$$Q_{CD} = nC_V(2T_i - 6T_i) = -6nRT_i$$
$$Q_{DA} = nC_P(T_i - 2T_i) = -2.50nRT_i$$

(a) Therefore, $Q_{\text{entering}} = |Q_h| = Q_{AB} + Q_{BC} = \boxed{10.5nRT_i}$

(b) $Q_{\text{leaving}} = |Q_c| = |Q_{CD} + Q_{DA}| = \boxed{8.50nRT_i}$

(c) Actual efficiency, $e = \frac{|Q_h| - |Q_c|}{|Q_h|} = \boxed{0.190}$

(d) Carnot efficiency, $e_c = 1 - \frac{T_c}{T_h} = 1 - \frac{T_i}{6T_i} = \boxed{0.833}$

P18.52 $\Delta S = \int_i^f \frac{dQ}{T} = \int_i^f \frac{nC_P dT}{T} = nC_P \int_i^f T^{-1} dT = nC_P \ln T\Big|_{T_i}^{T_f} = nC_P(\ln T_f - \ln T_i) = nC_P \ln\left(\frac{T_f}{T_i}\right)$

$\Delta S = nC_P \ln\left(\frac{PV_f}{nR}\frac{nR}{PV_i}\right) = \boxed{nC_P \ln 3}$

P18.53 (a) The ideal gas at constant temperature keeps constant internal energy. As it puts out energy by work in expanding it must take in an equal amount of energy by heat. Thus its entropy increases. Let P_i, V_i, T_i represent the state of the gas before the isothermal expansion. Let P_C, V_C, T_i represent the state after this process, so that $P_iV_i = P_CV_C$. Let P_i, $3V_i$, T_f represent the state after the adiabatic compression.

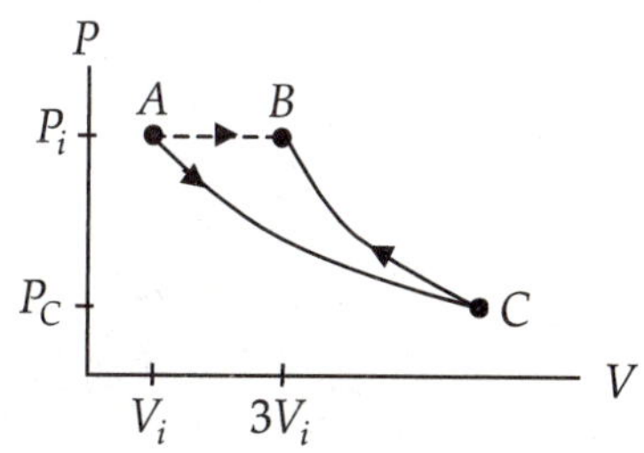

FIG. P18.53

Then $P_CV_C^\gamma = P_i(3V_i)^\gamma$

Substituting $P_C = \frac{P_iV_i}{V_C}$

continued on next page

gives $$P_iV_iV_C^{\gamma-1}=P_i\left(3^\gamma V_i^\gamma\right)$$

Then $$V_C^{\gamma-1}=3^\gamma V_i^{\gamma-1}\text{ and }\frac{V_C}{V_i}=3^{\gamma/(\gamma-1)}$$

The work output in the isothermal expansion is

$$W=\int_i^C PdV=nRT_i\int_i^C V^{-1}dV=nRT_i\ln\left(\frac{V_C}{V_i}\right)=nRT_i\ln\left(3^{\gamma/(\gamma-1)}\right)=nRT_i\left(\frac{\gamma}{\gamma-1}\right)\ln 3$$

This is also the input heat, so the entropy change is

$$\Delta S=\frac{Q}{T}=nR\left(\frac{\gamma}{\gamma-1}\right)\ln 3$$

Since $$C_P=\gamma C_V=C_V+R$$

we have $$(\gamma-1)C_V=R,\ C_V=\frac{R}{\gamma-1}$$

and $$C_P=\frac{\gamma R}{\gamma-1}$$

Then the result is $$\boxed{\Delta S=nC_P\ln 3}$$

(b) The pair of processes considered here carry the gas from the initial state in Problem 18.52 to the final state there. Entropy is a function of state. Entropy change does not depend on path. Therefore the entropy change in Problem 18.52 equals $\Delta S_{\text{isothermal}}+\Delta S_{\text{adiabatic}}$ in this problem. Since $\Delta S_{\text{adiabatic}}=0$, the answers to Problems 18.52 and 18.53 (a) must be the same.

***P18.54** Simply evaluate the maximum (Carnot) efficiency.

$$e_C=\frac{\Delta T}{T_h}=\frac{4.00\text{ K}}{277\text{ K}}=\boxed{0.014\,4}$$

The proposal does not merit serious consideration.

P18.55 (a) Use the equation of state for an ideal gas

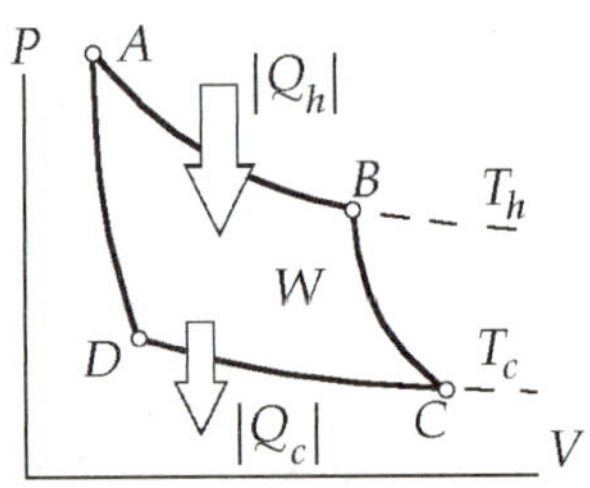

FIG. P18.55

$$V=\frac{nRT}{P}$$

$$V_A=\frac{1.00(8.314)(600)}{25.0\left(1.013\times10^5\right)}=\boxed{1.97\times10^{-3}\text{ m}^3}$$

$$V_C=\frac{1.00(8.314)(400)}{1.013\times10^5}=\boxed{32.8\times10^{-3}\text{ m}^3}$$

Since AB is isothermal, $$P_AV_A=P_BV_B$$

and since BC is adiabatic, $$P_BV_B^\gamma=P_CV_C^\gamma$$

continued on next page

Combining these expressions, $V_B = \left[\left(\frac{P_C}{P_A}\right)\frac{V_C^\gamma}{V_A}\right]^{1/(\gamma-1)} = \left[\left(\frac{1.00}{25.0}\right)\frac{\left(32.8\times10^{-3}\ \text{m}^3\right)^{1.40}}{1.97\times10^{-3}\ \text{m}^3}\right]^{(1/0.400)}$

$$V_B = \boxed{11.9\times10^{-3}\ \text{m}^3}$$

Similarly, $V_D = \left[\left(\frac{P_A}{P_C}\right)\frac{V_A^\gamma}{V_C}\right]^{1/(\gamma-1)} = \left[\left(\frac{25.0}{1.00}\right)\frac{\left(1.97\times10^{-3}\ \text{m}^3\right)^{1.40}}{32.8\times10^{-3}\ \text{m}^3}\right]^{(1/0.400)}$

or $V_D = \boxed{5.44\times10^{-3}\ \text{m}^3}$

Since AB is isothermal, $P_A V_A = P_B V_B$

and $P_B = P_A\left(\frac{V_A}{V_B}\right) = 25.0\ \text{atm}\left(\frac{1.97\times10^{-3}\ \text{m}^3}{11.9\times10^{-3}\ \text{m}^3}\right) = \boxed{4.14\ \text{atm}}$

Also, CD is an isothermal and $P_D = P_C\left(\frac{V_C}{V_D}\right) = 1.00\ \text{atm}\left(\frac{32.8\times10^{-3}\ \text{m}^3}{5.44\times10^{-3}\ \text{m}^3}\right) = \boxed{6.03\ \text{atm}}$

Solving part (c) before part (b):

(c) For this Carnot cycle, $e_c = 1 - \frac{T_c}{T_h} = 1 - \frac{400\ \text{K}}{600\ \text{K}} = \boxed{0.333}$

(b) Energy is added by heat to the gas during the process AB. For the isothermal process, $\Delta E_{\text{int}} = 0$.

and the first law gives $Q_{AB} = -W_{AB} = nRT_h \ln\left(\frac{V_B}{V_A}\right)$

or $|Q_h| = Q_{AB} = 1.00\ \text{mol}(8.314\ \text{J/mol}\cdot\text{K})(600\ \text{K})\ln\left(\frac{11.9}{1.97}\right) = 8.97\ \text{kJ}$

Then, from $e = \frac{W_{\text{eng}}}{|Q_h|}$

the net work done per cycle is $W_{\text{eng}} = e_c|Q_h| = 0.333(8.97\ \text{kJ}) = \boxed{2.99\ \text{kJ}}$.

P18.56 $\frac{|Q_c|}{W} = \text{COP}_C\ (\text{refrigerator}) = \frac{T_c}{T_h - T_c} = \frac{\frac{|Q_c|}{\Delta t}}{\frac{W}{\Delta t}}$

$$\frac{0.150\ \text{W}}{\frac{W}{\Delta t}} = \frac{260\ \text{K}}{40.0\ \text{K}}$$

$$\mathcal{P} = \frac{W}{\Delta t} = 0.150\ \text{W}\left(\frac{40.0\ \text{K}}{260\ \text{K}}\right) = \boxed{23.1\ \text{mW}}$$

P18.57 Define $T_1 = \text{Temp Cream} = 5.00°\text{C} = 278\text{ K}$. Define $T_2 = \text{Temp Coffee} = 60.0°\text{C} = 333\text{ K}$. The final temperature of the mixture is: $T_f = \frac{(20.0\text{ g})T_1 + (200\text{ g})T_2}{220\text{ g}} = 55.0°\text{C} = 328\text{ K}$. The entropy change due to this mixing is $\Delta S = (20.0\text{ g})\int_{T_1}^{T_f} \frac{c_V dT}{T} + (200\text{ g})\int_{T_2}^{T_f} \frac{c_V dT}{T}$

$$\Delta S = (84.0\text{ J/K})\ln\left(\frac{T_f}{T_1}\right) + (840\text{ J/K})\ln\left(\frac{T_f}{T_2}\right) = (84.0\text{ K/J})\ln\left(\frac{328}{278}\right) + (840\text{ J/K})\ln\left(\frac{328}{333}\right)$$

$$\Delta S = \boxed{+1.18\text{ J/K}}$$

P18.58 No energy is transferred by heat during the adiabatic processes. Energy Q_h is added by heat during process BC and energy $|Q_h|$ is exhausted during process DA. We have $Q_h = nC_V(T_C - T_B)$ and $|Q_C| = nC_V(T_D - T_A)$. So $e = 1 - \frac{|Q_C|}{Q_h} = 1 - \frac{T_D - T_A}{T_C - T_B}$. For the adiabatic processes, $P_C V_2^{\gamma} = P_D V_1^{\gamma}$ and $P_B V_2^{\gamma} = P_A V_1^{\gamma}$.

From the gas law $P_C = \frac{nRT_C}{V_2}$ $P_D = \frac{nRT_D}{V_1}$

$$P_B = \frac{nRT_B}{V_2} \qquad P_A = \frac{nRT_A}{V_1}$$

Then $nRT_C V_2^{\gamma-1} = nRT_D V_1^{\gamma-1}$ and $T_B V_2^{\gamma-1} = T_A V_1^{\gamma-1}$. Thus, $\frac{T_C}{T_D} = \left(\frac{V_1}{V_2}\right)^{\gamma-1} = r^{\gamma-1}$ and

$\frac{T_B}{T_A} = \left(\frac{V_1}{V_2}\right)^{\gamma-1} = r^{\gamma-1}$. Then $e = 1 - \frac{T_D - T_A}{T_D r^{\gamma-1} - T_A r^{\gamma-1}} = 1 - \frac{1}{r^{\gamma-1}} = 1 - r^{1-\gamma}$.

ANSWERS TO EVEN PROBLEMS

P18.2 (a) 29.4 L/h; (b) 185 hp; (c) 527 N·m; (d) 1.91×10^5 W

P18.4 13.7°C

P18.6 (a) 870 MJ; (b) 330 MJ

P18.8 546°C

P18.10 (a) 5.12%; (b) 5.27 TJ/h; (c) As conventional energy sources become more expensive, or as their true costs are recognized, alternative sources become economically viable.

P18.12 (a)

State	P (kPa)	V (L)	T (K)
A	1 400	10.0	720
B	875	16.0	720
C	445	24.0	549
D	712	15.0	549

(b)

Process	Q (kJ)	W (kJ)	ΔE_{int} (kJ)
$A \to B$	6.58	–6.58	0
$B \to C$	0	–4.98	–4.98
$C \to D$	–5.02	5.02	0
$D \to A$	0	4.98	4.98
$ABCDA$	1.56	–1.56	0

P18.14 77.8 W

P18.16 9.00

P18.18 11.8

P18.20 1.86

P18.22 (a) $\$1.00 \times 10^{-3}$/min; (b) $\$2.22 \times 10^{-4}$/min; (c) $\$2.78 \times 10^{-5}$/min

P18.24 –610 J/K

P18.26 244 J/K

P18.28 (a) 2 heads and 2 tails; (b) either all heads or all tails; (c) 2 heads and 2 tails

P18.30 3.27 J/K

P18.32 718 J/K

P18.34 5.76 J/K, no change in temperature

P18.36 9×10^{14} W

P18.38 (a) 214 J, 64.3 J;
(b) −35.7 J, −35.7 J. The net effect is the transport of energy by heat from the cold to the hot reservoir without expenditure of external work.
(c) 333 J, 233 J;
(d) 83.3 J, 83.3 J, 0. The net effect is converting energy, taken in as heat, entirely into energy output by work in a cyclic process.
(e) −0.111 J/K, the entropy of the Universe has decreased.

P18.40 8.36×10^{6} J/K

P18.42 (a) 39.4 J; (b) 65.4 rad/s; (c) 293 rad/s

P18.44 (a) 0.476 J/K; (b) 417 J;
(c) $W_{\text{net}} = T_1 \Delta S_U = 167$ J

P18.46 $\dfrac{m}{t} = \dfrac{\mathcal{P} T_c}{(T_h - T_c)c\Delta T}$

P18.48 (a) 3.19 cal/K; (b) 98.19°F, 2.59 cal/K

P18.50 (a) 8.48 kW; (b) 1.52 kW;
(c) 1.09×10^{4} J/K;
(d) COP drops by 20.0%

P18.52 $nC_P \ln 3$

P18.54 No, the device cannot meet its specifications, according to the second law of thermodynamics.

P18.56 23.1 mW

P18.58 see the solution

CONTEXT 5 CONCLUSION SOLUTIONS TO PROBLEMS

CC5.1 We take 214 K as the tropopause temperature and −6.5 °C/km as the lapse rate. Then

$$214\text{ K} - T_{\text{surface}} = (-6.5\,^\circ\text{C/km})13\text{ km} = -84^\circ\text{C}$$

$$T_{\text{surface}} = \boxed{298\text{ K}}$$

CC5.2 The lapse rate is

$$\frac{\Delta T}{\Delta y} = \frac{T_{\text{tropopause}} - T_{\text{surface}}}{y_{\text{tropopause}} - 0}$$

$$y_{\text{tropopause}} = \frac{T_{\text{tropopause}} - T_{\text{surface}}}{\frac{\Delta T}{\Delta y}} = \frac{200\text{ K} - 732\text{ K}}{-8.8\,^\circ\text{C/km}} = \boxed{60\text{ km}}$$

CC5.3 (a) The power balance equation for the planet as a whole is

$$\mathcal{P}_{\text{absorbed}} = \mathcal{P}_{\text{emitted}}: \quad e_{vis} I_s A_{\text{absorbing}} = e_{ir} \sigma A_{\text{emitting}} T_1^{\ 4}$$

$$0.70\left(1\ 370\ \text{W/m}^2\right)\pi R_E^{\ 2} = 1\left(5.67\times10^{-8}\ \text{W/m}^2\cdot\text{K}^4\right)\left(4\pi R_E^{\ 2}\right)T_1^{\ 4}$$

$$T_1 = \sqrt[4]{\frac{0.70\left(1370\ \text{W/m}^2\right)}{4\left(5.67\times10^{-8}\ \text{W/m}^2\cdot\text{K}^4\right)}} = \boxed{255\ \text{K}}$$

(b) The power balance equation for layer 1, required for the constancy of its temperature, is $e_{ir}\sigma\left(8\pi R_E^{\ 2}\right)T_1^{\ 4} = e_{ir}\sigma\left(4\pi R_E^{\ 2}\right)T_2^{\ 4}$ or $2T_1^{\ 4} = T_2^{\ 4}$. Similarly, for layer 2, $2T_2^{\ 4} = T_1^{\ 4} + T_3^{\ 4}$. Again, $2T_3^{\ 4} = T_2^{\ 4} + T_4^{\ 4}, \ldots\ 2T_N^{\ 4} = T_{N-1}^{\ \ 4} + T_S^{\ 4}$. Thus, $T_2 = \sqrt[4]{2}\,T_1$.

By substitution, $4T_1^{\ 4} = T_1^{\ 4} + T_3^{\ 4}$ $\quad T_3 = \sqrt[4]{3}\,T_1$

$6T_1^{\ 4} = 2T_1^{\ 4} + T_4^{\ 4}$ $\quad T_4 = \sqrt[4]{4}\,T_1, \ldots\ T_N = \sqrt[4]{N}\,T_1$

and $T_S = (N+1)^{1/4}T_1$

(c) From part (b), $T_S = (2+1)^{1/4}(255\ \text{K}) = \boxed{336\ \text{K}}$.

(d) The troposphere and stratosphere are too thick to be accurately modeled as having uniform temperatures.

(e) Venus is at 1.08×10^{11} m from the Sun, instead of 1.496×10^{11} m like the Earth. The intensity of solar radiation incident on Venus is then $\left(1\ 370\ \text{W/m}^2\right)\left(\frac{1.496}{1.08}\right)^2 = 2.63\times10^3\ \text{W/m}^2$. The overall power balance equation for Venus is $e_{vis} I_s A_{\text{absorbing}} = e_{ir}\sigma A_{\text{emitting}} T_1^{\ 4}$

$$(1-0.77)\left(2.63\times10^3\ \text{W/m}^2\right)\left(\pi R_V^{\ 2}\right) = 1\left(5.67\times10^{-8}\ \text{W/m}^2\cdot\text{K}^4\right)\left(4\pi R_V^{\ 2}\right)T_1^{\ 4}$$

$$T_1 = \sqrt[4]{\frac{0.23\left(2.63\times10^3\ \text{W/m}^2\right)}{4\left(5.67\times10^{-8}\ \text{W/m}^2\cdot\text{K}^4\right)}} = \boxed{227\ \text{K}}$$

(f) $T_S = (N+1)^{1/4}T_1$: $\quad N = \left(\frac{T_S}{T_1}\right)^4 - 1 = \left(\frac{732}{227}\right)^4 - 1 = \boxed{107}$

(g) The multi-layer model should be better for Venus than for Earth. There are many layers, so the temperature of each can be reasonably uniform.

ANSWERS TO EVEN CONTEXT 5 CONCLUSION PROBLEMS

CC5.2 60 km

Chapter 19

Electric Forces and Electric Fields

CHAPTER OUTLINE

ANSWERS TO QUESTIONS

Q19.1 A neutral atom is one that has no net charge. This means that it has the same number of electrons orbiting the nucleus as it has protons in the nucleus. A negatively charged atom has one or more excess electrons.

Q19.2 The clothes dryer rubs dissimilar materials together as it tumbles the clothes. Electrons are transferred from one kind of molecule to another. The charges on pieces of cloth, or on nearby objects charged by induction, can produce strong electric fields that promote the ionization process in the surrounding air that is necessary for a spark to occur. Then you hear or see the sparks.

Q19.3 To avoid making a spark. Rubber-soled shoes acquire a charge by friction with the floor and could discharge with a spark, possibly causing an explosion of any flammable material in the oxygen-enriched atmosphere.

Q19.4 Similarities: A force of gravity is proportional to the product of the intrinsic properties (masses) of two particles, and inversely proportional to the square of the separation distance. An electrical force exhibits the same proportionalities, with charge as the intrinsic property.

Differences: The electrical force can either attract or repel, while the gravitational force as described by Newton's law can only attract. The electrical force between elementary particles is vastly stronger than the gravitational force.

Q19.5 No. The balloon induces polarization of the molecules in the wall, so that a layer of positive charge exists near the balloon. This is just like the situation in Figure 19.5a, except that the signs of the charges are reversed. The attraction between these charges and the negative charges on the balloon is stronger than the repulsion between the negative charges on the balloon and the negative charges in the polarized molecules (because they are farther from the balloon), so that there is a net attractive force toward the wall. Ionization processes in the air surrounding the balloon provide ions to which excess electrons in the balloon can transfer, reducing the charge on the balloon and eventually causing the attractive force to be insufficient to support the weight of the balloon.

Q19.6 An electric field once established by a positive or negative charge extends in all directions from the charge. Thus, it can exist in empty space if that is what surrounds the charge. There is no material at point A in Figure 19.18(a), so there is no charge, nor is there a force. There would be a force if a charge were present at point A, however. A field does exist at point A.

Q19.7 If a charge distribution is small compared to the distance of a field point from it, the charge distribution can be modeled as a single particle with charge equal to the net charge of the distribution. Further, if a charge distribution is spherically symmetric, it will create a field at exterior points just as if all of its charge were a point charge at its center.

Q19.8 Both figures are drawn correctly. $\vec{\mathbf{E}}_1$ and $\vec{\mathbf{E}}_2$ are the electric fields separately created by the point charges q and $-q$ in Figure 19.11. The net electric field is the vector sum of $\vec{\mathbf{E}}_1$ and $\vec{\mathbf{E}}_2$, shown as $\vec{\mathbf{E}}$. Figure 19.16 shows only one electric field line at each point away from the charge. At the point location of an object modeled as a point charge, the direction of the field is undefined, and so is its magnitude.

Q19.9 No. Life would be no different if electrons were + charged and protons were – charged. Opposite charges would still attract, and like charges would repel. The naming of + and – charge is merely a convention.

Q19.10 At a point exactly midway between the two charges.

Q19.11 In special orientations the force between two dipoles can be zero or a force of repulsion. In general each dipole will exert a torque on the other, tending to align its axis with the field created by the first dipole. After this alignment, each dipole exerts a force of attraction on the other.

Q19.12 The negative charge will be drawn to the center of the positively charged ring. Since it will then have velocity, it will continue on, to an equidistant point on the opposite side of the ring. It will then start moving back and arrive again at point P. This periodic motion will continue. If x is much less than a, the motion can be shown from the solution for the electric field to be simple harmonic.

Q19.13 The surface must enclose a positive total charge.

Q19.14 The net flux through any gaussian surface is zero. We can argue it two ways. Any surface contains zero charge so Gauss's law says the total flux is zero. The field is uniform, so the field lines entering one side of the closed surface come out the other side and the net flux is zero.

Q19.15 Gauss's law cannot tell the different values of the electric field at different points on the surface. When E is an unknown number, then we can say $\int E\cos\theta dA = E\int \cos\theta dA$. When $E(x, y, z)$ is an unknown function, then there is no such simplification.

Q19.16 Inject some charge at arbitrary places within a conducting object. Every bit of the charge repels every other bit, so each bit runs away as far as it can, stopping only when it reaches the outer surface of the conductor.

Q19.17 If the person is uncharged, the electric field inside the sphere is zero. The interior wall of the shell carries no charge. The person is not harmed by touching this wall. If the person carries a (small) charge q, the electric field inside the sphere is no longer zero. Charge $-q$ is induced on the inner wall of the sphere. The person will get a (small) shock when touching the sphere, as all the charge on his body jumps to the metal.

Q19.18 The electric fields outside are identical. The electric fields inside are very different. We have $\vec{\mathbf{E}}=0$ everywhere inside the conducting sphere while E decreases gradually as you go below the surface of the sphere with uniform volume charge density.

Q19.19 There is zero force. The huge charged sheet creates a uniform field. The field can polarize the neutral sheet, creating in effect a film of opposite charge on the near face and a film with an equal amount of like charge on the far face of the neutral sheet. Since the field is uniform, the films of charge feel equal-magnitude forces of attraction and repulsion to the charged sheet. The forces add to zero.

SOLUTIONS TO PROBLEMS

Section 19.1 Historical Overview

No problems in this section

Section 19.2 Properties of Electric Charges

***P19.1** (a) The mass of an average neutral hydrogen atom is 1.007 9u. Losing one electron reduces its mass by a negligible amount, to

$$1.007\,9\left(1.660\times10^{-27}\text{ kg}\right)-9.11\times10^{-31}\text{ kg}=\boxed{1.67\times10^{-27}\text{ kg}}.$$

Its charge, due to loss of one electron, is

$$0-1\left(-1.60\times10^{-19}\text{ C}\right)=\boxed{+1.60\times10^{-19}\text{ C}}.$$

(b) By similar logic, charge $=\boxed{+1.60\times10^{-19}\text{ C}}$

$$\text{mass}=22.99\left(1.66\times10^{-27}\text{ kg}\right)-9.11\times10^{-31}\text{ kg}=\boxed{3.82\times10^{-26}\text{ kg}}$$

(c) charge of $\text{Cl}^-=\boxed{-1.60\times10^{-19}\text{ C}}$

$$\text{mass}=35.453\left(1.66\times10^{-27}\text{ kg}\right)+9.11\times10^{-31}\text{ kg}=\boxed{5.89\times10^{-26}\text{ kg}}$$

(d) charge of $\text{Ca}^{++}=-2\left(-1.60\times10^{-19}\text{ C}\right)=\boxed{+3.20\times10^{-19}\text{ C}}$

$$\text{mass}=40.078\left(1.66\times10^{-27}\text{ kg}\right)-2\left(9.11\times10^{-31}\text{ kg}\right)=\boxed{6.65\times10^{-26}\text{ kg}}$$

(e) charge of $\text{N}^{3-}=3\left(-1.60\times10^{-19}\text{ C}\right)=\boxed{-4.80\times10^{-19}\text{ C}}$

$$\text{mass}=14.007\left(1.66\times10^{-27}\text{ kg}\right)+3\left(9.11\times10^{-31}\text{ kg}\right)=\boxed{2.33\times10^{-26}\text{ kg}}$$

continued on next page

(f) charge of $N^{4+} = 4\left(1.60\times10^{-19}\text{ C}\right) = \boxed{+6.40\times10^{-19}\text{ C}}$

$$\text{mass} = 14.007\left(1.66\times10^{-27}\text{ kg}\right) - 4\left(9.11\times10^{-31}\text{ kg}\right) = \boxed{2.32\times10^{-26}\text{ kg}}$$

(g) We think of a nitrogen nucleus as a seven-times ionized nitrogen atom.

$$\text{charge} = 7\left(1.60\times10^{-19}\text{ C}\right) = \boxed{1.12\times10^{-18}\text{ C}}$$

$$\text{mass} = 14.007\left(1.66\times10^{-27}\text{ kg}\right) - 7\left(9.11\times10^{-31}\text{ kg}\right) = \boxed{2.32\times10^{-26}\text{ kg}}$$

(h) charge $= \boxed{-1.60\times10^{-19}\text{ C}}$

$$\text{mass} = \left[2(1.007\,9) + 15.999\right]1.66\times10^{-27}\text{ kg} + 9.11\times10^{-31}\text{ kg} = \boxed{2.99\times10^{-26}\text{ kg}}$$

P19.2 (a) $$N = \left(\frac{10.0\text{ grams}}{107.87\text{ grams/mol}}\right)\left(6.02\times10^{23}\ \frac{\text{atoms}}{\text{mol}}\right)\left(47\ \frac{\text{electrons}}{\text{atom}}\right) = \boxed{2.62\times10^{24}}$$

(b) $$\#\text{ electrons added} = \frac{Q}{e} = \frac{1.00\times10^{-3}\text{ C}}{1.60\times10^{-19}\text{ C/electron}} = 6.25\times10^{15}$$

or $\boxed{2.38\text{ electrons for every }10^9\text{ already present}}$.

Section 19.3 Insulators and Conductors

No problems in this section

Section 19.4 Coulomb's Law

P19.3 If each person has a mass of $\approx 70\text{ kg}$ and is (almost) composed of water, then each person contains

$$N = \left(\frac{70\,000\text{ grams}}{18\text{ grams/mol}}\right)\left(6.02\times10^{23}\ \frac{\text{molecules}}{\text{mol}}\right)\left(10\ \frac{\text{protons}}{\text{molecule}}\right) = 2.3\times10^{28}\text{ protons}.$$

With an excess of 1% electrons over protons, each person has a charge

$$q = 0.01\left(1.6\times10^{-19}\text{ C}\right)\left(2.3\times10^{28}\right) = 3.7\times10^{7}\text{ C}.$$

So $$F = k_e\frac{q_1q_2}{r^2} = \left(9\times10^9\right)\frac{\left(3.7\times10^7\right)^2}{0.6^2}\text{ N} = 4\times10^{25}\text{ N}\ \boxed{\sim10^{26}\text{ N}}.$$

This force is almost enough to lift a weight equal to that of the Earth:

$$Mg = 6\times10^{24}\text{ kg}\left(9.8\text{ m/s}^2\right) = 6\times10^{25}\text{ N} \sim 10^{26}\text{ N}.$$

P19.4 The force on one proton is $\vec{\mathbf{F}} = \frac{k_e q_1 q_2}{r^2}$ away from the other proton. Its magnitude is

$$\left(8.99\times10^{9}\ \text{N}\cdot\text{m/C}^2\right)\left(\frac{1.6\times10^{-19}\ \text{C}}{2\times10^{-15}\ \text{m}}\right)^2 = \boxed{57.5\ \text{N}}.$$

P19.5

$$F_1 = k_e\frac{q_1q_2}{r^2} = \frac{\left(8.99\times10^{9}\ \text{N}\cdot\text{m}^2/\text{C}^2\right)\left(7.00\times10^{-6}\ \text{C}\right)\left(2.00\times10^{-6}\ \text{C}\right)}{(0.500\ \text{m})^2} = 0.503\ \text{N}$$

$$F_2 = k_e\frac{q_1q_2}{r^2} = \frac{\left(8.99\times10^{9}\ \text{N}\cdot\text{m}^2/\text{C}^2\right)\left(7.00\times10^{-6}\ \text{C}\right)\left(4.00\times10^{-6}\ \text{C}\right)}{(0.500\ \text{m})^2} = 1.01\ \text{N}$$

$$F_x = 0.503\cos 60.0° + 1.01\cos 60.0° = 0.755\ \text{N}$$
$$F_y = 0.503\sin 60.0° - 1.01\sin 60.0° = -0.436\ \text{N}$$
$$\vec{\mathbf{F}} = (0.755\ \text{N})\hat{\mathbf{i}} - (0.436\ \text{N})\hat{\mathbf{j}} = \boxed{0.872\ \text{N at an angle of } 330°}$$

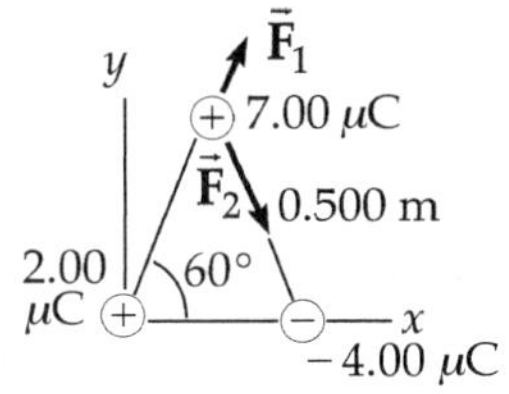

FIG. P19.5

***P19.6** In the first situation, $\vec{\mathbf{F}}_{A\text{ on }B,1} = \frac{k_e|q_A||q_B|}{r_1^2}\hat{\mathbf{i}}$. In the second situation, $|q_A|$ and $|q_B|$ are the same.

$$\vec{\mathbf{F}}_{B\text{ on }A,2} = -\vec{\mathbf{F}}_{A\text{ on }B} = \frac{k_e|q_A||q_B|}{r_2^2}\left(-\hat{\mathbf{i}}\right)$$

$$\frac{F_2}{F_1} = \frac{k_e|q_A||q_B|}{r_2^2}\frac{r_1^2}{k_e|q_A||q_B|}$$

$$F_2 = \frac{F_1r_1^2}{r_2^2} = 2.62\ \mu\text{N}\left(\frac{13.7\ \text{mm}}{17.7\ \text{mm}}\right)^2 = 1.57\ \mu\text{N}$$

Then $\vec{\mathbf{F}}_{B\text{ on }A,2} = \boxed{1.57\ \mu\text{N to the left}}$.

P19.7 (a) The force is one of $\boxed{\text{attraction}}$. The distance r in Coulomb's law is the distance between centers. The magnitude of the force is

$$F = \frac{k_eq_1q_2}{r^2} = \left(8.99\times10^{9}\ \text{N}\cdot\text{m}^2/\text{C}^2\right)\frac{\left(12.0\times10^{-9}\ \text{C}\right)\left(18.0\times10^{-9}\ \text{C}\right)}{(0.300\ \text{m})^2} = \boxed{2.16\times10^{-5}\ \text{N}}.$$

(b) The net charge of -6.00×10^{-9} C will be equally split between the two spheres, or -3.00×10^{-9} C on each. The force is one of $\boxed{\text{repulsion}}$, and its magnitude is

$$F = \frac{k_eq_1q_2}{r^2} = \left(8.99\times10^{9}\ \text{N}\cdot\text{m}^2/\text{C}^2\right)\frac{\left(3.00\times10^{-9}\ \text{C}\right)\left(3.00\times10^{-9}\ \text{C}\right)}{(0.300\ \text{m})^2} = \boxed{8.99\times10^{-7}\ \text{N}}.$$

***P19.8** Let the third bead have charge Q and be located distance x from the left end of the rod. This bead will experience a net force given by

$$\vec{\mathbf{F}} = \frac{k_e(3q)Q}{x^2}\hat{\mathbf{i}} + \frac{k_e(q)Q}{(d-x)^2}\left(-\hat{\mathbf{i}}\right).$$

The net force will be zero if $\frac{3}{x^2} = \frac{1}{(d-x)^2}$, or $d - x = \frac{x}{\sqrt{3}}$.

This gives an equilibrium position of the third bead of $x = \boxed{0.634d}$.

The equilibrium is $\boxed{\text{stable if the third bead has positive charge}}$.

P19.9 (a) $$F = \frac{k_e e^2}{r^2} = \left(8.99\times10^9\ \text{N}\cdot\text{m}^2/\text{C}^2\right)\frac{\left(1.60\times10^{-19}\ \text{C}\right)^2}{\left(0.529\times10^{-10}\ \text{m}\right)^2} = \boxed{8.22\times10^{-8}\ \text{N}}$$

(b) We have $F = \frac{mv^2}{r}$ from which

$$v = \sqrt{\frac{Fr}{m}} = \sqrt{\frac{8.22\times10^{-8}\ \text{N}\left(0.529\times10^{-10}\ \text{m}\right)}{9.11\times10^{-31}\ \text{kg}}} = \boxed{2.19\times10^{6}\ \text{m/s}}.$$

Section 19.5 Electric Fields

P19.10 For equilibrium, $\vec{\mathbf{F}}_e = -\vec{\mathbf{F}}_g$

or $q\vec{\mathbf{E}} = -mg\left(-\hat{\mathbf{j}}\right)$.

Thus, $\vec{\mathbf{E}} = \frac{mg}{q}\hat{\mathbf{j}}$.

(a) $$\vec{\mathbf{E}} = \frac{mg}{q}\hat{\mathbf{j}} = \frac{\left(9.11\times10^{-31}\ \text{kg}\right)\left(9.80\ \text{m/s}^2\right)}{\left(-1.60\times10^{-19}\ \text{C}\right)}\hat{\mathbf{j}} = \boxed{-\left(5.58\times10^{-11}\ \text{N/C}\right)\hat{\mathbf{j}}}$$

(b) $$\vec{\mathbf{E}} = \frac{mg}{q}\hat{\mathbf{j}} = \frac{\left(1.67\times10^{-27}\ \text{kg}\right)\left(9.80\ \text{m/s}^2\right)}{\left(1.60\times10^{-19}\ \text{C}\right)}\hat{\mathbf{j}} = \boxed{\left(1.02\times10^{-7}\ \text{N/C}\right)\hat{\mathbf{j}}}$$

P19.11 The point is designated in the sketch. The magnitudes of the electric fields, E_1, (due to the -2.50×10^{-6} C charge) and E_2 (due to the 6.00×10^{-6} C charge) are

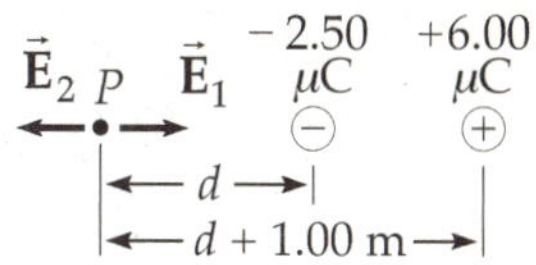

FIG. P19.11

$$E_1=\frac{k_eq}{r^2}=\frac{\left(8.99\times10^9\ \text{N}\cdot\text{m}^2/\text{C}^2\right)\left(2.50\times10^{-6}\ \text{C}\right)}{d^2} \quad (1)$$

$$E_2=\frac{k_eq}{r^2}=\frac{\left(8.99\times10^9\ \text{N}\cdot\text{m}^2/\text{C}^2\right)\left(6.00\times10^{-6}\ \text{C}\right)}{(d+1.00\ \text{m})^2} \quad (2)$$

Equate the right sides of (1) and (2)

to get $(d+1.00\ \text{m})^2=2.40d^2$

or $d+1.00\ \text{m}=\pm1.55d$

which yields $d=1.82$ m

or $d=-0.392$ m.

The negative value for d is unsatisfactory because that locates a point between the charges where both fields are in the same direction.

Thus, $d=\boxed{1.82\text{ m to the left of the }-2.50\ \mu\text{C charge}}$.

***P19.12** The first charge creates at the origin field $\frac{k_eQ}{a^2}$ to the right. Suppose the total field at the origin is to the right. Then q must be negative:

FIG. P19.12

$$\frac{k_eQ}{a^2}\hat{\mathbf{i}}+\frac{k_eq}{(3a)^2}\left(-\hat{\mathbf{i}}\right)=\frac{2k_eQ}{a^2}\hat{\mathbf{i}} \qquad \boxed{q=-9Q}.$$

In the alternative, the total field at the origin is to the left:

$$\frac{k_eQ}{a^2}\hat{\mathbf{i}}+\frac{k_eq}{9a^2}\left(-\hat{\mathbf{i}}\right)=\frac{2k_eQ}{a^2}\left(-\hat{\mathbf{i}}\right) \qquad \boxed{q=+27Q}.$$

P19.13 (a)

$$\vec{\mathbf{E}}_1=\frac{k_e|q_1|}{r_1^2}\left(-\hat{\mathbf{j}}\right)=\frac{\left(8.99\times10^9\right)\left(3.00\times10^{-9}\right)}{(0.100)^2}\left(-\hat{\mathbf{j}}\right)=-\left(2.70\times10^3\ \text{N/C}\right)\hat{\mathbf{j}}$$

$$\vec{\mathbf{E}}_2=\frac{k_e|q_2|}{r_2^2}\left(-\hat{\mathbf{i}}\right)=\frac{\left(8.99\times10^9\right)\left(6.00\times10^{-9}\right)}{(0.300)^2}\left(-\hat{\mathbf{i}}\right)=-\left(5.99\times10^2\ \text{N/C}\right)\hat{\mathbf{i}}$$

$$\vec{\mathbf{E}}=\vec{\mathbf{E}}_2+\vec{\mathbf{E}}_1=\boxed{-\left(5.99\times10^2\ \text{N/C}\right)\hat{\mathbf{i}}-\left(2.70\times10^3\ \text{N/C}\right)\hat{\mathbf{j}}}$$

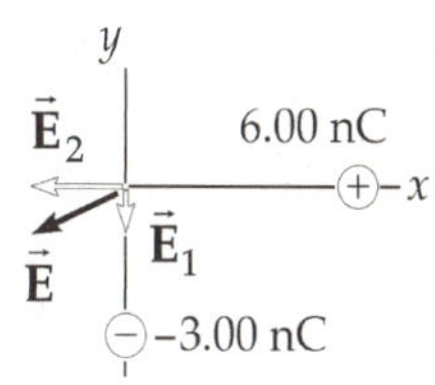

FIG. P19.13

(b)

$$\vec{\mathbf{F}}=q\vec{\mathbf{E}}=\left(5.00\times10^{-9}\ \text{C}\right)\left(-599\hat{\mathbf{i}}-2\,700\hat{\mathbf{j}}\right)\ \text{N/C}$$

$$\vec{\mathbf{F}}=\left(-3.00\times10^{-6}\hat{\mathbf{i}}-13.5\times10^{-6}\hat{\mathbf{j}}\right)\ \text{N}=\boxed{\left(-3.00\hat{\mathbf{i}}-13.5\hat{\mathbf{j}}\right)\ \mu\text{N}}$$

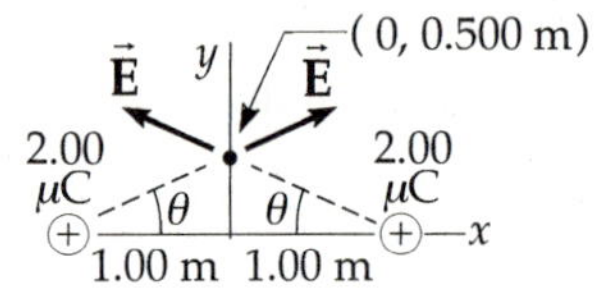

FIG. P19.14

P19.14 (a) $E=\frac{k_e q}{r^2}=\frac{\left(8.99\times10^9\right)\left(2.00\times10^{-6}\right)}{(1.12)^2}=14\,400\text{ N/C}$

$E_x=0$ and $E_y=2(14\,400)\sin 26.6°=1.29\times10^4\text{ N/C}$

so $\boxed{\vec{\mathbf{E}}=1.29\times10^4\,\hat{\mathbf{j}}\text{ N/C}}$.

(b) $\vec{\mathbf{F}}=q\vec{\mathbf{E}}=\left(-3.00\times10^{-6}\right)\left(1.29\times10^4\,\hat{\mathbf{j}}\right)=\boxed{-3.86\times10^{-2}\,\hat{\mathbf{j}}\text{ N}}$

P19.15 (a) $\vec{\mathbf{E}}=\frac{k_e q_1}{r_1^2}\hat{\mathbf{r}}_1+\frac{k_e q_2}{r_2^2}\hat{\mathbf{r}}_2+\frac{k_e q_3}{r_3^2}\hat{\mathbf{r}}_3=\frac{k_e(2q)}{a^2}\hat{\mathbf{i}}+\frac{k_e(3q)}{2a^2}\left(\hat{\mathbf{i}}\cos 45.0°+\hat{\mathbf{j}}\sin 45.0°\right)+\frac{k_e(4q)}{a^2}\hat{\mathbf{j}}$

$\vec{\mathbf{E}}=3.06\frac{k_e q}{a^2}\hat{\mathbf{i}}+5.06\frac{k_e q}{a^2}\hat{\mathbf{j}}=\boxed{5.91\frac{k_e q}{a^2}\text{ at }58.8°}$

(b) $\vec{\mathbf{F}}=q\vec{\mathbf{E}}=\boxed{5.91\frac{k_e q^2}{a^2}\text{ at }58.8°}$

P19.16 The electric field at any point x is

$$E=\frac{k_e q}{(x-a)^2}-\frac{k_e q}{(x-(-a))^2}=\frac{k_e q(4ax)}{\left(x^2-a^2\right)^2}.$$

When x is much, much greater than a, we find $E\approx\boxed{\frac{4a(k_e q)}{x^3}}$.

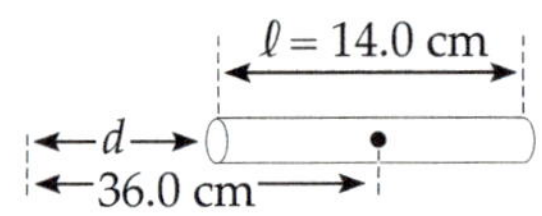

FIG. P19.17

P19.17 $E=\frac{k_e\lambda\ell}{d(\ell+d)}=\frac{k_e(Q/\ell)\ell}{d(\ell+d)}=\frac{k_e Q}{d(\ell+d)}=\frac{\left(8.99\times10^9\right)\left(22.0\times10^{-6}\right)}{(0.290)(0.140+0.290)}$

$\vec{\mathbf{E}}=\boxed{1.59\times10^6\text{ N/C, directed toward the rod.}}$

P19.18 $E=\int\frac{k_e dq}{x^2}$, where $dq=\lambda_0 dx$

$E=k_e\lambda_0\int_{x_0}^{\infty}\frac{dx}{x^2}=k_e\lambda_0\left(-\frac{1}{x}\right)\Big|_{x_0}^{\infty}=\boxed{\frac{k_e\lambda_0}{x_0}}$ $\boxed{\text{The direction is }-\hat{\mathbf{i}}\text{ or left for }\lambda_0>0}$

P19.19 $E=\frac{k_e xQ}{\left(x^2+a^2\right)^{3/2}}=\frac{\left(8.99\times10^9\right)\left(75.0\times10^{-6}\right)x}{\left(x^2+0.100^2\right)^{3/2}}=\frac{6.74\times10^5\,x}{\left(x^2+0.010\,0\right)^{3/2}}$

(a) At $x=0.010\,0\text{ m}$, $\vec{\mathbf{E}}=6.64\times10^6\,\hat{\mathbf{i}}\text{ N/C}=\boxed{6.64\hat{\mathbf{i}}\text{ MN/C}}$

(b) At $x=0.050\,0\text{ m}$, $\vec{\mathbf{E}}=2.41\times10^7\,\hat{\mathbf{i}}\text{ N/C}=\boxed{24.1\hat{\mathbf{i}}\text{ MN/C}}$

continued on next page

(c) At $x = 0.300$ m, $\vec{\mathbf{E}} = 6.40 \times 10^6 \hat{\mathbf{i}}$ N/C $= \boxed{6.40\hat{\mathbf{i}} \text{ MN/C}}$

(d) At $x = 1.00$ m, $\vec{\mathbf{E}} = 6.64 \times 10^5 \hat{\mathbf{i}}$ N/C $= \boxed{0.664\hat{\mathbf{i}} \text{ MN/C}}$

***P19.20** $E = \dfrac{k_e Q x}{\left(x^2 + a^2\right)^{3/2}}$

For a maximum, $\dfrac{dE}{dx} = Qk_e\left[\dfrac{1}{\left(x^2 + a^2\right)^{3/2}} - \dfrac{3x^2}{\left(x^2 + a^2\right)^{5/2}}\right] = 0$

$x^2 + a^2 - 3x^2 = 0$ or $x = \dfrac{a}{\sqrt{2}}$.

Substituting into the expression for E gives

$$E = \frac{k_e Q a}{\sqrt{2}\left(\frac{3}{2}a^2\right)^{3/2}} = \frac{k_e Q}{3\frac{\sqrt{3}}{2}a^2} = \boxed{\frac{2k_e Q}{3\sqrt{3}a^2}} = \boxed{\frac{Q}{6\sqrt{3}\pi \in_0 a^2}}.$$

P19.21 Due to symmetry $E_y = \int dE_y = 0$, and $E_x = \int dE \sin\theta = k_e \int \dfrac{dq \sin\theta}{r^2}$

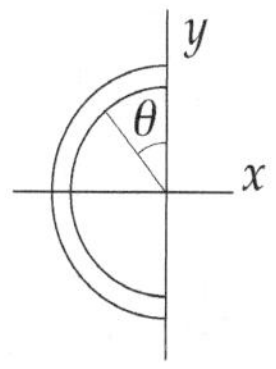

FIG. P19.21

where $dq = \lambda ds = \lambda r d\theta$,

so that, $E_x = \dfrac{k_e \lambda}{r}\displaystyle\int_0^{\pi} \sin\theta d\theta = \dfrac{k_e \lambda}{r}(-\cos\theta)\Big|_0^{\pi} = \dfrac{2k_e\lambda}{r}$

where $\lambda = \dfrac{q}{L}$ and $r = \dfrac{L}{\pi}$.

Thus, $E_x = \dfrac{2k_e q \pi}{L^2} = \dfrac{2\left(8.99 \times 10^9 \text{ N}\cdot\text{m}^2/\text{C}^2\right)\left(7.50 \times 10^{-6} \text{ C}\right)\pi}{(0.140 \text{ m})^2}$.

Solving, $E_x = 2.16 \times 10^7$ N/C.

Since the rod has a negative charge, $\vec{\mathbf{E}} = \left(-2.16 \times 10^7 \hat{\mathbf{i}}\right)$ N/C $= \boxed{-21.6\hat{\mathbf{i}} \text{ MN/C}}$.

P19.22 (a) The electric field at point P due to each element of length dx, is $dE = \frac{k_e dq}{x^2 + y^2}$ and is directed along the line joining the element to point P. By symmetry,

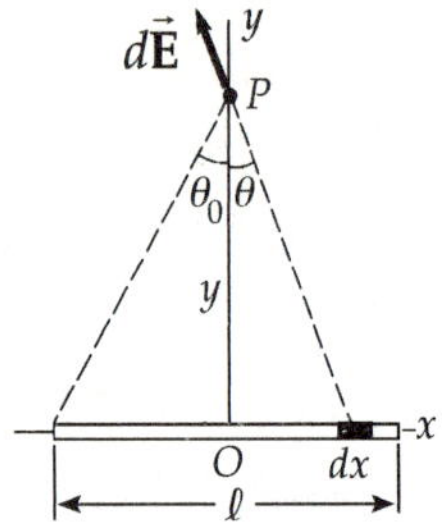

FIG. P19.22

$E_x = \int dE_x = 0$ and since $dq = \lambda dx$,

$E = E_y = \int dE_y = \int dE \cos\theta$ where $\cos\theta = \frac{y}{\sqrt{x^2 + y^2}}$.

Therefore, $E = 2k_e \lambda y \int_0^{\ell/2} \frac{dx}{\left(x^2 + y^2\right)^{3/2}} = \boxed{\frac{2k_e \lambda \sin\theta_0}{y}}$.

(b) For a bar of infinite length, $\theta_0 = 90°$ and $E_y = \boxed{\frac{2k_e \lambda}{y}}$.

P19.23 (a) The whole surface area of the cylinder is $A = 2\pi r^2 + 2\pi rL = 2\pi r(r + L)$.

$Q = \sigma A = \left(15.0 \times 10^{-9} \text{ C/m}^2\right) 2\pi(0.025\,0 \text{ m})\left[0.025\,0 \text{ m} + 0.060\,0 \text{ m}\right] = \boxed{2.00 \times 10^{-10} \text{ C}}$

(b) For the curved lateral surface only, $A = 2\pi rL$.

$Q = \sigma A = \left(15.0 \times 10^{-9} \text{ C/m}^2\right)\left[2\pi(0.025\,0 \text{ m})(0.060\,0 \text{ m})\right] = \boxed{1.41 \times 10^{-10} \text{ C}}$

(c) $Q = \rho V = \rho \pi r^2 L = \left(500 \times 10^{-9} \text{ C/m}^3\right)\left[\pi(0.025\,0 \text{ m})^2 (0.060\,0 \text{ m})\right] = \boxed{5.89 \times 10^{-11} \text{ C}}$

Section 19.6 Electric Field Lines

***P19.24** (a) $\frac{q_1}{q_2} = \frac{-6}{18} = \boxed{-\frac{1}{3}}$

(b) $\boxed{q_1 \text{ is negative, } q_2 \text{ is positive}}$

P19.25

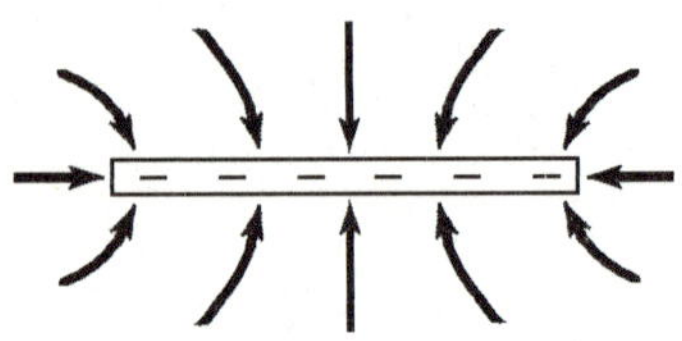

FIG. P19.25

P19.26 (a) The electric field has the general appearance shown. It is zero $\boxed{\text{at the center}}$, where (by symmetry) one can see that the three charges individually produce fields that cancel out.

In addition to the center of the triangle, the electric field lines in the second figure to the right indicate three other points near the middle of each leg of the triangle where $E = 0$, but they are more difficult to find mathematically.

(b) You may need to review vector addition in Chapter One. The electric field at point P can be found by adding the electric field vectors due to each of the two lower point charges: $\vec{\mathbf{E}} = \vec{\mathbf{E}}_1 + \vec{\mathbf{E}}_2$.

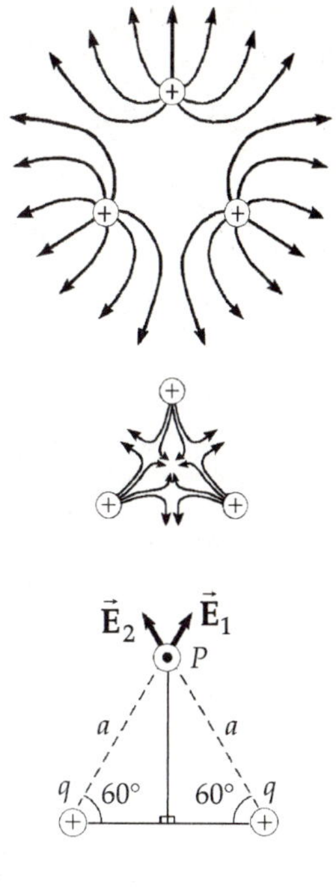

FIG. P19.26

The electric field from a point charge is $\vec{\mathbf{E}} = k_e \dfrac{q}{r^2}\hat{\mathbf{r}}$.

As shown in the solution figure at right,

$\vec{\mathbf{E}}_1 = k_e \dfrac{q}{a^2}$ to the right and upward at 60°

$\vec{\mathbf{E}}_2 = k_e \dfrac{q}{a^2}$ to the left and upward at 60°

$$\vec{\mathbf{E}} = \vec{\mathbf{E}}_1 + \vec{\mathbf{E}}_2 = k_e \frac{q}{a^2}\left[\left(\cos 60°\,\hat{\mathbf{i}} + \sin 60°\,\hat{\mathbf{j}}\right) + \left(-\cos 60°\,\hat{\mathbf{i}} + \sin 60°\,\hat{\mathbf{j}}\right)\right] = k_e \frac{q}{a^2}\left[2\left(\sin 60°\,\hat{\mathbf{j}}\right)\right]$$

$$= \boxed{1.73 k_e \frac{q}{a^2}\hat{\mathbf{j}}}$$

Section 19.7 Motion of Charged Particles in a Uniform Electric Field

P19.27 (a) $a = \dfrac{qE}{m} = \dfrac{1.60 \times 10^{-19}(640)}{1.67 \times 10^{-27}} = \boxed{6.13 \times 10^{10}\ \text{m/s}^2}$

(b) $v_f = v_i + at$ $\qquad 1.20 \times 10^6 = \left(6.13 \times 10^{10}\right)t$ $\qquad t = \boxed{1.95 \times 10^{-5}\ \text{s}}$

(c) $x_f - x_i = \dfrac{1}{2}\left(v_i + v_f\right)t$ $\qquad x_f = \dfrac{1}{2}\left(1.20 \times 10^6\right)\left(1.95 \times 10^{-5}\right) = \boxed{11.7\ \text{m}}$

(d) $K = \dfrac{1}{2}mv^2 = \dfrac{1}{2}\left(1.67 \times 10^{-27}\ \text{kg}\right)\left(1.20 \times 10^6\ \text{m/s}\right)^2 = \boxed{1.20 \times 10^{-15}\ \text{J}}$

P19.28 The required electric field will be $\boxed{\text{in the direction of motion}}$.

Work done $= \Delta K$

so, $-Fd = -\frac{1}{2}mv_i^2$ (since the final velocity $= 0$)

which becomes $eEd = K$

and $E = \boxed{\frac{K}{ed}}$.

P19.29 (a) $t = \frac{x}{v_x} = \frac{0.050\,0}{4.50 \times 10^5} = 1.11 \times 10^{-7}\ \text{s} = \boxed{111\ \text{ns}}$

(b) $a_y = \frac{qE}{m} = \frac{\left(1.602 \times 10^{-19}\right)\left(9.60 \times 10^3\right)}{\left(1.67 \times 10^{-27}\right)} = 9.21 \times 10^{11}\ \text{m/s}^2$

$y_f - y_i = v_{yi}t + \frac{1}{2}a_y t^2$: $y_f = \frac{1}{2}\left(9.21 \times 10^{11}\right)\left(1.11 \times 10^{-7}\right)^2 = 5.68 \times 10^{-3}\ \text{m} = \boxed{5.68\ \text{mm}}$

(c) $v_x = \boxed{4.50 \times 10^5\ \text{m/s}}$ $v_{yf} = v_{yi} + a_y t = \left(9.21 \times 10^{11}\right)\left(1.11 \times 10^{-7}\right) = \boxed{1.02 \times 10^5\ \text{m/s}}$

Section 19.8 Electric Flux

P19.30 $\Phi_E = EA\cos\theta = \left(2.00 \times 10^4\ \text{N/C}\right)\left(18.0\ \text{m}^2\right)\cos 10.0° = \boxed{355\ \text{kN} \cdot \text{m}^2/\text{C}}$

P19.31 $\Phi_E = EA\cos\theta$ $A = \pi r^2 = \pi(0.200)^2 = 0.126\ \text{m}^2$

$5.20 \times 10^5 = E(0.126)\cos 0°$ $E = 4.14 \times 10^6\ \text{N/C} = \boxed{4.14\ \text{MN/C}}$

Section 19.9 Gauss's Law

P19.32 (a) $E = \frac{k_e Q}{r^2}$: $8.90 \times 10^2 = \frac{\left(8.99 \times 10^9\right)Q}{(0.750)^2}$

But Q is negative since $\vec{\mathbf{E}}$ points inward. $Q = -5.57 \times 10^{-8}\ \text{C} = \boxed{-55.7\ \text{nC}}$

(b) The $\boxed{\text{negative}}$ charge has a $\boxed{\text{spherically symmetric}}$ charge distribution, concentric with the spherical shell.

P19.33 (a) With δ very small, all points on the hemisphere are nearly at a distance R from the charge, so the field everywhere on the curved surface is $\frac{k_e Q}{R^2}$ radially outward (normal to the surface). Therefore, the flux is this field strength times the area of half a sphere:

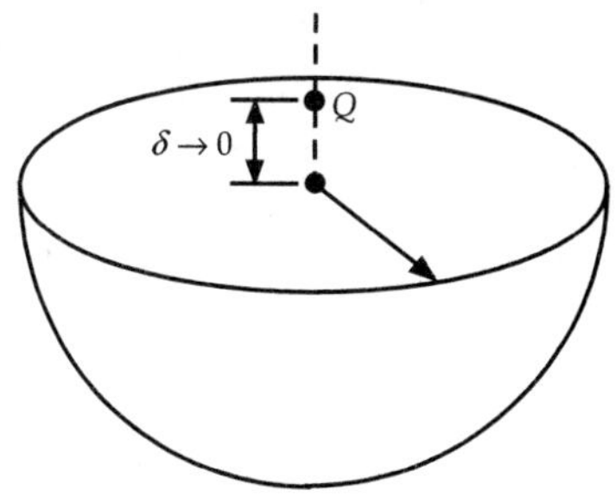

FIG. P19.33

$$\Phi_{\text{curved}} = \int \vec{\mathbf{E}} \cdot d\vec{\mathbf{A}} = E_{\text{local}} A_{\text{hemisphere}}$$

$$\Phi_{\text{curved}} = \left(k_e \frac{Q}{R^2}\right)\left(\frac{1}{2} 4\pi R^2\right) = \frac{1}{4\pi \epsilon_0} Q(2\pi) = \boxed{\frac{+Q}{2\epsilon_0}}$$

(b) The closed surface encloses zero charge so Gauss's law gives

$$\Phi_{\text{curved}} + \Phi_{\text{flat}} = 0 \qquad \text{or} \qquad \Phi_{\text{flat}} = -\Phi_{\text{curved}} = \boxed{\frac{-Q}{2\epsilon_0}}.$$

P19.34 $\Phi_E = \frac{q_{\text{in}}}{\epsilon_0} = \frac{170 \times 10^{-6}\ \text{C}}{8.85 \times 10^{-12}\ \text{C}^2/\text{N} \cdot \text{m}^2} = 1.92 \times 10^7\ \text{N} \cdot \text{m}^2/\text{C}$

(a) $(\Phi_E)_{\text{one face}} = \frac{1}{6}\Phi_E = \frac{1.92 \times 10^7\ \text{N} \cdot \text{m}^2/\text{C}}{6}$ $\qquad (\Phi_E)_{\text{one face}} = \boxed{3.20\ \text{MN} \cdot \text{m}^2/\text{C}}$

(b) $\Phi_E = \boxed{19.2\ \text{MN} \cdot \text{m}^2/\text{C}}$

(c) The answer to (a) would change because the flux through each face of the cube would not be equal with an asymmetric charge distribution. The sides of the cube nearer the charge would have more flux and the ones further away would have less. The answer to (b) would remain the same, since the overall flux would remain the same.

Section 19.10 Application of Gauss's Law to Symmetric Charge Distributions

P19.35 (a) $E = \frac{k_e Q r}{a^3} = \boxed{0}$

(b) $E = \frac{k_e Q r}{a^3} = \frac{(8.99 \times 10^9)(26.0 \times 10^{-6})(0.100)}{(0.400)^3} = \boxed{365\ \text{kN/C}}$

(c) $E = \frac{k_e Q}{r^2} = \frac{(8.99 \times 10^9)(26.0 \times 10^{-6})}{(0.400)^2} = \boxed{1.46\ \text{MN/C}}$

(d) $E = \frac{k_e Q}{r^2} = \frac{(8.99 \times 10^9)(26.0 \times 10^{-6})}{(0.600)^2} = \boxed{649\ \text{kN/C}}$

The direction for each electric field is $\boxed{\text{radially outward}}$.

***P19.36** $mg = qE = q\left(\dfrac{\sigma}{2\epsilon_0}\right) = q\left(\dfrac{Q/A}{2\epsilon_0}\right)$ $\qquad \dfrac{Q}{A} = \dfrac{2\epsilon_0\, mg}{q} = \dfrac{2\left(8.85\times10^{-12}\right)(0.01)(9.8)}{-0.7\times10^{-6}} = \boxed{-2.48\ \mu\text{C/m}^2}$

P19.37 (a) $E = \dfrac{2k_e\lambda}{r}$ $\qquad 3.60\times10^{4} = \dfrac{2\left(8.99\times10^{9}\right)(Q/2.40)}{0.190}$

$Q = +9.13\times10^{-7}\ \text{C} = \boxed{+913\ \text{nC}}$

(b) $\vec{\mathbf{E}} = \boxed{0}$

P19.38 (a) $\vec{\mathbf{E}} = \boxed{0}$

(b) $E = \dfrac{k_eQ}{r^2} = \dfrac{\left(8.99\times10^{9}\right)\left(32.0\times10^{-6}\right)}{(0.200)^2} = 7.19\ \text{MN/C}$ $\qquad \vec{\mathbf{E}} = \boxed{7.19\ \text{MN/C radially outward}}$

P19.39 If ρ is positive, the field must be radially outward. Choose as the gaussian surface a cylinder of length L and radius r, contained inside the charged rod. Its volume is πr^2L and it encloses charge $\rho\pi r^2L$. Because the charge distribution is long, no electric flux passes through the circular end caps; $\vec{\mathbf{E}}\cdot d\vec{\mathbf{A}} = EdA\cos 90.0° = 0$. The curved surface has $\vec{\mathbf{E}}\cdot d\vec{\mathbf{A}} = EdA\cos 0°$, and E must be the same strength everywhere over the curved surface.

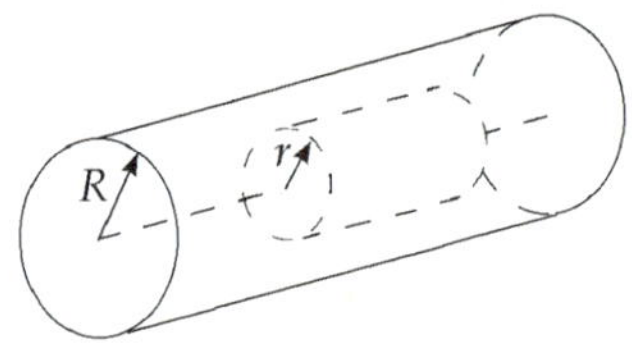

FIG. P19.39

Gauss's law, $\oint \vec{\mathbf{E}}\cdot d\vec{\mathbf{A}} = \dfrac{q}{\epsilon_0}$, $\qquad$ becomes $\qquad E\displaystyle\int_{\substack{\text{Curved}\\\text{Surface}}} dA = \dfrac{\rho\pi r^2L}{\epsilon_0}$.

Now the lateral surface area of the cylinder is $2\pi rL$:

$E(2\pi r)L = \dfrac{\rho\pi r^2L}{\epsilon_0}$. $\qquad$ Thus, $\qquad \vec{\mathbf{E}} = \boxed{\dfrac{\rho r}{2\epsilon_0}\ \text{radially away from the cylinder axis}}$.

***P19.40** The charge density is determined by $Q = \dfrac{4}{3}\pi a^3\rho$ $\qquad \rho = \dfrac{3Q}{4\pi a^3}$

(a) The flux is that created by the enclosed charge within radius r:

$$\Phi_E = \frac{q_{\text{in}}}{\epsilon_0} = \frac{4\pi r^3\rho}{3\epsilon_0} = \frac{4\pi r^3\, 3Q}{3\epsilon_0\, 4\pi a^3} = \boxed{\frac{Qr^3}{\epsilon_0\, a^3}}$$

(b) $\Phi_E = \boxed{\dfrac{Q}{\epsilon_0}}$. Note that the answers to parts (a) and (b) agree at $r = a$.

continued on next page

(c)

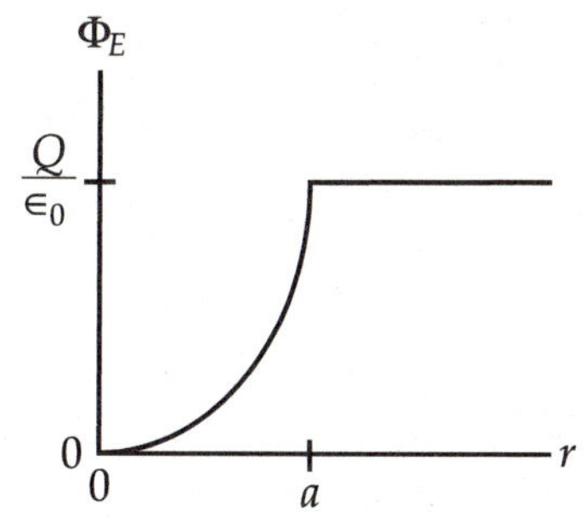

FIG. P19.40(c)

P19.41 The distance between centers is $2\times 5.90\times 10^{-15}$ m. Each produces a field as if it were a point charge at its center, and each feels a force as if all its charge were a point at its center.

$$F=\frac{k_e q_1 q_2}{r^2}=\left(8.99\times 10^9\ \text{N}\cdot\text{m}^2/\text{C}^2\right)\frac{(46)^2\left(1.60\times 10^{-19}\ \text{C}\right)^2}{\left(2\times 5.90\times 10^{-15}\ \text{m}\right)^2}=3.50\times 10^3\ \text{N}=\boxed{3.50\ \text{kN}}$$

Section 19.11 Conductors in Electrostatic Equilibrium

P19.42 $\oint E dA = E(2\pi r l)=\frac{q_{in}}{\epsilon_0}$ $\quad E=\frac{\frac{q_{in}}{l}}{2\pi\epsilon_0 r}=\frac{\lambda}{2\pi\epsilon_0 r}$

(a) $r=3.00$ cm $\quad \vec{\mathbf{E}}=\boxed{0}$

(b) $r=10.0$ cm $\quad \vec{\mathbf{E}}=\frac{30.0\times 10^{-9}}{2\pi\left(8.85\times 10^{-12}\right)(0.100)}=\boxed{5\,400\ \text{N/C, outward}}$

(c) $r=100$ cm $\quad \vec{\mathbf{E}}=\frac{30.0\times 10^{-9}}{2\pi\left(8.85\times 10^{-12}\right)(1.00)}=\boxed{540\ \text{N/C, outward}}$

P19.43 The fields are equal. The Equation 19.25 $E=\frac{\sigma_{\text{conductor}}}{\epsilon_0}$ for the field outside the aluminum looks different from Equation 19.24 $E=\frac{\sigma_{\text{insulator}}}{2\epsilon_0}$ for the field around glass. But its charge will spread out to cover both sides of the aluminum plate, so the density is $\sigma_{\text{conductor}}=\frac{Q}{2A}$. The glass carries charge only on area A, with $\sigma_{\text{insulator}}=\frac{Q}{A}$. The two fields are $\frac{Q}{2A\epsilon_0}$ the same in magnitude, and both are perpendicular to the plates, vertically upward if Q is positive.

P19.44 (a) $E=\frac{\sigma}{\epsilon_0}$ $\sigma=(8.00\times10^{4})(8.85\times10^{-12})=7.08\times10^{-7}\ \mathrm{C/m^2}$

$\sigma=\boxed{708\ \mathrm{nC/m^2}}$, positive on one face and negative on the other.

(b) $\sigma=\frac{Q}{A}$ $Q=\sigma A=(7.08\times10^{-7})(0.500)^2\ \mathrm{C}$

$Q=1.77\times10^{-7}\ \mathrm{C}=\boxed{177\ \mathrm{nC}}$, positive on one face and negative on the other.

P19.45 (a) $\vec{\mathbf{E}}=\boxed{0}$

(b) $E=\frac{k_eQ}{r^2}=\frac{(8.99\times10^{9})(8.00\times10^{-6})}{(0.030\,0)^2}=7.99\times10^{7}\ \mathrm{N/C}$ $\vec{\mathbf{E}}=\boxed{79.9\ \mathrm{MN/C\ radially\ outward}}$

(c) $\vec{\mathbf{E}}=\boxed{0}$

(d) $E=\frac{k_eQ}{r^2}=\frac{(8.99\times10^{9})(4.00\times10^{-6})}{(0.070\,0)^2}=7.34\times10^{6}\ \mathrm{N/C}$ $\vec{\mathbf{E}}=\boxed{7.34\ \mathrm{MN/C\ radially\ outward}}$

P19.46 The electric field on the surface of a conductor varies inversely with the radius of curvature of the surface. Thus, the field is most intense where the radius of curvature is smallest and vice-versa. The local charge density and the electric field intensity are related by

$$E=\frac{\sigma}{\epsilon_0}\qquad\text{or}\qquad\sigma=\epsilon_0 E.$$

(a) Where the radius of curvature is the greatest,

$$\sigma=\epsilon_0 E_{\text{min}}=(8.85\times10^{-12}\ \mathrm{C^2/N\cdot m^2})(2.80\times10^{4}\ \mathrm{N/C})=\boxed{248\ \mathrm{nC/m^2}}.$$

(b) Where the radius of curvature is the smallest,

$$\sigma=\epsilon_0 E_{\text{max}}=(8.85\times10^{-12}\ \mathrm{C^2/N\cdot m^2})(5.60\times10^{4}\ \mathrm{N/C})=\boxed{496\ \mathrm{nC/m^2}}.$$

P19.47 (a) Inside surface: consider a cylindrical surface within the metal. Since E inside the conducting shell is zero, the total charge inside the gaussian surface must be zero, so the inside charge/length $=-\lambda$.

$0=\lambda\ell+q_{\text{in}}$ so $\frac{q_{\text{in}}}{\ell}=\boxed{-\lambda}$

Outside surface: The total charge on the metal cylinder is $2\lambda\ell=q_{\text{in}}+q_{\text{out}}$

$q_{\text{out}}=2\lambda\ell+\lambda\ell$ so the outside charge/length is $\boxed{3\lambda}$.

(b) $E=\frac{2k_e(3\lambda)}{r}=\frac{6k_e\lambda}{r}=\boxed{\frac{3\lambda}{2\pi\epsilon_0 r}\ \text{radially outward}}$

***P19.48** (a) Consider a gaussian surface in the shape of a rectangular box with two faces perpendicular to the direction of the field. It encloses some charge, so the net flux out of the box is nonzero. The field must be stronger on one side than on the other. It cannot be uniform in magnitude.

(b) Now the volume contains no charge. The net flux out of the box is zero. The flux entering is equal to the flux exiting. The field magnitude is uniform.

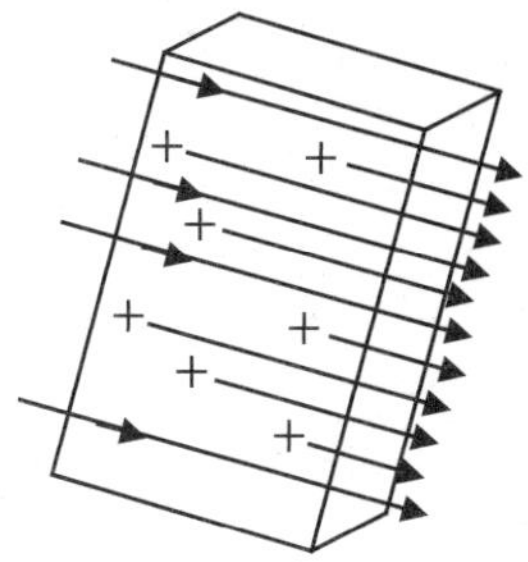

FIG. P19.48(a)

P19.49 (a) The charge density on each of the surfaces (upper and lower) of the plate is:

$$\sigma = \frac{1}{2}\left(\frac{q}{A}\right) = \frac{1}{2}\frac{\left(4.00\times10^{-8}\ \text{C}\right)}{(0.500\ \text{m})^2} = 8.00\times10^{-8}\ \text{C/m}^2 = \boxed{80.0\ \text{nC/m}^2}.$$

(b) $$\vec{\mathbf{E}} = \left(\frac{\sigma}{\epsilon_0}\right)\hat{\mathbf{k}} = \left(\frac{8.00\times10^{-8}\ \text{C/m}^2}{8.85\times10^{-12}\ \text{C}^2/\text{N}\cdot\text{m}^2}\right)\hat{\mathbf{k}} = \boxed{(9.04\ \text{kN/C})\hat{\mathbf{k}}}$$

(c) $$\vec{\mathbf{E}} = \boxed{(-9.04\ \text{kN/C})\hat{\mathbf{k}}}$$

Section 19.12 Context Connection—The Atmospheric Electric Field

P19.50 (a) $\vec{\mathbf{E}} = \dfrac{\sigma}{\epsilon_0}$ toward a $\boxed{\text{negative}}$ charge.

$$\sigma = E\epsilon_0 = (120\ \text{N/C})\left(8.85\times10^{-12}\ \text{C}^2/\text{N}\cdot\text{m}^2\right) = \boxed{1.06\times10^{-9}\ \text{C/m}^2}$$

(b) $$Q = \sigma A = \sigma\left(4\pi R^2\right) = \left(-1.06\times10^{-9}\ \text{C/m}^2\right)\left[4\pi\left(6.37\times10^6\ \text{m}\right)^2\right] = \boxed{-5.42\times10^5\ \text{C}}$$

$$-5.42\times10^5\ \text{C}\left(\frac{1\ \text{electron}}{-1.6\times10^{-19}\ \text{C}}\right) = \boxed{3.38\times10^{24}\ \text{excess electrons}}$$

P19.51 Consider as a gaussian surface a box with horizontal area A, lying between 500 and 600 m elevation.

$$\oint \vec{\mathbf{E}}\cdot d\vec{\mathbf{A}} = \frac{q}{\epsilon_0}: \qquad (+120\ \text{N/C})A + (-100\ \text{N/C})A = \frac{\rho A(100\ \text{m})}{\epsilon_0}$$

$$\rho = \frac{(20\ \text{N/C})\left(8.85\times10^{-12}\ \text{C}^2/\text{N}\cdot\text{m}^2\right)}{100\ \text{m}} = \boxed{1.77\times10^{-12}\ \text{C/m}^3}$$

The charge is $\boxed{\text{positive}}$, to produce the net outward flux of electric field.

P19.52 (a) The Moon would feel a force $\boxed{\text{away from Earth}}$ of magnitude

$$F=\frac{k_e q_1 q_2}{r^2}=\left(8.99\times10^9\ \text{N}\cdot\text{m}^2/\text{C}^2\right)\left(\frac{\left(-5\times10^5\ \text{C}\right)\left(-1.37\times10^5\ \text{C}\right)}{\left(3.84\times10^8\ \text{m}\right)^2}\right)=\boxed{4.18\times10^3\ \text{N}}.$$

(b) The gravitational force is

$$F=\frac{Gm_1 m_2}{r^2}=\frac{\left(6.67\times10^{-11}\ \text{N}\cdot\text{m}^2/\text{kg}^2\right)\left(5.98\times10^{24}\ \text{kg}\right)\left(7.36\times10^{22}\ \text{kg}\right)}{\left(3.84\times10^8\ \text{m}\right)^2}$$

$F=1.99\times10^{20}$ N toward Earth.

Thus, $\boxed{\text{the electric force is weaker by}}$

$$\frac{1.99\times10^{20}\ \text{N}}{4.18\times10^3\ \text{N}}=\boxed{4.77\times10^{16}\ \text{times and in the opposite direction}}.$$

Additional Problems

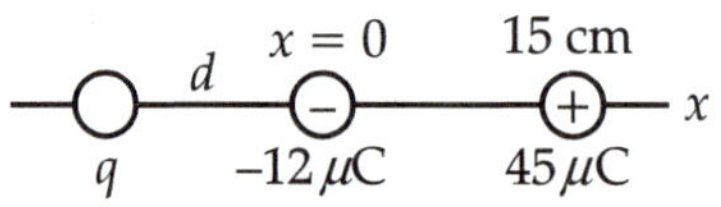

FIG. P19.53

***P19.53** The two given charges exert equal-size forces of attraction on each other. If a third charge, positive or negative, were placed between them they could not be in equilibrium. If the third charge were at a point $x>15$ cm, it would exert a stronger force on the 45 μC than on the −12 μC, and could not produce equilibrium for both. Thus the third charge must be at $x=-d<0$. Its equilibrium requires

$$\frac{k_e q(12\ \mu\text{C})}{d^2}=\frac{k_e q(45\ \mu\text{C})}{(15\ \text{cm}+d)^2}\qquad\left(\frac{15\ \text{cm}+d}{d}\right)^2=\frac{45}{12}=3.75$$

$$15\ \text{cm}+d=1.94d\qquad d=16.0\ \text{cm}.$$

The third charge is at $\boxed{x=-16.0\ \text{cm}}$. The equilibrium of the −12 μC requires

$$\frac{k_e q(12\ \mu\text{C})}{(16.0\ \text{cm})^2}=\frac{k_e(45\ \mu\text{C})12\ \mu\text{C}}{(15\ \text{cm})^2}\qquad\boxed{q=51.3\ \mu\text{C}}.$$

All six individual forces are now equal in magnitude, so we have equilibrium as required, and this is the only solution.

P19.54 From the free-body diagram shown,

$\sum F_y = 0$: $\quad T\cos 15.0° = 1.96\times 10^{-2}$ N.

So $\quad T = 2.03\times 10^{-2}$ N.

From $\sum F_x = 0$, we have $\quad qE = T\sin 15.0°$

or $$q = \frac{T\sin 15.0°}{E} = \frac{\left(2.03\times 10^{-2}\ \text{N}\right)\sin 15.0°}{1.00\times 10^{3}\ \text{N/C}} = 5.25\times 10^{-6}\ \text{C} = \boxed{5.25\ \mu\text{C}}.$$

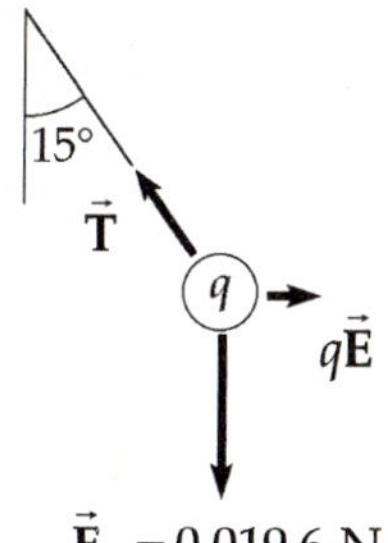

FIG. P19.54

P19.55 $F = \dfrac{k_e q_1 q_2}{r^2}$: $\quad \tan\theta = \dfrac{15.0}{60.0} \quad \theta = 14.0°$

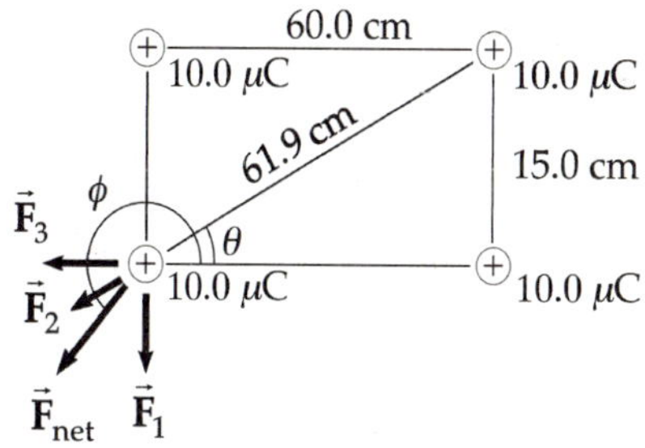

FIG. P19.55

$$F_1 = \frac{\left(8.99\times 10^{9}\right)\left(10.0\times 10^{-6}\right)^2}{(0.150)^2} = 40.0\ \text{N}$$

$$F_3 = \frac{\left(8.99\times 10^{9}\right)\left(10.0\times 10^{-6}\right)^2}{(0.600)^2} = 2.50\ \text{N}$$

$$F_2 = \frac{\left(8.99\times 10^{9}\right)\left(10.0\times 10^{-6}\right)^2}{(0.619)^2} = 2.35\ \text{N}$$

$$F_x = -F_3 - F_2\cos 14.0° = -2.50 - 2.35\cos 14.0° = -4.78\ \text{N}$$
$$F_y = -F_1 - F_2\sin 14.0° = -40.0 - 2.35\sin 14.0° = -40.6\ \text{N}$$

$$F_{\text{net}} = \sqrt{F_x^2 + F_y^2} = \sqrt{(4.78)^2 + (40.6)^2} = \boxed{40.9\ \text{N}}$$

$$\tan\phi = \frac{F_y}{F_x} = \frac{-40.6}{-4.78}$$

$$\phi = \boxed{263°}$$

P19.56 **FIG. P19.56(a)**: $d\cos 30.0° = 15.0 \text{ cm}$,

or $d = \dfrac{15.0 \text{ cm}}{\cos 30.0°}$

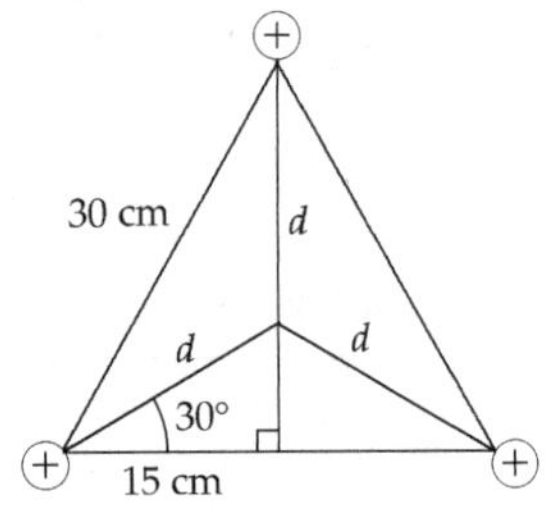

FIG. P19.56(a)

FIG. P19.56(b): $\theta = \sin^{-1}\left(\dfrac{d}{50.0 \text{ cm}}\right)$

$$\theta = \sin^{-1}\left(\frac{15.0 \text{ cm}}{50.0 \text{ cm}(\cos 30.0°)}\right) = 20.3°$$

$$\frac{F_q}{mg} = \tan\theta$$

or $F_q = mg\tan 20.3°$ (1)

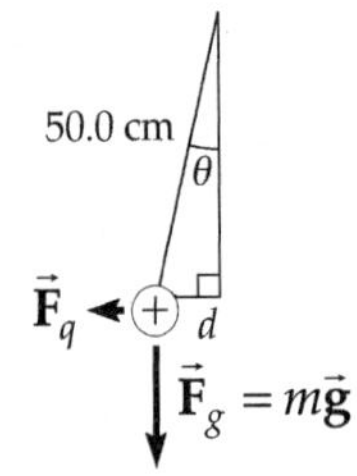

FIG. P19.56(b)

FIG. P19.56(c): $F_q = 2F\cos 30.0°$

$$F_q = 2\left[\frac{k_e q^2}{(0.300 \text{ m})^2}\right]\cos 30.0° \quad (2)$$

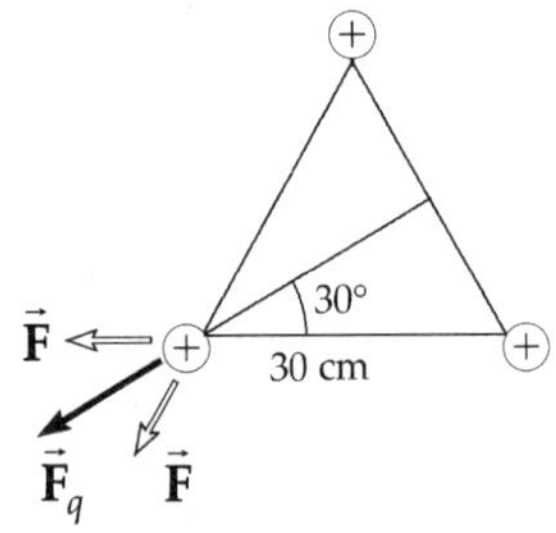

FIG. P19.56(c)

Combining equations (1) and (2),

$$2\left[\frac{k_e q^2}{(0.300 \text{ m})^2}\right]\cos 30.0° = mg\tan 20.3°$$

$$q^2 = \frac{mg(0.300 \text{ m})^2\tan 20.3°}{2k_e\cos 30.0°}$$

$$q^2 = \frac{(2.00\times10^{-3} \text{ kg})(9.80 \text{ m/s}^2)(0.300 \text{ m})^2\tan 20.3°}{2(8.99\times10^9 \text{ N}\cdot\text{m}^2/\text{C}^2)\cos 30.0°}$$

$$q = \sqrt{4.20\times10^{-14} \text{ C}^2} = 2.05\times10^{-7} \text{ C} = \boxed{0.205\ \mu\text{C}}$$

P19.57 Charge $\dfrac{Q}{2}$ resides on each block, which repel as point charges:

$$F = \frac{k_e\left(\frac{Q}{2}\right)\left(\frac{Q}{2}\right)}{L^2} = k(L - L_i)$$

$$Q = 2L\sqrt{\frac{k(L-L_i)}{k_e}} = 2(0.400 \text{ m})\sqrt{\frac{(100 \text{ N/m})(0.100 \text{ m})}{(8.99\times10^9 \text{ N}\cdot\text{m}^2/\text{C}^2)}} = \boxed{26.7\ \mu\text{C}}$$

P19.58 Charge $\frac{Q}{2}$ resides on each block, which repel as point charges: $F = \frac{k_e\left(\frac{Q}{2}\right)\left(\frac{Q}{2}\right)}{L^2} = k(L - L_i)$.

Solving for Q, $Q = \boxed{2L\sqrt{\frac{k(L-L_i)}{k_e}}}$.

Observing the spring extension gives a way of measuring the charge. If L is large compared to L_i, the charge is proportional to L to the $\frac{3}{2}$ power. The charge is proportional to the spring constant to the $\frac{1}{2}$ power. We demonstrate dimensional correctness by evaluating the units of the right hand side:

$$\mathrm{m}\left(\frac{\mathrm{N \cdot m \cdot C^2}}{\mathrm{m \cdot N \cdot m^2}}\right)^{1/2} = \mathrm{C}.$$

P19.59 (a) From the $2Q$ charge we have $F_e - T_2 \sin\theta_2 = 0$ and $mg - T_2\cos\theta_2 = 0$.

Combining these we find $\frac{F_e}{mg} = \frac{T_2 \sin\theta_2}{T_2\cos\theta_2} = \tan\theta_2$.

From the Q charge we have $F_e = T_1 \sin\theta_1 = 0$ and $mg - T_1\cos\theta_1 = 0$.

Combining these we find $\frac{F_e}{mg} = \frac{T_1 \sin\theta_1}{T_1\cos\theta_1} = \tan\theta_1$ or $\boxed{\theta_2 = \theta_1}$.

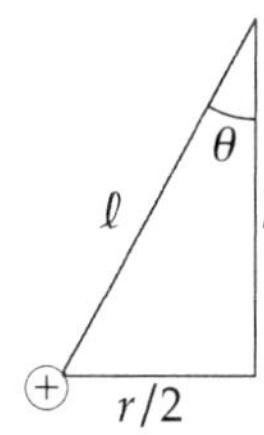

FIG. P19.59

(b) $F_e = \frac{k_e 2QQ}{r^2} = \frac{2k_e Q^2}{r^2}$

If we assume θ is small then $\tan\theta \approx \frac{r/2}{\ell}$.

Substitute expressions for F_e and $\tan\theta$ into either equation found in part (a) and solve for r.

$\frac{F_e}{mg} = \tan\theta$ then $\frac{2k_e Q^2}{r^2}\left(\frac{1}{mg}\right) \approx \frac{r}{2\ell}$ and solving for r we find $r \approx \left(\frac{4k_e Q^2 \ell}{mg}\right)^{1/3}$.

P19.60 At an equilibrium position, the net force on the charge Q is zero. The equilibrium position can be located by determining the angle θ corresponding to equilibrium.

In terms of lengths s, $\frac{1}{2}a\sqrt{3}$, and r, shown in Figure P19.60, the charge at the origin exerts an attractive force

$$\frac{k_e Qq}{\left(s + \frac{1}{2}a\sqrt{3}\right)^2}$$

The other two charges exert equal repulsive forces of magnitude $\frac{k_e Qq}{r^2}$. The horizontal components of the two repulsive forces add, balancing the attractive force,

continued on next page

$$F_{net} = k_e Qq\left[\frac{2\cos\theta}{r^2} - \frac{1}{\left(s+\frac{1}{2}a\sqrt{3}\right)^2}\right] = 0$$

From Figure P19.60 $\quad r = \dfrac{\frac{1}{2}a}{\sin\theta} \qquad s = \dfrac{1}{2}a\cot\theta$

The equilibrium condition, in terms of θ, is $\quad F_{net} = \left(\dfrac{4}{a^2}\right)k_e Qq\left(2\cos\theta\sin^2\theta - \dfrac{1}{\left(\sqrt{3}+\cot\theta\right)^2}\right) = 0.$

Thus the equilibrium value of θ satisfies $\quad 2\cos\theta\sin^2\theta\left(\sqrt{3}+\cot\theta\right)^2 = 1.$

One method for solving for θ is to tabulate the left side. To three significant figures a value of θ corresponding to equilibrium is 81.7°.

The distance from the vertical side of the triangle to the equilibrium position is

$$s = \frac{1}{2}a\cot 81.7° = \boxed{0.072\,9a}.$$

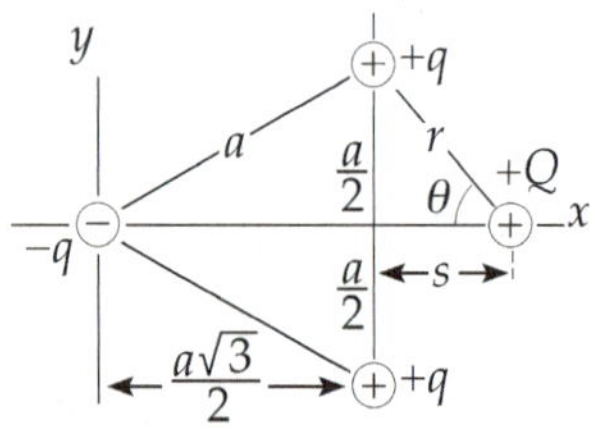

FIG. P19.60

θ	$2\cos\theta\sin^2\theta\left(\sqrt{3}+\cot\theta\right)^2$
60°	4
70°	2.654
80°	1.226
90°	0
81°	1.091
81.5°	1.024
81.7°	0.997

A second zero-field point is on the negative side of the x-axis, where $\theta = -9.16°$ and $s = -3.10a$.

P19.61 (a) Zero contribution from the same face due to symmetry, opposite face contributes

$4\left(\dfrac{k_e q}{r^2}\sin\phi\right)$ where $r = \sqrt{\left(\dfrac{s}{2}\right)^2 + \left(\dfrac{s}{2}\right)^2 + s^2} = \sqrt{1.5}s = 1.22s$

$\sin\phi = \dfrac{s}{r} \qquad E = 4\dfrac{k_e qs}{r^3} = \dfrac{4}{(1.22)^3}\dfrac{k_e q}{s^2} = \boxed{2.18\dfrac{k_e q}{s^2}}$

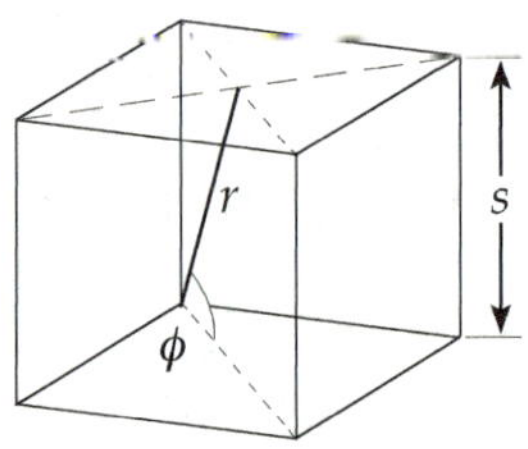

FIG. P19.61

(b) $\boxed{\text{The direction is the } \hat{\mathbf{k}} \text{ direction.}}$

P19.62 Consider the field due to a single sheet and let E_+ and E_- represent the fields due to the positive and negative sheets. The field at any distance from each sheet has a magnitude given by Equation 19.24:

$$|E_+| = |E_-| = \frac{\sigma}{2\epsilon_0}.$$

(a) To the left of the positive sheet, E_+ is directed toward the left and E_- toward the right and the net field over this region is $\vec{\mathbf{E}} = \boxed{0}$.

(b) In the region between the sheets, E_+ and E_- are both directed toward the right and the net field is

$$\vec{\mathbf{E}} = \boxed{\frac{\sigma}{\epsilon_0} \text{ to the right}}.$$

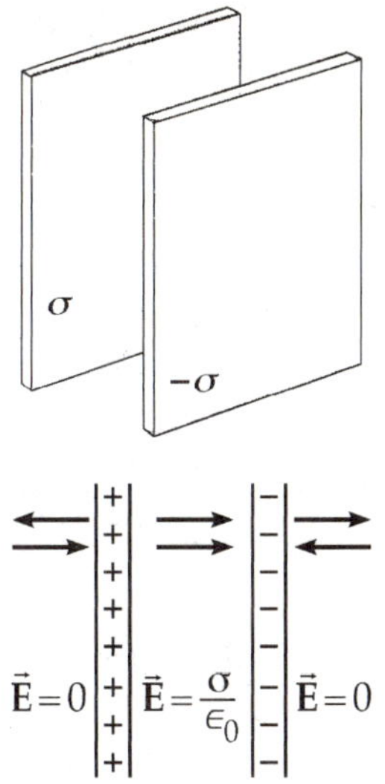

FIG. P19.62

(c) To the right of the negative sheet, E_+ and E_- are again oppositely directed and $\vec{\mathbf{E}} = \boxed{0}$.

P19.63 The magnitude of the field due to the each sheet given by Equation 19.24 is

$$E = \frac{\sigma}{2\epsilon_0} \text{ directed perpendicular to the sheet.}$$

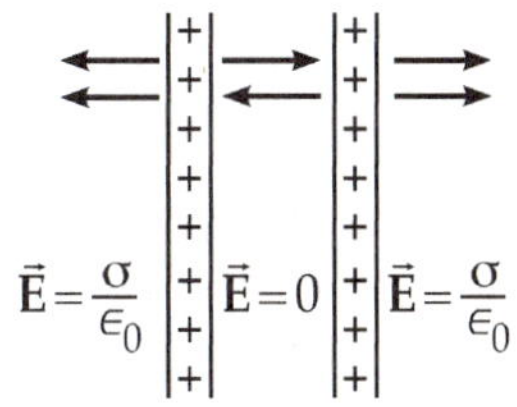

FIG. P19.63

(a) In the region to the left of the pair of sheets, both fields are directed toward the left and the net field is

$$\vec{\mathbf{E}} = \boxed{\frac{\sigma}{\epsilon_0} \text{ to the left}}.$$

(b) In the region between the sheets, the fields due to the individual sheets are oppositely directed and the net field is

$$\vec{\mathbf{E}} = \boxed{0}.$$

(c) In the region to the right of the pair of sheets, both fields are directed toward the right and the net field is

$$\vec{\mathbf{E}} = \boxed{\frac{\sigma}{\epsilon_0} \text{ to the right}}.$$

P19.64 $$d\vec{\mathbf{E}}=\frac{k_e dq}{x^2+(0.150\text{ m})^2}\left(\frac{-x\hat{\mathbf{i}}+0.150\text{ m}\hat{\mathbf{j}}}{\sqrt{x^2+(0.150\text{ m})^2}}\right)=\frac{k_e\lambda\left(-x\hat{\mathbf{i}}+0.150\text{ m}\hat{\mathbf{j}}\right)dx}{\left[x^2+(0.150\text{ m})^2\right]^{3/2}}$$

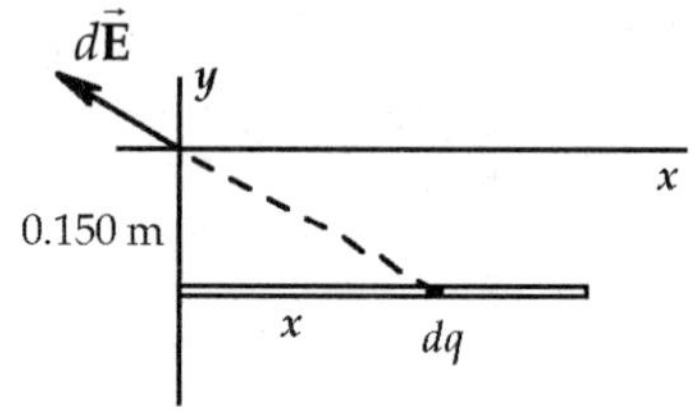

FIG. P19.64

$$\vec{\mathbf{E}}=\int_{\text{all charge}}d\vec{\mathbf{E}}=k_e\lambda\int_{x=0}^{0.400\text{ m}}\frac{\left(-x\hat{\mathbf{i}}+0.150\text{ m}\hat{\mathbf{j}}\right)dx}{\left[x^2+(0.150\text{ m})^2\right]^{3/2}}$$

$$\vec{\mathbf{E}}=k_e\lambda\left[\frac{+\hat{\mathbf{i}}}{\sqrt{x^2+(0.150\text{ m})^2}}\Bigg|_0^{0.400\text{ m}}+\frac{(0.150\text{ m})\hat{\mathbf{j}}x}{(0.150\text{ m})^2\sqrt{x^2+(0.150\text{ m})^2}}\Bigg|_0^{0.400\text{ m}}\right]$$

$$\vec{\mathbf{E}}=\left(8.99\times10^9\text{ N}\cdot\text{m}^2/\text{C}^2\right)\left(35.0\times10^{-9}\text{ C/m}\right)\left[\hat{\mathbf{i}}(2.34-6.67)\text{ m}^{-1}+\hat{\mathbf{j}}(6.24-0)\text{ m}^{-1}\right]$$

$$\vec{\mathbf{E}}=\left(-1.36\hat{\mathbf{i}}+1.96\hat{\mathbf{j}}\right)\times10^3\text{ N/C}=\boxed{\left(-1.36\hat{\mathbf{i}}+1.96\hat{\mathbf{j}}\right)\text{ kN/C}}$$

P19.65 (a) $\oint\vec{\mathbf{E}}\cdot d\vec{\mathbf{A}}=E\left(4\pi r^2\right)=\frac{q_{\text{in}}}{\epsilon_0}$

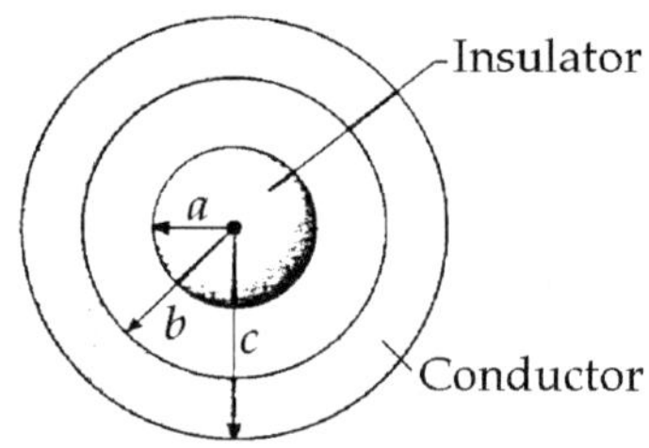

FIG. P19.65

For $r<a$, $q_{\text{in}}=\rho\left(\frac{4}{3}\pi r^3\right)$

so $E=\boxed{\frac{\rho r}{3\epsilon_0}}$.

For $a<r<b$ and $c<r$, $q_{\text{in}}=Q$.

So $E=\boxed{\frac{Q}{4\pi r^2\epsilon_0}}$.

For $b\le r\le c$, $E=0$, since $\boxed{E=0}$ inside a conductor.

(b) Let $q_1=$ induced charge on the inner surface of the hollow sphere. Since $E=0$ inside the conductor, the total charge enclosed by a spherical surface of radius $b\le r\le c$ must be zero.

Therefore, $q_1+Q=0$ and $\sigma_1=\frac{q_1}{4\pi b^2}=\boxed{\frac{-Q}{4\pi b^2}}$.

Let $q_2=$ induced charge on the outside surface of the hollow sphere. Since the hollow sphere is uncharged, we require

$q_1+q_2=0$ and $\sigma_2=\frac{q_1}{4\pi c^2}=\boxed{\frac{Q}{4\pi c^2}}$.

P19.66 The field on the axis of the ring is calculated in Example 19.5, $E = E_x = \dfrac{k_e x Q}{\left(x^2 + a^2\right)^{3/2}}$

The force experienced by a charge $-q$ placed along the axis of the ring is $F = -k_e Q q \left[\dfrac{x}{\left(x^2+a^2\right)^{3/2}}\right]$

and when $x \ll a$, this becomes $F = -\left(\dfrac{k_e Q q}{a^3}\right)x$

This expression for the force is in the form of Hooke's law, with an

effective spring constant of $k = \dfrac{k_e Q q}{a^3}$

Since $\omega = 2\pi f = \sqrt{\dfrac{k}{m}}$, we have $f = \boxed{\dfrac{1}{2\pi}\sqrt{\dfrac{k_e Q q}{m a^3}}}$.

P19.67 The resultant field within the cavity is the superposition of two fields, one $\vec{\mathbf{E}}_+$ due to a uniform sphere of positive charge of radius $2a$, and the other $\vec{\mathbf{E}}_-$ due to a sphere of negative charge of radius a centered within the cavity.

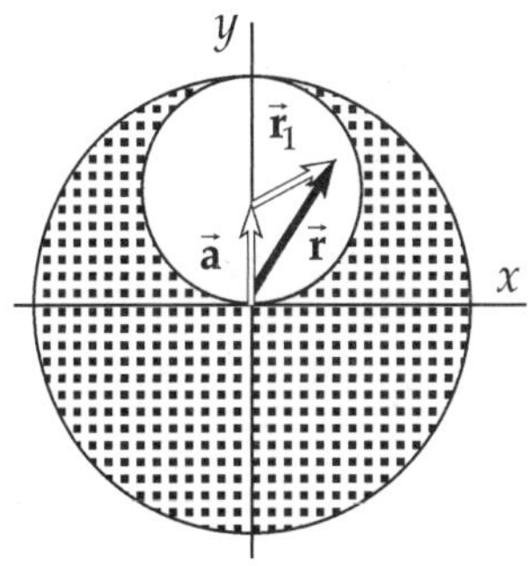

FIG. P19.67

$$\frac{4}{3}\left(\frac{\pi r^3 \rho}{\epsilon_0}\right) = 4\pi r^2 E_+ \qquad \text{so} \qquad \vec{\mathbf{E}}_+ = \frac{\rho r}{3\epsilon_0}\hat{\mathbf{r}} = \frac{\rho\vec{\mathbf{r}}}{3\epsilon_0}$$

$$-\frac{4}{3}\left(\frac{\pi r_1^3 \rho}{\epsilon_0}\right) = 4\pi r_1^2 E_- \qquad \text{so} \qquad \vec{\mathbf{E}}_- = \frac{\rho r_1}{3\epsilon_0}\left(-\hat{\mathbf{r}}_1\right) = \frac{-\rho}{3\epsilon_0}\vec{\mathbf{r}}_1.$$

Since $\vec{\mathbf{r}} = \vec{\mathbf{a}} + \vec{\mathbf{r}}_1$, $\quad \vec{\mathbf{E}}_- = \dfrac{-\rho\left(\vec{\mathbf{r}} - \vec{\mathbf{a}}\right)}{3\epsilon_0}$

$$\vec{\mathbf{E}} = \vec{\mathbf{E}}_+ + \vec{\mathbf{E}}_- = \frac{\rho\vec{\mathbf{r}}}{3\epsilon_0} - \frac{\rho\vec{\mathbf{r}}}{3\epsilon_0} + \frac{\rho\vec{\mathbf{a}}}{3\epsilon_0} = \frac{\rho\vec{\mathbf{a}}}{3\epsilon_0} = 0\hat{\mathbf{i}} + \frac{\rho a}{3\epsilon_0}\hat{\mathbf{j}}.$$

Thus, $\boxed{E_x = 0}$

and $\boxed{E_y = \dfrac{\rho a}{3\epsilon_0}}$ at all points within the cavity.

ANSWERS TO EVEN PROBLEMS

P19.2 (a) 2.62×10^{24} electrons; (b) 2.38 electrons for every 10^9 already present

P19.4 57.5 N away from the other proton

P19.6 1.57 μN to the left

P19.8 $0.634d$, stable if the third bead has positive charge

P19.10 (a) $-\left(5.58\times10^{-11}\ \text{N/C}\right)\hat{\mathbf{j}}$; (b) $\left(1.02\times10^{-7}\ \text{N/C}\right)\hat{\mathbf{j}}$

P19.12 $-9Q$ and $+27Q$

P19.14 (a) $\vec{\mathbf{E}}=1.29\times10^{4}\hat{\mathbf{j}}\ \text{N/C}$; (b) $-3.86\times10^{-2}\hat{\mathbf{j}}\ \text{N}$

P19.16 see the solution

P19.18 $\dfrac{k_e\lambda_0}{x_0}$, The direction is $-\hat{\mathbf{i}}$ or left for $\lambda_0>0$

P19.20 see the solution

P19.22 (a) see the solution; (b) see the solution

P19.24 (a) $-\dfrac{1}{3}$; (b) q_1 is negative and q_2 is positive

P19.26 (a) at the center; (b) $1.73k_e\dfrac{q}{a^2}\hat{\mathbf{j}}$

P19.28 in the direction of motion, $\dfrac{K}{ed}$

P19.30 $355\ \text{kN}\cdot\text{m}^2/\text{C}$

P19.32 (a) $-55.7\ \text{nC}$; (b) negative, spherically symmetric, and concentric with the shell

P19.34 (a) $3.20\ \text{MN}\cdot\text{m}^2/\text{C}$; (b) $19.2\ \text{MN}\cdot\text{m}^2/\text{C}$; (c) see the solution

P19.36 $-2.48\ \mu\text{C/m}^2$

P19.38 (a) 0; (b) 7.19 MN/C radially outward

P19.40 (a) $\dfrac{Qr^3}{\epsilon_0 a^3}$; (b) $\dfrac{Q}{\epsilon_0}$; (c) see the solution

P19.42 (a) 0; (b) 5 400 N/C, outward; (c) 540 N/C, outward

P19.44 (a) $708\ \text{nC/m}^2$ and $-708\ \text{nC/m}^2$; (b) 177 nC and −177 nC

P19.46 (a) $248\ \text{nC/m}^2$; (b) $496\ \text{nC/m}^2$

P19.48 (a) If the volume charge density is nonzero, the field cannot be uniform in magnitude. (b) The field must be uniform in magnitude.

P19.50 (a) negative, $1.06\times10^{-9}\ \text{C/m}^2$; (b) $-5.42\times10^{5}\ \text{C}$, 3.38×10^{24} excess electrons

P19.52 (a) away from Earth, $4.18\times10^{3}\ \text{N}$; (b) the electric force is weaker by 4.77×10^{16} times and in the opposite direction

P19.54 5.25 μC

P19.56 0.205 μC

P19.58 $2L\sqrt{\dfrac{k(L-L_i)}{k_e}}$

P19.60 0.072 9a

P19.62 (a) 0; (b) $\dfrac{\sigma}{\epsilon_0}$ to the right; (c) 0

P19.64 $\left(-1.36\hat{\mathbf{i}}+1.96\hat{\mathbf{j}}\right)\ \text{kN/C}$

P19.66 see the solution

Chapter 20

Electric Potential and Capacitance

CHAPTER OUTLINE

ANSWERS TO QUESTIONS

Q20.1 When one object B with electric charge is immersed in the electric field of another charge or charges A, the system possesses electric potential energy. The energy can be measured by seeing how much work the field does on the charge B as it moves to a reference location. We choose not to visualize A's effect on B as an action-at-a-distance, but as the result of a two-step process: Charge A creates electric potential throughout the surrounding space. Then the potential acts on B to inject the system with energy.

Q20.2 The potential energy increases. When an outside agent makes it move in the direction of the field, the charge moves to a region of lower electric potential. Then the product of its negative charge with a lower number of volts gives a higher number of joules. Keep in mind that a negative charge feels an electric force in the *opposite* direction to the field, while the potential is the work done on the charge to move it in a field per unit charge.

Q20.3 To move like charges together from an infinite separation, at which the potential energy of the system of two charges is zero, requires *work* to be done on the system by an outside agent. Hence energy is stored, and potential energy is positive. As charges with opposite signs move together from an infinite separation, energy is released, and the potential energy of the set of charges becomes negative.

Q20.4 If there were a potential difference between two points on the conductor, the free electrons in the conductor would move until the potential difference disappears.

Q20.5 A sharp point in a charged conductor would imply a large electric field in that region. An electric discharge could most easily take place at that sharp point.

Q20.6 Use a conductive box to shield the equipment. Any stray electric field will cause charges on the outer surface of the conductor to rearrange and cancel the stray field inside the volume it encloses.

Q20.7 The grounding wire can be touched equally well to any point on the sphere. Electrons will drain away into the ground and the sphere will be left positively charged. The ground, wire, and sphere are all conducting. They together form an equipotential volume at zero volts during the contact. However close the grounding wire is to the negative charge, electrons have no difficulty in moving within the metal through the grounding wire to ground. The ground can act as an infinite source or sink of electrons. In this case, it is an electron sink.

Q20.8 Nothing happens to the charge if the wires are disconnected. If the wires are connected to each other, charges in the single conductor that now exists move between the wires and the plates until the entire conductor is at a single potential and the capacitor is discharged.

Q20.9 The parallel-connected capacitors store more energy because they have higher equivalent capacitance.

Q20.10 Seventeen combinations:

Individual C_1, C_2, C_3

Parallel $C_1 + C_2 + C_3,\ C_1 + C_2,\ C_1 + C_3,\ C_2 + C_3$

Series-Parallel $\left(\frac{1}{C_1}+\frac{1}{C_2}\right)^{-1}+C_3,\ \left(\frac{1}{C_1}+\frac{1}{C_3}\right)^{-1}+C_2,\ \left(\frac{1}{C_2}+\frac{1}{C_3}\right)^{-1}+C_1$

$$\left(\frac{1}{C_1+C_2}+\frac{1}{C_3}\right)^{-1},\ \left(\frac{1}{C_1+C_3}+\frac{1}{C_2}\right)^{-1},\ \left(\frac{1}{C_2+C_3}+\frac{1}{C_1}\right)^{-1}$$

Series $\left(\frac{1}{C_1}+\frac{1}{C_2}+\frac{1}{C_3}\right)^{-1},\ \left(\frac{1}{C_1}+\frac{1}{C_2}\right)^{-1},\ \left(\frac{1}{C_2}+\frac{1}{C_3}\right)^{-1},\ \left(\frac{1}{C_1}+\frac{1}{C_3}\right)^{-1}$

Q20.11 Energy is proportional to voltage squared. It gets four times larger.

Q20.12 The work you do to pull the plates apart becomes additional electric potential energy stored in the capacitor. The charge is constant and the capacitance decreases but the potential difference increases to drive up the potential energy $\frac{1}{2}Q\Delta V$. The electric field between the plates is constant in strength but fills more volume as you pull the plates apart.

Q20.13 Let C = the capacitance of an individual capacitor, and C_s represent the equivalent capacitance of the group in series. While being charged in parallel, each capacitor receives charge

$Q = C\Delta V_{\text{charge}} = \left(5.00\times10^{-4}\ \text{F}\right)(800\ \text{V}) = 0.400\ \text{C}$.

While being discharged in series, $\Delta V_{\text{discharge}} = \frac{Q}{C_s} = \frac{Q}{C/10} = \frac{0.400\ \text{C}}{5.00\times10^{-5}\ \text{F}} = 8.00\ \text{kV}$

(or 10 times the original voltage).

Q20.14 Put a material with higher dielectric strength between the plates, or evacuate the space between the plates. At very high voltages, you may want to cool off the plates or choose to make them of a different chemically stable material, because atoms in the plates themselves can ionize, showing *thermionic emission* under high electric fields.

Q20.15 The primary choice would be the dielectric. You would want to chose a dielectric that has a large dielectric constant and dielectric strength, such as strontium titanate, where $\kappa \approx 233$ (Table 20.1). A convenient choice could be thick plastic or mylar. Secondly, geometry would be a factor. To maximize capacitance, one would want the individual plates as close as possible, since the capacitance is proportional to the inverse of the plate separation—hence the need for a dielectric with a high dielectric strength. Also, one would want to build, instead of a single parallel plate capacitor, several capacitors in parallel. This could be achieved through "stacking" the plates of the capacitor. For example, you can alternately lay down sheets of a conducting material, such as aluminum foil, sandwiched between your sheets of insulating dielectric. Making sure that none of the conducting sheets are in contact with their next neighbors, connect every other plate together. Figure Q20.15 illustrates this idea.

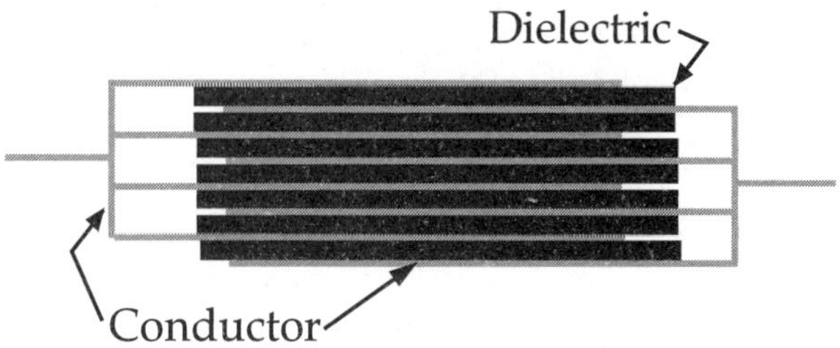

FIG. Q20.15

This technique is often used when "home-brewing" signal capacitors for radio applications, as they can withstand huge potential differences without flashover (without either discharge between plates around the dielectric or dielectric breakdown). One variation on this technique is to sandwich together flexible materials such as aluminum roof flashing and thick plastic, so the whole product can be rolled up into a "capacitor burrito" and placed in an insulating tube, such as a PVC pipe, and then filled with motor oil (again to prevent flashover).

Q20.16 The material of the dielectric may be able to support a larger electric field than air, without breaking down to pass a spark between the capacitor plates.

SOLUTIONS TO PROBLEMS

Section 20.1 Potential Difference and Electric Potential

P20.1 (a) Energy of the proton-field system is conserved as the proton moves from high to low potential, which can be defined for this problem as moving from 120 V down to 0 V.

$$K_i + U_i + \Delta E_{\text{mech}} = K_f + U_f \qquad 0 + qV + 0 = \frac{1}{2}mv_p^2 + 0$$

$$\left(1.60\times10^{-19}\text{ C}\right)(120\text{ V})\left(\frac{1\text{ J}}{1\text{ V}\cdot\text{C}}\right) = \frac{1}{2}\left(1.67\times10^{-27}\text{ kg}\right)v_p^2$$

$$v_p = \boxed{1.52\times10^5\text{ m/s}}$$

continued on next page

(b) The electron will gain speed in moving the other way,

from $V_i = 0$ to $V_f = 120$ V: $K_i + U_i + \Delta E_{\text{mech}} = K_f + U_f$

$$0 + 0 + 0 = \frac{1}{2}mv_e^2 + qV$$

$$0 = \frac{1}{2}\left(9.11\times10^{-31}\text{ kg}\right)v_e^2 + \left(-1.60\times10^{-19}\text{ C}\right)(120\text{ J/C})$$

$$v_e = \boxed{6.49\times10^{6}\text{ m/s}}$$

P20.2 $\Delta V = -14.0$ V and $Q = -N_A e = -\left(6.02\times10^{23}\right)\left(1.60\times10^{-19}\right) = -9.63\times10^{4}$ C

$\Delta V = \frac{W}{Q}$, so $W = Q\Delta V = \left(-9.63\times10^{4}\text{ C}\right)(-14.0\text{ J/C}) = \boxed{1.35\text{ MJ}}$

Section 20.2 Potential Differences in a Uniform Electric Field

P20.3 (a) We follow the path from (0, 0) to (20.0 cm, 0) to (20.0 cm, 50.0 cm).

$\Delta U = -$(work done)

$\Delta U = -$(work from origin to (20.0 cm, 0)) – (work from (20.0 cm, 0) to (20.0 cm, 50.0 cm))

Note that the last term is equal to 0 because the force is perpendicular to the displacement.

$$\Delta U = -(qE_x)\Delta x = -\left(12.0\times10^{-6}\text{ C}\right)(250\text{ V/m})(0.200\text{ m}) = \boxed{-6.00\times10^{-4}\text{ J}}$$

(b) $$\Delta V = \frac{\Delta U}{q} = -\frac{6.00\times10^{-4}\text{ J}}{12.0\times10^{-6}\text{ C}} = -50.0\text{ J/C} = \boxed{-50.0\text{ V}}$$

P20.4 $$E = \frac{|\Delta V|}{d} = \frac{25.0\times10^{3}\text{ J/C}}{1.50\times10^{-2}\text{ m}} = 1.67\times10^{6}\text{ N/C} = \boxed{1.67\text{ MN/C}}$$

P20.5 $$\Delta U = -\frac{1}{2}m\left(v_f^2 - v_i^2\right) = -\frac{1}{2}\left(9.11\times10^{-31}\text{ kg}\right)\left[\left(1.40\times10^{5}\text{ m/s}\right)^2 - \left(3.70\times10^{6}\text{ m/s}\right)^2\right] = 6.23\times10^{-18}\text{ J}$$

$\Delta U = q\Delta V$: $+6.23\times10^{-18} = \left(-1.60\times10^{-19}\right)\Delta V$

$\Delta V = \boxed{-38.9\text{ V. The origin is at highest potential.}}$

P20.6 (a) Arbitrarily choose $V=0$ at 0. Then at other points

$V=-Ex$ and $U_e=QV=-QEx$.

Between the endpoints of the motion,

$$(K+U_s+U_e)_i=(K+U_s+U_e)_f$$

$$0+0+0=0+\frac{1}{2}kx_{max}^2-QEx_{max} \text{ so } x_{max}=\boxed{\frac{2QE}{k}}.$$

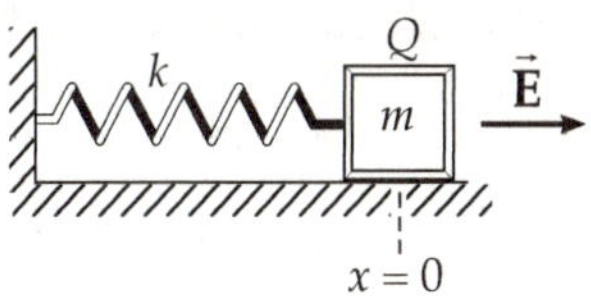

FIG. P20.6

(b) At equilibrium,

$\sum F_x=-F_s+F_e=0$ or $kx=QE$.

So the equilibrium position is at $x=\boxed{\frac{QE}{k}}$.

(c) The block's equation of motion is $\sum F_x=-kx+QE=m\frac{d^2x}{dt^2}$.

Let $x'=x-\frac{QE}{k}$, or $x=x'+\frac{QE}{k}$,

so the equation of motion becomes:

$$-k\left(x'+\frac{QE}{k}\right)+QE=m\frac{d^2(x+QE/k)}{dt^2}, \text{ or } \frac{d^2x'}{dt^2}=-\left(\frac{k}{m}\right)x'.$$

This is the equation for simple harmonic motion $a_{x'}=-\omega^2x'$

with $\omega=\sqrt{\frac{k}{m}}$.

The period of the motion is then $T=\frac{2\pi}{\omega}=\boxed{2\pi\sqrt{\frac{m}{k}}}$.

(d) $(K+U_s+U_e)_i+\Delta E_{mech}=(K+U_s+U_e)_f$

$$0+0+0-\mu_k mgx_{max}=0+\frac{1}{2}kx_{max}^2-QEx_{max}$$

$$x_{max}=\boxed{\frac{2(QE-\mu_k mg)}{k}}$$

Section 20.3 Electric Potential and Electric Potential Energy Due to Point Charges

P20.7 (a) The potential at 1.00 cm is $V_1 = k_e \frac{q}{r} = \frac{\left(8.99\times10^9\ \text{N}\cdot\text{m}^2/\text{C}^2\right)\left(1.60\times10^{-19}\ \text{C}\right)}{1.00\times10^{-2}\ \text{m}} = \boxed{1.44\times10^{-7}\ \text{V}}$.

(b) The potential at 2.00 cm is $V_2 = k_e \frac{q}{r} = \frac{\left(8.99\times10^9\ \text{N}\cdot\text{m}^2/\text{C}^2\right)\left(1.60\times10^{-19}\ \text{C}\right)}{2.00\times10^{-2}\ \text{m}} = 0.719\times10^{-7}\ \text{V}$.

Thus, the difference in potential between the two points is $\Delta V = V_2 - V_1 = \boxed{-7.19\times10^{-8}\ \text{V}}$.

(c) The approach is the same as above except the charge is $-1.60\times10^{-19}\ \text{C}$. This changes the sign of each answer, with its magnitude remaining the same.

That is, the potential at 1.00 cm is $\boxed{-1.44\times10^{-7}\ \text{V}}$.

The potential at 2.00 cm is $-0.719\times10^{-7}\ \text{V}$, so $\Delta V = V_2 - V_1 = \boxed{7.19\times10^{-8}\ \text{V}}$.

P20.8 (a) Since the charges are equal and placed symmetrically, $\boxed{F=0}$.

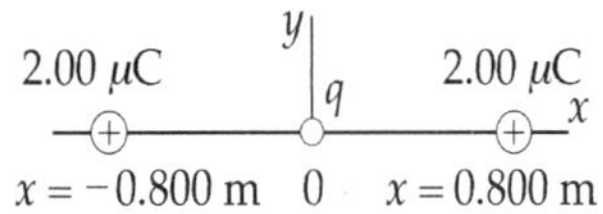

FIG. P20.8

(b) Since $F = qE = 0$, $\boxed{E=0}$.

(c) $V = 2k_e \frac{q}{r} = 2\left(8.99\times10^9\ \text{N}\cdot\text{m}^2/\text{C}^2\right)\left(\frac{2.00\times10^{-6}\ \text{C}}{0.800\ \text{m}}\right)$

$V = 4.50\times10^4\ \text{V} = \boxed{45.0\ \text{kV}}$

P20.9 (a) $E_x = \frac{k_e q_1}{x^2} + \frac{k_e q_2}{(x-2.00)^2} = 0$ becomes $E_x = k_e\left(\frac{+q}{x^2} + \frac{-2q}{(x-2.00)^2}\right) = 0$.

Dividing by k_e, $2qx^2 = q(x-2.00)^2$ $x^2 + 4.00x - 4.00 = 0$.

Therefore $E=0$ when $x = \frac{-4.00 \pm \sqrt{16.0+16.0}}{2} = \boxed{-4.83\ \text{m}}$.

(Note that the positive root does not correspond to a physically valid situation.)

(b) $V = \frac{k_e q_1}{x} + \frac{k_e q_2}{2.00 - x} = 0$ or $V = k_e\left(\frac{+q}{x} - \frac{2q}{2.00-x}\right) = 0$.

Again solving for *x*, $2qx = q(2.00 - x)$.

For $0 \le x \le 2.00$ $V = 0$ when $x = \boxed{0.667\ \text{m}}$

and $\frac{q}{|x|} = \frac{-2q}{|2-x|}$. For $x<0$ $x = \boxed{-2.00\ \text{m}}$.

P20.10 $U_e = q_4V_1 + q_4V_2 + q_4V_3 = q_4\left(\frac{1}{4\pi\epsilon_0}\right)\left(\frac{q_1}{r_1}+\frac{q_2}{r_2}+\frac{q_3}{r_3}\right)$

$$U_e = \left(10.0\times10^{-6}\text{ C}\right)^2\left(8.99\times10^9\text{ N}\cdot\text{m}^2/\text{C}^2\right)\left(\frac{1}{0.600\text{ m}}+\frac{1}{0.150\text{ m}}+\frac{1}{\sqrt{(0.600\text{ m})^2+(0.150\text{ m})^2}}\right)$$

$$U_e = \boxed{8.95\text{ J}}$$

P20.11 $V = \sum_i k\frac{q_i}{r_i}$

$$V = \left(8.99\times10^9\right)\left(7.00\times10^{-6}\right)\left[\frac{-1}{0.010\,0}-\frac{1}{0.010\,0}+\frac{1}{0.038\,7}\right]$$

$$V = \boxed{-1.10\times10^7\text{ V} = -11.0\text{ MV}}$$

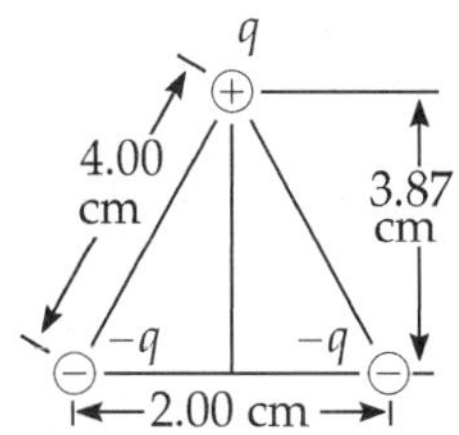

FIG. P20.11

P20.12 (a) $V = \frac{k_eq_1}{r_1}+\frac{k_eq_2}{r_2} = 2\left(\frac{k_eq}{r}\right)$

$$V = 2\left(\frac{\left(8.99\times10^9\text{ N}\cdot\text{m}^2/\text{C}^2\right)\left(2.00\times10^{-6}\text{ C}\right)}{\sqrt{(1.00\text{ m})^2+(0.500\text{ m})^2}}\right)$$

$$V = 3.22\times10^4\text{ V} = \boxed{32.2\text{ kV}}$$

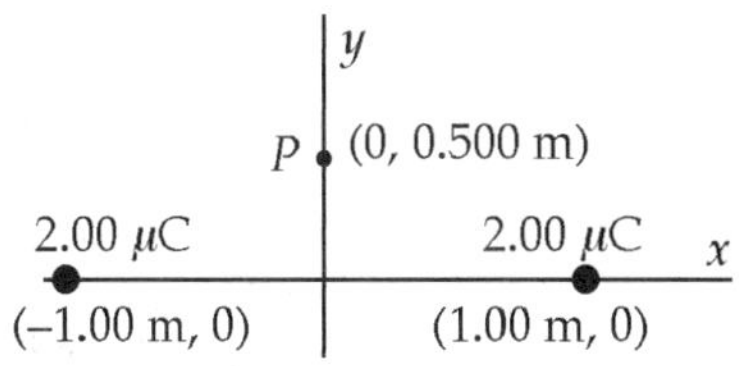

FIG. P20.12

(b) $U = qV = \left(-3.00\times10^{-6}\text{ C}\right)\left(3.22\times10^4\text{ J/C}\right) = \boxed{-9.65\times10^{-2}\text{ J}}$

P20.13 $U = U_1 + U_2 + U_3 + U_4$

$$U = 0 + U_{12} + (U_{13}+U_{23}) + (U_{14}+U_{24}+U_{34})$$

$$U = 0 + \frac{k_eQ^2}{s} + \frac{k_eQ^2}{s}\left(\frac{1}{\sqrt{2}}+1\right) + \frac{k_eQ^2}{s}\left(1+\frac{1}{\sqrt{2}}+1\right)$$

$$U = \frac{k_eQ^2}{s}\left(4+\frac{2}{\sqrt{2}}\right) = \boxed{5.41\frac{k_eQ^2}{s}}$$

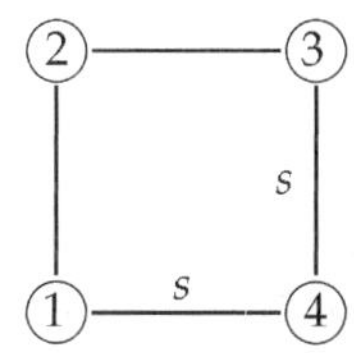

FIG. P20.13

***P20.14** (a) The first expression, with distances squared, describes an electric field. The second expression describes an electric potential. Then a $\boxed{\text{positive 7 nC charge}}$ is 7 cm from the origin. To create field that is to the left and downward, it must be in the first quadrant, with position vector $\boxed{\text{7 cm at } 70°}$. A $\boxed{\text{negative 8 nC charge}}$ 3 cm from the origin creates an upward electric field at the origin, so it must be at $\boxed{\text{3 cm at } 90°}$. We evaluate the given expressions:

$$\vec{\mathbf{E}} = -4.39\text{ kN/C}\,\hat{\mathbf{i}} + 67.8\text{ kN/C}\,\hat{\mathbf{j}}$$
$$V = -1.50\text{ kV}$$

continued on next page

(b) $\vec{\mathbf{F}} = q\vec{\mathbf{E}} = -16\times10^{-9}\ \text{C}\left(-4.39\hat{\mathbf{i}} + 67.8\hat{\mathbf{j}}\right)10^{3}\ \text{N/C} = \boxed{\left(7.03\hat{\mathbf{i}} - 109\hat{\mathbf{j}}\right)\times10^{-5}\ \text{N}}$

(c) $U_e = qV = -16\times10^{-9}\ \text{C}\left(-1.50\times10^{3}\ \text{J/C}\right) = \boxed{-2.40\times10^{-5}\ \text{J}}$

P20.15 Consider the two spheres as a system.

(a) Conservation of momentum: $0 = m_1 v_1\hat{\mathbf{i}} + m_2 v_2\left(-\hat{\mathbf{i}}\right)$ or $v_2 = \dfrac{m_1 v_1}{m_2}$

By conservation of energy, $0 = \dfrac{k_e(-q_1)q_2}{d} = \dfrac{1}{2}m_1 v_1^2 + \dfrac{1}{2}m_2 v_2^2 + \dfrac{k_e(-q_1)q_2}{r_1+r_2}$

and $\dfrac{k_e q_1 q_2}{r_1+r_2} - \dfrac{k_e q_1 q_2}{d} = \dfrac{1}{2}m_1 v_1^2 + \dfrac{1}{2}\dfrac{m_1^2 v_1^2}{m_2}$

$$v_1 = \sqrt{\frac{2m_2 k_e q_1 q_2}{m_1(m_1+m_2)}\left(\frac{1}{r_1+r_2} - \frac{1}{d}\right)}$$

$$v_1 = \sqrt{\frac{2(0.700\ \text{kg})\left(8.99\times10^{9}\ \text{N}\cdot\text{m}^2/\text{C}^2\right)\left(2\times10^{-6}\ \text{C}\right)\left(3\times10^{-6}\ \text{C}\right)}{(0.100\ \text{kg})(0.800\ \text{kg})}\left(\frac{1}{8\times10^{-3}\ \text{m}} - \frac{1}{1.00\ \text{m}}\right)}$$

$$= \boxed{10.8\ \text{m/s}}$$

$$v_2 = \frac{m_1 v_1}{m_2} = \frac{0.100\ \text{kg}(10.8\ \text{m/s})}{0.700\ \text{kg}} = \boxed{1.55\ \text{m/s}}$$

(b) If the spheres are metal, electrons will move around on them with negligible energy loss to place the centers of excess charge on the insides of the spheres. Then just before they touch, the effective distance between charges will be less than $r_1 + r_2$ and the spheres will really be moving $\boxed{\text{faster than calculated in (a)}}$.

P20.16 Consider the two spheres as a system.

(a) Conservation of momentum: $0 = m_1 v_1\hat{\mathbf{i}} + m_2 v_2\left(-\hat{\mathbf{i}}\right)$

or $v_2 = \dfrac{m_1 v_1}{m_2}$.

By conservation of energy, $0 = \dfrac{k_e(-q_1)q_2}{d} = \dfrac{1}{2}m_1 v_1^2 + \dfrac{1}{2}m_2 v_2^2 + \dfrac{k_e(-q_1)q_2}{r_1+r_2}$

and $\dfrac{k_e q_1 q_2}{r_1+r_2} - \dfrac{k_e q_1 q_2}{d} = \dfrac{1}{2}m_1 v_1^2 + \dfrac{1}{2}\dfrac{m_1^2 v_1^2}{m_2}$.

continued on next page

$$v_1 = \boxed{\sqrt{\frac{2m_2k_eq_1q_2}{m_1(m_1+m_2)}\left(\frac{1}{r_1+r_2}-\frac{1}{d}\right)}}$$

$$v_2 = \left(\frac{m_1}{m_2}\right)v_1 = \boxed{\sqrt{\frac{2m_1k_eq_1q_2}{m_2(m_1+m_2)}\left(\frac{1}{r_1+r_2}-\frac{1}{d}\right)}}$$

(b) If the spheres are metal, electrons will move around on them with negligible energy loss to place the centers of excess charge on the insides of the spheres. Then just before they touch, the effective distance between charges will be less than r_1+r_2 and the spheres will really be moving $\boxed{\text{faster than calculated in (a)}}$. We prove the equation for v_1 is dimensionally correct by evaluating the units of the right-hand side:

$$\left(\frac{\text{kg}\cdot\text{N}\cdot\text{m}^2\text{CC}}{\text{C}^2\text{kg}^2\cdot\text{m}}\right)^{1/2} = \left(\frac{\text{kg}\cdot\text{mm}}{\text{s}^2\text{kg}}\right)^{1/2} = \frac{\text{m}}{\text{s}}$$

as required. As either q_1 or q_2 increases, both v_1 and v_2 increase. The charge on either object affects the speeds of both by equal factors. As $m_1 \to \infty$, $v_1 \to 0$. As $m_2 \to \infty$, v_1 approaches $\sqrt{\frac{2K_eq_1q_2}{m_1}\left(\frac{1}{r_1+r_2}-\frac{1}{d}\right)}$. In this case, all of the potential energy becomes kinetic energy of just the lighter object. As $r_1+r_2 \to 0$, the speeds go to infinity in our classical approximation.

P20.17 (a) Each charge separately creates positive potential everywhere. The total potential produced by the three charges together is then the sum of three positive terms. There is $\boxed{\text{no point}}$ located at a finite distance from the charges, at which this total potential is zero.

(b) $V = \frac{k_eq}{a}+\frac{k_eq}{a} = \boxed{\frac{2k_eq}{a}}$

P20.18 (a) In an empty universe, the 20-nC charge can be placed at its location with no energy investment. At a distance of 4 cm, it creates a potential

$$V_1 = \frac{k_eq_1}{r} = \frac{\left(8.99\times10^9\ \text{N}\cdot\text{m}^2/\text{C}^2\right)\left(20\times10^{-9}\ \text{C}\right)}{0.04\ \text{m}} = 4.50\ \text{kV}.$$

To place the 10-nC charge there we must put in energy

$$U_{12} = q_2V_1 = \left(10\times10^{-9}\ \text{C}\right)\left(4.5\times10^3\ \text{V}\right) = 4.50\times10^{-5}\ \text{J}.$$

Next, to bring up the –20-nC charge requires energy

$$U_{23}+U_{13} = q_3V_2+q_3V_1 = q_3(V_2+V_1)$$

$$= -20\times10^{-9}\ \text{C}\left(8.99\times10^9\ \text{N}\cdot\text{m}^2/\text{C}^2\right)\left(\frac{10\times10^{-9}\ \text{C}}{0.04\ \text{m}}+\frac{20\times10^{-9}\ \text{C}}{0.08\ \text{m}}\right)$$

$$= -4.50\times10^{-5}\ \text{J} - 4.50\times10^{-5}\ \text{J}$$

continued on next page

The total energy of the three charges is

$$U_{12}+U_{23}+U_{13}=\boxed{-4.50\times10^{-5}\ \text{J}}.$$

(b) The three fixed charges create this potential at the location where the fourth is released:

$$V=V_1+V_2+V_3=\left(8.99\times10^{9}\ \text{N}\cdot\text{m}^2/\text{C}^2\right)\left(\frac{20\times10^{-9}}{\sqrt{0.04^2+0.03^2}}+\frac{10\times10^{-9}}{0.03}-\frac{20\times10^{-9}}{0.05}\right)\ \text{C/m}$$

$$V=3.00\times10^{3}\ \text{V}$$

Energy of the system of four charged objects is conserved as the fourth charge flies away:

$$\left(\frac{1}{2}mv^2+qV\right)_i=\left(\frac{1}{2}mv^2+qV\right)_f$$

$$0+\left(40\times10^{-9}\ \text{C}\right)\left(3.00\times10^{3}\ \text{V}\right)=\frac{1}{2}\left(2.00\times10^{-13}\ \text{kg}\right)v^2+0$$

$$v=\sqrt{\frac{2\left(1.20\times10^{-4}\ \text{J}\right)}{2\times10^{-13}\ \text{kg}}}=\boxed{3.46\times10^{4}\ \text{m/s}}$$

***P20.19** The original electrical potential energy is

$$U_e=qV=q\frac{k_eq}{d}.$$

In the final configuration we have mechanical equilibrium. The spring and electrostatic forces on each charge are $-k(2d)+q\frac{k_eq}{(3d)^2}=0$. Then $k=\frac{k_eq^2}{18d^3}$. In the final configuration the total potential energy is $\frac{1}{2}kx^2+qV=\frac{1}{2}\frac{k_eq^2}{18d^3}(2d)^2+q\frac{k_eq}{3d}=\frac{4}{9}\frac{k_eq^2}{d}$. The missing energy must have become internal energy, as the system is isolated: $\frac{k_eq^2}{d}=\frac{4k_eq^2}{9d}+\Delta E_{\text{int}}$

$$\boxed{\Delta E_{\text{int}}=\frac{5}{9}\frac{k_eq^2}{d}}.$$

P20.20 Using conservation of energy for the alpha particle-nucleus system,

we have $K_f+U_f=K_i+U_i$.

But $U_i=\frac{k_eq_\alpha q_{\text{gold}}}{r_i}$

and $r_i\approx\infty$.

Thus, $U_i=0$.

Also $K_f=0$ ($v_f=0$ at turning point),

so $U_f=K_i$

continued on next page

or $$\frac{k_e q_\alpha q_{\text{gold}}}{r_{\min}} = \frac{1}{2} m_\alpha v_\alpha^2$$

$$r_{\min} = \frac{2k_e q_\alpha q_{\text{gold}}}{m_\alpha v_\alpha^2} = \frac{2\left(8.99\times10^9 \text{ N}\cdot\text{m}^2/\text{C}^2\right)(2)(79)\left(1.60\times10^{-19} \text{ C}\right)^2}{\left(6.64\times10^{-27} \text{ kg}\right)\left(2.00\times10^7 \text{ m/s}\right)^2} = 2.74\times10^{-14} \text{ m} = \boxed{27.4 \text{ fm}}.$$

Section 20.4 Obtaining Electric Field from Electric Potential

P20.21 $V = a + bx = 10.0 \text{ V} + (-7.00 \text{ V/m})x$

(a) At $x = 0$, $V = \boxed{10.0 \text{ V}}$

At $x = 3.00$ m, $V = \boxed{-11.0 \text{ V}}$

At $x = 6.00$ m, $V = \boxed{-32.0 \text{ V}}$

(b) $E = -\frac{dV}{dx} = -b = -(-7.00 \text{ V/m}) = \boxed{7.00 \text{ N/C in the } +x \text{ direction}}$

P20.22 (a) For $r < R$ $V = \frac{k_e Q}{R}$

$$E_r = -\frac{dV}{dr} = \boxed{0}$$

(b) For $r \geq R$ $V = \frac{k_e Q}{r}$

$$E_r = -\frac{dV}{dr} = -\left(-\frac{k_e Q}{r^2}\right) = \boxed{\frac{k_e Q}{r^2}}$$

P20.23 $V = 5x - 3x^2y + 2yz^2$

Evaluate $\vec{\mathbf{E}}$ at $(1, 0, -2)$

$$E_x = -\frac{\partial V}{\partial x} = \boxed{-5 + 6xy} = -5 + 6(1)(0) = -5$$

$$E_y = -\frac{\partial V}{\partial y} = \boxed{+3x^2 - 2z^2} = 3(1)^2 - 2(-2)^2 = -5$$

$$E_z = -\frac{\partial V}{\partial z} = \boxed{-4yz} = -4(0)(-2) = 0$$

$$E = \sqrt{E_x^2 + E_y^2 + E_z^2} = \sqrt{(-5)^2 + (-5)^2 + 0^2} = \boxed{7.07 \text{ N/C}}$$

Section 20.5 Electric Potential Due to Continuous Charge Distributions

P20.24 $\Delta V = V_{2R} - V_0 = \frac{k_e Q}{\sqrt{R^2 + (2R)^2}} - \frac{k_e Q}{R} = \frac{k_e Q}{R}\left(\frac{1}{\sqrt{5}} - 1\right) = \boxed{-0.553\frac{k_e Q}{R}}$

P20.25 (a) $[\alpha] = \left[\frac{\lambda}{x}\right] = \frac{\text{C}}{\text{m}} \cdot \left(\frac{1}{\text{m}}\right) = \boxed{\frac{\text{C}}{\text{m}^2}}$

(b) $V = k_e \int \frac{dq}{r} = k_e \int \frac{\lambda dx}{r} = k_e \alpha \int_0^L \frac{x dx}{d + x} = \boxed{k_e \alpha \left[L - d \ln\left(1 + \frac{L}{d}\right)\right]}$

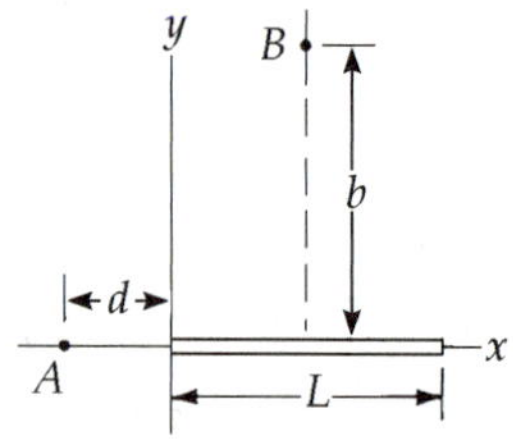

FIG. P20.25

P20.26 $V = \int \frac{k_e dq}{r} = k_e \int \frac{\alpha x dx}{\sqrt{b^2 + \left(\frac{L}{2} - x\right)^2}}$

Let $z = \frac{L}{2} - x$.

Then $x = \frac{L}{2} - z$, and $dx = -dz$

$$V = k_e \alpha \int \frac{\left(\frac{L}{2} - z\right)(-dz)}{\sqrt{b^2 + z^2}} = -\frac{k_e \alpha L}{2} \int \frac{dz}{\sqrt{b^2 + z^2}} + k_e \alpha \int \frac{z dz}{\sqrt{b^2 + z^2}} = -\frac{k_e \alpha L}{2} \ln\left(z + \sqrt{z^2 + b^2}\right) + k_e \alpha \sqrt{z^2 + b^2}$$

$$V = -\frac{k_e \alpha L}{2} \ln\left[\left(\frac{L}{2} - x\right) + \sqrt{\left(\frac{L}{2} - x\right)^2 + b^2}\right]_0^L + k_e \alpha \sqrt{\left(\frac{L}{2} - x\right)^2 + b^2}\,\Bigg|_0^L$$

$$V = -\frac{k_e \alpha L}{2} \ln\left[\frac{\frac{L}{2} - L + \sqrt{\left(\frac{L}{2}\right)^2 + b^2}}{\frac{L}{2} + \sqrt{\left(\frac{L}{2}\right)^2 + b^2}}\right] + k_e \alpha \left[\sqrt{\left(\frac{L}{2} - L\right)^2 + b^2} - \sqrt{\left(\frac{L}{2}\right)^2 + b^2}\right]$$

$$V = \boxed{-\frac{k_e \alpha L}{2} \ln\left[\frac{\sqrt{b^2 + \frac{L^2}{4}} - \frac{L}{2}}{\sqrt{b^2 + \frac{L^2}{4}} + \frac{L}{2}}\right]}$$

P20.27 $V = \int dV = \frac{1}{4\pi \in_0} \int \frac{dq}{r}$

All bits of charge are at the same distance from O.

So $V = \frac{1}{4\pi \in_0}\left(\frac{Q}{R}\right) = \left(8.99 \times 10^9 \text{ N} \cdot \text{m}^2/\text{C}^2\right)\left(\frac{-7.50 \times 10^{-6} \text{ C}}{0.140 \text{ m}/\pi}\right) = \boxed{-1.51 \text{ MV}}$.

Section 20.6 Electric Potential of a Charged Conductor

P20.28 Substituting given values into $V = \frac{k_e q}{r}$

$$7.50 \times 10^3 \text{ V} = \frac{\left(8.99 \times 10^9 \text{ N} \cdot \text{m}^2/\text{C}^2\right) q}{0.300 \text{ m}}.$$

Substituting $q = 2.50 \times 10^{-7}$ C,

$$N = \frac{2.50 \times 10^{-7} \text{ C}}{1.60 \times 10^{-19} \text{ C}/e^-} = \boxed{1.56 \times 10^{12} \text{ electrons}}.$$

P20.29 (a) $E = \boxed{0}$;

$$V = \frac{k_e q}{R} = \frac{\left(8.99 \times 10^9\right)\left(26.0 \times 10^{-6}\right)}{0.140} = \boxed{1.67 \text{ MV}}$$

(b) $$E = \frac{k_e q}{r^2} = \frac{\left(8.99 \times 10^9\right)\left(26.0 \times 10^{-6}\right)}{(0.200)^2} = \boxed{5.84 \text{ MN/C}} \text{ away}$$

$$V = \frac{k_e q}{R} = \frac{\left(8.99 \times 10^9\right)\left(26.0 \times 10^{-6}\right)}{0.200} = \boxed{1.17 \text{ MV}}$$

(c) $$E = \frac{k_e q}{R^2} = \frac{\left(8.99 \times 10^9\right)\left(26.0 \times 10^{-6}\right)}{(0.140)^2} = \boxed{11.9 \text{ MN/C}} \text{ away}$$

$$V = \frac{k_e q}{R} = \boxed{1.67 \text{ MV}}$$

P20.30 (a) Both spheres must be at the same potential according to $\frac{k_e q_1}{r_1} = \frac{k_e q_2}{r_2}$

where also $q_1 + q_2 = 1.20 \times 10^{-6}$ C.

Then $q_1 = \frac{q_2 r_1}{r_2}$

$$\frac{q_2 r_1}{r_2} + q_2 = 1.20 \times 10^{-6} \text{ C}$$

$$q_2 = \frac{1.20 \times 10^{-6} \text{ C}}{1 + 6 \text{ cm}/2 \text{ cm}} = 0.300 \times 10^{-6} \text{ C on the smaller sphere}$$

$$q_1 = 1.20 \times 10^{-6} \text{ C} - 0.300 \times 10^{-6} \text{ C} = 0.900 \times 10^{-6} \text{ C}$$

$$V = \frac{k_e q_1}{r_1} = \frac{\left(8.99 \times 10^9 \text{ N} \cdot \text{m}^2/\text{C}^2\right)\left(0.900 \times 10^{-6} \text{ C}\right)}{6 \times 10^{-2} \text{ m}} = \boxed{1.35 \times 10^5 \text{ V}}$$

continued on next page

(b) Outside the larger sphere,

$$\vec{\mathbf{E}}_1 = \frac{k_e q_1}{r_1^2}\hat{\mathbf{r}} = \frac{V_1}{r_1}\hat{\mathbf{r}} = \frac{1.35\times10^5\ \text{V}}{0.06\ \text{m}}\hat{\mathbf{r}} = \boxed{2.25\times10^6\ \text{V/m away}}.$$

Outside the smaller sphere,

$$\vec{\mathbf{E}}_2 = \frac{1.35\times10^5\ \text{V}}{0.02\ \text{m}}\hat{\mathbf{r}} = \boxed{6.74\times10^6\ \text{V/m away}}.$$

The smaller sphere carries less charge but creates a much stronger electric field than the larger sphere.

Section 20.7 Capacitance

P20.31 (a) $Q = C\Delta V = (4.00\times10^{-6}\ \text{F})(12.0\ \text{V}) = 4.80\times10^{-5}\ \text{C} = \boxed{48.0\ \mu\text{C}}$

(b) $Q = C\Delta V = (4.00\times10^{-6}\ \text{F})(1.50\ \text{V}) = 6.00\times10^{-6}\ \text{C} = \boxed{6.00\ \mu\text{C}}$

P20.32 (a) $C = \dfrac{Q}{\Delta V} = \dfrac{10.0\times10^{-6}\ \text{C}}{10.0\ \text{V}} = 1.00\times10^{-6}\ \text{F} = \boxed{1.00\ \mu\text{F}}$

(b) $\Delta V = \dfrac{Q}{C} = \dfrac{100\times10^{-6}\ \text{C}}{1.00\times10^{-6}\ \text{F}} = \boxed{100\ \text{V}}$

P20.33 $E = \dfrac{k_e q}{r^2}$: $\quad q = \dfrac{(4.90\times10^4\ \text{N/C})(0.210\ \text{m})^2}{(8.99\times10^9\ \text{N}\cdot\text{m}^2/\text{C}^2)} = 0.240\ \mu\text{C}$

(a) $\sigma = \dfrac{q}{A} = \dfrac{0.240\times10^{-6}}{4\pi(0.120)^2} = \boxed{1.33\ \mu\text{C/m}^2}$

(b) $C = 4\pi\epsilon_0 r = 4\pi(8.85\times10^{-12})(0.120) = \boxed{13.3\ \text{pF}}$

P20.34 With $\theta = \pi$, the plates are out of mesh and the overlap area is zero. With $\theta = 0$, the overlap area is that of a semi-circle, $\dfrac{\pi R^2}{2}$. By proportion, the effective area of a single sheet of charge is $\dfrac{(\pi-\theta)R^2}{2}$.

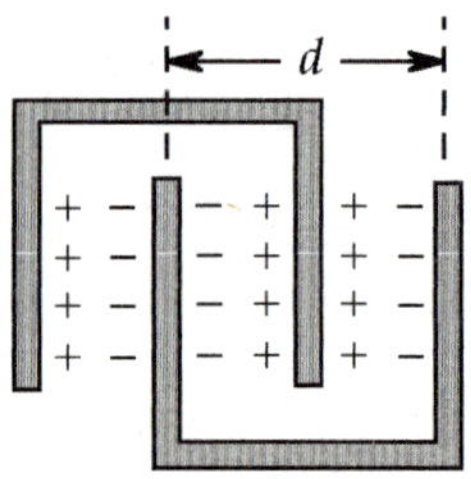

FIG. P20.34

When there are two plates in each comb, the number of adjoining sheets of positive and negative charge is 3, as shown in the sketch. When there are N plates on each comb, the number of parallel capacitors is $2N-1$ and the total capacitance is

$$C = (2N-1)\frac{\epsilon_0 A_{\text{effective}}}{\text{distance}} = \frac{(2N-1)\epsilon_0(\pi-\theta)\frac{R^2}{2}}{\frac{d}{2}} = \boxed{\frac{(2N-1)\epsilon_0(\pi-\theta)R^2}{d}}.$$

P20.35 (a) $\Delta V = Ed$

$$E = \frac{20.0\text{ V}}{1.80\times10^{-3}\text{ m}} = \boxed{11.1\text{ kV/m}}$$

(b) $E = \frac{\sigma}{\epsilon_0}$

$$\sigma = \left(1.11\times10^{4}\text{ N/C}\right)\left(8.85\times10^{-12}\text{ C}^2/\text{N}\cdot\text{m}^2\right) = \boxed{98.3\text{ nC/m}^2}$$

(c) $$C = \frac{\epsilon_0 A}{d} = \frac{\left(8.85\times10^{-12}\text{ C}^2/\text{N}\cdot\text{m}^2\right)\left(7.60\text{ cm}^2\right)\left(1.00\text{ m}/100\text{ cm}\right)^2}{1.80\times10^{-3}\text{ m}} = \boxed{3.74\text{ pF}}$$

(d) $\Delta V = \frac{Q}{C}$

$$Q = (20.0\text{ V})\left(3.74\times10^{-12}\text{ F}\right) = \boxed{74.7\text{ pC}}$$

P20.36 (a) $$C = \frac{\ell}{2k_e\ln\left(\frac{b}{a}\right)} = \frac{50.0}{2\left(8.99\times10^{9}\right)\ln\left(\frac{7.27}{2.58}\right)} = \boxed{2.68\text{ nF}}$$

(b) Method 1: $\Delta V = 2k_e\lambda\ln\left(\frac{b}{a}\right)$

$$\lambda = \frac{q}{\ell} = \frac{8.10\times10^{-6}\text{ C}}{50.0\text{ m}} = 1.62\times10^{-7}\text{ C/m}$$

$$\Delta V = 2\left(8.99\times10^{9}\right)\left(1.62\times10^{-7}\right)\ln\left(\frac{7.27}{2.58}\right) = \boxed{3.02\text{ kV}}$$

Method 2: $$\Delta V = \frac{Q}{C} = \frac{8.10\times10^{-6}}{2.68\times10^{-9}} = \boxed{3.02\text{ kV}}$$

P20.37 $\sum F_y = 0:$ $T\cos\theta - mg = 0$

$\sum F_x = 0:$ $T\sin\theta - Eq = 0$

Dividing, $\tan\theta = \frac{Eq}{mg}$

so $E = \frac{mg}{q}\tan\theta$

and $\Delta V = Ed = \boxed{\dfrac{mgd\tan\theta}{q}}$.

P20.38 (a) The electric field created by the charges points radially outward in the space between radius a and radius b. To find it, use a spherical gaussian surface of radius r. The magnitude of the field must be uniform over this surface, so $\oint \vec{\mathbf{E}} \cdot d\vec{\mathbf{A}} = \frac{q_{in}}{\epsilon_0}$ gives

$$E(4\pi r^2) = \frac{Q}{\epsilon_0}$$

$$E = \frac{Q}{4\pi \epsilon_0 r^2} = k_e \frac{Q}{r^2}$$

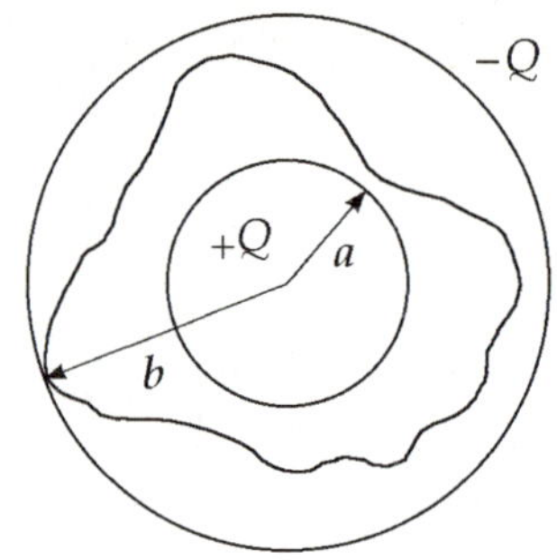

FIG. P20.38

Now to find the difference in electric potential between the conducting spheres, we use $V_b - V_a = -\int_a^b \vec{\mathbf{E}} \cdot d\vec{\mathbf{s}}$ along a radius line

$$V_b - V_a = -\int_{r=a}^{b} \frac{k_e Q}{r^2}(\cos 0)\, dr = -k_e Q \left.\frac{r^{-1}}{-1}\right|_a^b = k_e Q\left(\frac{1}{b} - \frac{1}{a}\right) = -k_e Q\left(\frac{1}{a} - \frac{1}{b}\right) = -k_e Q\left(\frac{b-a}{ab}\right)$$

$$V_a - V_b = +k_e Q\left(\frac{b-a}{ab}\right)$$

Then the capacitance is $C = \frac{Q}{V_a - V_b} = \boxed{\frac{ab}{k_e(b-a)}}$.

(b) As $b \to \infty$, $b - a \to b$, to give $C \to \frac{ab}{k_e b} = \boxed{\frac{a}{k_e}}$ with $k_e = \frac{1}{4\pi \epsilon_0}$, this is $C = 4\pi \epsilon_0 a$, in agreement with Equation 20.20.

Section 20.8 Combinations of Capacitors

P20.39 (a) Capacitors in parallel add. Thus, the equivalent capacitor has a value of

$$C_{eq} = C_1 + C_2 = 5.00\ \mu\text{F} + 12.0\ \mu\text{F} = \boxed{17.0\ \mu\text{F}}.$$

(b) The potential difference across each branch is the same and equal to the voltage of the battery.

$$\Delta V = \boxed{9.00\ \text{V}}$$

(c) $Q_5 = C\Delta V = (5.00\ \mu\text{F})(9.00\ \text{V}) = \boxed{45.0\ \mu\text{C}}$

and $Q_{12} = C\Delta V = (12.0\ \mu\text{F})(9.00\ \text{V}) = \boxed{108\ \mu\text{C}}$

P20.40 (a) In series capacitors add as

$$\frac{1}{C_{eq}}=\frac{1}{C_1}+\frac{1}{C_2}=\frac{1}{5.00\ \mu\text{F}}+\frac{1}{12.0\ \mu\text{F}}$$

and $C_{eq}=\boxed{3.53\ \mu\text{F}}$.

(c) The charge on the equivalent capacitor is $Q_{eq}=C_{eq}\Delta V=(3.53\ \mu\text{F})(9.00\ \text{V})=31.8\ \mu\text{C}$.

Each of the series capacitors has this same charge on it.

So $Q_1=Q_2=\boxed{31.8\ \mu\text{C}}$.

(b) The potential difference across each is $\Delta V_1=\frac{Q_1}{C_1}=\frac{31.8\ \mu\text{C}}{5.00\ \mu\text{F}}=\boxed{6.35\ \text{V}}$

and $\Delta V_2=\frac{Q_2}{C_2}=\frac{31.8\ \mu\text{C}}{12.0\ \mu\text{F}}=\boxed{2.65\ \text{V}}$.

P20.41 (a) $\frac{1}{C_s}=\frac{1}{15.0}+\frac{1}{3.00}$

$C_s=2.50\ \mu\text{F}$

$C_p=2.50+6.00=8.50\ \mu\text{F}$

$$C_{eq}=\left(\frac{1}{8.50\ \mu\text{F}}+\frac{1}{20.0\ \mu\text{F}}\right)^{-1}=\boxed{5.96\ \mu\text{F}}$$

(b) $Q=C\Delta V=(5.96\ \mu\text{F})(15.0\ \text{V})=\boxed{89.5\ \mu\text{C}}$ on $20.0\ \mu\text{F}$

$\Delta V=\frac{Q}{C}=\frac{89.5\ \mu\text{C}}{20.0\ \mu\text{F}}=4.47\ \text{V}$

$15.0-4.47=10.53\ \text{V}$

$Q=C\Delta V=(6.00\ \mu\text{F})(10.53\ \text{V})=\boxed{63.2\ \mu\text{C}}$ on $6.00\ \mu\text{F}$

$89.5-63.2=\boxed{26.3\ \mu\text{C}}$ on $15.0\ \mu\text{F}$ and $3.00\ \mu\text{F}$

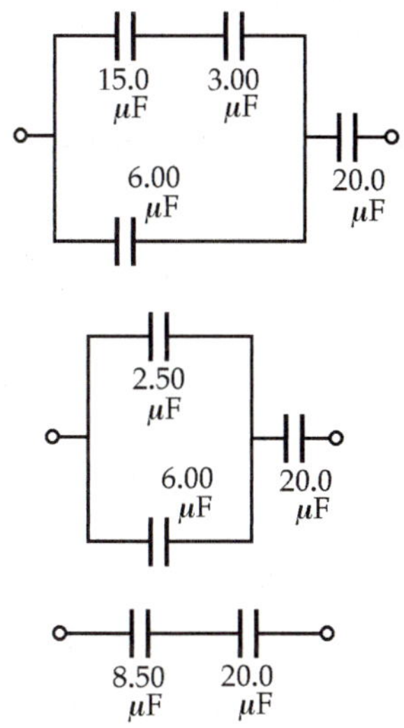

FIG. P20.41

P20.42 $C_p=C_1+C_2$

and $\frac{1}{C_s}=\frac{1}{C_1}+\frac{1}{C_2}$.

Substitute $C_2=C_p-C_1$: $\frac{1}{C_s}=\frac{1}{C_1}+\frac{1}{C_p-C_1}=\frac{C_p-C_1+C_1}{C_1(C_p-C_1)}$.

Simplifying, $C_1^2-C_1C_p+C_pC_s=0$

continued on next page

and $$C_1 = \frac{C_p \pm \sqrt{C_p^2 - 4C_pC_s}}{2} = \boxed{\frac{1}{2}C_p + \sqrt{\frac{1}{4}C_p^2 - C_pC_s}}$$

where the positive sign was arbitrarily chosen (choosing the negative sign gives the same values for the capacitances, with the names reversed).

Then, from $$C_2 = C_p - C_1$$

$$C_2 = \boxed{\frac{1}{2}C_p - \sqrt{\frac{1}{4}C_p^2 - C_pC_s}}\,.$$

P20.43 $C = \frac{Q}{\Delta V}$ so $6.00 \times 10^{-6} = \frac{Q}{20.0}$

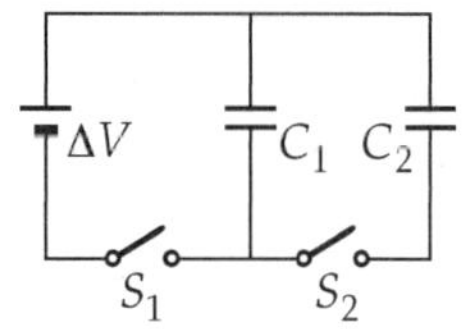

FIG. P20.43

and $Q = \boxed{120\ \mu\text{C}}$

$Q_1 = 120\ \mu\text{C} - Q_2$

and $\Delta V = \frac{Q}{C}$: $\frac{120 - Q_2}{C_1} = \frac{Q_2}{C_2}$

or $\frac{120 - Q_2}{6.00} = \frac{Q_2}{3.00}$

$(3.00)(120 - Q_2) = (6.00)Q_2$

$Q_2 = \frac{360}{9.00} = \boxed{40.0\ \mu\text{C}}$ $Q_1 = 120\ \mu\text{C} - 40.0\ \mu\text{C} = \boxed{80.0\ \mu\text{C}}$

***P20.44** (a) Capacitors 2 and 3 are in parallel and present equivalent capacitance 6C. This is in series with capacitor 1, so the battery sees capacitance $\left[\frac{1}{3C} + \frac{1}{6C}\right]^{-1} = \boxed{2C}$.

(b) If they were initially uncharged, C_1 stores the same charge as C_2 and C_3 together. With greater capacitance, C_3 stores more charge than C_2. Then $\boxed{Q_1 > Q_3 > Q_2}$.

(c) The $(C_2 \| C_3)$ equivalent capacitor stores the same charge as C_1. Since it has greater capacitance, $\Delta V = \frac{Q}{C}$ implies that it has smaller potential difference across it than C_1. In parallel with each other, C_2 and C_3 have equal voltages: $\boxed{\Delta V_1 > \Delta V_2 = \Delta V_3}$.

(d) If C_3 is increased, the overall equivalent capacitance increases. More charge moves through the battery and Q increases. As ΔV_1 increases, ΔV_2 must decrease so Q_2 decreases. Then Q_3 must increase even more: $\boxed{Q_3 \text{ and } Q_1 \text{ increase; } Q_2 \text{ decreases}}$.

***P20.45** (a) In $\boxed{\text{series}}$, to reduce the effective capacitance:

$$\frac{1}{32.0\ \mu\text{F}} = \frac{1}{34.8\ \mu\text{F}} + \frac{1}{C_s}$$

$$C_s = \frac{1}{2.51 \times 10^{-3}/\mu\text{F}} = \boxed{398\ \mu\text{F}}$$

(b) In $\boxed{\text{parallel}}$, to increase the total capacitance:

$$29.8\ \mu\text{F} + C_p = 32.0\ \mu\text{F}$$

$$C_p = \boxed{2.20\ \mu\text{F}}$$

P20.46 $C_s = \left(\frac{1}{5.00} + \frac{1}{7.00}\right)^{-1} = 2.92\ \mu\text{F}$

$$C_p = 2.92 + 4.00 + 6.00 = \boxed{12.9\ \mu\text{F}}$$

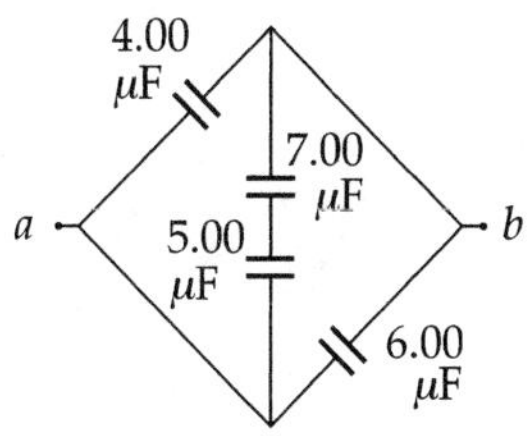

FIG. P20.46

Section 20.9 Energy Stored in a Charged Capacitor

P20.47 (a) $U = \frac{1}{2}C(\Delta V)^2 = \frac{1}{2}(3.00\ \mu\text{F})(12.0\ \text{V})^2 = \boxed{216\ \mu\text{J}}$

(b) $U = \frac{1}{2}C(\Delta V)^2 = \frac{1}{2}(3.00\ \mu\text{F})(6.00\ \text{V})^2 = \boxed{54.0\ \mu\text{J}}$

P20.48 $U = \frac{1}{2}C\Delta V^2$

$$\Delta V = \sqrt{\frac{2U}{C}} = \sqrt{\frac{2(300\ \text{J})}{30 \times 10^{-6}\ \text{C/V}}} = \boxed{4.47 \times 10^3\ \text{V}}$$

P20.49 $U = \frac{1}{2}C(\Delta V)^2$

The circuit diagram is shown at the right.

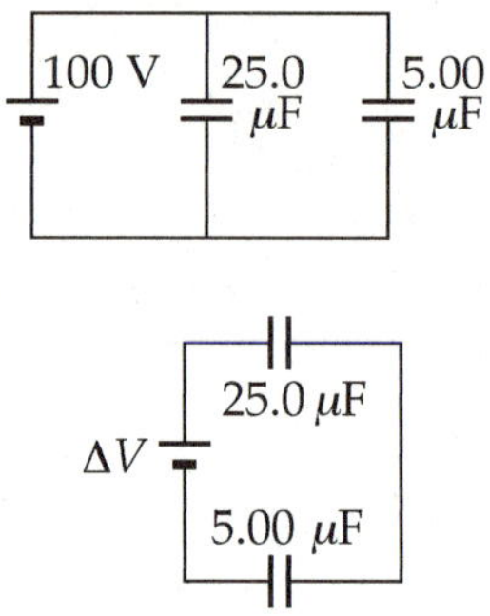

FIG. P20.49

(a) $C_p = C_1 + C_2 = 25.0\ \mu\text{F} + 5.00\ \mu\text{F} = 30.0\ \mu\text{F}$

$$U = \frac{1}{2}\left(30.0\times10^{-6}\right)(100)^2 = \boxed{0.150\ \text{J}}$$

(b) $$C_s = \left(\frac{1}{C_1} + \frac{1}{C_2}\right)^{-1} = \left(\frac{1}{25.0\ \mu\text{F}} + \frac{1}{5.00\ \mu\text{F}}\right)^{-1} = 4.17\ \mu\text{F}$$

$$U = \frac{1}{2}C(\Delta V)^2$$

$$\Delta V = \sqrt{\frac{2U}{C}} = \sqrt{\frac{2(0.150)}{4.17\times10^{-6}}} = \boxed{268\ \text{V}}$$

P20.50 (a) $Q = C\Delta V = \left(150\times10^{-12}\ \text{F}\right)\left(10\times10^{3}\ \text{V}\right) = \boxed{1.50\times10^{-6}\ \text{C}}$

(b) $$U = \frac{1}{2}C(\Delta V)^2$$

$$\Delta V = \sqrt{\frac{2U}{C}} = \sqrt{\frac{2\left(250\times10^{-6}\ \text{J}\right)}{150\times10^{-12}\ \text{F}}} = \boxed{1.83\times10^{3}\ \text{V}}$$

P20.51 $W = U = \int F dx$

$$\text{so } F = \frac{dU}{dx} = \frac{d}{dx}\left(\frac{Q^2}{2C}\right) = \frac{d}{dx}\left(\frac{Q^2 x}{2\epsilon_0 A}\right) = \boxed{\frac{Q^2}{2\epsilon_0 A}}$$

P20.52 $u = \frac{U}{V} = \frac{1}{2}\epsilon_0 E^2$

$$\frac{1.00\times10^{-7}}{V} = \frac{1}{2}\left(8.85\times10^{-12}\right)(3\,000)^2$$

$$V = \boxed{2.51\times10^{-3}\ \text{m}^3} = \left(2.51\times10^{-3}\ \text{m}^3\right)\left(\frac{1\,000\ \text{L}}{\text{m}^3}\right) = \boxed{2.51\ \text{L}}$$

Section 20.10 Capacitors with Dielectrics

P20.53 (a) $$C = \frac{\kappa\epsilon_0 A}{d} = \frac{2.10\left(8.85\times10^{-12}\ \text{F/m}\right)\left(1.75\times10^{-4}\ \text{m}^2\right)}{4.00\times10^{-5}\ \text{m}} = 8.13\times10^{-11}\ \text{F} = \boxed{81.3\ \text{pF}}$$

(b) $$\Delta V_{\text{max}} = E_{\text{max}}d = \left(60.0\times10^{6}\ \text{V/m}\right)\left(4.00\times10^{-5}\ \text{m}\right) = \boxed{2.40\ \text{kV}}$$

P20.54 $Q_{max} = C\Delta V_{max}$,

but $\Delta V_{max} = E_{max} d$.

Also, $C = \dfrac{\kappa \in_0 A}{d}$.

Thus, $Q_{max} = \dfrac{\kappa \in_0 A}{d}(E_{max} d) = \kappa \in_0 AE_{max}$.

(a) With air between the plates, $\kappa = 1.00$

and $E_{max} = 3.00\times10^6\ \text{V/m}$.

Therefore,

$$Q_{max} = \kappa \in_0 AE_{max} = (8.85\times10^{-12}\ \text{F/m})(5.00\times10^{-4}\ \text{m}^2)(3.00\times10^6\ \text{V/m}) = \boxed{13.3\ \text{nC}}.$$

(b) With polystyrene between the plates, $\kappa = 2.56$ and $E_{max} = 24.0\times10^6\ \text{V/m}$.

$$Q_{max} = \kappa \in_0 AE_{max} = 2.56(8.85\times10^{-12}\ \text{F/m})(5.00\times10^{-4}\ \text{m}^2)(24.0\times10^6\ \text{V/m}) = \boxed{272\ \text{nC}}$$

P20.55 $C = \dfrac{\kappa \in_0 A}{d}$

or $95.0\times10^{-9} = \dfrac{3.70(8.85\times10^{-12})(0.070\,0)\ell}{0.025\,0\times10^{-3}}$

$\ell = \boxed{1.04\ \text{m}}$

***P20.56** Consider two sheets of aluminum foil, each 40 cm by 100 cm, with one sheet of plastic between them. Suppose the plastic has $\kappa \cong 3$, $E_{max} \sim 10^7\ \text{V/m}$ and thickness 1 mil $= \dfrac{2.54\ \text{cm}}{1\,000}$. Then,

$$C = \frac{\kappa \in_0 A}{d} \sim \frac{3(8.85\times10^{-12}\ \text{C}^2/\text{N}\cdot\text{m}^2)(0.4\ \text{m}^2)}{2.54\times10^{-5}\ \text{m}} \sim \boxed{10^{-6}\ \text{F}}$$

$$\Delta V_{max} = E_{max} d \sim (10^7\ \text{V/m})(2.54\times10^{-5}\ \text{m}) \sim \boxed{10^2\ \text{V}}$$

P20.57 Originally, $C = \dfrac{\in_0 A}{d} = \dfrac{Q}{(\Delta V)_i}$.

(a) The charge is the same before and after immersion, with value $Q = \dfrac{\in_0 A(\Delta V)_i}{d}$.

$$Q = \frac{(8.85\times10^{-12}\ \text{C}^2/\text{N}\cdot\text{m}^2)(25.0\times10^{-4}\ \text{m}^2)(250\ \text{V})}{(1.50\times10^{-2}\ \text{m})} = \boxed{369\ \text{pC}}$$

continued on next page

(b) Finally,

$$C_f = \frac{\kappa \epsilon_0 A}{d} = \frac{Q}{(\Delta V)_f} \qquad C_f = \frac{80.0\left(8.85\times10^{-12}\ \text{C}^2/\text{N}\cdot\text{m}^2\right)\left(25.0\times10^{-4}\ \text{m}^2\right)}{\left(1.50\times10^{-2}\ \text{m}\right)} = \boxed{118\ \text{pF}}$$

$$(\Delta V)_f = \frac{Qd}{\kappa\epsilon_0 A} = \frac{\epsilon_0 A(\Delta V)_i d}{\kappa\epsilon_0 Ad} = \frac{(\Delta V)_i}{\kappa} = \frac{250\ \text{V}}{80.0} = \boxed{3.12\ \text{V}}.$$

(c) Originally, $U_i = \frac{1}{2}C(\Delta V)_i^2 = \frac{\epsilon_0 A(\Delta V)_i^2}{2d}$.

Finally, $U_f = \frac{1}{2}C_f(\Delta V)_f^2 = \frac{\kappa\epsilon_0 A(\Delta V)_i^2}{2d\kappa^2} = \frac{\epsilon_0 A(\Delta V)_i^2}{2d\kappa}$.

So, $\Delta U = U_f - U_i = \frac{-\epsilon_0 A(\Delta V)_i^2(\kappa - 1)}{2d\kappa}$

$$\Delta U = -\frac{\left(8.85\times10^{-12}\ \text{C}^2/\text{N}\cdot\text{m}^2\right)\left(25.0\times10^{-4}\ \text{m}^2\right)(250\ \text{V})^2(79.0)}{2\left(1.50\times10^{-2}\ \text{m}\right)(80.0)} = \boxed{-45.5\ \text{nJ}}.$$

Section 20.11 Context Connection—The Atmosphere as a Capacitor

P20.58 (a) $E_{max} = 3.00\times10^6\ \text{V/m} = \frac{k_eQ}{r^2} = \frac{k_eQ}{r}\left(\frac{1}{r}\right) = V_{max}\left(\frac{1}{r}\right)$

$V_{max} = E_{max}r = 3.00\times10^6(0.150) = \boxed{450\ \text{kV}}$

(b) $\frac{k_eQ_{max}}{r^2} = E_{max}$ $\left\{\text{or } \frac{k_eQ_{max}}{r} = V_{max}\right\}$ $Q_{max} = \frac{E_{max}r^2}{k_e} = \frac{3.00\times10^6(0.150)^2}{8.99\times10^9} = \boxed{7.51\ \mu\text{C}}$

P20.59 The energy transferred is $T_{ET} = \frac{1}{2}Q\Delta V = \frac{1}{2}(50.0\ \text{C})\left(1.00\times10^8\ \text{V}\right) = 2.50\times10^9\ \text{J}$

and 1% of this (or $\Delta E_{int} = 2.50\times10^7$ J) is absorbed by the tree. If m is the amount of water boiled away,

then $\Delta E_{int} = m(4\,186\ \text{J/kg}\cdot°\text{C})(100°\text{C} - 30.0°\text{C}) + m\left(2.26\times10^6\ \text{J/kg}\right) = 2.50\times10^7\ \text{J}$

giving $m = \boxed{9.79\ \text{kg}}$.

Additional Problems

***P20.60** (a) The two particles exert forces of repulsion on each other. As the projectile approaches the target particle, the projectile slows. The target starts to move in the x direction. As long as the projectile is moving faster than the second particle, the two will be approaching. Kinetic energy will be being converted into electric potential energy. When both particles move with equal speeds, the distance between them will momentarily not be changing: this is the instant of closest approach. Thereafter, the target particle, still feeling a forward force, will move faster than the projectile. The particles will separate again. The particles exert forces on each other but never touch. The particles will eventually be very far apart, with zero electric potential energy. All of the U_e they had at closest approach is converted back into kinetic energy. The whole process is an elastic collision. Compare this problem with Problem 8.46 in Chapter 8.

(b) Momentum is constant throughout the process. We equate it at the large-separation initial point and the point b of closest approach.

$$m_1\vec{\mathbf{v}}_{1i} + m_2\vec{\mathbf{v}}_{2i} = m_1\vec{\mathbf{v}}_{1b} + m_2\vec{\mathbf{v}}_{2b}$$

$$(2\text{ g})(21\hat{\mathbf{i}}\text{ m/s}) + 0 = (2\text{ g} + 5\text{ g})\vec{\mathbf{v}}_b$$

$$\vec{\mathbf{v}}_b = \boxed{6.00\hat{\mathbf{i}}\text{ m/s}}$$

(c) Energy conservation between the same two points:

$$\frac{1}{2}m_1v_{1i}^2 + \frac{1}{2}m_2v_{2i}^2 + 0 = \frac{1}{2}(m_1 + m_2)v_b^2 + \frac{k_eq_1q_2}{r_b}$$

$$\frac{1}{2}0.002\text{ kg}(21\text{ m/s})^2 + \frac{1}{2}0.005\text{ kg}(0)^2 + 0 = \frac{1}{2}0.007\text{ kg}(6\text{ m/s})^2 + \frac{8.99\times10^9\text{ Nm}^2}{\text{C}^2}\,\frac{15\times10^{-6}\text{ C }8.5\times10^{-6}\text{ C}}{r_b}$$

$$0.441\text{ J} - 0.126\text{ J} = \frac{1.15\text{ J}\cdot\text{m}}{r_b}$$

$$r_b = \frac{1.15\text{ m}}{0.315} = \boxed{3.64\text{ m}}$$

(d) The overall elastic collision is described by conservation of momentum:

$$m_1\vec{\mathbf{v}}_{1i} + m_2\vec{\mathbf{v}}_{2i} = m_1\vec{\mathbf{v}}_{1d} + m_2\vec{\mathbf{v}}_{2d}$$

$$(2\text{ g})(21\hat{\mathbf{i}}\text{ m/s}) + 0 = 2\text{ g}\vec{\mathbf{v}}_{1d}\hat{\mathbf{i}} + 5\text{ g}\vec{\mathbf{v}}_{2d}\hat{\mathbf{i}}$$

and by the relative velocity equation:

$$v_{1i} - v_{2i} = v_{2d} - v_{1d}$$

$$21\text{ m/s} - 0 = v_{2d} - v_{1d}$$

continued on next page

we substitute

$$v_{2d} = 21 \text{ m/s} + v_{1d}$$

$$42 \text{ g}\cdot\text{m/s} = 2 \text{ g}v_{1d} + 5 \text{ g}(21 \text{ m/s} + v_{1d}) = 2 \text{ g}v_{1d} + 105 \text{ g}\cdot\text{m/s} + 5 \text{ g}v_{1d}$$

$$-63 \text{ g}\cdot\text{m/s} = 7 \text{ g}v_{1d}$$

$$v_{1d} = -9.00 \text{ m/s}$$

$$\boxed{\vec{\mathbf{v}}_{1d} = -9.00\hat{\mathbf{i}} \text{ m/s}}$$

$$v_{2d} = 21 \text{ m/s} - 9 \text{ m/s} = 12.0 \text{ m/s}$$

$$\boxed{\vec{\mathbf{v}}_{2d} = 12.0\hat{\mathbf{i}} \text{ m/s}}$$

P20.61 $U = qV = k_e \dfrac{q_1 q_2}{r_{12}} = \left(8.99\times10^9\right)\dfrac{(38)(54)\left(1.60\times10^{-19}\right)^2}{(5.50+6.20)\times10^{-15}} = 4.04\times10^{-11} \text{ J} = \boxed{253 \text{ MeV}}$

P20.62 (a) $U = \dfrac{k_e q_1 q_2}{r} = \dfrac{-\left(8.99\times10^9\right)\left(1.60\times10^{-19}\right)^2}{0.0529\times10^{-9}} = -4.35\times10^{-18} \text{ J} = \boxed{-27.2 \text{ eV}}$

(b) $U = \dfrac{k_e q_1 q_2}{r} = \dfrac{-\left(8.99\times10^9\right)\left(1.60\times10^{-19}\right)^2}{2^2\left(0.0529\times10^{-9}\right)} = \boxed{-6.80 \text{ eV}}$

(c) $U = \dfrac{k_e q_1 q_2}{r} = \dfrac{-k_e e^2}{\infty} = \boxed{0}$

P20.63 $W = \displaystyle\int_0^Q V dq$

where $V = \dfrac{k_e q}{R}$.

Therefore, $W = \boxed{\dfrac{k_e Q^2}{2R}}$.

P20.64 (a) $V_B - V_A = -\int_A^B \vec{\mathbf{E}} \cdot d\vec{\mathbf{s}}$ and the field at distance r from a uniformly charged rod (where $r >$ radius of charged rod) is

$$E = \frac{\lambda}{2\pi \epsilon_0 r} = \frac{2k_e\lambda}{r}.$$

In this case, the field between the central wire and the coaxial cylinder is directed perpendicular to the line of charge so that

$$V_B - V_A = -\int_{r_a}^{r_b} \frac{2k_e\lambda}{r} dr = 2k_e\lambda \ln\left(\frac{r_a}{r_b}\right),$$

or $\boxed{\Delta V = 2k_e\lambda \ln\left(\frac{r_a}{r_b}\right)}$.

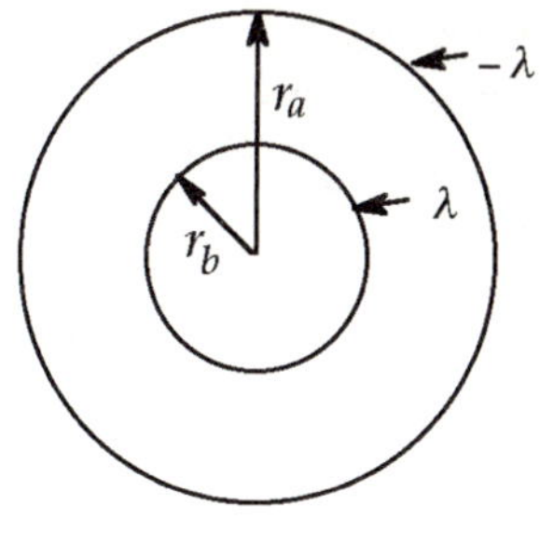

FIG. P20.64

(b) From part (a), when the outer cylinder is considered to be at zero potential, the potential at a distance r from the axis is

$$V = 2k_e\lambda \ln\left(\frac{r_a}{r}\right).$$

The field at r is given by

$$E = -\frac{\partial V}{\partial r} = -2k_e\lambda\left(\frac{r}{r_a}\right)\left(-\frac{r_a}{r^2}\right) = \frac{2k_e\lambda}{r}.$$

But, from part (a), $2k_e\lambda = \frac{\Delta V}{\ln\frac{r_a}{r_b}}$.

Therefore, $\boxed{E = \frac{\Delta V}{\ln\frac{r_a}{r_b}}\left(\frac{1}{r}\right)}$.

P20.65 $2\pi r\ell E = \frac{q_{\text{in}}}{\epsilon_0}$

so $E = \frac{\lambda}{2\pi r \epsilon_0}$

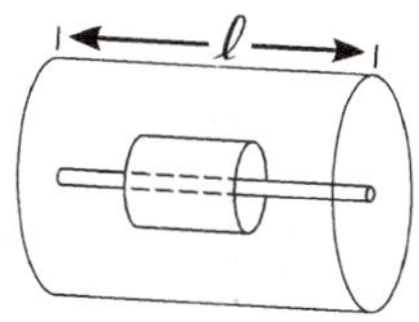

FIG. P20.65

$$\Delta V = -\int_{r_1}^{r_2} \vec{\mathbf{E}} \cdot d\vec{\mathbf{r}} = \int_{r_1}^{r_2} \frac{\lambda}{2\pi r \epsilon_0} dr = \frac{\lambda}{2\pi\epsilon_0}\ln\left(\frac{r_1}{r_2}\right)$$

$$\frac{\lambda_{\text{max}}}{2\pi\epsilon_0} = E_{\text{max}} r_{\text{inner}}$$

$$\Delta V = \left(1.20\times10^6 \text{ V/m}\right)\left(0.100\times10^{-3} \text{ m}\right)\ln\left(\frac{25.0}{0.200}\right)$$

$\Delta V_{\text{max}} = \boxed{579 \text{ V}}$

P20.66 (a) From Problem 64,

$$E = \frac{\Delta V}{\ln\left(\frac{r_a}{r_b}\right)} \frac{1}{r}.$$

We require just outside the central wire

$$5.50 \times 10^6 \text{ V/m} = \frac{50.0 \times 10^3 \text{ V}}{\ln\left(\frac{0.850 \text{ m}}{r_b}\right)} \left(\frac{1}{r_b}\right)$$

or $$\left(110 \text{ m}^{-1}\right) r_b \ln\left(\frac{0.850 \text{ m}}{r_b}\right) = 1.$$

We solve by homing in on the required value

r_b (m)	0.0 100	0.001 00	0.001 50	0.001 45	0.001 43	0.001 42
$\left(110 \text{ m}^{-1}\right) r_b \ln\left(\frac{0.850 \text{ m}}{r_b}\right)$	4.89	0.740	1.05	1.017	1.005	0.999

Thus, to three significant figures,

$$r_b = \boxed{1.42 \text{ mm}}.$$

(b) At r_a,

$$E = \frac{50.0 \text{ kV}}{\ln(0.850 \text{ m}/0.001\,42 \text{ m})} \left(\frac{1}{0.850 \text{ m}}\right) = \boxed{9.20 \text{ kV/m}}.$$

***P20.67** (a) $$V = \frac{m}{\rho} = \frac{1.00 \times 10^{-12} \text{ kg}}{1\,100 \text{ kg/m}^3} = \boxed{9.09 \times 10^{-16} \text{ m}^3}.$$

Since $V = \frac{4\pi r^3}{3}$, the radius is $r = \left[\frac{3V}{4\pi}\right]^{1/3}$, and the surface area is

$$A = 4\pi r^2 = 4\pi\left[\frac{3V}{4\pi}\right]^{2/3} = 4\pi\left[\frac{3\left(9.09 \times 10^{-16} \text{ m}^3\right)}{4\pi}\right]^{2/3} = \boxed{4.54 \times 10^{-10} \text{ m}^2}$$

(b) $$C = \frac{\kappa \epsilon_0 A}{d} = \frac{(5.00)\left(8.85 \times 10^{-12} \text{ C}^2/\text{N}\cdot\text{m}^2\right)\left(4.54 \times 10^{-10} \text{ m}^2\right)}{100 \times 10^{-9} \text{ m}} = \boxed{2.01 \times 10^{-13} \text{ F}}$$

(c) $$Q = C(\Delta V) = \left(2.01 \times 10^{-13} \text{ F}\right)\left(100 \times 10^{-3} \text{ V}\right) = \boxed{2.01 \times 10^{-14} \text{ C}},$$

and the number of electronic charges is

$$n = \frac{Q}{e} = \frac{2.01 \times 10^{-14} \text{ C}}{1.60 \times 10^{-19} \text{ C}} = \boxed{1.26 \times 10^5}.$$

***P20.68** Take the illustration presented with the problem as an initial picture. No external horizontal forces act on the set of four balls, so its center of mass stays fixed at the location of the center of the square. As the charged balls 1 and 2 swing out and away from each other, balls 3 and 4 move up with equal y-components of velocity. The maximum-kinetic-energy point is illustrated. System energy is conserved:

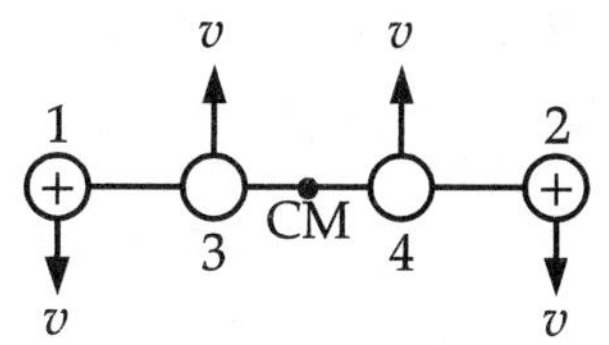

FIG. P20.68

$$\frac{k_e q^2}{a} = \frac{k_e q^2}{3a} + \frac{1}{2}mv^2 + \frac{1}{2}mv^2 + \frac{1}{2}mv^2 + \frac{1}{2}mv^2$$

$$\frac{2k_e q^2}{3a} = 2mv^2 \qquad \boxed{v = \sqrt{\frac{k_e q^2}{3am}}}$$

P20.69 From Example 20.5, the potential at the center of the ring is $V_i = \frac{k_e Q}{R}$ and the potential at an infinite distance from the ring is $V_f = 0$. Thus, the initial and final potential energies of the point charge-ring system are:

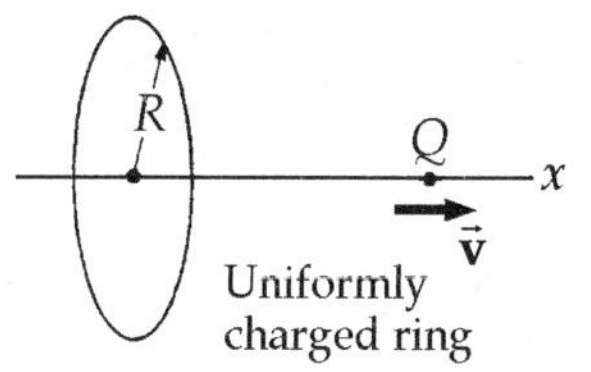

FIG. P20.69

$$U_i = QV_i = \frac{k_e Q^2}{R}$$

and $$U_f = QV_f = 0.$$

From conservation of energy,

$$K_f + U_f = K_i + U_i$$

or $$\frac{1}{2}Mv_f^2 + 0 = 0 + \frac{k_e Q^2}{R}$$

giving $$v_f = \boxed{\sqrt{\frac{2k_e Q^2}{MR}}}.$$

P20.70 (a) $$V = \frac{k_e q}{r_1} - \frac{k_e q}{r_2} = \frac{k_e q}{r_1 r_2}(r_2 - r_1)$$

From the figure, for $r >> a$, $r_2 - r_1 \approx 2a\cos\theta$.

Then $$V \approx \frac{k_e q}{r_1 r_2} 2a\cos\theta \cong \frac{k_e p\cos\theta}{r^2}.$$

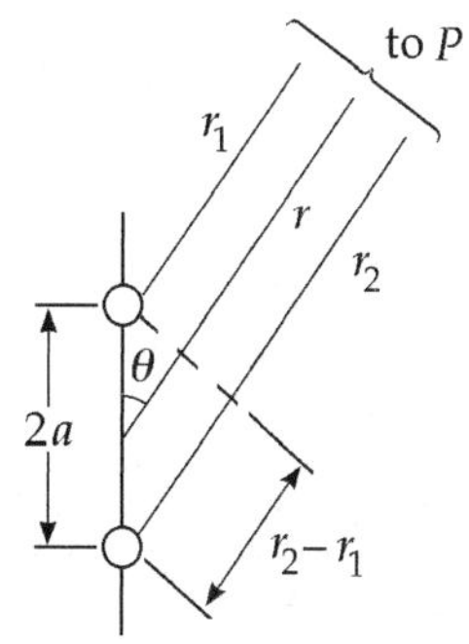

FIG. P20.70

(b) $$E_r = -\frac{\partial V}{\partial r} = \boxed{\frac{2k_e p\cos\theta}{r^3}}$$

In spherical coordinates, the θ component of the gradient is $\frac{1}{r}\left(\frac{\partial}{\partial\theta}\right)$.

Therefore, $$E_\theta = -\frac{1}{r}\left(\frac{\partial V}{\partial\theta}\right) = \boxed{\frac{k_e p\sin\theta}{r^3}}.$$

continued on next page

For $r >> a$ $\qquad E_r(0°) = \frac{2k_e p}{r^3}$

and $\qquad E_r(90°) = 0,$

$$E_\theta(0°) = 0$$

and $\qquad E_\theta(90°) = \frac{k_e p}{r^3}.$

These results are $\boxed{\text{reasonable for } r >> a}$. Their directions are as shown in Figure 20.8(c). However, for $\boxed{r \to 0,\ E(0) \to \infty. \text{ This is unreasonable,}}$ since r is not much greater than a if it is 0.

(c) $\boxed{V = \frac{k_e p y}{\left(x^2 + y^2\right)^{3/2}}}$

and $\boxed{E_x = -\frac{\partial V}{\partial x} = \frac{3k_e p x y}{\left(x^2 + y^2\right)^{5/2}}}$

$$\boxed{E_y = -\frac{\partial V}{\partial y} = \frac{k_e p\left(2y^2 - x^2\right)}{\left(x^2 + y^2\right)^{5/2}}}$$

***P20.71** Imagine the center plate is split along its midplane and pulled apart. We have two capacitors in parallel, supporting the same ΔV and carrying total charge Q. The upper has capacitance $C_1 = \frac{\epsilon_0 A}{d}$ and the lower $C_2 = \frac{\epsilon_0 A}{2d}$. Charge flows from ground onto each of the outside plates so that $Q_1 + Q_2 = Q \qquad \Delta V_1 = \Delta V_2 = \Delta V$.

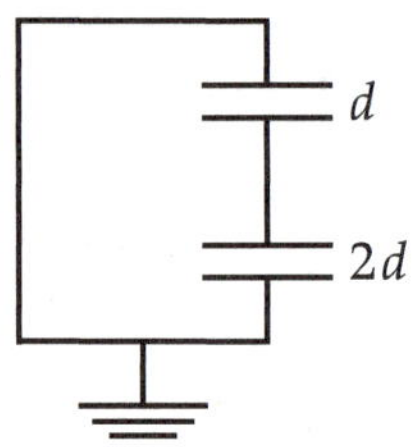

FIG. P20.71

Then $\quad \frac{Q_1}{C_1} = \frac{Q_2}{C_2} = \frac{Q_1 d}{\epsilon_0 A} = \frac{Q_2 2d}{\epsilon_0 A} \qquad Q_1 = 2Q_2 \qquad 2Q_2 + Q_2 = Q.$

(a) $Q_2 = \frac{Q}{3}$. $\boxed{\text{On the lower plate the charge is } -\frac{Q}{3}.}$

$Q_1 = \frac{2Q}{3}$. $\boxed{\text{On the upper plate the charge is } -\frac{2Q}{3}.}$

(b) $\Delta V = \frac{Q_1}{C_1} = \boxed{\frac{2Qd}{3\epsilon_0 A}}$

P20.72 (a) We use Equation 20.30 to find the potential energy of the capacitor. As we will see, the potential difference ΔV changes as the dielectric is withdrawn. The initial and final energies are $U_i = \frac{1}{2}\left(\frac{Q^2}{C_i}\right)$ and $U_f = \frac{1}{2}\left(\frac{Q^2}{C_f}\right)$.

But the initial capacitance (with the dielectric) is $C_i = \kappa C_f$. Therefore, $U_f = \frac{1}{2}\kappa\left(\frac{Q^2}{C_i}\right)$.

Since the work done by the external force in removing the dielectric equals the change in potential energy, we have $W = U_f - U_i = \frac{1}{2}\kappa\left(\frac{Q^2}{C_i}\right) - \frac{1}{2}\left(\frac{Q^2}{C_i}\right) = \frac{1}{2}\left(\frac{Q^2}{C_i}\right)(\kappa - 1)$.

To express this relation in terms of potential difference ΔV_i, we substitute $Q = C_i(\Delta V_i)$, and evaluate: $W = \frac{1}{2}C_i(\Delta V_i)^2(\kappa - 1) = \frac{1}{2}\left(2.00\times10^{-9}\text{ F}\right)(100\text{ V})^2(5.00 - 1.00) = \boxed{4.00\times10^{-5}\text{ J}}$.

The positive result confirms that the final energy of the capacitor is greater than the initial energy. The extra energy comes from the work done *on* the system by the external force that pulled out the dielectric.

(b) The final potential difference across the capacitor is $\Delta V_f = \frac{Q}{C_f}$.

Substituting $C_f = \frac{C_i}{\kappa}$ and $Q = C_i(\Delta V_i)$ gives $\Delta V_f = \kappa\Delta V_i = 5.00(100\text{ V}) = \boxed{500\text{ V}}$.

Even though the capacitor is isolated and its charge remains constant, the potential difference across the plates does increase in this case.

P20.73 $\kappa = 3.00$, $E_{\text{max}} = 2.00\times10^8\text{ V/m} = \frac{\Delta V_{\text{max}}}{d}$

For $C = \frac{\kappa\in_0 A}{d} = 0.250\times10^{-6}\text{ F}$

$$A = \frac{Cd}{\kappa\in_0} = \frac{C\Delta V_{\text{max}}}{\kappa\in_0 E_{\text{max}}} = \frac{\left(0.250\times10^{-6}\right)(4\,000)}{3.00\left(8.85\times10^{-12}\right)\left(2.00\times10^8\right)} = \boxed{0.188\text{ m}^2}$$

P20.74 The initial charge on the larger capacitor is

$$Q = C\Delta V = 10\ \mu\text{F}(15\text{ V}) = 150\ \mu\text{C}.$$

An additional charge q is pushed through the 50-V battery, giving the smaller capacitor charge q and the larger charge $150\ \mu\text{C} + q$.

Then $$50\text{ V} = \frac{q}{5\ \mu\text{F}} + \frac{150\ \mu\text{C} + q}{10\ \mu\text{F}}$$

$$500\ \mu\text{C} = 2q + 150\ \mu\text{C} + q$$

$$q = 117\ \mu\text{C}$$

So across the 5-μF capacitor $\Delta V = \frac{q}{C} = \frac{117\ \mu\text{C}}{5\ \mu\text{F}} = \boxed{23.3\text{ V}}$.

Across the 10-μF capacitor $\Delta V = \frac{150\ \mu\text{C} + 117\ \mu\text{C}}{10\ \mu\text{F}} = \boxed{26.7\text{ V}}$.

P20.75 (a) The portion of the device containing the dielectric has plate area ℓx and capacitance $C_1 = \dfrac{\kappa \in_0 \ell x}{d}$. The unfilled part has area $\ell(\ell - x)$ and capacitance $C_2 = \dfrac{\in_0 \ell(\ell - x)}{d}$. The total capacitance is $C_1 + C_2 = \boxed{\dfrac{\in_0}{d}\left[\ell^2 + \ell x(\kappa - 1)\right]}$.

(b) The stored energy is $U = \dfrac{1}{2}\dfrac{Q^2}{C} = \boxed{\dfrac{Q^2 d}{2\in_0\left(\ell^2 + \ell x(\kappa - 1)\right)}}$.

(c) $\vec{\mathbf{F}} = -\left(\dfrac{dU}{dx}\right)\hat{\mathbf{i}} = \boxed{\dfrac{Q^2 d\,\ell(\kappa - 1)}{2\in_0\left(\ell^2 + \ell x(\kappa - 1)\right)^2}\hat{\mathbf{i}}}$ When $x = 0$, the original value of the force is $\dfrac{Q^2 d(\kappa - 1)}{2\in_0 \ell^3}\hat{\mathbf{i}}$. As the dielectric slides in, the charges on the plates redistribute themselves. The force decreases to its final value $\dfrac{Q^2 d(\kappa - 1)}{2\in_0 \ell^3 \kappa^2}\hat{\mathbf{i}}$.

(d) At $x = \dfrac{\ell}{2}$, $\vec{\mathbf{F}} = \dfrac{2Q^2 d(\kappa - 1)}{\in_0 \ell^3 (\kappa + 1)^2}\hat{\mathbf{i}}$.

For the constant charge on the capacitor and the initial voltage we have the relationship

$$Q = C_0 \Delta V = \frac{\in_0 \ell^2 \Delta V}{d}.$$

Then the force is $\vec{\mathbf{F}} = \dfrac{2\in_0 \ell(\Delta V)^2(\kappa - 1)}{d(\kappa + 1)^2}\hat{\mathbf{i}}$

$$\vec{\mathbf{F}} = \frac{2\left(8.85 \times 10^{-12}\ \text{C}^2\right)0.05\ \text{m}(2\ 000\ \text{Nm})^2(4.5 - 1)}{\text{Nm}^2(0.002\ \text{m})\text{C}^2(4.5 + 1)^2}\hat{\mathbf{i}} = \boxed{205\,\mu\text{N}\ \hat{\mathbf{i}}}.$$

P20.76 The portion of the sandwich nearly filled by metal has

capacitance $\dfrac{\kappa \in_0 (\ell x)}{d} \to \infty$

and stored energy $\dfrac{Q^2}{2C} \to 0$.

The unfilled portion has

capacitance $\dfrac{\in_0 \ell(\ell - x)}{d}$.

The charge on this portion is $Q = \dfrac{(\ell - x)Q_0}{\ell}$.

continued on next page

(a) The stored energy is

$$U=\frac{Q^2}{2C}=\frac{\left[(\ell-x)\frac{Q_0}{\ell}\right]^2}{2\epsilon_0\,\ell\,\frac{\ell-x}{d}}=\boxed{\frac{Q_0^2(\ell-x)d}{2\epsilon_0\,\ell^3}}.$$

(b) $$F=-\frac{dU}{dx}=-\frac{d}{dx}\left(\frac{Q_0^2(\ell-x)d}{2\epsilon_0\,\ell^3}\right)=+\frac{Q_0^2 d}{2\epsilon_0\,\ell^3}$$

$$\vec{\mathbf{F}}=\boxed{\frac{Q_0^2 d}{2\epsilon_0\,\ell^3}\text{ to the right}}\text{ (into the pair of plates)}$$

(c) $$\text{Stress }=\frac{F}{\ell d}=\boxed{\frac{Q_0^2}{2\epsilon_0\,\ell^4}}$$

(d) $$u=\frac{1}{2}\epsilon_0 E^2=\frac{1}{2}\epsilon_0\left(\frac{\sigma}{\epsilon_0}\right)^2=\frac{1}{2}\epsilon_0\left(\frac{Q_0}{\epsilon_0\,\ell^2}\right)^2=\boxed{\frac{Q_0^2}{2\epsilon_0\,\ell^4}}$$

P20.77 By symmetry, the potential difference across 3C is zero, so the circuit reduces to

$$C_{eq}=\left(\frac{1}{2C}+\frac{1}{4C}\right)^{-1}=\frac{8}{6}C=\boxed{\frac{4}{3}C}.$$

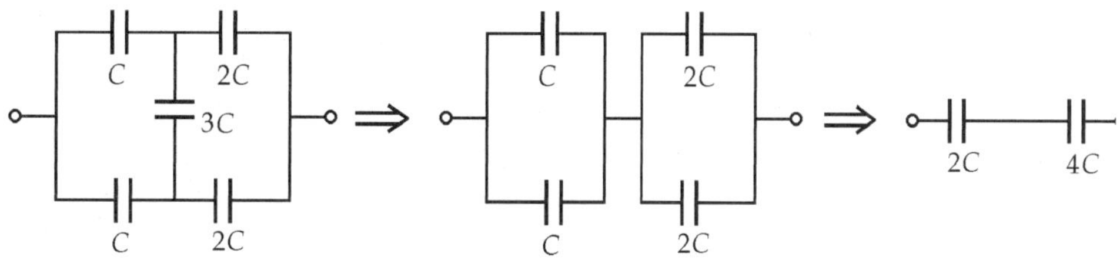

FIG. P20.77

ANSWERS TO EVEN PROBLEMS

P20.2 1.35 MJ

P20.4 1.67 MN/C

P20.6 (a) $\frac{2QE}{k}$; (b) $\frac{QE}{k}$; (c) $2\pi\sqrt{\frac{m}{k}}$; (d) $\frac{2(QE-\mu_k mg)}{k}$

P20.8 (a) $\vec{\mathbf{F}}=0$; (b) $\vec{\mathbf{E}}=0$; (c) 45.0 kV

P20.10 8.95 J

P20.12 (a) 32.2 kV ; (b) -9.65×10^{-2} J

P20.14 (a) +7.00 nC with position vector 7.00 cm at 70.0° and –8.00 nC with position vector 3.00 cm at 90.0°;

(b) $\left(0.070\,3\hat{\mathbf{i}}-1.09\hat{\mathbf{j}}\right)$ mN ; (c) $-24.0\ \mu$J

P20.16 (a) $v_1=\sqrt{\frac{2m_2k_eq_1q_2}{m_1(m_1+m_2)}\left(\frac{1}{r_1+r_2}-\frac{1}{d}\right)}$,

$v_2=\sqrt{\frac{2m_1k_eq_1q_2}{m_2(m_1+m_2)}\left(\frac{1}{r_1+r_2}-\frac{1}{d}\right)}$;

(b) Faster than calculated in (a)

P20.18 (a) $-45.0\ \mu$J ; (b) 34.6 km/s

P20.20 27.4 fm

P20.22 (a) 0; (b) $\frac{k_e Q}{r^2}$

P20.24 $-0.553\frac{k_e Q}{R}$

P20.26 $V=-\left(\frac{k_e \alpha L}{2}\right)\ln\left[\frac{\sqrt{(L^2/4)+b^2}-L/2}{\sqrt{(L^2/4)+b^2}+L/2}\right]$

P20.28 1.56×10^{12} electrons removed

P20.30 (a) 135 kV;
(b) 2.25 MV/m away for the large sphere and 6.74 MV/m away for the small sphere

P20.32 (a) $1.00\ \mu\text{F}$; (b) 100 V

P20.34 $\frac{(2N-1)(\pi-\theta)R^2 \in_0}{d}$

P20.36 (a) 2.68 nF; (b) 3.02 kV

P20.38 see the solution

P20.40 (a) $3.53\ \mu\text{F}$; (b) 6.35 V and 2.65 V;
(c) $31.8\ \mu\text{C}$ on each

P20.42 $\frac{1}{2}C_p+\sqrt{\frac{1}{4}C_p^2-C_pC_s}$

P20.44 (a) $2C$; (b) $Q_1>Q_3>Q_2$;
(c) $\Delta V_1>\Delta V_2=\Delta V_3$;
(d) Q_3 and Q_1 increase, Q_2 decreases

P20.46 $12.9\ \mu\text{F}$

P20.48 4.47 kV

P20.50 (a) $1.50\ \mu\text{C}$; (b) 1.83 kV

P20.52 $2.51\times10^{-3}\ \text{m}^3=2.51\ \text{L}$

P20.54 (a) 13.3 nC; (b) 272 nC

P20.56 $\sim10^{-6}$ F and $\sim10^{2}$ V for two 40-cm by 100-cm sheets of aluminum foil sandwiching a thin sheet of plastic

P20.58 (a) 450 kV; (b) $7.51\ \mu\text{C}$

P20.60 (a) The velocity of one particle relative to the other is first a velocity of approach, then zero at closest approach, and then a velocity of recession;
(b) $6.00\hat{\mathbf{i}}$ m/s;(c) 3.64 m;
(d) $-9.00\hat{\mathbf{i}}$ m/s for the incident particle and $12.0\hat{\mathbf{i}}$ m/s for the target particle.

P20.62 (a) –27.2 eV; (b) –6.80 eV; (c) 0

P20.64 see the solution

P20.66 (a) 1.42 mm; (b) 9.20 kV/m

P20.68 $\left(\frac{k_e q^2}{3am}\right)^{1/2}$

P20.70 (a) see the solution;
(b) $E_r=\frac{2k_e p\cos\theta}{r^3}$, $E_\theta=\frac{k_e p\sin\theta}{r^3}$, yes, no;
(c) $V=k_e py\left(x^2+y^2\right)^{-3/2}$,
$\vec{\mathbf{E}}=\frac{3k_e pxy\hat{\mathbf{i}}}{\left(x^2+y^2\right)^{5/2}}+\frac{k_e p\left(2y^2-x^2\right)\hat{\mathbf{j}}}{\left(x^2+y^2\right)^{5/2}}$

P20.72 (a) $40.0\ \mu\text{J}$; (b) 500 V

P20.74 23.3 V; 26.7 V

P20.76 (a) $\frac{Q_0^2 d(\ell-x)}{2\ell^3 \in_0}$; (b) $\frac{Q_0^2 d}{2\ell^3 \in_0}$ to the right;
(c) $\frac{Q_0^2}{2\ell^4 \in_0}$; (d) $\frac{Q_0^2}{2\ell^4 \in_0}$

Chapter 21

Current and Direct Current Circuits

CHAPTER OUTLINE

ANSWERS TO QUESTIONS

Q21.1 Consider all of the eastbound lanes on a highway, taken together. Individual vehicles correspond to bits of charge. The number of vehicles that pass a certain milepost, divided by the time during which they go past, corresponds to the value of the current.

Q21.2 Geometry and resistivity. In turn, the resistivity of the material depends on the temperature.

Q21.3 The radius of wire B is $\sqrt{3}$ times the radius of wire A, to make its cross–sectional area 3 times larger.

Q21.4 In a normal metal, suppose that we could proceed to a limit of zero resistance by lengthening the average time between collisions. The classical model of conduction then suggests that a constant applied voltage would cause constant acceleration of the free electrons, and a current steadily increasing in time.

On the other hand, we can actually switch to zero resistance by substituting a superconducting wire for the normal metal. In this case, the drift velocity of electrons is established by vibrations of atoms in the crystal lattice; the maximum current is limited; and it becomes impossible to establish a potential difference across the superconductor.

Q21.5 The amplitude of atomic vibrations increases with temperature. Atoms can then scatter electrons more efficiently.

Q21.6 A current will continue to exist in a superconductor without voltage because there is no resistance loss.

Q21.7 Because there are so many electrons in a conductor (approximately 10^{28} electrons/m^3) the average velocity of charges is very slow. When you connect a wire to a potential difference, you establish an electric field everywhere in the wire nearly instantaneously, to make electrons start drifting everywhere all at once.

Q21.8 The 25 W bulb has a higher resistance. The 100 W bulb carries more current.

Q21.9 One ampere–hour is 3 600 coulombs. The ampere–hour rating is the quantity of charge that the battery can lift though its nominal potential difference.

Q21.10 In series, the current is the same through each resistor. Without knowing individual resistances, nothing can be determined about potential differences or power.

Q21.11 In parallel, the potential difference is the same across each resistor. Without knowing individual resistances, nothing can be determined about current or power.

Q21.12 A short circuit can develop when the last bit of insulation frays away between the two conductors in a lamp cord. Then the two conductors touch each other, opening a low-resistance branch in parallel with the lamp. The lamp will immediately go out, carrying no current and presenting no danger. A very large current exists in the power supply, the house wiring, and the rest of the lamp cord up to the contact point. Before it blows the fuse or pops the circuit breaker, the large current can quickly raise the temperature in the short-circuit path.

Q21.13 The whole wire is very nearly at one uniform potential. There is essentially zero *difference* in potential between the bird's feet. Then negligible current goes through the bird. The resistance through the bird's body between its feet is much larger than the resistance through the wire between the same two points.

Q21.14 A wire or cable in a transmission line is thick and made of material with very low resistivity. Only when its length is very large does its resistance become significant. To transmit power over a long distance it is most efficient to use low current at high voltage, minimizing the I^2R power loss in the transmission line. Alternating current, as opposed to the direct current we study first, can be stepped up in voltage and then down again, with high-efficiency transformers at both ends of the power line.

Q21.15 The bulb will light up for a while immediately after the switch is closed. As the capacitor charges, the bulb gets progressively dimmer. When the capacitor is fully charged the current in the circuit is zero and the bulb does not glow at all. If the value of RC is small, this whole process might occupy a very short time interval.

Q21.16 Car headlights are in parallel. If they were in series, both would go out when the filament of one failed. An important safety factor would be lost.

Q21.17 Kirchhoff's junction rule expresses conservation of electric charge. If the total current into a point were different from the total current out, then charge would be continuously created or annihilated at that point.

Kirchhoff's loop rule expresses conservation of energy. For a single-loop circle with two resistors, the loop rule reads $+\mathcal{E} - IR_1 - IR_2 = 0$. This is algebraically equivalent to $q\mathcal{E} = qIR_1 + qIR_2$, where $q = I\Delta t$ is the charge passing a point in the loop in the time interval Δt. The equivalent equation states that the power supply injects energy into the circuit equal in amount to that which the resistors degrade into internal energy.

Q21.18 At their normal operating temperatures, from $\mathcal{P} = \dfrac{\Delta V^2}{R}$, the bulbs present resistances $R = \dfrac{\Delta V^2}{\mathcal{P}} = \dfrac{(120\text{ V})^2}{60\text{ W}} = 240\ \Omega$, and $\dfrac{(120\text{ V})^2}{75\text{ W}} = 190\ \Omega$, and $\dfrac{(120\text{ V})^2}{200\text{ W}} = 72\ \Omega$. The nominal 60 W lamp has greatest resistance. When they are connected in series, they all carry the same small current. Here the highest-resistance bulb glows most brightly and the one with lowest resistance is faintest. This is just the reverse of their order of intensity if they were connected in parallel, as they are designed to be.

Q21.19 Answer their question with a challenge. If the student is just looking at a diagram, provide the materials to build the circuit. If you are looking at a circuit where the second bulb really is fainter, get the student to unscrew them both and interchange them. But check that the student's understanding of potential has not been impaired: if you patch past the first bulb to short it out, the second gets brighter.

Q21.20 The hospital maintenance worker is right. A hospital room is full of electrical grounds, including the bed frame. If your grandmother touched the faulty knob and the bed frame at the same time, she could receive quite a jolt, as there would be a potential difference of 120 V across her. If the 120 V is DC, the shock could send her into ventricular fibrillation, and the hospital staff could use the defibrillator you read about in Example 21.10. If the 120 V is AC, which is most likely, the current could produce external and internal burns along the path of conduction. Likely no one got a shock from the radio back at home because her bedroom contained no electrical grounds—no conductors connected to zero volts. Just like the bird in Question 21.13, granny could touch the "hot" knob without getting a shock so long as there was no path to ground to supply a potential difference across her. A new appliance in the bedroom or a flood could make the radio lethal. Repair it or discard it. Enjoy the news from Lake Wobegon on the new plastic radio.

Q21.21 Suppose $\mathcal{E} = 12\text{ V}$ and each lamp has $R = 2\ \Omega$. Before the switch is closed the current is $\dfrac{12\text{ V}}{6\ \Omega} = 2\text{ A}$. The potential difference across each lamp is $(2\text{ A})(2\ \Omega) = 4\text{ V}$. The power of each lamp is $(2\text{ A})(4\text{ V}) = 8\text{ W}$, totaling 24 W for the circuit. Closing the switch makes the switch and the wires connected to it a zero-resistance branch. All of the current through A and B will go through the switch and (b) lamp C goes out, with zero voltage across it. With less total resistance, the (c) current in the battery $\dfrac{12\text{ V}}{4\ \Omega} = 3\text{ A}$ becomes larger than before and (a) lamps A and B get brighter. (d) The voltage across each of A and B is $(3\text{ A})(2\ \Omega) = 6\text{ V}$, larger than before. Each converts power $(3\text{ A})(6\text{ V}) = 18\text{ W}$, totaling 36 W, which is (e) an increase.

Q21.22 Two runs in series: . Three runs in parallel: 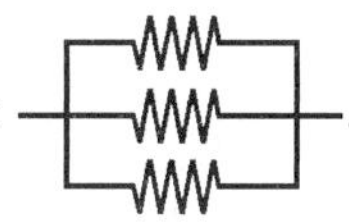. Junction of one lift and two runs: 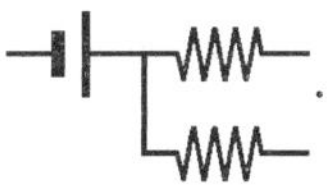.

Gustav Robert Kirchhoff, Professor of Physics at Heidelberg and Berlin, was master of the obvious. A junction rule: The number of skiers coming into any junction must be equal to the number of skiers leaving, so that if we count the outward bound skiers as negative the net accumulation of skiers is zero. A loop rule: the total change in altitude must be zero for any skier completing a closed path.

SOLUTIONS TO PROBLEMS

Section 21.1 Electric Current

P21.1 $I = \dfrac{\Delta Q}{\Delta t}$ $\qquad \Delta Q = I\Delta t = \left(30.0\times10^{-6}\text{ A}\right)(40.0\text{ s}) = 1.20\times10^{-3}\text{ C}$

$$N = \frac{Q}{e} = \frac{1.20\times10^{-3}\text{ C}}{1.60\times10^{-19}\text{ C/electron}} = \boxed{7.50\times10^{15}\text{ electrons}}$$

P21.2 The period of revolution for the sphere is $T=\frac{2\pi}{\omega}$, and the average current represented by this revolving charge is $I=\frac{q}{T}=\boxed{\frac{q\omega}{2\pi}}$.

P21.3 $Q(t)=\int_0^t I\,dt=I_0\tau\left(1-e^{-t/\tau}\right)$

(a) $Q(\tau)=I_0\tau\left(1-e^{-1}\right)=\boxed{(0.632)I_0\tau}$

(b) $Q(10\tau)=I_0\tau\left(1-e^{-10}\right)=\boxed{(0.999\,95)I_0\tau}$

(c) $Q(\infty)=I_0\tau\left(1-e^{-\infty}\right)=\boxed{I_0\tau}$

P21.4 $q=4t^3+5t+6$

$$A=\left(2.00\text{ cm}^2\right)\left(\frac{1.00\text{ m}}{100\text{ cm}}\right)^2=2.00\times10^{-4}\text{ m}^2$$

(a) $I(1.00\text{ s})=\left.\frac{dq}{dt}\right|_{t=1.00\text{ s}}=\left.\left(12t^2+5\right)\right|_{t=1.00\text{ s}}=\boxed{17.0\text{ A}}$

(b) $J=\frac{I}{A}=\frac{17.0\text{ A}}{2.00\times10^{-4}\text{ m}^2}=\boxed{85.0\text{ kA/m}^2}$

P21.5 We use $I=nqAv_d$. Here n is the number of charge carriers per unit volume, and is identical to the number of atoms per unit volume. We assume a contribution of 1 free electron per atom in the relationship above. For aluminum, which has a molar mass of 27, we know that Avogadro's number of atoms, N_A, has a mass of 27.0 g. Thus, the mass per atom is

$$\frac{27.0\text{ g}}{N_A}=\frac{27.0\text{ g}}{6.02\times10^{23}}=4.49\times10^{-23}\text{ g/atom}.$$

Thus, $$n=\frac{\text{density of aluminum}}{\text{mass per atom}}=\frac{2.70\text{ g/cm}^3}{4.49\times10^{-23}\text{ g/atom}}$$

$$n=6.02\times10^{22}\text{ atoms/cm}^3=6.02\times10^{28}\text{ atoms/m}^3.$$

Therefore, $$v_d=\frac{I}{nqA}=\frac{5.00\text{ A}}{\left(6.02\times10^{28}\text{ m}^{-3}\right)\left(1.60\times10^{-19}\text{ C}\right)\left(4.00\times10^{-6}\text{ m}^2\right)}=1.30\times10^{-4}\text{ m/s}$$

or, $v_d=\boxed{0.130\text{ mm/s}}$.

Section 21.2 Resistance and Ohm's Law

P21.6 $I=\frac{\Delta V}{R}=\frac{120\text{ V}}{240\ \Omega}=0.500\text{ A}=\boxed{500\text{ mA}}$

P21.7 $\Delta V=IR$

and $R=\frac{\rho\ell}{A}$: $A=(0.600\text{ mm})^2\left(\frac{1.00\text{ m}}{1\,000\text{ mm}}\right)^2=6.00\times10^{-7}\text{ m}^2$

$\Delta V=\frac{I\rho\ell}{A}$: $I=\frac{\Delta VA}{\rho\ell}=\frac{(0.900\text{ V})\left(6.00\times10^{-7}\text{ m}^2\right)}{\left(5.60\times10^{-8}\ \Omega\cdot\text{m}\right)(1.50\text{ m})}$

$I=\boxed{6.43\text{ A}}$

P21.8 (a) Given $M=\rho_dV=\rho_dA\ell$ where $\rho_d\equiv$ mass density,

we obtain: $A=\frac{M}{\rho_d\ell}$. Taking $\rho_r\equiv$ resistivity, $R=\frac{\rho_r\ell}{A}=\frac{\rho_r\ell}{M/\rho_d\ell}=\frac{\rho_r\rho_d\ell^2}{M}$.

Thus, $\ell=\sqrt{\frac{MR}{\rho_r\rho_d}}=\sqrt{\frac{\left(1.00\times10^{-3}\right)(0.500)}{\left(1.70\times10^{-8}\right)\left(8.92\times10^3\right)}}$ $\ell=\boxed{1.82\text{ m}}$.

(b) $V=\frac{M}{\rho_d}$, or $\pi r^2\ell=\frac{M}{\rho_d}$.

Thus, $r=\sqrt{\frac{M}{\pi\rho_d\ell}}=\sqrt{\frac{1.00\times10^{-3}}{\pi\left(8.92\times10^3\right)(1.82)}}$ $r=1.40\times10^{-4}\text{ m}$.

The diameter is twice this distance: diameter $=\boxed{280\ \mu\text{m}}$.

P21.9 (a) $\rho=\rho_0\left[1+\alpha(T-T_0)\right]=\left(2.82\times10^{-8}\ \Omega\cdot\text{m}\right)\left[1+3.90\times10^{-3}(30.0°)\right]=\boxed{3.15\times10^{-8}\ \Omega\cdot\text{m}}$

(b) $J=\frac{E}{\rho}=\frac{0.200\text{ V/m}}{3.15\times10^{-8}\ \Omega\cdot\text{m}}=\boxed{6.35\times10^6\text{ A/m}^2}$

(c) $I=JA=J\left(\frac{\pi d^2}{4}\right)=\left(6.35\times10^6\text{ A/m}^2\right)\left[\frac{\pi\left(1.00\times10^{-4}\text{ m}\right)^2}{4}\right]=\boxed{49.9\text{ mA}}$

continued on next page

(d) $$n = \frac{6.02 \times 10^{23} \text{ electrons}}{\left[26.98 \text{ g}/\left(2.70 \times 10^{6} \text{ g/m}^3\right)\right]} = 6.02 \times 10^{28} \text{ electrons/m}^3$$

$$v_d = \frac{J}{ne} = \frac{\left(6.35 \times 10^{6} \text{ A/m}^2\right)}{\left(6.02 \times 10^{28} \text{ electrons/m}^3\right)\left(1.60 \times 10^{-19} \text{ C}\right)} = \boxed{659 \ \mu\text{m/s}}$$

(e) $$\Delta V = E\ell = (0.200 \text{ V/m})(2.00 \text{ m}) = \boxed{0.400 \text{ V}}$$

P21.10 At the low temperature T_C we write $R_C = \frac{\Delta V}{I_C} = R_0\left[1 + \alpha(T_C - T_0)\right]$ where $T_0 = 20.0°\text{C}$.

At the high temperature T_h, $$R_h = \frac{\Delta V}{I_h} = \frac{\Delta V}{1 \text{ A}} = R_0\left[1 + \alpha(T_h - T_0)\right].$$

Then $$\frac{(\Delta V)/(1.00 \text{ A})}{(\Delta V)/I_C} = \frac{1 + \left(3.90 \times 10^{-3}\right)(38.0)}{1 + \left(3.90 \times 10^{-3}\right)(-108)}$$

and $$I_C = (1.00 \text{ A})\left(\frac{1.15}{0.579}\right) = \boxed{1.98 \text{ A}}.$$

P21.11 For aluminum,

$$\alpha_E = 3.90 \times 10^{-3} \text{ °C}^{-1} \qquad \text{(Table 21.1)}$$

$$\alpha = 24.0 \times 10^{-6} \text{ °C}^{-1} \qquad \text{(Table 16.1)}$$

$$R = \frac{\rho\ell}{A} = \frac{\rho_0(1 + \alpha_E\Delta T)\ell(1 + \alpha\Delta T)}{A(1 + \alpha\Delta T)^2} = R_0\frac{(1 + \alpha_E\Delta T)}{(1 + \alpha\Delta T)} = (1.234 \ \Omega)\left(\frac{1.39}{1.002\,4}\right) = \boxed{1.71 \ \Omega}$$

Section 21.3 Superconductors

No problems in this section

Section 21.4 A Structural Model for Electrical Conduction

P21.12 (a) n is $\boxed{\text{unaffected}}$

(b) $$|J| = \frac{I}{A} \propto I$$

so it $\boxed{\text{doubles}}$.

continued on next page

(c) $J = nev_d$

so v_d $\boxed{\text{doubles}}$.

(d) $\tau = \dfrac{m\sigma}{nq^2}$ is $\boxed{\text{unchanged}}$ as long as σ does not change due to a temperature change in the conductor.

P21.13 $\rho = \dfrac{m}{nq^2\tau}$

so $$\tau = \frac{m}{\rho nq^2} = \frac{(9.11\times10^{-31})}{(1.70\times10^{-8})(8.49\times10^{28})(1.60\times10^{-19})} = 2.47\times10^{-14}\text{ s}$$

$$v_d = \frac{qE}{m}\tau$$

so $$7.84\times10^{-4} = \frac{(1.60\times10^{-19})E(2.47\times10^{-14})}{9.11\times10^{-31}}$$

Therefore, $E = \boxed{0.181\text{ V/m}}$. We could also compute the field from $E = \rho J$ with $J = nqv_d$.

Section 21.5 Electric Energy and Power

P21.14 $I = \dfrac{\mathcal{P}}{\Delta V} = \dfrac{600\text{ W}}{120\text{ V}} = \boxed{5.00\text{ A}}$

and $R = \dfrac{\Delta V}{I} = \dfrac{120\text{ V}}{5.00\text{ A}} = \boxed{24.0\ \Omega}$.

P21.15 $\mathcal{P} = 0.800(1\,500\text{ hp})(746\text{ W/hp}) = 8.95\times10^5\text{ W}$

$\mathcal{P} = I\Delta V$ $\quad 8.95\times10^5 = I(2\,000)$ $\quad I = \boxed{448\text{ A}}$

***P21.16** The battery takes in energy by electric transmission

$$\mathcal{P}\Delta t = (\Delta V)I(\Delta t) = 2.3\text{ J/C}\left(13.5\times10^{-3}\text{ C/s}\right)4.2\text{ h}\left(\frac{3\,600\text{ s}}{1\text{ h}}\right) = 469\text{ J}.$$

It puts out energy by electric transmission

$$(\Delta V)I(\Delta t) = 1.6\text{ J/C}\left(18\times10^{-3}\text{ C/s}\right)2.4\text{ h}\left(\frac{3\,600\text{ s}}{1\text{ h}}\right) = 249\text{ J}.$$

(a) efficiency $= \dfrac{\text{useful output}}{\text{total input}} = \dfrac{249\text{ J}}{469\text{ J}} = \boxed{0.530}$

continued on next page

(b) The only place for the missing energy to go is into internal energy:

$$469\text{ J} = 249\text{ J} + \Delta E_{\text{int}}$$

$$\Delta E_{\text{int}} = \boxed{221\text{ J}}$$

(c) We imagine toasting the battery over a fire with 221 J of heat input:

$$Q = mc\Delta T$$

$$\Delta T = \frac{Q}{mc} = \frac{221\text{ J}}{0.015\text{ kg}}\,\frac{\text{kg}^\circ\text{C}}{975\text{ J}} = \boxed{15.1^\circ\text{C}}$$

P21.17 $$\frac{\mathcal{P}}{\mathcal{P}_0} = \frac{(\Delta V)^2/R}{(\Delta V_0)^2/R} = \left(\frac{\Delta V}{\Delta V_0}\right)^2 = \left(\frac{140}{120}\right)^2 = 1.361$$

$$\Delta\% = \left(\frac{\mathcal{P} - \mathcal{P}_0}{\mathcal{P}_0}\right)(100\%) = \left(\frac{\mathcal{P}}{\mathcal{P}_0} - 1\right)(100\%) = (1.361 - 1)100\% = \boxed{36.1\%}$$

***P21.18** The energy taken in by electric transmission for the fluorescent lamp is

$$\mathcal{P}\Delta t = 11\text{ J/s}(100\text{ h})\left(\frac{3\,600\text{ s}}{1\text{ h}}\right) = 3.96 \times 10^6\text{ J}$$

$$\text{cost} = 3.96 \times 10^6\text{ J}\left(\frac{\$0.08}{\text{kWh}}\right)\left(\frac{\text{k}}{1\,000}\right)\left(\frac{\text{W}\cdot\text{s}}{\text{J}}\right)\left(\frac{\text{h}}{3\,600\text{ s}}\right) = \$0.088$$

For the incandescent bulb,

$$\mathcal{P}\Delta t = 40\text{ W}(100\text{ h})\left(\frac{3\,600\text{ s}}{1\text{ h}}\right) = 1.44 \times 10^7\text{ J}$$

$$\text{cost} = 1.44 \times 10^7\text{ J}\left(\frac{\$0.08}{3.6 \times 10^6\text{ J}}\right) = \$0.32$$

$$\text{saving} = \$0.32 - \$0.088 = \boxed{\$0.232}$$

P21.19 At operating temperature,

(a) $$\mathcal{P} = I\Delta V = (1.53\text{ A})(120\text{ V}) = \boxed{184\text{ W}}$$

(b) Use the change in resistance to find the final operating temperature of the toaster.

$$R = R_0(1 + \alpha\Delta T) \qquad \frac{120}{1.53} = \frac{120}{1.80}\left[1 + \left(0.400 \times 10^{-3}\right)\Delta T\right]$$

$$\Delta T = 441^\circ\text{C} \qquad T = 20.0^\circ\text{C} + 441^\circ\text{C} = \boxed{461^\circ\text{C}}$$

P21.20 The total clock power is

$$\left(270\times10^{6}\text{ clocks}\right)\left(2.50\ \frac{\text{J/s}}{\text{clock}}\right)\left(\frac{3\,600\text{ s}}{1\text{ h}}\right)=2.43\times10^{12}\ \text{J/h}.$$

From $e=\dfrac{W_{\text{out}}}{Q_{\text{in}}}$, the power input to the generating plants must be:

$$\frac{Q_{\text{in}}}{\Delta t}=\frac{W_{\text{out}}/\Delta t}{e}=\frac{2.43\times10^{12}\ \text{J/h}}{0.250}=9.72\times10^{12}\ \text{J/h}$$

and the rate of coal consumption is

$$\text{Rate }=\left(9.72\times10^{12}\ \text{J/h}\right)\left(\frac{1.00\text{ kg coal}}{33.0\times10^{6}\text{ J}}\right)=2.95\times10^{5}\ \text{kg coal/h}=\boxed{295\ \text{metric ton/h}}.$$

P21.21 You pay the electric company for energy transferred in the amount $E=\mathcal{P}\Delta t$

(a) $$\mathcal{P}\Delta t=40\text{ W}(2\text{ weeks})\left(\frac{7\text{ d}}{1\text{ week}}\right)\left(\frac{86\,400\text{ s}}{1\text{ d}}\right)\left(\frac{1\text{ J}}{1\text{ W}\cdot\text{s}}\right)=48.4\text{ MJ}$$

$$\mathcal{P}\Delta t=40\text{ W}(2\text{ weeks})\left(\frac{7\text{ d}}{1\text{ week}}\right)\left(\frac{24\text{ h}}{1\text{ d}}\right)\left(\frac{\text{k}}{1\,000}\right)=13.4\text{ kWh}$$

$$\mathcal{P}\Delta t=40\text{ W}(2\text{ weeks})\left(\frac{7\text{ d}}{1\text{ week}}\right)\left(\frac{24\text{ h}}{1\text{ d}}\right)\left(\frac{\text{k}}{1\,000}\right)\left(\frac{0.12\ \$}{\text{kWh}}\right)=\boxed{\$1.61}$$

(b) $$\mathcal{P}\Delta t=970\text{ W}(3\text{ min})\left(\frac{1\text{ h}}{60\text{ min}}\right)\left(\frac{\text{k}}{1\,000}\right)\left(\frac{0.12\ \$}{\text{kWh}}\right)=\boxed{\$0.005\,82}=0.582¢$$

(c) $$\mathcal{P}\Delta t=5\,200\text{ W}(40\text{ min})\left(\frac{1\text{ h}}{60\text{ min}}\right)\left(\frac{\text{k}}{1\,000}\right)\left(\frac{0.12\ \$}{\text{kWh}}\right)=\boxed{\$0.416}$$

***P21.22** The heater should put out constant power

$$\mathcal{P}=\frac{Q}{\Delta t}=\frac{mc\left(T_f-T_i\right)}{\Delta t}=\frac{(0.250\text{ kg})(4\,186\text{ J})(100^\circ\text{C}-20^\circ\text{C})}{\text{kg}\cdot^\circ\text{C}(4\text{ min})}\left(\frac{1\text{ min}}{60\text{ s}}\right)=349\ \text{J/s}$$

Then its resistance should be described by

$$\mathcal{P}=(\Delta V)I=\frac{(\Delta V)(\Delta V)}{R}$$

$$R=\frac{(\Delta V)^2}{\mathcal{P}}=\frac{(120\ \text{J/C})^2}{349\ \text{J/s}}=41.3\ \Omega$$

continued on next page

Its resistivity at 100°C is given by

$$\rho = \rho_0\left[1+\alpha(T-T_0)\right] = \left(1.50\times10^{-6}\ \Omega\cdot\text{m}\right)\left[1+0.4\times10^{-3}(80)\right] = 1.55\times10^{-6}\ \Omega\cdot\text{m}$$

Then for a wire of circular cross section

$$R = \rho\frac{\ell}{A} = \rho\frac{\ell}{\pi r^2} = \rho\frac{\ell 4}{\pi d^2}$$

$$41.3\ \Omega = \left(1.55\times10^{-6}\ \Omega\cdot\text{m}\right)\frac{4\ell}{\pi d^2}$$

$$\frac{\ell}{d^2} = 2.09\times10^{+7}/\text{m} \qquad \text{or} \qquad d^2 = \left(4.77\times10^{-8}\ \text{m}\right)\ell$$

One possible choice is $\ell = 0.900$ m and $d = 2.07\times10^{-4}$ m. If ℓ and d are made too small, the surface area will be inadequate to transfer heat into the water fast enough to prevent overheating of the filament. To make the volume less than 0.5 cm^3, we want ℓ and d less than those described by $\frac{\pi d^2}{4}\ell = 0.5\times10^{-6}\ \text{m}^3$. Substituting $d^2 = \left(4.77\times10^{-8}\ \text{m}\right)\ell$ gives $\frac{\pi}{4}\left(4.77\times10^{-8}\ \text{m}\right)\ell^2 = 0.5\times10^{-6}\ \text{m}^3$, $\ell = 3.65$ m and $d = 4.18\times10^{-4}$ m. Thus our answer is: Any diameter d and length ℓ related by $d^2 = \left(4.77\times10^{-8}\ \text{m}\right)\ell$ would have the right resistance. One possibility is length 0.900 m and diameter 0.207 mm, but such a small wire might overheat rapidly if it were not surrounded by water. The volume can be less than 0.5 cm^3.

P21.23 (a) $\mathscr{P} = I\Delta V$

so $I = \frac{\mathscr{P}}{\Delta V} = \frac{8.00\times10^3\ \text{W}}{12.0\ \text{V}} = \boxed{667\ \text{A}}$.

(b) $\Delta t = \frac{\Delta U}{\mathscr{P}} = \frac{2.00\times10^7\ \text{J}}{8.00\times10^3\ \text{W}} = 2.50\times10^3\ \text{s}$

and $\Delta x = v\Delta t = (20.0\ \text{m/s})\left(2.50\times10^3\ \text{s}\right) = \boxed{50.0\ \text{km}}$.

P21.24 Consider a 400-W blow dryer used for ten minutes daily for a year. The energy transferred to the dryer is

$$\mathscr{P}\Delta t = (400\ \text{J/s})(600\ \text{s/d})(365\ \text{d}) \approx 9\times10^7\ \text{J}\left(\frac{1\ \text{kWh}}{3.6\times10^6\ \text{J}}\right) \approx 20\ \text{kWh}.$$

We suppose that electrically transmitted energy costs on the order of ten cents per kilowatt-hour. Then the cost of using the dryer for a year is on the order of

$$\text{Cost} \approx (20\ \text{kWh})(\$0.10/\text{kWh}) = \$2\ \boxed{\sim\$1}.$$

Section 21.6 Sources of emf

P21.25 (a) $\mathscr{P} = \dfrac{(\Delta V)^2}{R}$

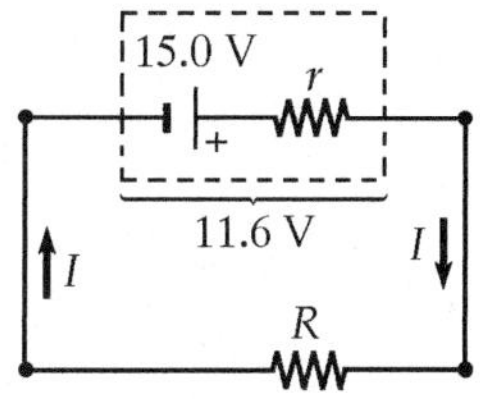

FIG. P21.25

becomes $20.0\text{ W} = \dfrac{(11.6\text{ V})^2}{R}$

so $R = \boxed{6.73\ \Omega}$.

(b) $\Delta V = IR$

so $11.6\text{ V} = I(6.73\ \Omega)$

and $I = 1.72\text{ A}$

$\mathcal{E} = IR + Ir$

so $15.0\text{ V} = 11.6\text{ V} + (1.72\text{ A})r$

$r = \boxed{1.97\ \Omega}$.

P21.26 The total resistance is $R = \dfrac{3.00\text{ V}}{0.600\text{ A}} = 5.00\ \Omega$.

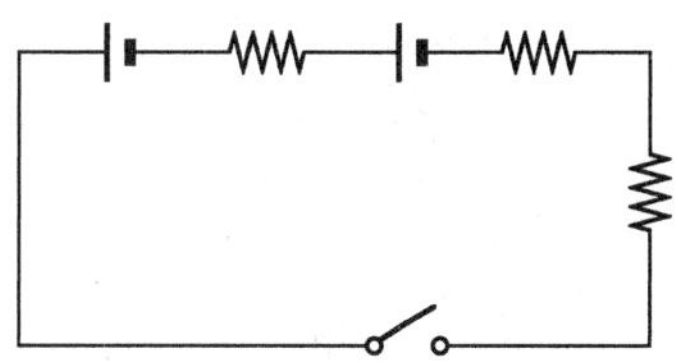

FIG. P21.26

(a) $R_{\text{lamp}} = R - r_{\text{batteries}} = 5.00\ \Omega - 0.408\ \Omega = \boxed{4.59\ \Omega}$

(b) $\dfrac{\mathscr{P}_{\text{batteries}}}{\mathscr{P}_{\text{total}}} = \dfrac{(0.408\ \Omega)I^2}{(5.00\ \Omega)I^2} = 0.081\,6 = \boxed{8.16\%}$

Section 21.7 Resistors in Series and in Parallel

P21.27 (a) $R_p = \dfrac{1}{(1/7.00\ \Omega) + (1/10.0\ \Omega)} = 4.12\ \Omega$

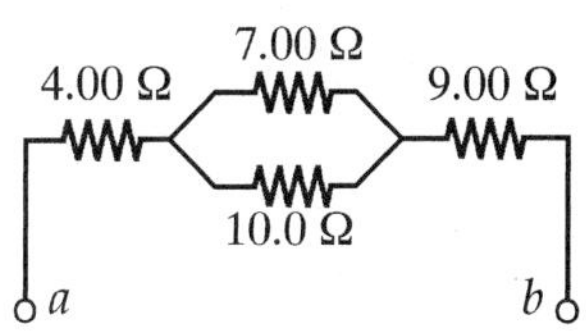

FIG. P21.27

$R_s = R_1 + R_2 + R_3 = 4.00 + 4.12 + 9.00 = \boxed{17.1\ \Omega}$

(b) $\Delta V = IR$

$34.0\text{ V} = I(17.1\ \Omega)$

$I = \boxed{1.99\text{ A}}$ for 4.00 Ω, 9.00 Ω resistors.

Applying $\Delta V = IR$, $(1.99\text{ A})(4.12\ \Omega) = 8.18\text{ V}$

$8.18\text{ V} = I(7.00\ \Omega)$

so $I = \boxed{1.17\text{ A}}$ for 7.00 Ω resistor

$8.18\text{ V} = I(10.0\ \Omega)$

so $I = \boxed{0.818\text{ A}}$ for 10.0 Ω resistor.

P21.28 We assume that the metal wand makes low-resistance contact with the person's hand and that the resistance through the person's body is negligible compared to the resistance R_{shoes} of the shoe soles. The equivalent resistance seen by the power supply is $1.00\ \text{M}\Omega + R_{\text{shoes}}$. The current through both resistors is $\dfrac{50.0\ \text{V}}{1.00\ \text{M}\Omega + R_{\text{shoes}}}$. The voltmeter displays

$$\Delta V = I(1.00\ \text{M}\Omega) = \frac{50.0\ \text{V}(1.00\ \text{M}\Omega)}{1.00\ \text{M}\Omega + R_{\text{shoes}}} = \Delta V.$$

(a) We solve to obtain $50.0\ \text{V}(1.00\ \text{M}\Omega) = \Delta V(1.00\ \text{M}\Omega) + \Delta V(R_{\text{shoes}})$

$$R_{\text{shoes}} = \frac{1.00\ \text{M}\Omega(50.0 - \Delta V)}{\Delta V}.$$

(b) With $R_{\text{shoes}} \to 0$, the current through the person's body is

$$\frac{50.0\ \text{V}}{1.00\ \text{M}\Omega} = 50.0\ \mu\text{A} \qquad \boxed{\text{The current will never exceed } 50\ \mu\text{A}}.$$

P21.29 If we turn the given diagram on its side, we find that it is the same as figure (a). The 20.0 Ω and 5.00 Ω resistors are in series, so the first reduction is shown in (b). In addition, since the 10.0 Ω, 5.00 Ω, and 25.0 Ω resistors are then in parallel, we can solve for their equivalent resistance as:

$$R_{\text{eq}} = \frac{1}{\left(\frac{1}{10.0\ \Omega} + \frac{1}{5.00\ \Omega} + \frac{1}{25.0\ \Omega}\right)} = 2.94\ \Omega.$$

This is shown in figure (c), which in turn reduces to the circuit shown in figure (d).

Next, we work backwards through the diagrams applying $I = \dfrac{\Delta V}{R}$ and $\Delta V = IR$ alternately to every resistor, real and equivalent. The 12.94 Ω resistor is connected across 25.0 V, so the current through the battery in every diagram is

$$I = \frac{\Delta V}{R} = \frac{25.0\ \text{V}}{12.94\ \Omega} = 1.93\ \text{A}.$$

In figure (c), this 1.93 A goes through the 2.94 Ω equivalent resistor to give a potential difference of:

$$\Delta V = IR = (1.93\ \text{A})(2.94\ \Omega) = 5.68\ \text{V}.$$

From figure (b), we see that this potential difference is the same across ΔV_{ab}, the 10 Ω resistor, and the 5.00 Ω resistor.

(b) Therefore, $\Delta V_{ab} = \boxed{5.68\ \text{V}}$.

(a) Since the current through the 20.0 Ω resistor is also the current through the 25.0 Ω line *ab*,

$$I = \frac{\Delta V_{ab}}{R_{ab}} = \frac{5.68\ \text{V}}{25.0\ \Omega} = 0.227\ \text{A} = \boxed{227\ \text{mA}}.$$

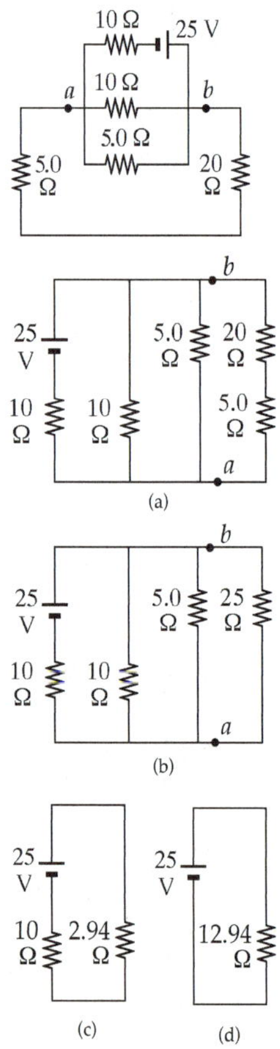

FIG. P21.29

P21.30 (a) Since all the current in the circuit must pass through the series 100 Ω resistor, $\mathcal{P} = I^2R$

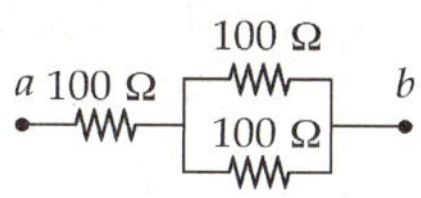

FIG. P21.30

$$\mathcal{P}_{\max} = RI_{\max}^2$$

so $$I_{\max} = \sqrt{\frac{\mathcal{P}}{R}} = \sqrt{\frac{25.0 \text{ W}}{100 \ \Omega}} = 0.500 \text{ A}$$

$$R_{eq} = 100 \ \Omega + \left(\frac{1}{100} + \frac{1}{100}\right)^{-1} \Omega = 150 \ \Omega$$

$$\Delta V_{\max} = R_{eq} I_{\max} = \boxed{75.0 \text{ V}}$$

(b) $\mathcal{P} = I\Delta V = (0.500 \text{ A})(75.0 \text{ V}) = \boxed{37.5 \text{ W}}$ total power

$$\mathcal{P}_1 = \boxed{25.0 \text{ W}}$$

$$\mathcal{P}_2 = \mathcal{P}_3 = RI^2 = (100 \ \Omega)(0.250 \text{ A})^2 = \boxed{6.25 \text{ W}}$$

P21.31 $$R_p = \left(\frac{1}{3.00} + \frac{1}{1.00}\right)^{-1} = 0.750 \ \Omega$$

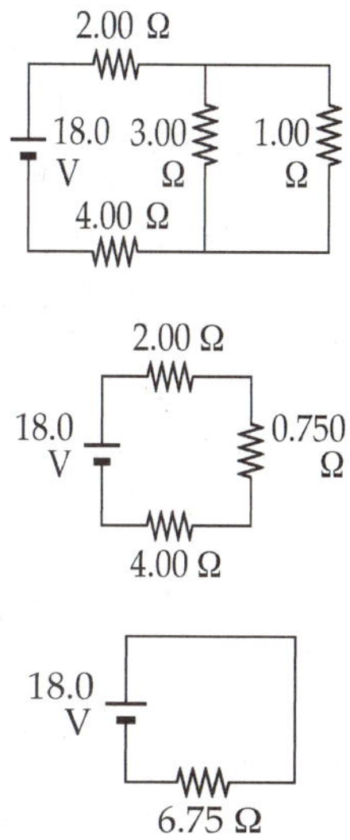

FIG. P21.31

$$R_s = (2.00 + 0.750 + 4.00) \ \Omega = 6.75 \ \Omega$$

$$I_{\text{battery}} = \frac{\Delta V}{R_s} = \frac{18.0 \text{ V}}{6.75 \ \Omega} = 2.67 \text{ A}$$

$\mathcal{P} = I^2R$: $$\mathcal{P}_2 = (2.67 \text{ A})^2(2.00 \ \Omega)$$

$$\mathcal{P}_2 = \boxed{14.2 \text{ W}} \text{ in } 2.00 \ \Omega$$

$$\mathcal{P}_4 = (2.67 \text{ A})^2(4.00 \text{ A}) = \boxed{28.4 \text{ W}} \text{ in } 4.00 \ \Omega$$

$$\Delta V_2 = (2.67 \text{ A})(2.00 \ \Omega) = 5.33 \text{ V},$$

$$\Delta V_4 = (2.67 \text{ A})(4.00 \ \Omega) = 10.67 \text{ V}$$

$$\Delta V_p = 18.0 \text{ V} - \Delta V_2 - \Delta V_4 = 2.00 \text{ V} (= \Delta V_3 = \Delta V_1)$$

$$\mathcal{P}_3 = \frac{(\Delta V_3)^2}{R_3} = \frac{(2.00 \text{ V})^2}{3.00 \ \Omega} = \boxed{1.33 \text{ W}} \text{ in } 3.00 \ \Omega$$

$$\mathcal{P}_1 = \frac{(\Delta V_1)}{R_1} = \frac{(2.00 \text{ V})^2}{1.00 \ \Omega} = \boxed{4.00 \text{ W}} \text{ in } 1.00 \ \Omega$$

***P21.32** (a) The resistors 2, 3, and 4 can be combined to a single $2R$ resistor. This is in series with resistor 1, with resistance R, so the equivalent resistance of the whole circuit is $3R$. In series, potential difference is shared in proportion to the resistance, so resistor 1 gets $\frac{1}{3}$ of the battery voltage and the 2-3-4 parallel combination get $\frac{2}{3}$ of the battery voltage. This is the potential difference across resistor 4, but resistors 2 and 3 must share this voltage. $\frac{1}{3}$ goes to 2 and $\frac{2}{3}$ to 3. The ranking by potential difference is $\boxed{\Delta V_4 > \Delta V_3 > \Delta V_1 > \Delta V_2}$.

continued on next page

(b) Based on the reasoning above the potential differences are

$$\boxed{\Delta V_1 = \frac{\mathcal{E}}{3},\ \Delta V_2 = \frac{2\mathcal{E}}{9},\ \Delta V_3 = \frac{4\mathcal{E}}{9},\ \Delta V_4 = \frac{2\mathcal{E}}{3}}.$$

(c) All the current goes through resistor 1, so it gets the most. The current then splits at the parallel combination. Resistor 4 gets more than half, because the resistance in that branch is less than in the other branch. Resistors 2 and 3 have equal currents because they are in series. The ranking by current is $\boxed{I_1 > I_4 > I_2 = I_3}$.

(d) Resistor 1 has a current of I. Because the resistance of 2 and 3 in series is twice that of resistor 4, twice as much current goes through 4 as through 2 and 3. The current through the resistors are $\boxed{I_1 = I,\ I_2 = I_3 = \frac{I}{3},\ I_4 = \frac{2I}{3}}$.

(e) Increasing resistor 3 increases the equivalent resistance of the entire circuit. The current in the circuit, which is the current through resistor 1, decreases. This decreases the potential difference across resistor 1, increasing the potential difference across the parallel combination. With a larger potential difference the current through resistor 4 is increased. With more current through 4, and less in the circuit to start with, the current through resistors 2 and 3 must decrease. To summarize, $\boxed{I_4 \text{ increases and } I_1,\ I_2, \text{ and } I_3 \text{ decrease}}$.

(f) If resistor 3 has an infinite resistance it blocks any current from passing through that branch, and the circuit effectively is just resistor 1 and resistor 4 in series with the battery. The circuit now has an equivalent resistance of $4R$. The current in the circuit drops to $\frac{3}{4}$ of the original current because the resistance has increased by $\frac{4}{3}$. All this current passes through resistors 1 and 4, and none passes through 2 or 3. Therefore $\boxed{I_1 = \frac{3I}{4},\ I_2 = I_3 = 0,\ I_4 = \frac{3I}{4}}$.

***P21.33** (a) We find the resistance intrinsic to the vacuum cleaner:

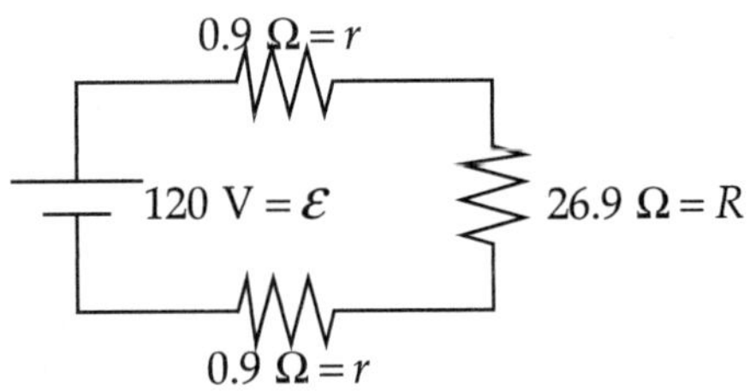

FIG. P21.33

$$\mathcal{P} = I\Delta V = \frac{(\Delta V)^2}{R}$$

$$R = \frac{(\Delta V)^2}{\mathcal{P}} = \frac{(120\ \text{V})^2}{535\ \text{W}} = 26.9\ \Omega$$

with the inexpensive cord, the equivalent resistance is $0.9\ \Omega + 26.9\ \Omega + 0.9\ \Omega = 28.7\ \Omega$ so the current throughout the circuit is

$$I = \frac{\mathcal{E}}{R_{\text{Tot}}} = \frac{120\ \text{V}}{28.7\ \Omega} = 4.18\ \text{A}$$

continued on next page

and the cleaner power is

$$\mathcal{P}_{\text{cleaner}} = I(\Delta V)_{\text{cleaner}} = I^2 R = (4.18 \text{ A})^2 (26.9 \ \Omega) = \boxed{470 \text{ W}}.$$

In symbols, $R_{\text{Tot}} = R + 2r$, $I = \dfrac{\mathcal{E}}{R+2r}$ and $\mathcal{P}_{\text{cleaner}} = I^2 R = \dfrac{\mathcal{E}^2 R}{(R+2r)^2}$.

(b) $$R + 2r = \left(\frac{\mathcal{E}^2 R}{\mathcal{P}_{\text{cleaner}}}\right)^{1/2} = 120 \text{ V}\left(\frac{26.9 \ \Omega}{525 \text{ W}}\right)^{1/2} = 27.2 \ \Omega$$

$$r = \frac{27.2 \ \Omega - 26.9 \ \Omega}{2} = 0.128 \ \Omega = \frac{\rho \ell}{A} = \frac{\rho \ell 4}{\pi d^2}$$

$$d = \left(\frac{4\rho\ell}{\pi r}\right)^{1/2} = \left(\frac{4(1.7 \times 10^{-8} \ \Omega \cdot \text{m})(15 \text{ m})}{\pi(0.128 \ \Omega)}\right)^{1/2} = \boxed{1.60 \text{ mm or more}}$$

(c) Unless the extension cord is a superconductor, it is impossible to attain cleaner power 535 W. To move from 525 W to 532 W will require a lot more copper, as we show here:

$$r = \frac{\mathcal{E}}{2}\left(\frac{R}{\mathcal{P}_{\text{cleaner}}}\right)^{1/2} - \frac{R}{2} = \frac{120 \text{ V}}{2}\left(\frac{26.9 \ \Omega}{532 \text{ W}}\right)^{1/2} - \frac{26.9 \ \Omega}{2} = 0.037\,9 \ \Omega$$

$$d = \left(\frac{4\rho\ell}{\pi r}\right)^{1/2} = \left(\frac{4(1.7 \times 10^{-8} \ \Omega \cdot \text{m})(15 \text{ m})}{\pi(0.037\,9 \ \Omega)}\right)^{1/2} = \boxed{2.93 \text{ mm or more}}$$

Section 21.8 Kirchhoff's Rules

P21.34 $+15.0 - (7.00)I_1 - (2.00)(5.00) = 0$

$5.00 = 7.00 I_1$ so $\boxed{I_1 = 0.714 \text{ A}}$

$I_1 + I_2 - 2.00 \text{ A} = 0$

$0.714 + I_2 = 2.00$ so $\boxed{I_2 = 1.29 \text{ A}}$

$+\mathcal{E} - 2.00(1.29) - 5.00(2.00) = 0$ $\boxed{\mathcal{E} = 12.6 \text{ V}}$

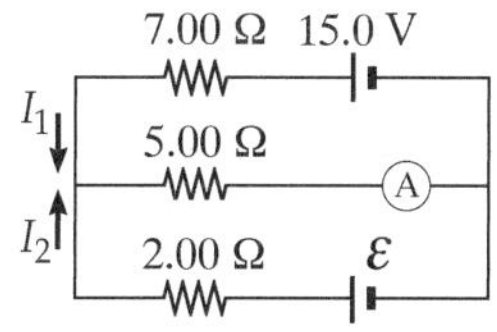

FIG. P21.34

P21.35 We name currents I_1, I_2, and I_3 as shown.

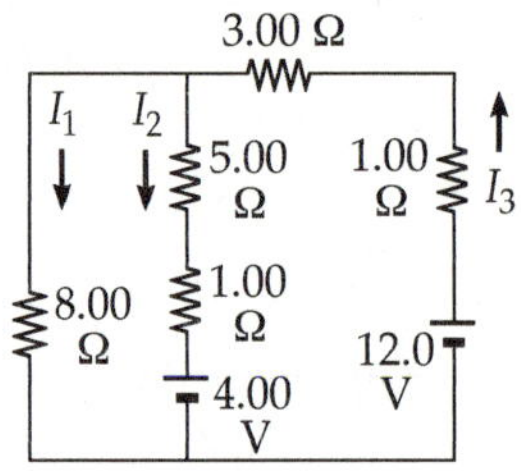

FIG. P21.35

From Kirchhoff's current rule, $I_3 - I_1 - I_2 = 0$.

Applying Kirchhoff's voltage rule to the loop containing I_2 and I_3,

$$12.0\text{ V} - (4.00)I_3 - (6.00)I_2 - 4.00\text{ V} = 0$$
$$8.00 = (4.00)I_3 + (6.00)I_2$$

Applying Kirchhoff's voltage rule to the loop containing I_1 and I_2,

$$-(6.00)I_2 - 4.00\text{ V} + (8.00)I_1 = 0 \qquad (8.00)I_1 = 4.00 + (6.00)I_2.$$

Solving the above linear system, we proceed to the pair of simultaneous equations:

$$\begin{cases} 8 = 4I_1 + 4I_2 + 6I_2 \\ 8I_1 = 4 + 6I_2 \end{cases} \quad \text{or} \quad \begin{cases} 8 = 4I_1 + 10I_2 \\ I_2 = 1.33I_1 - 0.667 \end{cases}$$

and to the single equation $8 = 4I_1 + 13.3I_1 - 6.67$

$$I_1 = \frac{14.7\text{ V}}{17.3\ \Omega} = 0.846\text{ A}. \qquad \text{Then} \qquad I_2 = 1.33(0.846\text{ A}) - 0.667$$

and $I_3 = I_1 + I_2$ give $\boxed{I_1 = 846\text{ mA},\ I_2 = 462\text{ mA},\ I_3 = 1.31\text{ A}}$.

All currents are in the directions indicated by the arrows in the circuit diagram.

P21.36 See Figure P21.36 for solution to this problem.

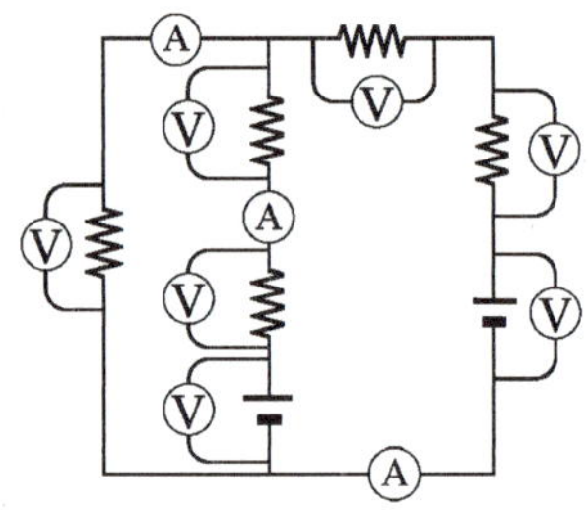

FIG. P21.36

P21.37 We use the results of Problem 21.35.

(a) By the 4.00-V battery: $\Delta U = (\Delta V)I\Delta t = (4.00\text{ V})(-0.462\text{ A})120\text{ s} = \boxed{-222\text{ J}}$.

By the 12.0-V battery: $(12.0\text{ V})(1.31\text{ A})120\text{ s} = \boxed{1.88\text{ kJ}}$.

(b) By the 8.00-Ω resistor: $I^2R\Delta t = (0.846\text{ A})^2(8.00\ \Omega)120\text{ s} = \boxed{687\text{ J}}$.

By the 5.00-Ω resistor: $(0.462\text{ A})^2(5.00\ \Omega)120\text{ s} = \boxed{128\text{ J}}$.

By the 1.00-Ω resistor: $(0.462\text{ A})^2(1.00\ \Omega)120\text{ s} = \boxed{25.6\text{ J}}$.

By the 3.00-Ω resistor: $(1.31\text{ A})^2(3.00\ \Omega)120\text{ s} = \boxed{616\text{ J}}$.

By the 1.00-Ω resistor: $(1.31\text{ A})^2(1.00\ \Omega)120\text{ s} = \boxed{205\text{ J}}$.

continued on next page

(c) We can count the energy twice, like a child counting his lunch money at ten o'clock and again at eleven.

$-222\text{ J}+1.88\text{ kJ}=\boxed{1.66\text{ kJ}}$ from chemical to electrically transmitted.

$687\text{ J}+128\text{ J}+25.6\text{ J}+616\text{ J}+205\text{ J}=1.66\text{ kJ}$ from electrically transmitted to internal.

***P21.38** (a) The first equation represents Kirchhoff's loop theorem. We choose to think of it as describing a clockwise trip around the left-hand loop in a circuit, see Figure (a). For the right-hand loop see Figure (b). The junctions must be between the 5.8 V and the 370 Ω and between the 370 Ω and the 150 Ω. Then we have Figure (c). This is consistent with the third equation,

$$I_1+I_3-I_2=0$$
$$I_2=I_1+I_3$$

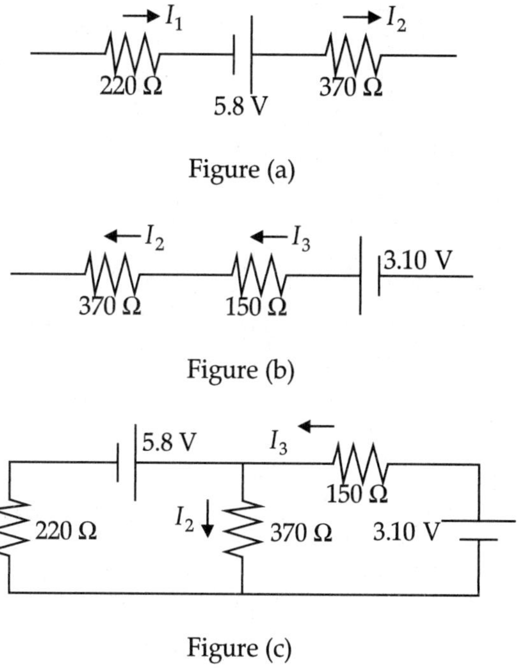

FIG. P21.38

(b) We substitute:

$$-220I_1+5.8-370I_1-370I_3=0$$
$$+370I_1+370I_3+150I_3-3.1=0$$

Next

$$I_3=\frac{5.8-590I_1}{370}$$
$$370I_1+\frac{520}{370}(5.8-590I_1)-3.1=0$$
$$370I_1+8.15-829I_1-3.1=0$$
$$I_1=\frac{5.05\text{ V}}{459\ \Omega}=\boxed{\text{11.0 mA in the 220-}\Omega\text{ resistor and out of the positive pole of the 5.8-V battery}}$$
$$I_3=\frac{5.8-590(0.011\,0)}{370}=-1.87\text{ mA}$$

$\boxed{\text{The current is 1.87 mA in the 150-}\Omega\text{ resistor and out of the negative pole of the 3.1-V battery}}$

$$I_2=11.0-1.87=\boxed{9.13\text{ mA in the 370-}\Omega\text{ resistor}}$$

P21.39 Label the currents in the branches as shown in the first figure. Reduce the circuit by combining the two parallel resistors as shown in the second figure.

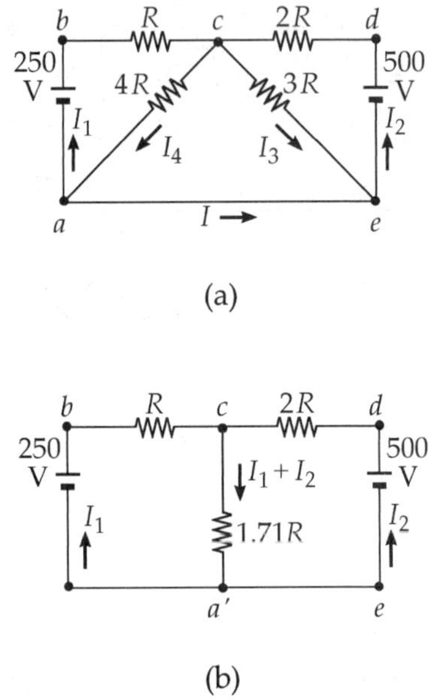

FIG. P21.39

Apply Kirchhoff's loop rule to both loops in Figure (b) to obtain:

$$(2.71R)I_1 + (1.71R)I_2 = 250$$

and $$(1.71R)I_1 + (3.71R)I_2 = 500.$$

With $R = 1\,000\ \Omega$, simultaneous solution of these equations yields:

$$I_1 = 10.0\text{ mA}$$

and $$I_2 = 130.0\text{ mA}.$$

From Figure (b), $$V_c - V_a = (I_1 + I_2)(1.71R) = 240\text{ V}.$$

Thus, from Figure (a), $$I_4 = \frac{V_c - V_a}{4R} = \frac{240\text{ V}}{4\,000\ \Omega} = 60.0\text{ mA}.$$

Finally, applying Kirchhoff's point rule at point a in Figure (a) gives:

$$I_4 - I_1 - I = 0$$
$$I = I_4 - I_1 = 60.0\text{ mA} - 10.0\text{ mA} = +50.0\text{ mA}$$

or $$I = \boxed{50.0\text{ mA from point } a \text{ to point } e}.$$

P21.40 Using Kirchhoff's rules,

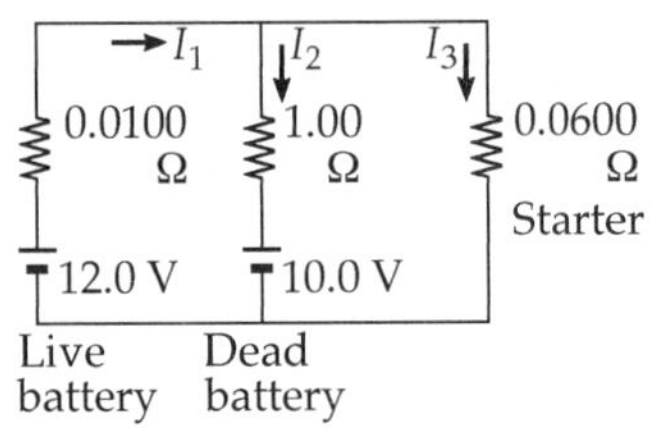

FIG. P21.40

$$12.0 - (0.010\,0)I_1 - (0.060\,0)I_3 = 0$$
$$10.0 + (1.00)I_2 - (0.060\,0)I_3 = 0$$

and $I_1 - I_2 - I_3 = 0$ or $I_1 = I_2 + I_3$

then $12.0 - (0.010\,0)I_2 - (0.070\,0)I_3 = 0$

and $I_2 = 0.06I_3 - 10$

Solving simultaneously, $12 - (0.01)(0.06I_3 - 10) - 0.07I_3 = 0$

$$I_2 = \boxed{0.283\text{ A downward}} \text{ in the dead battery}$$

and $$I_3 = \boxed{171\text{ A downward}} \text{ in the starter.}$$

The currents are forward in the live battery and in the starter, relative to normal starting operation. The current is backward in the dead battery, tending to charge it up.

Section 21.9 *RC* Circuits

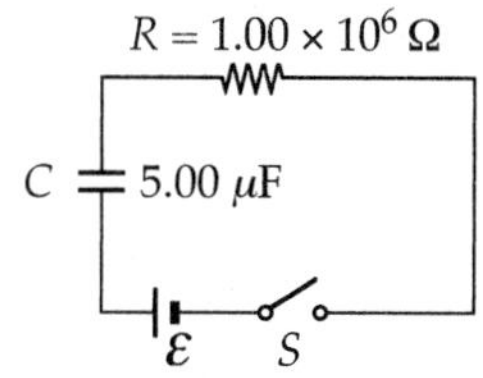

FIG. P21.41

P21.41 (a) $RC = (1.00\times10^6\ \Omega)(5.00\times10^{-6}\ \text{F}) = \boxed{5.00\ \text{s}}$

(b) $Q = C\mathcal{E} = (5.00\times10^{-6}\ \text{C})(30.0\ \text{V}) = \boxed{150\ \mu\text{C}}$

(c) $I(t) = \dfrac{\mathcal{E}}{R}e^{-t/RC} = \left(\dfrac{30.0}{1.00\times10^6}\right)\exp\left[\dfrac{-10.0}{(1.00\times10^6)(5.00\times10^{-6})}\right] = \boxed{4.06\ \mu\text{A}}$

P21.42 (a) $I(t) = -I_0e^{-t/RC}$

$$I_0 = \frac{Q}{RC} = \frac{5.10\times10^{-6}\ \text{C}}{(1\,300\ \Omega)(2.00\times10^{-9}\ \text{F})} = 1.96\ \text{A}$$

$$I(t) = -(1.96\ \text{A})\exp\left[\frac{-9.00\times10^{-6}\ \text{s}}{(1\,300\ \Omega)(2.00\times10^{-9}\ \text{F})}\right] = \boxed{-61.6\ \text{mA}}$$

(b) $q(t) = Qe^{-t/RC} = (5.10\ \mu\text{C})\exp\left[\dfrac{-8.00\times10^{-6}\ \text{s}}{(1\,300\ \Omega)(2.00\times10^{-9}\ \text{F})}\right] = \boxed{0.235\ \mu\text{C}}$

(c) The magnitude of the maximum current is $I_0 = \boxed{1.96\ \text{A}}$.

P21.43 (a) $\tau = RC = (1.50\times10^5\ \Omega)(10.0\times10^{-6}\ \text{F}) = \boxed{1.50\ \text{s}}$

(b) $\tau = (1.00\times10^5\ \Omega)(10.0\times10^{-6}\ \text{F}) = \boxed{1.00\ \text{s}}$

(c) The battery carries current $\dfrac{10.0\ \text{V}}{50.0\times10^3\ \Omega} = 200\ \mu\text{A}$.

The 100 kΩ carries current of magnitude $I = I_0e^{-t/RC} = \left(\dfrac{10.0\ \text{V}}{100\times10^3\ \Omega}\right)e^{-t/1.00\ \text{s}}$.

So the switch carries downward current $\boxed{200\ \mu\text{A} + (100\ \mu\text{A})e^{-t/1.00\ \text{s}}}$.

***P21.44** (a) We model the person's body and street shoes as shown. For the discharge to reach 100 V,

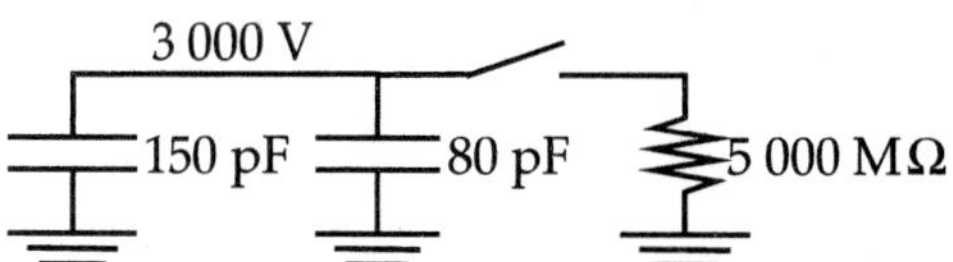

FIG. P21.44(a)

$$q(t) = Qe^{-t/RC} = C\Delta V(t) = C\Delta V_0e^{-t/RC}$$

$$\frac{\Delta V}{\Delta V_0} = e^{-t/RC} \qquad \frac{\Delta V_0}{\Delta V} = e^{+t/RC} \qquad \frac{t}{RC} = \ln\left(\frac{\Delta V_0}{\Delta V}\right)$$

$$t = RC\ln\left(\frac{\Delta V_0}{\Delta V}\right) = 5\,000\times10^6\ \Omega(230\times10^{-12}\ \text{F})\ln\left(\frac{3\,000}{100}\right) = \boxed{3.91\ \text{s}}$$

(b) $t = 1\times10^6\ \text{V/A}(230\times10^{-12}\ \text{C/V})\ln 30 = \boxed{782\ \mu\text{s}}$

P21.45 (a) Call the potential at the left junction V_L and at the right V_R. After a "long" time, the capacitor is fully charged.

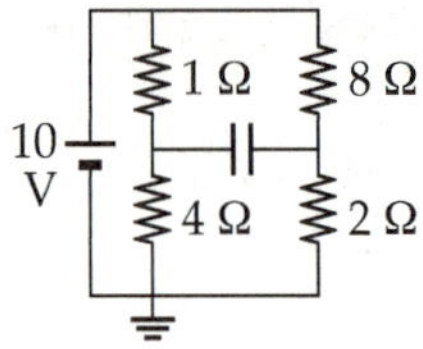

FIG. P21.45(a)

$V_L = 8.00$ V because of voltage divider:

$$I_L = \frac{10.0 \text{ V}}{5.00 \ \Omega} = 2.00 \text{ A}$$

$$V_L = 10.0 \text{ V} - (2.00 \text{ A})(1.00 \ \Omega) = 8.00 \text{ V}$$

Likewise, $$V_R = \left(\frac{2.00 \ \Omega}{2.00 \ \Omega + 8.00 \ \Omega}\right)(10.0 \text{ V}) = 2.00 \text{ V}$$

or $$I_R = \frac{10.0 \text{ V}}{10.0 \ \Omega} = 1.00 \text{ A}$$

$$V_R = (10.0 \text{ V}) - (8.00 \ \Omega)(1.00 \text{ A}) = 2.00 \text{ V}.$$

Therefore, $$\Delta V = V_L - V_R = 8.00 - 2.00 = \boxed{6.00 \text{ V}}.$$

(b) Redraw the circuit $$R = \frac{1}{(1/9.00 \ \Omega) + (1/6.00 \ \Omega)} = 3.60 \ \Omega$$

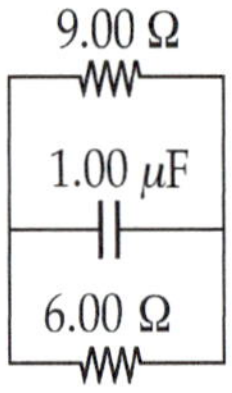

FIG. P21.45(b)

$$RC = 3.60 \times 10^{-6} \text{ s}$$

and $$e^{-t/RC} = \frac{1}{10}$$

so $$t = RC \ln 10 = \boxed{8.29 \ \mu\text{s}}.$$

P21.46 The potential difference across the capacitor $$\Delta V(t) = \Delta V_{\max}\left(1 - e^{-t/RC}\right).$$

Using 1 Farad $= 1$ s/Ω, $$4.00 \text{ V} = (10.0 \text{ V})\left[1 - e^{-(3.00 \text{ s})/\left[R\left(10.0\times10^{-6} \text{ s}/\Omega\right)\right]}\right].$$

Therefore, $$0.400 = 1.00 - e^{-\left(3.00\times10^{5} \ \Omega\right)/R}$$

or $$e^{-\left(3.00\times10^{5} \ \Omega\right)/R} = 0.600.$$

Taking the natural logarithm of both sides, $$-\frac{3.00 \times 10^5 \ \Omega}{R} = \ln(0.600)$$

and $$R = -\frac{3.00 \times 10^5 \ \Omega}{\ln(0.600)} = +5.87 \times 10^5 \ \Omega = \boxed{587 \text{ k}\Omega}.$$

Section 21.10 Context Connection—The Atmosphere as a Conductor

P21.47 $J = \sigma E$ so $$\sigma = \frac{J}{E} = \frac{6.00 \times 10^{-13} \text{ A/m}^2}{100 \text{ V/m}} = \boxed{6.00 \times 10^{-15} \ (\Omega \cdot \text{m})^{-1}}$$

P21.48 (a) $\mathcal{P} = \Delta V I = \left(300 \times 10^3 \text{ J/C}\right)\left(1.00 \times 10^3 \text{ C/s}\right) = \boxed{3.00 \times 10^8 \text{ W}}$

A large electric generating station, fed by a trainload of coal each day, converts energy faster.

(b) $I = \frac{\mathcal{P}}{A} = \frac{\mathcal{P}}{\pi r^2}$ $\qquad \mathcal{P} = I\left(\pi r^2\right) = \left(1\,370 \text{ W/m}^2\right)\left[\pi(6.37 \times 10^6 \text{ m})^2\right] = \boxed{1.75 \times 10^{17} \text{ W}}$

Terrestrial solar power is immense compared to lightning and compared to all human energy conversions.

Additional Problems

P21.49 (a) $I = \frac{\Delta V}{R}$ so $\mathcal{P} = I\Delta V = \frac{(\Delta V)^2}{R}$

$R = \frac{(\Delta V)^2}{\mathcal{P}} = \frac{(120 \text{ V})^2}{25.0 \text{ W}} = \boxed{576\ \Omega}$ and $R = \frac{(\Delta V)^2}{\mathcal{P}} = \frac{(120 \text{ V})^2}{100 \text{ W}} = \boxed{144\ \Omega}$

(b) $I = \frac{\mathcal{P}}{\Delta V} = \frac{25.0 \text{ W}}{120 \text{ V}} = 0.208 \text{ A} = \frac{Q}{\Delta t} = \frac{1.00 \text{ C}}{\Delta t}$

$\Delta t = \frac{1.00 \text{ C}}{0.208 \text{ A}} = \boxed{4.80 \text{ s}}$

The bulb takes in charge at high potential and puts out the same amount of charge at low potential. The charge in itself is identical.

(c) $\mathcal{P} = 25.0 \text{ W} = \frac{\Delta U}{\Delta t} = \frac{1.00 \text{ J}}{\Delta t}$ $\qquad \Delta t = \frac{1.00 \text{ J}}{25.0 \text{ W}} = \boxed{0.040\,0 \text{ s}}$

The bulb takes in energy by electrical transmission and puts out the same amount of energy by heat and light.

(d) $\Delta U = \mathcal{P}\Delta t = (25.0 \text{ J/s})(86\,400 \text{ s/d})(30.0 \text{ d}) = 64.8 \times 10^8 \text{ J}$

The electric company sells $\boxed{\text{energy}}$.

$$\text{Cost} = 64.8 \times 10^6 \text{ J}\left(\frac{\$0.070\,0}{\text{kWh}}\right)\left(\frac{\text{k}}{1\,000}\right)\left(\frac{\text{W} \cdot \text{s}}{\text{J}}\right)\left(\frac{\text{h}}{3\,600 \text{ s}}\right) = \boxed{\$1.26}$$

$$\text{Cost per joule} = \frac{\$0.070\,0}{\text{kWh}}\left(\frac{\text{kWh}}{3.60 \times 10^6 \text{ J}}\right) = \boxed{\$1.94 \times 10^{-8}/\text{J}}$$

P21.50 $\rho = \frac{RA}{\ell} = \frac{(\Delta V)}{I}\frac{A}{\ell}$

ℓ (m)	R (Ω)	ρ ($\Omega \cdot$m)
0.540	10.4	1.41×10^{-6}
1.028	21.1	1.50×10^{-6}
1.543	31.8	1.50×10^{-6}

$\bar{\rho} = \boxed{1.47 \times 10^{-6}\ \Omega \cdot \text{m}}$ (in agreement with tabulated value of $1.50 \times 10^{-6}\ \Omega \cdot \text{m}$ in Table 21.1)

P21.51 (a) $\vec{\mathbf{E}} = -\frac{dV}{dx}\hat{\mathbf{i}} = -\frac{(0-4.00)\text{ V}}{(0.500-0)\text{ m}} = \boxed{8.00\hat{\mathbf{i}}\text{ V/m}}$

(b) $R = \frac{\rho \ell}{A} = \frac{(4.00\times 10^{-8}\ \Omega\cdot\text{m})(0.500\text{ m})}{\pi(1.00\times 10^{-4}\text{ m})^2} = \boxed{0.637\ \Omega}$

(c) $I = \frac{\Delta V}{R} = \frac{4.00\text{ V}}{0.637\ \Omega} = \boxed{6.28\text{ A}}$

(d) $\vec{\mathbf{J}} = \frac{I}{A}\hat{\mathbf{i}} = \frac{6.28\text{ A}}{\pi(1.00\times 10^{-4}\text{ m})^2} = 2.00\times 10^{8}\hat{\mathbf{i}}\text{ A/m}^2 = \boxed{200\hat{\mathbf{i}}\text{ MA/m}^2}$

(e) $\rho\vec{\mathbf{J}} = (4.00\times 10^{-8}\ \Omega\cdot\text{m})(2.00\times 10^{8}\hat{\mathbf{i}}\text{ A/m}^2) = 8.00\hat{\mathbf{i}}\text{ V/m} = \vec{\mathbf{E}}$

P21.52 (a) $\vec{\mathbf{E}} = -\frac{dV(x)}{dx}\hat{\mathbf{i}} = \boxed{\frac{V}{L}\hat{\mathbf{i}}}$ Note that the field is independent of the material and the area.

(b) $R = \frac{\rho \ell}{A} = \boxed{\frac{4\rho L}{\pi d^2}}$ Independent of voltage, field, and current.

(c) $I = \frac{\Delta V}{R} = \boxed{\frac{V\pi d^2}{4\rho L}}$ This is how Georg Simon Ohm thought of his experimental results on all the things current depends on.

(d) $\vec{\mathbf{J}} = \frac{I}{A}\hat{\mathbf{i}} = \boxed{\frac{V}{\rho L}\hat{\mathbf{i}}}$ Independent of area.

(e) $\rho\vec{\mathbf{J}} = \frac{V}{L}\hat{\mathbf{i}} = \boxed{\vec{\mathbf{E}}}$ Independent of length and area. At each point within the material, vector field causes vector current density according to this relation. This is how an atom thinks of Ohm's law.

P21.53 (a) $\mathcal{P} = I\Delta V$: So for the Heater, $I = \frac{\mathcal{P}}{\Delta V} = \frac{1\,500\text{ W}}{120\text{ V}} = \boxed{12.5\text{ A}}$.

For the Toaster, $I = \frac{750\text{ W}}{120\text{ V}} = \boxed{6.25\text{ A}}$.

And for the Grill, $I = \frac{1\,000\text{ W}}{120\text{ V}} = \boxed{8.33\text{ A}}$.

(b) $12.5 + 6.25 + 8.33 = \boxed{27.1\text{ A}}$

The current draw is greater than 25.0 amps, so this circuit would not be sufficient.

P21.54 (a) A thin cylindrical shell of radius r, thickness dr, and length L contributes resistance

$$dR = \frac{\rho\, d\ell}{A} = \frac{\rho\, dr}{(2\pi r)L} = \left(\frac{\rho}{2\pi L}\right)\frac{dr}{r}.$$

The resistance of the whole annulus is the series summation of the contributions of the thin shells:

$$R = \frac{\rho}{2\pi L}\int_{r_a}^{r_b}\frac{dr}{r} = \boxed{\frac{\rho}{2\pi L}\ln\left(\frac{r_b}{r_a}\right)}.$$

(b) In this equation $\frac{\Delta V}{I} = \frac{\rho}{2\pi L}\ln\left(\frac{r_b}{r_a}\right)$

we solve for $\boxed{\rho = \frac{2\pi L \Delta V}{I\ln(r_b/r_a)}}$.

***P21.55** The set of four batteries boosts the electric potential of each bit of charge that goes through them by $4\times 1.50\text{ V} = 6.00\text{ V}$. The chemical energy they store is

$$\Delta U = q\Delta V = (240\text{ C})(6.00\text{ J/C}) = 1\,440\text{ J}.$$

The radio draws current $I = \frac{\Delta V}{R} = \frac{6.00\text{ V}}{200\ \Omega} = 0.030\,0\text{ A}.$

So, its power is $\mathcal{P} = (\Delta V)I = (6.00\text{ V})(0.030\,0\text{ A}) = 0.180\text{ W} = 0.180\text{ J/s}.$

Then for the time the energy lasts, we have $\mathcal{P} = \frac{E}{\Delta t}$: $\Delta t = \frac{E}{\mathcal{P}} = \frac{1\,440\text{ J}}{0.180\text{ J/s}} = 8.00\times 10^3\text{ s}.$

We could also compute this from $I = \frac{Q}{\Delta t}$: $\Delta t = \frac{Q}{I} = \frac{240\text{ C}}{0.030\,0\text{ A}} = 8.00\times 10^3\text{ s} = \boxed{2.22\text{ h}}.$

P21.56 $I = \frac{\mathcal{E}}{R+r}$, so $\mathcal{P} = I^2R = \frac{\mathcal{E}^2 R}{(R+r)^2}$ or $(R+r)^2 = \left(\frac{\mathcal{E}^2}{\mathcal{P}}\right)R.$

Let $x \equiv \frac{\mathcal{E}^2}{\mathcal{P}}$, then $(R+r)^2 = xR$ or $R^2 + (2r - x)R - r^2 = 0.$

With $r = 1.20\ \Omega$, this becomes $R^2 + (2.40 - x)R - 1.44 = 0,$

which has solutions of $R = \frac{-(2.40-x)\pm\sqrt{(2.40-x)^2 - 5.76}}{2}.$

(a) With $\mathcal{E} = 9.20\text{ V}$ and $\mathcal{P} = 12.8\text{ W}$, $x = 6.61$:

$$R = \frac{+4.21 \pm \sqrt{(4.21)^2 - 5.76}}{2} = \boxed{3.84\ \Omega}\text{ or }\boxed{0.375\ \Omega}.$$

continued on next page

(b) For $\mathcal{E} = 9.20$ V and $\mathcal{P} = 21.2$ W, $x \equiv \dfrac{\mathcal{E}^2}{\mathcal{P}} = 3.99$

$$R = \frac{+1.59 \pm \sqrt{(1.59)^2 - 5.76}}{2} = \frac{1.59 \pm \sqrt{-3.22}}{2}.$$

The equation for the load resistance yields a complex number, so $\boxed{\text{there is no resistance}}$ that will extract 21.2 W from this battery. The maximum power output occurs when $R = r = 1.20\ \Omega$, and that maximum is: $\mathcal{P}_{max} = \dfrac{\mathcal{E}^2}{4r} = 17.6$ W.

P21.57 The current in the simple loop circuit will be $I = \dfrac{\mathcal{E}}{R+r}$.

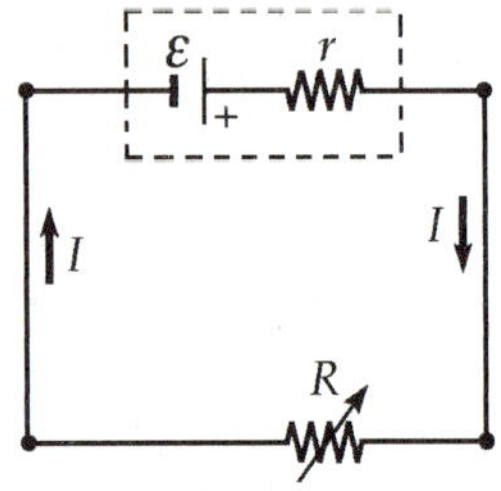

FIG. P21.57

(a) $\Delta V_{ter} = \mathcal{E} - Ir = \dfrac{\mathcal{E}R}{R+r}$ and $\Delta V_{ter} \to \mathcal{E}$ as $\boxed{R \to \infty}$.

(b) $I = \dfrac{\mathcal{E}}{R+r}$ and $I \to \dfrac{\mathcal{E}}{r}$ as $\boxed{R \to 0}$.

(c) $\mathcal{P} = I^2R = \mathcal{E}^2 \dfrac{R}{(R+r)^2}$ $\dfrac{d\mathcal{P}}{dR} = \dfrac{-2\mathcal{E}^2R}{(R+r)^3} + \dfrac{\mathcal{E}^2}{(R+r)^2} = 0$

Then $2R = R + r$ and $\boxed{R = r}$.

P21.58 The battery supplies energy at a changing rate $\dfrac{dE}{dt} = \mathcal{P} = \mathcal{E}I = \mathcal{E}\left(\dfrac{\mathcal{E}}{R}e^{-1/RC}\right)$.

Then the total energy put out by the battery is $\int dE = \int_{t=0}^{\infty} \dfrac{\mathcal{E}^2}{R}\exp\left(-\dfrac{t}{RC}\right)dt$

$$\int dE = \frac{\mathcal{E}^2}{R}(-RC)\int_0^{\infty} \exp\left(-\frac{t}{RC}\right)\left(-\frac{dt}{RC}\right) = -\mathcal{E}^2C\exp\left(-\frac{t}{RC}\right)\Big|_0^{\infty} = -\mathcal{E}^2C[0-1] = \mathcal{E}^2C.$$

The power delivered to the resistor is $\dfrac{dE}{dt} = \mathcal{P} = \Delta V_R I = I^2R = R\dfrac{\mathcal{E}^2}{R^2}\exp\left(-\dfrac{2t}{RC}\right)$.

So the total internal energy appearing in the resistor is $\int dE = \int_0^{\infty} \dfrac{\mathcal{E}^2}{R}\exp\left(-\dfrac{2t}{RC}\right)dt$

$$\int dE = \frac{\mathcal{E}^2}{R}\left(-\frac{RC}{2}\right)\int_0^{\infty}\exp\left(-\frac{2t}{RC}\right)\left(-\frac{2dt}{RC}\right) = -\frac{\mathcal{E}^2C}{2}\exp\left(-\frac{2t}{RC}\right)\Big|_0^{\infty} = -\frac{\mathcal{E}^2C}{2}[0-1] = \frac{\mathcal{E}^2C}{2}.$$

The energy finally stored in the capacitor is $U = \frac{1}{2}C(\Delta V)^2 = \frac{1}{2}C\mathcal{E}^2$. Thus, energy of the circuit is conserved $\mathcal{E}^2C = \frac{1}{2}\mathcal{E}^2C + \frac{1}{2}\mathcal{E}^2C$ and resistor and capacitor share equally in the energy from the battery.

P21.59 (a) $q = C\Delta V\left(1 - e^{-t/RC}\right)$

$$q = \left(1.00\times10^{-6}\text{ F}\right)(10.0\text{ V})\left[1 - e^{-10.0/\left[\left(2.00\times10^{6}\right)\left(1.00\times10^{-6}\right)\right]}\right] = \boxed{9.93\ \mu\text{C}}$$

(b) $I = \dfrac{dq}{dt} = \left(\dfrac{\Delta V}{R}\right)e^{-t/RC}$

$$I = \left(\frac{10.0\text{ V}}{2.00\times10^{6}\ \Omega}\right)e^{-5.00} = 3.37\times10^{-8}\text{ A} = \boxed{33.7\text{ nA}}$$

(c) $\dfrac{dU}{dt} = \dfrac{d}{dt}\left(\dfrac{1}{2}\dfrac{q^2}{C}\right) = \left(\dfrac{q}{C}\right)\dfrac{dq}{dt} = \left(\dfrac{q}{C}\right)I$

$$\frac{dU}{dt} = \left(\frac{9.93\times10^{-6}\text{ C}}{1.00\times10^{-6}\text{ C/V}}\right)\left(3.37\times10^{-8}\text{ A}\right) = 3.34\times10^{-7}\text{ W} = \boxed{334\text{ nW}}$$

(d) $\mathcal{P}_{\text{battery}} = I\mathcal{E} = \left(3.37\times10^{-8}\text{ A}\right)(10.0\text{ V}) = 3.37\times10^{-7}\text{ W} = \boxed{337\text{ nW}}$

P21.60 Start at the point when the voltage has just reached $\frac{2}{3}\Delta V$ and the switch has just closed. The voltage is $\frac{2}{3}\Delta V$ and is decaying towards 0 V with a time constant R_BC

$$\Delta V_C(t) = \left[\frac{2}{3}\Delta V\right]e^{-t/R_BC}.$$

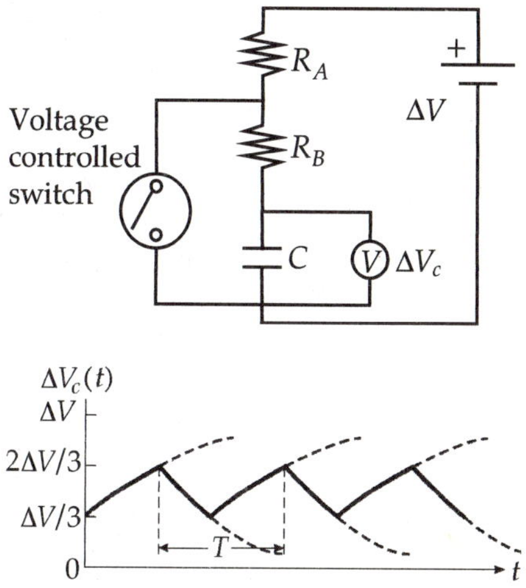

FIG. P21.60

We want to know when $\Delta V_C(t)$ will reach $\frac{1}{3}\Delta V$.

Therefore, $\dfrac{1}{3}\Delta V = \left[\dfrac{2}{3}\Delta V\right]e^{-t/R_BC}$

or $e^{-t/R_BC} = \dfrac{1}{2}$

or $t_1 = R_BC\ln 2$.

After the switch opens, the voltage is $\frac{1}{3}\Delta V$, increasing toward ΔV with time constant $(R_A + R_B)C$:

$$\Delta V_C(t) = \Delta V - \left[\frac{2}{3}\Delta V\right]e^{-t/(R_A+R_B)C}.$$

When $\Delta V_C(t) = \dfrac{2}{3}\Delta V$

$\dfrac{2}{3}\Delta V = \Delta V - \dfrac{2}{3}\Delta V e^{-t/(R_A+R_B)C}$ or $e^{-t/(R_A+R_B)C} = \dfrac{1}{2}$.

So $t_2 = (R_A + R_B)C\ln 2$ and $T = t_1 + t_2 = \boxed{(R_A + 2R_B)C\ln 2}$.

P21.61 (a) With the switch closed, current exists in a simple series circuit as shown. The capacitors carry no current. For R_2 we have

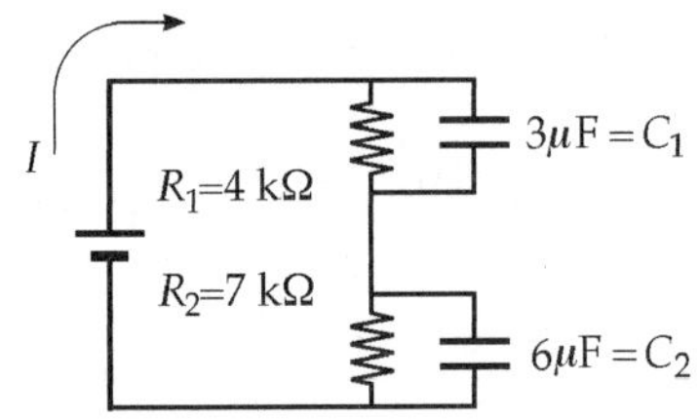

FIG. P21.61(a)

$$\mathcal{P} = I^2 R_2 \qquad I = \sqrt{\frac{\mathcal{P}}{R_2}} = \sqrt{\frac{2.40 \text{ V} \cdot \text{A}}{7\,000 \text{ V/A}}} = 18.5 \text{ mA}.$$

The potential difference across R_1 and C_1 is

$$\Delta V = IR_1 = (1.85 \times 10^{-2} \text{ A})(4\,000 \text{ V/A}) = 74.1 \text{ V}.$$

The charge on C_1 is

$$Q = C_1 \Delta V = (3.00 \times 10^{-6} \text{ C/V})(74.1 \text{ V}) = \boxed{222\ \mu\text{C}}.$$

The potential difference across R_2 and C_2 is

$$\Delta V = IR_2 = (1.85 \times 10^{-2} \text{ A})(7\,000\ \Omega) = 130 \text{ V}.$$

The charge on C_2 is

$$Q = C_2 \Delta V = (6.00 \times 10^{-6} \text{ C/V})(130 \text{ V}) = 778\ \mu\text{C}.$$

The battery emf is

$$IR_{eq} = I(R_1 + R_2) = 1.85 \times 10^{-2} \text{ A}(4\,000 + 7\,000) \text{ V/A} = 204 \text{ V}.$$

(b) In equilibrium after the switch has been opened, no current exists. The potential difference across each resistor is zero. The full 204 V appears across both capacitors. The new charge on C_2 is

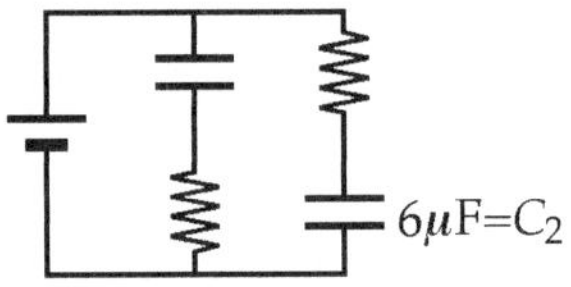

FIG. P21.61(b)

$$Q = C_2 \Delta V = (6.00 \times 10^{-6} \text{ C/V})(204 \text{ V}) = 1\,222\ \mu\text{C}$$

for a change of $1\,222\ \mu\text{C} - 778\ \mu\text{C} = \boxed{444\ \mu\text{C}}$.

P21.62 $\Delta V = \mathcal{E} e^{-t/RC}$

so $\ln\left(\frac{\mathcal{E}}{\Delta V}\right) = \left(\frac{1}{RC}\right)t.$

A plot of $\ln\left(\frac{\mathcal{E}}{\Delta V}\right)$ versus t should be a straight line with slope equal to $\frac{1}{RC}$.

Using the given data values:

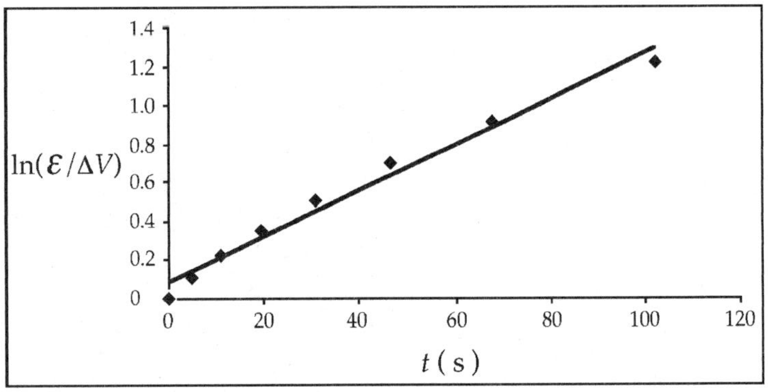

FIG. P21.62

continued on next page

(a) A least-square fit to this data yields the graph above.

$\sum x_i = 282$, $\sum x_i^2 = 1.86 \times 10^4$,

$\sum x_i y_i = 244$, $\sum y_i = 4.03$, $N = 8$

$t(s)$	$\Delta V(V)$	$\ln(\mathcal{E}/\Delta V)$
0	6.19	0
4.87	5.55	0.109
11.1	4.93	0.228
19.4	4.34	0.355
30.8	3.72	0.509
46.6	3.09	0.695
67.3	2.47	0.919
102.2	1.83	1.219

$$\text{Slope} = \frac{N\left(\sum x_i y_i\right) - \left(\sum x_i\right)\left(\sum y_i\right)}{N\left(\sum x_i^2\right) - \left(\sum x_i\right)^2} = 0.011\,8$$

$$\text{Intercept} = \frac{\left(\sum x_i^2\right)\left(\sum y_i\right) - \left(\sum x_i\right)\left(\sum x_i y_i\right)}{N\left(\sum x_i^2\right) - \left(\sum x_i\right)^2} = 0.088\,2$$

The equation of the best fit line is: $\boxed{\ln\left(\frac{\mathcal{E}}{\Delta V}\right) = (0.011\,8)t + 0.088\,2}$.

(b) Thus, the time constant is $\tau = RC = \frac{1}{\text{slope}} = \frac{1}{0.011\,8} = \boxed{84.7\text{ s}}$

and the capacitance is $C = \frac{\tau}{R} = \frac{84.7\text{ s}}{10.0 \times 10^6\ \Omega} = \boxed{8.47\ \mu\text{F}}$.

P21.63 **An editorial in *The Physics Teacher* suggested the idea for this problem.**

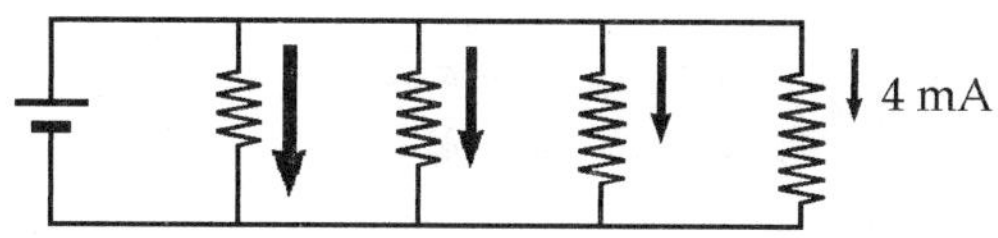

FIG. P21.63

The battery current is

$(150 + 45 + 14 + 4)\text{ mA} = 213\text{ mA}$.

(a) The resistor with highest resistance is that carrying 4 mA. Doubling its resistance will reduce the current it carries to 2 mA. Then the total current is

$(150 + 45 + 14 + 2)\text{ mA} = 211\text{ mA}$, nearly the same as before. The ratio is $\frac{211}{213} = \boxed{0.991}$.

(b) The resistor with least resistance carries 150 mA. Doubling its resistance changes this current to 75 mA and changes the total to

$(75 + 45 + 14 + 4)\text{ mA} = 138\text{ mA}$. The ratio is $\frac{138}{213} = \boxed{0.648}$, representing a much larger reduction (35.2% instead of 0.9%).

(c) This problem is precisely analogous. As a battery maintained a potential difference in parts (a) and (b), a furnace maintains a temperature difference here. Energy flow by heat is analogous to current and takes place through thermal resistances in parallel. Each resistance can have its "*R*-value" increased by adding insulation. Doubling the thermal resistance of the attic door will produce only a negligible (0.9%) saving in fuel. Doubling the thermal resistance of $\boxed{\text{the ceiling}}$ will produce a much larger saving. The ceiling originally has the smallest thermal resistance.

***P21.64** (a) For the first measurement, the equivalent circuit is as shown in Figure 1.

$$R_{ab} = R_1 = R_y + R_y = 2R_y$$

so $$R_y = \frac{1}{2}R_1. \quad (1)$$

For the second measurement, the equivalent circuit is shown in Figure 2.

Thus, $$R_{ac} = R_2 = \frac{1}{2}R_y + R_x. \quad (2)$$

Substitute (1) into (2) to obtain:

$$R_2 = \frac{1}{2}\left(\frac{1}{2}R_1\right) + R_x, \text{ or } \boxed{R_x = R_2 - \frac{1}{4}R_1}.$$

(b) If $R_1 = 13.0\ \Omega$ and $R_2 = 6.00\ \Omega$, then $\boxed{R_x = 2.75\ \Omega}$.

The antenna is inadequately grounded since this exceeds the limit of $2.00\ \Omega$.

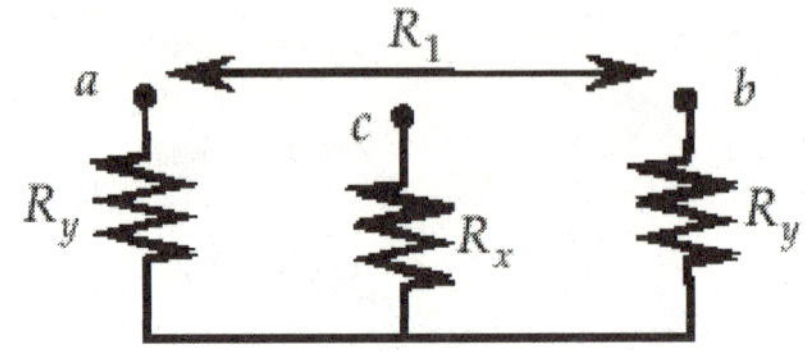

Figure 1

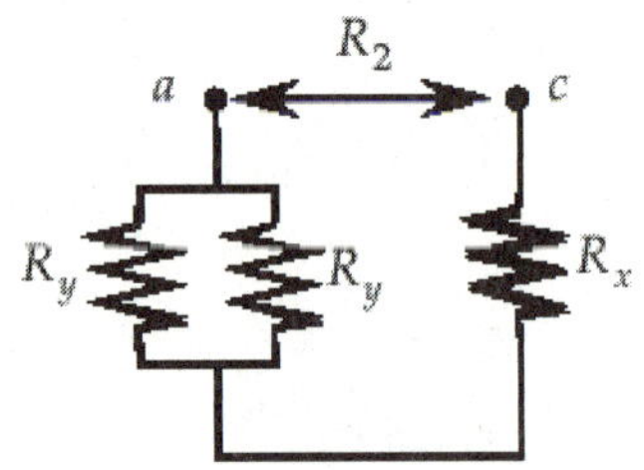

Figure 2

FIG. P21.64

ANSWERS TO EVEN PROBLEMS

P21.2 $\frac{q\omega}{2\pi}$

P21.4 (a) 17.0 A ; (b) 85.0 kA/m^2

P21.6 500 mA

P21.8 (a) 1.82 m; (b) 280 μm

P21.10 1.98 A

P21.12 (a) no change; (b) doubles; (c) doubles; (d) no change

P21.14 5.00 A; 24.0 Ω

P21.16 (a) 0.530; (b) 221 J; (c) 15.1°C

P21.18 \$0.232

P21.20 295 metric ton/h

P21.22 Any diameter d and length ℓ related by $d^2 = (4.77 \times 10^{-8}\ \text{m})\ell$, such as length 0.900 m and diameter 0.207 mm. Yes.

P21.24 ~\$1

P21.26 (a) 4.59 Ω; (b) 8.16%

P21.28 (a) see the solution; (b) no

P21.30 (a) 75.0 V;
(b) 25.0 W, 6.25 W, and 6.25 W; 37.5 W

P21.32 (a) $\Delta V_4 > \Delta V_3 > \Delta V_1 > \Delta V_2$;
(b) $\Delta V_1 = \frac{\mathcal{E}}{3}$, $\Delta V_2 = \frac{2\mathcal{E}}{9}$, $\Delta V_3 = \frac{4\mathcal{E}}{9}$, $\Delta V_4 = \frac{2\mathcal{E}}{3}$;
(c) $I_1 > I_4 > I_2 = I_3$;
(d) $I_1 = I$, $I_2 = I_3 = \frac{I}{3}$, $I_4 = \frac{2I}{3}$;
(e) I_4 increases while I_1, I_2, and I_3 decrease;
(f) $I_1 = \frac{3I}{4}$, $I_2 = I_3 = 0$, $I_4 = \frac{3I}{4}$

P21.34 0.714 A, 1.29 A, 12.6 V

P21.36 see the solution

P21.38 (a) see the solution;
(b) The current in the 220-Ω resistor and the 5.80-V battery is 11.0 mA out of the positive battery pole. The current in the 370-Ω resistor is 9.13 mA. The current in the 150-Ω resistor and the 3.10-V battery is 1.87 mA out of the negative battery pole.

P21.40 starter 171 A; battery 0.283 A

P21.42 (a) –61.6 mA; (b) 0.235 μC; (c) 1.96 A

P21.44 (a) 3.91 s; (b) 0.782 ms

P21.46 587 kΩ

P21.48 (a) 300 MW; (b) 175 PW

P21.50 Experimental resistivity $= 1.47\ \mu\Omega\cdot\text{m} \pm 4\%$, in agreement with 1.50 $\mu\Omega\cdot\text{m}$

P21.52 (a) $\frac{V}{L}\hat{\mathbf{i}}$; (b) $\frac{4\rho L}{\pi d^2}$; (c) $\frac{V\pi d^2}{4\rho L}$; (d) $\frac{V}{\rho L}\hat{\mathbf{i}}$;
(e) see the solution

P21.54 (a) $R = \frac{\rho}{2\pi L}\ln\left(\frac{r_b}{r_a}\right)$; (b) $\rho = \frac{2\pi L\Delta V}{I\ln\left(\frac{r_b}{r_a}\right)}$

P21.56 (a) either 3.84 Ω or 0.375 Ω;
(b) impossible

P21.58 see the solution

P21.60 $(R_A + 2R_B)C\ln 2$

P21.62 (a) $\ln\left(\frac{\mathcal{E}}{\Delta V}\right) = 0.011\,8t + 0.088\,2$;
(b) 84.7 s, 8.47 μF

P21.64 (a) $R_x = R_2 - \frac{R_1}{4}$;
(b) $R_x = 2.75\ \Omega$, the station is inadequately grounded

CONTEXT 6 CONCLUSION SOLUTIONS TO PROBLEMS

CC6.1 The resistance of the spherical shell of air carrying radial electric current is

$$R = \frac{\rho L}{A} = \frac{\left(2\times 10^{13}\ \Omega\cdot\text{m}\right)(4\,000\ \text{m})}{4\pi\left(6.37\times 10^{6}\ \text{m}\right)^2} = 157\ \Omega.$$

The time constant for discharge is $RC = (157\ \text{V/A})(0.8\ \text{C/V}) = 126\ \text{s}$.

(a) $q = Qe^{-t/RC}$ describes the discharge

$$\frac{q}{Q} = e^{-t/RC}$$

$$e^{+t/RC} = \frac{Q}{q}$$

$$\frac{t}{RC} = \ln\left(\frac{Q}{q}\right)$$

$$t = RC\ln\left(\frac{Q}{q}\right)$$

$$t = 126\ \text{s}\left(\ln\left(\frac{4\times 10^4\ \text{C}}{2\times 10^4\ \text{C}}\right)\right) = \boxed{87.0\ \text{s}}$$

continued on next page

(b) $$t = 126\text{ s}\left(\ln\left(\frac{4\times10^4\text{ C}}{5\times10^3\text{ C}}\right)\right) = \boxed{261\text{ s}}$$

(c) $$t = 126\text{ s}\left(\ln\left(\frac{4\times10^4\text{ C}}{0}\right)\right)\ \boxed{\to\infty}$$

The exponential decay goes on forever, as modeled mathematically.

CC6.2 (a) $$I = \frac{Q}{\Delta t}$$

$$\Delta t = \frac{Q}{I} = \frac{25\text{ C}}{2000\text{ C/s}} = 1.25\times10^{-2}\text{ s} \approx \boxed{0.01\text{ s}}$$

(b) $$\text{number} = 1\text{ d}\left(\frac{86\,400\text{ s}}{1\text{ d}}\right)\left(\frac{1}{1.25\times10^{-2}\text{ s}}\right) \approx \boxed{7\times10^6}$$

CC6.3 (a) For radial currents, the two layers of the atmosphere are resistors in series, presenting equivalent resistance

$$\frac{\rho_1 L}{A} + \frac{\rho_2 L}{A} = \frac{2\times10^{13}\ \Omega\cdot\text{m}\left(2.5\times10^3\text{ m}\right)}{4\pi\left(6.37\times10^6\text{ m}\right)^2} + \frac{0.2\times10^{13}\ \Omega\cdot\text{m}\left(2.5\times10^3\text{ m}\right)}{4\pi\left(6.37\times10^6\text{ m}\right)^2} = 108\ \Omega.$$

Then the time constant for discharge is

$$RC = 0.9\text{ F}(108\ \Omega) = 97.1\text{ s}.$$

According to the argument of the context conclusion, we consider six time constants as the time for 2×10^4 lightning strikes. Then the number in a day is

$$\frac{2\times10^4}{6RC} = \frac{2\times10^4}{6(97.1\text{ s})}\left(\frac{86\,400\text{ s}}{1\text{ d}}\right) = \boxed{3\times10^6}.$$

(b) For radial currents through the atmosphere, the air in the two hemispheres presents resistances in parallel, with equivalent resistance

$$R_{eq} = \frac{1}{\left(\frac{\rho_1 L}{A}\right)^{-1} + \left(\frac{\rho_2 L}{A}\right)^{-1}} = \left(\frac{2\pi\left(6.37\times10^6\text{ m}\right)^2}{2\times10^{13}\ \Omega\cdot\text{m}\left(5\times10^3\text{ m}\right)} + \frac{2\pi\left(6.37\times10^6\text{ m}\right)^2}{0.2\times10^{13}\ \Omega\cdot\text{m}\left(5\times10^3\text{ m}\right)}\right)^{-1} = 35.7\ \Omega$$

$$\text{number of daily strokes} = \frac{2\times10^4}{6RC} = \frac{2\times10^4}{6(35.7\ \Omega)(0.9\text{ F})}\left(\frac{86\,400\text{ s}}{1\text{ d}}\right) = \boxed{9\times10^6}$$

ANSWERS TO EVEN CONTEXT 6 CONCLUSION PROBLEMS

CC6.2 (a) 0.01 s; (b) 7×10^6

Chapter 22

Magnetic Forces and Magnetic Fields

CHAPTER OUTLINE

ANSWERS TO QUESTIONS

Q22.1 If they are projected in the same direction into the same magnetic field, the charges are of opposite sign.

Q22.2 Similarities: Both can alter the velocity of a charged particle moving through the field. Both exert forces directly proportional to the charge of the particle feeling the force. Positive and negative charges feel forces in opposite directions. Differences: The direction of the electric force is parallel or antiparallel to the direction of the electric field, but the direction of the magnetic force is perpendicular to the magnetic field and to the velocity of the charged particle. Electric forces can accelerate a charged particle from rest or stop a moving particle, but magnetic forces cannot.

Q22.3 (a) The $q\vec{\mathbf{v}}\times\vec{\mathbf{B}}$ force on each electron is down. Since electrons are negative, $\vec{\mathbf{v}}\times\vec{\mathbf{B}}$ must be up. With $\vec{\mathbf{v}}$ to the right, $\vec{\mathbf{B}}$ must be into the page, away from you.

(b) Reversing the current in the coils would reverse the direction of $\vec{\mathbf{B}}$, making it toward you. Then $\vec{\mathbf{v}}\times\vec{\mathbf{B}}$ is in the direction **right** $\times$ **toward you** = **down**, and $q\vec{\mathbf{v}}\times\vec{\mathbf{B}}$ will make the electron beam curve up.

Q22.4 If the current is in a direction *parallel* or *antiparallel* to the magnetic field, then there is no force.

Q22.5 Yes. If the magnetic field is perpendicular to the plane of the loop, then it exerts no torque on the loop.

Q22.6 If you can hook a spring balance to the particle and measure the force on it in a known electric field, then $q=\frac{F}{E}$ will tell you its charge. You cannot hook a spring balance to an electron. Measuring the acceleration of small particles by observing their deflection in known electric and magnetic fields can tell you the charge-to-mass ratio, but not separately the charge or mass. Both an acceleration produced by an electric field and an acceleration caused by a magnetic field depend on the properties of the particle only by being proportional to the ratio $\frac{q}{m}$.

Q22.7 The Earth's magnetic field exerts force on a charged incoming cosmic ray, tending to make it spiral around a magnetic field line. If the particle energy is low enough, the spiral will be tight enough that the particle will first hit some matter as it follows a field line down into the atmosphere or to the surface at a high geographic latitude.

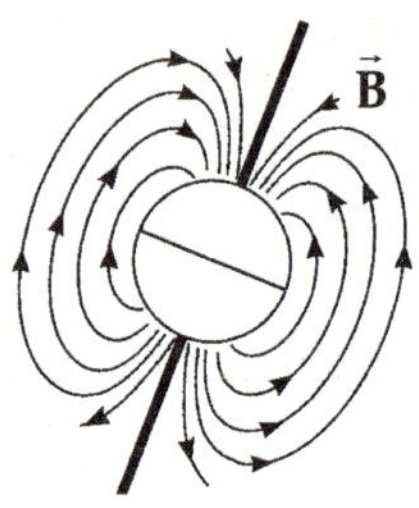

FIG. Q22.7

Q22.8 The spiral tracks are left by charged particles gradually losing kinetic energy. A straight path might be left by an uncharged particle that managed to leave a trail of bubbles, or it might be the imperceptibly curving track of a very fast charged particle, or of a very massive charged particle.

Q22.9 The magnetic field created by wire 1 at the position of wire 2 is into the paper. Hence, the magnetic force on wire 2 is in direction down × into the paper = to the right, away from wire 1. Now wire 2 creates a magnetic field into the page at the location of wire 1, so wire 1 feels force up × into the paper = left, away from wire 2.

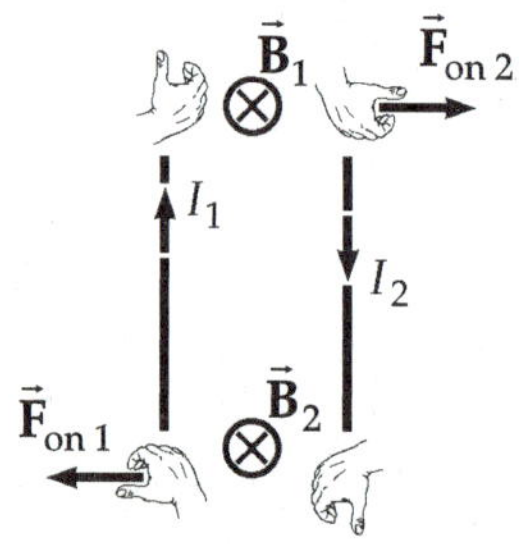

FIG. Q22.9

Q22.10 No total force, but a torque. Let wire 1 carry current in the y direction, toward the top of the page. Let wire 2 be a millimeter above the plane of the paper and carry current to the right, in the x direction. On the left-hand side of wire 1, wire 1 creates magnetic field in the z direction, which exerts force in the $\hat{\mathbf{i}}\times\hat{\mathbf{k}}=-\hat{\mathbf{j}}$ direction on wire 2. On the right-hand side, wire 1 produces magnetic field in the $-\hat{\mathbf{k}}$ direction and makes a $\hat{\mathbf{i}}\times\left(-\hat{\mathbf{k}}\right)=+\hat{\mathbf{j}}$ force of equal magnitude act on wire 2. If wire 2 is free to move, its center section will twist counterclockwise and then be attracted to wire 1.

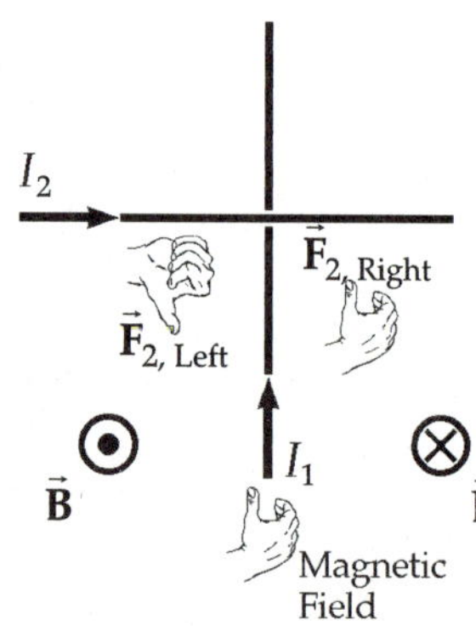

FIG. Q22.10

Q22.11 Apply Ampère's law to the circular path labeled 1 in the picture. Since there is no current inside this path, the magnetic field inside the tube must be zero. On the other hand, the current through path 2 is the current carried by the conductor. Therefore the magnetic field outside the tube is nonzero.

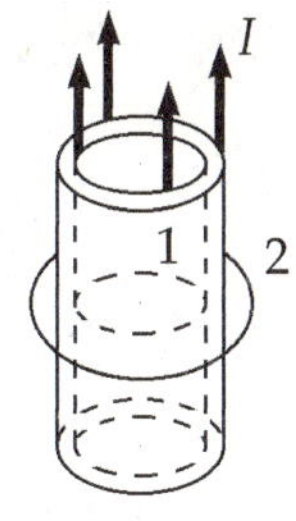

FIG. Q22.11

Q22.12 The magnetic field inside a long solenoid is given by $B = \frac{\mu_0 NI}{\ell}$.

(a) If the length ℓ is doubled, the field is cut in half.

(b) If N is doubled, the magnetic field is doubled.

Q22.13 Magnetic domain alignment creates a stronger external magnetic field. The field of one piece of iron in turn can align domains in another iron sample. A nonuniform magnetic field exerts a net force of attraction on magnetic dipoles aligned with the field.

Q22.14 The geographic North Pole is a magnetic South Pole.

Q22.15 The shock misaligns the domains. Heating will also decrease magnetism.

Q22.16 The medium for any magnetic recording should be a hard ferromagnetic substance, so that thermal vibrations and stray magnetic fields will not rapidly erase the information.

Q22.17 (a) The magnets repel each other with a force equal to the weight of one of them.

(b) The pencil prevents motion to the side and prevents the magnets from rotating under their mutual torques. Its constraint changes unstable equilibrium into stable.

(c) Most likely, the disks are magnetized perpendicular to their flat faces, making one face a north pole and the other a south pole. One disk has its north pole on the top side and the other has its north pole on the bottom side.

(d) Then if either were inverted they would attract each other and stick firmly together.

SOLUTIONS TO PROBLEMS

Section 22.1 Historical Overview

No problems in this section

Section 22.2 The Magnetic Field

P22.1 (a) up

(b) out of the page, since the charge is negative.

(c) no deflection

(d) into the page

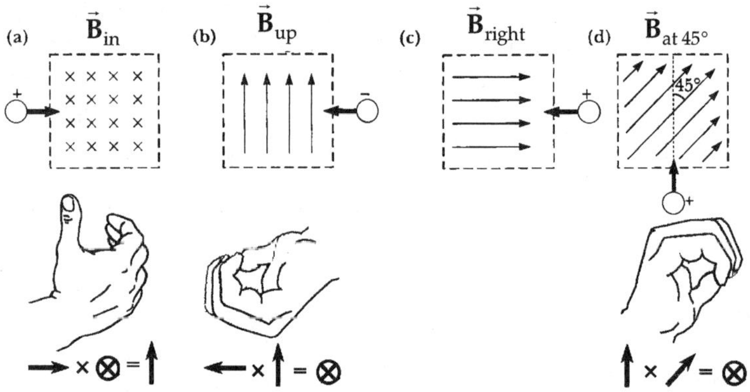

FIG. P22.1

***P22.2** At the equator, the Earth's magnetic field is horizontally north. Because an electron has negative charge, $\vec{F} = q\vec{v} \times \vec{B}$ is opposite in direction to $\vec{v} \times \vec{B}$. Figures are drawn looking down.

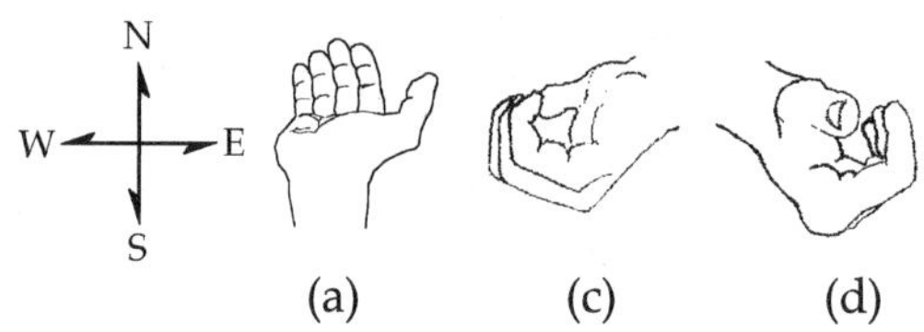

FIG. P22.2

(a) Down × North = East, so the force is directed $\boxed{\text{West}}$.

(b) North × North $= \sin 0° = 0$: $\boxed{\text{Zero deflection}}$.

(c) West × North = Down, so the force is directed $\boxed{\text{Up}}$.

(d) Southeast × North = Up, so the force is $\boxed{\text{Down}}$.

P22.3 (a) $F_B = qvB\sin\theta = \left(1.60\times10^{-19}\ \text{C}\right)\left(3.00\times10^{6}\ \text{m/s}\right)\left(3.00\times10^{-1}\ \text{T}\right)\sin 37.0°$

$$F_B = \boxed{8.67\times10^{-14}\ \text{N}}$$

(b) $a = \dfrac{F}{m} = \dfrac{8.67\times10^{-14}\ \text{N}}{1.67\times10^{-27}\ \text{kg}} = \boxed{5.19\times10^{13}\ \text{m/s}^2}$

P22.4 First find the speed of the electron.

$$\Delta K = \frac{1}{2}mv^2 = e\Delta V = \Delta U: \qquad v = \sqrt{\frac{2e\Delta V}{m}} = \sqrt{\frac{2\left(1.60\times10^{-19}\ \text{C}\right)\left(2\,400\ \text{J/C}\right)}{\left(9.11\times10^{-31}\ \text{kg}\right)}} = 2.90\times10^{7}\ \text{m/s}$$

(a) $F_{B,\,\max} = qvB = \left(1.60\times10^{-19}\ \text{C}\right)\left(2.90\times10^{7}\ \text{m/s}\right)(1.70\ \text{T}) = \boxed{7.90\times10^{-12}\ \text{N}}$

(b) $F_{B,\,\min} = \boxed{0}$ occurs when $\vec{v}$ is either parallel to or anti-parallel to $\vec{B}$.

P22.5 Gravitational force: $F_g = mg = \left(9.11\times10^{-31}\text{ kg}\right)\left(9.80\text{ m/s}^2\right) = \boxed{8.93\times10^{-30}\text{ N down}}$.

Electric force: $F_e = qE = \left(-1.60\times10^{-19}\text{ C}\right)(100\text{ N/C down}) = \boxed{1.60\times10^{-17}\text{ N up}}$.

Magnetic force: $\vec{\mathbf{F}}_B = q\vec{\mathbf{v}}\times\vec{\mathbf{B}} = \left(-1.60\times10^{-19}\text{ C}\right)\left(6.00\times10^{6}\text{ m/s }\hat{\mathbf{E}}\right)\times\left(50.0\times10^{-6}\text{ N}\cdot\text{s/C}\cdot\text{m }\hat{\mathbf{N}}\right)$.

$\vec{\mathbf{F}}_B = -4.80\times10^{-17}\text{ N up} = \boxed{4.80\times10^{-17}\text{ N down}}$.

P22.6 $\vec{\mathbf{F}}_B = q\vec{\mathbf{v}}\times\vec{\mathbf{B}}$

$$\vec{\mathbf{v}}\times\vec{\mathbf{B}} = \begin{vmatrix}\hat{\mathbf{i}} & \hat{\mathbf{j}} & \hat{\mathbf{k}}\\ +2 & -4 & +1\\ +1 & +2 & -3\end{vmatrix} = (12-2)\hat{\mathbf{i}} + (1+6)\hat{\mathbf{j}} + (4+4)\hat{\mathbf{k}} = 10\hat{\mathbf{i}} + 7\hat{\mathbf{j}} + 8\hat{\mathbf{k}}$$

$$\left|\vec{\mathbf{v}}\times\vec{\mathbf{B}}\right| = \sqrt{10^2+7^2+8^2} = 14.6\text{ T}\cdot\text{m/s}$$

$$\left|\vec{\mathbf{F}}_B\right| = q\left|\vec{\mathbf{v}}\times\vec{\mathbf{B}}\right| = \left(1.60\times10^{-19}\text{ C}\right)(14.6\text{ T}\cdot\text{m/s}) = \boxed{2.34\times10^{-18}\text{ N}}$$

Section 22.3 Motion of a Charged Particle in a Uniform Magnetic Field

***P22.7** For each electron, $|q|vB\sin 90.0° = \dfrac{mv^2}{r}$ and $v = \dfrac{eBr}{m}$.

The electrons have no internal structure to absorb energy, so the collision must be perfectly elastic:

$$K = \frac{1}{2}mv_{1i}^2 + 0 = \frac{1}{2}mv_{1f}^2 + \frac{1}{2}mv_{2f}^2$$

$$K = \frac{1}{2}m\left(\frac{e^2B^2R_1^2}{m^2}\right) + \frac{1}{2}m\left(\frac{e^2B^2R_2^2}{m^2}\right) = \frac{e^2B^2}{2m}\left(R_1^2+R_2^2\right)$$

$$K = \frac{e\left(1.60\times10^{-19}\text{ C}\right)\left(0.044\,0\text{ N}\cdot\text{s/C}\cdot\text{m}\right)^2}{2\left(9.11\times10^{-31}\text{ kg}\right)}\left[(0.010\,0\text{ m})^2 + (0.024\,0\text{ m})^2\right] = \boxed{115\text{ keV}}$$

P22.8 (a) We begin with $qvB = \dfrac{mv^2}{R}$

or $qRB = mv$.

But $L = mvR = qR^2B$.

Therefore, $R = \sqrt{\dfrac{L}{qB}} = \sqrt{\dfrac{4.00\times10^{-25}\text{ J}\cdot\text{s}}{\left(1.60\times10^{-19}\text{ C}\right)\left(1.00\times10^{-3}\text{ T}\right)}} = 0.050\,0\text{ m} = \boxed{5.00\text{ cm}}$.

(b) Thus, $v = \dfrac{L}{mR} = \dfrac{4.00\times10^{-25}\text{ J}\cdot\text{s}}{\left(9.11\times10^{-31}\text{ kg}\right)(0.050\,0\text{ m})} = \boxed{8.78\times10^{6}\text{ m/s}}$.

P22.9 $E=\frac{1}{2}mv^2=e\Delta V$

and $evB\sin 90°=\frac{mv^2}{R}$

$$B=\frac{mv}{eR}=\frac{m}{eR}\sqrt{\frac{2e\Delta V}{m}}=\frac{1}{R}\sqrt{\frac{2m\Delta V}{e}}$$

$$B=\frac{1}{5.80\times10^{10}\text{ m}}\sqrt{\frac{2\left(1.67\times10^{-27}\text{ kg}\right)\left(10.0\times10^{6}\text{ V}\right)}{1.60\times10^{-19}\text{ C}}}=\boxed{7.88\times10^{-12}\text{ T}}$$

Section 22.4 Applications Involving Charged Particles Moving in a Magnetic Field

P22.10 $F_B=F_e$

so $qvB=qE$

where $v=\sqrt{\frac{2K}{m}}$ and K is kinetic energy of the electron.

$$E=vB=\sqrt{\frac{2K}{m}}B=\sqrt{\frac{2(750)\left(1.60\times10^{-19}\right)}{9.11\times10^{-31}}}(0.015\,0)=\boxed{244\text{ kV/m}}$$

P22.11 In the velocity selector: $v=\frac{E}{B}=\frac{2\,500\text{ V/m}}{0.035\,0\text{ T}}=7.14\times10^{4}\text{ m/s}$.

In the deflection chamber: $r=\frac{mv}{qB}=\frac{\left(2.18\times10^{-26}\text{ kg}\right)\left(7.14\times10^{4}\text{ m/s}\right)}{\left(1.60\times10^{-19}\text{ C}\right)(0.035\,0\text{ T})}=\boxed{0.278\text{ m}}$.

P22.12 Note that the "cyclotron frequency" is an angular speed. The motion of the proton is described by $\sum F=ma$:

$$|q|vB\sin 90°=\frac{mv^2}{r}$$

$$|q|B=m\frac{v}{r}=m\omega$$

(a) $$\omega=\frac{|q|B}{m}=\frac{\left(1.60\times10^{-19}\text{ C}\right)(0.8\text{ N}\cdot\text{s/C}\cdot\text{m})}{\left(1.67\times10^{-27}\text{ kg}\right)}\left(\frac{\text{kg}\cdot\text{m}}{\text{N}\cdot\text{s}^2}\right)=\boxed{7.66\times10^{7}\text{ rad/s}}$$

(b) $$v=\omega r=\left(7.66\times10^{7}\text{ rad/s}\right)(0.350\text{ m})\left(\frac{1}{1\text{ rad}}\right)=\boxed{2.68\times10^{7}\text{ m/s}}$$

(c) $$K=\frac{1}{2}mv^2=\frac{1}{2}\left(1.67\times10^{-27}\text{ kg}\right)\left(2.68\times10^{7}\text{ m/s}\right)^2\left(\frac{1\text{ eV}}{1.6\times10^{-19}\text{ J}}\right)=\boxed{3.76\times10^{6}\text{ eV}}$$

continued on next page

(d) The proton gains 600 eV twice during each revolution, so the number of revolutions is

$$\frac{3.76\times10^{6}\text{ eV}}{2(600\text{ eV})}=\boxed{3.13\times10^{3}\text{ revolutions}}.$$

(e) $\theta=\omega t$ $$t=\frac{\theta}{\omega}=\frac{3.13\times10^{3}\text{ rev}}{7.66\times10^{7}\text{ rad/s}}\left(\frac{2\pi\text{ rad}}{1\text{ rev}}\right)=\boxed{2.57\times10^{-4}\text{ s}}$$

P22.13 $\theta=\tan^{-1}\left(\frac{25.0}{10.0}\right)=68.2°$ and $R=\frac{1.00\text{ cm}}{\sin 68.2°}=1.08\text{ cm}$.

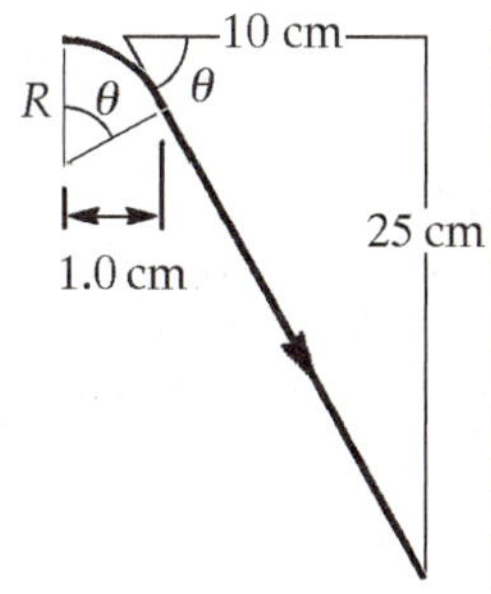

FIG. P22.13

Ignoring relativistic correction, the kinetic energy of the electrons is

$\frac{1}{2}mv^2=q\Delta V$ so $v=\sqrt{\frac{2q\Delta V}{m}}=1.33\times10^{8}\text{ m/s}$.

From Newton's second law $\frac{mv^2}{R}=qvB$, we find the magnetic field

$$B=\frac{mv}{|q|R}=\frac{\left(9.11\times10^{-31}\text{ kg}\right)\left(1.33\times10^{8}\text{ m/s}\right)}{\left(1.60\times10^{-19}\text{ C}\right)\left(1.08\times10^{-2}\text{ m}\right)}=\boxed{70.1\text{ mT}}.$$

P22.14 (a) If the charge carriers are negative, to carry current in the x direction they move with drift velocity v_d in the $-x$ direction. The magnetic force $q\vec{\mathbf{v}}\times\vec{\mathbf{B}}$ is in the $-(-\hat{\mathbf{i}})\times\hat{\mathbf{j}}=\hat{\mathbf{k}}$ direction, so the negative charges are deflected to the top of the ribbon, point c. Some accumulate there to make V_c negative with respect to V_a, until ...

(b) ... an upward electric field $\vec{\mathbf{E}}=\frac{|V_c-V_a|}{d}\hat{\mathbf{k}}$ exerts a downward force on the other charge carriers to let them drift in equilibrium according to

$\Sigma F_z=0$: $$|q|\frac{|V_c-V_a|}{d}(-\hat{\mathbf{k}})+|q||\vec{\mathbf{v}}_d||\vec{\mathbf{B}}|\hat{\mathbf{k}}=0$$

$$v_d=\frac{|\Delta V_H|}{dB}.$$

Since the measured current is $$I=n|q|v_d t d$$

we have $$I=n|q|\frac{|\Delta V_H|}{dB}td$$

$$n=\frac{IB}{|q||\Delta V_H|t}.$$

Since we have shown that ΔV_H is negative if q is negative, this expression simplifies to

$$\boxed{n=\frac{IB}{q\Delta V_H t}}.$$

Section 22.5 Magnetic Force on a Current-Carrying Conductor

P22.15 $\vec{\mathbf{F}}_B = I\vec{\ell}\times\vec{\mathbf{B}} = (2.40\text{ A})(0.750\text{ m})\hat{\mathbf{i}}\times(1.60\text{ T})\hat{\mathbf{k}} = \boxed{\left(-2.88\hat{\mathbf{j}}\right)\text{ N}}$

P22.16 (a) $F_B = ILB\sin\theta = (5.00\text{ A})(2.80\text{ m})(0.390\text{ T})\sin 60.0° = \boxed{4.73\text{ N}}$

(b) $F_B = (5.00\text{ A})(2.80\text{ m})(0.390\text{ T})\sin 90.0° = \boxed{5.46\text{ N}}$

(c) $F_B = (5.00\text{ A})(2.80\text{ m})(0.390\text{ T})\sin 120° = \boxed{4.73\text{ N}}$

P22.17 The magnetic force on each bit of ring is $Id\vec{\mathbf{s}}\times\vec{\mathbf{B}} = IdsB$ radially inward and upward, at angle θ above the radial line. The radially inward components tend to squeeze the ring but all cancel out as forces. The upward components $IdsB\sin\theta$ all add to $\boxed{I2\pi rB\sin\theta\text{ up}}$.

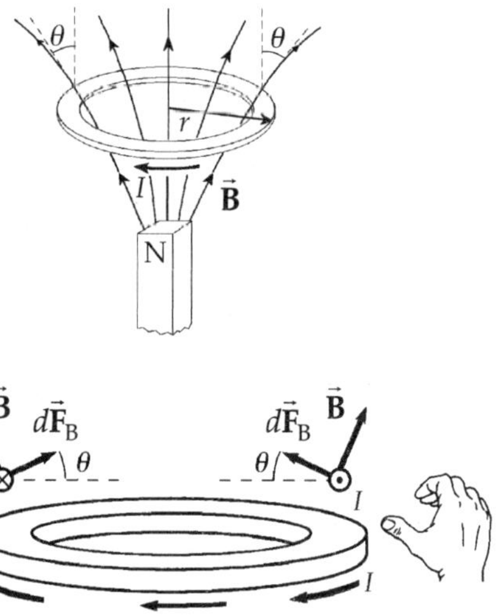

FIG. P22.17

P22.18 For each segment, $I = 5.00\text{ A}$ and $\vec{\mathbf{B}} = 0.020\,0\text{ N/A}\cdot\text{m}\,\hat{\mathbf{j}}$.

Segment	$\vec{\ell}$	$\vec{\mathbf{F}}_B = I\left(\vec{\ell}\times\vec{\mathbf{B}}\right)$
ab	$-0.400\text{ m}\,\hat{\mathbf{j}}$	$\boxed{0}$
bc	$0.400\text{ m}\,\hat{\mathbf{k}}$	$\boxed{(40.0\text{ mN})\left(-\hat{\mathbf{i}}\right)}$
cd	$-0.400\text{ m}\,\hat{\mathbf{i}} + 0.400\text{ m}\,\hat{\mathbf{j}}$	$\boxed{(40.0\text{ mN})\left(-\hat{\mathbf{k}}\right)}$
da	$0.400\text{ m}\,\hat{\mathbf{i}} - 0.400\text{ m}\,\hat{\mathbf{k}}$	$\boxed{(40.0\text{ mN})\left(\hat{\mathbf{k}}+\hat{\mathbf{i}}\right)}$

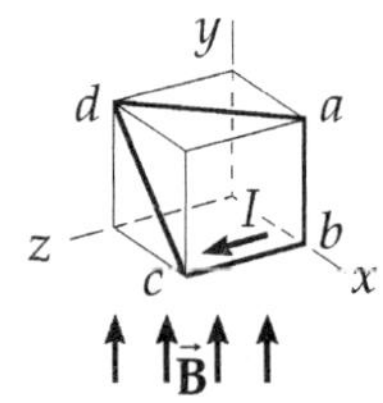

FIG. P22.18

Section 22.6 Torque on a Current Loop in a Uniform Magnetic Field

P22.19 (a) $2\pi r = 2.00\text{ m}$

so $r = 0.318\text{ m}$

$$\mu = IA = \left(17.0\times10^{-3}\text{ A}\right)\left[\pi(0.318)^2\text{ m}^2\right] = \boxed{5.41\text{ mA}\cdot\text{m}^2}$$

(b) $\vec{\tau} = \vec{\mu}\times\vec{\mathbf{B}}$

so $\tau = \left(5.41\times10^{-3}\text{ A}\cdot\text{m}^2\right)(0.800\text{ T}) = \boxed{4.33\text{ mN}\cdot\text{m}}$

P22.20 Choose $U = 0$ when the dipole moment is at $\theta = 90.0°$ to the field. The field exerts torque of magnitude $\mu B\sin\theta$ on the dipole, tending to turn the dipole moment in the direction of decreasing θ. According to Equations 7.15 and 10.28, the potential energy of the dipole-field system is given by

$$U - 0 = \int_{90.0°}^{\theta}\mu B\sin\theta\,d\theta = \mu B(-\cos\theta)\Big|_{90.0°}^{\theta} = -\mu B\cos\theta + 0 \quad\text{or}\quad \boxed{U = -\vec{\mu}\cdot\vec{\mathbf{B}}}.$$

P22.21 $\tau = NBAI\sin\phi$

$\tau = 100(0.800\text{ T})\left(0.400\times0.300\text{ m}^2\right)(1.20\text{ A})\sin60°$

$\tau = \boxed{9.98\text{ N}\cdot\text{m}}$

Note that ϕ is the angle between the magnetic moment and the $\vec{\mathbf{B}}$ field. The loop will rotate so as to align the magnetic moment with the $\vec{\mathbf{B}}$ field. Looking down along the y-axis, the loop will rotate in a $\boxed{\text{clockwise}}$ direction.

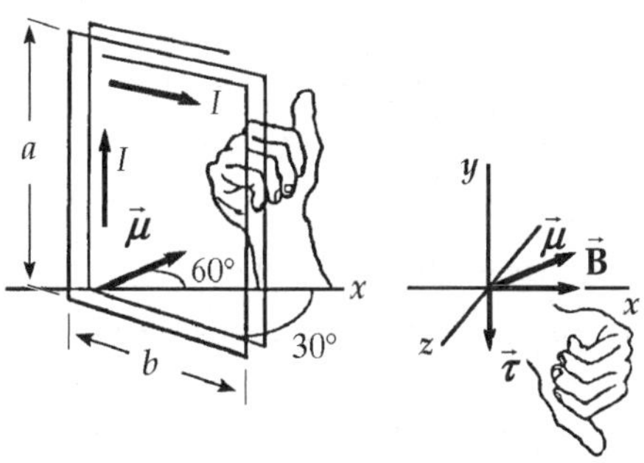

FIG. P22.21

***P22.22** (a) $|\vec{\tau}| = |\vec{\mu}\times\vec{\mathbf{B}}| = NIAB\sin\theta$

$$\tau_{\text{max}} = 80\left(10^{-2}\text{ A}\right)(0.025\text{ m}\cdot0.04\text{ m})(0.8\text{ N/A}\cdot\text{m})\sin90° = \boxed{6.40\times10^{-4}\text{ N}\cdot\text{m}}$$

(b) $$\mathcal{P}_{\text{max}} = \tau_{\text{max}}\omega = 6.40\times10^{-4}\text{ N}\cdot\text{m}(3\,600\text{ rev/min})\left(\frac{2\pi\text{ rad}}{1\text{ rev}}\right)\left(\frac{1\text{ min}}{60\text{ s}}\right) = \boxed{0.241\text{ W}}$$

(c) In one half revolution the work is

$$W = U_{\text{max}} - U_{\text{min}} = -\mu B\cos180° - (-\mu B\cos0°) = 2\mu B$$
$$= 2NIAB = 2\left(6.40\times10^{-4}\text{ N}\cdot\text{m}\right) = 1.28\times10^{-3}\text{ J}$$

In one full revolution, $W = 2\left(1.28\times10^{-3}\text{ J}\right) = \boxed{2.56\times10^{-3}\text{ J}}$.

(d) $$\mathcal{P}_{\text{avg}} = \frac{W}{\Delta t} = \frac{2.56\times10^{-3}\text{ J}}{(1/60)\text{ s}} = \boxed{0.154\text{ W}}$$

The peak power in (b) is greater by the factor $\frac{\pi}{2}$.

Section 22.7 The Biot-Savart Law

P22.23 $B = \dfrac{\mu_0 I}{2R} = \dfrac{\mu_0 q(v/2\pi R)}{2R} = \boxed{12.5\ \text{T}}$

P22.24 $B = \dfrac{\mu_0 I}{2\pi R} = \dfrac{(4\pi\times10^{-7}\ \text{T}\cdot\text{m/A})(1.00\times10^{4}\ \text{A})}{2\pi(100\ \text{m})} = 2.00\times10^{-5}\ \text{T} = \boxed{20.0\ \mu\text{T}}$

P22.25 For leg 1, $d\vec{\mathbf{s}}\times\hat{\mathbf{r}} = 0$, so there is no contribution to the field from this segment. For leg 2, the wire is only semi-infinite; thus,

$$B = \frac{1}{2}\left(\frac{\mu_0 I}{2\pi x}\right) = \boxed{\frac{\mu_0 I}{4\pi x}\ \text{into the paper}}.$$

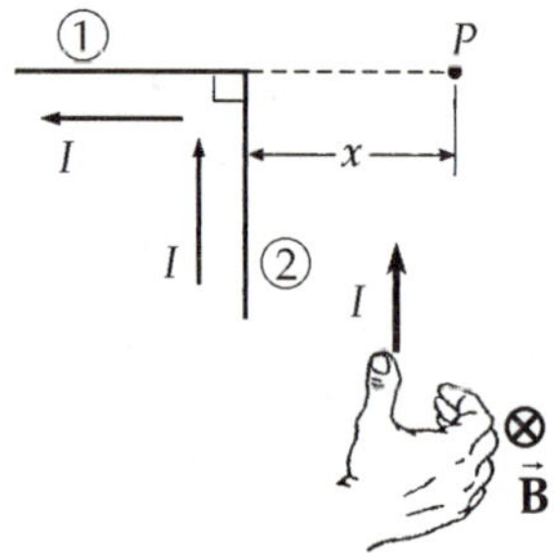

FIG. P22.25

P22.26 $B = \dfrac{\mu_0 I}{2\pi r} = \dfrac{(4\pi\times10^{-7})(1.00\ \text{A})}{2\pi(1.00\ \text{m})} = \boxed{2.00\times10^{-7}\ \text{T}}$

P22.27 We can think of the total magnetic field as the superposition of the field due to the long straight wire (having magnitude $\dfrac{\mu_0 I}{2\pi R}$ and directed into the page) and the field due to the circular loop (having magnitude $\dfrac{\mu_0 I}{2R}$ and directed into the page). The resultant magnetic field is:

$$\boxed{\vec{\mathbf{B}} = \left(1+\frac{1}{\pi}\right)\frac{\mu_0 I}{2R}\ \text{(directed into the page)}}.$$

P22.28 Along the axis of a circular loop of radius R,

$$B = \frac{\mu_0 I R^2}{2(x^2+R^2)^{3/2}}$$

or $$\frac{B}{B_0} = \left[\frac{1}{(x/R)^2+1}\right]^{3/2}$$

where $$B_0 \equiv \frac{\mu_0 I}{2R}.$$

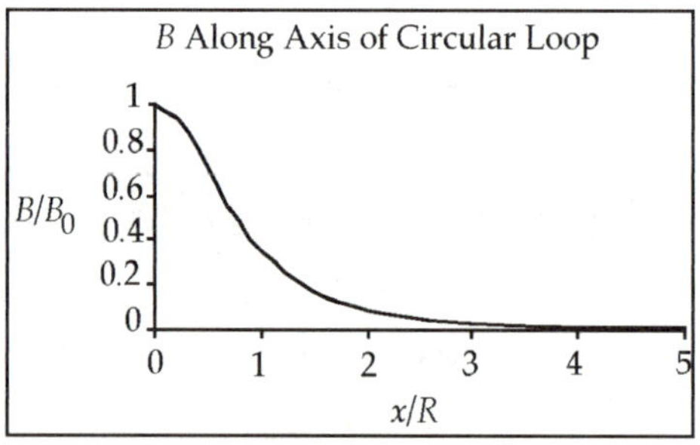

FIG. P22.28

x/R	B/B_0
0.00	1.00
1.00	0.354
2.00	0.0894
3.00	0.0316
4.00	0.0143
5.00	0.00754

***P22.29** Wire 1 creates at the origin magnetic field

$\vec{\mathbf{B}}_1 = \dfrac{\mu_0 I}{2\pi r}$ right hand rule $= \dfrac{\mu_0 I_1}{2\pi a}$  $= \dfrac{\mu_0 I_1}{2\pi a}\hat{\mathbf{j}}$

(a) If the total field at the origin is $\dfrac{2\mu_0 I_1}{2\pi a}\hat{\mathbf{j}} = \dfrac{\mu_0 I_1}{2\pi a}\hat{\mathbf{j}} + \vec{\mathbf{B}}_2$ then the second wire must create field according to $\vec{\mathbf{B}}_2 = \dfrac{\mu_0 I_1}{2\pi a}\hat{\mathbf{j}} = \dfrac{\mu_0 I_2}{2\pi(2a)}$ .

Then $I_2 = \boxed{2I_1 \text{ out of the paper}}$.

(b) The other possibility is $\vec{\mathbf{B}}_1 + \vec{\mathbf{B}}_2 = \dfrac{2\mu_0 I_1}{2\pi a}(-\hat{\mathbf{j}}) = \dfrac{\mu_0 I_1}{2\pi a}\hat{\mathbf{j}} + \vec{\mathbf{B}}_2$. Then

$\vec{\mathbf{B}}_2 = \dfrac{3\mu_0 I_1}{2\pi a}(-\hat{\mathbf{j}}) = \dfrac{\mu_0 I_2}{2\pi(2a)}$  $\qquad I_2 = \boxed{6I_1 \text{ into the paper}}$

P22.30 (a) Above the pair of wires, the field out of the page of the 50 A current will be stronger than the $(-\hat{\mathbf{k}})$ field of the 30 A current, so they cannot add to zero. Between the wires, both produce fields into the page. They can only add to zero below the wires, at coordinate $y = -|y|$. Here the total field is

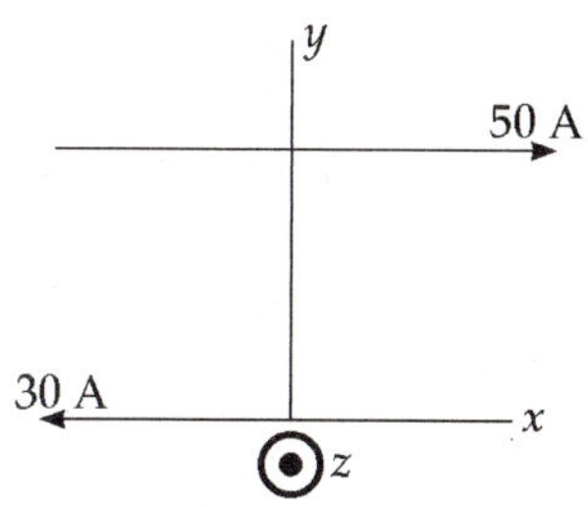

FIG. P22.30

$\vec{\mathbf{B}} = \dfrac{\mu_0 I}{2\pi r}$  $+ \dfrac{\mu_0 I}{2\pi r}$  :

$$0 = \frac{\mu_0}{2\pi}\left[\frac{50 \text{ A}}{(|y| + 0.28 \text{ m})}(-\hat{\mathbf{k}}) + \frac{30 \text{ A}}{|y|}(\hat{\mathbf{k}})\right]$$

$$50|y| = 30(|y| + 0.28 \text{ m})$$

$$50(-y) = 30(0.28 \text{ m} - y)$$

$$-20y = 30(0.28 \text{ m}) \boxed{\text{at } y = -0.420 \text{ m}}$$

(b) At $y = 0.1$ m the total field is $\vec{\mathbf{B}} = \dfrac{\mu_0 I}{2\pi r}$  $+ \dfrac{\mu_0 I}{2\pi r}$  :

$$\vec{\mathbf{B}} = \frac{4\pi \times 10^{-7} \text{ T}\cdot\text{m/A}}{2\pi}\left(\frac{50 \text{ A}}{(0.28 - 0.10) \text{ m}}(-\hat{\mathbf{k}}) + \frac{30 \text{ A}}{0.10 \text{ m}}(-\hat{\mathbf{k}})\right) = 1.16 \times 10^{-4} \text{ T}(-\hat{\mathbf{k}}).$$

The force on the particle is

$$\vec{\mathbf{F}} = q\vec{\mathbf{v}} \times \vec{\mathbf{B}} = (-2 \times 10^{-6} \text{ C})(150 \times 10^6 \text{ m/s})(\hat{\mathbf{i}}) \times (1.16 \times 10^{-4} \text{ N}\cdot\text{s/C}\cdot\text{m})(-\hat{\mathbf{k}}) = \boxed{3.47 \times 10^{-2} \text{ N}(-\hat{\mathbf{j}})}.$$

(c) We require $\vec{\mathbf{F}}_e = 3.47 \times 10^{-2} \text{ N}(+\hat{\mathbf{j}}) = q\vec{\mathbf{E}} = (-2 \times 10^{-6} \text{ C})\vec{\mathbf{E}}$.

So $\vec{\mathbf{E}} = \boxed{-1.73 \times 10^4 \hat{\mathbf{j}} \text{ N/C}}$.

P22.31 We use the Biot-Savart law. For bits of wire along the straight-line sections, $d\vec{\mathbf{s}}$ is at 0° or 180° to $\hat{\mathbf{r}}$, so $d\vec{\mathbf{s}}\times\hat{\mathbf{r}}=0$. Thus, only the curved section of wire contributes to $\vec{\mathbf{B}}$ at P. Hence, $d\vec{\mathbf{s}}$ is tangent to the arc and $\hat{\mathbf{r}}$ is radially inward; so $d\vec{\mathbf{s}}\times\hat{\mathbf{r}}=|ds|\ell\sin 90°=|ds|\otimes$. All points along the curve are the same distance $r=0.600$ m from the field point, so

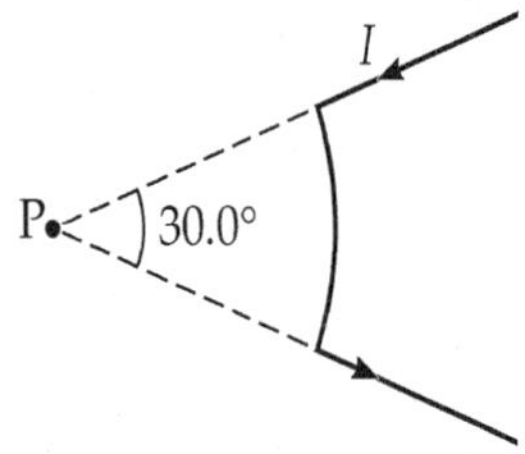

FIG. P22.31

$$B=\int_{\text{all current}}\left|d\vec{\mathbf{B}}\right|=\int\frac{\mu_0}{4\pi}\frac{I\left|d\vec{\mathbf{s}}\times\hat{\mathbf{r}}\right|}{r^2}=\frac{\mu_0}{4\pi}\frac{I}{r^2}\int|ds|=\frac{\mu_0}{4\pi}\frac{I}{r^2}s$$

where s is the arc length of the curved wire,

$$s=r\theta=(0.600\text{ m})(30.0°)\left(\frac{2\pi}{360°}\right)=0.314\text{ m}$$

Then, $B=\left(10^{-7}\text{ T}\cdot\text{m/A}\right)\dfrac{(3.00\text{ A})}{(0.600\text{ m})^2}(0.314\text{ m})$ $\qquad B=\boxed{261\text{ nT into the page}}$

P22.32 Label the wires 1, 2, and 3 as shown in Figure (a) and let the magnetic field created by the currents in these wires be $\vec{\mathbf{B}}_1$, $\vec{\mathbf{B}}_2$, and $\vec{\mathbf{B}}_3$, respectively.

(a) At point A: $B_1=B_2=\dfrac{\mu_0 I}{2\pi\left(a\sqrt{2}\right)}$ and $B_3=\dfrac{\mu_0 I}{2\pi(3a)}$.

The directions of these fields are shown in Figure (b). Observe that the horizontal components of $\vec{\mathbf{B}}_1$ and $\vec{\mathbf{B}}_2$ cancel while their vertical components both add to $\vec{\mathbf{B}}_3$.

Therefore, the net field at point A is:

$$B_A=B_1\cos 45.0°+B_2\cos 45.0°+B_3$$
$$=\frac{\mu_0 I}{2\pi a}\left[\frac{2}{\sqrt{2}}\cos 45.0°+\frac{1}{3}\right]$$
$$B_A=\frac{\left(4\pi\times10^{-7}\text{ T}\cdot\text{m/A}\right)(2.00\text{ A})}{2\pi\left(1.00\times10^{-2}\text{ m}\right)}\left[\frac{2}{\sqrt{2}}\cos 45°+\frac{1}{3}\right]$$

$B_A=\boxed{53.3\ \mu\text{T toward the bottom of the page}}$

(b) At point B: $\vec{\mathbf{B}}_1$ and $\vec{\mathbf{B}}_2$ cancel, leaving

$$B_B=B_3=\frac{\mu_0 I}{2\pi(2a)}$$
$$B_B=\frac{\left(4\pi\times10^{-7}\text{ T}\cdot\text{m/A}\right)(2.00\text{ A})}{2\pi(2)\left(1.00\times10^{-2}\text{ m}\right)}$$

$=\boxed{20.0\ \mu\text{T toward the bottom of the page}}$

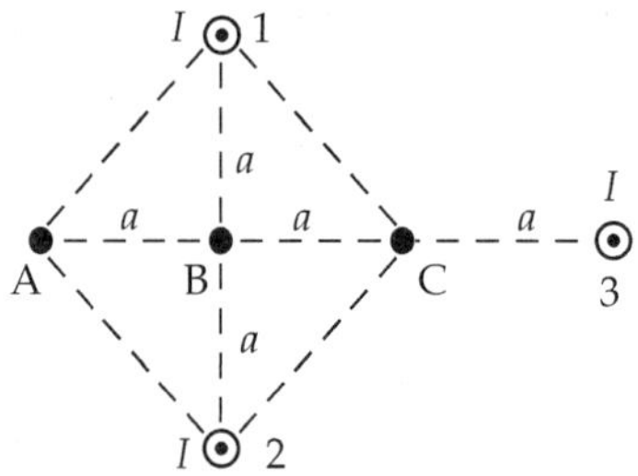

Figure (a)

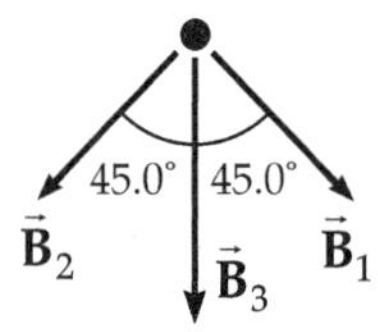

Figure (b)

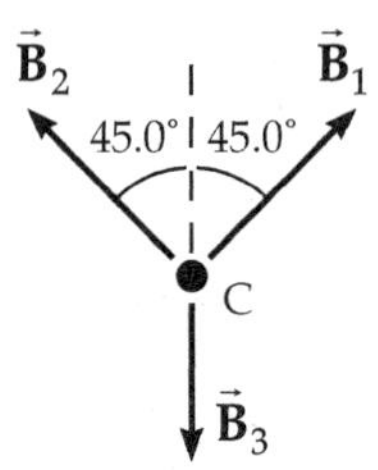

Figure (c)

FIG. P22.32

continued on next page

(c) At point C: $B_1 = B_2 = \dfrac{\mu_0 I}{2\pi\left(a\sqrt{2}\right)}$ and $B_3 = \dfrac{\mu_0 I}{2\pi a}$ with the directions shown in Figure (c). Again, the horizontal components of $\vec{\mathbf{B}}_1$ and $\vec{\mathbf{B}}_2$ cancel. The vertical components both oppose $\vec{\mathbf{B}}_3$ giving

$$B_C = 2\left[\frac{\mu_0 I}{2\pi\left(a\sqrt{2}\right)}\cos 45.0°\right] - \frac{\mu_0 I}{2\pi a} = \frac{\mu_0 I}{2\pi a}\left[\frac{2}{\sqrt{2}}\cos 45.0° - 1\right] = \boxed{0}.$$

***P22.33** (a) Each coil separately produces field given by $B = \dfrac{N\mu_0 I R^2}{2\left(R^2 + x^2\right)^{3/2}}$ at the point halfway between them. Together they produce field

$$2B = \frac{N\mu_0 I R^2}{\left(R^2 + x^2\right)^{3/2}} = 4.5 \times 10^{-5}\ \text{T} = \frac{50\left(4\pi \times 10^{-7}\ \text{Tm}\right) I (0.012\ \text{m})^2}{\text{A}\left[(0.012\ \text{m})^2 + (0.011\ \text{m})^2\right]^{3/2}} = \frac{9.05 \times 10^{-9}\ \text{Tm}^3}{4.31 \times 10^{-6}\ \text{m}^3\text{A}} I$$

$$I = \frac{4.5 \times 10^{-5}\ \text{T A}}{2.10 \times 10^{-3}\ \text{T}} = \boxed{21.5\ \text{mA}}$$

(b) $\Delta V = IR = (0.021\,5\ \text{A})(210\ \Omega) = \boxed{4.51\ \text{V}}$

(c) $\mathcal{P} = (\Delta V)I = (4.51\ \text{V})(0.021\,5\ \text{A}) = \boxed{96.7\ \text{mW}}$

Section 22.8 The Magnetic Force Between Two Parallel Conductors

P22.34 Let both wires carry current in the x direction, the first at $y = 0$ and the second at $y = 10.0$ cm.

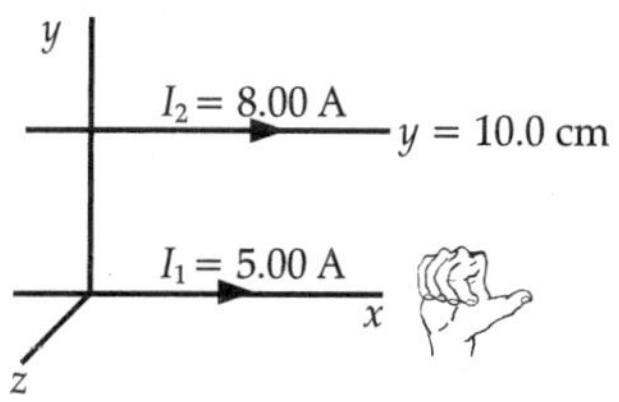

FIG. P22.34(a)

(a) $$\vec{\mathbf{B}} = \frac{\mu_0 I}{2\pi r}\hat{\mathbf{k}} = \frac{\left(4\pi \times 10^{-7}\ \text{T}\cdot\text{m/A}\right)(5.00\ \text{A})}{2\pi(0.100\ \text{m})}\hat{\mathbf{k}}$$

$$\vec{\mathbf{B}} = \boxed{1.00 \times 10^{-5}\ \text{T out of the page}}$$

(b) $$\vec{\mathbf{F}}_B = I_2 \vec{\ell} \times \vec{\mathbf{B}} = (8.00\ \text{A})\left[(1.00\ \text{m})\hat{\mathbf{i}} \times \left(1.00 \times 10^{-5}\ \text{T}\right)\hat{\mathbf{k}}\right] = \left(8.00 \times 10^{-5}\ \text{N}\right)\left(-\hat{\mathbf{j}}\right)$$

$$\boxed{\vec{\mathbf{F}}_B = 8.00 \times 10^{-5}\ \text{N toward the first wire}}$$

(c) $$\vec{\mathbf{B}} = \frac{\mu_0 I}{2\pi r}\left(-\hat{\mathbf{k}}\right) = \frac{\left(4\pi \times 10^{-7}\ \text{T}\cdot\text{m/A}\right)(8.00\ \text{A})}{2\pi(0.100\ \text{m})}\left(-\hat{\mathbf{k}}\right) = \left(1.60 \times 10^{-5}\ \text{T}\right)\left(-\hat{\mathbf{k}}\right)$$

$$\vec{\mathbf{B}} = \boxed{1.60 \times 10^{-5}\ \text{T into the page}}$$

continued on next page

(d) $\vec{\mathbf{F}}_B = I_1\vec{\ell}\times\vec{\mathbf{B}} = (5.00\text{ A})\left[(1.00\text{ m})\hat{\mathbf{i}}\times\left(1.60\times10^{-5}\text{ T}\right)\left(-\hat{\mathbf{k}}\right)\right] = \left(8.00\times10^{-5}\text{ N}\right)\left(+\hat{\mathbf{j}}\right)$

$\vec{\mathbf{F}}_B = \boxed{8.00\times10^{-5}\text{ N towards the second wire}}$

P22.35 By symmetry, we note that the magnetic forces on the top and bottom segments of the rectangle cancel. The net force on the vertical segments of the rectangle is (using Equation 22.27)

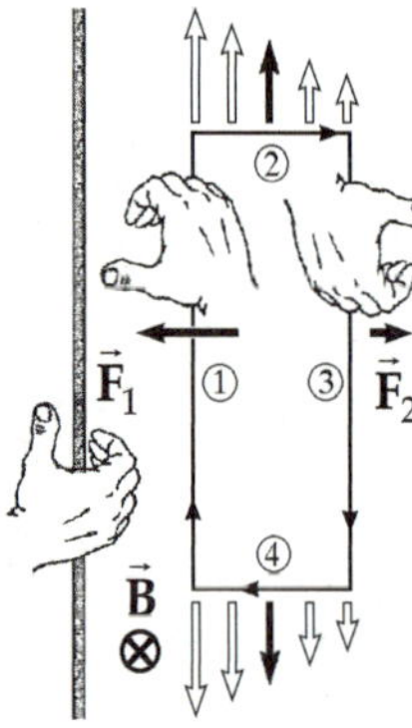

FIG. P22.35

$$\vec{\mathbf{F}} = \vec{\mathbf{F}}_1 + \vec{\mathbf{F}}_2 = \frac{\mu_0 I_1 I_2 \ell}{2\pi}\left(\frac{1}{c+a} - \frac{1}{c}\right)\hat{\mathbf{i}} = \frac{\mu_0 I_1 I_2 \ell}{2\pi}\left(\frac{-a}{c(c+a)}\right)\hat{\mathbf{i}}$$

$$\vec{\mathbf{F}} = \frac{\left(4\pi\times10^{-7}\text{ N/A}^2\right)(5.00\text{ A})(10.0\text{ A})(0.450\text{ m})}{2\pi}\left(\frac{-0.150\text{ m}}{(0.100\text{ m})(0.250\text{ m})}\right)\hat{\mathbf{i}}$$

$$\vec{\mathbf{F}} = \left(-2.70\times10^{-5}\hat{\mathbf{i}}\right)\text{ N}$$

or $\vec{\mathbf{F}} = \boxed{2.70\times10^{-5}\text{ N toward the left}}$.

***P22.36** Carrying oppositely directed currents, wires 1 and 2 repel each other. If wire 3 were between them, it would have to repel either 1 or 2, so the force on that wire could not be zero. If wire 3 were to the right of wire 2, it would feel a larger force exerted by 2 than that exerted by 1, so the total force on 3 could not be zero. Therefore wire 3 must be to the left of both other wires as shown. It must carry downward current so that it can attract wire 2.

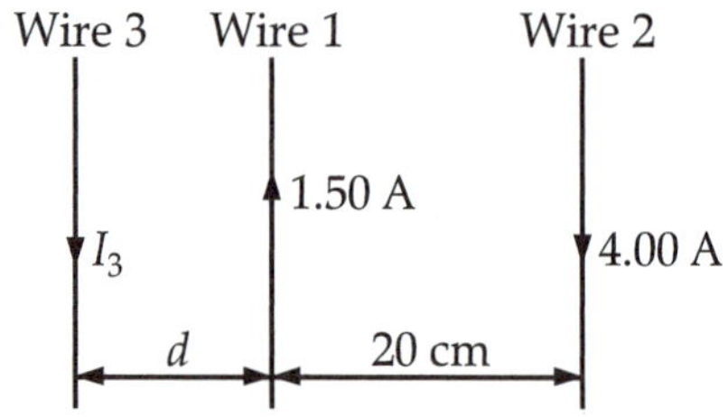

FIG. P22.36

(a) For the equilibrium of wire 3 we have

$$F_{1\text{ on }3} = F_{2\text{ on }3} \qquad \frac{\mu_0(1.50\text{ A})I_3}{2\pi d} = \frac{\mu_0(4\text{ A})I_3}{2\pi(20\text{ cm}+d)}$$

$$1.5(20\text{ cm}+d) = 4d \qquad d = \frac{30\text{ cm}}{2.5} = \boxed{12.0\text{ cm to the left of wire 1}}.$$

(b) For the equilibrium of wire 1,

$$\frac{\mu_0 I_3(1.5\text{ A})}{2\pi(12\text{ cm})} = \frac{\mu_0(4\text{ A})(1.5\text{ A})}{2\pi(20\text{ cm})} \qquad I_3 = \frac{12}{20}4\text{ A} = \boxed{2.40\text{ A down}}.$$

We know that wire 2 must be in equilibrium because the forces on it are equal in magnitude to the forces that it exerts on wires 1 and 3, which are equal because they both balance the equal-magnitude forces that 1 exerts on 3 and that 3 exerts on 1.

Section 22.9 Ampère's Law

P22.37 Each wire is distant from P by

$(0.200\ \text{m})\cos 45.0° = 0.141\ \text{m}.$

Each wire produces a field at P of equal magnitude:

$$B_A = \frac{\mu_0 I}{2\pi a} = \frac{(2.00\times10^{-7}\ \text{T}\cdot\text{m/A})(5.00\ \text{A})}{(0.141\ \text{m})} = 7.07\ \mu\text{T}.$$

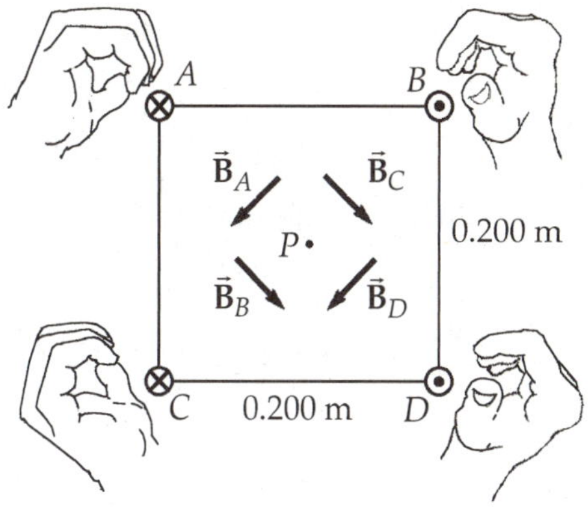

FIG. P22.37

Carrying currents into the page, A produces at P a field of 7.07 μT to the left and down at –135°, while B creates a field to the right and down at –45°. Carrying currents toward you, C produces a field downward and to the right at –45°, while D's contribution is downward and to the left. The total field is then

$4(7.07\ \mu\text{T})\sin 45.0° = \boxed{20.0\ \mu\text{T}}$ toward the bottom of the page.

P22.38 Let the current I be to the right. It creates a field $B = \frac{\mu_0 I}{2\pi d}$ at the proton's location. And we have a balance between the weight of the proton and the magnetic force

$$mg(-\hat{\mathbf{j}}) + qv(-\hat{\mathbf{i}})\times\frac{\mu_0 I}{2\pi d}(\hat{\mathbf{k}}) = 0 \text{ at a distance } d \text{ from the wire}$$

$$d = \frac{qv\mu_0 I}{2\pi mg} = \frac{(1.60\times10^{-19}\ \text{C})(2.30\times10^{4}\ \text{m/s})(4\pi\times10^{-7}\ \text{T}\cdot\text{m/A})(1.20\times10^{-6}\ \text{A})}{2\pi(1.67\times10^{-27}\ \text{kg})(9.80\ \text{m/s}^2)} = \boxed{5.40\ \text{cm}}.$$

P22.39 (a) One wire feels force due to the field of the other ninety-nine.

$$B = \frac{\mu_0 I_0 r}{2\pi R^2} = \frac{(4\pi\times10^{-7}\ \text{T}\cdot\text{m/A})(99)(2.00\ \text{A})(0.200\times10^{-2}\ \text{m})}{2\pi(0.500\times10^{-2}\ \text{m})^2} = 3.17\times10^{-3}\ \text{T}$$

This field points tangent to a circle of radius 0.200 cm and exerts force $\vec{\mathbf{F}} = I\vec{\ell}\times\vec{\mathbf{B}}$ toward the center of the bundle, on the single hundredth wire:

$$\frac{F}{\ell} = IB\sin\theta = (2.00\ \text{A})(3.17\times10^{-3}\ \text{T})\sin 90° - 6.34\ \text{mN/m}$$

$$\frac{F_B}{\ell} = \boxed{6.34\times10^{-3}\ \text{N/m inward}}.$$

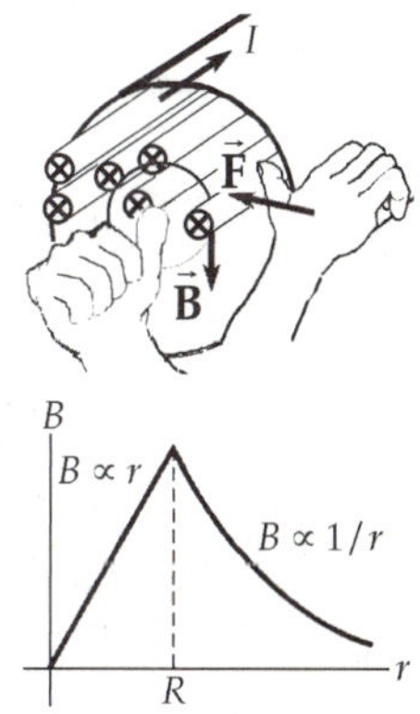

FIG. P22.39

(b) $B \propto r$, so B is greatest at the outside of the bundle. Since each wire carries the same current, F is $\boxed{\text{greatest at the outer surface}}$.

***P22.40** (a) In $B=\dfrac{\mu_0 I}{2\pi r}$, the field will be one-tenth as large at a ten-times larger distance: $\boxed{400\text{ cm}}$

(b) $$\vec{\mathbf{B}}=\frac{\mu_0 I}{2\pi r_1}\hat{\mathbf{k}}+\frac{\mu_0 I}{2\pi r_2}\left(-\hat{\mathbf{k}}\right)\text{ so }B=\frac{4\pi\times10^{-7}\text{ T}\cdot\text{m}(2.00\text{ A})}{2\pi\text{ A}}\left(\frac{1}{0.398\,5\text{ m}}-\frac{1}{0.401\,5\text{ m}}\right)=\boxed{7.50\text{ nT}}$$

(c) Call r the distance from cord center to field point and $2d=3.00$ mm the distance between conductors.

$$B=\frac{\mu_0 I}{2\pi}\left(\frac{1}{r-d}-\frac{1}{r+d}\right)=\frac{\mu_0 I}{2\pi}\frac{2d}{r^2-d^2}$$

$$7.50\times10^{-10}\text{ T}=\left(2.00\times10^{-7}\text{ T}\cdot\text{m/A}\right)(2.00\text{ A})\frac{\left(3.00\times10^{-3}\text{ m}\right)}{r^2-2.25\times10^{-6}\text{ m}^2}\quad\text{so }r=\boxed{1.26\text{ m}}$$

The field of the two-conductor cord is weak to start with and falls off rapidly with distance.

(d) The cable creates $\boxed{\text{zero}}$ field at exterior points, since a loop in Ampère's law encloses zero total current. Shall we sell coaxial-cable power cords to people who worry about biological damage from weak magnetic fields?

P22.41 (a) $$B_{\text{inner}}=\frac{\mu_0 NI}{2\pi r}=\frac{\left(4\pi\times10^{-7}\text{ T}\cdot\text{m/A}\right)(900)\left(14.0\times10^3\text{ A}\right)}{2\pi(0.700\text{ m})}=\boxed{3.60\text{ T}}$$

(b) $$B_{\text{outer}}=\frac{\mu_0 NI}{2\pi r}=\frac{\left(2\times10^{-7}\text{ T}\cdot\text{m/A}\right)(900)\left(14.0\times10^3\text{ A}\right)}{1.30\text{ m}}=\boxed{1.94\text{ T}}$$

P22.42 We assume the current is vertically upward.

(a) Consider a circle of radius r slightly less than R. It encloses no current so from

$$\oint\vec{\mathbf{B}}\cdot d\vec{\mathbf{s}}=\mu_0 I_{\text{inside}}\qquad B(2\pi r)=0$$

we conclude that the magnetic field is $\boxed{\text{zero}}$.

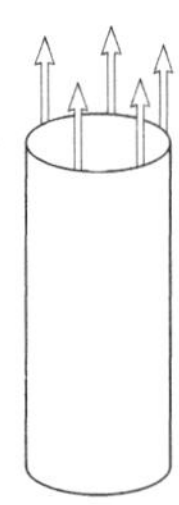

FIG. P22.42(a)

(b) Now let the r be barely larger than R. Ampère's law becomes $B(2\pi R)=\mu_0 I$,

so $\boxed{B=\dfrac{\mu_0 I}{2\pi R}}$.

The field's direction is 

$\boxed{\text{tangent to the wall of the cylinder in a counterclockwise sense}}$.

continued on next page

(c) Consider a strip of the wall of width dx and length ℓ. Its width is so small compared to $2\pi R$ that the field at its location would be essentially unchanged if the current in the strip were turned off.

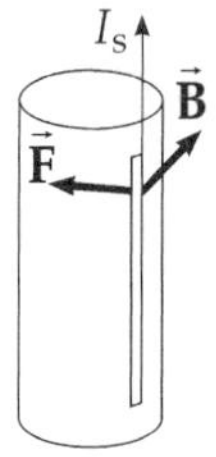

FIG. P22.42(c)

The current it carries is $I_s = \dfrac{Idx}{2\pi R}$ up.

The force on it is

$$\vec{\mathbf{F}} = I_s \vec{\ell} \times \vec{\mathbf{B}} = \frac{Idx}{2\pi R}\left(\ell \frac{\mu_0 I}{2\pi R}\right) \textbf{up} \times \textbf{into page} = \frac{\mu_0 I^2 \ell dx}{4\pi^2 R^2} \textbf{radially inward}.$$

The pressure on the strip and everywhere on the cylinder is

$$P = \frac{F}{A} = \frac{\mu_0 I^2 \ell dx}{4\pi^2 R^2 \ell dx} = \boxed{\frac{\mu_0 I^2}{(2\pi R)^2} \text{ inward}}.$$

The pinch effect makes an effective demonstration when an aluminum can crushes itself as it carries a large current along its length.

P22.43 From $\oint \vec{\mathbf{B}} \cdot d\vec{\ell} = \mu_0 I$, $\quad I = \dfrac{2\pi r B}{\mu_0} = \dfrac{2\pi\left(1.00 \times 10^{-3}\right)(0.100)}{4\pi \times 10^{-7}} = \boxed{500 \text{ A}}$.

Section 22.10 The Magnetic Field of a Solenoid

***P22.44** The field produced by the solenoid in its interior is given by

$$\vec{\mathbf{B}} = \mu_0 n I\left(-\hat{\mathbf{i}}\right) = \left(4\pi \times 10^{-7} \text{ T}\cdot\text{m/A}\right)\left(\frac{30.0}{10^{-2} \text{ m}}\right)(15.0 \text{ A})\left(-\hat{\mathbf{i}}\right)$$

$$\vec{\mathbf{B}} = -\left(5.65 \times 10^{-2} \text{ T}\right)\hat{\mathbf{i}}$$

The force exerted on side AB of the square current loop is

$$\left(\vec{\mathbf{F}}_B\right)_{AB} = I\vec{\ell} \times \vec{\mathbf{B}} = (0.200 \text{ A})\left[\left(2.00 \times 10^{-2} \text{ m}\right)\hat{\mathbf{j}} \times \left(5.65 \times 10^{-2} \text{ T}\right)\left(-\hat{\mathbf{i}}\right)\right]$$

$$\left(\vec{\mathbf{F}}_B\right)_{AB} = \left(2.26 \times 10^{-4} \text{ N}\right)\hat{\mathbf{k}}$$

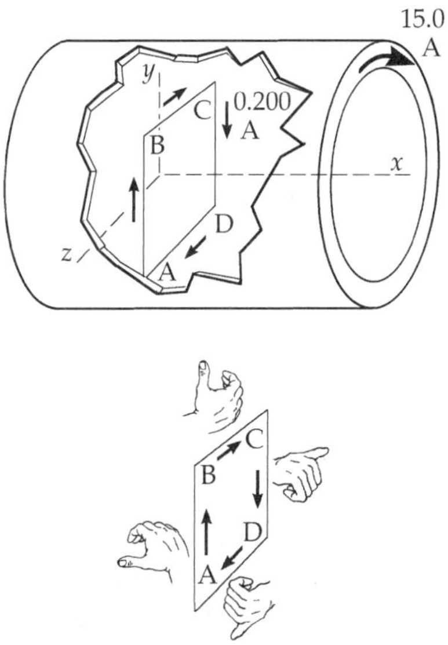

FIG. P22.44

Similarly, each side of the square loop experiences a force, lying in the plane of the loop, of $\boxed{226\ \mu\text{N directed away from the center}}$. From the above result, it is seen that the net torque exerted on the square loop by the field of the solenoid should be zero. More formally, the magnetic dipole moment of the square loop is given by

$$\vec{\mu} = I\vec{\mathbf{A}} = (0.200 \text{ A})\left(2.00 \times 10^{-2} \text{ m}\right)^2\left(-\hat{\mathbf{i}}\right) = -80.0\ \mu\text{A}\cdot\text{m}^2\hat{\mathbf{i}}$$

The torque exerted on the loop is then $\vec{\tau} = \vec{\mu} \times \vec{\mathbf{B}} = \left(-80.0\ \mu\text{A}\cdot\text{m}^2\hat{\mathbf{i}}\right) \times \left(-5.65 \times 10^{-2} \text{ T}\hat{\mathbf{i}}\right) = \boxed{0}$

P22.45 $B = \mu_0 \dfrac{N}{\ell} I$ so $I = \dfrac{B}{\mu_0 n} = \dfrac{(1.00 \times 10^{-4}\ \text{T})0.400\ \text{m}}{(4\pi \times 10^{-7}\ \text{T}\cdot\text{m/A})1\,000} = \boxed{31.8\ \text{mA}}$

P22.46 Let the axis of the solenoid lie along the y–axis from $y = 0$ to $y = \ell$. We will determine the field at $y = a$. This point will be inside the solenoid if $0 < a < \ell$ and outside if $a < 0$ or $a > \ell$. We think of solenoid as formed of rings, each of thickness dy. Now I is the symbol for the current in each turn of wire and the number of turns per length is $\left(\dfrac{N}{\ell}\right)$. So the number of turns in the ring is $\left(\dfrac{N}{\ell}\right)dy$ and the current in the ring is $I_{\text{ring}} = I\left(\dfrac{N}{\ell}\right)dy$. Now we use the result of Example 22.6 for the field created by one ring:

$$B_{\text{ring}} = \frac{\mu_0 I_{\text{ring}} R^2}{2\left(x^2 + R^2\right)^{3/2}}$$

where x is the name of the distance from the center of the ring, at location y, to the field point $x = a - y$. Each ring creates field in the same direction, along our y–axis, so the whole field of the solenoid is

$$B = \sum_{\text{all rings}} B_{\text{ring}} = \sum \frac{\mu_0 I_{\text{ring}} R^2}{2\left(x^2 + R^2\right)^{3/2}} = \int_0^{\ell} \frac{\mu_0 I (N/\ell) dy R^2}{2\left((a-y)^2 + R^2\right)^{3/2}} = \frac{\mu_0 I N R^2}{2\ell} \int_0^{\ell} \frac{dy}{2\left((a-y)^2 + R^2\right)^{3/2}}.$$

To perform the integral we change variables to $u = a - y$.

$$B = \frac{\mu_0 I N R^2}{2\ell} \int_a^{a-\ell} \frac{-du}{\left(u^2 + R^2\right)^{3/2}}$$

and then use the table of integrals in the appendix:

(a) $$B = \frac{\mu_0 I N R^2}{2\ell} \left. \frac{-u}{R^2 \sqrt{u^2 + R^2}} \right|_a^{a-\ell} = \boxed{\frac{\mu_0 I N}{2\ell}\left[\frac{a}{\sqrt{a^2 + R^2}} - \frac{a-\ell}{\sqrt{(a-\ell)^2 + R^2}}\right]}$$

(b) If ℓ is much larger than R and $a = 0$,

we have $B \approx \dfrac{\mu_0 I N}{2\ell}\left[0 - \dfrac{-\ell}{\sqrt{\ell^2}}\right] = \dfrac{\mu_0 I N}{2\ell}$.

This is just half the magnitude of the field deep within the solenoid. We would get the same result by substituting $a = \ell$ to describe the other end.

***P22.47** The number of turns is $N = \dfrac{75 \text{ cm}}{0.1 \text{ cm}} = 750$. We assume that the solenoid is long enough to qualify as a long solenoid. Then the field within it (not close to the ends) is $B = \dfrac{N\mu_0 I}{\ell}$ so

$$I = \frac{B\ell}{N\mu_0} = \frac{\left(8\times10^{-3}\text{ T}\right)(0.75\text{ m}\cdot\text{A})}{750\left(4\pi\times10^{-7}\text{ T}\cdot\text{m}\right)} = 6.37\text{ A}.$$

The resistance of the wire is

$$R = \frac{\rho\ell_{\text{wire}}}{A} = \frac{\left(1.7\times10^{-8}\ \Omega\cdot\text{m}\right)2\pi(0.05\text{ m})750}{\pi\left(0.05\times10^{-2}\text{ m}\right)^2} = 5.10\ \Omega.$$

The power delivered is

$$\mathscr{P} = I\Delta V = I^2R = (6.37\text{ A})^2(5.10\ \Omega) = \boxed{207\text{ W}}.$$

The power required would be smaller if wire were wrapped in several layers.

Section 22.11 Magnetism in Matter

P22.48 (a) $I = \dfrac{ev}{2\pi r}$ $\qquad \mu = IA = \left(\dfrac{ev}{2\pi r}\right)\pi r^2 = \boxed{9.27\times10^{-24}\text{ A}\cdot\text{m}^2}$

FIG. P22.48

The Bohr model predicts the correct magnetic moment. However, the "planetary model" is seriously deficient in other regards.

(b) Because the electron is (–), its [conventional] current is clockwise, as seen from above, and μ points $\boxed{\text{downward}}$.

P22.49 (a) Number of unpaired electrons $= \dfrac{8.00\times10^{22}\text{ A}\cdot\text{m}^2}{9.27\times10^{-24}\text{ A}\cdot\text{m}^2} = \boxed{8.63\times10^{45}}$.

Each iron atom has two unpaired electrons, so the number of iron atoms required is $\dfrac{1}{2}\left(8.63\times10^{45}\right)$.

(b) $$\text{Mass} = \frac{\left(4.31\times10^{45}\text{ atoms}\right)\left(7\,900\text{ kg/m}^3\right)}{\left(8.50\times10^{28}\text{ atoms/m}^3\right)} = \boxed{4.01\times10^{20}\text{ kg}}$$

Section 22.12 Context Connection—The Attractive Model for Magnetic Levitation

P22.50 Let N be the number of charges. For the vehicle we want

$$\Sigma F_y = 0: \qquad -mg + Nqv B \sin 90° = 0$$

$$N = \frac{mg}{qvB} = \frac{5\times 10^4 \text{ kg}(9.8 \text{ m/s}^2)}{10^{-6} \text{ C}(400 \text{ km/h})(1\,000/\text{k})(1 \text{ h}/3\,600 \text{ s})(0.1 \text{ N}\cdot\text{s/C}\cdot\text{m})} = \boxed{4\times 10^{10}}$$

P22.51 The energy per distance is the effective force required to propel the vehicle:

$$\frac{W}{\Delta x} = F = \frac{W/t}{\Delta x/t} = \frac{\mathcal{P}}{v}$$

(a) $$\frac{\mathcal{P}}{v} = \frac{10^2(10^3 \text{ J/s})(3\,600 \text{ s/1 h})}{400\times 10^3 \text{ m/h}} = 900 \text{ N} = (900 \text{ J/m})\left(\frac{1\,609 \text{ m}}{1 \text{ mi}}\right) = \boxed{1.4\times 10^6 \text{ J/mi}}$$

(b) We call 20 mi/gal the fuel economy. Then 1 gal/20 mi is the measure of energy use in which we are interested:

$$\frac{W}{\Delta x} = \frac{1 \text{ gal}}{20 \text{ mi}}\left(\frac{3.786 \text{ L}}{1 \text{ gal}}\right)\left(\frac{10^{-3} \text{ m}^3}{1 \text{ L}}\right)\left(\frac{754 \text{ kg}}{\text{m}^3}\right)\left(\frac{40\times 10^6 \text{ J}}{\text{kg}}\right) = \boxed{5.7\times 10^6 \text{ J/mi}}.$$

(c) One automobile passenger uses chemical energy at the rate 5.7×10^6 J/mi. One Transrapid passenger uses electric energy at the rate

$$\frac{1.4\times 10^6 \text{ J/mi}}{100} = 1.4\times 10^4 \text{ J/mi}, \boxed{\text{smaller by 400 times}}.$$

If we suppose that electric energy must be generated by burning a fossil fuel with limited efficiency, then a fair comparison is between the output energies propelling the vehicles, namely $0.20(5.7\times 10^6 \text{ J/mi}) = 1.1\times 10^6$ J/mi for the car and 1.4×10^4 J/mi for the maglev vehicle. The latter is 80 times smaller.

The dependence of American society on gasoline is a dangerous and destructive addiction that we cannot continue in the long run.

Additional Problems

P22.52 (a) The element $d\vec{s}$ is a distance r from P. The direction of the field at P due to this element is out of the page, since $d\vec{s} \times \hat{r}$ is out of the page. In fact, *all* elements give contributions directly out of the page at P. Therefore we have only to determine the magnitude of the field.

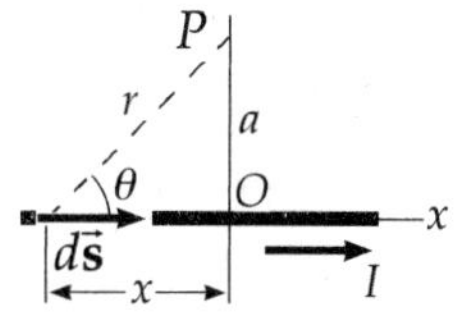

FIG. P22.52(a)

Since $d\vec{s} = \hat{i}dx$ in this case,

we see that $\left|d\vec{s} \times \hat{r}\right| = dx\sin\theta$.

Using this in the Biot-Savart law

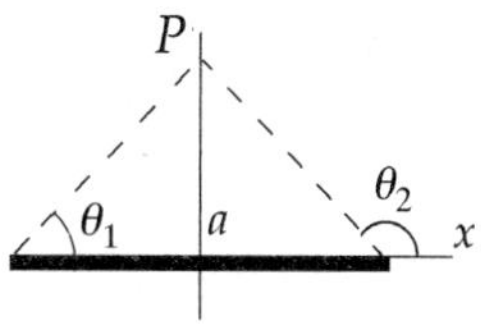

FIG. P22.52(b)

we get
$$dB = \frac{\mu_0 I}{4\pi}\frac{dx\sin\theta}{r^2}. \quad (1)$$

In order to integrate this expression, we must relate the variables θ, x, and r. One approach is to express x and r in terms of θ. From the geometry in Figure P22.52a and some simple differentiation, we obtain the following relationships:

$$r = \frac{a}{\sin\theta} = a\csc\theta. \quad (2)$$

Since $\tan\theta = -\frac{a}{x}$ from the right triangle in Figure 22.52a, we have $x = -a\cot\theta$, so

$$dx = a\csc^2\theta\,d\theta. \quad (3)$$

Substitution of (2) and (3) into (1) gives

$$dB = \frac{\mu_0 I}{4\pi a}\sin\theta\,d\theta. \quad (4)$$

Thus, we have reduced the expression to one involving only the variable θ. We can now obtain the total field at P by integrating (4) over all elements that subtend angles ranging from θ_1 to θ_2, as defined in Figure 22.52. This gives

$$B = \frac{\mu_0 I}{4\pi a}\int_{\theta_1}^{\theta_2}\sin\theta\,d\theta = \frac{\mu_0 I}{4\pi a}(\cos\theta_1 - \cos\theta_2).$$

We can apply this result to find the magnetic field of any straight wire—if we know the geometry and the hence the angles θ_1 and θ_2.

(b) Consider the point P a distance a from the wire. Then for an infinite wire,

$$\theta_1 \to 0$$

and $\theta_2 \to -\pi$ radians

and $B = \frac{\mu_0 I}{2\pi a}$.

P22.53 $J_s = \dfrac{I}{\ell}$

From Ampère's Law, $\oint \vec{\mathbf{B}} \cdot d\vec{\mathbf{s}} = \mu_0 I$

$$B \cdot 2\ell = \mu_0 J_s \ell \quad \therefore \quad \boxed{B = \frac{\mu_0 J_s}{2}}$$

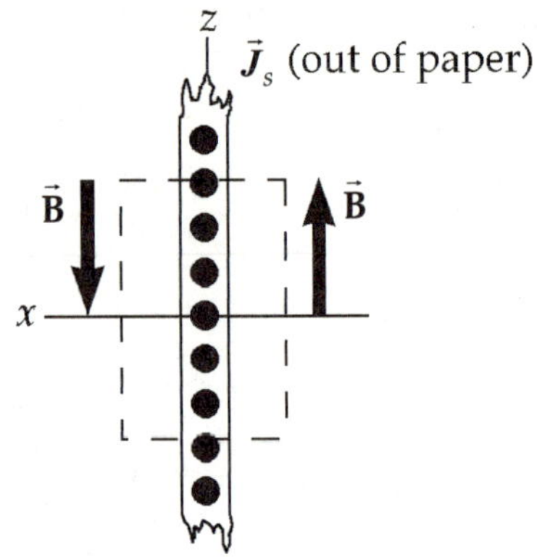

FIG. P22.53

P22.54 (a) The boundary between a region of strong magnetic field and a region of zero field cannot be perfectly sharp, but we ignore the thickness of the transition zone. In the field the electron moves on an arc of a circle:

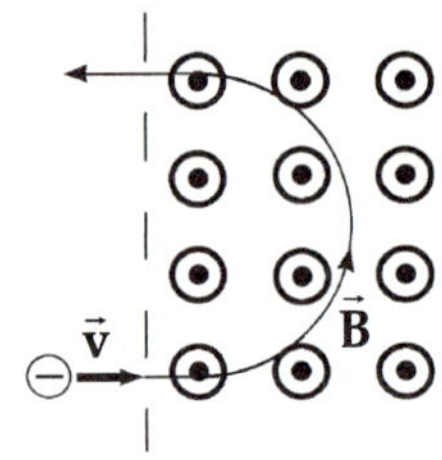

FIG. P22.54(a)

$\sum F = ma$:

$$|q|vB \sin 90° = \frac{mv^2}{r}$$

$$\frac{v}{r} = \omega = \frac{|q|B}{m} = \frac{\left(1.60 \times 10^{-19}\ \text{C}\right)\left(10^{-3}\ \text{N} \cdot \text{s/C} \cdot \text{m}\right)}{\left(9.11 \times 10^{-31}\ \text{kg}\right)} = 1.76 \times 10^{8}\ \text{rad/s}$$

The time for one half revolution is,

from $\Delta\theta = \omega \Delta t$

$$\Delta t = \frac{\Delta\theta}{\omega} = \frac{\pi\,\text{rad}}{1.76 \times 10^{8}\ \text{rad/s}} = \boxed{1.79 \times 10^{-8}\ \text{s}}.$$

(b) The maximum depth of penetration is the radius of the path.

Then $v = \omega r = \left(1.76 \times 10^{8}\ \text{s}^{-1}\right)(0.02\ \text{m}) = 3.51 \times 10^{6}\ \text{m/s}$

and

$$K = \frac{1}{2}mv^2 = \frac{1}{2}\left(9.11 \times 10^{-31}\ \text{kg}\right)\left(3.51 \times 10^{6}\ \text{m/s}\right)^2 = 5.62 \times 10^{-18}\ \text{J} = \frac{5.62 \times 10^{-18}\ \text{J} \cdot \text{e}}{1.60 \times 10^{-19}\ \text{C}}$$

$$= \boxed{35.1\ \text{eV}}.$$

P22.55 (a) Define vector $\vec{\mathbf{h}}$ to have the downward direction of the current, and vector $\vec{\mathbf{L}}$ to be along the pipe into the page as shown. The electric current experiences a magnetic force.

$I(\vec{\mathbf{h}} \times \vec{\mathbf{B}})$ in the direction of $\vec{\mathbf{L}}$.

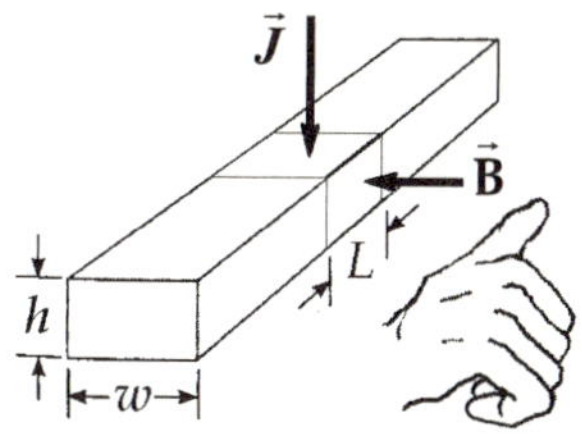

FIG. P22.55

(b) The sodium, consisting of ions and electrons, flows along the pipe transporting no net charge. But inside the section of length L, electrons drift upward to constitute downward electric current $J \times (\text{area}) = JLw$.

The current then feels a magnetic force $I|\vec{\mathbf{h}} \times \vec{\mathbf{B}}| = JLwhB \sin 90°$.

This force along the pipe axis will make the fluid move, exerting pressure

$$\frac{F}{\text{area}} = \frac{JLwhB}{hw} = \boxed{JLB}.$$

(c) Charge moves within the fluid inside the length L, but charge does not accumulate: the fluid is not charged after it leaves the pump. It is not current-carrying and it is not magnetized.

P22.56 $\sum F_y = 0$: $\quad +n - mg = 0$

$\sum F_x = 0$: $\quad -\mu_k n + IBd \sin 90.0° = 0$

$$B = \frac{\mu_k mg}{Id} = \frac{0.100(0.200 \text{ kg})(9.80 \text{ m/s}^2)}{(10.0 \text{ A})(0.500 \text{ m})} = \boxed{39.2 \text{ mT}}$$

P22.57 (a) The net force is the Lorentz force given by

$$\vec{\mathbf{F}} = q\vec{\mathbf{E}} + q\vec{\mathbf{v}} \times \vec{\mathbf{B}} = q(\vec{\mathbf{E}} + \vec{\mathbf{v}} \times \vec{\mathbf{B}})$$

$$\vec{\mathbf{F}} = (3.20 \times 10^{-19})\left[(4\hat{\mathbf{i}} - 1\hat{\mathbf{j}} - 2\hat{\mathbf{k}}) + (2\hat{\mathbf{i}} + 3\hat{\mathbf{j}} - 1\hat{\mathbf{k}}) \times (2\hat{\mathbf{i}} + 4\hat{\mathbf{j}} + 1\hat{\mathbf{k}})\right] \text{N}$$

Carrying out the indicated operations, we find:

$$\vec{\mathbf{F}} = \boxed{(3.52\hat{\mathbf{i}} - 1.60\hat{\mathbf{j}}) \times 10^{-18} \text{ N}}.$$

(b) $$\theta = \cos^{-1}\left(\frac{F_x}{F}\right) = \cos^{-1}\left(\frac{3.52}{\sqrt{(3.52)^2 + (1.60)^2}}\right) = \boxed{24.4°}$$

P22.58 The magnetic force on each proton, $\vec{\mathbf{F}}_B = q\vec{\mathbf{v}} \times \vec{\mathbf{B}} = qvB \sin 90°$ downward perpendicular to velocity, causes centripetal acceleration, guiding it into a circular path of radius r, with

$$qvB = \frac{mv^2}{r}$$

and $$r = \frac{mv}{qB}.$$

We compute this radius by first finding the proton's speed:

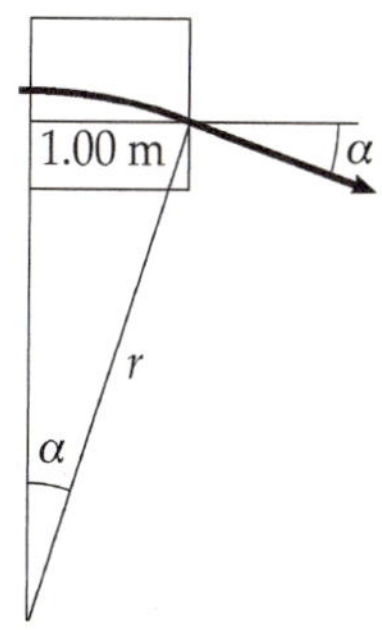

FIG. P22.58

continued on next page

$$K = \frac{1}{2}mv^2$$

$$v = \sqrt{\frac{2K}{m}} = \sqrt{\frac{2(5.00\times10^6\text{ eV})(1.60\times10^{-19}\text{ J/eV})}{1.67\times10^{-27}\text{ kg}}} = 3.10\times10^7\text{ m/s}.$$

$$\text{Now, } r = \frac{mv}{qB} = \frac{(1.67\times10^{-27}\text{ kg})(3.10\times10^7\text{ m/s})}{(1.60\times10^{-19}\text{ C})(0.050\,0\text{ N}\cdot\text{s/C}\cdot\text{m})} = 6.46\text{ m}.$$

(b) From the figure, observe that

$$\sin\alpha = \frac{1.00\text{ m}}{r} = \frac{1\text{ m}}{6.46\text{ m}}$$

$$\boxed{\alpha = 8.90°}$$

(a) The magnitude of the proton momentum stays constant, and its final y component is

$$-(1.67\times10^{-27}\text{ kg})(3.10\times10^7\text{ m/s})\sin 8.90° = \boxed{-8.00\times10^{-21}\text{ kg}\cdot\text{m/s}}.$$

P22.59 Suppose the input power is

$120\text{ W} = (120\text{ V})I$: $\boxed{I \sim 1\text{ A} = 10^0\text{ A}}$.

Suppose $\omega = 2\,000\text{ rev/min}\left(\frac{1\text{ min}}{60\text{ s}}\right)\left(\frac{2\pi\text{ rad}}{1\text{ rev}}\right) \sim 200\text{ rad/s}$

and the output power is $20\text{ W} = \tau\omega = \tau(200\text{ rad/s})$ $\boxed{\tau \sim 10^{-1}\text{ N}\cdot\text{m}}$.

Suppose the area is about $(3\text{ cm})\times(4\text{ cm})$, or $\boxed{A \sim 10^{-3}\text{ m}^2}$.

Suppose that the field is $\boxed{B \sim 10^{-1}\text{ T}}$.

Then, the number of turns in the coil may be found from $\tau \approx NIAB$:

$$0.1\text{ N}\cdot\text{m} \sim N(1\text{ C/s})(10^{-3}\text{ m}^2)(10^{-1}\text{ N}\cdot\text{s/C}\cdot\text{m})$$

giving $\boxed{N \sim 10^3}$.

P22.60 $\sum F = ma$ or $qvB\sin 90.0° = \frac{mv^2}{r}$

$\therefore$ the angular frequency for each ion is $\frac{v}{r} = \omega = \frac{qB}{m}$ and

$$\Delta\omega = \omega_{12} - \omega_{14} = qB\left(\frac{1}{m_{12}} - \frac{1}{m_{14}}\right) = \frac{(1.60\times10^{-19}\text{ C})(2.40\text{ T})}{(1.66\times10^{-27}\text{ kg/u})}\left(\frac{1}{12.0\text{ u}} - \frac{1}{14.0\text{ u}}\right)$$

$$\Delta\omega = \omega_{12} - \omega_{14} = 2.75\times10^6\text{ s}^{-1} = \boxed{2.75\text{ Mrad/s}}$$

P22.61 Let v_x and $v_\perp$ be the components of the velocity of the positron parallel to and perpendicular to the direction of the magnetic field.

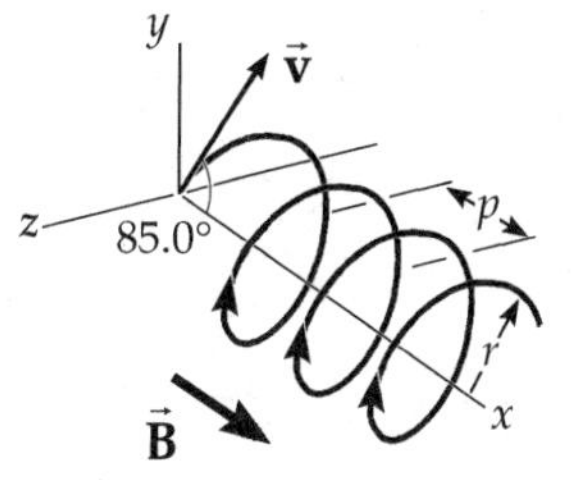

FIG. P22.61

(a) The pitch of trajectory is the distance moved along x by the positron during each period, T

$$p = v_x T = (v\cos 85.0°)\left(\frac{2\pi m}{Bq}\right)$$

$$p = \frac{(5.00\times10^6)(\cos 85.0°)(2\pi)(9.11\times10^{-31})}{0.150(1.60\times10^{-19})} = \boxed{1.04\times10^{-4}\ \text{m}}$$

(b) From Equation 22.3, $r = \frac{mv_\perp}{Bq} = \frac{mv\sin 85.0°}{Bq}$

$$r = \frac{(9.11\times10^{-31})(5.00\times10^6)(\sin 85.0°)}{(0.150)(1.60\times10^{-19})} = \boxed{1.89\times10^{-4}\ \text{m}}$$

P22.62 (a) The magnetic force acting on ions in the blood stream will deflect positive charges toward point A and negative charges toward point B. This separation of charges produces an electric field directed from A toward B. At equilibrium, the electric force caused by this field must balance the magnetic force, so

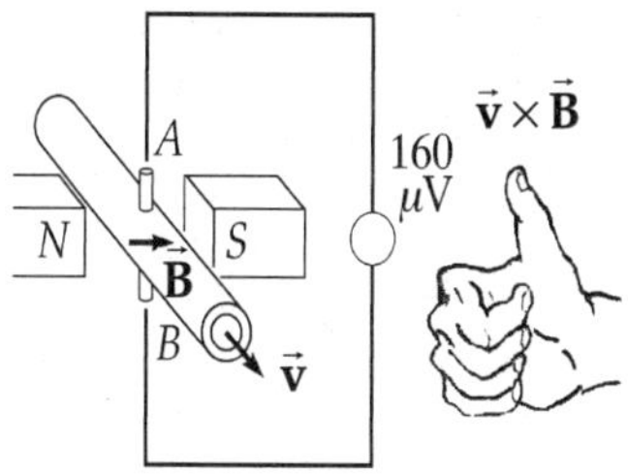

FIG. P22.62

$$qvB = qE = q\left(\frac{\Delta V}{d}\right)$$

or $$v = \frac{\Delta V}{Bd} = \frac{(160\times10^{-6}\ \text{V})}{(0.040\,0\ \text{T})(3.00\times10^{-3}\ \text{m})} = \boxed{1.33\ \text{m/s}}.$$

(b) $\boxed{\text{No}}$. Negative ions moving in the direction of v would be deflected toward point B, giving A a higher potential than B. Positive ions moving in the direction of v would be deflected toward A, again giving A a higher potential than B. Therefore, the sign of the potential difference does not depend on whether the ions in the blood are positively or negatively charged.

P22.63 Consider a longitudinal filament of the strip of width dr as shown in the sketch. The contribution to the field at point P due to the current dI in the element dr is

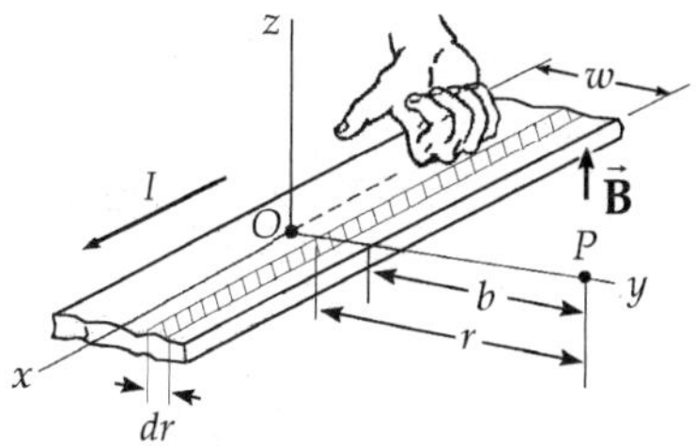

FIG. P22.63

$$dB = \frac{\mu_0 dI}{2\pi r}$$

where $$dI = I\left(\frac{dr}{w}\right)$$

$$\vec{\mathbf{B}} = \int d\vec{\mathbf{B}} = \int_b^{b+w} \frac{\mu_0 I dr}{2\pi w r}\hat{\mathbf{k}} = \boxed{\frac{\mu_0 I}{2\pi w}\ln\left(1+\frac{w}{b}\right)\hat{\mathbf{k}}}.$$

P22.64 $B = \frac{\mu_0 I R^2}{2(R^2 + R^2)^{3/2}} = \frac{\mu_0 I}{2^{5/2} R}$ $\quad I = \frac{2^{5/2} BR}{\mu_0} = \frac{2^{5/2}(7.00 \times 10^{-5}\ \text{T})(6.37 \times 10^6\ \text{m})}{(4\pi \times 10^{-7}\ \text{T} \cdot \text{m/A})}$

so $\boxed{I = 2.01 \times 10^9\ \text{A}}$ toward the west

P22.65 On the axis of a current loop, the magnetic field is given by $B = \frac{\mu_0 I R^2}{2(x^2 + R^2)^{3/2}}$

where in this case $I = \frac{q}{(2\pi/\omega)}$. Therefore, $B = \frac{\mu_0 \omega R^2 q}{4\pi(x^2 + R^2)^{3/2}}$

when $x = \frac{R}{2}$. then $B = \frac{\mu_0 \omega R^2 q}{4\pi\left(\frac{5}{4}R^2\right)^{3/2}} = \boxed{\frac{\mu_0 q \omega}{2.5\sqrt{5}\pi R}}$.

P22.66 (a) Use Equation 22.33 twice: $B_x = \frac{\mu_0 I R^2}{2(x^2 + R^2)^{3/2}}$

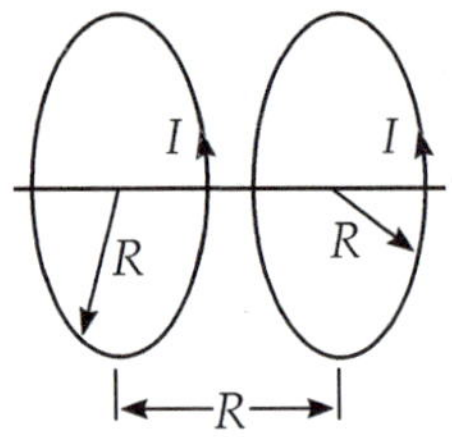

FIG. P22.66

If each coil has N turns, the field is just N times larger.

$$B = B_{x1} + B_{x2} = \frac{N\mu_0 I R^2}{2}\left[\frac{1}{(x^2 + R^2)^{3/2}} + \frac{1}{\left[(R - x)^2 + R^2\right]^{3/2}}\right]$$

$$\boxed{B = \frac{N\mu_0 I R^2}{2}\left[\frac{1}{(x^2 + R^2)^{3/2}} + \frac{1}{(2R^2 + x^2 - 2xR)^{3/2}}\right]}$$

(b) $$\frac{dB}{dx} = \frac{N\mu_0 I R^2}{2}\left[-\frac{3}{2}(2x)(x^2 + R^2)^{-5/2} - \frac{3}{2}(2R^2 + x^2 - 2xR)^{-5/2}(2x - 2R)\right]$$

Substituting $x = \frac{R}{2}$ and canceling terms, $\boxed{\frac{dB}{dx} = 0}$.

$$\frac{d^2B}{dx^2} = \frac{-3N\mu_0 I R^2}{2}\left[(x^2 + R^2)^{-5/2} - 5x^2(x^2 + R^2)^{-7/2} + (2R^2 + x^2 - 2xR)^{-5/2} - 5(x - R)^2(2R^2 + x^2 - 2xR)^{-7/2}\right]$$

Again substituting $x = \frac{R}{2}$ and canceling terms, $\boxed{\frac{d^2B}{dx^2} = 0}$.

P22.67 Model the two wires as straight parallel wires (!)

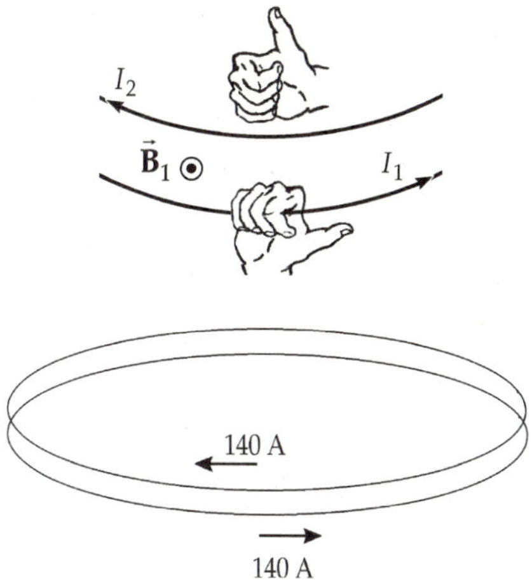

FIG. P22.67

(a) $$F_B = \frac{\mu_0 I^2 \ell}{2\pi a} \text{ (Equation 22.27)}$$

$$F_B = \frac{\left(4\pi\times10^{-7}\right)(140)^2(2\pi)(0.100)}{2\pi\left(1.00\times10^{-3}\right)}$$

$$= \boxed{2.46\text{ N}} \text{ upward}$$

(b) $$a_{\text{loop}} = \frac{2.46\text{ N} - m_{\text{loop}}g}{m_{\text{loop}}} = \boxed{107\text{ m/s}^2} \text{ upward}$$

P22.68 (a) $$B = \frac{\mu_0 I}{2\pi r} = \frac{\left(4\pi\times10^{-7}\text{ T}\cdot\text{m/A}\right)(24.0\text{ A})}{2\pi(0.017\,5\text{ m})} = \boxed{2.74\times10^{-4}\text{ T}}$$

(b) At point C, conductor AB produces a field $\frac{1}{2}\left(2.74\times10^{-4}\text{ T}\right)\left(-\hat{\mathbf{j}}\right)$,  conductor DE produces a field of $\frac{1}{2}\left(2.74\times10^{-4}\text{ T}\right)\left(-\hat{\mathbf{j}}\right)$,  BD produces no field, and AE produces negligible field. The total field at C is $\boxed{2.74\times10^{-4}\text{ T}\left(-\hat{\mathbf{j}}\right)}$.

(c) $$\vec{\mathbf{F}}_B = I\vec{\ell}\times\vec{\mathbf{B}} = (24.0\text{ A})\left(0.035\,0\text{ m}\hat{\mathbf{k}}\right)\times\left[5\left(2.74\times10^{-4}\text{ T}\right)\left(-\hat{\mathbf{j}}\right)\right] = \boxed{\left(1.15\times10^{-3}\text{ N}\right)\hat{\mathbf{i}}}$$ 

(d) $$\vec{\mathbf{a}} = \frac{\sum\vec{\mathbf{F}}}{m} = \frac{\left(1.15\times10^{-3}\text{ N}\right)\hat{\mathbf{i}}}{3.0\times10^{-3}\text{ kg}} = \boxed{\left(0.384\text{ m/s}^2\right)\hat{\mathbf{i}}}$$

(e) The bar is already so far from AE that it moves through nearly constant magnetic field. The force acting on the bar is constant, and therefore the bar's $\boxed{\text{acceleration is constant}}$.

(f) $$v_f^2 = v_i^2 + 2ax = 0 + 2\left(0.384\text{ m/s}^2\right)(1.30\text{ m}),\text{ so } \vec{\mathbf{v}}_f = \boxed{(0.999\text{ m/s})\hat{\mathbf{i}}}$$

ANSWERS TO EVEN PROBLEMS

P22.2 (a) west; (b) no deflection; (c) up; (d) down

P22.4 (a) 7.90 pN; (b) 0

P22.6 2.34 aN

P22.8 (a) 5.00 cm; (b) 8.78×10^6 m/s

P22.10 244 kV/m

P22.12 (a) 7.66×10^7 rad/s; (b) 26.8 Mm/s; (c) 3.76 MeV; (d) 3.13×10^3 rev; (e) 257 μs

P22.14 (a) see the solution; (b) $n = \dfrac{IB}{qt\Delta V_H}$

P22.16 (a) 4.73 N; (b) 5.46 N; (c) 4.73 N

P22.18 *ab*: 0, *bc*: $-40.0\text{ mN}\hat{\mathbf{i}}$, *cd*: $-40.0\text{ mN}\hat{\mathbf{k}}$, *da*: $(40.0\hat{\mathbf{i}}+40.0\hat{\mathbf{k}})\text{ mN}$

P22.20 see the solution

P22.22 (a) 640 μN·m; (b) 241 mW; (c) 2.56 mJ; (d) 154 mW

P22.24 20.0 μT

P22.26 200 nT

P22.28 see the solution

P22.30 (a) along the line $(y=-0.420\text{ m},\ z=0)$; (b) $-34.7\hat{\mathbf{j}}$ mN; (c) $-17.3\hat{\mathbf{j}}$ kN/C

P22.32 (a) 53.3 μT toward the bottom of the page; (b) 20.0 μT toward the bottom of the page; (c) zero

P22.34 (a) 10.0 μT out of the page; (b) 80.0 μN toward the first wire; (c) 16.0 μT into the page; (d) 80.0 μN toward the second wire

P22.36 (a) 12.0 cm to the left of W_1; (b) 2.40 A, downward

P22.38 5.40 cm

P22.40 (a) 400 cm; (b) 7.50 nT; (c) 1.26 m; (d) zero

P22.42 (a) 0; (b) $\dfrac{\mu_0 I}{2\pi R}$ tangent to the wall in a counterclockwise sense; (c) $\dfrac{\mu_0 I^2}{(2\pi R)^2}$ inward

P22.44 226 μN away from the center of the loop, 0

P22.46 (a) $\dfrac{\mu_0 IN}{2\ell}\left(\dfrac{a}{\sqrt{a^2+R^2}}-\dfrac{a-\ell}{\sqrt{(a-\ell)^2+R^2}}\right)$; (b) see the solution

P22.48 (a) $9.27\times10^{-24}\text{ A}\cdot\text{m}^2$; (b) down

P22.50 4×10^{10}

P22.52 see the solution

P22.54 (a) 17.9 ns; (b) 35.1 eV

P22.56 39.2 mT

P22.58 (a) $-8.00\times10^{-21}\text{ kg}\cdot\text{m/s}$; (b) 8.90°

P22.60 2.75 Mrad/s

P22.62 (a) 1.33 m/s; (b) Positive ions moving toward you in magnetic field to the right feel upward magnetic force, and migrate upward in the blood vessel. Negative ions moving toward you feel downward magnetic force and accumulate at the bottom of this section of vessel. Thus both species can participate in the generation of the emf.

P22.64 2.01 GA toward the west

P22.66 see the solution

P22.68 (a) 274 μT; (b) $-274\hat{\mathbf{j}}\ \mu$T; (c) $1.15\hat{\mathbf{i}}$ mN; (d) $0.384\hat{\mathbf{i}}\text{ m/s}^2$; (e) acceleration is constant; (f) $0.999\hat{\mathbf{i}}$ m/s

Chapter 23

Faraday's Law and Inductance

CHAPTER OUTLINE

ANSWERS TO QUESTIONS

Q23.1 The magnetic flux is $\Phi_B = BA\cos\theta$. Therefore the flux is maximum when $\vec{\mathbf{B}}$ is perpendicular to the loop of wire and zero when there is no component of magnetic field perpendicular to the loop. The flux is zero when the loop is turned so that the field lies in the plane of its area.

Q23.2 (a) The south pole of the magnet produces an upward magnetic field that increases as the magnet approaches. The loop opposes change by making its own downward magnetic field; it carries current clockwise, which goes to the left through the resistor.

(b) The north pole of the magnet produces an upward magnetic field. The loop sees decreasing upward flux as the magnet falls away, and tries to make an upward magnetic field of its own by carrying current counterclockwise, to the right in the resistor.

Q23.3 The force on positive charges in the bar is $\vec{\mathbf{F}} = q\left(\vec{\mathbf{v}} \times \vec{\mathbf{B}}\right)$. If the bar is moving to the left, positive charge will move downward and accumulate at the bottom end of the bar, so that an electric field will be established upward.

Q23.4 No. The magnetic force acts within the bar, but has no influence on the forward motion of the bar.

Q23.5 By the magnetic force law $\vec{\mathbf{F}} = q\left(\vec{\mathbf{v}} \times \vec{\mathbf{B}}\right)$: the positive charges in the moving bar will flow downward and therefore clockwise in the circuit. If the bar is moving to the left, the positive charge in the bar will flow upward and therefore counterclockwise in the circuit.

Q23.6 We ignore mechanical friction between the bar and the rails. Moving the conducting bar through the magnetic field will force charges to move around the circuit to constitute clockwise current. The downward current in the bar feels a magnetic force to the left. Then a counterbalancing applied force to the right is required to maintain the motion.

Q23.7 As water falls, it gains speed and kinetic energy. It then pushes against turbine blades, transferring its energy to the rotor coils of a large AC generator. The rotor of the generator turns within a strong magnetic field. Because the rotor is spinning, the magnetic flux through its turns changes in time as $\Phi_B = BA\cos\omega t$. Generated in the rotor is an induced emf of $\mathcal{E} = \frac{-Nd\Phi_B}{dt}$. This induced emf is the voltage driving the current in our electric power lines.

Q23.8 Yes. The induced eddy currents on the surface of the aluminum will slow the descent of the aluminum. It may fall very slowly.

Q23.9 The increasing counterclockwise current in the solenoid coil produces an upward magnetic field that increases rapidly. The increasing upward flux of this field through the ring induces an emf to produce clockwise current in the ring. The magnetic field of the solenoid has a radially outward component at each point on the ring. This field component exerts upward force on the current in the ring there. The whole ring feels a total upward force larger than its weight.

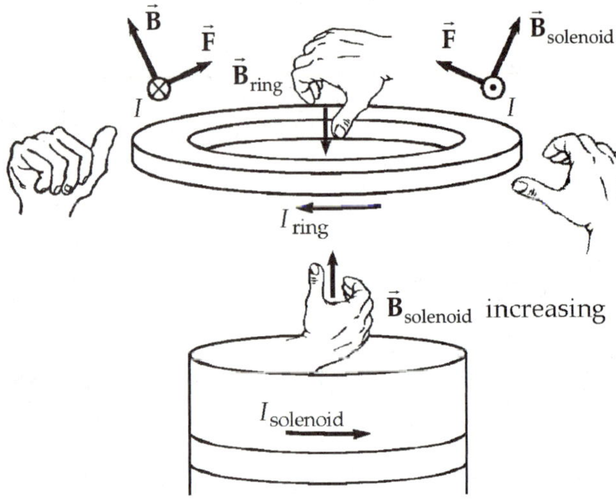

FIG. Q23.9

Q23.10 Oscillating current in the solenoid produces an always-changing magnetic field. Vertical flux through the ring, alternately increasing and decreasing, produces current in it with a direction that is alternately clockwise and counterclockwise. The current through the ring's resistance produces internal energy at the rate I^2R.

Q23.11 (a) The battery makes counterclockwise current I_1 in the primary coil, so its magnetic field $\vec{B}_1$ is to the right and increasing just after the switch is closed. The secondary coil will oppose the change with a leftward field $\vec{B}_2$, which comes from an induced clockwise current I_2 that goes to the right in the resistor.

(b) At steady state the primary magnetic field is unchanging, so no emf is induced in the secondary.

(c) The primary's field is to the right and decreasing as the switch is opened. The secondary coil opposes this decrease by making its own field to the right, carrying counterclockwise current to the left in the resistor.

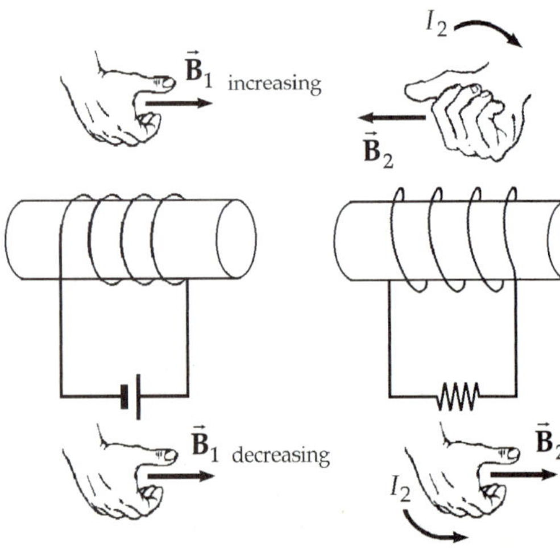

FIG. Q23.11

Q23.12 The motional emf between the wingtips cannot be used to run a light bulb. To connect the light, an extra insulated wire would have to be run out along each wing, making contact with the wing tip. The wings with the extra wires and the bulb constitute a single-loop circuit. As the plane flies through a uniform magnetic field, the magnetic flux through this loop is constant and zero emf is generated. On the other hand, if the magnetic field is not uniform, a large loop towed through it will generate pulses of positive and negative emf. This phenomenon has been demonstrated with a cable unreeled from the Space Shuttle.

Q23.13 As Section 23.4 describes, the electric field lines form closed loops for an electric field created by a changing magnetic field. If a positively charged object is carried around one of these loops in the direction of $\vec{\mathbf{E}}$, the electric field does positive work on it every step of the way. The work adds up to a nonzero positive total. This fact proves that the force is nonconservative. The work in fact is $W = \oint \vec{\mathbf{F}} \cdot d\vec{\mathbf{s}} = \oint q\vec{\mathbf{E}} \cdot d\vec{\mathbf{s}} = -q\frac{d}{dt}\int \vec{\mathbf{B}} \cdot d\vec{\mathbf{A}}$. If going around once extracts energy from the changing magnetic field, going around several lines is better. This is why generators and transformers have coils with many turns.

Q23.14 The inductance of a coil is determined by (a) the geometry of the coil and (b) the "contents" of the coil. This is similar to the parameters that determine the capacitance of a capacitor and the resistance of a resistor. With an inductor, the most important factor in the geometry is the number of turns of wire, or turns per unit length. By the "contents" we refer to the material in which the inductor establishes a magnetic field, notably the magnetic properties of the core around which the wire is wrapped.

Q23.15 The current decreases not instantaneously but over some span of time. The faster the decrease in the current, the larger will be the emf generated in the inductor. A spark can appear at the switch as it is opened because the self-induced voltage is a maximum at this instant. The voltage can therefore briefly cause dielectric breakdown of the air between the contacts.

Q23.16 A physicist's list of constituents of the universe in 1829 might include matter, light, heat, the stuff of stars, charge, momentum, and several other entries. Our list today might include the quarks, electrons, muons, tauons, and neutrinos of matter; gravitons of gravitational fields; photons of electric and magnetic fields; W and Z particles; gluons; energy; momentum; angular momentum; charge; baryon number; three different lepton numbers; upness; downness; strangeness; charm; topness; and bottomness. Alternatively, the relativistic interconvertability of mass and energy, and of electric and magnetic fields, can be used to make the list look shorter. Some might think of the conserved quantities energy, momentum, … bottomness as properties of matter, rather than as things with their own existence. The idea of a field is not due to Henry, but rather to Faraday, to whom Henry personally demonstrated self-induction. Still the thesis stated in the question has an important germ of truth. Henry precipitated a basic change if he did not cause it. The biggest difference between the two lists is that the 1829 list does not include fields and today's list does.

Q23.17 The energy stored in the magnetic field of an inductor is proportional to the square of the current. Doubling I makes $U = \frac{1}{2}LI^2$ get four times larger.

Q23.18 The energy stored in a capacitor is proportional to the square of the electric field, and the energy stored in an induction coil is proportional to the square of the magnetic field. The capacitor's energy is proportional to its capacitance, which depends on its geometry and the dielectric material inside. The coil's energy is proportional to its inductance, which depends on its geometry and the core material. On the other hand, we can think of Henry's discovery of self-inductance as fundamentally new. Before a certain school vacation at the Albany Academy about 1830, one could visualize the universe as consisting of only one thing, matter. All the forms of energy then known (kinetic, gravitational, elastic, internal, electrical) belonged to chunks of matter. But the energy that temporarily maintains a current in a coil after the battery is removed is not energy that belongs to any bit of matter. This energy is vastly larger than the kinetic energy of the drifting electrons in the wires. This energy belongs to the magnetic field around the coil. Beginning in 1830, Nature has forced us to admit that the universe consists of matter and also of fields, massless and invisible, known only by their effects.

Q23.19 The inductance of the series combination of inductor L_1 and inductor L_2 is $L_1 + L_2 + M_{12}$, where M_{12} is the mutual inductance of the two coils. It can be defined as the emf induced in coil two when the current in coil one changes at one ampere per second, due to the magnetic field of coil one producing flux through coil two. The coils can be arranged to have large mutual inductance, as by winding them onto the same core. The coils can be arranged to have negligible mutual inductance, as separate toroids do.

Q23.20 An object cannot exert a net force on itself. An object cannot create momentum out of nothing. A coil can induce an emf in itself. When it does so, the actual forces acting on charges in different parts of the loop add as vectors to zero. The term electromotive force does not refer to a force, but to a voltage.

SOLUTIONS TO PROBLEMS

Section 23.1 Faraday's Law of Induction

P23.1 $$|\mathcal{E}| = \left|\frac{\Delta\Phi_B}{\Delta t}\right| = \frac{\Delta(\vec{\mathbf{B}}\cdot\vec{\mathbf{A}})}{\Delta t} = \frac{(2.50\text{ T} - 0.500\text{ T})(8.00\times10^{-4}\text{ m}^2)}{1.00\text{ s}}\left(\frac{1\text{ N}\cdot\text{s}}{1\text{ T}\cdot\text{C}\cdot\text{m}}\right)\left(\frac{1\text{ V}\cdot\text{C}}{1\text{ N}\cdot\text{m}}\right)$$

$$|\mathcal{E}| = 1.60\text{ mV and } I_{\text{loop}} = \frac{\mathcal{E}}{R} = \frac{1.60\text{ mV}}{2.00\ \Omega} = \boxed{0.800\text{ mA}}$$

P23.2 $$\mathcal{E} = -N\frac{\Delta BA\cos\theta}{\Delta t} = -NB\pi r^2\left(\frac{\cos\theta_f - \cos\theta_i}{\Delta t}\right) = -25.0\left(50.0\times10^{-6}\text{ T}\right)\left[\pi(0.500\text{ m})^2\right]\left(\frac{\cos 180° - \cos 0°}{0.200\text{ s}}\right)$$

$$\mathcal{E} = \boxed{+9.82\text{ mV}}$$

P23.3 Noting unit conversions from $\vec{\mathbf{F}} = q\vec{\mathbf{v}}\times\vec{\mathbf{B}}$ and $U = qV$, the induced voltage is

$$\mathcal{E} = -N\frac{d(\vec{\mathbf{B}}\cdot\vec{\mathbf{A}})}{dt} = -N\left(\frac{0 - B_iA\cos\theta}{\Delta t}\right) = \frac{+200(1.60\text{ T})(0.200\text{ m}^2)\cos 0°}{20.0\times10^{-3}\text{ s}}\left(\frac{1\text{ N}\cdot\text{s}}{1\text{ T}\cdot\text{C}\cdot\text{m}}\right)\left(\frac{1\text{ V}\cdot\text{C}}{\text{N}\cdot\text{m}}\right) = 3\,200\text{ V}$$

$$I = \frac{\mathcal{E}}{R} = \frac{3\,200\text{ V}}{20.0\ \Omega} = \boxed{160\text{ A}}$$

P23.4 $$|\mathcal{E}| = \frac{d(BA)}{dt} = 0.500\mu_0 nA\frac{dI}{dt} = 0.500\mu_0 n\pi r_2^2\frac{\Delta I}{\Delta t}$$

(a) $$I_{\text{ring}} = \frac{\mathcal{E}}{R} = \boxed{\frac{\mu_0 n\pi r_2^2}{2R}\frac{\Delta I}{\Delta t}}$$

(b) $$B = \frac{\mu_0 I}{2r_1} = \boxed{\frac{\mu_0^2 n\pi r_2^2}{4r_1 R}\frac{\Delta I}{\Delta t}}$$

(c) The coil's field points downward, and is increasing, so $\boxed{B_{\text{ring}}\text{ points upward}}$.

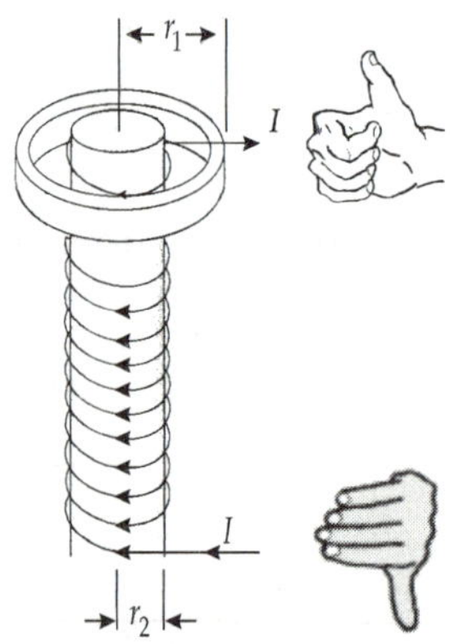

FIG. P23.4

P23.5 (a) $d\Phi_B = \vec{\mathbf{B}} \cdot d\vec{\mathbf{A}} = \frac{\mu_0 I}{2\pi x} L dx$: $\Phi_B = \int_h^{h+w} \frac{\mu_0 IL}{2\pi} \frac{dx}{x} = \boxed{\frac{\mu_0 IL}{2\pi} \ln\left(\frac{h+w}{h}\right)}$

(b) $$\mathcal{E} = -\frac{d\Phi_B}{dt} = -\frac{d}{dt}\left[\frac{\mu_0 IL}{2\pi}\ln\left(\frac{h+w}{h}\right)\right] = -\left[\frac{\mu_0 L}{2\pi}\ln\left(\frac{h+w}{h}\right)\right]\frac{dI}{dt}$$

$$\mathcal{E} = -\frac{\left(4\pi\times10^{-7}\ \text{T}\cdot\text{m/A}\right)(1.00\ \text{m})}{2\pi}\ln\left(\frac{1.00+10.0}{1.00}\right)(10.0\ \text{A/s}) = \boxed{-4.80\ \mu\text{V}}$$

The long wire produces magnetic flux into the page through the rectangle, shown by the first hand in the figure to the right.

As the magnetic flux increases, the rectangle produces its own magnetic field out of the page, which it does by carrying $\boxed{\text{counterclockwise}}$ current (second hand in the figure).

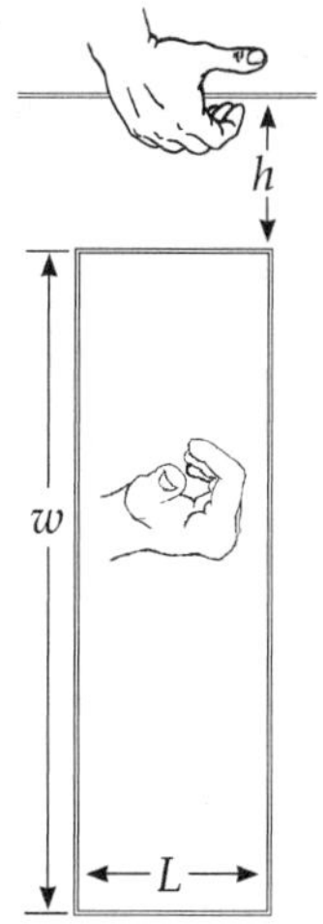

FIG. P23.5

P23.6 $\Phi_B = (\mu_0 n I) A_{\text{solenoid}}$

$$\mathcal{E} = -N\frac{d\Phi_B}{dt} = -N\mu_0 n\left(\pi r_{\text{solenoid}}^2\right)\frac{dI}{dt}$$

$$\mathcal{E} = -15.0\left(4\pi\times10^{-7}\ \text{T}\cdot\text{m/A}\right)\left(1.00\times10^3\ \text{m}^{-1}\right)\pi(0.020\,0\ \text{m})^2(600\ \text{A/s})\cos(120t)$$

$$\boxed{\mathcal{E} = -14.2\cos(120t)\ \text{mV}}$$

P23.7 $|\mathcal{E}| = \left|\frac{\Delta\Phi_B}{\Delta t}\right| = N\left(\frac{dB}{dt}\right)A = N(0.010\,0 + 0.080\,0t)A$

At $t = 5.00$ s, $|\mathcal{E}| = 30.0(0.410\ \text{T/s})\left[\pi(0.040\,0\ \text{m})^2\right] = \boxed{61.8\ \text{mV}}$

***P23.8** (a) Each coil has a pulse of voltage tending to produce counterclockwise current as the projectile approaches, and then a pulse of clockwise voltage as the projectile recedes.

(b) $v = \frac{d}{t} = \frac{1.50\ \text{m}}{2.40\times10^{-3}\ \text{s}} = \boxed{625\ \text{m/s}}$

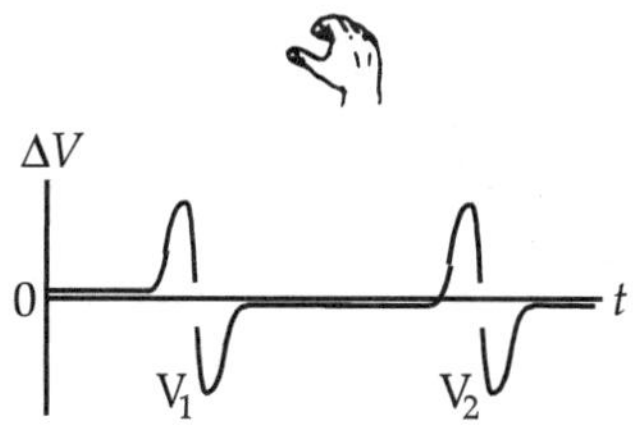

FIG. P23.8

P23.9 (a) Suppose, first, that the central wire is long and straight. The enclosed current of unknown amplitude creates a circular magnetic field around it, with the magnitude of the field given by Ampère's Law.

$$\oint \vec{\mathbf{B}}\cdot d\vec{\mathbf{s}} = \mu_0 I: \quad B = \frac{\mu_0 I_{\max}\sin\omega t}{2\pi R}$$

at the location of the Rogowski coil, which we assume is centered on the wire. This field passes perpendicularly through each turn of the toroid, producing flux

$$\vec{\mathbf{B}}\cdot\vec{\mathbf{A}} = \frac{\mu_0 I_{\max} A\sin\omega t}{2\pi R}.$$

The toroid has $2\pi Rn$ turns. As the magnetic field varies, the emf induced in it is

$$\mathcal{E} = -N\frac{d}{dt}\vec{\mathbf{B}}\cdot\vec{\mathbf{A}} = -2\pi Rn\frac{\mu_0 I_{\max} A}{2\pi R}\frac{d}{dt}\sin\omega t = -\mu_0 I_{\max} nA\omega\cos\omega t.$$

This is an alternating voltage with amplitude $\mathcal{E}_{\max} = \mu_0 nA\omega I_{\max}$. Measuring the amplitude determines the size $I_{\max}$ of the central current. Our assumptions that the central wire is long and straight and passes perpendicularly through the center of the Rogowski coil are all unnecessary.

(b) If the wire is not centered, the coil will respond to stronger magnetic fields on one side, but to correspondingly weaker fields on the opposite side. The emf induced in the coil is proportional to the line integral of the magnetic field around the circular axis of the toroid. Ampère's Law says that this line integral depends only on the amount of current the coil encloses. It does not depend on the shape or location of the current within the coil, or on any currents outside the coil.

***P23.10** The upper loop has area $\pi(0.05\text{ m})^2 = 7.85\times10^{-3}\text{ m}^2$. The induced emf in it is

$$\mathcal{E} = -N\frac{d}{dt}BA\cos\theta = -1A\cos 0°\frac{dB}{dt} = -7.85\times10^{-3}\text{ m}^2(2\text{ T/s}) = -1.57\times10^{-2}\text{ V}.$$

The minus sign indicates that it tends to produce counterclockwise current, to make its own magnetic field out of the page. Similarly, the induced emf in the lower loop is

$$\mathcal{E} = -NA\cos\theta\frac{dB}{dt} = -\pi(0.09\text{ m})^2 2\text{ T/s} = -5.09\times10^{-2}\text{ V} = +5.09\times10^{-2}\text{ V to produce}$$

counterclockwise current in the lower loop, which becomes clockwise current in the upper loop.

The net emf for current in this sense around the figure 8 is

$5.09\times10^{-2}\text{ V} - 1.57\times10^{-2}\text{ V} = 3.52\times10^{-2}\text{ V}$.

It pushes current in this sense through series resistance $[2\pi(0.05\text{ m}) + 2\pi(0.09\text{ m})]3\ \Omega/\text{m} = 2.64\ \Omega$.

The current is $I = \dfrac{\mathcal{E}}{R} = \dfrac{3.52\times10^{-2}\text{ V}}{2.64\ \Omega} = \boxed{13.3\text{ mA}}$.

Section 23.2 Motional emf

Section 23.3 Lenz's Law

P23.11 (a) For maximum induced emf, with positive charge at the top of the antenna,

$\vec{\mathbf{F}}_+ = q_+(\vec{\mathbf{v}} \times \vec{\mathbf{B}})$, so the auto must move $\boxed{\textit{east}}$, since east × north = up.

(b) $$\mathcal{E} = B\ell v = (5.00 \times 10^{-5}\ \text{T})(1.20\ \text{m})\left(\frac{65.0 \times 10^3\ \text{m}}{3\ 600\ \text{s}}\right)\cos 65.0° = \boxed{4.58 \times 10^{-4}\ \text{V}}$$

P23.12 $I = \frac{\mathcal{E}}{R} = \frac{B\ell v}{R}$

$\boxed{v = 1.00\ \text{m/s}}$

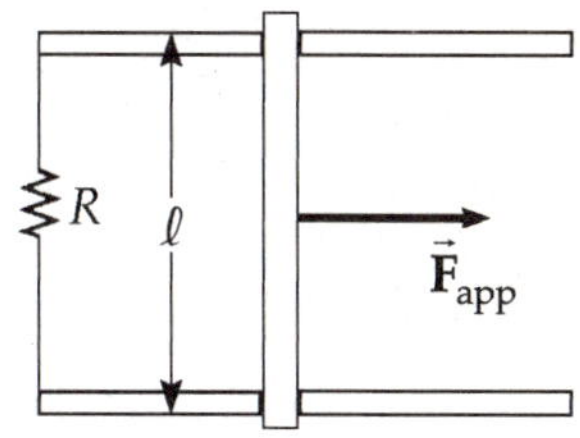

FIG. P23.12

P23.13 (a) $\left|\vec{\mathbf{F}}_B\right| = I\left|\vec{\ell} \times \vec{\mathbf{B}}\right| = I\ell B$

When $I = \frac{\mathcal{E}}{R}$

and $\mathcal{E} = B\ell v$

we get $F_B = \frac{B\ell v}{R}(\ell B) = \frac{B^2\ell^2 v}{R} = \frac{(2.50)^2(1.20)^2(2.00)}{6.00} = 3.00\ \text{N}$.

The applied force is $\boxed{3.00\ \text{N to the right}}$.

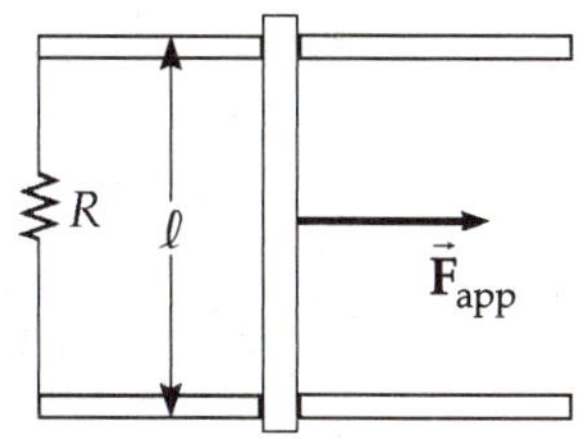

FIG. P23.13

(b) $\mathcal{P} = I^2R = \frac{B^2\ell^2v^2}{R} = 6.00\ \text{W}$ or $\mathcal{P} = Fv = \boxed{6.00\ \text{W}}$

P23.14 $F_B = I\ell B$ and $\mathcal{E} = B\ell v$

$I = \frac{\mathcal{E}}{R} = \frac{B\ell v}{R}$ so $B = \frac{IR}{\ell v}$

(a) $F_B = \frac{I^2\ell R}{\ell v}$ and $I = \sqrt{\frac{F_B v}{R}} = \boxed{0.500\ \text{A}}$

(b) $I^2R = \boxed{2.00\ \text{W}}$

(c) For constant force, $\mathcal{P} = \vec{\mathbf{F}} \cdot \vec{\mathbf{v}} = (1.00\ \text{N})(2.00\ \text{m/s}) = \boxed{2.00\ \text{W}}$.

***P23.15** With v representing the initial speed of the bar, let u represent its speed at any later time. The motional emf induced in the bar is $\mathcal{E} = B\ell u$. The induced current is $I = \frac{\mathcal{E}}{R} = \frac{B\ell u}{R}$. The magnetic force on the bar is backward $F = -I\ell B = -\frac{B^2\ell^2 u}{R} = \frac{m\,du}{dt}$.

Method one: To find u as a function of time, we separate variables thus:

$$-\frac{B^2\ell^2}{Rm}dt = \frac{du}{u}$$

$$\int_0^t -\frac{B^2\ell^2}{Rm}dt = \int_v^u \frac{du}{u}$$

$$-\frac{B^2\ell^2}{Rm}(t-0) = \ln u - \ln v = \ln\frac{u}{v}$$

$$e^{-B^2\ell^2 t/Rm} = \frac{u}{v} \qquad u = ve^{-B^2\ell^2 t/Rm} = \frac{dx}{dt}.$$

The distance traveled is given by

$$\int_0^{x_{\max}} dx = \int_0^{\infty} ve^{-B^2\ell^2 t/Rm}dt = v\left(-\frac{Rm}{B^2\ell^2}\right)\int_0^{\infty} e^{-B^2\ell^2 t/Rm}\left(-\frac{-B^2\ell^2 dt}{Rm}\right)$$

$$x_{\max} - 0 = -\frac{Rmv}{B^2\ell^2}\left[e^{-\infty} - e^{-0}\right] = \boxed{\frac{Rmv}{B^2\ell^2}}.$$

Method two: Newton's second law is

$$-\frac{B^2\ell^2 u}{R} = -\frac{B^2\ell^2}{R}\frac{dx}{dt} = m\frac{du}{dt}$$

$$m\,du = -\frac{B^2\ell^2}{R}dx.$$

Direct integration from the initial to the stopping point gives

$$\int_v^0 m\,du = \int_0^{x_{\max}} -\frac{B^2\ell^2}{R}dx$$

$$m(0-v) = -\frac{B^2\ell^2}{R}(x_{\max} - 0)$$

$$x_{\max} = \frac{mvR}{B^2\ell^2}.$$

***P23.16** Model the magnetic flux inside the metallic tube as constant as it shrinks from radius R to radius r:

$$2.50\text{ T}\left(\pi R^2\right) = B_f \pi r^2$$

$$B_f = 2.50\text{ T}\left(\frac{R}{r}\right)^2 = 2.50\text{ T}(12)^2 = \boxed{360\text{ T}}$$

P23.17 Observe that the homopolar generator has no commutator and produces a voltage constant in time: DC with no ripple. In time dt, the disk turns by angle $d\theta = \omega dt$. The outer brush slides over distance $rd\theta$.

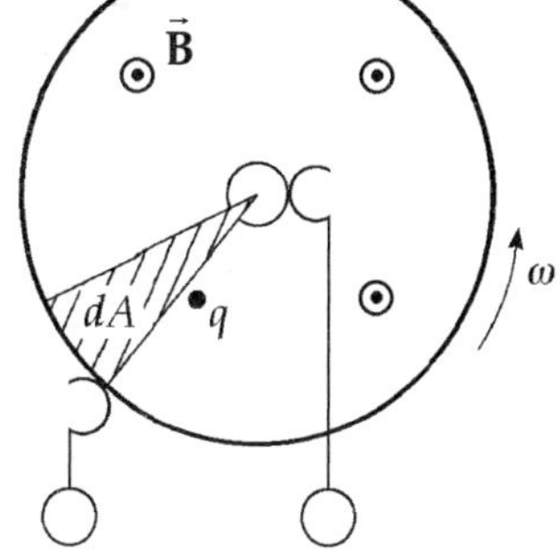

FIG. P23.17

The radial line to the outer brush sweeps over area

$$dA = \frac{1}{2} r r d\theta = \frac{1}{2} r^2 \omega dt.$$

The emf generated is $\qquad \mathcal{E} = -N \frac{d}{dt} \vec{\mathbf{B}} \cdot \vec{\mathbf{A}}$

$$\mathcal{E} = -(1) B \cos 0° \frac{dA}{dt} = -B\left(\frac{1}{2} r^2 \omega\right).$$

(We could think of this as following from the result of Example 23.3.)

The magnitude of the emf is

$$|\mathcal{E}| = B\left(\frac{1}{2} r^2 \omega\right) = (0.9 \text{ N}\cdot\text{s/C}\cdot\text{m})\left[\frac{1}{2}(0.4 \text{ m})^2 (3\,200 \text{ rev/min})\right]\left(\frac{2\pi \text{ rad/rev}}{60 \text{ s/min}}\right)$$

$$|\mathcal{E}| = \boxed{24.1 \text{ V}}.$$

A free positive charge q shown, turning with the disk, feels a magnetic force $q\vec{\mathbf{v}} \times \vec{\mathbf{B}}$ radially outward. Thus the $\boxed{\text{outer contact is positive}}$.

***P23.18** The speed of waves on the wire is

$$v = \sqrt{\frac{T}{\mu}} = \sqrt{\frac{267 \text{ N}\cdot\text{m}}{3\times 10^{-3} \text{ kg}}} = 298 \text{ m/s}.$$

In the simplest standing-wave vibration state,

$$d_{NN} = 0.64 \text{ m} = \frac{\lambda}{2} \qquad\qquad \lambda = 1.28 \text{ m}$$

and $f = \dfrac{v}{\lambda} = \dfrac{298 \text{ m/s}}{1.28 \text{ m}} = 233 \text{ Hz}$.

(a) The changing flux of magnetic field through the circuit containing the wire will drive current to the left in the wire as it moves up and to the right as it moves down. The emf will have this same frequency of $\boxed{233 \text{ Hz}}$.

(b) The vertical coordinate of the center of the wire is described by

$x = A\cos\omega t = (1.5 \text{ cm})\cos(2\pi 233\, t/\text{s})$.

Its velocity is $v = \dfrac{dx}{dt} = -(1.5 \text{ cm})(2\pi\, 233/\text{s})\sin(2\pi 233 \ t/\text{s})$.

Its maximum speed is $1.5 \text{ cm}(2\pi)233/\text{s} = 22.0 \text{ m/s}$.

The induced emf is $\mathcal{E} = -B\ell v$, with amplitude

$$\mathcal{E}_{\max} = B\ell v_{\max} = 4.50\times 10^{-3} \text{ T}(0.02 \text{ m})22 \text{ m/s} = \boxed{1.98\times 10^{-3} \text{ V}}.$$

P23.19 $\omega = (2.00 \text{ rev/s})(2\pi \text{ rad/rev}) = 4.00\pi \text{ rad/s}$

$$\mathcal{E} = \frac{1}{2}B\omega\ell^2 = \boxed{2.83 \text{ mV}}$$

P23.20 (a) $\vec{\mathbf{B}}_{ext} = B_{ext}\hat{\mathbf{i}}$ and B_{ext} decreases; therefore, the induced field is $\vec{\mathbf{B}}_0 = B_0\hat{\mathbf{i}}$ (to the right) and the current in the resistor is directed $\boxed{\text{to the right}}$.

(b) $\vec{\mathbf{B}}_{ext} = B_{ext}(-\hat{\mathbf{i}})$ increases; therefore, the induced field $\vec{\mathbf{B}}_0 = B_0(+\hat{\mathbf{i}})$ is to the right, and the current in the resistor is directed $\boxed{\text{to the right}}$.

(c) $\vec{\mathbf{B}}_{ext} = B_{ext}(-\hat{\mathbf{k}})$ into the paper and B_{ext} decreases; therefore, the induced field is $\vec{\mathbf{B}}_0 = B_0(-\hat{\mathbf{k}})$ into the paper, and the current in the resistor is directed $\boxed{\text{to the right}}$.

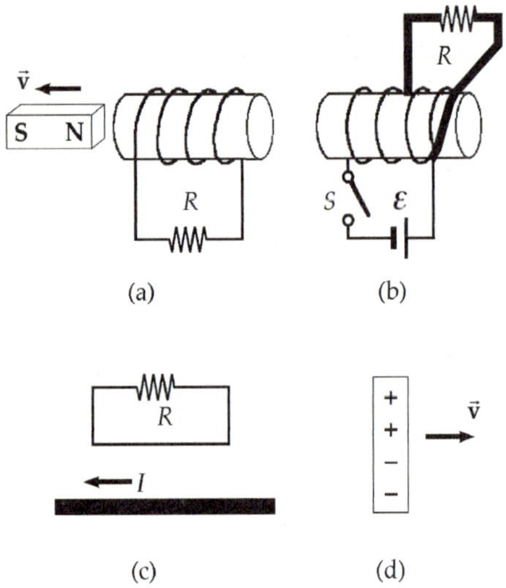

FIG. P23.20

(d) By the magnetic force law, $\vec{\mathbf{F}}_B = q(\vec{\mathbf{v}} \times \vec{\mathbf{B}})$. Therefore, a positive charge will move to the top of the bar if $\vec{\mathbf{B}}$ is $\boxed{\text{into the paper}}$.

P23.21 (a) At terminal speed,

$$Mg = F_B = IwB = \left(\frac{\mathcal{E}}{R}\right)wB = \left(\frac{Bwv_T}{R}\right)wB = \frac{B^2w^2v_T}{R}$$

or $$\boxed{v_T = \frac{MgR}{B^2\omega^2}}.$$

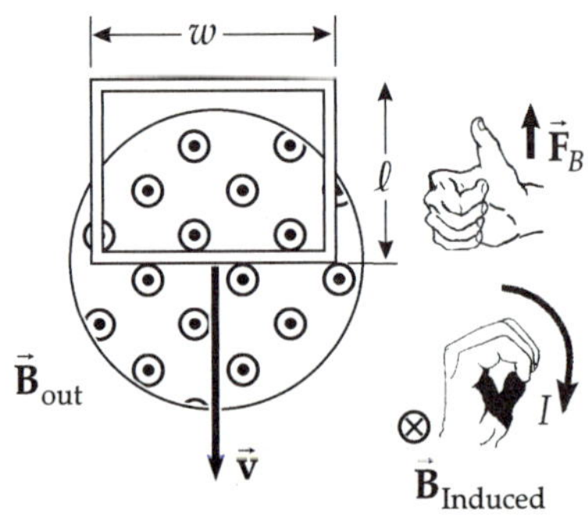

FIG. P23.21

(b) The emf is directly proportional to v_T, but the current is inversely proportional to R. A large R means a small current at a given speed, so the loop must travel faster to get $F_B = mg$.

(c) At a given speed, the current is directly proportional to the magnetic field. But the force is proportional to the product of the current and the field. For a small B, the speed must increase to compensate for both the small B and also the current, so $v_T \propto \frac{1}{B^2}$.

P23.22 (a) The force on the side of the coil entering the field (consisting of N wires) is

$$F = N(ILB) = N(IwB).$$

The induced emf in the coil is

$$|\mathcal{E}| = N\frac{d\Phi_B}{dt} = N\frac{d(Bwx)}{dt} = NBwv.$$

so the current is $I = \frac{|\mathcal{E}|}{R} = \frac{NBwv}{R}$ counterclockwise.

The force on the leading side of the coil is then:

$$F = N\left(\frac{NBwv}{R}\right)wB = \boxed{\frac{N^2B^2w^2v}{R} \text{ to the left}}.$$

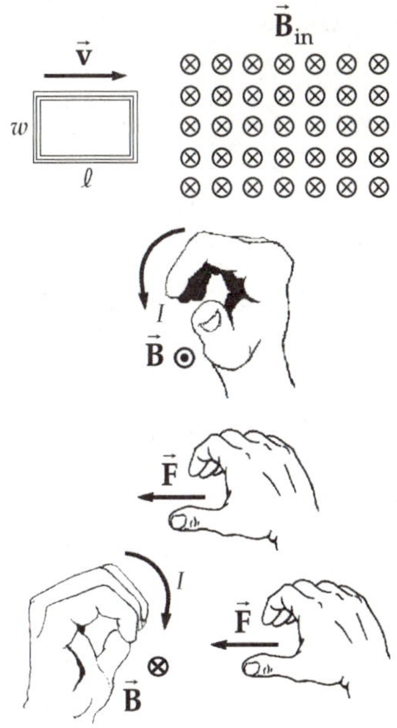

FIG. P23.22

(b) Once the coil is entirely inside the field, $\Phi_B = NBA = \text{constant}$,

so $\mathcal{E} = 0$, $I = 0$, and $F = \boxed{0}$.

(c) As the coil starts to leave the field, the flux *decreases* at the rate Bwv, so the magnitude of the current is the same as in part (a), but now the current is clockwise. Thus, the force exerted on the trailing side of the coil is:

$$F = \boxed{\frac{N^2B^2w^2v}{R} \text{ to the left again}}.$$

P23.23 (a) $\mathcal{E}_{max} = NAB\omega = (1\,000)(0.100)(0.200)(120\pi) = \boxed{7.54 \text{ kV}}$

(b) $\mathcal{E}(t) = NBA\omega \sin\omega t = NBA\omega \sin\theta$

$|\mathcal{E}|$ is maximal when $|\sin\theta| = 1$

or $\theta = \pm\frac{\pi}{2}$

so the $\boxed{\text{plane of coil is parallel to } \vec{\mathbf{B}}}$.

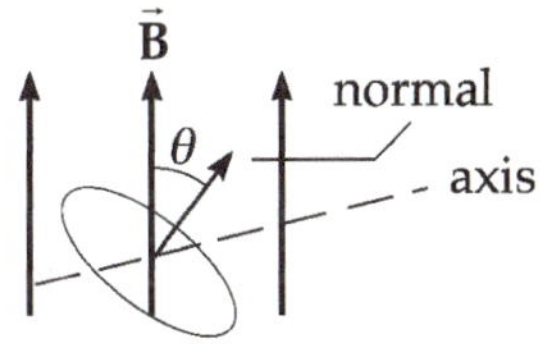

FIG. P23.23

***P23.24** $B = \mu_0 nI = (4\pi \times 10^{-7} \text{ T}\cdot\text{m/A})(200 \text{ m}^{-1})(15.0 \text{ A}) = 3.77 \times 10^{-3} \text{ T}$

For the small coil, $\Phi_B = N\vec{\mathbf{B}}\cdot\vec{\mathbf{A}} = NBA\cos\omega t = NB(\pi r^2)\cos\omega t$.

Thus, $\mathcal{E} = -\frac{d\Phi_B}{dt} = NB\pi r^2 \omega \sin\omega t$

$$\mathcal{E} = (30.0)(3.77 \times 10^{-3} \text{ T})\pi(0.080\,0 \text{ m})^2(4.00\pi \text{ s}^{-1})\sin(4.00\pi t) = \boxed{(28.6 \text{ mV})\sin(4.00\pi t)}.$$

Section 23.4 Induced emfs and Electric Fields

P23.25 $\frac{dB}{dt} = 0.060\,0t$ $\qquad |\mathcal{E}| = \frac{d\Phi_B}{dt} = \pi r_1^2 \frac{dB}{dt} = 2\pi r_1 E$

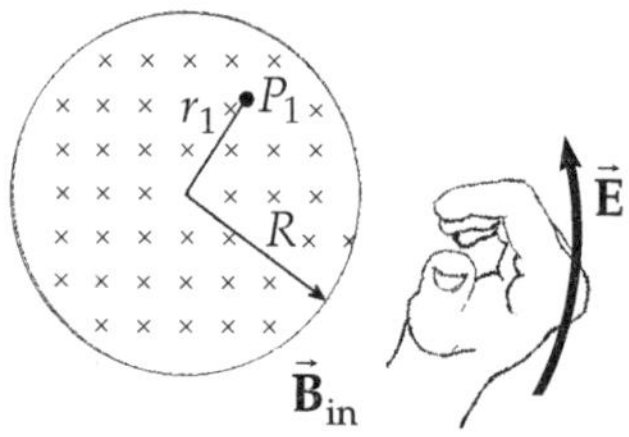

FIG. P23.25

At $t = 3.00$ s,

$$E = \left(\frac{\pi r_1^2}{2\pi r_1}\right)\frac{dB}{dt} = \frac{0.020\,0\text{ m}}{2}\left(0.060\,0\text{ T/s}^2\right)(3.00\text{ s})\left(\frac{1\text{ N}\cdot\text{s}}{1\text{ T}\cdot\text{C}\cdot\text{m}}\right)$$

$$\vec{\mathbf{E}} = \boxed{1.80\times10^{-3}\text{ N/C perpendicular to } r_1 \text{ and counterclockwise}}$$

P23.26 (a) $\frac{dB}{dt} = 6.00t^2 - 8.00t$ $\qquad |\mathcal{E}| = \frac{d\Phi_B}{dt}$

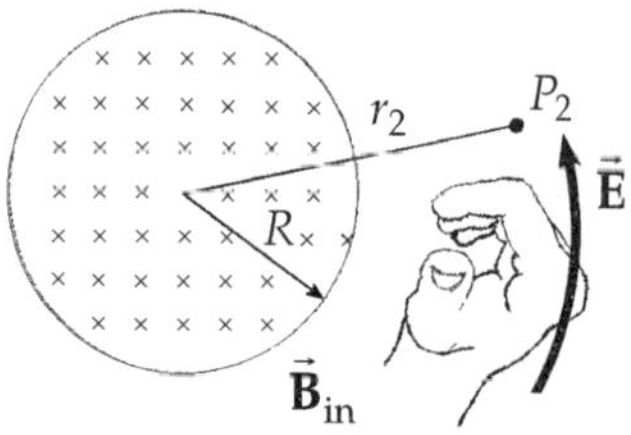

FIG. P23.26

At $t = 2.00$ s, $\qquad E = \frac{\pi R^2(dB/dt)}{2\pi r_2} = \frac{8.00\pi(0.025\,0)^2}{2\pi(0.050\,0)}$

$$F = qE = \boxed{8.00\times10^{-21}\text{ N}}$$

(b) When $6.00t^2 - 8.00t = 0$, at $t = \boxed{1.33\text{ s}}$ and at $t = \boxed{0}$.

Section 23.5 Self-Inductance

P23.27 $|\mathcal{E}| = L\frac{\Delta I}{\Delta t} = \left(3.00\times10^{-3}\text{ H}\right)\left(\frac{1.50\text{ A} - 0.200\text{ A}}{0.200\text{ s}}\right) = 1.95\times10^{-2}\text{ V} = \boxed{19.5\text{ mV}}$

P23.28 Treating the telephone cord as a solenoid, we have:

$$L = \frac{\mu_0 N^2 A}{\ell} = \frac{\left(4\pi\times10^{-7}\text{ T}\cdot\text{m/A}\right)(70.0)^2\pi\left(6.50\times10^{-3}\text{ m}\right)^2}{0.600\text{ m}} = \boxed{1.36\ \mu\text{H}}.$$

P23.29 $\mathcal{E} = -L\frac{dI}{dt} = -L\frac{d}{dt}\left(I_{max}\sin\omega t\right) = -L\omega I_{max}\cos\omega t = -\left(10.0\times10^{-3}\right)(120\pi)(5.00)\cos\omega t$

$$\mathcal{E} = -(6.00\pi)\cos(120\pi t) = \boxed{-(18.8\text{ V})\cos(377t)}$$

P23.30 From $|\mathcal{E}| = L\left(\frac{\Delta I}{\Delta t}\right)$, we have $\quad L = \frac{\mathcal{E}}{(\Delta I/\Delta t)} = \frac{24.0\times10^{-3}\text{ V}}{10.0\text{ A/s}} = 2.40\times10^{-3}\text{ H}.$

From $L = \frac{N\Phi_B}{I}$, we have $\quad \Phi_B = \frac{LI}{N} = \frac{\left(2.40\times10^{-3}\text{ H}\right)(4.00\text{ A})}{500} = \boxed{19.2\ \mu\text{T}\cdot\text{m}^2}.$

P23.31 $|\mathcal{E}| = L\frac{dI}{dt} = \left(90.0\times10^{-3}\right)\frac{d}{dt}\left(t^2 - 6t\right)$ V

(a) At $t = 1.00$ s, $\mathcal{E} = \boxed{360 \text{ mV}}$

(b) At $t = 4.00$ s, $\mathcal{E} = \boxed{180 \text{ mV}}$

(c) $\mathcal{E} = \left(90.0\times10^{-3}\right)(2t-6) = 0$

when $\boxed{t = 3.00 \text{ s}}$.

P23.32 $L = \frac{N\Phi_B}{I} = \frac{NBA}{I} \approx \frac{NA}{I}\cdot\frac{\mu_0 NI}{2\pi R} = \boxed{\frac{\mu_0 N^2 A}{2\pi R}}$

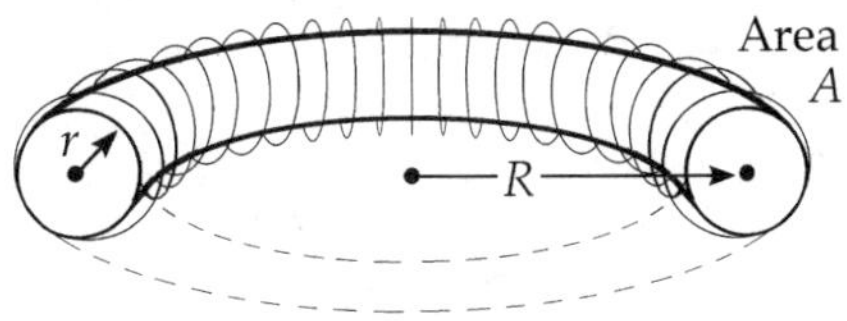

FIG. P23.32

Section 23.6 *RL* Circuits

P23.33 (a) At time t, $I(t) = \frac{\mathcal{E}\left(1 - e^{-t/\tau}\right)}{R}$

where $\tau = \frac{L}{R} = 0.200$ s.

After a long time, $I_{max} = \frac{\mathcal{E}\left(1 - e^{-\infty}\right)}{R} = \frac{\mathcal{E}}{R}$.

At $I(t) = 0.500 I_{max}$ $(0.500)\frac{\mathcal{E}}{R} = \frac{\mathcal{E}\left(1 - e^{-t/0.200\text{ s}}\right)}{R}$

so $0.500 = 1 - e^{-t/0.200\text{ s}}$.

Isolating the constants on the right, $\ln\left(e^{-t/0.200\text{ s}}\right) = \ln(0.500)$

and solving for t, $-\frac{t}{0.200\text{ s}} = -0.693$

or $t = \boxed{0.139 \text{ s}}$.

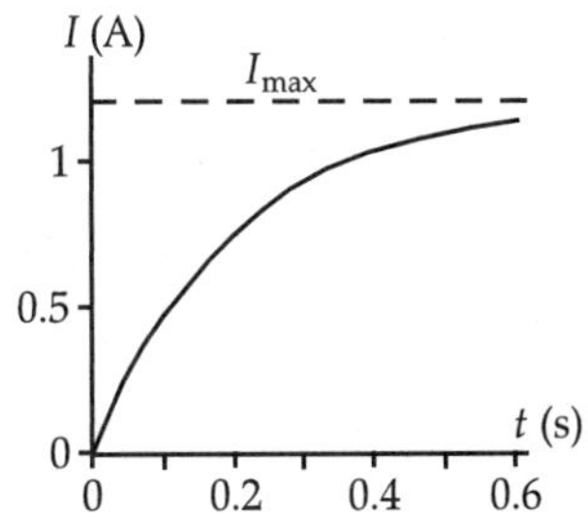

FIG. P23.33

(b) Similarly, to reach 90% of I_{max}, $0.900 = 1 - e^{-t/\tau}$

and $t = -\tau\ln(1 - 0.900)$.

Thus, $t = -(0.200\text{ s})\ln(0.100) = \boxed{0.461 \text{ s}}$.

P23.34 Taking $\tau = \frac{L}{R}$, $I = I_0 e^{-t/\tau}$: $\frac{dI}{dt} = I_0 e^{-t/\tau}\left(-\frac{1}{\tau}\right)$

$IR + L\frac{dI}{dt} = 0$ will be true if $I_0 R e^{-t/\tau} + L\left(I_0 e^{-t/\tau}\right)\left(-\frac{1}{\tau}\right) = 0$.

Because $\tau = \frac{L}{R}$, we have agreement with $0 = 0$.

P23.35 (a) $\tau = \frac{L}{R} = 2.00 \times 10^{-3}\text{ s} = \boxed{2.00\text{ ms}}$

(b) $I = I_{\max}\left(1 - e^{-t/\tau}\right) = \left(\frac{6.00\text{ V}}{4.00\ \Omega}\right)\left(1 - e^{-0.250/2.00}\right) = \boxed{0.176\text{ A}}$

(c) $I_{\max} = \frac{\mathcal{E}}{R} = \frac{6.00\text{ V}}{4.00\ \Omega} = \boxed{1.50\text{ A}}$

(d) $0.800 = 1 - e^{-t/2.00\text{ ms}} \rightarrow t = -(2.00\text{ ms})\ln(0.200) = \boxed{3.22\text{ ms}}$

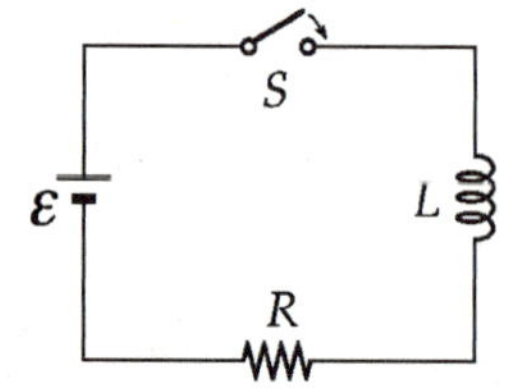

FIG. P23.35

P23.36 (a) $\Delta V_R = IR = (8.00\ \Omega)(2.00\text{ A}) = 16.0\text{ V}$

and $\Delta V_L = \mathcal{E} - \Delta V_R = 36.0\text{ V} - 16.0\text{ V} = 20.0\text{ V}$.

Therefore, $\frac{\Delta V_R}{\Delta V_L} = \frac{16.0\text{ V}}{20.0\text{ V}} = \boxed{0.800}$.

(b) $\Delta V_R = IR = (4.50\text{ A})(8.00\ \Omega) = 36.0\text{ V}$

$\Delta V_L = \mathcal{E} - \Delta V_R = \boxed{0}$

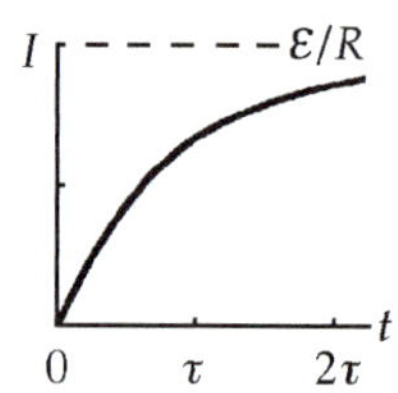

FIG. P23.36

***P23.37** For the increasing current $I = \frac{\mathcal{E}}{R}\left(1 - e^{-Lt/R}\right)$. The final value is $\frac{\mathcal{E}}{R}$, so the condition on Δt is

$$0.8\frac{\mathcal{E}}{R} = \frac{\mathcal{E}}{R}\left(1 - e^{-L\Delta t/R}\right)$$
$$e^{-L\Delta t/R} = 0.2$$
$$e^{+L\Delta t/R} = 5$$
$$\frac{L\Delta t}{R} = \ln 5$$
$$\Delta t = \frac{R\ln 5}{L}$$

At the moment when the battery is removed, the current in the coil is quite precisely $\frac{\mathcal{E}}{R}$. During the decrease, $I = I_i e^{-Lt/R} = \frac{\mathcal{E}}{R}e^{-Lt/R}$.

(a) at $t = \Delta t = \frac{R\ln 5}{L}$, $\frac{I}{I_i} = e^{-L\Delta t/R} = 0.200 = \boxed{20.0\%}$

(b) at $t = 2\Delta t$, $\frac{I}{I_i} = e^{-L2\Delta t/R} = \left(e^{-L\Delta t/R}\right)^2 = (0.2)^2 = 0.04 = \boxed{4.00\%}$

P23.38 $I = I_{max}\left(1 - e^{-t/\tau}\right)$: $\quad 0.980 = 1 - e^{-3.00\times10^{-3}/\tau}$

$$0.020\,0 = e^{-3.00\times10^{-3}/\tau}$$

$$\tau = -\frac{3.00\times10^{-3}}{\ln(0.020\,0)} = 7.67\times10^{-4}\text{ s}$$

$\tau = \dfrac{L}{R}$, so $\quad L = \tau R = \left(7.67\times10^{-4}\right)(10.0) = \boxed{7.67\text{ mH}}$

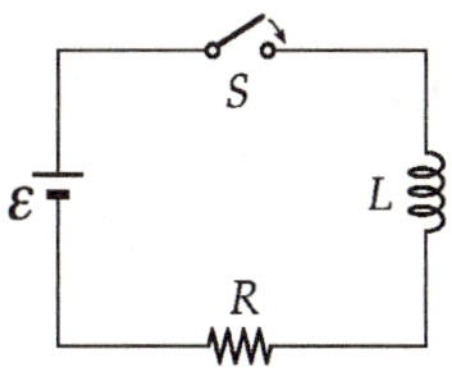

FIG. P23.38

P23.39 Name the currents as shown. By Kirchhoff's laws:

$$+I_1 - I_2 - I_3 = 0 \quad (1)$$

$$+10.0\text{ V} - 4.00I_1 - 4.00I_2 = 0 \quad (2)$$

$$+10.0\text{ V} - 4.00I_1 - 8.00I_3 - (1.00)\frac{dI_3}{dt} = 0 \quad (3)$$

From (1) and (2), $\quad +10.0 - 4.00I_1 - 4.00I_1 + 4.00I_3 = 0$

and $\quad I_1 = 0.500I_3 + 1.25\text{ A}$.

Then (3) becomes $\quad 10.0\text{ V} - 4.00(0.500I_3 + 1.25\text{ A}) - 8.00I_3 - (1.00)\dfrac{dI_3}{dt} = 0$

$$(1.00\text{ H})\left(\frac{dI_3}{dt}\right) + (10.0\ \Omega)I_3 = 5.00\text{ V}.$$

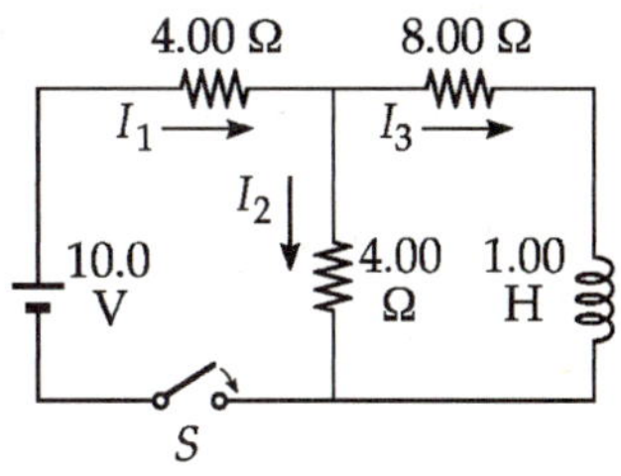

FIG. P23.39

We solve the differential equation using Equations 23.13 and 23.14:

$$I_3(t) = \left(\frac{5.00\text{ V}}{10.0\ \Omega}\right)\left[1 - e^{-(10.0\ \Omega)t/1.00\text{ H}}\right] = \boxed{(0.500\text{ A})\left[1 - e^{-10t/\text{s}}\right]}$$

$$I_1 = 1.25 + 0.500I_3 = \boxed{1.50\text{ A} - (0.250\text{ A})e^{-10t/\text{s}}}$$

P23.40 (a) $\quad I = \dfrac{\mathcal{E}}{R} = \dfrac{12.0\text{ V}}{12.0\ \Omega} = \boxed{1.00\text{ A}}$

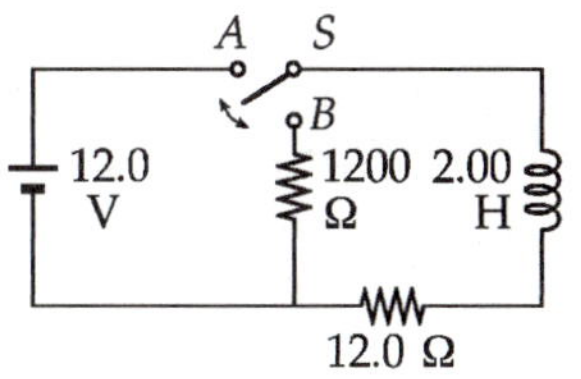

FIG. P23.40

(b) Initial current is 1.00 A: $\Delta V_{12} = (1.00\text{ A})(12.00\ \Omega) = \boxed{12.0\text{ V}}$

$$\Delta V_{1\,200} = (1.00\text{ A})(1\,200\ \Omega) = \boxed{1.20\text{ kV}}$$

$$\Delta V_L = \boxed{1.21\text{ kV}}.$$

(c) $I = I_{max}e^{-Rt/L}$: $\quad \dfrac{dI}{dt} = -I_{max}\dfrac{R}{L}e^{-Rt/L}$

and $\quad -L\dfrac{dI}{dt} = \Delta V_L = I_{max}Re^{-Rt/L}$.

Solving $\quad 12.0\text{ V} = (1\,212\text{ V})e^{-1\,212t/2.00}$

so $\quad 9.90\times10^{-3} = e^{-606t}$.

Thus, $\quad \boxed{t = 7.62\text{ ms}}$.

P23.41 $\tau = \frac{L}{R} = \frac{0.140}{4.90} = 28.6 \text{ ms}$

$I_{max} = \frac{\mathcal{E}}{R} = \frac{6.00 \text{ V}}{4.90\ \Omega} = 1.22 \text{ A}$

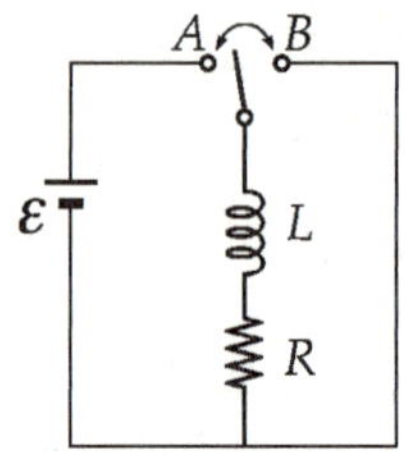

FIG. P23.41

(a) $I = I_{max}\left(1 - e^{-t/\tau}\right)$ so $0.220 = 1.22\left(1 - e^{-t/\tau}\right)$

$e^{-t/\tau} = 0.820:$ $t = -\tau \ln(0.820) = \boxed{5.66 \text{ ms}}$

(b) $I = I_{max}\left(1 - e^{-10.0/0.028\,6}\right) = (1.22 \text{ A})\left(1 - e^{-350}\right) = \boxed{1.22 \text{ A}}$

(c) $I = I_{max} e^{-t/\tau}$ and $0.160 = 1.22 e^{-t/\tau}$

so $t = -\tau \ln(0.131) = \boxed{58.1 \text{ ms}}$.

Section 23.7 Energy Stored in a Magnetic Field

P23.42 (a) The magnetic energy density is given by

$$u = \frac{B^2}{2\mu_0} = \frac{(4.50 \text{ T})^2}{2\left(1.26 \times 10^{-6} \text{ T} \cdot \text{m/A}\right)} = \boxed{8.06 \times 10^6 \text{ J/m}^3}.$$

(b) The magnetic energy stored in the field equals u times the volume of the solenoid (the volume in which B is non-zero).

$$U = uV = \left(8.06 \times 10^6 \text{ J/m}^3\right)\left[(0.260 \text{ m})\pi(0.031\,0 \text{ m})^2\right] = \boxed{6.32 \text{ kJ}}$$

P23.43 $L = \mu_0 \frac{N^2 A}{\ell} = \mu_0 \frac{(68.0)^2\left[\pi\left(0.600 \times 10^{-2}\right)^2\right]}{0.080\,0} = 8.21\ \mu\text{H}$

$U = \frac{1}{2} L I^2 = \frac{1}{2}\left(8.21 \times 10^{-6} \text{ H}\right)(0.770 \text{ A})^2 = \boxed{2.44\ \mu\text{J}}$

P23.44 (a) $U = \frac{1}{2} L I^2 = \frac{1}{2}(4.00 \text{ H})(0.500 \text{ A})^2$ $U = \boxed{0.500 \text{ J}}$

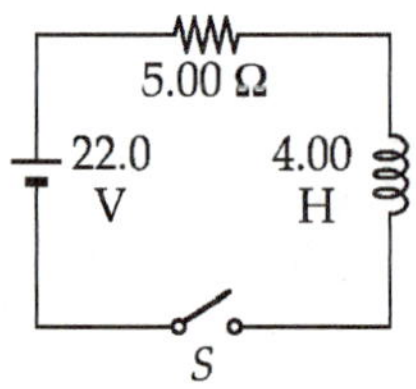

FIG. P23.44

(b) When the current is 1.00 A,

Kirchhoff's loop rule reads $+22.0 \text{ V} - (1.00 \text{ A})(5.00\ \Omega) - \Delta V_L = 0$.

Then $\Delta V_L = 17.0 \text{ V}$.

The power being stored in the inductor is

$I\Delta V_L = (1.00 \text{ A})(17.0 \text{ V}) = \boxed{17.0 \text{ W}}$.

(c) $\mathcal{P} = I\Delta V = (0.500 \text{ A})(22.0 \text{ V})$ $\mathcal{P} = \boxed{11.0 \text{ W}}$

P23.45 $u = \epsilon_0 \dfrac{E^2}{2} = \boxed{44.2\ \text{nJ/m}^3}$ $\qquad u = \dfrac{B^2}{2\mu_0} = \boxed{995\ \mu\text{J/m}^3}$

Section 23.8 Context Connection—The Repulsive Model for Magnetic Levitation

P23.46 The induced emf in the leading edge of the loop is

$$\mathcal{E} = B\ell v = B(0.2\ \text{m})(400\ \text{km/h})\left(\frac{1\,000\ \text{m}}{\text{km}}\right)\left(\frac{\text{h}}{3\,600\ \text{s}}\right) = \left(22.2\ \text{m}^2/\text{s}\right)B$$

The induced current is $\qquad I = \dfrac{\mathcal{E}}{R} = \dfrac{\left(22.2\ \text{m}^2\right)B}{25\ \Omega\cdot\text{s}}$

The magnetic force on the lower section of one loop is $\vec{\mathbf{F}}_B = I\vec{\ell}\times\vec{\mathbf{B}}$:

$$\vec{\mathbf{F}}_B = \left(\frac{22.2\ \text{m}^2\text{A}}{25\ \text{V}\cdot\text{s}}\right)B(0.1\ \text{m})B\sin 90°$$  $$= \left(0.0889\ \text{m}^3\cdot\text{C}^2/\text{J}\cdot\text{s}^2\right)B^2\ \textbf{up}$$

We require $\Sigma F_y = 0$: $\qquad 100F_B - mg = 0$

$$B^2\left(8.89\ \text{m}^2\cdot\text{C}^2/\text{N}\cdot\text{s}^2\right) = 5\times10^4(9.80\ \text{N})$$

$$B = 235\ \text{N}\cdot\text{s/C}\cdot\text{m} = \boxed{2\times10^2\ \text{T}}$$

P23.47 (a) $B = \dfrac{\mu_0 NI}{\ell} = \dfrac{\left(4\pi\times10^{-7}\ \text{T}\cdot\text{m/A}\right)(1\,400)(2.00\ \text{A})}{1.20\ \text{m}} = \boxed{2.93\times10^{-3}\ \text{T (upward)}}$

(b) $u = \dfrac{B^2}{2\mu_0} = \dfrac{\left(2.93\times10^{-3}\ \text{T}\right)^2}{2\left(4\pi\times10^{-7}\ \text{T}\cdot\text{m/A}\right)} = \left(3.42\ \text{J/m}^3\right)\left(\dfrac{1\ \text{N}\cdot\text{m}}{1\ \text{J}}\right) = 3.42\ \text{N/m}^2 = \boxed{3.42\ \text{Pa}}$

(c) To produce a downward magnetic field, the surface of the superconductor must carry a $\boxed{\text{clockwise}}$ current.

(d) The vertical component of the field of the solenoid exerts an inward force on the superconductor. The total horizontal force is zero. Over the top end of the solenoid, its field diverges and has a radially outward horizontal component. This component exerts upward force on the clockwise superconductor current. The total force on the core is $\boxed{\text{upward}}$. You can think of it as a force of repulsion between the solenoid with its north end pointing up, and the core, with its north end pointing down.

(e) $F = PA = (3.42\ \text{Pa})\left[\pi\left(1.10\times10^{-2}\ \text{m}\right)^2\right] = \boxed{1.30\times10^{-3}\ \text{N}}$

Note that we have not proven that energy density is pressure. In fact, it is not in some cases; Equation 16.13 shows that the pressure is two-thirds of the translational energy density in an ideal gas.

Additional Problems

P23.48 (a) Doubling the number of turns.

Amplitude doubles: period unchanged

(b) Doubling the angular velocity.

doubles the amplitude: cuts the period in half

(c) Doubling the angular velocity while reducing the number of turns to one half the original value.

Amplitude unchanged: cuts the period in half

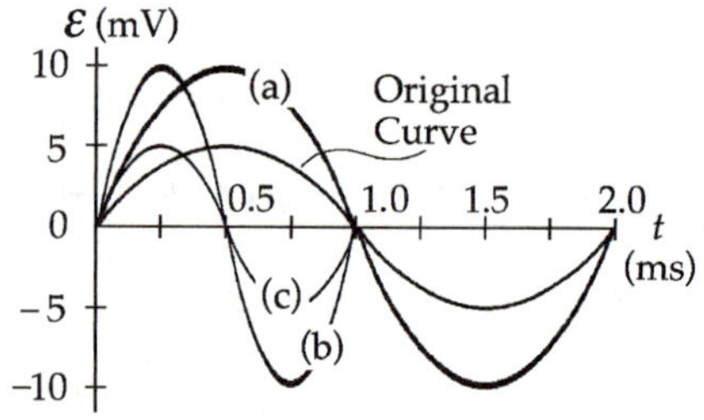

FIG. P23.48

P23.49 $\mathcal{E}=-N\frac{d}{dt}(BA\cos\theta)=-N(\pi r^2)\cos 0°\left(\frac{dB}{dt}\right)$

$$\mathcal{E}=-(30.0)\left[\pi(2.70\times10^{-3}\text{ m})^2\right](1)\frac{d}{dt}\left[50.0\text{ mT}+(3.20\text{ mT})\sin\left(2\pi\left[523t\text{ s}^{-1}\right]\right)\right]$$

$$\mathcal{E}=-(30.0)\left[\pi(2.70\times10^{-3}\text{ m})^2\right](3.20\times10^{-3}\text{ T})\left[2\pi(523\text{ s}^{-1})\cos\left(2\pi\left[523t\text{ s}^{-1}\right]\right)\right]$$

$$\mathcal{E}=\boxed{-(7.22\times10^{-3}\text{ V})\cos\left[2\pi(523t\text{ s}^{-1})\right]}$$

***P23.50** $\mathcal{E}=-N\frac{\Delta}{\Delta t}(BA\cos\theta)=-N(\pi r^2)\cos 0°\frac{\Delta B}{\Delta t}=-1(0.005\,00\text{ m}^2)(1)\left(\frac{1.50\text{ T}-5.00\text{ T}}{20.0\times10^{-3}\text{ s}}\right)=0.875\text{ V}$

(a) $I=\frac{\mathcal{E}}{R}=\frac{0.875\text{ V}}{0.020\,0\ \Omega}=\boxed{43.8\text{ A}}$

(b) $\mathcal{P}=\mathcal{E}I=(0.875\text{ V})(43.8\text{ A})=\boxed{38.3\text{ W}}$

***P23.51** Suppose we wrap twenty turns of wire into a flat compact circular coil of diameter 3 cm. Suppose we use a bar magnet to produce field 10^{-3} T through the coil in one direction along its axis. Suppose we then flip the magnet to reverse the flux in 10^{-1} s. The average induced emf is then

$$\bar{\mathcal{E}}=-N\frac{\Delta\Phi_B}{\Delta t}=-N\frac{\Delta[BA\cos\theta]}{\Delta t}=-NB(\pi r^2)\left(\frac{\cos180°-\cos0°}{\Delta t}\right)$$

$$\bar{\mathcal{E}}=-(20)(10^{-3}\text{ T})\pi(0.015\,0\text{ m})^2\left(\frac{-2}{10^{-1}\text{ s}}\right)\boxed{\sim10^{-4}\text{ V}}$$

P23.52 $I=\dfrac{\mathcal{E}+\mathcal{E}_{\text{induced}}}{R}$ and $\mathcal{E}_{\text{induced}}=-\dfrac{d}{dt}(BA)$

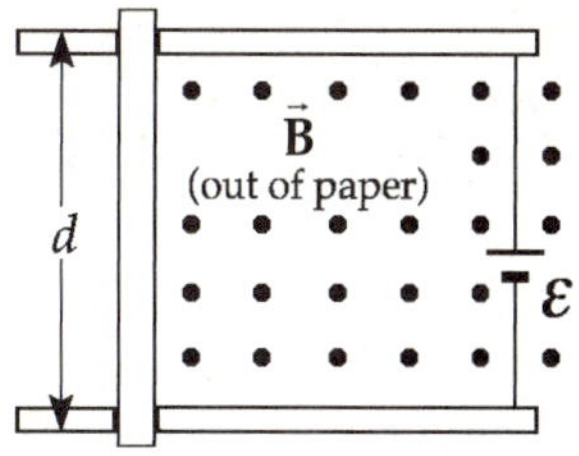

FIG. P23.52

$$F=m\frac{dv}{dt}=IBd$$

$$\frac{dv}{dt}=\frac{IBd}{m}=\frac{Bd}{mR}(\mathcal{E}+\mathcal{E}_{\text{induced}})$$

$$\frac{dv}{dt}=\frac{Bd}{mR}(\mathcal{E}-Bvd)$$

To solve the differential equation, let $u=\mathcal{E}-Bvd$

$$\frac{du}{dt}=-Bd\frac{dv}{dt}$$

$$-\frac{1}{Bd}\frac{du}{dt}=\frac{Bd}{mR}u$$

so $$\int_{u_0}^{u}\frac{du}{u}=-\int_0^t\frac{(Bd)^2}{mR}dt.$$

Integrating from $t=0$ to $t=t$, $$\ln\frac{u}{u_0}=-\frac{(Bd)^2}{mR}t$$

or $$\frac{u}{u_0}=e^{-B^2d^2t/mR}.$$

Since $v=0$ when $t=0$, $u_0=\mathcal{E}$

and $u=\mathcal{E}-Bvd$

$$\mathcal{E}-Bvd=\mathcal{E}e^{-B^2d^2t/mR}.$$

Therefore, $$\boxed{v=\frac{\mathcal{E}}{Bd}\left(1-e^{-B^2d^2t/mR}\right)}.$$

P23.53 The enclosed flux is $\Phi_B=BA=B\pi r^2$.

The particle moves according to $\sum\vec{\mathbf{F}}=m\vec{\mathbf{a}}$: $qvB\sin 90°=\dfrac{mv^2}{r}$

$$r=\frac{mv}{qB}.$$

Then $$\Phi_B=\frac{B\pi m^2v^2}{q^2B^2}.$$

(a) $$v=\sqrt{\frac{\Phi_Bq^2B}{\pi m^2}}=\sqrt{\frac{(15\times10^{-6}\text{ T}\cdot\text{m}^2)(30\times10^{-9}\text{ C})^2(0.6\text{ T})}{\pi(2\times10^{-16}\text{ kg})^2}}=\boxed{2.54\times10^5\text{ m/s}}$$

continued on next page

(b) Energy for the particle-electric field system is conserved in the firing process:

$U_i = K_f$: $\quad q\Delta V = \frac{1}{2}mv^2$

$$\Delta V = \frac{mv^2}{2q} = \frac{\left(2\times10^{-16}\text{ kg}\right)\left(2.54\times10^5\text{ m/s}\right)^2}{2\left(30\times10^{-9}\text{ C}\right)} = \boxed{215\text{ V}}.$$

P23.54 (a) Consider an annulus of radius r, width dr, height b, and resistivity ρ. Around its circumference, a voltage is induced according to

$$\mathcal{E} = -N\frac{d}{dt}\vec{\mathbf{B}}\cdot\vec{\mathbf{A}} = -1\frac{d}{dt}B_{\max}(\cos\omega t)\pi r^2 = +B_{\max}\pi r^2\omega\sin\omega t.$$

The resistance around the loop is $\frac{\rho\ell}{A_x} = \frac{\rho(2\pi r)}{bdr}$.

The eddy current in the ring is $dI = \frac{\mathcal{E}}{\text{resistance}} = \frac{B_{\max}\pi r^2\omega(\sin\omega t)bdr}{\rho(2\pi r)} = \frac{B_{\max}rb\omega dr\sin\omega t}{2\rho}$.

The instantaneous power is $d\mathcal{P}_i = \mathcal{E}dI = \frac{B_{\max}^2\pi r^3 b\omega^2 dr\sin^2\omega t}{2\rho}$.

The time average of the function $\sin^2\omega t = \frac{1}{2} - \frac{1}{2}\cos 2\omega t$ is $\frac{1}{2} - 0 = \frac{1}{2}$

so the time-averaged power delivered to the annulus is

$$d\mathcal{P} = \frac{B_{\max}^2\pi r^3 b\omega^2 dr}{4\rho}.$$

The power delivered to the disk is $\mathcal{P} = \int d\mathcal{P} = \int_0^R \frac{B_{\max}^2\pi b\omega^2}{4\rho}r^3 dr$

$$\mathcal{P} = \frac{B_{\max}^2\pi b\omega^2}{4\rho}\left(\frac{R^4}{4} - 0\right) = \boxed{\frac{\pi B_{\max}^2 R^4 b\omega^2}{16\rho}}.$$

(b) When $B_{\max}$ gets two times larger, $B_{\max}^2$ and $\mathcal{P}$ get $\boxed{4}$ times larger.

(c) When f and $\omega = 2\pi f$ double, ω^2 and $\mathcal{P}$ get $\boxed{4}$ times larger.

(d) When R doubles, R^4 and $\mathcal{P}$ become $2^4 = \boxed{16}$ times larger.

P23.55 We are given $\Phi_B = \left(6.00t^3 - 18.0t^2\right)\text{ T}\cdot\text{m}^2$

and $\mathcal{E} = -\frac{d\Phi_B}{dt} = -18.0t^2 + 36.0t$.

Maximum $\mathcal{E}$ occurs when $\frac{d\mathcal{E}}{dt} = -36.0t + 36.0 = 0$

which gives $t = 1.00\text{ s}$.

Therefore, the maximum current (at $t = 1.00\text{ s}$) is $I = \frac{\mathcal{E}}{R} = \frac{(-18.0+36.0)\text{ V}}{3.00\ \Omega} = \boxed{6.00\text{ A}}$.

***P23.56** (a) $\mathcal{E}=-N\frac{d}{dt}BA\cos\theta=-1\frac{d}{dt}B\frac{\theta a^2}{2}\cos 0°=-\frac{Ba^2}{2}\frac{d\theta}{dt}=-\frac{1}{2}Ba^2\omega=-\frac{1}{2}(0.5\text{ T})(0.5\text{ m})^2 2\text{ rad/s}$

$=-0.125\text{ V}=\boxed{0.125\text{ V clockwise}}$

The – sign indicates that the induced emf produces clockwise current, to make its own magnetic field into the page.

(b) At this instant $\theta=\omega t=2\text{ rad/s}(0.25\text{ s})=0.5\text{ rad}$. The arc PQ has length $r\theta=(0.5\text{ rad})(0.5\text{ m})=0.25\text{ m}$. The length of the circuit is $0.5\text{ m}+0.5\text{ m}+0.25\text{ m}=1.25\text{ m}$ its resistance is $1.25\text{ m}(5\ \Omega/\text{m})=6.25\ \Omega$. The current is $\frac{0.125\text{ V}}{6.25\ \Omega}=\boxed{0.020\,0\text{ A clockwise}}$.

P23.57 Suppose the field is vertically down. When an electron is moving away from you, the force on it is in the direction given by

$q\vec{\mathbf{v}}\times\vec{\mathbf{B}}_c$ as $-(\text{away})\times\text{down}=-$ 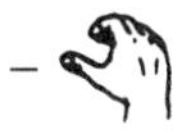 $=-\text{left}=\text{right}$.

Therefore, the electrons circulate clockwise.

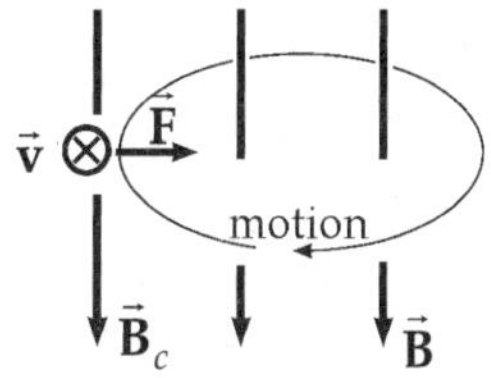

FIG. P23.57

(a) As the downward field increases, an emf is induced to produce some current that in turn produces an upward field. This current is directed  counterclockwise, carried by negative electrons moving clockwise. Therefore the original electron motion speeds up.

(b) At the circumference, we have $\sum F_c=ma_c$: $\quad |q|vB_c\sin 90°=\frac{mv^2}{r}$

$mv=|q|rB_c$.

The increasing magnetic field $\vec{\mathbf{B}}_{\text{av}}$ in the area enclosed by the orbit produces a tangential electric field according to

$$\left|\oint\vec{\mathbf{E}}\cdot d\vec{\mathbf{s}}\right|=\left|-\frac{d}{dt}\vec{\mathbf{B}}_{av}\cdot\vec{\mathbf{A}}\right| \qquad E(2\pi r)=\pi r^2\frac{dB_{av}}{dt} \qquad E=\frac{r}{2}\frac{dB_{av}}{dt}.$$

An electron feels a tangential force according to $\sum F_t=ma_t$: $\quad |q|E=m\frac{dv}{dt}$.

Then $\quad |q|\frac{r}{2}\frac{dB_{av}}{dt}=m\frac{dv}{dt} \qquad |q|\frac{r}{2}B_{av}=mv=|q|rB_c$

and $\quad B_{av}=2B_c$.

P23.58 $\mathcal{E}=-N\frac{d\Phi_B}{dt}=-N\frac{d}{dt}(BA\cos\theta)$

$$\mathcal{E}=-NB\cos\theta\left(\frac{\Delta A}{\Delta t}\right)=-200\left(50.0\times10^{-6}\text{ T}\right)(\cos 62.0°)\left(\frac{39.0\times10^{-4}\text{ m}^2}{1.80\text{ s}}\right)=\boxed{-10.2\ \mu\text{V}}$$

P23.59 The magnetic field produced by the current in the straight wire is perpendicular to the plane of the coil at all points within the coil. The magnitude of the field is $B = \frac{\mu_0 I}{2\pi r}$. Thus, the flux linkage is

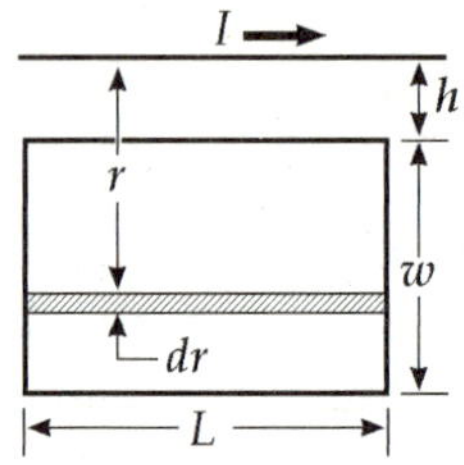

FIG. P23.59

$$N\Phi_B = \frac{\mu_0 NIL}{2\pi}\int_h^{h+w}\frac{dr}{r} = \frac{\mu_0 NI_{max}L}{2\pi}\ln\left(\frac{h+w}{h}\right)\sin(\omega t+\phi).$$

Finally, the induced emf is

$$\mathcal{E} = -\frac{\mu_0 NI_{max}L\omega}{2\pi}\ln\left(1+\frac{w}{h}\right)\cos(\omega t+\phi)$$

$$\mathcal{E} = -\frac{\left(4\pi\times10^{-7}\right)(100)(50.0)(0.200\text{ m})\left(200\pi\text{ s}^{-1}\right)}{2\pi}\ln\left(1+\frac{5.00\text{ cm}}{5.00\text{ cm}}\right)\cos(\omega t+\phi)$$

$$\mathcal{E} = \boxed{-(87.1\text{ mV})\cos(200\pi t+\phi)}.$$

The term $\sin(\omega t+\phi)$ in the expression for the current in the straight wire does not change appreciably when ωt changes by 0.10 rad or less. Thus, the current does not change appreciably during a time interval

$$\Delta t < \frac{0.10}{\left(200\pi\text{ s}^{-1}\right)} = 1.6\times10^{-4}\text{ s}.$$

We define a critical length, $c\Delta t = \left(3.00\times10^8\text{ m/s}\right)\left(1.6\times10^{-4}\text{ s}\right) = 4.8\times10^4\text{ m}$ equal to the distance to which field changes could be propagated during an interval of 1.6×10^{-4} s. This length is so much larger than any dimension of the coil or its distance from the wire that, although we consider the straight wire to be infinitely long, we can also safely ignore the field propagation effects in the vicinity of the coil. Moreover, the phase angle can be considered to be constant along the wire in the vicinity of the coil.

If the angular frequency ω were much larger, say, $200\pi\times10^5\text{ s}^{-1}$, the corresponding critical length would be only 48 cm. In this situation propagation effects would be important and the above expression for $\mathcal{E}$ would require modification. As a "rule of thumb" we can consider field propagation effects for circuits of laboratory size to be negligible for frequencies, $f = \frac{\omega}{2\pi}$, that are less than about 10^6 Hz.

P23.60 $B = \frac{\mu_0 NI}{2\pi r}$

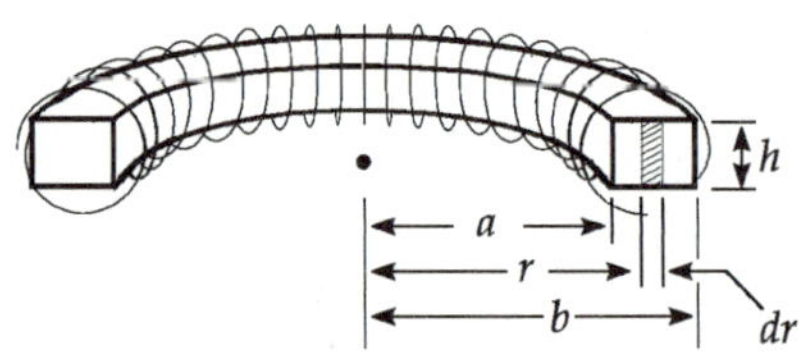

FIG. P23.60

(a) $\Phi_B = \int B\,dA = \int_a^b \frac{\mu_0 NI}{2\pi r}h\,dr = \frac{\mu_0 NIh}{2\pi}\int_a^b\frac{dr}{r} = \frac{\mu_0 NIh}{2\pi}\ln\left(\frac{b}{a}\right)$

$$L = \frac{N\Phi_B}{I} = \boxed{\frac{\mu_0 N^2 h}{2\pi}\ln\left(\frac{b}{a}\right)}$$

continued on next page

(b) $$L = \frac{\mu_0(500)^2(0.010\,0)}{2\pi}\ln\left(\frac{12.0}{10.0}\right) = \boxed{91.2\ \mu\text{H}}$$

(c) $$L_{\text{appx}} = \frac{\mu_0 N^2}{2\pi}\left(\frac{A}{R}\right) = \frac{\mu_0(500)^2}{2\pi}\left(\frac{2.00\times10^{-4}\ \text{m}^2}{0.110}\right) = \boxed{90.9\ \mu\text{H}}$$, only 0.3% different.

P23.61 (a) At the center, $$B = \frac{N\mu_0 IR^2}{2\left(R^2+0^2\right)^{3/2}} = \frac{N\mu_0 I}{2R}.$$

So the coil creates flux through itself $$\Phi_B = BA\cos\theta = \frac{N\mu_0 I}{2R}\pi R^2\cos0° = \frac{\pi}{2}N\mu_0 IR.$$

When the current it carries changes, $$\mathcal{E}_L = -N\frac{d\Phi_B}{dt} \approx -N\left(\frac{\pi}{2}\right)N\mu_0 R\frac{dI}{dt} = -L\frac{dI}{dt}$$

so $$\boxed{L \approx \frac{\pi}{2}N^2\mu_0 R}.$$

(b) $2\pi r = 3(0.3\ \text{m})$ so $r \approx 0.14\ \text{m}$

$$L \approx \frac{\pi}{2}\left(1^2\right)\left(4\pi\times10^{-7}\ \text{T}\cdot\text{m/A}\right)(0.14\ \text{m}) = 2.8\times10^{-7}\ \text{H}$$

$$\boxed{L \sim 100\ \text{nH}}$$

(c) $$\frac{L}{R} = \frac{2.8\times10^{-7}\ \text{V}\cdot\text{s/A}}{270\ \text{V/A}} = 1.0\times10^{-9}\ \text{s} \qquad \boxed{\frac{L}{R} \sim 1\ \text{ns}}$$

P23.62 When the switch is closed, as shown in figure (a), the current in the inductor is I:

$12.0 - 7.50I - 10.0 = 0 \rightarrow I = 0.267\ \text{A}$.

When the switch is opened, the initial current in the inductor remains at 0.267 A.

$IR = \Delta V$: $(0.267\ \text{A})R \le 80.0\ \text{V}$

$$\boxed{R \le 300\ \Omega}$$

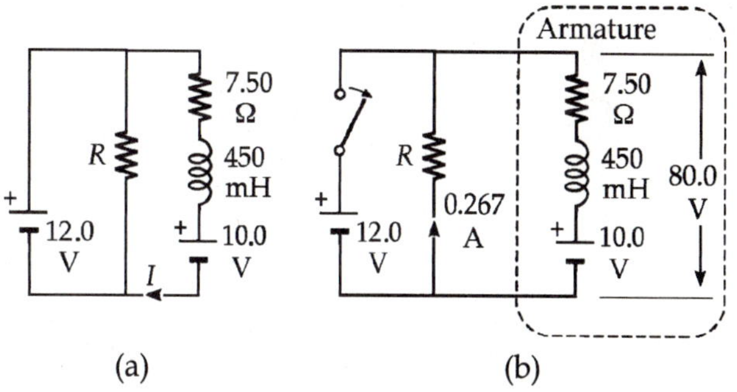

FIG. P23.62

P23.63 For an RL circuit,

$$I(t) = I_{\max} e^{-(R/L)t}: \qquad \frac{I(t)}{I_{\max}} = 1 - 10^{-9} = e^{-(R/L)t} \approx 1 - \frac{R}{L}t$$

$$\frac{R}{L}t = 10^{-9} \quad \text{so} \quad R_{\max} = \frac{\left(3.14\times10^{-8}\right)\left(10^{-9}\right)}{(2.50\text{ yr})\left(3.16\times10^{7}\text{ s/yr}\right)} = \boxed{3.97\times10^{-25}\ \Omega}.$$

(If the ring were of purest copper, of diameter 1 cm, and cross-sectional area 1 mm^2, its resistance would be at least $10^{-6}\ \Omega$.)

P23.64 (a) $U_B = \frac{1}{2}LI^2 = \frac{1}{2}(50.0\text{ H})\left(50.0\times10^{3}\text{ A}\right)^2 = \boxed{6.25\times10^{10}\text{ J}}$

(b) Two adjacent turns are parallel wires carrying current in the same direction. Since the loops have such large radius, a one-meter section can be regarded as straight.

Then one wire creates a field of $B = \frac{\mu_0 I}{2\pi r}$.

This causes a force on the next wire of $F = I\ell B\sin\theta$

giving $F = I\ell\frac{\mu_0 I}{2\pi r}\sin 90° = \frac{\mu_0 \ell I^2}{2\pi r}$.

Evaluating the force, $F = \left(4\pi\times10^{-7}\text{ N/A}^2\right)\frac{(1.00\text{ m})\left(50.0\times10^{3}\text{ A}\right)^2}{2\pi(0.250\text{ m})} = \boxed{2\,000\text{ N}}$.

P23.65 $\mathscr{P} = I\Delta V \qquad I = \frac{\mathscr{P}}{\Delta V} = \frac{1.00\times10^{9}\text{ W}}{200\times10^{3}\text{ V}} = 5.00\times10^{3}\text{ A}$

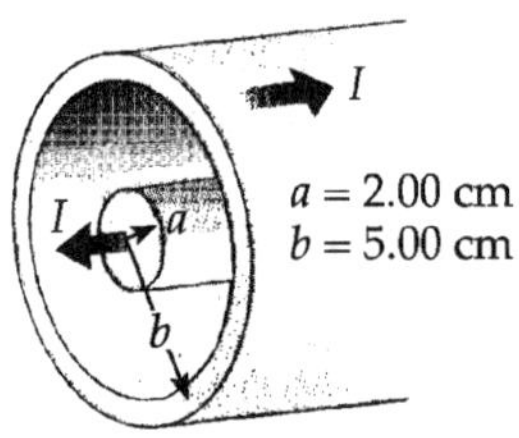

FIG. P23.65

From Ampère's law, $B(2\pi r) = \mu_0 I_{\text{enclosed}}$ or $B = \frac{\mu_0 I_{\text{enclosed}}}{2\pi r}$.

(a) At $r = a = 0.020\,0\text{ m}$, $I_{\text{enclosed}} = 5.00\times10^{3}\text{ A}$

and $B = \frac{\left(4\pi\times10^{-7}\text{ T}\cdot\text{m/A}\right)\left(5.00\times10^{3}\text{ A}\right)}{2\pi(0.020\,0\text{ m})} = 0.050\,0\text{ T} = \boxed{50.0\text{ mT}}$.

(b) At $r = b = 0.050\,0\text{ m}$, $I_{\text{enclosed}} = I = 5.00\times10^{3}\text{ A}$

and $B = \frac{\left(4\pi\times10^{-7}\text{ T}\cdot\text{m/A}\right)\left(5.00\times10^{3}\text{ A}\right)}{2\pi(0.050\,0\text{ m})} = 0.020\,0\text{ T} = \boxed{20.0\text{ mT}}$.

continued on next page

(c) $$U=\int u\,dV=\int_a^b \frac{[B(r)]^2(2\pi r\ell dr)}{2\mu_0}=\frac{\mu_0 I^2\ell}{4\pi}\int_a^b \frac{dr}{r}=\frac{\mu_0 I^2\ell}{4\pi}\ln\left(\frac{b}{a}\right)$$

$$U=\frac{(4\pi\times10^{-7}\ \text{T}\cdot\text{m/A})(5.00\times10^{3}\ \text{A})^2(1\,000\times10^{3}\ \text{m})}{4\pi}\ln\left(\frac{5.00\ \text{cm}}{2.00\ \text{cm}}\right)=2.29\times10^{6}\ \text{J}=\boxed{2.29\ \text{MJ}}$$

(d) The magnetic field created by the inner conductor exerts a force of repulsion on the current in the outer sheath. The strength of this field, from part (b), is 20.0 mT. Consider a small rectangular section of the outer cylinder of length ℓ and width w.

It carries a current of $$(5.00\times10^{3}\ \text{A})\left(\frac{w}{2\pi(0.050\,0\ \text{m})}\right)$$

and experiences an outward force $$F=I\ell B\sin\theta=\frac{(5.00\times10^{3}\ \text{A})w}{2\pi(0.050\,0\ \text{m})}\ell(20.0\times10^{-3}\ \text{T})\sin 90.0°.$$

The pressure on it is $$P=\frac{F}{A}=\frac{F}{w\ell}=\frac{(5.00\times10^{3}\ \text{A})(20.0\times10^{-3}\ \text{T})}{2\pi(0.050\,0\ \text{m})}=\boxed{318\ \text{Pa}}.$$

ANSWERS TO EVEN PROBLEMS

P23.2 9.82 mV

P23.4 (a) $\frac{\mu_0 n\pi r_2^2}{2R}\left(\frac{\Delta I}{\Delta t}\right)$; (b) $\frac{\mu_0^2 n\pi r_2^2}{4r_1 R}\left(\frac{\Delta I}{\Delta t}\right)$; (c) upward

P23.6 $-(14.2\ \text{mV})\cos(120t)$

P23.8 (a) see the solution; (b) 625 m/s

P23.10 13.3 mA counterclockwise in the lower loop and clockwise in the upper loop

P23.12 1.00 m/s

P23.14 (a) 0.500 A; (b) 2.00 W; (c) 2.00 W

P23.16 360 T

P23.18 (a) 233 Hz; (b) 1.98 mV

P23.20 (a) to the right; (b) to the right; (c) to the right; (d) into the paper

P23.22 (a) $F=\frac{N^2B^2w^2v}{R}$ to the left; (b) 0; (c) $F=\frac{N^2B^2w^2v}{R}$ to the left

P23.24 $(28.6\ \text{mV})\sin(4\pi t)$

P23.26 (a) 8.00×10^{-21} N, clockwise; (b) 1.33 s and 0

P23.28 1.36 μH

P23.30 19.2 $\mu\text{T}\cdot\text{m}^2$

P23.32 see the solution

P23.34 see the solution

P23.36 (a) 0.800; (b) 0

P23.38 7.67 mH

P23.40 (a) 1.00 A; (b) 12.0 V, 1.20 kV, 1.21 kV; (c) 7.62 ms

P23.42 (a) $8.06 \times 10^6 \ \text{J/m}^3$; (b) 6.32 kJ

P23.44 (a) 0.500 J; (b) 17.0 W; (c) 11.0 W

P23.46 2×10^2 T

P23.48 see the solution
(a) doubling N doubles amplitude;
(b) doubling ω doubles amplitude and halves period;
(c) doubling ω and halving N leaves the amplitude the same and cuts the period in half

P23.50 (a) 43.8 A; (b) 38.3 W

P23.52 see the solution

P23.54 (a) $\dfrac{\pi B_{\text{max}}^2 R^4 \omega^2 b}{16\rho}$; (b) 4; (c) 4; (d) 16

P23.56 (a) 0.125 V to produce clockwise current; (b) 20.0 mA clockwise

P23.58 $-10.2\ \mu$V

P23.60 (a) see the solution; (b) 91.2 μH; (c) 90.9 μH

P23.62 300 Ω

P23.64 (a) 62.5 GJ; (b) 2 000 N

CONTEXT 7 CONCLUSION SOLUTIONS TO PROBLEMS

CC7.1 The alternating current must complete a cycle of reversing and reversing again as the vehicle travels forward by $2(10 \text{ m}) = 20 \text{ m}$. The period of alternation must be

$$T = \frac{\Delta x}{v} = \frac{20 \text{ m} \cdot \text{h}}{400 \text{ km}} \left(\frac{\text{km}}{1\,000 \text{ m}} \right) \left(\frac{3\,600 \text{ s}}{\text{h}} \right) = 0.180 \text{ s}.$$

So the frequency is $f = \dfrac{1}{T} = \boxed{5.56 \text{ Hz}}$.

CC7.2 The current in the magnet creates an  upward magnetic field, so the N and S poles on the solenoid core are shown correctly. On the rail in front of the brake, the upward flux of $\vec{\mathbf{B}}$ increases as the coil approaches, so a current is induced here to create a downward magnetic field. This is  clockwise current, so the S pole on the rail is shown correctly. On the rail behind the brake, the upward magnetic flux is decreasing. The induced current in the rail will produce upward magnetic field by being  counterclockwise as the picture correctly shows.

ANSWERS TO EVEN CONTEXT 7 CONCLUSION PROBLEMS

CC7.2 both are correct

Chapter 24

Electromagnetic Waves

CHAPTER OUTLINE

ANSWERS TO QUESTIONS

Q24.1 Radio waves move at the speed of light. They can travel around the curved surface of the Earth, bouncing between the ground and the ionosphere, which has an altitude that is small when compared to the radius of the Earth. The distance across the lower forty-eight states is approximately 5 000 km, requiring a transit time of $\frac{5\times10^6 \text{ m}}{3\times10^8 \text{ m/s}} \sim 10^{-2}$ s. To go halfway around the Earth takes only 0.07 s. In other words, a speech can be heard on the other side of the world before it is heard at the back of a large room.

Q24.2 Energy moves. No matter moves. You could say that electric and magnetic fields move, but it is nicer to say that the fields at one point stay at that point and oscillate. The fields vary in time, like sports fans in the grandstand when the crowd does the wave. The fields constitute the medium for the wave, and energy moves.

Q24.3 Acceleration of electric charge.

Q24.4 A wire connected to the terminals of a battery does not radiate electromagnetic waves. The battery establishes an electric field, which produces current in the wire. The current in the wire creates a magnetic field. Both fields are constant in time, so no electromagnetic induction or "magneto-electric induction" happens. Neither field creates a new cycle of the other field. No wave propagation occurs.

Q24.5 No. Static electricity is just that: static. Without acceleration of the charge, there can be no electromagnetic wave.

Q24.6

Sound	*Light*
The world of sound extends to the top of the atmosphere and stops there; sound requires a material medium. Sound propagates by a chain reaction of density and pressure disturbances recreating each other. Sound in air moves at hundreds of meters per second. Audible sound has frequencies over a range of three decades (ten octaves) from 20 Hz to 20 kHz. Audible sound has wavelengths of ordinary size (1.7 cm to 17 m). Sound waves are longitudinal.	The universe of light fills the whole universe. Light moves through materials, but faster in a vacuum. Light propagates by a chain reaction of electric and magnetic fields recreating each other. Light in air moves at hundreds of millions of meters per second. Visible light has frequencies over a range of less than one octave, from 430 to 750 **Tera**hertz. Visible light has wavelengths of very small size (400 nm to 700 nm). Light waves are transverse.

continued on next page

Sound and light can both be reflected, refracted, or absorbed to produce internal energy. Both have amplitude and frequency set by the source, speed set by the medium, and wavelength set by both source and medium. Sound and light both exhibit the Doppler effect, standing waves, beats, interference, diffraction, and resonance. Both can be focused to make images. Both are described by wave functions satisfying wave equations. **Both carry energy**. If the source is small, their intensities both follow an inverse-square law. Both are waves.

Q24.7 When the capacitor is fully discharged, the current in the circuit is a maximum. The inductance of the coil is making the current continue to flow. At this time the magnetic field of the coil contains all the energy that was originally stored in the charged capacitor. The current has just finished discharging the capacitor and is proceeding to charge it up again with the opposite polarity.

Q24.8 The Poynting vector $\vec{S}$ describes the energy flow associated with an electromagnetic wave. The direction of $\vec{S}$ is along the direction of propagation and the magnitude of $\vec{S}$ is the rate at which electromagnetic energy crosses a unit surface area perpendicular to the direction of $\vec{S}$.

Q24.9 Different stations have transmitting antennas at different locations. For best reception align your rabbit ears perpendicular to the straight-line path from your TV to the transmitting antenna. The transmitted signals are also polarized. The polarization direction of the wave can be changed by reflection from surfaces—including the atmosphere—and through Kerr rotation—a change in polarization axis when passing through an organic substance. In your home, the plane of polarization is determined by your surroundings, so antennas need to be adjusted to align with the polarization of the wave.

Q24.10 You become part of the receiving antenna! You are a big sack of salt water. Your contribution usually increases the gain of the antenna by a few tenths of a dB, enough to noticeably improve reception.

Q24.11 Consider a typical metal rod antenna for a car radio. The rod detects the electric field portion of the carrier wave. Variations in the amplitude of the incoming radio wave cause the electrons in the rod to vibrate with amplitudes emulating those of the carrier wave. Likewise, for frequency modulation, the variations of the frequency of the carrier wave cause constant-amplitude vibrations of the electrons in the rod but at frequencies that imitate those of the carrier.

Q24.12 The frequency of EM waves in a microwave oven, typically 2.45 GHz, is chosen to be in a band of frequencies absorbed by water molecules. The plastic and the glass contain no water molecules. Plastic and glass have very different absorption frequencies from water, so they may not absorb any significant microwave energy and remain cool to the touch.

Q24.13 Lightbulbs and the toaster shine brightly in the infrared. Somewhat fainter are the back of the refrigerator and the back of the television set, while the TV screen is dark. The pipes under the sink show the same weak sheen as the walls until you turn on the faucets. Then the pipe on the right turns very black while that on the left develops a rich glow that quickly runs up along its length. The food on your plate shines; so does human skin, the same color for all races. Clothing is dark as a rule, but your bottom glows like a monkey's rump when you get up from a chair, and you leave behind a patch of the same blush on the chair seat. Your face shows you are lit from within, like a jack-o-lantern: your nostrils and the openings of your ear canals are bright; brighter still are just the pupils of your eyes.

Q24.14 People of all the world's races have skin the same color in the infrared. When you blush or exercise or get excited, you stand out like a beacon in an infrared group picture. The brightest portions of your face show where you radiate the most. Your nostrils and the openings of your ear canals are bright; brighter still are just the pupils of your eyes.

Q24.15 Welding produces ultraviolet light, along with high intensity visible and infrared.

Q24.16 12.2 cm waves have a frequency of 2.46 GHz. If the Q value of the phone is low (namely if it is cheap), and your microwave oven is not well shielded (namely, if it is also cheap), the phone can likely pick up interference from the oven. If the phone is well constructed and has a high Q value, then there should be no interference at all.

Q24.17 Stimulated emission coerces atoms to emit photons along a specific axis, rather than in the random directions of spontaneously emitted photons. The photons that are emitted through stimulation can be made to accumulate over time. The fraction allowed to escape constitutes the intense, collimated, and coherent laser beam. If this process relied solely on spontaneous emission, the emitted photons would not exit the laser tube or crystal in the same direction. Neither would they be coherent with one another.

Q24.18 Photons carry momentum. Recalling what we learned in Chapter 8, the impulse imparted to a particle that bounces elastically is twice that imparted to an object that sticks to a massive wall. Similarly, the impulse, and hence the pressure exerted by a photon reflecting from a surface must be twice that exerted by a photon that is absorbed.

SOLUTIONS TO PROBLEMS

Section 24.1 Displacement Current and the Generalized Ampère's Law

P24.1 We use the extended form of Ampère's law, Equation 24.7. Since no moving charges are present, $I=0$ and we have

$$\oint \vec{\mathbf{B}}\cdot d\vec{\ell}=\mu_0\in_0\frac{d\Phi_E}{dt}.$$

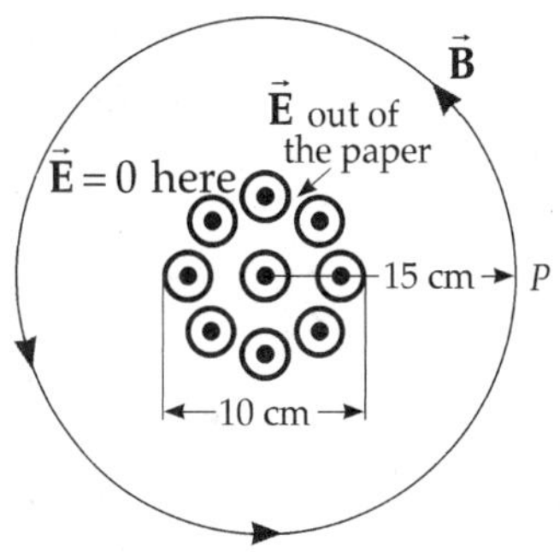

FIG. P24.1

In order to evaluate the integral, we make use of the symmetry of the situation. Symmetry requires that no particular direction from the center can be any different from any other direction. Therefore, there must be *circular symmetry* about the central axis. We know the magnetic field lines are circles about the axis. Therefore, as we travel around such a magnetic field circle, the magnetic field remains constant in magnitude. Setting aside until later the determination of the *direction* of $\vec{\mathbf{B}}$, we integrate $\oint \vec{\mathbf{B}}\cdot d\vec{\ell}$ around the circle

at $R=0.15$ m

to obtain $2\pi RB$.

Differentiating the expression $\Phi_E=AE$

we have $\dfrac{d\Phi_E}{dt}=\left(\dfrac{\pi d^2}{4}\right)\dfrac{dE}{dt}$.

Thus, $\oint \vec{\mathbf{B}}\cdot d\vec{\ell}=2\pi RB=\mu_0\in_0\left(\dfrac{\pi d^2}{4}\right)\dfrac{dE}{dt}$

continued on next page

Solving for B gives $\quad B=\dfrac{\mu_0 \in_0}{2\pi R}\left(\dfrac{\pi d^2}{4}\right)\dfrac{dE}{dt}.$

Substituting numerical values, $\quad B=\dfrac{\left(4\pi\times10^{-7}\ \text{H/m}\right)\left(8.85\times10^{-12}\ \text{F/m}\right)\left[\pi(0.10\ \text{m})^2\right](20\ \text{V/m}\cdot\text{s})}{2\pi(0.15\ \text{m})(4)}$

$$B=\boxed{1.85\times10^{-18}\ \text{T}}.$$

In Figure 24.1, the direction of the *increase* of the electric field is out the plane of the paper. By the right-hand rule, this implies that the direction of $\vec{\mathbf{B}}$ is *counterclockwise*. Thus, the direction of $\vec{\mathbf{B}}$ at P is upwards.

Section 24.2 Maxwell's Equations

P24.2 (a) The rod creates the same electric field that it would if stationary. We apply Gauss's law to a cylinder of radius $r=20$ cm and length ℓ:

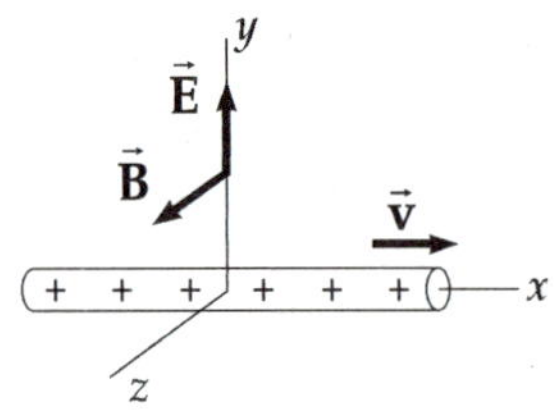

FIG. P24.2

$$\oint \vec{\mathbf{E}}\cdot d\vec{\mathbf{A}}=\frac{q_{\text{inside}}}{\in_0}$$

$$E(2\pi r l)\cos 0°=\frac{\lambda l}{\in_0}$$

$$\vec{\mathbf{E}}=\frac{\lambda}{2\pi\in_0 r}\ \text{radially outward}=\frac{\left(35\times10^{-9}\ \text{C/m}\right)\text{N}\cdot\text{m}^2}{2\pi\left(8.85\times10^{-12}\ \text{C}^2\right)(0.2\ \text{m})}\hat{\mathbf{j}}=\boxed{3.15\times10^3\hat{\mathbf{j}}\ \text{N/C}}.$$

(b) The charge in motion constitutes a current of $\left(35\times10^{-9}\ \text{C/m}\right)\left(15\times10^6\ \text{m/s}\right)=0.525\ \text{A}$. This current creates a magnetic field.

$$\vec{\mathbf{B}}=\frac{\mu_0 I}{2\pi r}$$  $$=\frac{\left(4\pi\times10^{-7}\ \text{T}\cdot\text{m/A}\right)(0.525\ \text{A})}{2\pi(0.2\ \text{m})}\hat{\mathbf{k}}=\boxed{5.25\times10^{-7}\hat{\mathbf{k}}\ \text{T}}$$

(c) The Lorentz force on the electron is $F=q\vec{\mathbf{E}}+q\vec{\mathbf{v}}\times\vec{\mathbf{B}}$

$$\vec{\mathbf{F}}=\left(-1.6\times10^{-19}\ \text{C}\right)\left(3.15\times10^3\hat{\mathbf{j}}\ \text{N/C}\right)+\left(-1.6\times10^{-19}\ \text{C}\right)\left(240\times10^6\hat{\mathbf{i}}\ \text{m/s}\right)\times\left(5.25\times10^{-7}\hat{\mathbf{k}}\ \frac{\text{N}\cdot\text{s}}{\text{C}\cdot\text{m}}\right)$$

$$\vec{\mathbf{F}}=5.04\times10^{-16}\left(-\hat{\mathbf{j}}\right)\text{N}+2.02\times10^{-17}\left(+\hat{\mathbf{j}}\right)\text{N}=\boxed{4.83\times10^{-16}\left(-\hat{\mathbf{j}}\right)\text{N}}$$

***P24.3** $\vec{\mathbf{F}} = m\vec{\mathbf{a}} = q\vec{\mathbf{E}} + q\vec{\mathbf{v}} \times \vec{\mathbf{B}}$

$$\vec{\mathbf{a}} = \frac{e}{m}\left[\vec{\mathbf{E}} + \vec{\mathbf{v}} \times \vec{\mathbf{B}}\right] \text{ where } \vec{\mathbf{v}} \times \vec{\mathbf{B}} = \begin{vmatrix} \hat{\mathbf{i}} & \hat{\mathbf{j}} & \hat{\mathbf{k}} \\ 200 & 0 & 0 \\ 0.200 & 0.300 & 0.400 \end{vmatrix} = -200(0.400)\hat{\mathbf{j}} + 200(0.300)\hat{\mathbf{k}}$$

$$\vec{\mathbf{a}} = \frac{1.60 \times 10^{-19}}{1.67 \times 10^{-27}}\left[50.0\hat{\mathbf{j}} - 80.0\hat{\mathbf{j}} + 60.0\hat{\mathbf{k}}\right] = 9.58 \times 10^{7}\left[-30.0\hat{\mathbf{j}} + 60.0\hat{\mathbf{k}}\right]$$

$$\vec{\mathbf{a}} = 2.87 \times 10^{9}\left[-\hat{\mathbf{j}} + 2\hat{\mathbf{k}}\right] \text{ m/s}^2 = \boxed{\left(-2.87 \times 10^{9}\hat{\mathbf{j}} + 5.75 \times 10^{9}\hat{\mathbf{k}}\right) \text{ m/s}^2}$$

Section 24.3 Electromagnetic Waves

P24.4 (a) Since the light from this star travels at 3.00×10^{8} m/s

the last bit of light will hit the Earth in $\dfrac{6.44 \times 10^{18} \text{ m}}{3.00 \times 10^{8} \text{ m/s}} = 2.15 \times 10^{10} \text{ s} = 680 \text{ years}$.

Therefore, it will disappear from the sky in the year $2\,004 + 680 = \boxed{2.68 \times 10^{3} \text{ C.E.}}$.

The star is 680 light-years away.

(b) $$\Delta t = \frac{\Delta x}{v} = \frac{1.496 \times 10^{11} \text{ m}}{3 \times 10^{8} \text{ m/s}} = \boxed{499 \text{ s}} = 8.31 \text{ min}$$

(c) $$\Delta t = \frac{\Delta x}{v} = \frac{2\left(3.84 \times 10^{8} \text{ m}\right)}{3 \times 10^{8} \text{ m/s}} = \boxed{2.56 \text{ s}}$$

(d) $$\Delta t = \frac{\Delta x}{v} = \frac{2\pi\left(6.37 \times 10^{6} \text{ m}\right)}{3 \times 10^{8} \text{ m/s}} = \boxed{0.133 \text{ s}}$$

(e) $$\Delta t = \frac{\Delta x}{v} = \frac{10 \times 10^{3} \text{ m}}{3 \times 10^{8} \text{ m/s}} = \boxed{3.33 \times 10^{-5} \text{ s}}$$

P24.5 $$v = \frac{1}{\sqrt{\kappa\mu_0 \in_0}} = \frac{1}{\sqrt{1.78}}c = 0.750c = \boxed{2.25 \times 10^{8} \text{ m/s}}$$

P24.6 $$\frac{E}{B} = c$$

or $$\frac{220}{B} = 3.00 \times 10^{8}$$

so $B = 7.33 \times 10^{-7} \text{ T} = \boxed{733 \text{ nT}}$.

P24.7 (a) $f\lambda = c$

or $f(50.0\text{ m}) = 3.00\times10^{8}\text{ m/s}$

so $\boxed{f = 6.00\times10^{6}\text{ Hz} = 6.00\text{ MHz}}$.

(b) $\dfrac{E}{B} = c$

or $\dfrac{22.0}{B_{\max}} = 3.00\times10^{8}$

so $\vec{\mathbf{B}}_{\max} = \boxed{-73.3\hat{\mathbf{k}}\text{ nT}}$.

(c) $k = \dfrac{2\pi}{\lambda} = \dfrac{2\pi}{50.0} = 0.126\text{ m}^{-1}$

and $\omega = 2\pi f = 2\pi\left(6.00\times10^{6}\text{ s}^{-1}\right) = 3.77\times10^{7}\text{ rad/s}$

$$\vec{\mathbf{B}} = \vec{\mathbf{B}}_{\max}\cos(kx-\omega t) = \boxed{-73.3\cos\left(0.126x - 3.77\times10^{7}t\right)\hat{\mathbf{k}}\text{ nT}}.$$

P24.8 (a) $B = \dfrac{E}{c} = \dfrac{100\text{ V/m}}{3.00\times10^{8}\text{ m/s}} = 3.33\times10^{-7}\text{ T} = \boxed{0.333\ \mu\text{T}}$

(b) $\lambda = \dfrac{2\pi}{k} = \dfrac{2\pi}{1.00\times10^{7}\text{ m}^{-1}} = \boxed{0.628\ \mu\text{m}}$

(c) $f = \dfrac{c}{\lambda} = \dfrac{3.00\times10^{8}\text{ m/s}}{6.28\times10^{-7}\text{ m}} = \boxed{4.77\times10^{14}\text{ Hz}}$

P24.9 $E = E_{\max}\cos(kx-\omega t)$

$$\frac{\partial E}{\partial x} = -E_{\max}\sin(kx-\omega t)(k)$$
$$\frac{\partial E}{\partial t} = -E_{\max}\sin(kx-\omega t)(-\omega)$$
$$\frac{\partial^2 E}{\partial x^2} = -E_{\max}\cos(kx-\omega t)\left(k^2\right)$$
$$\frac{\partial^2 E}{\partial t^2} = -E_{\max}\cos(kx-\omega t)(-\omega)^2$$

We must show: $\dfrac{\partial E}{\partial x^2} = \mu_0\in_0\dfrac{\partial^2 E}{\partial t^2}$.

That is, $-\left(k^2\right)E_{\max}\cos(kx-\omega t) = -\mu_0\in_0(-\omega)^2E_{\max}\cos(kx-\omega t)$.

But this is true, because $\dfrac{k^2}{\omega^2} = \left(\dfrac{1}{f\lambda}\right)^2 = \dfrac{1}{c^2} = \mu_0\in_0$.

The proof for the wave of magnetic field follows precisely the same steps.

P24.10 In the fundamental mode, there is a single loop in the standing wave between the plates. Therefore, the distance between the plates is equal to half a wavelength.

$$\lambda = 2L = 2(2.00\text{ m}) = 4.00\text{ m}$$

Thus, $$f = \frac{c}{\lambda} = \frac{3.00\times10^{8}\text{ m/s}}{4.00\text{ m}} = 7.50\times10^{7}\text{ Hz} = \boxed{75.0\text{ MHz}}.$$

P24.11 $d_{\text{A to A}} = 6\text{ cm}\pm5\% = \frac{\lambda}{2}$

$\lambda = 12\text{ cm}\pm5\%$

$v = \lambda f = (0.12\text{ m}\pm5\%)\left(2.45\times10^{9}\text{ s}^{-1}\right) = \boxed{2.9\times10^{8}\text{ m/s}\pm5\%}$

***P24.12** The orbital speed of the Earth is as described by $\sum F = ma$: $\frac{Gm_Sm_E}{r^2} = \frac{m_Ev^2}{r}$

$$v = \sqrt{\frac{Gm_S}{r}} = \sqrt{\frac{\left(6.67\times10^{-11}\text{ N}\cdot\text{m}^2/\text{kg}^2\right)\left(1.99\times10^{30}\text{ kg}\right)}{1.496\times10^{11}\text{ m}}} = 2.98\times10^{4}\text{ m/s}.$$

The maximum frequency received by the extraterrestrials is

$$f_{\text{obs}} = f_{\text{source}}\sqrt{\frac{1+v/c}{1-v/c}} = \left(57.0\times10^{6}\text{ Hz}\right)\sqrt{\frac{1+\left(2.98\times10^{4}\text{ m/s}\right)/\left(3.00\times10^{8}\text{ m/s}\right)}{1-\left(2.98\times10^{4}\text{ m/s}\right)/\left(3.00\times10^{8}\text{ m/s}\right)}} = 57.005\,66\times10^{6}\text{ Hz}.$$

The minimum frequency received is

$$f_{\text{obs}} = f_{\text{source}}\sqrt{\frac{1-v/c}{1+v/c}} = \left(57.0\times10^{6}\text{ Hz}\right)\sqrt{\frac{1-\left(2.98\times10^{4}\text{ m/s}\right)/\left(3.00\times10^{8}\text{ m/s}\right)}{1+\left(2.98\times10^{4}\text{ m/s}\right)/\left(3.00\times10^{8}\text{ m/s}\right)}} = 56.994\,34\times10^{6}\text{ Hz}.$$

The difference, which lets them figure out the speed of our planet, is

$$(57.005\,66 - 56.994\,34)\times10^{6}\text{ Hz} = \boxed{1.13\times10^{4}\text{ Hz}}.$$

***P24.13** (a) Let f_c be the frequency as seen by the car. Thus, $f_c = f_{\text{source}}\sqrt{\frac{c+v}{c-v}}$

and, if f is the frequency of the reflected wave, $f = f_c\sqrt{\frac{c+v}{c-v}}$.

Combining gives $\boxed{f = f_{\text{source}}\frac{(c+v)}{(c-v)}}$.

(b) Using the above result, $f(c-v) = f_{\text{source}}(c+v)$

which gives $(f - f_{\text{source}})c = (f + f_{\text{source}})v \approx 2f_{\text{source}}v$.

The beat frequency is then $f_{\text{beat}} = f - f_{\text{source}} = \frac{2f_{\text{source}}v}{c} = \boxed{\frac{2v}{\lambda}}$.

continued on next page

(c) $$f_{\text{beat}} = \frac{(2)(30.0 \text{ m/s})(10.0\times10^9 \text{ Hz})}{3.00\times10^8 \text{ m/s}} = \frac{(2)(30.0 \text{ m/s})}{(0.030\,0 \text{ m})} = 2\,000 \text{ Hz} = \boxed{2.00 \text{ kHz}}$$

$$\lambda = \frac{c}{f_{\text{source}}} = \frac{3.00\times10^8 \text{ m/s}}{10.0\times10^9 \text{ Hz}} = 3.00 \text{ cm}$$

(d) $$v = \frac{f_{\text{beat}}\lambda}{2} \quad \text{so} \quad \Delta v = \frac{\Delta f_{\text{beat}}\lambda}{2} = \frac{(5 \text{ Hz})(0.030\,0 \text{ m})}{2} = \boxed{0.075\,0 \text{ m/s} \approx 0.2 \text{ mi/h}}$$

P24.14 (a) When the source moves away from an observer, the observed frequency is

$$f_{\text{obs}} = f_{\text{source}}\left(\frac{c - v_s}{c + v_s}\right)^{1/2} \quad \text{where } v_s = v_{\text{source}}.$$

When $v_s << c$, the binomial expansion gives

$$\left(\frac{c - v_s}{c + v_s}\right)^{1/2} = \left[1 - \left(\frac{v_s}{c}\right)\right]^{1/2}\left[1 + \left(\frac{v_s}{c}\right)\right]^{-1/2} \approx \left(1 - \frac{v_s}{2c}\right)\left(1 - \frac{v_s}{2c}\right) \approx \left(1 - \frac{v_s}{c}\right).$$

So, $$f_{\text{obs}} \approx f_{\text{source}}\left(1 - \frac{v_s}{c}\right).$$

The observed wavelength is found from $c = \lambda_{\text{obs}} f_{\text{obs}} = \lambda f_{\text{source}}$:

$$\lambda_{\text{obs}} = \frac{\lambda f_{\text{source}}}{f_{\text{obs}}} \approx \frac{\lambda f_{\text{source}}}{f_{\text{source}}(1 - v_s/c)} = \frac{\lambda}{1 - v_s/c}$$

$$\Delta\lambda = \lambda_{\text{obs}} - \lambda = \lambda\left(\frac{1}{1 - v_s/c} - 1\right) = \lambda\left(\frac{v_s/c}{1 - v_s/c}\right).$$

Since $1 - \frac{v_s}{c} \approx 1$, $\boxed{\frac{\Delta\lambda}{\lambda} \approx \frac{v_{\text{source}}}{c}}$.

(b) $$v_{\text{source}} = c\left(\frac{\Delta\lambda}{\lambda}\right) = c\left(\frac{20.0 \text{ nm}}{397 \text{ nm}}\right) = \boxed{0.050\,4c}$$

***P24.15** For the light as observed

$$f_{\text{obs}} = \frac{c}{\lambda_{\text{obs}}} = \sqrt{\frac{1 + v/c}{1 - v/c}} f_{\text{source}} = \sqrt{\frac{1 + v/c}{1 - v/c}} \frac{c}{\lambda_{\text{source}}}$$

$$\sqrt{\frac{1 + v/c}{1 - v/c}} = \frac{\lambda_{\text{source}}}{\lambda_{\text{obs}}} = \frac{650 \text{ nm}}{520 \text{ nm}} \qquad \frac{1 + v/c}{1 - v/c} = 1.25^2 = 1.562$$

$$1 + \frac{v}{c} = 1.562 - 1.562\frac{v}{c} \qquad \frac{v}{c} = \frac{0.562}{2.562} = 0.220$$

$$v = \boxed{0.220c} = 6.59\times10^7 \text{ m/s}$$

***P24.16** (a) Consider the raindrops moving toward the station, at speed v. They receive radio waves with the Doppler-enhanced frequency $f' = f\sqrt{\frac{c+v}{c-v}}$ where $f = 2.85$ GHz. These raindrops reflect the waves at frequency f'. The waves are received by the station with another upward Doppler shift in frequency to

$$f'' = f'\sqrt{\frac{c+v}{c-v}} = f\sqrt{\frac{c+v}{c-v}}\sqrt{\frac{c+v}{c-v}}$$

$$2.85\times10^9\text{ Hz} + 254\text{ Hz} = 2.85\times10^9\text{ Hz}\left(\frac{c+v}{c-v}\right)$$

$$1 + 8.91\times10^{-8} = \frac{c+v}{c-v}$$

$$c + 8.91\times10^{-8}c - v - 8.91\times10^{-8}v = c + v$$

$$8.91\times10^{-8}c = 2.000\,000\,089v$$

$$v = \left(4.46\times10^{-8}\right)\left(3\times10^8\text{ m/s}\right) = \boxed{13.4\text{ m/s}}.$$

The same calculation with 254 Hz replaced by –254 Hz applies to the receding raindrops and given the same velocity magnitude. Thus the velocities are 13.4 m/s toward the station and 13.4 m/s away from the station.

(b) Radio waves travel to the rain and back again in 180 μs, so the one-way distance is $\frac{1}{2}\left(3\times10^8\text{ m/s}\right)\left(180\times10^{-6}\text{ s}\right) = 27\,000$ m. The diameter of the vortex is $s = r\theta = 27\,000\text{ m}(1°)\left(\frac{\pi\text{ rad}}{180°}\right) = 471$ m. So its radius is $\frac{1}{2}(471\text{ m}) = 236$ m and the angular speed of the rain is $\omega = \frac{v}{r} = \frac{13.4\text{ m/s}}{236\text{ m}} = \boxed{0.0567\text{ rad/s}}$. A Doppler weather radar computer performs a calculation like this to detect a "tornado vortex signature."

Section 24.4 Hertz's Discoveries

P24.17 This radio is a radiotelephone on a ship, according to frequency assignments made by international treaties, laws, and decisions of the National Telecommunications and Information Administration.

The resonance frequency is $f_0 = \frac{1}{2\pi\sqrt{LC}}$.

Thus, $C = \frac{1}{(2\pi f_0)^2 L} = \frac{1}{\left[2\pi\left(6.30\times10^6\text{ Hz}\right)\right]^2\left(1.05\times10^{-6}\text{ H}\right)} = \boxed{608\text{ pF}}$.

P24.18 $f = \frac{1}{2\pi\sqrt{LC}}$: $L = \frac{1}{(2\pi f)^2 C} = \frac{1}{[2\pi(120)]^2\left(8.00\times10^{-6}\right)} = \boxed{0.220\text{ H}}$

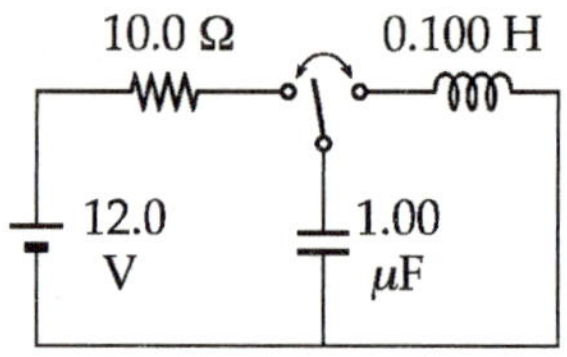

FIG. P24.19

P24.19 (a) $f=\frac{1}{2\pi\sqrt{LC}}=\frac{1}{2\pi\sqrt{(0.100\text{ H})\left(1.00\times10^{-6}\text{ F}\right)}}=\boxed{503\text{ Hz}}$

(b) $Q=C\mathcal{E}=\left(1.00\times10^{-6}\text{ F}\right)(12.0\text{ V})=\boxed{12.0\ \mu\text{C}}$

(c) $\frac{1}{2}C\mathcal{E}^2=\frac{1}{2}LI_{\max}^2$

$$I_{\max}=\mathcal{E}\sqrt{\frac{C}{L}}=12\text{ V}\sqrt{\frac{1.00\times10^{-6}\text{ F}}{0.100\text{ H}}}=\boxed{37.9\text{ mA}}$$

(d) At all times $U=\frac{1}{2}C\mathcal{E}^2=\frac{1}{2}\left(1.00\times10^{-6}\text{ F}\right)(12.0\text{ V})^2=\boxed{72.0\ \mu\text{J}}$.

Section 24.5 Energy Carried by Electromagnetic Waves

P24.20 $S=I=\frac{U}{At}=\frac{Uc}{V}=uc$ $\qquad \frac{\text{Energy}}{\text{Unit Volume}}=u=\frac{I}{c}=\frac{1\,000\text{ W/m}^2}{3.00\times10^8\text{ m/s}}=\boxed{3.33\ \mu\text{J/m}^3}$

P24.21 $r=(5.00\text{ mi})(1\,609\text{ m/mi})=8.04\times10^3\text{ m}$

$$S=\frac{\overline{\mathcal{P}}}{4\pi r^2}=\frac{250\times10^3\text{ W}}{4\pi\left(8.04\times10^3\text{ W}\right)^2}=\boxed{307\ \mu\text{W/m}^2}$$

P24.22 $S_{\text{av}}=\frac{\overline{\mathcal{P}}}{4\pi r^2}=\frac{4.00\times10^3\text{ W}}{4\pi(4.00\times1\,609\text{ m})^2}=7.68\ \mu\text{W/m}^2$

$$E_{\max}=\sqrt{2\mu_0cS_{\text{av}}}=0.076\,1\text{ V/m}$$

$$\Delta V_{\max}=E_{\max}L=(76.1\text{ mV/m})(0.650\text{ m})=\boxed{49.5\text{ mV (amplitude)}}\text{ or 35.0 mV (rms)}$$

P24.23 Power output = (power input)(efficiency).

Thus, $\qquad$ $\text{Power input}=\frac{\text{Power out}}{\text{eff}}=\frac{1.00\times10^6\text{ W}}{0.300}=3.33\times10^6\text{ W}$

and $\qquad$ $A=\frac{\mathcal{P}}{I}=\frac{3.33\times10^6\text{ W}}{1.00\times10^3\text{ W/m}^2}=\boxed{3.33\times10^3\text{ m}^2}$.

P24.24 (a) $$I=\frac{E_{max}^2}{2\mu_0 c}=\frac{\left(3\times10^6\text{ V/m}\right)^2}{2\left(4\pi\times10^{-7}\text{ T}\cdot\text{m/A}\right)\left(3\times10^8\text{ m/s}\right)}\left(\frac{\text{J}}{\text{V}\cdot\text{C}}\right)^2\left(\frac{\text{C}}{\text{A}\cdot\text{s}}\right)\left(\frac{\text{T}\cdot\text{C}\cdot\text{m}}{\text{N}\cdot\text{s}}\right)\left(\frac{\text{N}\cdot\text{m}}{\text{J}}\right)$$

$$I=\boxed{1.19\times10^{10}\text{ W/m}^2}$$

(b) $$\mathcal{P}=IA=\left(1.19\times10^{10}\text{ W/m}^2\right)\pi\left(\frac{5\times10^{-3}\text{ m}}{2}\right)^2=\boxed{2.34\times10^5\text{ W}}$$

P24.25 (a) $$\mathcal{P}=I^2R=150\text{ W}$$

$$A=2\pi rL=2\pi\left(0.900\times10^{-3}\text{ m}\right)(0.080\,0\text{ m})=4.52\times10^{-4}\text{ m}^2$$

$$S=\frac{\mathcal{P}}{A}=\boxed{332\text{ kW/m}^2}\text{ (points radially inward)}$$

(b) $$B=\frac{\mu_0 I}{2\pi r}=\frac{\mu_0(1.00)}{2\pi\left(0.900\times10^{-3}\right)}=\boxed{222\ \mu\text{T}}$$

$$E=\frac{\Delta V}{\Delta x}=\frac{IR}{L}=\frac{150\text{ V}}{0.080\,0\text{ m}}=\boxed{1.88\text{ kV/m}}$$

Note: $$S=\frac{EB}{\mu_0}=332\text{ kW/m}^2$$

P24.26 (a) $$\vec{\mathbf{E}}\bullet\vec{\mathbf{B}}=\left(80.0\hat{\mathbf{i}}+32.0\hat{\mathbf{j}}-64.0\hat{\mathbf{k}}\right)(\text{N/C})\bullet\left(0.200\hat{\mathbf{i}}+0.080\,0\hat{\mathbf{j}}+0.290\hat{\mathbf{k}}\right)\mu\text{T}$$

$$\vec{\mathbf{E}}\bullet\vec{\mathbf{B}}=(16.0+2.56-18.56)\text{ N}^2\cdot\text{s/C}^2\cdot\text{m}=\boxed{0}$$

(b) $$\vec{\mathbf{S}}=\frac{1}{\mu_0}\vec{\mathbf{E}}\times\vec{\mathbf{B}}=\frac{\left[\left(80.0\hat{\mathbf{i}}+32.0\hat{\mathbf{j}}-64.0\hat{\mathbf{k}}\right)\text{ N/C}\right]\times\left[\left(0.200\hat{\mathbf{i}}+0.080\,0\hat{\mathbf{j}}+0.290\hat{\mathbf{k}}\right)\mu\text{T}\right]}{4\pi\times10^{-7}\text{ T}\cdot\text{m/A}}$$

$$\vec{\mathbf{S}}=\frac{\left(6.40\hat{\mathbf{k}}-23.2\hat{\mathbf{j}}-6.40\hat{\mathbf{k}}+9.28\hat{\mathbf{i}}-12.8\hat{\mathbf{j}}+5.12\hat{\mathbf{i}}\right)\times10^{-6}\text{ W/m}^2}{4\pi\times10^{-7}}$$

$$\vec{\mathbf{S}}=\boxed{\left(11.5\hat{\mathbf{i}}-28.6\hat{\mathbf{j}}\right)\text{ W/m}^2}=30.9\text{ W/m}^2\text{ at }-68.2^\circ\text{ from the }+x\text{ axis.}$$

P24.27 $$\text{Power}=SA=\frac{E_{max}{}^2}{2\mu_0 c}\left(4\pi r^2\right)$$

Solving for r, $$r=\sqrt{\frac{\mathcal{P}\mu_0 c}{2\pi E_{max}{}^2}}=\sqrt{\frac{(100\text{ W})\mu_0 c}{2\pi(15.0\text{ V/m})^2}}=\boxed{5.16\text{ m}}$$

***P24.28** (a) We assume that the starlight moves through space without any of it being absorbed. The radial distance is

$$20\text{ ly}=20c(1\text{ yr})=20\left(3\times10^{8}\text{ m/s}\right)\left(3.16\times10^{7}\text{ s}\right)=1.89\times10^{17}\text{ m}$$

$$I=\frac{\mathcal{P}}{4\pi r^2}=\frac{4\times10^{28}\text{ W}}{4\pi\left(1.89\times10^{17}\text{ m}\right)^2}=\boxed{8.88\times10^{-8}\text{ W/m}^2}.$$

(b) The Earth presents the projected target area of a flat circle:

$$\mathcal{P}=IA=\left(8.88\times10^{-8}\text{ W/m}^2\right)\pi\left(6.37\times10^{6}\text{ m}\right)^2=\boxed{1.13\times10^{7}\text{ W}}.$$

Section 24.6 Momentum and Radiation Pressure

P24.29 $I=\dfrac{\mathcal{P}}{\pi r^2}=\dfrac{E_{max}^2}{2\mu_0 c}$

(a) $E_{max}=\sqrt{\dfrac{\mathcal{P}(2\mu_0 c)}{\pi r^2}}=\boxed{1.90\text{ kN/C}}$

(b) $\dfrac{15\times10^{-3}\text{ J/s}}{3.00\times10^{8}\text{ m/s}}(1.00\text{ m})=\boxed{50.0\text{ pJ}}$

(c) $p=\dfrac{U}{c}=\dfrac{5\times10^{-11}}{3.00\times10^{8}}=\boxed{1.67\times10^{-19}\text{ kg}\cdot\text{m/s}}$

P24.30 (a) The radiation pressure is $\dfrac{2\left(1\,370\text{ W/m}^2\right)}{3.00\times10^{8}\text{ m/s}^2}=9.13\times10^{-6}\text{ N/m}^2$.

Multiplying by the total area, $A=6.00\times10^{5}\text{ m}^2$ gives: $F=\boxed{5.48\text{ N}}$.

(b) The acceleration is: $a=\dfrac{F}{m}=\dfrac{5.48\text{ N}}{6\,000\text{ kg}}=\boxed{9.13\times10^{-4}\text{ m/s}^2}$.

(c) It will arrive at time t where $d=\dfrac{1}{2}at^2$

or $t=\sqrt{\dfrac{2d}{a}}=\sqrt{\dfrac{2\left(3.84\times10^{8}\text{ m}\right)}{\left(9.13\times10^{-4}\text{ m/s}^2\right)}}=9.17\times10^{5}\text{ s}=\boxed{10.6\text{ days}}$.

Section 24.7 The Spectrum of Electromagnetic Waves

P24.31 The time for the radio signal to travel 100 km is: $\Delta t_r = \dfrac{100\times10^3 \text{ m}}{3.00\times10^8 \text{ m/s}} = 3.33\times10^{-4} \text{ s}$.

The sound wave travels 3.00 m across the room in: $\Delta t_s = \dfrac{3.00 \text{ m}}{343 \text{ m/s}} = 8.75\times10^{-3} \text{ s}$.

Therefore, $\boxed{\text{listeners 100 km away}}$ will receive the news before the people in the newsroom by a total time difference of

$\Delta t = 8.75\times10^{-3} \text{ s} - 3.33\times10^{-4} \text{ s} = 8.41\times10^{-3} \text{ s}$.

P24.32 From the electromagnetic spectrum chart and accompanying text discussion, the following identifications are made:

Frequency, f	Wavelength, $\lambda = \dfrac{c}{f}$	Classification
$2 \text{ Hz} = 2\times10^0 \text{ Hz}$	150 Mm	Radio
$2 \text{ kHz} = 2\times10^3 \text{ Hz}$	150 km	Radio
$2 \text{ MHz} = 2\times10^6 \text{ Hz}$	150 m	Radio
$2 \text{ GHz} = 2\times10^9 \text{ Hz}$	15 cm	Microwave
$2 \text{ THz} = 2\times10^{12} \text{ Hz}$	150 μm	Infrared
$2 \text{ PHz} = 2\times10^{15} \text{ Hz}$	150 nm	Ultraviolet
$2 \text{ EHz} = 2\times10^{18} \text{ Hz}$	150 pm	X-ray
$2 \text{ ZHz} = 2\times10^{21} \text{ Hz}$	150 fm	Gamma ray
$2 \text{ YHz} = 2\times10^{24} \text{ Hz}$	150 am	Gamma ray

Wavelength, λ	Frequency, $f = \dfrac{c}{\lambda}$	Classification
$2 \text{ km} = 2\times10^3 \text{ m}$	$1.5\times10^5 \text{ Hz}$	Radio
$2 \text{ m} = 2\times10^0 \text{ m}$	$1.5\times10^8 \text{ Hz}$	Radio
$2 \text{ mm} = 2\times10^{-3} \text{ m}$	$1.5\times10^{11} \text{ Hz}$	Microwave
$2 \ \mu\text{m} = 2\times10^{-6} \text{ m}$	$1.5\times10^{14} \text{ Hz}$	Infrared
$2 \text{ nm} = 2\times10^{-9} \text{ m}$	$1.5\times10^{17} \text{ Hz}$	Ultraviolet/X-ray
$2 \text{ pm} = 2\times10^{-12} \text{ m}$	$1.5\times10^{20} \text{ Hz}$	X-ray/Gamma ray
$2 \text{ fm} = 2\times10^{-15} \text{ m}$	$1.5\times10^{23} \text{ Hz}$	Gamma ray
$2 \text{ am} = 2\times10^{-18} \text{ m}$	$1.5\times10^{26} \text{ Hz}$	Gamma ray

P24.33 $f = \dfrac{c}{\lambda} = \dfrac{3.00\times10^8 \text{ m/s}}{5.50\times10^{-7} \text{ m}} = \boxed{5.45\times10^{14} \text{ Hz}}$

P24.34 (a) $f = \dfrac{c}{\lambda} = \dfrac{3 \times 10^8 \text{ m/s}}{1.7 \text{ m}}$ $\boxed{\sim 10^8 \text{ Hz}}$ $\boxed{\text{radio wave}}$

(b) 1 000 pages, 500 sheets, is about 3 cm thick so one sheet is about 6×10^{-5} m thick.

$$f = \frac{3.00 \times 10^8 \text{ m/s}}{6 \times 10^{-5} \text{ m}} \boxed{\sim 10^{13} \text{ Hz}} \boxed{\text{infrared}}$$

P24.35 (a) $f\lambda = c$ gives $(5.00 \times 10^{19} \text{ Hz})\lambda = 3.00 \times 10^8 \text{ m/s}$: $\boxed{\lambda = 6.00 \times 10^{-12} \text{ m} = 6.00 \text{ pm}}$

(b) $f\lambda = c$ gives $(4.00 \times 10^9 \text{ Hz})\lambda = 3.00 \times 10^8 \text{ m/s}$: $\boxed{\lambda = 0.075 \text{ m} = 7.50 \text{ cm}}$

***P24.36** $\lambda = \dfrac{c}{f} = \dfrac{3.00 \times 10^8 \text{ m/s}}{27.33 \times 10^6 \text{ Hz}} = \boxed{11.0 \text{ m}}$

P24.37 For the proton, $\Sigma F = ma$: $qvB \sin 90.0° = \dfrac{mv^2}{R}$.

The period and frequency of the proton's circular motion are therefore:

$$T = \frac{2\pi R}{v} = \frac{2\pi m}{qB} = \frac{2\pi(1.67 \times 10^{-27} \text{ kg})}{(1.60 \times 10^{-19} \text{ C})(0.350 \text{ T})} = 1.87 \times 10^{-7} \text{ s}$$

$$f = 5.34 \times 10^6 \text{ Hz}.$$

The charge will radiate at this same frequency,

with $\lambda = \dfrac{c}{f} = \dfrac{3.00 \times 10^8 \text{ m/s}}{5.34 \times 10^6 \text{ Hz}} = \boxed{56.2 \text{ m}}$.

P24.38 (a) Channel 4: $f_{min} = 66$ MHz $\lambda_{max} = \boxed{4.55 \text{ m}}$

$f_{max} = 72$ MHz $\lambda_{min} = \boxed{4.17 \text{ m}}$

(b) Channel 6: $f_{min} = 82$ MHz $\lambda_{max} = \boxed{3.66 \text{ m}}$

$f_{max} = 88$ MHz $\lambda_{min} = \boxed{3.41 \text{ m}}$

(c) Channel 8: $f_{min} = 180$ MHz $\lambda_{max} = \boxed{1.67 \text{ m}}$

$f_{max} = 186$ MHz $\lambda_{min} = \boxed{1.61 \text{ m}}$

P24.39 (a) $\lambda = \dfrac{c}{f} = \dfrac{3.00 \times 10^8 \text{ m/s}}{1\,150 \times 10^3 \text{ s}^{-1}} = 261 \text{ m}$ so $\dfrac{180 \text{ m}}{261 \text{ m}} = \boxed{0.690 \text{ wavelengths}}$

(b) $\lambda = \dfrac{c}{f} = \dfrac{3.00 \times 10^8 \text{ m/s}}{98.1 \times 10^6 \text{ s}^{-1}} = 3.06 \text{ m}$ so $\dfrac{180 \text{ m}}{3.06 \text{ m}} = \boxed{58.9 \text{ wavelengths}}$

***P24.40** Time to reach object $=\frac{1}{2}(\text{total time of flight})=\frac{1}{2}\left(4.00\times10^{-4}\text{ s}\right)=2.00\times10^{-4}\text{ s}$.

Thus, $d=vt=\left(3.00\times10^{8}\text{ m/s}\right)\left(2.00\times10^{-4}\text{ s}\right)=6.00\times10^{4}\text{ m}=\boxed{60.0\text{ km}}$.

Section 24.8 Polarization

P24.41 $I=I_{\max}\cos^2\theta \quad\Rightarrow\quad \theta=\cos^{-1}\sqrt{\frac{I}{I_{\max}}}$

(a) $\frac{I}{I_{\max}}=\frac{1}{3.00} \quad\Rightarrow\quad \theta=\cos^{-1}\sqrt{\frac{1}{3.00}}=\boxed{54.7°}$

(b) $\frac{I}{I_{\max}}=\frac{1}{5.00} \quad\Rightarrow\quad \theta=\cos^{-1}\sqrt{\frac{1}{5.00}}=\boxed{63.4°}$

(c) $\frac{I}{I_{\max}}=\frac{1}{10.0} \quad\Rightarrow\quad \theta=\cos^{-1}\sqrt{\frac{1}{10.0}}=\boxed{71.6°}$

P24.42 The average value of the cosine-squared function is one-half, so the first polarizer transmits $\frac{1}{2}$ the light. The second transmits $\cos^2 30.0°=\frac{3}{4}$.

$$I_f=\frac{1}{2}\times\frac{3}{4}I_i=\boxed{\frac{3}{8}I_i}$$

P24.43 For incident unpolarized light of intensity $I_{\max}$:

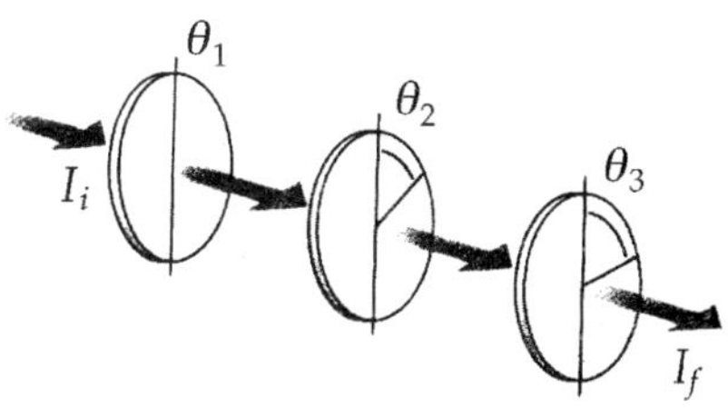

FIG. P24.43

After transmitting 1st disk: $I=\frac{1}{2}I_{\max}$.

After transmitting 2nd disk: $I=\frac{1}{2}I_{\max}\cos^2\theta$.

After transmitting 3rd disk: $I=\frac{1}{2}I_{\max}\cos^2\theta\cos^2(90°-\theta)$.

where the angle between the first and second disk is $\theta=\omega t$.

Using trigonometric identities $\cos^2\theta=\frac{1}{2}(1+\cos2\theta)$

and $\cos^2(90°-\theta)=\sin^2\theta=\frac{1}{2}(1-\cos2\theta)$

continued on next page

we have
$$I = \frac{1}{2} I_{max} \left[\frac{(1+\cos 2\theta)}{2} \right] \left[\frac{(1-\cos 2\theta)}{2} \right]$$

$$I = \frac{1}{8} I_{max} \left(1 - \cos^2 2\theta\right) = \frac{1}{8} I_{max} \left(\frac{1}{2}\right)(1 - \cos 4\theta) .$$

Since $\theta = \omega t$, the intensity of the emerging beam is given by $\boxed{I = \frac{1}{16} I_{max} (1 - \cos 4\omega t)}$.

***P24.44** $\mathscr{P} = \frac{(\Delta V)^2}{R}$ or $\mathscr{P} \propto (\Delta V)^2$

$\Delta V = (-)E_y \cdot \Delta y = E_y \cdot \ell \cos\theta$

$\Delta V \propto \cos\theta$ so $\mathscr{P} \propto \cos^2\theta$

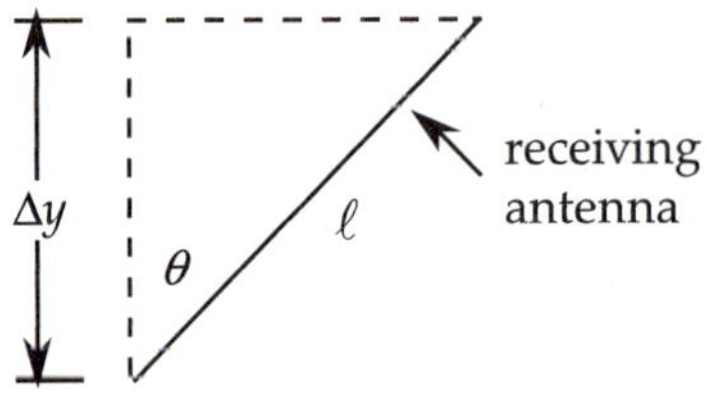

FIG. P24.44

(a) $\theta = 15.0°$: $\mathscr{P} = \mathscr{P}_{max} \cos^2(15.0°) = 0.933 \mathscr{P}_{max} = \boxed{93.3\%}$

(b) $\theta = 45.0°$: $\mathscr{P} = \mathscr{P}_{max} \cos^2(45.0°) = 0.500 \mathscr{P}_{max} = \boxed{50.0\%}$

(c) $\theta = 90.0°$: $\mathscr{P} = \mathscr{P}_{max} \cos^2(90.0°) = \boxed{0}$

P24.45 $\frac{I}{I_{max}} = \frac{1}{2}\left(\cos^2 45.0°\right)\left(\cos^2 45.0°\right) = \boxed{\frac{1}{8}}$

P24.46 Let the first sheet have its axis at angle θ to the original plane of polarization, and let each further sheet have its axis turned by the same angle.

The first sheet passes intensity $I_{max} \cos^2\theta$.

The second sheet passes $I_{max} \cos^4\theta$

and the nth sheet lets through $I_{max} \cos^{2n}\theta \geq 0.90 I_{max}$ where $\theta = \frac{45°}{n}$.

Try different integers to find $\cos^{2\times 5}\left(\frac{45°}{5}\right) = 0.885$ $\cos^{2\times 6}\left(\frac{45°}{6}\right) = 0.902$.

(a) So $n = \boxed{6}$

(b) $\theta = \boxed{7.50°}$

Section 24.9 Context Connection—The Special Properties of Laser Light

P24.47 The photon energy is $E_3^* - E_2 = (20.66 - 18.70)\text{ eV} = 1.96\text{ eV} = \dfrac{hc}{\lambda}$

$$\lambda = \frac{(6.626\times10^{-34}\text{ J}\cdot\text{s})(3.00\times10^{8}\text{ m/s})}{1.96(1.60\times10^{-19}\text{ J})} = \boxed{633\text{ nm}}.$$

P24.48 (a) $B_{\max} = \dfrac{E_{\max}}{c}$: $\quad B_{\max} = \dfrac{7.00\times10^{5}\text{ N/C}}{3.00\times10^{8}\text{ m/s}} = \boxed{2.33\text{ mT}}$

(b) $I = \dfrac{E_{\max}^2}{2\mu_0 c}$: $\quad I = \dfrac{(7.00\times10^{5})^2}{2(4\pi\times10^{-7})(3.00\times10^{8})} = \boxed{650\text{ MW/m}^2}$

(c) $I = \dfrac{\mathcal{P}}{A}$: $\quad \mathcal{P} = IA = (6.50\times10^{8}\text{ W/m}^2)\left[\dfrac{\pi}{4}(1.00\times10^{-3}\text{ m})^2\right] = \boxed{510\text{ W}}$

P24.49 (a) $I = \dfrac{(3.00\times10^{-3}\text{ J})}{(1.00\times10^{-9}\text{ s})\left[\pi(15.0\times10^{-6}\text{ m})^2\right]} = \boxed{4.24\times10^{15}\text{ W/m}^2}$

(b) $(3.00\times10^{-3}\text{ J})\dfrac{(0.600\times10^{-9}\text{ m})^2}{(30.0\times10^{-6}\text{ m})^2} = \boxed{1.20\times10^{-12}\text{ J}} = 7.50\text{ MeV}$

P24.50 $f = \dfrac{E}{h} = \dfrac{0.117\text{ eV}}{6.63\times10^{-34}\text{ J}\cdot\text{s}}\left(\dfrac{1.60\times10^{-19}\text{ C}}{e}\right)\left(\dfrac{1\text{ J}}{1\text{ V}\cdot\text{C}}\right) = \boxed{2.82\times10^{13}\text{ s}^{-1}}$

$\lambda = \dfrac{c}{f} = \dfrac{3.00\times10^{8}\text{ m/s}}{2.82\times10^{13}\text{ s}^{-1}} = \boxed{10.6\ \mu\text{m}}$, $\boxed{\text{infrared}}$

***P24.51** $E = \mathcal{P}\Delta t = (1.00\times10^{6}\text{ W})(1.00\times10^{-8}\text{ s}) = 0.010\,0\text{ J}$

$E_\gamma = hf = \dfrac{hc}{\lambda} = \dfrac{(6.626\times10^{-34})(3.00\times10^{8})}{694.3\times10^{-9}}\text{ J} = 2.86\times10^{-19}\text{ J}$

$N = \dfrac{E}{E_\gamma} = \dfrac{0.010\,0}{2.86\times10^{-19}} = \boxed{3.49\times10^{16}\text{ photons}}$

P24.52 (a) $\left(3.00\times10^{8}\ \text{m/s}\right)\left(14.0\times10^{-12}\ \text{s}\right)=\boxed{4.20\ \text{mm}}$

(b) $E=\dfrac{hc}{\lambda}=2.86\times10^{-19}\ \text{J}$

$$N=\frac{3.00\ \text{J}}{2.86\times10^{-19}\ \text{J}}=\boxed{1.05\times10^{19}\ \text{photons}}$$

(c) $V=(4.20\ \text{mm})\left[\pi(3.00\ \text{mm})^2\right]=119\ \text{mm}^3$

$$n=\frac{1.05\times10^{19}}{119}=\boxed{8.82\times10^{16}\ \text{mm}^{-3}}$$

***P24.53** (a) The distance between nodes is $\dfrac{\lambda}{2}$, so we require solutions to $35.124\,103\ \text{cm}=\dfrac{N}{2}\lambda$ where N is an integer and λ is in the required range. The midpoint of the range is 632.809 10 nm, giving $N_{\text{trial}}=\dfrac{2\left(35.124\,103\times10^{-2}\ \text{m}\right)}{632.809\,1\times10^{9}\ \text{m}}=1\,110\,101.07$. So we try $N=1\,110\,101$, 1 110 102, 1 110 100, 1 110 103, and so on:

$$\lambda_1=\frac{2\left(35.124\,103\times10^{-2}\ \text{m}\right)}{1\,110\,101}=\boxed{632.809\,14\ \text{nm}}$$

$$\lambda_2=\frac{2\left(35.124\,103\times10^{-2}\ \text{m}\right)}{1\,110\,102}=\boxed{632.808\,57\ \text{nm}}$$

$$\lambda_3=\frac{2\left(35.124\,103\times10^{-2}\ \text{m}\right)}{1\,110\,100}=\boxed{632.809\,71\ \text{nm}}$$

$\lambda_{\text{trial}}=\dfrac{2\left(35.124\,103\times10^{-2}\ \text{m}\right)}{1\,110\,103}=632.808\,00\ \text{nm}$, outside the range. Thus the laser light has just $\boxed{\text{three}}$ wavelength components.

(b) $\dfrac{1}{2}m_0v^2=\dfrac{3}{2}kT$. We use the periodic table.

$$v=\sqrt{\frac{3kT}{m_0}}=\sqrt{\frac{3\left(1.38\times10^{-23}\ \text{J/K}\right)(393\ \text{K})}{20.18\ \text{u}}\left(\frac{1\ \text{u}}{1.66\times10^{-27}\ \text{kg}}\right)}=\boxed{697\ \text{m/s}}$$

(c) For a neon atom moving toward one mirror at the rms speed as it emits, the Doppler shift is described by

$$f'=f\sqrt{\frac{c+v}{c-v}}=\frac{c}{\lambda'}=\frac{c}{\lambda}\sqrt{\frac{c+v}{c-v}}$$

$$\lambda'=\lambda\sqrt{\frac{c-v}{c+v}}=632.809\,1\ \text{nm}\sqrt{\frac{3\times10^{8}-697}{3\times10^{8}+697}}=632.807\,63\ \text{nm}.$$

This is outside the given range. Many atoms are moving faster than the rms speed, so we should expect still more Doppler broadening of the resonance amplification peak.

P24.54 (a) $$\frac{N_3}{N_2}=\frac{N_g e^{-E_3/(k_B\cdot 300\text{ K})}}{N_g e^{-E_2/(k_B\cdot 300\text{ K})}}=e^{-(E_3-E_2)/(k_B\cdot 300\text{ K})}=e^{-hc/\lambda(k_B\cdot 300\text{ K})}$$

where λ is the wavelength of light radiated in the $3\to 2$ transition.

$$\frac{N_3}{N_2}=e^{-(6.63\times 10^{-34}\text{ J}\cdot\text{s})(3\times 10^8\text{ m/s})/(632.8\times 10^{-9}\text{ m})(1.38\times 10^{-23}\text{ J/K})(300\text{ K})}$$

$$\frac{N_3}{N_2}=e^{-75.9}=\boxed{1.07\times 10^{-33}}$$

(b) $$\frac{N_u}{N_\ell}=e^{-(E_u-E_\ell)/k_BT}$$

where the subscript u refers to an upper energy state and the subscript ℓ to a lower energy state.

Since $E_u-E_\ell=E_{\text{photon}}=\dfrac{hc}{\lambda}$ $\quad\dfrac{N_u}{N_\ell}=e^{-hc/\lambda k_BT}$.

Thus, we require $\quad 1.02=e^{-hc/\lambda k_BT}$

or $$\ln(1.02)=-\frac{(6.63\times 10^{-34}\text{ J}\cdot\text{s})(3\times 10^8\text{ m/s})}{(632.8\times 10^{-9}\text{ m})(1.38\times 10^{-23}\text{ J/K})T}$$

$$T=-\frac{2.28\times 10^4}{\ln(1.02)}=\boxed{-1.15\times 10^6\text{ K}}.$$

A negative-temperature state is not achieved by cooling the system below 0 K, but by heating it above $T=\infty$, for as $T\to\infty$ the populations of upper and lower states approach equality.

(c) Because $E_u-E_\ell>0$, and in any real equilibrium state $T>0$,

$e^{-(E_u-E_\ell)/k_BT}<1$ and $N_u<N_\ell$.

Thus, a population inversion cannot happen in thermal equilibrium.

Additional Problems

P24.55 (a) $\mathscr{P}=SA$: $$\mathscr{P}=(1\,370\text{ W/m}^2)\left[4\pi(1.496\times 10^{11}\text{ m})^2\right]=\boxed{3.85\times 10^{26}\text{ W}}$$

(b) $S=\dfrac{cB_{\max}^2}{2\mu_0}$ so $$B_{\max}=\sqrt{\frac{2\mu_0 S}{c}}=\sqrt{\frac{2(4\pi\times 10^{-7}\text{ N/A}^2)(1\,370\text{ W/m}^2)}{3.00\times 10^8\text{ m/s}}}=\boxed{3.39\ \mu\text{T}}$$

$S=\dfrac{E_{\max}^2}{2\mu_0 c}$ so $$E_{\max}=\sqrt{2\mu_0 cS}=\sqrt{2(4\pi\times 10^{-7})(3.00\times 10^8)(1\,370)}=\boxed{1.02\text{ kV/m}}$$

P24.56 Suppose you cover a 1.7 m × 0.3 m section of beach blanket. Suppose the elevation angle of the Sun is 60°. Then the target area you fill in the Sun's field of view is

$$(1.7\text{ m})(0.3\text{ m})\cos 30° = 0.4\text{ m}^2 .$$

Now $I = \frac{\mathcal{P}}{A} = \frac{U}{At}$ $\quad U = IAt = \left(1\,370\text{ W/m}^2\right)\left[(0.6)(0.5)\left(0.4\text{ m}^2\right)\right](3\,600\text{ s})\boxed{\sim 10^6\text{ J}}.$

P24.57 (a) $\mathcal{E} = -\frac{d\Phi_B}{dt} = -\frac{d}{dt}(BA\cos\theta)$ $\qquad \mathcal{E} = -A\frac{d}{dt}(B_{max}\cos\omega t\cos\theta) = AB_{max}\omega(\sin\omega t\cos\theta)$

$\mathcal{E}(t) = 2\pi fB_{max}A\sin 2\pi ft\cos\theta$ $\qquad \mathcal{E}(t) = 2\pi^2r^2 fB_{max}\cos\theta\sin 2\pi ft$

Thus, $\boxed{\mathcal{E}_{max} = 2\pi^2r^2 f\,B_{max}\cos\theta}$

where θ is the angle between the magnetic field and the normal to the loop.

(b) If $\vec{\mathbf{E}}$ is vertical, $\vec{\mathbf{B}}$ is horizontal, so $\boxed{\text{the plane of the loop should be vertical}}$

and $\boxed{\text{the plane should contain the line of sight of the transmitter}}$.

***P24.58** (a) The power incident on the mirror is: $\mathcal{P}_I = IA = \left(1\,370\text{ W/m}^2\right)\left[\pi(100\text{ m})^2\right] = 4.30\times 10^7\text{ W}.$

The power reflected through the atmosphere is $\mathcal{P}_R = 0.746\left(4.30\times 10^7\text{ W}\right) = \boxed{3.21\times 10^7\text{ W}}.$

(b) $$S = \frac{\mathcal{P}_R}{A} = \frac{3.21\times 10^7\text{ W}}{\pi\left(4.00\times 10^3\text{ m}\right)^2} = \boxed{0.639\text{ W/m}^2}$$

(c) Noon sunshine in St. Petersburg produces this power-per-area on a horizontal surface:

$$\frac{\mathcal{P}_N}{A} = 0.746\left(1\,370\text{ W/m}^2\right)\sin 7.00° = 125\text{ W/m}^2 .$$

The radiation intensity received from the mirror is

$\left(\frac{0.639\text{ W/m}^2}{125\text{ W/m}^2}\right)100\% = \boxed{0.513\%}$ of that from the noon Sun in January.

P24.59 (a) $B_{\max}=\frac{E_{\max}}{c}=\boxed{6.67\times10^{-16}\ \text{T}}$

(b) $S_{\text{av}}=\frac{E_{\max}^2}{2\mu_0 c}=\boxed{5.31\times10^{-17}\ \text{W/m}^2}$

(c) $\mathcal{P}=S_{\text{av}}A=\boxed{1.67\times10^{-14}\ \text{W}}$

(d) $F=PA=\left(\frac{S_{\text{av}}}{c}\right)A=\boxed{5.56\times10^{-23}\ \text{N}}$ (≈ the weight of 3 000 H atoms!)

FIG. P24.59

P24.60 The area over which we model the antenna as radiating is the lateral surface of a cylinder,

$$A=2\pi r\ell=2\pi\left(4.00\times10^{-2}\ \text{m}\right)(0.100\ \text{m})=2.51\times10^{-2}\ \text{m}^2.$$

(a) The intensity is then: $S=\frac{\mathcal{P}}{A}=\frac{0.600\ \text{W}}{2.51\times10^{-2}\ \text{m}^2}=\boxed{23.9\ \text{W/m}^2}$.

(b) The standard is:

$$0.570\ \text{mW/cm}^2=0.570\left(\text{mW/cm}^2\right)\left(\frac{1.00\times10^{-3}\ \text{W}}{1.00\ \text{mW}}\right)\left(\frac{1.00\times10^{4}\ \text{cm}^2}{1.00\ \text{m}^2}\right)=5.70\ \text{W/m}^2.$$

While it is on, the telephone is over the standard by $\frac{23.9\ \text{W/m}^2}{5.70\ \text{W/m}^2}=\boxed{4.19\ \text{times}}$.

P24.61 $u=\frac{1}{2}\epsilon_0 E_{\max}^2$ $\qquad E_{\max}=\sqrt{\frac{2u}{\epsilon_0}}=\boxed{95.1\ \text{mV/m}}$

P24.62 (a) $B_{\max}=\frac{E_{\max}}{c}=\frac{175\ \text{V/m}}{3.00\times10^{8}\ \text{m/s}}=\boxed{5.83\times10^{-7}\ \text{T}}$

$k=\frac{2\pi}{\lambda}=\frac{2\pi}{0.015\,0\ \text{m}}=\boxed{419\ \text{rad/m}}$ $\qquad \omega=kc=\boxed{1.26\times10^{11}\ \text{rad/s}}$

Since $\vec{\mathbf{S}}$ is along x, and $\vec{\mathbf{E}}$ is along y, $\vec{\mathbf{B}}$ must be in $\boxed{\text{the } z \text{ direction}}$. (That is $\vec{\mathbf{S}}\propto\vec{\mathbf{E}}\times\vec{\mathbf{B}}$.)

(b) $S_{av}=\frac{E_{\max}B_{\max}}{2\mu_0}=40.6\ \text{W/m}^2$ $\qquad \vec{\mathbf{S}}_{av}=\boxed{\left(40.6\ \text{W/m}^2\right)\hat{\mathbf{i}}}$

(c) $P_r=\frac{2S}{c}=\boxed{2.71\times10^{-7}\ \text{N/m}^2}$

(d) $a=\frac{\sum F}{m}=\frac{PA}{m}=\frac{\left(2.71\times10^{-7}\ \text{N/m}^2\right)\left(0.750\ \text{m}^2\right)}{0.500\ \text{kg}}=4.06\times10^{-7}\ \text{m/s}^2$ $\qquad \vec{\mathbf{a}}=\boxed{\left(406\ \text{nm/s}^2\right)\hat{\mathbf{i}}}$

P24.63 The mirror intercepts power $\mathcal{P} = I_1 A_1 = \left(1.00\times10^3\ \text{W/m}^2\right)\left[\pi(0.500\ \text{m})^2\right] = 785\ \text{W}$.

In the image,

(a) $I_2 = \dfrac{\mathcal{P}}{A_2}$: $\quad I_2 = \dfrac{785\ \text{W}}{\pi(0.020\,0\ \text{m})^2} = \boxed{625\ \text{kW/m}^2}$

(b) $I_2 = \dfrac{E_{max}^2}{2\mu_0 c}$ so $\quad E_{max} = \sqrt{2\mu_0 c I_2} = \sqrt{2\left(4\pi\times10^{-7}\right)\left(3.00\times10^8\right)\left(6.25\times10^5\right)} = \boxed{21.7\ \text{kN/C}}$

$$B_{max} = \frac{E_{max}}{c} = \boxed{72.4\ \mu\text{T}}$$

(c) $0.400\mathcal{P}\Delta t = mc\Delta T$

$$0.400(785\ \text{W})\Delta t = (1.00\ \text{kg})(4\,186\ \text{J/kg}\cdot{}^\circ\text{C})(100^\circ\text{C} - 20.0^\circ\text{C})$$

$$\Delta t = \frac{3.35\times10^5\ \text{J}}{314\ \text{W}} = 1.07\times10^3\ \text{s} = \boxed{17.8\ \text{min}}$$

***P24.64** (a) $\lambda = \dfrac{c}{f} = \dfrac{3.00\times10^8\ \text{m/s}}{20.0\times10^9\ \text{s}^{-1}} = \boxed{1.50\ \text{cm}}$

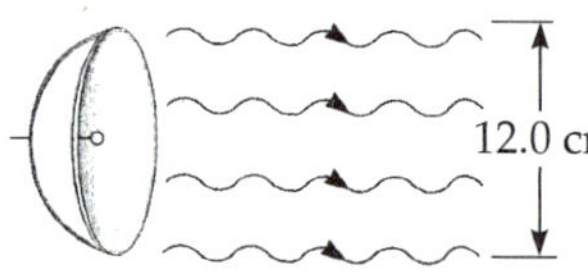

FIG. P24.64

(b) $U = \mathcal{P}(\Delta t) = \left(25.0\times10^3\ \text{J/s}\right)\left(1.00\times10^{-9}\ \text{s}\right) = 25.0\times10^{-6}\ \text{J}$

$= \boxed{25.0\ \mu\text{J}}$

(c) $u_{av} = \dfrac{U}{V} = \dfrac{U}{\left(\pi r^2\right)\ell} = \dfrac{U}{\left(\pi r^2\right)c(\Delta t)} = \dfrac{25.0\times10^{-6}\ \text{J}}{\pi(0.060\,0\ \text{m})^2\left(3.00\times10^8\ \text{m/s}\right)\left(1.00\times10^{-9}\ \text{s}\right)}$

$$u_{av} = 7.37\times10^{-3}\ \text{J/m}^3 = \boxed{7.37\ \text{mJ/m}^3}$$

(d) $E_{max} = \sqrt{\dfrac{2u_{av}}{\epsilon_0}} = \sqrt{\dfrac{2\left(7.37\times10^{-3}\ \text{J/m}^3\right)}{8.85\times10^{-12}\ \text{C}^2/\text{N}\cdot\text{m}^2}} = 4.08\times10^4\ \text{V/m} = \boxed{40.8\ \text{kV/m}}$

$$B_{max} = \frac{E_{max}}{c} = \frac{4.08\times10^4\ \text{V/m}}{3.00\times10^8\ \text{m/s}} = 1.36\times10^{-4}\ \text{T} = \boxed{136\ \mu\text{T}}$$

(e) $F = PA = \left(\dfrac{S}{c}\right)A = u_{av}A = \left(7.37\times10^{-3}\ \text{J/m}^3\right)\pi(0.060\,0\ \text{m})^2 = 8.33\times10^{-5}\ \text{N} = \boxed{83.3\ \mu\text{N}}$

P24.65 (a) On the right side of the equation, $\frac{C^2\left(m/s^2\right)^2}{\left(C^2/N\cdot m^2\right)\left(m/s\right)^3}=\frac{N\cdot m^2\cdot C^2\cdot m^2\cdot s^3}{C^2\cdot s^4\cdot m^3}=\frac{N\cdot m}{s}=\frac{J}{s}=W$.

(b) $F=ma=qE$ or $a=\frac{qE}{m}=\frac{\left(1.60\times10^{-19}\text{ C}\right)\left(100\text{ N/C}\right)}{9.11\times10^{-31}\text{ kg}}=\boxed{1.76\times10^{13}\text{ m/s}^2}$.

The radiated power is then: $\mathscr{P}=\frac{q^2a^2}{6\pi\in_0 c^3}=\frac{\left(1.60\times10^{-19}\right)^2\left(1.76\times10^{13}\right)^2}{6\pi\left(8.85\times10^{-12}\right)\left(3.00\times10^8\right)^3}=\boxed{1.75\times10^{-27}\text{ W}}$.

(c) $F=ma_c=m\left(\frac{v^2}{r}\right)=qvB$ so $v=\frac{qBr}{m}$.

The proton accelerates at $a=\frac{v^2}{r}=\frac{q^2B^2r}{m^2}=\frac{\left(1.60\times10^{-19}\right)^2(0.350)^2(0.500)}{\left(1.67\times10^{-27}\right)^2}=5.62\times10^{14}\text{ m/s}^2$.

The proton then radiates $\mathscr{P}=\frac{q^2a^2}{6\pi\in_0 c^3}=\frac{\left(1.60\times10^{-19}\right)^2\left(5.62\times10^{14}\right)^2}{6\pi\left(8.85\times10^{-12}\right)\left(3.00\times10^8\right)^3}=\boxed{1.80\times10^{-24}\text{ W}}$.

***P24.66** (a) $m=\rho V=\rho\frac{1}{2}\frac{4}{3}\pi r^3$

$$r=\left(\frac{6m}{\rho4\pi}\right)^{1/3}=\left(\frac{6(8.7\text{ kg})}{\left(990\text{ kg/m}^3\right)4\pi}\right)^{1/3}=\boxed{0.161\text{ m}}$$

(b) $A=\frac{1}{2}4\pi r^2=2\pi(0.161\text{ m})^2=\boxed{0.163\text{ m}^2}$

(c) $I=e\sigma T^4=0.970\left(5.67\times10^{-8}\text{ W/m}^2\cdot\text{K}^4\right)(304\text{ K})^4=\boxed{470\text{ W/m}^2}$

(d) $\mathscr{P}=IA=\left(470\text{ W/m}^2\right)0.163\text{ m}^2=\boxed{76.8\text{ W}}$

(e) $I=\frac{E_{max}^2}{2\mu_0 c}$

$$E_{max}=(2\mu_0 cI)^{1/2}=\left[\left(8\pi\times10^{-7}\text{ Tm/A}\right)\left(3\times10^8\text{ m/s}\right)\left(470\text{ W/m}^2\right)\right]^{1/2}=\boxed{595\text{ N/C}}$$

(f) $E_{max}=cB_{max}$

$$B_{max}=\frac{595\text{ N/C}}{3\times10^8\text{ m/s}}=\boxed{1.98\ \mu\text{T}}$$

continued on next page

(g) The sleeping cats are uncharged and nonmagnetic. They carry no macroscopic current. They are a source of infrared radiation. They glow not by visible-light emission but by infrared emission.

(h) Each kitten has radius $r_k = \left(\dfrac{6(0.8)}{990 \times 4\pi}\right)^{1/3} = 0.072\,8\text{ m}$ and radiating area

$2\pi(0.072\,8\text{ m})^2 = 0.033\,3\text{ m}^2$. Eliza has area $2\pi\left(\dfrac{6(5.5)}{990 \times 4\pi}\right)^{2/3} = 0.120\text{ m}^2$. The total glowing

area is $0.120\text{ m}^2 + 4(0.033\,3\text{ m}^2) = 0.254\text{ m}^2$ and has power output

$\mathcal{P} = IA = (470\text{ W/m}^2)0.254\text{ m}^2 = \boxed{119\text{ W}}$.

P24.67 (a) At steady state, $\mathcal{P}_{\text{in}} = \mathcal{P}_{\text{out}}$ and the power radiated out is $\mathcal{P}_{\text{out}} = e\sigma AT^4$.

Thus, $0.900(1\,000\text{ W/m}^2)A = 0.700(5.67 \times 10^{-8}\text{ W/m}^2 \cdot \text{K}^4)AT^4$

or $$T = \left[\frac{900\text{ W/m}^2}{0.700(5.67 \times 10^{-8}\text{ W/m}^2 \cdot \text{K}^4)}\right]^{1/4} = \boxed{388\text{ K}} = 115°\text{C}.$$

(b) The box of horizontal area A, presents projected area $A\sin 50.0°$ perpendicular to the sunlight. Then by the same reasoning,

$$0.900(1\,000\text{ W/m}^2)A\sin 50.0° = 0.700(5.67 \times 10^{-8}\text{ W/m}^2 \cdot \text{K}^4)AT^4$$

or $$T = \left[\frac{(900\text{ W/m}^2)\sin 50.0°}{0.700(5.67 \times 10^{-8}\text{ W/m}^2 \cdot \text{K}^4)}\right]^{1/4} = \boxed{363\text{ K}} = 90.0°\text{C}.$$

***P24.68** We take R to be the planet's distance from its star. The planet, of radius r, presents a $\boxed{\text{projected area } \pi r^2}$ perpendicular to the starlight. $\boxed{\text{It radiates over area } 4\pi r^2.}$

At steady-state, $\mathcal{P}_{\text{in}} = \mathcal{P}_{\text{out}}$: $eI_{\text{in}}(\pi r^2) = e\sigma(4\pi r^2)T^4$

$e\left(\dfrac{6.00 \times 10^{23}\text{ W}}{4\pi R^2}\right)(\pi r^2) = e\sigma(4\pi r^2)T^4$ so that $6.00 \times 10^{23}\text{ W} = 16\pi\sigma R^2 T^4$

$$R = \sqrt{\frac{6.00 \times 10^{23}\text{ W}}{16\pi\sigma T^4}} = \sqrt{\frac{6.00 \times 10^{23}\text{ W}}{16\pi(5.67 \times 10^{-8}\text{ W/m}^2 \cdot \text{K}^4)(310\text{ K})^4}} = \boxed{4.77 \times 10^9\text{ m} = 4.77\text{ Gm}}.$$

P24.69 (a) $P = \frac{F}{A} = \frac{I}{c}$ $\qquad F = \frac{IA}{c} = \frac{\mathscr{P}}{c} = \frac{100 \text{ J/s}}{3.00 \times 10^{8} \text{ m/s}} = 3.33 \times 10^{-7} \text{ N} = (110 \text{ kg})a$

$a = 3.03 \times 10^{-9} \text{ m/s}^2$ and $\qquad x = \frac{1}{2}at^2$

$$t = \sqrt{\frac{2x}{a}} = 8.12 \times 10^{4} \text{ s} = \boxed{22.6 \text{ h}}$$

(b) $0 = (107 \text{ kg})v - (3.00 \text{ kg})(12.0 \text{ m/s} - v) = (107 \text{ kg})v - 36.0 \text{ kg} \cdot \text{m/s} + (3.00 \text{ kg})v$

$v = \frac{36.0}{110} = 0.327 \text{ m/s}$ $\qquad t = \boxed{30.6 \text{ s}}$

ANSWERS TO EVEN PROBLEMS

P24.2 (a) $3.15\hat{\mathbf{j}}$ kN/C; (b) 525 nT$\hat{\mathbf{k}}$; (c) $-483\hat{\mathbf{j}}$ aN

P24.4 (a) 2.68×10^{3} A.D.; (b) 499 s; (c) 2.56 s; (d) 0.133 s; (e) 33.3 μs

P24.6 733 nT

P24.8 (a) 0.333 μT; (b) 0.628 μm; (c) 477 THz

P24.10 75.0 MHz

P24.12 11.3 kHz

P24.14 0.050 4c

P24.16 (a) 13.4 m/s toward the station and 13.4 m/s away from the station.
(b) 0.056 7 rad/s

P24.18 0.220 H

P24.20 3.33 $\mu\text{J/m}^3$

P24.22 49.5 mV

P24.24 (a) 11.9 GW/m^2; (b) 234 kW

P24.26 (a) $\vec{\mathbf{E}} \cdot \vec{\mathbf{B}} = 0$; (b) $(11.5\hat{\mathbf{i}} - 28.6\hat{\mathbf{j}})$ W/m^2

P24.28 (a) 88.8 nW/m^2; (b) 11.3 MW

P24.30 (a) 5.48 N; (b) 913 $\mu\text{m/s}^2$; (c) 10.6 d

P24.32 radio, radio, radio, radio or microwave, infrared, ultraviolet, x-ray, γ-ray, γ-ray; radio, radio, microwave, infrared, ultraviolet or x-ray, x- or γ-ray, γ-ray, γ-ray

P24.34 (a) $\sim 10^{8}$ Hz radio wave;
(b) $\sim 10^{13}$ Hz infrared light

P24.36 11.0 m

P24.38 (a) 4.17 m to 4.55 m; (b) 3.41 m to 3.66 m; (c) 1.61 m to 1.67 m

P24.40 60.0 km

P24.42 $\frac{3}{8}$

P24.44 (a) 93.3%; (b) 50.0%; (c) 0

P24.46 (a) 6; (b) 7.50°

P24.48 (a) 2.33 mT; (b) 650 MW/m^2; (c) 510 W

P24.50 28.2 THz, 10.6 μm, infrared

P24.52 (a) 4.20 mm; (b) 1.05×10^{19} photons; (c) $8.82 \times 10^{16}/\text{mm}^3$

P24.54 (a) 1.07×10^{-33}; (b) -1.15×10^{6} K; (c) no real T is below 0 K

P24.56 $\sim 10^6$ J

P24.58 (a) 32.1 MW; (b) 0.639 W/m^2; (c) 0.513%

P24.60 (a) 23.9 W/m^2; (b) It is 4.19 times the standard.

P24.62 (a) $\vec{\mathbf{B}}_{\text{max}} = 583\text{ nT}\hat{\mathbf{k}}$, $k = 419\text{ rad/m}$, $\omega = 12.6\text{ Trad/s}$; (b) $S_{\text{av}} = 40.6\text{ W/m}^2$; (c) 271 nPa; (d) 406 nm/s^2

P24.64 (a) 1.50 cm; (b) 25.0 μJ; (c) 7.37 mJ/m^3; (d) 40.8 kV/m, 136 μT; (e) 83.3 μN

P24.66 (a) 16.1 cm; (b) 0.163 m^2; (c) 470 W/m^2; (d) 76.8 W; (e) 595 N/C; (f) 1.98 μT; (g) The cats are nonmagnetic and carry no macroscopic charge or current. Oscillating charges within molecules make them emit infrared radiation. (h) 119 W

P24.68 4.77 Gm

Chapter 25

Reflection and Refraction of Light

CHAPTER OUTLINE

ANSWERS TO QUESTIONS

Q25.1 The ray approximation, predicting sharp shadows, is valid for $\lambda << d$. For $\lambda \sim d$ diffraction effects become important, and the light waves will spread out noticeably beyond the slit.

Q25.2 With a vertical shop window, streetlights and his own reflection can impede the window shopper's clear view of the display. The tilted shop window can put these reflections out of the way. Windows of airport control towers are also tilted like this, as are automobile windshields.

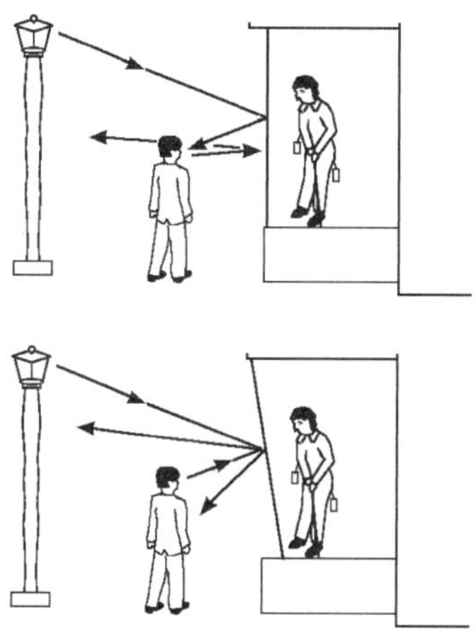

FIG. Q25.2

Q25.3 The right-hand fish image is light from the right side of the actual fish, refracted toward the observer, and the second image is light from the left side of the fish refracted toward the observer.

Q25.4 An echo is an example of the reflection of sound. Hearing the noise of a distant highway on a cold morning, when you cannot hear it after the ground warms up, is an example of acoustical refraction. You can use a rubber inner tube inflated with helium as an acoustical lens to concentrate sound in the way a lens can focus light. At your next party, see if you can experimentally find the approximate focal point!

Q25.5 Suppose the light moves into a medium of higher refractive index. Then its wavelength decreases. The frequency remains constant. The speed diminishes by a factor equal to the index of refraction.

Q25.6 If a laser beam enters a sugar solution with a concentration gradient (density and index of refraction increasing with depth) then the laser beam will be progressively bent downward (toward the normal) as it passes into regions of greater index of refraction.

Q25.7 Diamond has higher index of refraction than glass and consequently a smaller critical angle for total internal reflection. A brilliant-cut diamond is shaped to admit light from above, reflect it totally at the converging facets on the underside of the jewel, and let the light escape only at the top. Glass will have less light internally reflected.

Q25.8 The index of refraction of diamond varies with the frequency of the light. Different color-components of the white light are refracted off in different directions by the jewel. The diamond disperses light to form a spectrum, as any prism does.

Q25.9 A faceted diamond or a stone of cubic zirconia sparkles because the light entering the stone from above is totally internally reflected and the stone is cut so the light can only escape back out the top. If the diamond or the cubic zirconia is immersed in a high index of refraction liquid, then the total internal reflection is thwarted and the diamond loses its "sparkle". For an exact match of index of refraction between cubic zirconia and corn syrup, the cubic zirconia stone would be invisible.

Q25.10 Take a half-circular disk of plastic. Center it on a piece of polar-coordinate paper on a horizontal corkboard. Slowly move a pin around the curved side while you look for it, gazing at the center of the flat wall. When you can barely see the pin as your line of sight grazes the flat side of the block, the light from the pin is reaching the origin at the critical angle θ_c. You can conclude that the index of refraction of the plastic is $\dfrac{1}{\sin\theta_c}$.

Q25.11 The light with the greater change in speed will have the larger deviation. If the glass has a higher index than the surrounding medium, *X* travels slower in the glass.

Q25.12 Total internal reflection occurs only when light moving originally in a medium of high index of refraction falls on an interface with a medium of lower index of refraction. Thus, light moving from air ($n = 1$) to water ($n = 1.33$) cannot undergo total internal reflection.

Q25.13 Highly silvered mirrors reflect about 98% of the incident light. With a 2-mirror periscope, that results in approximately a 4% decrease in intensity of light as the light passes through the periscope. This may not seem like much, but in low-light conditions, that lost light may mean the difference between being able to distinguish an enemy armada or an iceberg from the sky beyond. Using prisms results in total internal reflection, meaning that 100% of the incident light is reflected by each prism. That is the "total" in total internal reflection. At the surfaces of entry into and exit from the prisms, antireflective coatings can minimize light loss.

Q25.14 Light from the lamps along the edges of the sheet enters the plastic. Then it is totally internally reflected by the front and back faces of the plastic, wherever the plastic has an interface with air. If the refractive index of the grease is intermediate between 1.55 and 1.00, some of this light can leave the plastic into the grease and leave the grease into the air. The surface of the grease is rough, so the grease can send out light in all directions. The customer sees the grease shining against a black background. The spotlight method of producing the same effect is much less efficient. With it, much of the light from the spotlight is absorbed by the blackboard. The refractive index of the grease must be less than 1.55. Perhaps the best choice would be $\sqrt{1.55 \times 1.00} = 1.24$.

Q25.15 At the altitude of the plane the surface of the Earth need not block off the lower half of the rainbow. Thus, the full circle can be seen. You can see such a rainbow by climbing on a stepladder above a garden sprinkler in the middle of a sunny day. Set the sprinkler for fine mist. Do not let the slippery children fall from the ladder.

Q25.16 A mirage occurs when light changes direction as it moves between batches of air having different indices of refraction because they have different densities at different temperatures. When the sun makes a blacktop road hot, an apparent wet spot is bright due to refraction of light from the bright sky. The light, originally headed a little below the horizontal, always bends up as it first enters and then leaves sequentially hotter, lower-density, lower-index layers of air closer to the road surface.

SOLUTIONS TO PROBLEMS

Section 25.1 The Nature of Light

No problems in this section

Section 25.2 The Ray Model in Geometric Optics

Section 25.3 The Wave Under Reflection

Section 25.4 The Wave Under Refraction

P25.1 (a) From geometry, $1.25 \text{ m} = d \sin 40.0°$

so $d = \boxed{1.94 \text{ m}}$.

(b) $\boxed{50.0° \text{ above the horizontal}}$

or parallel to the incident ray.

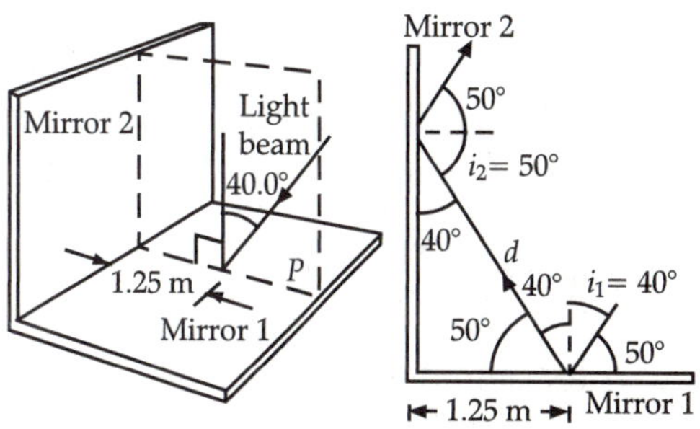

FIG. P25.1

P25.2 (a) **Method One:**

The incident ray makes angle $\alpha = 90° - \theta_1$

with the first mirror. In the picture, the law of reflection implies that

$$\theta_1 = \theta_1'.$$

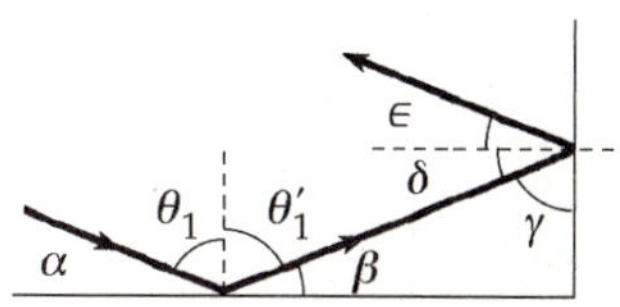

FIG. P25.2

Then $\beta = 90° - \theta_1' = 90 - \theta_1 = \alpha$.

In the triangle made by the mirrors and the ray passing between them,

$$\beta + 90° + \gamma = 180°$$
$$\gamma = 90° - \beta.$$

continued on next page

Further, $\delta = 90° - \gamma = \beta = \alpha$

and $\in = \delta = \alpha$.

Thus the final ray makes the same angle with the first mirror as did the incident ray. Its direction is opposite to the incident ray.

Method Two:

The vector velocity of the incident light has a component v_y perpendicular to the first mirror and a component v_x perpendicular to the second. The v_y component is reversed upon the first reflection, which leaves v_x unchanged. The second reflection reverses v_x and leaves v_y unchanged. The doubly reflected ray then has velocity opposite to the incident ray.

(b) The incident ray has velocity $v_x\hat{\mathbf{i}} + v_y\hat{\mathbf{j}} + v_z\hat{\mathbf{k}}$. Each reflection reverses one component and leaves the other two unchanged. After all the reflections, the light has velocity $-v_x\hat{\mathbf{i}} - v_y\hat{\mathbf{j}} - v_z\hat{\mathbf{k}}$, opposite to the incident ray.

P25.3 The incident light reaches the left-hand mirror at distance

$(1.00\text{ m})\tan 5.00° = 0.087\,5\text{ m}$

above its bottom edge. The reflected light first reaches the right-hand mirror at height

$2(0.087\,5\text{ m}) = 0.175\text{ m}$.

It bounces between the mirrors with this distance between points of contact with either.

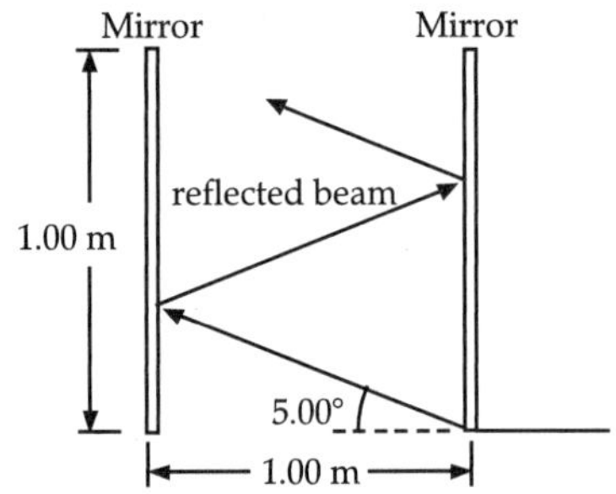

FIG. P25.3

Since $\dfrac{1.00\text{ m}}{0.175\text{ m}} = 5.72$

the light reflects $\boxed{\text{five times from the right-hand mirror and six times from the left}}$.

P25.4 Using Snell's law, $\sin\theta_2 = \dfrac{n_1}{n_2}\sin\theta_1$

$\theta_2 = \boxed{25.5°}$

$\lambda_2 = \dfrac{\lambda_1}{n_1} = \boxed{442\text{ nm}}$.

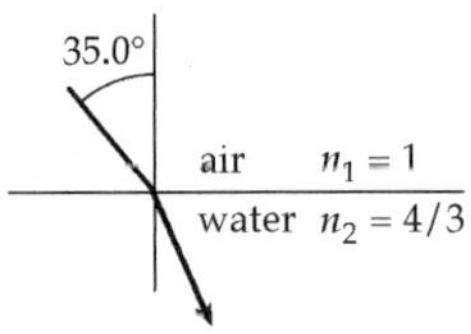

FIG. P25.4

***P25.5** The law of refraction $n_1 \sin\theta_1 = n_2 \sin\theta_2$ can be put into the more general form

$$\frac{c}{v_1}\sin\theta_1 = \frac{c}{v_2}\sin\theta_2$$
$$\frac{\sin\theta_1}{v_1} = \frac{\sin\theta_2}{v_2}$$

In this form it applies to all kinds of waves that move through space.

$$\frac{\sin 3.5°}{343\ \text{m/s}} = \frac{\sin\theta_2}{1\,493\ \text{m/s}}$$
$$\sin\theta_2 = 0.266$$
$$\theta_2 = \boxed{15.4°}$$

The wave keeps constant frequency in

$$f = \frac{v_1}{\lambda_1} = \frac{v_2}{\lambda_2}$$
$$\lambda_2 = \frac{v_2\lambda_1}{v_1} = \frac{1\,493\ \text{m/s}(0.589\ \text{m})}{343\ \text{m/s}} = \boxed{2.56\ \text{m}}$$

P25.6 (a) $f = \dfrac{c}{\lambda} = \dfrac{3.00\times10^8\ \text{m/s}}{6.328\times10^{-7}\ \text{m}} = \boxed{4.74\times10^{14}\ \text{Hz}}$

(b) $\lambda_{\text{glass}} = \dfrac{\lambda_{\text{air}}}{n} = \dfrac{632.8\ \text{nm}}{1.50} = \boxed{422\ \text{nm}}$

(c) $v_{\text{glass}} = \dfrac{c_{\text{air}}}{n} = \dfrac{3.00\times10^8\ \text{m/s}}{1.50} = 2.00\times10^8\ \text{m/s} = \boxed{200\ \text{Mm/s}}$

P25.7
$$n_1 \sin\theta_1 = n_2 \sin\theta_2$$
$$\sin\theta_1 = 1.333\sin 45°$$
$$\sin\theta_1 = (1.33)(0.707) = 0.943$$
$$\theta_1 = 70.5° \rightarrow \boxed{19.5°\ \text{above the horizon}}$$

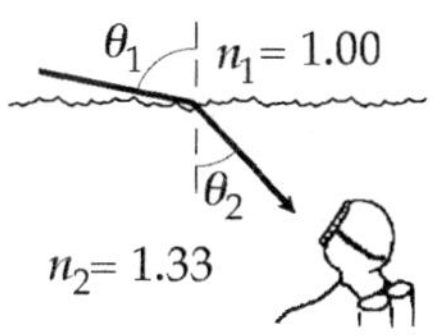

FIG. P25.7

***P25.8** We find the angle of incidence:

$$n_1 \sin\theta_1 = n_2 \sin\theta_2$$
$$1.333\sin\theta_1 = 1.52\sin 19.6°$$
$$\theta_1 = 22.5°$$

The angle of reflection of the beam in water is then also $\boxed{22.5°}$.

***P25.9** (a) As measured from the diagram, the incidence angle is 60°, and the refraction angle is 35°. From Snell's law, $\frac{\sin\theta_2}{\sin\theta_1}=\frac{v_2}{v_1}$, then $\frac{\sin 35°}{\sin 60°}=\frac{v_2}{c}$ and the speed of light in the block is $\boxed{2.0\times10^8 \text{ m/s}}$.

(b) The frequency of the light does not change upon refraction. Knowing the wavelength in a vacuum, we can use the speed of light in a vacuum to determine the frequency: $c=f\lambda$, thus $3.00\times10^8=f\left(632.8\times10^{-9}\right)$, so the frequency is $\boxed{474.1 \text{ THz}}$.

(c) To find the wavelength of light in the block, we use the same wave speed relation, $v=f\lambda$, so $2.0\times10^8=\left(4.741\times10^{14}\right)\lambda$, so $\lambda_{\text{glass}}=\boxed{420 \text{ nm}}$.

P25.10 (a) $n_1\sin\theta_1=n_2\sin\theta_2$

$1.00\sin 30.0°=n\sin 19.24°$

$n=\boxed{1.52}$

(c) $f=\frac{c}{\lambda}=\frac{3.00\times10^8 \text{ m/s}}{6.328\times10^{-7} \text{ m}}=\boxed{4.74\times10^{14} \text{ Hz}}$ in air and in syrup.

(d) $v=\frac{c}{n}=\frac{3.00\times10^8 \text{ m/s}}{1.52}=1.98\times10^8 \text{ m/s}=\boxed{198 \text{ Mm/s}}$

(b) $\lambda=\frac{v}{f}=\frac{1.98\times10^8 \text{ m/s}}{4.74\times10^{14}/\text{s}}=\boxed{417 \text{ nm}}$

P25.11 (a) Flint Glass: $v=\frac{c}{n}=\frac{3.00\times10^8 \text{ m/s}}{1.66}=1.81\times10^8 \text{ m/s}=\boxed{181 \text{ Mm/s}}$

(b) Water: $v=\frac{c}{n}=\frac{3.00\times10^8 \text{ m/s}}{1.333}=2.25\times10^8 \text{ m/s}=\boxed{225 \text{ Mm/s}}$

(c) Cubic Zirconia: $v=\frac{c}{n}=\frac{3.00\times10^8 \text{ m/s}}{2.20}=1.36\times10^8 \text{ m/s}=\boxed{136 \text{ Mm/s}}$

P25.12 $n_1\sin\theta_1=n_2\sin\theta_2$: $1.333\sin 37.0°=n_2\sin 25.0°$

$n_2=1.90=\frac{c}{v}$: $v=\frac{c}{1.90}=1.58\times10^8 \text{ m/s}=\boxed{158 \text{ Mm/s}}$

P25.13 $n_1 \sin\theta_1 = n_2 \sin\theta_2$: $\theta_2 = \sin^{-1}\left(\frac{n_1 \sin\theta_1}{n_2}\right)$

$$\theta_2 = \sin^{-1}\left\{\frac{1.00 \sin 30°}{1.50}\right\} = \boxed{19.5°}$$

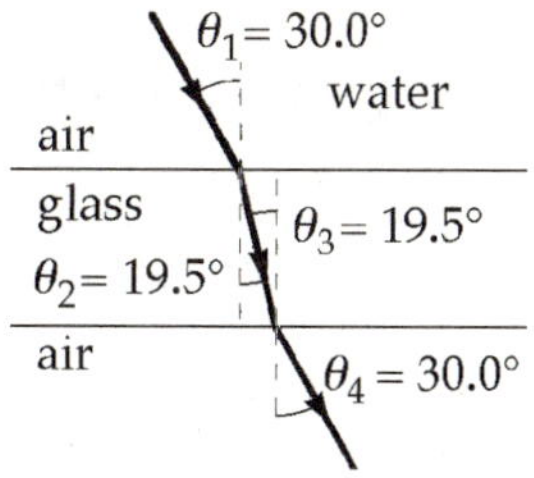

FIG. P25.13

θ_2 and θ_3 are alternate interior angles formed by the ray cutting parallel normals.

So, $\theta_3 = \theta_2 = \boxed{19.5°}$

$$1.50 \sin\theta_3 = 1.00 \sin\theta_4$$

$$\theta_4 = \boxed{30.0°}$$

P25.14 $\sin\theta_1 = n_w \sin\theta_2$

$$\sin\theta_2 = \frac{1}{1.333}\sin\theta_1 = \frac{1}{1.333}\sin(90.0°–28.0°) = 0.662$$

$$\theta_2 = \sin^{-1}(0.662) = 41.5°$$

$$h = \frac{d}{\tan\theta_2} = \frac{3.00\text{ m}}{\tan 41.5°} = \boxed{3.39\text{ m}}$$

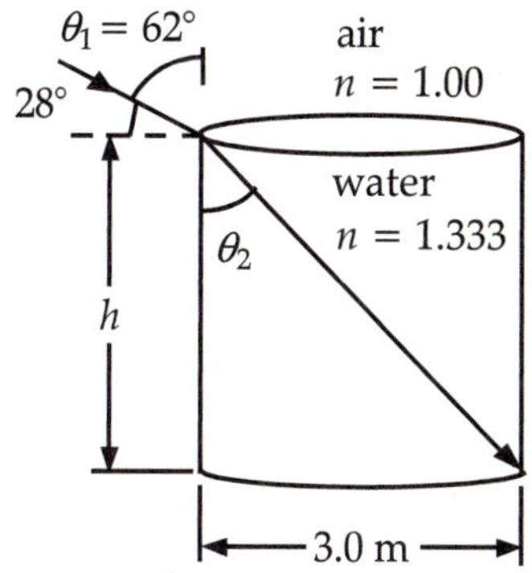

FIG. P25.14

P25.15 For $\alpha + \beta = 90°$

with $\theta_1' + \alpha + \beta + \theta_2 = 180°$

we have $\theta_1' + \theta_2 = 90°$.

Also, $\theta_1' = \theta_1$

and $1 \sin\theta_1 = n \sin\theta_2$.

Then, $\sin\theta_1 = n \sin(90 - \theta_1) = n \cos\theta_1$

$$\frac{\sin\theta_1}{\cos\theta_1} = n = \tan\theta_1 \qquad \boxed{\theta_1 = \tan^{-1} n}.$$

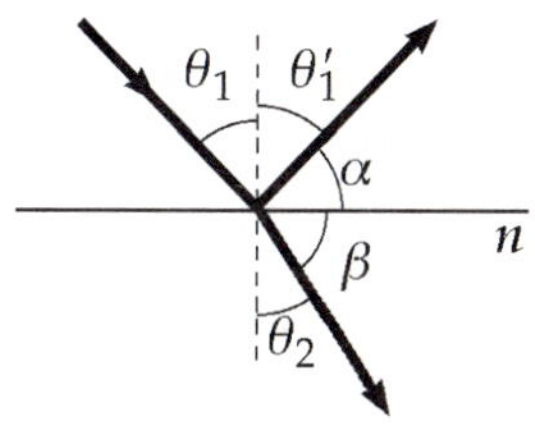

FIG. P25.15

***P25.16** From Snell's law, $\sin\theta = \left(\dfrac{n_{\text{medium}}}{n_{\text{liver}}}\right)\sin 50.0°$

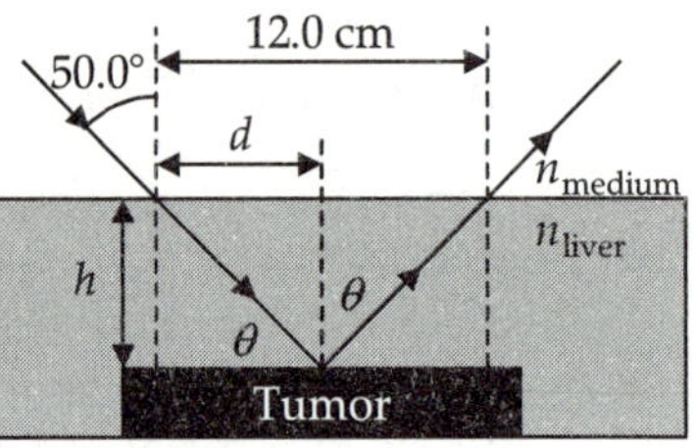

FIG. P25.16

But, $\dfrac{n_{\text{medium}}}{n_{\text{liver}}} = \dfrac{c/v_{\text{medium}}}{c/v_{\text{liver}}} = \dfrac{v_{\text{liver}}}{v_{\text{medium}}} = 0.900$,

so, $\theta = \sin^{-1}\left[(0.900)\sin 50.0°\right] = 43.6°$.

From the law of reflection,

$$d = \frac{12.0 \text{ cm}}{2} = 6.00 \text{ cm, and } h = \frac{d}{\tan\theta} = \frac{6.00 \text{ cm}}{\tan(43.6°)} = \boxed{6.30 \text{ cm}}$$

P25.17 At entry, $n_1 \sin\theta_1 = n_2 \sin\theta_2$

or $1.00 \sin 30.0° = 1.50 \sin\theta_2$

$\theta_2 = 19.5°$.

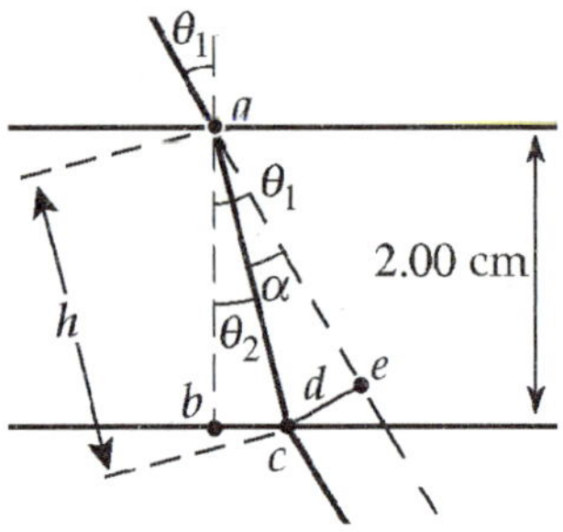

FIG. P25.17

The distance h the light travels in the medium is given by

$$\cos\theta_2 = \frac{2.00 \text{ cm}}{h}$$

or $h = \dfrac{2.00 \text{ cm}}{\cos 19.5°} = 2.12 \text{ cm}$.

The angle of deviation upon entry is $\alpha = \theta_1 - \theta_2 = 30.0° – 19.5° = 10.5°$.

The offset distance comes from $\sin\alpha = \dfrac{d}{h}$: $d = (2.21 \text{ cm})\sin 10.5° = \boxed{0.388 \text{ cm}}$.

P25.18 The distance, h, traveled by the light is $h = \dfrac{2.00 \text{ cm}}{\cos 19.5°} = 2.12 \text{ cm}$.

The speed of light in the material is $v = \dfrac{c}{n} = \dfrac{3.00 \times 10^8 \text{ m/s}}{1.50} = 2.00 \times 10^8 \text{ m/s}$.

Therefore, $t = \dfrac{h}{v} = \dfrac{2.12 \times 10^{-2} \text{ m}}{2.00 \times 10^8 \text{ m/s}} = 1.06 \times 10^{-10} \text{ s} = \boxed{106 \text{ ps}}$.

P25.19 Applying Snell's law at the air-oil interface,

$$n_{\text{air}} \sin\theta = n_{\text{oil}} \sin 20.0°$$

yields $\boxed{\theta = 30.4°}$.

Applying Snell's law at the oil-water interface

$$n_w \sin\theta' = n_{\text{oil}} \sin 20.0°$$

yields $\boxed{\theta' = 22.3°}$.

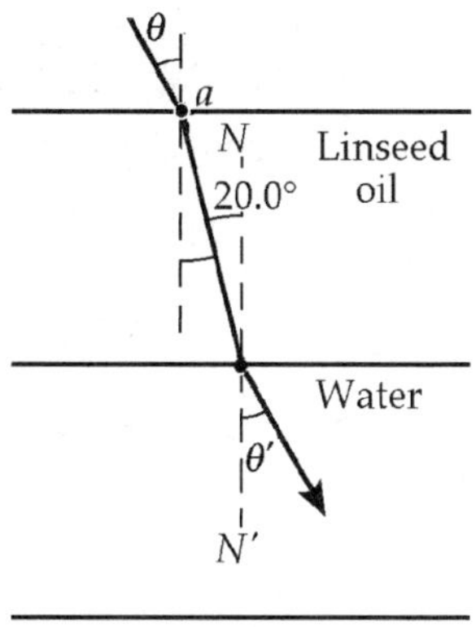

FIG. P25.19

***P25.20** From the figure we have $w = 2b + a$

$$b = \frac{w-a}{2} = \frac{700\ \mu\text{m} - 1\ \mu\text{m}}{2} = 349.5\ \mu\text{m}$$

$$\tan\theta_2 = \frac{b}{t} = \frac{349.5\ \mu\text{m}}{1\,200\ \mu\text{m}} = 0.291 \qquad \theta_2 = 16.2°$$

For refraction at entry,

$$n_1 \sin\theta_1 = n_2 \sin\theta_2$$

$$\theta_1 = \sin^{-1}\frac{n_2 \sin\theta_2}{n_1} = \sin^{-1}\frac{1.55 \sin 16.2°}{1.00} = \sin^{-1} 0.433 = \boxed{25.7°}$$

P25.21 Consider glass with an index of refraction of 1.5, which is 3 mm thick. The speed of light in the glass is

$$\frac{3\times10^{8}\ \text{m/s}}{1.5} = 2\times10^{8}\ \text{m/s}.$$

The extra travel time is $\dfrac{3\times10^{-3}\ \text{m}}{2\times10^{8}\ \text{m/s}} - \dfrac{3\times10^{-3}\ \text{m}}{3\times10^{8}\ \text{m/s}} \boxed{\sim 10^{-11}\ \text{s}}$.

For light of wavelength 600 nm in vacuum and wavelength $\dfrac{600\ \text{nm}}{1.5} = 400\ \text{nm}$ in glass,

the extra optical path, in wavelengths, is $\dfrac{3\times10^{-3}\ \text{m}}{4\times10^{-7}\ \text{m}} - \dfrac{3\times10^{-3}\ \text{m}}{6\times10^{-7}\ \text{m}} \boxed{\sim 10^{3}\ \text{wavelengths}}$.

P25.22 See the sketch showing the path of the light ray. α and γ are angles of incidence at mirrors 1 and 2.

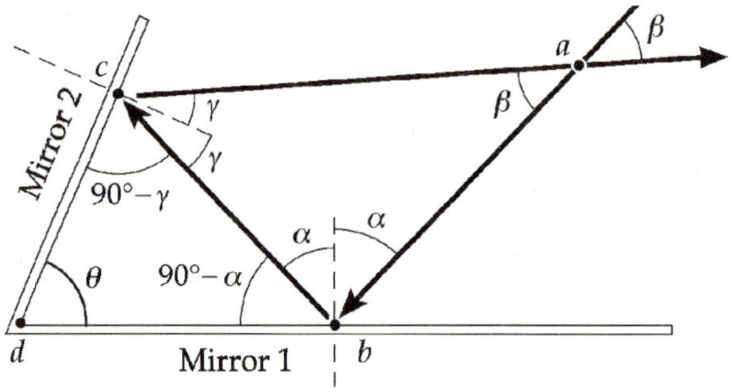

FIG. P25.22

For triangle *abca*,

$$2\alpha+2\gamma+\beta=180°$$

or $\beta=180°-2(\alpha+\gamma)$. (1)

Now for triangle *bcdb*,

$$(90.0°-\alpha)+(90.0°-\gamma)+\theta=180°$$

or $\theta=\alpha+\gamma$. (2)

Substituting Equation (2) into Equation (1) gives $\boxed{\beta=180°-2\theta}$.

Note: From Equation (2), $\gamma=\theta-\alpha$. Thus, the ray will follow a path like that shown only if $\alpha<\theta$. For $\alpha>\theta$, γ is negative and multiple reflections from each mirror will occur before the incident and reflected rays intersect.

Section 25.5 Dispersion and Prisms

P25.23 From the dispersion curve for fused quartz in the chapter (that is, the graph of its refractive index versus wavelength) we have

$$n_v=1.470 \text{ at } 400 \text{ nm} \quad \text{and} \quad n_r=1.458 \text{ at } 700 \text{ nm}.$$

Then $1.00\sin\theta=1.470\sin\theta_v$ and $1.00\sin\theta=1.458\sin\theta_r$

$$\delta_r-\delta_v=\theta_r-\theta_v=\sin^{-1}\left(\frac{\sin\theta}{1.458}\right)-\sin^{-1}\left(\frac{\sin\theta}{1.470}\right)$$

$$\Delta\delta=\sin^{-1}\left(\frac{\sin 30.0°}{1.458}\right)-\sin^{-1}\left(\frac{\sin 30.0°}{1.470}\right)=\boxed{0.171°}$$

P25.24 $n_1\sin\theta_1=n_2\sin\theta_2$: $\quad \theta_2=\sin^{-1}\left(\frac{n_1\sin\theta_1}{n_2}\right)$

$$\theta_2=\sin^{-1}\left(\frac{1.00\sin 30.0°}{1.50}\right)=\boxed{19.5°}$$

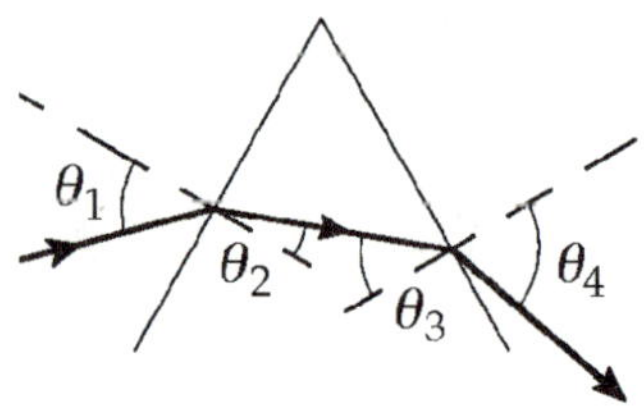

FIG. P25.24

The surface of entry, the surface of exit, and the ray within the prism form a triangle. Inside the triangle the angles must add up according to

$$90.0°-\theta_2+60.0°+90.0°-\theta_3=180°$$

$$\theta_3=\left(\left[(90.0°-19.5°)+60.0°\right]-180°\right)+90.0°=\boxed{40.5°}$$

$n_3\sin\theta_3=n_4\sin\theta_4$: $\quad \theta_4=\sin^{-1}\left(\frac{n_3\sin\theta_3}{n_4}\right)=\sin^{-1}\left(\frac{1.50\sin 40.5°}{1.00}\right)=\boxed{77.1°}$

P25.25 At the first refraction, $1.00\sin\theta_1 = n\sin\theta_2$.

The critical angle at the second surface is given by $n\sin\theta_3 = 1.00$:

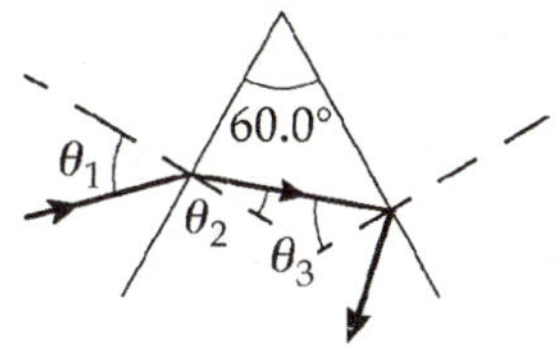

FIG. P25.25

or $$\theta_3 = \sin^{-1}\left(\frac{1.00}{1.50}\right) = 41.8°.$$

But, $$\theta_2 = 60.0° - \theta_3.$$

Thus, to avoid total internal reflection at the second surface (i.e., have $\theta_3 < 41.8°$)

it is necessary that $$\theta_2 > 18.2°.$$

Since $\sin\theta_1 = n\sin\theta_2$, this becomes $$\sin\theta_1 > 1.50\sin 18.2° = 0.468$$

or $$\theta_1 > \boxed{27.9°}.$$

P25.26 At the first refraction, $1.00\sin\theta_1 = n\sin\theta_2$.

The critical angle at the second surface is given by

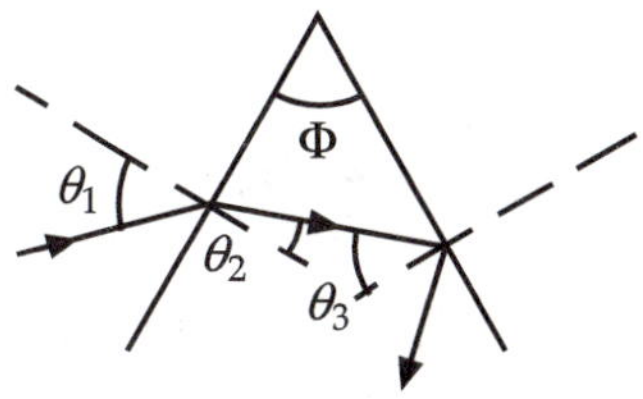

FIG. P25.26

$n\sin\theta_3 = 1.00$, or $$\theta_3 = \sin^{-1}\left(\frac{1.00}{n}\right).$$

But $$(90.0° - \theta_2) + (90.0° - \theta_3) + \Phi = 180°$$

which gives $$\theta_2 = \Phi - \theta_3.$$

Thus, to have $\theta_3 < \sin^{-1}\left(\frac{1.00}{n}\right)$ and avoid total internal reflection at the second surface,

it is necessary that $$\theta_2 > \Phi - \sin^{-1}\left(\frac{1.00}{n}\right).$$

Since $\sin\theta_1 = n\sin\theta_2$, this requirement becomes $$\sin\theta_1 > n\sin\left[\Phi - \sin^{-1}\left(\frac{1.00}{n}\right)\right]$$

or $$\theta_1 > \boxed{\sin^{-1}\left(n\sin\left[\Phi - \sin^{-1}\left(\frac{1.00}{n}\right)\right]\right)}.$$

Through the application of trigonometric identities, $$\theta_1 > \boxed{\sin^{-1}\left(\sqrt{n^2 - 1}\sin\Phi - \cos\Phi\right)}.$$

As n increases, θ_1 increases. As Φ increases, both of the terms inside the parentheses increase, so θ_1 increases. It becomes easier to produce total internal reflection. If Φ decreases below $\tan^{-1}\left(\left(n^2 - 1\right)^{-1/2}\right)$ it becomes necessary for θ_1 to be negative for total internal reflection to occur. That is, the incident ray must be on the other side of the normal to the first surface, but the equation is still true. If the quantity $\sqrt{n^2 - 1}\sin\Phi - \cos\Phi$ is greater than 1, total internal reflection happens for all values of θ_1. However, if this quantity is greater than n, the ray internal to the prism never reaches the second surface.

P25.27 For the incoming ray, $\sin\theta_2 = \frac{\sin\theta_1}{n}$.

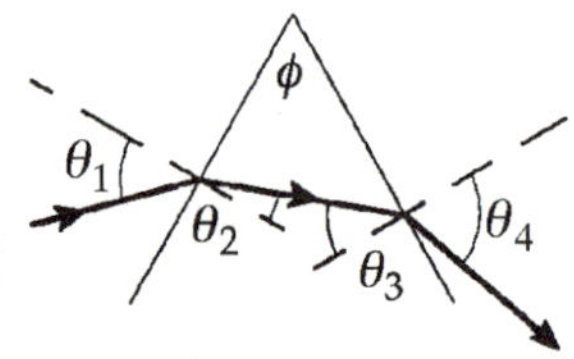

FIG. P25.27

Using the figure to the right, $(\theta_2)_{\text{violet}} = \sin^{-1}\left(\frac{\sin 50.0°}{1.66}\right) = 27.48°$

$(\theta_2)_{\text{red}} = \sin^{-1}\left(\frac{\sin 50.0°}{1.62}\right) = 28.22°$.

For the outgoing ray, $\theta_3 = 60.0° - \theta_2$

and $\sin\theta_4 = n\sin\theta_3$: $(\theta_4)_{\text{violet}} = \sin^{-1}[1.66\sin 32.52°] = 63.17°$

$(\theta_4)_{\text{red}} = \sin^{-1}[1.62\sin 31.78°] = 58.56°$.

The angular dispersion is the difference $\Delta\theta_4 = (\theta_4)_{\text{violet}} - (\theta_4)_{\text{red}} = 63.17° - 58.56° = \boxed{4.61°}$.

Section 25.6 Huygens's Principle

P25.28 (a) For the diagrams of contour lines and wave fronts and rays, see Figures (a) and (b) below. As the waves move to shallower water, the wave fronts bend to become more nearly parallel to the contour lines.

(b) For the diagrams of contour lines and wave fronts and rays, see Figures (c) and (d) below. We suppose that the headlands are steep underwater, as they are above water. The rays are everywhere perpendicular to the wave fronts of the incoming refracting waves. As shown, the rays bend toward the headlands and deliver more energy per length at the headlands.

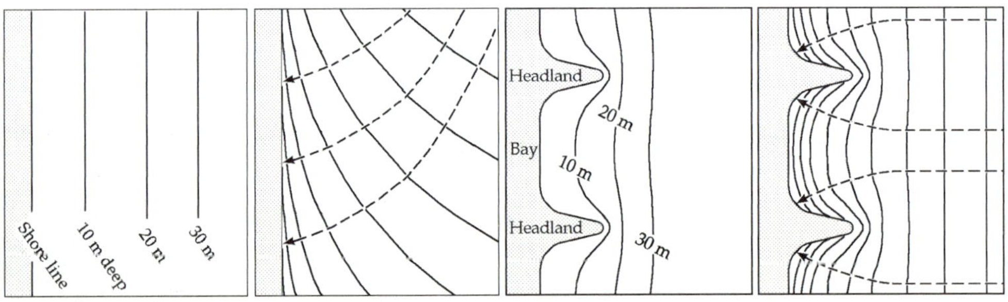

(a) Contour lines (b) Wave fronts and rays (c) Contour lines (d) Wave fronts and rays

FIG. P25.28

Section 25.7 Total Internal Reflection

P25.29 $n \sin\theta = 1$. From the table of refractive indices

(a) $$\theta = \sin^{-1}\left(\frac{1}{2.419}\right) = \boxed{24.4°}$$

(b) $$\theta = \sin^{-1}\left(\frac{1}{1.66}\right) = \boxed{37.0°}$$

(c) $$\theta = \sin^{-1}\left(\frac{1}{1.309}\right) = \boxed{49.8°}$$

P25.30 (a) $$\frac{\sin\theta_2}{\sin\theta_1} = \frac{v_2}{v_1}$$

and $\theta_2 = 90.0°$ at the critical angle

$$\frac{\sin 90.0°}{\sin\theta_c} = \frac{1\,850 \text{ m/s}}{343 \text{ m/s}}$$

so $\theta_c = \sin^{-1}(0.185) = \boxed{10.7°}$.

(b) Sound can be totally reflected if it is traveling in the medium where it travels slower: $\boxed{\text{air}}$.

(c) $\boxed{\text{Sound in air falling on the wall from most directions is 100\% reflected}}$, so the wall is a good mirror.

P25.31 $$\sin\theta_c = \frac{n_2}{n_1}$$

$$n_2 = n_1 \sin 88.8° = (1.000\,3)(0.999\,8) = \boxed{1.000\,08}$$

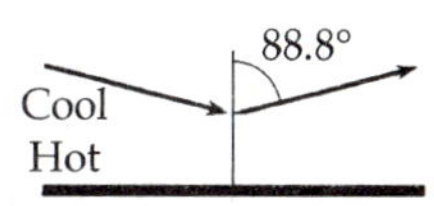

FIG. P25.31

P25.32 For plastic with index of refraction $\boxed{n \geq 1.42}$ surrounded by air, the critical angle for total internal reflection is given by

$$\theta_c = \sin^{-1}\left(\frac{1}{n}\right) \leq \sin^{-1}\left(\frac{1}{1.42}\right) = 44.8°.$$

In the gasoline gauge, skylight from above travels down the plastic. The rays close to the vertical are totally reflected from the sides of the slab and from both facets at the lower end of the plastic, where it is not immersed in gasoline. This light returns up inside the plastic and makes it look bright. Where the plastic is immersed in gasoline, with index of refraction about 1.50, total internal reflection should not happen. The light passes out of the lower end of the plastic with little reflected, making this part of the gauge look dark. To frustrate total internal reflection in the gasoline, the index of refraction of the plastic should be $\boxed{n < 2.12}$

since $\theta_c = \sin^{-1}\left(\frac{1.50}{2.12}\right) = 45.0°.$

Section 25.8 Context Connection—Optical Fibers

P25.33 $\sin\theta_c = \frac{n_{air}}{n_{pipe}} = \frac{1.00}{1.36} = 0.735$ $\theta_c = 47.3°$

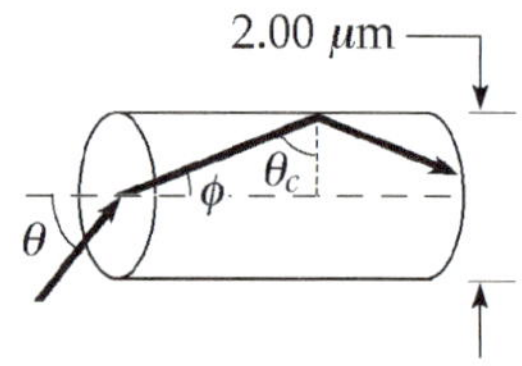

FIG. P25.33

Geometry shows that the angle of refraction at the end is

$\phi = 90.0° - \theta_c = 90.0° - 47.3° = 42.7°$.

Then, Snell's law at the end, $1.00\sin\theta = 1.36\sin 42.7°$

gives $\boxed{\theta = 67.2°}$.

The 2-μm diameter is unnecessary information.

P25.34 For total internal reflection, $n_1 \sin\theta_1 = n_2 \sin 90.0°$

$1.50 \sin\theta_1 = 1.33(1.00)$ or $\theta_1 = \boxed{62.4°}$

P25.35 As the beam enters the slab,

$$1.00\sin 50.0° = 1.48\sin\theta_2$$

giving $\theta_2 = 31.2°$.

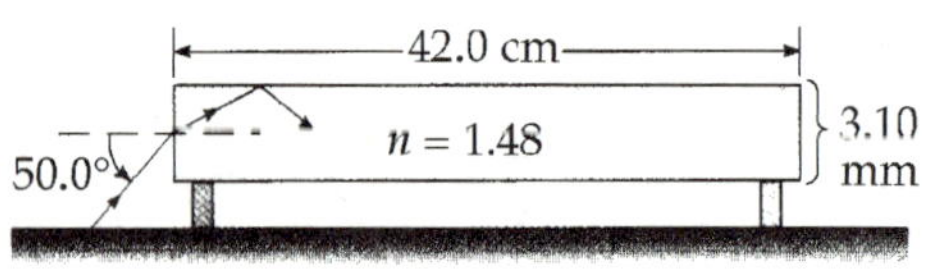

FIG. P25.35

The beam then strikes the top of the slab at $x_1 = \frac{1.55 \text{ mm}}{\tan 31.2°}$ from the left end. Thereafter, the beam strikes a face each time it has traveled a distance of $2x_1$ along the length of the slab. Since the slab is 420 mm long, the beam has an additional $420 \text{ mm} - x_1$ to travel after the first reflection. The number of additional reflections is

continued on next page

$$\frac{420\text{ mm}-x_1}{2x_1}=\frac{420\text{ mm}-1.55\text{ mm}/\tan 31.2°}{3.10\text{ mm}/\tan 31.2°}=81.5 \qquad \text{or 81 reflections}$$

since the answer must be an integer. The total number of reflections made in the slab is then $\boxed{82}$.

P25.36 (a) A ray along the inner edge will escape if any ray escapes. Its angle of incidence is described by $\sin\theta=\frac{R-d}{R}$ and by $n\sin\theta>1\sin 90°$. Then

$$\frac{n(R-d)}{R}>1 \qquad nR-nd>R \qquad nR-R>nd \qquad R>\boxed{\frac{nd}{n-1}}.$$

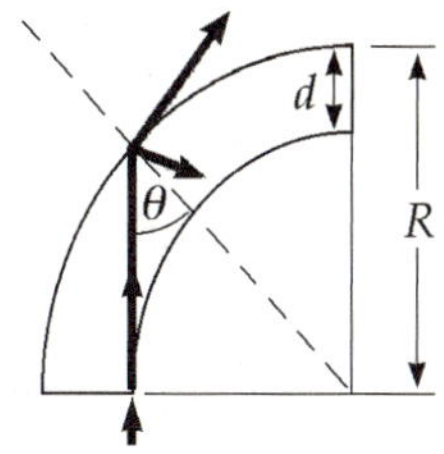

FIG. P25.36

(b) As $d\to 0$, $R_{\text{min}}\to 0$. This is reasonable.

As n increases, R_{min} decreases. This is reasonable.

As n decreases toward 1, R_{min} increases. This is reasonable.

(c) $$R_{\text{min}}=\frac{1.40\left(100\times 10^{-6}\text{ m}\right)}{0.40}=\boxed{350\times 10^{-6}\text{ m}}$$

Additional Problems

***P25.37** For sheets 1 and 2 as described,

$$n_1\sin 26.5°=n_2\sin 31.7°$$
$$0.849n_1=n_2$$

For the trial with sheets 3 and 2,

$$n_3\sin 26.5°=n_2\sin 36.7°$$
$$0.747n_3=n_2$$

Now

$$0.747n_3=0.849n_1$$
$$n_3=1.14n_1$$

For the third trial,

$$n_1\sin 26.5°=n_3\sin\theta_3=1.14n_1\sin\theta_3$$
$$\theta_3=\boxed{23.1°}$$

***P25.38** Scattered light leaves the center of the photograph (a) in all horizontal directions between $\theta_1 = 0°$ and $90°$ from the normal. When it immediately enters the water (b), it is gathered into a fan between $0°$ and $\theta_{2\,\max}$ given by

$$n_1 \sin\theta_1 = n_2 \sin\theta_2$$
$$1.00 \sin 90 = 1.333 \sin\theta_{2\,\max}$$
$$\theta_{2\,\max} = 48.6°$$

The light leaves the cylinder without deviation, so the viewer only receives light from the center of the photograph when he has turned by an angle less than 48.6°. When the paperweight is turned farther, light at the back surface undergoes total internal reflection (c). The viewer sees things outside the globe on the far side.

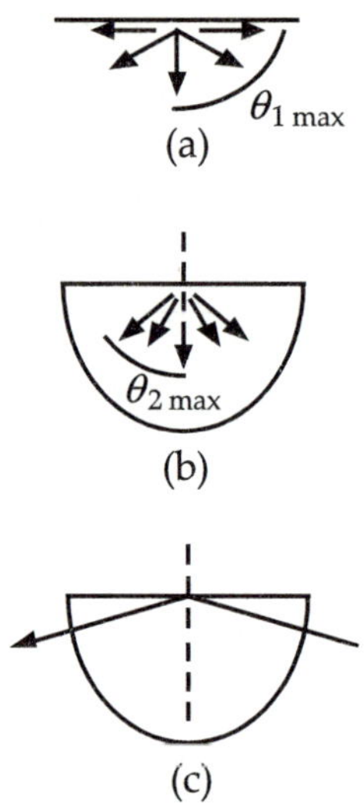

FIG. P25.38

P25.39 Let $n(x)$ be the index of refraction at distance x below the top of the atmosphere and $n(x = h) = n$ be its value at the planet surface.

Then,
$$n(x) = 1.000 + \left(\frac{n - 1.000}{h}\right)x.$$

(a) The total time interval required to traverse the atmosphere is

$$\Delta t = \int_0^h \frac{dx}{v} = \int_0^h \frac{n(x)}{c}dx: \qquad \Delta t = \frac{1}{c}\int_0^h \left[1.000 + \left(\frac{n - 1.000}{h}\right)x\right]dx$$

$$\Delta t = \frac{h}{c} + \frac{(n - 1.000)}{ch}\left(\frac{h^2}{2}\right) = \boxed{\frac{h}{c}\left(\frac{n + 1.000}{2}\right)}.$$

(b) The travel time in the absence of an atmosphere would be $\frac{h}{c}$.

Thus, the time in the presence of an atmosphere is $\boxed{\left(\frac{n + 1.000}{2}\right) \text{ times larger}}$.

P25.40 (a) $\theta_1' = \theta_1 = \boxed{30.0°}$ $\quad n_1 \sin\theta_1 = n_2 \sin\theta_2$

$$1.00 \sin 30.0° = 1.55 \sin\theta_2$$
$$\theta_2 = \boxed{18.8°}$$

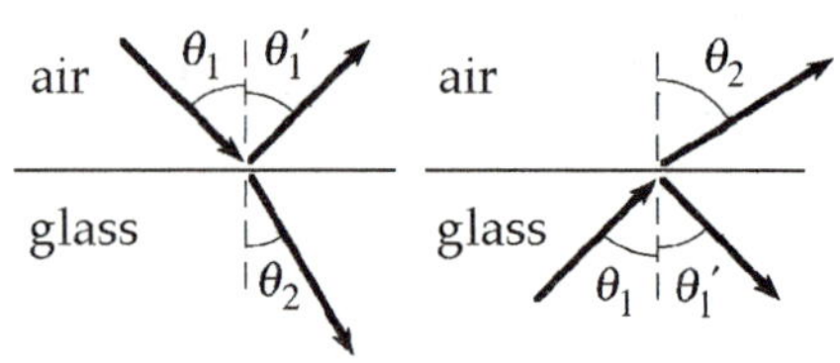

FIG. P25.40

(b) $\theta_1' = \theta_1 = \boxed{30.0°}$ $\quad \theta_2 = \sin^{-1}\left(\frac{n_1 \sin\theta_1}{n_2}\right)$

$$= \sin^{-1}\left(\frac{1.55 \sin 30.0°}{1}\right) = \boxed{50.8°}$$

continued on next page

(c), (d) The other entries are computed similarly, and are shown in the table below.

(c) air into glass, angles in degrees			(d) glass into air, angles in degrees		
incidence	reflection	refraction	incidence	reflection	refraction
0	0	0	0	0	0
10.0	10.0	6.43	10.0	10.0	15.6
20.0	20.0	12.7	20.0	20.0	32.0
30.0	30.0	18.8	30.0	30.0	50.8
40.0	40.0	24.5	40.0	40.0	85.1
50.0	50.0	29.6	50.0	50.0	none*
60.0	60.0	34.0	60.0	60.0	none*
70.0	70.0	37.3	70.0	70.0	none*
80.0	80.0	39.4	80.0	80.0	none*
90.0	90.0	40.2	90.0	90.0	none*

*total internal reflection

P25.41 For water, $\sin\theta_c = \frac{1}{4/3} = \frac{3}{4}$.

Thus $\theta_c = \sin^{-1}(0.750) = 48.6°$

and $d = 2\left[(1.00\text{ m})\tan\theta_c\right]$

$d = (2.00\text{ m})\tan 48.6° = \boxed{2.27\text{ m}}$.

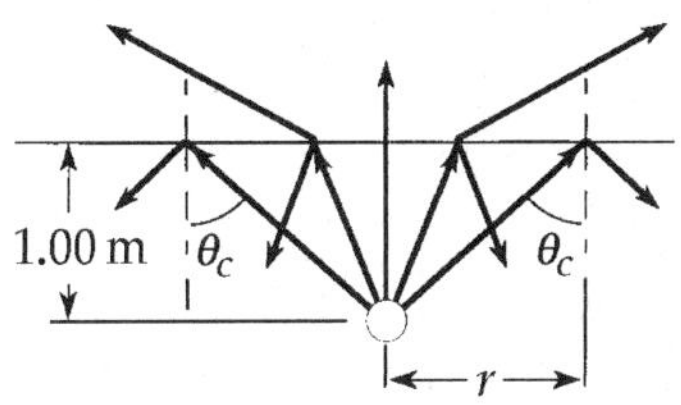

FIG. P25.41

P25.42 Call θ_1 the angle of incidence and of reflection on the left face and θ_2 those angles on the right face. Let α represent the complement of θ_1 and β be the complement of θ_2. Now $\alpha = \gamma$ and $\beta = \delta$ because they are pairs of alternate interior angles. We have

$$A = \gamma + \delta = \alpha + \beta$$

and $B = \alpha + A + \beta = \alpha + \beta + A = \boxed{2A}$.

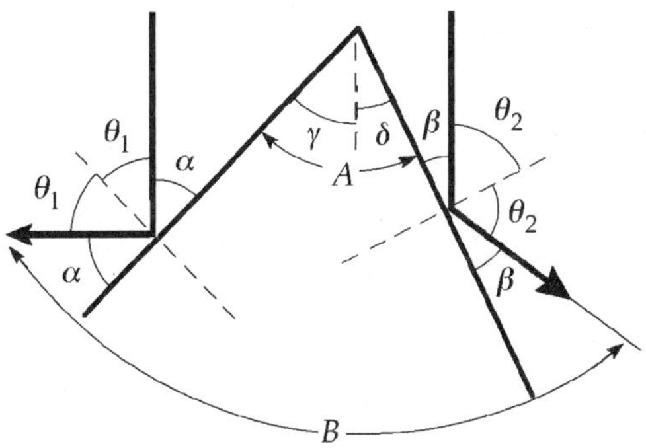

FIG. P25.42

P25.43 (a) We see the Sun moving from east to west across the sky. Its angular speed is

$$\omega = \frac{\Delta\theta}{\Delta t} = \frac{2\pi\,\text{rad}}{86\,400\text{ s}} = 7.27\times10^{-5}\text{ rad/s}.$$

The direction of sunlight crossing the cell from the window changes at this rate, moving on the opposite wall at speed

$$v = r\omega = (2.37\text{ m})\left(7.27\times10^{-5}\text{ rad/s}\right) = 1.72\times10^{-4}\text{ m/s} = \boxed{0.172\text{ mm/s}}.$$

continued on next page

(b) The mirror folds into the cell the motion that would occur in a room twice as wide:

$$v = r\omega = 2(0.174 \text{ mm/s}) = \boxed{0.345 \text{ mm/s}}.$$

(c), (d) As the Sun moves southward and upward at 50.0°, we may regard the corner of the window as fixed, and both patches of light move $\boxed{\text{northward and downward at } 50.0°}$.

P25.44 (a) $\boxed{45.0°}$ as shown in the first figure to the right.

(b) $\boxed{\text{Yes}}$

If grazing angle is halved, the number of reflections from the side faces is doubled.

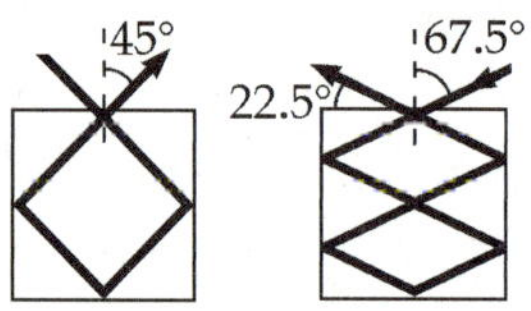

FIG. P25.44

P25.45 Horizontal light rays from the setting Sun pass above the hiker. The light rays are twice refracted and once reflected, as in Figure (b). The most intense light reaching the hiker, (the light that represents the visible rainbow,) is located between angles of 40° and 42° from the hiker's shadow.

The hiker sees a greater percentage of the violet inner edge, so we consider the red outer edge. The radius R of the circle of droplets is

$$R = (8.00 \text{ km})\sin 42.0° = 5.35 \text{ km}.$$

Then the angle ϕ, between the vertical and the radius where the bow touches the ground, is given by

$$\cos\phi = \frac{2.00 \text{ km}}{R} = \frac{2.00 \text{ km}}{5.35 \text{ km}} = 0.374$$

or $\phi = 68.1°$.

The angle filled by the visible bow is $360° - (2 \times 68.1°) = 224°$

so the visible bow is $\frac{224°}{360°} = \boxed{62.2\% \text{ of a circle}}$.

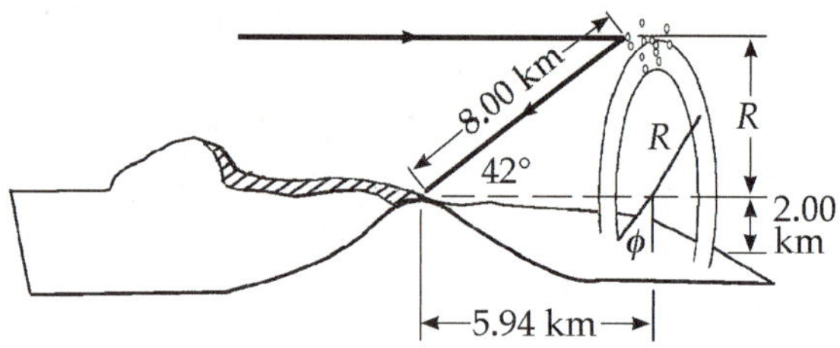

Figure (a)

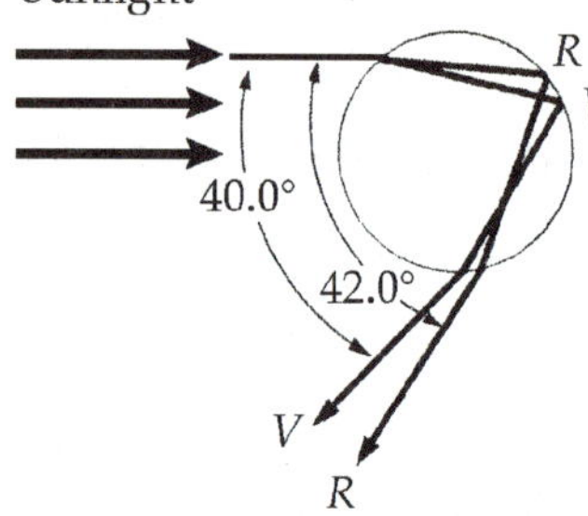

Figure (b)

FIG. P25.45

This striking view motivated Charles Wilson's 1906 invention of the cloud chamber, a standard tool of nuclear physics. The effect is mentioned in the Bible Ezekiel 1:29.

P25.46 Light passing the top of the pole makes an angle of incidence $\phi_1 = 90.0° - \theta$. It falls on the water surface at distance from the pole

$$s_1 = \frac{L-d}{\tan\theta}$$

and has an angle of refraction ϕ_2 from $1.00\sin\phi_1 = n\sin\phi_2$.

Then $s_2 = d\tan\phi_2$

and the whole shadow length is

$$s_1 + s_2 = \frac{L-d}{\tan\theta} + d\tan\left(\sin^{-1}\left(\frac{\sin\phi_1}{n}\right)\right)$$

$$s_1 + s_2 = \frac{L-d}{\tan\theta} + d\tan\left(\sin^{-1}\left(\frac{\cos\theta}{n}\right)\right)$$

$$= \frac{2.00\text{ m}}{\tan 40.0°} + (2.00\text{ m})\tan\left(\sin^{-1}\left(\frac{\cos 40.0°}{1.33}\right)\right) = \boxed{3.79\text{ m}}$$

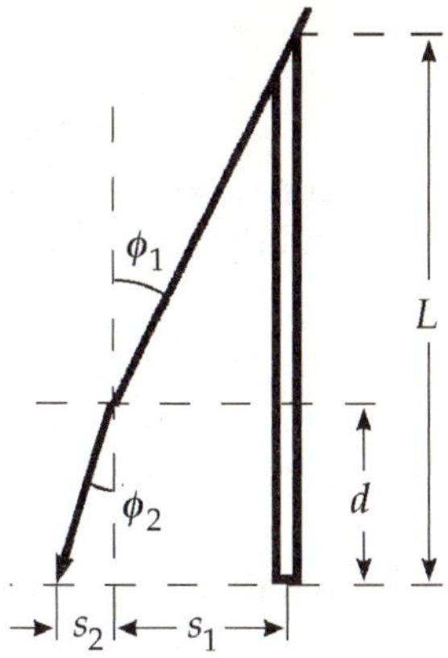

FIG. P25.46

P25.47 (a) $\frac{S_1'}{S_1} = \left[\frac{n_2 - n_1}{n_2 + n_1}\right]^2 = \left[\frac{1.52 - 1.00}{1.52 + 1.00}\right]^2 = \boxed{0.042\,6}$

(b) If medium 1 is glass and medium 2 is air, $\frac{S_1'}{S_1} = \left[\frac{n_2 - n_1}{n_2 + n_1}\right]^2 = \left[\frac{1.00 - 1.52}{1.00 + 1.52}\right]^2 = 0.042\,6$.

There is $\boxed{\text{no difference}}$.

P25.48 (a) With $n_1 = 1$

and $n_2 = n$

the reflected fractional intensity is $\frac{S_1'}{S_1} = \left(\frac{n-1}{n+1}\right)^2$.

The remaining intensity must be transmitted:

$$\frac{S_2}{S_1} = 1 - \left(\frac{n-1}{n+1}\right)^2 = \frac{(n+1)^2 - (n-1)^2}{(n+1)^2} = \frac{n^2 + 2n + 1 - n^2 + 2n - 1}{(n+1)^2} = \boxed{\frac{4n}{(n+1)^2}}.$$

(b) At entry, $\frac{S_2}{S_1} = 1 - \left(\frac{n-1}{n+1}\right)^2 = \frac{4(2.419)}{(2.419+1)^2} = 0.828$.

At exit, $\frac{S_3}{S_2} = 0.828$.

Overall, $\frac{S_3}{S_1} = \left(\frac{S_3}{S_2}\right)\left(\frac{S_2}{S_1}\right) = (0.828)^2 = 0.685$

or $\boxed{68.5\%}$.

P25.49 Define $T=\frac{4n}{(n+1)^2}$ as the transmission coefficient for one encounter with an interface. For diamond and air, it is 0.828, as in Problem P25.48.

As shown in the figure, the total amount transmitted is

$$T^2+T^2(1-T)^2+T^2(1-T)^4+T^2(1-T)^6$$
$$+\ldots+T^2(1-T)^{2n}+\ldots$$

We have $1-T=1-0.828=0.172$ so the total transmission is

$$(0.828)^2\left[1+(0.172)^2+(0.172)^4+(0.172)^6+\ldots\right].$$

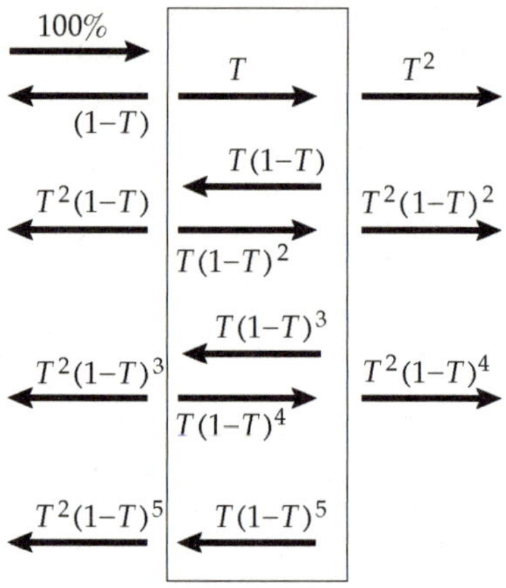

FIG. P25.49

To sum this series, define $F=1+(0.172)^2+(0.172)^4+(0.172)^6+\ldots$.

Note that $(0.172)^2F=(0.172)^2+(0.172)^4+(0.172)^6+\ldots$, and
$1+(0.172)^2F=1+(0.172)^2+(0.172)^4+(0.172)^6+\ldots=F$.

Then, $1=F-(0.172)^2F$ or $F=\frac{1}{1-(0.172)^2}$.

The overall transmission is then $\frac{(0.828)^2}{1-(0.172)^2}=0.706$ or $\boxed{70.6\%}$.

P25.50 (a) As the mirror turns through angle θ, the angle of incidence increases by θ and so does the angle of reflection. The incident ray is stationary, so the reflected ray turns through angle 2θ. The angular speed of the reflected ray is $2\omega_m$. The speed of the dot of light on the circular wall is $\boxed{2\omega_m R}$.

(b) The two angles marked θ in the figure to the right are equal because their sides are perpendicular, right side to right side and left side to left side.

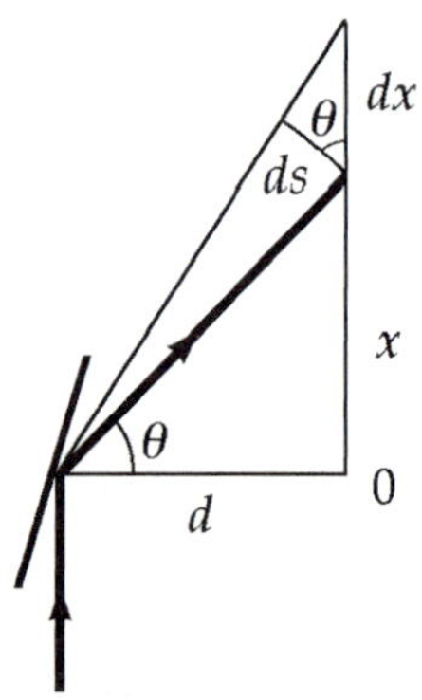

FIG. P25.50

We have $\cos\theta=\frac{d}{\sqrt{x^2+d^2}}=\frac{ds}{dx}$

and $\frac{ds}{dt}=2\omega_m\sqrt{x^2+d^2}$.

So $\frac{dx}{dt}=\frac{ds}{dt}\frac{\sqrt{x^2+d^2}}{d}=\boxed{2\omega_m\frac{x^2+d^2}{d}}$.

P25.51 Define n_1 to be the index of refraction of the surrounding medium and n_2 to be that for the prism material. We can use the critical angle of 42.0° to find the ratio $\frac{n_2}{n_1}$:

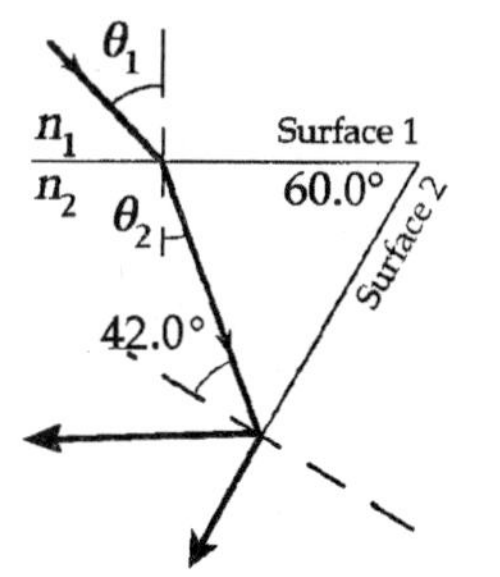

FIG. P25.51

$$n_2 \sin 42.0° = n_1 \sin 90.0° .$$

So, $$\frac{n_2}{n_1} = \frac{1}{\sin 42.0°} = 1.49 .$$

Call the angle of refraction θ_2 at the surface 1. The ray inside the prism forms a triangle with surfaces 1 and 2, so the sum of the interior angles of this triangle must be 180°.

Thus, $(90.0° - \theta_2) + 60.0° + (90.0° - 42.0°) = 180° .$

Therefore, $\theta_2 = 18.0° .$

Applying Snell's law at surface 1, $n_1 \sin \theta_1 = n_2 \sin 18.0°$

$\sin \theta_1 = \left(\frac{n_2}{n_1}\right) \sin \theta_2 = 1.49 \sin 18.0°$ $\boxed{\theta_1 = 27.5°}$.

***P25.52** The picture illustrates optical sunrise. At the center of the earth,

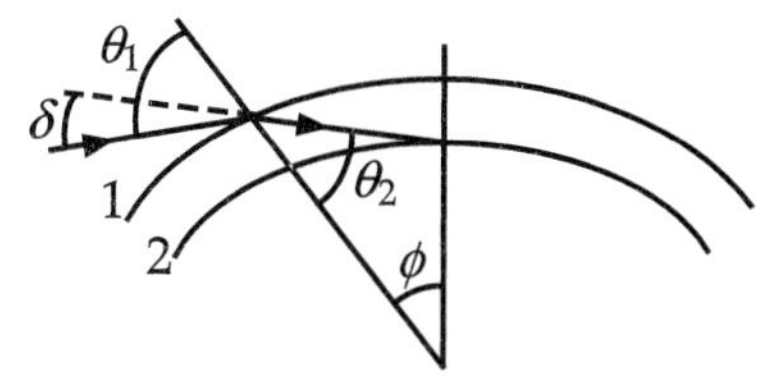

FIG. P25.52

$$\cos \phi = \frac{6.37 \times 10^6 \text{ m}}{6.37 \times 10^6 \text{ m} + 8\,614}$$

$\phi = 2.98°$

$\theta_2 = 90 - 2.98° = 87.0°$

At the top of the atmosphere

$n_1 \sin \theta_1 = n_2 \sin \theta_2$

$1 \sin \theta_1 = 1.000\,293 \sin 87.0°$

$\theta_1 = 87.4°$

Deviation upon entry is

$\delta = |\theta_1 - \theta_2|$

$\delta = 87.364° - 87.022° = 0.342°$

Sunrise of the optical day is before geometric sunrise by $0.342°\left(\frac{86\,400 \text{ s}}{360°}\right) = 82.2$ s. Optical sunset occurs later too, so the optical day is longer by $\boxed{164 \text{ s}}$.

P25.53 (a) For polystyrene *surrounded by air*, internal reflection requires

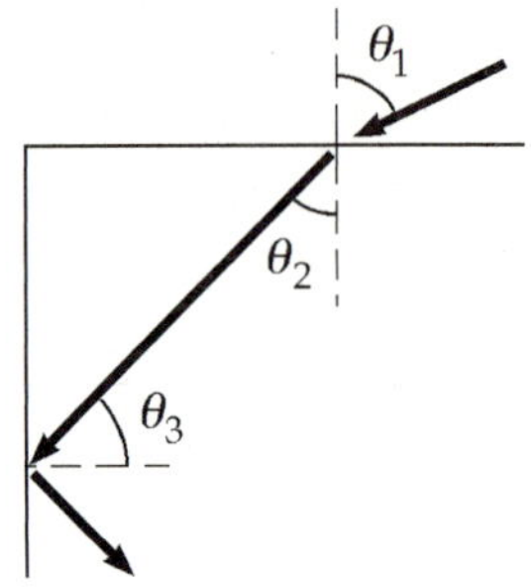

FIG. P25.53

$$\theta_3 = \sin^{-1}\left(\frac{1.00}{1.49}\right) = 42.2°.$$

Then from geometry, $\theta_2 = 90.0° - \theta_3 = 47.8°$.

From Snell's law, $\sin\theta_1 = 1.49\sin 47.8° = 1.10$.

This has no solution.

Therefore, total internal reflection $\boxed{\text{always happens}}$.

(b) For polystyrene *surrounded by water*, $\theta_3 = \sin^{-1}\left(\frac{1.33}{1.49}\right) = 63.2°$

and $\theta_2 = 26.8°$.

From Snell's law, $\theta_1 = \boxed{30.3°}$.

(c) $\boxed{\text{No internal refraction is possible}}$

since the beam is initially traveling in a medium of lower index of refraction.

P25.54 $\delta = \theta_1 - \theta_2 = 10.0°$ and $n_1 \sin\theta_1 = n_2 \sin\theta_2$

with $n_1 = 1$, $n_2 = \frac{4}{3}$.

Thus, $\theta_1 = \sin^{-1}(n_2 \sin\theta_2) = \sin^{-1}[n_2 \sin(\theta_1 - 10.0°)]$.

(You can use a calculator to home in on an approximate solution to this equation, testing different values of θ_1 until you find that $\theta_1 = \boxed{36.5°}$. Alternatively, you can solve for θ_1 exactly, as shown below.)

We are given that $\sin\theta_1 = \frac{4}{3}\sin(\theta_1 - 10.0°)$.

This is the sine of a difference, so $\frac{3}{4}\sin\theta_1 = \sin\theta_1 \cos 10.0° - \cos\theta_1 \sin 10.0°$.

Rearranging, $\sin 10.0° \cos\theta_1 = \left(\cos 10.0° - \frac{3}{4}\right)\sin\theta_1$

$\frac{\sin 10.0°}{\cos 10.0° - 0.750} = \tan\theta_1$ and $\theta_1 = \tan^{-1}(0.740) = \boxed{36.5°}$.

P25.55 $\tan\theta_1 = \dfrac{4.00\text{ cm}}{h}$

and $\tan\theta_2 = \dfrac{2.00\text{ cm}}{h}$

$$\tan^2\theta_1 = (2.00\tan\theta_2)^2 = 4.00\tan^2\theta_2$$

$$\frac{\sin^2\theta_1}{1-\sin^2\theta_1} = 4.00\left(\frac{\sin^2\theta_2}{1-\sin^2\theta_2}\right). \qquad (1)$$

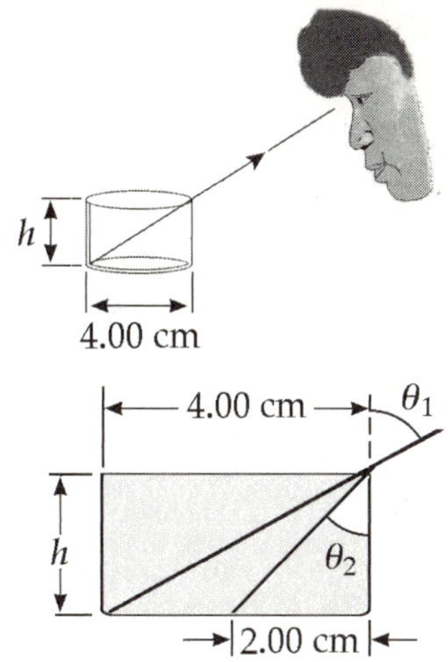

FIG. P25.55

Snell's law in this case is: $n_1\sin\theta_1 = n_2\sin\theta_2$

$$\sin\theta_1 = 1.333\sin\theta_2.$$

Squaring both sides, $\sin^2\theta_1 = 1.777\sin^2\theta_2. \qquad (2)$

Substituting (2) into (1), $\dfrac{1.777\sin^2\theta_2}{1-1.777\sin^2\theta_2} = 4.00\left(\dfrac{\sin^2\theta_2}{1-\sin^2\theta_2}\right).$

Defining $x = \sin^2\theta$, $\dfrac{0.444}{1-1.777x} = \dfrac{1}{1-x}.$

Solving for x, $0.444 - 0.444x = 1 - 1.777x$ and $x = 0.417.$

From x we can solve for θ_2: $\theta_2 = \sin^{-1}\sqrt{0.417} = 40.2°.$

Thus, the height is $h = \dfrac{2.00\text{ cm}}{\tan\theta_2} = \dfrac{2.00\text{ cm}}{\tan 40.2°} = \boxed{2.36\text{ cm}}.$

P25.56 Observe in the sketch that the angle of incidence at point P is γ, and using triangle OPQ:

$$\sin\gamma = \frac{L}{R}.$$

Also, $\cos\gamma = \sqrt{1-\sin^2\gamma} = \dfrac{\sqrt{R^2-L^2}}{R}.$

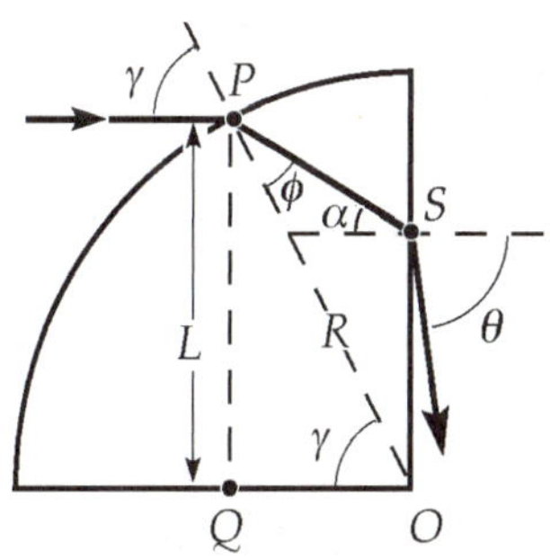

FIG. P25.56

Applying Snell's law at point P, $1.00\sin\gamma = n\sin\phi$.

Thus, $\sin\phi = \dfrac{\sin\gamma}{n} = \dfrac{L}{nR}$

and $\cos\phi = \sqrt{1-\sin^2\phi} = \dfrac{\sqrt{n^2R^2-L^2}}{nR}.$

From triangle OPS, $\phi + (\alpha + 90.0°) + (90.0° - \gamma) = 180°$ or the angle of incidence at point S is $\alpha = \gamma - \phi$. Then, applying Snell's law at point S

continued on next page

gives $\quad 1.00\sin\theta = n\sin\alpha = n\sin(\gamma-\phi)$

or $\quad \sin\theta = n[\sin\gamma\cos\phi - \cos\gamma\sin\phi] = n\left[\left(\frac{L}{R}\right)\frac{\sqrt{n^2R^2-L^2}}{nR} - \frac{\sqrt{R^2-L^2}}{R}\left(\frac{L}{nR}\right)\right]$

$$\sin\theta = \frac{L}{R^2}\left(\sqrt{n^2R^2-L^2} - \sqrt{R^2-L^2}\right)$$

and $\quad \theta = \boxed{\sin^{-1}\left[\frac{L}{R^2}\left(\sqrt{n^2R^2-L^2} - \sqrt{R^2-L^2}\right)\right]}$.

P25.57 As shown in the sketch, the angle of incidence at point A is:

$$\theta = \sin^{-1}\left(\frac{d/2}{R}\right) = \sin^{-1}\left(\frac{1.00\text{ m}}{2.00\text{ m}}\right) = 30.0°.$$

If the emerging ray is to be parallel to the incident ray, the path must be symmetric about the centerline CB of the cylinder. In the isosceles triangle ABC,

$$\gamma = \alpha \quad \text{and} \quad \beta = 180° - \theta.$$

Therefore, $\quad \alpha + \beta + \gamma = 180°$

becomes $\quad 2\alpha + 180° - \theta = 180°$

or $\quad \alpha = \frac{\theta}{2} = 15.0°.$

Then, applying Snell's law at point A,

$$n\sin\alpha = 1.00\sin\theta$$

or $\quad n = \frac{\sin\theta}{\sin\alpha} = \frac{\sin 30.0°}{\sin 15.0°} = \boxed{1.93}$.

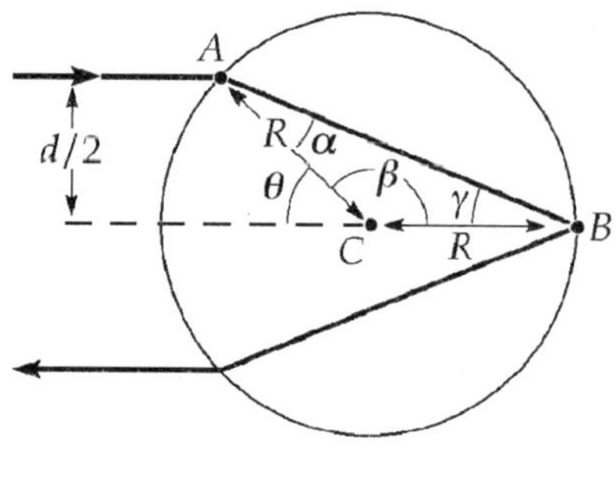

FIG. P25.57

P25.58

$\sin\theta_1$	$\sin\theta_2$	$\frac{\sin\theta_1}{\sin\theta_2}$
0.174	0.131	1.330 4
0.342	0.261	1.312 9
0.500	0.379	1.317 7
0.643	0.480	1.338 5
0.766	0.576	1.328 9
0.866	0.647	1.339 0
0.940	0.711	1.322 0
0.985	0.740	1.331 5

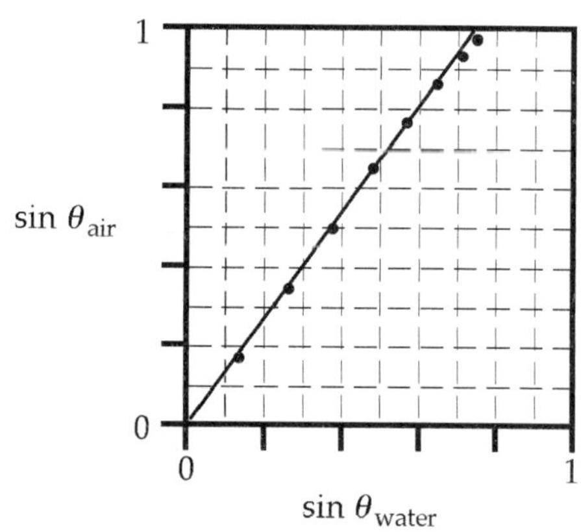

FIG. P25.58

The straightness of the graph line demonstrates Snell's proportionality.

The slope of the line is $\quad \bar{n} = 1.327\,6 \pm 0.01$

and $\quad n = \boxed{1.328 \pm 0.8\%}$.

P25.59 (a) Given that $\theta_1 = 45.0°$ and $\theta_2 = 76.0°$.

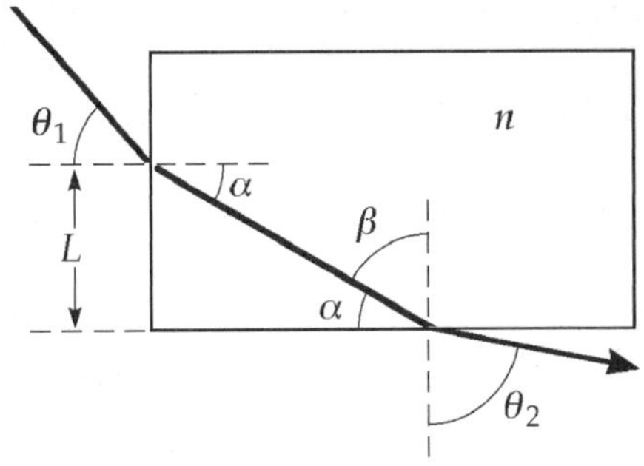

FIG. P25.59

Snell's law at the first surface gives

$$n \sin\alpha = 1.00 \sin 45.0°. \qquad (1)$$

Observe that the angle of incidence at the second surface is

$$\beta = 90.0° - \alpha.$$

Thus, Snell's law at the second surface yields

$$n \sin\beta = n \sin(90.0° - \alpha) = 1.00 \sin 76.0°$$

or $$n \cos\alpha = \sin 76.0°. \qquad (2)$$

Dividing Equation (1) by Equation (2), $$\tan\alpha = \frac{\sin 45.0°}{\sin 76.0°} = 0.729$$

or $$\alpha = 36.1°.$$

Then, from Equation (1), $$n = \frac{\sin 45.0°}{\sin\alpha} = \frac{\sin 45.0°}{\sin 36.1°} = \boxed{1.20}.$$

(b) From the sketch, observe that the distance the light travels in the plastic is $d = \frac{L}{\sin\alpha}$. Also, the speed of light in the plastic is $v = \frac{c}{n}$, so the time required to travel through the plastic is

$$\Delta t = \frac{d}{v} = \frac{nL}{c\sin\alpha} = \frac{1.20(0.500 \text{ m})}{(3.00\times10^{8} \text{ m/s})\sin 36.1°} = 3.40\times10^{-9} \text{ s} = \boxed{3.40 \text{ ns}}.$$

***P25.60** Consider an insulated box with the imagined one-way mirror forming one face, installed so that 90% of the electromagnetic radiation incident from the outside is transmitted to the inside and only a lower percentage of the electromagnetic waves from the inside make it through to the outside. Suppose the interior and exterior of the box are originally at the same temperature. Objects within and without are radiating and absorbing electromagnetic waves. They would all maintain constant temperature if the box had an open window. With the glass letting more energy in than out, the interior of the box will rise in temperature. But this is impossible, according to Clausius's statement of the second law. This reduction to a contradiction proves that it is impossible for the one-way mirror to exist.

ANSWERS TO EVEN PROBLEMS

P25.2 see the solution

P25.4 25.5°; 442 nm

P25.6 (a) 474 THz; (b) 422 nm; (c) 200 Mm/s

P25.8 22.5°

P25.10 (a) 1.52; (b) 417 nm; (c) 474 THz; (d) 198 Mm/s

P25.12 158 Mm/s

P25.14 3.39 m

P25.16 6.30 cm

P25.18 106 ps

P25.20 25.7°

P25.22 see the solution

P25.24 30.0° and 19.5° at entry, 40.5° and 77.1° at exit

P25.26 $\sin^{-1}\left(\sqrt{n^2-1}\sin\Phi-\cos\Phi\right)$

P25.28 see the solution

P25.30 (a) 10.7°; (b) air;
(c) sound in air falling on the wall from most directions is 100% reflected

P25.32 Skylight incident from the above travels down the plastic. If the index of refraction of the plastic is greater than 1.41, the rays close in direction to the vertical are totally reflected from the side walls of the slab and from both facets at the bottom of the plastic, where it is not immersed in gasoline. This light returns up inside the plastic and makes it look bright. Where the plastic is immersed in gasoline, total internal reflection is frustrated and the downward-propagating light passes from the plastic out into the gasoline. Little light is reflected up, and the gauge looks dark.

P25.34 62.4°

P25.36 (a) $\frac{dn}{n-1}$; (b) yes, yes, yes; (c) 350 μm

P25.38 Scattered light leaving the photograph in all forward horizontal directions in air is gathered by refraction into a fan in the water of half-angle 48.6°. At larger angles I see things on the other side of the globe, reflected by total internal reflection at the back surface of the cylinder.

P25.40 (a) see the solution, the angles are $\theta_1' = 30.0°$ and $\theta_2 = 18.8°$;
(b) see the solution, the angles are $\theta_1' = 30.0°$ and $\theta_2 = 50.8°$;
(c) see the solution; (d) see the solution

P25.42 see the solution

P25.44 (a) 45.0°;
(b) yes, such as 67.5°, see the solution

P25.46 3.79 m

P25.48 (b) 68.5%

P25.50 (a) $2R\omega_m$; (b) $\frac{2\omega_m\left(x^2+d^2\right)}{d}$

P25.52 164 s

P25.54 36.5°

P25.56 $\sin^{-1}\left[\frac{L}{R^2}\left(\sqrt{n^2R^2-L^2}-\sqrt{R^2-L^2}\right)\right]$

P25.58 see the solution; $n = \text{slope} = 1.328 \pm 0.8\%$

P25.60 see the solution

Chapter 26

Image Formation by Mirrors and Lenses

CHAPTER OUTLINE

ANSWERS TO QUESTIONS

Q26.1 With a concave spherical mirror, for objects beyond the focal length the image will be real and inverted. For objects inside the focal length, the image will be virtual, upright, and magnified. Try a shaving or makeup mirror as an example.

Q26.2 With a convex spherical mirror, all images of real objects are upright, virtual, and smaller than the object. As seen in Question 26.1, you only get a change of orientation when you pass the focal point—but the focal point of a convex mirror is on the non-reflecting side!

Q26.3 The mirror equation and the magnification equation apply to plane mirrors. A curved mirror is made flat by increasing its radius of curvature without bound, so that its focal length goes to infinity. From $\frac{1}{p}+\frac{1}{q}=\frac{1}{f}=0$ we have $\frac{1}{p}=-\frac{1}{q}$; therefore, $p=-q$. The virtual image is as far behind the mirror as the object is in front. The magnification is $M=-\frac{q}{p}=\frac{p}{p}=1$. The image is right side up and actual size.

Q26.4 Stones at the bottom of a clear stream always appear closer to the surface because light is refracted away from the normal at the surface. Example 26.5 in the textbook shows that its apparent depth is three quarters of its actual depth.

Q26.5 For definiteness, we consider real objects ($p>0$).

(a) For $M=-\frac{q}{p}$ to be negative, q must be positive. This will happen in $\frac{1}{q}=\frac{1}{f}-\frac{1}{p}$ if $p>f$, if the object is farther than the focal point.

(b) For $M=-\frac{q}{p}$ to be positive, q must be negative.

From $\frac{1}{q}=\frac{1}{f}-\frac{1}{p}$ we need $p<f$.

continued on next page

(c) For a real image, q must be positive.

As in part (a), it is sufficient for p to be larger than f.

(d) For $q<0$ we need $p<f$.

(e) For $|M|>1$, we consider separately $M<-1$ and $M>1$.

If $M=-\frac{q}{p}<-1$, we need $\frac{q}{p}>1$ or $q>p$

or $\frac{1}{q}<\frac{1}{p}$.

From $\frac{1}{p}+\frac{1}{q}=\frac{1}{f}$, $\frac{1}{p}+\frac{1}{p}>\frac{1}{f}$ or $\frac{2}{p}>\frac{1}{f}$

or $\frac{p}{2}<f$ or $p<2f$.

Now if $-\frac{q}{p}>1$ or $-q>p$ or $q<-p$

we may require $q<0$, since then $\frac{1}{p}=\frac{1}{f}-\frac{1}{q}$ with $\frac{1}{f}>0$

gives $\frac{1}{p}>-\frac{1}{q}$ as required or $-p>q$.

For $q<0$ in $\frac{1}{q}=\frac{1}{f}-\frac{1}{p}$ we need $p<f$.

Thus the overall condition for an enlarged image is simply $p<2f$.

(f) For $|M|<1$, we have the reverse of part (e), requiring $p>2f$.

Q26.6 Using the same analysis as in Question 26.5 except $f<0$.

(a) Never.

(b) Always.

(c) Never, for light rays passing through the lens will always diverge.

(d) Always.

(e) Never.

(f) Always.

Q26.7 We assume the lens has a refractive index higher than its surroundings. For the biconvex lens in Figure 26.20(a), $R_1 > 0$ and $R_2 < 0$. Then all terms in $(n-1)\left(\frac{1}{R_1}-\frac{1}{R_2}\right)$ are positive and $f > 0$. For the other two lenses in part (a) of the figure, R_1 and R_2 are both positive but R_1 is less than R_2. Then $\frac{1}{R_1} > \frac{1}{R_2}$ and the focal length is again positive.

For the biconcave lens and the plano-concave lens in Figure 26.20(b), $R_1 < 0$ and $R_2 > 0$. Then both terms are negative in $\frac{1}{R_1}-\frac{1}{R_2}$ and the focal length is negative. For the middle lens in part (b) of the figure, R_1 and R_2 are both positive but R_1 is greater than R_2. Then $\frac{1}{R_1} < \frac{1}{R_2}$ and the focal length is again negative.

Q26.8 Both words are inverted. However OXIDE has up-down symmetry whereas LEAD does not.

Q26.9 In the diagram, only two of the three principal rays have been used to locate images to reduce the amount of visual clutter. The upright shaded arrows are the objects, and the correspondingly numbered inverted arrows are the images. As you can see, object 2 is closer to the focal point than object 1, and image 2 is farther to the left than image 1.

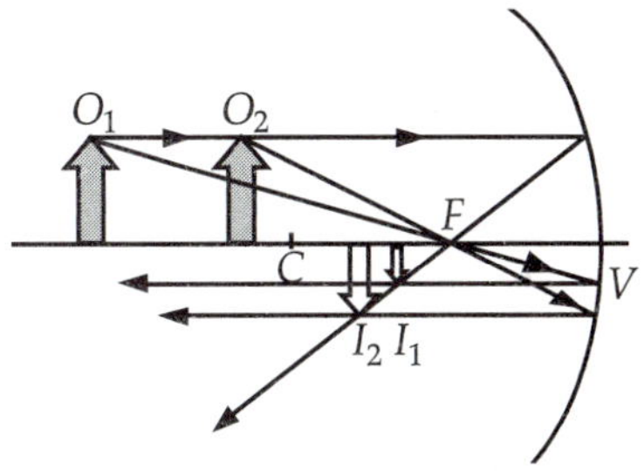

FIG. Q26.9

Q26.10 As in the diagram, let the center of curvature *C* of the fishbowl and the bottom of the fish define the optical axis, intersecting the fishbowl at vertex *V*. A ray from the top of the fish that reaches the bowl surface along a radial line through *C* has angle of incidence zero and angle of refraction zero. This ray exits from the bowl unchanged in direction. A ray from the top of the fish to *V* is refracted to bend away from the normal. Its extension back inside the fishbowl determines the location of the image and the characteristics of the image. The image is upright, virtual, and enlarged.

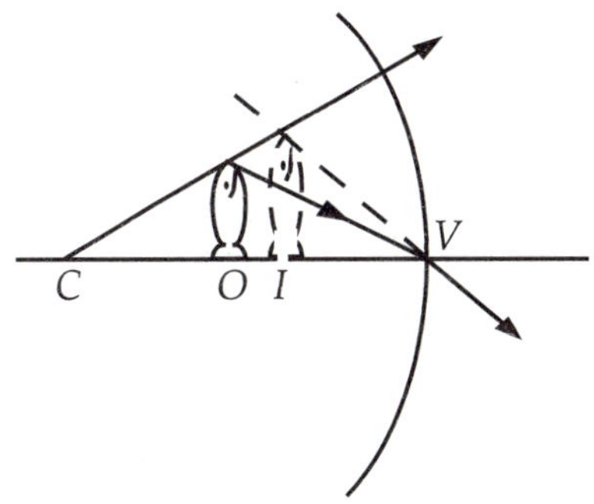

FIG. Q26.10

Q26.11 Because when you look at the ƎƆИA⅃UᙠMA in your rear view mirror, the apparent left-right inversion clearly displays the name of the AMBULANCE behind you. Do not jam on your brakes when a MIAMI city bus is right behind you.

Q26.12 With the meniscus design, when you direct your gaze near the outer circumference of the lens you receive a ray that has passed through glass with more nearly parallel surfaces of entry and exit. Thus, the lens minimally distorts the direction to the object you are looking at. If you wear glasses, turn them around and look through them the wrong way to maximize this distortion.

Q26.13 An infinite number. In general, an infinite number of rays leave each point of any object and travel in all directions. Note that the three principal rays that we use for imaging are just a subset of the infinite number of rays. All three principal rays can be drawn in a ray diagram, provided that we extend the plane of the lens as shown in Figure Q26.13.

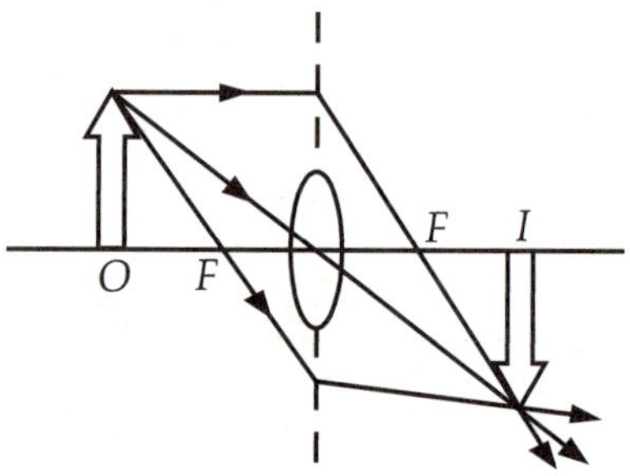

FIG. Q26.13

Q26.14 Absolutely. Only absorbed light, not transmitted light, contributes internal energy to a transparent object. A clear lens can stay ice-cold and solid as megajoules of light energy pass through it.

Q26.15 The focal point is defined as the location of the image formed by rays originally parallel to the axis. An object at a large but finite distance will radiate rays nearly but not exactly parallel. Infinite object distance describes the definite limiting case in which these rays become parallel. To measure the focal length of a converging lens, set it up to form an image of the farthest object you can see outside a window. The image distance will be equal to the focal length within one percent or better if the object distance is a hundred times larger or more.

Q26.16 Use a converging lens as the projection lens in a slide projector. Place the brightly illuminated slide slightly farther than its focal length away from it, so that the lens will produce a real, inverted, enlarged image on the screen.

Q26.17 Make the mirror an efficient reflector (shiny). Make it reflect to the image even rays far from the axis, by giving it a parabolic shape. Most important, make it large in diameter to intercept a lot of solar power. And you get higher temperature if the image is smaller, as you get with shorter focal length; and if the furnace enclosure is an efficient absorber (black).

Q26.18 The artist's statements are accurate, perceptive, and eloquent. The image you see is "almost one's whole surroundings," including things behind you and things farther in front of you than the globe is, but nothing eclipsed by the opaque globe or by your head. For example, we cannot see Escher's index and middle fingers or their reflections in the globe.

The point halfway between your eyes is indeed the focus in a figurative sense, but it is not an optical focus. The principal axis will always lie in a line that runs through the center of the sphere and the bridge of your nose. Outside the globe, you are at the center of your observable universe. If you wink at the ball, the center of the looking-glass world hops over to the location of the image of your open eye.

Q26.19 The three mirrors, two of which are shown as M and N in the figure to the right, reflect any incident ray back parallel to its original direction. When you look into the corner you see image I_3 of yourself.

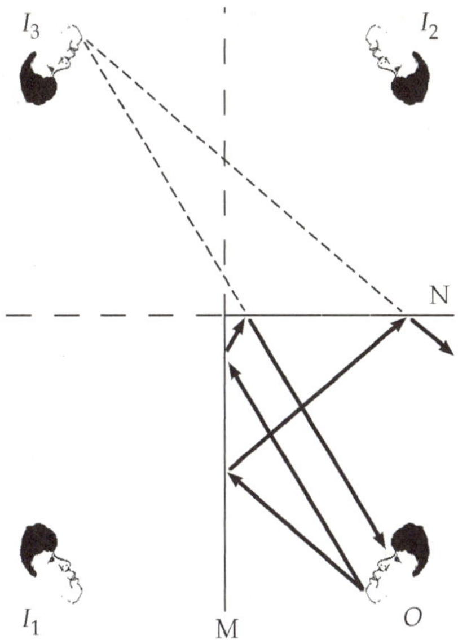

FIG. Q26.19

SOLUTIONS TO PROBLEMS

Section 26.1 Images Formed by Flat Mirrors

P26.1 I stand 40 cm from my bathroom mirror. I scatter light, which travels to the mirror and back to me in time

$$\frac{0.8 \text{ m}}{3\times10^8 \text{ m/s}} \boxed{\sim 10^{-9} \text{ s}}$$

showing me a view of myself as I was at that look-back time. I'm no Dorian Gray!

P26.2 The virtual image is as far behind the mirror as the choir is in front of the mirror. Thus, the image is 5.30 m behind the mirror. The image of the choir is $0.800 \text{ m}+5.30 \text{ m}=6.10 \text{ m}$ from the organist. Using similar triangles:

$$\frac{h'}{0.600 \text{ m}}=\frac{6.10 \text{ m}}{0.800 \text{ m}}$$

or $$h'=(0.600 \text{ m})\left(\frac{6.10 \text{ m}}{0.800 \text{ m}}\right)=\boxed{4.58 \text{ m}}.$$

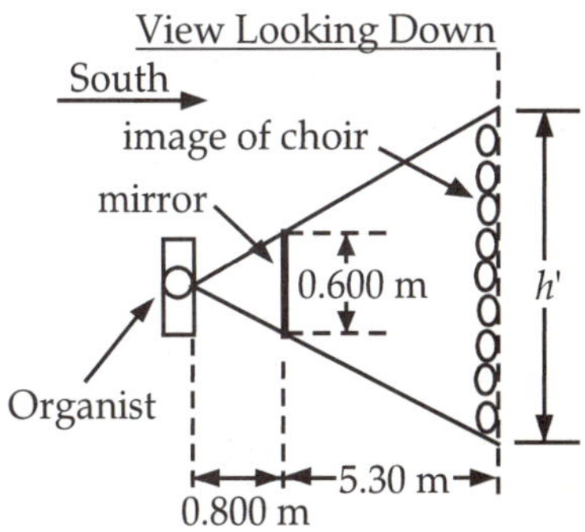

FIG. P26.2

P26.3 The flatness of the mirror is described

by $R = \infty,\ f = \infty$

and $\frac{1}{f} = 0$.

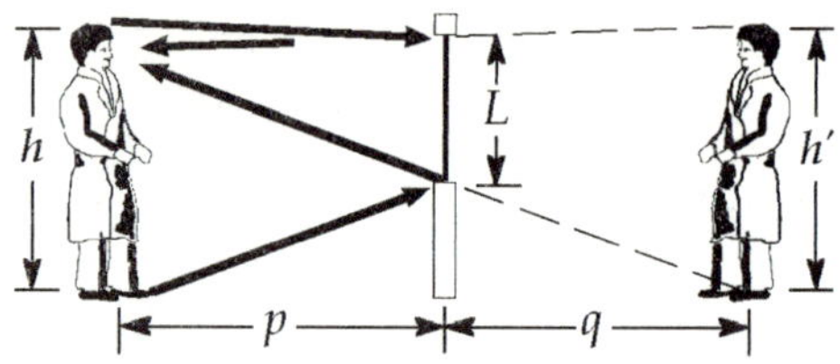

FIG. P26.3

By our general mirror equation,

$$\frac{1}{p} + \frac{1}{q} = \frac{1}{f} = 0$$

or $q = -p$.

Thus, the image is as far behind the mirror as the person is in front. The magnification is then

$$M = \frac{-q}{p} = 1 = \frac{h'}{h}$$

so $h' = h = 70.0$ inches .

The required height of the mirror is defined by the triangle from the person's eyes to the top and bottom of his image, as shown. From the geometry of the triangle, we see that the mirror height must be:

$$h'\left(\frac{p}{p-q}\right) = h'\left(\frac{p}{2p}\right) = \frac{h'}{2}.$$

Thus, the mirror must be $\boxed{\text{at least 35.0 inches high}}$.

P26.4 A graphical construction produces 5 images, with images I_1 and I_2 directly into the mirrors from the object O,

and (O, I_3, I_4)

and (I_2, I_1, I_5)

forming the vertices of equilateral triangles.

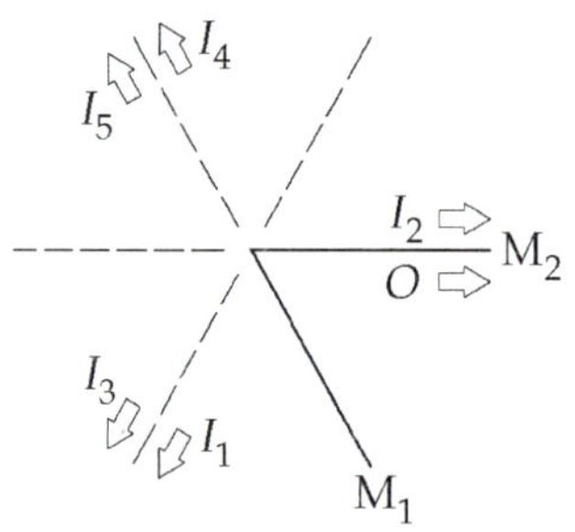

FIG. P26.4

P26.5 (1) The first image in the left mirror is 5.00 ft behind the mirror, or $\boxed{10.0 \text{ ft}}$ from the position of the person.

(2) The first image in the right mirror is located 10.0 ft behind the right mirror, but this location is 25.0 ft from the left mirror. Thus, the second image in the left mirror is 25.0 ft behind the mirror, or $\boxed{30.0 \text{ ft}}$ from the person.

(3) The first image in the left mirror forms an image in the right mirror. This first image is 20.0 ft from the right mirror, and, thus, an image 20.0 ft behind the right mirror is formed. This image in the right mirror also forms an image in the left mirror. The distance from this image in the right mirror to the left mirror is 35.0 ft. The third image in the left mirror is, thus, 35.0 ft behind the mirror, or $\boxed{40.0 \text{ ft}}$ from the person.

P26.6 (a) The flat mirrors have

$$R \to \infty$$

and $f \to \infty$.

The upper mirror M_1 produces a virtual, actual sized image I_1 according to

$$\frac{1}{p_1} + \frac{1}{q_1} = \frac{1}{f} = \frac{1}{\infty} = 0$$

$$q_1 = -p_1$$

with $M_1 = -\dfrac{q_1}{p_1} = +1$.

As shown, this image is above the upper mirror. It is the object for mirror M_2, at object distance

$$p_2 = p_1 + h.$$

The lower mirror produces a virtual, actual-size, right-side-up image according to

$$\frac{1}{p_2} + \frac{1}{q_2} = 0$$

$$q_2 = -p_2 = -(p_1 + h)$$

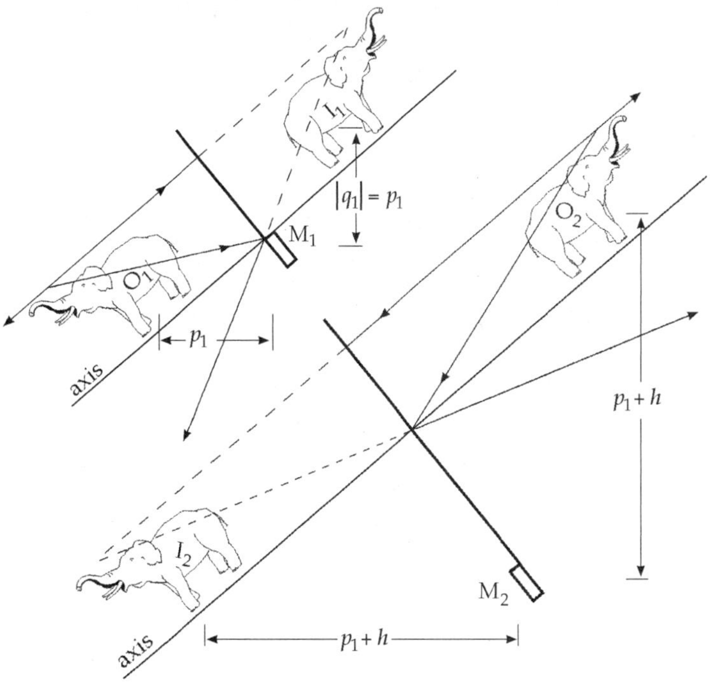

FIG. P26.6

with $M_2 = -\dfrac{q_2}{p_2} = +1$ and $M_{\text{overall}} = M_1 M_2 = 1$.

Thus the final image is at distance $\boxed{p_1 + h}$ behind the lower mirror.

(b) It is $\boxed{\text{virtual}}$.

(c) $\boxed{\text{Upright}}$

(d) With magnification $\boxed{+1}$.

(e) It $\boxed{\text{does not appear to be reversed}}$ left and right. In a top view of the periscope, parallel rays from the right and left sides of the object stay parallel and on the right and left.

Section 26.2 Images Formed by Spherical Mirrors

P26.7 For a concave mirror, both R and f are positive.

We also know that $$f=\frac{R}{2}=10.0\text{ cm}.$$

(a) $$\frac{1}{q}=\frac{1}{f}-\frac{1}{p}=\frac{1}{10.0\text{ cm}}-\frac{1}{40.0\text{ cm}}=\frac{3}{40.0\text{ cm}}$$

and $$\boxed{q=13.3\text{ cm}}$$

$$M=\frac{q}{p}=-\frac{13.3\text{ cm}}{40.0\text{ cm}}=\boxed{-0.333}.$$

The image is 13.3 cm in front of the mirror, $\boxed{\text{real, and inverted}}$.

(b) $$\frac{1}{q}=\frac{1}{f}-\frac{1}{p}=\frac{1}{10.0\text{ cm}}-\frac{1}{20.0\text{ cm}}=\frac{1}{20.0\text{ cm}}$$

and $$\boxed{q=20.0\text{ cm}}$$

$$M=\frac{q}{p}=-\frac{20.0\text{ cm}}{20.0\text{ cm}}=\boxed{-1.00}.$$

The image is 20.0 cm in front of the mirror, $\boxed{\text{real, and inverted}}$.

(c) $$\frac{1}{q}=\frac{1}{f}-\frac{1}{p}=\frac{1}{10.0\text{ cm}}-\frac{1}{10.0\text{ cm}}=0$$

Thus, q = infinity.

$\boxed{\text{No image is formed}}$. The rays are reflected parallel to each other.

***P26.8** $$\frac{1}{q}=\frac{1}{f}-\frac{1}{p}=-\frac{1}{0.275\text{ m}}-\frac{1}{10.0\text{ m}}$$ gives $\boxed{q=-0.267\text{ m}}$.

Thus, the image is $\boxed{\text{virtual}}$.

$$M=\frac{-q}{p}=-\frac{-0.267}{10.0\text{ m}}=\boxed{0.026\,7}$$

Thus, the image is $\boxed{\text{upright}}$ $(+M)$ and $\boxed{\text{diminished}}$ $(|M|<1)$.

P26.9 (a) $\frac{1}{p}+\frac{1}{q}=\frac{2}{R}$ gives $\frac{1}{30.0\text{ cm}}+\frac{1}{q}=\frac{2}{(-40.0\text{ cm})}$

$\frac{1}{q}=-\frac{2}{40.0\text{ cm}}-\frac{1}{30.0\text{ cm}}=-0.083\,3\text{ cm}^{-1}$ so $q=\boxed{-12.0\text{ cm}}$

$M=\frac{-q}{p}=-\frac{(-12.0\text{ cm})}{30.0\text{ cm}}=\boxed{0.400}$.

(b) $\frac{1}{p}+\frac{1}{q}=\frac{2}{R}$ gives $\frac{1}{60.0\text{ cm}}+\frac{1}{q}=\frac{2}{(-40.0\text{ cm})}$

$\frac{1}{q}=-\frac{2}{40.0\text{ cm}}-\frac{1}{60.0\text{ cm}}=-0.066\,6\text{ cm}^{-1}$ so $q=\boxed{-15.0\text{ cm}}$

$M=\frac{-q}{p}=-\frac{(-15.0\text{ cm})}{60.0\text{ cm}}=\boxed{0.250}$.

(c) Since $M>0$, the images are $\boxed{\text{upright}}$.

P26.10 With radius 2.50 m, the cylindrical wall is a highly efficient mirror for sound, with focal length

$$f=\frac{R}{2}=1.25\text{ m}.$$

In a vertical plane the sound disperses as usual, but that radiated in a horizontal plane is concentrated in a sound image at distance q from the back of the niche, where

$$\frac{1}{p}+\frac{1}{q}=\frac{1}{f} \quad \text{so} \quad \frac{1}{2.00\text{ m}}+\frac{1}{q}=\frac{1}{1.25\text{ m}}$$

$$q=\boxed{3.33\text{ m}}.$$

P26.11 (a) $\frac{1}{p}+\frac{1}{q}=\frac{2}{R}$ becomes $\frac{1}{q}=\frac{2}{60.0\text{ cm}}-\frac{1}{90.0\text{ cm}}$

$q=\boxed{45.0\text{ cm}}$ and $M=\frac{-q}{p}=-\frac{45.0\text{ cm}}{90.0\text{ cm}}=\boxed{-0.500}$.

(b) $\frac{1}{p}+\frac{1}{q}=\frac{2}{R}$ becomes $\frac{1}{q}=\frac{2}{60.0\text{ cm}}-\frac{1}{20.0\text{ cm}}$

$q=\boxed{-60.0\text{ cm}}$ and $M=\frac{-q}{p}=-\frac{(-60.0\text{ cm})}{(20.0\text{ cm})}=\boxed{3.00}$.

(c) The image (a) is real, inverted and diminished. That of (b) is virtual, upright, and enlarged. The ray diagrams are shown.

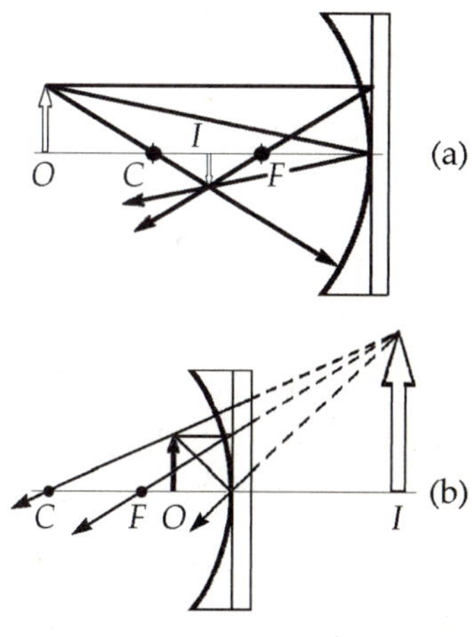

FIG. P26.11

***P26.12** (a) Since the object is in front of the mirror, $p>0$. With the image behind the mirror, $q<0$. The mirror equation gives the radius of curvature as $\frac{2}{R}=\frac{1}{p}+\frac{1}{q}=\frac{1}{1.00\text{ cm}}-\frac{1}{10.0\text{ cm}}=\frac{10-1}{10.0\text{ cm}}$, or

$$R=2\left(\frac{10.0\text{ cm}}{9}\right)=\boxed{+2.22\text{ cm}}.$$

(b) The magnification is $M=-\frac{q}{p}=-\frac{(-10.0\text{ cm})}{1.00\text{ cm}}=\boxed{+10.0}$.

***P26.13** The ball is a convex mirror with $R=-4.25$ cm and $f=\frac{R}{2}=-2.125$ cm. We have

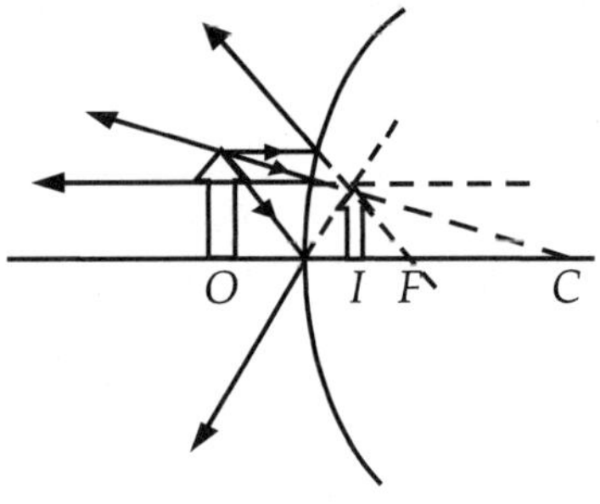

FIG. P26.13

$$M=\frac{3}{4}=-\frac{q}{p}$$

$$q=-\frac{3}{4}p$$

$$\frac{1}{p}+\frac{1}{q}=\frac{1}{f}$$

$$\frac{1}{p}+\frac{1}{-(3/4)p}=\frac{1}{-2.125\text{ cm}}$$

$$\frac{3}{3p}-\frac{4}{3p}=\frac{1}{-2.125\text{ cm}}$$

$$3p=2.125\text{ cm}$$

$\boxed{p=0.708\text{ cm}}$ in front of the sphere.

The image is upright, virtual, and diminished.

P26.14 (a) $M=-4=-\frac{q}{p}$ $\quad q=4p$

$q-p=0.60\text{ m}=4p-p$ $\quad p=0.2$ m $\quad q=0.8$ m

$\frac{1}{f}=\frac{1}{p}+\frac{1}{q}=\frac{1}{0.2\text{ m}}+\frac{1}{0.8\text{ m}}$ $\quad f=\boxed{160\text{ mm}}$

(b) $M=+\frac{1}{2}=-\frac{q}{p}$ $\quad p=-2q$

$|q|+p=0.20\text{ m}=-q+p=-q-2q$

$q=-66.7$ mm $\quad p=133$ mm

$\frac{1}{p}+\frac{1}{q}=\frac{2}{R}=\frac{1}{0.133\text{ m}}+\frac{1}{-0.0667\text{ m}}$ $\quad R=\boxed{-267\text{ mm}}$

P26.15 $M = -\frac{q}{p}$

$q = -Mp = -0.013(30 \text{ cm}) = -0.39 \text{ cm}$

$\frac{1}{p} + \frac{1}{q} = \frac{1}{f} = \frac{2}{R}$

$\frac{1}{30 \text{ cm}} + \frac{1}{-0.39 \text{ cm}} = \frac{2}{R}$

$R = \frac{2}{-2.53 \text{ m}^{-1}} = -0.790 \text{ cm}$

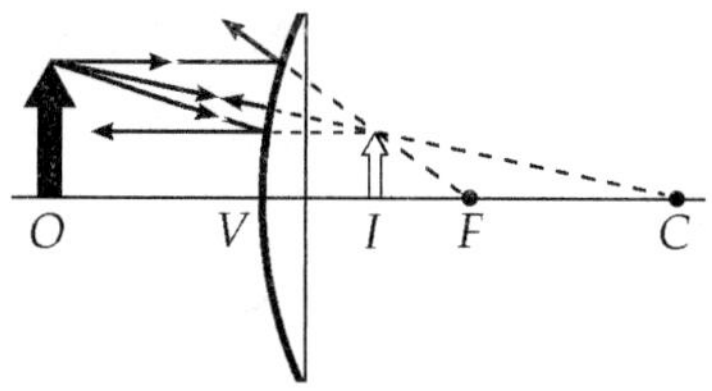

FIG. P26.15

The cornea is convex, with radius of curvature $\boxed{0.790 \text{ cm}}$.

***P26.16** With

$M = \frac{h'}{h} = \frac{+4.00 \text{ cm}}{10.0 \text{ cm}} = +0.400 = -\frac{q}{p}$

$q = -0.400p$

the image must be virtual.

(a) It is a $\boxed{\text{convex}}$ mirror that produces a diminished upright virtual image.

(b) We must have

$p + |q| = 42.0 \text{ cm} = p - q$

$p = 42.0 \text{ cm} + q$

$p = 42.0 \text{ cm} - 0.400p$

$p = \frac{42.0 \text{ cm}}{1.40} = 30.0 \text{ cm}$

The mirror is $\boxed{\text{at the 30.0 cm mark}}$.

(c) $\frac{1}{p} + \frac{1}{q} = \frac{1}{f} = \frac{1}{30 \text{ cm}} + \frac{1}{-0.4(30 \text{ cm})} = \frac{1}{f} = -0.050\,0/\text{cm}$ $\boxed{f = -20.0 \text{ cm}}$

The ray diagram looks like Figure 26.12(c) in the text.

P26.17 (a) $q = (p + 5.00 \text{ m})$ and, since the image must be real,

$M = -\frac{q}{p} = -5$ or $q = 5p$.

Therefore, $p + 5.00 \text{ m} = 5p$

or $p = 1.25 \text{ m}$ and $q = 6.25 \text{ m}$.

From $\frac{1}{p} + \frac{1}{q} = \frac{2}{R}$, $R = \frac{2pq}{p+q} = \frac{2(1.25)(6.25)}{1.25 + 6.25}$

$= \boxed{2.08 \text{ m(concave)}}$

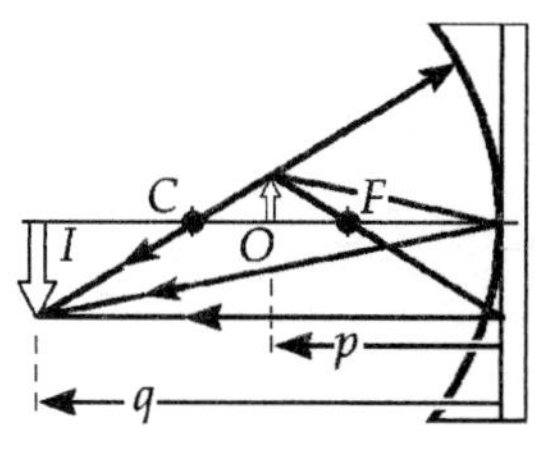

FIG. P26.17

(b) From part (a), $p = 1.25 \text{ m}$; the mirror should be $\boxed{1.25 \text{ m}}$ in front of the object.

P26.18 Assume that the object distance is the same in both cases (i.e., her face is the same distance from the hubcap regardless of which way it is turned). Also realize that the near image ($q=-10.0$ cm) occurs when using the convex side of the hubcap. Applying the mirror equation to both cases gives:

(concave side: $R=|R|$, $q=-30.0$ cm)

$$\frac{1}{p}-\frac{1}{30.0}=\frac{2}{|R|}$$

or

$$\frac{2}{|R|}=\frac{30.0\text{ cm}-p}{(30.0\text{ cm})p} \tag{1}$$

(convex side: $R=-|R|$, $q=-10.0$ cm)

$$\frac{1}{p}-\frac{1}{10.0}=-\frac{2}{|R|}$$

or

$$\frac{2}{|R|}=\frac{p-10.0\text{ cm}}{(10.0\text{ cm})p}. \tag{2}$$

(a) Equating Equations (1) and (2) gives:

$$\frac{30.0\text{ cm}-p}{3.00}=p-10.0\text{ cm}$$

or $p=15.0$ cm.

Thus, her face is $\boxed{15.0\text{ cm}}$ from the hubcap.

(b) Using the above result ($p=15.0$ cm) in Equation (1) gives:

$$\frac{2}{|R|}=\frac{30.0\text{ cm}-15.0\text{ cm}}{(30.0\text{ cm})(15.0\text{ cm})}$$

or

$$\frac{2}{|R|}=\frac{1}{30.0\text{ cm}}$$

and $|R|=60.0$ cm.

The radius of the hubcap is $\boxed{60.0\text{ cm}}$.

P26.19 (a) The flat mirror produces an image according to

$$\frac{1}{p}+\frac{1}{q}=\frac{1}{f}=\frac{2}{R} \qquad \frac{1}{24\text{ cm}}+\frac{1}{q}=\frac{1}{\infty}=0 \qquad q=-24.0\text{ m}.$$

The image is 24.0 m behind the mirror, distant from your eyes by

$$1.55\text{ m}+24.0\text{ m}=\boxed{25.6\text{ m}}.$$

(b) The image is the same size as the object, so $\theta=\frac{h}{d}=\frac{1.50\text{ m}}{25.6\text{ m}}=\boxed{0.058\,7\text{ rad}}$.

continued on next page

(c) $\frac{1}{p}+\frac{1}{q}=\frac{2}{R}$ $\frac{1}{24\text{ m}}+\frac{1}{q}=\frac{2}{(-2\text{ m})}$ $q=\frac{1}{-(1/1\text{ m})-(1/24\text{ m})}=-0.960\text{ m}$

This image is distant from your eyes by $1.55\text{ m}+0.960\text{ m}=\boxed{2.51\text{ m}}$.

(d) The image size is given by $M=\frac{h'}{h}=-\frac{q}{p}$ $h'=-h\frac{q}{p}=-1.50\text{ m}\left(\frac{-0.960\text{ m}}{24\text{ m}}\right)=0.060\,0\text{ m}$.

So its angular size at your eye is $\theta'=\frac{h'}{d}=\frac{0.06\text{ m}}{2.51\text{ m}}=\boxed{0.023\,9\text{ rad}}$.

(e) Your brain assumes that the car is 1.50 m high and calculate its distance as

$d'=\frac{h}{\theta'}=\frac{1.50\text{ m}}{0.023\,9}=\boxed{62.8\text{ m}}$.

***P26.20** (a) The image starts from a point whose height above the mirror vertex is given by

$\frac{1}{p}+\frac{1}{q}=\frac{1}{f}=\frac{2}{R}$ $\frac{1}{3.00\text{ m}}+\frac{1}{q}=\frac{1}{0.500\text{ m}}$. Therefore, $q=0.600\text{ m}$.

As the ball falls, p decreases and q increases. Ball and image pass when $q_1=p_1$. When this is true,

$\frac{1}{p_1}+\frac{1}{p_1}=\frac{1}{0.500\text{ m}}=\frac{2}{p_1}$ or $p_1=1.00\text{ m}$.

As the ball passes the focal point, the image switches from infinitely far above the mirror to infinitely far below the mirror. As the ball approaches the mirror from above, the virtual image approaches the mirror from below, reaching it together when $p_2=q_2=0$.

(b) The falling ball passes its real image when it has fallen

$3.00\text{ m}-1.00\text{ m}=2.00\text{ m}=\frac{1}{2}gt^2$, or when $t=\sqrt{\frac{2(2.00\text{ m})}{9.80\text{ m/s}^2}}=\boxed{0.639\text{ s}}$.

The ball reaches its virtual image when it has traversed

$3.00\text{ m}-0=3.00\text{ m}=\frac{1}{2}gt^2$, or at $t=\sqrt{\frac{2(3.00\text{ m})}{9.80\text{ m/s}^2}}=\boxed{0.782\text{ s}}$.

Section 26.3 Images Formed by Refraction

P26.21 $\frac{n_1}{p}+\frac{n_2}{q}=\frac{n_2-n_1}{R}=0$ and $R\to\infty$

$$q=-\frac{n_2}{n_1}p=-\frac{1}{1.309}(50.0\text{ cm})=-38.2\text{ cm}$$

Thus, the virtual image of the dust speck is $\boxed{38.2\text{ cm below the top surface}}$ of the ice.

P26.22 When $R \to \infty$, the equation describing image formation at a single refracting surface becomes $q = -p\left(\frac{n_2}{n_1}\right)$. We use this to locate the final images of the two surfaces of the glass plate. First, find the image the glass forms of the *bottom* of the plate.

$$q_{B1} = -\left(\frac{1.33}{1.66}\right)(8.00 \text{ cm}) = -6.41 \text{ cm}$$

This virtual image is 6.41 cm below the top surface of the glass of 18.41 cm below the water surface. Next, use this image as an object and locate the image the water forms of the bottom of the plate.

$$q_{B2} = -\left(\frac{1.00}{1.33}\right)(18.41 \text{ cm}) = -13.84 \text{ cm} \quad \text{or} \quad 13.84 \text{ cm below the water surface.}$$

Now find image the water forms of the *top* surface of the glass.

$$q_3 = -\left(\frac{1}{1.33}\right)(12.0 \text{ cm}) = -9.02 \text{ cm} \quad \text{or} \quad 9.02 \text{ cm below the water surface.}$$

Therefore, the apparent thickness of the glass is $\Delta t = 13.84 \text{ cm} - 9.02 \text{ cm} = \boxed{4.82 \text{ cm}}$.

P26.23 For refraction at a curved surface, $\frac{n_1}{p} + \frac{n_2}{q} = \frac{n_2 - n_1}{R}$.

Solve for q to find $q = \frac{n_2 R p}{p(n_2 - n_1) - n_1 R}$.

In this case, $n_1 = 1.50$, $n_2 = 1.00$, $R = -15.0$ cm

and $p = 10.0$ cm.

So $q = \frac{(1.00)(-15.0 \text{ cm})(10.0 \text{ cm})}{(10.0 \text{ cm})(1.00 - 1.50) - (1.50)(-15.0 \text{ cm})} = -8.57$ cm.

Therefore, the $\boxed{\text{apparent depth is 8.57 cm}}$.

P26.24 $\frac{n_1}{p} + \frac{n_2}{q} = \frac{n_2 - n_1}{R}$ so $\frac{1.00}{\infty} + \frac{1.40}{21.0 \text{ mm}} = \frac{1.40 - 1.00}{6.00 \text{ mm}}$

and $0.0667 = 0.0667$.

They agree. $\boxed{\text{The image is inverted, real and diminished.}}$

P26.25 $\frac{n_1}{p}+\frac{n_2}{q}=\frac{n_2-n_1}{R}$ becomes $\frac{1.00}{p}+\frac{1.50}{q}=\frac{1.50-1.00}{6.00\text{ cm}}=\frac{1}{12.0\text{ cm}}$

(a) $\frac{1.00}{20.0\text{ cm}}+\frac{1.50}{q}=\frac{1}{12.0\text{ cm}}$ or $q=\frac{1.50}{[(1.00/12.0\text{ cm})-(1.00/20.0\text{ cm})]}=\boxed{45.0\text{ cm}}$

(b) $\frac{1.00}{10.0\text{ cm}}+\frac{1.50}{q}=\frac{1}{12.0\text{ cm}}$ or $q=\frac{1.50}{[(1.00/12.0\text{ cm})-(1.00/10.0\text{ cm})]}=\boxed{-90.0\text{ cm}}$

(c) $\frac{1.00}{3.0\text{ cm}}+\frac{1.50}{q}=\frac{1}{12.0\text{ cm}}$ or $q=\frac{1.50}{[(1.00/12.0\text{ cm})-(1.00/3.0\text{ cm})]}=\boxed{-6.00\text{ cm}}$

***P26.26** For a plane surface, $\frac{n_1}{p}+\frac{n_2}{q}=\frac{n_2-n_1}{R}$ becomes $q=-\frac{n_2 p}{n_1}$.

Thus, the magnitudes of the rate of change in the image and object positions are related by

$$\left|\frac{dq}{dt}\right|=\frac{n_2}{n_1}\left|\frac{dp}{dt}\right|.$$

If the fish swims toward the wall with a speed of 2.00 cm/s, the speed of the image is given by

$$v_{\text{image}}=\left|\frac{dq}{dt}\right|=\frac{1.00}{1.33}(2.00\text{ cm/s})=\boxed{1.50\text{ cm/s}}.$$

Section 26.4 Thin Lenses

P26.27 (a) $\frac{1}{f}=(n-1)\left[\frac{1}{R_1}-\frac{1}{R_2}\right]=(0.440)\left[\frac{1}{12.0\text{ cm}}-\frac{1}{(-18.0\text{ cm})}\right]$

$f=\boxed{16.4\text{ cm}}$

(b) $\frac{1}{f}=(0.440)\left[\frac{1}{18.0\text{ cm}}-\frac{1}{(-12.0\text{ cm})}\right]$

$f=\boxed{16.4\text{ cm}}$

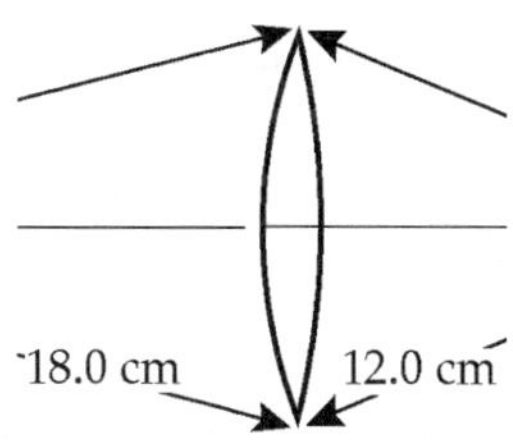

FIG. P26.27

P26.28 Let R_1 = outer radius and R_2 = inner radius

$$\frac{1}{f}=(n-1)\left[\frac{1}{R_1}-\frac{1}{R_2}\right]=(1.50-1)\left[\frac{1}{2.00\text{ m}}-\frac{1}{2.50\text{ cm}}\right]=0.0500\text{ cm}^{-1}$$

so $f=\boxed{20.0\text{ cm}}$.

P26.29 (a) $\frac{1}{q}=\frac{1}{f}-\frac{1}{p}=\frac{1}{25.0\text{ cm}}-\frac{1}{26.0\text{ cm}}$ $\boxed{q=650\text{ cm}}$

The image is $\boxed{\text{real, inverted, and enlarged}}$.

(b) $\frac{1}{q}=\frac{1}{f}-\frac{1}{p}=\frac{1}{25.0\text{ cm}}-\frac{1}{24.0\text{ cm}}$ $\boxed{q=-600\text{ cm}}$

The image is $\boxed{\text{virtual, upright, and enlarged}}$.

P26.30 For a converging lens, f is positive. We use $\frac{1}{p}+\frac{1}{q}=\frac{1}{f}$.

(a) $\frac{1}{q}=\frac{1}{f}-\frac{1}{p}=\frac{1}{20.0\text{ cm}}-\frac{1}{40.0\text{ cm}}=\frac{1}{40.0\text{ cm}}$ $\boxed{q=40.0\text{ cm}}$

$$M=-\frac{q}{p}=-\frac{40.0}{40.0}=\boxed{-1.00}$$

The image is $\boxed{\text{real, inverted}}$, and located 40.0 cm past the lens.

(b) $\frac{1}{q}=\frac{1}{f}-\frac{1}{p}=\frac{1}{20.0\text{ cm}}-\frac{1}{20.0\text{ cm}}=0$ $\boxed{q=\text{infinity}}$

$\boxed{\text{No image}}$ is formed. The rays emerging from the lens are parallel to each other.

(c) $\frac{1}{q}=\frac{1}{f}-\frac{1}{p}=\frac{1}{20.0\text{ cm}}-\frac{1}{10.0\text{ cm}}=-\frac{1}{20.0\text{ cm}}$ $\boxed{q=-20.0\text{ cm}}$

$$M=-\frac{q}{p}=-\frac{(-20.0)}{10.0}=\boxed{2.00}$$

The image is $\boxed{\text{upright, virtual}}$ and 20.0 cm in front of the lens.

P26.31 We are looking at an enlarged, upright, virtual image:

$M=\frac{h'}{h}=2=-\frac{q}{p}$ so $p=-\frac{q}{2}=-\frac{(-2.84\text{ cm})}{2}=+1.42\text{ cm}$

$\frac{1}{p}+\frac{1}{q}=\frac{1}{f}$ gives $\frac{1}{1.42\text{ cm}}+\frac{1}{(-2.84\text{ cm})}=\frac{1}{f}$

$\boxed{f=2.84\text{ cm}}$.

FIG. P26.31

P26.32 (a) $\frac{1}{p}+\frac{1}{q}=\frac{1}{f}$: $\qquad \frac{1}{32.0\text{ cm}}+\frac{1}{8.00\text{ cm}}=\frac{1}{f}$

so $\boxed{f=6.40\text{ cm}}$

(b) $M=-\frac{q}{p}=-\frac{8.00\text{ cm}}{32.0\text{ cm}}=\boxed{-0.250}$

(c) Since $f>0$, the lens is $\boxed{\text{converging}}$.

P26.33 $\frac{1}{p}+\frac{1}{q}=\frac{1}{f}$: $\qquad p^{-1}+q^{-1}=\text{constant}$

We may differentiate through with respect to p: $\qquad -1p^{-2}-1q^{-2}\frac{dq}{dp}=0$

$$\frac{dq}{dp}=-\frac{q^2}{p^2}=-M^2.$$

P26.34 The image is inverted: $\qquad M=\frac{h'}{h}=\frac{-1.80\text{ m}}{0.0240\text{ m}}=-75.0=\frac{-q}{p} \qquad q=75.0p$.

(b) $q+p=3.00\text{ m}=75.0p+p \qquad p=\boxed{39.5\text{ mm}}$

(a) $q=2.96\text{ m} \qquad \frac{1}{f}=\frac{1}{p}+\frac{1}{q}=\frac{1}{0.0395\text{ m}}+\frac{1}{2.96\text{ m}} \qquad f=\boxed{39.0\text{ mm}}$

P26.35 (a) $\frac{1}{p}+\frac{1}{q}=\frac{1}{f} \qquad \frac{1}{20.0\text{ cm}}+\frac{1}{q}=\frac{1}{(-32.0\text{ cm})}$

so $\qquad q=-\left(\frac{1}{20.0}+\frac{1}{32.0}\right)^{-1}=\boxed{-12.3\text{ cm}}$

The image is 12.3 cm to the left of the lens.

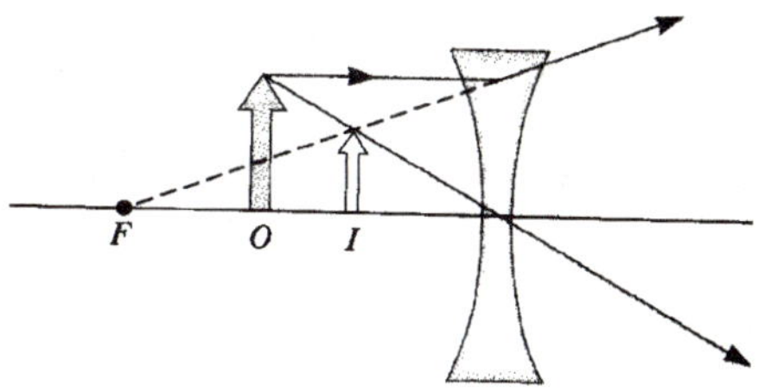

FIG. P26.35

(b) $M=-\frac{q}{p}=-\frac{(-12.3\text{ cm})}{20.0\text{ cm}}=\boxed{0.615}$

(c) See the ray diagram to the right.

***P26.36** Comparing $\frac{1}{p}+\frac{1}{q}=\frac{1}{f}$ with $\frac{1}{p}+\frac{1}{-3.5p}=\frac{1}{7.5\text{ cm}}$ we see $q=-3.5p$ and $f=7.50$ cm for a converging lens.

(a) To solve, we add the fractions:

$$\frac{-3.5+1}{-3.5p}=\frac{1}{7.5\text{ cm}}$$

$$\frac{3.5p}{2.5}=7.5\text{ cm}$$

$$p=\boxed{5.36\text{ cm}}$$

(b) $q=-3.5(5.36\text{ cm})=\boxed{-18.8\text{ cm}}$

$$M=-\frac{q}{p}=-\frac{-18.8\text{ cm}}{5.36\text{ cm}}=+3.50$$

(c)

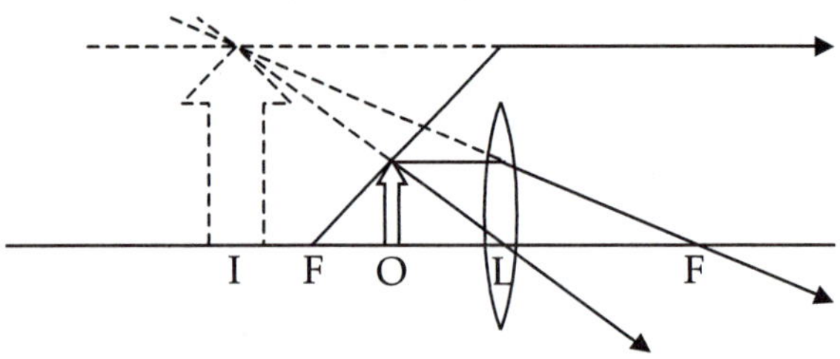

P26.36(c)

The image is $\boxed{\text{enlarged, upright, and virtual.}}$

(d) The lens is being used as a magnifying glass. Statement: A magnifying glass with focal length 7.50 cm is used to form an image of a stamp, enlarged 3.50 times. Find the object distance. Locate and describe the image.

***P26.37** In

$$\frac{1}{p}+\frac{1}{q}=\frac{1}{f}$$

$$p^{-1}+q^{-1}=\text{constant},$$

we differentiate with respect to time

$$-1\left(p^{-2}\right)\frac{dp}{dt}-1\left(q^{-2}\right)\frac{dq}{dt}=0$$

$$\frac{dq}{dt}=\frac{-q^2}{p^2}\frac{dp}{dt}.$$

continued on next page

We must find the momentary image location q:

$$\frac{1}{20\text{ m}}+\frac{1}{q}=\frac{1}{0.3\text{ m}}$$

$q=0.305\text{ m}$.

Now $\frac{dq}{dt}=-\frac{(0.305\text{ m})^2}{(20\text{ m})^2}5\text{ m/s}=-0.001\,16\text{ m/s}=\boxed{1.16\text{ mm/s toward the lens}}$.

P26.38 (a) $\frac{1}{f}=(n-1)\left[\frac{1}{R_1}-\frac{1}{R_2}\right]=(1.50-1)\left[\frac{1}{15.0\text{ cm}}-\frac{1}{(-12.0\text{ cm})}\right]$ or $\boxed{f=13.3\text{ cm}}$

(b)

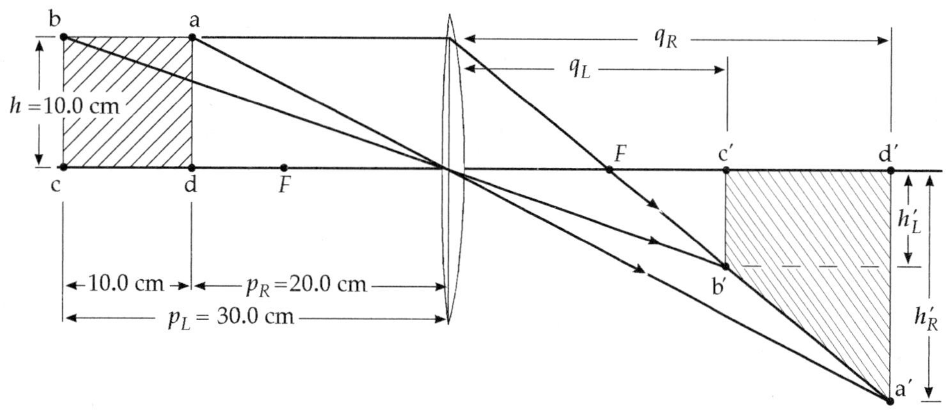

The square is imaged as a trapezoid.

FIG. P26.38(b)

(c) To find the area, first find q_R and q_L, along with the heights h'_R and h'_L, using the thin lens equation.

$\frac{1}{p_R}+\frac{1}{q_R}=\frac{1}{f}$ becomes $\frac{1}{20.0\text{ cm}}+\frac{1}{q_R}=\frac{1}{13.3\text{ cm}}$ or $q_R=40.0\text{ cm}$

$$h'_R=hM_R=h\left(\frac{-q_R}{p_R}\right)=(10.0\text{ cm})(-2.00)=-20.0\text{ cm}$$

$\frac{1}{30.0\text{ cm}}+\frac{1}{q_L}=\frac{1}{13.3\text{ cm}}$ or $q_L=24.0\text{ cm}$

$$h'_L=hM_L=(10.0\text{ cm})(-0.800)=-8.00\text{ cm}$$

Thus, the area of the image is: Area $=|q_R-q_L||h'_L|+\frac{1}{2}|q_R-q_L||h'_R-h'_L|=\boxed{224\text{ cm}^2}$.

P26.39 (a) The image distance is: $q = d - p$.

Thus, $\frac{1}{p} + \frac{1}{q} = \frac{1}{f}$ becomes $\frac{1}{p} + \frac{1}{d-p} = \frac{1}{f}$.

This reduces to a quadratic equation: $p^2 + (-d)p + fd = 0$

which yields: $\boxed{p = \frac{d \pm \sqrt{d^2 - 4fd}}{2} = \frac{d}{2} \pm \sqrt{\frac{d^2}{4} - fd}}$.

Since $f < \frac{d}{4}$, both solutions are meaningful and the two solutions are not equal to each other. Thus, there are two distinct lens positions that form an image on the screen.

(b) The smaller solution for p gives a larger value for q, with a $\boxed{\text{real, enlarged, inverted image}}$.

The larger solution for p describes a $\boxed{\text{real, diminished, inverted image}}$.

P26.40 To properly focus the image of a distant object, the lens must be at a distance equal to the focal length from the film ($q_1 = 65.0$ mm). For the closer object:

$$\frac{1}{p_2} + \frac{1}{q_2} = \frac{1}{f}$$

becomes $$\frac{1}{2\,000 \text{ mm}} + \frac{1}{q_2} = \frac{1}{65.0 \text{ mm}}$$

and $$q_2 = (65.0 \text{ mm})\left(\frac{2\,000}{2\,000 - 65.0}\right).$$

The lens must be moved $\boxed{\text{away from the film}}$ by a distance

$$D = q_2 - q_1 = (65.0 \text{ mm})\left(\frac{2\,000}{2\,000 - 65.0}\right) - 65.0 \text{ mm} = \boxed{2.18 \text{ mm}}.$$

Section 26.5 Context Connection—Medical Fiberscopes

P26.41 The image will be inverted. With $h = 6$ cm, we require $h' = -1$ mm.

(a) $M = \frac{h'}{h} = -\frac{q}{p}$ $\qquad q = -p\frac{h'}{h} = -50 \text{ mm}\frac{(-1 \text{ mm})}{60 \text{ mm}} = \boxed{0.833 \text{ mm}}$

(b) $\frac{1}{p} + \frac{1}{q} = \frac{1}{f} = \frac{1}{50 \text{ mm}} + \frac{1}{0.833 \text{ mm}}$ $\qquad f = \boxed{0.820 \text{ mm}}$

P26.42 (a) Light leaving the sphere refracts away from the normal, so light that travels toward the upper right comes from the $\boxed{\text{bottom half}}$ of the sphere.

(b) $\sin\theta_1 = \dfrac{1\text{ mm}}{R}$ $\qquad n_1 \sin\theta_1 = n_2 \sin\theta_2$

$1.50\left(\dfrac{1\text{ mm}}{R}\right) = 1\sin 90°$ $\qquad R = \boxed{1.50\text{ mm}}$

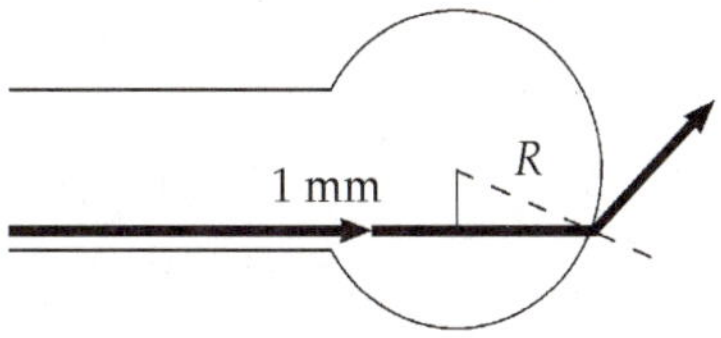

FIG. P26.42(b)

(c) $\delta = \left|\theta_2 - \theta_1\right| = \left|90° - \sin^{-1}\left(\dfrac{1}{1.5}\right)\right| = \boxed{48.2°}$

(d) No ray can have an angle of refraction larger than 90°, so the ray considered in parts (b) and (c) has the largest possible angle of refraction and then the largest possible deviation.

All other rays, at distances from the axis of less than 1 mm, will leave the sphere at smaller angles with the axis than 48.2°. The angular diameter of the cone of diverging light is $2 \times 48.2° = \boxed{96.4°}$.

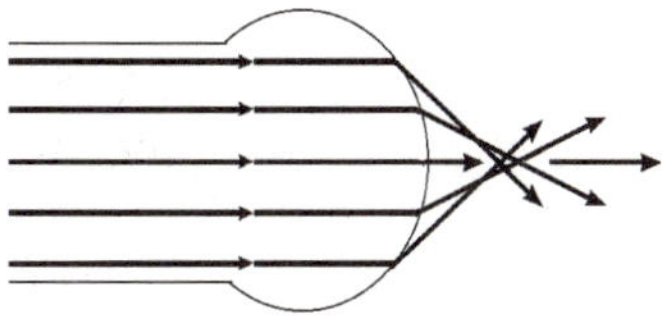

FIG. P26.42(d)

(e) Light can be absorbed by a coating on the sphere and reradiated in any direction. The sphere deviates from a perfectly spherical shape, probably macroscopically and surely at the scale of the wavelength of light. Light rays enter the sphere along directions not parallel to the axis of the fiber. Inhomogeneities within the sphere scatter light. Light reflects from the interior surface of the sphere.

Additional Problems

P26.43 If $M < 1$, the lens is diverging and the image is virtual. $\qquad d = p - |q| = p + q$

$M = -\dfrac{q}{p}$ so $q = -Mp$ and $d = p - Mp$

$p = \dfrac{d}{1-M}$: $\qquad \dfrac{1}{p} + \dfrac{1}{q} = \dfrac{1}{f} = \dfrac{1}{p} + \dfrac{1}{(-Mp)} = \dfrac{(-M+1)}{-Mp} = \dfrac{(1-M)^2}{-Md}$ $\qquad \boxed{f = \dfrac{-Md}{(1-M)^2}}$.

If $M > 1$, the lens is converging and the image is still virtual.

Now $\qquad d = -q - p$.

We obtain in this case $\qquad \boxed{f = \dfrac{Md}{(M-1)^2}}$.

P26.44 Start with the first pass through the lens.

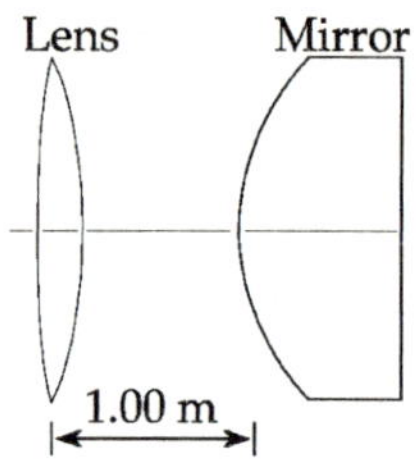

FIG. P26.44

$$\frac{1}{q_1}=\frac{1}{f_1}-\frac{1}{p_1}=\frac{1}{80.0\text{ cm}}-\frac{1}{100\text{ cm}}\qquad q_1=400\text{ cm to right of lens}$$

For the mirror, $p_2=-300\text{ cm}$

$$\frac{1}{q_2}=\frac{1}{f_2}-\frac{1}{p_2}=\frac{1}{(-50.0\text{ cm})}-\frac{1}{(-300\text{ cm})}\qquad q_2=-60.0\text{ cm}$$

For the second pass through the lens, $p_3=160\text{ cm}$

$$\frac{1}{q_3}=\frac{1}{f_1}-\frac{1}{p_3}=\frac{1}{80.0\text{ cm}}-\frac{1}{160\text{ cm}}\qquad q_3=\boxed{160\text{ cm to the left of lens}}$$

$$M_1=-\frac{q_1}{p_1}=-\frac{400\text{ cm}}{100\text{ cm}}=-4.00\qquad M_2=-\frac{q_2}{p_2}=-\frac{(-60.0\text{ cm})}{(-300\text{ cm})}=-\frac{1}{5}$$

$$M_3=-\frac{q_3}{p_3}=-\frac{160\text{ cm}}{160\text{ cm}}=-1\qquad M=M_1M_2M_3=\boxed{-0.800}$$

Since $M<0$ the final image is $\boxed{\text{inverted}}$.

P26.45 The real image formed by the concave mirror serves as a real object for the convex mirror with $p=50$ cm and $q=-10$ cm. Therefore,

$$\frac{1}{f}=\frac{1}{p}+\frac{1}{q}\qquad \frac{1}{f}=\frac{1}{50\text{ cm}}+\frac{1}{(-10\text{ cm})}$$

gives $f=-12.5$ cm and $R=2f=\boxed{-25.0\text{ cm}}$.

P26.46 With $n_1=1$ and $n_2=n$ for the lens material, the first refraction is described by

$$\frac{1}{\infty}+\frac{1}{q_1}=\frac{n-1}{R_1}$$

$$q_1=\frac{R_1}{n-1}$$

The real image formed by the first surface is a virtual object seen by the second, at $p_2=-q_1$. Then for the second refraction

$$\frac{1}{-q_1}+\frac{1}{q_2}=\frac{1-n}{R_2}$$

$$\frac{1-n}{R_1}+\frac{1}{q_2}=\frac{1-n}{R_2}$$

$$\frac{1}{q_2}=(n-1)\left(\frac{1}{R_1}-\frac{1}{R_2}\right)$$

This image distance for an object at infinity is by definition the focal length.

***P26.47** (a) $\frac{1}{q_1} = \frac{1}{f_1} - \frac{1}{p_1} = \frac{1}{5 \text{ cm}} - \frac{1}{7.5 \text{ cm}}$ $\therefore q_1 = 15 \text{ cm}$

$$M_1 = -\frac{q_1}{p_1} = -\frac{15 \text{ cm}}{7.5 \text{ cm}} = -2$$

$$M = M_1 M_2 \qquad \therefore 1 = (-2) M_2$$

$$\therefore M_2 = -\frac{1}{2} = -\frac{q_2}{p_2} \qquad \therefore p_2 = 2q_2$$

$$\frac{1}{p_2} + \frac{1}{q_2} = \frac{1}{f_2} \qquad \therefore \frac{1}{2q_2} + \frac{1}{q_2} = \frac{1}{10 \text{ cm}} \qquad \therefore q_2 = 15 \text{ cm}, \ p_2 = 30 \text{ cm}$$

$$p_1 + q_1 + p_2 + q_2 = 7.5 \text{ cm} + 15 \text{ cm} + 30 \text{ cm} + 15 \text{ cm} = \boxed{67.5 \text{ cm}}$$

(b) $\frac{1}{p_1'} + \frac{1}{q_1'} = \frac{1}{f_1} = \frac{1}{5 \text{ cm}}$

Solve for q_1' in terms of p_1': $q_1' = \frac{5p_1'}{p_1' - 5}$ (1)

$M_1' = -\frac{q_1'}{p_1'} = -\frac{5}{p_1' - 5}$, using (1).

$$M' = M_1' M_2' \qquad \therefore M_2' = \frac{M'}{M_1'} = -\frac{3}{5}(p_1' - 5) = -\frac{q_2'}{p_2'}$$

$\therefore q_2' = \frac{3}{5} p_2' (p_1' - 5)$ (2)

Substitute (2) into the lens equation $\frac{1}{p_2'} + \frac{1}{q_2'} = \frac{1}{f_2} = \frac{1}{10 \text{ cm}}$ and obtain p_2' in terms of p_1':

$p_2' = \frac{10(3p_1' - 10)}{3(p_1' - 5)}$. (3)

Substituting (3) in (2), obtain q_2' in terms of p_1':

$q_2' = 2(3p_1' - 10)$. (4)

Now, $p_1' + q_1' + p_2' + q_2' =$ a constant.

Using (1), (3) and (4), and the value obtained in (a):

$$p_1' + \frac{5p_1'}{p_1' - 5} + \frac{10(3p_1' - 10)}{3(p' - 5)} + 2(3p_1' - 10) = 67.5.$$

This reduces to the quadratic equation

$$21p_1'^2 - 322.5p_1' + 1\,212.5 = 0,$$

which has solutions $p_1' = 8.784$ cm and 6.573 cm.

continued on next page

Case 1: $p_1' = 8.784 \text{ cm}$

$\therefore p_1' - p_1 = 8.784 \text{ cm} - 7.5 \text{ cm} = 1.28 \text{ cm}.$

From (4): $q_2' = 32.7 \text{ cm}$

$\therefore q_2' - q_2 = 32.7 \text{ cm} - 15 \text{ cm} = 17.7 \text{ cm}.$

Case 2: $p_1' = 6.573 \text{ cm}$

$\therefore p_1' - p_1 = 6.573 \text{ cm} - 7.5 \text{ cm} = -0.927 \text{ cm}.$

From (4): $q_2' = 19.44 \text{ cm}$

$\therefore q_2' = q_2 = 19.44 \text{ cm} - 15 \text{ cm} = 4.44 \text{ cm}.$

From these results it is concluded that:

The lenses can be displaced in two ways. The first lens can be moved 1.28 cm farther from the object and the second lens 17.7 cm toward the object. Alternatively, the first lens can be moved 0.927 cm toward the object and the second lens 4.44 cm toward the object.

P26.48 $$\frac{1}{q_1} = \frac{1}{f_1} - \frac{1}{p_1} = \frac{1}{10.0 \text{ cm}} - \frac{1}{12.5 \text{ cm}}$$

so $q_1 = 50.0 \text{ cm}$ (to left of mirror).

This serves as an object for the lens (a virtual object), so

$$\frac{1}{q_2} = \frac{1}{f_2} - \frac{1}{p_2} = \frac{1}{(-16.7 \text{ cm})} - \frac{1}{(-25.0 \text{ cm})} \text{ and } q_2 = -50.3 \text{ cm},$$

meaning 50.3 cm to the right of the lens. Thus, the final image is located 25.3 cm to right of mirror.

$$M_1 = -\frac{q_1}{p_1} = -\frac{50.0 \text{ cm}}{12.5 \text{ cm}} = -4.00$$

$$M_2 = -\frac{q_2}{p_2} = -\frac{(-50.3 \text{ cm})}{(-25.0 \text{ cm})} = -2.01$$

$$M = M_1 M_2 = 8.05$$

Thus, the final image is virtual, upright, 8.05 times the size of object, and 25.3 cm to right of the mirror.

P26.49 A hemisphere is too thick to be described as a thin lens. The light is undeviated on entry into the flat face. We next consider the light's exit from the second surface, for which $R = -6.00$ cm.

The incident rays are parallel, so $p = \infty$.

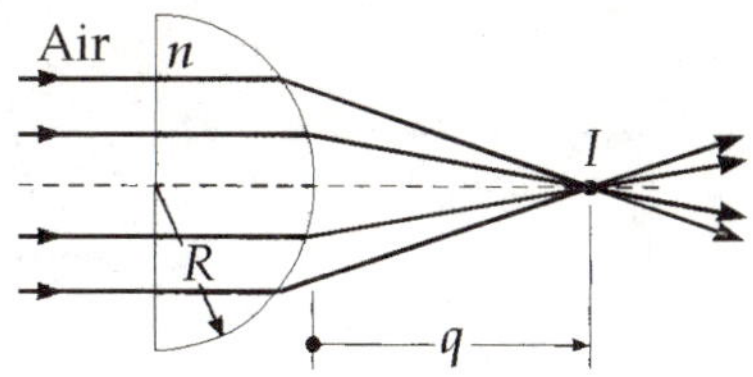

FIG. P26.49

Then, $$\frac{n_1}{p} + \frac{n_2}{q} = \frac{n_2 - n_1}{R}$$

becomes $$0 + \frac{1}{q} = \frac{1.00 - 1.56}{-6.00 \text{ cm}}$$

and $$\boxed{q = 10.7 \text{ cm}}.$$

P26.50 (a) $$I = \frac{\mathcal{P}}{4\pi r^2} = \frac{4.50 \text{ W}}{4\pi\left(1.60 \times 10^{-2} \text{ m}\right)^2} = \boxed{1.40 \text{ kW/m}^2}$$

(b) $$I = \frac{\mathcal{P}}{4\pi r^2} = \frac{4.50 \text{ W}}{4\pi(7.20 \text{ m})^2} = \boxed{6.91 \text{ mW/m}^2}$$

(c) $\frac{1}{p} + \frac{1}{q} = \frac{1}{f}$: $$\frac{1}{7.20 \text{ m}} + \frac{1}{q} = \frac{1}{0.350 \text{ m}}$$

so $$q = 0.368 \text{ m}$$

and $$M = \frac{h'}{3.20 \text{ cm}} = -\frac{q}{p} = -\frac{0.368 \text{ m}}{7.20 \text{ m}}$$

$$h' = \boxed{0.164 \text{ cm}}$$

(d) The lens intercepts power given by $$\mathcal{P} = IA = \left(6.91 \times 10^{-3} \text{ W/m}^2\right)\left[\frac{\pi}{4}(0.150 \text{ m})^2\right]$$

and puts it all onto the image where $$I = \frac{\mathcal{P}}{A} = \frac{\left(6.91 \times 10^{-3} \text{ W/m}^2\right)\left[\pi(15.0 \text{ cm})^2/4\right]}{\pi(0.164 \text{ cm})^2/4}$$

$$I = \boxed{58.1 \text{ W/m}^2}.$$

P26.51 From the thin lens equation, $$q_1 = \frac{f_1 p_1}{p_1 - f_1} = \frac{(-6.00 \text{ cm})(12.0 \text{ cm})}{12.0 \text{ cm} - (-6.00 \text{ cm})} = -4.00 \text{ cm}.$$

When we require that $q_2 \to \infty$, the thin lens equation becomes $p_2 = f_2$.

In this case, $p_2 = d - (-4.00 \text{ cm})$.

Therefore, $d + 4.00 \text{ cm} = f_2 = 12.0 \text{ cm}$ and $d = \boxed{8.00 \text{ cm}}$.

continued on next page

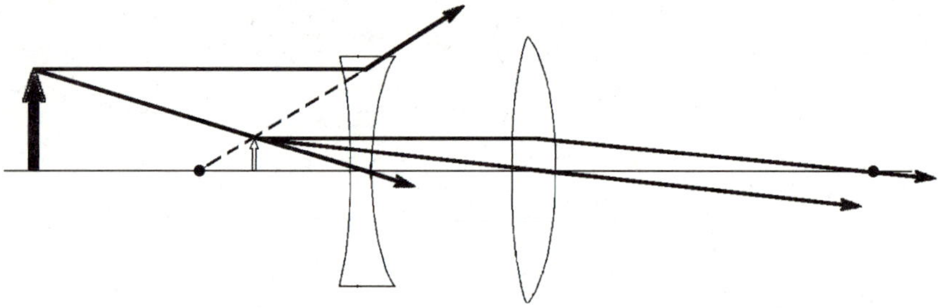

FIG. P26.51

P26.52 The inverted real image is formed by the lens operating on light directly from the object, on light that has not reflected from the mirror.

For this we have $M = -1.50 = -\frac{q}{p}$ $\quad q = 1.50p$

$$\frac{1}{p}+\frac{1}{q}=\frac{1}{f} \qquad \frac{1}{p}+\frac{1}{1.50p}=\frac{1}{10\text{ cm}}=\frac{2.50}{1.50p} \qquad p = 10\text{ cm}\left(\frac{2.5}{1.5}\right) = 16.7\text{ cm}$$

Then the object is distant from the mirror by $\quad 40.0\text{ cm} - 16.7\text{ cm} = 23.3\text{ cm}$.

The second image seen by the person is formed by light that first reflects from the mirror and then goes through the lens. For it to be in the same position as the inverted image, the lens must be receiving light from an image formed by the mirror at the same location as the physical object. The formation of this image is described by

$$\frac{1}{p}+\frac{1}{q}=\frac{1}{f} \qquad \frac{1}{23.3\text{ cm}}+\frac{1}{23.3\text{ cm}}=\frac{1}{f} \qquad f = \boxed{11.7\text{ cm}}.$$

P26.53 For the mirror, $f = \frac{R}{2} = +1.50\text{ m}$. In addition, because the distance to the Sun is so much larger than any other distances, we can take $p = \infty$.

The mirror equation, $\frac{1}{p}+\frac{1}{q}=\frac{1}{f}$, then gives $\quad q = f = \boxed{1.50\text{ m}}$.

Now, in $\quad M = -\frac{q}{p} - \frac{h'}{h}$

the magnification is nearly zero, but we can be more precise: $\frac{h}{p}$ is the angular diameter of the object.

Thus, the image diameter is

$$h' = -\frac{hq}{p} = (-0.533°)\left(\frac{\pi\text{ rad}}{180°}\right)(1.50\text{ m}) = -0.0140\text{ m} = \boxed{-1.40\text{ cm}}.$$

P26.54 (a) For the light the mirror intercepts,

$$\mathscr{P} = I_0 A = I_0 \pi R_a^2$$

$$350 \text{ W} = \left(1\,000 \text{ W/m}^2\right)\pi R_a^2$$

and $R_a = \boxed{0.334 \text{ m or larger}}$.

(b) In $\frac{1}{p} + \frac{1}{q} = \frac{1}{f} = \frac{2}{R}$

we have $p \to \infty$

so $q = \frac{R}{2}$

$$M = \frac{h'}{h} = -\frac{q}{p}$$

so $h' = -q\left(\frac{h}{p}\right) = -\left(\frac{R}{2}\right)\left[0.533°\left(\frac{\pi \text{ rad}}{180°}\right)\right] = -\left(\frac{R}{2}\right)(9.30 \text{ m rad})$

where $\frac{h}{p}$ is the angle the Sun subtends. The intensity at the image is

then $I = \frac{\mathscr{P}}{\pi h'^2/4} = \frac{4I_0 \pi R_a^2}{\pi h'^2} = \frac{4I_0 R_a^2}{(R/2)^2\left(9.30 \times 10^{-3} \text{ rad}\right)^2}$

$$120 \times 10^3 \text{ W/m}^2 = \frac{16\left(1\,000 \text{ W/m}^2\right)R_a^2}{R^2\left(9.30 \times 10^{-3} \text{ rad}\right)^2}$$

so $\boxed{\frac{R_a}{R} = 0.025\,5 \text{ or larger}}$.

P26.55 In the original situation, $p_1 + q_1 = 1.50 \text{ m}$.

In the final situation, $p_2 = p_1 + 0.900 \text{ m}$

and $q_2 = q_1 - 0.900 \text{ m} = 0.600 \text{ m} - p_1$.

Our lens equation is $\frac{1}{p_1} + \frac{1}{q_1} = \frac{1}{f} = \frac{1}{p_2} + \frac{1}{q_2}$.

Substituting, we have $\frac{1}{p_1} + \frac{1}{1.50 \text{ m} - p_1} = \frac{1}{p_1 + 0.900} + \frac{1}{0.600 - p_1}$.

Adding the fractions, $\frac{1.50 \text{ m} - p_1 + p_1}{p_1(1.50 \text{ m} - p_1)} = \frac{0.600 - p_1 + p_1 + 0.900}{(p_1 + 0.900)(0.600 - p_1)}$.

Simplified, this becomes $p_1(1.50 \text{ m} - p_1) = (p_1 + 0.900)(0.600 - p_1)$.

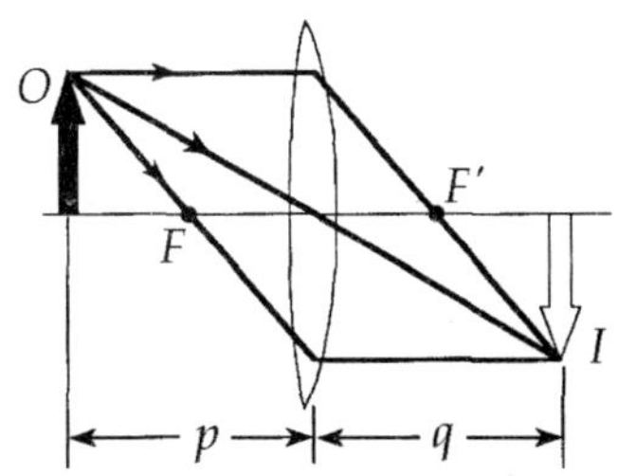

FIG. P26.55

continued on next page

(a) Thus, $p_1 = \frac{0.540}{1.80}\ \text{m} = \boxed{0.300\ \text{m}}$ $\qquad p_2 = p_1 + 0.900 = \boxed{1.20\ \text{m}}$

(b) $\frac{1}{f} = \frac{1}{0.300\ \text{m}} + \frac{1}{1.50\ \text{m} - 0.300\ \text{m}}$ and $f = \boxed{0.240\ \text{m}}$

(c) The second image is $\boxed{\text{real, inverted, and diminished}}$

with $M = -\frac{q_2}{p_2} = \boxed{-0.250}$.

P26.56 A telescope with an eyepiece decreases the diameter of a beam of parallel rays. When light is sent through the same device in the opposite direction, the beam expands. Send the light first through the diverging lens. It will then be diverging from a virtual image found like this:

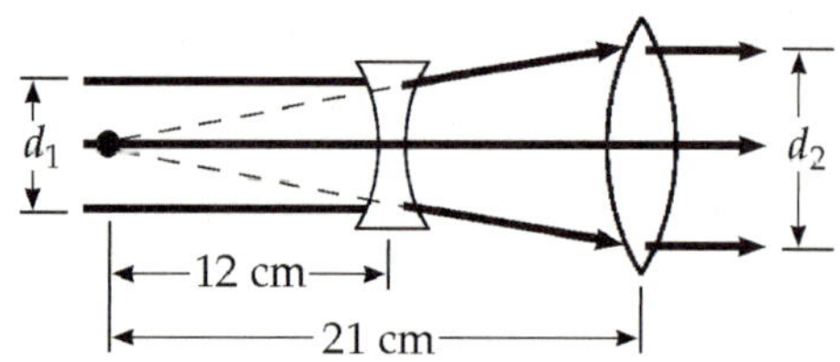

FIG. P26.56

$$\frac{1}{p} + \frac{1}{q} = \frac{1}{f} \qquad \frac{1}{\infty} + \frac{1}{q} = \frac{1}{-12\ \text{cm}}$$

$$q = -12\ \text{cm}.$$

Use this image as a real object for the converging lens, placing it at the focal point on the object side of the lens, at $p = 21$ cm. Then

$$\frac{1}{p} + \frac{1}{q} = \frac{1}{f} \qquad \frac{1}{21\ \text{cm}} + \frac{1}{q} = \frac{1}{21\ \text{cm}}$$

$$q = \infty.$$

The exiting rays will be parallel. The lenses must be $21.0\ \text{cm} - 12.0\ \text{cm} = 9.00\ \text{cm}$ apart.

By similar triangles, $\frac{d_2}{d_1} = \frac{21\ \text{cm}}{12\ \text{cm}} = \boxed{1.75\ \text{times}}$.

***P26.57** (a) For the lens in air,

$$\frac{1}{f} = (n-1)\left(\frac{1}{R_1} - \frac{1}{R_2}\right)$$

$$\frac{1}{79\ \text{cm}} = (1.55-1)\left(\frac{1}{R_1} - \frac{1}{R_2}\right)$$

For the same lens in water

$$\frac{1}{f'} = \left(\frac{n_2}{n_1} - 1\right)\left(\frac{1}{R_1} - \frac{1}{R_2}\right)$$

$$\frac{1}{f'} = \left(\frac{1.55}{1.33} - 1\right)\left(\frac{1}{R_1} - \frac{1}{R_2}\right)$$

continued on next page

By division,

$$\frac{\frac{1}{79\text{ cm}}}{\frac{1}{f'}} = \frac{0.55}{0.165} = \frac{f'}{79\text{ cm}}$$

$$f' = 79\text{ cm}(3.32) = \boxed{263\text{ cm}}$$

(b) The path of a reflected ray does not depend on the refractive index of the medium which the reflecting surface bounds. Therefore the focal length of a mirror does not change when it is put into a different medium: $f' = \frac{R}{2} = f = \boxed{79.0\text{ cm}}$.

P26.58 (a) The lens makers' equation, $\frac{1}{f} = (n-1)\left(\frac{1}{R_1} + \frac{1}{R_2}\right)$

becomes: $\frac{1}{5.00\text{ cm}} = (n-1)\left[\frac{1}{9.00\text{ cm}} - \frac{1}{(-11.0\text{ cm})}\right]$ giving $n = \boxed{1.99}$.

(b) As the light passes through the lens for the first time, the thin lens equation

$$\frac{1}{p_1} + \frac{1}{q_1} = \frac{1}{f}$$

becomes:

$$\frac{1}{8.00\text{ cm}} + \frac{1}{q_1} = \frac{1}{5.00\text{ cm}}$$

or $q_1 = 13.3\text{ cm}$, and $M_1 = -\frac{q_1}{p_1} = -\frac{13.3\text{ cm}}{8.00\text{ cm}} = -1.67$.

This image becomes the object for the concave mirror with:

$$p_m = 20.0\text{ cm} - q_1 = 20.0\text{ cm} - 13.3\text{ cm} = 6.67\text{ cm}$$

and

$$f = \frac{R}{2} = +4.00\text{ cm}.$$

The mirror equation becomes:

$$\frac{1}{6.67\text{ cm}} + \frac{1}{q_m} = \frac{1}{4.00\text{ cm}}$$

giving

$$q_m = 10.0\text{ cm}$$

and

$$M_2 = -\frac{q_m}{p_m} = -\frac{10.0\text{ cm}}{6.67\text{ cm}} = -1.50.$$

The image formed by the mirror serves as a real object for the lens on the second pass of the light through the lens with:

$$p_3 = 20.0\text{ cm} - q_m = +10.0\text{ cm}.$$

The thin lens equation yields:

$$\frac{1}{10.0\text{ cm}} + \frac{1}{q_3} = \frac{1}{5.00\text{ cm}}$$

continued on next page

or $$q_3 = 10.0\text{ cm}$$

and $$M_3 = -\frac{q_3}{p_3} = -\frac{10.0\text{ cm}}{10.0\text{ cm}} = -1.00\,.$$

The final image is a real image located $\boxed{\text{10.0 cm to the left of the lens}}$.

The overall magnification is $M_{\text{total}} = M_1 M_2 M_3 = \boxed{-2.50}$.

(c) Since the total magnification is negative, this final image is $\boxed{\text{inverted}}$.

P26.59 The object is located at the focal point of the upper mirror. Thus, the upper mirror creates an image at infinity (i.e., parallel rays leave this mirror).

The lower mirror focuses these parallel rays at its focal point, located at the hole in the upper mirror.

Thus, the $\boxed{\text{image is real, inverted, and actual size}}$.

For the upper mirror:

$$\frac{1}{p} + \frac{1}{q} = \frac{1}{f}: \qquad \frac{1}{7.50\text{ cm}} + \frac{1}{q_1} = \frac{1}{7.50\text{ cm}} \qquad q_1 = \infty\,.$$

For the lower mirror:

$$\frac{1}{\infty} + \frac{1}{q_2} = \frac{1}{7.50\text{ cm}} \qquad q_2 = 7.50\text{ cm}.$$

Light directed into the hole in the upper mirror reflects as shown, to behave as if it were reflecting from the hole.

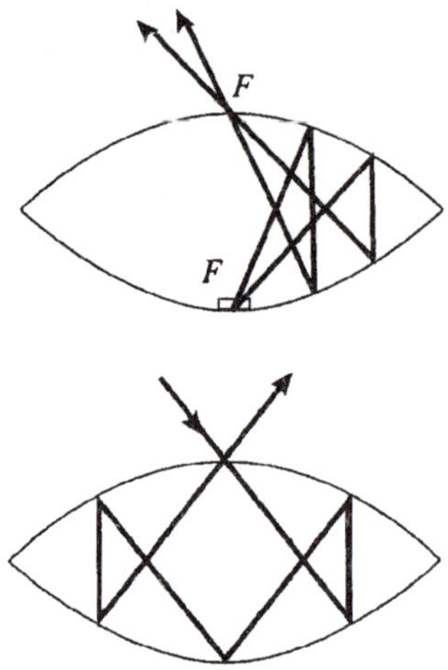

FIG. P26.59

P26.60 (a) For lens one, as shown in the first figure,

$$\frac{1}{40.0\text{ cm}}+\frac{1}{q_1}=\frac{1}{30.0\text{ cm}}$$

$$q_1=120\text{ cm}$$

$$M_1=-\frac{q_1}{p_1}=-\frac{120\text{ cm}}{40.0\text{ cm}}=-3.00$$

This real image $I_1=O_2$ is a virtual object for the second lens. That is, it is *behind* the lens, as shown in the second figure. The object distance is

$$p_2=110\text{ cm}-120\text{ cm}=-10.0\text{ cm}$$

$$\frac{1}{-10.0\text{ cm}}+\frac{1}{q_2}=\frac{1}{-20.0\text{ cm}}:$$

$$q_2=\boxed{20.0\text{ cm}}$$

$$M_2=-\frac{q_2}{p_2}=-\frac{20.0\text{ cm}}{(-10.0\text{ cm})}=+2.00$$

$$M_{\text{overall}}=M_1M_2=\boxed{-6.00}$$

(b) $M_{\text{overall}}<0$, so final image is $\boxed{\text{inverted}}$.

(c) If lens two is a converging lens (third figure):

$$\frac{1}{-10.0\text{ cm}}+\frac{1}{q_2}=\frac{1}{20.0\text{ cm}}$$

$$q_2=\boxed{6.67\text{ cm}}$$

$$M_2=-\frac{6.67\text{ cm}}{(-10.0\text{ cm})}=+0.667$$

$$M_{\text{overall}}=M_1M_2=\boxed{-2.00}$$

Again, $M_{\text{overall}}<0$ and the final image is $\boxed{\text{inverted}}$.

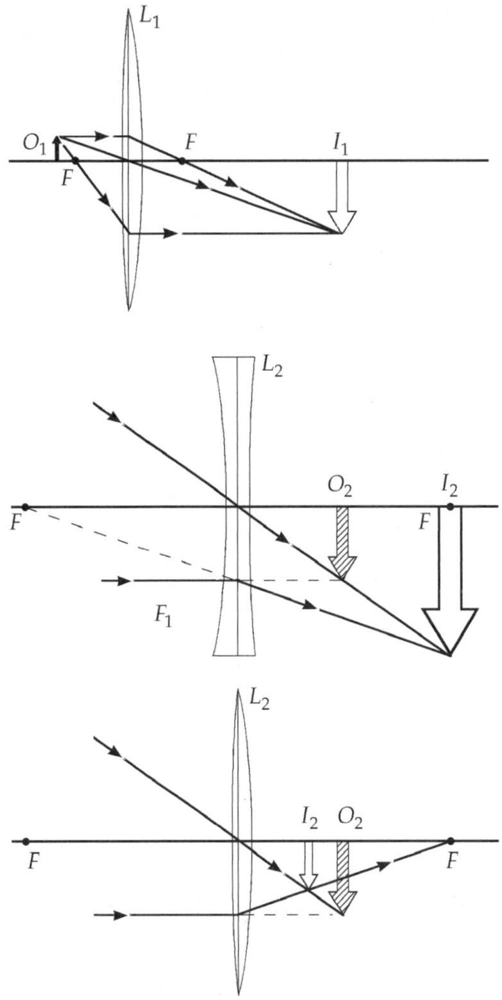

FIG. P26.60

ANSWERS TO EVEN PROBLEMS

P26.2 4.58 m

P26.4 see the solution

P26.6 (a) $-(p_1+h)$; (b) virtual; (c) upright; (d) 1.00; (e) no

P26.8 at $q=-0.267$ m virtual upright diminished with $M=0.0267$

P26.10 behind the worshiper, 3.33 m from the deepest point in the niche

P26.12 (a) 2.22 cm; (b) 10.0

P26.14 (a) 160 mm; (b) –267 mm

P26.16 (a) convex; (b) at the 30.0 cm mark; (c) –20.0 cm

P26.18 (a) 15.0 cm; (b) 60.0 cm

P26.20 (a) see the solution; (b) at 0.639 s and 0.782 s

P26.22 4.82 cm

P26.24 see the solution, the image is real, inverted, and diminished

P26.26 1.50 cm/s

P26.28 20.0 cm

P26.30 (a) $q = 40.0$ cm, $M = -1.00$, real and inverted; (b) $q \to \infty$, no image is formed; (c) $q = -20.0$ cm, $M = +2.00$, virtual and upright

P26.32 (a) 6.40 cm; (b) –0.250; (c) converging

P26.34 (a) $f = 39.0$ mm; (b) $p = 39.5$ mm

P26.36 (a) 5.36 cm; (b) –18.8 cm; (c) virtual, right side up, enlarged; (d) A magnifying glass with focal length 7.50 cm is used to form an image of a stamp, enlarged 3.50 times. Find the object distance. Locate and describe the image.

P26.38 (a) $f = 13.3$ cm; (b) see the solution; (c) 224 cm^2

P26.40 2.18 mm away from the film plane

P26.42 (a) the bottom half; (b) 1.50 mm; (c) 48.2°; (d) see the solution, 96.4°; (e) see the solution

P26.44 160 cm to the left of the lens, inverted, $M = -0.800$

P26.46 see the solution

P26.48 25.3 cm to the right of the mirror, virtual, upright, enlarged 8.05 times

P26.50 (a) 1.40 kW/m^2; (b) 6.91 mW/m^2; (c) 0.164 cm; (d) 58.1 W/m^2

P26.52 11.7 cm

P26.54 (a) 0.334 m or larger; (b) R_a must be at least 0.025 5 R

P26.56 Align the lenses on the same axis and 9.00 cm apart. Let the light pass first through the diverging lens and then through the converging lens. The diameter increases by a factor of 1.75.

P26.58 (a) 1.99; (b) 10.0 cm to the left of the lens, –2.50; (c) inverted

P26.60 (a) 20.0 cm to the right of the second lens, –6.00; (b) inverted; (c) 6.67 to the right of the second lens, –2.00, inverted

Chapter 27

Wave Optics

CHAPTER OUTLINE

ANSWERS TO QUESTIONS

Q27.1 (a) Two waves interfere constructively if their path difference is zero, or an integral multiple of the wavelength, according to $\delta = m\lambda$, with $m = 0, 1, 2, 3, \ldots$.

(b) Two waves interfere destructively if their path difference is a half wavelength, or an odd multiple of $\frac{\lambda}{2}$, described by $\delta = \left(m + \frac{1}{2}\right)\lambda$, with $m = 0, 1, 2, 3, \ldots$.

Q27.2 The light from the flashlights consists of many different wavelengths (that's why it's white) with random time differences between the light waves. There is no *coherence* between the two sources. The light from the two flashlights does not maintain a constant phase relationship over time. These three equivalent statements mean no possibility of an interference pattern.

Q27.3 Every color produces its own pattern, with a spacing between the maxima that is characteristic of the wavelength. With several colors, the patterns are superimposed and it can be difficult to pick out a single maximum. Using monochromatic light can eliminate this problem.

Q27.4 Underwater, the wavelength of the light would decrease, $\lambda_{\text{water}} = \frac{\lambda_{\text{air}}}{n_{\text{water}}}$. Since the positions of light and dark bands are proportional to λ, (according to Equations 27.2 and 27.3), the underwater fringe separations will decrease.

Q27.5 The threads that are woven together to make the cloth have small meshes between them. These bits of space act as pinholes through which the light diffracts. Since the cloth is a grid of such pinholes, an interference pattern is formed, as when you look through a diffraction grating.

Q27.6 If the oil film is brightest where it is thinnest, then $n_{\text{air}} < n_{\text{oil}} < n_{\text{water}}$. With this condition, light reflecting from both the top and the bottom surface of the oil film will undergo phase reversal. Then these two beams will be in phase with each other where the film is very thin. This is the condition for constructive interference as the thickness of the oil film decreases toward zero.

Q27.7 As water evaporates from the 'soap' bubble, the thickness of the bubble wall approaches zero. Since light reflecting from the front of the water surface is phase-shifted 180° and light reflecting from the back of the soap film is phase-shifted 0°, the reflected light meets the conditions for a minimum. Thus the soap film appears black, as in the illustration accompanying textbook Example 27.4, "Interference in a Wedge-Shaped Film."

Q27.8 If the film is more than a few wavelengths thick, the interference fringes are so close together that you cannot resolve them.

Q27.9 Suppose the coating is intermediate in index of refraction between vacuum and the glass. When the coating is very thin, light reflected from its top and bottom surfaces will interfere constructively, so you see the surface white and brighter. As the thickness reaches one quarter of the wavelength of violet light in the coating, destructive interference for violet will make the surface look red or perhaps orange. Next to interfere destructively are blue, green, yellow, orange, and red, making the surface look red, purple, and then blue. As the coating gets still thicker, we can get constructive interference for violet and then for other colors in spectral order. Still thicker coating will give constructive and destructive interference for several visible wavelengths, so the reflected light will start to look white again.

Q27.10 The wavelength of light is extremely small in comparison to the dimensions of your hand, so the diffraction of light around an obstacle the size of your hand is totally negligible. However, sound waves have wavelengths that are comparable to the dimensions of the hand or even larger. Therefore, significant diffraction of sound waves occurs around hand-sized obstacles.

Q27.11 Audible sound has wavelengths on the order of meters or centimeters, while visible light has a wavelength on the order of half a micrometer. In this world of breadbox-sized objects, $\frac{\lambda}{a}$ is large for sound. When sound goes through a doorway, the sound diffracts around behind the adjoining walls. But $\frac{\lambda}{a}$ is a tiny fraction for visible light passing ordinary-size objects or apertures, so light changes its direction by only very small angles when it diffracts.

Another way of phrasing the answer: We can see by a small angle around a small obstacle or around the edge of a small opening. The side fringes in Figure 27.12 and the Arago spot in the center of Figure 27.13 show this diffraction. We cannot always hear around corners. Out-of-doors, away from reflecting surfaces, have someone a few meters distant face away from you and whisper. The high-frequency, short-wavelength, information-carrying components of the sound do not diffract around his head enough for you to understand his words.

Q27.12 Yes, but no diffraction effects are observed because the separation distance between adjacent ribs is so much greater than the wavelength of x-rays. Diffraction does not limit the resolution of an x-ray image. Diffraction might sometimes limit the resolution of an ultrasonogram.

Q27.13 The intensity of the light coming through the slit decreases, as you would expect. The central maximum increases in width as the width of the slit decreases. In the condition $\sin\theta = \frac{\lambda}{a}$ for destructive interference on each side of the central maximum, θ increases as a decreases.

Q27.14 It is shown in the correct orientation. If the horizontal width of the opening is equal to or less than the wavelength of the sound, then the equation $a\sin\theta = (1)\lambda$ has the solution $\theta = 90°$, or has no solution. The central diffraction maximum covers the whole seaward side. If the vertical height of the opening is large compared to the wavelength, then the angle in $a\sin\theta = (1)\lambda$ will be small, and the central diffraction maximum will form a thin horizontal sheet.

Q27.15 The fine hair blocks off light that would otherwise go through a fine slit and produce a diffraction pattern on a distant screen. The width of the central maximum in the pattern is inversely proportional to the distance across the slit. When the hair is in place, it subtracts the same diffraction pattern from the projected disk of laser light. The hair produces a diffraction minimum that crosses the bright circle on the screen. The width of the minimum is inversely proportional to the diameter of the hair. The central minimum is flanked by narrower maxima and minima. Measure the width $2y$ of the central minimum between the maxima bracketing it, and use Equation 27.13 in the form $\frac{y}{L} = \frac{\lambda}{a}$ to find the width a of the hair.

Q27.16 The condition for constructive interference is that the three radio signals arrive at the city in phase. We know the speed of the waves (it is the speed of light c), the angular bearing θ of the city east of north from the broadcast site, and the distance d between adjacent towers. The wave from the westernmost tower must travel an extra distance $2d\sin\theta$ to reach the city, compared to the signal from the eastern tower. For each cycle of the carrier wave, the western antenna would transmit first, the center antenna after a time delay $\frac{d\sin\theta}{c}$, and the eastern antenna after an additional equal time delay.

SOLUTIONS TO PROBLEMS

Section 27.1 Conditions for Interference

Section 27.2 Young's Double-Slit Experiment

Section 27.3 Light Waves in Interference

P27.1 $y_{\text{bright}} = \frac{\lambda L}{d} m$

For $m = 1$, $\lambda = \frac{yd}{L} = \frac{(3.40 \times 10^{-3}\text{ m})(5.00 \times 10^{-4}\text{ m})}{3.30\text{ m}} = \boxed{515\text{ nm}}$

P27.2 $\lambda = \frac{v}{f} = \frac{354\text{ m/s}}{2\,000\text{ s}^{-1}} = 0.177\text{ m}$

(a) $d\sin\theta = m\lambda$ so $(0.300\text{ m})\sin\theta = 1(0.177\text{ m})$ and $\theta = \boxed{36.2°}$

(b) $d\sin\theta = m\lambda$ so $d\sin 36.2° = 1(0.030\,0\text{ m})$ and $d = \boxed{5.08\text{ cm}}$

continued on next page

(c) $(1.00\times10^{-6}\text{ m})\sin 36.2° = (1)\lambda$ so $\lambda = 590\text{ nm}$

$$f = \frac{c}{\lambda} = \frac{3.00\times10^{8}\text{ m/s}}{5.90\times10^{-7}\text{ m}} = \boxed{508\text{ THz}}$$

P27.3 Note, with the conditions given, the small angle approximation **does not work well.** That is, $\sin\theta$, $\tan\theta$, and θ are significantly different. We treat the interference as a Fraunhofer pattern.

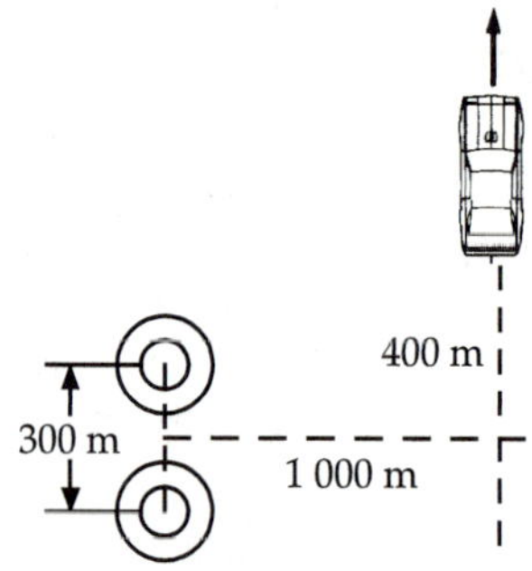

FIG. P27.3

(a) At the $m = 2$ maximum, $\tan\theta = \dfrac{400\text{ m}}{1\,000\text{ m}} = 0.400$

$$\theta = 21.8°$$

so $$\lambda = \frac{d\sin\theta}{m} = \frac{(300\text{ m})\sin 21.8°}{2} = \boxed{55.7\text{ m}}.$$

(b) The next minimum encountered is the $m = 2$ minimum;

and at that point, $$d\sin\theta = \left(m + \frac{1}{2}\right)\lambda$$

which becomes $$d\sin\theta = \frac{5}{2}\lambda$$

or $$\sin\theta = \frac{5}{2}\frac{\lambda}{d} = \frac{5}{2}\left(\frac{55.7\text{ m}}{300\text{ m}}\right) = 0.464$$

and $$\theta = 27.7°$$

so $$y = (1\,000\text{ m})\tan 27.7° = 524\text{ m}.$$

Therefore, the car must travel an additional $\boxed{124\text{ m}}$.

If we considered Fresnel interference, we would more precisely find (a) $\lambda = \frac{1}{2}\left(\sqrt{550^2 + 1\,000^2} - \sqrt{250^2 + 1\,000^2}\right) = 55.2\text{ m}$ and (b) 123 m.

P27.4 $$\lambda = \frac{340\text{ m/s}}{2\,000\text{ Hz}} = 0.170\text{ m}$$

Maxima are at $d\sin\theta = m\lambda$:

$m = 0$ gives $\theta = 0°$

$m = 1$ gives $\sin\theta = \dfrac{\lambda}{d} = \dfrac{0.170\text{ m}}{0.350\text{ m}}$ $\theta = 29.1°$

$m = 2$ gives $\sin\theta = \dfrac{2\lambda}{d} = 0.971$ $\theta = 76.3°$

$m = 3$ gives $\sin\theta = 1.46$ No solution.

continued on next page

Minima are at $d\sin\theta = \left(m + \frac{1}{2}\right)\lambda$:

$m = 0$ gives $\sin\theta = \frac{\lambda}{2d} = 0.243$ $\theta = 14.1°$

$m = 1$ gives $\sin\theta = \frac{3\lambda}{2d} = 0.729$ $\theta = 46.8°$

$m = 2$ gives $\sin\theta = 1.21$ No solution.

$\boxed{\text{So we have maxima at } 0°, 29.1°, \text{ and } 76.3°; \text{ minima at } 14.1° \text{ and } 46.8°}$.

P27.5 In the equation $d\sin\theta = \left(m + \frac{1}{2}\right)\lambda$.

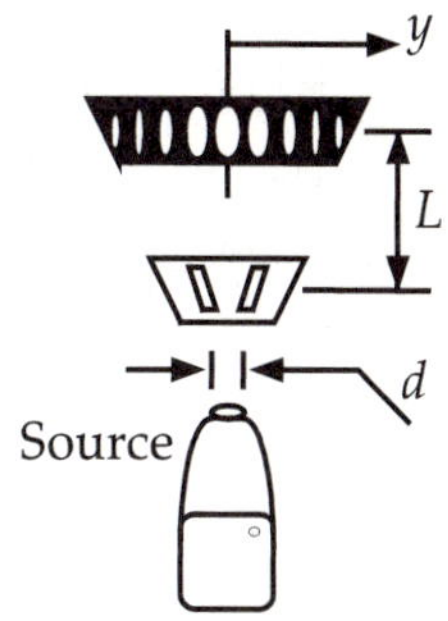

FIG. P27.5

The first minimum is described by $m = 0$

and the tenth by $m = 9$: $\sin\theta = \frac{\lambda}{d}\left(9 + \frac{1}{2}\right)$.

Also, $\tan\theta = \frac{y}{L}$

but for small θ, $\sin\theta \approx \tan\theta$.

Thus, $d = \frac{9.5\lambda}{\sin\theta} = \frac{9.5\lambda L}{y}$

$$d = \frac{9.5(5\,890 \times 10^{-10} \text{ m})(2.00 \text{ m})}{7.26 \times 10^{-3} \text{ m}} = 1.54 \times 10^{-3} \text{ m} = \boxed{1.54 \text{ mm}}.$$

***P27.6** Location of A = central maximum,

Location of B = first minimum.

So, $\Delta y = [y_{\text{min}} - y_{\text{max}}] = \frac{\lambda L}{d}\left(0 + \frac{1}{2}\right) - 0 = \frac{1}{2}\frac{\lambda L}{d} = 20.0 \text{ m}$.

Thus, $d = \frac{\lambda L}{2(20.0 \text{ m})} = \frac{(3.00 \text{ m})(150 \text{ m})}{40.0 \text{ m}} = \boxed{11.3 \text{ m}}$.

P27.7 At $30.0°$, $d\sin\theta = m\lambda$

$(3.20 \times 10^{-4} \text{ m})\sin 30.0° = m(500 \times 10^{-9} \text{ m})$ so $m = 320$

There are 320 maxima to the right, 320 to the left, and one for $m = 0$ straight ahead.

There are $\boxed{641 \text{ maxima}}$.

P27.8 Observe that the pilot must not only home in on the airport, but must be headed in the right direction when she arrives at the end of the runway.

(a) $\lambda = \frac{c}{f} = \frac{3\times 10^8 \text{ m/s}}{30\times 10^6 \text{ s}^{-1}} = \boxed{10.0 \text{ m}}$

(b) The first side maximum is at an angle given by $d\sin\theta = (1)\lambda$.

$(40 \text{ m})\sin\theta = 10 \text{ m}$ $\qquad \theta = 14.5°$ $\qquad \tan\theta = \frac{y}{L}$

$y = L\tan\theta = (2\,000 \text{ m})\tan 14.5° = \boxed{516 \text{ m}}$

(c) The signal of 10-m wavelength in parts (a) and (b) would show maxima at 0°, 14.5°, 30.0°, 48.6°, and 90°. A signal of wavelength 11.23-m would show maxima at 0°, 16.3°, 34.2°, and 57.3°. The only value in common is 0°. If λ_1 and λ_2 were related by a ratio of small integers (a just musical consonance!) in $\frac{\lambda_1}{\lambda_2} = \frac{n_1}{n_2}$, then the equations $d\sin\theta = n_2\lambda_1$ and $d\sin\theta = n_1\lambda_2$ would both be satisfied for the same nonzero angle. The pilot could come flying in with that inappropriate bearing, and run off the runway immediately after touchdown.

P27.9 $\phi = \frac{2\pi}{\lambda} d\sin\theta = \frac{2\pi}{\lambda} d\left(\frac{y}{L}\right)$

(a) $\phi = \frac{2\pi}{(5.00\times 10^{-7} \text{ m})}(1.20\times 10^{-4} \text{ m})\sin(0.500°) = \boxed{13.2 \text{ rad}}$

(b) $\phi = \frac{2\pi}{(5.00\times 10^{-7} \text{ m})}(1.20\times 10^{-4} \text{ m})\left(\frac{5.00\times 10^{-3} \text{ m}}{1.20 \text{ m}}\right) = \boxed{6.28 \text{ rad}}$

(c) If $\phi = 0.333 \text{ rad} = \frac{2\pi d\sin\theta}{\lambda}$ $\qquad \theta = \sin^{-1}\left(\frac{\lambda\phi}{2\pi d}\right) = \sin^{-1}\left[\frac{(5.00\times 10^{-7} \text{ m})(0.333 \text{ rad})}{2\pi(1.20\times 10^{-4} \text{ m})}\right]$

$\theta = \boxed{1.27\times 10^{-2} \text{ deg}}$.

(d) If $d\sin\theta = \frac{\lambda}{4}$ $\qquad \theta = \sin^{-1}\left(\frac{\lambda}{4d}\right) = \sin^{-1}\left[\frac{5\times 10^{-7} \text{ m}}{4(1.20\times 10^{-4} \text{ m})}\right]$

$\theta = \boxed{5.97\times 10^{-2} \text{ deg}}$.

***P27.10** The path difference between rays 1 and 2 is: $\delta = d\sin\theta_1 - d\sin\theta_2$.

For constructive interference, this path difference must be equal to an integral number of wavelengths: $d\sin\theta_1 - d\sin\theta_2 = m\lambda$, or

$\boxed{d(\sin\theta_1 - \sin\theta_2) = m\lambda}$.

P27.11 $I_{av} = I_{\max}\cos^2\left(\frac{\pi d\sin\theta}{\lambda}\right)$

For small θ, $\sin\theta \approx \frac{y}{L}$

and $I_{av} = 0.750 I_{\max}$

$$y = \frac{\lambda L}{\pi d}\cos^{-1}\sqrt{\frac{I_{av}}{I_{\max}}}$$

$$y = \frac{(6.00\times10^{-7})(1.20\text{ m})}{\pi(2.50\times10^{-3}\text{ m})}\cos^{-1}\sqrt{\frac{0.750 I_{\max}}{I_{\max}}} = \boxed{48.0\ \mu\text{m}}$$

P27.12 (a) The intensity of the sum of two coherent waves is described by $\frac{I}{I_{\max}} = \cos^2\left(\frac{\phi}{2}\right)$. Therefore,

$$\phi = 2\cos^{-1}\sqrt{\frac{I}{I_{\max}}} = 2\cos^{-1}\sqrt{0.640} = \boxed{1.29\text{ rad}}.$$

(b) $\delta = \frac{\lambda\phi}{2\pi} = \frac{(486.1\text{ nm})(1.29\text{ rad})}{2\pi} = \boxed{99.6\text{ nm}}$

P27.13 $I = I_{\max}\cos^2\left(\frac{\pi y d}{\lambda L}\right)$

$$\frac{I}{I_{\max}} = \cos^2\left[\frac{\pi(6.00\times10^{-3}\text{ m})(1.80\times10^{-4}\text{ m})}{(656.3\times10^{-9}\text{ m})(0.800\text{ m})}\right] = \boxed{0.968}$$

Section 27.4 Change of Phase Due to Reflection

Section 27.5 Interference in Thin Films

P27.14 Light reflecting from the first surface suffers phase reversal. Light reflecting from the second surface does not, but passes twice through the thickness t of the film. So, for constructive interference, we require

$$\frac{\lambda_n}{2} + 2t = \lambda_n$$

where $\lambda_n = \frac{\lambda}{n}$ is the wavelength in the material.

Then $2t = \frac{\lambda_n}{2} = \frac{\lambda}{2n}$

$$\lambda = 4nt = 4(1.33)(115\text{ nm}) = \boxed{612\text{ nm}}.$$

P27.15 (a) The light reflected from the top of the oil film undergoes phase reversal. Since $1.45 > 1.33$, the light reflected from the bottom undergoes no reversal. For constructive interference of reflected light, we then have

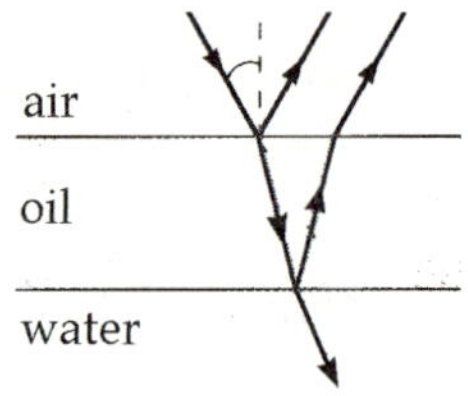

FIG. P27.15

$$2nt = \left(m + \frac{1}{2}\right)\lambda$$

or $$\lambda_m = \frac{2nt}{m + (1/2)} = \frac{2(1.45)(280\text{ nm})}{m + (1/2)}.$$

Substituting for m gives: $m = 0$, $\lambda_0 = 1\,620$ nm (infrared)

$m = 1$, $\lambda_1 = 541$ nm (green)

$m = 2$, $\lambda_2 = 325$ nm (ultraviolet).

Both infrared and ultraviolet light are invisible to the human eye, so the dominant color in reflected light is **green**.

(b) The dominant wavelengths in the transmitted light are those that produce destructive interference in the reflected light. The condition for destructive interference upon reflection is

$$2nt = m\lambda$$

or $$\lambda_m = \frac{2nt}{m} = \frac{812\text{ nm}}{m}.$$

Substituting for m gives: $m = 1$, $\lambda_1 = 812$ nm (near infrared)

$m = 2$, $\lambda_2 = 406$ nm (violet)

$m = 3$, $\lambda_3 = 271$ nm (ultraviolet).

Of these, the only wavelength visible to the human eye (and hence the dominate wavelength observed in the transmitted light) is 406 nm. Thus, the dominant color in the transmitted light is **violet**.

P27.16 Treating the anti-reflectance coating like a camera-lens coating,

$$2t = \left(m + \frac{1}{2}\right)\frac{\lambda}{n}.$$

Let $m = 0$: $$t = \frac{\lambda}{4n} = \frac{3.00\text{ cm}}{4(1.50)} = \boxed{0.500\text{ cm}}.$$

This anti-reflectance coating could be easily countered by changing the wavelength of the radar—to 1.50 cm—now creating maximum reflection!

P27.17 $$2nt = \left(m + \frac{1}{2}\right)\lambda \quad \text{so} \quad t = \left(m + \frac{1}{2}\right)\frac{\lambda}{2n}$$

Minimum $$t = \left(\frac{1}{2}\right)\frac{(500\text{ nm})}{2(1.30)} = \boxed{96.2\text{ nm}}.$$

***P27.18** (a) For maximum transmission, we want destructive interference in the light reflected from the front and back surfaces of the film.

If the surrounding glass has refractive index greater than 1.378, light reflected from the front surface suffers no phase reversal and light reflected from the back does undergo phase reversal. This effect by itself would produce destructive interference, so we want the distance down and back to be one whole wavelength in the film: $2t = \frac{\lambda}{n}$.

$$t = \frac{\lambda}{2n} = \frac{656.3\text{ nm}}{2(1.378)} = \boxed{238\text{ nm}}$$

(b) The filter will expand. As t increases in $2nt = \lambda$, so does $\boxed{\lambda \text{ increase}}$.

(c) Destructive interference for reflected light happens also for λ in $2nt = 2\lambda$,

or $\lambda = 1.378(238\text{ nm}) = \boxed{328\text{ nm}}$ (near ultraviolet).

P27.19 For destructive interference in the air,

$2t = m\lambda$.

For 30 dark fringes, including the one where the plates meet,

$$t = \frac{29(600\text{ nm})}{2} = 8.70 \times 10^{-6}\text{ m}.$$

Therefore, the *radius* of the wire is

$$r = \frac{t}{2} = \frac{8.70\ \mu\text{m}}{2} = \boxed{4.35\ \mu\text{m}}.$$

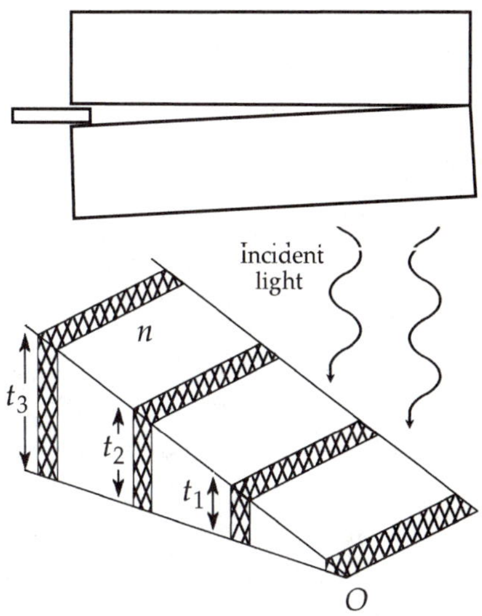

FIG. P27.19

Section 27.6 Diffraction Patterns

P27.20 $\sin\theta = \frac{\lambda}{a} = \frac{6.328 \times 10^{-7}}{3.00 \times 10^{-4}} = 2.11 \times 10^{-3}$

$$\frac{y}{1.00\text{ m}} = \tan\theta \approx \sin\theta = \theta \text{ (for small } \theta)$$

$$2y = \boxed{4.22\text{ mm}}$$

P27.21 $\frac{y}{L} \approx \sin\theta = \frac{m\lambda}{a}$ $\Delta y = 3.00 \times 10^{-3}$ nm

$\Delta m = 3 - 1 = 2$ and $a = \frac{\Delta m \lambda L}{\Delta y}$

$$a = \frac{2\left(690 \times 10^{-9} \text{ m}\right)(0.500 \text{ m})}{\left(3.00 \times 10^{-3} \text{ m}\right)} = \boxed{2.30 \times 10^{-4} \text{ m}}$$

P27.22 The positions of the first-order minima are $\frac{y}{L} \approx \sin\theta = \pm\frac{\lambda}{a}$. Thus, the spacing between these two minima is $\Delta y = 2\left(\frac{\lambda}{a}\right)L$ and the wavelength is

$$\lambda = \left(\frac{\Delta y}{2}\right)\left(\frac{a}{L}\right) = \left(\frac{4.10 \times 10^{-3} \text{ m}}{2}\right)\left(\frac{0.550 \times 10^{-3} \text{ m}}{2.06 \text{ m}}\right) = \boxed{547 \text{ nm}}.$$

P27.23 For destructive interference,

$$\sin\theta = m\frac{\lambda}{a} = \frac{\lambda}{a} = \frac{5.00 \text{ cm}}{36.0 \text{ cm}} = 0.139$$

and $\theta = 7.98°$

$$\frac{d}{L} = \tan\theta$$

gives $d = L\tan\theta = (6.50 \text{ m})\tan 7.98° = 0.912 \text{ m}$

$$d = \boxed{91.2 \text{ cm}}.$$

P27.24 If the speed of sound is 340 m/s, $\lambda = \frac{v}{f} = \frac{340 \text{ m/s}}{650 \text{ s}^{-1}} = 0.523 \text{ m}$.

Diffraction minima occur at angles described by $a\sin\theta = m\lambda$

$(1.10 \text{ m})\sin\theta_1 = 1(0.523 \text{ m})$ $\theta_1 = 28.4°$

$(1.10 \text{ m})\sin\theta_2 = 2(0.523 \text{ m})$ $\theta_2 = 72.0°$

$(1.10 \text{ m})\sin\theta_3 = 3(0.523 \text{ m})$ θ_3 nonexistent

Maxima appear straight ahead at $\boxed{0°}$ and left and right at an angle given approximately by

$(1.10 \text{ m})\sin\theta_x = 1.5(0.523 \text{ m})$ $\theta_x \boxed{\approx 46°}$.

There is no solution to $a\sin\theta = 2.5\lambda$, so our answer is already complete, with $\boxed{\text{three}}$ sound maxima.

***P27.25** The rectangular patch on the wall is wider than it is tall. The aperture will be taller than it is wide. For horizontal spreading we have

$$\tan\theta_{\text{width}} = \frac{y_{\text{width}}}{L} = \frac{0.110 \text{ m}/2}{4.5 \text{ m}} = 0.012\,2$$

$$a_{\text{width}} \sin\theta_{\text{width}} = 1\lambda$$

$$a_{\text{width}} = \frac{632.8\times10^{-9} \text{ m}}{0.012\,2} = \boxed{5.18\times10^{-5} \text{ m}}$$

For vertical spreading, similarly

$$\tan\theta_{\text{height}} = \frac{0.006 \text{ m}/2}{4.5 \text{ m}} = 0.000\,667$$

$$a_{\text{height}} = \frac{1\lambda}{\sin\theta_h} = \frac{632.8\times10^{-9} \text{ m}}{0.000\,667} = \boxed{9.49\times10^{-4} \text{ m}}$$

Section 27.7 Resolution of Single-Slit and Circular Apertures

P27.26 $\sin\theta = \frac{\lambda}{a} = \frac{5.00\times10^{-7} \text{ m}}{5.00\times10^{-4}} = \boxed{1.00\times10^{-3} \text{ rad}}$

P27.27 Undergoing diffraction from a circular opening, the beam spreads into a cone of half-angle

$$\theta_{\text{min}} = 1.22\frac{\lambda}{D} = 1.22\left(\frac{632.8\times10^{-9} \text{ m}}{0.005\,00 \text{ m}}\right) = 1.54\times10^{-4} \text{ rad}.$$

The radius of the beam ten kilometers away is, from the definition of radian measure,

$$r_{\text{beam}} = \theta_{\text{min}}\left(1.00\times10^{4} \text{ m}\right) = 1.544 \text{ m}$$

and its diameter is $d_{\text{beam}} = 2r_{\text{beam}} = \boxed{3.09 \text{ m}}$.

***P27.28** By Rayleigh's criterion: $\theta_{\text{min}} = \frac{d}{L} = 1.22\frac{\lambda}{D}$, where θ_{min} is the smallest angular separation of two objects for which they are resolved by an aperture of diameter D, d is the separation of the two objects, and L is the maximum distance of the aperture from the two objects at which they can be resolved.

Two objects can be resolved if their angular separation is greater than θ_{min}. Thus, θ_{min} should be as small as possible. Therefore, the smaller of the two given wavelengths is easier to resolve, i.e. $\boxed{\text{violet}}$.

$$L = \frac{Dd}{1.22\lambda} = \frac{\left(5.20\times10^{-3} \text{ m}\right)\left(2.80\times10^{-2} \text{ m}\right)}{1.22\lambda} = \frac{1.193\times10^{-4} \text{ m}^2}{\lambda}$$

Thus $L = 186$ m for $\lambda = 640$ nm, and $L = 271$ m for $\lambda = 440$ nm. The viewer can resolve adjacent tubes of violet in the range $\boxed{186 \text{ m to } 271 \text{ m}}$, but cannot resolve adjacent tubes of red in this range.

P27.29 By Rayleigh's criterion, two dots separated center-to-center by 2.00 mm would overlap

when $\theta_{\min} = \dfrac{d}{L} = 1.22\dfrac{\lambda}{D}$.

Thus, $L = \dfrac{dD}{1.22\lambda} = \dfrac{\left(2.00\times10^{-3}\text{ m}\right)\left(4.00\times10^{-3}\text{ m}\right)}{1.22\left(500\times10^{-9}\text{ m}\right)} = \boxed{13.1\text{ m}}$.

P27.30 The concave mirror of the spy satellite is probably about 2 m in diameter, and is surely not more than 5 m in diameter. That is the size of the largest piece of glass successfully cast to a precise shape, for the mirror of the Hale telescope on Mount Palomar. If the spy satellite had a larger mirror, its manufacture could not be kept secret, and it would be visible from the ground. Outer space is probably closer than your state capitol, but the satellite is surely above 200-km altitude, for reasonably low air friction. We find the distance between barely resolvable objects at a distance of 200 km, seen in yellow light through a 5-m aperture:

$$\frac{y}{L} = 1.22\frac{\lambda}{D}$$

$$y = (200\,000\text{ m})(1.22)\left(\frac{6\times10^{-7}\text{ m}}{5\text{ m}}\right) = 3\text{ cm}$$

(Considering atmospheric seeing caused by variations in air density and temperature, the distance between barely resolvable objects is more like $(200\,000\text{ m})(1\text{ s})\left(\dfrac{1^\circ}{3\,600\text{ s}}\right)\left(\dfrac{\pi\text{ rad}}{180^\circ}\right) = 97\text{ cm}$.) Thus the snooping spy satellite cannot see the difference between III and II or IV on a license plate. It cannot count coins spilled on a sidewalk, much less read the date on them.

P27.31 $1.22\dfrac{\lambda}{D} = \dfrac{d}{L}$ $\qquad \lambda = \dfrac{c}{f} = 0.020\,0\text{ m}$

$D = 2.10\text{ m}$ $\qquad L = 9\,000\text{ m}$

$$d = 1.22\frac{(0.020\,0\text{ m})(9\,000\text{ m})}{2.10\text{ m}} = \boxed{105\text{ m}}$$

Section 27.8 The Diffraction Grating

P27.32 The principal maxima are defined by

$d\sin\theta = m\lambda$ $\qquad m = 0, 1, 2, \ldots.$

For $m = 1$, $\qquad \lambda = d\sin\theta$

where θ is the angle between the central ($m = 0$) and the first order ($m = 1$) maxima. The value of θ can be determined from the information given about the distance between maxima and the grating-to-screen distance. From the figure,

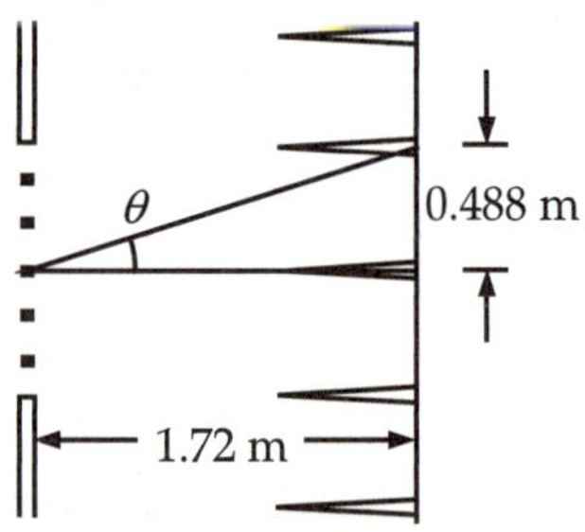

FIG. P27.32

$$\tan\theta = \frac{0.488\text{ m}}{1.72\text{ m}} = 0.284$$

continued on next page

so $\theta = 15.8°$

and $\sin\theta = 0.273$.

The distance between grating "slits" equals the reciprocal of the number of grating lines per centimeter

$$d = \frac{1}{5\,310\text{ cm}^{-1}} = 1.88\times10^{-4}\text{ cm} = 1.88\times10^{3}\text{ nm}.$$

The wavelength is $\lambda = d\sin\theta = (1.88\times10^{3}\text{ nm})(0.273) = \boxed{514\text{ nm}}$.

P27.33 The grating spacing is $d = \frac{1.00\times10^{-2}\text{ m}}{4\,500} = 2.22\times10^{-6}\text{ m}$.

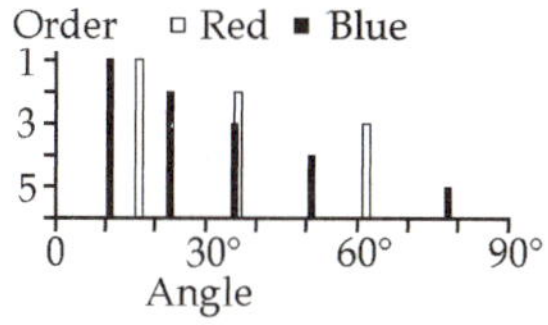

FIG. P27.33

In the 1st-order spectrum, diffraction angles are given by

$\sin\theta = \frac{\lambda}{d}$: $\sin\theta_1 = \frac{656\times10^{-9}\text{ m}}{2.22\times10^{-6}\text{ m}} = 0.295$

so that for red $\theta_1 = 17.17°$

and for blue $\sin\theta_2 = \frac{434\times10^{-9}\text{ m}}{2.22\times10^{-6}\text{ m}} = 0.195$

so that $\theta_2 = 11.26°$.

The angular separation is in first-order, $\Delta\theta = 17.17° - 11.26° = \boxed{5.91°}$.

In the second-order spectrum, $\Delta\theta = \sin^{-1}\left(\frac{2\lambda_1}{d}\right) - \sin^{-1}\left(\frac{2\lambda_2}{d}\right) = \boxed{13.2°}$.

Again, in the third order, $\Delta\theta = \sin^{-1}\left(\frac{3\lambda_1}{d}\right) - \sin^{-1}\left(\frac{3\lambda_2}{d}\right) = \boxed{26.5°}$.

Since the red does not appear in the fourth-order spectrum, the answer is complete.

P27.34 $\sin\theta = 0.350$: $d = \frac{\lambda}{\sin\theta} = \frac{632.8\text{ nm}}{0.350} = 1.81\times10^{3}\text{ nm}$

Line spacing $= \boxed{1.81\ \mu\text{m}}$

P27.35 $d = \frac{1.00\times10^{-3}\text{ m/mm}}{250\text{ lines/mm}} = 4.00\times10^{-6}\text{ m} = 4\,000\text{ nm}$ $\quad d\sin\theta = m\lambda \Rightarrow m = \frac{d\sin\theta}{\lambda}$

(a) The number of times a complete order is seen is the same as the number of orders in which the long wavelength limit is visible.

$$m_{\max} = \frac{d\sin\theta_{\max}}{\lambda} = \frac{(4\,000\text{ nm})\sin 90.0°}{700\text{ nm}} = 5.71$$

or $\boxed{\text{5 orders is the maximum}}$.

continued on next page

(b) The highest order in which the violet end of the spectrum can be seen is:

$$m_{\max} = \frac{d\sin\theta_{\max}}{\lambda} = \frac{(4\,000\text{ nm})\sin 90.0°}{400\text{ nm}} = 10.0$$

or $\boxed{\text{10 orders in the short-wavelength region}}$.

***P27.36** $\sin\theta = \frac{m\lambda}{d}$

Therefore, taking the ends of the visible spectrum to be $\lambda_v = 400$ nm and $\lambda_r = 750$ nm, the ends the different order spectra are:

End of second order: $\sin\theta_{2r} = \frac{2\lambda_r}{d} = \frac{1\,500\text{ nm}}{d}$.

Start of third order: $\sin\theta_{3v} = \frac{3\lambda_v}{d} = \frac{1\,200\text{ nm}}{d}$.

Thus, it is seen that $\boxed{\theta_{2r} > \theta_{3v} \text{ and these orders must overlap}}$ regardless of the value of the grating spacing *d*.

***P27.37** The sound has wavelength $\lambda = \frac{v}{f} = \frac{343\text{ m/s}}{37.2\times 10^3/\text{s}} = 9.22\times 10^{-3}$ m. Each diffracted beam is described by

$$d\sin\theta = m\lambda, \quad m = 0, 1, 2, \ldots$$

The zero-order beam is at $m = 0$, $\theta = 0$. The beams in the first order of interference are to the left and right at $\theta = \sin^{-1}\frac{1\lambda}{d} = \sin^{-1}\frac{9.22\times 10^{-3}\text{ m}}{1.3\times 10^{-2}\text{ m}} = \sin^{-1}0.709 = 45.2°$. For a second-order beam we would need $\theta = \sin^{-1}\frac{2\lambda}{d} = \sin^{-1}2(0.709) = \sin^{-1}1.42$. No angle, smaller or larger than 90° has a sine greater than 1. Then a diffracted beam does not exist for the second order or any higher order. The whole answer is then, $\boxed{\text{three beams, at } 0° \text{ and at } 45.2° \text{ to the right and to the left}}$.

Section 27.9 Diffraction of X-Rays by Crystals

P27.38 $2d\sin\theta = m\lambda$: $\lambda = \frac{2d\sin\theta}{m} = \frac{2(0.353\times 10^{-9}\text{ m})\sin 7.60°}{1} = 9.34\times 10^{-11}\text{ m} = \boxed{0.093\,4\text{ nm}}$

P27.39 $2d\sin\theta = m\lambda$: $\sin\theta = \frac{m\lambda}{2d} = \frac{1(0.140\times 10^{-9}\text{ m})}{2(0.281\times 10^{-9}\text{ m})} = 0.249$

and $\boxed{\theta = 14.4°}$

P27.40 Figure 27.28 of the text shows the situation.

$$2d\sin\theta = m\lambda \qquad \text{or} \qquad \lambda = \frac{2d\sin\theta}{m}$$

$m = 1$: $\lambda_1 = \dfrac{2(2.80\text{ m})\sin 80.0°}{1} = \boxed{5.51\text{ m}}$

$m = 2$: $\lambda_2 = \dfrac{2(2.80\text{ m})\sin 80.0°}{2} = \boxed{2.76\text{ m}}$

$m = 3$: $\lambda_3 = \dfrac{2(2.80\text{ m})\sin 80.0°}{3} = \boxed{1.84\text{ m}}$

Section 27.10 Context Connection—Holography

P27.41 (a) The several narrow parallel slits make a diffraction grating. The zeroth- and first- order maxima are separated according to

$$d\sin\theta = (1)\lambda \qquad \sin\theta = \frac{\lambda}{d} = \frac{632.8\times 10^{-9}\text{ m}}{1.2\times 10^{-3}\text{ m}}$$

$$\theta = \sin^{-1}(0.000\,527) = 0.000\,527\text{ rad}$$

$$y = L\tan\theta = (1.40\text{ m})(0.000\,527) = \boxed{0.738\text{ mm}}.$$

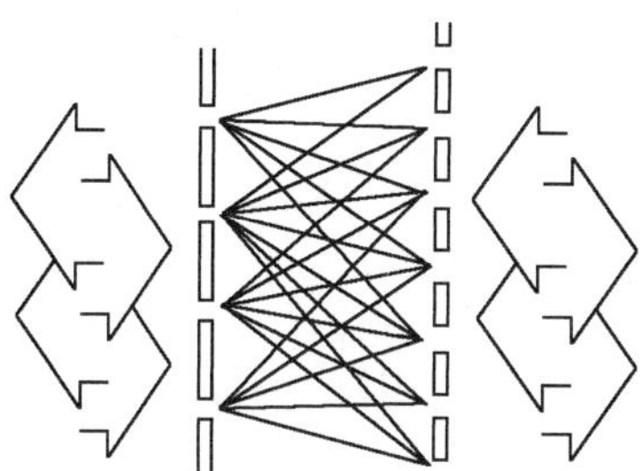

FIG. P27.41

(b) Many equally spaced transparent lines appear on the film. It is itself a diffraction grating. When the same light is sent through the film, it produces interference maxima separated according to

$$d\sin\theta = (1)\lambda \qquad \sin\theta = \frac{\lambda}{d} = \frac{632.8\times 10^{-9}\text{ m}}{0.738\times 10^{-3}\text{ m}} = 0.000\,857$$

$$y = L\tan\theta = (1.40\text{ m})(0.000\,857) = 1.20\text{ mm}$$

An image of the original set of slits appears on the screen. If the screen is removed, light diverges from the real images with the same wave fronts reconstructed as the original slits produced. Reasoning from the mathematics of Fourier transforms, Gabor showed that light diverging from any object, not just a set of slits, could be used. In the picture, the slits or maxima on the left are separated by 1.20 mm. The slits or maxima on the right are separated by 0.738 mm. The length difference between any pair of lines is an integer number of wavelengths. Light can be sent through equally well toward the right or toward the left.

P27.42 (a) The light in the cavity is incident perpendicularly on the mirrors, although the diagram shows a large angle of incidence for clarity. We ignore the variation of the index of refraction with wavelength. To minimize reflection at a vacuum wavelength of 632.8 nm, the net phase difference between rays (1) and (2) should be 180°. There is automatically a 180° shift in one of the two rays upon reflection, so the extra distance traveled by ray (2) should be one whole wavelength:

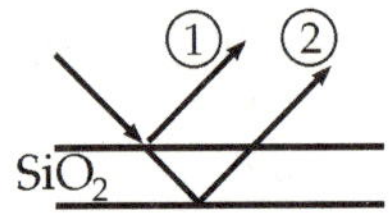

FIG. P27.42

$$2t = \frac{\lambda}{n}$$

$$t = \frac{\lambda}{2n} = \frac{632.8 \text{ nm}}{2(1.458)} = \boxed{217 \text{ nm}}$$

(b) The total phase difference should be 360°, including contributions of 180° by reflection and 180° by extra distance traveled:

$$2t = \frac{\lambda}{2n}$$

$$t = \frac{\lambda}{4n} = \frac{543 \text{ nm}}{4(1.458)} = \boxed{93.1 \text{ nm}}$$

Additional Problems

***P27.43** Along the line of length d joining the source, two identical waves moving in opposite directions add to give a standing wave. An antinode is halfway between the sources. If $\frac{d}{2} > \frac{\lambda}{2}$, there is space for two more antinodes for a total of three. If $\frac{d}{2} > \lambda$, there will be at least five antinodes, and so on. To repeat, if $\frac{d}{\lambda} > 0$, the number of antinodes is 1 or more. If $\frac{d}{\lambda} > 1$, the number of antinodes is 3 or more. If $\frac{d}{\lambda} > 2$, the number of antinodes is 5 or more. In general,

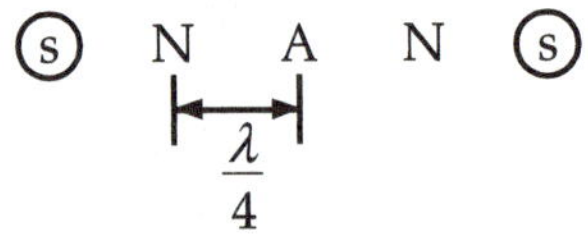

FIG. P27.43

$$\boxed{\text{The number of antinodes is 1 plus 2 times the greatest integer less than or equal to } \frac{d}{\lambda}.}$$

If $\frac{d}{2} < \frac{\lambda}{4}$, there will be no nodes. If $\frac{d}{2} > \frac{\lambda}{4}$, there will be space for at least two nodes, as shown in the picture. If $\frac{d}{2} > \frac{3\lambda}{4}$, there will be at least four nodes. If $\frac{d}{2} > \frac{5\lambda}{4}$ six or more nodes will fit in, and so on. To repeat, if $2d < \lambda$ the number of nodes is 0. If $2d > \lambda$ the number of nodes is 2 or more. If $2d > 3\lambda$ the number of nodes is 4 or more. If $2d > 5\lambda$ the number of nodes is 6 or more. Again, if $\left(\frac{d}{\lambda} + \frac{1}{2}\right) > 1$, the number of nodes is at least 2. If $\left(\frac{d}{\lambda} + \frac{1}{2}\right) > 2$, the number of nodes is at least 4. If $\left(\frac{d}{\lambda} + \frac{1}{2}\right) > 3$, the number of nodes is at least 6. In general,

continued on next page

the number of nodes is 2 times the greatest nonzero integer less than $\left(\frac{d}{\lambda}+\frac{1}{2}\right)$.

Next, we enumerate the zones of constructive interference. They are described by $d\sin\theta = m\lambda$, $m = 0, 1, 2, \ldots$ with θ counted as positive both left and right of the maximum at $\theta = 0$ in the center. The number of side maxima on each side is the greatest integer satisfying $\sin\theta \le 1$, $d1 \ge m\lambda$, $m \le \frac{d}{\lambda}$. So the total

number of bright fringes is one plus 2 times the greatest integer less than or equal to $\frac{d}{\lambda}$.

It is equal to the number of antinodes on the line joining the sources.

The interference minima are to the left and right at angles described by $d\sin\theta = \left(m+\frac{1}{2}\right)\lambda$, $m = 0, 1, 2, \ldots$. With $\sin\theta < 1$, $d1 > \left(m_{\max}+\frac{1}{2}\right)\lambda$, $m_{\max} < \frac{d}{\lambda}-\frac{1}{2}$ or $m_{\max}+1 < \frac{d}{\lambda}+\frac{1}{2}$. Let $n = 1, 2, 3, \ldots$. Then the number of side minima is the greatest integer n less than $\frac{d}{\lambda}+\frac{1}{2}$. Counting both left and right, the number of dark fringes is two times the greatest positive integer less than $\left(\frac{d}{\lambda}+\frac{1}{2}\right)$. It is equal to the number of nodes in the standing wave between the sources.

P27.44 My middle finger has width $d = 2$ cm.

(a) Two adjacent directions of constructive interference for 600-nm light are described by

$$d\sin\theta = m\lambda$$
$$\theta_0 = 0$$
$$\left(2\times10^{-2}\text{ m}\right)\sin\theta_1 = 1\left(6\times10^{-7}\text{ m}\right)$$

Thus, $\theta_1 = 2\times10^{-3}$ degree

and $\theta_1 - \theta_0 \boxed{\sim 10^{-3}\text{ degree}}$.

(b) Choose $\theta_1 = 20°$

$$\left(2\times10^{-2}\text{ m}\right)\sin 20° = (1)\lambda$$
$$\lambda = 7\text{ mm}$$

Millimeter waves are microwaves.

$f = \frac{c}{\lambda}$: $\quad f = \frac{3\times10^{8}\text{ m/s}}{7\times10^{-3}\text{ m}} \boxed{\sim 10^{11}\text{ Hz}}$

P27.45 No phase shift occurs upon reflection from the upper surface (glass to air) of the film, but there will be a shift of $\frac{\lambda}{2}$ due to the reflection at the lower surface of the film (air to metal). The total phase difference in the two reflected beams is

then $$\delta = 2nt + \frac{\lambda}{2}.$$

For constructive interference, $\delta = m\lambda$

or $$2(1.00)t + \frac{\lambda}{2} = m\lambda.$$

Thus, the film thickness for the mth order bright fringe is

$$t_m = \left(m - \frac{1}{2}\right)\frac{\lambda}{2} = m\left(\frac{\lambda}{2}\right) - \frac{\lambda}{4}$$

and the thickness for the $m-1$ bright fringe is:

$$t_{m-1} = (m-1)\left(\frac{\lambda}{2}\right) - \frac{\lambda}{4}.$$

Therefore, the change in thickness required to go from one bright fringe to the next is

$$\Delta t = t_m - t_{m-1} = \frac{\lambda}{2}.$$

To go through 200 bright fringes, the change in thickness of the air film must be:

$$200\left(\frac{\lambda}{2}\right) = 100\lambda.$$

Thus, the increase in the length of the rod is

$$\Delta L = 100\lambda = 100\left(5.00 \times 10^{-7}\text{ m}\right) = 5.00 \times 10^{-5}\text{ m}.$$

From $$\Delta L = L_i \alpha \Delta T$$

we have: $$\alpha = \frac{\Delta L}{L_i \Delta T} = \frac{5.00 \times 10^{-5}\text{ m}}{(0.100\text{ m})(25.0°\text{C})} = \boxed{20.0 \times 10^{-6}\ °\text{C}^{-1}}.$$

***P27.46** The central bright fringe is wider than the side bright fringes, so the light must have been diffracted by a $\boxed{\text{single slit}}$. For precision, we measure to the third minimum from the center

$$y = 4.0\text{ cm}$$

$$\tan\theta = \frac{y}{L} = \frac{0.04\text{ m}}{2.6\text{ m}} = 0.015\ 4$$

$$\theta = 0.881° = 0.015\ 4\text{ rad}$$

$$a\sin\theta = m\lambda$$

$$a = \frac{m\lambda}{\sin\theta} = \frac{3\left(632.8 \times 10^{-9}\text{ m}\right)}{\sin 0.881°} = \frac{3\left(632.8 \times 10^{-9}\text{ m}\right)}{0.015\ 4} = \boxed{1.23 \times 10^{-4}\text{ m}}$$

P27.47 For destructive interference, the path length must differ by $m\lambda$. We may treat this problem as a double slit experiment if we remember the light undergoes a $\frac{\pi}{2}$-phase shift at the mirror. The second slit is the mirror image of the source, 1.00 cm below the mirror plane. Modifying the equation for Young's experiment, $\frac{dy}{L} = m\lambda$ we have

$$y_{\text{dark}} = \frac{m\lambda L}{d} = \frac{1(5.00\times10^{-7}\text{ m})(100\text{ m})}{(2.00\times10^{-2}\text{ m})} = \boxed{2.50\text{ mm}}.$$

P27.48 $2\sqrt{(15.0\text{ km})^2 + h^2} = 30.175\text{ km}$

$(15.0\text{ km})^2 + h^2 = 227.63$

$h = \boxed{1.62\text{ km}}$

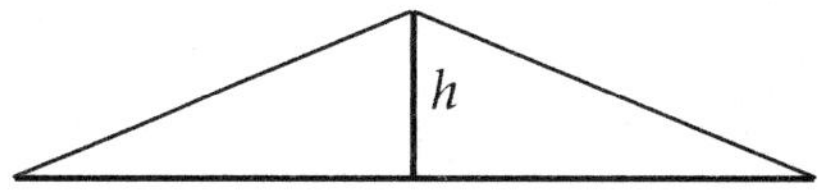

FIG. P27.48

P27.49 For dark fringes, $2nt = m\lambda$

and at the edge of the wedge, $t = \frac{84(500\text{ nm})}{2}$.

When submerged in water, $2nt = m\lambda$

$$m = \frac{2(1.33)(42)(500\text{ nm})}{500\text{ nm}}$$

so $m + 1 = \boxed{113\text{ dark fringes}}$.

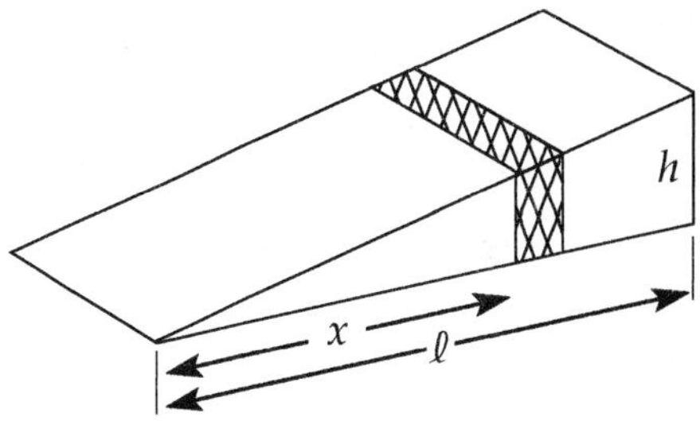

FIG. P27.49

P27.50 (a) Minimum: $2nt = m\lambda_2$ for $m = 0, 1, 2, \ldots$

Maximum: $2nt = \left(m' + \frac{1}{2}\right)\lambda_1$ for $m' = 0, 1, 2, \ldots$

for $\lambda_1 > \lambda_2$, $\left(m' + \frac{1}{2}\right) < m$

so $m' = m - 1$.

Then $2nt = m\lambda_2 = \left(m - \frac{1}{2}\right)\lambda_1$

$2m\lambda_2 = 2m\lambda_1 - \lambda_1$

so $\boxed{m = \frac{\lambda_1}{2(\lambda_1 - \lambda_2)}}$.

continued on next page

(b) $$m = \frac{500}{2(500 - 370)} = 1.92 \rightarrow 2 \text{ (wavelengths measured to } \pm 5\text{ nm)}$$

Minimum: $2nt = m\lambda_2$

$2(1.40)t = 2(370\text{ nm})$ $\qquad t = 264\text{ nm}$

Maximum: $2nt = \left(m - 1 + \frac{1}{2}\right)\lambda = 1.5\lambda$

$2(1.40)t = 1.5(500\text{ nm})$ $\qquad t = 268\text{ nm}$

Film thickness $= \boxed{266\text{ nm}}$.

P27.51 The shift between the two reflected waves is $\delta = 2na - b - \frac{\lambda}{2}$ where a and b are as shown in the ray diagram, n is the index of refraction, and the term $\frac{\lambda}{2}$ is due to phase reversal at the top surface. For constructive interference, $\delta = m\lambda$ where m has integer values. This condition becomes

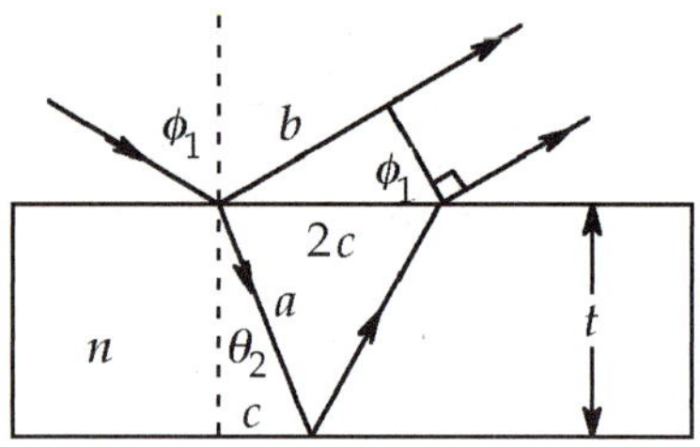

FIG. P27.51

$$2na - b = \left(m + \frac{1}{2}\right)\lambda \qquad (1)$$

From the figure's geometry, $a = \dfrac{t}{\cos\theta_2}$

$$c = a\sin\theta_2 = \frac{t\sin\theta_2}{\cos\theta_2}$$

$$b = 2c\sin\phi_1 = \frac{2t\sin\theta_2}{\cos\theta_2}\sin\phi_1$$

Also, from Snell's law, $\sin\phi_1 = n\sin\theta_2$.

Thus, $b = \dfrac{2nt\sin^2\theta_2}{\cos\theta_2}$.

With these results, the condition for constructive interference given in Equation (1) becomes:

$$2n\left(\frac{t}{\cos\theta_2}\right) - \frac{2nt\sin^2\theta_2}{\cos\theta_2} = \frac{2nt}{\cos\theta_2}\left(1 - \sin^2\theta_2\right) = \left(m + \frac{1}{2}\right)\lambda$$

or $$\boxed{2nt\cos\theta_2 = \left(m + \frac{1}{2}\right)\lambda}.$$

P27.52 (a) Bright bands are observed when $2nt = \left(m + \frac{1}{2}\right)\lambda$.

Hence, the first bright band ($m = 0$) corresponds to $nt = \frac{\lambda}{4}$.

Since $\frac{x_1}{x_2} = \frac{t_1}{t_2}$

we have $x_2 = x_1\left(\frac{t_2}{t_1}\right) = x_1\left(\frac{\lambda_2}{\lambda_1}\right) = (3.00\text{ cm})\left(\frac{680\text{ nm}}{420\text{ nm}}\right) = \boxed{4.86\text{ cm}}$.

(b) $t_1 = \frac{\lambda_1}{4n} = \frac{420\text{ nm}}{4(1.33)} = \boxed{78.9\text{ nm}}$

$t_2 = \frac{\lambda_2}{4n} = \frac{680\text{ nm}}{4(1.33)} = \boxed{128\text{ nm}}$

(c) $\theta \approx \tan\theta = \frac{t_1}{x_1} = \frac{78.9\text{ nm}}{3.00\text{ cm}} = \boxed{2.63 \times 10^{-6}\text{ rad}}$

P27.53 The first minimum is at $a\sin\theta = (1)\lambda$.

This has no solution if $\frac{\lambda}{a} > 1$.

or if $a < \lambda = \boxed{632.8\text{ nm}}$.

P27.54 Consider vocal sound moving at 340 m/s and of frequency 3 000 Hz. Its wavelength is

$$\lambda = \frac{v}{f} = \frac{340\text{ m/s}}{3\,000\text{ Hz}} = 0.113\text{ m}.$$

If your mouth, for horizontal dispersion, behaves similarly to a slit 6.00 cm wide, then $a\sin\theta = m\lambda$ predicts no diffraction minima. You are a nearly isotropic source of this sound. It spreads out from you nearly equally in all directions. On the other hand, if you use a megaphone with width 60.0 cm at its wide end, then $a\sin\theta = m\lambda$ predicts the first diffraction minimum at

$$\theta = \sin^{-1}\left(\frac{m\lambda}{a}\right) = \sin^{-1}\left(\frac{0.113\text{ m}}{0.600\text{ m}}\right) = 10.9°.$$

This suggests that the sound is radiated mostly toward the front into a diverging beam of angular diameter only about 20°. With less sound energy wasted in other directions, more is available for your intended auditors. We could check that a distant observer to the side or behind you receives less sound when a megaphone is used.

P27.55 $d = \dfrac{1}{400\ \text{mm}^{-1}} = 2.50\times10^{-6}\ \text{m}$

(a) $d\sin\theta = m\lambda$ $\qquad \theta_a = \sin^{-1}\left(\dfrac{2\times541\times10^{-9}\ \text{m}}{2.50\times10^{-6}\ \text{m}}\right) = \boxed{25.6°}$

(b) $\lambda = \dfrac{541\times10^{-9}\ \text{m}}{1.33} = 4.07\times10^{-7}\ \text{m}$ $\qquad \theta_b = \sin^{-1}\left(\dfrac{2\times4.07\times10^{-7}\ \text{m}}{2.50\times10^{-6}\ \text{m}}\right) = \boxed{19.0°}$

(c) $d\sin\theta_a = 2\lambda$ $\qquad d\sin\theta_b = \dfrac{2\lambda}{n}$

$$n\sin\theta_b = (1)\sin\theta_a$$

P27.56 (a) $\lambda = \dfrac{v}{f}$: $\qquad \lambda = \dfrac{3.00\times10^{8}\ \text{m/s}}{1.40\times10^{9}\ \text{s}^{-1}} = 0.214\ \text{m}$

$\theta_{\min} = 1.22\dfrac{\lambda}{D}$: $\qquad \theta_{\min} = 1.22\left(\dfrac{0.214\ \text{m}}{3.60\times10^{4}\ \text{m}}\right) = \boxed{7.26\ \mu\text{rad}}$

$$\theta_{\min} = 7.26\ \mu\text{rad}\left(\frac{180\times60\times60\ \text{s}}{\pi}\right) = \boxed{1.50\ \text{arc seconds}}$$

(b) $\theta_{\min} = \dfrac{d}{L}$: $\qquad d = \theta_{\min}L = \left(7.26\times10^{-6}\ \text{rad}\right)(26\,000\ \text{ly}) = \boxed{0.189\ \text{ly}}$

(c) $\theta_{\min} = 1.22\dfrac{\lambda}{D}$ $\qquad \theta_{\min} = 1.22\left(\dfrac{500\times10^{-9}\ \text{m}}{12.0\times10^{-3}\ \text{m}}\right) = \boxed{50.8\ \mu\text{rad}}$ (10.5 seconds of arc)

(d) $d = \theta_{\min}L = \left(50.8\times10^{-6}\ \text{rad}\right)(30.0\ \text{m}) = 1.52\times10^{-3}\ \text{m} = \boxed{1.52\ \text{mm}}$

P27.57 (a) $d\sin\theta = m\lambda$

or $d = \dfrac{m\lambda}{\sin\theta} = \dfrac{3\left(500\times10^{-9}\ \text{m}\right)}{\sin 32.0°} = 2.83\ \mu\text{m}$

Therefore, lines per unit length $= \dfrac{1}{d} = \dfrac{1}{2.83\times10^{-6}\ \text{m}}$

or lines per unit length $= 3.53\times10^{5}\ \text{m}^{-1} = \boxed{3.53\times10^{3}\ \text{cm}^{-1}}$.

continued on next page

(b) $\sin\theta = \frac{m\lambda}{d} = \frac{m(500\times10^{-9}\text{ m})}{2.83\times10^{-6}\text{ m}} = m(0.177)$

For $\sin\theta \le 1.00$, we must have $\quad m(0.177) \le 1.00$

or $\quad m \le 5.66$.

Therefore, the highest order observed is $\quad m = 5$.

Total number of primary maxima observed is $\quad 2m+1 = \boxed{11}$.

***P27.58** (a) Bragg's law applies to the space lattice of melanin rods. Consider the planes $d = 0.25\ \mu\text{m}$ apart. For light at near-normal incidence, strong reflection happens for the wavelength given by $2d\sin\theta = m\lambda$. The longest wavelength reflected strongly corresponds to $m = 1$:

$2(0.25\times10^{-6}\text{ m})\sin 90° = 1\lambda \qquad \lambda = 500\text{ nm}$. This is the blue-green color.

(b) For light incident at grazing angle 60°, $2d\sin\theta = m\lambda$ gives $1\lambda = 2(0.25\times10^{-6}\text{ m})\sin 60° = 433\text{ nm}$. This is violet.

(c) Your two eyes receive light reflected from the feather at different angles, so they receive light incident at different angles and containing different colors reinforced by constructive interference.

(d) The longest wavelength that can be reflected with extra strength by these melanin rods is the one we computed first, 500 nm blue-green.

(e) If the melanin rods were farther apart (say $0.32\ \mu\text{m}$) they could reflect red with constructive interference.

***P27.59** A central maximum and side maxima in seven orders of interference appear. If the seventh order is just at 90°,

$d\sin\theta = m\lambda \qquad d1 = 7(654\times10^{-9}\text{ m}) \qquad d = 4.58\ \mu\text{m}$.

If the seventh order is at less than 90°, the eighth order might be nearly ready to appear according to

$d1 = 8(654\times10^{-9}\text{ m}) \qquad d = 5.23\ \mu\text{m}$.

Thus $\boxed{4.58\ \mu\text{m} < d < 5.23\ \mu\text{m}}$.

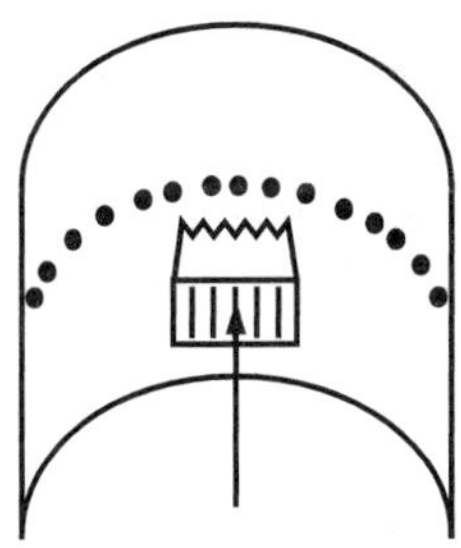

FIG. P27.59

P27.60 (a) We require $\quad \theta_{\min} = 1.22\frac{\lambda}{D} = \frac{\text{radius of diffraction disk}}{L} = \frac{D}{2L}$.

Then $\quad \boxed{D^2 = 2.44\lambda L}$.

(b) $D = \sqrt{2.44(500\times10^{-9}\text{ m})(0.150\text{ m})} = \boxed{428\ \mu\text{m}}$

P27.61 $d\sin\theta = m\lambda$

and, differentiating, $d(\cos\theta)d\theta = md\lambda$

or $d\sqrt{1-\sin^2\theta}\Delta\theta \approx m\Delta\lambda$

$$d\sqrt{1-\frac{m^2\lambda^2}{d^2}}\Delta\theta \approx m\Delta\lambda$$

so $\boxed{\Delta\theta \approx \dfrac{\Delta\lambda}{\sqrt{(d^2/m^2)-\lambda^2}}}$.

ANSWERS TO EVEN PROBLEMS

P27.2 (a) 36.2°; (b) 5.08 cm; (c) 508 THz

P27.4 maxima at 0°, 29.1°, and 76.3°; minima at 14.1° and 46.8°

P27.6 11.3 m

P27.8 (a) 10.0 m;
(b) 516 m;
(c) Only the runway centerline is a maximum for the interference patterns for both frequencies. If the frequencies were related by a ratio of small integers $\frac{k}{\ell}$, the plane could by mistake fly along the k-th side maximum of one signal where it coincides with the ℓ-th side maximum of the other.

P27.10 see the solution

P27.12 (a) 1.29 rad; (b) 99.6 nm

P27.14 612 nm

P27.16 0.500 cm

P27.18 (a) 238 nm; (b) λ increase; (c) 328 nm

P27.20 4.22 mm

P27.22 547 nm

P27.24 three maxima, at 0° and near 46° on both sides

P27.26 1.00 mrad

P27.28 she can resolve only the violet tubes if she is between 186 m and 271 m away

P27.30 neither, it can resolve objects no closer than several centimeters apart

P27.32 514 nm

P27.34 1.81 μm1.81 m

P27.36 see the solution

P27.38 93.4 pm

P27.40 5.51 m, 2.76 m, 1.84 m

P27.42 (a) 217 nm; (b) 93.1 nm

P27.44 (a) $\sim 10^{-3}$ degree;
(b) $\sim 10^{11}$ Hz, microwave

P27.46 one slit 0.123 mm wide

P27.48 1.62 km

P27.50 (a) see the solution; (b) 266 nm

P27.52 (a) 4.86 cm from the top;
(b) 78.9 nm and 128 nm;
(c) 2.63×10^{-6} rad

P27.54 see the solution

P27.56 (a) 7.26 μrad (1.50 seconds of arc);
(b) 0.189 ly;
(c) 50.8 μrad (10.5 seconds of arc);
(d) 1.52 mm

P27.58 see the solution

P27.60 (a) $D^2 = 2.44\lambda L$; (b) 0.428 mm

CONTEXT 8 CONCLUSION SOLUTIONS TO PROBLEMS

CC8.1 Depth = one-quarter of the wavelength in plastic.

$$t = \frac{\lambda}{4n} = \frac{780 \text{ nm}}{4(1.50)} = \boxed{130 \text{ nm}}$$

CC8.2 For a side maximum, $\tan\theta = \frac{y}{L} = \frac{0.4\ \mu\text{m}}{6.9\ \mu\text{m}}$

$$\theta = 3.32°$$

$d\sin\theta = m\lambda$ $\quad d = \frac{(1)(780 \times 10^{-9} \text{ m})}{\sin 3.32°} = 13.5\ \mu\text{m}.$

The number of grooves per millimeter $= \frac{1 \times 10^{-3} \text{ m}}{13.5 \times 10^{-6} \text{ m}} = \boxed{74.2}$.

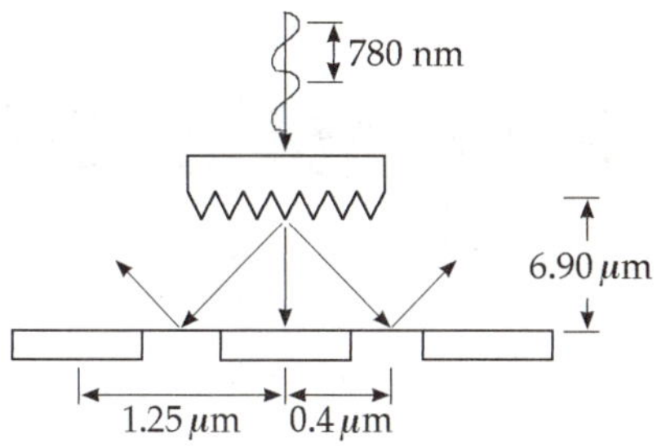

FIG. CC8.2

***CC8.3** In one second, a 1.3-m length of audio track passes the laser. This length includes 705 600 bits of information. The average length per bit is $\frac{1.3 \text{ m}}{705\ 600 \text{ bits}} = 1.8 \times 10^{-6} \text{ m/bit} = \boxed{1.8\mu\text{m/bit}}$. The average length per bit of *total* information on a CD is smaller than this because the disc includes additional information, such as error correction codes, song numbers, and timing codes. The shortest length per bit of total information is actually about $0.8\mu\text{m/bit}$.

***CC8.4** We convert each bit to a power of 2 and add them up:

$$\begin{array}{llll} 1\times 2^{15} = 32\ 768 & 0\times 2^{14} = 0 & 1\times 2^{13} = 8\ 192 & 1\times 2^{12} = 4\ 096 \\ 1\times 2^{11} = 2\ 048 & 0\times 2^{12} = 0 & 1\times 2^{9} = 512 & 1\times 2^{8} = 256 \\ 1\times 2^{7} = 128 & 0\times 2^{6} = 0 & 1\times 2^{5} = 32 & 1\times 2^{4} = 16 \\ 1\times 2^{3} = 8 & 0\times 2^{2} = 0 & 1\times 2^{1} = 2 & 1\times 2^{0} = 1 \end{array}$$

$$\text{sum} = \boxed{48\ 059}$$

The CD player converts this number into a voltage, representing one of the 44 100 values that will be used to build one second of the electronic waveform that represents the recorded sound.

CC8.5 The volume of the cylinder is $\pi r^2 h = \pi(10^{-6} \text{ m})^2 (10^{-6} \text{ m}) \approx 3 \times 10^{-18} \text{ m}^3$

The mass of the magnetic material is $\rho V = (2 \times 10^3 \text{ kg/m}^3)(3 \times 10^{-18} \text{ m}^3) \approx 6 \times 10^{-15} \text{ kg}$

We model the disk as absorbing all of the incident light. To identify the amount of energy transfer T_{ER} by electromagnetic radiation, we compute the heat that would have the same effect:

continued on next page

$$\text{“}Q\text{”} = mc\Delta T = \left(6\times10^{-15}\ \text{kg}\right)\left(\frac{300\ \text{J}}{\text{kg}\cdot{}^\circ\text{C}}\right)(600\ \text{K} - 300\ \text{K})\left(\frac{1^\circ\text{C}}{1\ \text{K}}\right)$$

$$\text{“}Q\text{”} \approx 6\times10^{-10}\ \text{J}$$

The pit moves past in a time interval of $\Delta t = \dfrac{2\times10^{-6}\ \text{m}}{1\ \text{m/s}} = 2\times10^{-6}\ \text{s}$.

So, the intensity is $I = \dfrac{\mathscr{P}}{A} = \dfrac{\text{“}Q\text{”}}{A\,\Delta t}$: $I \approx \dfrac{6\times10^{-10}\ \text{J}}{\left[\pi\left(10^{-6}\ \text{m}\right)^2\right]\left(2\times10^{-6}\ \text{s}\right)} = 9\times10^{7}\ \text{W/m}^2$

$$I \boxed{\sim 10^{8}\ \text{W/m}^2}$$

ANSWERS TO EVEN CONTEXT 8 CONCLUSION PROBLEMS

CC8.2 74.2 grooves/mm

CC8.4 48 059

Chapter 28

Quantum Physics

CHAPTER OUTLINE

ANSWERS TO QUESTIONS

Q28.1 Planck made two new assumptions: (1) molecular energy is quantized and (2) molecules emit or absorb energy in discrete irreducible packets. These assumptions contradict the classical idea of energy as continuously divisible. They also imply that an atom must have a definite structure—it cannot just be a soup of electrons orbiting the nucleus.

Q28.2 (c) UV light has the highest frequency of the three, and hence each photon delivers more energy to a skin cell. This explains why you can become sunburned on a cloudy day: clouds block visible light and infrared, but not much ultraviolet. You usually do not become sunburned through window glass, even though you can see the visible light from the Sun coming through the window, because the glass absorbs much of the ultraviolet and reemits it as infrared.

Q28.3 No. The second metal may have a larger work function than the first, in which case the incident photons may not have enough energy to eject photoelectrons.

Q28.4 The Compton effect describes the *scattering* of photons from electrons, while the photoelectric effect predicts the ejection of electrons due to the *absorption* of photons by a material.

Q28.5 Wave theory predicts that the photoelectric effect should occur at any frequency, provided the light intensity is high enough. However, as seen in the photoelectric experiments, the light must have a sufficiently high frequency for the effect to occur.

Q28.6 A few photons would only give a few dots of exposure, apparently randomly scattered.

Q28.7 The x-ray photon transfers some of its energy to the electron. Thus, its frequency must decrease.

Q28.8 Light has both classical-wave and classical-particle characteristics. In single- and double-slit experiments light behaves like a wave. In the photoelectric effect light behaves like a particle. Light may be characterized as an electromagnetic wave with a particular wavelength or frequency, yet at the same time light may be characterized as a stream of photons, each carrying a discrete energy, hf. Since light displays *both* wave and particle characteristics, perhaps it would be fair to call light a "wavicle". It is customary to call a photon a quantum particle, different from a classical particle.

Q28.9 An electron has both classical-wave and classical-particle characteristics. In single- and double-slit diffraction and interference experiments, electrons behave like classical waves. An electron has mass and charge. It carries kinetic energy and momentum in parcels of definite size, as classical particles do. At the same time it has a particular wavelength and frequency. Since an electron displays characteristics of both classical waves and classical particles, it is neither a classical wave nor a classical particle. It is customary to call it a quantum particle, but another invented term, such as "wavicle", could serve equally well.

Q28.10 The discovery of electron diffraction by Davisson and Germer was a fundamental advance in our understanding of the motion of material particles. Newton's laws fail to properly describe the motion of an object with small mass. It moves as a wave, not as a classical particle. Proceeding from this recognition, the development of quantum mechanics made possible describing the motion of electrons in atoms; understanding molecular structure and the behavior of matter at the atomic scale, including electronics, photonics, and engineered materials; accounting for the motion of nucleons in nuclei; and studying elementary particles.

Q28.11 Any object of macroscopic size—including a grain of dust—has an undetectably small wavelength and does not exhibit quantum behavior.

Q28.12 If we set $\frac{p^2}{2m} = q\Delta V$, which is the same for both particles, then we see that the electron has the smaller momentum and therefore the longer wavelength $\left(\lambda = \frac{h}{p}\right)$.

Q28.13 The *intensity* of electron waves in some small region of space determines the *probability* that an electron will be found in that region.

Q28.14 (a) The slot is blacker than any black material or pigment. Any radiation going in through the hole will be absorbed by the walls or the contents of the box, perhaps after several reflections. Essentially none of that energy will come out through the hole again. Figure 28.1 in the text shows this effect if you imagine the beam getting weaker at each reflection.

(b) The open slots between the glowing tubes are brightest. When you look into a slot, you receive direct radiation emitted by the wall on the far side of a cavity enclosed by the fixture; and you also receive radiation that was emitted by other sections of the cavity wall and has bounced around a few or many times before escaping through the slot. In Figure 28.1 in the text, reverse all of the arrowheads and imagine the beam getting stronger at each reflection. Then the figure shows the extra efficiency of a cavity radiator. Here is the conclusion of Kirchhoff's thermodynamic argument: ... energy radiated. A poor reflector—a good absorber—avoids rising in temperature by being an efficient emitter. Its emissivity is equal to its absorptivity: $e = a$. The slot in the box in part (a) of the question is a black body with reflectivity zero and absorptivity 1, so it must also be the most efficient possible radiator, to avoid rising in temperature above its surroundings in thermal equilibrium. Its emissivity in Stefan's law is $100\% = 1$, higher than perhaps 0.9 for black paper, 0.1 for light-colored paint, or 0.04 for shiny metal. Only in this way can the material objects underneath these different surfaces maintain equal temperatures after they come to thermal equilibrium and continue to exchange energy by electromagnetic radiation. By considering one blackbody facing another, Kirchhoff proved logically that the material forming the walls of the cavity made no difference to the radiation. By thinking about inserting color filters between two cavity radiators, he proved that the spectral distribution of blackbody radiation must be a universal function of wavelength, the same for all materials and depending only on the temperature. Blackbody radiation is a fundamental connection between the matter and the energy that physicists had previously studied separately.

Q28.15 The motion of the quantum particle does not consist of moving through successive points. The particle has no definite position. It can sometimes be found on one side of a node and sometimes on the other side, but never at the node itself. There is no contradiction here, for the quantum particle is moving as a wave. It is not a classical particle. In particular, the particle does not speed up to infinite speed to cross the node.

SOLUTIONS TO PROBLEMS

Section 28.1 Blackbody Radiation and Planck's Theory

***P28.1** This is an example of Stefan's law.

At the lower temperature, $\mathcal{P}_1 = \sigma A e T_1^4$.

At the higher , $\mathcal{P}_2 = \sigma A e T_2^4$.

Then $\dfrac{\mathcal{P}_2}{\mathcal{P}_1} = \left(\dfrac{T_2}{T_1}\right)^4 = \left(\dfrac{273+38.3}{273+37}\right)^4 = (1.004\ 19)^4 = 1.016\ 9$.

The percentage increase in power is $(1.016\ 9 - 1) = 0.016\ 9 = \boxed{1.69\%}$, four times larger than the percentage increase in temperature.

P28.2 (a) $\mathcal{P} = eA\sigma T^4$, so

$$T = \left(\frac{\mathcal{P}}{eA\sigma}\right)^{1/4} = \left[\frac{3.85\times10^{26}\ \text{W}}{1\left[4\pi\left(6.96\times10^{8}\ \text{m}\right)^2\right]\left(5.67\times10^{-8}\ \text{W/m}^2\cdot\text{K}^4\right)}\right]^{1/4} = \boxed{5.78\times10^{3}\ \text{K}}$$

(b) $$\lambda_{\max} = \frac{2.898\times10^{-3}\ \text{m}\cdot\text{K}}{T} = \frac{2.898\times10^{-3}\ \text{m}\cdot\text{K}}{5.78\times10^{3}\ \text{K}} = 5.01\times10^{-7}\ \text{m} = \boxed{501\ \text{nm}}$$

***P28.3** The peak radiation occurs at approximately 560 nm wavelength. From Wien's displacement law,

$$T = \frac{0.289\ 8\times10^{-2}\ \text{m}\cdot\text{K}}{\lambda_{\max}} = \frac{0.289\ 8\times10^{-2}\ \text{m}\cdot\text{K}}{560\times10^{-9}\ \text{m}} \approx \boxed{5\ 200\ \text{K}}.$$

Clearly, a firefly is not at this temperature, so $\boxed{\text{this is not blackbody radiation}}$.

P28.4 (a) $$E = hf = \left(6.626\times10^{-34}\ \text{J}\cdot\text{s}\right)\left(620\times10^{12}\ \text{s}^{-1}\right)\left(\frac{1.00\ \text{eV}}{1.60\times10^{-19}\ \text{J}}\right) = \boxed{2.57\ \text{eV}}$$

(b) $$E = hf = \left(6.626\times10^{-34}\ \text{J}\cdot\text{s}\right)\left(3.10\times10^{9}\ \text{s}^{-1}\right)\left(\frac{1.00\ \text{eV}}{1.60\times10^{-19}\ \text{J}}\right) = \boxed{1.28\times10^{-5}\ \text{eV}}$$

(c) $$E = hf = \left(6.626\times10^{-34}\ \text{J}\cdot\text{s}\right)\left(46.0\times10^{6}\ \text{s}^{-1}\right)\left(\frac{1.00\ \text{eV}}{1.60\times10^{-19}\ \text{J}}\right) = \boxed{1.91\times10^{-7}\ \text{eV}}$$

continued on next page

(d) $\lambda = \frac{c}{f} = \frac{3.00 \times 10^8 \text{ m/s}}{620 \times 10^{12} \text{ Hz}} = 4.84 \times 10^{-7} \text{ m} = \boxed{484 \text{ nm, visible light (blue)}}$

$$\lambda = \frac{c}{f} = \frac{3.00 \times 10^8 \text{ m/s}}{3.10 \times 10^9 \text{ Hz}} = 9.68 \times 10^{-2} \text{ m} = \boxed{9.68 \text{ cm, radio wave}}$$

$$\lambda = \frac{c}{f} = \frac{3.00 \times 10^8 \text{ m/s}}{46.0 \times 10^6 \text{ Hz}} = \boxed{6.52 \text{ m, radio wave}}$$

P28.5 Each photon has an energy $E = hf = \left(6.626 \times 10^{-34}\right)\left(99.7 \times 10^6\right) = 6.61 \times 10^{-26} \text{ J}$.

This implies that there are $\frac{150 \times 10^3 \text{ J/s}}{6.61 \times 10^{-26} \text{ J/photon}} = \boxed{2.27 \times 10^{30} \text{ photons/s}}$.

P28.6 Energy of a single 500-nm photon:

$$E_\gamma = hf = \frac{hc}{\lambda} = \frac{\left(6.626 \times 10^{-34} \text{ J} \cdot \text{s}\right)\left(3.00 \times 10^8 \text{ m/s}\right)}{\left(500 \times 10^{-9} \text{ m}\right)} = 3.98 \times 10^{-19} \text{ J}.$$

The energy entering the eye each second

$$E = \mathcal{P}\Delta t = IA\Delta t = \left(4.00 \times 10^{-11} \text{ W/m}^2\right)\left[\frac{\pi}{4}\left(8.50 \times 10^{-3} \text{ m}\right)^2\right](1.00 \text{ s}) = 2.27 \times 10^{-15} \text{ J}.$$

The number of photons required to yield this energy

$$n = \frac{E}{E_\gamma} = \frac{2.27 \times 10^{-15} \text{ J}}{3.98 \times 10^{-19} \text{ J/photon}} = \boxed{5.71 \times 10^3 \text{ photons}}.$$

P28.7 We take $\theta = 0.030\,0$ radians. Then the pendulum's total energy is

$$E = mgh = mg(L - L\cos\theta)$$
$$E = (1.00 \text{ kg})\left(9.80 \text{ m/s}^2\right)(1.00 - 0.999\,5) = 4.41 \times 10^{-3} \text{ J}$$

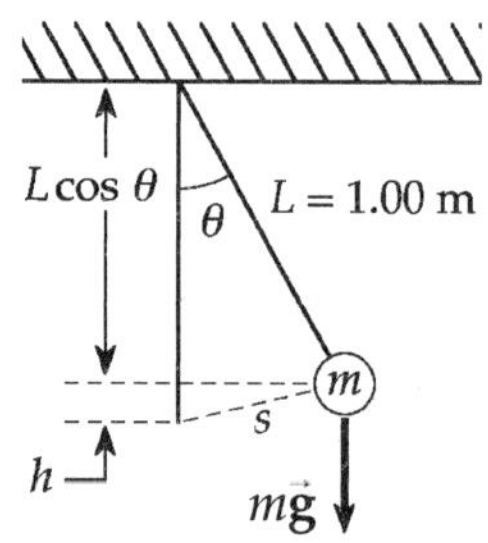

FIG. P28.7

The frequency of oscillation is $f = \frac{\omega}{2\pi} = \frac{1}{2\pi}\sqrt{\frac{g}{L}} = 0.498 \text{ Hz}$.

The energy is quantized, $E = nhf$.

Therefore, $n = \frac{E}{hf} = \frac{4.41 \times 10^{-3} \text{ J}}{\left(6.626 \times 10^{-34} \text{ J} \cdot \text{s}\right)\left(0.498 \text{ s}^{-1}\right)}$

$$= \boxed{1.34 \times 10^{31}}$$

***P28.8** (a) $V = \frac{4}{3}\pi r^3 = \frac{4}{3}\pi(0.02 \text{ m})^3 = 3.35 \times 10^{-5} \text{ m}^3$

$$m = \rho V = 7.86 \times 10^3 \text{ kg/m}^3 \; 3.35 \times 10^{-5} \text{ m}^3 = \boxed{0.263 \text{ kg}}$$

continued on next page

(b) $A = 4\pi r^2 = 4\pi(0.02\text{ m})^2 = 5.03\times10^{-3}\text{ m}^2$

$\mathscr{P} = \sigma AeT^4 = 5.67\times10^{-8}\text{ W/m}^2\text{K}^4\ 5.03\times10^{-3}\text{ m}^2\ 0.86(293\text{ K})^4 = \boxed{1.81\text{ W}}$

(c) It emits but does not absorb radiation, so its temperature must drop according to

$$Q = mc\Delta T = mc\left(T_f - T_i\right) \qquad \frac{dQ}{dt} = mc\frac{dT_f}{dt}$$

$$\frac{dT_f}{dt} = \frac{\frac{dQ}{dt}}{mc} = \frac{-\mathscr{P}}{mc} = \frac{-1.81\text{ J/s}}{0.263\text{ kg }448\text{ J/kg}\cdot\text{C}^\circ} = \boxed{-0.015\ 3\ ^\circ\text{C/s}} = -0.919\ ^\circ\text{C/min}$$

(d) $\lambda_{\max}T = 2.898\times10^{-3}\text{ m}\cdot\text{K}$

$\lambda_{\max} = \dfrac{2.898\times10^{-3}\text{ m}\cdot\text{K}}{293\text{ K}} = \boxed{9.89\times10^{-6}\text{ m}}$ infrared

(e) $E = hf = \dfrac{hc}{\lambda} = \dfrac{6.63\times10^{-34}\text{ Js }3\times10^{8}\text{ m/s}}{9.89\times10^{-6}\text{ m}} = \boxed{2.01\times10^{-20}\text{ J}}$

(f) The energy output each second is carried by photons according to

$$\mathscr{P} = \left(\frac{N}{\Delta t}\right)E$$

$$\frac{N}{\Delta t} = \frac{\mathscr{P}}{E} = \frac{1.81\text{ J/s}}{2.01\times10^{-20}\text{ J/photon}} = \boxed{8.98\times10^{19}\text{ photon/s}}$$

Matter is coupled to radiation, quite strongly, in terms of photon numbers.

Section 28.2 The Photoelectric Effect

P28.9 (a) $\lambda_c = \dfrac{hc}{\phi} = \dfrac{\left(6.626\times10^{-34}\text{ J}\cdot\text{s}\right)\left(3.00\times10^{8}\text{ m/s}\right)}{(4.20\text{ eV})\left(1.60\times10^{-19}\text{ J/eV}\right)} = \boxed{296\text{ nm}}$

$f_c = \dfrac{c}{\lambda_c} = \dfrac{3.00\times10^{8}\text{ m/s}}{296\times10^{-9}\text{ m}} = \boxed{1.01\times10^{15}\text{ Hz}}$

(b) $\dfrac{hc}{\lambda} = \phi + e\Delta V_S$: $\dfrac{\left(6.626\times10^{-34}\right)\left(3.00\times10^{8}\right)}{180\times10^{-9}} = (4.20\text{ eV})\left(1.60\times10^{-19}\text{ J/eV}\right) + \left(1.60\times10^{-19}\right)\Delta V_S$

Therefore, $\boxed{\Delta V_S = 2.71\text{ V}}$

P28.10 $K_{max} = \frac{1}{2}mv_{max}^2 = \frac{1}{2}(9.11\times10^{-31})(4.60\times10^5)^2 = 9.64\times10^{-20}\text{ J} = 0.602\text{ eV}$

(a) $\phi = E - K_{max} = \frac{1\,240\text{ eV}\cdot\text{nm}}{625\text{ nm}} - 0.602\text{ nm} = \boxed{1.38\text{ eV}}$

(b) $f_c = \frac{\phi}{h} = \frac{1.38\text{ eV}}{6.626\times10^{-34}\text{ J}\cdot\text{s}}\left(\frac{1.60\times10^{-19}\text{ J}}{1\text{ eV}}\right) = \boxed{3.34\times10^{14}\text{ Hz}}$

P28.11 (a) $e\Delta V_S = \frac{hc}{\lambda} - \phi \rightarrow \phi = \frac{1\,240\text{ nm}\cdot\text{eV}}{546.1\text{ nm}} - 0.376\text{ eV} = \boxed{1.90\text{ eV}}$

(b) $e\Delta V_S = \frac{hc}{\lambda} - \phi = \frac{1\,240\text{ nm}\cdot\text{eV}}{587.5\text{ nm}} - 1.90\text{ eV} \rightarrow \boxed{\Delta V_S = 0.216\text{ V}}$

P28.12 The energy needed is $E = 1.00\text{ eV} = 1.60\times10^{-19}\text{ J}$.

The energy absorbed in time interval Δt is $E = \mathcal{P}\Delta t = IA\Delta t$

so $\Delta t = \frac{E}{IA} = \frac{1.60\times10^{-19}\text{ J}}{(500\text{ J/s}\cdot\text{m}^2)\left[\pi(2.82\times10^{-15}\text{ m})^2\right]} = 1.28\times10^7\text{ s} = \boxed{148\text{ days}}$.

The gross failure of the classical theory of the photoelectric effect contrasts with the success of quantum mechanics.

P28.13 Ultraviolet photons will be absorbed to knock electrons out of the sphere with maximum kinetic energy $K_{max} = hf - \phi$,

or $K_{max} = \frac{(6.626\times10^{-34}\text{ J}\cdot\text{s})(3.00\times10^8\text{ m/s})}{200\times10^{-9}\text{ m}}\left(\frac{1.00\text{ eV}}{1.60\times10^{-19}\text{ J}}\right) - 4.70\text{ eV} = 1.51\text{ eV}$.

The sphere is left with positive charge and so with positive potential relative to $V = 0$ at $r = \infty$. As its potential approaches 1.51 V, no further electrons will be able to escape, but will fall back onto the sphere. Its charge is then given by

$V = \frac{k_e Q}{r}$ or $Q = \frac{rV}{k_e} = \frac{(5.00\times10^{-2}\text{ m})(1.51\text{ N}\cdot\text{m/C})}{8.99\times10^9\text{ N}\cdot\text{m}^2/\text{C}^2} = \boxed{8.41\times10^{-12}\text{ C}}$.

Section 28.3 The Compton Effect

P28.14 $E = \frac{hc}{\lambda} = \frac{(6.626\times10^{-34}\text{ J}\cdot\text{s})(3.00\times10^8\text{ m/s})}{700\times10^{-9}\text{ m}} = 2.84\times10^{-19}\text{ J} = \boxed{1.78\text{ eV}}$

$p = \frac{h}{\lambda} = \frac{6.626\times10^{-34}\text{ J}\cdot\text{s}}{700\times10^{-9}\text{ m}} = \boxed{9.47\times10^{-28}\text{ kg}\cdot\text{m/s}}$

P28.15 (a) $\Delta\lambda = \frac{h}{m_e c}(1-\cos\theta)$: $\Delta\lambda = \frac{6.626\times10^{-34}}{(9.11\times10^{-31})(3.00\times10^{8})}(1-\cos 37.0°) = \boxed{4.88\times10^{-13}\text{ m}}$

(b) $E_0 = \frac{hc}{\lambda_0}$: $\quad (300\times10^{3}\text{ eV})(1.60\times10^{-19}\text{ J/eV}) = \frac{(6.626\times10^{-34})(3.00\times10^{8}\text{ m/s})}{\lambda_0}$

$$\lambda_0 = 4.14\times10^{-12}\text{ m}$$

and $\quad \lambda' = \lambda_0 + \Delta\lambda = 4.63\times10^{-12}\text{ m}$

$$E' = \frac{hc}{\lambda'} = \frac{(6.626\times10^{-34}\text{ J}\cdot\text{s})(3.00\times10^{8}\text{ m/s})}{4.63\times10^{-12}\text{ m}} = 4.30\times10^{-14}\text{ J} = \boxed{268\text{ keV}}$$

(c) $K_e = E_0 - E' = 300\text{ keV} - 268.5\text{ keV} = \boxed{31.5\text{ keV}}$

P28.16 This is Compton scattering through 180°:

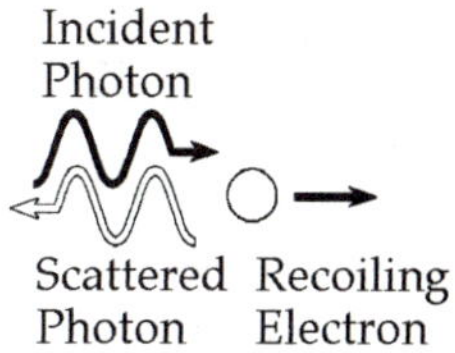

FIG. P28.16

$$E_0 = \frac{hc}{\lambda_0} = \frac{(6.626\times10^{-34}\text{ J}\cdot\text{s})(3.00\times10^{8}\text{ m/s})}{(0.110\times10^{-9}\text{ m})(1.60\times10^{-19}\text{ J/eV})} = 11.3\text{ keV}$$

$$\Delta\lambda = \frac{h}{m_e c}(1-\cos\theta) = (2.43\times10^{-12}\text{ m})(1-\cos180°) = 4.86\times10^{-12}\text{ m}$$

$\lambda' = \lambda_0 + \Delta\lambda = 0.115\text{ nm}$ so $\quad E' = \frac{hc}{\lambda'} = 10.8\text{ keV}$.

By conservation of momentum for the photon-electron system, $\frac{h}{\lambda_0}\hat{\mathbf{i}} = \frac{h}{\lambda'}(-\hat{\mathbf{i}}) + p_e\hat{\mathbf{i}}$

and $\quad p_e = h\left(\frac{1}{\lambda_0} + \frac{1}{\lambda'}\right)$

$$p_e = (6.626\times10^{-34}\text{ J}\cdot\text{s})\left(\frac{(3.00\times10^{8}\text{ m/s})/c}{1.60\times10^{-19}\text{ J/eV}}\right)\left(\frac{1}{0.110\times10^{-9}\text{ m}} + \frac{1}{0.115\times10^{-9}\text{ m}}\right) = \boxed{\frac{22.1\text{ keV}}{c}}.$$

By conservation of system energy, $\quad 11.3\text{ keV} = 10.8\text{ keV} + K_e$

so that $\quad \boxed{K_e = 478\text{ eV}}$.

Check: $E^2 = p^2c^2 + m_e^2c^4$ or $\quad (m_ec^2 + K_e)^2 = (pc)^2 + (m_ec^2)^2$

$$(511\text{ keV} + 0.478\text{ keV})^2 = (22.1\text{ keV})^2 + (511\text{ keV})^2$$
$$2.62\times10^{11} = 2.62\times10^{11}$$

P28.17 With $K_e = E'$, $K_e = E_0 - E'$ gives $E' = E_0 - E'$

$$E' = \frac{E_0}{2} \text{ and } \lambda' = \frac{hc}{E'} \qquad \lambda' = \frac{hc}{E_0/2} = 2\frac{hc}{E_0} = 2\lambda_0$$

$$\lambda' = \lambda_0 + \lambda_C(1 - \cos\theta) \qquad 2\lambda_0 = \lambda_0 + \lambda_C(1 - \cos\theta)$$

$$1 - \cos\theta = \frac{\lambda_0}{\lambda_C} = \frac{0.001\,60}{0.002\,43} \qquad \boxed{\theta = 70.0°}$$

P28.18 (a) $K = \frac{1}{2}m_e v^2$: $\quad K = \frac{1}{2}\left(9.11\times10^{-31}\text{ kg}\right)\left(1.40\times10^{6}\text{ m/s}\right)^2 = 8.93\times10^{-19}\text{ J} = 5.58\text{ eV}$

$$E_0 = \frac{hc}{\lambda_0} = \frac{1\,240\text{ eV}\cdot\text{nm}}{0.800\text{ nm}} = 1\,550\text{ eV}$$

$E' = E_0 - K$ and $\quad \lambda' = \frac{hc}{E'} = \frac{1\,240\text{ eV}\cdot\text{nm}}{1\,550\text{ eV} - 5.58\text{ eV}} = 0.803\text{ nm}$

$$\Delta\lambda = \lambda' - \lambda_0 = 0.002\,88\text{ nm} = \boxed{2.88\text{ pm}}$$

(b) $\Delta\lambda = \lambda_C(1 - \cos\theta)$: $\quad \cos\theta = 1 - \frac{\Delta\lambda}{\lambda_C} = 1 - \frac{0.002\,88\text{ nm}}{0.002\,43\text{ nm}} = -0.189$,

so $\quad \boxed{\theta = 101°}$

Section 28.4 Photons and Electromagnetic Waves

***P28.19** With photon energy $10.0\text{ eV} = hf$

$$f = \frac{10.0\left(1.6\times10^{-19}\text{ J}\right)}{6.63\times10^{-34}\text{ J}\cdot\text{s}} = 2.41\times10^{15}\text{ Hz}.$$

Any electromagnetic wave with frequency higher than 2.41×10^{15} Hz counts as ionizing radiation. This includes far ultraviolet light, x-rays, and gamma rays.

Section 28.5 The Wave Properties of Particles

P28.20 $\lambda = \frac{h}{p} = \frac{h}{mv} = \frac{6.626\times10^{-34}\text{ J}\cdot\text{s}}{\left(1.67\times10^{-27}\text{ kg}\right)\left(1.00\times10^{6}\text{ m/s}\right)} = \boxed{3.97\times10^{-13}\text{ m}}$

P28.21 (a) Electron: $\lambda = \frac{h}{p}$ and $K = \frac{1}{2}m_e v^2 = \frac{m_e^2 v^2}{2m_e} = \frac{p^2}{2m_e}$ so $p = \sqrt{2m_e K}$

and $\lambda = \frac{h}{\sqrt{2m_e K}} = \frac{6.626\times10^{-34}\text{ J}\cdot\text{s}}{\sqrt{2(9.11\times10^{-31}\text{ kg})(3.00)(1.60\times10^{-19}\text{ J})}}$

$\lambda = 7.09\times10^{-10}\text{ m} = \boxed{0.709\text{ nm}}.$

(b) Photon: $\lambda = \frac{c}{f}$ and $E = hf$ so $f = \frac{E}{h}$

and $\lambda = \frac{hc}{E} = \frac{(6.626\times10^{-34}\text{ J}\cdot\text{s})(3.00\times10^{8}\text{ m/s})}{3(1.60\times10^{-19}\text{ J})} = 4.14\times10^{-7}\text{ m} = \boxed{414\text{ nm}}.$

P28.22 From the condition for Bragg reflection,

$$m\lambda = 2d\sin\theta = 2d\cos\left(\frac{\phi}{2}\right).$$

But $d = a\sin\left(\frac{\phi}{2}\right)$

where a is the lattice spacing.

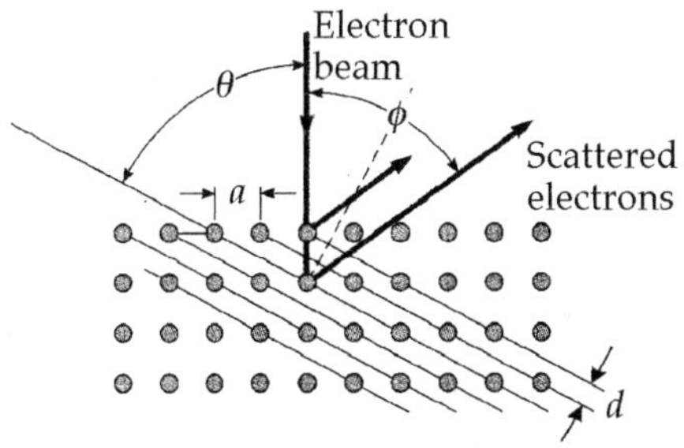

FIG. P28.22

Thus, with $m = 1$, $\lambda = 2a\sin\left(\frac{\phi}{2}\right)\cos\left(\frac{\phi}{2}\right) = a\sin\phi$

$\lambda = \frac{h}{p} = \frac{h}{\sqrt{2m_e K}}$ $\lambda = \frac{6.626\times10^{-34}\text{ J}\cdot\text{s}}{\sqrt{2(9.11\times10^{-31}\text{ kg})(54.0\times1.60\times10^{-19}\text{ J})}} = 1.67\times10^{-10}\text{ m}.$

Therefore, the lattice spacing is $a = \frac{\lambda}{\sin\phi} = \frac{1.67\times10^{-10}\text{ m}}{\sin 50.0°} = 2.18\times10^{-10} = \boxed{0.218\text{ nm}}.$

P28.23 (a) $\lambda \sim 10^{-14}$ m or less. $p = \frac{h}{\lambda} \sim \frac{6.6\times10^{-34}\text{ J}\cdot\text{s}}{10^{-14}\text{ m}} = 10^{-19}\text{ kg}\cdot\text{m/s}$ or more.

The energy of the electron is $E = \sqrt{p^2c^2 + m_e^2c^4} \sim \sqrt{(10^{-19})^2(3\times10^8)^2 + (9\times10^{-31})^2(3\times10^8)^4}$

or $E \sim 10^{-11}\text{ J} \sim 10^{8}\text{ eV}$ or more,

so that $K = E - m_ec^2 \sim 10^8\text{ eV} - (0.5\times10^6\text{ eV})\boxed{\sim 10^8\text{ eV}}$ or more.

(b) The electric potential energy of the electron-nucleus system would be

$$U_e = \frac{k_e q_1 q_2}{r} \sim \frac{(9\times10^9\text{ N}\cdot\text{m}^2/\text{C}^2)(10^{-19}\text{ C})(-e)}{10^{-14}\text{ m}} \sim -10^5\text{ eV}.$$

With its $K + U_e >> 0$, $\boxed{\text{the electron would immediately escape the nucleus}}$.

P28.24 (a) The wavelength of the student is $\lambda = \frac{h}{p} = \frac{h}{mv}$. If w is the width of the diffracting aperture,

then we need $$w \le 10.0\lambda = 10.0\left(\frac{h}{mv}\right)$$

so that $$v \le 10.0\frac{h}{mw} = 10.0\left(\frac{6.626\times10^{-34}\text{ J}\cdot\text{s}}{(80.0\text{ kg})(0.750\text{ m})}\right) = \boxed{1.10\times10^{-34}\text{ m/s}}.$$

(b) Using $\Delta t = \frac{d}{v}$ we get: $$\Delta t \ge \frac{0.150\text{ m}}{1.10\times10^{-34}\text{ m/s}} = \boxed{1.36\times10^{33}\text{ s}}.$$

(c) $\boxed{\text{No}}$. The minimum time to pass through the door is over 10^{15} times the age of the Universe.

P28.25 $\lambda = \frac{h}{p}$ $$p = \frac{h}{\lambda} = \frac{6.626\times10^{-34}\text{ J}\cdot\text{s}}{1.00\times10^{-11}\text{ m}} = 6.63\times10^{-23}\text{ kg}\cdot\text{m/s}$$

(a) electrons: $$K_e = \frac{p^2}{2m_e} = \frac{\left(6.63\times10^{-23}\right)^2}{2\left(9.11\times10^{-31}\right)}\text{ J} = 15.1\text{ keV}$$

The relativistic answer is more precisely correct:

$$K_e = \sqrt{p^2c^2 + m_e^2c^4} - m_ec^2 = \boxed{14.9\text{ keV}}.$$

(b) photons: $$E_\gamma = pc = \left(6.63\times10^{-23}\right)\left(3.00\times10^8\right) = \boxed{124\text{ keV}}$$

Section 28.6 The Quantum Particle

***P28.26** $E = K = \frac{1}{2}mu^2 = hf$ and $\lambda = \frac{h}{mu}$.

$$v_{\text{phase}} = f\lambda = \frac{mu^2}{2h}\frac{h}{mu} = \boxed{\frac{u}{2} = v_{\text{phase}}}.$$

This is different from the speed u at which the particle transports mass, energy, and momentum.

P28.27 As a bonus, we begin by proving that the phase speed $v_p = \frac{\omega}{k}$ is not the speed of the particle.

$$v_p = \frac{\omega}{k} = \frac{\sqrt{p^2c^2 + m^2c^4}\,\hbar}{\hbar\gamma mv} = \frac{\sqrt{\gamma^2m^2v^2c^2 + m^2c^4}}{\sqrt{\gamma^2m^2v^2}} = c\sqrt{1 + \frac{c^2}{\gamma^2v^2}} = c\sqrt{1 + \frac{c^2}{v^2}\left(1 - \frac{v^2}{c^2}\right)} = c\sqrt{1 + \frac{c^2}{v^2} - 1} = \frac{c^2}{v}$$

In fact, the phase speed is larger than the speed of light. A point of constant phase in the wave function carries no mass, no energy, and no information.

Now for the group speed:

continued on next page

$$v_g = \frac{d\omega}{dk} = \frac{d\hbar\omega}{d\hbar k} = \frac{dE}{dp} = \frac{d}{dp}\sqrt{m^2c^4 + p^2c^2}$$

$$v_g = \frac{1}{2}\left(m^2c^4 + p^2c^2\right)^{-1/2}\left(0 + 2pc^2\right) = \sqrt{\frac{p^2c^4}{p^2c^2 + m^2c^4}}$$

$$v_g = c\sqrt{\frac{\gamma^2m^2v^2}{\gamma^2m^2v^2 + m^2c^2}} = c\sqrt{\frac{v^2/\left(1 - v^2/c^2\right)}{v^2/\left(1 - v^2/c^2\right) + c^2}} = c\sqrt{\frac{v^2/\left(1 - v^2/c^2\right)}{\left(v^2 + c^2 - v^2\right)/\left(1 - v^2/c^2\right)}} = v$$

It is this speed at which mass, energy, and momentum are transported.

Section 28.7 The Double-Slit Experiment Revisited

P28.28 Consider the first bright band away from the center:

$d\sin\theta = m\lambda$ $\qquad$ $\left(6.00\times10^{-8}\text{ m}\right)\sin\left(\tan^{-1}\left[\frac{0.400}{200}\right]\right) = (1)\lambda = 1.20\times10^{-10}\text{ m}$

$\lambda = \frac{h}{m_e v}$ so $\qquad$ $m_e v = \frac{h}{\lambda}$

and $\qquad$ $K = \frac{1}{2}m_e v^2 = \frac{m_e^2v^2}{2m_e} = \frac{h^2}{2m_e\lambda^2} = e\Delta V$

$\Delta V = \frac{h^2}{2em_e\lambda^2}$ $\qquad$ $\Delta V = \frac{\left(6.626\times10^{-34}\text{ J}\cdot\text{s}\right)^2}{2\left(1.60\times10^{-19}\text{ C}\right)\left(9.11\times10^{-31}\text{ kg}\right)\left(1.20\times10^{-10}\text{ m}\right)^2} = \boxed{105\text{ V}}$.

P28.29 (a) $\lambda = \frac{h}{mv} = \frac{6.626\times10^{-34}\text{ J}\cdot\text{s}}{\left(1.67\times10^{-27}\text{ kg}\right)(0.400\text{ m/s})} = \boxed{9.92\times10^{-7}\text{ m}}$

(b) For destructive interference in a multiple-slit experiment, $d\sin\theta = \left(m + \frac{1}{2}\right)\lambda$, with $m = 0$ for the first minimum.

Then, $\qquad$ $\theta = \sin^{-1}\left(\frac{\lambda}{2d}\right) = 0.028\,4°$

so $\frac{y}{L} = \tan\theta$ $\qquad$ $y = L\tan\theta = (10.0\text{ m})(\tan 0.028\,4°) = \boxed{4.96\text{ mm}}$.

(c) We cannot say the neutron passed through one slit. We can only say it passed through the slits.

***P28.30** We find the speed of each electron from energy conservation in the firing process:

$$0 = K_f + U_f = \frac{1}{2}mv^2 - eV$$

$$v = \sqrt{\frac{2eV}{m}} = \sqrt{\frac{2 \times 1.6 \times 10^{-19}\ \text{C}(45\ \text{V})}{9.11 \times 10^{-31}\ \text{kg}}} = 3.98 \times 10^6\ \text{m/s}$$

The time of flight is $\Delta t = \frac{\Delta x}{v} = \frac{0.28\ \text{m}}{3.98 \times 10^6\ \text{m/s}} = 7.04 \times 10^{-8}\ \text{s}$. The current when electrons are 28 cm apart is $I = \frac{q}{t} = \frac{e}{\Delta t} = \frac{1.6 \times 10^{-19}\ \text{C}}{7.04 \times 10^{-8}\ \text{s}} = \boxed{2.27 \times 10^{-12}\ \text{A}}$.

Section 28.8 The Uncertainty Principle

P28.31 For the electron, $\Delta p = m_e \Delta v = (9.11 \times 10^{-31}\ \text{kg})(500\ \text{m/s})(1.00 \times 10^{-4}) = 4.56 \times 10^{-32}\ \text{kg} \cdot \text{m/s}$

$$\Delta x = \frac{h}{4\pi \Delta p} = \frac{6.626 \times 10^{-34}\ \text{J} \cdot \text{s}}{4\pi(4.56 \times 10^{-32}\ \text{kg} \cdot \text{m/s})} = \boxed{1.16\ \text{mm}}.$$

For the bullet, $\Delta p = m \Delta v = (0.020\,0\ \text{kg})(500\ \text{m/s})(1.00 \times 10^{-4}) = 1.00 \times 10^{-3}\ \text{kg} \cdot \text{m/s}$

$$\Delta x = \frac{h}{4\pi \Delta p} = \boxed{5.28 \times 10^{-32}\ \text{m}}.$$

P28.32 (a) $\Delta p \Delta x = m \Delta v \Delta x \geq \frac{\hbar}{2}$ so $\Delta v \geq \frac{h}{4\pi m \Delta x} = \frac{2\pi\ \text{J} \cdot \text{s}}{4\pi(2.00\ \text{kg})(1.00\ \text{m})} = \boxed{0.250\ \text{m/s}}$.

(b) The duck might move by $(0.25\ \text{m/s})(5\ \text{s}) = 1.25\ \text{m}$. With original position uncertainty of 1.00 m, we can think of Δx growing to $1.00\ \text{m} + 1.25\ \text{m} = \boxed{2.25\ \text{m}}$.

P28.33 $\frac{\Delta y}{x} = \frac{\Delta p_y}{p_x}$ and $d \Delta p_y \geq \frac{h}{4\pi}$.

Eliminate Δp_y and solve for x.

$$x = 4\pi p_x (\Delta y)\frac{d}{h}: \qquad x = 4\pi(1.00 \times 10^{-3}\ \text{kg})(100\ \text{m/s})(1.00 \times 10^{-2}\ \text{m})\frac{(2.00 \times 10^{-3}\ \text{m})}{(6.626 \times 10^{-34}\ \text{J} \cdot \text{s})}$$

The answer, $x = \boxed{3.79 \times 10^{28}\ \text{m}}$, is 190 times greater than the diameter of the observable Universe!

P28.34 From the uncertainty principle $\Delta E \Delta t \geq \frac{\hbar}{2}$

or $\Delta\left(mc^2\right)\Delta t = \frac{\hbar}{2}$.

Therefore,

$$\frac{\Delta m}{m} = \frac{h}{4\pi c^2(\Delta t)m} = \frac{h}{4\pi(\Delta t)E_R}$$

$$\frac{\Delta m}{m} = \frac{6.626\times10^{-34}\ \text{J}\cdot\text{s}}{4\pi\left(8.70\times10^{-17}\ \text{s}\right)(135\ \text{MeV})}\left(\frac{1\ \text{MeV}}{1.60\times10^{-13}\ \text{J}}\right) = \boxed{2.81\times10^{-8}}.$$

P28.35 (a) At the top of the ladder, the woman holds a pellet inside a small region Δx_i. Thus, the uncertainty principle requires her to release it with typical horizontal momentum $\Delta p_x = m\Delta v_x = \frac{\hbar}{2\Delta x_i}$. It falls to the floor in a travel time given by $H = 0 + \frac{1}{2}gt^2$ as $t = \sqrt{\frac{2H}{g}}$, so the total width of the impact points is

$$\Delta x_f = \Delta x_i + (\Delta v_x)t = \Delta x_i + \left(\frac{\hbar}{2m\Delta x_i}\right)\sqrt{\frac{2H}{g}} = \Delta x_i + \frac{A}{\Delta x_i}$$

where $A = \frac{\hbar}{2m}\sqrt{\frac{2H}{g}}$.

To minimize Δx_f, we require $\frac{d(\Delta x_f)}{d(\Delta x_i)} = 0$ or $1 - \frac{A}{\Delta x_i^2} = 0$

so $\Delta x_i = \sqrt{A}$.

The minimum width of the impact points is

$$\left(\Delta x_f\right)_{\min} = \left(\Delta x_i + \frac{A}{\Delta x_i}\right)\Bigg|_{\Delta x_i = \sqrt{A}} = 2\sqrt{A} = \boxed{\sqrt{\frac{2\hbar}{m}}\left(\frac{2H}{g}\right)^{1/4}}.$$

(b)

$$\left(\Delta x_f\right)_{\min} = \left[\frac{2\left(1.0546\times10^{-34}\ \text{J}\cdot\text{s}\right)}{5.00\times10^{-4}\ \text{kg}}\right]^{1/2}\left[\frac{2(2.00\ \text{m})}{9.80\ \text{m/s}^2}\right]^{1/4} = \boxed{5.19\times10^{-16}\ \text{m}}$$

Section 28.9 An Interpretation of Quantum Mechanics

P28.36 Probability

$$P = \int_{-a}^{a}|\psi(x)|^2 = \int_{-a}^{a}\frac{a}{\pi\left(x^2+a^2\right)}dx = \left(\frac{a}{\pi}\right)\left(\frac{1}{a}\right)\tan^{-1}\left(\frac{x}{a}\right)\Bigg|_{-a}^{a}$$

$$P = \frac{1}{\pi}\left[\tan^{-1}1 - \tan^{-1}(-1)\right] = \frac{1}{\pi}\left[\frac{\pi}{4} - \left(-\frac{\pi}{4}\right)\right] = \boxed{\frac{1}{2}}$$

P28.37 (a) $\psi(x) = Ae^{i(5.00\times10^{10}x)} = A\cos(5\times10^{10}x) + Ai\sin(5\times10^{10}x) = A\cos(kx) + Ai\sin(kx)$ goes through a full cycle when x changes by λ and when kx changes by 2π. Then $k\lambda = 2\pi$ where $k = 5.00\times10^{10}\ \text{m}^{-1} = \frac{2\pi}{\lambda}$. Then $\lambda = \frac{2\pi\ \text{m}}{(5.00\times10^{10})} = \boxed{1.26\times10^{-10}\ \text{m}}$.

(b) $p = \frac{h}{\lambda} = \frac{6.626\times10^{-34}\ \text{J}\cdot\text{s}}{1.26\times10^{-10}\ \text{m}} = \boxed{5.27\times10^{-24}\ \text{kg}\cdot\text{m/s}}$

(c) $m_e = 9.11\times10^{-31}\ \text{kg}$

$$K = \frac{m_e^2v^2}{2m_e} = \frac{p^2}{2m} = \frac{(5.27\times10^{-24}\ \text{kg}\cdot\text{m/s})^2}{(2\times9.11\times10^{-31}\ \text{kg})} = 1.52\times10^{-17}\ \text{J} = \frac{1.52\times10^{-17}\ \text{J}}{1.60\times10^{-19}\ \text{J/eV}} = \boxed{95.5\ \text{eV}}$$

Section 28.10 A Particle in a Box

P28.38 For an electron wave to "fit" into an infinitely deep potential well, an integral number of half-wavelengths must equal the width of the well.

$$\frac{n\lambda}{2} = 1.00\times10^{-9}\ \text{m} \qquad \text{so} \qquad \lambda = \frac{2.00\times10^{-9}}{n} = \frac{h}{p}$$

FIG. P28.38

(a) Since $K = \frac{p^2}{2m_e} = \frac{(h^2/\lambda^2)}{2m_e} = \frac{h^2}{2m_e}\frac{n^2}{(2\times10^{-9})^2} = (0.377n^2)\ \text{eV}$

For $K \approx 6\ \text{eV}$ $\qquad \boxed{n = 4}$

(b) With $n = 4$, $\qquad \boxed{K = 6.03\ \text{eV}}$

P28.39 (a) We can draw a diagram that parallels our treatment of standing mechanical waves. In each state, we measure the distance d from one node to another (N to N), and base our solution upon that:

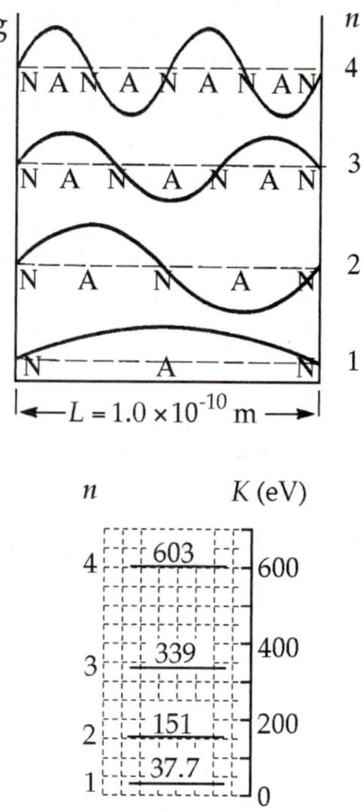

FIG. P28.39

Since $\quad d_{\text{N to N}} = \dfrac{\lambda}{2}$ and $\lambda = \dfrac{h}{p}$

$$p = \frac{h}{\lambda} = \frac{h}{2d}.$$

Next, $\quad K = \dfrac{p^2}{2m_e} = \dfrac{h^2}{8m_e d^2} = \dfrac{1}{d^2}\left[\dfrac{\left(6.626\times10^{-34}\ \text{J}\cdot\text{s}\right)^2}{8\left(9.11\times10^{-31}\ \text{kg}\right)}\right].$

Evaluating, $\quad K = \dfrac{6.02\times10^{-38}\ \text{J}\cdot\text{m}^2}{d^2} \qquad K = \dfrac{3.77\times10^{-19}\ \text{eV}\cdot\text{m}^2}{d^2}.$

In state 1, $\quad d = 1.00\times10^{-10}\ \text{m} \qquad K_1 = 37.7\ \text{eV}.$

In state 2, $\quad d = 5.00\times10^{-11}\ \text{m} \qquad K_2 = 151\ \text{eV}.$

In state 3, $\quad d = 3.33\times10^{-11}\ \text{m} \qquad K_3 = 339\ \text{eV}.$

In state 4, $\quad d = 2.50\times10^{-11}\ \text{m} \qquad K_4 = 603\ \text{eV}.$

(b) When the electron falls from state 2 to state 1, it puts out energy

$$E = 151\ \text{eV} - 37.7\ \text{eV} = 113\ \text{eV} = hf = \frac{hc}{\lambda}$$

into emitting a photon of wavelength

$$\lambda = \frac{hc}{E} = \frac{\left(6.626\times10^{-34}\ \text{J}\cdot\text{s}\right)\left(3.00\times10^{8}\ \text{m/s}\right)}{(113\ \text{eV})\left(1.60\times10^{-19}\ \text{J/eV}\right)} = 11.0\ \text{nm}.$$

The wavelengths of the other spectral lines we find similarly:

Transition	$4\to3$	$4\to2$	$4\to1$	$3\to2$	$3\to1$	$2\to1$
E(eV)	264	452	565	188	302	113
λ(nm)	4.71	2.75	2.20	6.60	4.12	11.0

P28.40 The confined proton can be described in the same way as a standing wave on a string. At level 1, the node-to-node distance of the standing wave is 1.00×10^{-14} m, so the wavelength is twice this distance: $\frac{h}{p}=2.00\times10^{-14}$ m.

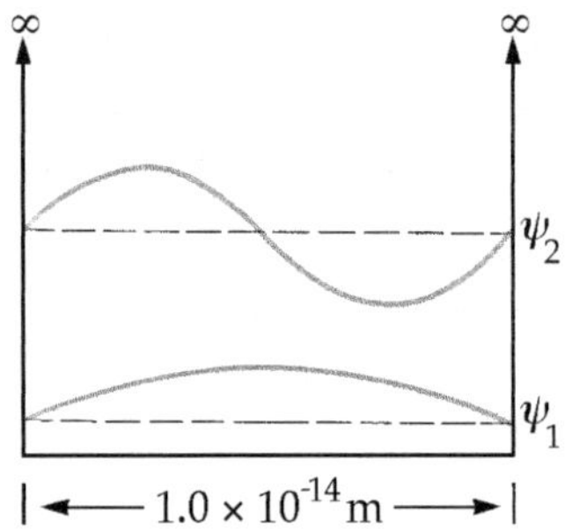

FIG. P28.40

The proton's kinetic energy is

$$K=\frac{1}{2}mv^2=\frac{p^2}{2m}=\frac{h^2}{2m\lambda^2}=\frac{\left(6.626\times10^{-34}\ \text{J}\cdot\text{s}\right)^2}{2\left(1.67\times10^{-27}\ \text{kg}\right)\left(2.00\times10^{-14}\ \text{m}\right)^2}$$

$$=\frac{3.29\times10^{-13}\ \text{J}}{1.60\times10^{-19}\ \text{J/eV}}=2.05\ \text{MeV}$$

In the first excited state, level 2, the node-to-node distance is half as long as in state 1. The momentum is two times larger and the energy is four times larger: $K=8.22$ MeV.

The proton has mass, has charge, moves slowly compared to light in a standing wave state, and stays inside the nucleus. When it falls from level 2 to level 1, its energy change is

$2.05\ \text{MeV}-8.22\ \text{MeV}=-6.16\ \text{MeV}$.

Therefore, we know that a photon (a traveling wave with no mass and no charge) is emitted at the speed of light, and that it has an energy of $\boxed{+6.16\ \text{MeV}}$.

Its frequency is $$f=\frac{E}{h}=\frac{\left(6.16\times10^6\ \text{eV}\right)\left(1.60\times10^{-19}\ \text{J/eV}\right)}{6.626\times10^{-34}\ \text{J}\cdot\text{s}}=1.49\times10^{21}\ \text{Hz}.$$

And its wavelength is $$\lambda=\frac{c}{f}=\frac{3.00\times10^8\ \text{m/s}}{1.49\times10^{21}\ \text{s}^{-1}}=\boxed{2.02\times10^{-13}\ \text{m}}.$$

This is a gamma ray, according to the electromagnetic spectrum chart in Chapter 24.

***P28.41** (a) The energies of the confined electron are $E_n=\frac{h^2}{8m_eL^2}n^2$. Its energy gain in the quantum jump from state 1 to state 4 is $\frac{h^2}{8m_eL^2}\left(4^2-1^2\right)$ and this is the photon energy:

$$\frac{h^2 15}{8m_eL^2}=hf=\frac{hc}{\lambda}.\ \text{Then } 8m_ecL^2=15h\lambda \text{ and } \boxed{L=\left(\frac{15h\lambda}{8m_ec}\right)^{1/2}}.$$

(b) Let λ' represent the wavelength of the photon emitted: $\frac{hc}{\lambda'}=\frac{h^2}{8m_eL^2}4^2-\frac{h^2}{8m_eL^2}2^2=\frac{12h^2}{8m_eL^2}$.

Then $\frac{hc}{\lambda}\frac{\lambda'}{hc}=\frac{h^2 15\left(8m_eL^2\right)}{8m_eL^2 12h^2}=\frac{5}{4}$ and $\boxed{\lambda'=1.25\lambda}$.

Section 28.11 The Quantum Particle Under Boundary Conditions

Section 28.12 The Schrödinger Equation

P28.42 $\psi(x) = A\cos kx + B\sin kx$ $\qquad \dfrac{\partial\psi}{\partial x} = -kA\sin kx + kB\cos kx$

$\dfrac{\partial^2\psi}{\partial x^2} = -k^2 A\cos kx - k^2 B\sin kx$ $\qquad -\dfrac{2m}{\hbar}(E-U)\psi = -\dfrac{2mE}{\hbar^2}(A\cos kx + B\sin kx)$

Therefore the Schrödinger equation is satisfied if

$\dfrac{\partial^2\psi}{\partial x^2} = \left(-\dfrac{2m}{\hbar^2}\right)(E-U)\psi$ or $\qquad -k^2(A\cos kx + B\sin kx) = \left(-\dfrac{2mE}{\hbar^2}\right)(A\cos kx + B\sin kx).$

This is true as an identity (functional equality) for all x if $\boxed{E = \dfrac{\hbar^2 k^2}{2m}}$.

P28.43 We have $\quad \psi = Ae^{i(kx-\omega t)} \quad \dfrac{\partial\psi}{\partial x} = ik\psi \quad$ and $\quad \dfrac{\partial^2\psi}{\partial x^2} = -k^2\psi.$

Schrödinger's equation: $\qquad \dfrac{\partial^2\psi}{\partial x^2} = -k^2\psi = -\dfrac{2m}{\hbar^2}(E-U)\psi.$

Since $\quad k^2 = \dfrac{(2\pi)^2}{\lambda^2} = \dfrac{(2\pi p)^2}{h^2} = \dfrac{p^2}{\hbar^2} \quad$ and $\quad E - U = \dfrac{p^2}{2m}.$

$\boxed{\text{Thus this equation balances.}}$

P28.44 (a) With $\quad \psi(x) = Ae^{ikx}$

$$\frac{d}{dx}Ae^{ikx} = Aikxe^{ikx} \text{ and } \frac{d^2\psi}{dx^2} = -Ak^2e^{ikx}.$$

Then $\quad -\dfrac{\hbar^2}{2m}\dfrac{d^2\psi}{dx^2} = +\dfrac{\hbar^2k^2}{2m}Ae^{ikx} = \dfrac{h^2}{4\pi^2}\dfrac{4\pi^2}{\lambda^2 2m}\psi = \dfrac{p^2}{2m}\psi = \dfrac{m^2v^2}{2m}\psi = \dfrac{1}{2}mv^2\psi = K\psi.$

(b) With $\quad \psi(x) = \sqrt{\dfrac{2}{L}}\sin\left(\dfrac{n\pi x}{L}\right) = A\sin kx, \ \dfrac{d\psi}{dx} = Ak\cos kx$ and $\dfrac{d^2\psi}{dx^2} = -Ak^2\sin kx$

Then $\quad -\dfrac{\hbar^2}{2m}\dfrac{d^2\psi}{dx^2} = +\dfrac{\hbar^2}{2m}Ak^2\sin kx = \dfrac{h^2 4\pi^2}{4\pi^2\lambda^2 2m}\psi = \dfrac{p^2}{2m}\psi = K\psi.$

P28.45 (a) $\quad \langle x\rangle = \displaystyle\int_0^L x\frac{2}{L}\sin^2\left(\frac{2\pi x}{L}\right)dx = \frac{2}{L}\int_0^L x\left(\frac{1}{2} - \frac{1}{2}\cos\frac{4\pi x}{L}\right)dx$

$$\langle x\rangle = \frac{1}{L}\frac{x^2}{2}\Bigg|_0^L - \frac{1}{L}\frac{L^2}{16\pi^2}\left[\frac{4\pi x}{L}\sin\frac{4\pi x}{L} + \cos\frac{4\pi x}{L}\right]_0^L = \boxed{\frac{L}{2}}$$

continued on next page

(b) $$\text{Probability} = \int_{0.490L}^{0.510L} \frac{2}{L}\sin^2\left(\frac{2\pi x}{L}\right)dx = \left[\frac{1}{L}x - \frac{1}{L}\frac{L}{4\pi}\sin\frac{4\pi x}{L}\right]_{0.490L}^{0.510L}$$

$$\text{Probability} = 0.020 - \frac{1}{4\pi}(\sin 2.04\pi - \sin 1.96\pi) = \boxed{5.26\times10^{-5}}$$

(c) $$\text{Probability}\left[\frac{x}{L} - \frac{1}{4\pi}\sin\frac{4\pi x}{L}\right]_{0.240L}^{0.260L} = \boxed{3.99\times10^{-2}}$$

(d) In the $n = 2$ graph in Figure 28.23 (b), it is more probable to find the particle either near $x = \frac{L}{4}$ or $x = \frac{3L}{4}$ than at the center, where the probability density is zero.

Nevertheless, the symmetry of the distribution means that the average position is $\frac{L}{2}$.

P28.46 Normalization requires

$$\int_{\text{all space}} |\psi|^2 dx = 1 \quad \text{or} \quad \int_0^L A^2 \sin^2\left(\frac{n\pi x}{L}\right)dx = 1$$

$$\int_0^L A^2 \sin^2\left(\frac{n\pi x}{L}\right)dx = A^2\left(\frac{L}{2}\right) = 1 \quad \text{or} \quad \boxed{A = \sqrt{\frac{2}{L}}}.$$

P28.47 The desired probability is $$P = \int_0^{L/4} |\psi|^2 dx = \frac{2}{L}\int_0^{L/4} \sin^2\left(\frac{2\pi x}{L}\right)dx$$

where $$\sin^2\theta = \frac{1-\cos 2\theta}{2}.$$

Thus, $$P = \left(\frac{x}{L} - \frac{1}{4\pi}\sin\frac{4\pi x}{L}\right)\Bigg|_0^{L/4} = \left(\frac{1}{4} - 0 - 0 + 0\right) = \boxed{0.250}.$$

P28.48 (a) $$\psi(x) = A\left(1 - \frac{x^2}{L^2}\right) \qquad \frac{d\psi}{dx} = -\frac{2Ax}{L^2} \qquad \frac{d^2\psi}{dx^2} = -\frac{2A}{L^2}$$

Schrödinger's equation $$\frac{d^2\psi}{dx^2} = -\frac{2m}{\hbar^2}(E-U)\psi$$

becomes $$-\frac{2A}{L^2} = -\frac{2m}{\hbar^2}EA\left(1 - \frac{x^2}{L^2}\right) + \frac{2m}{\hbar^2}\frac{\left(-\hbar^2x^2\right)A\left(1 - x^2/L^2\right)}{mL^2\left(L^2 - x^2\right)}$$

$$-\frac{1}{L^2} = -\frac{mE}{\hbar^2} + \frac{mEx^2}{\hbar^2L^2} - \frac{x^2}{L^4}.$$

This will be true for all x if both $$\frac{1}{L^2} = \frac{mE}{\hbar^2}$$

and $$\frac{mE}{\hbar^2L^2} - \frac{1}{L^4} = 0$$

both these conditions are satisfied for a particle of energy $\boxed{E = \frac{\hbar^2}{L^2m}}$.

continued on next page

(b) For normalization,
$$1=\int_{-L}^{L}A^2\left(1-\frac{x^2}{L^2}\right)^2 dx=A^2\int_{-L}^{L}\left(1-\frac{2x^2}{L^2}+\frac{x^4}{L^4}\right)dx$$

$$1=A^2\left[x-\frac{2x^3}{3L^2}+\frac{x^5}{5L^4}\right]_{-L}^{L}=A^2\left[L-\frac{2}{3}L+\frac{L}{5}+L-\frac{2}{3}L+\frac{L}{5}\right]=A^2\left(\frac{16L}{15}\right)\qquad \boxed{A=\sqrt{\frac{15}{16L}}}.$$

(c) $$P=\int_{-L/3}^{L/3}\psi^2 dx=\frac{15}{16L}\int_{-L/3}^{L/3}\left(1-\frac{2x^2}{L^2}+\frac{x^4}{L^4}\right)dx=\frac{15}{16L}\left[x-\frac{2x^3}{3L^2}+\frac{x^5}{5L^5}\right]_{-L/3}^{L/3}=\frac{30}{16L}\left[\frac{L}{3}-\frac{2L}{81}+\frac{L}{1\,215}\right]$$

$$P=\frac{47}{81}=\boxed{0.580}$$

Section 28.13 Tunneling Through a Potential Energy Barrier

P28.49 $T=e^{-2CL}$ where $C=\dfrac{\sqrt{2m(U-E)}}{\hbar}$

$$2CL=\frac{2\sqrt{2\left(9.11\times10^{-31}\right)\left(8.00\times10^{-19}\right)}}{1.055\times10^{-34}}\left(2.00\times10^{-10}\right)=4.58$$

(a) $T=e^{-4.58}=\boxed{0.010\,3}$, a 1% chance of transmission.

(b) $R=1-T=\boxed{0.990}$, a 99% chance of reflection.

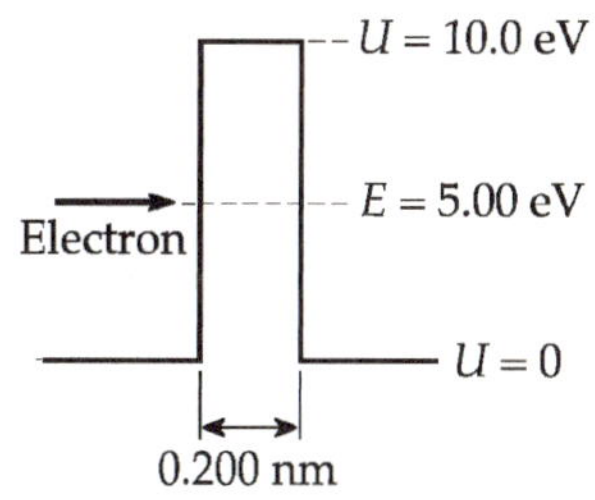

FIG. P28.49

P28.50 $$C=\frac{\sqrt{2\left(9.11\times10^{-31}\right)(5.00-4.50)\left(1.60\times10^{-19}\right)}\ \text{kg}\cdot\text{m/s}}{1.055\times10^{-34}\ \text{J}\cdot\text{s}}=3.62\times10^{9}\ \text{m}^{-1}$$

$$T=e^{-2CL}=\exp\left[-2\left(3.62\times10^{9}\ \text{m}^{-1}\right)\left(950\times10^{-12}\ \text{m}\right)\right]=\exp(-6.88)$$

$$T=\boxed{1.03\times10^{-3}}$$

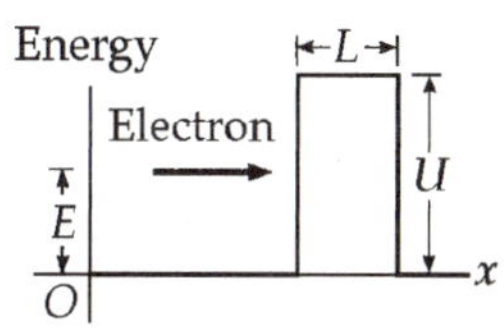

FIG. P28.50

***P28.51** The original tunneling probability is $T=e^{-2CL}$ where

$$C=\frac{\left(2m(U-E)\right)^{1/2}}{\hbar}=\frac{2\pi\left(2\times9.11\times10^{-31}\ \text{kg}(20-12)1.6\times10^{-19}\ \text{J}\right)^{1/2}}{6.626\times10^{-34}\ \text{J}\cdot\text{s}}=1.448\,1\times10^{10}\ \text{m}^{-1}.$$

The photon energy is $hf=\dfrac{hc}{\lambda}=\dfrac{1\,240\ \text{eV}\cdot\text{nm}}{546\ \text{nm}}=2.27\ \text{eV}$, to make the electron's new kinetic energy $12+2.27=14.27\ \text{eV}$ and its decay coefficient inside the barrier

$$C'=\frac{2\pi\left(2\times9.11\times10^{-31}\ \text{kg}(20-14.27)1.6\times10^{-19}\ \text{J}\right)^{1/2}}{6.626\times10^{-34}\ \text{J}\cdot\text{s}}=1.225\,5\times10^{10}\ \text{m}^{-1}.$$

Now the factor of increase in transmission probability is

$$\frac{e^{-2C'L}}{e^{-2CL}}=e^{2L(C-C')}=e^{2\times10^{-9}\ \text{m}\times0.223\times10^{10}\ \text{m}^{-1}}=e^{4.45}=\boxed{85.9}.$$

Section 28.14 Context Connection—The Cosmic Temperature

P28.52 The radiation wavelength of $\lambda' = 500$ nm that is observed by observers on Earth is not the true wavelength, λ, emitted by the star because of the Doppler effect. The true wavelength is related to the observed wavelength using:

$$\frac{c}{\lambda'} = \frac{c}{\lambda}\sqrt{\frac{1-(v/c)}{1+(v/c)}}: \qquad \lambda = \lambda'\sqrt{\frac{1-(v/c)}{1+(v/c)}} = (500 \text{ nm})\sqrt{\frac{1-(0.280)}{1+(0.280)}} = 375 \text{ nm}.$$

The temperature of the star is given by $\lambda_{max} T = 2.898 \times 10^{-3}$ m·K:

$$T = \frac{2.898 \times 10^{-3} \text{ m}\cdot\text{K}}{\lambda_{max}}: \qquad T = \frac{2.898 \times 10^{-3} \text{ m}\cdot\text{K}}{375 \times 10^{-9}} = \boxed{7.73 \times 10^3 \text{ K}}.$$

P28.53 (a) Wien's law: $\lambda_{max} T = 2.898 \times 10^{-3}$ m·K.

Thus, $$\lambda_{max} = \frac{2.898 \times 10^{-3} \text{ m}\cdot\text{K}}{T} = \frac{2.898 \times 10^{-3} \text{ m}\cdot\text{K}}{2.73 \text{ K}} = 1.06 \times 10^{-3} \text{ m} = \boxed{1.06 \text{ mm}}.$$

(b) This is a $\boxed{\text{microwave}}$.

P28.54 We suppose that the fireball of the Big Bang is a black body.

$$I = e\sigma T^4 = (1)\left(5.67 \times 10^{-8} \text{ W/m}^2\cdot\text{K}^4\right)(2.73 \text{ K})^4 = \boxed{3.15 \times 10^{-6} \text{ W/m}^2}$$

As a bonus, we can find the current power of direct radiation from the Big Bang in the section of the universe observable to us. If it is fifteen billion years old, the fireball is a perfect sphere of radius fifteen billion light years, centered at the point halfway between your eyes:

$$\mathcal{P} = IA = I(4\pi r^2) = \left(3.15 \times 10^{-6} \text{ W/m}^2\right)(4\pi)\left(15 \times 10^9 \text{ ly}\right)^2\left(\frac{3 \times 10^8 \text{ m/s}}{1 \text{ ly/yr}}\right)^2\left(3.156 \times 10^7 \text{ s/yr}\right)^2$$

$$\mathcal{P} = 7.98 \times 10^{47} \text{ W}.$$

Additional Problems

***P28.55** The condition on electric power delivered to the filament is $\mathcal{P} = I\Delta V = \frac{(\Delta V)^2}{R} = \frac{(\Delta V)^2 A}{\rho\ell} = \frac{(\Delta V)^2 \pi r^2}{\rho\ell}$

so $r = \left(\frac{\mathcal{P}\rho\ell}{\pi(\Delta V)^2}\right)^{1/2}$. Here $\mathcal{P} = 75$ W, $\rho = 7.13 \times 10^{-7}\,\Omega\cdot\text{m}$, and $\Delta V = 120$ V. As the filament radiates in steady state, it must emit all of this power through its lateral surface area $\mathcal{P} = \sigma e A T^4 = \sigma e 2\pi r \ell T^4$. We combine the conditions by substitution:

continued on next page

$$\mathcal{P} = \sigma e 2\pi \left(\frac{\mathcal{P}\rho \ell}{\pi(\Delta V)^2} \right)^{1/2} \ell T^4$$

$$(\Delta V)\mathcal{P}^{1/2} = \sigma e 2\pi^{1/2} \rho^{1/2} \ell^{3/2} T^4$$

$$\ell = \left(\frac{(\Delta V)\mathcal{P}^{1/2}}{\sigma e 2\pi^{1/2} \rho^{1/2} T^4} \right)^{2/3} = \left(\frac{120 \text{ V}(75 \text{ W})^{1/2} \text{ m}^2\text{K}^4}{5.67\times10^{-8} \text{ W } 0.450(2)\pi^{1/2}\left(7.13\times10^{-7}\,\Omega\cdot\text{m}\right)^{1/2}(2\,900 \text{ K})^4} \right)^{2/3}$$

$$= \left(0.192 \text{ m}^{3/2}\right)^{2/3} = \boxed{0.333 \text{ m} = \ell}$$

and $r = \left(\frac{\mathcal{P}\rho\ell}{\pi(\Delta V)^2} \right)^{1/2} = \left(\frac{75 \text{ W } 7.13\times10^{-7}\,\Omega\cdot\text{m } 0.333 \text{ m}}{\pi(120 \text{ V})^2} \right)^{1/2} = \boxed{r = 1.98\times10^{-5} \text{ m}}$.

P28.56 $\Delta V_S = \left(\frac{h}{e}\right) f - \frac{\phi}{e}$

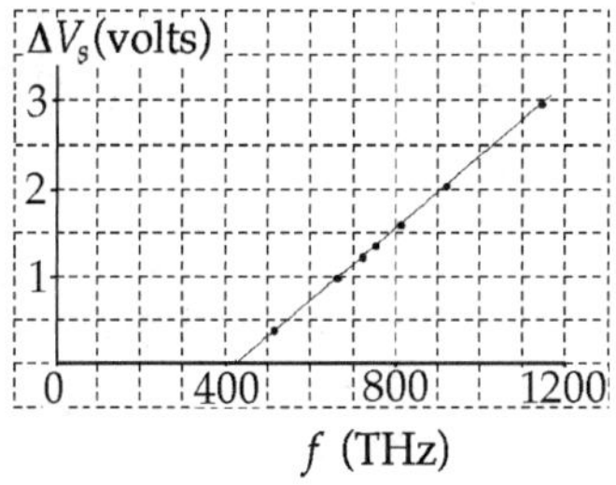

FIG. P28.56

From two points on the graph $0 = \left(\frac{h}{e}\right)\left(4.1\times10^{14} \text{ Hz}\right) - \frac{\phi}{e}$

and $3.3 \text{ V} = \left(\frac{h}{e}\right)\left(12\times10^{14} \text{ Hz}\right) - \frac{\phi}{e}$.

Combining these two expressions we find:

(a) $\phi = \boxed{1.7 \text{ eV}}$

(b) $\frac{h}{e} = \boxed{4.2\times10^{-15} \text{ V}\cdot\text{s}}$

(c) At the cutoff wavelength $\frac{hc}{\lambda_c} = \phi = \left(\frac{h}{e}\right)\frac{ec}{\lambda_c}$

$$\lambda_c = \left(4.2\times10^{-15} \text{ V}\cdot\text{s}\right)\left(1.6\times10^{-19} \text{ C}\right)\frac{\left(3\times10^8 \text{ m/s}\right)}{(1.7 \text{ eV})\left(1.6\times10^{-19} \text{ J/eV}\right)} = \boxed{730 \text{ nm}}$$

P28.57 We want an Einstein plot of K_{max} versus $f = \frac{c}{\lambda}$

λ, nm	f, 10^{14} Hz	K_{max}, eV
588	5.10	0.67
505	5.94	0.98
445	6.74	1.35
399	7.52	1.63

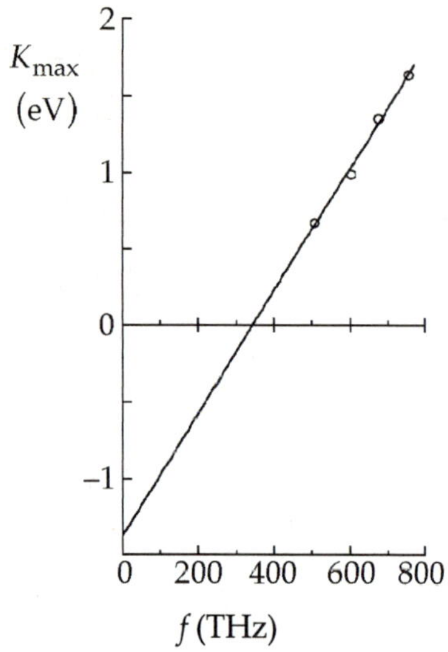

FIG. P28.57

(a) $\text{slope} = \frac{0.402 \text{ eV}}{10^{14} \text{ Hz}} \pm 8\%$

(b) $e\Delta V_S = hf - \phi$

$$h = (0.402)\left(\frac{1.60 \times 10^{-19} \text{ J} \cdot \text{s}}{10^{14}}\right) = \boxed{6.4 \times 10^{-34} \text{ J} \cdot \text{s} \pm 8\%}$$

(c) $K_{max} = 0$

at $f \approx 344 \times 10^{12}$ Hz

$$\phi = hf = 2.32 \times 10^{-19} \text{ J} = \boxed{1.4 \text{ eV}}$$

P28.58 From the path the electrons follow in the magnetic field, the maximum kinetic energy is seen to be:

$$K_{max} = \frac{e^2B^2R^2}{2m_e}.$$

From the photoelectric equation, $K_{max} = hf - \phi = \frac{hc}{\lambda} - \phi.$

Thus, the work function is $\phi = \frac{hc}{\lambda} - K_{max} = \boxed{\frac{hc}{\lambda} - \frac{e^2B^2R^2}{2m_e}}.$

P28.59 (a) $mgy_i = \frac{1}{2}mv_f^2$

$$v_f = \sqrt{2gy_i} = \sqrt{2(9.80 \text{ m/s}^2)(50.0 \text{ m})} = 31.3 \text{ m/s}$$

$$\lambda = \frac{h}{mv} = \frac{6.626 \times 10^{-34} \text{ J} \cdot \text{s}}{(75.0 \text{ kg})(31.3 \text{ m/s})} = \boxed{2.82 \times 10^{-37} \text{ m}} \quad \text{(not observable)}$$

(b) $\Delta E \Delta t \geq \frac{\hbar}{2}$

$$\text{so } \Delta E \geq \frac{6.626 \times 10^{-34} \text{ J} \cdot \text{s}}{4\pi(5.00 \times 10^{-3} \text{ s})} = \boxed{1.06 \times 10^{-32} \text{ J}}$$

(c) $$\frac{\Delta E}{E} = \frac{1.06 \times 10^{-32} \text{ J}}{(75.0 \text{ kg})(9.80 \text{ m/s}^2)(50.0 \text{ m})} = \boxed{2.87 \times 10^{-35}\%}$$

P28.60 (a) $\lambda = 2L = \boxed{2.00\times 10^{-10}\text{ m}}$

(b) $p = \frac{h}{\lambda} = \frac{6.626\times 10^{-34}\text{ J}\cdot\text{s}}{2.00\times 10^{-10}\text{ m}} = \boxed{3.31\times 10^{-24}\text{ kg}\cdot\text{m/s}}$

(c) $E = \frac{p^2}{2m} = \boxed{0.172\text{ eV}}$

P28.61 (a) See the figure. (b) See the figure.

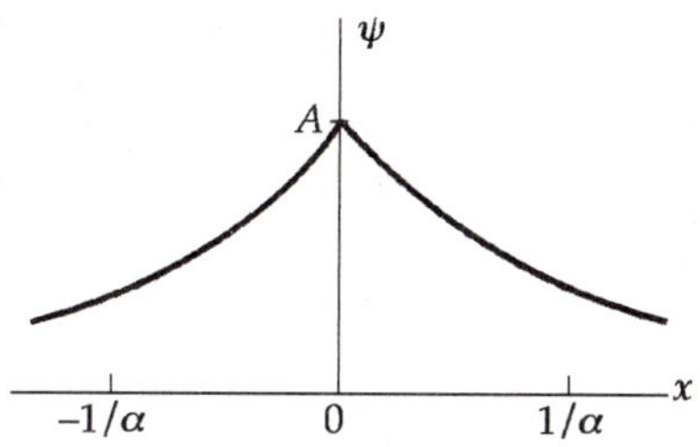

FIG. P28.61(a)

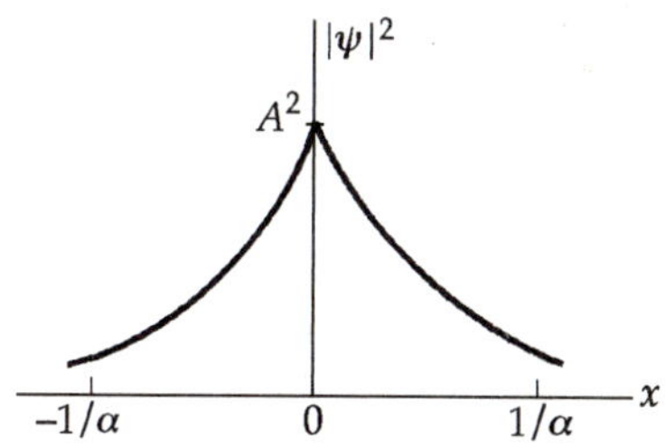

FIG. P28.61(b)

(c) ψ is continuous and $\psi \to 0$ as $x \to \pm\infty$. The function can be normalized. It describes a particle bound near $x = 0$, by a very deep, very narrow well of potential energy.

(d) Since ψ is symmetric,

$$\int_{-\infty}^{\infty} |\psi|^2 dx = 2\int_0^{\infty} |\psi|^2 dx = 1$$

or $$2A^2\int_0^{\infty} e^{-2\alpha x}dx = \left(\frac{2A^2}{-2\alpha}\right)\left(e^{-\infty} - e^0\right) = 1.$$

This gives $\boxed{A = \sqrt{\alpha}}$.

(e) $$P_{(-1/2\alpha)\to(1/2\alpha)} = 2\left(\sqrt{a}\right)^2\int_{x=0}^{1/2\alpha} e^{-2\alpha x}dx = \left(\frac{2\alpha}{-2\alpha}\right)\left(e^{-2\alpha/2\alpha} - 1\right) = \left(1 - e^{-1}\right) = \boxed{0.632}$$

P28.62 (a) Use Schrödinger's equation

$$\frac{\partial^2\psi}{\partial x^2} = -\frac{2m}{\hbar^2}(E - U)\psi$$

with solutions

$$\psi_1 = Ae^{ik_1x} + Be^{-ik_1x} \qquad \text{[region } I\text{]}$$

$$\psi_2 = Ce^{ik_2x} \qquad \text{[region } II\text{]}.$$

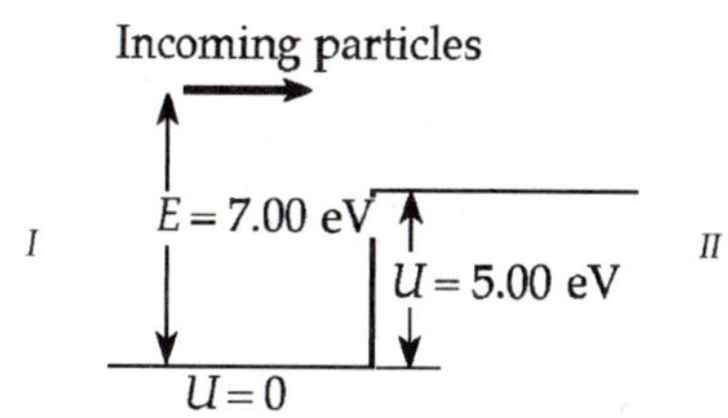

FIG. P28.62(a)

continued on next page

Where $k_1 = \frac{\sqrt{2mE}}{\hbar}$

and $k_2 = \frac{\sqrt{2m(E-U)}}{\hbar}$.

Then, matching functions and derivatives at $x = 0$

$(\psi_1)_0 = (\psi_2)_0$ gives $A + B = C$

and $\left(\frac{d\psi_1}{dx}\right)_0 = \left(\frac{d\psi_2}{dx}\right)_0$ gives $k_1(A - B) = k_2 C$.

Then $B = \frac{1 - k_2/k_1}{1 + k_2/k_1} A$

and $C = \frac{2}{1 + k_2/k_1} A$.

Incident wave Ae^{ikx} reflects Be^{-ikx}, with probability $R = \frac{B^2}{A^2} = \frac{(1 - k_2/k_1)^2}{(1 + k_2/k_1)^2} = \boxed{\frac{(k_1 - k_2)^2}{(k_1 + k_2)^2}}$.

(b) With $E = 7.00\text{ eV}$

and $U = 5.00\text{ eV}$

$$\frac{k_2}{k_1} = \sqrt{\frac{E - U}{E}} = \sqrt{\frac{2.00}{7.00}} = 0.535.$$

The reflection probability is $R = \frac{(1 - 0.535)^2}{(1 + 0.535)^2} = \boxed{0.092\,0}$.

The probability of transmission is $T = 1 - R = \boxed{0.908}$.

P28.63 $\langle x^2 \rangle = \int_{-\infty}^{\infty} x^2 |\psi|^2 dx$

For a one-dimensional box of width L, $\psi_n = \sqrt{\frac{2}{L}} \sin\left(\frac{n\pi x}{L}\right)$.

Thus, $\langle x^2 \rangle = \frac{2}{L}\int_0^L x^2 \sin^2\left(\frac{n\pi x}{L}\right) dx = \boxed{\frac{L^2}{3} - \frac{L^2}{2n^2\pi^2}}$ (from integral tables).

***P28.64** (a) The requirement that $\frac{n\lambda}{2} = L$ so $p = \frac{h}{\lambda} = \frac{nh}{2L}$ is still valid.

$$E = \sqrt{(pc)^2 + \left(mc^2\right)^2} \Rightarrow E_n = \sqrt{\left(\frac{nhc}{2L}\right)^2 + \left(mc^2\right)^2}$$

$$K_n = E_n - mc^2 = \boxed{\sqrt{\left(\frac{nhc}{2L}\right)^2 + \left(mc^2\right)^2} - mc^2}$$

(b) Taking $L = 1.00\times10^{-12}$ m, $m = 9.11\times10^{-31}$ kg, and $n = 1$, we find $K_1 = \boxed{4.69\times10^{-14}\text{ J}}$.

$$\text{Nonrelativistic, } E_1 = \frac{h^2}{8mL^2} = \frac{\left(6.626\times10^{-34}\text{ J}\cdot\text{s}\right)^2}{8\left(9.11\times10^{-31}\text{ kg}\right)\left(1.00\times10^{-12}\text{ m}\right)^2} = 6.02\times10^{-14}\text{ J}.$$

Comparing this to K_1, we see that this value is too large by $\boxed{28.6\%}$.

P28.65 For a particle with wave function

$$\psi(x) = \sqrt{\frac{2}{a}}e^{-x/a} \qquad \text{for } x > 0$$

and 0 for $x < 0$.

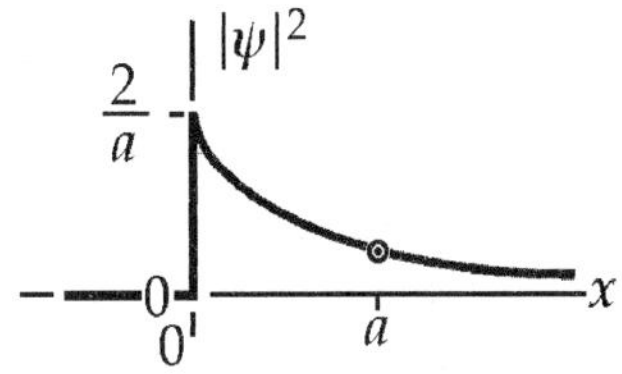

FIG. P28.65

(a) $|\psi(x)|^2 = 0,\ x < 0$ and $\left|\psi^2(x)\right| = \frac{2}{a}e^{-2x/a},\ x > 0$

(b) $\text{Prob}(x < 0) = \int_{-\infty}^{0}|\psi(x)|^2 dx = \int_{-\infty}^{0}(0)dx = \boxed{0}$

(c) Normalization $\int_{-\infty}^{\infty}|\psi(x)|^2 dx = \int_{-\infty}^{0}|\psi|^2 dx + \int_{0}^{\infty}|\psi|^2 dx = 1$

$$\int_{-\infty}^{0}0dx + \int_{0}^{\infty}\left(\frac{2}{a}\right)e^{-2x/a}dx = 0 - e^{-2x/a}\Big|_0^{\infty} = -\left(e^{-\infty} - 1\right) = 1$$

$$\text{Prob}(0 < x < a) = \int_0^a|\psi|^2 dx = \int_0^a\left(\frac{2}{a}\right)e^{-2x/a}dx = -e^{-2x/a}\Big|_0^a = 1 - e^{-2} = \boxed{0.865}$$

This wave function would require the potential energy to be $+\infty$ for $x < 0$ and $-\infty$ for $x = 0$. This potential energy function cannot be realistic in detail.

ANSWERS TO EVEN PROBLEMS

P28.2 (a) 5.78×10^3 K; (b) 501 nm

P28.4 (a) 2.57 eV; (b) 12.8 μeV; (c) 191 neV; (d) 484 nm visible, 9.68 cm and 6.52 m radio waves

P28.6 5.71×10^3 photons

P28.8 (a) 0.263 kg; (b) 1.81 W; (c) $-0.015\,3$ °C/s $= -0.919$ °C/min; (d) 9.89 μm; (e) 2.01×10^{-20} J; (f) 8.98×10^{19} photon/s

P28.10 (a) 1.38 eV; (b) 334 THz

P28.12 148 days, absurdly large

P28.14 1.78 eV, 9.47×10^{-28} kg·m/s

P28.16 $\dfrac{22.1 \text{ keV}}{c}$, 478 eV

P28.18 (a) 2.88 pm; (b) 101°

P28.20 397 fm

P28.22 0.218 nm

P28.24 (a) 1.10×10^{-34} m/s; (b) 1.36×10^{33} s; (c) no, the time is over 10^{15} times the age of the universe

P28.26 $v_{\text{phase}} = \dfrac{u}{2}$

P28.28 105 V

P28.30 2.27 pA

P28.32 (a) 0.250 m/s; (b) 2.25 m

P28.34 2.81×10^{-8}

P28.36 $\dfrac{1}{2}$

P28.38 (a) $n = 4$; (b) 6.03 eV

P28.40 6.16 MeV, 202 fm, a gamma ray

P28.42 see the solution, $E = \dfrac{\hbar^2 k^2}{2m}$

P28.44 see the solution

P28.46 see the solution

P28.48 (a) $E = \dfrac{\hbar^2}{mL^2}$; (b) requiring $\int_{-L}^{L} A^2\left(1 - \dfrac{x^2}{L^2}\right)^2 dx = 1$ gives $A = \left(\dfrac{15}{16L}\right)^{1/2}$; (c) $\dfrac{47}{81} = 0.580$

P28.50 1.03×10^{-3}

P28.52 7.73×10^3 K

P28.54 3.15 μW/m^2

P28.56 (a) 1.7 eV; (b) 4.2 fV·s; (c) 730 nm

P28.58 $\dfrac{hc}{\lambda} - \dfrac{e^2 B^2 R^2}{2m_e}$

P28.60 (a) 2.00×10^{-10} m; (b) 3.31×10^{-24} kg·m/s; (c) 0.172 eV

P28.62 (a) see the solution; (b) $R = 0.092\,0$, $T = 0.908$

P28.64 (a) $\sqrt{\left(\dfrac{nhc}{2L}\right)^2 + m^2c^4} - mc^2$; (b) 46.9 fJ, 28.6%

Chapter 29

Atomic Physics

CHAPTER OUTLINE

ANSWERS TO QUESTIONS

Q29.1 Bohr modeled the electron as moving in a perfect circle, with zero uncertainity in its radial coordinate. Then its radial velocity is always zero with zero uncertainty. Bohr's theory violates the uncertainty principle by making the uncertainty product $\Delta r \Delta p_r$ be zero, less than the minimum allowable $\frac{\hbar}{2}$.

Q29.2 Fundamentally, three quantum numbers describe an orbital wave function because we live in three-dimensional space. They arise mathematically from boundary conditions on the wave function, expressed as a product of a function of r, a function of θ, and a function of ϕ.

Q29.3 Bohr's theory pictures the electron as moving in a flat circle like a classical particle described by $\sum F = ma$. Schrödinger's theory pictures the electron as a cloud of probability amplitude in the three-dimensional space around the hydrogen nucleus, with its motion described by a wave equation. In the Bohr model, the ground-state angular momentum is $1\hbar$; in the Schrödinger model the ground-state angular momentum is zero. Both models predict that the electron's energy is limited to discrete energy levels, given by $\frac{-13.606 \text{ eV}}{n^2}$ with $n = 1, 2, 3$.

Q29.4 The term *electron cloud* refers to the unpredictable location of an electron around an atomic nucleus. It is a cloud of probability amplitude. An electron in an *s* subshell has a spherically symmetric probability distribution. Electrons in *p*, *d*, and *f* subshells have directionality to their distribution. The shape of these electron clouds influences how atoms form molecules and chemical compounds.

Q29.5 The direction of the magnetic moment due to an orbiting charge is given by the right hand rule, but assumes a *positive* charge. Since the electron is negatively charged, its magnetic moment is in the opposite direction to its angular momentum.

Q29.6 The deflecting force on an atom with a magnetic moment is proportional to the *gradient* of the magnetic field. Thus, atoms with oppositely directed magnetic moments would be deflected in *opposite* directions in an inhomogeneous magnetic field.

Q29.7 Practically speaking, no. Ions have a net charge and the magnetic force $q(\vec{\mathbf{v}} \times \vec{\mathbf{B}})$ would deflect the beam, making it difficult to separate the atoms with different orientations of magnetic moments.

Q29.8 The Stern-Gerlach experiment with hydrogen atoms shows that the component of an electron's spin angular momentum along an applied magnetic field can have only one of two allowed values. So does electron spin resonance on atoms with one unpaired electron.

Q29.9 If the exclusion principle were not valid, the elements and their chemical behavior would be grossly different because every electron would end up in the lowest energy level of the atom. All matter would be nearly alike in its chemistry and composition, since the shell structures of all elements would be identical. Most materials would have a much higher density. The spectra of atoms and molecules would be very simple, and there would be very little color in the world.

Q29.10 An atom is a quantum system described by a wave function. The electric force of attraction to the nucleus imposes a constraint on the electrons. The physical constraint implies mathematical boundary conditions on the wave functions, with consequent quantization so that only certain wave functions are allowed to exist. The Schrödinger equation assigns a definite energy to each allowed wave function. Each wave function is spread out in space, describing an electron with no definite position. If you like analogies, think of a classical standing wave on a string fixed at both ends. Its position is spread out to fill the whole string, but its frequency is one of a certain set of quantized values.

Q29.11 The three elements have similar electronic configurations. Each has filled inner shells plus one electron in an s orbital. Their single outer electrons largely determine their chemical interactions with other atoms.

Q29.12 In a neutral helium atom, one electron can be modeled as moving in an electric field created by the nucleus and the other electron. According to Gauss's law, if the electron is above the ground state it moves in the electric field of a net charge of $+2e-1e=+1e$ We say the nuclear charge is *screened* by the inner electron. The electron in a He^+ ion moves in the field of the unscreened nuclear charge of 2 protons. Then the potential energy function for the electron is about double that of one electron in the neutral atom.

Q29.13 At low density, the gas consists of essentially separate atoms. As the density increases, the atoms interact with each other. This has the effect of giving different atoms levels at slightly different energies, at any one instant. The collection of atoms can then emit photons in lines or bands, narrower or wider, depending on the density.

Q29.14 Each of the electrons must have at least one quantum number different from the quantum numbers of each of the other electrons. They can differ (in m_s) by being spin-up or spin-down. They can also differ (in ℓ) in angular momentum and in the general shape of the wave function. Those electrons with $\ell=1$ can differ (in m_ℓ) in orientation of angular momentum—look at Figure Q29.14.

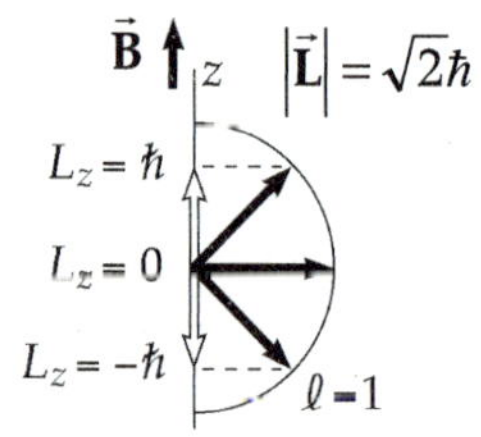

FIG. Q29.14

Q29.15 The Mosely graph shows that the reciprocal square root of the wavelength of K_α characteristic x-rays is a linear function of atomic number. Then measuring this wavelength for a new chemical element reveals its location on the graph, including its atomic number.

Q29.16 (a) The terms "I define" and "this part of the universe" seem vague, in contrast to the precision of the rest of the statement. But the statement is true in the sense of being experimentally verifiable. The way to test the orientation of the magnetic moment of an electron is to apply a magnetic field to it. When that is done for any electron, it has precisely a 50% chance of being either spin-up or spin-down. Its spin magnetic moment vector must make one of two allowed angles with the applied magnetic field. They are given by $\cos\theta = \frac{S_z}{S} = \frac{\frac{1}{2}}{\frac{\sqrt{3}}{2}}$ and $\cos\theta = \frac{-\frac{1}{2}}{\frac{\sqrt{3}}{2}}$. You can calculate as many digits of the two angles allowed by "space quantization" as you wish.

(b) This statement may be true. There is no reason to suppose that an ant can comprehend the cosmos, and no reason to suppose that a human can comprehend all of it. Our experience with macroscopic objects does not prepare us to understand quantum particles. On the other hand, what seems strange to us now may be the common knowledge of tomorrow. Looking back at the past 150 years of physics, great strides in understanding the Universe—from the quantum to the galactic scale—have been made. Think of trying to explain the photoelectric effect using Newtonian mechanics. What seems strange sometimes just has an underlying structure that has not yet been described fully. On the other hand still, it has been demonstrated that a "hidden-variable" theory, that would model quantum uncertainty as caused by some determinate but fluctuating quantity, cannot agree with experiment.

Q29.17 People commonly say that science is difficult to learn. Many other branches of knowledge are constructed by humans. Learning them can require getting your mind to follow the path of someone else's mind. A well-developed science does not mirror human patterns of thought, but the way nature works. Thus we agree with the view stated in the problem. A scientific discovery can be like a garbled communication in the sense that it does not explain itself. It can seem fragmentary and in need of context. It can seem unfriendly in the sense that it has no regard for ease of human comprehension. Evolution has adapted the human mind for catching rabbits and outwitting bison and leopards, but not necessarily for understanding atoms. Education in science gives the student opportunities for understanding a wide variety of ideas; for evaluating new information; and for recognizing that extraordinary evidence must be adduced for extraordinary claims.

SOLUTIONS TO PROBLEMS

Section 29.1 Early Structural Models of the Atom

P29.1 (a) For a classical atom, the centripetal acceleration is

$$a = \frac{v^2}{r} = \frac{1}{4\pi \epsilon_0} \frac{e^2}{r^2 m_e}$$

$$E = -\frac{e^2}{4\pi \epsilon_0 r} + \frac{m_e v^2}{2} = -\frac{e^2}{8\pi \epsilon_0 r}$$

so $$\frac{dE}{dt} = \frac{e^2}{8\pi \epsilon_0 r^2}\frac{dr}{dt} = \frac{-1}{6\pi \epsilon_0}\frac{e^2 a^2}{c^3} = \frac{-e^2}{6\pi \epsilon_0 c^3}\left(\frac{e^2}{4\pi \epsilon_0 r^2 m_e}\right)^2.$$

Therefore, $$\frac{dr}{dt} = -\frac{e^4}{12\pi^2 \epsilon_0^2 r^2 m_e^2 c^3}.$$

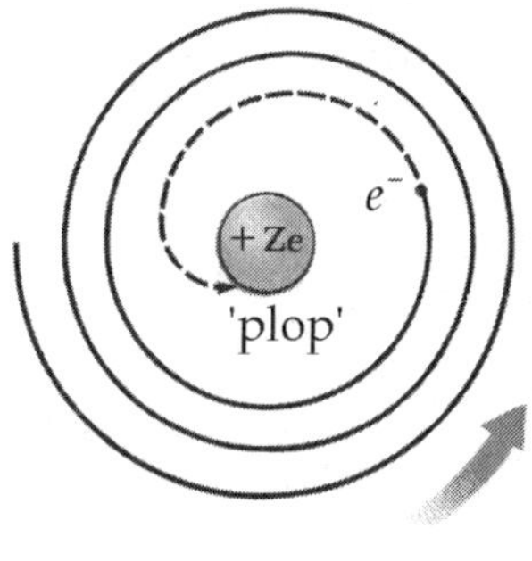

FIG. P29.1

continued on next page

(b) $-\int_{2.00\times10^{-10}\text{ m}}^{0} 12\pi^2 \epsilon_0^2 r^2 m_e^2 c^3 dr = e^4 \int_0^T dt \qquad \frac{12\pi^2 \epsilon_0^2 m_e^2 c^3}{e^4} \left.\frac{r^3}{3}\right|_0^{2.00\times10^{-10}} = T = \boxed{8.46\times10^{-10}\text{ s}}$

Since atoms last a lot longer than 0.8 ns, the classical laws (fortunately!) do not hold for systems of atomic size.

P29.2 (a) The point of closest approach is found when

$$E = K + U = 0 + \frac{k_e q_\alpha q_{\text{Au}}}{r}$$

or $$r_{\min} = \frac{k_e(2e)(79e)}{E}$$

$$r_{\min} = \frac{(8.99\times10^9\text{ N}\cdot\text{m}^2/\text{C}^2)(158)(1.60\times10^{-19}\text{ C})^2}{(4.00\text{ MeV})(1.60\times10^{-13}\text{ J/MeV})} = \boxed{5.68\times10^{-14}\text{ m}}.$$

(b) The maximum force exerted on the alpha particle is

$$F_{\max} = \frac{k_e q_\alpha q_{\text{Au}}}{r_{\min}^2} = \frac{(8.99\times10^9\text{ N}\cdot\text{m}^2/\text{C}^2)(158)(1.60\times10^{-19}\text{ C})^2}{(5.68\times10^{-14}\text{ m})^2} = \boxed{11.3\text{ N}}$$ away from the nucleus.

***P29.3** (a) The speed of the moon in its orbit is $v = \frac{2\pi r}{T} = \frac{2\pi(3.84\times10^8\text{ m})}{2.36\times10^6\text{ s}} = 1.02\times10^3\text{ m/s}$.

So, $$L = mvr = (7.36\times10^{22}\text{ kg})(1.02\times10^3\text{ m/s})(3.84\times10^8\text{ m}) = \boxed{2.89\times10^{34}\text{ kg}\cdot\text{m}^2/\text{s}}.$$

(b) We have $L = n\hbar$

or $$n = \frac{L}{\hbar} = \frac{2.89\times10^{34}\text{ kg}\cdot\text{m}^2/\text{s}}{1.055\times10^{-34}\text{ J}\cdot\text{s}} = \boxed{2.74\times10^{68}}.$$

(c) We have $n\hbar = L = mvr = m\left(\frac{GM_e}{r}\right)^{1/2} r$,

so $$r = \frac{\hbar^2}{m^2 G M_e} n^2 = Rn^2 \text{ and } \frac{\Delta r}{r} = \frac{(n+1)^2 R - n^2 R}{n^2 R} = \frac{2n+1}{n^2}$$

which is approximately equal to $\frac{2}{n} = \boxed{7.30\times10^{-69}}$.

***P29.4** (a) The fifth excited state must lie above the second excited state by the photon energy

$$E_{52} = hf = \frac{hc}{\lambda} = \frac{6.63\times10^{-34}\ \text{J}\cdot\text{s}\ 3\times10^{8}\ \text{m/s}}{520\times10^{-9}\ \text{m}} = 3.82\times10^{-19}\ \text{J}.$$

The sixth excited state exceeds the second in energy by

$$E_{62} = \frac{6.63\times10^{-34}\ \text{J}\cdot\text{s}\ 3\times10^{8}\ \text{m/s}}{410\times10^{-9}\ \text{m}} = 4.85\times10^{-19}\ \text{J}.$$

Then the sixth excited state is above the fifth by $(4.85-3.82)10^{-19}\ \text{J} = 1.03\times10^{-19}\ \text{J}$. In the 6 to 5 transition the atom emits a photon with the infrared wave length

$$\lambda = \frac{hc}{E_{62}} = \frac{6.63\times10^{-34}\ \text{J}\cdot\text{s}\ 3\times10^{8}\ \text{m/s}}{1.03\times10^{-19}\ \text{J}} = \boxed{1.94\times10^{-6}\ \text{m}}.$$

(b) The same steps solve the symbolic problem.

$$E_{CA} = E_{CB} + E_{BA}$$
$$\frac{hc}{\lambda_{CA}} = \frac{hc}{\lambda_{CB}} + \frac{hc}{\lambda_{BA}}$$
$$\frac{1}{\lambda_{CB}} = \frac{1}{\lambda_{CA}} - \frac{1}{\lambda_{BA}}$$
$$\boxed{\lambda_{CB} = \left(\frac{1}{\lambda_{CA}} - \frac{1}{\lambda_{BA}}\right)^{-1}}$$

Section 29.2 The Hydrogen Atom Revisited

P29.5 (a) Longest wavelength implies lowest frequency and smallest energy:

the atom falls from $n = 3$

to $n = 2$

losing energy $-\frac{13.6\ \text{eV}}{3^2} + \frac{13.6\ \text{eV}}{2^2} = \boxed{1.89\ \text{eV}}$

The photon frequency is $f = \frac{\Delta E}{h}$

and its wavelength is

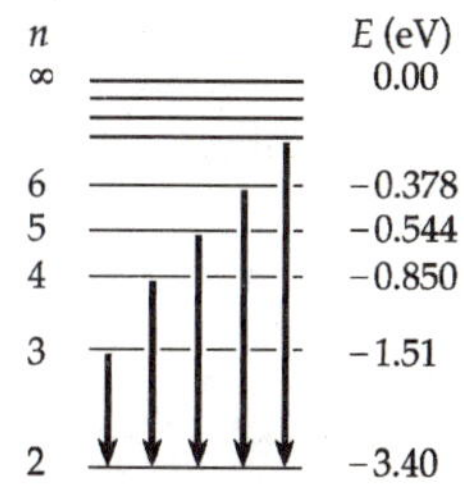

Balmer Series

FIG. P29.5

$$\lambda = \frac{c}{f} = \frac{hc}{\Delta E} = \frac{(6.626\times10^{-34}\ \text{J}\cdot\text{s})(3.00\times10^{8}\ \text{m/s})}{(1.89\ \text{eV})}\left(\frac{\text{eV}}{1.60\times10^{-19}\ \text{J}}\right)$$
$$\lambda = \boxed{656\ \text{nm}}$$

continued on next page

(b) The biggest energy loss is for an atom to fall from an ionized configuration,

$$n=\infty$$

to the $n=2$ state.

It loses energy $-\dfrac{13.6\text{ eV}}{\infty}+\dfrac{13.6\text{ eV}}{2^2}=\boxed{3.40\text{ eV}}$

to emit light of wavelength $\lambda=\dfrac{hc}{\Delta E}=\dfrac{(6.626\times10^{-34}\text{ J}\cdot\text{s})(3.00\times10^{8}\text{ m/s})}{(3.40\text{ eV})(1.60\times10^{-19}\text{ J/eV})}=\boxed{365\text{ nm}}$.

P29.6 (a) The photon has energy 2.28 eV.

And $\dfrac{13.6\text{ eV}}{2^2}=3.40\text{ eV}$ is required to ionize a hydrogen atom from state $n=2$. So while the photon cannot ionize a hydrogen atom pre-excited to $n=2$, it can ionize a hydrogen atom in the $n=\boxed{3}$ state, with energy $-\dfrac{13.6\text{ eV}}{3^2}=-1.51\text{ eV}$.

(b) The electron thus freed can have kinetic energy $K_e=2.28\text{ eV}-1.51\text{ eV}=0.769\text{ eV}=\dfrac{1}{2}m_e v^2$.

Therefore, $v=\sqrt{\dfrac{2(0.769)(1.60\times10^{-19})\text{ J}}{9.11\times10^{-31}\text{ kg}}}=\boxed{520\text{ km/s}}$.

P29.7 The reduced mass of positronium is **less** than hydrogen, so the photon energy will be **less** for positronium than for hydrogen. This means that the wavelength of the emitted photon will be **longer** than 656.3 nm. On the other hand, helium has about the same reduced mass but more charge than hydrogen, so its transition energy will be **larger**, corresponding to a wavelength **shorter** than 656.3 nm.

All the factors in the given equation are constant for this problem except for the reduced mass and the nuclear charge. Therefore, the wavelength corresponding to the energy difference for the transition can be found simply from the ratio of mass and charge variables.

For hydrogen, $\mu=\dfrac{m_p m_e}{m_p+m_e}\approx m_e$. The photon energy is $\Delta E=E_3-E_2$.

Its wavelength is $\lambda=656.3\text{ nm}$, where $\lambda=\dfrac{c}{f}=\dfrac{hc}{\Delta E}$.

(a) For positronium, $\mu=\dfrac{m_e m_e}{m_e+m_e}=\dfrac{m_e}{2}$

so the energy of each level is one half as large as in hydrogen, which we could call "protonium". The photon energy is inversely proportional to its wavelength , so for positronium,

$$\lambda_{32}=2(656.3\text{ nm})=\boxed{1.31\ \mu\text{m}}\text{ (in the infrared region).}$$

continued on next page

(b) For He^+, $\mu \approx m_e$, $q_1 = e$, and $q_2 = 2e$,

so the transition energy is $2^2 = 4$ times larger than hydrogen.

Then, $\lambda_{32} = \left(\frac{656}{4}\right)\text{nm} = \boxed{164\text{ nm}}$ (in the ultraviolet region).

P29.8 (a) For a particular transition from n_i to n_f, $\Delta E_H = -\frac{\mu_H k_e^2 e^4}{2\hbar^2}\left(\frac{1}{n_f^2} - \frac{1}{n_i^2}\right) = \frac{hc}{\lambda_H}$ and

$\Delta E_D = -\frac{\mu_D k_e^2 e^4}{2\hbar^2}\left(\frac{1}{n_f^2} - \frac{1}{n_i^2}\right) = \frac{hc}{\lambda_D}$ where $\mu_H = \frac{m_e m_p}{m_e + m_p}$ and $\mu_D = \frac{m_e m_D}{m_e + m_D}$. By division,

$\frac{\Delta E_H}{\Delta E_D} = \frac{\mu_H}{\mu_D} = \frac{\lambda_D}{\lambda_H}$ or $\lambda_D = \left(\frac{\mu_H}{\mu_D}\right)\lambda_H$. Then, $\boxed{\lambda_H - \lambda_D = \left(1 - \frac{\mu_H}{\mu_D}\right)\lambda_H}$.

(b) $$\frac{\mu_H}{\mu_D} = \left(\frac{m_e m_p}{m_e + m_p}\right)\left(\frac{m_e + m_D}{m_e m_D}\right) = \frac{(1.007\ 276\text{ u})(0.000\ 549\text{ u} + 2.013\ 553\text{ u})}{(0.000\ 549\text{ u} + 1.007\ 276\text{ u})(2.013\ 553\text{ u})} = 0.999\ 728$$

$\lambda_H - \lambda_D = (1 - 0.999\ 728)(656.3\text{ nm}) = \boxed{0.179\text{ nm}}$

***P29.9** (a) $\Delta x \Delta p \geq \frac{\hbar}{2}$ so if $\Delta x = r$, $\Delta p \geq \boxed{\frac{\hbar}{2r}}$.

(b) Arbitrarily choosing $\Delta p \approx \frac{\hbar}{r}$, we find $K = \frac{p^2}{2m_e} \approx \frac{(\Delta p)^2}{2m_e} = \boxed{\frac{\hbar^2}{2m_e r^2}}$

$U = \frac{-k_e e^2}{r}$, so $E = K + U \approx \boxed{\frac{\hbar^2}{2m_e r^2} - \frac{k_e e^2}{r}}$.

(c) To minimize E,

$$\frac{dE}{dr} = -\frac{\hbar^2}{m_e r^3} + \frac{k_e e^2}{r^2} = 0 \to r = \boxed{\frac{\hbar^2}{m_e k_e e^2} = a_0} \text{ (the Bohr radius).}$$

Then, $E = \frac{\hbar^2}{2m_e}\left(\frac{m_e k_e e^2}{\hbar^2}\right)^2 - k_e e^2\left(\frac{m_e k_e e^2}{\hbar^2}\right) = -\frac{m_e k_e^2 e^4}{2\hbar^2} = \boxed{-13.6\text{ eV}}$.

Section 29.3 The Wave Functions for Hydrogen

P29.10 $\psi_{1s}(r) = \dfrac{1}{\sqrt{\pi a_0^3}} e^{-r/a_0}$ (Eq. 29.3)

$P_{1s}(r) = \dfrac{4r^2}{a_0^3} e^{-2r/a_0}$ (Eq. 29.7)

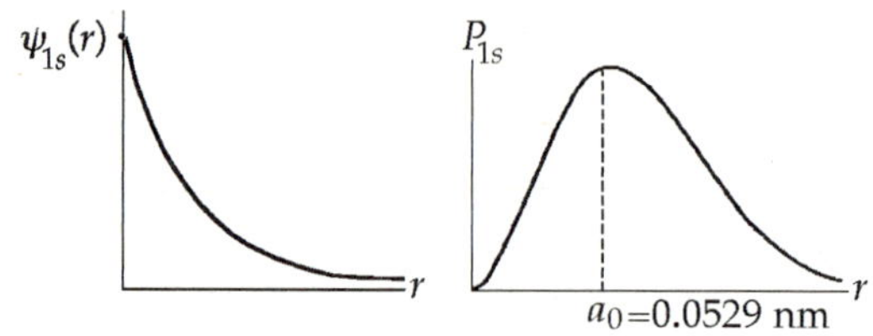

FIG. P29.10

P29.11 (a) $\int |\psi|^2 dV = 4\pi \int_0^\infty |\psi|^2 r^2 dr = 4\pi \left(\dfrac{1}{\pi a_0^3}\right) \int_0^\infty r^2 e^{-2r/a_0} dr$

Using integral tables, $\int |\psi|^2 dV = -\dfrac{2}{a_0^2}\left[e^{-2r/a_0}\left(r^2 + a_0 r + \dfrac{a_0^2}{2}\right)\right]_0^\infty = \left(-\dfrac{2}{a_0^2}\right)\left(-\dfrac{a_0^2}{2}\right) = \boxed{1}$

so the wave function as given is normalized.

(b) $P_{a_0/2 \to 3a_0/2} = 4\pi \int_{a_0/2}^{3a_0/2} |\psi|^2 r^2 dr = 4\pi \left(\dfrac{1}{\pi a_0^3}\right) \int_{a_0/2}^{3a_0/2} r^2 e^{-2r/a_0} dr$

Again, using integral tables,

$$P_{a_0/2 \to 3a_0/2} = -\frac{2}{a_0^2}\left[e^{-2r/a_0}\left(r^2 + a_0 r + \frac{a_0^2}{2}\right)\right]_{a_0/2}^{3a_0/2} = -\frac{2}{a_0^2}\left[e^{-3}\left(\frac{17a_0^2}{4}\right) - e^{-1}\left(\frac{5a_0^2}{4}\right)\right] = \boxed{0.497}.$$

P29.12 $\psi = \dfrac{1}{\sqrt{3}} \dfrac{1}{(2a_0)^{3/2}} \dfrac{r}{a_0} e^{-r/2a_0}$

so $\quad P_r = 4\pi r^2 \left|\psi^2\right| = 4\pi r^2 \dfrac{r^2}{24a_0^5} e^{-r/a_0}.$

Set $\quad \dfrac{dP}{dr} = \dfrac{4\pi}{24a_0^5}\left[4r^3 e^{-r/a_0} + r^4\left(-\dfrac{1}{a_0}\right)e^{-r/a_0}\right] = 0.$

Solving for r, this is a maximum at $\boxed{r = 4a_0}$.

P29.13 $\psi = \frac{1}{\sqrt{\pi a_0^3}} e^{-r/a_0}$ $\quad \frac{2}{r}\frac{d\psi}{dr} = \frac{-2}{r\sqrt{\pi a_0^5}} e^{-r/a_0} = -\frac{2}{r a_0}\psi$

$$\frac{d^2\psi}{dr^2} = \frac{1}{\sqrt{\pi a_0^7}} e^{-r/a_0} = \frac{1}{a_0^2}\psi \qquad -\frac{\hbar^2}{2m_e}\left(\frac{1}{a_0^2} - \frac{2}{r a_0}\right)\psi - \frac{e^2}{4\pi \in_0 r}\psi = E\psi$$

But $\quad a_0 = \frac{\hbar^2 (4\pi \in_0)}{m_e e^2}$

so $\quad -\frac{e^2}{8\pi \in_0 a_0} = E$

or $\quad \boxed{E = -\frac{k_e e^2}{2a_0}}$.

This is true, so the Schrödinger equation is satisfied.

P29.14 The hydrogen ground-state radial probability density is

$$P(r) = 4\pi r^2 |\psi_{1s}|^2 = \frac{4r^2}{a_0^3}\exp\left(-\frac{2r}{a_0}\right).$$

The number of observations at $2a_0$ is, by proportion

$$N = 1\,000\frac{P(2a_0)}{P\left(\frac{a_0}{2}\right)} = 1\,000\frac{(2a_0)^2 e^{-4a_0/a_0}}{\left(\frac{a_0}{2}\right)^2 e^{-a_0/a_0}} = 1\,000(16)e^{-3} = \boxed{797 \text{ times}}.$$

Section 29.4 Physical Interpretation of the Quantum Numbers

Note: Problems 22.20 and 22.17 in Chapter 22 can be assigned with Section 29.4.

P29.15 (a) In the 3*d* subshell, $n = 3$ and $\ell = 2$,

we have n	3	3	3	3	3	3	3	3	3	3
ℓ	2	2	2	2	2	2	2	2	2	2
m_ℓ	+2	+2	+1	+1	0	0	–1	–1	–2	–2
m_s	+1/2	–1/2	+1/2	–1/2	+1/2	–1/2	+1/2	–1/2	+1/2	–1/2

(A total of 10 states)

(b) In the 3*p* subshell, $n = 3$ and $\ell = 1$,

we have n	3	3	3	3	3	3
ℓ	1	1	1	1	1	1
m_ℓ	+1	+1	+0	+0	–1	–1
m_s	+1/2	–1/2	+1/2	–1/2	+1/2	–1/2

(A total of 6 states)

P29.16 (a) For the d state, $\ell = 2$, $L = \boxed{\sqrt{6}\hbar} = 2.58 \times 10^{-34}\ \text{J}\cdot\text{s}$.

(b) For the f state, $\ell = 3$, $L = \sqrt{\ell(\ell+1)}\hbar = \boxed{\sqrt{12}\hbar} = 3.65 \times 10^{-34}\ \text{J}\cdot\text{s}$.

P29.17 $L = \sqrt{\ell(\ell+1)}\hbar$:

$$4.714 \times 10^{-34} = \sqrt{\ell(\ell+1)}\left(\frac{6.626 \times 10^{-34}}{2\pi}\right)$$

$$\ell(\ell+1) = \frac{\left(4.714 \times 10^{-34}\right)^2 (2\pi)^2}{\left(6.626 \times 10^{-34}\right)^2} = 1.998 \times 10^1 \approx 20 = 4(4+1)$$

so $\boxed{\ell = 4}$.

P29.18 The 5th excited state has $n = 6$, energy $\dfrac{-13.6\ \text{eV}}{36} = -0.378\ \text{eV}$.

The atom loses this much energy: $\dfrac{hc}{\lambda} = \dfrac{\left(6.626 \times 10^{-34}\ \text{J}\cdot\text{s}\right)\left(3.00 \times 10^8\ \text{m/s}\right)}{\left(1\,090 \times 10^{-9}\ \text{m}\right)\left(1.60 \times 10^{-19}\ \text{J/eV}\right)} = 1.14\ \text{eV}$

to end up with energy $-0.378\ \text{eV} - 1.14\ \text{eV} = -1.52\ \text{eV}$

which is the energy in state 3: $-\dfrac{13.6\ \text{eV}}{3^3} = -1.51\ \text{eV}$.

While $n = 3$, ℓ can be as large as 2, giving angular momentum $\sqrt{\ell(\ell+1)}\hbar = \boxed{\sqrt{6}\hbar}$.

P29.19 (a) $n = 1$: For $n = 1$, $\ell = 0$, $m_\ell = 0$, $m_s = \pm\dfrac{1}{2}$

n	ℓ	m_ℓ	m_s
1	0	0	−1/2
1	0	0	+1/2

Yields 2 sets; $2n^2 = 2(1)^2 = \boxed{2}$

(b) $n = 2$: For $n = 2$,

we have

n	ℓ	m_ℓ	m_s
2	0	0	±1/2
2	1	−1	±1/2
2	1	0	±1/2
2	1	1	±1/2

yields 8 sets; $2n^2 = 2(2)^2 = \boxed{8}$

continued on next page

Note that the number is twice the number of m_ℓ values. Also, for each ℓ there are $(2\ell+1)$ different m_ℓ values. Finally, ℓ can take on values ranging from 0 to $n-1$.

So the general expression is $\quad \text{number} = \sum_0^{n-1} 2(2\ell+1)$.

The series is an arithmetic progression: $\quad 2+6+10+14\ldots$

the sum of which is $\quad \text{number} = \frac{n}{2}\left[2a+(n-1)d\right]$

where $a=2$, $d=4$: $\quad \text{number} = \frac{n}{2}\left[4+(n-1)4\right] = 2n^2$.

(c) $n=3$: $\quad 2(1)+2(3)+2(5)=2+6+10=18 \quad 2n^2 = 2(3)^2 = \boxed{18}$

(d) $n=4$: $\quad 2(1)+2(3)+2(5)+2(7)=32 \quad 2n^2 = 2(4)^2 = \boxed{32}$

(e) $n=5$: $\quad 32+2(9)=32+18=50 \quad 2n^2 = 2(5)^2 = \boxed{50}$

P29.20 For a 3*d* state, $\quad n=3$ and $\ell=2$.

Therefore, $\quad L=\sqrt{\ell(\ell+1)}\hbar = \boxed{\sqrt{6}\hbar} = 2.58\times10^{-34}\ \text{J}\cdot\text{s}$

m_ℓ can have the values $\quad -2, -1, 0, 1$, and 2

so $\quad \boxed{L_z \text{ can have the values } -2\hbar,\ -\hbar,\ 0,\ \hbar \text{ and } 2\hbar}$.

Using the relation $\quad \cos\theta = \frac{L_z}{L}$

we find the possible values of θ $\quad \boxed{145°,\ 114°,\ 90.0°,\ 65.9°, \text{ and } 35.3°}$.

P29.21 (a) Density of a proton: $\quad \rho = \frac{m}{V} = \frac{1.67\times10^{-27}\ \text{kg}}{(4/3)\pi\left(1.00\times10^{-15}\ \text{m}\right)^3} = \boxed{3.99\times10^{17}\ \text{kg/m}^3}$.

(b) Size of model electron: $\quad r = \left(\frac{3m}{4\pi\rho}\right)^{1/3} = \left(\frac{3\left(9.11\times10^{-31}\ \text{kg}\right)}{4\pi\left(3.99\times10^{17}\ \text{kg/m}^3\right)}\right)^{1/3} = \boxed{8.17\times10^{-17}\ \text{m}}$.

continued on next page

(c) Moment of inertia: $I=\frac{2}{5}mr^2=\frac{2}{5}(9.11\times10^{-31}\text{ kg})(8.17\times10^{-17}\text{ m})^2=2.43\times10^{-63}\text{ kg}\cdot\text{m}^2$

$$L_z=I\omega=\frac{\hbar}{2}=\frac{Iv}{r}.$$

Therefore, $v=\frac{\hbar r}{2I}=\frac{(6.626\times10^{-34}\text{ J}\cdot\text{s})(8.17\times10^{-17}\text{ m})}{2\pi(2\times2.43\times10^{-63}\text{ kg}\cdot\text{m}^2)}=\boxed{1.77\times10^{12}\text{ m/s}}.$

(d) This is $\boxed{5.91\times10^3 \text{ times larger}}$ than the speed of light.

P29.22 In the N shell, $n=4$. For $n=4$, ℓ can take on values of 0, 1, 2, and 3. For each value of ℓ, m_ℓ can be $-\ell$ to ℓ in integral steps. Thus, the maximum value for m_ℓ is 3. Since $L_z=m_\ell\hbar$, the maximum value for L_z is $L_z=\boxed{3\hbar}$.

P29.23 The 3*d* subshell has $\ell=2$, and $n=3$. Also, we have $s=1$.

Therefore, we can have $\boxed{n=3,\ \ell=2;\ m_\ell=-2,\ -1,\ 0,\ 1,\ 2;\ s=1;\text{ and }m_s=-1,\ 0,\ 1}$

leading to the following table:

n	3	3	3	3	3	3	3	3	3	3	3	3	3	3	3
ℓ	2	2	2	2	2	2	2	2	2	2	2	2	2	2	2
m_ℓ	–2	–2	–2	–1	–1	–1	0	0	0	1	1	1	2	2	2
s	1	1	1	1	1	1	1	1	1	1	1	1	1	1	1
m_s	–1	0	1	–1	0	1	–1	0	1	–1	0	1	–1	0	1

Section 29.5 The Exclusion Principle and the Periodic Table

P29.24 (a) $\boxed{1s^2 2s^2 2p^4}$

(b) For the 1*s* electrons, $n=1$, $\ell=0$, $m_\ell=0$, $m_s=+\frac{1}{2}$ and $-\frac{1}{2}$.

For the two 2*s* electrons, $n=2$, $\ell=0$, $m_\ell=0$, $m_s=+\frac{1}{2}$ and $-\frac{1}{2}$.

For the four 2*p* electrons, $n=2$; $\ell=1$; $m_\ell=-1, 0,$ or 1; and $m_s=+\frac{1}{2}$ or $-\frac{1}{2}$.

P29.25 The $\boxed{\text{4s subshell fills first}}$, for potassium and calcium, before the 3*d* subshell starts to fill for scandium through zinc. Thus, we would first suppose that $[\text{Ar}]3d^4 4s^2$ would have lower energy than $[\text{Ar}]3d^5 4s^1$. But the latter has more unpaired spins, six instead of four, and Hund's rule suggests that this could give the latter configuration lower energy. In fact it must, for $[\text{Ar}]3d^5 4s^1$ is the ground state for chromium.

P29.26 Electronic configuration: Sodium to Argon

$\left[1s^2 2s^2 2p^6\right]$	$+3s^1$	$\rightarrow$	Na^{11}
	$+3s^2$	$\rightarrow$	Mg^{12}
	$+3s^2 3p^1$	$\rightarrow$	Al^{13}
	$+3s^2 3p^2$	$\rightarrow$	Si^{14}
	$+3s^2 3p^3$	$\rightarrow$	P^{15}
	$+3s^2 3p^4$	$\rightarrow$	S^{16}
	$+3s^2 3p^5$	$\rightarrow$	Cl^{17}
	$+3s^2 3p^6$	$\rightarrow$	Ar^{18}
$\left[1s^2 2s^2 2p^6 3s^2 3p^6\right] 4s^1$		$\rightarrow$	K^{19}

P29.27 In the table of electronic configurations in the text, or on a periodic table, we look for the element whose last electron is in a $3p$ state and which has three electrons outside a closed shell. Its electron configuration then ends in $3s^2 3p^1$. The element is $\boxed{\text{aluminum}}$.

P29.28 (a) For electron one and also for electron two, $n = 3$ and $\ell = 1$. The possible states are listed here in columns giving the other quantum numbers:

electron one	m_ℓ	1	1	1	1	1	1	1	1	1	1	0	0	0	0	0
	m_s	$\frac{1}{2}$	$\frac{1}{2}$	$\frac{1}{2}$	$\frac{1}{2}$	$\frac{1}{2}$	$-\frac{1}{2}$	$-\frac{1}{2}$	$-\frac{1}{2}$	$-\frac{1}{2}$	$-\frac{1}{2}$	$\frac{1}{2}$	$\frac{1}{2}$	$\frac{1}{2}$	$\frac{1}{2}$	$\frac{1}{2}$
electron two	m_ℓ	1	0	0	−1	−1	1	0	0	−1	−1	1	1	0	−1	−1
	m_s	$-\frac{1}{2}$	$\frac{1}{2}$	$-\frac{1}{2}$	$\frac{1}{2}$	$-\frac{1}{2}$	$\frac{1}{2}$	$\frac{1}{2}$	$-\frac{1}{2}$	$\frac{1}{2}$	$-\frac{1}{2}$	$\frac{1}{2}$	$-\frac{1}{2}$	$-\frac{1}{2}$	$\frac{1}{2}$	$-\frac{1}{2}$

electron one	m_ℓ	0	0	0	0	0	−1	−1	−1	−1	−1	−1	−1	−1	−1	−1
	m_s	$-\frac{1}{2}$	$-\frac{1}{2}$	$-\frac{1}{2}$	$-\frac{1}{2}$	$-\frac{1}{2}$	$\frac{1}{2}$	$\frac{1}{2}$	$\frac{1}{2}$	$\frac{1}{2}$	$\frac{1}{2}$	$-\frac{1}{2}$	$-\frac{1}{2}$	$-\frac{1}{2}$	$-\frac{1}{2}$	$-\frac{1}{2}$
electron two	m_ℓ	1	1	0	−1	−1	1	1	0	0	−1	1	1	0	0	−1
	m_s	$\frac{1}{2}$	$-\frac{1}{2}$	$\frac{1}{2}$	$\frac{1}{2}$	$-\frac{1}{2}$	$\frac{1}{2}$	$-\frac{1}{2}$	$\frac{1}{2}$	$-\frac{1}{2}$	$-\frac{1}{2}$	$\frac{1}{2}$	$-\frac{1}{2}$	$\frac{1}{2}$	$-\frac{1}{2}$	$\frac{1}{2}$

There are thirty allowed states, since electron one can have any of three possible values for m_ℓ for both spin up and spin down, amounting to six states, and the second electron can have any of the other five states.

(b) Were it not for the exclusion principle, there would be $\boxed{36}$ possible states, six for each electron independently.

P29.29 (a)

$n+\ell$	1	2	3	4	5	6	7
subshell	$1s$	$2s$	$2p, 3s$	$3p, 4s$	$3d, 4p, 5s$	$4d, 5p, 6s$	$4f, 5d, 6p, 7s$

(b) $Z = 15$:

Filled subshells: $1s, 2s, 2p, 3s$ (12 electrons)

Valence subshell: 3 electrons in $3p$ subshell

Prediction: Valence $= +5$ or -3

Element is phosphorus, Valence $= +5$ or -3 (Prediction correct)

$Z = 47$:

Filled subshells: $1s, 2s, 2p, 3s, 3p, 4s, 3d, 4p, 5s$ (38 electrons)

Outer subshell: 9 electrons in $4d$ subshell

Prediction: Valence $= -1$

Element is silver, (Prediction fails) Valence is $+1$

$Z = 86$:

Filled subshells: $1s, 2s, 2p, 3s, 3p, 4s, 3d, 4p, 5s, 4d, 5p, 6s, 4f, 5d, 6p$ (86 electrons)

Prediction: Outer subshell is full: inert gas

Element is radon, inert (Prediction correct)

P29.30 Listing subshells in the order of filling, we have for element 110,

$1s^2 2s^2 2p^6 3s^2 3p^6 4s^2 3d^{10} 4p^6 5s^2 4d^{10} 5p^6 6s^2 4f^{14} 5d^{10} 6p^6 7s^2 5f^{14} 6d^8$.

In order of increasing principal quantum number, this is

$$\boxed{1s^2 2s^2 2p^6 3s^2 3p^6 3d^{10} 4s^2 4p^6 4d^{10} 4f^{14} 5s^2 5p^6 5d^{10} 5f^{14} 6s^2 6p^6 6d^8 7s^2}.$$

P29.31 In the ground state of sodium, the outermost electron is in an s state. This state is spherically symmetric, so it generates no magnetic field by orbital motion, and has the same energy no matter whether the electron is spin-up or spin-down. The energies of the states $3p\uparrow$ and $3p\downarrow$ above 3s are $hf_1 = \frac{hc}{\lambda}$ and $hf_2 = \frac{hc}{\lambda_2}$.

The energy difference is

$$2\mu_B B = hc\left(\frac{1}{\lambda_1} - \frac{1}{\lambda_2}\right)$$

continued on next page

so $B = \frac{hc}{2\mu_B}\left(\frac{1}{\lambda_1} - \frac{1}{\lambda_2}\right) = \frac{(6.63\times10^{-34}\text{ J}\cdot\text{s})(3\times10^8\text{ m/s})}{2(9.27\times10^{-24}\text{ J/T})}\left(\frac{1}{588.995\times10^{-9}\text{ m}} - \frac{1}{589.592\times10^{-9}\text{ m}}\right)$

$B = \boxed{18.4\text{ T}}$.

Section 29.6 More on Atomic Spectra: Visible and X-Ray

P29.32 (a) $\boxed{n=3,\ \ell=0,\ m_\ell=0}$

$\boxed{n=3,\ \ell=1,\ m_\ell=-1,\ 0,\ 1}$

For $\boxed{n=3,\ \ell=2,\ m_\ell=-2,\ -1,\ 0,\ 1,\ 2}$

(b) ψ_{300} corresponds to $E_{300} = -\frac{Z^2E_0}{n^2} = -\frac{2^2(13.6)}{3^2} = \boxed{-6.05\text{ eV}}$.

ψ_{31-1}, ψ_{310}, ψ_{311} have the same energy since n is the same.

ψ_{32-2}, ψ_{32-1}, ψ_{320}, ψ_{321}, ψ_{322} have the same energy since n is the same.

All states are degenerate.

P29.33 $E = \frac{hc}{\lambda} = e\Delta V$: $\quad \frac{(6.626\times10^{-34}\text{ J}\cdot\text{s})(3.00\times10^8\text{ m/s})}{(10.0\times10^{-9}\text{ m})} = (1.60\times10^{-19})\Delta V$

$\Delta V = \boxed{124\text{ V}}$

P29.34 Some electrons can give all their kinetic energy $K_e = e\Delta V$ to the creation of a single photon of x-radiation, with

$$hf = \frac{hc}{\lambda} = e\Delta V$$

$$\lambda = \frac{hc}{e\Delta V} = \frac{(6.626\,1\times10^{-34}\text{ J}\cdot\text{s})(2.997\,9\times10^8\text{ m/s})}{(1.602\,2\times10^{-19}\text{ C})\Delta V} = \boxed{\frac{1\,240\text{ nm}\cdot\text{V}}{\Delta V}}$$

P29.35 Following Example 29.8 $\quad E_\gamma = \frac{3}{4}(42-1)^2(13.6\text{ eV}) = 1.71\times10^4\text{ eV} = 2.74\times10^{-15}\text{ J}$

$f = 4.14\times10^{18}\text{ Hz}$

and $\quad \lambda = \boxed{0.072\,5\text{ nm}}$.

P29.36 $E=\frac{hc}{\lambda}=\frac{1\,240\text{ eV}\cdot\text{nm}}{\lambda}=\frac{1.240\text{ keV}\cdot\text{nm}}{\lambda}$

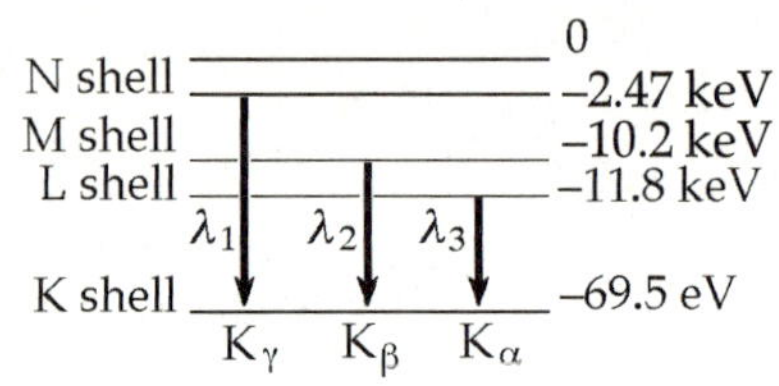

FIG. P29.36

For $\lambda_1=0.018\,5\text{ nm}$, $E=67.03\text{ keV}$

$\lambda_2=0.020\,9\text{ nm}$, $E=59.3\text{ keV}$

$\lambda_3=0.021\,5\text{ nm}$, $E=57.7\text{ keV}$

The ionization energy for the K shell is 69.5 keV, so the ionization energies for the other shells are:

$\boxed{\text{L shell}=11.8\text{ keV}}$ $\boxed{\text{M shell}=10.2\text{ keV}}$ $\boxed{\text{N shell}=2.47\text{ keV}}$.

P29.37 The K_β x-rays are emitted when there is a vacancy in the ($n=1$) K shell and an electron from the ($n=3$) M shell falls down to fill it. Then this electron is shielded by nine electrons originally and by one in its final state.

$$\frac{hc}{\lambda}=-\frac{13.6(Z-9)^2}{3^2}\text{ eV}+\frac{13.6(Z-1)^2}{1^2}\text{ eV}$$

$$\frac{\left(6.626\times10^{-34}\text{ J}\cdot\text{s}\right)\left(3.00\times10^{8}\text{ m/s}\right)}{\left(0.152\times10^{-9}\text{ m}\right)\left(1.60\times10^{-19}\text{ J/eV}\right)}=(13.6\text{ eV})\left(-\frac{Z^2}{9}+\frac{18Z}{9}-\frac{81}{9}+Z^2-2Z+1\right)$$

$$8.17\times10^{3}\text{ eV}=(13.6\text{ eV})\left(\frac{8Z^2}{9}-8\right)$$

so $$601=\frac{8Z^2}{9}-8$$

and $Z=26$ $\boxed{\text{Iron}}$.

Section 29.7 Context Connection—Atoms in Space

P29.38 (a) The configuration we may model as $\boxed{\text{SN}}\ \boxed{\text{NS}}$ has higher energy than $\boxed{\text{SN}}\ \boxed{\text{SN}}$. The higher energy state has antiparallel magnetic moments, so it has $\boxed{\text{parallel spins}}$ of the oppositely charged particles.

(b) $E=\frac{hc}{\lambda}=9.42\times10^{-25}\text{ J}=\boxed{5.89\ \mu\text{eV}}$

(c) $\Delta E\Delta t\approx\frac{\hbar}{2}$ so $\Delta E\approx\frac{1.055\times10^{-34}\text{ J}\cdot\text{s}}{2\left(10^{7}\text{ yr}\right)\left(3.16\times10^{7}\text{ s/yr}\right)}\left(\frac{1.00\text{ eV}}{1.60\times10^{-19}\text{ J}}\right)=\boxed{1.04\times10^{-30}\text{ eV}}$

P29.39 Section 24.3 says $f_{\text{observer}} = f_{\text{source}}\sqrt{\dfrac{1+v_a/c}{1-v_a/c}}$.

The velocity of approach, v_a, is the negative of the velocity of mutual recession: $v_a = -v$.

Then, $\dfrac{c}{\lambda'} = \dfrac{c}{\lambda}\sqrt{\dfrac{1-v/c}{1+v/c}}$ and $\boxed{\lambda' = \lambda\sqrt{\dfrac{1+v/c}{1-v/c}}}$.

***P29.40** (a) The energy of the ground state is: $E_1 = -\dfrac{hc}{\lambda_{\text{series limit}}} = -\dfrac{1\,240\text{ eV}\cdot\text{nm}}{152.0\text{ nm}} = \boxed{-8.16\text{ eV}}$.

From the wavelength of the Lyman α line: $E_2 - E_1 = \dfrac{hc}{\lambda} = \dfrac{1\,240\text{ nm}\cdot\text{eV}}{202.6\text{ nm}} = 6.12\text{ eV}$

$$E_2 = E_1 + 6.12\text{ eV} = \boxed{-2.04\text{ eV}}.$$

The wavelength of the Lyman β line gives: $E_3 - E_1 = \dfrac{1\,240\text{ nm}\cdot\text{eV}}{170.9\text{ nm}} = 7.26\text{ eV}$

so $E_3 = \boxed{-0.902\text{ eV}}$.

Next, using the Lyman γ line gives: $E_4 - E_1 = \dfrac{1\,240\text{ nm}\cdot\text{eV}}{162.1\text{ nm}} = 7.65\text{ eV}$

and $E_4 = \boxed{-0.508\text{ eV}}$.

From the Lyman δ line, $E_5 - E_1 = \dfrac{1\,240\text{ nm}\cdot\text{eV}}{158.3\text{ nm}} = 7.83\text{ eV}$

so $E_5 = \boxed{-0.325\text{ eV}}$.

(b) For the Balmer series, $\dfrac{hc}{\lambda} = E_i - E_2$, or $\lambda = \dfrac{1\,240\text{ nm}\cdot\text{eV}}{E_i - E_2}$.

For the α line, $E_i = E_3$ and so $\lambda_a = \dfrac{1\,240\text{ nm}\cdot\text{eV}}{(-0.902\text{ eV})-(-2.04\text{ eV})} = \boxed{1\,090\text{ nm}}$.

Similarly, the wavelengths of the β line, γ line, and the short wavelength limit are found to be: $\boxed{811\text{ nm}}$, $\boxed{724\text{ nm}}$, and $\boxed{609\text{ nm}}$.

(c) Computing 60.0% of the wavelengths of the spectral lines shown on the energy-level diagram gives:

$0.600(202.6\text{ nm}) = \boxed{122\text{ nm}}$, $0.600(170.9\text{ nm}) = \boxed{103\text{ nm}}$, $0.600(162.1\text{ nm}) = \boxed{97.3\text{ nm}}$, $0.600(158.3\text{ nm}) = \boxed{95.0\text{ nm}}$, and $0.600(152.0\text{ nm}) = \boxed{91.2\text{ nm}}$

These are seen to be the wavelengths of the α, β, γ, and δ lines as well as the short wavelength limit for the Lyman series in Hydrogen.

continued on next page

(d) The observed wavelengths could be the result of Doppler shift when the source moves away from the Earth. The required speed of the source is found from

$$\frac{f'}{f}=\frac{\lambda}{\lambda'}=\sqrt{\frac{1-(v/c)}{1+(v/c)}}=0.600 \qquad \text{yielding} \qquad \boxed{v=0.471c}.$$

P29.41 (a) $\lambda'=\lambda\sqrt{\dfrac{1+v/c}{1-v/c}}$ $\qquad 510\text{ nm}=434\text{ nm}\sqrt{\dfrac{1+v/c}{1-v/c}}$

$$1.18^2=\frac{1+v/c}{1-v/c}=1.381$$

$$1+\frac{v}{c}=1.381-1.381\frac{v}{c} \qquad 2.38\frac{v}{c}=0.381$$

$$\frac{v}{c}=0.160 \qquad \text{or} \qquad v=\boxed{0.160c}=4.80\times10^{7}\text{ m/s}$$

(b) $v=HR$: $\qquad R=\dfrac{v}{H}=\dfrac{4.80\times10^{7}\text{ m/s}}{17\times10^{-3}\text{ m/s}\cdot\text{ly}}=\boxed{2.82\times10^{9}\text{ ly}}$

Additional Problems

***P29.42** (a) The energy emitted by the atom is $\Delta E=E_4-E_2=-13.6\text{ eV}\left(\dfrac{1}{4^2}-\dfrac{1}{2^2}\right)=2.55\text{ eV}$. The wavelength of the photon produced is then

$$\lambda=\frac{hc}{E_\gamma}=\frac{hc}{\Delta E}=\frac{(6.626\times10^{-34}\text{ J}\cdot\text{s})(3.00\times10^{8}\text{ m/s})}{(2.55\text{ eV})(1.60\times10^{-19}\text{ J/eV})}=4.87\times10^{-7}\text{ m}=\boxed{487\text{ nm}}$$

(b) Since momentum must be conserved, the photon and the atom go in opposite directions with equal magnitude momenta. Thus, $p=m_{\text{atom}}v=\dfrac{h}{\lambda}$ or

$$v=\frac{h}{m_{\text{atom}}\lambda}=\frac{6.626\times10^{-34}\text{ J}\cdot\text{s}}{(1.67\times10^{-27}\text{ kg})(4.87\times10^{-7}\text{ m})}=\boxed{0.814\text{ m/s}}.$$

P29.43 (a) $$\Delta E=\frac{e\hbar B}{m_e}=\frac{1.60\times10^{-19}\text{ C}(6.63\times10^{-34}\text{ J}\cdot\text{s})(5.26\text{ T})}{2\pi(9.11\times10^{-31}\text{ kg})}\left(\frac{\text{N}\cdot\text{s}}{\text{T}\cdot\text{C}\cdot\text{m}}\right)\left(\frac{\text{kg}\cdot\text{m}}{\text{N}\cdot\text{s}^2}\right)=9.75\times10^{-23}\text{ J}$$

$$=\boxed{609\ \mu\text{eV}}$$

(b) $k_BT=(1.38\times10^{-23}\text{ J/K})(80\times10^{-3}\text{ K})=1.10\times10^{-24}\text{ J}=\boxed{6.90\ \mu\text{eV}}$

continued on next page

(c) $$f = \frac{\Delta E}{h} = \frac{9.75\times10^{-23}\text{ J}}{6.63\times10^{-34}\text{ J}\cdot\text{s}} = \boxed{1.47\times10^{11}\text{ Hz}}$$

$$\lambda = \frac{c}{f} = \frac{3\times10^{8}\text{ m/s}}{1.47\times10^{11}\text{ Hz}} = \boxed{2.04\times10^{-3}\text{ m}}$$

P29.44 (a) The energy difference between these two states is equal to the energy that is absorbed.

Thus, $$E = E_2 - E_1 = \frac{(-13.6\text{ eV})}{4} - \frac{(-13.6\text{ eV})}{1} = 10.2\text{ eV} = \boxed{1.63\times10^{-18}\text{ J}}.$$

(b) $E = \frac{3}{2}k_BT$ or $$T = \frac{2E}{3k_B} = \frac{2\left(1.63\times10^{-18}\text{ J}\right)}{3\left(1.38\times10^{-23}\text{ J/K}\right)} = \boxed{7.88\times10^{4}\text{ K}}.$$

P29.45 $$r_{av} = \int_0^\infty rP(r)\,dr = \int_0^\infty \left(\frac{4r^3}{a_0{}^3}\right)\left(e^{-2r/a_0}\right)dr$$

Make a change of variables with $\frac{2r}{a_0} = x$ and $dr = \frac{a_0}{2}dx$.

Then $$r_{av} = \frac{a_0}{4}\int_0^\infty x^3e^{-x}\,dx = \frac{a_0}{4}\left[-x^3e^{-x} + 3\left(-x^2e^{-x} + 2e^{-x}(-x-1)\right)\right]_0^\infty = \boxed{\frac{3}{2}a_0}.$$

***P29.46** The average squared separation distance is

$$\langle r^2\rangle = \int_{\text{all space}} \psi_{1s}^* r^2 \psi_{1s}\,dV = \int_{r=0}^\infty \frac{1}{\sqrt{\pi a_0^3}}e^{-r/a_0}r^2\frac{1}{\sqrt{\pi a_0^3}}e^{-r/a_0}4\pi r^2\,dr = \frac{4\pi}{\pi a_0^3}\int_0^\infty r^4e^{-2r/a_0}\,dr.$$

We use $\int_0^\infty x^ne^{-ax}\,dx = \frac{n!}{a^{n+1}}$ from Table B.6.

$$\langle r^2\rangle = \frac{4}{a_0^3}\frac{4!}{\left(\frac{2}{a_0}\right)^5} = \frac{a_0^2\,96}{32} = 3a_0^2$$

$$\Delta r = \sqrt{\langle r^2\rangle - \langle r\rangle^2} = \left(3a_0^2 - \left(\frac{3a_0}{2}\right)^2\right)^{1/2} = \left(3a_0^2 - \frac{9a_0^2}{4}\right)^{1/2} = \boxed{\left(\frac{3}{4}\right)^{1/2}a_0}$$

P29.47 $$hf = \Delta E = \frac{4\pi^2mk_e^2e^4}{2h^2}\left(\frac{1}{(n-1)^2} - \frac{1}{n^2}\right)$$

$$f = \frac{2\pi^2mk_e^2e^4}{h^3}\left(\frac{2n-1}{(n-1)^2n^2}\right)$$

As n approaches infinity, we have f approaching $$\frac{2\pi^2mk_e^2e^4}{h^3}\frac{2}{n^3}$$

continued on next page

The classical frequency is $$f = \frac{v}{2\pi r} = \frac{1}{2\pi}\sqrt{\frac{k_e e^2}{m}}\frac{1}{r^{3/2}}$$

where $$r = \frac{n^2 h^2}{4\pi m k_e e^2}$$

Using this equation to eliminate r from the expression for f, $$f = \frac{2\pi^2 m k_e^2 e^4}{h^3}\frac{2}{n^3}$$

***P29.48** (a) $$\text{Probability} = \int_r^\infty P_{1s}(r')dr' = \frac{4}{a_0^3}\int_r^\infty r'^2 e^{-2r'/a_0}dr' = \left[-\left(\frac{2r'^2}{a_0^2} + \frac{2r'}{a_0} + 1\right)e^{-2r'/a_0}\right]_r^\infty,$$

using integration by parts,

$$= \boxed{\left(\frac{2r^2}{a_0^2} + \frac{2r}{a_0} + 1\right)e^{-2r/a_0}}$$

(b)

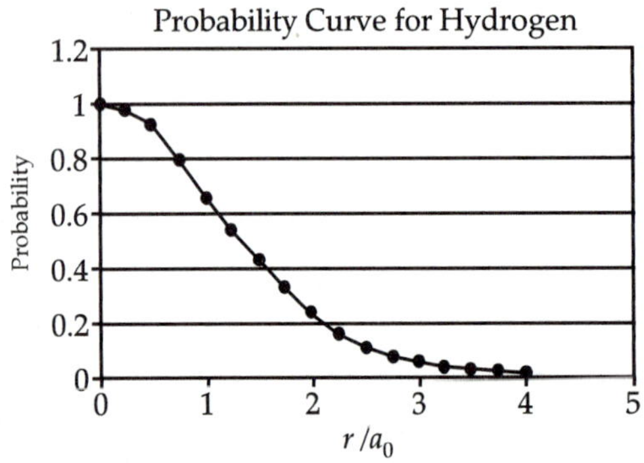

FIG. P29.48(b)

(c) The probability of finding the electron inside or outside the sphere of radius r is $\frac{1}{2}$.

$$\therefore \left(\frac{2r^2}{a_0^2} + \frac{2r}{a_0} + 1\right)e^{-2r/a_0} = \frac{1}{2} \text{ or } z^2 + 2z + 2 = e^z \text{ where } z = \frac{2r}{a_0}$$

One can home in on a solution to this transcendental equation for r on a calculator, the result being $r = \boxed{1.34a_0}$ to three digits.

P29.49 The wave function for the 2s state is given by Eq. 29.8: $\psi_{2s}(r) = \frac{1}{4\sqrt{2\pi}}\left(\frac{1}{a_0}\right)^{3/2}\left[2 - \frac{r}{a_0}\right]e^{-r/2a_0}$.

(a) Taking $r = a_0 = 0.529 \times 10^{-10}$ m

we find $$\psi_{2s}(a_0) = \frac{1}{4\sqrt{2\pi}}\left(\frac{1}{0.529 \times 10^{-10}\text{ m}}\right)^{3/2}[2-1]e^{-1/2} = \boxed{1.57 \times 10^{14}\text{ m}^{-3/2}}.$$

continued on next page

(b) $\left|\psi_{2s}(a_0)\right|^2 = \left(1.57\times10^{14}\ \text{m}^{-3/2}\right)^2 = \boxed{2.47\times10^{28}\ \text{m}^{-3}}$

(c) Using Equation 29.6 and the results to (b) gives $P_{2s}(a_0) = 4\pi a_0^2\left|\psi_{2s}(a_0)\right|^2 = \boxed{8.69\times10^{8}\ \text{m}^{-1}}$.

P29.50 From Figure 29.12, a typical ionization energy is 8 eV. For internal energy to ionize most of the atoms we require

$$\frac{3}{2}k_BT = 8\ \text{eV}: \qquad T = \frac{2\times8\left(1.60\times10^{-19}\ \text{J}\right)}{3\left(1.38\times10^{-23}\ \text{J/K}\right)}\boxed{\sim\text{ between }10^4\text{ K and }10^5\text{ K}}.$$

P29.51 The fermions are described by the exclusion principle. Two of them, one spin-up and one spin-down, will be in the ground energy level, with

$$d_{\text{NN}} = L = \frac{1}{2}\lambda,\ \lambda = 2L = \frac{h}{p},\text{ and } p = \frac{h}{2L} \qquad K = \frac{1}{2}mv^2 = \frac{p^2}{2m} = \frac{h^2}{8mL^2}.$$

The third must be in the next higher level, with

$$d_{\text{NN}} = \frac{L}{2} = \frac{\lambda}{2},\ \lambda = L,\text{ and } p = \frac{h}{L} \qquad K = \frac{p^2}{2m} = \frac{h^2}{2mL^2}.$$

The total energy is then $\frac{h^2}{8mL^2} + \frac{h^2}{8mL^2} + \frac{h^2}{2mL^2} = \boxed{\frac{3h^2}{4mL^2}}$.

P29.52 $$\Delta z = \frac{at^2}{2} = \frac{1}{2}\left(\frac{F_z}{m_0}\right)t^2 = \frac{\mu_z\left(dB_z/dz\right)}{2m_0}\left(\frac{\Delta x}{v}\right)^2 \quad\text{and}\quad \mu_z = \frac{e\hbar}{2m_e}$$

$$\frac{dB_z}{dz} = \frac{2m_0(\Delta z)v^2(2m_e)}{\Delta x^2 e\hbar} = \frac{2(108)\left(1.66\times10^{-27}\ \text{kg}\right)\left(10^{-3}\ \text{m}\right)\left(10^4\ \text{m}^2/\text{s}^2\right)\left(2\times9.11\times10^{-31}\ \text{kg}\right)}{\left(1.00\ \text{m}^2\right)\left(1.60\times10^{-19}\ \text{C}\right)\left(1.05\times10^{-34}\ \text{J}\cdot\text{s}\right)}$$

$$\frac{dB_z}{dz} = \boxed{0.389\ \text{T/m}}$$

P29.53 We use $$\psi_{2s}(r) = \frac{1}{4}\left(2\pi a_0^3\right)^{-1/2}\left(2 - \frac{r}{a_0}\right)e^{-r/2a_0}.$$

By Equation 29.6, $$P(r) = 4\pi r^2\psi^2 = \frac{1}{8}\left(\frac{r^2}{a_0^3}\right)\left(2 - \frac{r}{a_0}\right)^2 e^{-r/a_0}.$$

(a) $$\frac{dP(r)}{dr} = \frac{1}{8}\left[\frac{2r}{a_0^3}\left(2 - \frac{r}{a_0}\right)^2 - \frac{2r^2}{a_0^3}\left(\frac{1}{a_0}\right)\left(2 - \frac{r}{a_0}\right) - \frac{r^2}{a_0^3}\left(2 - \frac{r}{a_0}\right)^2\left(\frac{1}{a_0}\right)\right]e^{-r/a_0} = 0$$

$$\text{or } \frac{1}{8}\left(\frac{r}{a_0^3}\right)\left(2 - \frac{r}{a_0}\right)\left[2\left(2 - \frac{r}{a_0}\right) - \frac{2r}{a_0} - \frac{r}{a_0}\left(2 - \frac{r}{a_0}\right)\right]e^{-r/a_0} = 0.$$

The roots of $\frac{dP}{dr} = 0$ at $r = 0$, $r = 2a_0$ and $r = \infty$ are minima with $P(r) = 0$.

continued on next page

Therefore we require $$[.....]=4-\left(\frac{6r}{a_0}\right)+\left(\frac{r}{a_0}\right)^2=0$$

with solutions $$r=\left(3\pm\sqrt{5}\right)a_0.$$

We substitute the last two roots into $P(r)$ to determine the most probable value:

When $r=\left(3-\sqrt{5}\right)a_0=0.763\,9a_0$, $\qquad P(r)=\dfrac{0.051\,9}{a_0}$.

When $r=\left(3+\sqrt{5}\right)a_0=5.236a_0$, $\qquad P(r)=\dfrac{0.191}{a_0}$.

Therefore, the most probable value of r is $\qquad \left(3+\sqrt{5}\right)a_0=\boxed{5.236a_0}$.

(b) $$\int_0^\infty P(r)dr=\int_0^\infty \frac{1}{8}\left(\frac{r^2}{a_0^3}\right)\left(2-\frac{r}{a_0}\right)^2 e^{-r/a_0}dr$$

Let $u=\dfrac{r}{a_0}$, $dr=a_0du$,

$$\int_0^\infty P(r)dr=\int_0^\infty \frac{1}{8}u^2\left(4-4u+u^2\right)e^{-u}dr=\int_0^\infty \frac{1}{8}\left(u^4-4u^3+4u^2\right)e^{-u}du=-\frac{1}{8}\left(u^4+4u^2+8u+8\right)e^{-u}\Big|_0^\infty=1$$

This is as required for normalization.

***P29.54** (a) The system of a separated proton and electron puts out energy 13.606 eV to become a hydrogen atom in its ground state. This decrease in its rest energy appears also as a decrease in mass: the mass is smaller.

(b) $$|\Delta m|=\frac{|E|}{c^2}=\frac{13.6\text{ eV}}{\left(3\times10^8\text{ m/s}\right)^2}\left(\frac{1.6\times10^{-19}\text{ J}}{1\text{ eV}}\right)=2.42\times10^{-35}\text{ kg}$$

$$=2.42\times10^{-35}\text{ kg}\left(\frac{1\text{ u}}{1.66\times10^{-27}\text{ kg}}\right)=\boxed{1.46\times10^{-8}\text{ u}}$$

(c) $$\frac{1.46\times10^{-8}\text{ u}}{1.007\,825\text{ u}}=1.45\times10^{-8}=\boxed{1.45\times10^{-6}\%}$$

(d) Table A.3 lists 1.007 825 u as the atomic mass of hydrogen. This correction of 0.000 000 01 u is on the order of 100 times too small to affect the values listed.

P29.55 With one vacancy in the K shell, excess energy

$$\Delta E \approx -(Z-1)^2(13.6\text{ eV})\left(\frac{1}{2^2}-\frac{1}{1^2}\right)=5.40\text{ keV}.$$

We suppose the outermost 4*s* electron is shielded by 22 electrons inside its orbit:

$$E_{\text{ionization}} \approx \frac{2^2(13.6\text{ eV})}{4^2}=3.40\text{ eV}.$$

Note the experimental ionization energy is 6.76 eV.

$$K=\Delta E-E_{\text{ionization}} \approx \boxed{5.39\text{ keV}}.$$

P29.56 $E=\frac{hc}{\lambda}=\frac{1\,240\text{ eV}\cdot\text{nm}}{\lambda}=\Delta E$

$\lambda_1=310\text{ nm},$ so $\Delta E_1=4.00\text{ eV}$

$\lambda_2=400\text{ nm},$ $\Delta E_2=3.10\text{ eV}$

$\lambda_3=1\,378\text{ nm},$ $\Delta E_3=0.900\text{ eV}$

and the ionization energy $=4.10\text{ eV}$.

The energy level diagram having the fewest levels and consistent with these energies is shown at the right.

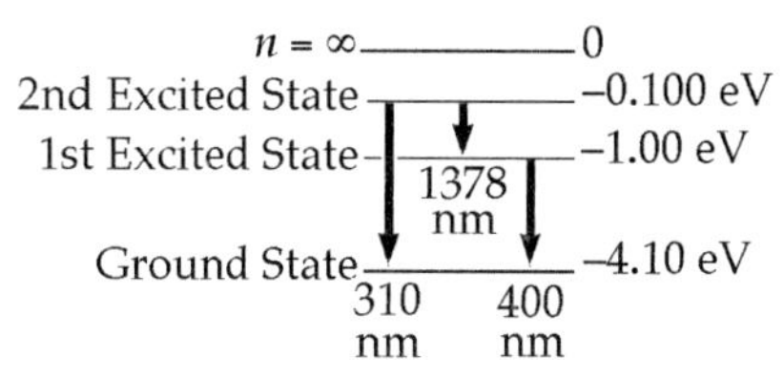

FIG. P29.56

P29.57 $P=\int_{2.50a_0}^{\infty}\frac{4r^2}{a_0^3}e^{-2r/a_0}dr=\frac{1}{2}\int_{5.00}^{\infty}z^2e^{-z}dz$ where $z\equiv\frac{2r}{a_0}$

$$P=-\frac{1}{2}\left(z^2+2z+2\right)e^{-z}\Big|_{5.00}^{\infty}=-\frac{1}{2}[0]+\frac{1}{2}(25.0+10.0+2.00)e^{-5}=\left(\frac{37}{2}\right)(0.006\,74)=\boxed{0.125}$$

P29.58 (a) One molecule's share of volume

$$\text{Al: } V=\frac{\text{mass per molecule}}{\text{density}}=\left(\frac{27.0\text{ g/mol}}{6.02\times10^{23}\text{ molecules/mol}}\right)\left(\frac{1.00\times10^{-6}\text{ m}^3}{2.70\text{ g}}\right)=1.66\times10^{-29}\text{ m}^3$$

$$\sqrt[3]{V}=\boxed{2.55\times10^{-10}\text{ m}\sim10^{-1}\text{ nm}}.$$

$$\text{U: } V=\left(\frac{238\text{ g}}{6.02\times10^{23}\text{ molecules}}\right)\left(\frac{1.00\times10^{-6}\text{ m}^3}{18.9\text{ g}}\right)=2.09\times10^{-29}\text{ m}^3$$

$$\sqrt[3]{V}=\boxed{2.76\times10^{-10}\text{ m}\sim10^{-1}\text{ nm}}.$$

(b) The outermost electron in any atom sees the nuclear charge screened by all the electrons below it. If we can visualize a single outermost electron, it moves in the electric field of net charge, $+Ze-(Z-1)e=+e$, the charge of a single proton, as felt by the electron in hydrogen. So the Bohr radius sets the scale for the outside diameter of every atom. An innermost electron, on the other hand, sees the nuclear charge unscreened, and the scale size of its (K-shell) orbit is $\frac{a_0}{Z}$.

P29.59 $\Delta E = 2\mu_B B = hf$

so $$2\left(9.27\times10^{-24}\text{ J/T}\right)(0.350\text{ T}) = \left(6.626\times10^{-34}\text{ J}\cdot\text{s}\right)f$$

and $$f = \boxed{9.79\times10^{9}\text{ Hz}}.$$

P29.60 $$\psi_{2s} = \frac{1}{4}(2\pi)^{-1/2}\left(\frac{1}{a_0}\right)^{3/2}\left(2-\frac{r}{a_0}\right)e^{-r/2a_0} = A\left(2-\frac{r}{a_0}\right)e^{-r/2a_0} \qquad \frac{d\psi}{dr} = Ae^{-r/2a_0}\left(-\frac{2}{a_0}+\frac{r}{2a_0^2}\right)$$

$$\frac{d^2\psi}{dr^2} = \left(\frac{Ae^{-r/2a_0}}{a_0^2}\right)\left(\frac{3}{2}-\frac{r}{4a_0}\right)$$

Substituting into Schrödinger's equation and dividing by $Ae^{-r/2a_0}$, we will have a solution if

$$-\frac{5}{4}\frac{\hbar^2}{m_e a_0^2}+\frac{k_e e^2}{a_0}+\frac{\hbar^2 r}{8m_e a_0^3}+\frac{2\hbar^2}{m_e a_0 r}-\frac{2k_e e^2}{r} = 2E-\frac{Er}{a_0}.$$

Now with $a_0 = \frac{\hbar^2}{m_e e^2 k_e}$, this reduces to

$$-\frac{m_e e^4 k_e^2}{8\hbar^2}\left(2-\frac{r}{a_0}\right) = E\left(2-\frac{r}{a_0}\right).$$

This is true, so ψ_{2s} is a solution to the Schrödinger equation, provided $E = \frac{1}{4}E_1 = -3.40\text{ eV}$.

P29.61 (a) Suppose the atoms move in the $+x$ direction. The absorption of a photon by an atom is a completely inelastic collision, described by

$$mv_i\hat{\mathbf{i}}+\frac{h}{\lambda}\left(-\hat{\mathbf{i}}\right) = mv_f\hat{\mathbf{i}} \qquad \text{so} \qquad v_f - v_i = -\frac{h}{m\lambda}.$$

This happens promptly every time an atom has fallen back into the ground state, so it happens every $10^{-8}\text{ s} = \Delta t$. Then,

$$a = \frac{v_f - v_i}{\Delta t} = -\frac{h}{m\lambda\Delta t} \sim -\frac{6.626\times10^{-34}\text{ J}\cdot\text{s}}{\left(10^{-25}\text{ kg}\right)\left(500\times10^{-9}\text{ m}\right)\left(10^{-8}\text{ s}\right)} \sim \boxed{-10^{6}\text{ m/s}^2}.$$

(b) With constant average acceleration,

$$v_f^2 = v_i^2 + 2a\Delta x \qquad 0 \sim \left(10^3\text{ m/s}\right)^2 + 2\left(-10^6\text{ m/s}^2\right)\Delta x$$

so $$\Delta x \sim \frac{\left(10^3\text{ m/s}\right)^2}{10^6\text{ m/s}^2} \boxed{\sim 1\text{ m}}.$$

***P29.62** $$\left\langle \frac{1}{r} \right\rangle = \int_0^\infty \frac{4r^2}{a_0^3} e^{-2r/a_0} \frac{1}{r} dr = \frac{4}{a_0^3} \int_0^\infty r e^{-(2/a_0)r} dr = \frac{4}{a_0^3} \frac{1}{\left(\frac{2}{a_0}\right)^2} = \boxed{\frac{1}{a_0}}$$

We compare this to $\frac{1}{\langle r \rangle} = \frac{1}{\frac{3a_0}{2}} = \frac{2}{3a_0}$, and find that the average reciprocal value is $\boxed{\text{NOT}}$ the reciprocal of the average value.

ANSWERS TO EVEN PROBLEMS

P29.2 (a) 56.8 fm; (b) 11.3 N

P29.4 (a) 1.94 μm; (b) $\lambda_{CB} = \frac{1}{\frac{1}{\lambda_{CA}} - \frac{1}{\lambda_{BA}}}$

P29.6 (a) 3; (b) 520 km/s

P29.8 (a) see the solution; (b) 0.179 nm

P29.10 see the solution

P29.12 $4a_0$

P29.14 797 times

P29.16 (a) $\sqrt{6}\,\hbar$; (b) $\sqrt{12}\,\hbar$

P29.18 $\sqrt{6}\,\hbar$

P29.20 $L = \sqrt{6}\,\hbar$, $L_z = -2\hbar, -\hbar, 0, \hbar, 2\hbar$, $\theta = 145°$, 114°, 90.0°, 65.9°, and 35.3°

P29.22 $3\,\hbar$

P29.24 (a) $1s^2 2s^2 p^4$;

(b)

n	ℓ	m_ℓ	m_s
1	0	0	1/2
1	0	0	–1/2
2	0	0	1/2
2	0	0	–1/2
2	1	1	1/2
2	1	0	1/2
2	1	–1	1/2
2	1	–1	–1/2

P29.26 see the solution

P29.28 (a) see the solution; (b) 36

P29.30 $1s^2\, 2s^2\, 2p^6\, 3s^2\, 3p^6\, 3d^{10}\, 4s^2\, 4p^6\, 4d^{10}\, 4f^{14}$ $5s^2\, 5p^6\, 5d^{10}\, 5f^{14}\, 6s^2\, 6p^6\, 6d^8\, 7s^2$

P29.32 (a) $\ell = 0$ and $m_\ell = 0$, or $\ell = 1$ and $m_\ell = -1$, 0, or 1, or $\ell = 2$ and $m_\ell = -2, -1$, 0, 1, or 2;
(b) –6.05 eV

P29.34 see the solution

P29.36 L shell = 11.8 keV, M shell = 10.2 keV, N shell = 2.47 keV, see the solution

P29.38 (a) parallel spins, with antiparallel magnetic moments;
(b) 5.89 μeV; (c) 1.04×10^{-30} eV

P29.40 (a) –8.16 eV, –2.04 eV, –0.902 eV, –0.508 eV, –0.325 eV;
(b) 1090 nm, 811 nm, 724 nm, 609 nm;
(c) see the solution;
(d) The spectrum could be that of hydrogen, Doppler shifted by motion away from us at speed 0.471c.

P29.42 (a) 487 nm; (b) 0.814 m/s

P29.44 (a) 1.63×10^{-18} J; (b) 7.88×10^4 K

P29.46 $\sqrt{\frac{3}{4}}a_0 = 0.866a_0$

P29.48 (a) Probability $= \left(\frac{2r^2}{a_0^2} + \frac{2r}{a_0} + 1\right)e^{-2r/a_0}$;
(b) see the solution; (c) $1.34a_0$

P29.50 between 10^4 K and 10^5 K

P29.52 0.389 T/m

P29.54 (a) smaller; (b) 1.46×10^{-8} u;
(c) $1.45 \times 10^{-6}\%$; (d) no

P29.56 see the solution

P29.58 (a) For Al, about 0.255 nm ~ 0.1 nm. For U, about 0.276 nm ~ 0.1 nm. For an outer electron, the nuclear charge is screened by the interior electrons.
(b) For an inner-shell electron, the nuclear charge is unscreened. The distance scale of the wave function, representing the orbital size, is proportional to $\frac{a_0}{Z}$.

P29.60 see the solution

P29.62 $\frac{1}{a_0}$, no

Chapter 30

Nuclear Physics

CHAPTER OUTLINE

ANSWERS TO QUESTIONS

Q30.1 Because of electrostatic repulsion between the positively-charged nucleus and the $+2e$ alpha particle. To drive the α-particle into the nucleus would require extremely high kinetic energy.

Q30.2 Nuclei with more nucleons than bismuth-209 are unstable because the electrical repulsion forces among all of the protons is stronger than the nuclear attractive force between nucleons.

Q30.3 Extra neutrons are required to overcome the increasing electrostatic repulsion of the protons. The neutrons participate in the net attractive effect of the nuclear force, but feel no Coulomb repulsion.

Q30.4 The nuclear force favors the formation of neutron-proton pairs, so a stable nucleus cannot be too far away from having equal numbers of protons and neutrons. This effect sets the upper boundary of the zone of stability on the neutron-proton diagram. All of the protons repel one another electrically, so a stable nucleus cannot have too many protons. This effect sets the lower boundary of the zone of stability.

Q30.5 ^4He, ^{16}O, ^{40}Ca, and ^{208}Pb.

Q30.6 If one half the number of radioactive nuclei decay in one year, then one half the remaining number will decay in the second year. Three quarters of the original nuclei will be gone, and one quarter will remain.

Q30.7 Since the samples are of the same radioactive isotope, their half-lives are the same. When prepared, sample A has twice the activity (number of radioactive decays per second) of sample B. After 5 half-lives, the activity of sample A is decreased by a factor of 2^5, and after 5 half-lives the activity of sample B is decreased by a factor of 2^5. So after 5 half-lives, the ratio of activities is still 2:1.

Q30.8 The statement is false. Both patterns show monotonic decrease over time, but with very different shapes. For radioactive decay, maximum activity occurs at time zero. Cohorts of people now living will be dying most rapidly perhaps forty years from now. Everyone now living will be dead within less than two centuries, while the mathematical model of radioactive decay tails off exponentially forever. A radioactive nucleus never gets old. It has constant probability of decay however long it has existed.

Q30.9 The alpha particle and the daughter nucleus carry equal amounts of momentum in opposite directions. Since kinetic energy can be written as $\frac{p^2}{2m}$, the small-mass alpha particle has much more of the decay energy than the recoiling nucleus.

Q30.10 Yes. The daughter nucleus can be left in its ground state or sometimes in one of a set of excited states. If the energy carried by the alpha particle is mysteriously low, the daughter nucleus can quickly emit the missing energy in a gamma ray.

Q30.11 The samples would have started with more carbon-14 than we first thought. We would increase our estimates of their ages.

Q30.12 In a heavy nucleus each nucleon is strongly bound to its momentary neighbors. Even if the nucleus could step down in energy by shedding an individual proton or neutron, one individual nucleon is never free to escape. Instead, the nucleus can decay when two protons and two neutrons, strongly bound to one another but not to their neighbors, happen momentarily to have a lot of kinetic energy, to lie at the surface of the nucleus, to be headed outward, and to tunnel successfully through the potential energy barrier they encounter.

Q30.13 The decay of a radioactive nucleus at one particular moment instead of at another instant cannot be predicted and has no cause. Natural events are not just like a perfect clockworks. In history, the idea of a determinate mechanical Universe arose temporarily from an unwarranted wild extrapolation of Isaac Newton's account of planetary motion. Before Newton's time [really you can blame Pierre Simon de Laplace] and again now, no one thought of natural events as just like a perfect row of falling dominos. We can and do use the word "cause" more loosely to describe antecedent enabling events. One gear turning another is intelligible. So is the process of a hot dog getting toasted over a campfire, even though random molecular motion is at the essence of that process. In summary, we say that the future is not determinate. All natural events have causes in the ordinary sense of the word, but not necessarily in the contrived sense of a cause operating infallibly and predictably in a way that can be calculated. We have better reason now than ever before to think of the Universe as intelligible. First describing natural events, and second determining their causes form the basis of science, including physics but also scientific medicine and scientific bread-baking. The evidence alone of the past hundred years of discoveries in physics, finding causes of natural events from the photoelectric effect to x-rays and jets emitted by black holes, suggests that human intelligence is a good tool for figuring out how things go. Even without organized science, humans have always been searching for the causes of natural events, with explanations ranging from "the will of the gods" to Schrödinger's equation. We depend on the principle that things are intelligible as we make significant strides towards understanding the Universe. To hope that our search is not futile is the best part of human nature.

Q30.14 Fusion of light nuclei to a heavier nucleus releases energy. Fission of a heavy nucleus to lighter nuclei releases energy. Both processes are steps towards greater stability on the curve of binding energy, Figure 30.10. The energy release per nucleon is typically greater for fusion, and this process is harder to control.

Q30.15 Sometimes the references are oblique indeed. Some must serve for more than one form of energy or mode of transfer. Here is one list:

kinetic: ocean currents
rotational kinetic: Earth turning
gravitational: water lifted up; you on the perilous bridge
elastic: Elastic energy is necessary for sound, listed below.
internal: infernos, lava
chemical: corrosive smoke
sound: thunder
electrical transmission: lightning
electromagnetic radiation: heavens blazing; lightning
atomic electronic: In the blazing heavens, stars have different colors because of different predominant energy losses by atoms at their surfaces.
nuclear: The blaze of the heavens is produced by nuclear reactions in the cores of stars.

Remarkably, the word "energy" in this translation is an anachronism. Goethe wrote the song a few years before Thomas Young coined the term.

SOLUTIONS TO PROBLEMS

Section 30.1 Some Properties of Nuclei

P30.1 An iron nucleus (in hemoglobin) has a few more neutrons than protons, but in a typical water molecule there are eight neutrons and ten protons.

So protons and neutrons are nearly equally numerous in your body, each contributing mass (say) 35 kg:

$$35\text{ kg}\left(\frac{1\text{ nucleon}}{1.67\times10^{-27}\text{ kg}}\right)\boxed{\sim10^{28}\text{ protons}}$$

and $\boxed{\sim10^{28}\text{ neutrons}}$.

The electron number is precisely equal to the proton number, $\boxed{\sim10^{28}\text{ electrons}}$.

P30.2 $\frac{1}{2}mv^2=q\Delta V$ and $\frac{mv^2}{r}=qvB$

$2m\Delta V=qr^2B^2$: $r=\sqrt{\frac{2m\Delta V}{qB^2}}=\sqrt{\frac{2(1\,000\text{ V})}{(1.60\times10^{-19}\text{ C})(0.200\text{ T})^2}}\sqrt{m}$

$$r=\left(5.59\times10^{11}\text{ m}/\sqrt{\text{kg}}\right)\sqrt{m}$$

(a) For ^{12}C, $m=12$ u and $r=\left(5.59\times10^{11}\text{ m}/\sqrt{\text{kg}}\right)\sqrt{12\left(1.66\times10^{-27}\text{ kg}\right)}$

$$r=0.078\,9\text{ m}=\boxed{7.89\text{ cm}}.$$

For ^{13}C: $r=\left(5.59\times10^{11}\text{ m}/\sqrt{\text{kg}}\right)\sqrt{13\left(1.66\times10^{-27}\text{ kg}\right)}$

$$r=0.082\,1\text{ m}=\boxed{8.21\text{ cm}}.$$

continued on next page

(b) With $r_1 = \sqrt{\dfrac{2m_1\Delta V}{qB^2}}$ and $r_2 = \sqrt{\dfrac{2m_2\Delta V}{qB^2}}$

the ratio gives $\boxed{\dfrac{r_1}{r_2} = \sqrt{\dfrac{m_1}{m_2}}}$

$$\frac{r_1}{r_2} = \frac{7.89 \text{ cm}}{8.21 \text{ cm}} = 0.961$$

and $\sqrt{\dfrac{m_1}{m_2}} = \sqrt{\dfrac{12 \text{ u}}{13 \text{ u}}} = 0.961$

so they do agree.

***P30.3** $E_\alpha = 7.70 \text{ MeV} = \dfrac{1}{2}mv^2 = k_e \dfrac{2eZe}{d_{\min}}$

(a) $d_{\min} = \dfrac{4k_e Ze^2}{mv^2} = \dfrac{2k_e Ze^2}{E_\alpha} = \dfrac{2(8.99\times 10^9)(79)(1.60\times 10^{-19})^2}{7.70(1.60\times 10^{-13})} = 29.5\times 10^{-15} \text{ m} = \boxed{29.5 \text{ fm}}$

(b) The de Broglie wavelength of the α is

$$\lambda = \frac{h}{m_\alpha v_\alpha} = \frac{h}{\sqrt{2m_\alpha E_\alpha}} = \frac{6.626\times 10^{-34}}{\sqrt{2(6.64\times 10^{-27})7.70(1.60\times 10^{-13})}} = 5.18\times 10^{-15} \text{ m} = \boxed{5.18 \text{ fm}}.$$

(c) Since $\boxed{\lambda \text{ is much less than the distance of closest approach}}$, the α may be considered a particle.

P30.4 (a) $r = r_0 A^{1/3} = (1.20\times 10^{-15} \text{ m})(4)^{1/3} = \boxed{1.90\times 10^{-15} \text{ m}}$

(b) $r = r_0 A^{1/3} = (1.20\times 10^{-15} \text{ m})(238)^{1/3} = \boxed{7.44\times 10^{-15} \text{ m}}$

P30.5 The number of nucleons in a star of two solar masses is

$$A = \frac{2(1.99\times 10^{30} \text{ kg})}{1.67\times 10^{-27} \text{ kg/nucleon}} = 2.38\times 10^{57} \text{ nucleons}.$$

Therefore $r = r_0 A^{1/3} = (1.20\times 10^{-15} \text{ m})(2.38\times 10^{57})^{1/3} = \boxed{16.0 \text{ km}}.$

P30.6 $V=\frac{4}{3}\pi r^4=\frac{4}{3}\pi(0.021\,5\text{ m})^3=4.16\times10^{-5}\text{ m}^3$

We take the nuclear density from Example 30.1

$$m=\rho V=\left(2.3\times10^{17}\text{ kg/m}^3\right)\left(4.16\times10^{-5}\text{ m}^3\right)=9.57\times10^{12}\text{ kg}$$

and $F=G\frac{m_1m_2}{r^2}=\left(6.67\times10^{-11}\text{ N}\cdot\text{m}^2/\text{kg}^2\right)\frac{\left(9.57\times10^{12}\text{ kg}\right)^2}{(1.00\text{ m})^2}$

$F=\boxed{6.11\times10^{15}\text{ N}}$ toward the other ball.

P30.7 (a) $$f_n=\frac{|2\mu B|}{h}=\frac{2(1.913\,5)\left(5.05\times10^{-27}\text{ J/T}\right)(1.00\text{ T})}{6.626\times10^{-34}\text{ J}\cdot\text{s}}=\boxed{29.2\text{ MHz}}$$

(b) $$f_p=\frac{2(2.792\,8)\left(5.05\times10^{-27}\text{ J/T}\right)(1.00\text{ T})}{6.626\times10^{-34}\text{ J}\cdot\text{s}}=\boxed{42.6\text{ MHz}}$$

(c) In the Earth's magnetic field,

$$f_p=\frac{2(2.792\,8)\left(5.05\times10^{-27}\right)\left(50.0\times10^{-6}\right)}{6.626\times10^{-34}}=\boxed{2.13\text{ kHz}}.$$

Section 30.2 Binding Energy

P30.8 Using atomic masses as given in Table A.3,

(a) For ${}^2_1\text{H}$: $$\frac{-2.014\,102+1(1.008\,665)+1(1.007\,825)}{2}$$

$$\frac{E_b}{A}=(0.001\,194\text{ u})\left(\frac{931.5\text{ MeV}}{\text{u}}\right)=\boxed{1.11\text{ MeV/nucleon}}.$$

(b) For ${}^4_2\text{He}$: $$\frac{2(1.008\,665)+2(1.007\,825)-4.002\,603}{4}$$

$$\frac{E_b}{A}=0.007\,59\text{ u}c^2=\boxed{7.07\text{ MeV/nucleon}}.$$

(c) For ${}^{56}_{26}\text{Fe}$: $30(1.008\,665)+26(1.007\,825)-55.934\,942=0.528\text{ u}$

$$\frac{E_b}{A}=\frac{0.528}{56}=0.009\,44\text{ u}c^2=\boxed{8.79\text{ MeV/nucleon}}.$$

(d) For ${}^{238}_{92}\text{U}$: $146(1.008\,665)+92(1.007\,825)-238.050\,783=1.934\,2\text{ u}$

$$\frac{E_b}{A}=\frac{1.934\,2}{238}=0.008\,13\text{ u}c^2=\boxed{7.57\text{ MeV/nucleon}}.$$

P30.9 The binding energy of a nucleus is $E_b(\text{MeV}) = \left[ZM(\text{H}) + Nm_n - M\left({}^A_Z\text{X}\right)\right](931.494\ \text{MeV/u})$.

For ${}^{15}_{8}\text{O}$: $E_b = [8(1.007\,825\ \text{u}) + 7(1.008\,665\ \text{u}) - 15.003\,065\ \text{u}](931.494\ \text{MeV/u}) = 111.96\ \text{MeV}$.

For ${}^{15}_{7}\text{N}$: $E_b = [7(1.007\,825\ \text{u}) + 8(1.008\,665\ \text{u}) - 15.000\,109\ \text{u}](931.494\ \text{MeV/u}) = 115.49\ \text{MeV}$.

Therefore, $\boxed{\text{the binding energy of } {}^{15}_{7}\text{N is larger by 3.54 MeV}}$.

P30.10 (a) The neutron-to-proton ratio $\dfrac{A-Z}{Z}$ is greatest for $\boxed{{}^{139}_{55}\text{Cs}}$ and is equal to 1.53.

(b) $\boxed{{}^{139}\text{La}}$ has the largest binding energy per nucleon of 8.378 MeV.

(c) ${}^{139}\text{Cs}$ with a mass of 138.913 u. We locate the nuclei carefully on Figure 30.4, the neutron–proton plot of stable nuclei. $\boxed{\text{Cesium}}$ appears to be farther from the center of the zone of stability. Its instability means extra energy and extra mass.

P30.11 $\Delta E_b = E_{bf} - E_{bi}$

For $A = 200,\ \dfrac{E_b}{A} = 7.8\ \text{MeV}$

so $E_{bi} = 200(7.8\ \text{MeV}) = 1\,560\ \text{MeV}$.

For $A \approx 100,\ \dfrac{E_b}{A} = 8.7\ \text{MeV}$

so $E_{bf} = 2(100)(8.7\ \text{MeV}) = 1\,740\ \text{MeV}$.

$\Delta E_b = E_{bf} - E_{bi}$: $E_b = 1\,740\ \text{MeV} - 1\,560\ \text{MeV} \approx \boxed{200\ \text{MeV}}$

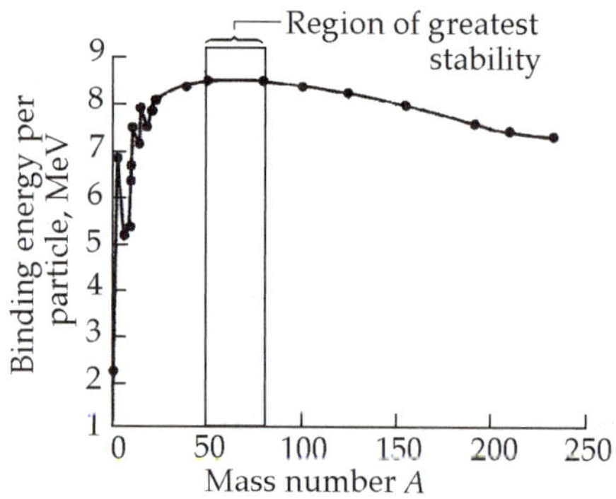

FIG. P30.11

Section 30.3 Radioactivity

P30.12 $R = R_0 e^{-\lambda t} = (6.40\ \text{mCi})e^{-(\ln 2/8.04\ \text{d})(40.2\ \text{d})} = (6.40\ \text{mCi})\left(e^{-\ln 2}\right)^5 = (6.40\ \text{mCi})\left(\dfrac{1}{2^5}\right) = \boxed{0.200\ \text{mCi}}$

P30.13 (a) From $R = R_0 e^{-\lambda t}$,

$$\lambda = \frac{1}{t}\ln\left(\frac{R_0}{R}\right) = \left(\frac{1}{4.00\ \text{h}}\right)\ln\left(\frac{10.0}{8.00}\right) = 5.58\times10^{-2}\ \text{h}^{-1} = \boxed{1.55\times10^{-5}\ \text{s}^{-1}}$$

$$T_{1/2} = \frac{\ln 2}{\lambda} = \boxed{12.4\ \text{h}}$$

(b) $$N_0 = \frac{R_0}{\lambda} = \frac{10.0\times10^{-3}\ \text{Ci}}{1.55\times10^{-5}/\text{s}}\left(\frac{3.70\times10^{10}/\text{s}}{1\ \text{Ci}}\right) = \boxed{2.39\times10^{13}\ \text{atoms}}$$

(c) $$R = R_0 e^{-\lambda t} = (10.0\ \text{mCi})\exp\left(-5.58\times10^{-2}\times30.0\right) = \boxed{1.88\ \text{mCi}}$$

P30.14 $\frac{dN}{dt} = -\lambda N$

so $\lambda = \frac{1}{N}\left(-\frac{dN}{dt}\right) = \left(1.00\times 10^{-15}\right)\left(6.00\times 10^{11}\right) = 6.00\times 10^{-4}\ \text{s}^{-1}$

$$T_{1/2} = \frac{\ln 2}{\lambda} = \boxed{1.16\times 10^{3}\ \text{s}}\ (= 19.3\ \text{min})$$

P30.15 $R = R_0 e^{-\lambda t}$ where $\lambda = \frac{\ln 2}{26.0\ \text{h}} = 0.026\,6\ \text{h}^{-1}$

$\frac{R}{R_0} = 0.100 = e^{-\lambda t}$ so $\ln(0.100) = -\lambda t$

$2.30 = \left(\frac{0.026\,6}{\text{h}}\right)t$ $t = \boxed{86.4\ \text{h}}$

P30.16 The number of nuclei which decay during the interval will be $N_1 - N_2 = N_0\left(e^{-\lambda t_1} - e^{-\lambda t_2}\right)$.

First we find λ: $\lambda = \frac{\ln 2}{T_{1/2}}$

so $e^{-\lambda t} = e^{\ln 2\left(-t/T_{1/2}\right)} = 2^{-t/T_{1/2}}$

and $N_0 = \frac{R_0}{\lambda} = \frac{R_0 T_{1/2}}{\ln 2}$.

Substituting in these values $N_1 - N_2 = \frac{R_0 T_{1/2}}{\ln 2}\left(e^{-\lambda t_1} - e^{-\lambda t_2}\right) = \boxed{\frac{R_0 T_{1/2}}{\ln 2}\left(2^{-t_1/T_{1/2}} - 2^{-t_2/T_{1/2}}\right)}$.

***P30.17** We have all this information: $N_x(0) = 2.50 N_y(0)$

$N_x(3\text{d}) = 4.20 N_y(3\text{d})$

$N_x(0)e^{-\lambda_x 3\text{d}} = 4.20 N_y(0)e^{-\lambda_y 3\text{d}} = 4.20\frac{N_x(0)}{2.50}e^{-\lambda_y 3\text{d}}$

$e^{3\text{d}\lambda_x} = \frac{2.5}{4.2}e^{3\text{d}\lambda_y}$

$3\text{d}\lambda_x = \ln\left(\frac{2.5}{4.2}\right) + 3\text{d}\lambda_y$

$3\text{d}\frac{0.693}{T_{1/2\,x}} = \ln\left(\frac{2.5}{4.2}\right) + 3\text{d}\frac{0.693}{1.60\ \text{d}} = 0.781$

$T_{1/2\,x} = \boxed{2.66\ \text{d}}$

***P30.18** (a) $\frac{dN_2}{dt}$ = rate of change of N_2

= rate of production of N_2 – rate of decay of N_2

= rate of decay of N_1 – rate of decay of N_2

$= \lambda_1 N_1 - \lambda_2 N_2$

(b) From the trial solution

$$N_2(t) = \frac{N_{10}\lambda_1}{\lambda_1 - \lambda_2}\left(e^{\lambda_2 t} - e^{-\lambda_1 t}\right)$$

$$\therefore \frac{dN_2}{dt} = \frac{N_{10}\lambda_1}{\lambda_1 - \lambda_2}\left(-\lambda_2 e^{-\lambda_2 t} + \lambda_1 e^{-\lambda_1 t}\right) \quad (1)$$

$$\therefore \frac{dN_2}{dt} + \lambda_2 N_2 = \frac{N_{10}\lambda_1}{\lambda_1 \ \ \lambda_2}\left(-\lambda_2 e^{-\lambda_2 t} + \lambda_1 e^{-\lambda_1 t} + \lambda_2 e^{-\lambda_2 t} - \lambda_2 e^{-\lambda_1 t}\right)$$

$$= \frac{N_{10}\lambda_1}{\lambda_1 - \lambda_2}(\lambda_1 - \lambda_2)e^{-\lambda_1 t}$$

$$= \lambda_1 N_1$$

So $\frac{dN_2}{dt} = \lambda_1 N_1 - \lambda_2 N_2$ as required.

(c) The functions to be plotted are

$$N_1(t) = 1\,000 e^{-\left(0.223\,6\ \text{min}^{-1}\right)t}$$

$$N_2(t) = 1\,130.8\left[e^{-\left(0.223\,6\ \text{min}^{-1}\right)t} - e^{-\left(0.025\,9\ \text{min}^{-1}\right)t}\right]$$

From the graph: $t_m \approx \boxed{10.9\text{ min}}$

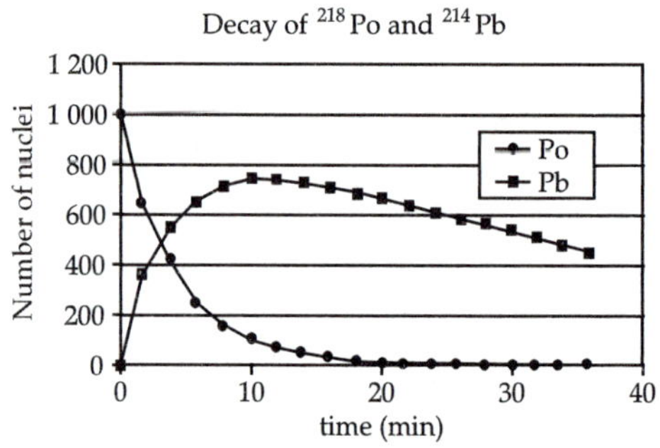

FIG. P30.18(c)

(d) From (1), $\frac{dN_2}{dt} = 0$ if $\lambda_2 e^{-\lambda_2 t} = \lambda_1 e^{-\lambda_1 t}$. $\therefore e^{(\lambda_1 - \lambda_2)t} = \frac{\lambda_1}{\lambda_2}$. Thus, $t = \boxed{t_m = \frac{\ln(\lambda_1/\lambda_2)}{\lambda_1 - \lambda_2}}$.

With $\lambda_1 = 0.223\,6\ \text{min}^{-1}$, $\lambda_2 = 0.025\,9\ \text{min}^{-1}$, this formula gives $t_m = \boxed{10.9\text{ min}}$, in agreement with the result of part (c).

Section 30.4 The Radioactive Decay Processes

P30.19 $Q = \left(M_{\text{U-238}} - M_{\text{Th-234}} - M_{\text{He-4}}\right)(931.5\ \text{MeV/u})$

$Q = (238.050\,783 - 234.043\,596 - 4.002\,603)\ \text{u}(931.5\ \text{MeV/u}) = \boxed{4.27\ \text{MeV}}$

P30.20 (a) A gamma ray has zero charge and it contains no protons or neutrons. So for a gamma ray $Z=0$ and $A=0$. Keeping the total values of Z and A for the system conserved then requires $Z=28$ and $A=65$ for X. With this atomic number it must be nickel, and the nucleus must be in an exited state, so it is $\boxed{{}^{65}_{28}\text{Ni}^*}$.

(b) $\alpha={}^4_2\text{He}$ has $Z=2$ and $A=4$

so for X we require $Z=84-2=82$

for Pb and $A=215-4=211$, $X=\boxed{{}^{211}_{82}\text{Pb}}$.

(c) A positron $e^+={}^0_1e$ has charge the same as a nucleus with $Z=1$. A neutrino ${}^0_0\nu$ has no charge. Neither contains any protons or neutrons. So X must have by conservation $Z=26+1=27$. It is Co. And $A=55+0=55$. It is $\boxed{{}^{55}_{27}\text{Co}}$.

Similar reasoning about balancing the sums of Z and A across the reaction reveals:

(d) $\boxed{{}^{0}_{-1}e}$

(e) $\boxed{{}^1_1\text{H (or p)}}$. Note that this process is a nuclear reaction, rather than radioactive decay. We can solve it from the same principles, which are fundamentally conservation of charge and conservation of baryon number.

P30.21 $N_C=\left(\dfrac{0.021\,0\text{ g}}{12.0\text{ g/mol}}\right)\left(6.02\times10^{23}\text{ molecules/mol}\right)$

$\left(N_C=1.05\times10^{21}\text{ carbon atoms}\right)$ of which 1 in 7.70×10^{11} is a ${}^{14}\text{C}$ atom

$$\left(N_0\right)_{\text{C-14}}=1.37\times10^9,\ \lambda_{\text{C-14}}=\frac{\ln 2}{5\,730\text{ yr}}=1.21\times10^{-4}\text{ yr}^{-1}=3.83\times10^{-12}\text{ s}^{-1}$$

$$R=\lambda N=\lambda N_0e^{-\lambda t}$$

At $t=0$, $$R_0=\lambda N_0=\left(3.83\times10^{-12}\text{ s}^{-1}\right)\left(1.37\times10^9\right)\left[\frac{7(86\,400\text{ s})}{1\text{ week}}\right]=3.17\times10^3\text{ decays/week}.$$

At time t, $$R=\frac{837}{0.88}=951\text{ decays/week}.$$

Taking logarithms, $\ln\dfrac{R}{R_0}=-\lambda t$ so $t=\dfrac{-1}{\lambda}\ln\left(\dfrac{R}{R_0}\right)$

$$t=\frac{-1}{1.21\times10^{-4}\text{ yr}^{-1}}\ln\left(\frac{951}{3.17\times10^3}\right)=\boxed{9.96\times10^3\text{ yr}}.$$

P30.22 $^3_1\text{H nucleus} \rightarrow ^3_2\text{He nucleus} + e^- + \bar{\nu}$

becomes $^3_1\text{H nucleus} + e^- \rightarrow ^3_2\text{He nucleus} + 2e^- + \bar{\nu}$.

Ignoring the slight difference in ionization energies,

we have $\boxed{^3_1\text{H atom} \rightarrow ^3_2\text{He atom} + \bar{\nu}}$

$$3.016\,049\text{ u} = 3.016\,029\text{ u} + 0 + \frac{Q}{c^2}$$

$$Q = (3.016\,049\text{ u} - 3.016\,029\text{ u})(931.5\text{ MeV/u}) = 0.018\,6\text{ MeV} = \boxed{18.6\text{ keV}}$$

P30.23 (a) $\boxed{e^- + p \rightarrow n + \nu}$

(b) For nuclei, $^{15}\text{O} + e^- \rightarrow ^{15}\text{N} + \nu$.
Add seven electrons to both sides to obtain $\boxed{^{15}_8\text{O atom} \rightarrow ^{15}_7\text{N atom} + \nu}$.

(c) From Table A.3, $m(^{15}\text{O}) = m(^{15}\text{N}) + \frac{Q}{c^2}$

$$\Delta m = 15.003\,065\text{ u} - 15.000\,109\text{ u} = 0.002\,956\text{ u}$$

$$Q = (931.5\text{ MeV/u})(0.002\,956\text{ u}) = \boxed{2.75\text{ MeV}}$$

P30.24

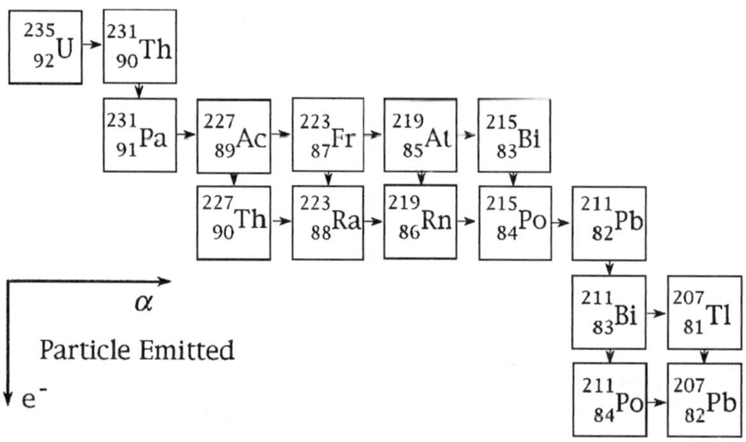

FIG. P30.24

P30.25 (a) $$4.00\text{ pCi/L} = \left(\frac{4.00 \times 10^{-12}\text{ Ci}}{1\text{ L}}\right)\left(\frac{3.70 \times 10^{10}\text{ Bq}}{1\text{ Ci}}\right)\left(\frac{1.00 \times 10^3\text{ L}}{1\text{ m}^3}\right) = \boxed{148\text{ Bq/m}^3}$$

(b) $$N = \frac{R}{\lambda} = R\left(\frac{T_{1/2}}{\ln 2}\right) = \left(148\text{ Bq/m}^3\right)\left(\frac{3.82\text{ d}}{\ln 2}\right)\left(\frac{86\,400\text{ s}}{1\text{ d}}\right) = \boxed{7.05 \times 10^7\text{ atoms/m}^3}$$

continued on next page

(c) $$\text{mass} = \left(7.05\times10^{7}\ \text{atoms/m}^3\right)\left(\frac{1\ \text{mol}}{6.02\times10^{23}\ \text{atoms}}\right)\left(\frac{222\ \text{g}}{1\ \text{mol}}\right) = 2.60\times10^{-14}\ \text{g/m}^3$$

Since air has a density of 1.20 kg/m^3, the fraction consisting of radon is

$$\text{fraction} = \frac{2.60\times10^{-14}\ \text{g/m}^3}{1\,200\ \text{g/m}^3} = \boxed{2.17\times10^{-17}}$$

Section 30.5 Nuclear Reactions

P30.26 (a) For X, $A = 24+1-4 = 21$

and $Z = 12+0-2 = 10$, so X is $\boxed{{}^{21}_{10}\text{Ne}}$.

(b) $A = 235+1-90-2 = 144$

and $Z = 92+0-38-0 = 54$, so X is $\boxed{{}^{144}_{54}\text{Xe}}$.

(c) $A = 2-2 = 0$

and $Z = 2-1 = +1$, so X must be a positron.

As it is ejected, so is a neutrino: $\boxed{X={}^{0}_{1}\text{e}^{+}}$ and $\boxed{X'={}^{0}_{0}\nu}$.

P30.27 (a) $${}^{197}_{79}\text{Au} + {}^{1}_{0}\text{n} \rightarrow {}^{198}_{79}\text{Au}^{*} \rightarrow {}^{198}_{80}\text{Hg} + {}^{0}_{-1}\text{e} + \overline{\nu}$$

(b) Consider adding 79 electrons:

$${}^{197}_{79}\text{Au atom} + {}^{1}_{0}\text{n} \rightarrow {}^{198}_{80}\text{Hg atom} + \overline{\nu} + Q$$

$$Q = \left[M_{{}^{197}\text{Au}} + m_n - M_{{}^{198}\text{Hg}}\right]c^2$$

$$Q = \left[196.966\,552 + 1.008\,665 - 197.966\,752\right]\text{u}(931.5\ \text{MeV/u}) = \boxed{7.89\ \text{MeV}}$$

P30.28 Neglect recoil of product nucleus, (i.e., do not require momentum conservation for the system of colliding particles). The energy balance gives $K_{\text{emerging}} = K_{\text{incident}} + Q$. To find Q:

$$Q = \left[(M_{\text{H}} + M_{\text{Al}}) - (M_{\text{Si}} + m_n)\right]c^2$$

$$Q = \left[(1.007\,825 + 26.981\,539) - (26.986\,705 + 1.008\,665)\right]\text{u}(931.5\ \text{MeV/u}) = -5.59\ \text{MeV}$$

Thus, $K_{\text{emerging}} = 6.61\ \text{MeV} - 5.59\ \text{MeV} = \boxed{1.02\ \text{MeV}}$.

P30.29 The available energy to do work is 0.200 times the energy content of the fuel.

$$(1.00 \text{ kg fuel})(0.034\,0\ {}^{235}\text{U/fuel})\left(\frac{1\,000 \text{ g}}{1 \text{ kg}}\right)\left(\frac{1 \text{ mol}}{235 \text{ g}}\right)(6.02\times10^{23}/\text{mol})\left(\frac{(208)(1.60\times10^{-13}\text{ J})}{\text{fission}}\right)=2.90\times10^{12}\text{ J}$$

$$(2.90\times10^{12}\text{ J})(0.200)=5.80\times10^{11}\text{ J}=(1.00\times10^{5}\text{ N})\Delta r$$

$$\Delta r=5.80\times10^{6}\text{ m}=\boxed{5.80 \text{ Mm}}$$

P30.30 (a)

$$Q=(\Delta m)c^2=\left[m_n+M_{\text{U235}}-M_{\text{Ba141}}-M_{\text{Kr92}}-3m_n\right]c^2$$

$$\Delta m=\left[(1.008\,665+235.043\,923)-(140.914\,4+91.926\,2+3\times1.008\,665)\right]\text{u}=0.185\,993\text{ u}$$

$$Q=(0.185\,993\text{ u})(931.5\text{ MeV/u})=\boxed{173\text{ MeV}}$$

(b)

$$f=\frac{\Delta m}{m_i}=\frac{0.185993\text{ u}}{236.05\text{ u}}=7.88\times10^{-4}=\boxed{0.078\,8\%}$$

P30.31

$$\text{mass of }{}^{235}\text{U available}\approx(0.007)(10^{9}\text{ metric tons})\left(\frac{10^{6}\text{ g}}{1\text{ metric ton}}\right)=7\times10^{12}\text{ g}$$

$$\text{number of nuclei}\approx\left(\frac{7\times10^{12}\text{ g}}{235\text{ g/mol}}\right)(6.02\times10^{23}\text{ nuclei/mol})=1.8\times10^{34}\text{ nuclei}$$

The energy available from fission (at 208 MeV/event) is

$$E\approx(1.8\times10^{34}\text{ events})(208\text{ MeV/event})(1.60\times10^{-13}\text{ J/MeV})=6.0\times10^{23}\text{ J}.$$

This would last for a time interval of

$$\Delta t=\frac{E}{\mathcal{P}}\approx\frac{6.0\times10^{23}\text{ J}}{7.0\times10^{12}\text{ J/s}}=(8.6\times10^{10}\text{ s})\left(\frac{1\text{ yr}}{3.16\times10^{7}\text{ s}}\right)\approx\boxed{3\,000\text{ yr}}.$$

P30.32 (a)

$$V=(317\times10^{6}\text{ mi}^3)\left(\frac{1\,609\text{ m}}{1\text{ mi}}\right)^3=1.32\times10^{18}\text{ m}^3$$

$$m_{\text{water}}=\rho V=(10^{3}\text{ kg/m}^3)(1.32\times10^{18}\text{ m}^3)=1.32\times10^{21}\text{ kg}$$

$$m_{\text{H}_2}=\left(\frac{M_{\text{H}_2}}{M_{\text{H}_2\text{O}}}\right)m_{\text{H}_2\text{O}}=\left(\frac{2.016}{18.015}\right)(1.32\times10^{21}\text{ kg})=1.48\times10^{20}\text{ kg}$$

$$m_{\text{Deuterium}}=(0.030\,0\%)m_{\text{H}_2}=(0.030\,0\times10^{-2})(1.48\times10^{20}\text{ kg})=4.43\times10^{16}\text{ kg}$$

The number of deuterium nuclei in this mass is

$$N=\frac{m_{\text{Deuterium}}}{m_{\text{Deuteron}}}=\frac{4.43\times10^{16}\text{ kg}}{(2.014\text{ u})(1.66\times10^{-27}\text{ kg/u})}=1.33\times10^{43}.$$

Since two deuterium nuclei are used per fusion, ${}^2_1\text{H}+{}^2_1\text{H}\rightarrow{}^4_2\text{He}+Q$, the number of events is $\frac{N}{2}=6.63\times10^{42}$.

continued on next page

The energy released per event is

$$Q=\left[M_{^2\text{H}}+M_{^2\text{H}}-M_{^4\text{He}}\right]c^2=\left[2(2.014\,102)-4.002\,603\right]\text{u}(931.5\text{ MeV/u})=23.8\text{ MeV}.$$

The total energy available is then

$$E=\left(\frac{N}{2}\right)Q=\left(6.63\times10^{42}\right)(23.8\text{ MeV})\left(\frac{1.60\times10^{-13}\text{ J}}{1\text{ MeV}}\right)=\boxed{2.53\times10^{31}\text{ J}}.$$

(b) The time this energy could possibly meet world requirements is

$$\Delta t=\frac{E}{\mathcal{P}}=\frac{2.53\times10^{31}\text{ J}}{100\left(7.00\times10^{12}\text{ J/s}\right)}=\left(3.61\times10^{16}\text{ s}\right)\left(\frac{1\text{ yr}}{3.16\times10^{7}\text{ s}}\right)=\boxed{1.14\times10^{9}\text{ yr}}\sim1\text{ billion years}.$$

Section 30.6 Context Connection—The Engine of the Stars

P30.33 Add two electrons to both sides of the given reaction.

Then $4\ {}^1_1\text{H atom}\rightarrow{}^4_2\text{He atom}+Q$

where $$Q=(\Delta m)c^2=\left[4(1.007\,825)-4.002603\right]\text{u}(931.5\text{ MeV/u})=26.7\text{ MeV}$$

or $$Q=(26.7\text{ MeV})\left(1.60\times10^{-13}\text{ J/MeV}\right)=4.28\times10^{-12}\text{ J}.$$

The proton fusion rate is then

$$\text{rate}=\frac{\text{power output}}{\text{energy per proton}}=\frac{3.85\times10^{26}\text{ J/s}}{\left(4.28\times10^{-12}\text{ J}\right)/(4\text{ protons})}=\boxed{3.60\times10^{38}\text{ protons/s}}.$$

P30.34 (a) The solar-core temperature of 15 MK gives particles enough kinetic energy to overcome the Coulomb-repulsion barrier to ${}^1_1\text{H}+{}^3_2\text{He}\rightarrow{}^4_2\text{He}+\text{e}^++\nu$, estimated as $\frac{k_e(e)(2e)}{r}$. The Coulomb barrier to Bethe's fifth and eight reactions is like $\frac{k_e(e)(7e)}{r}$, larger by $\frac{7}{2}$ times, so the required temperature can be estimated as $\frac{7}{2}\left(15\times10^6\text{ K}\right)\approx\boxed{5\times10^7\text{ K}}$.

(b) For ${}^{12}\text{C}+{}^1\text{H}\rightarrow{}^{13}\text{N}+Q$,

$$Q_1=(12.000\,000+1.007\,825-13.005\,739)(931.5\text{ MeV})=\boxed{1.94\text{ MeV}}$$

For the second step, add seven electrons to both sides to have: ${}^{13}\text{N atom}\rightarrow{}^{13}\text{C atom}+\text{e}^++\text{e}^-+Q$

continued on next page

$$Q_2 = [13.005\,739 - 13.003\,355 - 2(0.000\,549)](931.5 \text{ MeV}) = \boxed{1.20 \text{ MeV}}$$

$$Q_3 = Q_7 = 2(0.000\,549)(931.5 \text{ MeV}) = \boxed{1.02 \text{ MeV}}$$

$$Q_4 = [13.003\,355 + 1.007\,825 - 14.003\,074](931.5 \text{ MeV}) = \boxed{7.55 \text{ MeV}}$$

$$Q_5 = [14.003\,074 + 1.007\,825 - 15.003\,065](931.5 \text{ MeV}) = \boxed{7.30 \text{ MeV}}$$

$$Q_6 = [15.003\,065 - 15.000\,109 - 2(0.000\,549)](931.5 \text{ MeV}) = \boxed{1.73 \text{ MeV}}$$

$$Q_8 = [15.000\,109 + 1.007\,825 - 12 - 4.002\,603](931.5 \text{ MeV}) = \boxed{4.97 \text{ MeV}}$$

The sum is $\boxed{26.7 \text{ MeV}}$, the same as for the proton-proton cycle.

(c) Not all of the energy released appears as internal energy in the star. When a neutrino is created, it will likely fly directly out of the star without interacting with any other particle.

P30.35 (a) $Q_{\text{I}} = [M_A + M_B - M_C - M_E]c^2$, and $\quad Q_{\text{II}} = [M_C + M_D - M_F - M_G]c^2$

$$Q_{\text{net}} = Q_{\text{I}} + Q_{\text{II}} = [M_A + M_B - M_C - M_E + M_C + M_D - M_F - M_G]c^2$$

$$Q_{\text{net}} = Q_{\text{I}} + Q_{\text{II}} = [M_A + M_B + M_D - M_E - M_F - M_G]c^2$$

Thus, reactions may be added. Any product like C used in a subsequent reaction does not contribute to the energy balance.

(b) Adding all five reactions gives

$${}^1_1\text{H} + {}^1_1\text{H} + {}^{\;0}_{-1}\text{e} + {}^1_1\text{H} + {}^1_1\text{H} + {}^{\;0}_{-1}\text{e} \to {}^4_2\text{He} + 2\nu + Q_{\text{net}}$$

or $\quad 4\,{}^1_1\text{H} + 2\,{}^{\;0}_{-1}\text{e} \to {}^4_2\text{He} + 2\nu + Q_{\text{net}}.$

Adding two electrons to each side $\quad 4\,{}^1_1\text{H atom} \to {}^4_2\text{He atom} + Q_{\text{net}}.$

Thus, $\quad Q_{\text{net}} = \left[4M_{{}^1_1\text{H}} - M_{{}^4_2\text{He}}\right]c^2 = [4(1.007\,825) - 4.002\,603]\text{u}(931.5 \text{ MeV/u}) = \boxed{26.7 \text{ MeV}}.$

P30.36 (a) Add two electrons to both sides of the reaction to have it in energy terms:

$4\,{}^1_1\text{H atom} \to {}^4_2\text{He atom} + Q \qquad Q = \Delta mc^2 = \left[4M_{{}^1_1\text{H}} - M_{{}^4_2\text{He}}\right]c^2$

$$Q = [4(1.007\,825 \text{ u}) - 4.002\,603 \text{ u}](931.5 \text{ MeV/u})\left(\frac{1.60 \times 10^{-13} \text{ J}}{1 \text{ MeV}}\right) = \boxed{4.28 \times 10^{-12} \text{ J}}$$

(b) $$N = \frac{1.99 \times 10^{30} \text{ kg}}{1.67 \times 10^{-27} \text{ kg/atom}} = \boxed{1.19 \times 10^{57} \text{ atoms}} = 1.19 \times 10^{57} \text{ protons}$$

(c) The energy that could be created by this many protons in this reaction is:

$$(1.19 \times 10^{57} \text{ protons})\left(\frac{4.28 \times 10^{-12} \text{ J}}{4 \text{ protons}}\right) = 1.27 \times 10^{45} \text{ J}$$

$\mathcal{P} = \dfrac{E}{\Delta t} \quad$ so $\quad \Delta t = \dfrac{E}{\mathcal{P}} = \dfrac{1.27 \times 10^{45} \text{ J}}{3.85 \times 10^{26} \text{ W}} = 3.31 \times 10^{18} \text{ s} = \boxed{105 \text{ billion years}}.$

P30.37 (a) At 6×10^8 K, the average kinetic energy of a carbon atom is

$$\frac{3}{2}k_BT=(1.5)\left(8.62\times10^{-5}\ \text{eV/K}\right)\left(6\times10^8\ \text{K}\right)=\boxed{8\times10^4\ \text{eV}}$$

Note that 6×10^8 K is about $6^2=36$ times larger than 1.5×10^7 K, the core temperature of the Sun. This factor corresponds to the higher potential-energy barrier to carbon fusion compared to hydrogen fusion. It could be misleading to compare it to the temperature $\sim10^8$ K required for fusion in a low-density plasma in a fusion reactor.

(b) The energy released is

$$E=\left[2m\left(\text{C}^{12}\right)-m\left(\text{Ne}^{20}\right)-m\left(\text{He}^4\right)\right]c^2$$

$$E=(24.000\,000-19.992\,440-4.002\,603)(931.5)\ \text{MeV}=\boxed{4.62\ \text{MeV}}$$

In the second reaction,

$$E=\left[2m\left(\text{C}^{12}\right)-m\left(\text{Mg}^{24}\right)\right](931.5)\ \text{MeV/u}$$

$$E=(24.000\,000-23.985\,042)(931.5)\ \text{MeV}=\boxed{13.9\ \text{MeV}}$$

(c) The energy released is the energy of reaction of the number of carbon nuclei in a 2.00-kg sample, which corresponds to

$$\Delta E=\left(2.00\times10^3\ \text{g}\right)\left(\frac{6.02\times10^{23}\ \text{atoms/mol}}{12.0\ \text{g/mol}}\right)\left(\frac{4.62\ \text{MeV/fusion event}}{2\ \text{nuclei/fusion event}}\right)\left(\frac{1\ \text{kWh}}{2.25\times10^{19}\ \text{MeV}}\right)$$

$$\Delta E=\frac{\left(1.00\times10^{26}\right)(4.62)}{2\left(2.25\times10^{19}\right)}\text{kWh}=\boxed{1.03\times10^7\ \text{kWh}}$$

P30.38 We have $N_{235}=N_{0,\,235}e^{-\lambda_{235}t}$

and $N_{238}=N_{0,\,238}e^{-\lambda_{238}t}$

$$\frac{N_{235}}{N_{238}}=0.007\,25=e^{\left(-(\ln 2)t/T_{h,\,235}+(\ln 2)t/T_{h,\,238}\right)}.$$

Taking logarithms,

$$-4.93=\left(-\frac{\ln 2}{0.704\times10^9\ \text{yr}}+\frac{\ln 2}{4.47\times10^9\ \text{yr}}\right)t$$

or

$$-4.93=\left(-\frac{1}{0.704\times10^9\ \text{yr}}+\frac{1}{4.47\times10^9\ \text{yr}}\right)(\ln 2)t$$

$$t=\frac{-4.93}{\left(-1.20\times10^{-9}\ \text{yr}^{-1}\right)\ln 2}=\boxed{5.94\times10^9\ \text{yr}}.$$

Additional Problems

***P30.39** (a) $$Q = \left[M_{^9\text{Be}} + M_{^4\text{He}} - M_{^{12}\text{C}} - m_n\right]c^2$$

$$Q = [9.012\,182\text{ u} + 4.002\,603\text{ u} - 12.000\,000\text{ u} - 1.008\,665\text{ u}](931.5\text{ MeV/u}) = \boxed{5.70\text{ MeV}}$$

(b) $$Q = \left[2M_{^2\text{H}} - M_{^3\text{He}} - m_n\right]$$

$$Q = [2(2.014\,102) - 3.016\,029 - 1.008\,665]\text{ u}(931.5\text{ MeV/u}) = \boxed{3.27\text{ MeV (exothermic)}}$$

***P30.40** (a) With m_n and v_n as the mass and speed of the neutrons, Eq. 8.25 of the text becomes, after making appropriate notational changes, for the two collisions $v_1 = \left(\frac{2m_n}{m_n + m_1}\right)v_n$,

$$v_2 = \left(\frac{2m_n}{m_n + m_2}\right)v_n$$

$$\therefore (m_n + m_2)v_2 = (m_n + m_1)v_1 = 2m_n v_n$$

$$\therefore m_n(v_2 - v_1) = m_1 v_1 - m_2 v_2$$

$$\therefore m_n = \frac{m_1 v_1 - m_2 v_2}{v_2 - v_1}$$

(b) $$m_n = \frac{(1\text{ u})(3.30\times10^7\text{ m/s}) - (14\text{ u})(4.70\times10^6\text{ m/s})}{4.70\times10^6\text{ m/s} - 3.30\times10^7\text{ m/s}} = \boxed{1.16\text{ u}}$$

P30.41 (a) At threshold, the particles have no kinetic energy relative to each other. That is, they move like two particles that have suffered a perfectly inelastic collision. Therefore, in order to calculate the reaction threshold energy, we can use the results of a perfectly inelastic collision. Initially, the projectile M_a moves with velocity v_a while the target M_X is at rest. We have from momentum conservation for the projectile-target system:

$$M_a v_a = (M_a + M_X)v_c.$$

The initial energy is: $E_i = \frac{1}{2}M_a v_a^2$.

The final kinetic energy is:

$$E_f = \frac{1}{2}(M_a + M_X)v_c^2 = \frac{1}{2}(M_a + M_X)\left[\frac{M_a v_a}{M_a + M_X}\right]^2 = \left[\frac{M_a}{M_a + M_X}\right]E_i.$$

From this, we see that E_f is always less than E_i and the change in energy, $E_f - E_i$, is given by

$$E_f - E_i = \left[\frac{M_a}{M_a + M_X} - 1\right]E_i = -\left[\frac{M_X}{M_a + M_X}\right]E_i.$$

This loss of kinetic energy in the isolated system corresponds to an increase in mass-energy during the reaction. Thus, the absolute value of this kinetic energy change is equal to $-Q$ (remember that Q is negative in an endothermic reaction). The initial kinetic energy E_i is the threshold energy E_{th}.

continued on next page

Therefore, $$-Q = \left[\frac{M_X}{M_a + M_X}\right] E_{th}$$

or $$E_{th} = -Q\left[\frac{M_X + M_a}{M_X}\right] = \boxed{-Q\left[1 + \frac{M_a}{M_X}\right]}.$$

(b) First, calculate the Q-value for the reaction: $Q = \left[M_{\text{N-14}} + M_{\text{He-4}} - M_{\text{O-17}} - M_{\text{H-1}}\right]c^2$

$$Q = [14.003\,074 + 4.002\,603 - 16.999\,132 - 1.007\,825]\text{u}(931.5 \text{ MeV/u}) = -1.19 \text{ MeV}.$$

Then, $$E_{th} = -Q\left[\frac{M_X + M_a}{M_X}\right] = -(-1.19 \text{ MeV})\left[1 + \frac{4.002\,603}{14.003\,074}\right] = \boxed{1.53 \text{ MeV}}.$$

P30.42 (a) $r = r_0 A^{1/3} = 1.20 \times 10^{-15} A^{1/3}$ m.

When $A = 12$, $r = \boxed{2.75 \times 10^{-15} \text{ m}}$.

(b) $$F = \frac{k_e(Z-1)e^2}{r^2} = \frac{\left(8.99 \times 10^9 \text{ N}\cdot\text{m}^2/\text{C}^2\right)(Z-1)\left(1.60 \times 10^{-19} \text{ C}\right)^2}{r^2}$$

When $Z = 6$ and $r = 2.75 \times 10^{-15}$ m, $F = \boxed{152 \text{ N}}$.

(c) $$U = \frac{k_e q_1 q_2}{r} = \frac{k_e(Z-1)e^2}{r} = \frac{\left(8.99 \times 10^9\right)(Z-1)\left(1.6 \times 10^{-19}\right)^2}{r}$$

When $Z = 6$ and $r = 2.75 \times 10^{-15}$ m, $U = 4.19 \times 10^{-13} \text{ J} = \boxed{2.62 \text{ MeV}}$.

(d) $A = 238$; $Z = 92$, $r = \boxed{7.44 \times 10^{-15} \text{ m}}$ $F = \boxed{379 \text{ N}}$

and $U = 2.82 \times 10^{-12} \text{ J} = \boxed{17.6 \text{ MeV}}$.

P30.43 (a) Because the reaction $p \to n + e^+ + \nu$ would violate the law of $\boxed{\text{conservation of energy}}$

$m_p = 1.007\,276$ u $m_n = 1.008\,665$ u $m_{e^+} = 5.49 \times 10^{-4}$ u.

Note that $m_n + m_{e^+} > m_p$.

(b) The $\boxed{\text{required energy can come from the electrostatic repulsion}}$ of protons in the nucleus.

(c) Add seven electrons to both sides of the reaction for nuclei ${}^{13}_{7}\text{N} \to {}^{13}_{6}\text{C} + e^+ + \nu$

to obtain the reaction for neutral atoms ${}^{13}_{7}\text{N atom} \to {}^{13}_{6}\text{C atom} + e^+ + e^- + \nu$

$$Q = c^2\left[m\left({}^{13}\text{N}\right) - m\left({}^{13}\text{C}\right) - m_{e^+} - m_{e^-} - m_\nu\right]$$

$$Q = (931.5 \text{ MeV/u})\left[13.005\,739 - 13.003\,355 - 2\left(5.49 \times 10^{-4}\right) - 0\right]\text{u}$$

$$Q = (931.5 \text{ MeV/u})\left(1.286 \times 10^{-3} \text{ u}\right) = \boxed{1.20 \text{ MeV}}$$

P30.44 (a) A least-square fit to the graph yields:

$$\lambda = -\text{slope} = -\left(-0.250\ \text{h}^{-1}\right) = 0.250\ \text{h}^{-1}$$

and

$$\ln(\text{cpm})\big|_{t=0} = \text{intercept} = 8.30.$$

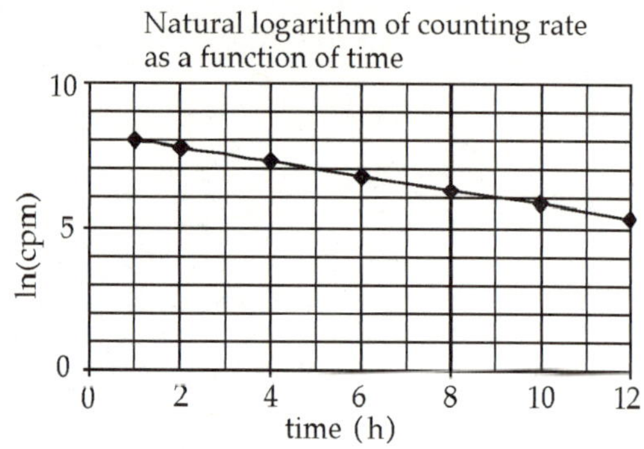

FIG. P30.44

(b) $$\lambda = 0.250\ \text{h}^{-1}\left(\frac{1\ \text{h}}{60.0\ \text{min}}\right) = \boxed{4.17\times10^{-3}\ \text{min}^{-1}}$$

$$T_{1/2} = \frac{\ln 2}{\lambda} = \frac{\ln 2}{4.17\times10^{-3}\ \text{min}^{-1}} = 166\ \text{min} = \boxed{2.77\ \text{h}}$$

(c) From (a), $\text{intercept} = \ln(\text{cpm})_0 = 8.30.$

Thus, $(\text{cpm})_0 = e^{8.30}\ \text{counts/min} = \boxed{4.02\times10^{3}\ \text{counts/min}}.$

(d) $$N_0 = \frac{R_0}{\lambda} = \frac{1}{\lambda}\frac{(\text{cpm})_0}{\text{Eff}} = \frac{4.02\times10^{3}\ \text{counts/min}}{\left(4.17\times10^{-3}\ \text{min}^{-1}\right)(0.100)} = \boxed{9.65\times10^{6}\ \text{atoms}}$$

P30.45 (a) If ΔE is the energy difference between the excited and ground states of the nucleus of mass M, and hf is the energy of the emitted photon, conservation of energy for the nucleus-photon system gives

$$\Delta E = hf + E_r. \tag{1}$$

Where E_r is the recoil energy of the nucleus, which can be expressed as

$$E_r = \frac{Mv^2}{2} = \frac{(Mv)^2}{2M}. \tag{2}$$

Since system momentum must also be conserved, we have

$$Mv = \frac{hf}{c}. \tag{3}$$

Hence, E_r can be expresses as $E_r = \dfrac{(hf)^2}{2Mc^2}.$

When $hf << Mc^2$

we can make the approximation that $hf \approx \Delta E$

so $E_r \approx \boxed{\dfrac{(\Delta E)^2}{2Mc^2}}.$

continued on next page

(b) $E_r = \frac{(\Delta E)^2}{2Mc^2}$ where $\Delta E = 0.014\,4\text{ MeV}$

and $Mc^2 = (57\text{ u})(931.5\text{ MeV/u}) = 5.31\times10^4\text{ MeV}$.

Therefore, $E_r = \frac{\left(1.44\times10^{-2}\text{ MeV}\right)^2}{2\left(5.31\times10^4\text{ MeV}\right)} = \boxed{1.94\times10^{-3}\text{ eV}}$.

P30.46 (a) One liter of milk contains this many ^{40}K nuclei:

$$N = (2.00\text{ g})\left(\frac{6.02\times10^{23}\text{ nuclei/mol}}{39.1\text{ g/mol}}\right)\left(\frac{0.011\,7}{100}\right) = 3.60\times10^{18}\text{ nuclei}$$

$$\lambda = \frac{\ln 2}{T_{1/2}} = \frac{\ln 2}{1.28\times10^9\text{ yr}}\left(\frac{1\text{ yr}}{3.156\times10^7\text{ s}}\right) = 1.72\times10^{-17}\text{ s}^{-1}$$

$$R = \lambda N = \left(1.72\times10^{-17}\text{ s}^{-1}\right)\left(3.60\times10^{18}\right) = \boxed{61.8\text{ Bq}}$$

(b) For the iodine, $R = R_0 e^{-\lambda t}$ with $\lambda = \frac{\ln 2}{8.04\text{ d}}$

$$t = \frac{1}{\lambda}\ln\left(\frac{R_0}{R}\right) = \frac{8.04\text{ d}}{\ln 2}\ln\left(\frac{2\,000}{61.8}\right) = \boxed{40.3\text{ d}}$$

P30.47 (a) For cobalt–56,

$$\lambda = \frac{\ln 2}{T_{1/2}} = \frac{\ln 2}{77.1\text{ d}}\left(\frac{365.25\text{ d}}{1\text{ yr}}\right) = 3.28\text{ yr}^{-1}.$$

The elapsed time from July 1054 to July 2005 is 951 yr.

$$R = R_0 e^{-\lambda t}$$

implies $\frac{R}{R_0} = e^{-\lambda t} = e^{-\left(3.28\text{ yr}^{-1}\right)(951\text{ yr})} = e^{-3\,123} = e^{-(\ln 10)1\,356} = \boxed{\sim 10^{-1\,356}}$.

(b) For carbon–14,

$$\lambda = \frac{\ln 2}{5\,730\text{ yr}} = 1.21\times10^{-4}\text{ yr}^{-1}$$

$$\frac{R}{R_0} = e^{-\lambda t} = e^{-\left(1.21\times10^{-4}\text{ yr}^{-1}\right)(951\text{ yr})} = e^{-0.115} = \boxed{0.891}$$

P30.48 (a) If we assume all the ^{87}Sr came from ^{87}Rb,

then $N = N_0 e^{-\lambda t}$

yields $t = \frac{-1}{\lambda}\ln\left(\frac{N}{N_0}\right) = \frac{T_{1/2}}{\ln 2}\ln\left(\frac{N_0}{N}\right)$

where $N = N_{\text{Rb-87}}$

continued on next page

and $N_0 = N_{\text{Sr-87}} + N_{\text{Rb-87}}$

$$t = \frac{(4.75\times10^{10}\text{ yr})}{\ln 2}\ln\left(\frac{1.82\times10^{10}+1.07\times10^{9}}{1.82\times10^{10}}\right) = \boxed{3.91\times10^{9}\text{ yr}}.$$

(b) It could be $\boxed{\text{no older}}$. The rock could be younger if some ^{87}Sr were originally present.

P30.49 (a) Let us assume that the parent nucleus (mass M_p) is initially at rest, and let us denote the masses of the daughter nucleus and alpha particle by M_d and M_α, respectively. Applying the equations of conservation of momentum and energy for the alpha decay process gives

$$M_d v_d = M_\alpha v_\alpha \quad (1)$$

$$M_p c^2 = M_d c^2 + M_\alpha c^2 + \frac{1}{2}M_\alpha v_\alpha^2 + \frac{1}{2}M_d v_d^2 \quad (2)$$

The disintegration energy Q is given by

$$Q = (M_p - M_d - M_\alpha)c^2 = \frac{1}{2}M_\alpha v_\alpha^2 + \frac{1}{2}M_d v_d^2 \quad (3)$$

Eliminating v_d from Equations (1) and (3) gives

$$Q = \frac{1}{2}M_d\left(\frac{M_\alpha}{M_d}v_\alpha\right)^2 + \frac{1}{2}M_\alpha v_\alpha^2 = \frac{1}{2}\frac{M_\alpha^2}{M_d}v_\alpha^2 + \frac{1}{2}M_\alpha v_\alpha^2 = \frac{1}{2}M_\alpha v_\alpha^2\left(1+\frac{M_\alpha}{M_d}\right)$$

$$= \boxed{K_\alpha\left(1+\frac{M_\alpha}{M_d}\right)}$$

(b) $$K_\alpha = \frac{Q}{1+\frac{M_\alpha}{M_d}} = \frac{4.87\text{ MeV}}{1+\frac{4}{222}} = \boxed{4.78\text{ MeV}}$$

P30.50 $R = R_0 \exp(-\lambda t)$ lets us write $\ln R = \ln R_0 - \lambda t$

which is the equation of a straight line with $|\text{slope}| = \lambda$.

The logarithmic plot shown in Figure P30.50 is fitted by

$\ln R = 8.44 - 0.262t$.

If t is measured in minutes, then decay constant λ is 0.262 per minute. The half-life is

$$T_{1/2} = \frac{\ln 2}{\lambda} = \frac{\ln 2}{0.262/\text{min}} = \boxed{2.64\text{ min}}.$$

The reported half-life of ^{137}Ba is 2.55 min. The difference reflects experimental uncertainties.

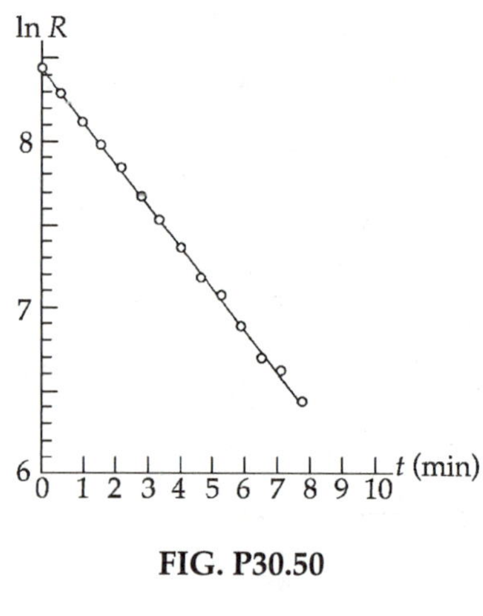

FIG. P30.50

P30.51 $N_0 = \dfrac{\text{mass present}}{\text{mass of nucleus}} = \dfrac{5.00\text{ kg}}{(89.907\,7\text{ u})(1.66\times10^{-27}\text{ kg/u})} = 3.35\times10^{25}\text{ nuclei}$

$$\lambda = \frac{\ln 2}{T_{1/2}} = \frac{\ln 2}{29.1\text{ yr}} = 2.38\times10^{-2}\text{ yr}^{-1} = 4.53\times10^{-8}\text{ min}^{-1}$$

$$R_0 = \lambda N_0 = (4.53\times10^{-8}\text{ min}^{-1})(3.35\times10^{25}) = 1.52\times10^{18}\text{ counts/min}$$

$$\frac{R}{R_0} = e^{-\lambda t} = \frac{10.0\text{ counts/min}}{1.52\times10^{18}\text{ counts/min}} = 6.59\times10^{-18}$$

and $\lambda t = -\ln(6.59\times10^{-18}) = 39.6$

giving $t = \dfrac{39.6}{\lambda} = \dfrac{39.6}{2.38\times10^{-2}\text{ yr}^{-1}} = \boxed{1.66\times10^{3}\text{ yr}}$.

***P30.52** (a) $I = I_0 e^{-\mu x}$, so $x = \dfrac{1}{\mu}\ln\left(\dfrac{I_0}{I}\right)$

With $\mu = 1.59\text{ cm}^{-1}$, the thickness when $I = \dfrac{I_0}{2}$ is $x = \dfrac{1}{1.59\text{ cm}^{-1}}\ln(2) = \boxed{0.436\text{ cm}}$.

(b) When $\dfrac{I_0}{I} = 1.00\times10^{4}$, $x = \dfrac{1}{1.59\text{ cm}^{-1}}\ln(1.00\times10^{4}) = \boxed{5.79\text{ cm}}$.

***P30.53** (a) With $I(x) = \dfrac{1}{2}I_0$, $I(x) = I_0 e^{-\mu x}$ becomes

$$\frac{1}{2}I_0 = I_0 e^{-0.72x/\text{mm}} \qquad 2 = e^{+0.72x/\text{mm}}$$

$$\ln 2 = 0.72\,x/\text{mm} \qquad x = \frac{\ln 2\text{ mm}}{0.72} = \boxed{0.963\text{ mm}}$$

(b) $I(0.8\text{ mm}) = I_0 e^{-0.72(0.8)} = 0.562 I_0$

$I(0.7\text{ mm}) = I_0 e^{-0.72(0.7)} = 0.604 I_0$

$$\text{fractional change} = \frac{0.604I_0 - 0.562I_0}{0.562I_0} = 0.074\,7 = \boxed{7.47\%}$$

***P30.54** The initial specific activity of ^{59}Fe in the steel,

$$(R/m)_0 = \frac{20.0\ \mu\text{Ci}}{0.200\text{ kg}} = \frac{100\ \mu\text{Ci}}{\text{kg}}\left(\frac{3.70\times10^{4}\text{ Bq}}{1\ \mu\text{Ci}}\right) = 3.70\times10^{6}\text{ Bq/kg}.$$

After 1 000 h, $\dfrac{R}{m} = \left(\dfrac{R}{m}\right)_0 e^{-\lambda t} = (3.70\times10^{6}\text{ Bq/kg})e^{-(6.40\times10^{-4}\text{ h}^{-1})(1\,000\text{ h})} = 1.95\times10^{6}\text{ Bq/kg}$.

The activity of the oil, $R_{\text{oil}} = \left(\dfrac{800}{60.0}\text{ Bq/liter}\right)(6.50\text{ liters}) = 86.7\text{ Bq}$.

Therefore, $m_{\text{in oil}} = \dfrac{R_{\text{oil}}}{(R/m)} = \dfrac{86.7\text{ Bq}}{1.95\times10^{6}\text{ Bq/kg}} = 4.45\times10^{-5}\text{ kg}$.

So that wear rate is $\dfrac{4.45\times10^{-5}\text{ kg}}{1\,000\text{ h}} = \boxed{4.45\times10^{-8}\text{ kg/h}}$.

P30.55 (a) Starting with $N = 0$ radioactive atoms at $t = 0$, the rate of increase is (production – decay)

$$\frac{dN}{dt} = R - \lambda N \qquad \text{so} \qquad dN = (R - \lambda N)dt.$$

The variables are separable.

$$\int_0^N \frac{dN}{R - \lambda N} = \int_0^t dt: \qquad -\frac{1}{\lambda}\ln\left(\frac{R - \lambda N}{R}\right) = t$$

so $$\ln\left(\frac{R - \lambda N}{R}\right) = -\lambda t \qquad \text{and} \qquad \left(\frac{R - \lambda N}{R}\right) = e^{-\lambda t}.$$

Therefore, $$1 - \frac{\lambda}{R}N = e^{-\lambda t} \qquad N = \boxed{\frac{R}{\lambda}\left(1 - e^{-\lambda t}\right)}.$$

(b) The maximum number of radioactive nuclei would be $\boxed{\frac{R}{\lambda}}$.

P30.56 The Japanese call it the *original child bomb.*

(a) Suppose each ^{235}U fission releases 208 MeV of energy. Then, the number of nuclei that must have undergone fission is

$$N = \frac{\text{total release}}{\text{energy per nuclei}} = \frac{5 \times 10^{13}\ \text{J}}{(208\ \text{MeV})\left(1.60 \times 10^{-13}\ \text{J/MeV}\right)} = \boxed{1.5 \times 10^{24}\ \text{nuclei}}.$$

(b) $$\text{mass} = \left(\frac{1.5 \times 10^{24}\ \text{nuclei}}{6.02 \times 10^{23}\ \text{nuclei/mol}}\right)(235\ \text{g/mol}) \approx \boxed{0.6\ \text{kg}}$$

P30.57 The complete fissioning of 1.00 gram of U^{235} releases

$$Q = \frac{(1.00\ \text{g})}{235\ \text{grams/mol}}\left(6.02 \times 10^{23}\ \text{atoms/mol}\right)(200\ \text{MeV/fission})\left(1.60 \times 10^{-13}\ \text{J/MeV}\right) = 8.20 \times 10^{10}\ \text{J}.$$

If all this energy could be utilized to convert m kilograms of 20.0°C water to 400°C steam (see Chapter 17 of text for values),

then $$Q = mc_w\,\Delta T + mL_v + mc_s\,\Delta T$$

$$Q = m\left[(4186\ \text{J/kg °C})(80.0\ \text{°C}) + 2.26 \times 10^6\ \text{J/kg} + (2010\ \text{J/kg °C})(300\ \text{°C})\right].$$

Therefore $$m = \frac{8.20 \times 10^{10}\ \text{J}}{3.20 \times 10^6\ \text{J/kg}} = \boxed{2.56 \times 10^4\ \text{kg}}.$$

P30.58 (a) $$\Delta V = 4\pi r^2 \Delta r = 4\pi\left(14.0 \times 10^3\ \text{m}\right)^2(0.05\ \text{m}) = 1.23 \times 10^8\ \text{m}^3 \ \boxed{\sim 10^8\ \text{m}^3}$$

(b) The force on the next layer is determined by atmospheric pressure.

$$W = P\Delta V = \left(1.013 \times 10^5\ \text{N/m}^2\right)\left(1.23 \times 10^8\ \text{m}^3\right) = 1.25 \times 10^{13}\ \text{J} \ \boxed{\sim 10^{13}\ \text{J}}$$

continued on next page

(c) $1.25\times10^{13}\text{ J}=\frac{1}{10}(\text{yield})$, so yield $=1.25\times10^{14}\text{ J}$ $\boxed{\sim10^{14}\text{ J}}$

(d) $$\frac{1.25\times10^{14}\text{ J}}{4.2\times10^{9}\text{ J/ton TNT}}=2.97\times10^{4}\text{ ton TNT}\sim10^{4}\text{ ton TNT}$$

or $\boxed{\sim10\text{ kilotons}}$

P30.59 (a) The number of molecules in 1.00 liter of water (mass = 1 000 g) is

$$N=\left(\frac{1.00\times10^{3}\text{ g}}{18.0\text{ g/mol}}\right)\left(6.02\times10^{23}\text{ molecules/mol}\right)=3.34\times10^{25}\text{ molecules}.$$

The number of deuterium nuclei contained in these molecules is

$$N'=\left(3.34\times10^{25}\text{ molecules}\right)\left(\frac{1\text{ deuteron}}{3300\text{ molecules}}\right)=1.01\times10^{22}\text{ deuterons}.$$

Since 2 deuterons are consumed per fusion event, the number of events possible is $\frac{N'}{2}=5.07\times10^{21}$ reactions, and the energy released is

$$E_{\text{fusion}}=\left(5.07\times10^{21}\text{ reactions}\right)(3.27\text{ MeV/reaction})=1.66\times10^{22}\text{ MeV}$$

$$E_{\text{fusion}}=\left(1.66\times10^{22}\text{ MeV}\right)\left(1.60\times10^{-13}\text{ J/MeV}\right)=\boxed{2.65\times10^{9}\text{ J}}.$$

(b) In comparison to burning 1.00 liter of gasoline, the energy from the fusion of deuterium is

$$\frac{E_{\text{fusion}}}{E_{\text{gasoline}}}=\frac{2.65\times10^{9}\text{ J}}{3.40\times10^{7}\text{ J}}=\boxed{78.0\text{ times larger}}.$$

P30.60 The number of nuclei in 0.155 kg of ^{210}Po is

$$N_0=\left(\frac{155\text{ g}}{209.98\text{ g/mol}}\right)\left(6.02\times10^{23}\text{ nuclei/mol}\right)=4.44\times10^{23}\text{ nuclei}.$$

The half-life of ^{210}Po is 138.38 days, so the decay constant is given by

$$\lambda=\frac{\ln 2}{T_{1/2}}=\frac{\ln 2}{(138.38\text{ d})\left(8.64\times10^{4}\text{ s/d}\right)}=5.80\times10^{-8}\text{ s}^{-1}.$$

The initial activity is

$$R_0=\lambda N_0=\left(5.80\times10^{-8}\text{ s}^{-1}\right)\left(4.44\times10^{23}\text{ nuclei}\right)=2.58\times10^{16}\text{ Bq}.$$

The energy released in each $^{210}_{84}\text{Po}\to{}^{206}_{82}\text{Pb}+{}^{4}_{2}\text{He}$ reaction is

$$Q=\left[M_{^{210}_{84}\text{Po}}-M_{^{206}_{82}\text{Pb}}-M_{^{4}_{2}\text{He}}\right]c^2:$$

$$Q=[209.982\,857-205.974\,449-4.002\,603]\text{u}(931.5\text{ MeV/u})=5.41\text{ MeV}.$$

Thus, assuming a conversion efficiency of 1.00%, the initial power output of the battery is

$$\mathcal{P}=(0.010\,0)R_0Q=(0.010\,0)\left(2.58\times10^{16}\text{ decays/s}\right)(5.41\text{ MeV/decay})\left(1.60\times10^{-13}\text{ J/MeV}\right)=\boxed{223\text{ W}}.$$

***P30.61** (a) $V=\ell^3=\dfrac{m}{\rho}$, so $\ell=\left(\dfrac{m}{\rho}\right)^{1/3}=\left(\dfrac{70.0\text{ kg}}{18.7\times10^3\text{ kg/m}^3}\right)^{1/3}=\boxed{0.155\text{ m}}$

(b) Add 92 electrons to both sides of the given nuclear reaction. Then it becomes
$^{238}_{92}\text{U atom}\rightarrow 8\ ^{4}_{2}\text{He atom}+\ ^{206}_{82}\text{Pb atom}+Q_{\text{net}}$.

$$Q_{\text{net}}=\left[M_{^{238}_{92}\text{U}}-8M_{^4_2\text{He}}-M_{^{206}_{82}\text{Pb}}\right]c^2=\left[238.050\,783-8(4.002\,603)-205.974\,449\right]\text{u }(931.5\text{ MeV/u})$$

$$Q_{\text{net}}=\boxed{51.7\text{ MeV}}$$

(c) If there is a single step of decay, the number of decays per time is the decay rate R and the energy released in each decay is Q. Then the energy released per time is $\boxed{\mathscr{P}=QR}$. If there is a series of decays in steady state, the equation is still true, with Q representing the net decay energy.

(d) The decay rate for all steps in the radioactive series in steady state is set by the parent uranium:

$$N=\left(\frac{7.00\times10^4\text{ g}}{238\text{ g/mol}}\right)\left(6.02\times10^{23}\text{ nuclei/mol}\right)=1.77\times10^{26}\text{ nuclei}$$

$$\lambda=\frac{\ln 2}{T_{1/2}}=\frac{\ln 2}{4.47\times10^9\text{ yr}}=1.55\times10^{-10}\ \frac{1}{\text{yr}}$$

$$R=\lambda N=\left(1.55\times10^{-10}\ \frac{1}{\text{yr}}\right)\left(1.77\times10^{26}\text{ nuclei}\right)=2.75\times10^{16}\text{ decays/yr},$$

so $\mathscr{P}=QR=(51.7\text{ MeV})\left(2.75\times10^{16}\ \dfrac{1}{\text{yr}}\right)\left(1.60\times10^{-13}\text{ J/MeV}\right)=\boxed{2.27\times10^5\text{ J/yr}}$.

(e) The allowed whole-body dose is then $(70.0\text{ kg})\left(\dfrac{4.55\times10^{-3}\text{ J}}{\text{kg}\cdot\text{yr}}\right)=\boxed{3.18\text{ J/yr}}$.

***P30.62** From Table A.3, the half-life of ^{32}P is 14.26 d. Thus, the decay constant is

$$\lambda=\frac{\ln 2}{T_{1/2}}=\frac{\ln 2}{14.26\text{ d}}=0.048\,6\text{ d}^{-1}=5.63\times10^{-7}\text{ s}^{-1}.$$

$$N_0=\frac{R_0}{\lambda}=\frac{5.22\times10^6\text{ decay/s}}{5.63\times10^{-7}\text{ s}^{-1}}=9.28\times10^{12}\text{ nuclei}$$

At $t=10.0$ days, the number remaining is

$$N=N_0e^{-\lambda t}=\left(9.28\times10^{12}\text{ nuclei}\right)e^{-(0.048\,6\text{ d}^{-1})(10.0\text{ d})}=5.71\times10^{12}\text{ nuclei}$$

so the number of decays has been $N_0-N=3.57\times10^{12}$ and the energy released is

$$E=\left(3.57\times10^{12}\right)(700\text{ keV})\left(1.60\times10^{-16}\text{ J/keV}\right)=\boxed{0.400\text{ J}}.$$

***P30.63** The decay constant is $\lambda = \frac{\ln 2}{T_{1/2}} = \frac{\ln 2}{17\text{ d}} = 0.040\,8/\text{d}$. The number of nuclei remaining after 30 d is $N = N_0 e^{-\lambda T} = N_0 e^{-0.040\,8(30)} = 0.294 N_0$. The number decayed is $N_0 - N = N_0(1 - 0.294) = 0.706 N_0$. Then the energy release is

$$2.12\text{ J} = 0.706 N_0 \left(21.0 \times 10^3\text{ eV}\right)\left(\frac{1.6 \times 10^{-19}\text{ J}}{1\text{ eV}}\right)$$

$$N_0 = \frac{2.12\text{ J}}{2.37 \times 10^{-15}\text{ J}} = 8.94 \times 10^{14}$$

(a) $R_0 = \lambda N_0 = \frac{0.040\,8}{\text{d}}\left(8.94 \times 10^{14}\right)\left(\frac{1\text{ d}}{86\,400\text{ s}}\right) = \boxed{4.22 \times 10^8\text{ Bq}}$

(b) $\text{original sample mass} = m = N_{\text{original}} m_{\text{one atom}} = 8.94 \times 10^{14}(103\text{ u})\left(\frac{1.66 \times 10^{-27}\text{ kg}}{1\text{ u}}\right)$

$$= \boxed{1.53 \times 10^{-10}\text{ kg}} = 1.53 \times 10^{-7}\text{ g} = 153\text{ ng}$$

P30.64 (a) The number of Pu nuclei in 1.00 kg $= \frac{6.02 \times 10^{23}\text{ nuclei/mol}}{239.05\text{ g/mol}}(1\,000\text{ g})$.

The total energy $= \left(25.2 \times 10^{23}\text{ nuclei}\right)(200\text{ MeV}) = 5.04 \times 10^{26}\text{ MeV}$

$$E = \left(5.04 \times 10^{26}\text{ MeV}\right)\left(4.44 \times 10^{-20}\text{ kWh/MeV}\right) = \boxed{2.24 \times 10^7\text{ kWh}}$$

or 22 million kWh.

(b) $E = \Delta m c^2 = (3.016\,049\text{ u} + 2.014\,102\text{ u} - 4.002\,603\text{ u} - 1.008\,665\text{ u})(931.5\text{ MeV/u})$

$E = \boxed{17.6\text{ MeV for each D-T fusion}}$

(c) $E_n = (\text{Total number of D nuclei})(17.6)\left(4.44 \times 10^{-20}\right)$

$$E_n = \left(6.02 \times 10^{23}\right)\left(\frac{1\,000}{2.014}\right)(17.6)\left(4.44 \times 10^{-20}\right) = \boxed{2.34 \times 10^8\text{ kWh}}$$

(d) $E_n = \text{the number of C atoms in 1.00 kg} \times 4.20\text{ eV}$

$$E_n = \left(\frac{6.02 \times 10^{26}}{12}\right)\left(4.20 \times 10^{-6}\text{ MeV}\right)\left(4.44 \times 10^{-20}\right) = \boxed{9.36\text{ kWh}}$$

(e) Coal is cheap at this moment in human history. We hope that safety and waste disposal problems can be solved so that nuclear energy can be affordable before scarcity drives up the price of fossil fuels.

ANSWERS TO EVEN PROBLEMS

P30.2 (a) 7.89 cm for ^{12}C, 8.21 cm for ^{13}C
(b) see the solution

P30.4 (a) 1.90 fm; (b) 7.44 fm

P30.6 6.11×10^{15} N toward the other ball

P30.8 (a) 1.11 MeV/nucleon;
(b) 7.07 MeV/nucleon;
(c) 8.79 MeV/nucleon;
(d) 7.57 MeV/nucleon

P30.10 (a) Cs; (b) La; (c) Cs

P30.12 0.200 mCi

P30.14 1.16 ks

P30.16 $\dfrac{R_0 T_{1/2}}{\ln 2}\left(2^{-t_1/T_{1/2}} - 2^{-t_2/T_{1/2}}\right)$

P30.18 (a) see the solution; (b) see the solution;
(c) see the solution; 10.9 min;
(d) $t_m = \dfrac{\ln(\lambda_1/\lambda_2)}{\lambda_1 - \lambda_2}$, yes

P30.20 (a) $^{65}_{28}\text{Ni}^*$; (b) $^{211}_{82}\text{Pb}$; (c) $^{55}_{27}\text{Co}$; (d) $^{0}_{-1}\text{e}$;
(e) $^{1}_{1}\text{H}$ (or p)

P30.22 $^{3}_{1}\text{H arom} \rightarrow {}^{3}_{2}\text{He atom} + \bar{\nu}$, 18.6 keV

P30.24 see the solution

P30.26 (a) $^{21}_{10}\text{Ne}$; (b) $^{144}_{54}\text{Xe}$; (c) $^{0}_{1}\text{e}^+ + ^{0}_{0}\nu$

P30.28 1.02 MeV

P30.30 (a) 173 MeV; (b) 0.078 8%

P30.32 (a) 2.53×10^{31} J; (b) 1.14×10^{9} yr

P30.34 (a) 5×10^{7} K;
(b) 1.94 MeV, 1.20 MeV, 1.02 MeV, 7.55 MeV, 7.30 MeV, 1.73 MeV, 1.02 MeV, 4.97 MeV, 26.7 MeV;
(c) Most of the neutrinos leave the star directly after their creation, without interacting with any other particles.

P30.36 (a) 4.28 pJ; (b) 1.19×10^{57} atoms;
(c) 105 Gyr

P30.38 5.94 Gyr

P30.40 (a) see the solution; (b) 1.16 u

P30.42 (a) 2.75 fm; (b) 152 N; (c) 2.62 MeV;
(d) 7.44 fm, 379 N, 17.6 MeV

P30.44 (a) see the solution;
(b) 4.17×10^{-3} min^{-1}, 2.77 h;
(c) 4.02×10^{3} counts/min;
(d) 9.65×10^{6} atoms

P30.46 (a) 61.8 Bq; (b) 40.3 d

P30.48 (a) 3.91 Gyr;
(b) No. We must make some assumption about the original quantity of radioactive material. In part (a) we assumed that the rock originally contained no strontium.

P30.50 2.64 min

P30.52 (a) 0.436 cm; (b) 5.79 cm

P30.54 4.45×10^{-8} kg/h

P30.56 (a) 1.5×10^{24} nuclei; (b) 0.6 kg

P30.58 (a) $\sim 10^{8}$ m^3; (b) $\sim 10^{13}$ J; (c) $\sim 10^{14}$ J;
(d) ~10 kilotons

P30.60 223 W

P30.62 0.400 J

P30.64 (a) 2.24×10^{7} kWh; (b) 17.6 MeV;
(c) 2.34×10^{8} kWh; (d) 9.36 kWh;
(e) Coal is cheap at this moment in human history. We hope that safety and waste disposal problems can be solved so that nuclear energy can be affordable before scarcity drives up the price of fossil fuels.

Chapter 31

Particle Physics

CHAPTER OUTLINE

ANSWERS TO QUESTIONS

Q31.1 Strong Force—Mediated by gluons.

Electromagnetic Force—Mediated by photons.

Weak Force—Mediated by W^+, W^-, and Z^0 bosons.

Gravitational Force—Mediated by gravitons.

Q31.2 In the quark model, all hadrons are composed of smaller units called quarks. Quarks have a fractional electric charge and a baryon number of $\frac{1}{3}$. There are 6 types of quarks: up, down, strange, charmed, top, and bottom. Further, all *baryons* contain 3 quarks, and all *mesons* contain one quark and one anti-quark. *Leptons* are thought to be fundamental particles.

Q31.3 Hadrons are massive particles with structure and size. There are two classes of hadron: mesons and baryons. Hadrons are composed of quarks. Hadrons interact via the strong force.

Leptons are light particles with no structure or size. It is believed that leptons are fundamental particles. Leptons interact via the weak force.

Q31.4 Baryons are heavy hadrons with spin $\frac{1}{2}$ or $\frac{3}{2}$, are composed of three quarks, and have long lifetimes. Mesons are light hadrons with spin 0 or 1, are composed of a quark and an antiquark, and have short lifetimes.

Q31.5 Resonances are hadrons. They decay into strongly interacting particles such as protons, neutrons, and pions, all of which are hadrons.

Q31.6 The baryon number of a proton or neutron is one. Since baryon number is conserved, the baryon number of the kaon must be zero.

Q31.7 Yes, protons interact via the weak interaction; but the strong interaction predominates.

Q31.8 Decays by the weak interaction typically take 10^{-10} s or longer to occur. This is slow in particle physics.

Q31.9 The decays of the muon, tau, charged pion, kaons, neutron, lambda, charged sigmas, xis, and omega occur by the weak interaction. All have lifetimes longer than 10^{-13} s. Several produce neutrinos; none produce photons. Several violate strangeness conservation.

Q31.10 The decays of the neutral pion, eta, and neutral sigma occur by the electromagnetic interaction. These are three of the shortest lifetimes in Table 31.2. All produce photons, which are the quanta of the electromagnetic force. All conserve strangeness.

Q31.11 You can think of a conservation law as a superficial regularity which we happen to notice, as a person who does not know the rules of chess might observe that one player's two bishops are always on squares of opposite colors. Alternatively, you can think of a conservation law as identifying some stuff of which the universe is made. In classical physics one can think of both matter and energy as fundamental constituents of the world. We buy and sell both of them. In classical physics you can also think of linear momentum, angular momentum, and electric charge as basic stuffs of which the universe is made. In relativity we learn that matter and energy are not conserved separately, but are both aspects of the conserved quantity *relativistic total energy*. Discovered more recently, four conservation laws appear equally general and thus equally fundamental: Conservation of baryon number, conservation of electron-lepton number, conservation of tau-lepton number, and conservation of muon-lepton number. Processes involving the strong force and the electromagnetic force follow conservation of strangeness, charm, bottomness, and topness, while the weak interaction can alter the total S, C, B and T quantum numbers of an isolated system.

Q31.12 No. Antibaryons have baryon number –1, mesons have baryon number 0, and baryons have baryon number +1. The reaction cannot occur because it would not conserve baryon number, unless so much energy is available that a baryon-antibaryon pair is produced.

Q31.13 The Standard Model consists of quantum chromodynamics (to describe the strong interaction) and the electroweak theory (to describe the electromagnetic and weak interactions). The Standard Model is our most comprehensive description of nature. It fails to unify the two theories it includes, and fails to include the gravitational force. It pictures matter as made of six quarks and six leptons, interacting by exchanging gluons, photons, and W and Z bosons.

Q31.14 All baryons and antibaryons consist of three quarks. All mesons and antimesons consist of two quarks. Since quarks have spin quantum number $\frac{1}{2}$ and can be spin-up or spin-down, it follows that the three-quark baryons must have a half-integer spin, while the two-quark mesons must have spin 0 or 1.

Q31.15 Each flavor of quark can have colors, designated as red, green and blue. Antiquarks are colored antired, antigreen, and antiblue. A baryon consists of three quarks, each having a different color. By analogy to additive color mixing we call it colorless. A meson consists of a quark of one color and antiquark with the corresponding anticolor, making it colorless as a whole.

Q31.16 In 1961 Gell-Mann predicted the omega-minus particle, with quark composition sss. Its discovery in 1964 confirmed the quark theory.

Q31.17 The Ξ^- particle has, from Table 31.2, charge $-e$, spin $\frac{1}{2}$, $B=1$, $L_e = L_\mu = L_\tau = 0$, and strangeness –2. All of these are described by its quark composition dss (Table 31.5). The properties of the quarks from Table 31.3 let us add up charge: $-\frac{1}{3}e - \frac{1}{3}e - \frac{1}{3}e = -e$; spin $+\frac{1}{2} - \frac{1}{2} + \frac{1}{2} = \frac{1}{2}$, supposing one of the quarks is spin-down relative to the other two; baryon number $\frac{1}{3} + \frac{1}{3} + \frac{1}{3} = 1$; lepton numbers, charm, bottomness, and topness zero; and strangeness $0 - 1 - 1 = -2$.

Q31.18 The electroweak theory of Glashow, Salam, and Weinberg predicted the W^+, W^-, and Z particles. Their discovery in 1983 confirmed the electroweak theory.

Q31.19 Hubble determined experimentally that all galaxies outside the Local Group are moving away from us, with speed directly proportional to the distance of the galaxy from us.

Q31.20 Before that time, the Universe was too hot for the electrons to remain in any sort of stable orbit around protons. The thermal motion of both protons and electrons was too rapid for them to be in close enough proximity for the Coulomb force to dominate.

Q31.21 The photon and the graviton must both have zero mass. If either had any nonzero mass m, then exchange of a virtual particle would require an energy fluctuation mc^2, lasting for time $\Delta t < \frac{\hbar}{2mc^2}$, and acting over a range no larger than $c\Delta t = \frac{\hbar}{2mc}$. The force could not follow an inverse square law but would have to cut off at a finite range.

Q31.22 The experiment described is analogous to Rutherford's scattering experiment described in Section 29.1. In the Rutherford experiment, alpha particles were scattered from atoms. Their behavior demonstrated that most of the mass of the atom was in a small structure containing the positive charge, namely the nucleus.

Q31.23 No. The Big Bang happened here, but it also happened everywhere else. We are at the center of the spherical section of the universe that we can observe, but the universe has no center. An observer in any other galaxy cluster would observe galaxies moving away from her in all directions, as described by the same Hubble's law.

SOLUTIONS TO PROBLEMS

Note: Problem 8.3 may be useful as review to prepare for decay problems such as Problems 31.11, 31.20, 31.55, 31.60, and 31.62. Problem 22.7 may be useful as review for Problems 31.28 and 31.57. Problems 8.46 and 12.46 may be useful as review for Problems 31.30, 31.58, and 31.65.

Section 31.1 The Fundamental Forces in Nature

Section 31.2 Positrons and Other Antiparticles

P31.1 Assuming that the proton and antiproton are left nearly at rest after they are produced, the energy E of the photon must be

$$E = 2E_0 = 2(938.3\text{ MeV}) = 1\,876.6\text{ MeV} = 3.00\times 10^{-10}\text{ J}.$$

Thus, $E = hf = 3.00\times 10^{-10}\text{ J}$

$$f = \frac{3.00\times 10^{-10}\text{ J}}{6.626\times 10^{-34}\text{ J}\cdot\text{s}} = \boxed{4.53\times 10^{23}\text{ Hz}}$$

$$\lambda = \frac{c}{f} = \frac{3.00\times 10^{8}\text{ m/s}}{4.53\times 10^{23}\text{ Hz}} = \boxed{6.62\times 10^{-16}\text{ m}}.$$

***P31.2** The half-life of ^{14}O is 70.6 s, so the decay constant is $\lambda = \dfrac{\ln 2}{T_{1/2}} = \dfrac{\ln 2}{70.6\text{ s}} = 0.009\,82\text{ s}^{-1}$.

The ^{14}O nuclei remaining after five min is $N = N_0 e^{-\lambda t} = \left(10^{10}\right) e^{-\left(0.009\,82\text{ s}^{-1}\right)(300\text{ s})} = 5.26 \times 10^{8}$.

The number of these in one cubic centimeter of blood is

$$N' = N\left(\frac{1.00\text{ cm}^3}{\text{total vol. of blood}}\right) = \left(5.26\times 10^{8}\right)\left(\frac{1.00\text{ cm}^3}{2000\text{ cm}^3}\right) = 2.63\times 10^{5}$$

and their activity is $R = \lambda N' = \left(0.009\,82\text{ s}^{-1}\right)\left(2.63\times 10^{5}\right) = 2.58\times 10^{3}\text{ Bq}\,\boxed{\sim 10^{3}\text{ Bq}}$.

***P31.3** (a) The rest energy of a total of 6.20 g of material is converted into energy of electromagnetic radiation:

$$E = mc^2 = 6.20\times 10^{-3}\text{ kg}\left(3\times 10^{8}\text{ m/s}\right)^2 = \boxed{5.58\times 10^{14}\text{ J}}$$

(b) $5.58\times 10^{14}\text{ J} = 5.58\times 10^{14}\text{ J}\left(\dfrac{\$0.14}{\text{kWh}}\right)\left(\dfrac{\text{k}}{1\,000}\right)\left(\dfrac{\text{W}}{\text{J/s}}\right)\left(\dfrac{1\text{ h}}{3\,600\text{ s}}\right) = \boxed{\$2.17\times 10^{7}}$

All from two cents' worth of stuff.

P31.4 The minimum energy is released, and hence the minimum frequency photons are produced, when the proton and antiproton are at rest when they annihilate.

That is, $E = E_0$ and $K = 0$. To conserve momentum, each photon must carry away one-half the energy.

Thus, $E_{\min} = \dfrac{2E_0}{2} = E_0 = 938.3\text{ MeV} = hf_{\min}$.

Thus, $f_{\min} = \dfrac{(938.3\text{ MeV})\left(1.60\times 10^{-13}\text{ J/MeV}\right)}{\left(6.626\times 10^{-34}\text{ J}\cdot\text{s}\right)} = \boxed{2.27\times 10^{23}\text{ Hz}}$

$$\lambda = \frac{c}{f_{\min}} = \frac{3.00\times 10^{8}\text{ m/s}}{2.27\times 10^{23}\text{ Hz}} = \boxed{1.32\times 10^{-15}\text{ m}}.$$

P31.5 In $\gamma \to p^+ + p^-$,

we start with energy 2.09 GeV

we end with energy 938.3 MeV + 938.3 MeV + 95.0 MeV + K_2

where K_2 is the kinetic energy of the second proton.

Conservation of energy for the creation process gives $\boxed{K_2 = 118\text{ MeV}}$.

Section 31.3 Mesons and the Beginning of Particle Physics

P31.6 The reaction is $\mu^+ + e^- \to \nu + \nu$

muon-lepton number before reaction: $(-1)+(0)=-1$

electron-lepton number before reaction: $(0)+(1)=1$.

Therefore, after the reaction, the muon-lepton number must be –1. Thus, one of the neutrinos must be the anti-neutrino associated with muons, and one of the neutrinos must be the neutrino associated with electrons:

$$\boxed{\bar{\nu}_\mu} \quad \text{and} \quad \boxed{\nu_e}.$$

Then $\mu^+ + e^- \to \bar{\nu}_\mu + \nu_e$.

P31.7 The creation of a virtual Z^0 boson is an energy fluctuation $\Delta E = 91.2\times10^9$ eV. It can last no longer than $\Delta t = \dfrac{\hbar}{2\Delta E}$ and move no farther than

$$c(\Delta t) = \frac{hc}{4\pi\Delta E} = \frac{(6.626\times10^{-34}\text{ J}\cdot\text{s})(3.00\times10^8\text{ m/s})}{4\pi(91.2\times10^9\text{ eV})}\left(\frac{1\text{ eV}}{1.60\times10^{-19}\text{ J}}\right) = 1.08\times10^{-18}\text{ m} = \boxed{\sim10^{-18}\text{ m}}.$$

P31.8 A proton has rest energy 938.3 MeV. The time interval during which a virtual proton could exist is at most Δt in $\Delta E\Delta t = \dfrac{\hbar}{2}$. The distance it could move is at most

$$c\Delta t = \frac{\hbar c}{2\Delta E} = \frac{(1.055\times10^{-34}\text{ J}\cdot\text{s})(3\times10^8\text{ m/s})}{2(938.3)(1.6\times10^{-13}\text{ J})}\boxed{\sim10^{-16}\text{ m}}.$$

According to Yukawa's line of reasoning, this distance is the range of a force that could be associated with the exchange of virtual protons between high-energy particles.

P31.9 By Table 31.2, $M_{\pi^0} = 135\text{ MeV}/c^2$.

Therefore, $E_\gamma = \boxed{67.5\text{ MeV}}$ for each photon

$$p = \frac{E_\gamma}{c} = \boxed{67.5\text{ MeV}/c}$$

and

$$f = \frac{E_\gamma}{h} = \boxed{1.63\times10^{22}\text{ Hz}}.$$

P31.10 The time interval for a particle traveling with the speed of light to travel a distance of 3×10^{-15} m is

$$\Delta t = \frac{d}{v} = \frac{3\times10^{-15}\text{ m}}{3\times10^8\text{ m/s}} = \boxed{\sim10^{-23}\text{ s}}.$$

P31.11 (a) $\Delta E = \left(m_n - m_p - m_e\right)c^2$

From Table A-3, $\Delta E = (1.008\,665 - 1.007\,825)(931.5) = \boxed{0.782 \text{ MeV}}$.

(b) Assuming the neutron at rest, momentum conservation for the decay process implies $p_p = p_e$. Relativistic energy for the system is conserved

$$\sqrt{\left(m_p c^2\right)^2 + p_p^2 c^2} + \sqrt{\left(m_e c^2\right)^2 + p_e^2 c^2} = m_n c^2.$$

Since $p_p = p_e$, $\sqrt{(938.3)^2 + (pc)^2} + \sqrt{(0.511)^2 + (pc)^2} = 939.6 \text{ MeV}$.

Solving the algebra, $pc = 1.19 \text{ MeV}$.

If $p_e c = \gamma m_e v_e c = 1.19 \text{ MeV}$, then $\dfrac{\gamma v_e}{c} = \dfrac{1.19 \text{ MeV}}{0.511 \text{ MeV}} = \dfrac{x}{\sqrt{1-x^2}} = 2.33$ where $x = \dfrac{v_e}{c}$.

Solving, $x^2 = \left(1 - x^2\right)5.43$ and $x = \dfrac{v_e}{c} = 0.919$

$$\boxed{v_e = 0.919c}.$$

Then $m_p v_p = \gamma_e m_e v_e$: $v_p = \dfrac{\gamma_e m_e v_e c}{m_p c} = \dfrac{(1.19 \text{ MeV})\left(1.60 \times 10^{-13} \text{ J/MeV}\right)}{\left(1.67 \times 10^{-27}\right)\left(3.00 \times 10^8 \text{ m/s}\right)}$

$$v_p = 3.80 \times 10^5 \text{ m/s} = \boxed{380 \text{ km/s}}.$$

(c) $\boxed{\text{The electron is relativistic, the proton is not.}}$

Section 31.4 Classification of Particles

P31.12 In $? + p^+ \to n + \mu^+$, charge conservation requires the unknown particle to be neutral. Baryon number conservation requires baryon number = 0. The muon-lepton number of ? must be –1.

So the unknown particle must be $\boxed{\overline{\nu}_\mu}$.

P31.13 $\Omega^+ \to \overline{\Lambda}^0 + K^+$ $\quad \overline{K}_S^0 \to \pi^+ + \pi^-$ $\quad$ (or $\pi^0 + \pi^0$)

$\overline{\Lambda}^0 \to \overline{p} + \pi^+$ $\quad \overline{n} \to \overline{p} + e^+ + \nu_e$

Section 31.5 Conservation Laws

P31.14 (a) $p+\bar{p}\to\mu^{+}+e^{-}$ $\boxed{L_e}$ $0+0\to0+1$

and $\boxed{L_\mu}$ $0+0\to-1+0$

(b) $\pi^{-}+p\to p+\pi^{+}$ $\boxed{\text{charge}}$ $-1+1\to+1+1$

(c) $p+p\to p+\pi^{+}$ $\boxed{\text{baryon number}}$: $1+1\to1+0$

(d) $p+p\to p+p+n$ $\boxed{\text{baryon number}}$: $1+1\to1+1+1$

(e) $\gamma+p\to n+\pi^{0}$ $\boxed{\text{charge}}$ $0+1\to0+0$

P31.15 (a) Baryon number and charge are conserved, with values of $0+1=0+1$

and $1+1=1+1$ in both reactions.

(b) $\boxed{\text{Strangeness is } \textit{not} \text{ conserved}}$ in the second reaction.

P31.16 Baryon number conservation $\boxed{\text{allows the first and forbids the second}}$.

P31.17 (a) $\pi^{-}\to\mu^{-}+\boxed{\bar{\nu}_\mu}$ L_μ: $0\to1-1$

(b) $K^{+}\to\mu^{+}+\boxed{\nu_\mu}$ L_μ: $0\to-1+1$

(c) $\boxed{\bar{\nu}_e}+p^{+}\to n+e^{+}$ L_e: $-1+0\to0-1$

(d) $\boxed{\nu_e}+n\to p^{+}+e^{-}$ L_e: $1+0\to0+1$

(e) $\boxed{\nu_\mu}+n\to p^{+}+\mu^{-}$ L_μ: $1+0\to0+1$

(f) $\mu^{-}\to e^{-}+\boxed{\bar{\nu}_e}+\boxed{\nu_\mu}$ L_μ: $1\to0+0+1$ and L_e: $0\to1-1+0$

P31.18 Momentum conservation for the decay requires the pions to have equal speeds.

The total energy of each is $\dfrac{497.7\text{ MeV}}{2}$

so $E^2=p^2c^2+\left(mc^2\right)^2$ gives

$(248.8\text{ MeV})^2=(pc)^2+(139.6\text{ MeV})^2$.

continued on next page

Solving, $$pc = 206\text{ MeV} = \gamma mvc = \frac{mc^2}{\sqrt{1-(v/c)^2}}\left(\frac{v}{c}\right)$$

$$\frac{pc}{mc^2} = \frac{206\text{ MeV}}{139.6\text{ MeV}} = \frac{1}{\sqrt{1-(v/c)^2}}\left(\frac{v}{c}\right) = 1.48$$

$$\frac{v}{c} = 1.48\sqrt{1-\left(\frac{v}{c}\right)^2}$$

and $$\left(\frac{v}{c}\right)^2 = 2.18\left[1-\left(\frac{v}{c}\right)^2\right] = 2.18 - 2.18\left(\frac{v}{c}\right)^2$$

$$3.18\left(\frac{v}{c}\right)^2 = 2.18$$

so $$\frac{v}{c} = \sqrt{\frac{2.18}{3.18}} = 0.828$$

and $\boxed{v = 0.828c}$.

P31.19 (a) $p^+ \to \pi^+ + \pi^0$ — $\boxed{\text{Baryon number}}$: $1 \to 0+0$

(b) $p^+ + p^+ \to p^+ + p^+ + \pi^0$ — This reaction $\boxed{\text{can occur}}$.

(c) $p^+ + p^+ \to p^+ + \pi^+$ — $\boxed{\text{Baryon number}}$ is violated: $1+1 \to 1+0$

(d) $\pi^+ \to \mu^+ + \nu_\mu$ — This reaction $\boxed{\text{can occur}}$.

(e) $n^0 \to p^+ + e^- + \bar{\nu}_e$ — This reaction $\boxed{\text{can occur}}$.

(f) $\pi^+ \to \mu^+ + n$ — Violates $\boxed{\text{baryon number}}$: $0 \to 0+1$

Violates $\boxed{\text{muon-lepton number}}$: $0 \to -1+0$

P31.20 (a) $p \to e^+ + \gamma$

Baryon number: $+1 \to 0+0$

$\Delta B \neq 0$, so baryon number conservation is violated.

(b) From conservation of momentum for the decay: $p_e = p_\gamma$.

Then, for the positron, $E_e^2 = (p_e c)^2 + E_{0,e}^2$

becomes $E_e^2 = (p_\gamma c)^2 + E_{0,e}^2 = E_\gamma^2 + E_{0,e}^2$.

continued on next page

From conservation of energy for the system: $E_{0,\,p} = E_e + E_\gamma$

or $E_e = E_{0,\,p} - E_\gamma$

so $E_e^2 = E_{0,\,p}^2 - 2E_{0,\,p}E_\gamma + E_\gamma^2.$

Equating this to the result from above gives $E_\gamma^2 + E_{0,\,e}^2 = E_{0,\,p}^2 - 2E_{0,\,p}E_\gamma + E_\gamma^2$

or $$E_\gamma = \frac{E_{0,\,p}^2 - E_{0,\,e}^2}{2E_{0,\,p}} = \frac{(938.3\text{ MeV})^2 - (0.511\text{ MeV})^2}{2(938.3\text{ MeV})} = \boxed{469\text{ MeV}}.$$

Thus, $$E_e = E_{0,\,p} - E_\gamma = 938.3\text{ MeV} - 469\text{ MeV} = \boxed{469\text{ MeV}}.$$

Also, $$p_\gamma = \frac{E_\gamma}{c} = \boxed{\frac{469\text{ MeV}}{c}}$$

and $$p_e = p_\gamma = \boxed{\frac{469\text{ MeV}}{c}}.$$

(c) The total energy of the positron is $E_e = 469\text{ MeV}$.

But, $$E_e = \gamma E_{0,\,e} = \frac{E_{0,\,e}}{\sqrt{1-(v/c)^2}}$$

so $$\sqrt{1-\left(\frac{v}{c}\right)^2} = \frac{E_{0,\,e}}{E_e} = \frac{0.511\text{ MeV}}{469\text{ MeV}} = 1.09 \times 10^{-3}$$

which yields: $\boxed{v = 0.999\,999\,4c}$.

P31.21 The relevant conservation laws are: $\Delta L_e = 0$

$\Delta L_\mu = 0$

and $\Delta L_\tau = 0$.

(a) $\pi^+ \to \pi^0 + e^+ + ?$ L_e: $0 \to 0 - 1 + L_e$ implies $L_e = 1$ and we have a $\boxed{\nu_e}$

(b) $? + p \to \mu^- + p + \pi^+$ L_μ: $L_\mu + 0 \to +1 + 0 + 0$ implies $L_\mu = 1$ and we have a $\boxed{\nu_\mu}$

(c) $\Lambda^0 \to p + \mu^- + ?$ L_μ: $0 \to 0 + 1 + L_\mu$ implies $L_\mu = -1$ and we have a $\boxed{\bar{\nu}_\mu}$

(d) $\tau^+ \to \mu^+ + ? + ?$ L_μ: $0 \to -1 + L_\mu$ implies $L_\mu = 1$ and we have a $\boxed{\nu_\mu}$

L_τ: $-1 \to 0 + L_\tau$ implies $L_\tau = -1$ and we have a $\boxed{\bar{\nu}_\tau}$

Conclusion for (d): $L_\mu = 1$ for one particle, and $L_\tau = -1$ for the other particle.

We have $\boxed{\nu_\mu}$

and $\boxed{\bar{\nu}_\tau}$.

Section 31.6 Strange Particles and Strangeness

P31.22 The $\rho^0 \to \pi^+ + \pi^-$ decay must occur via the strong interaction.

The $K_S^0 \to \pi^+ + \pi^-$ decay must occur via the weak interaction.

P31.23 (a) $\Lambda^0 \to p + \pi^-$ Strangeness: $-1 \to 0 + 0$ (strangeness is $\boxed{\text{not conserved}}$)

(b) $\pi^- + p \to \Lambda^0 + K^0$ Strangeness: $0 + 0 \to -1 + 1$ ($0 = 0$ and strangeness is $\boxed{\text{conserved}}$)

(c) $\bar{p} + p \to \bar{\Lambda}^0 + \Lambda^0$ Strangeness: $0 + 0 \to +1 - 1$ ($0 = 0$ and strangeness is $\boxed{\text{conserved}}$)

(d) $\pi^- + p \to \pi^- + \Sigma^+$ Strangeness: $0 + 0 \to 0 - 1$ ($0 \neq -1$: strangeness is $\boxed{\text{not conserved}}$)

(e) $\Xi^- \to \Lambda^0 + \pi^-$ Strangeness: $-2 \to -1 + 0$ ($-2 \neq -1$ so strangeness is $\boxed{\text{not conserved}}$)

(f) $\Xi^0 \to p + \pi^-$ Strangeness: $-2 \to 0 + 0$ ($-2 \neq 0$ so strangeness is $\boxed{\text{not conserved}}$)

P31.24 (a) $\mu^- \to e^- + \gamma$ L_e: $0 \to 1 + 0$,

and L_μ: $1 \to 0$

(b) $n \to p + e^- + \nu_e$ L_e: $0 \to 0 + 1 + 1$

(c) $\Lambda^0 \to p + \pi^0$ Strangeness: $-1 \to 0 + 0$,

and charge: $0 \to +1 + 0$

(d) $p \to e^+ + \pi^0$ Baryon number: $+1 \to 0 + 0$

(e) $\Xi^0 \to n + \pi^0$ Strangeness: $-2 \to 0 + 0$

P31.25 (a) $\pi^- + p \to 2\eta$ violates conservation of baryon number as $0 + 1 \to 0$, $\boxed{\text{not allowed}}$.

(b) $K^- + n \to \Lambda^0 + \pi^-$

Baryon number, $0 + 1 \to 1 + 0$

Charge, $-1 + 0 \to 0 - 1$

Strangeness, $-1 + 0 \to -1 + 0$

Lepton number, $0 \to 0$

The interaction may occur via the $\boxed{\text{strong interaction}}$ since all are conserved.

continued on next page

(c) $K^- \rightarrow \pi^- + \pi^0$

Strangeness,	$-1 \rightarrow 0+0$
Baryon number,	$0 \rightarrow 0$
Lepton number,	$0 \rightarrow 0$
Charge,	$-1 \rightarrow -1+0$

Strangeness is violated by one unit, but everything else is conserved. Thus, the reaction can occur via the **weak interaction**, but not the strong or electromagnetic interaction.

(d) $\Omega^- \rightarrow \Xi^- + \pi^0$

Baryon number,	$1 \rightarrow 1+0$
Lepton number,	$0 \rightarrow 0$
Charge,	$-1 \rightarrow -1+0$
Strangeness,	$-3 \rightarrow -2+0$

May occur by **weak interaction**, but not by strong or electromagnetic.

(e) $\eta \rightarrow 2\gamma$

Baryon number,	$0 \rightarrow 0$
Lepton number,	$0 \rightarrow 0$
Charge,	$0 \rightarrow 0$
Strangeness,	$0 \rightarrow 0$

No conservation laws are violated, but photons are the mediators of the electromagnetic interaction. Also, the lifetime of the η is consistent with the **electromagnetic interaction**.

P31.26 (a) $\Xi^- \rightarrow \Lambda^0 + \mu^- + \nu_\mu$

Baryon number:	$+1 \rightarrow +1+0+0$	Charge:	$-1 \rightarrow 0-1+0$
L_e:	$0 \rightarrow 0+0+0$	L_μ:	$0 \rightarrow 0+1+1$
L_τ:	$0 \rightarrow 0+0+0$	Strangeness:	$-2 \rightarrow -1+0+0$

Conserved quantities are: **B, charge, L_e, and L_τ**

(b) $K_S^0 \rightarrow 2\pi^0$

Baryon number:	$0 \rightarrow 0$	Charge:	$0 \rightarrow 0$
L_e:	$0 \rightarrow 0$	L_μ:	$0 \rightarrow 0$
L_τ:	$0 \rightarrow 0$	Strangeness:	$+1 \rightarrow 0$

Conserved quantities are: **B, charge, L_e, L_μ, and L_τ**

continued on next page

(c) $K^- + p \rightarrow \Sigma^0 + n$

Baryon number:	$0+1 \rightarrow 1+1$	Charge:	$-1+1 \rightarrow 0+0$
L_e:	$0+0 \rightarrow 0+0$	L_μ:	$0+0 \rightarrow 0+0$
L_τ:	$0+0 \rightarrow 0+0$	Strangeness:	$-1+0 \rightarrow -1+0$

Conserved quantities are: $\boxed{S, \text{ charge}, L_e, L_\mu, \text{ and } L_\tau}$

(d) $\Sigma^0 + \Lambda^0 + \gamma$

Baryon number:	$+1 \rightarrow 1+0$	Charge:	$0 \rightarrow 0$
L_e:	$0 \rightarrow 0+0$	L_μ:	$0 \rightarrow 0+0$
L_τ:	$0 \rightarrow 0+0$	Strangeness:	$-1 \rightarrow -1+0$

Conserved quantities are: $\boxed{B, S, \text{ charge}, L_e, L_\mu, \text{ and } L_\tau}$

(e) $e^+ + e^- \rightarrow \mu^+ + \mu^-$

Baryon number:	$0+0 \rightarrow 0+0$	Charge:	$+1-1 \rightarrow +1-1$
L_e:	$-1+1 \rightarrow 0+0$	L_μ:	$0+0 \rightarrow +1-1$
L_τ:	$0+0 \rightarrow 0+0$	Strangeness:	$0+0 \rightarrow 0+0$

Conserved quantities are: $\boxed{B, S, \text{ charge}, L_e, L_\mu, \text{ and } L_\tau}$

(f) $\overline{p} + n \rightarrow \overline{\Lambda}^0 + \Sigma^-$

Baryon number:	$-1+1 \rightarrow -1+1$	Charge:	$-1+0 \rightarrow 0-1$
L_e:	$0+0 \rightarrow 0+0$	L_μ:	$0+0 \rightarrow 0+0$
L_τ:	$0+0 \rightarrow 0+0$	Strangeness:	$0+0 \rightarrow +1-1$

Conserved quantities are: $\boxed{B, S, \text{ charge}, L_e, L_\mu, \text{ and } L_\tau}$

P31.27 (a) $K^+ + p \rightarrow \underline{?} + p$

The strong interaction conserves everything.

Baryon number,	$0+1 \rightarrow B+1$	so	$B=0$
Charge,	$+1+1 \rightarrow Q+1$	so	$Q=+1$
Lepton numbers,	$0+0 \rightarrow L+0$	so	$L_e = L_\mu = L_\tau = 0$
Strangeness,	$+1+0 \rightarrow S+0$	so	$S=1$

The conclusion is that the particle must be positively charged, a non-baryon, with strangeness of +1. Of particles in Table 31.2, it can only be the $\boxed{K^+}$. Thus, this is an elastic scattering process.

The weak interaction conserves all but strangeness, and $\Delta S = \pm 1$.

continued on next page

(b) $\Omega^- \to \underline{?} + \pi^-$

Baryon number,	$+1 \to B + 0$	so	$B = 1$
Charge,	$-1 \to Q - 1$	so	$Q = 0$
Lepton numbers,	$0 \to L + 0$	so	$L_e = L_\mu = L_\tau = 0$
Strangeness,	$-3 \to S + 0$	so	$\Delta S = 1 : S = -2$

The particle must be a neutral baryon with strangeness of –2. Thus, it is the $\boxed{\Xi^0}$.

(c) $K^+ \to \underline{?} + \mu^+ + \nu_\mu$

Baryon number,	$0 \to B + 0 + 0$	so	$B = 0$
Charge,	$+1 \to Q + 1 + 0$	so	$Q = 0$
Lepton numbers,	$L_e,\ 0 \to L_e + 0 + 0$	so	$L_e = 0$
	$L_\mu,\ 0 \to L_\mu - 1 + 1$	so	$L_\mu = 0$
	$L_\tau,\ 0 \to L_\tau + 0 + 0$	so	$L_\tau = 0$
Strangeness,	$1 \to S + 0 + 0$	so	$\Delta S = \pm 1$
			(for weak interaction): $S = 0$

The particle must be a neutral meson with strangeness $= 0 \Rightarrow \boxed{\pi^0}$.

Section 31.7 Measuring Particle Lifetimes

***P31.28** (a)

$$p_{\Sigma^+} = eBr_{\Sigma^+} = \frac{(1.602\,177 \times 10^{-19}\ \text{C})(1.15\ \text{T})(1.99\ \text{m})}{5.344\,288 \times 10^{-22}\ (\text{kg} \cdot \text{m/s})/(\text{MeV}/c)} = \boxed{686\ \text{MeV}/c}$$

$$p_{\pi^+} = eBr_{\pi^+} = \frac{(1.602\,177 \times 10^{-19}\ \text{C})(1.15\ \text{T})(0.580\ \text{m})}{5.344\,288 \times 10^{-22}\ (\text{kg} \cdot \text{m/s})/(\text{MeV}/c)} = \boxed{200\ \text{MeV}/c}$$

(b) Let φ be the angle made by the neutron's path with the path of the Σ^+ at the moment of decay. By conservation of momentum:

$$p_n \cos\varphi + (199.961\,581\ \text{MeV}/c)\cos 64.5° = 686.075\,081\ \text{MeV}/c$$

$$\therefore p_n \cos\varphi = 599.989\,401\ \text{MeV}/c \qquad (1)$$

$$p_n \sin\varphi = (199.961\,581\ \text{MeV}/c)\sin 64.5° = 180.482\,380\ \text{MeV}/c \qquad (2)$$

From (1) and (2): $p_n = \sqrt{(599.989\,401\ \text{MeV}/c)^2 + (180.482\,380\ \text{MeV}/c)^2} = \boxed{627\ \text{MeV}/c}$

continued on next page

(c) $$E_{\pi^+} = \sqrt{(p_{\pi^+}c)^2 + (m_{\pi^+}c^2)^2} = \sqrt{(199.961\,581\text{ MeV})^2 + (139.6\text{ MeV})^2} = \boxed{244\text{ MeV}}$$

$$E_n = \sqrt{(p_n c)^2 + (m_n c^2)^2} = \sqrt{(626.547\,022\text{ MeV})^2 + (939.6\text{ MeV})^2} = \boxed{1\,130\text{ MeV}}$$

$$E_{\Sigma^+} = E_{\pi^+} + E_n = 243.870\,445\text{ MeV} + 1\,129.340\,219\text{ MeV} = \boxed{1\,370\text{ MeV}}$$

(d) $$m_{\Sigma^+}c^2 = \sqrt{E_{\Sigma^+}^2 - (p_{\Sigma^+}c)^2} = \sqrt{(1\,373.210\,664\text{ MeV})^2 - (686.075\,081\text{ MeV})^2} = 1\,190\text{ MeV}$$

$$\therefore m_{\Sigma^+} = \boxed{1\,190\text{ MeV}/c^2}$$

$$E_{\Sigma^+} = \gamma m_{\Sigma^+}c^2, \text{ where } \gamma = \left(1 - \frac{v^2}{c^2}\right)^{-1/2} = \frac{1\,373.210\,664\text{ MeV}}{1\,189.541\,303\text{ MeV}} = 1.154\,4$$

Solving for v, $v = \boxed{0.500c}$.

P31.29 Time-dilated lifetime:

$$T = \gamma T_0 = \frac{0.900 \times 10^{-10}\text{ s}}{\sqrt{1 - v^2/c^2}} = \frac{0.900 \times 10^{-10}\text{ s}}{\sqrt{1 - (0.960)^2}} = 3.214 \times 10^{-10}\text{ s}$$

$$\text{distance} = 0.960\left(3.00 \times 10^8\text{ m/s}\right)\left(3.214 \times 10^{-10}\text{ s}\right) = \boxed{9.26\text{ cm}}.$$

***P31.30** (a) Let $E_{\min}$ be the minimum total energy of the bombarding particle that is needed to induce the reaction. At this energy the product particles all move with the same velocity. The product particles are then equivalent to a single particle having mass equal to the total mass of the product particles, moving with the same velocity as each product particle. By conservation of energy:

$$E_{\min} + m_2c^2 = \sqrt{(m_3c^2)^2 + (p_3c)^2}. \qquad (1)$$

By conservation of momentum: $p_3 = p_1$

$$\therefore (p_3c)^2 = (p_1c)^2 = E_{\min}^2 - (m_1c^2)^2. \qquad (2)$$

Substitute (2) in (1): $E_{\min} + m_2c^2 = \sqrt{(m_3c^2)^2 + E_{\min}^2 - (m_1c^2)^2}$.

continued on next page

Square both sides:

$$E_{\min}^2 + 2E_{\min} m_2 c^2 + \left(m_2 c^2\right)^2 = \left(m_3 c^2\right)^2 + E_{\min}^2 - \left(m_1 c^2\right)^2$$

$$\therefore E_{\min} = \frac{\left(m_3^2 - m_1^2 - m_2^2\right)c^2}{2m_2}$$

$$\therefore K_{\min} = E_{\min} - m_1 c^2 = \frac{\left(m_3^2 - m_1^2 - m_2^2 - 2m_1 m_2\right)c^2}{2m_2} = \frac{\left[m_3^2 - \left(m_1 + m_2\right)^2\right]c^2}{2m_2}$$

Refer to Table 31.2 for the particle masses.

(b) $$K_{\min} = \frac{\left[4(938.3)\right]^2 \text{ MeV}^2/c^2 - \left[2(938.3)\right]^2 \text{ MeV}^2/c^2}{2\left(938.3 \text{ MeV}/c^2\right)} = \boxed{5.63 \text{ GeV}}$$

(c) $$K_{\min} = \frac{(497.7 + 1\,115.6)^2 \text{ MeV}^2/c^2 - (139.6 + 938.3)^2 \text{ MeV}^2/c^2}{2(938.3) \text{ MeV}/c^2} = \boxed{768 \text{ MeV}}$$

(d) $$K_{\min} = \frac{\left[2(938.3) + 135\right]^2 \text{ MeV}^2/c^2 - \left[2(938.3)\right]^2 \text{ MeV}^2/c^2}{2(938.3) \text{ MeV}/c^2} = \boxed{280 \text{ MeV}}$$

(e) $$K_{\min} = \frac{\left[\left(91.2 \times 10^3\right)^2 - \left[(938.3 + 938.3)^2\right] \text{ MeV}^2/c^2\right]}{2(938.3) \text{ MeV}/c^2} = \boxed{4.43 \text{ TeV}}$$

Section 31.8 Finding Patterns in the Particles

Section 31.9 Quarks

Section 31.10 Colored Quarks

Section 31.11 The Standard Model

P31.31 (a) The number of protons

$$N_p = 1\,000 \text{ g}\left(\frac{6.02 \times 10^{23} \text{ molecules}}{18.0 \text{ g}}\right)\left(\frac{10 \text{ protons}}{\text{molecule}}\right) = 3.34 \times 10^{26} \text{ protons}$$

and there are $$N_n = 1\,000 \text{ g}\left(\frac{6.02 \times 10^{23} \text{ molecules}}{18.0 \text{ g}}\right)\left(\frac{8 \text{ neutrons}}{\text{molecule}}\right) = 2.68 \times 10^{26} \text{ neutrons}.$$

So there are for electric neutrality $\boxed{3.34 \times 10^{26} \text{ electrons}}$.

The up quarks have number $2\left(3.34 \times 10^{26}\right) + 2.68 \times 10^{26} = \boxed{9.36 \times 10^{26} \text{ up quarks}}$

and there are $2\left(2.68 \times 10^{26}\right) + 3.34 \times 10^{26} = \boxed{8.70 \times 10^{26} \text{ down quarks}}$.

continued on next page

(b) Model yourself as 65 kg of water. Then you contain:

$$65\left(3.34\times10^{26}\right)\boxed{\sim10^{28}\text{ electrons}}$$

$$65\left(9.36\times10^{26}\right)\boxed{\sim10^{29}\text{ up quarks}}$$

$$65\left(8.70\times10^{26}\right)\boxed{\sim10^{29}\text{ down quarks}}.$$

Only these fundamental particles form your body. You have no strangeness, charm, topness or bottomness.

P31.32 (a)

	proton	u	u	d	total
strangeness	0	0	0	0	0
baryon number	1	1/3	1/3	1/3	1
charge	e	$2e/3$	$2e/3$	$-e/3$	e

(b)

	neutron	u	d	d	total
strangeness	0	0	0	0	0
baryon number	1	1/3	1/3	1/3	1
charge	0	$2e/3$	$-e/3$	$-e/3$	0

P31.33 Quark composition of proton = uud and of neutron = udd.

Thus, if we neglect binding energies, we may write

$$m_{\text{p}}=2m_{\text{u}}+m_{\text{d}} \tag{1}$$

and

$$m_{\text{n}}=m_{\text{u}}+2m_{\text{d}}. \tag{2}$$

Solving simultaneously,

we find $$m_{\text{u}}=\frac{1}{3}\left(2m_{\text{p}}-m_{\text{n}}\right)=\frac{1}{3}\left[2\left(938\text{ MeV}/c^2\right)-939.6\text{ meV}/c^2\right]=\boxed{312\text{ MeV}/c^2}$$

and from either (1) or (2), $m_{\text{d}}=\boxed{314\text{ MeV}/c^2}$.

P31.34 (a)

	K^0	d	$\bar{\text{s}}$	total
strangeness	1	0	1	1
baryon number	0	1/3	–1/3	0
charge	0	$-e/3$	$e/3$	0

(b)

	Λ^0	u	d	s	total
strangeness	–1	0	0	–1	–1
baryon number	1	1/3	1/3	1/3	1
charge	0	$2e/3$	$-e/3$	$-e/3$	0

P31.35 (a) $\pi^- + p \to K^0 + \Lambda^0$

In terms of constituent quarks: $\boxed{\bar{u}d + uud \to d\bar{s} + uds}$

up quarks:	$-1+2 \to 0+1,$	or	$1 \to 1$
down quarks:	$1+1 \to 1+1,$	or	$2 \to 2$
strange quarks:	$0+0 \to -1+1,$	or	$0 \to 0$

(b) $\pi^+ + p \to K^+ + \Sigma^+$ $\boxed{\bar{d}u + uud \to u\bar{s} + uus}$

up quarks:	$1+2 \to 1+2,$	or	$3 \to 3$
down quarks:	$-1+1 \to 0+0,$	or	$0 \to 0$
strange quarks:	$0+0 \to -1+1,$	or	$0 \to 0$

(c) $K^- + p \to K^+ + K^0 + \Omega^-$ $\boxed{\bar{u}s + uud \to u\bar{s} + d\bar{s} + sss}$

up quarks:	$-1+2 \to 1+0+0,$	or	$1 \to 1$
down quarks:	$0+1 \to 0+1+0,$	or	$1 \to 1$
strange quarks:	$1+0 \to -1-1+3,$	or	$1 \to 1$

(d) $p + p \to K^0 + p + \pi^+ + \underline{?}$ $uud + uud \to d\bar{s} + uud + u\bar{d} + \underline{?}$

The quark combination of ? must be such as to balance the last equation for up, down, and strange quarks.

up quarks:	$2+2 = 0+2+1+?$	(has 1 u quark)
down quarks:	$1+1 = 1+1-1+?$	(has 1 d quark)
strange quarks:	$0+0 = -1+0+0+?$	(has 1 s quark)

quark composition $= uds = \boxed{\Lambda^0 \text{ or } \Sigma^0}$

P31.36 In the first reaction, $\pi^- + p \to K^0 + \Lambda^0$, the quarks in the particles are: $\bar{u}d + uud \to d\bar{s} + uds$. There is a net of 1 up quark both before and after the reaction, a net of 2 down quarks both before and after, and a net of zero strange quarks both before and after. Thus, the reaction conserves the net number of each type of quark.

In the second reaction, $\pi^- + p \to K^0 + n$, the quarks in the particles are: $\bar{u}d + uud \to d\bar{s} + udd$. In this case, there is a net of 1 up and 2 down quarks before the reaction but a net of 1 up, 3 down, and 1 anti-strange quark after the reaction. Thus, the reaction does not conserve the net number of each type of quark.

P31.37 $\Sigma^0 + p \to \Sigma^+ + \gamma + X$

dds + uud → uds + 0 + ?

The left side has a net 3d, 2u and 1s. The right-hand side has 1d, 1u, and 1s leaving 2d and 1u missing.

$\boxed{\text{The unknown particle is a neutron, udd.}}$

Baryon and strangeness numbers are conserved.

P31.38 Compare the given quark states to the entries in Tables 31.4 and 31.5:

(a) $suu = \boxed{\Sigma^+}$

(b) $\bar{u}d = \boxed{\pi^-}$

(c) $\bar{s}d = \boxed{K^0}$

(d) $ssd = \boxed{\Xi^-}$

P31.39 (a) $\overline{u}\overline{u}\overline{d}$: charge $= \left(-\frac{2}{3}e\right) + \left(-\frac{2}{3}e\right) + \left(\frac{1}{3}e\right) = \boxed{-e}$. This is the $\boxed{\text{antiproton}}$.

(b) $\overline{u}\overline{d}\overline{d}$: charge $= \left(-\frac{2}{3}e\right) + \left(\frac{1}{3}e\right) + \left(\frac{1}{3}e\right) = \boxed{0}$. This is the $\boxed{\text{antineutron}}$.

Section 31.12 Context Connection—Investigating the Smallest System to Understand the Largest

***P31.40** We let r in Hubble's law represent any distance.

(a) $$v = Hr = 17\times10^{-3}\frac{\text{m}}{\text{s}\cdot\text{ly}}1.85\text{ m}\left(\frac{1\text{ ly}}{c\cdot1\text{ yr}}\right)\left(\frac{c}{3\times10^8\text{ m/s}}\right)\left(\frac{1\text{ yr}}{3.16\times10^7\text{ s}}\right)$$

$$= 17\times10^{-3}\frac{\text{m}}{\text{s}\cdot9.47\times10^{15}\text{ m}}1.85\text{ m} = 1.80\times10^{-18}\frac{1}{\text{s}}1.85\text{ m} = \boxed{3.32\times10^{-18}\text{ m/s}}$$

This is unobservably small.

(b) $$v = Hr = 1.80\times10^{-18}\frac{1}{\text{s}}3.84\times10^8\text{ m} = \boxed{6.90\times10^{-10}\text{ m/s}}$$

P31.41 $v = HR$ $$H = \frac{(1.7\times10^{-2}\ \text{m/s})}{\text{ly}}$$

(a) $v(2.00\times10^{6}\ \text{ly}) = 3.4\times10^{4}\ \text{m/s}$ $$\lambda' = \lambda\sqrt{\frac{1+v/c}{1-v/c}} = 590(1.000\,113\,3) = \boxed{590.07\ \text{nm}}$$

(b) $v(2.00\times10^{8}\ \text{ly}) = 3.4\times10^{6}\ \text{m/s}$ $$\lambda' = 590\sqrt{\frac{1+0.011\,33}{1-0.011\,33}} = \boxed{597\ \text{nm}}$$

(c) $v(2.00\times10^{9}\ \text{ly}) = 3.4\times10^{7}\ \text{m/s}$ $$\lambda' = 590\sqrt{\frac{1+0.113\,3}{1-0.113\,3}} = \boxed{661\ \text{nm}}$$

P31.42 (a) $$\lambda_n' = \lambda_n\sqrt{\frac{1+v/c}{1-v/c}} = (Z+1)\,\lambda_n \qquad \frac{1+v/c}{1-v/c} = (Z+1)^2$$

$$1+\frac{v}{c} = (Z+1)^2 - \left(\frac{v}{c}\right)(Z+1)^2 \qquad \left(\frac{v}{c}\right)\left(Z^2+2Z+2\right) = Z^2+2Z$$

$$v = \boxed{c\left(\frac{Z^2+2Z}{Z^2+2Z+2}\right)}$$

(b) $$R = \frac{v}{H} = \boxed{\frac{c}{H}\left(\frac{Z^2+2Z}{Z^2+2Z+2}\right)}$$

***P31.43** (a) The volume of the sphere bounded by the Earth's orbit is

$$V = \frac{4}{3}\pi r^3 = \frac{4}{3}\pi\left(1.496\times10^{11}\ \text{m}\right)^3 = 1.40\times10^{34}\ \text{m}^3$$
$$m = \rho V = 6\times10^{-28}\ \text{kg/m}^3\ 1.40\times10^{34}\ \text{m}^3 = \boxed{8.41\times10^{6}\ \text{kg}}$$

(b) By Gauss's law, the dark matter would create a gravitational field acting on the Earth to accelerate it toward the Sun. It would shorten the duration of the year in the same way that 8.41×10^{6} kg of extra material in the Sun would. This has the fractional effect $\frac{8.41\times10^{6}\ \text{kg}}{1.99\times10^{30}\ \text{kg}} = 4.23\times10^{-24}$. It is $\boxed{\text{immeasurably small}}$.

P31.44 The density of the Universe is

$$\rho = 1.20\rho_c = 1.20\left(\frac{3H^2}{8\pi G}\right).$$

Consider a remote galaxy at distance r. The mass interior to the sphere below it is

$$M = \rho\left(\frac{4}{3}\pi r^3\right) = 1.20\left(\frac{3H^2}{8\pi G}\right)\left(\frac{4}{3}\pi r^3\right) = \frac{0.600H^2r^3}{G}$$

both now and in the future when it has slowed to rest from its current speed $v = Hr$. The energy of this galaxy-sphere system is constant as the galaxy moves to apogee distance R:

$$\frac{1}{2}mv^2 - \frac{GmM}{r} = 0 - \frac{GmM}{R} \qquad \text{so} \qquad \frac{1}{2}mH^2r^2 - \frac{Gm}{r}\left(\frac{0.600H^2r^3}{G}\right) = 0 - \frac{Gm}{R}\left(\frac{0.600H^2r^3}{G}\right)$$

$$-0.100 = -0.600\frac{r}{R} \qquad \text{so} \qquad R = 6.00r.$$

The Universe will expand by a factor of $\boxed{6.00}$ from its current dimensions.

P31.45 (a) $k_B T \approx 2m_pc^2$

so $$T \approx \frac{2m_pc^2}{k_B} = \frac{2(938.3\text{ MeV})}{\left(1.38\times10^{-23}\text{ J/K}\right)}\left(\frac{1.60\times10^{-13}\text{ J}}{1\text{ MeV}}\right)\boxed{\sim 10^{13}\text{ K}}$$

(b) $k_B T \approx 2m_ec^2$

so $$T \sim \frac{2m_ec^2}{k_B} = \frac{2(0.511\text{ MeV})}{\left(1.38\times10^{-23}\text{ J/K}\right)}\left(\frac{1.60\times10^{-13}\text{ J}}{1\text{ MeV}}\right)\boxed{\sim 10^{10}\text{ K}}$$

P31.46 (a) The Hubble constant is defined in $v = HR$. The distance R between any two far-separated objects opens at constant speed according to $R = vt$. Then the time t since the Big Bang is found from

$$v = Hvt \qquad 1 = Ht \qquad t = \frac{1}{H}.$$

(b) $$\frac{1}{H} = \frac{1}{17\times10^{-3}\text{ m/s}\cdot\text{ly}}\left(\frac{3\times10^8\text{ m/s}}{1\text{ ly/yr}}\right) = \boxed{1.76\times10^{10}\text{ yr}} = 17.6\text{ billion years}$$

P31.47 (a) Consider a sphere around us of radius R large compared to the size of galaxy clusters. If the matter M inside the sphere has the critical density, then a galaxy of mass m at the surface of the sphere is moving just at escape speed v according to

$$K+U_g=0 \qquad \frac{1}{2}mv^2-\frac{GMm}{R}=0.$$

The energy of the galaxy-sphere system is conserved, so this equation is true throughout the history of the Universe after the Big Bang, where $v=\frac{dR}{dt}$. Then

$$\left(\frac{dR}{dt}\right)^2=\frac{2GM}{R} \qquad \frac{dR}{dt}=R^{-1/2}\sqrt{2GM} \qquad \int_0^R\sqrt{R}\,dR=\sqrt{2GM}\int_0^T dt$$

$$\left.\frac{R^{3/2}}{3/2}\right|_0^R=\sqrt{2GM}\,t\Big|_0^T \qquad \frac{2}{3}R^{3/2}=\sqrt{2GM}\,T$$

$$T=\frac{2}{3}\frac{R^{3/2}}{\sqrt{2GM}}=\frac{2}{3}\frac{R}{\sqrt{2GM/R}}.$$

From above, $\sqrt{\frac{2GM}{R}}=v$

so $T=\frac{2}{3}\frac{R}{v}$.

Now Hubble's law says $v=HR$.

So $T=\frac{2}{3}\frac{R}{HR}=\frac{2}{3H}$.

(b) $$T=\frac{2}{3\left(17\times10^{-3}\ \text{m/s}\cdot\text{ly}\right)}\left(\frac{3\times10^8\ \text{m/s}}{1\ \text{ly/yr}}\right)=\boxed{1.18\times10^{10}\ \text{yr}}=11.8\text{ billion years}$$

P31.48 In our frame of reference, Hubble's law is exemplified by $\vec{\mathbf{v}}_1=H\vec{\mathbf{R}}_1$ and $\vec{\mathbf{v}}_2=H\vec{\mathbf{R}}_2$. From these we may form the equations $-\vec{\mathbf{v}}_1=-H\vec{\mathbf{R}}_1$ and $\vec{\mathbf{v}}_2-\vec{\mathbf{v}}_1=H\left(\vec{\mathbf{R}}_2-\vec{\mathbf{R}}_1\right)$. These equations express Hubble's law as seen by the observer in the first galaxy cluster, as she looks at us to find $-\vec{\mathbf{v}}_1=H\left(-\vec{\mathbf{R}}_1\right)$ and as she looks at cluster two to find $\vec{\mathbf{v}}_2-\vec{\mathbf{v}}_1=H\left(\vec{\mathbf{R}}_2-\vec{\mathbf{R}}_1\right)$.

Additional Problems

P31.49 We find the number N of neutrinos:

$$10^{46}\text{ J} = N(6\text{ MeV}) = N\left(6\times 1.60\times 10^{-13}\text{ J}\right)$$

$$N = 1.0\times 10^{58}\text{ neutrinos}$$

The intensity at our location is

$$\frac{N}{A} = \frac{N}{4\pi r^2} = \frac{1.0\times 10^{58}}{4\pi\left(1.7\times 10^{5}\text{ ly}\right)^2}\left(\frac{1\text{ ly}}{\left(3.00\times 10^{8}\text{ m/s}\right)\left(3.16\times 10^{7}\text{ s}\right)}\right)^2 = 3.1\times 10^{14}\text{ m}^{-2}.$$

The number passing through a body presenting $5\,000\text{ cm}^2 = 0.50\text{ m}^2$

is then $\left(3.1\times 10^{14}\ \frac{1}{\text{m}^2}\right)\left(0.50\text{ m}^2\right) = 1.5\times 10^{14}$

or $\boxed{\sim 10^{14}}$.

P31.50 A photon travels the distance from the Large Magellanic Cloud to us in 170 000 years. The hypothetical massive neutrino travels the same distance in 170 000 years plus 10 seconds:

$$c(170\,000\text{ yr}) = v(170\,000\text{ yr} + 10\text{ s})$$

$$\frac{v}{c} = \frac{170\,000\text{ yr}}{170\,000\text{ yr} + 10\text{ s}} = \frac{1}{1+\left\{10\text{ s}/\left[\left(1.7\times 10^{5}\text{ yr}\right)\left(3.156\times 10^{7}\text{ s/yr}\right)\right]\right\}} = \frac{1}{1+1.86\times 10^{-12}}$$

For the neutrino we want to evaluate mc^2 in $E = \gamma mc^2$:

$$mc^2 = \frac{E}{\gamma} = E\sqrt{1-\frac{v^2}{c^2}} = 10\text{ MeV}\sqrt{1-\frac{1}{\left(1+1.86\times 10^{-12}\right)^2}} = 10\text{ MeV}\sqrt{\frac{\left(1+1.86\times 10^{-12}\right)^2-1}{\left(1+1.86\times 10^{-12}\right)^2}}$$

$$mc^2 \approx 10\text{ MeV}\sqrt{\frac{2\left(1.86\times 10^{-12}\right)}{1}} = 10\text{ MeV}\left(1.93\times 10^{-6}\right) = 19\text{ eV}.$$

Then the upper limit on the mass is

$$m = \boxed{\frac{19\text{ eV}}{c^2}}$$

$$m = \frac{19\text{ eV}}{c^2}\left(\frac{\text{u}}{931.5\times 10^{6}\text{ eV}/c^2}\right) = 2.1\times 10^{-8}\text{ u}.$$

P31.51 The total energy in neutrinos emitted per second by the Sun is:

$$(0.4)\left[4\pi\left(1.5\times10^{11}\right)^2\right]\text{W}=1.1\times10^{23}\ \text{W}.$$

Over 10^9 years, the Sun emits 3.6×10^{39} J in neutrinos. This represents an annihilated mass

$$mc^2=3.6\times10^{39}\ \text{J}$$

$$m=4.0\times10^{22}\ \text{kg}.$$

About $\boxed{\text{1 part in 50 000 000}}$ of the Sun's mass, over 10^9 years, has been lost to neutrinos.

P31.52 (a) $\pi^-+\text{p}\rightarrow\Sigma^++\pi^0$ is forbidden by $\boxed{\text{charge conservation}}$.

(b) $\mu^-\rightarrow\pi^-+\nu_e$ is forbidden by $\boxed{\text{energy conservation}}$.

(c) $\text{p}\rightarrow\pi^++\pi^++\pi^-$ is forbidden by $\boxed{\text{baryon number conservation}}$.

P31.53 $K=\frac{1}{2}mv^2$, so $v=\sqrt{\frac{2K}{m}}=\sqrt{\frac{2(0.0400\ \text{eV})\left(1.60\times10^{-19}\ \text{J/eV}\right)}{1.67\times10^{-27}\ \text{kg}}}=2.77\times10^3\ \text{m/s}$

The time for the trip is $\Delta t=\frac{\Delta x}{v}=\frac{1.00\times10^4\ \text{m}}{2.77\times10^3\ \text{m/s}}=3.61\ \text{s}$.

The number of neutrons finishing the trip is given by $N=N_0e^{-\lambda t}$.

The fraction decaying is $1-\frac{N}{N_0}=1-e^{-(\ln 2)t/T_{1/2}}=1-e^{-(\ln 2)(3.61\ \text{s}/614\ \text{s})}=0.004\ 00=\boxed{0.407\%}$.

P31.54 $\text{p}+\text{p}\rightarrow\text{p}+\pi^++X$

We suppose the protons each have 70.4 MeV of kinetic energy. From conservation of momentum for the collision, particle X has zero momentum and thus zero kinetic energy. Conservation of system energy then requires

$$M_\text{p}c^2+M_\pi c^2+M_Xc^2=\left(M_\text{p}c^2+K_\text{p}\right)+\left(M_\text{p}c^2+K_\text{p}\right)$$

$$M_Xc^2=M_\text{p}c^2+2K_\text{p}-M_\pi c^2=938.3\ \text{MeV}+2(70.4\ \text{MeV})-139.6\ \text{MeV}=939.5\ \text{MeV}$$

X must be a neutral baryon of rest energy 939.5 MeV. Thus X is a $\boxed{\text{neutron}}$.

P31.55 $m_\Lambda c^2=1115.6\ \text{MeV}$ $\Lambda^0\rightarrow\text{p}+\pi^-$

$m_pc^2=938.3\ \text{MeV}$ $m_\pi c^2=139.6\ \text{MeV}$

The difference between starting rest energy and final rest energy is the kinetic energy of the products.

$$K_p+K_\pi=37.7\ \text{MeV}\quad\text{and}\quad p_p=p_\pi=p$$

continued on next page

Applying conservation of relativistic energy to the decay process, we have

$$\left[\sqrt{(938.3)^2+p^2c^2}-938.3\right]+\left[\sqrt{(139.6)^2+p^2c^2}-139.6\right]=37.7\text{ MeV}.$$

Solving the algebra yields

$$p_\pi c = p_p c = 100.4\text{ MeV}.$$

Then, $K_p=\sqrt{\left(m_pc^2\right)^2+(100.4)^2}-m_pc^2=\boxed{5.35\text{ MeV}}$

$K_\pi=\sqrt{(139.6)^2+(100.4)^2}-139.6=\boxed{32.3\text{ MeV}}$.

***P31.56** (a) If $2N$ particles are annihilated, the energy released is $2Nmc^2$. The resulting photon momentum is $p=\frac{E}{c}=\frac{2Nmc^2}{c}=2Nmc$. Since the momentum of the system is conserved, the rocket will have momentum $2Nmc$ directed opposite the photon momentum.

$\boxed{p=2Nmc}$

(b) Consider a particle that is annihilated and gives up its rest energy mc^2 to another particle which also has initial rest energy mc^2 (but no momentum initially).

$E^2=p^2c^2+\left(mc^2\right)^2$

Thus $\left(2mc^2\right)^2=p^2c^2+\left(mc^2\right)^2$.

Where p is the momentum the second particle acquires as a result of the annihilation of the first particle. Thus $4\left(mc^2\right)^2=p^2c^2+\left(mc^2\right)^2$, $p^2=3\left(mc^2\right)^2$. So $p=\sqrt{3}mc$.

This process is repeated N times (annihilate $\frac{N}{2}$ protons and $\frac{N}{2}$ antiprotons). Thus the total momentum acquired by the ejected particles is $\sqrt{3}Nmc$, and this momentum is imparted to the rocket.

$\boxed{p=\sqrt{3}Nmc}$

Method (a) produces greater speed since $2Nmc>\sqrt{3}Nmc$.

P31.57 Momentum of proton is $qBr=\left(1.60\times10^{-19}\text{ C}\right)(0.250\text{ kg/C}\cdot\text{s})(1.33\text{ m})$

$p_p=5.32\times10^{-20}\text{ kg}\cdot m/s$ $\quad cp_p=1.60\times10^{-11}\text{ kg}\cdot\text{m}^2/\text{s}^2=1.60\times10^{-11}\text{ J}=99.8\text{ MeV}$.

Therefore, $p_p=99.8\text{ MeV}/c$.

The total energy of the proton is $E_p=\sqrt{E_0^2+(cp)^2}=\sqrt{(938.3)^2+(99.8)^2}=944\text{ MeV}$.

For pion, the momentum qBr is the same (as it must be from conservation of momentum in a 2-particle decay).

continued on next page

$p_\pi = 99.8\ \text{MeV}/c$ $\qquad E_{0\pi} = 139.6\ \text{MeV}$

$$E_\pi = \sqrt{E_0^2 + (cp)^2} = \sqrt{(139.6)^2 + (99.8)^2} = 172\ \text{MeV}$$

Thus, $\qquad E_{\text{total after}} = E_{\text{total before}} = \text{Rest energy}$.

Rest Energy of unknown particle $= 944\ \text{MeV} + 172\ \text{MeV} = 1\,116\ \text{MeV}$ (This is a Λ^0 particle!)

Mass $= \boxed{1\,116\ \text{MeV}/c^2}$.

P31.58 By relativistic energy conservation in the reaction, $\qquad E_\gamma + m_e c^2 = \dfrac{3m_e c^2}{\sqrt{1 - v^2/c^2}}$. (1)

By relativistic momentum conservation for the system, $\qquad \dfrac{E_\gamma}{c} = \dfrac{3m_e v}{\sqrt{1 - v^2/c^2}}$. (2)

Dividing (2) by (1), $\qquad X = \dfrac{E_\gamma}{E_\gamma + m_e c^2} = \dfrac{v}{c}$.

Subtracting (2) from (1), $\qquad m_e c^2 = \dfrac{3m_e c^2}{\sqrt{1 - X^2}} - \dfrac{3m_e c^2 X}{\sqrt{1 - X^2}}$.

Solving, $1 = \dfrac{3 - 3X}{\sqrt{1 - X^2}}$ and $X = \dfrac{4}{5}$ so $E_\gamma = 4m_e c^2 = \boxed{2.04\ \text{MeV}}$.

P31.59 $p + p \to p + n + \pi^+$

The total momentum is zero before the reaction. Thus, all three particles present after the reaction may be at rest and still conserve system momentum. This will be the case when the incident protons have minimum kinetic energy. Under these conditions, conservation of energy for the reaction gives

$$2\left(m_p c^2 + K_p\right) = m_p c^2 + m_n c^2 + m_\pi c^2$$

so the kinetic energy of each of the incident protons is

$$K_p = \frac{m_n c^2 + m_\pi c^2 - m_p c^2}{2} = \frac{(939.6 + 139.6 - 938.3)\ \text{MeV}}{2} = \boxed{70.4\ \text{MeV}}.$$

P31.60 $\Sigma^0 \to \Lambda^0 + \gamma$

From Table 31.2, $m_\Sigma = 1\,192.5\ \text{MeV}/c^2$ and $\qquad m_\Lambda = 1115.6\ \text{MeV}/c^2$.

Conservation of energy in the decay requires

$E_{0,\Sigma} = \left(E_{o,\Lambda} + K_\Lambda\right) + E_\gamma$ $\qquad$ or $\qquad 1\,192.5\ \text{MeV} = \left(1\,115.6\ \text{MeV} + \dfrac{p_\Lambda^{\ 2}}{2m_\Lambda}\right) + E_\gamma$.

continued on next page

System momentum conservation gives $|p_\Lambda| = |p_\gamma|$, so the last result may be written as

$$1\,192.5 \text{ MeV} = \left(1\,115.6 \text{ MeV} + \frac{p_\gamma^{\,2}}{2m_\Lambda}\right) + E_\gamma$$

or

$$1\,192.5 \text{ MeV} = \left(1\,115.6 \text{ MeV} + \frac{p_\gamma^{\,2}c^2}{2m_\Lambda c^2}\right) + E_\gamma.$$

Recognizing that $m_\Lambda c^2 = 1\,115.6 \text{ MeV}$ and $p_\gamma c = E_\gamma$

we now have

$$1\,192.5 \text{ MeV} = 1\,115.6 \text{ MeV} + \frac{E_\gamma^{\,2}}{2(1\,115.6 \text{ MeV})} + E_\gamma.$$

Solving this quadratic equation, $E_\gamma = \boxed{74.4 \text{ MeV}}$.

P31.61 The expression $e^{-E/k_BT}dE$ gives the fraction of the photons that have energy between E and $E + dE$. The fraction that have energy between E and infinity is

$$\frac{\int_E^\infty e^{-E/k_BT}dE}{\int_0^\infty e^{-E/k_BT}dE} = \frac{\int_E^\infty e^{-E/k_BT}(-dE/k_BT)}{\int_0^\infty e^{-E/k_BT}(-dE/k_BT)} = \frac{e^{-E/k_BT}\Big|_E^\infty}{e^{-E/k_BT}\Big|_0^\infty} = e^{-E/k_BT}.$$

We require T when this fraction has a value of 0.0100 (i.e., 1.00%)

and $E = 1.00 \text{ eV} = 1.60 \times 10^{-19} \text{ J}$.

Thus, $0.010\,0 = e^{-(1.60\times10^{-19} \text{ J})/(1.38\times10^{-23} \text{ J/K})T}$

or $\ln(0.010\,0) = -\dfrac{1.60\times10^{-19} \text{ J}}{(1.38\times10^{-23} \text{ J/K})T} = -\dfrac{1.16\times10^4 \text{ K}}{T}$ giving $T = \boxed{2.52\times10^3 \text{ K}}$.

P31.62 $\pi^- \to \mu^- + \bar{\nu}_\mu$: From the conservation laws for the decay,

$$m_\pi c^2 = 139.6 \text{ MeV} = E_\mu + E_\nu \qquad (1)$$

and $p_\mu = p_\nu$, $E_\nu = p_\nu c$:

$$E_\mu^2 = (p_\mu c)^2 + (105.7 \text{ MeV})^2 = (p_\nu c)^2 + (105.7 \text{ MeV})^2$$

or

$$E_\mu^2 - E_\nu^2 = (105.7 \text{ MeV})^2. \qquad (2)$$

Since

$$E_\mu + E_\nu = 139.6 \text{ MeV} \qquad (1)$$

and

$$(E_\mu + E_\nu)(E_\mu - E_\nu) = (105.7 \text{ MeV})^2 \qquad (2)$$

then

$$E_\mu - E_\nu = \frac{(105.7 \text{ MeV})^2}{139.6 \text{ MeV}} = 80.0. \qquad (3)$$

Subtracting (3) from (1), $2E_\nu = 59.6 \text{ MeV}$ and $\boxed{E_\nu = 29.8 \text{ MeV}}$.

P31.63 (a) The mediator of this weak interaction is a $\boxed{Z^0 \text{ boson}}$.

(b) The Feynman diagram shows a down quark and its antiparticle annihilating each other. They can produce a particle carrying energy, momentum, and angular momentum, but zero charge, zero baryon number, and, it may be, no color charge. In this case the product particle is a $\boxed{\text{photon}}$.

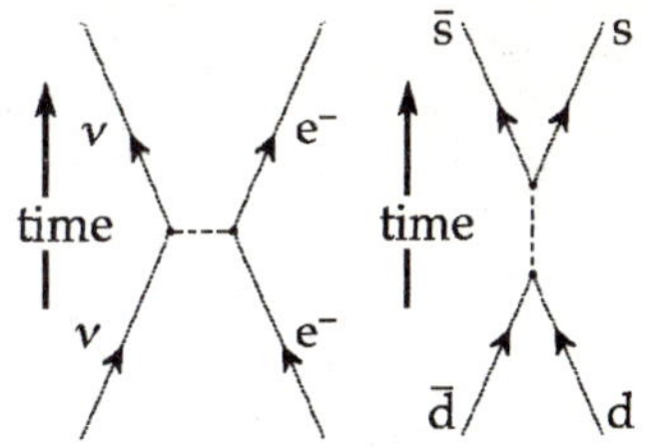

FIG. P31.63

For conservation of both energy and momentum in the collision we would expect two photons; but momentum need not be strictly conserved, according to the uncertainty principle, if the photon travels a sufficiently short distance before producing another matter-antimatter pair of particles, as shown in Figure P31.63. Depending on the color charges of the d and $\bar{\text{d}}$ quarks, the ephemeral particle could also be a $\boxed{\text{gluon}}$, as suggested in the discussion of Figure P31.13(b).

P31.64 (a) This diagram represents the annihilation of an electron and an antielectron. From charge and lepton-number conservation at either vertex, the exchanged particle must be an electron, $\boxed{e^-}$.

(b) This is the tough one. A neutrino collides with a neutron, changing it into a proton with release of a muon. This is a weak interaction. The exchanged particle has charge $+e$ and is a $\boxed{W^+}$.

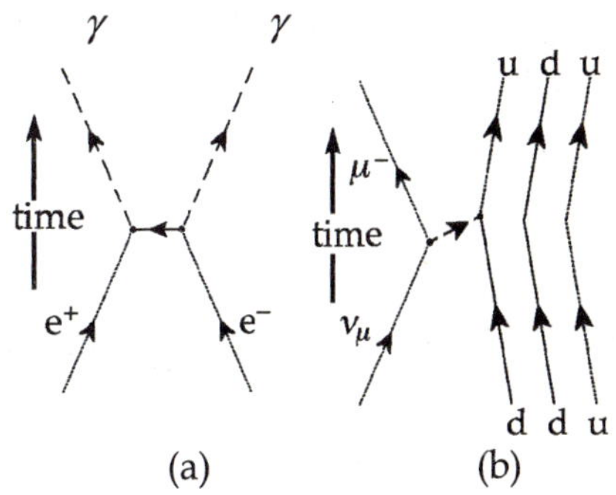

FIG. P31.64

***P31.65** (a) At threshold, we consider a photon and a proton colliding head-on to produce a proton and a pion at rest, according to $p+\gamma \rightarrow p+\pi^0$. Energy conservation gives

$$\frac{m_p c^2}{\sqrt{1-u^2/c^2}}+E_\gamma = m_p c^2 + m_\pi c^2.$$

Momentum conservation gives $\dfrac{m_p u}{\sqrt{1-u^2/c^2}}-\dfrac{E_\gamma}{c}=0.$

Combining the equations, we have

$$\frac{m_p c^2}{\sqrt{1-u^2/c^2}}+\frac{m_p c^2}{\sqrt{1-u^2/c^2}}\frac{u}{c}=m_p c^2+m_\pi c^2$$

$$\frac{938.3\text{ MeV}(1+u/c)}{\sqrt{(1-u/c)(1+u/c)}}=938.3\text{ MeV}+135.0\text{ MeV}$$

so $\dfrac{u}{c}=0.134$

and $E_\gamma = \boxed{127\text{ MeV}}$.

continued on next page

(b) $\lambda_{max}T = 2.898\text{ mm}\cdot\text{K}$

$$\lambda_{max} = \frac{2.898\text{ mm}\cdot\text{K}}{2.73\text{ K}} = \boxed{1.06\text{ mm}}$$

(c) $$E_\gamma = hf = \frac{hc}{\lambda} = \frac{1\,240\text{ eV}\cdot 10^{-9}\text{ m}}{1.06\times 10^{-3}\text{ m}} = \boxed{1.17\times 10^{-3}\text{ eV}}$$

(d) In the primed reference frame, the proton is moving to the right at $\frac{u'}{c} = 0.134$ and the photon is moving to the left with $hf' = 1.27\times 10^{8}$ eV. In the unprimed frame, $hf = 1.17\times 10^{-3}$ eV. Using the Doppler effect equation from Section 24.3, we have for the speed of the primed frame

$$1.27\times 10^{8} = \sqrt{\frac{1+v/c}{1-v/c}}\,1.17\times 10^{-3}$$

$$\frac{v}{c} = 1 - 1.71\times 10^{-22}$$

Then the speed of the proton is given by

$$\frac{u}{c} = \frac{u'/c + v/c}{1 + u'v/c^2} = \frac{0.134 + 1 - 1.71\times 10^{-22}}{1 + 0.134\left(1 - 1.71\times 10^{-22}\right)} = 1 - 1.30\times 10^{-22}.$$

And the energy of the proton is

$$\frac{m_pc^2}{\sqrt{1-u^2/c^2}} = \frac{938.3\text{ MeV}}{\sqrt{1-\left(1-1.30\times 10^{-22}\right)^2}} = 6.19\times 10^{10}\times 938.3\times 10^{6}\text{ eV} = \boxed{5.81\times 10^{19}\text{ eV}}.$$

ANSWERS TO EVEN PROBLEMS

P31.2 $\sim 10^{3}$ Bq

P31.4 2.27×10^{23} Hz, 1.32×10^{-15} m

P31.6 $\bar{\nu}_\mu$, ν_e

P31.8 $\sim 10^{-16}$ m

P31.10 $\sim 10^{-23}$ s

P31.12 $\bar{\nu}_\mu$

P31.14 (a) L_μ and L_e; (b) charge; (c) baryon number; (d) baryon number; (e) charge

P31.16 the second reaction does not conserve baryon number

P31.18 0.828 c

P31.20 (a) see the solution;
(b) $E_e = E_\gamma = 469$ MeV, $p_e = p_\gamma = \frac{469\text{ MeV}}{c}$
(c) 0.999 999 4 c

P31.22 The ρ^0 decays via the strong interaction; the K_s^0 must decay via the weak interaction.

P31.24 (a) electron and muon lepton numbers;
(b) electron lepton number;
(c) charge and strangeness;
(d) baryon number; (e) strangeness

P31.26 (a) baryon number, charge, L_e and L_τ;
(b) baryon number, charge, L_e, L_μ and L_τ;
(c) strangeness, charge, L_e, L_μ and L_τ;
(d) baryon number, strangeness, charge, L_e, L_μ and L_τ;
(e) baryon number, strangeness, charge, L_e, L_μ and L_τ;
(f) baryon number, strangeness, charge, L_e, L_μ and L_τ

P31.28 (a) 686 MeV/c and 200 MeV/c;
(b) 627 MeV/c;
(c) 244 MeV, 1 130 MeV, 1 370 MeV;
(d) 1 190 MeV/c^2, 0.500 c

P31.30 (a) see the solution; (b) 5.63 GeV;
(c) 768 MeV; (d) 280 MeV; (e) 4.43 TeV

P31.32 see the solution

P31.34 see the solution

P31.36 see the solution

P31.38 (a) Σ^+; (b) π^-; (c) K^0; (d) Ξ^-

P31.40 (a) 3.32 am/s; (b) 0.690 nm/s

P31.42 (a) $c\left(\dfrac{Z^2+2Z}{Z^2+2Z+2}\right)$; (b) $\dfrac{c}{H}\left(\dfrac{Z^2+2Z}{Z^2+2Z+2}\right)$

P31.44 6.00

P31.46 (a) see the solution; (b) 17.6 Gyr

P31.48 see the solution

P31.50 $\dfrac{19\text{ eV}}{c^2}$

P31.52 (a) charge conservation;
(b) energy conservation;
(c) baryon number conservation

P31.54 a neutron

P31.56 (a) $2Nmc$; (b) $\sqrt{3}Nmc$, method (a)

P31.58 2.04 MeV

P31.60 74.4 MeV

P31.62 29.8 MeV

P31.64 (a) electron-positron annihilation, e^-;
(b) a neutrino collides with a neutron, W^+

CONTEXT 9 CONCLUSION ANSWERS TO QUESTIONS

QCC9.1 The Universe is vast and could on its own terms get along very well without us. But as the cosmos is immense, life appears to be immensely scarce, and therefore precious. We must do our work, growing corn to feed the hungry while preserving our planet for future generations. One person has singular abilities and opportunities for effort, faithfulness, generosity, honor, curiosity, understanding, and wonder. His or her place is to use those abilities and opportunities, unique in all the Universe.

CONTEXT 9 CONCLUSION SOLUTIONS TO PROBLEMS

CC9.1 (a) $$L=\sqrt{\frac{\hbar G}{c^3}}=\sqrt{\frac{(1.055\times10^{-34}\text{ J}\cdot\text{s})(6.67\times10^{-11}\text{ N}\cdot\text{m}^2/\text{kg}^2)}{(3.00\times10^8\text{ m/s})^3}}=\boxed{1.61\times10^{-35}\text{ m}}$$

continued on next page

(b) This time is given as $T = \frac{L}{c} = \frac{1.61 \times 10^{-35} \text{ m}}{3.00 \times 10^{8} \text{ m/s}} = \boxed{5.38 \times 10^{-44} \text{ s}}$,

which is approximately equal to the ultra-hot epoch.

(c) Yes.